Karl Baedeker

Switzerland and the adjacent portions of Italy, Savoy, and Tyrol

Karl Baedeker

Switzerland and the adjacent portions of Italy, Savoy, and Tyrol

ISBN/EAN: 9783741149320

Manufactured in Europe, USA, Canada, Australia, Japa

Cover: Foto ©berggeist007 / pixelio.de

Manufactured and distributed by brebook publishing software (www.brebook.com)

Karl Baedeker

Switzerland and the adjacent portions of Italy, Savoy, and Tyrol

SWITZERLAND

AND THE ADJACENT PORTIONS OF

ITALY, SAVOY, AND TYROL

HANDBOOK FOR TRAVELLERS

BY

KARL BAEDEKER

WITH 47 MAPS, 12 PLANS, AND 12 PANORAMAS

SIXTEENTH EDITION

LEIPSIC: KARL BAEDEKER, PUBLISHER.
LONDON: DULAU AND CO., 37 SOHO SQUARE, W.
1895

'Go, little book, God send thee good passage,
And specially let this be thy prayere
Unto them all that thee will read or hear,
Where thou art wrong, after their help to call,
Thee to correct in any part or all.'

PREFACE.

The object of the Handbook for Switzerland is to supply the traveller with all needful information, to point out the most interesting places and the best way of reaching them, to render him comparatively independent of the services of guides and others, and thus to enable him thoroughly to enjoy his tour in this magnificent country.

With improved facilities for travel, the number of visitors to Switzerland has greatly increased of late years, and mountaineering ambition has been proportionally stimulated. Summits once deemed well-nigh inaccessible are now scaled annually by travellers from all parts of the world. The achievements of the modern Alpine clubs have dimmed the memory of De Saussure, Auldjo, and the other pioneers of these icy regions, and even ladies now frequently vie with the stronger sex in their deeds of daring.

The Handbook is based on the Editor's personal acquaintance with the places described, most of which he has carefully and repeatedly explored. This edition, which corresponds with the twenty-sixth German edition, has been thoroughly revised, and furnished with the latest information obtainable. Its contents are divided into SEVEN SECTIONS (I. North Switzerland; II. Lake of Lucerne and Environs, and St. Gotthard; III. Bernese Oberland; IV. South-Western Switzerland, Lake of Geneva, Lower Rhone Valley; V. Savoy, the Valais, and the adjacent Italian Alps; VI. South-Eastern Switzerland, Grisons; VII. Lakes of North Italy), each of which may be separately removed from the book by the mountaineer or pedestrian who desires to minimize the bulk of his luggage. To each section is prefixed a list of the routes it contains, so that each forms an approximately complete volume apart from the general table of contents or the general index.

The Editor will highly appreciate any corrections or suggestions with which travellers may favour him. The information already received from numerous correspondents, which he gratefully acknowledges, has in many instances proved most serviceable.

The MAPS and PLANS, on which special care has been bestowed, are based on the *Topographical Atlas of Switzer-*

land and on *Dufour's Map* (pp. xxii, **xxiii**), **and revised with the aid of other recent authorities and from the Editor's own experiences.**

TIME TABLES. The best Swiss **publications are the** '*Kursbücher*' (time-tables) of *Bürkli* **of Zürich and** *Krüsi* of Bâle (50 c. each), sold at most of the **railway-stations.**

HEIGHTS are given in the text in **English** feet, on the maps in mètres (1 Engl. ft. = 0.3048 mètre; 1 mètre = 3.281 Engl. ft., or about 3 ft. 3 1/3 in.). Comp. p. xxx. — DISTANCES on high-roads and railways are given in English miles; while those on bridle-paths and mountain-routes are expressed by the time which they usually take. The number of miles at the beginning of a paragraph denotes the distance from the starting-point, while the distances from place to place are generally stated within brackets; but on railway-routes the mileage is always reckoned from the starting-point.

HOTELS. Besides the first-class hotels, the Handbook mentions a number of the more modest inns also. The usual charges are stated in accordance with the Editor's **own** experience, **or** from the bills furnished to him by **travellers**. Hotel-charges, like carriage-fares and **fees** to guides, generally have an upward tendency, but **an approximate statement of these** items will enable the traveller to form an estimate **of his** probable expenditure. The value of the asterisks, which are used as marks of commendation, is relative only, signifying that the houses are good of their class. The Editor has distributed these asterisks as fully and impartially as his knowledge warrants, but there are doubtless many equally deserving houses among those not starred **or** even mentioned.

To hotel-keepers, tradesmen, and others **the** Editor begs to intimate that a character for fair dealing towards travellers forms the sole passport to his commendation, and that advertisements of every kind are strictly excluded from his Handbooks. Hotel-keepers are also warned against persons representing themselves as agents for Baedeker's Handbooks.

CONTENTS.

		Page
I.	Plan of Tour, etc.	xii
II.	Travelling Expenses. Money	xvii
III.	Hotels and Pensions	xvii
IV.	Passports. Custom House	xix
V.	Walking Tours	xix
VI.	Maps	xxi
VII.	Guides	xxii
VIII.	Carriages and Horses	xxiii
IX.	Diligences, Post Office, Telegraph	xxiii
X.	Railways	xxv
XI.	History. Statistics	xxvi
XII.	Metrical Measures. Thermometer	xxx

I. Northern Switzerland.

Route		
1.	Bâle	2
2.	From Bâle to Bienne and Bern through the Münster-Thal	9
3.	From Bâle to Bienne viâ Olten and Soleure	13
4.	From Bâle to Bern viâ Herzogenbuchsee	17
5.	From Bâle to Lucerne	18
6.	From Bâle to Zürich	19
7.	From Olten to Waldshut via Aarau and Brugg	22
8.	From Bâle to Schaffhausen and Constance	23
9.	The Falls of the Rhine	26
10.	From Friedrichshafen to Constance. Lake of Constance	28
11.	From Rorschach to Constance and Winterthur (Zürich)	31
12.	From Schaffhausen to Zürich	32
13.	Zürich and the Uetliberg	33
14.	From Zürich to Coire. Lakes of Zürich and Walenstadt	40
15.	From Zürich to Romanshorn and Friedrichshafen	48
16.	From Zürich to St. Gallen, Rorschach, and Lindau	49
17.	The Canton of Appenzell	53
18.	From Rorschach to Coire	60
19.	From Wyl through the Toggenburg to Buchs on the Rhine	62
20.	Ragatz and Pfäfers	64
21.	From Zürich to Glarus and Linthal	66
22.	From Stachelberg to Altdorf. Klausen	71
23.	From Schwyz to Glarus over the Pragel	72
24.	From Glarus to Coire through the Sernf-Thal	75

II. Lake of Lucerne and Environs. The St. Gotthard.

25.	From Zürich to Zug and Lucerne	78
26.	Lucerne	81
27.	Lake of Lucerne	86

CONTENTS.

Route	Page
28. The Rigi	94
29. From Lucerne to Alpnach-Stad. Pilatus	100
30. From Zug and Lucerne to Arth	103
31. From Zürich viâ Wädensweil to Arth-Goldau. From Biberbrücke to Einsiedeln	105
32. From Lucerne to Bellinzona. St. Gotthard Railway	108
33. From Göschenen to Airolo over the St. Gotthard	117
34. The Maderaner-Thal	122
35. From Göschenen to the Rhone Glacier. The Furka	124
36. From Lucerne to Altdorf viâ Stans and Engelberg. The Surenen Pass	127
37. From Lucerne over the Brünig to Meiringen and Brienz (Interlaken)	131
38. From Meiringen to Engelberg. Engstlen-Alp. Joch Pass	134
39. From Meiringen to Wasen. Susten Pass	136
40. From Lucerne to Bern. Entlebuch. Emmen-Thal	138
41. From Lucerne to Lenzburg (Aarau). The Seethal Railway	140

III. The Bernese Oberland.

42. Bern	144
43. From Bern to Thun	151
44. The Niesen	153
45. From Thun to Interlaken. Lake of Thun	155
46. Interlaken and Environs	158
47. The Lauterbrunnen Valley and Mürren	164
48. From Interlaken to Grindelwald	171
49. The Faulhorn	178
50. From Meiringen to Interlaken. Lake of Brienz	181
51. From Meiringen to Grindelwald	185
52. From Meiringen to the Rhone Glacier. Grimsel	187
53. From Spiez to Leuk over the Gemmi	191
54. The Adelboden Valley	196
55. From Gampel to Kandersteg. Lötschen Pass	197
56. From Thun to Sion over the Rawyl	199
57. From Thun to Saanen through the Simmen-Thal	202

IV. Western Switzerland. Lake of Geneva. Lower Valley of the Rhone.

58. From Bern to Neuchâtel	206
59. From Neuchâtel to Chaux-de-Fonds and Locle	209
60. From Neuchâtel to Pontarlier through the Val de Travers	211
61. From Neuchâtel to Lausanne	213
62. From Bern to Lausanne (Vevey)	215
63. From Lausanne to Payerne and Lyss	219
64. From Lausanne to Vallorbes and Pontarlier	220
65. Geneva and its Environs	221

CONTENTS.

Route	Page
66. From Geneva to Martigny viâ Lausanne and Villeneuve. Lake of Geneva *(North Bank)*	233
67. From Saanen to Aigle over the Col de Pillon	250
68. From Bulle to Château-d'Oex and Aigle	252
69. From Bex to Sion. Pas de Cheville	255
70. From Geneva to St. Maurice viâ Bouveret. Lake of Geneva *(South Bank)*. Val d'Illiez	256

V. Savoy, the Valais, and the adjacent Italian Alps.

Route	Page
71. From Geneva viâ Culoz and Aix-les-Bains to Chambéry, and back viâ Annecy	264
72. From Geneva to Chamonix	271
73. Chamonix and Environs	276
74. From Chamonix to Martigny over the Tête-Noire, or to Vernayaz viâ Triquent and Salvan	282
75. From Martigny to Chamonix. Col de Balme	286
76. From Chamonix to Courmayeur over the Col du Bonhomme and the Col de la Seigne. Tour du Mont Blanc	288
77. From Courmayeur to Aosta and Ivrea	293
78. From Martigny to Aosta. Great St. Bernard	298
79. From Martigny to Aosta over the Col de Fenêtre. Val de Bagnes	304
80. From Martigny to Domodossola over the Simplon	306
81. From the Rhone Glacier to Brig. The Eggishorn	314
82. From Ulrichen to Domodossola. Gries Pass. Falls of the Tosa. Val Formazza	319
83. The S. Valleys of the Valais, between Sion and Turtmann (Val d'Hérens, Val d'Anniviers, Turtmann Valley)	321
84. From Visp to Zermatt	331
85. From Visp to Saas and Mattmark	340
86. From Piedimulera to Macugnaga, and over the Monte Moro Pass to Mattmark	343
87. From Macugnaga to Zermatt round Monte Rosa	346
88. From Châtillon to Valtournanche and over the Théodule Pass to Zermatt	349

VI. South-Eastern Switzerland. The Grisons.

Route	Page
89. Coire	354
90. From Landquart to Davos through the Prätigau and to Schuls over the Flüela Pass	356
91. From Davos to Coire viâ Lenz (Landwasser Route)	360
92. From Coire to Davos through the Schanfigg-Thal. Arosa	364
93. From Coire to Göschenen. Oberalp	366
94. From Disentis to Biasca. The Lukmanier	375
95. From Coire viâ Thusis to Tiefenkasten (Schyn Road) or Splügen (Via Mala)	377

CONTENTS.

Route	Page
96. From Splügen to the Lake of Como	383
97. From Splügen to Bellinzona. Bernardino	385
98. From Coire to the Engadine over the Albula Pass	388
99. From Coire to the Engadine over the Julier	390
100. The Upper Engadine from the Maloja to Samaden	394
101. Pontresina and Environs	402
102. From Samaden to **Nauders**. Lower Engadine	412
103. From Samaden-Pontresina over the Bernina to Tirano and through the Valtellina to Colico	419
104. From the Maloja to Chiavenna. Val Bregaglia	422
105. From Tirano to **Nauders** over the Stelvio	425
106. From Nauders to **Bregenz** over the Arlberg	429

VII. The Italian Lakes.

107. From Bellinzona **to** Lugano and Como (Milan)	433
108. From Bellinzona **to** Locarno. Val Maggia	440
109. **Lago Maggiore**. The Borromean Islands	443
110. From Domodossola to Novara. Lake of Orta	451
111. From Luino on Lago Maggiore to Menaggio on the **Lake of Como**. Lake of Lugano	455
112. **The Lake** of Como	457
113. From Como to Milan	465
Index	469

List of Maps.

(Comp. Key Map after the General Index.)

1. MAP OF SWITZERLAND (1 : 1,000,000), before the title-page.
2. DISTRICT BETWEEN SCHAFFHAUSEN AND CONSTANCE (1 : 250,000); between pp. 26, 27.
3. ENVIRONS OF SCHAFFHAUSEN (1 : 35,000); p. 26.
4. LAKE OF CONSTANCE (1 : 250,000); between pp. 28, 29.
5. LAKES OF ZÜRICH AND ZUG (1 : 250,000); between pp. 40, 41.
6. CANTON OF APPENZELL (1 : 250,000); between pp. 54, 55.
7. CANTON OF GLARUS (1 : 250,000); between pp. 66, 67.
8. TÖDI DISTRICT (1 : 150,000); between pp. 68, 69.
9. LAKE OF LUCERNE (1 : 250,000); between pp. 86, 87.
10. PILATUS (1 : 100,000); p. 87.
11. RIGI (1 : 100,000); between pp. 94, 95.
12. ENVIRONS OF THE ST. GOTTHARD (1 : 250,000); between pp. 112, 113.
13. LOOP-TUNNELS OF THE ST. GOTTHARD RAILWAY (1 : 25,000); p. 113.
14. **TRIFT** DISTRICT (1 : 150,000); between pp. 118, 119.
15. ENVIRONS OF ENGELBERG (1 : 150,000); between pp. 126, 127.
16. ENVIRONS OF THUN (1 : 26,000); p. 152.
17. **BERNESE** OBERLAND (1 : 250,000); between pp. 152, 153.
18. ENVIRONS OF INTERLAKEN (1 : 26,000); p. 160.
19. ENVIRONS OF GRINDELWALD (1 : 150,000); between pp. 160, 161.
20. UPPER LAUTERBRUNNEN VALLEY (1 : 150,000); p. 161.
21. ENVIRONS OF KANDERSTEG (1 : 150,000); between pp. 192, 193.
22. ENVIRONS OF GENEVA (1 : 150,000); p. 223.
23. LAKE OF GENEVA (1 : 250,000); between pp. 234, 235.
24. ENVIRONS OF MONTREUX (1 : 50,000); p. 240.
25. ORMONT VALLEYS (1 : 150,000); between pp. 250, 251.

LIST OF MAPS.   xi

26. ENVIRONS OF CHAMONIX, SIXT, AND COURMAYEUR (1 : 250,000); between pp. 272, 273.
27. MONT BLANC DISTRICT (1 : 150,000); between pp. 276, 277.
28. ENVIRONS OF THE GREAT ST. BERNARD, from Martigny to Aosta (1 : 250,000); between pp. 298, 299.
29. LOWER VALLEY OF THE RHONE, from the Lake of Geneva to the Lötschen-Thal (1 : 250,000); between pp. 308, 309.
30. THE UPPER VALAIS (1 : 250,000; between pp. 310, 311.
31. ALETSCH DISTRICT (1 : 150,000); between pp. 316, 317.
32. VALAISIAN ALPS (1 : 250,000); between pp. 322, 323.
33. ENVIRONS OF ZERMATT (1 : 150,000); between pp. 332, 333.
34. ENVIRONS OF RAGATZ, THE PRÄTIGAU AND MONTAFON (1 : 250,000); between pp. 356, 357.
35. CENTRAL GRISONS ALPS, from Coire and Davos to Samaden (1 : 250,000); between pp. 360, 361.
36. VORDER-RHEINTHAL (1 : 250,000); between pp. 368, 369.
37. DISTRICT FROM THE LUKMANIER TO THE MALOJA (1 : 250,000); between pp. 382, 383.
38. THE ENGADINE AND VALTELLINA (1 : 500,000); between pp. 394, 395.
39. ENVIRONS OF PONTRESINA (1 : 150,000); between pp. 402, 403.
40. THE LOWER ENGADINE (1 : 250,000); between pp. 412, 413.
41. ENVIRONS OF LUGANO (1 : 150,000); p. 434.
42. ENVIRONS OF COMO (1 : 28,000); p. 435.
43. LAGO MAGGIORE (1 : 250,000); between pp. 448, 449.
44. ENVIRONS OF PALLANZA (1 : 65,000); p. 448.
45. ENVIRONS OF STRESA (1 : 65,000); p. 449.
46. LAKES OF COMO AND LUGANO (1 : 250,000); between pp. 456, 457.
47. KEY MAP OF SWITZERLAND (1 : 1,000,000), after the Index.

Panoramas and Views.

1. From the RIGI-KULM, between pp. 96, 97.
2. From the PILATUS, between pp. 102, 103.
3. From BERN, p. 145.
4. From the NIESEN, p. 153.
5. From the HEIMWEHFLUH, p. 161.
6. From MÜRREN, p. 169.
7. From the FAULHORN, between pp. 178, 179.
8. From the FLEGERE, between pp. 278, 279.
9. From the EGGISHORN, between pp. 316, 317.
10. From the GORNER GRAT, between pp. 334, 335.
11. From the PIZ LANGUARD, between pp. 406, 407.
12. From the MONTE GENEROSO, between pp. 438, 439.

Plans of Towns.

BÂLE, p. 2; CONSTANCE, p. 27; ZÜRICH, p. 82; RAGATZ, p. 66; LUCERNE, p. 86; BERN, p. 144; NEUCHÂTEL, p. 207; GENEVA, p. 222; LAUSANNE, p. 236; COIRE; p. 354; LUGANO, p. 434; MILAN, p. 464.

Abbreviations.

R. = Room, Route. N. = North, northern. min. = Minute.
B. = Breakfast.   S. = South, southern. carr. = Carriage.
D. = Dinner.   E. = East, eastern.   S.A.C. = Swiss Alpine Club.
L. = Light.   W. = West, western.   C.A.I. = Italian Alpine Club
A. = Attendance.   r. = Right.   S.B.G.H. = Société des
M. = English mile. l. = Left.   Bibliothèques des Grands
ft. (') = Engl. foot. hr. = Hour.   Hôtels (see p. xviii).

Asterisks are used as marks of commendation.
With regard to distances, see Preface.

I. Plan of Tour.

Season of the Year. Distribution of Time.

The traveller will save both time and money by planning his tour carefully before leaving home. The Handbook will help him to select the most interesting routes and the pleasantest resting-places, and point out how each day may be disposed of to the best advantage, provided the weather be favourable.

Season. The great majority of tourists visit Switzerland between the middle of July and the end of September; but to those who wish to see the scenery, the vegetation, and particularly the Alpine flowers in perfection June is recommended as the most charming month in the year. For expeditions among the higher Alps the month of August is the best season; but above a height of 6500 ft. snow-storms may occur at any time except in thoroughly settled weather. In ordinary seasons the snow disappears from the Rigi and the more frequented routes through the Bernese Oberland at the beginning of June. On the other hand snow sometimes lies throughout the whole season on the Furka, the Grimsel, the Gemmi, etc. The most loftily situated hotels are generally closed till the end of June.

Distribution of Time. ONE MONTH, as the annexed plan shows, suffices for a glimpse at the most interesting parts of Switzerland. Bâle, where the scenery is least interesting, is a good starting-point, but the traveller may find it more convenient to begin with Geneva or Neuchâtel.

	Days
By railway from *Bâle* to *Neuhausen;* visit the *Falls of the Rhine;* by railway from *Dachsen* to *Zürich* (RR. 1, 8, 9, 12)	1
Zürich and the *Uetliberg* (R. 13)	1
From Zürich by railway to *Zug;* by steamboat to *Arth;* by railway to the *Rigi-Kulm* (RR. 25, 30, 28)	1
From the Rigi by railway to *Vitznau* (or on foot to *Wäggis*); by steamboat to *Lucerne,* and one day at Lucerne (RR. 28, 27, 26)	1
By steamer on the *Lake of Lucerne* to *Brunnen;* visit the *Rütli, Axenstein,* etc. (R. 27)	1
By steamer from Brunnen to *Flüelen* (or by steamer to the *Tells-Platte* and thence on foot by the Axen-Strasse to Flüelen); by the St. Gotthard Railway to *Göschenen;* by omnibus or on foot to *Andermatt* (RR. 27, 32, 33)	1
By carriage or on foot over the *Furka* to the *Rhone Glacier* (R. 35); walk over the *Grimsel* to the *Grimsel Hospice* (R. 52)	
Drive or walk down the *Hasli-Thal* (Handegg Fall) to *Meiringen* (RR. 52, 50)	1
Walk from Meiringen (Falls of the Reichenbach) through the *Bernese Oberland,* by the *Scheidegg,* to *Grindelwald,* with ascent of the *Faulhorn* (RR. 51, 49)	1-2
By railway from Grindelwald over the *Kleine Scheidegg* (on foot to the *Wengern-Alp*) to *Lauterbrunnen* (Staubbach; R. 49) and *Mürren* (R. 47)	1
Walk viâ the *Obere Steinberg* to *Trachsellauenen* and back to Lauterbrunnen; by railway to *Interlaken* (R. 47)	1

I. PLAN OF TOUR.

	Days
Excursions from *Interlaken* (*St. Beatenberg, Schynige Platte, Brienzer Rothhorn*, etc.; RR. 46, 45, 50)	2
By railway or steamer to *Spiez*; drive or walk to *Kandersteg* (R. 53)	1
(Excursions from Kandersteg to the *Oeschinensee, Gastern-Thal*, etc.)	(1)
Walk from Kandersteg over the *Gemmi* to *Bad Leuk* (R. 53)	1
Drive to *Leuk* station (R. 53); by railway to *Visp* (R. 80) and *Zermatt* (R. 84)	1
Excursions from **Zermatt** (*Riffelhaus, Gorner Grat, Schwarzsee*, etc.; R. 84)	2
Railway to **Visp** (R. 84) and *Martigny* (R. 80)	1
To *Chamonix* by the *Col de Balme*, the *Tête-Noire*, or *Salvan* (RR. 75, 74)	1
Chamonix (R. 73)	1-2
By omnibus to *Geneva* (R. 72)	1
Geneva and Environs (R. 65)	1
By steamboat on the *Lake of Geneva* (R. 66) to *Montreux* (*Chillon, Glion*, etc.)	1-2
By railway to *Lausanne*; several hours at Lausanne; by railway in the afternoon to *Freiburg* (RR. 66, 62)	1
By railway to *Bern* (R. 62); at Bern (R. 42)	1
By railway to *Bâle* (R. 4); at Bâle (R. 1)	1

A fortnight additional may be pleasantly spent in **Eastern Switzerland** (Appenzell, Bad Pfäfers, Via Mala, Upper Engadine), whence the **Italian Lakes** are easily visited.

	Days
From *Rorschach* or *Zürich* to *Pfäfers* and *Coire* (RR. 14, 18, 20, 89)	1
Diligence to *Thusis*; visit the *Via Mala* as far as the third bridge, and return to Thusis (R. 95); walk or drive by the *Schyn Road* to *Tiefenkasten* (R. 95)	1
Diligence over the *Julier* to *Silvaplana* (R. 99) and *St. Moritz* (R. 100)	1
Drive to the *Maloja* and back (R. 100); in the afternoon to *Pontresina* (R. 101)	1
Pontresina (*Morteratsch* and *Roseg Glaciers*; ascent of the *Piz Languard*, etc.; R. 101)	2-3
Diligence over the *Bernina* to *Tirano* and *Sondrio* (R. 103); railway to *Colico* (R. 103); steamer to *Bellagio* (R. 112)	1½
Bellagio (*Villa Serbelloni, Villa Carlotta*, etc.); then **viâ** *Menaggio* and *Porlezza* to *Lugano* (RR. 112, 111)	1
Environs of Lugano (*Mte. S. Salvatore* or *Mte. Generoso*; R. 107)	1-1½
Steamboat to *Ponte Tresa*, railway to *Luino* (R. 111); steamer to the *Borromean Islands* and to *Pallanza* or *Stresa* (R. 109)	1
Steamboat to *Laveno*, and back by the St. Gotthard **Railway** to *Lucerne*	1
Or by railway and diligence over the **Simplon** to *Brieg* (R. 80)	

So comprehensive a tour as the above is of course rarely undertaken; but it will enable the traveller to plan an excursion of suitable length, such as one of the following: —

I. EIGHT DAYS FROM BÂLE.
(*Rigi, Bernese Oberland, Rhone Glacier, St. Gotthard Route.*)

1st. From *Bâle* (or *Constance* or *Romanshorn*) to *Zürich*. **Uetliberg**.
2nd. To *Zug*, *Arth*, the *Rigi*, and *Lucerne*.
3rd. By the *Brünig Railway* to *Meiringen* (*Gorge of the Aare; Pilatus* or *Brienzer Rothhorn* ½-1 day extra) and *Brienz*; by steamboat to the *Giessbach* and *Interlaken*.

4th. Railway to *Lauterbrunnen, Mürren*, and over the **Wengern-Alp** to *Grindelwald* (better partly on foot, taking another day).
5th. Over the *Great Scheidegg* to *Im Hof*.
6th. Through the *Hasli-Thal (Handegg Fall)* to the *Grimsel Hospice*.
7th. By the *Grimsel*, the *Rhone Glacier*, and the *Furka* to *Andermatt* or *Göschenen*.
8th. To *Flüelen, Lucerne*, and *Bâle*.

II. TWELVE OR FOURTEEN DAYS FROM BÂLE.
(Rigi, Bernese Oberland, Zermatt, Gemmi.)

1st-6th. As in Tour I.
7th. Over the *Grimsel* to the **Rhone Glacier**. Drive to *Fiesch*; walk or ride to the *Hôtel Jungfrau*.
8th. Ascend the *Eggishorn*; **walk viâ** the *Riederalp* to *Mörel*, drive to *Brig*. [Additional day: walk from the Riederalp to the *Belalp*; ascend the *Sparrenhorn*.]
9th. By railway to *Visp* and *Zermatt*.
10th. Ascend the *Riffelberg* and *Gorner Grat*, etc.
11th. Railway to *Visp* and *Loèche*; walk or drive to *Bad Leuk*.
12th. Over the *Gemmi* to *Kandersteg*; drive to *Spiez*; train to *Bern*.

III. SIXTEEN DAYS FROM BÂLE.
(Rigi, Bernese Oberland, Zermatt, Chamonix, **Lake of** *Geneva.)*

1st-9th. As in Tour II.
10th. By train to *Visp* and *Martigny*.
11th. Over the *Tête-Noire* or the *Col de Balme* to *Chamonix*.
12th. Excursions from *Chamonix*.
13th. By *Salvan* to *Vernayaz*; by train to *Montreux*.
14th, **15th**. To *Glion (Naye), Vevey, Lausanne*, and *Geneva*.
16th. To *Freiburg, Bern*, and *Bâle* (or from Bern to *Neuchâtel*).

IV. SEVENTEEN TO TWENTY DAYS FROM BÂLE.
(Rigi, Bernese **Oberland, Southern Valais***, Chamonix.)*

1st-8th. As in Tour II.
9th. Ascend the *Gorner Grat* and return **to** *St. Niklaus*.
10th. Cross the *Augstbord Pass* (ascent of *Schwarzhorn*) to *Gruben*.
11th. Cross the *Meiden* **Pass (ascent of *Bella* Tola**) to *St. Luc, Vissoye*, or *Zinal*.
12th. At *Zinal* (visit the *Alp Arpitetta*, etc.).
13th. Cross the *Col de Torrent* to *Evolena*.
14th, 15th. At *Evolena (Arolla* and *Ferpècle)*, and return to *Sion*.
16th, 17th. Cross the *Gemmi* to *Kandersteg* and *Thun* (or by railway to *Lausanne, Freiburg*, and *Bern*).
(Or: 15th. From *Evolena* to *Sion* and *Martigny*. 16th-20th. To *Chamonix, Geneva*, etc., as in Tour III.)

V. SEVEN DAYS FROM BÂLE.
(Bernese *Oberland*, **Rigi**, *St. Gotthard Railway*, *Italian* **Lakes.)**

1st. From *Bâle* to **Bern** and *Interlaken*.
2nd. To *Lauterbrunnen, Mürren*, and over the *Wengern-Alp* to **Grindelwald**.
3rd. Over the *Great Scheidegg* to *Meiringen*.
4th. Over the *Brünig* to *Alpnach-Stad* (ascent of *Pilatus*) and *Lucerne*.
5th. By the *St. Gotthard Railway* to *Laveno*; steamboat to *Stresa* (*Borromean Islands*).
6th. By *Luino* and *Lugano* to *Bellagio*.
7th. Steamer to *Como*; St. Gotthard Railway to Lucerne, etc.

VI. EIGHT OR TEN DAYS FROM BÂLE.
(Rigi, Lake of Lucerne, St. Gotthard, Italian Lakes. Splügen.)

1st. From *Bâle* to *Lucerne*, and by railway to the *Rigi-Kulm*.
2nd. Descend to *Vitznau*; steamer to *Brunnen (Axenstein, Rütli*, etc.).

(One or two additional days: visit the *Maderaner-Thal* from **Amsteg**, and return by the *Staffeln*. By train or carriage to *Göschenen*.)
- 3rd. By the *St. Gotthard* Line to *Locarno*.
- 4th. To the *Borromean Islands*, *Luino*, and *Lugano*.
- 5th. By *Como*, or by *Porlezza*, to *Bellagio*.
- 6th. Walks at Bellagio; steamer to *Colico*; drive to *Chiavenna*.
- 7th. Cross the *Splügen* to *Coire*.
- 8th. To **Zürich** and *Neuchâtel* (or to the *Falls of the Rhine* and *Bâle*).

VII. TWELVE TO FOURTEEN DAYS FROM BÂLE.

(Same as Tour VI., with the addition of the *Upper Engadine*.)
- 1st–5th. As in Tour VI.
- 6th. To *Chiavenna* and through the *Val Bregaglia* to **Casaccia**.
- 7th. Cross the *Maloja* to *St. Moritz* and *Pontresina*.
- 8th, 9th. At Pontresina (*Piz Languard*, etc.).
- 10th. Cross the *Albula* to *Tiefenkasten*.
- 11th. Through the *Schyn Pass* to *Thusis* (*Via Mala*) and *Coire*.
- 12th. To *Ragatz* (*Pfäfers*) and *Zürich*.

VIII. SIXTEEN TO EIGHTEEN DAYS FROM BÂLE.

(Same as Tour VII., with the addition of the *Valtellina* and *Lower Engadine*.)
- 1st–8th. As in Tour VII.
- 9th. Cross the *Bernina* to *Tirano*.
- 10th. Through the *Valtellina* to *Bormio*.
- 11th. Cross the *Wormser Joch* (*Piz Umbrail*) to *St. Maria* in the *Münster-Thal* (or cross the *Stelvio* to *Trafoi* and *Spondinig*).
- 12th. Over the *Ofen Pass* to *Zernets* (or drive by *Nauders* and *Martinsbruck* to *Schuls*).
- 13th. Cross the **Flüela Pass** to *Davos*.
- 14th. *Landwasser Route* to *Tiefenkasten*.
- 15th, 16th. As 11th and 12th of Tour VII.

IX. ONE MONTH FROM GENEVA.

(*Chamonix*, *Courmayeur*, *Zermatt*, *Macugnaga*, *Simplon*, *Upper Rhone Valley*, *Tosa Fall*, *St. Gotthard*, *Lake of Lucerne*, *Rigi*, *Bernese Oberland*.)
- 1st. From *Geneva* by steamer to *Chillon*, and by train to *Aigle*.
- 2nd. Drive to *Champéry*.
- 3rd. Cross the *Col de Coux* and *Col de Golèse* to **Samoëns** and **Sixt**.
- 4th. Cross the *Col d'Anterne* to *Chamonix*.
- 5th, 6th. At Chamonix; excursions.
- 7th. Cross the *Col de Voza* to *Contamines*.
- 8th. Cross the *Col de Bonhomme* and the *Col des Fours* to *Mottets*.
- 9th. Cross the *Col de la Seigne* to *Courmayeur* and *Aosta*.
- 10th. Railway to *Châtillon* and walk or ride to *Val Tournanche*.
- 11th. Cross the *Théodule Pass* to *Zermatt*.
- 12th, 13th. At Zermatt; excursions.
- 14th. To *Saas* and *Mattmark*.
- 15th. To *Macugnaga* by the *Monte Moro*.
- 16th. Walk or ride to *Piedimulera* (and thence, if time permit, devote a couple of days or more to the Italian Lakes).
- 17th. Cross the *Simplon* to *Brig*.
- 18th. Drive to *Fiesch*; ascend the *Eggishorn*.
- 19th. Drive to *Obergestelen* (perhaps visit the *Rhone Glacier* thence) and cross the *Gries Pass* to the *Fall of the Tosa*.
- 20th. Cross the *S. Giacomo Pass* to *Airolo*.
- 21st. By train to *Flüelen*; steamboat to *Vitznau*.
- 22nd. *Rigi*.
- 23rd. To *Lucerne*.
- 24th. Cross the *Brünig* to *Meiringen*.
- 25th. To *Rosenlaui* and *Grindelwald*.
- 26th. Cross the *Wengern-Alp* to *Lauterbrunnen* and *Mürren*.

27th. To *Interlaken*; visit *Giessbach*, etc.
28th. To *Thun*, *Bern*, and *Bâle*.

All the above tours are adapted for moderate walkers, and may of course be varied at pleasure.

Lastly, to travellers who are disinclined for a prolonged tour, the following notes may be acceptable: —

Famous Points of View.

1. In the **Jura** (with the Alps in **the distance, the** lower Swiss hills in the foreground, and, from **the westernmost** points, the lakes of Bienne, Neuchâtel, and Geneva) *Hôtel Schweizerhof* (p. 27), by the Falls of the Rhine; the *Weissenstein* (p. 16), near Soleure; the *Frohburg* (p. 14), near Olten; the *Chaumont* (p. 209) and the *Tête de Rang* (p. 210), in Canton Neuchâtel; the *Signal de Chexbres* (p. 218), the *Signal de Bougy* (p. 236), the *Dôle* (p. 235), **and the *Dent de Vaulion*** (p. 221), in the Canton de Vaud.

2. Nearer the Alps, or among the Lower Alps:

(a). On the N. side of the Alps: **the** *Kaien* (**p. 55**), *Hohe Kasten* (p. 59), and *Sentis* (p. 57), in Canton Appenzell; the *Uetliberg* (**p. 39) and** *Bachtel* (p. 44), near Zürich; the *Speer* (p. 45), near Weesen; the *Alvier* (**p. 47**), near Sargans; the *Rigi* (p. 94), *Pilatus* (p. 102), *Stanserhorn* (**p. 127**), *Myten* (p. 110), *Niederbauen* (p. 88), and *Pronalpstock* (p. 91), near **the Lake** of Lucerne; the *Napf* (**p. 139**), in the Emmen-Thal; the *Schänzli* (**p. 150**) and the *Gurten* (p. 151), near Bern; the *Moléson* (p. 253) **and *Jaman*** (**p. 254**), in Canton Freiburg; the *Salève* (p. 232) and the *Voirons* (**p. 233**), in Savoy, near Geneva; the *Rochers de Naye* (p. 248), near Glion; the *Chamossaire* (p. 246), near Villars.

(b). **On the** S. side of the Alps: *Monte Generoso* (p. 433), *Monte S. Salvatore* (p. 436), and *Monte Bré* (p. 437), near the Lake of Lugano; *Monte Mottarone* (**p. 450**) and *Monte Nudo* (p. 446), on Lago Maggiore; the *Monte S. Primo* (**p. 461**), near the Lake of Como; the *Becca di Nona* (p. 295), near Aosta; **the *Crammont* (p. 293)**, near Pré-St-Didier.

3. Among the **High Alps**: *Niesen* (**p. 154**), *Amisbühl* (p. 158), *Heimwehfluh* (p. 162), *Schynige Platte* (p. 164), *Mürren* (p. 168), *Schilthorn* (p. 168), *Obere Steinberg* (p. 167), *Wengern-Alp* (p. 173), *Lauberhorn* (p. 174), *Männlichen* (p. 177), *Faulhorn* (p. 179), *Brienzer Rothhorn* (p. 182), *Kleine Siedelhorn* (p. 189), and *Gemmi* (p. 194), **in the Bernese Oberland**; the *Pizzo Centrale* (p. 121), on the St. Gotthard; the *Furkahorn* (p. 126), *Eggishorn* (p. 316), *Sparrhorn* (p. 309), *Torrenthorn* (p. 195), *Pierre à Voir* (p. 250), *Mont Brûlé* (p. 300), *Gornergrat* (p. 334), *Schwarzhorn* (p. 331), *Bella Tola* (p. 328), and *Pic d'Arsinol* (p. 323), in the Valais; the *Col de Balme* (p. 287), *Flégère* (p. 279), and *Brévent* (p. 279), near Chamonix; *Piz Umbrail* (p. 426), on the Stelvio route; *Muottas Muraigl* (p. 406), *Schafberg* (p. 406), *Piz Languard* (p. 407), *Piz Ot* (p. 402), *Schwarzhorn* (p. 360), *Stätzerhorn* (p. 390), *Piz Mundaun* (p. 369), and *Piz Muraun* (p. 372), in the Grisons.

Principal Alpine Passes.

Pre-eminent in point of scenery is the *St. Gotthard* (RR. 32, 33), rendered easily accessible by the railway across it; but it need hardly be said that its attractions are not seen to advantage from the windows of a train. Next to it ranks the *Splügen* (RR. 95, 96), particularly on the N. side, where it coincides with the *Bernardino Route* (R. 97). The finest approach to the Engadine is by the *Schyn Road* (p. 379) and the *Albula Pass* (R. 98); and the beautiful *Maloja Pass* (RR. 100, 104) leads thence to the Lake of Como. From the Engadine the interesting *Bernina Pass* (R. 103) crosses to the somewhat monotonous Valtellina, the journey through which has, however, been much facilitated by the railway from Sondrio to Colico (p. 421). In Western Switzerland the *Simplon* (R. 80) is justly a favourite pass, though inferior to several of the above, while the famous *Great St. Bernard* (R. 78), apart from its hospice, is undoubtedly the least interesting of the series. Many of the grandest, and also easiest passes are comprised in the 9th of the above Tours.

II. TRAVELLING EXPENSES. MONEY.

Headquarters for Mountaineering.

The most important are *Grindelwald* (p. 175), *Zermatt* (p. 333), *Chamonix* (p. 276), *Courmayeur* (p. 292), *Macugnaga* (p. 344), and *Pontresina* (p. 402), at all of which experienced guides abound.

Health Resorts.

Switzerland can boast of few mineral springs, but 'Luftkurorte' ('air-cure places') and summer pensions abound in every part of the country. A few of the most important only need be mentioned here.

MINERAL BATHS. *Tarasp*, in the Lower Engadine (p. 416); *St. Moritz*, in the Upper Engadine (p. 398); *Ragatz* (p. 64); *Stachelberg* (p. 69); *Weissenburg* (p. 202); *Lenk* (p. 199); *Leuk* or *Loëche* (p. 194); the saline baths of *Bex* and *Aigle* (pp. 247, 216); *St. Gervais* (p. 278).

WINTER RESORTS for invalids: *Davos* (p. 361); *Montreux* (p. 242).

SUMMER RESORTS, see p. xviii.

Alpine Glow (*Alpenglühen*) is the name given to the rich glow seen on the snowy peaks and rocky summits of the Alps a few minutes after the setting sun has disappeared from view, while the valleys are already in twilight.

II. Travelling Expenses. Money.

Expenses. The cost of a tour in Switzerland depends of course upon the habits and tastes of the traveller. The pedestrian's daily expenditure, exclusive of guides, may be estimated at 12-15 fr., or even less, if he selects the more modest inns. The traveller, on the other hand, who prefers driving and riding to walking, who always goes to the best hotels, and never makes an ascent without a guide, must be prepared to spend at least twice the above sum; while the mountaineer's expenses will often amount to several pounds for a single glacier-expedition (comp. p. 281).

Money. The Swiss monetary system was assimilated to that of France in 1851. In gold there are coins of 20 fr., in silver of 5, 2, 1, and $^{1}/_{2}$ fr. (Those of 1859-63, with the sitting figure of Helvetia, which have been called in, Italian pieces of 2, 1, and $^{1}/_{2}$ fr., and Papal 1 fr. and $^{1}/_{2}$ fr. pieces should be declined; placards showing these illegal coins are hung up in every post-office.) In plated copper 20, 10, and 5 centimes (or 'Rappen'), and in copper 2 and 1 c. pieces. A few cantonal banks issue legal tender notes of 100 fr. and 50 fr. One franc = 100 c. = (in German money) 80 pfennigs = $9^{3}/_{4}d$. Twenty-franc-pieces are the most convenient money, and English sovereigns (25 fr.) and banknotes are received almost everywhere at the full value; but the circular notes of 10*l*. issued by many of the English banks, are safer for carrying large sums. German gold and banknotes also realize their full value (20 marks = 24 fr. 50-60 c.). — For *Savoy* (Chamonix) gold pieces or French banknotes are requisite. — In *Italy* the paper currency is much depreciated, and, as this is not always taken into account at hotels and railway-stations, it is advisable to provide oneself at a money-changer's with a supply of notes.

BAEDEKER, Switzerland. 16th Edition.   b

III. Hotels and Pensions.

Hotels. Switzerland is famous for its hotels. The large modern establishments at Geneva, Vevey, Zürich, Lucerne, Interlaken, etc., are models of organisation; the smaller hotels are often equally well conducted, and indeed a really bad inn is rarely met with in French or German Switzerland.

The ordinary charges at the first-class hotels are: bed-room from 2½ fr., candle 1 fr., service 1 fr.; breakfast (tea or coffee, bread, butter, and honey) 1½ fr. in the public room, 2 fr. in the traveller's apartment; luncheon ('déjeuner', 'Gabelfrühstück'), 3-3½ fr.; table d'hôte dinner ('dîner') 4-5 fr.; supper generally à la carte. Absence from table d'hôte is apt to be looked at askance. At the large hotels the best accommodation is generally reserved for families and parties, while the solitary traveller is consigned to the inferior rooms at equally high charges.

At the second-class inns the average charges are: bed-room from 1½-2 fr., breakfast 1-1¼ fr., table d'hôte 2-3 fr., service discretionary, and no charge for 'bougies'. In many of the more remote mountain-inns, however, the prices are higher owing to the difficulty and cost of the transport of supplies. The sensible traveller will easily make allowance for this; and he will generally find the entertainment remarkably good under the circumstances. Previous enquiry as to charges is quite customary.

Opinions regarding hotels often differ; but travellers will rarely have much cause to complain if they endeavour to comply with the customs of the country, restrict their luggage to a moderate quantity (p. xxvi), and learn enough of the language to make themselves intelligible.

If a prolonged stay is made at a hotel, the bill should be asked for every three or four days, in order that errors, whether accidental or designed, may more easily be detected. When an early departure is contemplated, the bill should be obtained over-night. It is not an uncommon practice to withhold the bill till the last moment, when the hurry and confusion of starting render overcharges less liable to discovery.

In the height of the season the hotels at the favourite resorts of travellers are often crowded. To prevent disappointment rooms should be telegraphed for (p. xxvi).

Most travellers err in giving too large Gratuities. When attendance is charged in the bill, nothing more need be given except to the boots and porter. In any case the amount of the fees should never exceed 5-10 per cent of the bill. In some of the best hotels the servants are forbidden to accept gratuities.

Many of the large hotels of Switzerland contain depots of the *Société des Bibliothèques des Grands Hôtels* (S. B. G. H.), a company formed for the sale of books (English, French, German) and maps in places not possessing a regular bookseller.

Pensions. Boarding-houses or 'pensions' abound at Lucerne, Geneva, Interlaken, and in many other parts of Switzerland; and most of the hotels also make pension arrangements with guests who stay for 4-5 days and upwards. The charge for board and lodging varies from 4½ to 10 fr. or more, and at some of the most famous

III. HOTELS AND PENSIONS.

health-resorts and watering-places sometimes amounts to 20 fr. per day. As the word 'pension' is sometimes used to signify board only, the traveller should ascertain whether rooms are included in the charge or not. It is always advantageous, when possible, to make arrangements for 'pension' in advance by writing to the landlord on a 'reply post-card'.

Among the Swiss Summer Resorts may be mentioned:—

In NORTHERN SWITZERLAND: The *Weissenstein* (4220'; p. 16), near Soleure; *Langenbruck* (2855'; p. 13) and *Frenkendorf* (1120'; p. 13), near Liestal; the *Frohburg* (2772'; p. 14), near Olten; the *Chaumont* (3845'; p. 209), near Neuchâtel; *Zürich* (1345'; p. 33) and the *Uetliberg* (2864'; p. 39); *Wädensweil* (1348'; p. 42) and other places on the Lake of Zürich (1342'); *Schönfels* and *Felsenegg* (3085'; p. 79), near Zug; *Ageri-Thal* (2380'; p. 80); *Weesen* (1410'; p. 45) and *Mürg* (p. 46), on the Walensee; *Obstalden* (2237'; p. 46), *Stachelberg* (2178'; p. 69), *Vorauen* (2640'), and *Richisau* (3590'), in the Klönthal (p. 74); the *Heinrichsbad* (2300'; p. 58), near Herisau; *Rorschach* (1312'; p. 52), *Walzenhausen* (2207'; p. 60), *Heiden* (2645'; p. 54), *Gais* (3075'; p. 56), and *Weissbad* (2080'; p. 56), in Appenzell.

On the LAKE OF LUCERNE (1435'): *Lucerne* (p. 81); *Meggen* (p. 104); *Hertenstein* (p. 87); *Weggis* (p. 87); *Beckenried* (p. 88); *Vitznau* (p. 87); *Gersau* (p. 89); *Brunnen* (p. 90); *Axenstein* (2460') and *Axenfels* (2065'; p. 91); *Seelisberg* (2628'; p. 89); *Bürgenstock* (2855'; p. 101); *Stoos* (4232'; p. 91); *Rigi-Klösterli* (4262'; p. 96), *Kaltbad* (4700'), *First* (4750'), *Staffel* (5262'), and *Scheidegg* (6405').

In CANTON LUCERNE: *Sonnenberg* (2560'; p. 86); *Schwarzenberg* (2760'; p. 138). In the EMMEN-THAL: *Rüttihubelbad* (2414'; p. 140). In UNTERWALDEN: *Engelberg* (3315'; p. 128); *Nieder-Rickenbach* (3830'; p. 128); *Melchsee-Frutt* (6115'; p. 132). In URI: *Amsteg* (1760'; p. 112); the *Maderaner-Thal* (4790'; p. 122); *Unterschächen* (3345'; p. 72); *Andermatt* (4738'; p. 119); *Hospenthal* (4800'; p. 120); *St. Gotthard* (6867'; p. 121).

In the BERNESE OBERLAND: *Bern* (1765'; p. 144); *Thun* (1844'; p. 151); *Oberhofen* (p. 156), *Gunten* (p. 156), *Spiez* (p. 156), and *Faulenseebad* (p. 157) on the Lake of Thun (1837'); *Aeschi* (2818'; p. 156); *Gurnigelbad* (3783'; p. 153); *Interlaken* (1863'; p. 158); *St. Beatenberg* (3775'; p. 158); *Abendberg* (3737'; p. 163); the *Giessbach* (1857'; p. 183); *Mürren* (5350'; p. 168); *Wengen* (4327'; p. 172); *Grindelwald* (3468'; p. 175); *Rosenlauibad* (4363'; p. 185); *Meiringen* (1968'; p. 181); *Engstlen-Alp* (6033'; p. 135); *Adelboden* (4450'; p. 196); *Kandersteg* (3840'; p. 190); *Lenk* (3527'; p. 199).

On the LAKE OF GENEVA, in the RHONE VALLEY, etc.: *Geneva* (1243'; p. 221); *Ouchy* (p. 236); *Lausanne* (p. 236); *Vevey* (p. 239); *Montreux* (p. 242); *Glion* (2254'; p. 242); *Aigle* (1375'; p. 246); *Bex* (1427'; p. 247); *Villars* (4166'; p. 246); the *Ormonts* (3815'; p. 251); *Gryon* (3632'; p. 250); *Château d'Oex* (3498'; p. 254); *Champéry* (3450'; p. 259); *Montana* (4045'; p. 308); *Fiesch* (3458'; p. 316); *Belalp* (7153'; p. 309); *Riederalp* (6315'; p. 317); *Eggishorn* (7195'; p. 316); *Berisal* (5005'; p. 311); *Zermatt* (5315'; p. 333); *Riffelalp* (7305'; p. 334), and *Riffelberg* (8430'; p. 334); *Saas in Grund* (5125'; p. 340); *Saas-Fee* (5900'; p. 341); *St. Luc* (5495'; p. 329); *Hôtel Weisshorn* (7550'; p. 328); *Zinal* (5505'; p. 328); *Evolena* (4520'; p. 322); *Chamonix* (3445'; p. 276).

In the GRISONS: *Samaden* (5670'; p. 401); *Pontresina* (5915'; p. 395); *St. Moritz* (6090'; p. 399); *Silvaplana* (5958'; p. 397); *Sils-Maria* (5895'; p. 396); *Maloja* (5960'; p. 394); *Zuoz* (5548'; p. 413); *Schuls* (3970'; p. 416); *Davos* (5115'; p. 361); *Arosa* (6035'; p. 365); *Klosters* (3966'; p. 358); *Seewis* (2885'; p. 357); *Waldhäuser* (3615'; p. 368), near Flims; *Thusis* (2448'; p. 378); *Disentis* (3773'; p. 372); *Wiesen* (4720'; p. 368); *Churwalden* (3976'; p. 390); *Parpan* (4956'; p. 396).

On the SOUTH SIDE OF THE ALPS: *Airolo* (3755'; p. 114), *Hôtel Piora* (6000'; p. 115), *Faido* (2485'; p. 116), and *Bignasco* (1424'; p. 442), in Ticino; *Macugnaga* (5115'; p. 344), *Alagna* (3955'; p. 346); *Gressoney* (5370'; p. 347); *Lugano* (982'; p. 434); *Bellagio* (p. 460), *Cadenabbia*, *Menaggio*, etc., on the Lake of Como (700'); *Locarno* (p. 440), *Pallanza* (p. 447), *Baveno* (p. 448), and *Stresa* (p. 449), on the Lago Maggiore (646'); *Monte Generoso* (3960'; p. 438) and *Lanzo d'Intelvi* (3117'; p. 456), near the Lake of Lugano.

b *

IV. Passports. Custom House.

Passports. In Switzerland passports are unnecessary, but as they must be shown in order to obtain delivery of registered letters, and are sometimes of service in proving the traveller's identity, it is unwise not to be provided with one. The principal passport-agents in London are: Lee and Carter, 440 West Strand; E. Stanford, 26 Cockspur St., Charing Cross; W. J. Adams, 59 Fleet Street; C. Smith & Son, 63 Charing Cross (charge 2s.; agent's fee 1s. 6d.).

Custom House. Luggage undergoes a slight examination at the Swiss frontier. The duty on cigars is 1½ fr. per 100. At the French, Italian, and Austrian frontiers the examination is sometimes strict, and tobacco and cigars pay a heavy duty, but at the German frontier the *visite* is usually lenient. As a rule the traveller should restrict his belongings as far as possible to wearing apparel and articles for personal use.

V. Walking Tours.

In a mountainous country like Switzerland it is to pedestrians alone that many of the finest points are accessible, and even where driving or riding is practicable, walking is often more enjoyable.

Disposition of Time. The first golden rule for the walker is to start early. If strength permits, and a suitable halting-place is to be met with, a walk of one or two hours may be accomplished before breakfast. At noon a moderate luncheon is preferable to a table d'hôte dinner. Rest should be taken during the hottest hours (12-3), and the journey then continued till 5 or 6 p.m. (comp. p. xviii), when a substantial meal (evening table d'hôte at the principal hotels) may be partaken of.

Equipment. A superabundance of luggage infallibly increases the delays, annoyances, and expenses of travel. To be provided with enough and no more, may be considered the second golden rule for the traveller. A light 'gibecière' or game-bag, which is far less irksome to carry than a knapsack, suffices to contain all that is necessary for a week's excursion. A change of flannel shirts and worsted stockings, a few pocket-handkerchiefs, a pair of slippers, and the 'objets de toilette' may, with a little practice, be carried with hardly a perceptible increase of fatigue. A pocket-knife with a corkscrew, a leathern drinking-cup, a spirit-flask, stout gloves, and a piece of green crape or coloured spectacles to protect the eyes from the glare of the snow, should not be forgotten. Useful, though less indispensable, are an opera-glass or small telescope, sewing materials, a supply of strong cord, sticking plaster, a small compass, a pocket-lantern, a thermometer, and an aneroid barometer. Special attention should be paid to the boots, which must be strong, well-tried, and thoroughly comfortable, as the slightest tendency to rub or blister

V. WALKING TOURS.

may seriously mar the enjoyment of the walk. For glacier-tours and mountain-ascents the soles must be supplied with nails, which, however, may be added on reaching the mountainous district. The traveller's reserve of clothing should be contained in a portmanteau of moderate size, which he can easily wield himself when necessary, and which may be forwarded from town to town by post.

The mountaineer should have a well-tried *Alpenstock* of seasoned ash, 5-6' long, shod with a steel point, and strong enough, when placed horizontally, with the ends supported, to bear the whole weight of the body. For the more difficult ascents an *Ice-Axe* and *Rope* are also necessary; the former may usually be borrowed at the hotel and the latter is generally furnished by the guide. The best ropes, light and strong, are made of silk or Manilla hemp. In crossing a glacier the precaution of using the rope should never be neglected. It should be securely tied round the waist of each member of the party, leaving a length of about 10' between each pair. Ice-axes are made in various forms, and are usually furnished with a spike at the end of the handle, so that they can in some measure be used like an Alpenstock. — Requisites for Alpine travelling may be obtained in London from *Carter*, 295 Oxford Street, or from *Adams & Sons*, 59 Fleet Street.

General Hints. The traveller's ambition often exceeds his powers of endurance, and if his strength be once overtaxed he will sometimes be incapacitated altogether for several days. At the outset, therefore, the walker's performances should be moderate; and even when he is in good training, they should rarely exceed 10 hrs. a day. When a mountain has to be breasted, the pedestrian should avoid 'spurts', and pursue the 'even tenor of his way' at a steady and moderate pace (*'chi va piano va sano; chi va sano va lontano'*). As another golden maxim for his guidance, the traveller should remember that — 'When fatigue begins, enjoyment ceases'.

To prevent the feet from blistering during a protracted walking tour, they may be rubbed morning and evening with brandy and tallow. A warm foot-bath with bran will be found soothing after a long day's march. Soaping the inside of the stocking is another well-known safeguard against abrasion of the skin.

Mountaineering among the higher Alps should not be attempted before the middle or end of July, nor at any period after a long continuance of rain or snow. Glaciers should be traversed as early in the morning as possible, before the sun softens the crust of ice formed during the night over the crevasses. Experienced guides are indispensable for such excursions.

The traveller is cautioned against sleeping in chalets, unless absolutely necessary. Whatever poetry there may be theoretically in 'a fragrant bed of hay', the cold night-air piercing abundant apertures, the ringing of the cow-bells, the grunting of the pigs, and the undiscarded garments, hardly conduce to refreshing slumber.

As a rule, therefore, the night previous to a mountain-expedition should be spent either at an inn or at one of the club-huts which the Swiss, German, and Italian Alpine Clubs have recently erected for the convenience of travellers.

Mountaineers should provide themselves with fresh meat, bread, and **wine** or spirits for long expeditions. The chalets usually afford nothing but Alpine fare (milk, cheese, and stale bread). Glacier-water should not be drunk except in small quantities, mixed with wine or cognac. Cold milk is also safer when qualified with **spirits**. One of the best beverages for quenching the thirst is cold tea.

Over all the **movements of the pedestrian** the weather holds despotic sway. The barometer and weather-wise natives should be consulted when an opportunity offers. The blowing down of the wind from the mountains into the valleys in the evening, **the melting away of the clouds, the fall of fresh snow on the mountains, and the ascent of the cattle to the higher parts of their pasture are all signs of fine weather.** On the other hand **it is a bad sign if the distant mountains are dark blue in colour and very distinct in outline, if the wind blows up the mountains, and if the dust rises in eddies on the roads.** West winds also usually bring rain.

Health. Tincture of arnica is a good remedy for *bruises*, and moreover has **a** bracing and invigorating effect if rubbed on the limbs after much fatigue; but it should never be applied to broken skin, as it is apt to produce erysipelas. Saturnine ointment or oxide of zinc ointment is beneficial in cases of inflammation of the skin, an inconvenience frequently caused by exposure to the glare of the sun on the snow. Cold cream, and, for the lips especially, vaseline or glycerine, are also recommended.

For *diarrhœa* 15 drops of **a** mixture of equal parts of tincture of opium and aromatic tincture may be safely taken every two hours until relief is afforded. The homœopathic tincture of camphor (5 drops on a lump of sugar every half-hour **or** so) is also **a** good remedy. The homœopathic camphor-globules **are** convenient, but are more apt to lose their strength.

VI. Maps.

1. MAPS ON A LARGE SCALE: —

Topographische Atlas der Schweiz, on **the scale of the** original drawings (flat districts 1 : 25,000, mountains 1 : 50,000), published by the Federal Staff Office **under** the superintendence of *Col. Siegfried* and known as the 'Siegfried Atlas'. The conformation of the ground is indicated by contour-lines at intervals of 10 and 30 mètres. Price, 1 fr. per sheet; four sheets in one, lithographed, 2 fr., mounted 3 fr. 30 c. Some of the more important districts are published in a special edition, in which the **system** of contour-lines is combined with **graduated** colouring (price 5 fr., mounted 6 fr. 30 c.). Key-plans, showing **the** extent of the **different** sheets, may be obtained gratis on application to Schmid, Franke, & Co. of Bern, Georg & Co. of Geneva, and other booksellers.

The four-sheet lithographs include Zürich and environs, Bern and environs, Thun and environs, Thun with the Stockhorn and Niesen district, Stockhorn chain and Jann-Thun, *Bernese Oberland I and II, Thun-Interlaken, Brienz-Guttannen, Jungfrau and Upper Valais, Gemmi and Blümlisalp, Evolena-Zermatt-Mte-Rosa, *Upper Engadine, *Albula district, and the *St. Gotthard.

Older than the above is the *Topographische Karte der Schweiz*, also from surveys made by order of the Federal authorities (under the superintendence of *General Dufour*); scale 1:100,000; 25 sheets, each 1 to 2 fr. (not mounted).

For Chamonix, *Reilly's* Map of Mont Blanc, and *Mieulet's* Massif du Montblanc (1:40,000).

2. MAPS ON A SMALLER SCALE: —

Generalkarte der Schweiz (1:250,000), reduced from Gen. Dufour's map (see above); four sheets at 2 fr., mounted 3 fr. 30 c.

Leusinger's Neue Karte der Schweiz (1:400,000); mounted, 8 fr.

Leusinger's Reise-Reliefkarte der Schweiz (1 630,000); mounted, 5 fr.

Müllhaupt's Karte (1:300,000); two sheets at 4 fr.

Distanzkarte der Schweiz in Marschstunden (1:500,000), 3 fr. 50 c.

The *Alpine Club Map of Switzerland*, published by R. C. Nichols (1:250,000); four sheets, 42s.

VII. Guides.

On well-trodden routes like those of the Rigi, Pilatus, Wengern-Alp, Faulhorn, Scheidegg, Grimsel, Gemmi, etc., the services of a guide are unnecessary in good weather; the maps and directions of the Handbook will be found entirely sufficient. The traveller may engage the first urchin he meets to carry his bag or knapsack for a trifling gratuity. Guides are, however, indispensable for expeditions among the higher mountains, especially on those which involve the passage of glaciers. The novice alone undervalues their services and forgets that snow-storms or mist may at any moment change security to danger. As a class, the Swiss guides will be found to be intelligent and respectable men, well versed in their duties, and acquainted with the people and resources of the country.

The great stations for guides are Interlaken, Lauterbrunnen, Grindelwald, Meiringen, Martigny, Chamonix, Courmayeur, Zermatt, and Pontresina, while for the principal passes guides are always to be found at the neighbouring villages. The traveller should select one of the *certificated* guides, who have passed an examination, and are furnished with legal certificates of character and qualifications. The usual pay of a guide is 6-8 fr. for a day of 8 hrs.; he is bound to carry 15-18 pounds of baggage, and to hold himself at the entire disposition of his employers. If dismissed at a distance from home, he is entitled to 6 fr. a day for the return-journey; but he is bound to return by the shortest practicable route.

Although a guide adds considerably to the traveller's expenses, the outlay will seldom be regretted. A good guide points out many objects which the best maps fail to indicate; he furnishes interesting information about manners and customs, battle-fields, and historical incidents; and when the traveller reaches his hotel, wearied with the fatigues of the day, his guide often renders him valuable service. It need hardly be said that a certain amount of good fellowship and confidence should subsist between the traveller and the man who is perhaps to be his sole companion for several days, and upon whose skill and **experience** his very life not unfrequently depends.

Divided among a party, the expense of a guide is of course greatly **diminished**; but **where** there is much luggage to carry, it is often better to **hire a horse** or mule, the attendant of which will serve as a guide on the ordinary routes.

Adult porters are entitled to 75 cent. or 1 fr. an hour, when not engaged by the day, **return** included. In every case it is advisable to **make** a distinct **bargain beforehand.**

VIII. Carriages and Horses.

Carriages. The ordinary charge for a carriage with one horse is 15-20 fr., with two horses 25-30 fr. per day, and the driver expects 10 per cent of the fare as a gratuity. In the height of summer the charges are slightly increased. In most cases there is now an official tariff, which also fixes the amount to be paid as the return-fare to the place where the driver was engaged. When this is not fixed, the driver is entitled to claim the full rate for his return-journey by the shortest route, a day being reckoned as 12 hrs.' driving. On the most frequented routes carriages may generally be ordered at the hotels, but it is usually more advantageous to deal personally with the driver. The carriage and horses should be inspected before the conclusion of the bargain. When the bargain is made for a future day the driver usually deposits a sum with his employer as earnest-money *(arrhes, caparra)*, afterwards to be added to the account. The hirer selects the hotels at which the nights are to be passed. Private posting, or the **system** of changing horses, is forbidden by law. Return-vehicles may sometimes be obtained for 10 to 15 fr. per day, but the use of them is in some places prohibited.

Horses. A horse or mule costs 10-12 fr. per day, **and the** attendant expects a gratuity of 1-2 fr. in addition; but in some places, **as at Chamonix, as** much is charged for the attendant as for the **animal.** If he cannot return home with his horse on the same day, the following day must be paid for. Walking, however, is preferable. A prolonged ascent on horseback is fatiguing, and the descent of a steep hill is disagreeable. Even ladies may easily ascend some of the finest points of view on foot, but if unequal to the task they may either ride or **engage** 'chaises-à-porteurs'. In the Bernese Oberland, however, the numerous mountain-railways make horses and chaises-à-porteurs alike superfluous.

IX. Diligences, Post Office, Telegraph.

Diligences. The Swiss coaching system is well organised. The diligences are generally well fitted up, the drivers and guards are respectable, and the fares moderate. These vehicles consist of the *coupé*, or first-class compartment in front, with 2-3 seats, the *intérieur*, or second-class compartment at the back, with 4-6 seats, which affords little or no view, and the *banquette* (used in summer only) for 2 passengers on the outside. In some cases there is only one outside-seat, which is reserved for the *conducteur*, or guard, but will be ceded by him on payment of the difference between the ordinary and the coupé fare. At the most important places, but not at all the intermediate stations, the traveller has a right to insist on transportation; and 'Beiwagen', or supplementary carriages, are supplied when the diligence is full. When there are many passengers it is advisable to keep an eye on one's luggage (see below), especially at a change of carriage.

On important routes the coupé is generally engaged several days beforehand. This may be done by letter or telegraph, giving the traveller's name, and the day and hour of departure. The fare must also be forwarded.

The *coupé* or *banquette* fare is on ordinary routes 20 c. per kilomètre (about 32 c. per Engl. M.), on Alpine passes 30 c. per kilom. (about 48 c. per Engl. M.); fare in the *intérieur* or *cabriolet* 15 or 25 c. per kilomètre (24 or 40 per Engl. M.). Children of 2-7 years of age pay half-fare. Each passenger is allowed 33 lbs. of luggage on ordinary routes, but 22 lbs. only on the high Alpine routes. Overweight is charged for at the ordinary postal tariff. Small articles may be taken into the carriage, but heavy luggage should be booked one hour before starting. The average speed of these sedate mail-coaches of Switzerland is about 6 M. per hour on level ground, and 4 M. per hour on mountain-routes.

Extra-Post. This is the term applied to the Swiss system of posting, managed by government, private posting being prohibited. The charge for each horse is 1/2 fr. per kilomètre (80 c. per M.); for a carriage with 2-3 seats 20 c. per kilom. (32 c. per M.), for one with 6 seats 25 c. per kilom. (40 c. per M.), for one with 7 or more seats 30 c. per kilom. (48 c. per M.). Besides these charges, a booking-fee of 2-4 fr. must be paid according to the size of the carriage. If the same vehicle is required for a journey of several stages, double-carriage-money is exacted. The postilions are strictly forbidden to demand gratuities. Extra-post may be ordered at the principal post-offices on the mountain-routes at one hour's notice. The fare must be paid in advance.

Letters of 250 grammes (about $8^1/_2$ oz.), prepaid, to any part of Switzerland 10 c.; if within a radius of 10 kilomètres, 5 c.; letters of 15 grammes (about $1/_2$ oz.) to all countries in the postal union 25 c., and 25 c. for each 15 gr. more. Registration-fee for Switzerland 10 c., for other countries 25 c. — Post-cards for Switzerland 5 c., for other countries 10 c. — Printed matter under 50 gr. for Switzerland 2 c., for other countries 5 c.

Post Office Orders within Switzerland must not exceed 1000 fr. for the larger, and 500 fr. for the smaller towns. The charge for an order not exceeding 20 fr. is 15 c., for 100 fr. 20 c., for each additional 100 fr. 10 c. more. Money-orders for foreign countries 25 c. for every 100 fr. (with a minimum fee of 50 c.). Money-orders, up to 200 fr., may also be transmitted by telegraph, at the ordinary money-order rate plus the cost of the telegram and a small extra fee.

Parcel Post. The rate of postage for an inland parcel from any post-office in Switzerland to any other is 15 c. for a weight not exceeding 500 grammes (1^1/$_{10}$ lb.); 25 c. from 500 to 2500 gr.; 40 c. from 2500 gr. to 5 kilogrammes (11 lb.); 70 c. from 5 to 10 kgr.; 1 fr. from 10 to 15 kgr.; 1 fr. 50 c. from 15 to 20 kgr. The tariff for parcels exceeding 20 kgr. varies according to the distance from 30 c. to 1 fr. 20 c. for every 5 kgr. Luggage can often be sent by post much more cheaply than by other means.

The **Telegraph System** of Switzerland is very complete, the aggregate length of the wires being at present greater than in any other country in proportion to the population. There are now upwards of 1000 offices; those in the large towns are open from 6 or 7 a.m. till 11 or 10 p.m. according to the season. The tariff for a telegram within Switzerland is 30 c., together with 2^1/$_2$ c. for each word; to Germany 50 c. and 10 c. for each word; to England 29 c. for each word; to France 10 c. for each word; to Italy 10 c. per word for telegrams to the frontier, or 17 c. for greater distances; to Austria 10 c. (Tyrol or Vorarlberg 7 c.) per word; to the United States from 1 fr. 50 c. per word. The rates for other foreign telegrams may be ascertained at the offices. For telegrams handed in at railway-stations an additional charge of 50 c. is made. Telegrams may be handed in at any post-office, from which, if not itself a telegraph-office, they are transmitted without delay to the nearest. In such cases the fee for the telegram is paid by affixing stamps of the requisite value. If in an envelope, the word 'telegram' should be added to the address.

X. Railways.

The **Carriages** on most of the Swiss lines are constructed on the American plan, holding 32-72 passengers, and furnished at each end with steps of easy access. Through each carriage, and indeed through the whole train, runs a passage, on each side of which the seats are disposed. This arrangement enables the traveller to change his seat at pleasure, and to see the scenery to advantage, unless the carriage is very full. Tickets are examined and collected in the carriages. — In French Switzerland passengers' tickets are checked as they leave the waiting-room before starting, and given up at the '*Sortie*' on their arrival.

Luggage must be booked and paid for after the traveller has obtained his own ticket, but small portmanteaus and travelling-bags may generally be taken into the carriage without objection. Indeed the forbearance of the Swiss railway officials in this respect is shamefully abused by inconsiderate travellers. Travellers with through-tickets from the German to the Swiss railways, or vice versâ, should see that their luggage is safe on reaching the frontier (Bâle, Geneva, Neuchâtel, Friedrichshafen, **Lindau**, Rorschach, Romanshorn, etc.). Where a frontier has to be crossed, ordinary luggage should never be sent by **goods-train**.

The enormous weight of the large trunks and boxes used by some travellers causes not only great labour but not infrequently serious and even lifelong injury to the railway and hotel porters who have to handle them. Heavy articles should be placed in the smaller packages, and only the lightest articles in the larger trunks.

Circular Tickets and return-tickets are issued at reduced rates on most of the Swiss lines, and also by **the German and French railways to Switzerland**. Information regarding them will be found in the time-tables; but they are apt to hamper the traveller's movements and to deprive him of the independence essential to enjoyment.

XI. History. Statistics.

The limits of this work preclude more than a brief historical sketch of the interesting country the traveller is now visiting, whose inhabitants have ever been noted for their spirit of freedom and independence.

Switzerland is believed to have been first peopled by the *Rhaeti*, who were driven from the plains to the mountains by the *Helvetii*, a Celtic tribe. The latter were conquered by the *Romans*, B. C. 58, and the Rhaeti were subdued in B. C. 15. The Romans made good military roads over the Great St. Bernard (p. 301) to Bâle, and over the Julier (p. 398), Septimer (p. 392), and Splügen (p. 383) to Bregenz (p. 432), and thence to Bâle. The chief settlements were *Aventicum* (Avenches, p. 219) in the Canton of Vaud, *Vindonissa* (p. 20) at the confluence of the Aare, Reuss, and Limmat, *Augusta Rauracorum* (Augst, p. 19) near Bâle, and *Curia Rhaetorum* (Coire, p. 354) in the Grisons. E. Switzerland as far as Pfyn (*ad fines*) in Thurgau, and *Pfyn* (p. 308) in the Upper Valais, belonged to the province of Rhaetia, while W. Switzerland formed part of Gaul. The name Helvetii had become extinct even before the time of Constantine. Under the Roman sway Helvetia enjoyed a flourishing trade, which covered the land with cities and villages. A trace of that period exists in the Romanic dialect, which is still spoken in some parts of Switzerland.

About A.D. 400 a great irruption of barbarians swept through the peaceful valleys of the Alps, and Huns, Burgundians, Alemanni, and Ostrogoths in succession settled in different parts of the country. The *Alemanni* occupied the whole of N. Switzerland, where German is now spoken; the *Burgundians* the W. part, where French is spoken; and the *Ostrogoths* S. Switzerland, where Italian and Romansch are now spoken. These races were gradually subdued by the *Franks*, who, however, did not take possession of the country themselves, but governed it by their officers. During this period Christianity was introduced, the monasteries of *Disentis* (p. 372), *St. Gallen* (p. 50), *Einsiedeln* (p. 106), and *Beromünster* were founded, and dukes and counts were appointed as vicegerents of the Franconian kings.

After the dissolution of the great Franconian empire, the E. half of Switzerland, the boundary of which extended from Eglisau over the Albis to Lucerne and the Grimsel, was united with the duchy of *Alemannia* or *Swabia*, and the W. part with the kingdom of *Burgundy* (912). After the downfall of the latter (1032) the *German Emperors* took possession of the country, and governed it by their vicegerents the dukes of Zæhringen (p. 145), who were perpetually at enmity with the Burgundian nobles and therefore favoured the inhabitants of the towns, and were themselves the founders of several new towns, such as Freiburg, Bern, and Burgdorf.

As the power of the emperors declined, and the nobles, spiritual and temporal, became more ambitious of independence, and more eager to fill their coffers at the expense of their neighbours, the Swiss towns and the few country-people who had succeeded in preserving their freedom from serfdom were compelled to consult their safety by entering into treaties with the feudal lords of the soil. Thus the inhabitants of Zürich placed themselves under the protection of the then unimportant *Counts of Hapsburg*, with whom the 'Three Cantons' of Uri, Schwyz, and Unterwalden were also allied. In 1231 and 1240 letters of independence were granted by Emperor Frederick II. to Uri and Schwyz, and after *Count Rudolph of Hapsburg* had become emperor he confirmed the privileges of the former in 1274, while Schwyz and Unterwalden still continued subject to the Hapsburg supremacy.

After the emperor's death in 1291 the Forest Cantons **formed their first league** for mutual safety and the protection of their liberty against the growing power of the House of Hapsburg. Rudolph's son *Albert* in particular endeavoured to rear the limited rights he enjoyed in these districts into absolute sovereignty, and to incorporate them with his empire.

The ancient cantons therefore embraced the cause of the rival monarch *Adolph of Nassau*, **who** confirmed their privileges. Victory, however, favoured Albert, **who** again deprived the cantons of their privileges, but does not appear to have treated them with much severity. To this period belongs the romantic but unfounded tradition of William Tell.†

After the assassination of Albert by John of Swabia in 1308, *Emperor Henry VII.*, who was also an opponent of the Hapsburgers, conferred **a charter** of independence on the Forest Cantons. The House of Hapsburg regarded this as an infringement of their rights, and sent a powerful army against these cantons, which after the death of Henry had declared their adherence to Lewis the Bavarian, the opponent of Frederick the Handsome. This army was destroyed at *Morgarten* (p. 80) in 1315. Subsequent attempts to subject the country to the supremacy of the House of Hapsburg were frustrated by the victories of the Swiss at *Sempach* (p. 19) in 1386, at *Näfels* (p. 67) in 1388, and at the *Stoss* (p. 56) in 1405.

In the Burgundian parts **of the** country too the nobility were jealous of the increasing importance of the towns, and therefore attempted to conquer Bern, but were defeated by the citizens at *Laupen* (p. 215) in 1339.

In **1354 a** confederacy was formed by eight independent districts and **towns**, which soon **became** powerful enough to assume the offensive, and **at length** actually wrested the hereditary domain **of** Hapsburg from the dukes of Austria, who tried in vain to recover **it.**

Even *Charles the Bold*, Duke of Burgundy, **the** mightiest prince of his time, **was defeated by** the Swiss at the three battles of *Grandson* (1476, p. 214), *Morat* (**1476, p.** 219), and *Nancy*, while at **an** earlier period a large body of irregular French and other troops, which had been made over to Austria by the **King of** France, sustained **a severe** check from the confederates at *St. Jacob* on the Birs (1444, p. 8).

In the Swabian **war** (1499) the bravery and unity of the Swiss achieved another triumph **in the** victory of *Dornach* (p. 10). At that period their independence of **the emperor was** formally recognised, but they continued nominally attached to **the empire** down to 1648.

The last-named victory formed a fitting termination to a successful career of two centuries, the most glorious in the history of Switzerland. At the beginning of the **16th** century a period of decline set in. The **enormous** booty captured **in the** Burgundian war had begotten a taste for wealth and luxury, the demoralising practice of serving as mercenary troops in foreign lands **began to** prevail, and a foundation was laid for the reproachful proverb, '**Pas** d'argent, pas de Suisses!'

† **The** legend of the national hero of Switzerland, as well as the story of the expulsion of the Austrian bailiffs in 1308, is destitute of historical foundation. No trace of such a person is to be found in the work of John of Winterthur (Vitodurunus, 1349) or that of Conrad Justinger of Bern (1420), the earliest Swiss historians. Mention is made of him for the first time in the Sarner Chronik of 1470, and the myth was subsequently embellished by Ægidius Tschudi of Glarus (d. 1342), and still more by Johann v. Müller (d. 1809), while Schiller's famous play has finally secured to the hero a world-wide celebrity. Similar traditions are met with among various northern nations, such as the Danes and Icelanders.

The cause of the Reformation under the auspices of Zwingli was zealously embraced by a large proportion of the population of Switzerland about the beginning of the 16th century; but the bitter jealousies thus sown between the Roman Catholic and the Reformed Cantons were attended with most disastrous consequences, and in the civil **wars** which ensued bloody battles were fought at *Kappel* (p. 79) in 1531, at *Villmergen* in 1656, and during the Toggenburg war (p. 62) in 1712.

Traces of unflinching bravery and of a noble spirit of **self-sacrifice in the cause of** conscience are observable in individual instances even at the close **of** the 18th century, as exemplified by the affairs of *Rothenthurm* (p. **107**) and *Stans* (p. 127), but the national vigour was gone. The resistance **of** individuals to the invasion of the French republicans proved fruitless, and the *Helvetian Republic* was founded on the ruins of the ancient liberties of the nation. In 1803 Napoleon restored the cantonal system, and in accordance with resolutions passed by the Congress **of** Vienna in 1815 the constitution was remodelled. The changes introduced **in** consequence of the revolution of July, 1830, were unhappily the **forerunners of** the civil war of the Sonderbund, or Separate League, in **November, 1847;** but this was of short duration, and on 12th September, **1848, a new** federal constitution was inaugurated. Since that period the **public tranquillity** has been undisturbed, and the prosperity and harmony **which now** prevail throughout the country **are** not unworthy of the glorious **traditions** of the past.

Area and Population
according to the census of 1st Dec., 1888.

Cantons.	Sq. Leagues	Confession.				Totals.
		Rom.Cath.	Prot.	Jews	Sects	
1. *Zürich*	74,8	40,408	294,356	1,416	2,960	339,014
2. *Bern*	294	65,246	468,120	1,245	1,604	539,305
3. *Lucerne*	54	127,583	7,939	215	93	135,780
4. *Uri*	47	16,892	378	3	11	17,284
5. *Schwyz*	40	49,289	1,067	2	8	50,396
6. *Unterwald*	33,5	27,096	457	—	3	27,556
7. *Glarus*	20,8	7,790	25,935	15	60	33,800
8. *Zug*	10,2	21,696	1,394	18	12	23,120
9. *Freiburg*	71,1	100,426	18,869	127	42	119,562
10. *Soleure*	34,5	63,539	21,898	154	125	85,720
11. *Bâle-ville*	1,5	22,402	50,326	1,078	441	74,247
Bâle-camp.	18,5	12,961	48,847	165	160	62,133
12. *Schaffhausen*	12,9	4,813	32,587	26	150	37,576
13. *Appenzell*						
(Rhodes ext.)	10,7	4,502	49,555	26	117	54,200
(Rhodes int.)	7,3	12,906	697	—	3	12,906
14. *St. Gallen*	87,7	135,706	93,705	575	355	229,441
15. *Grisons*	304,1	43,320	52,842	43	86	96,291
16. *Aargau*	60,4	85,962	100,414	1,064	394	193,834
17. *Thurgau*	42,3	30,337	74,782	61	411	105,091
18. *Ticino*	121,6	125,622	1,079	13	434	127,148
19. *Vaud*	138,7	22,428	227,475	638	755	251,296
20. *Valais*	226,5	100,925	805	3	44	101,887
21. *Neuchâtel*	34,7	12,689	95,040	774	584	109,087
22. *Geneva*	12,2	52,692	51,532	723	1,791	106,738
Total	1769,8	1,189,662	1,724,569	8,384	10,697	2,933,612
Census of 1880	—	1,161,055	1,666,984	7,380	10,683	2,846,102
Increase	—	28,607	57,385	1,004	14	117,510

XII. Comparative Tables of Measures.

Engl. Feet	Mètres	Mètres	Engl. Feet	Engl. Miles	Kilomètres	Kilomètres	Engl. Miles	Acres	Hectares	Hectares	Acres
1	0,30	1	3,28	1	1,61	1	0,62	1	0,40	1	2,47
2	0,61	2	6,56	2	3,22	2	1,24	2	0,81	2	4,94
3	0,91	3	9,84	3	4,83	3	1,86	3	1,21	3	7,41
4	1,22	4	13,12	4	6,44	4	2,48	4	1,61	4	9,88
5	1,52	5	16,40	5	8,04	5	3,10	5	2,02	5	12,35
6	1,83	6	19,69	6	9,65	6	3,73	6	2,42	6	14,82
7	2,13	7	22,97	7	11,26	7	4,35	7	2,83	7	17,30
8	2,44	8	26,25	8	12,87	8	4,97	8	3,23	8	19,77
9	2,74	9	29,53	9	14,58	9	5,59	9	3,63	9	22,24
10	3,04	10	32,81	10	16,09	10	6,21	10	4,04	10	24,71
11	3,35	11	36,09	11	17,70	11	6,83	11	4,44	11	27,19
12	3,66	12	39,37	12	19,31	12	7,45	12	4,85	12	29,65
13	3,96	13	42,65	13	20,92	13	8,07	13	5,25	13	32,12
14	4,27	14	45,93	14	22,53	14	8,69	14	5,66	14	34,59
15	4,57	15	49,21	15	24,13	15	9,31	15	6,06	15	37,05
16	4,88	16	52,49	16	25,74	16	9,93	16	6,46	16	39,58
17	5,18	17	55,78	17	27,35	17	10,55	17	6,87	17	42,00
18	5,49	18	59,06	18	28,96	18	11,18	18	7,27	18	44,47
19	5,79	19	62,34	19	30,67	19	11,80	19	7,67	19	46,95
20	6,10	20	65,62	20	32,18	20	12,42	20	8,08	20	49,42

Thermometric Scales.

Réaumur	Fahrenheit	Celsius	Réaumur	Fahrenheit	Celsius	Réaumur	Fahrenheit	Celsius	Réaumur	Fahrenheit	Celsius
+30,22	+100	+37,78	+21,78	+81	+27,22	+13,33	+62	+16,67	+4,89	+43	+6,11
29,78	99	37,22	21,33	80	26,67	12,89	61	16,11	4,44	42	5,56
29,33	98	36,67	20,89	79	26,11	12,44	60	15,56	4,00	41	5,00
28,89	97	36,11	20,44	78	25,56	12,00	59	15,00	3,56	40	4,44
28,44	96	35,56	20,00	77	25,00	11,56	58	14,44	3,11	39	3,89
28,00	95	35,00	19,56	76	24,44	11,11	57	13,89	2,67	38	3,33
27,56	94	34,44	19,11	75	23,89	10,67	56	13,33	2,22	37	2,78
27,11	93	33,89	18,67	74	23,33	10,22	55	12,78	1,78	36	2,22
26,67	92	33,33	18,22	73	22,78	9,78	54	12,22	1,33	35	1,61
26,22	91	32,78	17,78	72	22,22	9,33	53	11,67	0,89	34	1,11
25,78	90	32,22	17,33	71	21,67	8,89	52	11,11	0,44	33	0,56
25,33	89	31,67	16,89	70	21,11	8,44	51	10,56	0,00	32	0,00
24,89	88	31,11	16,44	69	20,56	8,00	50	10,00	−0,44	31	−0,56
24,44	87	30,56	16,00	68	20,00	7,56	49	9,44	0,89	30	1,11
24,00	86	30,00	15,56	67	19,44	7,11	48	8,89	1,33	29	1,67
23,56	85	29,44	15,11	66	18,89	6,67	47	8,33	1,78	28	2,22
23,11	84	28,89	14,67	65	18,33	6,22	46	7,78	2,22	27	2,78
22,67	83	28,33	14,22	64	17,78	5,78	45	7,22	2,67	26	3,33
22,22	82	27,78	13,78	63	17,22	5,33	44	6,67	3,11	25	3,89

1. NORTHERN SWITZERLAND.

1. Bâle 2
   From Bâle through the Birsigthal to **Flühen.** Landskron; Mariastein; Blauen, 9.
2. From Bâle to Bienne and Bern through the Münster-Thal . . . 9
   From Delémont to Porrentruy, 10. — Ascent of the Weissenstein from Münster. From Bévilard over the Montoz to Reuchenette, 11. — The Taubenloch-Schlucht. Macolin. Ascent of the Chasseral, 12.
3. From Bâle to Bienne viâ Olten and Soleure . . . 13
   From Liestal to Waldenburg. Langenbruck. The Schafmatt, 13. — Eptingen; Frohburg; Neu-Wartburg; Lostorf; Friedau, 14. — From Soleure to the Weissenstein, 16. — From Soleure to Burgdorf; to Lyss, 17.
4. From Bâle to **Bern** viâ Herzogenbuchsee 17
   From Herzogenbuchsee to Soleure, 17. — From Burgdorf to Langnau, 18.
5. **From** Bâle to Lucerne 18
   From Zofingen **to Suhr,** 18.
6. From Bâle to Zürich 19
   From Stein to Coblenz. Königsfelden; Vindonissa. From Brugg to Wohlen, 20. — From Wettingen to Oerlikon, 21.
7. **From Olten to** Waldshut viâ Aarau and Brugg . . . 22
   From Aarau to Baden, 22. — The Habsburg, 23.
8. **From** Bâle to Schaffhausen and Constance . . . 23
   From Singen to Etzweilen. The Island of Reichenau. Steamboat from Schaffhausen to Constance, 25.
9. The Falls of the Rhine 26
10. From Friedrichshafen to Constance. **Lake of Constance** 28
    The Mainau, 31.
11. From Rorschach to Constance and Winterthur (Zürich) 31
    From Etzweilen to Feuerthalen (Schaffhausen), 32.
12. **From** Schaffhausen **to** Zürich 32
13. Zürich and **the** Uetliberg 33
14. From Zürich to Coire. Lakes of Zürich **and Walenstadt** 40
    a. N.E. Railway from Zürich **to Meilen and Rapperswil** (Right Bank) 41
    The Pfannenstiel, 41.
    b. **N.E.** Railway from **Zürich to Ziegelbrücke (Left Bank)** 42
    The Wäggithal, 43.
    c. Railway from Zürich **to** Rapperswil and Sargans 43
    The Bachtel; Rieden, 44. — Biberlikopf; Amden; Speer, 45. — From Mühlehorn over the Kerenzenberg to Mollis. Mürtschenstock; the Murgthal; the Rothhor, 46. — The Widerstein-Furkel and Murgsee-Furkel. From Walenstadt over the Käserruck to Wildhaus in the Toggenburg. The Alvier. From Mels through the Weisstannen-Thal and Kalfeisen-Thal to Vättis, 47.

BAEDEKER, Switzerland. 16th Edition. 1

15. From Zürich to Romanshorn and Friedrichshafen . 48
From Oerlikon to Dielsdorf. Regensberg. From Winterthur to Waldshut. From Winterthur to Rüti (Tösstbal Railway), 48. — From Frauenfeld to Wyl. From Sulgen to Gossau, 49.

16. From Zürich to St. Gallen, Rorschach, and Lindau . 49
From Winkeln to Appenzell, 50. — Excursions from St. Gallen: Freudenberg; Rosenberg; Falkenburg, etc., 51. — Excursions from Rorschach: the Martinstobel; the Möttelischloss; Weinburg; Horn, 52. — Excursions from Lindau, 53.

17. The Canton of Appenzell . . 53
Chapel of St. Anthony; the Kaien. Vögelisegg, 55. — Gäbris; Stoss, 56. — The Wildkirchli and Ebenalp. The Sentis, 57. — From the Weissbad to Wildhaus, 58. — Altmann. From the Weissbad over the Hohe Kasten to the Valley of the Rhine. Teufen; Frölichsegg, 59.

18. From Rorschach to Coire 60
Thal; Walzenhausen; Meldegg; Berneck, 60. — Alvier; Gonzen, 61. — St. Luziensteig; Falknis, 62.

19. From Wyl through the Toggenburg to Buchs in the Rhine Valley 62
Ascent of the Speer from Ebnat or Nesslau. From Nesslau over the Kräzern Pass to Urnäsch, 63.

20. Ragatz and Pfäfers 64
Excursions from Ragatz: Guschenkopf; Pizalun; Vasanenkopf; Monteluna; Graue Hörner; Kunkels Pass, 66.

21. From Zürich to Glarus and Lintthal 66
The Rautispitz; Obersee; the Scheye, 67. — Schild; Fronalpstock; Schwändi, 68. — Oberblegi-See; Saasberg and Kärpfstock. Excursions from Stachelberg, 69. — The Pantenbrücke, Uellalp, Upper Sandalp, Tödi, etc., 70. — From Lintthal over the Kisten Pass to Ilanz, 71.

22. **From** Stachelberg to Altdorf. Klausen 71

23. **From** Schwyz to Glarus over **the** Pragel 72
From Muottathal to Altdorf over the Kinzig Pass, and to Stachelberg by the Bisithal, 73. — The Glärnisch, 74.

24. From Glarus to Coire through the Sernf-Thal . . . 75
From Elm over the Segnes Pass to Flims; over the Panixer Pass or the Sether Furka to Ilanz, 75. — From Elm over the Ramin Pass to Weisstannen. From Elm over the Sardona Pass, the Scheibe Pass, or the Muttenthaler Grat to Vättis. **From** Elm over the Richetli Pass to Lintthal, 76.

1. Bâle.

Railway Stations. The BADEN STATION (Pl. F, 1; *Restaurant*), at Klein-Basel, is on the right bank of the Rhine. — The Alsace and the Swiss lines both start from the CENTRAL STATION (Pl. D, E, 6; *Restaurant*, B. 1 fr.), in Bâle, on the S. side of the town. These two stations are connected by a *Junction Line* (10 min.; fares 1 fr., 70 c., 50 c.), and also by an *Omnibus* (20 c.), crossing the Old Rhine Bridge.

Hotels. *TROIS ROIS (Pl. a; D, 2, 3), on the Rhine, R., L., & A. 4-4½, B. 1½, lunch 3½, D. 5, pens. 12½, omn. 1 fr. — At the Central Station, to the right: *HÔTEL NATIONAL (Pl. d; E, 6), R., L., & A. from 4, lunch 3½, D.

4 fr.; *Hôtel Suisse (Pl. c; E, 6), R. & A. 4-4½, D. 4-5 fr., these two of the first class; *Hôtel Victoria (Pl. e; E, 6), R., L., & A. from 3, lunch 3, D. 4 fr.; *Hôtel St. Gotthard, R., L., & A. 3, D. 3, pens. 7-8 fr. To the left of the station: *Hôtel Euler (Pl. b; D, 6), R., L., & A. from 4½, omnibus 1 fr., first-class; Hôtel Hofer (Pl. f; D, 6), R., L., & A. 3-3½, B. 1¼ fr.; *Hôtel du Jura, R., L., & A. from 2, B. 1, D. 2½ fr.; Hôt. Gehrig, R. 2½, B. 1 fr., the last three with restaurants. — In the town: Faucon (Pl. g; D, 6), corner of the Elisabethen-Str., R. 2-3, B. 1 fr.; *Métropole (Pl. h; D, 4), R., L., & A. 3, B. 1¼, D. 3, pens. 7-8 fr.; Hôt. Central zum Wilden Mann (Pl. i; D, 4), well spoken of; *Cigogne (Pl. k; D, 3), R., L., & A. 3, D. 3, pens. 7-8 fr.; Couronne (Pl. l; D, 3), R., L., & A. 2, B. 1, D. 2, pens. 5-6 fr.; *Bellevue (Pl. m; D, 3), R., L., & A. 2-2½, B. 1, D. 3, pens. 7-9 fr., both on the Rhine; *Post (Pl. n; D, 3, 4), R. from 1½, B. 1, D. 2½ fr. — At *Klein-Basel:* *Hôtel Krafft (Pl. p; E, 3), R., L., & A. 3, B. 1¼, D. 3 fr.; *Croix Blanche (Pl. q; E, 3), R. & A. 2½-3 fr., both on the Rhine; Hôtel de Bâle (Pl. r; F, 2), R. & A. 3, B. 1¼ fr.; Hôtel Schrieder (Pl. s; F, 1), opposite the Baden Station, R., L., & A. 3, B. 1¼ fr

Cafés. *Casino*, Barfüsser-Platz, corner of the Steinenberg; *Trois Rois*, adjoining the hotel of that name (p. 2); *National*, in Klein-Basel, by the old bridge, with a terrace overlooking the Rhine; these all restaurants also. — **Confectioners** (who sell 'Basler Leckerli'): *Koch*, near the old bridge; *Kissling-Kuentzy*, Freie-Str. 19; *Burckhardt*, Schneidergasse; *Speiser*, Freie-Str. 61.

Restaurants (all with Markgräfler wine on draught). **Casino* (see above); **Zum Safran*, in the guild-house of that name; **Veltliner-Halle*, Freie-Str. 25; **Restaur. Kunsthalle* (p. 8); *Pschorrbräu*, Freie-Str. 49 (Munich beer); *Bühler's Bierhalle*, close to the Casino (in summer, *Bühler's Bier-Garten*, in the Sternengässlein). — In Klein-Basel: *Burgvogtei*, Rebgasse 14, with garden; *Warteck Brewery*, near the Baden station, corner of the Clara-Str. — **Sommer-Casino* (Pl. F, 6), near the St. Jacob Monument (p. 8), with a pleasant garden, music on Wed. and Frid. at 7.30, on Sun. at 6 p.m. (50 c.); *Schützenhaus* (Pl. B, 4), built in 1651 and restored in 1881-83, with old and new stained glass, good wine.

Cabs. For ¼ hr., 1-2 persons, 80 c.; second ¼ hr. 60, each additional ¼ hr. 50 c.; 3-4 pers. 1 fr. 20 c., the second ¼ hr. 90, each additional ¼ hr. 70 c. From either station into the town, 1-2 pers. 1 fr. 20 c., 3-4 pers. 1 fr. 80 c.; from one station to the other 1-2 pers. 1½, 3-4 pers. 2½ fr., each box 20 c. extra. At night (10 p. m. to 6 a. m.) 3 fr. for the first ½ hr. and 1 fr. for each additional ¼ hr., and 10 c. per ¼ hr. for lights.

Post and Telegraph Offices (Pl. D, 4) in the Freie-Str.; at the railway-stations; in the Johannes suburb; and at the Schützengraben.

Baths in the Rhine (Pl. E, 3, 4), entered from the Pfalz (p. 5), 1 fr. Warm Baths: Martinsgasse 20; Leonhard-Str. 12, etc.

The **Oeffentliche Verkehrsbureau** *(Public Enquiry Office)*, Schifflände 7, near the Old Bridge, gives information of all kinds (open 9-12 and 2-5).

Picture Gallery in the *Kunsthalle* (p. 8; open 9 to 12 and 2 to 6; adm. 50 c.; another at *Lang's*, Freie-Str.

English Church Service in a chapel at the Hôtel des Trois Rois (10.30 and 3). — **United States Consul**, *Mr. George Gifford.*

Bâle, or *Basel* (870'), the capital of the half-canton Bâle-Ville or Basel-Stadt (pop. 82,431), is first mentioned in the year 374 under the name of *Basilĕa*, having probably been founded by the Roman armies, when they fell back on the Rhine, near the old *Colonia Augusta Rauracorum*, which had been established in B. C. 27 by L. Munatius Plancus (now *Basel-Augst*, 5½ M. to the E., see p. 19). In the middle ages Bâle was a free town of the Empire and it has been a member of the Swiss Confederation since 1501. A university was founded here in 1460 by Pope Pius II. (Æneas Sylvius). The city lies on both banks of the *Rhine*, which here receives the waters of the *Birs* and the *Birsig* on the S. and of the *Wiese* on the N. On the left bank of the Rhine lies *Gross-Basel*, on two hills divided by the valley of the Birsig, through which run the **Freie-Strasse** and Gerber-Strasse, the ancient channels of the main traffic of the city. On the right bank lies *Klein-Basel*, with numerous manufactories.

Three Bridges cross the river, all affording admirable views. The wooden *Alte Brücke* (Pl. D, E, 3), 165 yds. in length, 16 yds. in breadth, and partly supported by stone piers, was originally built in 1225. In the middle of the bridge rise a chapel of the 16th cent. and a column with a barometer and weather-cock. Above the old bridge the river is crossed by the iron *Wettstein-Brücke* (Pl. F, 4), completed in 1879, with three spans, 200 ft. in width. At each end of the bridge are two basilisks, the heraldic symbol of Bâle. Below the old bridge is the five-arched *Johanniter-Brücke* (Pl. D, 1), completed in 1882.

The *****Münster** (Pl. E, 4), a picturesque edifice of red sandstone, with a brilliantly coloured new roof and two slender towers, is conspicuous in every view of the city. Down to the Reformation (1529) it was the cathedral of the old see of Bâle. Its foundation is ascribed to Emp. Henry II. (1010-1019), but the oldest parts now extant belong to a new building of 1185, which was sadly damaged in 1356 by an earthquake and a fire. It was at once rebuilt in the Gothic style and reconsecrated in 1365. Of the Romanesque structure the N. portal, or *St. Gallus Gateway* (built about 1200), still exists, and is adorned with statues of the Evangelists and John the Baptist; over the church-door is a relief representing the wise and foolish virgins; at the sides in six niches are the works of charity, and at the top Christ on the judgment-seat and the angels at the Last Day. The exterior of the *Choir*, with its round-arched arcades, is also Romanesque. The *W. Front*, with the towers, the principal portal, and two side-entrances, belongs entirely to the Gothic period. The tasteful *N. Tower* is 212' high; the *S. Tower*, completed in 1500, is 207' high. The sculptures on the façade represent the Virgin and Child, and under them the Emp. Henry, with a model of the church, and the Empress Kunigunde; on the two side-entrances are two knights, on the left St. George and the dragon, and on the right St. Martin. The whole building underwent a thorough restoration in 1852-56 and 1880-90.

The Interior is open to the public on Wed., 2-4 p.m.; at other times 50 c. for 1-2 pers. and 25 c. for each addit. person. The sacristan lives in the Münster Platz No. 13, but in summer he is generally to be found in the church (knock). The church, which is 213' long and 107' wide, originally consisted of nave and aisles, but is now provided with double aisles owing to the inclusion of the chapels. The general effect is very imposing, especially when seen from the galleries. The stained-glass windows are modern. The beautiful rood-loft of 1381 serves to support the large and excellent organ. The pulpit dates from 1486. In the left outer aisle are some monuments of the 13-15th cent. and (farther on) two reliefs with the martyrdom of St. Vincent and of St. Lawrence. The font is of 1465; on the pillar opposite is the tombstone of the learned Erasmus of Rotterdam (d. 1536), with a long Latin inscription. The right outer aisle contains a relief of six Apostles (11th cent.). In the transept are late-Gothic choir-stalls, with satirical representations (15th cent.). In the retro-choir are monuments of the Empress Anna (d. 1281), consort of Rudolph of Hapsburg and mother of Albert I., and of her youngest son Charles. The crypt is now occupied by the furnaces used in heating the church. — In 1431 the great *Council* began to sit in the Münster. It consisted of upwards of 500 clerics, including many great dignitaries, whose ostensible task was a 'reformation of the Church in head and members'; but after having disputed for years without any result, and having been excommunicated by Pope Eugene IV., it was at last dissolved in 1448.

On the S. side of the choir are extensive *Cloisters, at the entrance to which from the Ritter-Strasse stands a statue of *John Œcolampadius* (d. 1531), the Reformer. The vaulting of the cloisters is partly Romanesque, and partly late-Gothic (1470-90). They were restored in 1869-73, and used until 1850 as family burial-places. They extend to the **Pfalz**, a terrace behind the Münster, 65 ft. above the Rhine, planted with chestnuts, and affording a pleasing survey of the green river and the hills of the Black Forest. In the neighbourhood (Bäumleingasse 18) is the house of *Froben* the printer, in which Erasmus died in 1536.

In the Augustinergasse, which descends from the Münster-Platz towards the N.W. to the bridge, is the **Museum** (Pl. E, 3), constructed in 1843-49. On the groundfloor is the *University Library* (open on week-days 10-12 and 2-4), which contains 200,000 vols., including many incunabula, and 5000 MSS., mainly from the time of the Council of Bâle and the Reformation. Adjacent are two rooms containing an *Ethnographical and Prehistoric Collection* (lacustrine remains). On the staircase are three frescoes by *Böcklin* (1866-71), representing Gæa, Flora, and Apollo. The first floor contains the *Aula of the University*, with portraits of 107 scholars of Bâle, and the *Natural History Collections*. In the ante-room are marble busts of ten recent professors of the university. — The second floor is occupied by the *Picture Gallery (director, *Dr. Daniel Burckhardt*), which is chiefly interesting for its paintings and drawings by *Hans Holbein the Younger* (b. at Augsburg 1497, d. in London 1543), who lived at Bâle in 1515-26 and 1528-32. It is open free on Sun. 10.30-12.30, and in summer on Wed. 2-4; at other times fee 50 c.; no adm. between 12.30 and 1.30.

The staircase from the first to the second floor is adorned with cartoons by *Cornelius, Schnorr,* and *Steinle,* and by a painting by *Benner* (No. 178) of a Street in Capri. — ANTE-ROOM. Seven fragments of *Holbein's* obliterated frescoes in the Council Chamber and old and modern copies from them; painted organ-shutters from the Münster, by *Holbein*. — ROOM TO THE LEFT. MODERN SWISS MASTERS. To the left: *Arnold Böcklin* (b at Bâle in 1827), 10. Lady with a green veil, 15. Life a dream, *11. Pietà, 15b. Head of Medusa, 15a. Portrait of himself, 14. Naiads, *12. Battle of Centaurs; 27. *Ed. Girardet.* Fortune-teller; *21. *Zünd,* Forest-landscape with the Prodigal Son; *43. *Steffan,* Forest-landscape; *Böcklin,* 13. Sacred grove, 9. Diana hunting; 20. *Zünd,* Harvest; 37. *Borzaghi-Caitaneo,* Tasso and Leonora; *Diethelm Meyer,* 44. Girl of the Haslithal, 45. Girl of the Valais; 26. *Ed. Girardet,* Wounded Turcos; 49. *Staebli,* River scene; 54. *Rü-*

disohli, Marshy ground; 75. *E. de Pury*, Among the Lagoons; *A. van Muyden*, 29. Roman street-scene; 30. Italian woman with child; *1. *A. Calame*, Evening landscape; *Koller*, 32. 33. Cows at water, 31. Horses on a road through a dale; 25. *E. Girardet*, Barber's shop; 74. *Arthur Calame*, Nile landscape by moonlight; 57. *Castan*, Harvest; 18. *Anker*, Children's breakfast; *Vautier*, 16. Rustic debtor compelled by a rich neighbour and his agent to sell his property, 17. The involuntary confession; *23. *Zünd*, Noon; 24. *Ed. Girardet*, Snow-balling; *Stückelberg*, 7. The painter's children, 6. Marionettes, 5. Pilgrimage among the Sabine Mts.; 59. *S. Durand*, Wayfarers; *1, 3. *Calame*, Forest-landscapes; 19. *Anker*, Quack; 36. *Gleyre*, Nymph. We now return to the ante-room and enter the —

ROOM OF THE DRAWINGS. These include, partly on the walls and partly in cabinets, admirable examples of *Hans Holbein the Elder* (15-27a). *Albrecht Dürer* (30-32), and **Hans Holbein the Younger* (61-138, 142). Among the last may be specially mentioned: 111. Family of Sir Thomas More, 113. Combat of foot-soldiers, 114. Samuel and Saul (these two sketches for the lost pictures in the Council Chamber), 125-128. Women's costumes of Bâle, 91-100. The Passion. Between the first and second window are the original of *Holbein's* Praise of Folly (Laus stultitiæ) and drawings by other German masters of the 16th century. — We next enter the —

LARGE SALOON. **Here we turn to** the left, pass *Imhof's* statue of Rebecca, cross the old-German room, pass between the so-called *Steinhäuser Apollo* and the replica of the *Farnese Hercules* (two ancient heads), and reach the NORTH ANTE-ROOM, with a continuation of the pictures by Swiss masters. No. 64. *Veillon*, Lagoons of Venice; 76. *Frölicher*, Spring landscape; 48. *Grob*, Pestalozzi; 63. *Bosshardt*, Hans von Hallwyl at the battle of Murat; 89. *Burnagd-Cattaneo*, Fiesco; *Buchser*, 62. Capuchins and worldlings, 61. Rapids of Sault Ste. Marie, Canada; 82. *Sandreuter*, Heroic landscape; 69. *Bocion*, The harbour of Ouchy; 65. *Humbert*, Cattle watering. We return to the FIRST SECTION of the Large Saloon. To the right: *H. Holbein the Younger*, 6a. and 6b. Schoolmaster's signboard of 1516; *7. Erasmus; 10. The burgomaster Jacob Meyer and his wife (1516); *11. Last Supper; 12. Adam and Eve; 13. Ecce Homo; *14. The Passion, in eight separate scenes, formerly in the Rathhaus; *15. The dead body of Christ, of startling realism (1521); *16. Boniface Amerbach (1519); 17. Erasmus; *18. Lais Corinthiaca, the portrait of a lady of the noble family of Offenburg (1526); 19. The same lady with Cupid; *20. Wife and children of the painter (1528); 21. A London merchant; 23, 24. *Ambrose Holbein*, Portraits of boys; *M. Grunewald*, 32. Crucifixion, 33. Resurrection; *Hans Baldung Grien*, 34. Crucifixion, 35. Nativity, *36, *37. Pictures with figures of Death; 41-43. *N. Manuel Deutsch*; 58, 59. *Tob. Stimmer*, Full-length portraits of Jac. Schwytzer and his wife (1564). — SECOND SECTION. In the centre, marble statue of Jason, by *Schlöth*. Nos. 65-72. School of *Gerrit van St. Jans*; *Dutch Master* of the 15th cent., 73. Pius Joachim, 74. Coronation of the Virgin; 90. *Strigel*, St. Anna; 102. *Lucas Cranach the Elder*, Luther and Catharine von Bora; 107. *Schoreel*, The Anabaptist David Joris; 108-111. In the style of *H. met de Bles*, Altar-piece. — At the entrance to the next section, to the right, 166a. *Bronzino*, Portrait of a man; to the left, 73a. *Early French School*, Jacques de Savoie, Count of Romont. — THIRD SECTION. On the wall, ancient Greek head of a youth. To the left, 140. *Fr. Mieris the Elder*, Fishmonger; 146. *S. Ruysdael*, Landscape; 131. *Teniers the Younger*, Dutch interior; 139. *C. Dusart*, Peasants; 137. *C. du Jardin*, Before the inn; above, 121. *Peter Thys*, Pieta; 192. *Teniers the Younger*, Boors; 138. *Berchem*, Cattle crossing a stream; *119. *Rubens*, Bearing of the Cross (sketch); 136. *Wouwerman*, Horses and ass; 145. *Decker*, Landscape; 133. *Teniers the Younger*, Tavern music; 183a. *Matt. Merian*, Portrait of G. J. Muller (1687); *156. *Dutch Master*, Forest-scene; 125. *Dirk van Sandvoort*, Strolling singers; 145. *J. van Rombouts*, Forest-scene; 165. Old copy of Raphael's Joanna of Aragon. — FOURTH SECTION. Marble statuette of a runner, by *Kissling*, and a bust of S. Birman. To the left, 213. *Ph. de Champaigne*, Portrait; 218. *Meucheron*, Landscape; to the right, 237. *Teniers the Younger*, Smoker; 208. *N. Poussin*, Bacchus. — FIFTH SECTION.

Marble statue of Psyche, by *Schlöth*. To the left, *Leopold Robert*, 288. Wounded bandit and his wife, 289. Bandits' wives in flight, 289a. Palm Sunday; 305. *Landerer*, Federal representatives entering Bâle in 1501 to administer the federal oath to the town; 302. *Lud. Burckhardt*, Canine family; 292-297. Landscapes by *J. Frey*, of Bâle; 300. *Fr. Diday*, Lake of Brienz; 306. *Lessing*, Forest-landscape; 280, 281. *J. Schraudolph*, Angels; 266-268. *J. A. Koch*, Landscapes; 307. *Feuerbach*, Idyl. — The SOUTH ANTE-ROOM contains German drawings of the first half of the 19th century. In the centre a *Relief of the Jungfrau on the scale of 1:10,000, by *S. Simon*.

The **Rathhaus** (Pl. D, 3), or Town Hall, in the Markt-Platz (No. 13), was erected in the Burgundian late-Gothic style in 1508-21 and restored in 1824-28. By the flight of steps in the court is a *Statue of Munatius Plancus* (p. 3), erected here in 1580. The handsome *Council Hall* in the interior is adorned with fine panelling and stained glass. — Adjacent are the *Bâle Bank* and the *Geltenzunfthaus*, the latter a building of 1578, now used as a beer-saloon. — The late-Gothic *Fischmarkt-Brunnen* (Pl. D, 3) dates from 1467. — The *Post Office* (Pl. D, 4), at the corner of the Freie-Str. and the Gerbergasse, was formerly the exchange; the addition of 1880 is also in the late-Gothic style.

The *Barfüsser-Kirche* (Pl. D, E, 4), a huge building of the beginning of the 14th cent., with an unusually lofty choir, now contains the *Historical Museum, the chief collection of the kind in Switzerland (Sun. 10.30-12.30 and 2-4, and Wed. 2-4, free; other days 8-6 in summer, 10-4 in winter, fee 50 c.; director, Prof. *Ad. Burckhardt-Finsler*).

NAVE. Architectural fragments and sculptures from the churches and secular edifices of Bâle. *St. Martin*, from the Minster. To the left, the so-called *Holbein Fountain* (p. 9). Above St. Martin, the 'Lällenkönig', a curious piece of mechanism, formerly on the exterior of the tower (removed in 1839) of the Rhine bridge; when the clock struck the head stuck out its tongue and rolled its eyes. — The adjoining *Waffensammlung* or *Collection of Weapons* contains the chief curiosities of the arsenal of Bâle, interesting cannon (in the middle a finely ornamented twelve-pounder of 1514), Bâle uniforms, trophies of war (in the case to the right, hauberk supposed to have belonged to Charles the Bold), handsome weapons (in the case to the left, three daggers with silver-gilt sheaths of the 16th cent.), tent, guild-banners, etc. Next come some *State Sleighs* and fine specimens of *Smith's and Locksmith's Work*. — To the right and left of the nave and in the aisles is a series of rooms intended to exhibit the development of the furnishing and adorning of dwelling-houses from the 15th cent, onwards. To the right of the entrance: "1. *Room from the Spiesshof* (1601), with panelling and a large bed; 2. *Room from the Spiesshof* (1580), with fine cabinets and doors and the old Bâle council-table; 3. *Room from the Strassburger Hof* (1600), with a large bed, cabinet, and chests; *4. *Dining-Room of Councillor Iselin* (1607), with beautiful panelling; 5. *Room from Schwyz* (1650), with heavy coffered ceiling; 6. *State Room from the Haus zum Cardinal* (1540). — We now cross to the other side of the nave. 7. *Old Kitchen*, with large chimney-piece; 8. *Schönau Room* from the Château of Oeschgen (17th cent.); 9. *Gothic Room* (15th cent.), with a large bedstead of 1510 and other Gothic furniture; 10. *Gobelins Room* (1780); 11. *Neustück Room* (1787), with a collection of models of gates of Bâle and of neighbouring castles. *Room 12* (at the entrance to the church) contains the *Collection of Coins*, including coins, medals, and dies of Bâle and other Swiss towns, and also a few ancient coins and vases.

8   I. Route 1.   BÂLE.   Hist. Museum.

The Choir contains ecclesiastical antiquities. To the left, Fragments of the famous *Death Dance* of Bâle, a fresco which once adorned the wall of the Dominican burial-ground (taken down in 1805), painted early in the 15th century; bells of the 15th cent.; fine choir-stalls of 1598; Carved Altars of the 15-16th centuries. On the high-altar, *Altar of St. Maria Calanca*, in the Grisons (1512); to the right, *Votive Tablet of the Duchess Isabella of Burgundy* (1433), in enamelled bronze; above the last, kneeling figure of the knight Hügelin von Schönegg (1378); farther on, winged altar-piece from the church of Baden in the Aargau (15th cent.). — To the left is the entrance to the TREASURY, which contains reliquaries, monstrances, crosses, and chalices of the 13-18th cent.; cups and goblets belonging to the University (16-17th cent.); handsome plate of the guilds and trade-companies of Bâle. To the left of the entrance, cast of the golden antependium presented to the Cathedral of Bâle by Emp. Henry II. (beginning of the 11th cent.), which, along with other objects of value, was assigned to Bâle-Campagne at the division of the canton in 1833 and forthwith sold (now in the Musée de Cluny at Paris).
    We now return to the nave and ascend the staircase to the right to the GALLERIES of the aisles, in which the smaller objects of the collection are exhibited. *Roman, Alemannian,* and *Burgundian Antiquities,* found at Augst (p. 19) and elsewhere. — Bâle *Looms* and specimens of *Ribbon Weaving* at Bâle in the 17-18th centuries. — Embroidery, fans; Bâle and other Swiss *Costumes* of the 17-18th centuries. — On the old organ-screen (above the entrance), *Musical Antiquities,* showing in particular the development of the piano, wooden wind-instruments, and musical notation; also *Stained Glass.* — Farther on, *Small Works of Art.* Wood-carvings (in a case to the right, Adam and Eve, box-wood figures of 1500), ivory carvings, enamels, book-bindings, goldsmiths' models, small bronzes. — *Domestic Utensils:* porcelain, faïence, glass, pottery, tin-ware, works in leather, book-bindings, toys, clocks and watches, armorial windows. — *Government and Judicial Antiquities:* weights and measures of the 14-18th cent.; staves for the officers of justice, judicial swords, executioner's dress. — We now descend to the nave and enter, from the end of the right aisle, the —
    COURT, which contains stone monuments of the Roman, mediæval, and Renaissance periods, gates in hammered iron, and other objects.
    Near the Historical Museum, in the Steinenberg, is the **Kunsthalle** (Pl. E, 5), built by Stehlin in 1870-72. The staircase is adorned with a fresco by *Stückelberg* (the Awakening of Art), and on the garden-façade are caricature heads by *Böcklin.* Adjacent is a restaurant. — In the Elisabethen-Str. is the handsome **St. Elisabethenkirche** (Pl. E, 5), built in the Gothic style in 1857-65, with stained-glass windows from Munich and an open-work tower, 232' high.
    The S.E. SUBURBS are occupied by the well-to-do classes of the inhabitants. From the *St. Albans-Thor* (Pl. G, 5), in this quarter, the promenades of the St. Alban-Anlage and of the Æschengraben extend on the site of the former ramparts to the railway-station. In the Æschen-Platz (Pl. E, F, 5) is a large fountain (jet 148' high), which, however, seldom plays. The old *St. Alban's Convent* (Pl. F, 4) possesses fine Romanesque cloisters. The *Monument of St. Jacob* (Pl. F, 6), by *F. Schlöth,* completed in 1872, commemorates the heroism and death of 1300 Confederates who opposed the Armagnac invaders under the Dauphin (afterwards Louis XI.) in 1444. Beyond, to the right, is the *Sommer-Casino* (p. 3). — In the promenades, near the station, is the *Strassburg Monument,* a marble group in memory of the reception of the Strassburg fugitives on Swiss soil in 1870, by Bartholdi of Paris (1895).

In the W. Quarter, in the Spalen Suburb (Pl. C, 3, 4), is the *Spalen Fountain*, with a relief of dancing peasants (after Holbein) and the figure of a bag-piper, restored in 1887. The **Spalen-Thor** (Pl. C, 3), erected about 1400, is the handsomest of the remaining gates of Bâle. Near it are two modern buildings belonging to the University: the *Vesalianum* (Pl. C, 3), or institute for anatomy and physiology, and the *Bernoullianum* (Pl. C, 2, 3), for the study of physics, chemistry, and astronomy. In the vestibule of the latter are busts of the famous mathematicians of Bâle, James and John Bernouilli (d. 1705 and 1748). In the Hebel-Str. is the house (tablet) where the Alemannian poet *Hebel* (1760-1826) was born, — The *Mission House* (Pl. B, 3) contains an ethnographical collection, mainly from the countries (E. Indies, China, W. Africa) served by the Bâle missionaries (catalogue 1 fr.).

The **Zoological Garden** (Pl. B, C, 6) contains good examples of Swiss animals (adm. 50 c.). Concerts are frequently given on Sun. afternoons. — About 1 M. from the Old Rhine Bridge, to the N. of the Baden Station (Pl. F, 1), on the *Wiese*, is the **Erlen-Park**, much frequented on Sun. (Restaurant).

From Bâle to Flühen, 8 M., narrow-gauge railway ('Birsigthalbahn') in 50 min. (fares 1 fr. 30, 95 c.). The train, starting from the Steinenthor-Str. (Pl. D, 5), passes the Zoological Garden (see above), and traverses the fertile valley of the *Birsig*. Stations: 1¼ M. *Binningen* (Hirsch; Bär), a large village with 4700 inhab. and the church of *St. Margaret*, commanding a good view; 1³/₄ M. *Bottminger-Mühle*; 2¹/₂ M. *Bottmingen*, with the *Bottminger Schlösschen* (Inn and pretty park); 3 M. *Oberwyl* (Krone), with an extensive parquetry-factory; 4¹/₄ M. *Therwil* (*Rössli*), a substantial village in the *Leimen-Thal*; 5¹/₂ M. *Ettingen* (Badhaus), with a chalybeate spring. The line then skirts the hills to the W. viâ *Witterswyl* and *Bättwyl* to (8 M.) **Flühen** (1250'; Inn and Baths), a small village with a chalybeate spring, prettily situated in a defile at the foot of the *Blauen*, close to the frontier of Alsace. Interesting excursion hence viâ *Tannwald* to the (1¹/₂ M.) well-preserved ruin of *Landskron (1890 ft.), the tower of which commands a wide view (key at the last house in Tannwald). — A road leads to the S. from Flühen to (1¹/₂ M.) **Mariastein** (1685'; *Kreuz; Post*), formerly a Benedictine abbey, with a frequented pilgrimage-church, picturesquely situated on a steep crag. A spacious rock-cavern beneath the church contains the chapel of *Maria im Stein*. From Mariastein the *Landskron* may be reached viâ Tannwald in 25 minutes. — The road goes on beyond Mariastein to *Metzerlen* and (2¹/₄ M.) *Burg* (1735'; *Inn), a charmingly-situated village with a mineral spring and a château commanding fine views. — The *Blauen* (2690'), which may be ascended from Ettingen (see above) or Mariastein in 1¹/₂ hr., commands a wide prospect, extending on the S.E. to the Bernese Alps.

2. From Bâle to Bienne and Bern through the Münster-Thal.

77 M. Railway (*Jura-Simplon Line*) to Bienne (56 M.) in 3-4 hrs. (fares 9 fr. 30, 6 fr. 65, 4 fr. 75 c.); from Bienne to Bern (21 M.) in 1-1¹/₄ hr. (fares 3 fr. 55, 2 fr. 50, 1 fr. 80 c.). [Railway from Bienne to Neuchâtel (20 M.) in ³/₄-1¹/₄ hr.; to Geneva (102 M.) in 5¹/₄-7¹/₄ hrs.; from Bâle to Geneva, express in 7³/₄ hrs. Through-carriages to Geneva and St. Maurice.]

Bâle (870'), see p. 2. Leaving the Central Station, the train soon diverges from the Central Line (p. 13) to the right, passes the cemetery, and near (3 M.) *Mönchenstein*, the scene of a terrible railway accident in 1891, crosses the *Birs*. On the hills to the left are several ruined castles. — 5 M. *Dornach-Arlesheim* (Munzinger's Restaurant). On a wooded hill, 1½ M. to the E., near *Arlesheim* (1130 ft. ; *Löwe; Ochs), rises *Schloss Birseck*, once a château of the bishops of Bâle, with a pleasant park, interesting grottoes, and a hermitage. (Apply to the gardener at the foot of the hill.)

The train follows the right bank of the Birs. On the left is the village of *Dornach*, with its picturesque ruined castle. 7 M. *Aesch* (Ochs), a village on the left bank. The valley contracts. The train passes through a tunnel under the modernised château of *Angenstein*, and enters the **canton** of Bern. On a hill to the right is the picturesque ruin of *Pfeffingen* (1850'). On the right, near (9¼ M.) *Grellingen* (*Bär), are several factories. The train passes through a deep cutting and crosses the Birs twice; the valley then expands. *Schloss Zwingen*, on the right, was formerly the seat of the episcopal governors of the district.

14½ M. **Laufen** (1155'; *Hôt. Jura; Sonne*) lies at the confluence of the *Lützel* and Birs. The train traverses a narrow, wooded valley. Beyond (16 M.) *Bärschwyl* it passes through two tunnels and crosses the Birs twice. 18½ M. *Liesberg*. At (22½ M.) *Saugern*, Fr. *Soyhières* (Hôtel de la Gare), the language changes from German to French. On the right is the ruined castle of that name. At the rocky egress of the valley, before its expansion into a broad plain, lies *Bellerive*, on the left, now a factory. On a hill to the right is the ruin of *Vorburg*.

24½ M. **Delémont**, Ger. *Delsberg* (1430'; *Faucon; Lion d'Or; Hôtel Lachat*, at the station; *Rail. Restaurant*) is an old town (3638 inhab.) on the *Sorne*, with a château of the former Bishops of Bâle.

From Delémont to Porrentruy, 18 M., railway in ¾-1¼ hr. (fares 3 fr. 5, 2 fr. 15, 1 fr. 50 c.). The line traverses the grassy valley of the *Sorne*. Stations *Courtetelle, Courfaivre, Bassecourt*, and (7½ M.) *Glovelier*. We next cross the large viaduct of *Combe-Maran*, and beyond a tunnel, 3200 yds. in length, and two others, reach (11 M.) *Ste. Ursanne* (*Deux Clefs; Bœuf), a picturesque old town in the romantic valley of the *Doubs* (p. 210), with a ruined château on a lofty rock. Another tunnel pierces the *Mont Terrible*, Stat. *Courgenay*. Then (18 M.) **Porrentruy**, Ger. *Pruntrut* (1455'; *Hôt. National*, near the rail. station; *Cheval Blanc*), a considerable old town (6500 inhab.) with a château, once the residence of the Bishops of Bâle. At *Réclère*, 7 M. to the W. of Porrentruy, near the French frontier, a large stalactite grotto has been discovered and made accessible. — The line leads hence to *Delle*, the French frontier-station, *Belfort*, and *Paris*.

Beyond (26½ M.) *Courrendlin*, Ger. *Rennendorf* (Cerf) the train enters the *Münster-Thal, Fr. Val Moutier*, a wild, romantic ravine of the Birs, flanked with huge limestone rocks. In the Roman period it was traversed by the road from **Aventicum** (p. 219) to Augusta Rauracorum (p. 3). The line is carried through these '*Gorges de Moutier*' by means of a series of tunnels, galleries, and cuttings. (A walk from Courrendlin to Münster is recommended.) — Above

(28½ M.) *Choindez*, and opposite the *Glass Works of Roche*, which lie on the right bank of the stream, we traverse a tunnel, 100 yds. in length, and reach (30 M.) *Roche* (1650'; *Cheval Blanc, moderate). The train threads nine short tunnels in rapid succession, crosses the Birs by a lofty bridge, and then, at the mouth of the defile, the *Rausbach*.

32 M. **Münster**, Fr. *Moutier* (1730'; *Hôtel de la Gare*, moderate). The thriving village (1750'; *Cerf; Couronne; Cheval*, well spoken of), with 2346 inhab. and a new Protestant church, is prettily situated in a green dale, on the left bank of the Birs.

ASCENT OF THE WEISSENSTEIN FROM MÜNSTER (3½ hrs.; comp. p. 16). About 10 min. to the N.E. of Münster, or 6 min. from the station, at the *Restaurant Sperisen* (good beer), a road (diligence to St. Joseph daily in 1 hr.) ascends to the right to (2 M.) *Granfelden* (Fr. *Grandval*, 2010') and (¾ M.) *Crémine* (2065'; Croix). It next ascends the gorge of the *Raus* to (2 M.) *St. Joseph am Gänsbrunnen* (2450'; Inn), at the N. base of the *Weissenstein*, the Curhaus on which (p. 16) may easily be reached hence by a narrow road in 1¾-2 hrs. The footpath to the left is shorter (1½ hr.). Carriage from Münster to the Weissenstein 25 fr., there and back 30 fr.; from Gänsbrunnen 15 fr.

The line traverses another very picturesque gorge, the *Roches de Court*, high above the Birs, and beyond a longer and two short tunnels reaches (35½ M.) **Court** (2200'; *Ours; Couronne*).

From Court, or better from *Bévilard* (see below), a steep path leads over the **Montoz** (4370') to (3 hrs.) *Reuchenette* (see below; guide advisable). View similar to that from the Weissenstein.

We traverse pleasant grassy dales, pass *Sorvilier*, *Malleray-Bévilard*, and *Reconvilier*, and reach —

43 M. **Tavannes**, Ger. *Dachsfelden* (2500'; *Hôtel de la Gare*, poor; *Brasserie*, with restaurant and rooms), a large village near the source of the Birs (branch-line in 35 min. to *Tramelan*). The train ascends slightly, and passes by means of a tunnel (1500 yds.) under the *Pierre Pertuis*, a natural opening in the rock, fortified in the Roman times, through which the high-road passes. It then descends the slope to the right, describes a sharp curve between *Sombeval* and *Corgémont*, and crosses the *Suze*, or *Schüss*.

47½ M. **Sonceboz** (2150'; *Couronne; Cerf*, well spoken of), the junction for *La Chaux-de-Fonds* (see p. 211).

The train again crosses the Suze, and passes through a tunnel under the S.W. spur of the *Montoz* (see above). The stream is crossed several times in its beautiful wooded valley. 50½ M. *La Heutte*; 53 M. *Reuchenette* (1940'; Hôtel de la Truite). The line now suddenly turns towards the S., and enters the narrow passage which the Suze has forced through the last heights of the Jura range. Four tunnels between this point and Bienne. On the right beyond the first tunnel is a fall of the Suze, and on the hill is the ruined château of *Rondchâtel*. Two more tunnels. Pleasant view of the green valley of *Orvin* (Ger. *Ilfingen*) to the right. Beyond another long tunnel the train crosses the deep and wild ravine of the Suze (*Taubenloch*, see p. 12) by a lofty bridge, and quits the ravine. We now obtain a striking view of the rich plains of Bienne, with the whole of the Alpine chain from the mountains of Unterwalden to Mont Blanc in the distance. We then descend vine-clad slopes and thread a short tunnel.

56 M. **Bienne**, Ger. *Biel* (1445'; *Couronne*, R. from 2, D. 3, S. 2½ fr.; *Hôtel de Bienne*, near the station, R., L., & A. 2½-4, B. 1¼, lunch 3, D. 3½ fr.; *Victoria*, at the station; *Hôt. Suisse*, R. 2½, B. 1 fr.; *Croix*; *Hôt. de la Gare*, near the station, well spoken of; *Rail. Restaurant*), an ancient and thriving town (18,000 inhab.). The *Museum Schwab*, founded by Col. Schwab and presented by him to the town, is an interesting collection of antiquities from the lake-villages, Celtic and Roman weapons, implements, coins, etc. (open on Sun. and Thurs. 2-4, 50 c.; at other times on application). The beautiful avenues enclosing the town stretch to the (1/2 M.) *Lake of Bienne* (p. 206; lake-baths).

Tramway from the station into the town, to Nidau, and to the N. to (20 min.) *Bözingen*, Fr. *Boujean* (Hirsch; Rössli). An attractive walk leads hence through the picturesque *Taubenloch-Schlucht, watered by the copious Schüss, to the (½ hr.) hamlet of *Friedliswart*, Fr. *Frinvillier* (Restaurants des Gorges and de la Truite, good trout), and thence past the ruin of *Rondchâtel* to (¾ hr.) the station of *Reuchenette* (p. 11).

A WIRE-ROPE RAILWAY (station 10 min. to the N.W. of the railway station at Bienne, where an omnibus is waiting) ascends in ¼ hr. (1 fr.; return 1½ fr.) to the Curhaus of *Macolin*, Ger. *Magglingen* (2860'; R., L., & A. from 4, D. 4, pens. 8-12 fr.), splendidly situated on the slopes of the Jura, 1¼ hr. above Bienne, and visited as a health-resort. Large wooded grounds, and fine view of the Alps from the Sentis to Mont Blanc. English Church Service in August. About ½ M. lower down is the unpretending *Restaurant & Pension Magglingen* (pens. 3-4 fr.).

A very pleasant round of 3 hrs. is as follows: by cable-railway to Macolin, thence viâ the (25 min.) prettily-situated village of *Leubringen*, Fr. *Evilard* (*Curhaus*; *Drei Tannen*, well spoken of), through magnificent pine-woods, or viâ *Orvin* (p. 11) to *Frinvillier* and by the *Taubenloch-Schlucht* to *Bözingen* (tramway to Bienne). — The ascent of the Chasseral (5280') takes about 4 hrs. from Macolin. From the hotel a good path crosses the hill to the S.W. to *Lamboing*, *Diesse*, and *Nods*, at the N.E. foot of the mountain, whence a steep and stony path leads to the top (descent to *St. Imier*, see p. 211). — From Macolin to *Twannberg* (p. 206), a pleasant walk of 1½ hr.

From **Bienne** to *Soleure*, see p. 17; to *Neuchâtel* and *Geneva*, see RR. 58, 61.

The RAILWAY FROM BIENNE TO BERN crosses the *Zihl* near (58½ M.) *Brügg*, and the *Aare* before (61 M.) *Busswyl*.

63 M. Lyss (*Hirsch*; *Restaurant de la Poste*, *Ritter*, at the rail. station) is the junction of the lines to *Payerne* on the S. (p. 220) and to *Soleure* on the N. (p. 17). — 64½ M. *Suberg*; 68 M. *Schüpfen*; 71 M. *München-Buchsee* (*Hôt. Küch; Krone; Bär). On the right, the Bernese Alps from the Jungfrau to the Balmhorn become visible, but soon disappear. — 73 M. *Zollikofen*, a station on the Central Line (Bâle-Herzogenbuchsee-Bern). Thence to (77 M.) *Bern*, see p. 18.

3. From Bâle to Bienne viâ Olten and Soleure.

63 M. RAILWAY in 3-4 hrs. (fares 10 fr. 65, 7 fr. 45, 5 fr. 35 c.).

Bâle, see p. 2. The train crosses the *Birs*. 3 M. *Muttenz*. 5 M. *Pratteln*, the junction for Zürich (p. 19). On the Rhine, $1^1/_2$ M. to the N.W. (branch-railway in 10 min.), are the well-equipped salt-baths of *Schweizerhalle*.

The line leaves the valley of the Rhine, enters the Jura Mts., and follows the left bank of the *Ergolz*. Near ($7^1/_2$ M.) *Nieder-Schönthal*, on a hill to the right, lies *Frenkendorf* (1120'; Wilder Mann; Löwe), a pretty summer-resort. A good road leads from Nieder-Schönthal to ($2^1/_4$ M.) Bad Schauenburg (see below).

9 M. Liestal (1033'; 4927 inhab.; *Falke*, with salt-baths and garden, pens. $4^1/_2$–$5^1/_2$ fr.; *Schlüssel; Engel; Sonne; Hôt. de la Gare*), prettily situated on the Ergolz, is the seat of government of the half-canton of Basel-Land or Bâle-Campagne. In the town-hall are a collection of coins and the cup of Charles the Bold, found in his tent after the battle of Nancy (1477).

Bienenberg (*Curhaus*, with salt-baths), $1^1/_2$ M. to the N.W. of Liestal, is a pleasant summer-resort, and about $1^1/_2$ M. beyond it is *Bad Schauenburg* (1580'), below the ruin of the same name (1975'; *View). Road to Nieder-Schönthal, see above.

To WALDENBURG, $8^1/_2$ M., narrow-gauge railway in 1 hr., through the pretty *Frenkenthal*. $2^1/_2$ M. *Bad Bubendorf*, with mineral and salt baths. (The village with its ruined castle lies 1 M. to the right.) 4 M. *Lampenberg*; $5^1/_2$ M. *Hölstein*, in a narrow part of the valley, with manufactories of silk ribbon. Passing *Niederdorf* and *Oberdorf*, we reach ($8^1/_2$ M.) Waldenburg (1713'; *Löwe; Schlüssel*), a little town with a ruined castle and a pretty church. A good road leads hence (diligence 4 times daily in 50 min.) to (3 M.) Langenbruck (*Curhaus*, pens. $5^1/_2$-8 fr.; *Ochsen*, pens. 5 fr.; *Pens. Bider*, etc.), situated on the pass of the *Obere Hauenstein* (2855'), a quiet and pleasant hill-sanatorium. — A high-road leads from Langenbruck to the S.E. to *Friedau* and (5 M.) *Egerkingen* (p. 14); another to the S.W. viâ *Holderbank* and the picturesque ruin of *Falkenstein* to *Balsthal*, and through the *Œnsinger Klus*, a defile formerly fortified, with the rebuilt château of *Blauenstein*, to ($10^1/_2$ M.) *Œnsingen* (p. 15). On the hill to the left is the restored château of *Bechburg*.

11 M. *Lausen*. Near (13 M.) *Sissach* (1233'; Löwe), a thriving village, we pass (r.) the small château and park of *Ebenrain*. Fine view from the *Sissacher Fluh* (2400'), 1 hr. to the N.

FROM SISSACH OVER THE SCHAFMATT TO AARAU ($13^1/_2$ M.). Branch-line viâ *Böckten* in $^1/_4$ hr. to ($2^1/_2$ M.) *Gelterkinden* (1370'; *Rössli), a manufacturing village; thence road through a picturesque valley past the *Hanggiessen* waterfall to ($1^1/_2$ M.) *Tecknau* (1440'); ($1^1/_2$ M.) *Wenslingen* (1860'); ($1^1/_2$ M.) *Oltingen* (1940'; Ochs), with a mineral spring. The path ascending the ($1/_2$ hr.) *Schafmatt* (2515') diverges close to the 'Ochs', and is easily found, being provided with finger-posts. The summit commands an extensive panorama of the Jura and the Alps, which we enjoy until we reach a point overlooking the deep valley of *Rohr*. Turning to the left here, we reach the upper part of a meadow, at the foot of which ($1/_2$ hr. from the top) lies a chalet and whey-cure establishment. From this point we enjoy a view of the environs of the Lake of Lucerne, the Rigi, Pilatus, etc., framed by the mountains between which we stand. From the chalet to *Aarau* (p. 22) in $1^1/_4$ hr., past the *Laurenzenbad* (p. 22), situated in a side-valley to the left, and *Erlisbach*.

To the S. of Sissach lies (7 M.; diligence twice daily in $1^1/_4$ hr.

14 *I. Route 3.*   OLTEN.   *From Bâle*

viâ *Zunzgen, Tenniken,* and *Diegten*) Eptingen or *Ruch-Eptingen* (1873'; *Curhaus,* with saline and mineral baths, pens. 4-5 fr.), situated in a narrow valley at the base of the *Hauenstein* (footpath to *Läufelfingen,* see below, 1 hr.; to *Langenbruck,* **see** p. 13, 1¹/₄ hr.).

The train turns to the S. into the narrow *Homburger-Thal,* and beyond (16 M.) *Sommerau* passes through two tunnels. — 19¹/₂ M. *Läufelfingen* (2010'; Sonne), at the foot of the *Hauenstein.*
On the summit of the Hauenstein, ascended in ³/₄ hr. from stat. Läufelfingen viâ *Reisen* and *Erlimoos* (each of which has a Curhaus), is situated the *Frohburg (2770'; **Hotel & Pension,* R. 2¹/₂, B. 1¹/₄, pens. 6-7 fr.), commanding a beautiful view of the Alps, from the Sentis to Mont Blanc; in the foreground the Wartburg (see below) and the Wigger-Thal with the railway to Lucerne; on the right rises Pilatus, on the left the Rigi. About 10 min. from the inn are some scanty ruins of a castle destroyed by an earthquake. Descent by *Trimbach* in 1 hr. to *Olten.*

The train now enters the *Hauenstein Tunnel,* 2970 yds. long and traversed in five minutes. Beyond it we observe on a hill to the right the small château of *Neu - Wartburg* (see below), to the right of which, farther on, the Bernese Alps gradually become visible from the Wetterhorn to the Doldenhorn, with the Jungfrau in the middle (comp. the Panorama, p. 145). The train descends by a long curve to the *Aare,* crosses it, and ascends on the right bank to —

24¹/₂ M. **Olten.** — *Hôtel Suisse, R. 2¹/₂, B. 1 fr.; St. Gotthard, unpretending, both at the station; Halbmond, well spoken of. — **Rail. Restaurant.*

Carriages generally changed here. Detention of ¹/₄-¹/₂ hr. As we leave the waiting-rooms, the trains for Bâle and Zürich are to the *left,* those for Lucerne and Bern to the *right.* Pocket-picking not uncommon here.

Olten (1295'; 4936 inhab.), the second town in the canton of Soleure, prettily situated on the *Aare,* is the junction of the lines to Aarau **and** Brugg (R. 7), to Aarburg and Lucerne (R. 5), to Bern (R. 4), **and** to Soleure and Neuchâtel (see below). The *Parish Church* contains an Ascension by Disteli, and the *Capuchin Church* a Madonna **by Deschwanden.** Extensive railway-workshops and large shoe-manufactories.

To the S.E. of Olten, on an isolated hill on the right bank of the Aare, rises the Neu-Wartburg or *Sälischloss* (2235'; *Restaurant*), a small château with a fine view of the Alps from the Sentis to the Jungfrau. Good paths from Olten and from Aarburg to the top in ³/₄ hr.

About 4¹/₂ M. to the N.E. of Olten (diligence twice daily in summer in 1¹/₄ hr.) are the sulphur-baths of **Lostorf** (**Curhaus,* moderate, pens. 5 fr.), prettily situated at the foot of the Jura. On a cliff above (¹/₄ hr.) rises the small château of *Wartenfels* (2060'), with a fine view.

Beyond Olten the train diverges to the right from the Bern and Lucerne line (p. 17), crosses the Aare, and traverses the plain watered by the *Dünnern,* at the base of the Jura. To the left the view **of the Alps** from the Glärnisch to the Altels is gradually unfolded. 26 M. *Olten-Hammer;* 27¹/₂ M. *Wangen;* 29 M. *Hägendorf.* — 31 M. *Egerkingen* (Kreuz).

Diligence twice daily in 40 min. to Friedau (2300'; **Curhaus,* pens. 5¹/₂-6 fr.), situated on the slope of the Jura, and well fitted up. Beautiful view of the Alps from Sentis to Mont Blanc. Shady grounds and extensive wood-walks. The road also leads to *Langenbruck,* 3 M. farther on (see p. 13; diligence in summer daily).

32 M. *Oberbuchsitten*; 36 M. *Œnsingen* (diligence twice daily in 1³/₄ hr. to *Langenbruck*, p. 13); 37 M. *Niederbipp* (to the right of which is *Oberbipp*, with a handsome modern château). At (41 M.) *Wangen* the train crosses the Aare. Beyond *Deitingen* and *Luterbach* we obtain a view of Soleure with the minster of St. Ours; to the right are the Röthi and the Curhaus on the Weissenstein (p. 16). The train crosses the *Grosse Emme*, not far from its confluence with the Aare. — 47 M. *Neu-Solothurn*.

Soleure. — Soleure has two RAILWAY STATIONS: *Neu-Solothurn*, on the right bank of the Aare, for the lines to Olten, Herzogenbuchsee, Burgdorf, Lyss, and Bienne, and *Alt-Solothurn*, on the left bank, to the W. of the town, for the line to Bienne.

Hotels. *KRONE, with café-restaurant, R., L., & A. 3, B. 1¹/₄, D. 3 fr.; *STORCH; HIRSCH; THURM; SCHWAN, well spoken of.

Soleure, or *Solothurn* (1425'; 8462 inhab.), on the *Aare*, a quiet place, the capital of Canton Soleure, was incorporated with the Confederation in 1481, and claims to be the oldest town on this side of the Alps next to Trèves. ('*In Celtis nihil est Salodoro antiquius, unis exceptis Treviris, quarum ego dicta soror*', is the inscription on the clock-tower.) It was the Roman *Salodurum*.

The CATHEDRAL OF ST. OURS, a cathedral of the Bishopric of Bâle (p. 4), was built in 1762-73 on the site of an edifice of 1050, in the form of a cross, surmounted with a dome and two half-domes. A flight of 36 steps leads to the façade. One of the adjoining fountains is adorned with a statue of Moses striking the rock, the other with a figure of Gideon wringing the dew from the fleece. The ten large altar-pieces, dating from the latter half of the 18th cent., are unimportant. The treasury, in the sacristy, contains some good artistic work in metal and textile fabrics of the 14-18th centuries.

The *ARSENAL, not far from the cathedral, contains (on the second floor) a collection of ancient armour and weapons. Among the curiosities is a mitrailleuse of the 15th century. A large plastic group close to the entrance represents the reconciliation of the Confederates effected at the Diet of Stans in 1481 by Brother Klaus (p. 132).

The oldest building in Soleure is the CLOCK TOWER, recently restored, which is said to have been erected in the 4th cent. B.C., but really an early Burgundian building of the 5th or 6th cent. A.D. The figures and mechanism of the clock resemble those at Bern (p. 146).

The *Natural History Cabinet*, in the suburb on the right bank of the Aare, contains valuable collections of zoology and palæontology. In the Cantonal School are a number of *Roman and Mediaeval Antiquities* and the *Cantonal Library*. The *Town Library* contains about 40,000 vols. and 200 incunabula, besides coins and medals. The *Municipal Picture Gallery* possesses a *Virgin and Child, with SS. Ours and Martin of Tours, one of the chief works of *Holbein the Younger* (1522), much restored. — A tablet on No. 5 Gurzelngasse marks the house in which *Thaddeus Kosciuszko* died (1817).

WEISSENSTEIN.

The *Weissenstein (4220'), 3 hours' walk or drive to the N. of Soleure, is deservedly a very favourite point of view. It is reached either by the carriage-road viâ *Längendorf* and *Oberdorf* (two-horse carr. in 2½ hrs., up 20, down 10, there and back 25 fr. and fee), or (preferable) by the footpath (guide or porter 4-5 fr.) ascending the Verena-Thal. Taking the latter, we pass the cathedral of St. Ours, and through the handsome Bâle gate, and then bear to the left towards the *Villa Cartier* with its two towers, where we turn to the right. Farther on we enter the avenue to the left, at the end of which we turn to the right towards the church of *St. Nicholas*. Before reaching the church our route passes the *Restaurant Wengistein* and turns to the left into the °St. Verena-Thal (1 M. from Soleure), a narrow, cool, and shady ravine, ½ M. in length. The path to the left, at the beginning of the gorge, leads to the Wengistein (see below). At the other end of the valley are quarries of Portland limestone, where interesting fossils are found. The blocks of granite on the neighbouring slopes are believed by geologists to have been deposited by ancient Alpine glaciers. This gorge is now converted into a promenade.

At the N. end of the ravine is the **Hermitage of St. Verena**. On the right are the hermit's dwelling and a chapel; on the left is a rock-hewn chapel, reached by a broad flight of steps, and containing a representation of the Holy Sepulchre with life-size figures. We may now ascend by the chapel to the crosses, pass near the large quarries (with 'Gletscherschliffe', or rocks worn by the action of the glaciers), and traverse the wood to the **Wengistein**, the view from which is similar to that from the Weissenstein, though on a smaller scale. A huge granite boulder here bears a Latin inscription recording two memorable events in the history of Soleure.

From the restaurant beyond the hermitage we take to the right, in the direction of the Weissenstein; at (10 min.) the village of *Widlisbach* we turn to the left and cross the hill to (12 min.) the hamlet of *Failern* (1827'), at the foot of the Weissenstein. Above it we enter the wood to the left by a finger-post, ascend gradually, and then in steep zigzags to the (40 min.) first bench, above which there are several others. The path soon quits the wood and ascends an abrupt rocky gully, partly by means of steps. Farther up, the ascent is through wood and more gradual. In 40 min. we regain the road (to the left) above the *Nesselboden Alp* (3447'), and, following it, reach in 40 min. more the °**Curhaus** on the *Vordere Weissenstein* (4220'; R., L., & A. 3-4, B. 1¼, D. 3½, S. 2½, pension 7-10 fr.; telephone to Soleure), a sanatorium surrounded by woods and pastures, and much resorted to in summer (English Church service). The footpath, diverging to the right at the end of the wide curve, 8 min. from the Nesselboden Alp, and then ascending abruptly to the left at the post on the top, is a short-cut.

The *VIEW is less picturesque, but more extensive than that from the **Rigi**; and no spot commands a better view of the whole Alpine chain from Tyrol to Mont Blanc. To the E. are distinguished the Sentis, the Glärnisch, with the Rigi in the foreground, the Tödi between the Rigi and Pilatus, the lofty saddle of Titlis, and the Sustenhorn; beyond Soleure are the Wetterhorn and Schreckhorn, the Finsteraarhorn, Eiger, Mönch, Jungfrau, Blümlisalp, **and** Doldenhorn; then the Balmhorn, Altels, Wildstrubel, Wildhorn, Diablerets, and to the S., Mont Blanc. To the S.W. glitter the lakes of **Bienne**, Morat, and Neuchâtel; the Aare winds to the S. through the fertile **plains**, and the Grosse Emme flows into it at the foot of the mountain.

Pleasant walk to the W. through the wood **to** the (10 min.) *Käuzeli* (4093'). — The **Röthi** (4590'), 1½ hr. to the E. of the hotel, commands an extensive view to the N. and E. of the Black Forest and Vosges, which are hidden from the Weissenstein, and of the picturesque mountains and valleys of the Jura. — Towards the W. the view is concealed by the *Hasenmatt (4745'), 1¼ hr. from the hotel, whence an uninterrupted panorama may be enjoyed. The path to it (white marks) leads across the pastures to the W. to (25 min.) the *Hintere Weissenstein* (4027'; Inn). A pleasanter route leads by the shady footpath, which enters the woods to the right above the pastures, but which must be quitted as soon as it begins

BURGDORF. *I. Route 4.* 17

to ascend more steeply. Shortly before reaching the Hintere Weissenstein we descend a little to the left and cross the ridge to (20 min.) the end of the meadows; then descend for 1/4 hr. in the *Kesselwald*, and ascend across pastures to (20 min.) the chalet of *Althüsli* (4375'; simple rfmts.), on the saddle, with a good spring. An easy path leads hence to the summit in 20 min. (the path, diverging to the left, 10 min. before the chalet, is shorter but steeper). — We may descend from the Hasenmatt or the chalet on the S. side, pass *Lommiswyl*, and regain Soleure, or the nearer station of Selzach (see below). Those returning from the Curhaus to Soleure follow the road from Fallern (p. 16) to (1/2 M.) a sign-post with four arms, whence a path between pine-woods and large quarries brings them in 1/2 hr. to the N.W. gate of Soleure. Carriages may also be directed to return by a route affording an opportunity of visiting the St. Verena gorge.
From *Soleure to Herzogenbuchsee*, see below.

FROM SOLEURE TO BURGDORF (13 M.) by the Emmenthal railway in 40-50 minutes. The principal station is (7 M.) *Utzensdorf*, the largest village in the lower Emmen-Thal. *Burgdorf*, see below.

FROM SOLEURE TO LYSS (15 M.) by railway, skirting the right bank of the Aare, in 1-1 1/2 hour. The chief intermediate station is (10 M.) *Büren* (Krone), a small town with an old château. *Lyss*, see p. 12.

The Bienne line crosses the Aare. 48 M. *Alt-Solothurn* (p. 15); 51 M. *Selzach*; 54 1/2 M. *Grenchen* or *Granges*; 57 M. *Pieterlen*.

63 M. *Bienne*, see p. 12.

4. From Bâle to Bern viâ Herzogenbuchsee.

66 M. RAILWAY in 3 1/4-4 3/4 hrs. (fares 11 fr. 50, 8 fr. 5, 5 fr. 75 c.).

To (24 1/2 M.) *Olten*, see pp. 13, 14. The line skirts the right bank of the Aare; to the left, the château of *Neu-Wartburg* (p. 14).

27 M. Aarburg (1285'; *Krone; Bär*), a thriving little town (2079 inhab.), picturesquely situated on the Aare (junction for *Lucerne*, p. 18). The old castle on a hill, built in 1660, is now a factory.

As we proceed we have glimpses of the Alps, first to the right and then to the left. The line continues on the right bank of the Aare. 30 M. *Rothrist*; 33 M. *Murgenthal*, where we cross the *Murg*; 35 M. *Roggwyl*; 37 1/2 M. *Langenthal* (*Bär; Löwe*), a thriving village with a busy timber-trade (branch-line in 40 min. to *Huttwil*); 39 1/2 M. *Bützberg*.

42 M. Herzogenbuchsee (1500'; 2316 inhab.; *Sonne; Hôt. de la Gare*) is a considerable place, with a loftily situated church.

To SOLEURE (9 1/2 M.) railway in 40 minutes. 2 1/2 M. *Inkwyl*; 5 1/2 M. *Subigen*; 7 M. *Derendingen*, beyond which we cross the *Grosse Emme* to *Neu-Solothurn* (p. 15).

Near (45 1/2 M.) *Riedwyl* we enter a grassy valley with wooded slopes. Beyond (48 M.) *Wynigen* a long tunnel (1 min.). The train crosses the *Grosse Emme* to —

52 1/2 M. Burgdorf, Fr. *Berthoud* (1863'; *Hôt. Gugglsberg, Hôt. de la Gare*, both at the station; *Maison de Ville; Ours*), a busy town (6876 inhab.), picturesquely situated. The substantially built houses are flanked with 'Lauben', or arcades, as at Bern. The public buildings, the hospital, schools, orphanage, and technical institute testify to the wealth and taste of the community. In the château of

Burgdorf, in 1798, Pestalozzi established his famous school, which in 1804 he removed to Yverdon (p. 214). Beautiful views from the church and château; finer from the *Lueg* (2885'), 2 hrs. to the E.

FROM BURGDORF TO LANGNAU, 14 M., railway in ³/₄-1 hr. The line ascends the fertile *Emmen-Thal*. 2¹/₂ M. *Oberburg*; 4¹/₂ M. *Hasle-Rüegsau*. From Rüegsau, 1¹/₂ M. to the N.E. of the railway, the *Rachisberg* (2768'; fine view of the Alps and the Jura) may be ascended in ¹/₂ hr. — 6 M. *Lützelflüh-Goldbach*. Lützelflüh was the home of the pastor Albert Bitzius (d. 1854), a popular author well-known under the name of Jeremias Gotthelf. — 7¹/₂ M. *Ramsey-Sumiswald* (the latter lying 3 M. to the N.); 9 M. *Zollbrück*; 14 M. *Langnau* (p. 140).

From Burgdorf to *Soleure*, see p. 17.

54¹/₂ M. **Lyssach**. Beyond (56 M.) *Hindelbank* **a monument**, to the left of the railway, commemorates the battle between the Bernese and the French in the *Grauholz*, March 15th, 1798. — 59 M. *Schönbühl*. Beyond (61¹/₂ M.) *Zollikofen* (junction for *Bienne*, p. 12) the train crosses the iron *Worblaufen Bridge* (below, to the right, the handsome *Tiefenau Bridge* over the Aare) and then ascends through a cutting to the *Wyler Feld*, whence, to the left, we obtain a magnificent view of the Bernese Alps (comp. Panorama, p. 145). To the right is the suburb *Lorraine*, beyond which we cross the Aare and enter the station of Bern. The *Bridge, 200 yds. long and 142' high, has a roadway for ordinary traffic below the railway. — 66 M. *Bern*, see p. 144.

5. From Bâle to Lucerne.

59 M. RAILWAY in 2¹/₂-4¹/₂ hrs. (fares 10 fr. 25, 7 fr. 15. 5 fr. 10 c.).

To (27 M.) *Aarburg*, the junction for *Bern* (R. 4), see p. 17. The Lucerne line traverses the broad grassy *Wiggerthal*.

30 M. Zofingen (1430'; pop. 4496; *Rössli*; *Ochs*), a busy little town. The library in the Town Hall contains a collection of coins, autographs of Swiss reformers, and the album of the society of Swiss artists, founded in the year 1806, which formerly met at Zofingen annually. On the branches of the fine old lime-trees near the *Schützenhaus* two 'ball-rooms' have been constructed. In the *Bleichegut*, near the town, are the remains of a Roman bath.

FROM ZOFINGEN TO SUHR, railway in 36 minutes. Stations *Safenwyl*, *Kölliken*, *Entfelden*, well-to-do villages, and (10¹/₂ M.) *Suhr*, the junction for Aarau and Baden (p. 22).

33 M. *Reiden*, an old lodge of the knights of Malta, now a parsonage. 35 M. *Dagmersellen*; 37 M. *Nebikon* (diligence daily in 3 hrs., viâ *Willisau*, to *Wohlhausen* in the Entlebuch, p. 198). To the right appear the Bernese Alps; in the centre the Jungfrau, with the Mönch and Eiger to the left of it and the Altels to the right. Beyond (39¹/₂ M.) *Wauwyl* the little *Mauensee*, with its island and castle, lies on the right.

43¹/₂ M **Sursee** (1690'; pop. 2135; *Sonne*; *Hirsch*), an old town, over whose gates the double eagle of Hapsburg is still enthroned. The *Town Hall* recalls the Burgundian style.

RHEINFELDEN. *I. Route 6.* 19

Near (46 M.) *Nottwyl* we approach the *Lake of Sempach* (1663'), 5 M. long, 1½ M. broad, and abounding in fish. On a hill to the right rises *Schloss Wartensee*.

49½ M. **Sempach.** The small town (pop. 1097; *Kreuz*; *Adler*, moderate) lies 1½ M. to the N., on the S.E. bank of the lake. Near Sempach Duke Leopold of Austria was signally defeated on 9th July, 1386, by the Swiss Confederates, owing, according to the story, to the noble self-sacrifice of Arnold von Winkelried. The duke himself and 263 of his knights were slain. A column surmounted by a lion was erected beside the church in 1886 on the 500th anniversary of the victory.

A CHAPEL (2064'), 1½ M. to the N.E. of Sempach, marks the spot where Leopold fell. His uncle, another Duke Leopold, had been defeated by the Swiss 71 years before at Morgarten (p. 80). The anniversary is still kept.

The train intersects plantations of firs. On the right appear the precipitous cliffs and peaks of Pilatus; on the left the long crest of the Rigi; between these tower the snowy Alps (see p. 83); the isolated mountain adjacent to Pilatus, rising above the lake, is the Titlis. 53 M. *Rothenburg*; 56 M. *Emmenbrücke* (Hôtel Emmenbrücke; Restaurant Seethal), the junction of the 'Seethal' line to Lenzburg (p. 140). The line crosses the *Emme*, a little above its junction with the *Reuss*, and follows the latter, being joined on the left by the Zürich and Lucerne line (p. 81), and on the right by the Bern and Lucerne line (p. 138). Lastly we pass through a tunnel under the *Gütsch* (p. 85).

59 M. *Lucerne*, see p. 81.

6. From Bâle to Zürich.

56 M. RAILWAY in 2¼-3¼ hrs. (fares 9 fr. 40, 6 fr. 60, 4 fr. 75 c.).

To (5 M.) *Pratteln*, see p. 13. Near (7½ M.) *Augst*, picturesquely situated, we cross the *Ergolz* and approach the Rhine. On the left is *Kaiser-Augst*, with salt-works and an old church; opposite, on the left bank of the Ergolz is the hamlet of *Basel-Augst* (p. 3).

10½ M. **Rheinfelden.** — *GRAND HÔTEL DES SALINES, 5 min. above the town, pension 8-12 fr.; *HÔTEL DIETSCHY ZUR KRONE, with terrace on the Rhine; *HÔT.-PENS. ZUM SCHÜTZEN, R. & A. 1½-6, B. 1, D. 2½, pens. 6-7½, omn. ½ fr.; *DREIKÖNIG, pens. 5 fr.; SCHIFF, R., L., & A. 1½-2, B. 1, D. 2½, pens. 5 fr., all with salt-baths; *BELLEVUE, well situated on the right bank of the Rhine, R. 1½-2, B. 1, D. 2½, pens. 5-6 fr.; beer at the *Salmen*. — ENGLISH CHURCH SERVICE in summer.

Rheinfelden (873'), an old town with 2400 inhab., once strongly fortified, with walls and towers still partly preserved, was one of the outposts of the Holy Roman Empire. After repeated sieges it was razed to the ground by the French in 1744. Since 1801 it has belonged to Switzerland. The foaming river here dashes over the rocks, forming the *Höllenhaken* rapids. Near the town are extensive salt-works on the Rhine.

We quit the Rhine, which here describes a bend to the N., pass (13 M.) *Möhlin* and (17 M.) *Mumpf* (*Soolbad zur Sonne; Güntert),

2 *

and then return to the river for a short distance. — 18½ M. *Stein* (990'; *Löwe), connected by a covered bridge with *Säckingen* (p. 23).

FROM STEIN TO COBLENZ, 16 M., railway in 48 min. (2 fr. 80, 2 fr., 1 fr. 40 c.). The line skirts the left bank of the Rhine; stations *Sisseln*, *Laufenburg* (p. 23), *Sulz*, *Etzgen*, *Schwaderloch*, *Leibstatt*, *Felsenau*; then across the Aare to *Coblenz* (p. 23).

We quit the Rhine, and at (20½ M.) *Eiken* enter the fertile *Sisseln-Thal*. 23 M. *Frick* (1120'; Adler; Engel), a considerable village. The train ascends in a long curve to (26 M.) *Hornussen* (1275'). 28½ M. *Effingen* (1425'), the highest point on the line. Then a tunnel, 2697 yds. long (4 min.), under the **Bötzberg** (1945'), the **Mons Vocetius** of the Romans. 31 M. *Bötzenegg* is the station for the village of *Schinznach* (p. 23). The train gradually descends, affording a magnificent view of the valley of the Aare to the right, and, in clear weather, of the St. Gall, Glarus, and Schwyz Alps, and crosses the **Aare** by a bridge 259 yds. long and 104' high.

36 M. **Brugg** (1096'; pop. 1572; *Rothes Haus*; *Rössli*; *Hôt. Bahnhof*; *Rail. Restaurant*), an antiquated little town, the junction for *Aarau* and *Waldshut* (R. 7), is best surveyed from the bridge over the Aare, here hemmed in by rocks. The 'Schwarze Thurm', by the bridge, dates from the later Roman Empire; the upper part was rebuilt in the 15th century.

The ancient Abbey of **Königsfelden** (¾ M. to the S.E. of Brugg), formerly a convent of Minorites, was founded in 1310 by the Empress Elizabeth and her daughter, Queen Agnes of Hungary, on the spot where **Albert of** Austria, husband of the former, had been murdered two years before (1308) by John of Swabia and his accomplices. It was secularised in 1528; the building was converted into an hospital, and in 1872 into a lunatic asylum. Of the old buildings there now remain the S. part only, the church, and the dwelling of Queen Agnes, which last now contains a collection of antiquities. The stained-glass "Windows in the choir, opposite the door, are of the 14th cent. and portray the history of Agnes, etc. Part of the choir, with the tomb of Duke Leopold (p. 19), is now a cart-shed. On the walls are portraits of the chief knights who fell at Sempach (painted soon after the battle, but now much damaged).

On the tongue of land formed by the Reuss and the Aare once stood the considerable Helvetian town of **Vindonissa**, which in the early centuries of the Christian era was the headquarters of a Roman legion with its Rhaetian cohorts, as is proved by inscriptions. The position of the amphitheatre is recognisable; and the well of the Abbey of Königsfelden is fed by a subterranean Roman conduit. The town was destroyed in the 5th cent., and there is now no trace of its extensive edifices; but the name still survives in that of the village of *Windisch*, 1 M. to the E. of Brugg.

FROM BRUGG TO WOHLEN, 11 M., railway in 40 minutes. — 3 M. *Bierfeld*; 5½ M. *Othmarsingen* (junction for Wettingen and Aarau, p. 22); 7½ M. *Hendschikon* (p. 22); 8½ M. *Dottikon-Dintikon* (p. 22); 11 M. *Wohlen-Villmergen*. (To *Rothkreuz*, see p. 22.)

We cross the *Reuss* near its union with the Aare, and beyond (38 M.) *Turgi* (p. 23; Buffet), reach the **Limmat** and follow its left bank. The steep slopes are clad with vines.

42 M. **Baden** (1257'; pop. 3887; *Hôtel de la Gare*, R., L., & A. 1½-2, B. 1, D. 2½, pens. 6 fr.; *Hôtel de la Balance*, R. 1½-2, B. 1, D. 2, pens. 6½ fr.) was much visited even in Roman times for

the sake of its mineral springs *(Aquae Helvetiae)*. In the reign of Nero, according to Tacitus (Hist. i. 67), it had all the appearance of a town (*'in modum municipii exstructus locus, amoeno salubrium aquarum usu frequens'*). In the middle ages Baden was a fortress, and down to the beginning of the 15th cent. was often the residence of the counts of Hapsburg. The extensive ruins of the fortress *Stein zu Baden* (1505'), destroyed in 1415 and again in 1712, rise above the town (1/4 hr. from the station); pretty view from the top and from the adjacent *Café Belvedere*.

The hot mineral springs (98°-126° Fahr.) lie in the narrow valley of the Limmat (1190'), 5 min. to the N. of the station, 1/2 M. from the town. The '*Small Baths*' *(Adler; Engel; Hirsch; Rebstock; Schwan)*, in *Ennetbaden*, on the right bank of the Limmat, are chiefly frequented by the neighbouring peasantry; the '*Great Baths*' (*Neue Curanstalt Baden*, or *Grand Hôtel*, pens. 8-12 fr.; *Schiff*; *Verenahof*, 8 fr.; *Blume; Schweizerhof*, 6-7 fr.; *Freihof*; *Limmathof; Ochs; *Bär)* lie on the left bank. The Bahnhof-Str. leads from the station to the handsome *Cursaal*, with its pleasant grounds (*Restaurant; music several times daily) and to the **Curanstalt** (see above). Good view from the lower Limmat bridge (1175'); opposite, on the right bank, is the *Café Brunner*, with a garden. From the upper bridge a footpath leads to the left to (1/2 M.) the *Restaurant Schartenfels*, which commands a fine view.

From Baden to *Aarau*, see p. 22; station on the S.W. side of the upper town, 1 M. from the baths.

We pass through a short tunnel under the Stein zu Baden (see above), and cross the Limmat to (43 M.) *Wettingen*. The village lies on the left, at the foot of the vine-clad *Lägerngebirge* (2830'); on the right, surrounded by the Limmat, are the extensive buildings and gardens of the former Cistercian **Abbey of Wettingen**, now a seminary for teachers. The church contains a sarcophagus in which the remains of the Emp. Albert (see p. 20) reposed for 15 months before their removal to Spires. Stained-glass windows of the 16th and 17th cent., carved stalls of the 17th

FROM WETTINGEN TO OERLIKON, 13 1/2 M., railway in 1 1/4 hr. — 2 1/2 M. *Würenlos;* 4 1/2 M. *Otelfingen-Daenikon* (branch-line by *Buchs* and *Niederglatt* to *Bülach*, p. 48); 6 M. *Buchs-Daellikon*; 8 1/2 M. *Regensdorf-Watt*, a little to the E. of which is the small *Katzensee* (°Inn); 10 1/2 M. *Affoltern*; 12 1/2 M. *Seebach*; 13 1/2 M. *Oerlikon* (p. 48).

The train again crosses the deep bed of the Limmat and follows its left bank to Zürich. 46 M. *Killwangen*. — 49 M. **Dietikon** (1285'; *Löwe*). It was here that Masséna effected his famous passage of the Limmat, 24th Sept., 1799, after which he repulsed the Russians and took Zürich. — 51 M. *Schlieren;* 53 1/2 M. *Altstetten* (p. 78). To the right stretches the long ridge of the Uetli with its inn (p. 39). We now cross the *Sihl* and enter the station of —

56 M. **Zürich**, see p. 33.

7. From Olten to Waldshut viâ Aarau and Brugg.

32½ M. Railway in 2 hrs. (fares 5 fr. 60, 4 fr., 3 fr. 85 c.).
Olten, see p. 14. The train runs near the *Aare* as far as Brugg. To the left rise the picturesque Jura Mts.

4 M. *Dänikon*; 5½ M. *Schönenwerth;* on the opposite bank of the Aare is *Schloss Gösgen*, with a ruined tower. A tunnel now carries us under the loftily situated town of —

8½ M. **Aarau** (1263'; pop. 6809; **Rössli; *Ochs; *Löwe; *Wilder Mann* R. 2 fr.; U.S. Consular Agent), a manufacturing place, the capital of Canton Aargau, on the *Aare* (which is crossed by a suspension-bridge, constructed in 1850), and at the foot of the Jura, the slopes of which at places are planted with the vine. The *Gross-Rathsgebäude* contains fine stained glass (from the Abbey of Muri, 16th cent.) and the Cantonal Library (60,000 vols.). The Geographical and Commercial Society of Central Switzerland has here founded an interesting **Ethnographical Industrial Museum*. A house in the Rathhaus-Platz (No. 882) contains interesting antiquities from Vindonissa (p. 20). The historian Heinrich Zschokke (d. 1848) once lived here; his house, the '*Blumenhalde*', is passed on the pleasant walk across the suspension-bridge to the (¼ hr.) **Alpenzeiger* on the *Hungerberg* (Restaurant, with fine view; pens. 4 fr.).

Above the town, to the N., rises the *Wasserfluh* (2850'), and to the N.E. the *Giselafluh* (2540'), over which a path, with a view of the lakes of **Hallwyl** and **Baldegg**, leads to the Baths of Schinznach. — Pleasant road from Aarau by *Erlisbach* (p. 13) to the (4 M.) *°Laurenzenbad*, prettily situated in the Jura. — About 6 M. to the W of Aarau are the sulphur-baths of *Lostorf* (p. 14), the road to which passes Erlisbach and *Stüsslingen*. — From Aarau to *Sissach* over the *Schafmatt*, see p. 13.

From Aarau to Rothkreuz, 29½ M., railway in 1½-2 hrs. — 4 M. *Ruppersweil* (see below); 6 M. *Lenzburg* (p. 142); 8 M. *Hendschikon*; 10 M. *Dottikon-Dintikon;* 12½ M. *Wohlen-Villmergen*, two considerable villages (junction for *Brugg* and Bâle, p. 20). Branch-line hence to the E. to (5 M.) *Bremgarten* (Drei Könige; Kreuz), a small town on the Reuss, with a château. — Then (16 M.) *Boswyl-Bünzen* and the (18 M.) charmingly situated **Muri** (1590'; *°Löwe*, with salt and mineral baths, pens. 4-5½ fr.; *Adler*), with a former Benedictine Abbey burned down in 1889, but to be rebuilt. Near the town is the picturesque wooded *Mühltobel* with several waterfalls. On a hill, 1½ hr. to the S.E., is *°Schloss Horben* (2625'; pens. from 4 fr.), with extensive wood-walks and a beautiful view. — 20½ M. *Benzenschwyl;* 22½ M. *Mühlau*, on the Reuss; 25 M. *Sins;* 27 M. *Oberrüti*. We then cross the Reuss to (29½ M.) *Rothkreuz* (pp. 80, 109).

From Aarau to Baden, 17½ M., railway in 1½ hr. — 3 M. *Suhr* (branch-line to *Zofingen*, p. 18); 5½ M. *Hunzenschwyl* (on the right rises the *Staufberg*, see below). 7½ M. *Lenzburg* (p. 142; 'Seethalbahn' to Lucerne, see R. 41), where the *Aa* is crossed. 10½ M. *Othmarsingen*, junction for **Brugg** and Wohlen (p. 20). Near (11 M.) *Mägenwyl*, on a spur of the *Kestenberg*, to the left, rises *Schloss Brauuegg*. The train crosses the *Reuss*. 13½ M. *Mellingen;* 15½ M. *Dättwyl;* 17½ M. *Baden* (the station lies to the S.W. of the upper town, see p. 20).

On the left, beyond the Aare, at the foot of the Giselafluh, lies **Biberstein**, with an old castle. 13 M. *Ruppersweil;* to the right the **Staufberg** and the château of *Lenzburg* (p. 142). — 15 M. *Wildegg* (Aarhof), with a castle of that name, at the foot of the *Wülpelsberg*, has mineral springs containing iodine and bromine, the water of

SÄCKINGEN. *I. Route 8.* **23**

which is exported. On a hill beyond the Aare rises *Schloss Wildenstein*. — 17½ M. Stat. *Schinznach* lies ½ M. from **Bad Schinznach** (1203'), on the right bank of the Aare, with sulphur-baths, frequented by French visitors (physician, *Dr. Amsler*; R. in the *Neues Bad* from 4, board 8, bath 2, music ½ fr. per day; in the *Altes Bad*, frequented by Swiss visitors, about half as much).

The baths lie at the foot of the *Wülpelsberg* (1686'), on the top of which (½ hr.) are the ruins of the **Habsburg**, the cradle of the imperial family of Austria, erected by Count Radbod von Altenburg about 1020. The tower, with walls 8' thick, is the only part now standing. The adjoining house is occupied by a farmer. The view embraces the entire dominions of the ancient counts of Hapsburg, and the valleys of the Aare, Reuss, and Limmat, bounded on the S. by the Alps. — The village of *Schinznach* lies about 2½ M. to the S.W., on the left bank of the Aare. The nearest station is *Bötzenegg* (p. 20).

19½ M. **Brugg**, and thence to (22 M.) *Turgi*, see p. 20. The Waldshut train crosses the *Limmat* near its influx into the Aare, passes stat. *Siggenthal*, and traverses the broad valley of the Aare, which it approaches near (28 M.) *Döttingen-Klingnau*. It then describes a wide curve, passes through a tunnel, and crosses the Rhine near (30½ M.) *Coblenz*, above the mouth of the Aare.

32½ M. *Waldshut*, see p. 24.

8. From Bâle to Schaffhausen and Constance.

89 M. BADEN RAILWAY in 5 hrs. (to Schaffhausen 9 fr. 50, 6 fr. 30, 4 fr. 5 c.; to Constance 14 fr. 50, 9 fr. 65, 6 fr. 20 c.). *Neuhausen* (p. 24) is the station for the Falls of the Rhine (R. 9). Views to the *right*. — STEAMER from Schaffhausen to Constance in 3½-4 hrs. (descending in 3 hrs.), pleasant if time and weather permit (see p. 25; fares 4 fr., 1 fr. 95 c.).

Bâle (Baden station), see p. 2. We traverse a fertile plain between the S. spurs of the Black Forest and the Rhine. 3 M. *Grenzach*; 5 M. *Wyhlen* (Hôtel Bilmaier); 7½ M. *Herthen*. At (10 M.) *Bei Rheinfelden* (*Bellevue), opposite Rheinfelden (p. 19), the line approaches the *Rhine*, which here dashes over rocks. The left bank is precipitous and wooded. — 12 M. *Beuggen*; to the right are a large reformatory and a seminary, formerly a Teutonic lodge. 15 M. *Niederschwörstadt*. To the left of (17 M.) **Brennet** opens the *Wehrathal (see *Baedeker's Rhine*).

20 M. **Säckingen** (957'; *Soolbad* or *Löwe*; *Schütze*), a considerable town, has a large abbey-church with two towers. The castle on the Rhine, which figures in Scheffel's poem 'Der Trompeter von Säckingen', is now the property of Hr. Balli. Pretty grounds.

24 M. *Murg* (Zum Murgthal), where we cross the Murg. Opposite (25½ M.) *Laufenburg* (*Post) is the Swiss town of Laufenburg (980'; *Rheinsoolbad; Adler*), very picturesquely placed on the left bank, with its lofty church, ruined castle, and old watch-towers (railway-station, see p. 20). The Rhine here forms impetuous rapids called the '*Laufen*'.

A long tunnel; then, beyond (29 M.) *Albert-Hauenstein*, a lofty viaduct. At intervals we approach the river. Near (30 M.) *Albbruck* (Zum Albthal) the *Alb* is crossed. 32 M. *Dogern*.

35 M. **Waldshut** (1122'; *Hôtel Schätzle*, at the station; *Hôtel Blume; Rebstock*, in the town) lies high above the river. — Railway to *Turgi* (for Zürich), see p. 23; to *Winterthur*, see p. 48.

Beyond Waldshut a tunnel; to the right, occasional glimpses of the Alps. Before (38 M.) *Thiengen* (Krone) we cross the *Schlücht*, and at (40½ M.) *Oberlauchringen* the *Wutach*. To the right, on a wooded height, is the ruin of *Küssenberg*. Stations *Griessen*, *Ersingen*, *Wilchingen-Hallau*, *Neunkirch*, *Beringen*, and (57½ M.) *Neuhausen*, the station for the *Falls of the Rhine* (p. 26).

59 M. **Schaffhausen.** — Hôt. Müller, R., L., & A. from 2½, B. 1¼, déj. 2½. D. 3; pens. from 6 fr.; Rheinischer Hof, similar charges; *Riese*, R., L., & A. 2-2½. B. 1¼, lunch 2, D. 2½. pens. 7 fr., all three at the station; *Post*, in the Herrenacker, 3 min. from the station; *Schwan*, R., L., & A. 2-2½. B. 1, D. 2-2½ pens. 8 fr.; Krone, R., L., & A. 2-6, B. 1-1¼, lunch 1½-2, D. from 2, pens. 5-7 fr.; *Tanne*, plain, R., L., & A. 1¼-2, B. 1, D. 1½. pens. 4½ fr.; Schiff, on the Rhine, unpretending. — *Restaurant Rebmann*, at the station; *Rail. Restaurant.* — Baths in the Rhine, at the upper end of the town. 6-1 and 5-8, for ladies 2-5.

Schaffhausen (1415'; pop. 12,400), the capital of the canton of that name, still retains some of the features of a Swabian **town of the empire**. It presents a most picturesque appearance when seen from the village of *Feuerthalen*, on the left bank of the Rhine, or from the *Villa Charlottenfels* (1385') on the right bank. Hr. Moser (d. 1874), the late proprietor of the villa, originated the imposing **Waterworks** in the Rhine (outside the Mühlenthor), by means of which the factories of the town are supplied with water-power.

The Cathedral, once an abbey-church, an early-Romanesque basilica, was erected in **1052-1101**. Interior lately restored. The Gothic cloisters are tolerably **preserved**. The inscription on the great bell, cast in 1486: *Vivos voco, mortuos plango, fulgura frango*, suggested Schiller's beautiful 'Lied von der Glocke'. The late-Gothic *Church of St. John* contains an excellent new organ.

The castle of Munot (properly *Unnot*), built in 1564-82 and recently restored, commands the town. It consists of a round tower containing a winding **inclined plane instead of a staircase**, with walls 16' thick (fine view from the top).

The *Imthurneum*, in the Herrenacker, erected and presented to the town by Hr. Imthurn (d. 1881), a native of Schaffhausen and a London banker, contains a theatre, a music-school, and concert **rooms**. Opposite is the *Museum*, with natural history specimens and antiquities (including those found in the Kesslerloch near Thayingen), and the town-library. In the neighbouring government buildings is preserved a large ancient **onyx**, representing a goddess of peace (adm. 11-12 gratis; at other times 1 fr.).

In the pretty *Fäsenstaub Promenade* is a bust of the Swiss historian Johannes v. Müller (b. at Schaffhausen, 1752; d. at Cassel, 1809). The lofty terrace affords a fine view of the Rhine and the Alps.

From Schaffhausen to the *Falls of the Rhine* (2 M.), see p. 27. Carriage with one horse to the Schlösschen Wörth, and back from Neuhausen to Schaffhausen, including stay of 1 hr., 7 fr. Omnibus from the Schaffhausen station 12 times daily, see p. 27. — Pretty walk through the *Mühlen-Thal* to the *Seckelamtshüsli*, with a view of the Alps, and back to Schaffhausen by the *Hochfluh* (another fine point of view) and the suburb of *Steig* (1½ hr. in all). Other fine views may be obtained from the *Beringer Randen* (belvedere), 4 M. to the W. (to Beringen station in 20 min., see p. 24), and from the *Hohe Randen* (2855'), 10½ M. to the N.W., reached viâ *Hemmenstadt* or *Merishausen*. — From Feuerthalen to *Etzweilen*, see p. 32.

Stations *Herblingen*, *Thayingen*, and *Gottmadingen*. — 71 M. **Singen** (*Krone; *Adler; Ekkehard; Rail. Restaurant*), the junction for the Black Forest Railway. About 3 M. to the N.W. rises the *Hohentwiel* (2245'), with grand ruins and a noble view (see *Baedeker's Southern Germany*).

FROM SINGEN TO ETZWEILEN, railway in ½ hr. (1 fr. 30, 90, 65 c.). Stations *Rielasingen*, *Ramsen*. We cross the Rhine between *Hemishofen* and *Rheinklingen* (see below). 9 M. *Etzweilen* (p. 32).

75½ M. *Rickelshausen*. — 77½ M. **Radolfzell** (*Schiff; Krone; *Sonne*), an old town on the *Untersee*, with a Gothic church of 1436. Near it, on the lake, is the *Villa Seehalde*, with a monument to the poet Victor v. Scheffel (d. 1886). — 78 M. *Markelfingen*; 82 M. *Allensbach*. — 86 M. *Reichenau* is the station for the island of that name, situated to the right in the Untersee and connected with the shore by an embankment.

The island of **Reichenau** (3 M. long, 1¾ M. wide), now belonging to Baden, was formerly the seat of a celebrated Benedictine Abbey, founded in 724 and secularized in 1799. The Schaffhausen and Constance steamers touch at the island twice daily (see below). The road from the shore leads past the ruined tower of the castle of *Schopeln*, which was destroyed as early as 1284. The former collegiate church of St. George, near the houses of *Oberzell*, is a Romanesque basilica of the 11th and 12th cent., with interesting frescoes of the 10th century. — In the centre of the island lies its chief village, *Mittelzell* (Mohren; Bär), with 1000 inhabitants. The parish church, or Münster, is the former abbey church, which was consecrated in 806, and contains the remains of Charles the Fat, great-grandson of Charlemagne, who was dethroned in 887. The present edifice is a basilica, the 11th and 12th cent., borne by columns, with two transepts and a late-Gothic choir of 1448-54; the treasury, in the sacristy, contains several fine reliquaries. — The church of *Unterzell*, on the N.W. side of the island, is another basilica of the 9-12th centuries.

The train passes the large barracks of *Petershausen* and crosses the Rhine to (89 M.) **Constance** (p. 29), by an iron bridge embellished with statues.

STEAMBOAT FROM SCHAFFHAUSEN TO CONSTANCE. Charts of the journey are sold for 30 c. on board the steamboats. Below the stations are indicated with daggers. Pier above the bridge, near *Schloss Munot* (p. 24), opposite *Feuerthalen*. — Right: *Paradies*, formerly a nunnery.

† Left: *Büsingen*, with an old church.

R. *St. Catharinenthal*, formerly a nunnery, now a hospital for incurables; opposite (left) *Villa Rheinburg*.

† R. *Diessenhofen* (1325'; *Adler; Löwe; Hirsch*), the Roman *Gunodurum*. The Rhine is crossed here by a covered wooden bridge, below which the steamer lowers its funnel.

R. *Rheinklingen*; left, *Bibern*. We now pass under the handsome bridge of the 'Nordostbahn' (see p. 26). L. *Hemishofen*, with the **ruin** of *Wolkenstein* above. R. *Wagenhausen*.

† L. Stein (*°Sonne*; *°Rabe*), a picturesque old town, connected with the village of *Burg* (Wasserfels) by a new wooden bridge, and a station on the Winterthur railway (p. 32). The suppressed monastery of *St. George* contains a hall with a vaulted wooden roof, erected in 1515, and embellished with frescoes. The *Rathhaus* contains stained glass, old weapons, etc. The old château of *Hohenklingen* (1945¹), on a hill to the N. of the town, affords an admirable view.

Above Stein is the island of *St. Othmar*, with the **chapel of that name**. The Rhine widens, the steamer enters **the Untersee**. **R.** *Eschens* (p. 32); on the hill above it the château of *Freudenfels*.

† L. *Oberstaad*, an old mansion with a square tower, now **occupied** by dye-works; beyond it the suppressed monastery of *Oehningen*.

† R. *Mammern* (p. 32); in the **wood**, the ruin of *Neuburg*; **on the** bank, the house of *Glarisegg*.

† L. *Wangen* and the château of *Murbach* (now a hydropathic).
† R. *Steckborn* (p. 32). Below it, the former nunnery of *Feldbach*.
† R. *Berlingen* (p. 32). The lake expands, and we now see the island of Reichenau (p. 25). On the hill to the right is the château of *Eugensberg*, erected by Eugène Beauharnais, vice-king of Italy, and now the property of Count Reichenbach-Lessonitz.

† R. *Mannenbach* (p. 32), charmingly situated, above which is the handsome pinnacled château of *Salenstein*; then, on a wooded hill, *Arenaberg* (1052²), once the residence of Queen Hortense (d. 1837) and her son Napoleon III. (d. 1873), now the property of the ex-Empress Eugénie.

† L. *Reichenau*, on the island of Reichenau (p. 25).
† R. *Ermatingen* (p. 31), prettily situated on a promontory; on the hill above it, *Schloss Wolfsberg* (1690¹; "Hôtel-Pension, pens. 3½-6 fr.). The neighbouring *Schloss Hard*, with its beautiful garden, is not visible.

We now enter the narrow arm of the Rhine connecting the Untersee with the Lake of Constance.

† R. *Gottlieben* (Krone), with a château, restored by Napoleon III., in which Huss and Jerome of Prague, and afterwards Pope John XXII. were confined by order of the Council. Baron Scherer's château of *Castel*, on the hill at the back of the village, was built by Tafel of Stuttgart and is sumptuously fitted up (Alhambra room, frescoes by Häberlin, etc.). Beautiful retrospect of the Untersee, with the peaks of the Höhgau in the distance.

The banks now become flat, and at places marshy. We thread our way through reedy shallows (l. *Petershausen*, with large barracks), and at length pass under the handsome railway-bridge of **Constance** (p. 29). Passengers are landed at the pier with a lighthouse at its E. end.

9. The Falls of the Rhine.

Hotels. On the hill on the *right* bank, near stat. Neuhausen (p. 24): *SCHWEIZERHOF, 8 min. **from the** railway-station, R., L., & A. from 5, B. 1½. lunch 3, D. 5, pens. from 9 fr., omn. 75 c., with extensive grounds and the finest view of the **Falls and the Alps**; *BELLEVUE, at the rail. station, R., L., & A. 3, B. 1½, lunch 3, D. 4, pens. 8½ fr. — At Neuhausen *HÔTEL RHEINFALL, R., L., & A. 2-2½. B. 1¼, lunch 2½, D. 3, pens. 5-7, omn. ½ fr. — On the *left* bank, above the Falls, *HÔT. SCHLOSS LAUFEN, ½ M. from Dachsen station (p. 32), R., L., & A. 2½-4, B. 1¼, lunch 2½, D. 3½, pens. 6-7, omn. 1 fr.; *HÔT. WITZIG, at stat. Dachsen, ½ M. from the Falls (p. 32). Illumination of the Falls with electric and Bengal lights every evening in summer, for which ¾-1 fr. is charged in the hotel-bill. — *English Church* in the 'Schweizerhof' grounds.

The station for the Falls on the right bank is *Neuhausen* (p. 24) on the Baden Railway, that on the left bank *Dachsen* (p. 32) on the Swiss line. The best way to see the Falls is to start from Neuhausen and follow the route described below (cross the bridge to *Schloss Laufen*, descend to

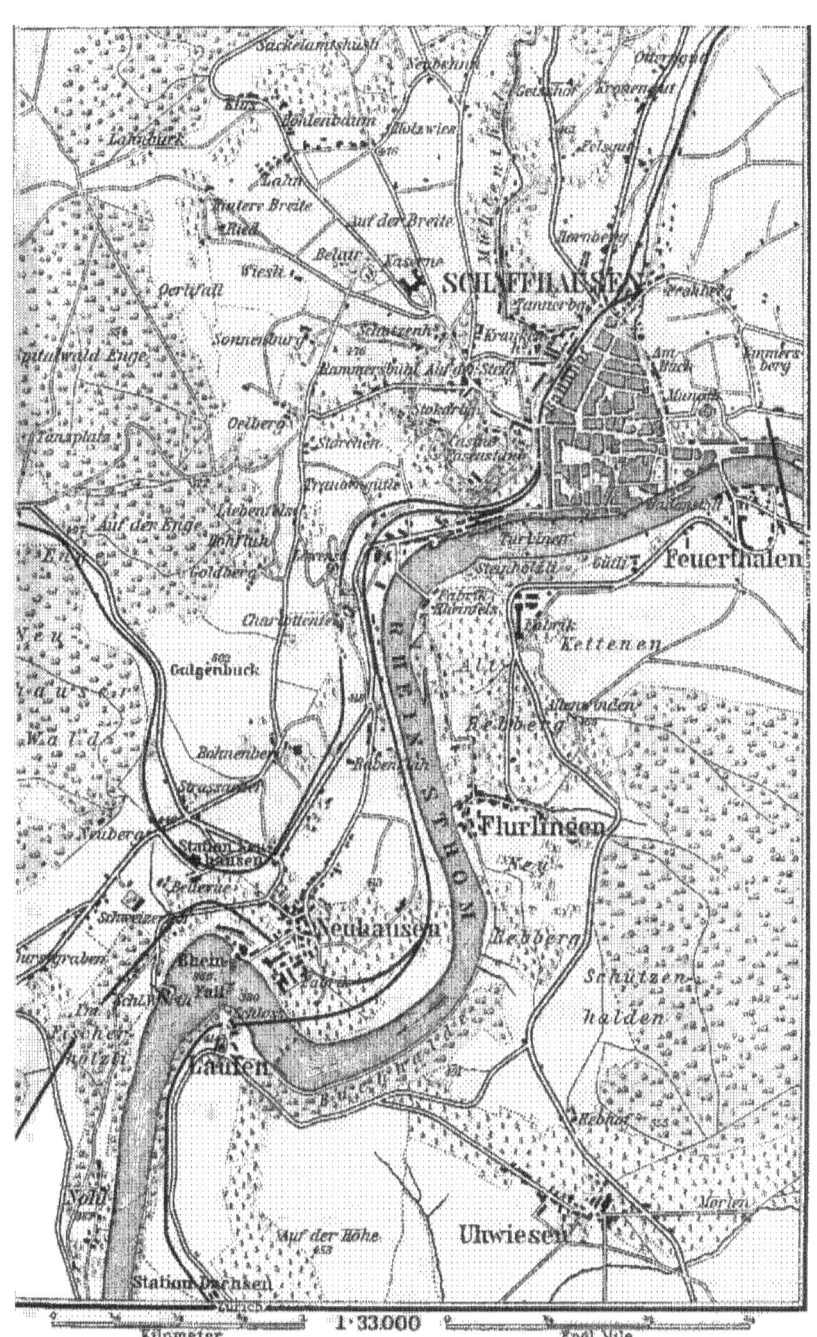

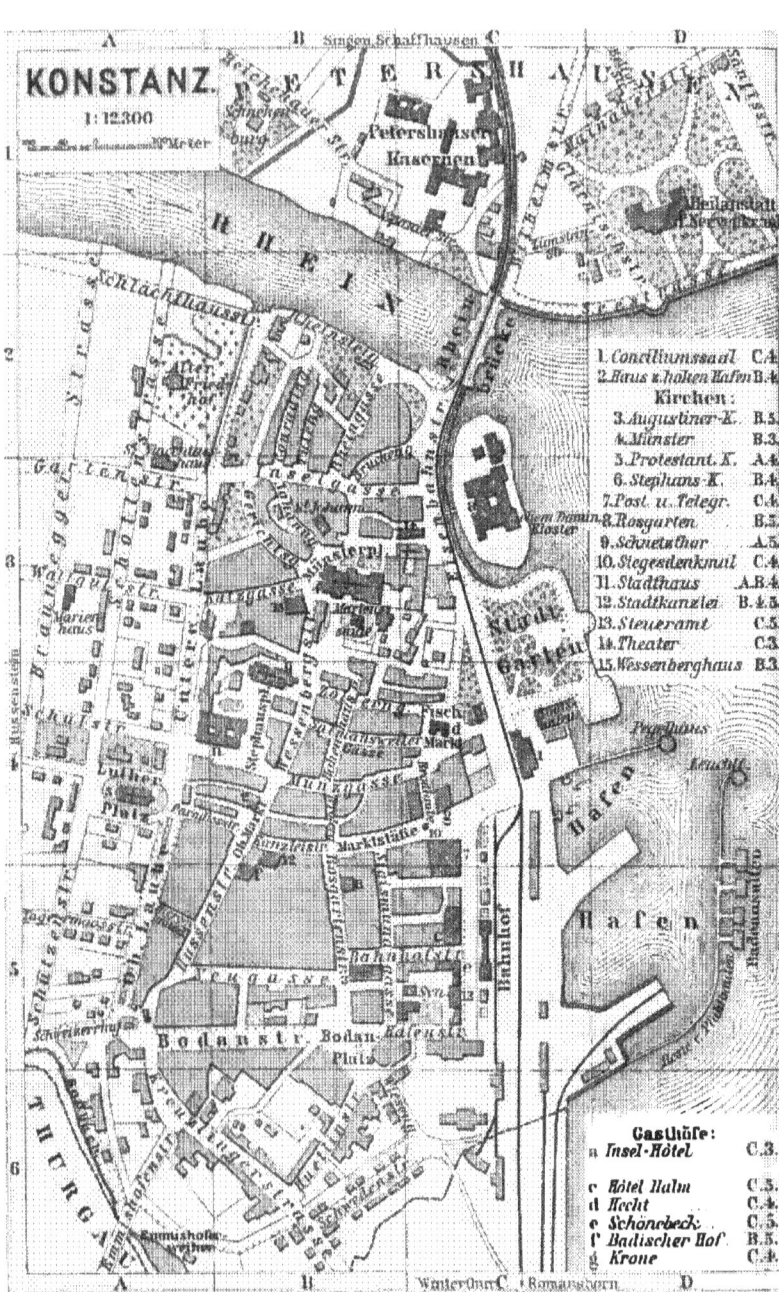

FALLS OF THE RHINE.   *I. Route 9.*

the *Fischetz*, cross to the *Schlösschen Wörth*, and return through the grounds, 1½ hr. in all). This round is often taken in the reverse direction, but as the Fischetz, the most striking point of all, is then visited first, the other points lose much of their impressiveness. — Travellers who desire to combine a visit to the falls with the journey to or from Switzerland alight at stat. *Dachsen*, walk or drive (omnibus there and back 1 fr.) to (½ M.) *Laufen*, descend through the grounds to the Fischetz, cross to the *Schlösschen Wörth*, and return to Schloss Laufen by the Rheinfall-Brücke; or descend from Wörth by the road on the right bank to the (⅔ M.) village of *Nohl*, cross the river (ferry 15-20 c.), and regain Dachsen in a few minutes. — The pleasantest way to visit the Falls from *Schaffhausen* (p. 24) is to drive in an open carriage, viâ Feuerthalen, to Schloss Laufen. Or the traveller may walk to Neuhausen and cross the railway-bridge to the Schloss (2 M.). Omnibuses ply from the railway-station at Schaffhausen to Neuhausen (Falls of the Rhine) in summer 12 times daily in 20 min. (40 c., the last two trips, at 8.30 and 10 p.m., 50 c.). — All the points of view should if possible be visited, as the traveller's impression of the Falls will otherwise be imperfect.

The **Falls of the Rhine** are in point of volume the grandest in Europe. The Rhine is precipitated in three leaps over an irregular rocky ledge, which on the side next the left bank is about 60' in height, and on the right bank about 48'. Above the Falls the river is about 125 yds. in width. If the rapids and the cataracts a few hundred paces farther up be taken into account, the total height of the Falls may be estimated at nearly 100'. (Level of the Rhine below the falls 1180'.) In June and July the river is much swelled by melting snow. Before 8 a.m. and after 3 p.m. numberless rainbows are formed by the sunshine in the clouds of silvery spray. The spectacle is also very impressive by moonlight.

Of the four limestone-rocks which rise above the Falls, that nearest the left bank has been worn by the action of the water to one-third of its original thickness, but has lately been buttressed with masonry. When viewed from a boat below, the rocks seem to tremble. The central and highest rock, surmounted by a small pavilion, may be reached by boat, and ascended by a path protected by a railing. The Falls are here surveyed to the best advantage. The passage, which only occupies a few minutes, is unattended with danger (1-2 pers. 3 fr. and fee; each additional person 1 fr.). — It is a curious fact that no mention of the Falls of the Rhine occurs in history before the year 960. It has therefore been supposed that they did not exist until about a thousand years ago, and that, while the bed of the river below the falls has gradually been deepened by erosion, the deepening process above the falls has been retarded by the hardness of the rocky barrier above mentioned.

Leaving the *Neuhausen Station* (p. 24), we follow the road to the left, and after a few paces descend by a path to the right to the village. At the Hôtel Rheinfall we descend to the right by a fingerpost, and after 100 paces take the shady path to the left, passing the *Gun and Waggon Factory* to the (¼ hr.) **Rheinfall-Brücke**, 210 yds. long, which carries the 'Nordostbahn' over the Rhine a little above the Falls (p. 32). The nine arches vary in span (42-66'), as it was difficult to obtain foundations for the piers. The footway on the upper side of the bridge affords an interesting view of the rocky bed of the river, the rapids, and the falls below.

On the left bank a path ascends to the left in 5 min. to the **Schloss Laufen** (1360'), picturesquely situated on a wooded rock

28   *1. Route 10.*   FRIEDRICHSHAFEN.

immediately above the Falls (adm. 1 fr.; no other fees). The balcony and a jutting pavilion with stained-glass windows command a good survey of the falls and the environs. Camera obscura, 50 c.

Footpaths descend through the grounds to the chief points of view: an iron *Pavilion, the wooden *Känzeli, and lastly the *Fischetz, an iron platform projecting over the foaming abyss. The scene here is stupendous. The vast emerald-green volume of water descends with a roar like thunder, apparently threatening to overwhelm the spectator, and bedewing him with its spray (waterproof overcoats are let to visitors; small fee).

Boats are in readiness here to ferry us across (50 c., return-fare 80 c.) to the **Schlösschen Wörth** (*Inn*, R. 1^3/$_4$ fr.; camera obscura 50 c.), on an island opposite the Falls, which is connected with the right bank by a bridge. This point commands the finest general *VIEW of the Falls. (Boat to the central rock, see p. 27.) We may now return to the Neuhausen station or to the Schweizerhof. To the W. of the hotel is the *Fischerhölzli*, with shady grounds and picturesque views. Or we may follow the path on the right bank, ascending the river (benches at intervals, commanding splendid views) and passing an *Aluminium Factory* (left), to the road, where we descend slightly to the right to a stone parapet near the sluices, affording another good survey of the Falls. The road thence to the left ascends through the village of Neuhausen to the station (see p. 27).

10. From Friedrichshafen to Constance. Lake of Constance.

STEAMBOAT six times daily in summer (twice direct, in 1^1/$_4$ hr.; four times viâ Meersburg in 1^1/$_2$-1^3/$_4$ hr.). Between the chief places on the lake, *Friedrichshafen, Lindau, Bregenz, Rorschach, Romanshorn, Constance, Meersburg, Ueberlingen*, and *Ludwigshafen*, the steamers (about 26 in number) ply at least once daily, and on the chief routes (Friedrichshafen-Constance 1^1/$_2$ hr., Friedrichshafen-Romanshorn 1 hr., Friedrichshafen-Rorschach 1^3/$_4$ hr., Lindau-Romanshorn 1^1/$_4$ hr., Rorschach-Lindau 1 hr., Constance-Lindau 3 hrs.) 2-6 times daily. Good restaurants on board. The lake being neutral, luggage is liable to custom-house examination on arriving in Germany or Austria from Switzerland, and nominally in the reverse case also. Passengers from one German port to another may avoid these formalities by obtaining on embarkation a custom-house ticket for their luggage (gratis).

The **Lake of Constance** (1305'; Ger. *Bodensee*, Lat. *Lacus Brigantinus*), an immense reservoir of the Rhine, 210 sq. M. in area, is, from Bregenz to the influx of the Stockach, 40 M. long, about 7^1/$_2$ M. wide, and between Friedrichshafen and Utweil 835' deep. In beauty of scenery the Bodensee cannot vie with the other Swiss lakes; but its broad expanse of water, its picturesque banks and green hills, the chain of the Appenzell Alps in the distance, the snow-clad Sentis in particular, and several snow-peaks of the Vorarlberg Alps, visible in clear weather, combine to present a very pleasing scene. In rough weather sea-sickness is sometimes experienced. The best fish are '*Felchen*' and trout, and the best wine grown on the banks is the '*Meersburger*'.

Friedrichshafen (*Deutsches Haus; Drei Könige*, well spoken of; *Krone; Sonne; Müller's Restaurant*), the S. terminus of the

CONSTANCE. *I. Route 10.* 29

Würtemberg Railway (to Stuttgart 6-7½ hrs.), is a busy place in summer. Its lake-baths attract many visitors, especially from Swabia, and it boasts of a *Curhalle* with pleasant grounds on the lake. The *Harbour* with its *Lighthouse* is 1 M. from the railway-station.

Travellers about to continue their journey by steamer may keep their seats until the train reaches the harbour-terminus, near the quay (Restaurant, with view-terrace). Those arriving by steamer may procure tickets immediately on landing, and step into the train at once.

The Constance steamer directs its course to the W. On the N. bank are the village of *Immenstaad*, the châteaux of *Herrsberg* and *Kirchberg*; then the village of *Hagnau*. On the N.W. arm of the lake, the *Ueberlinger See*, we next observe the picturesque little town of *Meersburg*; then the island of *Mainau* (p. 31), and in the distance *Ueberlingen*. The steamer passes the promontory which separates the Ueberlinger See from the bay of Constance, and reaches (1½ hr.) —

Constance (comp. Plan, p. 27). — *INSEL-HÔTEL (Pl. a; C, 3), formerly a Dominican monastery (p. 30), on the lake, with a garden and fine view, R., L., & A. 3-6, pens. 7-10 ℳ; HÔTEL HALM (Pl. c; C, 5), opposite the railway-station, R. 2-3, B. 1, D. 3, pens. 7-8 ℳ; *HECHT (Pl. d; C, 4), R., L., & A. 3, B. 1, D. 3 ℳ; *HÔTEL SCHÖNEBECK (Pl. e; C, 5), opposite the railway-station, R., L., & A. 2-2½, B. 1, D. 2½, pens. from 6 ℳ; BADISCHER HOF (Pl. f; A, 5); *KRONE (Pl. g; C, 4), ANKER, SCHIFF, BARBAROSSA, BODAN, FALKE, LAMM, *SCHNETZER, in the market-place, second-class, moderate; KATHOLISCHES VEREINSHAUS ST. JOHANN, near the Cathedral, with restaurant. — *Schönebeck Restaurant (see above), Victoria (beer), opposite the station; *Engler's Biergarten*, near the public park; *Café Maximilian*, Bahnhof-Str. — *Post Office* (Pl. 7; C, 4), near the station. — *Baths in the lake* (Pl. D, 4, 5), well fitted up (bath 40 pf.; ferry 10 pf.). — *English Church Service* in summer. — The former Constanzer Hof (Pl. D, 1), on the lake, is now an *Institute for Nervous Patients* (Dr. G. Fischer).

Constance (1335'; pop. 17,000), a free town of the Empire down to 1548, after the Reformation subject to Austria, and since the Peace of Pressburg in 1805 a town of Baden, lies at the N.W end of the Lake of Constance, at the efflux of the *Rhine*. The episcopal see, founded in 781, and held by 87 bishops in succession, was deprived of its temporalities in 1802, and suppressed in 1827.

The *CATHEDRAL (Pl. 4; B, 3), founded in 1052, originally a cruciform Romanesque edifice, was rebuilt in its present form in 1435 and 1680. The Gothic tower, designed by Hübsch, was erected in 1850-57; the open spire, with a platform on each side, commands an excellent survey of the town and lake (adm. 20 pf.).

INTERIOR. On the doors of the chief portal are *Reliefs in 20 sections, from the life of Christ, carved in oak by Simon Haider in 1470. *Choir-stalls, with satirical sculptures, of the same date. The organ-loft was enriched in the Renaissance style in 1680. In the nave, which is borne by 16 monolith columns (28' high, 3' thick), sixteen paces from the entrance, is a large stone slab, with a white spot which always remains dry when the rest is damp. On this spot Huss is said to have stood on 6th July, 1415, when the Council sentenced him to be burned at the stake. The N. chapel adjoining the choir contains a *Death of the Virgin, in stone, date 1460. In the left aisle is the monument of *J. H. v. Wessenberg* (p. 30).

The TREASURY (verger ½-1 ℳ) contains missals of 1426 with miniatures. On the E. side of the church is a CRYPT, containing the *Chapel of the Se-*

pulchre, a representation of the Holy Sepulchre in stone, 20' high (13th cent.). Adjoining the church on the N. stand two sides of the once handsome *Cloisters, erected about 1480 in the Gothic style.

The WESSENBERG-HAUS (Pl. 15; B, 3), once the residence of the benevolent HT. v. Wessenberg (d. 1860), who for many years was the administrator of the bishopric, contains a collection of pictures, engravings, and books, bequeathed by him to the town, and a number of paintings and sketches left by M. Ellenrieder (d. 1863), a lady-artist.

The late-Gothic church of ST. STEPHEN (Pl. 6; B, 4), of the 15th cent., with its slender tower, but disfigured externally, contains interesting sculptures in wood and stone. — The Wessenberg-Str. leads hence to the *Obere Markt*, at the corner of which is the house 'Zum Hohen Hafen' (Pl. 2; B, 4), where, according to the modern inscription, Frederick, Burgrave of Nuremberg, was invested with the March of Brandenburg by Emp. Sigismund on 18th April, 1417. Adjacent is an old house (now the *Hôtel Barbarossa*), styled by the inscription *Curia Pacis*, in which Emp. Frederick I. concluded peace with the Lombard towns in 1183.

The STADT-KANZLEI, or Town Hall (Pl. 12; B, 4, 5), erected in 1593 in the Renaissance style, and embellished in 1864 on the façade with frescoes relating to the history of Constance, contains the *Municipal Archives* in the lower rooms (2800 charters, chiefly from the Reformation period). Handsome inner court. — In the marketplace stands a *Victory*, by Baur (Pl. 10), erected in memory of the war of 1870-71.

The ROSGARTEN (Pl. 8; B, 5), the old guild-house of the butchers, contains the *Rosgarten Museum*, a fine collection of prehistoric remains, antiquities of Constance, and natural history specimens (adm. 40 pf.).

The KAUFHAUS (Pl. 1; C, 4), on the lake, erected in 1388, contains the large hall, 52 yds. long, 35 yds. wide, and borne by ten massive oaken pillars, where the conclave of cardinals met at the time of the Great Council (1414-18). The hall has been restored and adorned in 1875 with frescoes by Pecht and Schwörer from the history of the town (adm. 20 pf.). Upstairs a collection of Indian and Chinese curiosities, the property of the castellan (30 pf.).

The DOMINICAN MONASTERY (Pl. a; C, 3), in which Huss was confined, on an island, has been partly converted into a hotel ('Insel-Hôtel', p. 29). The well-preserved Romanesque cloisters (with frescoes by Häberlin, illustrating the history of the convent) and the finely-vaulted dining-room (formerly the church) are worthy of a visit.

Pleasant promenade in the *Stadtgarten* on the lake, with a marble bust of Emp. William I. and a charming view.

The house in which Huss was arrested, in the Husen-Strasse near the Schnetzthor (Pl. A, 5), is indicated by a tablet with a portrait of the Reformer in relief, put up in 1878. Adjoining it is an old relief, of 1415, with derisive verses. Behind it, in the 'Obere Laube', a bronze tablet with an inscription designates the spot where Jerome of Prague was imprisoned in 1415-16. In the *Brühl*, ½ M. to the W. of the town, a large boulder with inscriptions ('Husenstein') marks the spot where these illustrious Reformers suffered martyrdom.

Fine view of the lake and the Vorarlberg and Appenzell Alps from the *Allmannshöhe* (¾ hr.), with belvedere (Restaurant), 5 min. above the village of *Allmannsdorf*, on the road to the Mainau. — Pleasant walks to the *Loretto-Kapelle* (½ hr.); the *Jacob*, a restaurant with a fine view (½ hr.); and the *Kleine Rigi*, above Münsterlingen (Inn; 1 hr.).

In the N.W. arm of the Lake of Constance (*Ueberlinger See*, p. 29), 4½ M. from Constance, lies the pretty island of *Mainau*, formerly the seat of a commandery of the Teutonic Order, as is indicated by a cross on the S. side of the château, which was built in 1746. The island, 1½ M. in circumference, is connected with the mainland by an iron bridge 650 paces long. Since 1853 it has been the property of the Grand Duke of Baden, and is laid out in pleasure-grounds, where cypresses and other semi-tropical plants flourish in the open air. Near the château is a small inn. Steamboat from Constance in 55 min.; small boat (a pleasant trip of 1 hr.) 5 ℳ and gratuity; one-horse carr. 5-6, two-horse 8 ℳ. Walkers take a shorter route, partly through pleasant woods (1 hr.).

11. From Rorschach to Constance and Winterthur (Zürich).

Comp. Maps, pp. 25, 26.

60 M. RAILWAY (*Nordostbahn*) in 4¼-5¾ hrs. (fares 9 fr. 75, 6 fr. 85, 4 fr. 80 c.).

Rorschach, see p. 52. The line skirts the lake of Constance, of which it affords pretty glimpses. Rising conspicuously above the woods on the N. bank is *Heiligenberg* (1065' above the lake), a château of Prince Fürstenberg. Stations **Horn** (p. 52), *Arbon* (*Bär; Engel; Kreuz; Pens. Seebad*), a small town on the site of the Roman *Arbor Felix*. — 7½ M. *Egnach*.

9½ M. **Romanshorn**, see p. 49. — 12 M. *Uttwyl*; 13 M. *Kesswyl* (Bär; Pens. Seethal), well-to-do villages. To the right, on the lake, the *Moosburg* is visible. — 95 M. *Güttingen*, with a château; 16 M. *Altnau*; 18½ M. *Münsterlingen* (Pens. Schelling), with a lunatic asylum. — 21 M. *Kreuzlingen* (*Helvetia; Löwe; *Pens. Besmer*), a pleasant little town with the old Augustinian abbey of that name, at present a seminary for teachers. The church contains a curious piece of wood-carving of the 18th cent., with about 1000 small figures.

22 M. **Constance** (a terminus station), see p. 29. The train backs out and runs towards the W. through a fertile district. 23 M. *Emmishofen-Egelshofen*, 25 M. *Tägerweilen*, thriving villages; on the Rhine, to the right, *Gottlieben* (p. 26). Near (28 M.) **Ermatingen** (*Adler*) we approach the green *Untersee*, which we now skirt. Charming views; in the distance, to the N.W., rise the peaks of the Höhgau (p. 26). Near Ermatingen, on the height to the left, are the châ-

teaux of *Wolfsberg* (p. 26) and *Hard*; then *Arenaberg* (p. 26), and near (28½ M.) *Mannenbach* (*Pens. Schiff, 4-5 fr.) the handsome *Salenstein* (p. 26). To the right, in the lake, the large island of *Reichenau* (p. 25); on the left, *Schloss Eugensberg* (p. 26). At (30½ M.) *Berlingen* the Untersee attains its greatest width (5 M.), after which it divides into two branches.

32 M. Steckborn (*Krone; Sonne*), a small town with a castellated 'Kaufhaus', lately restored. Below it, on the right, the ironfoundry of *Feldbach*, once a nunnery. On the right, farther on, the mansion of *Glarisegg*; to the left, in the wood, the ruin of *Neuburg*. On the opposite (N.) bank are *Wangen* and the hydropathic establishment of *Marbach* (p. 26).

36 M. *Mammern* (Ochs, at the station), with a château, used as a Hydropathic Establishment(pens.). Then, on the right bank, *Oberstaad*, and on the hill the abbey of *Oehningen* (p. 26). At (37 M.) *Eschenz* the Untersee again narrows into the *Rhine* (p. 26). We follow the left bank to the station for (39 M.) Stein (p. 26; right bank), commanded by the castle of *Hohenklingen*; and then turn to the left to (41 M.) *Etzweilen* (Rail. Restaurant), the junction for *Singen* (p. 25).

FROM ETZWEILEN TO FEUERTHALEN, 10 M., railway in 35 minutes. — 2½ M. *Schlattingen*; 4½ M. *Diessenhofen* (p. 25); 7½ M. *Schlatt*. — 8½ M. *Langwiesen-Feuerthalen*, on the left bank of the Rhine, opposite *Schaffhausen* (see p. 21). A bridge is being built. — 10 M. *Feuerthalen*, see p. 24.

From Etzweilen to *Singen* (and *Stuttgart*), see p. 25.

On the left, as we proceed to the S., is the vine-clad and wooded *Stammheimer Berg* (1716'). 43½ M. *Stammheim*; 48½ M. Ossingen. We now cross the *Thur* by a bold iron bridge, 148' high, borne by seven iron buttresses. 53 M. *Thalheim-Altikon*; 54½ M. *Dynhard*; 56 M. *Seuzach*; 58½ M. *Ober-Winterthur*, a small town with an old Romanesque church (tower modern), the Roman *Vitodurum*.

60 M. *Winterthur* and thence to (76 M.) *Zürich*, see p. 48.

12. From Schaffhausen to Zürich.
Comp. Maps, pp. 26, 40.

35 M. RAILWAY (*Nordostbahn*) in 2 hrs.: to Winterthur 1 hr., to Zurich 1 hr. (fares 5 fr. 95, 4 fr. 20 c., 3 fr.). Views on the *right*.

Schaffhausen, see p. 24. The line skirts the lofty Fäsenstaub Promenade (p. 24), and passes below the *Villa Charlottenfels* (p. 24). On the right, high above, is the Waldshut railway (p. 24), which passes through a tunnel under Charlottenfels. Immediately beyond a long cutting we cross the *Rheinfall-Brücke* (see p. 27), obtaining a glimpse of the falls to the right, and enter a tunnel, 71 yds. long, under *Schloss Laufen* (p. 27). On emerging, and looking back to the right, we obtain another beautiful glance at the falls.

3 M. Dachsen (1296'; *Hôtel Witzig*, R. & B. 2 fr. 75, B. 1 fr. 30 c.) lies 1 M. to the S. of Schloss Laufen (comp. p. 26). As the train proceeds, it affords pleasing views at intervals of the bluish-green Rhine in its deep and narrow channel, enclosed by wooded banks.

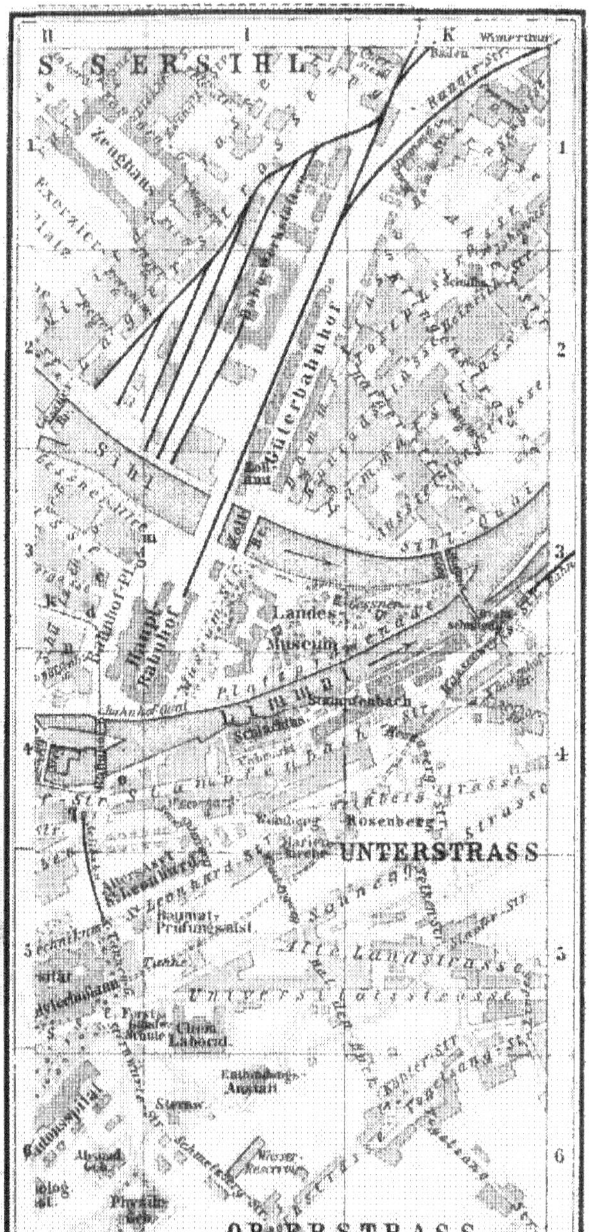

Geograph. Anst. v. Wagner & Debes, Leipzig

ZÜRICH. *I Route 13* 33

5½ M. *Marthalen.* The valley of (10½ M.) **Andelfingen** (1298'; *Löwe*) soon begins to open, and that thriving village appears in the distance to the right, on the steep bank of the *Thur*. We approach it by a wide curve, and cross the Thur above the village by an iron bridge 113' high. We then skirt the river for a short distance, and reach Andelfingen on the S. side. The site of the station has been excavated in an ancient moraine.

The route is now less interesting. 13 M. *Henggart,* ½ M. to the N.W. of which is the château of *Goldenberg* (pension, moderate). 14 M. **Hettlingen**. The vine-clad slopes of *Neftenbach,* to the right, produce the best wines in N. Switzerland, the finest of which is *Gallenspitz*. Near Winterthur the broad valley of the *Töss* is entered.

19 M. **Winterthur**, and thence to (35 M.) *Zürich*, see p. 48.

13. Zürich and the Uetliberg.

Railway Stations. *Central Station* (Pl. H, I, 3, 4; *Restaurant*), at the N. end of the town, ¾ M. from the lake (hotel-omn. ¾-1 fr., each box 20 c.; cab for 1-2 pers. 80 c.). — *Enge Station* (Pl. D, 2), on the left bank of the lake (p. 42). — *Uetli Station* (Pl. F, 1), also for the *Sihlthalbahn* (p. 40). — **Steamboats** (see pp. 34, 40) start from the Stadthaus-Platz (Pl. E, 4).

Hotels. *Hôtel Baur au Lac* (Pl. a; E, 3), with a pretty garden and delightful view, R. from 4, L. ¾, A. 1, B. 1½, luncheon 3½, D. 5-6, pens. 10-15, omn. 1 fr.; *Gr.-Hôt. Bellevue* (Pl. b; E, 4), on the lake, with fine view, R., L., & A. from 4, B. 1½, lunch 3½, D. 4, pens. from 10 fr.; *Grand Hôt. National* (Pl. d; H, 3), *Gr. Hôt. Victoria* (Pl. c; H, 3), both opposite the station, R., L., & A. from 4, B. 1½, lunch 3½, D. 4 (National 5), pens. from 9 fr.; Hôt. de l'Epée (Pl. e; G, 4), by the lower bridge, R. & L. from 3, D. 3-3½ fr.; *Hôtel Baur en Ville* (Pl. f; F, 3), R., L., & A. from 3, D. 4 fr.; Kupfer's Hôtel Habis (Pl. g; H, 3), near the station, R., L., & A. 3-4, B. 1¼, D. 3½, pens. from 7 fr.; *Hôtel de Zürich* (Pl. h; E. 5), R., L., & A. 2½-4, D. 3½ fr.; Cigogne (Pl. i; F, 4), commercial; *St. Gotthard* (Pl. k; H, 3) and *Wanner's Hôtel Garni* (Pl. l; H, 3), Bahnhof-Str.; Hôtel Bahnhof (Pl. m; H, 3) and Stadthof (Pl. n; H, 3, 4), R., L., & A. 3¼, B. 1½, D. 3 fr., both near the station; *Hôt. Central* (Pl. o; H, 4), on the right bank of the Limmat, near the station, D. incl. wine 3 fr.; *Schweizerhof* (Pl. p; G, 4), R., L., & A. from 2, B. 1¼, D. 3, pens. 8½ fr.; *Limmathof* (Pl. q; H, 4), R., L., & A. 2-2½, B. 1, D. 2½ fr.; Hôtel Jura, R., L., & A. 1½-2, B. 1, D. 1½-2, pens. from 5 fr., the last three on the Limmat-Quai; *Pfauen* (Pl. t; F, 6), next the Summer Theatre (p. 34) R., L., & A. 2, B. ½-1, D. 1½-2 fr.; Schwarzer Adler, Niederdorf-Str. 9, moderate; Rothes Haus (Pl. r; F, 4) and Seehof (Pl. s; F, 4, 5), on the Sonnen-Quai, moderate; Hôt.-Pens. Säntis, Seefeld-Str.; Weisses Kreuz, Krone, Hirsch, Lamm, Löwe, Schiff, etc., unpretending; *Hôt. Widder* (Evangelisches Vereinshaus), Rennweg 1, R. 1½-2, B. ½, D. 1¼, pens. 3¼-4½ fr. Visitors are received at all these hotels *en pension*, the charges being reduced in spring and autumn. — **Pensions**. *Pension Neptun*, in the Seefeld, 6-7 fr.; *Tiefenau*, Steinwies-Str., pens. 5-6 fr.; *Beau-Site*, Dufour-Str., near the Alpen-Quai, pens. from 5 fr.; *Villa Schanzenberg* (Frau Hepp), Schönberg-Str. 2 (5½-8 fr.); *Fortuna*, Mühlebach Str. 59, near the theatre (5-7 fr.); Pens. Internationale, Lavater-Str. 55, Enge (5-7 fr.); Merz, Tannen-Str. 15, Oberstrass; Karolinenburg Forster, and Plattenhof, at *Fluntern*, on the hill, 1½ M. to the E. of Zürich; Sonnenberg, Zürichberg (5 fr.).

Restaurants and Cafés. *Métropole*, Stadthaus-Quai; *Wanner*, Bahnhof-Str. (good Valais wine); *Orsini* (Munich Beer), *Zunfthaus zur Wang*, both in the Frau-Münster-Platz; *Café Baur*, Post-Str.; *Cafés National* and *Habis*, both at the station; *Dufour*, Schützengasse 17, near the rail. station; *Stahl*, Schiffländo 26; *Wiener Café*, Bahnhof-Str.; *Café Central*, Centralhof. On the right bank: *Kronenhalle*, D. at 12.30 p.m. 2 fr.; *Saffran*, opposite the Rathhaus; *Limmatburg*, Limmat-Quai. — Beer. *Kropf*, in Gassen (Pl. F, 3, 4), Munich beer; *Café Orsini* (p. 38); *Stadtkeller*, behind the Limmathof; *Metzgerbräu*, Beatengasse; *Franziskaner*, corner of Stüssihofstatt and Niederdorf-Str.; *Meyerei*, etc. — *Drahtschmidli*, with garden on the Limmat (p. 88). — Wine. Valtellina wine at the *Velltiner-Keller*; *Walliser Weinhalle*, near the Schweizerhof; *Wanner* (p. 33); *Gorgot*, Münstergasse 15 (Spanish wines). — Confectioners, *Sprüngli*, Parade-Platz; *Bourry*, *Untere* Kirchgasse, on the Sonnen-Quai.

Baths in the lake at the Stadthaus-Platz (Pl. E, 4), at the suburb of Enge (Pl. C, 3), at the Uto-Quai (Pl. C, 5), and, for ladies, at the Mythen-Quai (Pl. B, 2), the Uto-Quai, and in the Limmat below the Bauschanze (Pl. E, F, 4). *Neumünster Baths* (Pl. F, 5), at the S. end of the town. — *Warm Baths* (vapour, etc.) at *Treichler's*, at the *Werdmühle* in the Bahnhof-Str., and at *Stocker's*, Mühlebach-Str. (also pension).

Post and Telegraph Office (Pl. F, 3), Bahnhof-Strasse; branch-offices in various parts of the town.

Cabs. Drive within the town, or not exceeding ¼ hr., 1-2 pers. 80 c., 3-4 pers. 1 fr. 20 c., each box 20 c.; in the evening 10 c. extra for the lamps; from 10 p.m. to 6 a.m. double fares. For ½ hr., 1 fr. 50 or 2 fr. 50 c.; ¾ hr., 2 fr. or 2 fr. 90 c.; 1 hr., 2 fr. 50 or 3 fr. 60 c., etc. The cabmen are very apt to overcharge.

Tramway from the Station through the Bahnhof-Str. to the suburb of *Enge*; across the Bahnhofbrücke and by the Limmat-Quai, Tonhalle-Str., and Seefeld-Str. to *Riesbach* and *Tiefenbrunnen* (p. 41); and from the Parade-Platz northwards to the cemetery of *Aussersihl*. — Electric Tramway from the Quai-Brücke to the Kreuzplatz and Burgwies and from the Quai-Brücke to the Pfauen, Römerhof, and Kreuzplatz.

Cable Tramway (*Zürichberg-Drahtseilbahn*) from the Limmat-Quai to the *Polytechnic* (Pl. H, 4, 5), every 5-6 min. from 7 a.m. to 9 p.m. (in summer from 6 a.m. to 9.30 or 10 p.m.; fare, in either direction, 10 c.; journey 2½ min.). — Electric Cable Tramway (*Central-Zürichbergbahn*) every 6 min. from the Quai-Brücke to the Pfauen, Platte, and the church of Fluntern.

Steam Launches ('Dampfschwalben') ply on the lake-front of the city hourly (fares 10-50 c.). Stations on the right bank: *Stadthaus-Platz* (Pl. C, 4); *Theatre* (Pl. D, 5); *Mainaustrasse*; *Zürichhorn*; *Zollikon*; and *Küsnacht*. Stations on the left bank: *Enge* (at the Schloss and Belvoir); *Wollishofen*; *Bendlikon*; and *Thalweil* (p. 42).

Rowing Boats for 1-2 pers. 50 c. per hour; for 3 or more pers. 20 c. each per hour; each rower 60 c. per hour.

Theatre, Dufour-Platz, Uto-Quai (Pl. D, 4); performances from Sept. 15th to May 1st. — Panorama of the Battle of Morat (1476), by Prof. L. Braun, on the Uto-Quai (Pl. C, 4; open daily, from 7 a.m. till dusk; adm. 1 fr.).

Popular Resorts. *Tonhalle* (Pl. E, 3), near the lake, with restaurant, concerts every evening in summer; *Belvoir*, a beautiful park at the S. extremity of the Alpen-Quai (Pl. D, 3), with restaurant; adm. 20 c., concerts 50 c., free on Sun. and Wed. (tramway Paradeplatz-Enge); *Zürichhorn* (Pl. A, 6), park with restaurant and Nägeli's Museum of Stuffed Alpine Animals (50 c.), station of the steam-launches (see above); *Pfauen Summer Theatre* (Pl. F, 6), operettas, etc.; *Platten-Garten* (Pl. G, 6), adjoining the Polytechnic (exhibitions of animals; concerts). The *Waid* on the *Käferberg*, 3 M. to the N.W. of the town (pleasant route viâ Drahtschmidli, see p. 38); *Jakobsburg* (Munich beer), above Oberstrass; *Dolder Restaurant* (cable-tramway, see above), *Sonnenberg Restaurant* (p. 83), both on the slope of the *Zürichberg*, above Hottingen. The *Uetliberg* is the finest point in the environs (by railway in ½ hr.; see p. 39). — Information as to excursions, objects of interest, etc., at the *Official Enquiry Office*, on the ground-floor of the Exchange Buildings (Pl. E, 3; 9-12 and 2-5).

English Church Service in the *Chapel of St. Anne* (Pl. E, 3), near the Pelikan-Str., at 10.30 & 5 o'clock (comp. p. 37). — **Presbyterian Service** (Church of Scotland) in summer.

British Consul, *Henry Angst, Esq.*, 11 Bleicherweg; office-hours 9½-11½.
United States Consul, *Eug. Germain, Esq.*, Stadthaus-Quai 3, 9-12 and 2-4 p.m.

Permanent Exhibition of the Zurich Art Society in the 'Künstlerhaus', corner of Börsen-Str. and Thalgasse, next door to the Hôt. Baur au Lac (Swiss and foreign works of art), daily, 10-8, 1 fr. — Permanent Exhibition at *Staub & Co's*., Parade-Platz (gratis). — *Anglo-American Pharmacy*, Dr. C. Dünnenberg, Tonhalle-Platz.

Zürich (1345'; pop. 120,497, including the eleven recently incorporated 'Ausgemeinden' or suburbs), the capital of the canton, lies **at the N.** end of the lake, on the green and rapid *Limmat*, which divides it into the '*Grosse Stadt*' on the right, and the '*Kleine Stadt*' on the left bank. On the W. side flows the *Sihl*, an unimportant stream except in spring, which falls into the Limmat below the town. Zürich is one of the busiest manufacturing towns in Switzerland, silk being the staple product, while its cotton-mills, machine-works, and iron-foundries are also important.

Lacustrine remains prove that the site of Zürich was occupied in prehistoric times. In B. C. 58 Zürich (Turicum), along with the other towns of the Helvetii, passed under the sway of the Romans. It owed its early prosperity in the middle ages to the favour of the Carlovingians. In 1292 it united with Uri and Schwyz, and in 1351 it became a member of the Swiss Confederation. From an early date Zürich was the intellectual leader of Switzerland. As the home of Zwingli (1519) it was the focus of the Reformation, and its schools have for centuries sent forth men of distinction, such as Bodmer, Hottinger, Orelli, Gessner, Lavater, Hess, Pestalozzi, Heidegger, Horner, Hirzel, Henry Meyer, the friend of Goethe, and many others.

The SITUATION OF ZÜRICH is very beautiful. Both banks of the clear, pale-green lake are enlivened with villages, orchards, and vineyards, scattered over a highly cultivated country. In the background rise the snow-capped Alps; to the left is the crest of the *Glärnisch*, then the perpendicular sides of the *Griesetstock* (9200'), near it on the right the *Pfannstock*, and farther on, the *Drusberg*, the ice-clad *Bifertenstock*, and the *Tödi* (the highest of the group, the last two rising above the Linttthal); in front of these the *Clariden*, with their westernmost point the *Kammlistock* (10,610'); between this and the double-peaked *Scheerhorn* lies the *Gries Glacier*; then on the N. side of the *Schächen-Thal* the long *Rossstock Chain* with its fantastic peaks; the broad *Windgälle*; between this and the Scheerhorn appears the dark summit of the lower *Myten* near Schwyz; above the depression between the wooded *Kaiserstock* and the *Rossberg* towers the pyramidal *Bristenstock*, near Amsteg on the St. Gotthard route; then, if we occupy a commanding position, the *Blackenstock* and *Uri-Rothstock*, and part of the snow-mountains of the *Engelberger-Thal*, appearing above the *Albis*, to the right, the northernmost point of which is the *Uetliberg*, with the hotel on its summit.

In the BAHNHOF-PLATZ (Pl. H, 4) a fountain with a bronze *Statue of Alfred Escher* (d. 1882), the **statesman and founder of** the St. Gotthard Railway, by Kissling, was erected in 1889. The BAHNHOF-STRASSE (Pl. H, J, 3), nearly ¾ M. long, leads hence to the S. to the lake. It passes on the right, in the Linth-Escher-Platz (Pl. H, 3), the *Linth-Escher School*; then, on the right, the *Post Office* and the *Credit-Anstalt* (Pl. F, 3); on the left the *Centralhof*, a block of houses with tempting shops, and the *Kappeler Hof*; and on the right the *Zürich Cantonal Bank* and the *Exchange* (Pl. E, 3), the latter with an ethnographical collection on the fourth floor (9–12 and 2–5; adm. 50 c.). — Side-streets lead to the left to the tree-planted *Lindenhof* (Pl. G, 3, 4), 123' above the Limmat, which was fortified at the earliest period and afterwards became an imperial palace; to the late-Gothic *Augustine Church* (Pl. G, 3), now used by the Old Catholics, with paintings by Deschwanden; and to *St. Peter's Church* (Pl. F, 4), with its massive tower and large electric clock (dials 29' in diameter), where Lavater (d. 1801) was pastor for 23 years (grave on the N. side of the church).

The STADTHAUS-PLATZ is adjoined by a Terrace on the lake (Pl. E, 4), commanding a beautiful view; to the right is the steamboat-quay, to the left, a bathing-establishment (p. 34). — The broad *Alpen-Quai*, with its pleasant promenades and fine views of the lake and the Alps, skirts the lake to the right, passing the handsome new *Tonhalle* (Pl. E, 3; opened in Oct., 1895) and extending to the *Belvoir Park*, to the S. of the suburb of *Enge* (p. 34).

To the E. of the Stadthaus-Platz the handsome **Quai-Brücke** (Pl. E, 4; 180 yds. long), constructed in 1882–83, crosses the Limmat near its issue from the lake. Below the bridge, on the left bank of the Limmat, is the *Bauschanze*, a small pentagonal island, shaded with trees, and connected with the Stadthaus-Quai by a bridge. On the right bank of the lake also new promenades (*Uto-Quai* and *Seefeld-Quai*), with charming views, extend past the handsome new *Theatre* (Pl. D, 5) and the *Panorama* (Pl. C, 5) as far as the park of Zürichhorn (p. 34).

The next bridge below the Quai-**Brücke is the** four-arched **Münster-Brücke** (Pl. F, 4). Adjacent are the *Frau-Münsterkirche* of the 12–13th cent., with its high red-roofed tower, on the left bank, and the former *Wasserkirche* (1479–84), on the right bank. Over these rises the Gross-Münster (p. 37), the whole forming a quaint picture of old Zürich.

The old **Wasserkirche** now harbours the **Town Library** (Pl. F, 4), which contains 130,000 vols. and over 5000 MSS. (open on weekdays 9–12 and **4–6**; fee 50 c., for **a party** 1 fr.; entr. in the open vestibule adjoining the bridge).

A letter of *Zwingli* (p. 37) to his wife; Zwingli's Greek Bible with Hebrew annotations in his own handwriting; autograph letter of *Henry IV.* of France and a cast of his features; three autograph Latin letters of *Lady Jane Grey* to Antistes Bullinger; letter of *Frederick the Great*, dated 1784, to Prof. Müller; portraits of burgomasters and scholars of Zürich, including *Zwingli*; marble bust of *Lavater* by Dannecker; marble bust of *Pestalozzi* by Imhof; eight panes of stained glass of 1506. *Müller's Relief* of part of Switzerland, and one of the Engelberger-Thal on a much larger scale, are executed with great care and accuracy.

The *Helmhaus* (14th cent.), adjoining the Wasserkirche, contains the *Antiquarian Museum (adm. daily, 8–12 and 2–6, fee 50 c.; free on Sun. 10.30–1), including a large and excellent collection of relics from the ancient Swiss lake-villages, coins, etc.

The steps opposite the E. end of the Münster-Brücke lead to the **Gross-Münster** (Pl. F, 4), erected in the Romanesque style of the 11-13th centuries. The upper stories of the towers are Gothic, and in 1799 they were crowned with helmet-shaped tops with gilded flowers. On the W. tower is enthroned Charlemagne with gilded crown and sword, in recognition of donations made by him to the church. The choir contains three large modern stained-glass windows representing Christ, St. Peter, and St. Paul. The church and the adjoining *Cloisters*, of the beginning of the 13th cent., are open daily in summer from 11 to 12 (adm. 20 c., ascent of tower 30 c.; sacristan, Kirchgasse 13).

On the quay to the S. of the choir of the Gross-Münster is a bronze statue, by Natter, of *Zwingli*, who was incumbent of the Gross-Münster from 1519 till his death in 1531. — To the N. of the Münster-Brücke, on the Rathhaus-Quai, is the *Rüden*, restored in the German Renaissance style, containing the Swiss educational exhibition and the Pestalozzi cabinet. At the *Marktbrücke* or *Gemüsebrücke* (Pl. G, 4) we see on one side the *Rathhaus* (Pl. F, G, 4), a massive building of 1699, on the other the handsome *Fleischhalle*, or meat-market (Pl. G, 4).

From the Quai-Brücke we now ascend the RÄMI-STRASSE (Pl. E-H, 5, 6) to the E., then turn to the right to the **Hohe Promenade** (Pl. E, 5, 6), a loftily situated avenue of lime-trees. Beautiful view (best by morning-light) from the platform with the *Monument of Nägeli* (d. 1836), a favourite vocal composer. Adjacent is the *Old Cemetery*, where an *English Church* is now being built (see p. 35). — From the Hohe Promenade a road passing the N. side of the cemetery rejoins the Rämi-Strasse, where in the Kantonsschul-Platz (to the left) is the marble monument of *Ignaz Heim* (d. 1883), the composer. The street ascends to the Cantonal School (Pl. G, 6), and then bends to the N. To the left are the *Physical and Physiological Institute* of the University and the new *Ophthalmic Institute*, to the right are the *Cantonal Hospital* (Pl. H, 6), beyond it the *Physical Institute* of the Polytechnic, the *School of Forestry and Agriculture*, and the *Chemical Laboratory* (Pl I, 5).

At No. 15 Schönberggasse, behind the Physical Institute, *Jacob Bodmer* lived from 1739 till his death in 1783. — Lower down, on the slope, is the **Künstler-Gütli** (Pl. G, 5), containing the *Picture Gallery of the Zürich Artists' Union* (open in summer on Sat. 2-4, Sun. 10-12, free; at other times, 50 c.; catalogue 50 c.).

Large Room. To the right, 26. *Delachaux*, Choir-boys; 213. *Siemiradzki*, Venetian gondola; 227. *Stückelberg*, Charcoal-burners in the Jura; 2. *Anker*, Pestalozzi; 20. *Buchser*, Italian herdsmen; 29. *F. Diday*, Scene in the Valais; 60. *E. Girardet*, The sick child; 138. *Koller*, Alp in the Engelberg Valley; 270. *Zünd*, Chapel on the battle-field of Sempach; 238. *Ulrich*, Storm; 16. *Bosshardt*, Arrest of Canon Hämmerlin; 21. *A. Calame*, Lake of Lucerne; 1. *A. Achenbach*, Storm; 12. *Bodmer*, Stags; 22. *Carolus Duran*, Female figure; 174. *Ott*, Walensee; 140. *Koller*, Midday repose; 218. *Steffan*, Mountain torrent; 28. *Castan*, Winter-scene; 217. *Stauffer*, Portrait of a lady; *245. *Vautier*, The gallant professor; *142. *Koller*, Cattle at a lake; *Gg. Grob*, The artist on his travels; 198. *Sandreuter*, Charmey; 219. *Steffan*, Mountain-lake; 218. *Stückelberg*, Pilgrims; 271. *Zünd*, Oak-wood; 31. *Diday*, On the Handeck; *Böcklin*, 11. Arbour, *13. Spring; 246. *Veillon*, Evening on the Lake of Lucerne; 245. *Tobler*, Wedding in the Amperthal; 192. *Ritz*, Engineers among the mountains. — The smaller rooms contain portraits, water-colours, etc.

The handsome *Polytechnic (Pl. II, 5), to the left, designed by *G. Semper* (d. 1879), and erected in 1861-64, is the seat of the *University of Zürich* (600 students, 88 professors and lecturers) and of the federal *Polytechnic School* (800 students, 107 professors and lecturers). The sgraffito decorations of the N. façade were executed from Semper's designs by Schönherr and Walther.

The MAIN ENTRANCE is on the W. side. In the vestibule and on the staircase are busts of *Kopp* and *Bolley*, the chemists. On the ground-floor is the *Archaeological Collection* (casts, Greek vases, *Terracottas from Tanagra, etc.; Sun. 10-12, Tues. and Frid. 2-4). On the FIRST FLOOR are busts of *G. Semper* (see above) and *C. Culmann* (d. 1881), the engineer, and the *Mineralogical and Palaeontological Collection* (Thurs. 8-12 and 2-6, free; at other times 50c.). On the SECOND FLOOR are the *Zoological Collection* (open as above) and the *Aula*, handsomely decorated, with mythological ceiling-paintings by Bin of Paris and a marble bust of Orelli (d. 1819), the celebrated philologist, by Meili. Splendid view from the balcony. — The custodian, who opens the Aula, conducts visitors also to the TERRACE on the top of the building, which commands the best survey of the town and its beautiful environs.

The *Collection of Engineering* is shown only to professional engineers. The *Mechanical and Technical Collection* is open daily, 8-12 and 2-6 (adm. 50c.).

We may now return to the station by the *Cable Tramway* (Pl. II, 5, 4; p. 34), which ends opposite the *Bahnhof-Brücke*; or we may descend from the Künstler-Gütli by the Sempersteig to the Limmat-Quai, passing the handsome new *Girls' School* and the *Predigerkirche*.

The **Platz-Promenade** (Pl. I, K, 3, 4), so called from the former Schützen-Platz, an avenue of fine trees to the N. of the railway-station, between the Sihl and Limmat, affords a cool and pleasant walk. In this promenade are the new *Swiss National Museum*, a large building in the mediæval style from Gull's designs (to be opened in autumn, 1896), and the simple monuments of the idyllic poet *Salomon Gessner* (d. 1788), the minnesinger *Joh. Hadlaub*, and the composer *W. Baumgartner* (d. 1867). It terminates in the 'Platzspitz', a point of land formed by the junction of the Sihl with the Limmat. A bridge crosses the Limmat to the **Drahtschmidli** (Pl. K, 3), a beer-garden on the right bank; and this is also the pleasantest route to the *Waid* (p. 34; ascend the flight of steps, behind the Drahtschmidli to the right, to the upper road).

In *Aussersihl*, a new workmen's quarter on the left bank of the *Sihl*, is the *Military Depôt* of Canton Zürich, including barracks and an arsenal. The **Collection of Arms** in the arsenal (Pl. II, I, 1; open on week-days 8-12 and 1.30-6) contains battle-axes, halberds, armour, flags, and cross-bows, among which last is one of the many that claim to have belonged to Tell. *Zwingli's Battle-axe*, taken by the Lucerners at Kappel (p. 79), and once kept at Lucerne, was transferred hither after the War of the Separate League in 1847, and is now preserved here with his sword, coat-of-mail, and helmet.

The **Botanic Garden** (Pl. F, 2), well stocked with Alpine and other plants, contains bronze busts of A. P. de Candolle (d. 1841) and Conrad Gessner (d. 1565), and marble busts of H. Zollinger, a Swiss botanist (d. in Java, 1859), and Oswald Heer (d. 1883), the naturalist. In the garden rises the *Katz*, an old bastion, forming a lofty platform planted with trees.

To the E. of the Botanic Garden a bridge crosses the **Schanzengraben** (the old moat) to the suburb of *Selnau*. Immediately to the left is the **Gewerbe-Museum** (Pl. F, 2), containing industrial collections (including a *Room from a patrician house of the 17th cent., with fine panelling and stove) and a permanent exhibition (daily 8-12 and 2-5, except Mon.; on Sun., 10-12 and 2-5). Beyond it, towards the Sihl, is the *Uetliberg Station* (Pl. F, 1; see below).

The Uetliberg.

RAILWAY to the top in 1/2 hr. (fare, 1st class 3 fr. 50 c., 2nd cl. 2 fr.; return-ticket, 5 and 3 fr.; on Sun. and holidays by excursion-trains 1 fr., return-fare 1 1/2 fr.; season-tickets at reduced fares). This line, 5 1/2 M. long, with a maximum gradient of 7 : 100, is constructed in the ordinary way, but, as on the Rigi Railway, the locomotives are placed behind the trains. The station is in the suburb of *Selnau* (see above; Pl. F, 1), not far from the Botanic Garden, on the Sihl, 1/4 hr. from the Central Station and 12 min. from that of Enge.

The train (best views to the right) skirts the Sihl for a short distance and crosses it to (5 min.) stat. *Zürich-Binz* (1390'), where the ascent begins. At first we traverse an open slope, with a pleasant view of Zürich and the valley of the Limmat, and then ascend through wood to (17 min.) Stat. *Waldegg* (2040'; Inn). The train then describes a long curve on the slope of the hill and reaches the terminus (2677'). About 5 min. above the station is the large *Hôt.-Pens. Uetliberg* (R., L., & A. 3-5, B. 1 1/2, D. 4 fr.), and 3 min. higher, at the top of the hill, are the *Restaurant Uto-Kulm* and a view-tower 100' high (167 steps; adm. 50 c.). Pleasant shady walks in the woods near the hotel. On the S side, about 1/4 hr. from the top, is the *Hôtel Uto-Staffel* (pens. 5 fr.).

The *Uetliberg (2865'), the northernmost point of the Albis range, is the finest point in the environs of Zürich. The view, though inferior in grandeur to those from heights nearer the Alps, surpasses them in beauty. It embraces the Lake of Zürich and the valley of the Limmat; the Alps from the Sentis to the Jungfrau and the Stockhorn on the Lake of Thun, with the Rigi and Pilatus in the foreground; to the W. the Jura, from the Chasseral on the Lake of Bienne to its spurs near Aarau, over which appear some of the Vosges Mts.; farther to the N. are the Feldberg and Belchen in the Black Forest, and the volcanic peaks of the Höhgau, Hohentwiel, Hohenhöwen, and Hohenstoffeln. Baden with its old castle (p. 20) is also prominent. Good panorama by Keller. — On the Uto-Kulm is a

marble obelisk with a **bust of the** Zürich statesman *Jakob Dubs*
(d. 1879).

WALK TO THE UETLIBERG (2 hrs.). The road leads from the Parade-
platz (Pl. F, 3) viâ the Bleicher-Weg, the Beder-Strasse, and the Uto-
Strasse. After 1 M. we cross the Sihl, turn to the left viâ the Giesshübel-
Strasse, and reach (³/₄ M.) the *Albisgütli* (tavern; cab to this point 2-3 fr.).
We now turn to the right and ascend by a well-trodden path, winding
somewhat steeply up the valley, to the *Hôtel Uto-Staffel* (p. 39), on the
brow of the hill, where a view of the Rigi, Pilatus, and the Bernese
Alps is disclosed. To the summit 20 min. more.

FROM THE UETLIBERG TO THE ALBIS-HOCHWACHT, a beautiful walk of
3 hrs., ascending and descending on the Albis range, and chiefly through
wood. A few minutes' walk beyond the Hôtel Uto-Staffel (see above), at
the fork, we follow the road to the right, which alternates with a footpath, keeping nearer the E. margin of the hill and affording beautiful
views. Beyond *Baltern* (Inn) we reach (1¼ hr.) the *Felsenegg* (Restaurant;
view). To the left is the ravine of the Sihl, beyond it the blue lake with
its thousand glittering dwellings, to the right the pretty Türler See, and
farther off a fertile hilly tract, with the Alps rising in the distance. —
1 hr. *Nieder-Albis* (2600'; Hirsch; Windegg Restaurant); 20 min. *Albis-
Hochwacht* (2887'), with a pavilion and a splendid **view** of the **Lake** of Zug,
the Rigi, Pilatus, etc. At (¼ hr.) a fork **we may ascend to the right**
to the (¾ hr.) *Albishorn* (3010'), or descend to the left, through woods, to
(½ hr.) the forester's house of *Unter-Sihlwald* (good quarters), on the Sihl,
whence we may reach Zürich by the Sihlthal Line in ¾ hr.

SIHLTHAL RAILWAY from Zürich to *Sihlwald*, 8½ M. in ¾ hr., viâ *Ad-
liswil* and *Langnau-Gattikon*. Near the station of *Gontenbach* (½ hr. by rail)
is the *Langenberg*, a park 1½ M. in length, belonging to the town of Zürich
and stocked with deer, chamois, etc. (Restaurant).

14. From Zürich to Coire. Lakes of Zürich and Walenstadt.

Comp. Maps, pp. 54, 66.

RAILWAYS. — *N.E. Railway (Nordostbahn;* line on the right bank) from
Zürich viâ Küsnacht, Meilen, and Stäfa to Rapperswil. 22½ M. in 1½-2 hrs.
(fares 3 fr. 75, 2 fr. 65, 1 fr. 90 c.). — *N.E. Railway* (line on the left bank)
viâ Richterswil to Ziegelbrucke (p. 45, junction for Weesen and Sargans),
36 M., in 1½-2 hrs. (6 fr. 5, 4 fr. 25, 3 fr. 5 c.); to Glarus, 43 M., in 2-2½ hrs.
(7 fr. 20, 5 fr. 5, 3 fr. 60 c.). Comp. R. 21. — *United Swiss Railways (Ver-
einigte Schweizerbahnen*) viâ Wallisellen, Rapperswil, Weesen, and Sargans
to Coire, 79 M., in 3¾-4¾ hrs. (fares 12 fr. 45, 8 fr. 75, 6 fr. 25 c.). The
train does not approach the Lake of Zürich till it reaches Rapperswil.

STEAMBOAT from Zürich to Wädensweil and back 5 times daily in
summer in 1 hr. (each way), touching at Erlenbach, Herrliberg, and Obermeilen on the right bank, and at Thalweil and Horgen on the left bank).
Also from Zürich to Rapperswil twice daily in 2 hrs.

The *Lake of Zürich (1340'), 25 M. long, 2½ M. broad at its widest
part, and 470' deep, is fed by the *Lint* and drained by the *Limmat*.
Its scenery, though **with no pretension** to grandeur, is scarcely
equalled in beauty by that of any other **Swiss** lake. The banks rise
in gentle slopes, at the base of which are **meadows and** arable land;
above these is a belt of vineyards and **orchards, and** on the E. side
the hills, here about 2500' high, **are wooded.** Being sprinkled for
a long distance with houses, villages, and manufactories, the banks
are sometimes not unaptly termed the suburbs of Zürich. In the
background rises a long chain of snow-clad Alps (see p. 35).

a. **N.E. Railway from Zürich to Meilen and Rapperswil** (Right Bank). On leaving the *Central Railway Station* (p. 33) the train sweeps round to the N.E. (to the left the viaduct of the line to Winterthur, p. 48) and crosses the *Limmat*. 2 M. *Zürich-Letten*, with the pumping works for the Zürich water-supply (interesting for engineers; adm. free). The train ascends the right bank of the Limmat for a short time, beyond the Drahtschmidli passes under the *Zürichberg* by a tunnel (2288 yds. long), and reaches (3½ M.) *Zürich-Stadelhofen*, on the square of that name (Pl. E, 5), **near the** Uto-Quai. The line then passes under the suburb of *Neumünster* by another tunnel (1463 yds. long) and finally reaches **the light of day** at (5 M.) *Zürich-Tiefenbrunnen*, with its villas and gardens (tramway to Zürich, see p. 34). About ¾ M. to the W. is the Zürichhorn Park (p. 34). We now skirt the vine-clad bank of the lake. On the other side rises the long ridge of the Albis; in front are the Alps of Uri and Glarus. 6 M. *Zollikon*; the village, with its slender church-tower, lies above, to the left. — 7½ M. **Küsnacht** (*Sonne*, on the lake, with garden; *Seegarten Restaurant*), a large village (2750 inhab.), with a seminary for teachers. — 9 M. *Erlenbach* (Pension Seehof), beautifully **situated**. The train passes through cuttings and a short tunnel, then runs high above the lake (views). — 10½ M. *Herrliberg-Feldmeilen* (Hôt. Raben), opposite Horgen (p. 42). — 12½ M. **Meilen** (*Löwe*, on the lake; *Sonne*; *Rail. Restaurant*; *Bellevue*), **a large village** (2860 inhab.) **with an old** church, at the base of the *Pfannenstiel*. At *Obermeilen* (Hirsch), ¾ M. to the E., **the** first discovery of lake-dwellings was made in 1854.

The Pfannenstiel (*Okenshöhe*, 2418'), to which a good path ascends from Meilen in 1 hr., affords a charming view of the lakes of Zürich and Greifen and of the Alps from Sentis to Pilatus (panorama by Keller). At the top a monument to L. Oken (d. 1851), the naturalist, and a refreshment-pavilion.

Steamboat from Meilen to *Horgen* (p. 42) direct or viâ *Feldmeilen* 8-10 times daily in 12-15 minutes.

14½ M. *Uetikon* (Krone; Rail. Restaurant), **with a large manu**factory **of sulphuric acid.** — 15 M. *Männedorf* (*Wildenmann*, Löwe, both on the lake), a large village (2600 inhab.), with the Zeller Institute ('faith cure'). The high-lying churchyard commands an extensive view. — 17 M. **Stäfa** (pop. 3845; *Sonne*; *Rössli*), **the** largest village on the N. bank. The lake now attains its greatest breadth (2½ M.). To the **E.**, in the background, rises the Speer (p. 45); to the left of it the Sentis, beyond which tower the Toggenburg Mts.; to the right, above the lake, the wooded *Hohe Rhonen* (4040'). Steamboats to Wädensweil and Richtersweil, see p. 43. — 18 M. *Uerikon*. — 20 M. *Feldbach-Hombrechtikon* (Rössli; Feldbach Brewery, with restaurant).

To the right, in the lake (reached by small boat from Rapperswil in ½ hr.), are the small islands of *Lützelau* and *Ufnau*, in front of the wooded Etzel. *Ufnau*, the property of the abbey of Einsiedeln, contains a farm-house, and a church and chapel consecrated in 1141. *Ulrich von Hutten*, the Reformer, one of the boldest and most independent men of his time, sought refuge here when pursued by his enemies in 1523, and died a fortnight after his arrival, at the age of 36. His remains repose in the little churchyard, but the exact spot is unknown.

22½ M. **Rapperswil** (*Hôtel du Lac*, R. 2-3, B. 1, D. with wine 3, pens. 5-6 fr.; *Cygne*, R. 2-3, B. 1¼, D. 2½, pens. 5-7 fr.; *Bellevue*, all three on the lake; *Poste*, at the station, with garden; *Freihof*, in the town; *Restaurant Speer*, at the rail. station, with garden), a picturesquely situated town (2800 inhab.), lies at the foot of the *Lindenhof*, a hill planted with limes (fine view). The old *Schloss* contains a black marble column with the Polish eagle, erected in memory of the beginning of the hundred years' struggle of the Poles for independence, and the *Polish National Museum*, founded by Count R. Plater and including pictures, sculptures, engravings, gems, antiquities, coins, and a library (adm. 1 fr.; splendid view from the tower). The *Parish Church*, re-erected since a fire in 1881, contains valuable sacred vessels. At the foot of the Lindenhof on the lake are shady promenades, to which also flights of steps lead down from the Schloss and from the terrace in front. In 1878 the old wooden bridge connecting Rapperswil with (1 M.) *Hurden* (Adler; Rössli) and *Pfäffikon* (p. 43) was replaced by the *Seedamm*, a viaduct 1024 yds. long, with an iron swing-bridge 46' long (railway from Rapperswil viâ Pfäffikon to Samstagern-Einsiedeln, see p. 43).

From Rapperswil to **Weesen** and **Coire**, see p. 44.

---

b. N.E. Railway from Zürich to Ziegelbrücke (Left Bank). The train describes a wide curve round the town, crossing the *Sihl* twice, passes under the Uetliberg line, and at (3 M.) *Enge* (p. 33) approaches the lake, which it skirts all the way to Lachen, affording beautiful views to the left. 3½ M. *Wollishofen*, a pleasantly situated village; 5½ M. *Bendlikon-Kilchberg*, the latter situated on the hill above. Above (7 M.) *Rüschlikon* is the rustic *Nidelbad* (1 M. by road), with a chalybeate spring and pleasant walks. — 8 M. **Thalweil** (*Adler*, near the church, moderate; *Krone*, on the lake), a large village, charmingly situated. *View of the lake from the church, or better from the tower. — 9¼ M. *Oberrieden*. — 10½ M. **Horgen** (*Löwe; Meyerhof; Schützenhaus*, a café on the lake; W. F. *Kemmler*, U.S. Consul), with 5520 inhab. and handsome houses chiefly belonging to silkmanufacturers, pleasantly situated amidst vineyards and orchards.

Steamboat to *Meilen* (p. 41) 8-10 times daily in 12-15 min., to *Küsnacht* 7-9 times daily in ¾-1 hr. — About 1½ M. above Horgen is the *Curhaus Bocken* (p. 81). *Zimmerberg* (1 hr.), see p. 81. — To *Zug* diligence daily in 2½ hrs., see p. 81.

Near (13 M.) *Au* the peninsula of that name, with its orchards and meadows, projects far into the lake (*Hôt.-Pens. Au, 5 fr.). — 15½ M. **Wädensweil** (1348'; *Engel*, facing the quay, R., L., & A. 2-3, B. 1¼, D. 2½, pens. from 6 fr.; *Hôt. du Lac; Bellevue Restaurant*, well spoken of) is the largest village on the lake (6350 inhab.).

Railway to *Einsiedeln*, see R. 31; diligence twice daily in 1³/₄ hr. viâ *Schönenberg* to *Hütten* (p. 105).

17¹/₂ M. **Richtersweil** (pop. 3881; *Drei Könige*, or *Post*, R. 2-2¹/₂, B. 1, D. with wine 2¹/₂, pens. 5-7 fr.; *Engel*, R. 2, D. with wine 2¹/₂, pens. 5 fr.), another thriving village, prettily situated.

STEAMBOAT from Richtersweil viâ Wädensweil to *Stäfa* (p. 41) 12 times daily in 30-45 min.; to *Männedorf* (p. 41) 10-12 times daily in 27-50 minutes.

The lake attains its greatest width here (see p. 41). Towards the E. rise the mountains of the Toggenburg and Appenzell. To the left, farther on, are the islands of *Ufnau* and *Lützelau* (p. 41). — 21 M. *Pfäffikon* (*Hôt. Höfe).

Railway across the lake to *Rapperswil*, see p. 42; railway viâ *Wollerau* to *Samstagern* (Einsiedeln, etc.), see p. 105. Pleasant walk viâ the air-cure resort of (¹/₂ hr.) *Lugeten* (2130'; *Hôt.-Pension, 4-5 fr.) to (¹/₂ hr.) *Feusisberg* (p. 105) and (³/₄ hr.) *Schindellegi* (p. 105). Ascent of the Etzel, see p. 106.

The line now reaches the *Upper Lake*. On the slope to the right, above *Altendorf*, are the chapel of *St. Johann* (1656') and the *Johannisburg Pension & Restaurant* (pens. 4-5 fr.), with a fine view.

25 M. **Lachen** (1350'; *Bär*; *Ochs*; Hôtel Bahnhof, well spoken of), a considerable village with a pretty rococo church, on a bay near the mouth of the *Wäggithaler Aa*. About 2 M. to the N.E. is the small *Bad Nuolen*, pleasantly situated at the base of the *Untere Buchberg*, with mineral and lake baths. — The train leaves the lake and near (27¹/₂ M.) *Siebnen-Wangen* crosses the Aa.

Wäggithal. The road from *Siebnen* (°Rabe) follows first the left and then the right bank of the deep channel of the *Aa* to (4 M.) *Vorder-Wäggithal* (2400'; *Rössli, plain), pleasantly situated in a green basin. It then leads through the defile of *Stockerli*, between the *Grosse Auberig* (5585') on the right and the *Gugelberg* (3780') on the left, to (4 M.) *Hinter-Wäggithal*, or *Innerthal* (3800'; *Schäfli, unpretending). Pleasant excursions to the *Au* (20 min.); E. to the *Fläschenlochquelle* (¹/₄ hr.); to the *Aaberli-Alp* (3545'), ¹/₂ hr.; *Hohfläschen-Alp* (4725'), 1¹/₂ hr. — The *Grosse Auberig* (5585'), ascended by the *Bärlaut-Alp* in 3 hrs., and the *Fluhberg* or *Diethelm* (6873'), by the *Fläschli-Alp* in 4 hrs., are good points of view and present no difficulty (guide desirable). — From Innerthal to the *Klönthal* a pleasant route (to Richisau 3¹/₂ hrs.; guide advisable). Skirting the *Aabach*, the path ascends, past the *Aabern-Alp* (3565'), to the (2¹/₂ hrs.) *Schweinalp Pass* (5150'), and then descends by the *Brüsch-Alp* and the *Schwein-Alp* to (1 hr.) *Richisau* (p. 73).

We now traverse a somewhat marshy plain to (31 M.) *Reichenburg*. On the right rise the Glarus Mts., on the left the Untere and Obere Buchberg (p. 44), and above them the Speer (p. 45). 34¹/₂ M. Bilten (Hirsch); in the 'Herrenstube' is a handsome apartment with artistic wood-carving of the 17th century. We cross the *Lint Canal* (p. 44) to the Rapperswil and Coire railway at (36 M.) *Ziegelbrücke* (p. 45). Thence to (43 M.) **Glarus**, see p. 67.

---

c. RAILWAY FROM ZÜRICH TO RAPPERSWIL AND SARGANS. From Zürich to (5¹/₂ M.) *Wallisellen*, see p. 48. The line traverses a flat district, near the right bank of the *Glatt*, which flows out of the neighbouring *Greifensee* (1440'). Stations *Dübendorf*, *Schwer-*

senbach, and *Nänikon*. — 14 M. **Uster** (1530'; *Usterhof; Stern; Kreuz*), a large manufacturing village (7042 inhab.). On the right are the church with its pointed spire, and the loftily situated old castle with its massive tower, now the seat of the district court (*Restaurant*; fine view). In the vicinity are several large cotton-mills, driven by the *Aa*, a brook near the railway. Beyond (16 M.) *Aathal* the Alps of Glarus and Schwyz form the S. background. From (18 M.) *Wetzikon* (Schweizerhof) branch-lines lead to the N.W. to *Pfäffikon* and *Effretikon* (p. 48), and to the S.E. (10 min.) to *Hinweil* (Hirsch; Kreuz), at the N.W. base of the Bachtel (see below). Near (21 M.) *Bubikon* (Löwe, plain) the line attains its highest level (1800'). — 22½ M. **Rüti** (Pfau), with manufactures of machines and silk, is the junction of the *Tössthal Line* (p. 48).

The *Bachtel (3670'; *Inn; view-tower, 100' high), 2 hrs. to the N.E. of Rüti, commands a fine view to the N.W. over the district of Uster, sprinkled with factories, and the lakes of Greifen and Pfäffikon; to the S. the Lake of Zürich from Wädensweil to the Lint Canal, the Lint Valley as far as the bridge of Mollis, and the Alps from the Sentis to the Bernese Oberland. Consult *Keller's Panorama*, at the inn. It is most conveniently ascended from *Gibswyl* (p. 49; 8½ M. to the N. of Rüti) in 1 hr., from *Wald* (p. 49; 1½ M.) in 1½ hr., or from *Hinweil* (see above; small carriage to the top 7 fr.), in 1½ hr.

Beyond a tunnel the train descends, chiefly through wood. Near *Jona* (Schlüssel), a manufacturing village almost adjoining Rapperswil, we descry the Alps of Schwyz to the S., and farther on, the Mürtschenstock, Schäniser Berg, Speer, and Sentis on the left.

27 M. **Rapperswil**, see p. 42. The station is a terminus, from which the train backs out on its departure. Views to the *right* as far as Weesen. The line crosses the *Jona*, passes the nunnery and girls' school of *Wurmspach* on the right, and returns to the bank of the lake near *Bollingen*, with its large quarries.

33 M. **Schmerikon** *(*Gasthof zum Bad; *Rössli; Seehof; Adler)*, at the upper end of the lake, near the mouth of the *Lint*. We now enter a broad valley traversed by that river (see below). To the right, on the N.E. spur of the *Untere Buchberg* (p. 43), stands the ancient *Schloss* **Grynau**, with a frowning square tower.

35 M. **Utznach** *(Linthof)*, a manufacturing village (1378'; **Ochs; Falke*), lies on a hill to the left, overlooked by its church. (Diligence to *Wattwyl* 4 times daily in 2¼ hrs., p. 63.) To the left, on the hill, the monastery of *Sion* (2317'). 36½ M. *Kaltbrunn-Benken*. The wooded range on the right is the *Obere Buchberg* (2020').

A carriage-road leads from the station of Kaltbrunn-Benken or Utznach to (3 M.) *Rieden* (2360'; **Inn & Curhaus zum Rössli*, moderate), a health-resort, commanding charming views. Excursions may be made thence to the top of the *Speer* (p. 45), in 3½ hrs.; viâ *Alp Breitenau* to (2 hrs.) *Ebnat-Kappel* (p. 63), etc.

Beyond (39½ M.) **Schänis** (1450'; **Hirsch; Löwe*), another industrial place, the ancient frontier of Rhætia, we approach the *Lint Canal*, constructed in 1807-22 to connect the Lake of Zürich with the Walensee, and draining, in conjunction with the *Escher Canal*, a once dismal and swampy region. The canal runs parallel with the railway at the foot of the *Schäniser Berg* (5470'); to the right a striking view of the Valley of Glarus with its snow-mountains.

On the opposite bank of the Lint Canal is the *Lint-Colonie*, originally a colony of poor people, now an agricultural institution. 42½ M. *Ziegelbrücke* (Hôtel Berger) is the junction of the Glarus line, which soon diverges to the right (p. 67). The Weesen line rounds the *Biberlikopf* (see below), the extreme spur of the Schäniser Berg. To the right tower the beautiful Rautispitz and the Glärnisch (pp. 67, 74).

45½ M. **Weesen**. — Hotels. *Hôtel Speer, at the station, ⅛ M. from the lake, with fine view, R., L., & A. 2-3, B. 1¼, lunch 2½, D. 3, pens. 6-7 fr.; *Schwert. on the lake, R., L., & A. 2½, B. 1¼, lunch 2, D. 3, pens. from 5 fr.; *Hôt. Mariahalden, in an elevated situation; *Rössli, R., L., & A. 1½-2, B. 1, D. 2¼, pens. 4-5 fr. — *Rail. Restaurant.* — *English Church Service* in summer.

Weesen (1410'), a favourite summer-resort, lies in a sheltered situation at the W. end of the Walensee. The *Klosterberg* yields good wine.

Excursions. Shady paths ascend to the (20 min.) *Kapfenberg*, which affords a charming survey. — Pleasant walk (from the station ¾ hr., or from stat. Ziegelbrücke 20 min.) to the top of the Biberlikopf (1895), fine view of the Walensee and of the Linthal up to Netstall and down to the Buchberg. — A very attractive excursion may be made by boat across the lake to (¾ hr.) the hamlet of *Betlis*, prettily situated beside the ruin of *Strahlegg* at the foot of the Leistkamm. Fine view of Mühlehorn, the Mürtschenstock, etc. From Betlis, we may walk to the ruined *Serenmühle* and the *Falls of the Serenbach* (p. 46), or we may ascend to (1 hr.) *Amden*.

A new road (diligence from the rail. station twice daily) with fine views of the lake, but destitute of shade, ascends from Weesen to (1¼ hr.) Amden or *Ammon* (2875'; *Hirsch*), loftily situated on sunny pastures. Beautiful view of the lake from the (½ hr.) *Gyregarti*. — From Amden to the top of the *Leistkamm* (6890'), 3½ hrs., with guide (Thoma of Amden), interesting and not difficult. — From Amden to *Starkenbach* or *Stein* in the Toggenburg (p. 63) over the *Amdener Berg* (5065'), a route of 5 hrs., with beautiful views, but fatiguing on account of the stone pavement.

The *Speer (6417'), an admirable point of view, 4½-5 hrs. (guide unnecessary for experts). At the church we turn to the left, and ascend for the first ½ hr. over rough pavement of conglomerate (pleasant retrospects of the lake). Then a steep ascent through woods and meadows; 2 hrs. *Untere Bütz-Alp* (3563'); ¾ hr. *Unter-Käsern Alp* (4337'); 1 hr. *Ober-Käsern Alp* (5104'; *Inn Zum Hohen Speer). Thence to the top a steep ascent of ¾ hr. more. Beautiful view, especially of E. and N.E. Switzerland. From *Ebnat* or *Nesslau* (p. 63) the Speer is easily ascended in 3½-4 hrs.

The *Walensee, or *Lake of Walenstadt* (1395'), 9¼ M. long, 1¼ M. wide, and 495' deep, is hardly inferior to the Lake of Lucerne in mountainous grandeur. The N. bank consists of almost perpendicular precipices, 2000' to 3000' high, above which rise the barren peaks of the seven *Curfirsten* (*Leistkamm* 6890', *Selun* 7240', *Frümsel* 7434', *Brisi* 7477', *Zustoll* 7336', *Scheibenstoll* 7556', and *Hinterruck* 7523'). The hamlet of *Quinten* alone has found a site on the N. bank. On the S. bank also the rocks, pierced by nine tunnels, are very precipitous at places. At the mouths of the small torrents which descend from the *Mürtschenstock* (8012') lie several villages. The names of the hamlets, *Primsch*, *Gunz*, *Terzen*, *Quarten*, *Quinten*, and that of the lake itself, indicate that the inhabitants are of Rhætian or Latin, and not Germanic origin.

Beyond Weesen we cross the Lint Canal (to the right the Glarus line, see R. 21), and farther on the *Escher Canal* (p. 67) near its influx into the Walensee, and pass through two tunnels with apertures in the side next the lake. Beyond them we observe the *Bayerbach* waterfall on the opposite bank, and the village of **Amden** on the hill above; then the falls of the *Serenbach*, which sometimes disappear in summer. Three more tunnels, between which we obtain pleasant glimpses of the lake and the waterfalls and precipices opposite. — 50 M. **Mühlehorn** (*Zur Mühle*, *Tellsplatte*, both unpretending). To the right rises the bald *Mürtschenstock* (see below).

FROM MÜHLEHORN TO MOLLIS OVER THE KERENZENBERG (3 hrs.), an interesting walk. The road (diligence to Obstalden thrice daily in 1 hr.) ascends in wide curves (short-cuts for walkers) to *Voglingen* and (3 M.) Obstalden (2237'; *Hirsch*, with a shady garden, pens. 5½-6½ fr.; *Stern*; *Sonne*), a charmingly situated summer-resort, affording a fine view of the **Walensee**. A pleasant excursion may be made hence, or from Filzbach (see below), to the (1½ hr.) *Thalalp-See* (6309'). Thence vià the *Spannegg* and the *Platten-Alp* to Glarus, see p. 68; from the Spannegg to the *Mürtschen-Alp* and over the *Murgsee-Furkel* to the *Murgsee*, see p. 68. The Mürtschenstock (8012') may be ascended from Obstalden vià the *Meeren-Alp* (4920') in 5 hrs. (toilsome and for adepts only; guide, Jac. Heussi, 20 fr.). — Beyond Obstalden the road skirts the *Satterntobel*. 1¼ M. Filzbach (2338'; *Hôt. Mürtschenstock*; *Rössli*, plain), a village also frequented as a summer-resort. From the *Britterhöhe* (2920'), reached in ½ hr. by ascending to the left from the Hôt. Mürtschenstock (finger-post), we enjoy an admirable view of the Walensee and the mountains of Toggenburg and Glarus; a more extensive view is obtained from the *Neuenkamm* (6253'), reached vià *Habergschwend* in 3½ hrs. (guide desirable). — The road now ascends for a short distance, and then descends steadily. In 20 min. we reach a point (right), affording a good view of the head of the Walensee, the valley of the Lint Canal, bounded on the left by the *Hirzli* (5387'), and the Wiggis chain. Farther on we pass through the *Britterwald*. Near (3 M.) *Beglingen* we get a glimpse of the Glärnisch and the Tödi, and then descend in windings (avoided by short-cuts) to (1 M.) *Mollis* (p. 67).

A fine new road (recommended to pedestrians) leads from Mühlehorn vià (¾ M.) *Tiefenwinkel* (brewery) and (1½ M.) *Murg* to (2 M.) *Unter-Terzen* and (3½ M.) *Walenstadt*.

Two more tunnels (to the left, **Quinten**, see above).

51 M. **Murg** (*Schiffli*, *Rössli*, pens. at both 4 fr.; *Kreuz*, all rustic), charmingly situated at the mouth of the *Murgthal*, with factories and spinning-mills.

A visit to the "Murgthal, a valley 10 M. long, is recommended (guide unnecessary). The path ascends rapidly, past the Rössli, as far as (20 min.) a *Waterfall below a bridge, which we do not cross (or we may cross the bridge and return to Murg by the pleasant path on the other side). In 20 min. more we reach another bridge, and cross it. After a steep ascent of ¾ hr. on the left bank the path returns to the Murg and crosses it by a third bridge at the (1½ hr.) beginning of the *Merlen-Alp* (3640'). [To the right diverges the route to the *Mürtschen-Alp* (p. 47).] It then ascends on the right bank, through meadows and wood, to the (2½ hrs.) three *Murgseen* (5490', 5955', and 5980'). From the highest lake the *Roththor (8250') may be ascended in 2 hrs. (guide desirable; the fisherman or a herdsman); striking view (W. the Glärnisch, S.W. the Tödi, S.E. the Calanda, E. the Scesaplana, N. the Sentis and Curfirsten, N.W. the hill-country of Zürich). — From the highest lake a rough path crosses the Widerstein-Furkel (6607′) to the *Mühlebach-Thal* and (2½ hrs.) *Engi* in the Sernfthal (p. 75); another (guide required) leads over the Murgsee-Furkel (6570′) to the *Mürtschen-Alp* (6060′), past the *Mürtschenstock* and *Fronalpstock*, to the *Heuboden-Alp* (p. 68) and (5 hrs.) *Glarus*. Or, from the Mürtschen-Alp we may proceed viâ the *Spannegg* (p. 68) to the *Thalalp-See* and to (4½ hrs.) *Obstalden* or *Fïlzbach* (p. 46).

Beyond Murg another tunnel; above, to the right, the village of *Quarten* (1760′) with a new church (*Curhaus Quarten, with hydropathic, prettily situated about 1 M. from Unter-Terzen; pens. from 4 fr.). — 53½ M. *Unter-Terzen* (Freieck; Zur Blumenau). On the steep rocks of the opposite bank several waterfalls are visible; to the right, the village of *Mols*. Then a tunnel and a bridge across the *Seez Canal*.

56 M. **Walenstadt** (1395′; *Hôtel Churfirsten*, at the station, R., L., & A. 2-3, B. 1, lunch 1½, D. 2, pens. 5-6 fr.; *Hirsch*, in the village, moderate) lies ½ M. from the E. end of the lake (*Hôt.-Pens. Seehof*, on the lake).

EXCURSION (with guide) from Walenstadt by a steep path through wood to the (2 hrs.) *Alp Lösis*; then, nearly level, to the *Alp Büls* and (¾ hr.) the *Tschingeln-Alp* (5040′; milk); follow the slopes of the Curfirsten, with a series of beautiful views, to the (1¼ hr.) *Alp Schwaldis* (4775′) and return by *Alp Schrinen* (4205′) to (1½ hr.) Walenstadt; or proceed from Alp Schwaldis to the *Säls-Alp* (4800′), descend by the *Stäfeli* to the (1 hr.) *Laubegg Alp* (4505′) and thence by a steep path, but free from danger, to (1½ hr.) *Quinten* (p. 46), whence the lake is crossed by boat to *Murg*. — To AMDEN viâ the *Leistkamm*, 10 hrs. with guide, very attractive (comp. p. 45). — To WILDHAUS in the Toggenburg (p. 63) a rough path, with splendid views, crosses the *Käserruck* (7435′; 6 hrs.; guide necessary).

We now ascend the broad valley of the Seez. On a rock to the right, the ruins of *Gräplang* (Romansh *Crap Long*), or *Langenstein*; to the left, on a rocky height above *Bärschis*, the pilgrimage-church of *St. Georgen*. 58 M. **Flums** (1475′; Hôtel Bahnhof; Löwe). — Near (64 M.) **Mels** (1637′; *Melserhof*, at the station; *Frohsinn*) the Seez descends from the *Weisstannen-Thal*, a valley to the S.W.[1]

The *Alvier* (7758′), an admirable point of view, may be ascended hence in 5 hrs. (guide unnecessary for adepts). The path ascends steeply from the station to the right to the (3 hrs.) *Alp Palfries* (4850′; Curhaus, plain), traverses steep and rocky slopes, and (2½ hrs.) reaches the summit through a narrow cleft by steps cut in the rock (club-hut, dilapidated). The magnificent view embraces the Rhine Valley, the Rhætikon, and the Vorarlberg, Appenzell, and Glarus Mts. (good panorama by Simon). Good paths ascend from Flums, Sevelen, Buchs, and Trübbach (comp. p. 61).

FROM MELS TO VÄTTIS, through the *Weisstannen-Thal* and *Kalfeisen-Thal* (10-11 hrs.). Road to (8 M.) Weisstannen (3270′; *Alpenhof; Gamsli*). Thence (with guide), by *Unter-Lavtina* (4325′) and *Vattüsch* (5940′), in 4 hrs. to the Heidel Pass (7305′), between the *Seesberg* and the *Heidelspitz* (7980′), where we have a fine view of the huge Sardona Glacier, the Trinserhorn, and Ringelspitz. Descent into the *Kalfeisen-Thal*, to the Tamina bridge near *St. Martin* (4430′) 2 hrs., and to *Vättis* (p. 86) 2 hrs. more. — From Weisstannen to *Elm* by the *Foo* or *Ramin Pass*, see p. 76.

At (65 M.) **Sargans** (1590′; *Hôtel Thoma*, at the station, R. 2, B. 1 fr.; *Rail. Restaurant*; *Krone, Löwe*, in the town) we reach the Rhine Valley and the Rorschach and Coire line. The little town, ¾ M. to the N.W., rebuilt since a fire in 1811, lies picturesquely at the foot of the *Gonzen* (p. 61), and is commanded by an old castle (still habitable) of the former Counts of Toggenburg. Railway from Sargans viâ *Ragatz* to (79 M.) *Coire*, see R. 18.

15. From Zürich to Romanshorn and Friedrichshafen (Lindau).

Comp. *Maps*, pp. *20*, *26*, *28*.

RAILWAY to Romanshorn (51 M.) in 3 hrs. (8 fr. 65, 6 fr. 5, 4 fr. 35 c.). STEAMBOAT thence to Friedrichshafen in 1 hr. (1 ℳ 20 or 80 pf.); to Lindau in 1½ hr (2 ℳ 25 or 1 ℳ 50 pf.; see p. 28).

The train crosses the *Sihl*, ascends in a wide curve, crosses the *Limmat*, and passes under the *Käferberg* by a tunnel 1020 yds. long. — 3 M. *Oerlikon* (1443'; Sonne; Rail. Restaurant).

FROM OERLIKON TO DIELSDORF, 12 M., railway in 35 minutes. Stations *Glattbrugg*, *Rümlang*, and (8½ M.) *Oberglatt*, the junction for *Niederglatt* and (4½ M.) *Bülach* (see below). Then (10½ M.) *Niederhasli* and (12 M.) *Dielsdorf* (1410'; Sonne; Post), the terminus of the line, 1½ M. below the prettily situated old town of Regensberg (2024'; *Krone), on the E. spur of the *Lägern-Gebirge* (p. 21). Fine view from the tower of the old castle (now an institution for boys of weak intellect); still more extensive from the *Hochwacht* (2830'), 1 hr. farther on.

The line crosses the *Glatt*. At (5½ M.) *Wallisellen* (Linde) the *Rapperswil* line diverges to the right (see p. 43). Fine view of the Glarus Alps. 7½ M. *Dietlikon*; 10½ M. *Effretikon* (branchline to *Wetzikon* and *Hinweil*, p. 44); 13 M. *Kemptthal*. Near Winterthur the *Töss* is crossed. On a hill to the left, the ruins of *Hoch-Wülflingen* (1962').

16 M. Winterthur (1447'; pop. 15,985; *Goldner Löwe, R., L., & A. 2½-3, B. 1¼, D. 3, pens. 7-8 fr.; *Krone, R. & A. 2½ fr.; *Adler, R., L., & A. 1½-2, B. 1, D. 1½-3, pens. 5-8 fr.; *Rail. Restaurant; Rheinfels and *Walhalla Restaurants*; H. *Langsdorf*, U.S. Con. Agent), on the *Eulach*, is an industrial and wealthy town and an important railway-junction. The handsome *Stadthaus* was designed by Semper. The large *School* (with statues of Zwingli, Gessner, Pestalozzi, and Sulzer) contains the town-library and a few small Roman antiquities found near Ober-Winterthur (p. 23). In the *Kunsthalle* are some good Swiss paintings The *Panorama of the Rigi* near the Polytechnicum is worth seeing.

FROM WINTERTHUR TO WALDSHUT, 32 M., railway in 2 hrs. The line traverses the *Tössthal*. Stat. *Töss*, *Wülflingen*, *Pfungen-Neftenbach*, *Embrach-Rorbas*. The train leaves the Töss and passes through a tunnel (1980 yds.). 10½ M. Bülach (1874'; *Kopf*; *Kreuz*), a small town near the *Glatt*, once fortified (branch-line to *Oberglatt* and *Otelfingen*, p. 21). The line runs through the *Hardwald* to the N. to *Glattfelden* and (13½ M.) *Eglisau*; the latter (Löwe; Hirsch), with its castle, lies on the right bank of the *Rhine*. We now follow the left bank of the Rhine and cross the Glatt. Stat. *Zweidlen*; 19 M. *Weiach-Kaiserstuhl*, an old town with a massive tower; on the right bank *Schloss Röteln*, and farther on, the ruins of *Weiss-Wasserstelz*. Stat. *Rümikon*, *Reckingen*, *Zurzach*, and (30½ M.) *Coblenz*, where the Rhine is crossed to (32 M.) *Waldshut* (p. 24). Viâ *Laufenburg* to *Stein-Säckingen*, see pp. 21, 23.

FROM WINTERTHUR TO RÜTI, 29½ M., in 2-3 hrs., by the *Tössthalbahn*, Stations *Grüze* and *Seen*. Near (5 M.) *Sennhof* (25 min. to the S.W. of which is the old château of *Kyburg*, commanding a fine view) we enter the pretty *Tössthal*. Stations *Kollbrunn*, *Rikon*, *Zell* (10 M.) *Turbenthal* (Bär), *Wyla* (with a picturesquely situated church), *Saland*, (16 M.) *Bauma* (Tanne), all thriving industrial places. About 2¼ M. to the E. of Zell, on the slope of the *Schauenberg*, is the frequented *Gyrenbad*, with an alkaline spring (see below). Then *Steg*, *Fischenthal*, *Gibswyl-Ried*. From the last, situated on the watershed, the Bachtel may be ascended in 1 hr. Then through the picturesque valley of the *Jona* to (25 M.) *Wald* (2037'; *Löwe; Rössli*), at the S.E. foot of the Bachtel (p. 44). Passing the waterfall of *Hohe Lauf*, we join the Zürich and Rapperswil line at (29½ M.) *Rüti* (p. 44).

From Winterthur to *Schaffhausen*, see R. 12; to *St. Gallen* and *Rorschach*, see R. 16; to *Constance*, see R. 11.

The Romanshorn line traverses the green and fertile *Thurgau*. 20 M. *Wiesendangen*; 24 M. *Islikon*.

26 M. **Frauenfeld** (1340'; pop. 6087; *Falke; *Hôtel Bahnhof*, at both R. 1½-2, B. 1, D. with wine 2½ fr.; *Krone*), on the *Murg*, with large cotton-factories, is the capital of the Thurgau. The handsome *Schloss* on an ivy-clad rock is said to have been built by a Count of Kyburg in the 11th century.

From Frauenfeld to Wyl, 11 M., steam-tramway in 1-1¼ hr. (fares 1 fr. 80, 1 fr. 30 c.). Stations: *Murkart*, *Mazingen*, *Jakobsthal*, *Wängi*, *Rosenthal*, *Münchweilen*, and *Wyl* (see below).

29 M. *Felben*. Near (32¼ M.) *Müllheim* the train crosses the *Thur*. 35 M. *Märstetten*; 37½ M. *Weinfelden* (1463'). To the left *Schloss Weinfelden* (1850'; view), on the vine-clad *Ottenberg*. 39½ M. *Bürglen*. — 41 M. *Sulgen* (1584'; Helvetia; Schweizerhof).

From Sulgen to Gossau, 14½ M., railway in 67 min. (1 fr. 65, 1 fr. 15 c.). The line traverses the pretty valley of the *Thur*. Stations *Kradolf*, *Sitterthal*. 6 M. *Bischofzell* (1659'; *Linde; Thurbad*), a small town at the confluence of the Thur and *Sitter*. Then *Hauptweil*, *Arnegg*, *Gossau* (see p. 50).

Stations *Erlen* (Hôt. Bahnhof), *Amriswil*, and (51 M.) **Romanshorn** (1322'; *Hôtel Bodan*, R., L., & A. 2-3, B. 1 fr.; *Falke; Jäger; *Rail. Restaurant*), on a promontory on the *Lake of Constance*. Thence to *Friedrichshafen*, or *Lindau*, see p. 28.

16. From Zürich to St. Gallen, Rorschach, and Lindau.

Comp. Maps, pp. 40, 54, 28.

Railway to *St. Gallen* (52½ M.) in 3 hrs. (8 fr. 80, 6 fr. 20, 4 fr. 40 c.); to *Rorschach* (62 M.) in 3¾ hrs. (10 fr. 35, 7 fr. 45, 5 fr. 30 c.). — Steamboat from Rorschach to Lindau in 1¼ hr. (1 ℳ 65 or 1 ℳ 10 pf.).

From Zürich to (16 M.) *Winterthur*, see p. 48. The St. Gallen railway is unattractive. The Curfirsten gradually appear to the S., and the Appenzell Mts. to the S.E. — 20½ M. *Räterschen*; 24 M. *Elgg* (2012'; Ochs; Löwe). To the S. (4 M.) is the *Schauenberg* (2930'; fine view), on the S.W. slope of which lies the *Gyrenbad* (see above). Stations *Aadorf* (Linde), *Eschlikon*, *Sirnach*. — 34½ M. **Wyl** (1936'; *Hôtel Bahnhof*), a pleasant old town (3507 inhab.). Branch-line to *Ebnat*, see p. 63; steam-tramway to *Frauenfeld*, see above.

The train crosses the *Thur* by an iron bridge, near the old castle of *Schwarzenbach*. 39½ M. *Utzwyl*, the station for *Nieder-Utzwyl* on the left, and *Ober-Utzwyl* on the right. (Near the former, 1¼ M. from the station, is the hydropathic of *Buchenthal*.) — 43 M. **Flawyl** (2020'; *Rössli*; *Post*), a large manufacturing village. The *Glatt* is crossed. 46 M. *Gossau* (Hôt. Bahnhof; branch-line to *Sulgen*, see p. 49). — 48½ M. *Winkeln* (Kreuz).

From Winkeln to Appenzell, 16 M., in 1½ hr., by the narrow-gauge *Appenzell Railway*. The line passes the *Heinrichsbad* (*Curhaus, with chalybeate spring). 3 M. **Herisau** (2550'; 12,937 inhab.; *Löwe, R. 2¼, D. 3, pens. 7-8 fr.; *Storch*), a thriving town with extensive muslin-factories and a clock-tower attributed to the 7th century. — 5 M. *Wylen*; 5½ M. *Waldstatt* (2700'; *Hirsch; Pens. Sentishlick), with a chalybeate spring and whey-cure. Then through the *Urnäsch Valley*, by *Zürchersmühle*, to (9¼ M.) **Urnäsch** (2746'; *Krone; Bahnhof*). About ½ M. above Urnäsch is the primitive spa of *Rosenhügel* (2392'). — Beyond Urnäsch the train passes the (11½ M.) *Jacobsbad* (to the E.), with its mineral spring (good quarters), and goes on to (13 M.) **Gonten** (2970'; *Löwe; Krone; Bär*) and (14 M.) *Gontenbad* (2925'), a well-managed establishment, with a chalybeate spring (pens. 5-6 fr.). It then crosses the deep valley of the *Kaubach* to (16 M.) *Appenzell* (p. 56). — **Ascent** of the *Säntis* from Urnäsch, see p. 58. Over the *Kräzern Pass* to *Neu St. Johann*, see p. 63.

We now cross the deep valley of the *Sitter* by an imposing iron bridge, 207 yds. long, and 174' above the river. A little lower down is the *Kräzernbrücke*, with its two stone arches, built in 1810. — 50 M. *Bruggen*.

52½ M. **St. Gallen**. — Hotels. *Hecht, Theater-Platz, R., L., & A. 2½-4, D., incl. wine, 3½ fr., good cuisine; *Linde, Leonhard-Str., with café-restaurant; *Hirsch, in the market-place, R. & A. 2-2½, D. 3 fr.; *Walhalla, opposite the station, R., L., & A. 2½-3½, B. 1¼, D. 3, pens. 7-9 fr.; *Schiff, Ochs moderate. — Cafés. *Linde; Pavillon; Trischli; Hörnli; Rail. Restaurant. — Baths at the *Löchlibad*, *Tobler's* (St. Magnibalden), and *Seifert's* (Rorschacher-Str.); in summer, river-baths at *Dreilinden* (p. 51). — Cabs: ¼ hr., 1-2 persons 80 c., 3-4 pers. 1 fr. 20, ½ hr. 1 fr. 20 and 1 fr. 80, ¾ hr. 1 fr. 60 and 2 fr. 40, 1 hr. 2 fr. and 3 fr., luggage 20 c.; double fares at night. — U.S. Consul-General, *J. B. Richman, Esq*.

St. Gallen (2205'), one of the highest-lying of the larger towns of Europe, the capital of the canton of that name, and an episcopal see (since 1846), is one of the chief industrial towns in Switzerland, embroidered cotton goods being its staple product. Pop. 28,037.

From the railway-station we proceed to the left through the Post-Strasse or the Bahnhof-Strasse to the Market-Place, the central point of the crowded Old Town. The busy Marktgasse leads hence to the S. to the Protestant *Church of St. Lawrence*, rebuilt in the Gothic style in 1849-54 and provided with a lofty tower. Adjacent is the N. entrance to the Klosterhof ('Stiftseinfang'), containing the Benedictine Abbey, founded in the 7th cent. by St. Gallus, an Irish monk, rebuilt in the 18th cent., and suppressed in 1805, one of the most famous seats of learning in Europe from the 8th to the 10th century. The extensive buildings now accommodate the Cantonal offices, the bishop's residence, and the celebrated *Abbey Library*. The last (open on Mon., Wed., and Sat. 9-12 and 2-4, for strangers at other times also) contains 30,000 vols. (1358 incunabula) and many valuable MSS. (a psalter of Notker Labeo of the 10th cent. and a Nibelungenlied of the 13th cent.); of those mentioned in a catalogue of the year 823 about 400 still exist. — The *Abbey Church*, rebuilt in 1755-66 in the rococo style, contains finely carved choir-stalls and a beautiful iron choir-screen.

Behind the abbey flows the *Steinach*. — To the E., beyond the moat skirting this part of the old town, is the large *Cantonal School House*, containing the *Town Library* ('*Bibliotheca Vadiana*'; open **Tues.**, Thurs., and Sat., 2-4), which boasts of valuable MSS., chiefly of the Reformation period, and the collections of the *Geographical & Commercial Society* (open Sun. 10-12 & 1-3; Wed. and Sat. 1-3). — Near it, in the Museums-Str., by the Grosse Brühl, is the *Museum*, containing the municipal collections. On the groundfloor are the *Natural History Collections* (open Sun. 10-12 and 1-3, Wed. and Frid. 1-3), and on the first floor the *Picture Gallery of the Kunstverein* (open Sun. 10-12 & **1-3**, Wed. 1-4; works by Koller, Diday, Makart, A. Feuerbach, Ritz, Schirmer, and others), and the collections of the *Historical Society* (open Sun., 10-12 and Wed. 1-4). Behind the museum is the *Public Park*, prettily laid out.

The *Industrial Museum*, with a school of design, is in the Vadian-Strasse (open Sun. 10-12; on other days, except Mon., 9-12 and 1-5). — From the S. end of the town a cable-tramway (3 min.; fare 15, down 10 c.) ascends through the steep gorge of the Steinach to the suburb of *Mühlegg* (2440'; Restaurant). On the **other side** of the Steinach, at the base of the Freudenberg (see below), ½ M. to the E., are the open-air baths of *Dreilinden*, much frequented in summer.

EXCURSIONS. The *Freudenberg (2910'; Inn), 1½ M. to the E. of the town and ¾ M. from Mühlegg (see above; carriage for 1-2 pers. 7 fr., 3-4 pers. 12 fr.), commands a charming view of the Lake of Constance as far as Lindau; in the foreground lie St. Gallen and the surrounding country, dotted with houses, to the S. the Sentis chain, the Glärnisch, Tödi, etc. — The *Vögelisegg (4½ M.; carr. 6 or 10 fr.; p. 55) and the *Frölichsegg (4 M.; p. 58) also afford fine views. — The nunnery of *Notkersegg* (2580') and the *Kurzegg Inn* (2735'), both on the road to Vögelisegg, command fine views of the Bodensee. — To the **Rosenberg** (2170'; carriage 2 fr., 3 fr.), with the *Kurzenburg*, a deaf-and-dumb institution, and numerous villas; the route runs viâ *Rotmonten*, on the saddle, to the (1 hr.) inn of *SS. Peter and Paul* (2580'), with a large deer-park. — Through the Gemeindsböden or viâ Mühlegg (see above) to the **Falkenburg** (2560'), which commands the best view of the town. We then cross the wooded *Bernegg* to the *Vogelherd*, with a charming view and a monument to the poet Scheffel, to the (¾ hr.) *Im Nest Inn*, and to the (10 min.) *Solitude* (2690'; views). Then back by the Teufen road (2 M.). — **Kronbühl** (2035'; Inn; carriage 3 fr., 5 fr.), on the Arbon road, with a view of the Lake of Constance. — *Waid*, a health-resort, 3 M. to the N.E., with splendid view of the Lake of Constance (carriage 4 fr., 6 fr.; diligence from St. Fiden, see below). — *Bruggen* and the *Sitterbrücke* (**p. 50**), by rail in 8 minutes. — *Martinstobel* and *Mötteliechloss*, see p. 52. — Tramway to *Gais*, see p. 56.

From St. Gallen the line descends through a long cutting to (53½ M.) *St. Fiden* (2126'; **Hôt.** National), and enters the wild valley of the *Steinach*. Embankments and cuttings are traversed in rapid succession. Nearly the whole Lake of Constance is frequently

52   *I. Route 16.*   RORSCHACH.   *From Zürich*

visible, with Friedrichshafen on its N. bank. — Turning to the right, the line crosses the *Goldach* by a bridge of five arches near (56½ M.) *Mörschwil* (1778'; *Pens. Gallusberg, near the station).

62 M. **Rorschach.** — *Town Railway Station*, ½ M. to the E. of the pier, where the lines from St. Gallen and Romanshorn join that from Coire; *Lake Railway Station*, at the pier, not called at by all trains.

Hotels. °ANKER, R., L., & A. 2-4, B. 1¼, D. 8, pens. 6-7 fr.; °SEEHOF, with garden; HIRSCH, moderate; BADHOF; HÔTEL BODAN; HÔT. STIERLIN; SCHIFF, R. 1½, B. 1, D. 1½, pens. 5-6 fr.; HÔTEL BAHNHOF, POST, R. 2, D. 2½ fr., these two near the station; SCHÄFLE, with garden, moderate; RÖSSLE, R. 1-1¾, pens. 3½ fr.; ZUR ILGE; GRÜNER BAUM, R., L., & A. 1½-2, B. 1, D. 2-2½, pens. 5-7 fr., well spoken of; OCHS, with brewery. — *°Rail. Restaurant*, with a balcony and view of the lake. *Beer* at *Stierlin's*, behind the station, and at the *Falke* (with rooms to let). — Private apartments reasonable. — *Baths* at *Notter's*, on the lake; *Lake Baths* ¼ M. to the W. (bath with towel 35 c.).

Rorschach (1310'; pop. 5867), a busy town on the Lake of Constance, chiefly important for its corn-trade, is also a summer-resort.

EXCURSIONS. Above Rorschach rises the old abbey of **Marienberg**, with handsome cloisters, now a school. The view from the *Rorschacher Berg*, the green orchard-like hill behind the town, embraces the whole lake, with the Vorarlberg Mts. and the Rhætikon chain. Its summit, the *Rossbühel (Inn)*, may be reached in 1¼ hr. from Rorschach (boy to show the way desirable). The whole hillside is intersected by roads, which afford a great many pleasant walks. Good inns at (½ hr.) the *Sulzberg* and (½ hr.) the *Hohrain*. — The **St. Anna Schloss**, since 1449 the property of the Abbots of St. Gallen, has been partly restored (°Restaurant); fine view from the upper rooms. The road, which is steep towards the end, takes about ¾ hr. from the station. The view from the *Jägerhaus*, ½ hr. farther up, is still more extensive (Inn, good wine).

To the Martinstobel and Mötteliechloss and back, 3 hours. By the St. Gallen railway to *St. Fiden*, **see p. 51**. Below the station we take the road to *Neudorf* (brewery on the left), descend the high-road, and diverge to the right by the Heiden road into the **Martinstobel**, the gorge of the *Goldach*, spanned by an iron bridge 100' high. Here, at the beginning of the 10th cent., the monk Notker composed his '*Media vita in morte sumus*', upon seeing a man accidentally killed. Beyond the bridge we ascend the road to the left, passing the débris of a landslip which took place in 1845, to *Untereggen* (Schäfle), and thence descend the Goldach road as far as a road leading through a grassy dale to the right to the **Mötteliechloss**. This was formerly the seat of the Barons of Sulzberg, of whom it was purchased by the wealthy *Mötteli* family of St. Gallen, and after various vicissitudes it has now fallen into disrepair. °View from the platform on the top (gratuity), one of the finest near the lake. Pleasant walk back to Rorschach through the *Wilholz* (½ hr.). — To Tübach, surrounded by fruit-trees, and the *Castle of Steinach*, about 1 hr. — By the 'Obere Weg', with fine views, to (1 hr.) Wylen (°*Inn*), near the Duke of Parma's château of *Wartegg*, with its beautiful park. — By *Staad* (p. 60) to (1¼ hr.) Schloss Weinburg, the summer-residence of the Prince of Hohenzollern (visitors admitted to the fine park); splendid view from the *Steinerne Tisch*, above the château (return via *Thal* and *Rheinegg*, p. 60). — To *Walzenhausen* and the *°Meldegg*, see p. 60.

At Horn (on the lake, 1½ M. to the N W.; railway, see p. 31) there are a large *°Hotel & Bath-House* (pension 6 fr.), and the *Hirsch Inn*. Near Horn, to the left, is the château of the Landgrave of Hessen-Philippsthal.

Railway to *Coire*, see **p. 60**; to *Bregenz* and *Lindau*, see p. 432; to *Heiden*, see p. 54; to *Constance*, see p. 31.

To Lindau by steamer (1¼ hr.; D. 2½ *M.*, mediocre), comp. p. 28. To the S.E. is Bregenz, at the foot of the Pfänder; in the background the Rhætikon chain; to the S. rise the Appenzell Mts. and the Sentis.

Lindau. — *Bayrischer Hof, R., L., & A. 2½-4, B. 1 ℳ 20 pf., D. 3, pens. 6-8 ℳ; *Krone, *Hôtel Reutemann, *Lindauer Hof, R. 1½-2½, D. 2½ ℳ; Helvetia, R. 1¼-1½ ℳ, all on the lake; Sonne, in the Reichsplatz; Gärtchen auf der Mauer, a pension on the mainland. — Restaurants: *Seegarten*, near the Bayrischer Hof (also rooms); *Schützengarten*, a restaurant on the old bastion, near the Roman tower, with view; adjacent to it, *Rupflin* (wine); *Rail. Restaurant*. — *Lake Baths* on the N.W. side of the town, in the inner arm of the lake.

Lindau (5400 inhab.), the terminus of the Bavarian S.W. Railway (express to Augsburg 5, to Munich 5½ hrs.), once an imperial town and fortress, and in the middle ages a thriving commercial place, lies on an island in the Lake of Constance, connected with the mainland by a railway-embankment and by a wooden bridge, 356 yds. long. Lindau is said to have been the site of an ancient Roman fort, to which the venerable tower near the bridge perhaps belonged. On the quay is a monument to *King Max II.* (d. 1864), in bronze, designed by Halbig. At the end of the S. pier, on a granite pedestal 33' high, is placed an imposing lion in marble, 20' in height, also by Halbig; opposite, on the N. pier, is a *Lighthouse*. The harbour is adjoined to the S. by the *Alte Schanz*, which commands a view of the Alps from the Scesaplana to the Sentis (mountain indicator). In the Reichsplatz are the *Rathhaus*, erected in 1422-36 and restored in 1885-87, with painted façades and an interesting collection of antiquities (open 11-12, Sun. 2-5), and the handsome *Reichsbrunnen*, with a bronze figure of 'Lindauia' and other allegorical figures, erected in 1884.

Excursions. Pleasant walk on the N. bank of the lake towards the left (cross the railway embankment and turn to the left), passing the villas of *Näher*, *Lotzbeck* (pretty park), *Giebelbach*, *Lingg* (*Frescoes by Naue), and others, to the (2¼ M.) *Schachenbad* (Pens. Freihof) and the (¾ M.) Lindenhof (or Villa Gruber), with its beautiful grounds and hot-houses (adm. on Frid. gratis; at other times 1 ℳ, tickets at the Schachenbad; closed on Sun.). About ½ M. farther on is the château of *Alwind*. — Beautiful view from the (½ hr.) vine-clad *Hoierberg (1496'), which is reached by a path skirting the railway and passing the village of *Hoiren*, or to the left viâ *Entisweiler* (*Schmid's Restaurant) and *Schachen* (Zum Schlössle). The road from the Landthor leads viâ *Aeschach* (Schlatter). Two inns and a belvedere on the top. — To *Bregenz*, see p. 432.

17. The Canton of Appenzell.

The Canton of Appenzell cannot vie in grandeur with many other parts of Switzerland, but it includes within a small space most of the characteristics of the country. It boasts of one of Switzerland's largest lakes, of an almost southern vegetation, of great industrial prosperity, of the richest pastures, and even of lofty snow-mountains. The finest points are *Heiden*, *St. Antoni*, *Wildkirchli*, *Ebenalp*, the *Hohe Kasten*, and the *Sentis*.

This canton, which is entirely surrounded by that of St. Gallen, was divided after the religious wars of 1597 into two half-cantons, **Ausser-Rhoden** and **Inner-Rhoden**, and to this day party-feeling on religious questions is very strong. Inner-Rhoden, which consists of pasture-land and is 63 sq. M. in area, is almost exclusively Roman Catholic, and down to 1848 permitted no Protestants to settle within its limits; even Roman Catholics who were not natives of the canton were strictly excluded. This restriction was nominally rescinded by an article of the Federal consitution in 1818, but little change has practically taken place. Population 12,900, of whom about 700 only are Protestants. The inhabitants generally occupy scattered cottages and huts; they are, according to *Merian* (1650), 'a rough, hardy, homely, and pious folk'; their costume is picturesque and primitive, and cattle-breeding and cheese-making are their chief pursuits. — AUSSER-RHODEN (90 sq. M., 54,200 inhab., 3500 Rom. Cath.) belongs to the Reformed Church; one-fourth of its population is engaged in the cotton and silk manufacture, chiefly for firms at St. Gallen. Almost every house has its loom, the products of which often exhibit extraordinary taste and skill, and were objects of admiration at the London and Paris Industrial Exhibitions.

Railway from *Winkeln* to *Appenzell* in 1^1/$_2$-2 hrs.; from *St. Gallen* to *Gais* in 1^1/$_4$ hr.; from *Rorschach* to *Heiden* in 55 minutes. — Diligence from *Rheineck* to *Heiden* twice daily in 1^3/$_4$ hr.; from *Au* to *Heiden* viâ *Berneck*, once daily in 3 hrs.; from *Heiden* viâ *Trogen* and *Speicher* to *Teufen* twice daily in 2^3/$_4$ hrs.; from *Altstätten* to *Gais* daily in 2 hrs.; from *Gais* to *Appenzell* five times daily in 35 min.; from *St. Gallen* viâ *Speicher* to *Trogen* thrice daily in 1^3/$_4$ hr. — Carriage from St. Gallen to Trogen 6 fr. (3-4 pers. 10 fr.), to Appenzell 9 or 16, Weissbad 10 or 16^1/$_2$ fr.; half-fare more for the return.

The RAILWAY FROM RORSCHACH TO HEIDEN, 4^1/$_3$ M. long, is constructed on the rack-and-pinion system (maximum gradient 1 : 11). The train starts from the harbour station (p. 50), stops at the outer station, where the toothed rail begins, and then ascends through orchards, affording charming glimpses of the lake (best views to the left). On the left, below, is the picturesque château of *Wartegg*, on the right, above, *Wartensee*. We then cross a ravine, pass through a cutting, and traverse wood. Near (2^1/$_2$ M.) stat. *Wienacht-Tobel* (2025') are large quarries of fossiliferous sandstone. We then skirt the deep *Wienachter Tobel*, obtaining to the left a beautiful view of the rich valley, with the mountains of the Bregenzer Wald beyond, and the mouth of the Rhine below, while Heiden appears on the hill to the right. Beyond (3 M.) stat. *Schwendi* (2217') we cross the gorge by a lofty viaduct and ascend over pastures and through wood.

4^1/$_3$ M. Heiden. — *FREIHOF, R., L., & A. 3-4, B. 1^1/$_2$, D. 4, S. 2^1/$_2$. pens. 8^1/$_2$ fr.; *SCHWEIZERHOF, R., L., & A. 3^1/$_2$, B. 1^1/$_2$, D. 3^1/$_2$, S. 2^1/$_2$ fr.; *KRONE, R., L., & A. 1^1/$_2$-2^1/$_2$, B. 1, D. 2^1/$_2$, pens. 5^1/$_2$-7 fr.; *SONNEN-HÜGEL, at the upper end of the village, near the Curhalle, with baths and garden, R., L., & A. 2-3, B. 1^1/$_4$, D. 3, pens. 6-8 fr.; PENS. DIETRICH, with restaurant; LINDE; *ZUM PARADIES; LÖWE; *ZUR FROHEN AUSSICHT, R., L., & A. 2-2^1/$_2$, pens. 6-7 fr; PENS. BLUMENTHAL. Lodgings at *Arnold's* (view) and at *Tobler's*, the postmaster. Baths in the *Quellenhof*. — *Visitors' Tax* for a stay of several days 1 fr. 20 c. — *English Church Service* in summer.

Heiden (2655'; pop. 3453), a thriving village with substantial houses, rebuilt since a fire in 1838, lies in the midst of sunny and sheltered meadows, and is a favourite air-cure resort. Mineral water may also be procured. At the upper end is a tasteful *Curhalle*. The gallery of the church-tower and the grounds of the Freihof (see above) afford fine panoramic views.

WALKS. To the *Bellevue (2865'), a hill 25 min. to the S.E., on the right bank of the *Gstaldenbach*, with inn, belvedere (30 c.), and a beautiful view of Heiden and the Lake of Constance, and in 20 min. more to the *Seulishlick*; W. to the *Hasenbühl*, *Benzenrüti*, and *Steinli*, with a *of Appenzell.* TROGEN. *I. Route 17.* **55**

pavilion and charming view; S. to *Bischofsberg* (see below). To the W., below the Grub road (see below), the *Krähenwald* (pleasant grounds); N.W (¾ hr.) the *Rossbühel* above Grub (2925'; tavern, good wine).

A road affording picturesque views leads from Heiden to the N.E. viâ *Wolfhalden* (2350'; Friedberg) to (4½ M.) *Rheinegg* (p. 60; diligence twice daily in ¾ hr.); another attractive road to the W. viâ *Grub, Eggersried*, and the *Martinstobel* (p. 52) to (8 M.) *St. Gallen* (p. 50).

The *Chapel of St. Anthony* ('*St. Antönibild*'; 3640'), 1¼ hr. to the S. of Heiden, affords a famous view of the Rhine Valley (preferable to that from the Kaien), Bregenz, Lindau, part of the Lake of Constance, and the Vorarlberg and Appenzell Mts. One route to the chapel is by *Oberegg*; another, shorter, leads by the orphan-houses and the *Bischofsberg* (see above; both routes denoted by blue marks). From the chapel to *Altstätten* (p. 61) 1½ hr.

The *Kaien*, 1¼ hr. to the S.W. of Heiden, is also frequently ascended (guide, not indispensable, 1½ fr.). We follow the Trogen road for ¼ M. and then diverge to the right beyond a small bridge (finger-post 'Steinli, Kaien') and ascend by a good, red-marked path to the (1-1¼ hr.) summit of the *Kaien* (3612'). The view embraces a great part of the Lake of Constance and Canton Thurgau, the embouchures of the Rhine and the Bregenzer Ach, the Vorarlberg and Liechtenstein Mts., with the white chain of the Rhätikon and the Scesaplana above them to the S.E. To the S. it affords a characteristic glimpse of the Appenzell district: the Kamor and Hohe Kasten, the five peaks of the Furgglen-First and Kanzel, the double-peaked Altmann, the snow-fields of the Sentis, and the Tödi farther distant; in the foreground woods, meadows, and the thriving villages of Wald, Trogen, and Speicher; to the left above Trogen rises the Gäbris (see below); to the right, near Speicher, the Vögelisegg (see below); to the left, above Speicher, in the distance, the Pilatus and the Rigi. — The Kaien is 1½ hr. from Speicher, and 2½ hrs. from St. Gallen. Trogen seems almost within a stone's-throw, though really 3 M. distant. The path descends to the right by the *Gupf* (3545'; Inn) and *Rehetobel* (3140'; "Hirsch"), a village almost wholly burnt down in 1890, beyond which the road to Trogen is visible in the wooded ravine far below. Near the bridge, in the valley below, is a rustic tavern 'Am Goldach'.

The GÄBRIS (see p. 54) may be ascended from Heiden direct (avoiding Trogen): to *St. Anthony's Chapel* (see above) 1¼ hr.; then along the arête, with a charming survey of the Rhine Valley and the Sentis, to the *Landmark* (3265'; Inn, comp. p. 61), on the road from Altstätten to Trogen, and the summit of the *Gäbris*, a beautiful walk of 2 hrs. About 8 min. below the summit the St. Antoni route is joined by that from Trogen (finger-post 'Gais, Trogen, Speicher').

The road to Trogen (6½ M.) ascends the E. slope of the Kaien (see above) to the (2¼ M.) *Langenegg* (3185'; Inn) and then leads up and down hill, past *Rehetobel* (see above), situated beyond the deep valley of the Goldach on the right, and (2¼ M.) *Wald* (3150'; Sonne), to (2 M.) —

Trogen (2975'; pop. 2578; *Krone*; *Pens. Lindenbühl*), a prosperous village, pleasantly situated and visited as a summer-resort.

Road over the *Landmark* to (7 M.) *Altstätten*, see p. 61. — FROM ST. GALLEN TO TROGEN (6 M.), diligence thrice daily in 1¾ hr. The road leads past the nunnery of *Notkersegg* and the inn of *Kurzegg* (p. 51), to the (4 M.) *Vögelisegg* (1358'; *Hôtel-Pension*), which affords a fine view of the Lake of Constance, the populous and rich pasture-lands of Speicher and Trogen, and the Vorarlberg and Appenzell Mts. A point in front of the hotel commands a specially fine prospect of the Sentis. Descent to (¾ M.) *Speicher* (3070'; Löwe; Krone) and across the *Bachtobel* to (1¼ M.) *Trogen*. — From Trogen to (4¾ M.) *Teufen*, diligence twice daily in 1 hr. Steam-tramway from St. Gallen to Gais viâ Teufen, see p. 59.

From the church at Trogen a road leads viâ (3½ M.) *Bühler* (p. 59) to (1¾ M.) *Gais*, but the path over the *Gäbris (4100') is shorter and far more attractive.

The traveller coming from the Kaien follows the Trogen and Bühler road to the (½ hr.) top of the hill (3187'; view of the Sentis); a fingerpost here indicates the path to the left to Gais over the Gäbris. Those who come from Vögelisegg should not go on to Trogen, but quit the highroad beyond the *Bachtobel* (see above) by a flight of steps to the right. A small valley lies immediately on the right, and the path ascends gradually across meadows. After ¼ hr. (from Speicher) this path reaches the road from Trogen to Bühler a few hundred paces from the finger-post. About 5 min. beyond the latter we reach two houses. Where the ascent begins, 5 min. farther on, we keep to the left. Farther on the road skirts a wood (at the beginning of which the descent to the left is to be avoided). At the point (12 min.) where a row of old pine-trees flanks the road on the right, a footpath between two of these ascends, chiefly through wood, in 20 min. to the summit. The point first attained is the *Signalhöhe* (4110'), the view from which is obstructed by wood. A few min. farther on is an *Inn (4100'), whence a charming prospect is enjoyed (1½ hr. from Speicher). Hence to Gais a descent of ½ hour. Walkers in the reverse direction find fingerposts at doubtful points. Numerous benches.

Gais (3075'; pop. 2495; *Krone, R. & A. 2½-3½, pens. 7 fr.; *Ochs, *Adler, Hirsch, Gäbris, Hecht*, etc., plain; *Hackerbräu*, at the station), a trim-looking village, in the midst of green meadows, is the oldest of the Appenzell whey-resorts, having been in vogue since 1749. Fine view of the Sentis from the *Curgarten*.

Steam-tramway to *St. Gallen*, see p. 59. — The ROAD FROM GAIS TO ALTSTÄTTEN (6 M., diligence daily in 1¼ hr., from Altstätten to Gais in 1¾ hr.) is level for the first 1½ M., and then descends uninterruptedly from the point where it diverges from the old road and winds round the mountain. The old road, preferable for pedestrians, leads to the left viâ the (1¼ hr.) *Stoss (3180'; Pension Stoss)*, a chapel on the pass, with a celebrated view of the Rhine Valley, the Vorarlberg, and the Grisons. Here, on 17th June, 1405, 400 Appenzellers under Rudolf von Werdenberg signally defeated 3000 troops of the Archduke Frederick and the Abbot of St. Gallen. The shorter old road crosses the new immediately below the Stoss, and descends direct, partly through wood, to Altstätten (p. 61).

A **road** traversing meadows leads from Gais to (3½ M.) Appenzell, while a shorter **footpath** to the Weissbad diverges to the left halfway to Appenzell and crosses the *Guggerloch* (3084').

Appenzell (2550'; *Hecht, *Löwe, *Hirsch*, all moderate; beer at the *Krone*), the **capital of Canton Inner-Rhoden**, on the *Sitter*, a large village (4480 inhab.) consisting chiefly of old wooden houses. It contains two monasteries, and was formerly a country-seat of the Abbots of St. Gallen, Appenzell being a corruption of '*Abbatis Cella*'. The *Hospital*, the *Church*, erected in 1826, and the *Landes-Archiv*, containing interesting charters, are worthy of note. Shady promenades on the Sitter. — Railway to *Urnäsch* and *Winkeln*, see p. 50.

A **road** leads from Appenzell (also a footpath from the station; omnibus to and from the station, five times daily, 70 c.; carr. 4, with two horses 6 fr.) to the S.E., crossing the *Sitter* and passing the *Hôtel Steinegg*, to the (2 M.) *Weissbad (2680'), a summer and health resort (R. & A. 2-4, B. 1¼, D. 3, S. 2 fr., cheaper for a longer stay; also river-baths), pleasantly situated at the base of the Appenzell Mts., and a good starting-point for excursions. Besides the Curhaus there are two hotels, the *Weissbadbrücke* and the *Gemse*.

Guides' Fees *(Huber, Jac.,* and *Joh. Koster, Joh. Bapt. Rusch)*: Wildkirchli 5, Ebenalp 5, Sentis 10, over the Sentis to Wildhaus 20, Altmann 15, Hohe Kasten 6, over the latter into the Rhine Valley 10 fr. — Horse to Wildkirchli, Ebenalp, Seealp, or Ruhsitz 12 fr.

The favourite walk from the Weissbad is to the WILDKIRCHLI, 1¾ hr. to the S. (guide 5 fr., unnecessary). Following the road to Brülisau (p. 59) for 100 paces, we ascend to the right; 8 min. a house, whence the bridle-track diverges to the left, while the good footpath leads straight on through a gate, crossing the bridlepath at (¼ hr.) a double gate; we then cross the meadow, in the direction of the Ebenalp, to (40 min.) the depression between it and the wooded *Bommen-Alp* (to the left). Hence we ascend in windings through wood to the right, and in 10 min. reach a fingerpost showing the direct path to the Ebenalp (to the right; see below). The route to the Wildkirchli turns to the left and (10 min.) approaches the foot of the precipitous rocks which descend from the Ebenalp to the Seealp-Thal (see p. 56). Near the (¼ hr.) *Zum Escher Inn* (4790'; R. 1½-2 fr.; *View*) we ascend to the right by a narrow but safe path, skirting the perpendicular rocks, to the (2 min.) *Wildkirchli* (4845'), formerly a hermitage, founded in 1656, with a chapel dedicated to St. Michael, situated in a grotto (33' wide; tavern). On the patron-saint's day (at the beginning of July) and on St. Michael's Day (29th Sept.) solemn services are conducted here, and the grotto and the Ebenalp attract numerous visitors. View of the deep Seealp-Thal (with the path to the Sentis opposite, see p. 58), and, to the left, of the Lake of Constance.

A dark passage in the rock, **150 paces** long, closed by a door (opened by the landlord, who provides a light, ½ fr.), leads from the grotto to the *Ebenalp*, where an entirely new Alpine view is disclosed. The (25 min.) summit (5390'; *Inn*, 6 beds), **commands a superb view of the Sentis, Altmann, Lake of Constance, etc.** — We may descend direct to the (25 min.) saddle to the N. of the *Bommen-Alp* (see above; guide useful to the beginning of the distinct path).

Pleasant walk from Weissbad viâ *Schwendi* and (50 min.) *Wasserauen* (p. 58), crossing to the left bank of the Schwendibach, 4 min. farther on, passing the *Escherstein*, and ascending through a pretty wooded ravine to the (¾ hr.) Seealp-See (3735'; *Inn*, trout), very picturesquely situated in a basin between the *Gloggeren* and *Alten-Alp* (see p. 58). From the *Escher* (see above) a steep path descends to the Seealp-See in 1 hr. From the Seealp-See to the *Megglis-Alp* (see below) 1 hr., path recently improved (wire-rope at giddy points). The path unites with that from the Weissbad about 20 min. from the Megglis-Alp. — To the **Leuer Fall** (3185'), 1½ hr., also interesting; the path ascends the *Weissbach-Thal* (guide-post beyond the Weissbad), the last part through beautiful wood.

The snow-clad *Sentis (8215'), the highest mountain in the canton, is most conveniently ascended from the Weissbad (6 hrs.; guide 10 fr.; one-horse carr. to Wasserauen 4 fr.). A road diverges to the right from the road to Brülisau beyond the (3 min.)

bridge over the *Schwendibach*, and ascends on the right bank of the brook to (1/4 hr.) *Schwendi* (2790'; *Inn Zur Felsenburg, on the left bank), and to the (35 min.) *Wasserauen Inn*, where the road ceases. The ascent now commences *(Katzensteig)*, following the right side of a ravine through which a brook is precipitated; (40 min.) chalets of the *Hütten-Alp* (3940'; milk). The narrow, but well-defined path now skirts the *Schrennen*, the shelving pastures of the *Gloggeren* (below which are perpendicular rocks), affording beautiful glimpses of the *Seealp-See* far below, the Sentis and Altmann, and the Wildkirchli to the right. In 3/4 hr. we pass a refuge-hut, and in 3/4 hr. more we reach the *Megglis-Alp* (4985'; *Inn*), in a picturesque basin. The path ascends hence rather steeply on the left side of the valley and skirts the base of the *Kühmaad*, being frequently hewn in steps (the telegraph stakes commencing 10 min. from the Megglis-Alp may be followed). After 1 3/4 hr., at the *Wagenlucke* (6785'), the inn on the Sentis becomes visible. The path leaves the snow on the left and ascends, gradually becoming steeper and crossing large masses of rock (wire-rope), to (1 1/4 hr.) the *Inn* (8087'; bed 3-5 fr., mattress in the garret 1 1/2 fr.; food dear; often crowded; early arrival desirable). On the summit of the SENTIS, to which we finally mount by a path protected by a railing in 5 min. more, is a meteorological station (adm 30 c.). The **VIEW (see Heim's excellent Panorama) extends over N.E. and E. Switzerland, embracing the Lake of Constance, Swabia and Bavaria, the Tyrolese Mts., the Grisons, and the Alps of Glarus and Bern. — The N. peak, separated from the S. by the '*Blaue Schnee*' (not to be tried without a guide; see p. 59) is named the *Girespitz* (7766').

From the Sentis we may descend, at first over snow, and then by a path which is **very** steep at first, over the *Schafboden* (5660') and the *Flis-Alp* to (3 1/2-4 hrs.) in the reverse direction 6 hrs.) *Wildhaus* or *Unterwasser* in the Toggenburg (p. 63; guide desirable). — The usual ROUTE FROM THE WEISSBAD TO WILDHAUS (7 1/2-8 hrs.) leads by *Brülisau* and through the *Brüttobel* to the *Sämtis-See* (3965'), passes the *Fählen-See* (4750'; chalets), and ascends to the *Zwingli Pass* (6630'), between the *Altmann* (see below) on the right, and the *Kraialpfirst* (6910') on the left. We descend by the *Krai-Alp* (5933'), and the *Tesel-Alp* (4575') to *Wildhaus*. This route, however, is rough, and the Sentis route (not much longer) is preferable.

Mountaineers may combine a visit to the Wildkirchli (p. 57) with the ascent of the Sentis (7-8 hrs.; guide necessary, 15 fr.) by leaving the valley of **the** Seealp-See to the left. The path leads high above the Seealp-See **and at** the base of the *Zünster* and *Schäfler*, viâ the *Allen-Alp* and the **Oehrli,** to the *Muschelenberg* (numerous fossils); hence either to the left **across** the valley to the *Wagenlucke* (6785') by the path which ascends from **the** Megglis-Alp (see above), or (1 hr. shorter) across the *Blaue Schnee* (caution on account of the crevasses), past the base of the *Girespitz*, and over the *Platten* direct to the summit (7-8 hrs. in all). — A path, constructed by the S. A. C., ascends to the summit on the W. side also (6 hrs.; with guide). It starts from the *Gemeinen-Wesen Alp* (4210'; reached from *Urnäsch* or *Nesslau* in 2 hrs.), ascends over stony slopes, and mounts a steep rocky slope in zigzags to the first mountain-terrace. The ascent is then more gradual, over rock and pasture, to the *Fliesbordkamm* and the (2 1/2 hrs.) *Club-Hut* on the *Thierwies* (6835'). We next traverse rocks and débris on the *Grauhopf* (7255'), and ascend in zigzags to the arête between the *Girespitz* and the *Sentis*. Lastly we mount the *Platten* by a flight of steps 140 yds. long, protected by a wire railing, and reach the (1½ hr.) summit.

The Altmann (9000'; 7 hrs., with guide; toilsome), is ascended from the *Weissbad* viâ the *Fählen-Alp* and *Zwingli Pass* (see above); descent through the *Löchlibetter* to the *Mögglis-Alp* (p. 55).

FROM WEISSBAD TO THE RHINE VALLEY. The direct route by the HOHE KASTEN (5½ hrs.) leads to the S.E. through (½ hr.) *Brüllisau* (3030'; Krone, rustic); by the church we follow the paved path, past the first house, as far as a barn, and ascend the meadows as far as the last group of houses, ½ hr.; then straight on (not by the beaten path), through the enclosure on the right, to the Inn '*Ruhsitz*' (4495'; ½ hr., bridle-path thus far), at the S.W. base of the *Kamor* (5215'). From the inn a steep but good path ascends to (1¼ hr.) the summit of the *Hohe Kasten (5900'; *Inn*), which slopes precipitously on the E. towards the Rhine Valley. Splendid view of the Sentis group, with its three spurs on the N.E., which is nowhere seen to such advantage; in the other direction we see the Rhine Valley, stretching as far as the Lake of Constance, and the Alps of the Vorarlberg and Grisons. We may now descend by a steep and stony path to (3 hrs.) stat. *Sennwald-Salez* (p. 61). It diverges from the Weissbad path to the left, just below the saddle between the Kamor and Hohe Kasten, skirts the W. and S. slopes of the latter, and descends in zigzags (no possibility of mistake; several finger-posts lower down). Traversing wood for the last hour, we at length reach the village of *Sennwald* and the station.

Railway from Appenzell to Winkeln, viâ *Urnäsch* and *Hérisau*, see p. 48. — It is preferable, however, to drive viâ Gais and Teufen to St. Gallen (to Gais, 3½ M., **diligence** five times daily in 1 hr.; thence to St. Gallen, 8½ M., steam-tramway in 1¼ hr.). To (3½ M.) *Gais*, see p. 56. Thence the steam-tramway (rack-and-pinion line at the steeper places; pretty route) descends viâ *Zweibrücken*, where the road to Appenzell diverges to the left (p. 56), and along the *Rothbach* to (1¾ M.) the prettily situated village of *Bühler* ('2735'; Rössli, etc.), and beyond the *Rose* and **Linde** inns (good; pens. 4-5 fr.) ascends to (4½ M.) **Teufen** (2750'; pop. 4629; *Hecht*), a wealthy industrial village, picturesquely situated, with a fine view of the Sentis chain. It then skirts the W. slope of the *Teuferegg*, through meadows and wood, passing the stations of *Sternen*, *Niederteufen*, *Lustmühle*, and *Riethäusle*, and finally descends in sharp curves to (8½ M.) *St. Gallen* (p. 50).

The FOOTPATH FROM TEUFEN TO ST. GALLEN (1½ hr.) diverges from the high-road near the 'Hecht' inn, and immediately ascends to (¼ hr.) the *Schäfle's-Egg* (3185'; Inn); it then descends to (¾ hr.) *St. Georgen*, where it joins the high-road to (1½ M.) St. Gallen. — About 10 min. to the W. of the Schäfle's-Egg is the *Fröhlichsegg (3290'; *Inn*), which commands an admirable view; Teufen in the foreground, the green Alpine valley sprinkled with dwellings, and the Appenzell Mts., beginning with the Fähnern, on the left, the Kamor, the Hohe Kasten about the middle of the chain, the green Ebenalp below the snow, more to the right the Altmann and the Sentis with its snow-fields, then in the distance the Glärnisch and Speer; to the W., the railway and road to Wyl, and to the N., part of the Lake of Constance. Hence to St. Gallen, 3 M.

18. From Rorschach to Coire.

Comp. Maps, pp. 28, 54.

57 M. RAILWAY in 3¹/₄-4¹/₂ hrs. (9 fr. 75, 6 fr. 85, 4 fr. 90 c.; see Introd. X. as to circular-tickets, etc.).

Rorschach, see p. 52. The train skirts the lake for a short way. To the right is the castle of *Wartegg* (p. 52). 2¹/₂ M. *Staad* (Anker; good swimming and other baths), a picturesque place with quarries of white sandstone. *Heiden* (p. 54) is seen on the hill to the right. Farther on we have a glimpse of the *Weinburg* (p. 52), at the foot of the vine-clad **Buchberg**. The train traverses a delta, very fertile at places, which has been formed by the deposits of the Rhine. — 5¹/₂ M. **Rheinegg** (1320'; *Post; Rössli; Hecht*), a village at the foot of vineyards.

Omnibuses ply in 12 min. from the station to (1¹/₄ M.) **Thal** (1344'; *Ochs*), an industrial place with 3319 inhab., picturesquely situated at the foot of the *Buchberg* (to the *Steinerne Tisch*, 25 min., see p. 52).

A diligence runs thrice daily in 1 hr. 5 min. from Rheinegg to (3 M.) **Walzenhausen** (2225'; *Curhaus; *Hôt.-Pens. Rheinburg*, by the church, pens. 6-8¹/₂ fr.), a large village and health-resort, pleasantly situated and commanding beautiful views. The road (shorter footpath, ascending to the right beyond the Rhine bridge) runs from the church along the hill-side, affording charming views of the Rhine valley and traversing woods, to the (1 M.) *Convent of Grimmenstein* (2485'; *Löwe). About ¹/₂ M. farther on, **near** the Inn 'Zur Maldegg', the road to (3 M.) '*Au* (see below) diverges to the left. About ¹/₃ M. farther on, where the road makes its last ascent and bends to the right before descending to *Bernegg* (see below), a footpath, skirting the ridge to the left, leads to (10 min.) the *Meldegg (2415'; Inn in summer), a rocky promontory at the angle of the Rhine valley, commanding a splendid view of the valley, the Vorarlberg and Appenzell Alps, and the Lake of Constance. We then descend to (¹/₂ hr.) *Au* (see below) or (³/₄ hr.) *St. Margrethen* (see below).

Diligence from Rheinegg to *Wolfhalden* and **Heiden** twice daily in 1³/₄ hr., see p. 54.

Walzenhausen (see above) is visible on the hill to the right for a **short time**. At (3 M.) **St. Margrethen** (1330'; *Linde; Ochs; Sonne*) the **line** to Bregenz (p. 432) diverges to the left (to the *Meldegg*, 1 hr., see above).

We now cross the Rhine, the boundary between Switzerland and the Austrian **Vorarlberg**, by means of a timber-bridge. The *Rhine Valley*, formerly called the *Upper Rheingau*, was, like Ticino and Thurgau, governed down to 1798 by Swiss bailiffs. Part of its bottom is marshy and exposed to inundation when the water is high. Maize is largely cultivated. The train skirts the hills, which are covered with vineyards and orchards, and from *Heldsberg* to *Monstein* passes between the river and abrupt rocks. 9¹/₂ M. **Au** (1338'; *Schiff*, good wine; *Rössli; Railway Restaurant*), prettily situated at the foot of the *Meldegg* (see above). To the left rises the snow-clad Scesaplana and farther away the Drei Schwestern; to the right the **Hohe Kasten** with its inn (p. 59).

Road to (4 M.) *Walzenhausen*, see above. Ascent to the *Meldegg (³/₄-1 hr.), see above. — About 2 M. to the W., in a fertile, wine-growing basin, lies **Berneck** (1330'; *Drei Eidgenossen; Ochs; Pens. Tigelberg*), a pleasant village (2232 inhab.), with well-equipped public baths.

ALTSTÄTTEN. *I. Route 18.* **61**

12 M. *Heerbrugg;* 14 M. *Rebstein-Marbach.*
16½ M. **Altstätten** (1540'; pop. 8416; **Drei Könige*, moderate; *Freihof; Landhaus; Löwe*), a quaint little town. Through a gorge to the right is seen the Sentis (p. 58) and beside it the Fähnern. By the railway, to the right, is the Convent of the Good Shepherd (orphanage), with a large new domed church.

Roads lead hence viâ the *Landmark* (3265'; Inn) to (8 M.) *Trogen*, and viâ the *Stoss* (3135') to (6 M.) *Gais* (p. 56), and a pleasant path in 3 hrs. by the *Chapel of St. Anthony* to *Heiden* (p. 54).

19½ M. *Oberriet* (1387'; Sonne). On the E. slope of a wooded rock to the right is the square tower of the castle of *Blatten.*

22½ M. *Rüti* (Zum Bahnhof). — 27 M. *Saletz-Sennwald* (Restaurant by the station). To the left are the *Drei Schwestern* (6880').

Ascent of the *Hohe Kasten* (5900', 4½ hrs., without guide), see p. 59. — To the WEISSBAD (6 hrs.), a pleasant walk, by *Sax* and the *Saxer Lucke* (5430'), passing the *Fählen* and *Sämbtis* lakes (comp. p. 58).

29 M. *Haag-Gams* (*Kreuz), where the line crosses the Toggenburg and Feldkirch road (p. 59). Above (31 M.) **Buchs** (**Rail. Restaurant; *Rhaetia; Zum Arlberg*, both at the station) rises the well-preserved château of *Werdenberg.*

Railway to *Feldkirch*, see p. 432; custom-house examination at Buchs for travellers to or from Austria. — On a height, on the opposite bank of the Rhine, lies *Vaduz* (1525'; Engel; *Löwe), with the white château of *Liechtenstein* on a lofty rock, the capital of the principality of Liechtenstein, at the foot of the *Drei Schwestern* (see above).

Beyond the large village of (34½ M.) **Sevelen** (**Traube)* rises the ruined château of *Wartau* (2185'). On a height to the left, beyond the Rhine, near *Balzers*, is the extensive ruined castle of *Guttenberg*, where the ascent of the Luziensteig begins (p. 62). Beyond (39 M.) *Trübbach* (1585'; Löwe) the rocks of the *Schollberg* have been blasted to make way for the road and the railway.

The *Alvier* (7758'), an admirable point, ascended from Buchs, Sevelen, or Trübbach in 5-5½ hrs., see p. 47. The route from Trübbach is by *Atzmoos, Malans*, and past the ruin of *Wartau*, to (¾ hr.) *Oberschan* and (4½ hrs.) the top; descent 3 hrs. — The *Gonzen* (6014'), from Trübbach in 4½ hrs., is also easy and interesting.

42 M. **Sargans** (1590'; **Hôtel Thoma*, at the station), the junction of the Weesen (Glarus) and Zürich line (p. 48). Carriages sometimes changed here. The scenery becomes grander and more picturesque; to the N.W. appears the long serrated chain of the *Curfirsten* (p. 45), to the E. the *St. Luzienberg* or *Fläscherberg* (3730'; p. 62) and the grey pyramid of the *Falknis* (p. 62). To the right, near *Vilters*, is the *Untere Sar Fall*, fine after rain.

45 M. **Ragatz**, see p. 64. To the right is the ruin of *Freudenberg* (p. 64); farther on, to the left, are the pension and the ruined castle of *Wartenstein* (p. 65). Below the influx of the *Tamina* the train crosses the Rhine by a wooden bridge.

46 M. **Maienfeld** (1725'; pop. 1227; *Hôt.-Pens. Vilan*, at the station; *Hirsch; Zum Falknis; Rössli*, good wine) is an old and thriving little town. The tower (restaurant; fine view from the top) is said to have been erected in the 4th cent. by the Roman Emp. Constantius.

The **St. Luzienssteig** (2230'; Inn, good wine), a fortified defile between the *Fläscherberg* (3730') and the *Falknis*, through which the road to Vaduz and Feldkirch leads, is 2 M. from Maienfeld and is frequently visited from Ragatz. Fine view from the highest block-house (now destroyed), on the top of the Fläscherberg, 1¼ hr. farther to the W., and also on the return. — The Falknis (8420'), ascended from the Luzienssteig through the *Glecktobel* and by the *Sarina-Alp* or *Fläscher-Alp* (6 hrs.; with guide), is fatiguing but interesting. (Better from Maienfeld, with the guide Fortunat Enderlin, via *Jenins*, the *Vordere Alp*, *Sarina Alp*, and through the *Fläscher-Thal*.)

On the vine-clad slopes to the left lie the villages of *Jenins* (above it the ruins of *Wyneck* and *Aspermont*) and *Malans* (p. 356). The train crosses the *Landquart*, near its influx **into** the Rhine. 49½ M. **Landquart** (1730'; *Rail. Restaurant; *Hôt. Landquart*, near the station, R., L., & A. **4**, D. with wine 3½ fr.), on the road to the Prätigau, **and the junction of the line to Davos** (see p. 356). To the W., in the background, rise the barren *Graue Hörner* (p. 66).

The district between Maienfeld and Coire, **with its numerous castles**, is remarkable for its fertility. Its central **point** is (52 M.) **Zizers** (1854'; *Krone; Zum Bahnhof*), an ancient **little** borough. To the left, at the foot of the hills, are *Molinära*, a summer-residence of the Bishop of Coire, and the village of *Trimmis*. On the right tower the peaks of **the** *Calanda* (356'); at its base are the ruined castles of *Liechtenstein, Grottenstein*, and *Haldenstein*, at the foot of which last lies the village of the same name, with a well-preserved walled château.

57 M. **Coire**, see p. 354.

19. From Wyl through the Toggenburg to Buchs in the Rhine Valley.

Comp. Map, p. 54.

RAILWAY from *Wyl* to *Ebnat*, 15½ M., in 1 hr. 5 min. (1 fr. 95, 1 fr. 40 c.; 2nd and 3rd cl. only). — From *Ebnat* to *Buchs*, 24 M., diligence thrice daily in 5¼ hrs. (5 fr. 70 c.); also several times daily to Nesslau in 1 hr., and to Alt St. Johann in 2⅔ hrs. — Carriage with one horse from Wildhaus to *Gams* 8 fr. (carriages in Gams to be had at the 'Kreuz' inn); to *Buchs* 9 fr.; to *Ebnat* 14 fr.

Wyl, on the Winterthur and St. Gallen line, see p. 49. The train traverses the *Toggenburg*, the busy and populous valley of the *Thur*.

When the Counts of Toggenburg became extinct (1436), the County was purchased by the Abbots of St. Gallen, who at the same time secured to the inhabitants their ancient rights and privileges. In the course of centuries, however, a great part of the population having embraced Protestantism, the abbots violated their contract, which resulted in their expulsion at the beginning of the 18th century. This gave rise to the *Toggenburg War*, a violent feud in which the Roman Catholic cantons espoused the cause of St. Gallen, while the Protestants took the part of the Toggenburgers. No fewer than 150,000 men were thus gradually brought into the field. In July, 1712, the Roman Catholics were at length defeated at Villmergen in the Aargau; and a general peace was concluded, which secured to the Toggenburgers full enjoyment of all their ancient liberties, though they were still to belong to the Canton of St. Gallen.

4½ M. *Batzenheid*; opposite is *Jonswyl*, with a new church. Opposite (6 M.) *Lütisburg* we cross the *Guggerloch* by a viaduct 170 yds. long, and 190' high. 8 M. *Bütschwyl*; 9½ M. *Dietfurt*.

10½ M. **Lichtensteig** (pop. 1529; *Krone*), a pleasant town on a rocky height, with a modern Gothic church. On a hill to the E. (1¼ hr.) is the ruin of *Neu-Toggenburg* (3565'), a fine point of view.

12½ M. **Wattwyl** (2027'; *Ross;* *Toggenburg*), a charming village, with 5260 inhab. and a new church. (Diligence to Utznach, 4 times daily in 1¾ hr., see p. 44.) On a hill to the right is the nunnery of *St. Maria der Engeln*, and above it the ruin of *Yberg*. The last station is (15½ M.) *Ebnat-Kappel*. The village of Ebnat (2106'; *Krone;* *Adler;* *Rosenbühl*, a restaurant with view) is a thriving place; 1 M. to the N. W. is *Kappel* (Traube; Stern).

The *Speer* (6417') may be ascended through the *Steinthal* in 5 hrs. (not difficult for experts, but near the top rather trying; comp. p. 45); or from *Neu St. Johann*, or from *Nesslau* (see below), by the *Alp im Laad* and the *Herren-Alp* in 5 hrs. (guide 7 fr.).

The road, commanding a view of the Curfirsten opposite, and, near Neu St. Johann, of the Sentis on the left, ascends slightly on the right bank of the Thur, to *Krummenau* (2385'), where the '*Sprung*', a natural rock-bridge, crosses the stream, *Neu St. Johann* (Schäfle), with an old Benedictine abbey, and (4½ M.) —

20 M. **Nesslau** (2470'; *Krone; Traube; Stern*), with a pretty church.

To URNÄSCH OVER THE KRÄZERN PASS (4½ hrs.), interesting. A road ascends from Neu St. Johann through the *Lauterthal*, viâ *Ennetbühl* and the *Riedbad* or *Ennetbühler-Bad*, to the (1½ hr.) *Alp Bernhalden* (3102'); a path to the left then ascends through the *Kräzernwald* to the **Kräzern Pass** (3936'), and crosses the pastures of *Kräzern* to the (2 hrs.) *Rossfall-Alp* (Inn), whence a road leads to (1 hr.) *Urnäsch* (p. 50). — Ascent of the *Sentis* (p. 57) from Nesslau, 6¼ hrs.: from (1½ hr.) *Bernhalden* (see above) in ¾ hr. to the *Alp Gemeinen-Wesen* (4210'); new path thence to the (4 hrs.) top (p. 58). — Ascent of the *Speer*, see above.

The scenery becomes bleaker. The road leads past a fine fall of the *Weisse Thur* to (2¼ M.) *Stein* (Krone) and (2¼ M.) *Starkenbach* (Drei Eidgenossen), a straggling village. To the right is the ruin of *Starkenstein*. (Over the *Amdener Berg* to *Weesen*, see p. 45; guide to the pass advisable.) Passing (1½ M.) *Alt St. Johann* (2920'; *Rössli) and (¾ M.) *Unterwasser* (Stern; Traube), prettily situated at the sources of the Thur, we ascend to (3¾ M.) —

30½ M. **Wildhaus** (3600', pens. 5 fr.; *Hirsch; Sonne; Tell*). A little before the village, on the right, is the wooden house, blackened with age, in which *Zwingli* was born in 1484. Behind the village, which lies at the foot of the *Schafberg* (7820'), we obtain a survey of the seven Curfirsten (p. 45); or still better from the (¾ hr.) *Sommerikopf* (4317').

Ascent of the *Sentis* from Wildhaus or Alt St. Johann (viâ the *Flis-Alp* and the *Schafboden* in 6 hrs., with guide; toilsome), see p. 58. — *To Weissbad* by the *Krayalp*, the *Fählensee*, and *Sämblis-See* (7 hrs.), see p. 58. — *To Walenstadt* over the *Käserruck*, 6 hrs., see p. 47.

The road descends, finally describing a long bend (short-cut for walkers to the right), to (6 M.) *Gams* (1575'; *Kreuz*), in the Rhine Valley, and then leads straight to (1½ M.) *Haag* (p. 61), while a road to the right leads viâ *Grabs* and *Werdenberg* to (3½ M.) —

39½ M. *Buchs* (p. 61).

20. Ragatz and Pfäfers.

Comp. **Plan**, *p. 66*, and *Map, p. 346.*

Hotels (most of them open during the season only). "QUELLENHOF (Pl. a), R., L., & A. from 6, B. 1½, lunch 4, D. 5, pens. 12-18 fr.; °HOF RAGATZ (Pl. b), R., L., & A. from 6, B. 1½, D. 5, pens. 10-15 fr.; "HÔTEL TAMINA (Pl. c), R., L., & A. 3-5, D. 4, pens. 8-10 fr.; °SCHWEIZERHOF (Pl. d), R. 2½-3½, B. 1¼, D. 3, pens. from 7 fr.; °HOT.-PENS. LATTMANN (Pl. f), R., L., & A. 2-3, B. 1, D. 2½, pens. from 6 fr., good cuisine (open in winter also); °KRONE (Pl. e), R., L., & A. 2½-3, B. 1¼, D. 3, S. 2½ fr. (open in winter also); °HÔT.-PENS. SCHOLL (Pl. f), R. 2-3, pens. 6 fr.; VILLA LOUISA; °HÔT.-PENS. FRIEDTHAL (Pl. h), R., L., & A. 1-2, B. 1, D. 3, pens. 5-6 fr.; °FREIECK (Pl. g); °HÔT. NATIONAL (Pl. l), R., L., & A. 2-3, D. 3, pens. 6-8 fr.; °POST, R., L., & A. 2, B. 1, D. 2, pens 6 fr.; °OCHSE, LÖWE, unpretending. — Near the station: °ROSENGARTEN, R. & A. 2½, B. 1¼, D. 3, pens. from 7 fr. (open in winter also). — °PENS. VILLA FLORA, with large garden, on the road to the Freudenberg; °PENS. HOME-VILLA; °PENS. WARTENSTEIN (p. 65).

Restaurants. *Cursaal*, see below; good Munich beer at the *Schweizerhof* and *Scholl's* (see above); *Rheinvilla*, Bahnhof-Str.; *Nussbaum*, Churer-Str.; *Löwe* and *Kreuz*, with gardens. *Felsenkeller*, ¼ M. from the town, on the way to the Freudenberg (see below).

Post Office (Pl. 6), near the Dorfbad. — **Telegraph Office** (Pl. 7), opposite the Krone.

Omnibus from the station to the village of Ragatz 75 c., trunk 25 c. — **Carriage**, with one horse from Ragatz to Bad Pfäfers and back, with halt of 2 hrs., for 1-2 pers. 7, 3-4 pers. 10 fr., and fee; to Wartenstein and Dorf Pfäfers 8 or 14, Vättis 18 or 25, Malenfeld 6 or 10. Luziensteig 10 or 15 fr.

Baths. Properties of the water, see p. 344. The *Mühlbad* (Pl. 4), *Neubad* (Pl. 2), and *Helenenbad* (Pl. 3) are near the Curhaus; the *Dorfbad* (Pl. 5), with Trinkhalle, in the Eisenbahn-Strasse, between the Schweizerhof and the Tamina Hotel. The Neubad contains a large swimming-bath (84° Fahr.; 2 fr. in the morning, 1 fr. in the afternoon; ladies 9.30-11.30 a.m. and 4-6 p.m.) and single baths (2-2½ fr.). Tickets at the office, to the left of the Hof Ragatz.

Visitors' Tax, in June and Sept. 2, in July and Aug. 3 fr. per week for each person. **Music** in the morning, **afternoon**, and evening, alternately in the Cur-Garten (or Cursaal), the **Badhalle at the Dorfbad**, or in the Hof Ragatz.

Ragatz (1710'; pop. 1932), prettily situated on the impetuous *Tamina*, which falls into the Rhine lower down, is a famous watering-place **and one of the** most frequented places in Switzerland (50,000 visitors annually, passing travellers included). The chief rallying-points are the *Cursaal*, with the *Cur-Garten*, and the *Baths* (see above), which **receive the** mineral **water** from Pfäfers by a conduit, 2½ M. long. Music, see above. The open colonnade on the E. side of the Cursaal **affords a** pleasing survey of the Rhine Valley.

In the *Cemetery* is the monument of **the** philosopher *Schelling* (d. at Ragatz in 1854), 20' high, with his bust. By the last houses (1 M.) on the road from the cemetery to Sargans, a path ascends to the left through vineyards to (½ M.) the ruined castle of *Freudenberg* (915'), with a fine view of the Rheinthal. We return by a road on the hillside, between houses and gardens

*****Bad Pfäfers** or *Pfävers* (50 min.) is one of the most curious spots in Switzerland. It lies in the narrow gorge of the *Tamina*, a glacier-**torrent**, **on the brink** of which the good but narrow road (walking

PFÄFERS.   *I. Route 20.*   65

recommended) gradually ascends, flanked by sombre limestone cliffs, 500 to 800' high. A little before the (1½ M.) *Schwattenfall Restaurant* a footpath leads to the left across the Tamina to *Valurgut* and (½ hr.) *Wartenstein* (see below). About ½ M. farther on, a few paces before the road passes through a rocky gateway, is another path (shady and picturesque but steep), leading to (¾ hr.) the village of Pfäfers. Both these routes are miry in wet weather.

The monastic-looking *Bath-House* (3240'), built in 1704, lies between precipices 600' high, and enjoys sunshine in the height of summer from 10 till 4 o'clock only. Accommodation good, but plain (R., L., & A. 2-3½, B. 1¼ fr.). Very pleasant baths (1 fr.; temp. 98°, at Ragatz 95°), chiefly frequented by the less wealthy classes, and by invalids who prefer taking the waters near their source.

The copious hot springs (99-102°), clear as crystal, and free from taste and smell, are impregnated with carbonate of lime, chloride of sodium, and magnesia, resembling those of Gastein and Wildbad in their composition. They rise about ¼ M. above the bath-house in the narrow and gloomy *Tamina Gorge (30-50' wide). Tickets for the gorge and the springs (1 fr. each; umbrellas advisable) are sold in the principal corridor of the bath-house, to the right. The wooden pathway to the springs, resting on the rock or on masonry, 30-40' above the torrent, passes under the 'Beschluss' (see below). In 6 min. we reach a small terrace, on the E. side of which the guide opens a locked door. Laying aside hat and overcoat, we enter a narrow shaft, filled with clouds of vapour. After 40 paces this shaft expands to a cavern, where the spring rises in a deep cavity protected by a parapet. — From the Ragatz station to the springs and back, 3 hrs. on foot, or 2 hrs. by carriage (p 64).

From the Baths to the Village of Pfäfers (1¼ hr.). The path ascends in windings on the left bank of the Tamina; after ¼ hr., by a finger-post, where the path to the right leads to Valens (see below; 10 min. from the Bad is the 'Calandaschau'), we descend to the left and (5 min.) cross the Tamina by a natural bridge, called the '*Beschluss*', 280' perpendicularly above the springs. We now ascend the steep path on the right bank, cut in steps, and slippery in rainy weather, to a (20 min.) meadow, whence we may either ascend (finger-post) to (10 min.) an auberge on the road leading to the right to Vättis (p. 66) and to the left to the village of Pfäfers; or (preferable) ascend by the footpath to the left, through meadows and wood, to the (¾ hr.) road, 2¼ M. from the village of Pfäfers.

A Cable Tramway ascends from behind the Hôtel Hof Ragatz in 10 min. at a gradient of 27:100 (2nd class 1 fr., third class 60 c.; return-ticket 1 fr. 30, 80 c.) to the *Hôtel-Pension Wartenstein (2463'; R., L., & A. 2½-5½, B. 1¼, D. 3, pens. 7-10 fr.), a health-resort with a garden-restaurant, affording a splendid view of the Rhine Valley as far as the Curfirsten to the N.W. (p. 45). Below are the ruin of *Wartenstein* and the *Chapel of St. George* (2453'). — The **Village of Pfäfers** (2696'; *Adler; Löwe*) lies ¾ M. farther up, on the top of the hill (road from Ragatz, 2½ M.). The once rich and powerful Benedictine Abbey of Pfäfers was converted into a

Baedeker, Switzerland. 16th Edition.   5

lunatic asylum *(St. Pirminsberg)* in 1838. The *Tabor* (2765'), a rocky hill 1/4 hr. to the N. of the abbey, also affords a fine view.

EXCURSIONS FROM RAGATZ. (Guides: *Füh*, of Ragatz; *Joh.* and *Gust. Rupp*, of Valens; *Wilh.* and *Dav. Kohler*, and *J. A. Sprecher*, of Vättis.) Ruin of *Freudenberg*, p. 64. — The top of the **Guschenkopf** (2468'), a wooded hill to the W. of Ragatz, on the right of the entrance to the Tamina Gorge, may be reached in 40 min., either by a path on the S. side, passing the *Bild* (a chapel), or by one on the W. side (diverging to the left from the road to Freudenberg, before the 'Felsenkeller'). Fine view of Ragatz, the Rhine Valley, the Appenzell and Prätigau Mts., the Graue Hörner, and the Calanda. — To *Maienfeld* (1 1/2 M.; by the road crossing the new Rhine bridge), see p. 61; *St. Luziensteig* (direct path by the railway-bridge 3 M., road via Maienfeld 4 1/2 M.), see p. 62. — The *Prätigau* (*Seewis, Valzeina*, etc.), see R. 92. — *Coire, Via Mala*, etc., see pp. 354, 380.

*Pizalun (4860'; 3 hrs.; guide from St. Margretenberg advisable for novices) a splendid point of view. From (25 min.) Dorf Pfäfers through wood to the pastures of (1 hr.) *St. Margretenberg* (4130'), thence to the end of the village 1/2 hr., then to the left, and lastly by steps in the rock to the (1/2 hr.) top.

To **Valens** (3018'; *Zum Frohsinn*) from Bad Pfäfers, 1/2 hr. (to the right at the finger-post mentioned above). On leaving the wood, the point of view called the *Calandaschau* affords a striking view of the Tamina Valley, with the Calanda in the background to the left, and the Monteluna and the Graue Hörner to the right. Below the church a path crosses the deep *Mühletobel* to (1/2 hr.) Vasön (3045'), amid sunny pastures, and the (1/4 hr.) road to Vättis (see below). — Ascent of the *Vasanenkopf (6675'), from Valens, easy (3 1/2 hrs.; with guide). Across pastures to the *Lusa-Alp* (6145'; small Inn) 3 hrs.; thence to the right to the top 1/2 hr. (wide view; still finer from the *Schlösslikopf*, 7295', 3/4 hr. farther on). Rich flora. — *Monteluna (7955') 4 hrs., from Valens by Vasön and the *Alp Vindels* (5410'), also easy and interesting. — The ascent of *Pizol* (9345'), the highest of the Graue Hörner, is grand and interesting, but trying (3 hrs. from the Lasa-Alp, see above).

FROM RAGATZ TO REICHENAU OVER THE KUNKELS PASS (7-8 hrs.). To (10 M.) *Vättis* a road (diligence from Ragatz daily in 3 hrs., 2 fr. 65 c.; two-horse carr. there and back in 2 1/4 hrs., 25 fr.); thence to Reichenau a mule-track. The road leads from the village of Pfäfers on the right side of the deep Tamina Valley, of which picturesque glimpses are obtained. After 1/2 hr. the path to the Baths of Pfäfers diverges to the right (p. 65); farther on the road passes the hamlets of *Ragol* (opposite Valens) and *Vadura* (opposite *Vasön*, at the foot of the *Monteluna*, see above), and skirts the precipitous slopes of the *Calanda*. The valley expands near (10 M.) **Vättis** (3120'; *Hôt. Tamina*, moderate; *Zur Lerche*), a sequestered village near the mouth of the *Kalfeisen-Thal* (p. 76), from which the Tamina issues. (Viâ *St. Martin* to the *Sardona-Alp*, 4 hrs., see p. 76.) The road ends here. The bridle-path (which is practicable for vehicles to the top of the pass) quits the Tamina, crosses the *Görbs* three times, and ascends, generally on the E. side of the valley. The chalets of the upper valley are collectively called *Kunkels*. On reaching the (2 hrs.) **Kunkels** or **Foppa Pass** (4438'), we turn to the **left** of the conduit and enter the defile of *La Foppa*. (About 5 min. to the **right** of the path a superb view of the Rhine Valley may be obtained.) Then a steep and stony descent to *Tamins* and (1 1/2 hr.) *Reichenau* (p. 367). — The Ringelspitz or Piz Bargias (10,667') may be ascended from Vättis viâ *Kunkels*, the *Hinteralp*, and the *Taminser Glacier* in about 8 hrs. (difficult, for experts only; guide 40 fr.).

21. From Zürich to Glarus and Lintthal.

53 M. RAILWAY (*Nordostbahn*) to Glarus (43 M.) in 2 1/2 hrs. (7 fr. 20, 5 fr. 5, 3 fr. 60 c.); from Glarus to Lintthal (10 M.) in 40-50 min. (1 fr. 60 c., 1 fr. 15 c., 80 c.). (From Weesen to Glarus, 7 1/2 M., in 25 min.; 1 fr. 25 c., 90 c., 65 c.) Carriages are usually changed at Glarus.

Railway on the left bank from Zürich to (36 M.) *Ziegelbrücke*, see pp. 43-45. The train again crosses the Lint Canal (p. 44); on the right the Wiggis and Glärnisch (see below). 37 M. *Nieder-* and *Ober-Urnen*; 39 M. *Näfels-Mollis*, junction for (1¼ M.) *Weesen* (p. 45).

Näfels (1434'; *Schwert; National; Schlüssel; Landolt Restaurant*, near the rail. station) and Ober-Urnen are the only Roman Catholic villages in Canton Glarus. The church is the finest in the canton. The restored *Freuler Palace*, now a poor-house, contains some exquisite panelling (adm. 50 c.). On 9th April, 1388, the canton here shook off the Austrian yoke. In the *Rautifelder*, where eleven attacks took place, stand eleven memorial **stones** (monument in the *Sändlen*). On the second Thursday of April the natives flock to Näfels to celebrate the anniversary. — On the right bank of the *Escher Canal* lies **Mollis** (1470'; *Bär, Löwe*, both moderate; *Pens. Halttli*), an industrial village. (Over the *Kerenzenberg* to *Mühlehorn*, see p. 46.)

Excursions (guide, *M. Hauser*). The **Rautispitz** (7493'), the summit of the *Wiggis Chain* (see below), rising abruptly to the S.W., is ascended from Näfels in 5½-6 hrs. (interesting; no difficulty; guide 12 fr.). On the right bank of the *Rautibach*, with its numerous falls, we ascend in zigzags, cross the *Thrängibach*, and reach a road through wood. Passing above the (1 hr.) *Hastensee* (2460'), we reach the (¾ hr.) charming *Obersee* (3225'; Curhaus, plain), skirt the lake to the left, ascend through wood to the *Grappli-Alp* (4730') and (2 hrs.) *Rauti-Alp* (5400'), and in 1½ hr. more to the summit, which slopes gradually on the W. side (beautiful view). — A rocky arête 1 hr. in length, traversed by a path which should not be attempted by those subject to dizziness, connects the Rautispitz with the Scheye (7420'), the second peak of the Wiggis. The Scheye is also ascended from Vorauen (p. 74) by the *Langenegg-Alp* (4½ hrs.), or from the Klönthaler See (p. 74) by the *Herberig* and the *Deyen-Alp* (4 hrs.), or from Netstall by the *Auern-Alp* (5 hrs.; guide 8 fr.).

41 M. **Netstall** *(St. Fridolin; Bär; Rabe; Schwert)*, a large village (pop. 2326), lies at the E. base of the Wiggis. The *Löntsch*, descending from the *Klönthal* (p. 74), falls into the Lint here (road to *Vorauen*, see p. 74).

43 M. **Glarus.** — *Glarner Hof*, at the station, R. & A. 3½, B. 1½, D. 4 fr.; *Drei Eidgenossen*, R., L., & A. 2, B. 1 fr.; Löwe; Sonne; Blume; Schweizerhof. — Beer at the *Café Tobias*, opposite the station, at the *Raben*, etc.; *Restaurant* (plain) on the *Bergli* (1885'), 20 min. to the W. of the town, an admirable point of view.

Glarus (1490'; pop. 5400), Fr. *Glaris*, the capital of the canton, with busy industries, lies at the N.E. base of the precipitous and imposing *Vorder-Glärnisch* (7648'), at the W. base of the *Schild* (7503'), and at the S.E. base of the *Wiggis* (see above), the barren, grey summits of which form a striking contrast to the fresh green on their slopes. The *Hausstock* (10,355') forms the background to the S.; to the left the *Kärpfstock* (9180'), to the right the *Ruchi* (10,190'). In 1861, during a violent 'Föhn' (S. wind), the greater part of the town was burned down. The new Romanesque church is used by the Roman Catholics and the Protestants in common. In 1506-12 the reformer Zwingli was pastor at the old church, on the site of which the law-courts now stand. The two grassy spaces in front represent the old cemetery. The *Law Courts* contain the Can-

5*

tonal Archives, the public Library, and collections of antiquities and natural curiosities (fine fossils). In the *Government & Postal Buildings* is an excellent relief-model of the canton of Glarus by Becker (adm. free). In the art department is a small *Picture Gallery*, containing chiefly works by Swiss artists. The *Public Gardens*, in front of the Glarner Hof, are embellished with a handsome fountain, and contain memorial stones to the statesmen J. Heer (d. 1879) and J. J. Blumer (d. 1876), both natives of Glarus. — On the opposite bank of the Lint lies the busy manufacturing village of *Ennenda* (Schützenhof, **Neues Bad**).

EXCURSIONS (guides, see p. 69). Pretty walk (road) viâ *Schweizerhaus* to (3½ M.) *Schwändi* (see below). — The **Schild** (7500') is a fine point (5½ hrs.; guide 8 fr.). The path from Glarus leads through wood and pastures, and over the *Ennetberge*, to the (3 hrs.) *Heuboden-Alp* (4770') and thence to the right, without difficulty, to the top in 2½ hrs. more. Admirable view of the Mürtschenstock, Tödi, and Glärnisch. — The **Fronalpstock** (6980'; similar view) is easily ascended by the Ennetberge and the *Fronalp* in 5 hrs. (guide 7 fr.). — To THE MURGTHAL from the Heuboden-Alp, by the *Mürtschen-Alp* (*Oberstafel*, 6063'), see p. 47 (to the *Merlen-Alp* direct, 2 hrs.; over the *Murgseefurkel* to the *Murgseen*, 2½ hrs.). — To OBSTALDEN (8 hrs.; guide unnecessary for experts), a fine route: we cross the *Fronalp* (*Mittlere* 5193', *Obere* 6039'), pass between the Fronalpstock and Föhristock to the (5 hrs.) *Spannegg* (5108'), skirt the little *Spannegg-See* (4757'; with the *Mürtschenstock* on our right, p. 46), and descend over the *Platten-Alp* to the *Thalalp-See* (3610') and (3 hrs.) *Obstalden* (p. 46). — The Vorder-Glärnisch (7648'), from Glarus viâ *Sackberg* and through the *Gleiterschlucht*, 5½-6 hrs. (guide 13 fr.), laborious and adapted for experts only; steep descent viâ *Mittelgruppen* to (2½ hrs.) *Schwändi* (see below).

The *Klönthal* (p. 74) deserves a visit. Good road to the *Klönthaler See* 4½ M., thence to *Vorauen* 4½ M., to *Richisau* 6 M. more (one-horse carr. there and back 14, two-horse carr. 20-25 fr.).

From Glarus over the *Pragel* to *Schwyz*, see R. 23.

The railway to Linttbal crosses the Lint six times. 44 M. *Ennenda* (see above). Near (45½ M.) *Mitlödi* (1665'; Hirsch), and again beyond it, we obtain a superb view of the Tödi and its neighbours, which are not visible beyond Schwanden. On the right bank lies *Ennetlint*. The scenery is picturesque, the fertile valley with its factories contrasting pleasantly with the rocky and wooded slopes and the snow-mountains at its head. Pedestrians, who will also find this valley attractive, follow the right bank of the Lint, viâ *Ennenda*, *Ennetlint*, *Sool*, and *Haslen*, to *Hätzingen* (p. 69).

47 M. **Schwanden** (1718'; *Rail. Restaurant;* *Schwandner Hof;* *Freihof;* *Adler*, pens. 5-6 fr.), with large factories, lies at the junction of the *Sernf-Thal* or *Klein-Thal* with the Lint-Thal or Gross-Thal.

A charming walk (road viâ *Thon* 1½ M., direct footpath 25 min.) may be taken to Schwändi (2300'; *Inn*), which commands a splendid view of the Tödi and Selbsanft. — From Schwändi to the *Oberblegi-See* (p. 69) viâ the *Guppen-Alp* (5508') and *Guppen-Seeli* 4 hrs.

We cross the Lint below the influx of the Sernf and traverse the village of Schwanden. 48 M. *Nidfurn-Haslen;* to the E., 2 M. higher up, is the plain *Curhaus Tannenberg*. Farther on is *Leuggelbach*, with a fine waterfall on the right. — 50 M. *Luchsingen-Hätzingen*, two well-to-do villages, one on each bank of the Lint.

to Lintthal. STACHELBERG. *I. Route 21.* 69

From Luchsingen a pleasant excursion may be made to the (2½ hrs.) **Oberblegi-See** (4680'), at the foot of the Büchistock (p. 74), with descent viâ the *Bösbächi-Alp* and *Braunwald* to (3 hrs.) *Stachelberg*. Fine view of the Freiberge, Tödi group, etc.

We cross the Lint to (51 M.) *Betschwanden-Diesbach* (1958'); on the left, the picturesque fall of the *Diesbach*.

The **Saasberg** (6167'), a spur of the *Freiberg Range*, easily ascended from Betschwanden, Rüti, or Stachelberg in 3¼-4 hrs., commands a striking view of the Tödi group, etc. — **Kärpfstock** (*Hochkärpf*, 9180'), the highest of the Freiberge, laborious, and suitable for experts only (7-8 hrs. from Betschwanden or Rüti, viâ *Bodmen-Alp* and *Kühthal*; guide 15 fr.).

Beyond stat. *Rüti* we cross the Lint for the last time. 53 M. **Lintthal**, the terminus, lies on the left bank. About ¼ M. to the N. are the favourite *****Baths of Stachelberg** (2178'; **Glarner's Hotel*, R., L., & A. 3½-4, B. 1 fr. 30 c., D. 4, S. 2½ fr., board 6½ fr., visitors' tax 1 fr. per week; dépendance at the 'Seggen', on the right bank), beautifully situated. The powerful sulphureous alkaline water drops from a cleft in the *Braunwaldberg*, 1½ M. distant. The *View of the head of the valley is very striking: in the centre is the *Selbsanft* (9938'), to the right the *Kammerstock* (6975'), and adjoining it part of the *Tödi* (11,887') to the left; between the latter and the *Bifertenstock* (11,240') lies the *Biferten Glacier*. Pleasant walks have been laid out on the wooded hillside. — **English Church Service** at the hotel in summer.

Above the rail. station, on the left bank of the Lint, is *Ennetlint* (Schweizerhof, at the station; Klausen, both plain), with large spinning-mills. On the right lies (¾ M.) **Lintthal** (2238'; **Bär* or *Post; *Rabe; Drei Eidgenossen*, well spoken of), a considerable village with 2230 inhabitants.

EXCURSIONS (guides: *Fritz Stüssi* of Glarus, *Heinrich Streiff* of Seerüti, *Abr. Stüssi* at the Glärnisch-Hütte, *Salomon* and *Fritz Zweifel*, *Heinrich Schiesser*, *Rob. Hämig*, *Thom. Wichser*, and *Fritz Vögeli* of Lintthal; *Peter Elmer* and *Hilarius Rhyner* of Elm; high charges). To the **Fätschbach Fall* (½ hr.): we traverse the village of Lintthal and ascend the Thierfehd road (see below), passing the church; at the finger-post we diverge to the right (opposite, on the left bank, the new Klausen road with its tunnels) and reach a good point of view opposite the fine fall, in its wooded gorge. — To the **Pantenbrücke*, **Ueli-Alp*, and *Sandalp*, see p. 70; also to the (1½ hr.) **Braunwaldberge* (4920', Niederschlacht and Rubschen inns), a mountain hamlet with a magnificent view of the Tödi, best from beside the school, 1½ M. farther on; to the *Oberblegi-See* (see above), etc. — **Kammerstock** (6975'), by the *Kammer-Alp*, 4 hrs., repaying, and not difficult. — **Ortstock** or **Silberstock** (8908'), by the *Alp Bräch* and the *Furkel*, 6 hrs., laborious; splendid view (guide 15 fr.). — **Grieset** or **Faulen** (8940'), by the *Braunwaldberge*, 6 hrs., attractive, and not difficult (guide 18 fr.). The **Böse Faulen** (9200'), the N. and higher peak of the Grieset, is difficult (6½-7 hrs.; guide 30 fr.). These peaks afford an interesting survey of the stony wilderness around. Other fine points are the *Pfannenstock* (8410'; 6 hrs.) and the *Kirchberg* (*Hoher Thurm*; 8761'; 7 hrs., with guide). From the Faulen viâ the *Drecklochalp* (5560') to the *Glärnisch-Hütte* (p. 74), 4½ hrs. — **Gemsfayrenstock** (9758'), from the Upper Sandalp (see p. 70), by the *Beckenen* and the *Claridenglacier* in 3½ hrs., not difficult (guide 20 fr.). The descent may be made by the *Gemsfayer-Alp* to the *Urner-Boden* (p. 71).

A road, at first ascending and then level, leads from Lintthal (one-horse carr. from Stachelberg 8 fr. for ½ day, two-horse 12 fr.;

whole day 12 or 20 fr.) by the *Auengüter* (Inn 'Im Auen') **to** the (3½ M.) **Thierfehd** (2680'; **Hôtel Tödi*, R. & B. 3½, D. 3, pens. 5½ fr.), a green pasture surrounded by lofty mountains. During the latter part of the route we have a view of the **Schreienbach Waterfall* (230' high), which the morning sun tints with rainbow **hues.** Fine view of the gorge of the Lint from the **Känzeli*, ¾ M. from the inn.

A few paces beyond the hotel a bridge crosses the Lint, beyond which the stony path ascends for ½ hour. A slab on a large rock on the left is to the memory of Dr. Wislicenus, who perished on the Grünhorn in 1866. The path then descends a little towards the ravine, turns a corner, and reaches (¼ hr.) the **Pantenbrücke* (3212'), 160' above **the** Lint, **in the midst** of imposing scenery. On the right bank, a path ascends the grassy slope **straight to the** (¼ hr.) ***Üeli-Alp** (3612'), where **we** enjoy a superb view of the Tödi.

Thence we may either return by the same path to the Hôtel Tödi; or we may retrace our steps for about 80 yds. and then ascend to the E. by a somewhat unobvious forest-path to the (1¼ hr.) **Lower Baumgarten-Alp** (5285'), which lies on the right bank of the valley above the Thierfehd and presents a magnificent view. We next descend by a narrow and dizzy path (guide desirable, but not always to be obtained at the Alp, which is usually empty in summer), skirting the precipice of the *Tritt*, turning to the left, 5 min. beyond the Baumgarten-Alp, to *Obort* (3425'; Curhaus, plain, pens. 3½ fr.), and thence to the right viâ the Auengüter to (1 hr.) Linttal. For persons subject to giddiness this excursion is preferable in the opposite direction: Linttal, Auengüter, Obort, Baumgarten-Alp, Üeli-Alp, Pantenbrücke. — A steep path leads to the E. from the Baumgarten-Alp along precipitous grassy slopes to (1¼ hr.) the rocks of the *Thor* (6755'), where it becomes easier and bends to the right to (¾ hr.) the *Nüschen-Alp* (7270'), thence skirting the *Muttenwändli* to (1¼ hr.) the club-hut on the romantically situated Muttensee (8200'), the loftiest lake among the Swiss Alps. The hut, which has accommodation for 20 persons, is the starting-point for the ascents of the *Nüschenstock* (9500'), *Rüchi* (9855'), *Scheidstöckli* (9220'), *Ruchi* (10,100'), *Hausstock* (10,340'), *Muttenstock* (10,140'), *Piz dä Dartgas* (9135'), *Bifertenstock* (11,240', see below), *Selbsanft* (9935'), and other peaks. Over the *Kisten Pass* to *Ilanz*, see below.

The **Upper Sandalp* (6358'), 3½ hrs. above the Pantenbrücke, is frequently visited on account of its grand situation. The path ascends beyond the Pantenbrücke to the right (that in a straight direction leads to the Üelialp, see above), crosses the *Limmern-Bach*, which descends from a narrow ravine, and the *Sand-Bach*, and ascends on the left bank to the (1 hr.) *Vordere Sandalp* (4400'; rfmts.). The path now returns to the right bank. By the *Hintere Sandalp* (4330') it crosses the *Biferten-Bach*, and then ascends the steep and fatiguing slope of the *Ochsenblauken*, 2000' in height, where the Sandbach forms a fine cascade. Lastly we recross to the left bank, where the brook forces its passage through a gorge, and soon reach the (2 hrs.) **chalets** of the *Upper Sandalp* (Alpine fare and hay-beds in July and August). **The best** point of view is ½ hr. beyond the chalets.

The Lint **Valley** is terminated by a magnificent group of snow-mountains. The giant of this group is the **Tödi* or **Piz Rusein** (11,887'; from Linttal 10-11 hrs.; fit for experts only; guide 35 fr.; two guides required for one traveller, **or** one guide for two travellers), with its brilliant snowy crest, the most conspicuous mountain of N.E. Switzerland, ascended for the first time in 1837. The route from the Hintere Sandalp leads through the *Biferten-Thal* viâ the *Märenblanken* to the (4½ hrs. from Thierfehd) *Fridolin Hut* of the S. A. C. (6824') on the *Biferten-Alpeli*. We thence ascend by a fair path over stones and the moraine of the *Hinter Röthi Firn* to the (1¼-1½ hr.) old *Grünhorn Hut* (8040') and along the left side of the *Biferten*

Glacier, crossing the *Schneerunse,* a gully exposed to ice-avalanches in the afternoon, and the *Gelbwändli,* to the upper snows of the glacier, and to the (3½-4½ hrs.) summit. Magnificent view. We may descend by the *Porta da Spescha* (8860'), between the *Piz Mellen* (11,055') and *Stockgron* (11,215'), to the *Val Rusein* and (6 hrs.) *Disentis* (p. 372; guide 50 fr.); or by the *Gliemsfortz* (10,925'), between the Stockgron and the *Piz Urlaun,* to the *Gliems Glacier;* then over the *Puntaiglas Pass* to the *Puntaiglas Glacier* and down the Val Puntaiglas to *Truns* (comp. p. 371). — The **Bifertenstock** or **Piz Durgin** (11,240'), the second peak of the Tödi group, may be ascended from the Muttensee Club-Hut (p. 70) via the Kisten Pass (see below) and the *Furggle* in 6-7 hrs. (difficult; for expert climbers only; guide 30 fr.).

PASSES. From the Upper Sandalp a fatiguing route crosses the *Sandfirn* and the SANDALP PASS (9120') to Disentis in 6-7 hrs. (p. 372; guide 30 fr.); another, laborious but interesting, crosses (8 hrs.) the CLARIDEN PASS (9818') to the Maderaner-Thal (p. 124; guide 36 fr.).

FROM LINTTHAL OVER THE KISTEN PASS TO ILANZ, 13 hrs. (guide 30 fr.), fatiguing. Ascent by the (3 hrs.) *Baumgarten-Alp* to the (3 hrs.) *Muttensee Club-Hut* (p. 70). Thence viâ the *Mutten-Alp,* the *Lattenfirn,* and the *Kistenband,* high above the *Limmern-Thal* and opposite the *Selbsanft* and *Bifertenstock* (with the *Gries* and *Limmern* glaciers), to the (1½ hr.) Kisten Pass (8200'), lying to the N. of the *Kistenstöckli* (9020'). Descent by the *Alp Rubi* in the *Val Frisal* to (3 hrs.) *Brigels* (p. 371) and thence to the left to (2½ hrs.) *Ilanz* (p. 369), or to the right viâ *Schlans* to (2 hrs.) *Truns* (p. 371).

From Stachelberg by the *Bisithal* to *Muotathal,* see p. 73.

22. From Stachelberg to Altdorf. Klausen.

Comp. Maps, pp. 68, 86.

10 hrs. Bridle-path to Unterschächen (road in progress): from Stachelberg to Spitelrüti 3¼, Klausen 2, Unterschächen 2¼ hrs.; road thence to (7 M.) Altdorf (diligence daily in 1½ hr.; 3 fr. 5 c.; one-horse carr. 10, from Altorf to Unterschächen 15 fr.). Guide unnecessary (to Unterschächen 10, to Altdorf 15 fr.); horse to Unterschächen 28, to Altdorf 35 fr.

The new *Klausen Road,* the use of which is not yet permitted, ascends from *Ennetlint* in windings along the slope of the *Frutberg,* passing through tunnels and galleries. The old Klausen route has here been destroyed. The route at present used crosses the Lint below the Fätschbach Fall (p. 69) and ascends the *Fätschberg* on the right bank of the *Fätschbach.* Between the *Lower* and *Upper Fätschbach Falls* it crosses the stream. Near the (1 hr.) *Curhaus Frutberg* (3385'; unpretending) we reach the old bridle-path, which ascends through wood, at first rapidly, then more gradually, to the (1¾ hr.) wall and gate forming the boundary between Glarus and Uri, at the point where the *Scheidbächli* (4290') descends from the right.

The **Urner Boden** (2¼ hrs. from Stachelberg), a broad grassy and at places marshy valley, with a few groups of chalets, about 4 M. long and ½ M. broad, now begins. It is bounded on the N. by the jagged ridge of the *Jägernstöcke* and *Märenberge,* culminating in the *Ortstock* (8908'), and on the S. by the glaciers and snow-fields of the *Clariden* (10,728'). About ½ hr. from the frontier of Glarus we pass the inn *Zur Sonne,* and then (25 min.) the chalets of *Spitelrüti,* with a chapel on a hill (4560').

The path traverses the pasture for ½ hr. more, and then ascends a stony slope, passing (¾ hr.) an excellent spring to the left, to the

(¹/₄ hr.) *Klausen-Alp* and the (¹/₂ hr.) **Klausen Pass** (6437'). On the W. side we descend the gentle slopes of the beautifully situated *Bödmer Alp* (to the left, the *Grosse Scheerhorn*, 10,815'). After ¹/₂ hr., where the path divides, we turn to the left to the (5 min.) chalets of the *Lower Balm* (5600'; Inn zum Klausenpass, small) and cross the brook to a rocky cleft, forming the approach to the **Balmwand**, which here descends precipitously to the Schächen-Thal. The steep but well-kept path descends in zigzags (to the right the new Klausen road) to the (¹/₂ hr.) hamlet Im **Aesch** (4173'; *Hôt. Stäubi*, plain). To the left, the discharge of the *Gries Glacier*, on the N. side of the Scheerhorn, forms the magnificent *Stäuber Waterfall*.

We now descend the wooded **Schächen-Thal**, on the left bank of the turbulent *Schächenbach*. On the right bank (35 min.) the *Chapel of St. Anna*; 10 min., we cross the stream; ¹/₄ hr., **Unterschächen** (3345'; *Hôtel Klausen*, R., L., & A. 1¹/₂-2¹/₂, B. 1¹/₄, D. 3, pens. 6 fr.; *Alpenrose*, unpretending), finely situated near the mouth of the *Brunni-Thal*, at the head of which rises the *Grosse Ruchen* (10,295') with its glaciers. (Over the *Ruchkehlen Pass* to the Maderaner Thal, see p. 124.) About ³/₄ M. to the S. of the village is a small bath-house, with a mineral spring. To the N. rises the *Schächenthaler Windgälle* (9052'), and farther to the W. the *Kinzig Pass* (see p. 73).

A road descends the pretty valley, by *Spiringen*, *Weiterschwanden*, and *Trudelingen*, to (5 M.) a stone bridge over the Schächenbach, and thence to (1 M.) *Bürglen* (p. 111) and (1 M.) *Altdorf* (see p. 111).

23. From Schwyz to Glarus over the Pragel.

Comp. Maps, pp. 86, 66.

11 hrs. DILIGENCE from Schwyz to (6 M.) Muotathal twice daily in 1¹/₂ hr. (1 fr. 55 c.); carriage with one horse 9, with two horses 14 fr. From Muotathal over the Pragel to (4¹/₄ hrs.) Richisau, a bridle-path, unattractive; guide advisable, especially early and late in the season when the pass is covered with snow (18 fr.; *Melchior Bürgler*, *Jos. Gwerder*, or *Xav. Hediger* of Muotathal). No inn between Muotathal and Richisau. The pass being uninteresting, it is preferable to visit the *Muota-Thal*, as far as the Suvoroff bridge, from Schwyz or Brunnen, and the *Klönthal* from Glarus (see p. 68).

Schwyz, see p. 109. The road ascends to the S. through orchards and meadows (view of the Lake of Lucerne to the right), and in a wooded ravine at the foot of the *Giebel* (3010') reaches the *Muota*, which flows through a deep rocky channel. Opposite, to the right, is *Ober-Schönenbuch*, upon which the French were driven **back by Suvoroff in 1799**. Farther up the Muota ravine (2¹/₂ M.), but not visible from the road, is the *Suvoroff Bridge*, which was contested by the Russians and the French for two days. (At a sharp bend in the road, 2¹/₂ M. from Schwyz, a road descends to the right to this bridge in 3 min.; we may then return to Schwyz through wood and pastures on the left bank, a pleasant walk of 2 hrs. in all.) Beyond (2¹/₂ M.) **Ried** (1855' *Adler*), on the left, is the pretty fall of the *Gstübtbach*, at first descending perpendicularly, and then gliding over the rock. At (1 M.) *Föllmis* (1900') we cross the Muota and pass the *Mettelbach Fall* in the *Kesseltobel*. Then (2 M.)—

8 M. **Muotathal** (1995'; pop. 2015; *Kreuz*; *Hirsch*, moderate; *Krone*), the capital of the valley, with the *Franciscan Nunnery of St. Joseph*, founded in 1280, in which Suvoroff had his headquarters in 1799. Fine rock scenery and waterfalls in the vicinity.

OVER THE KINZIG PASS TO ALTDORF, 8 hrs., somewhat fatiguing (guide not indispensable). After following the Pragel route for 1/4 hr., we diverge by the Muota bridge to the right, and ascend the *Huri-Thal*, passing the chalets of *Lipplisbühl* and *Wängi*, to the (31/2 hrs.) Kinzig Pass (*Kinzigkulm* or *Kinzerkulm*; 6790'), lying to the S.E. of the *Faulen* (8130'). Limited view. Then a rapid descent to the *Schächen-Thal* (p. 72), *Wetterschwanden*, and *Bürglen* (p. 111). The Kinzig Pass is famous for the masterly retreat of Suvoroff, who, when cut off from the Lake of Lucerne by the French in Sept., 1799, marched with his army through the Schächen-Thal to the Muota-Thal, thence over the Pragel to Glarus, and lastly over the Panixer Pass to Coire.

THROUGH THE BISITHAL TO STACHELBERG, 10 hrs., rough but attractive; guide necessary. Good path (at first a road) through the narrow Bisithal, watered by the Muota, to (21/2 hrs.) *Schwarzenbach* (3153'; *Inn*), with a fine fall of the Muota; steep ascent thence to the left to the (3 hrs.) *Alp Melchberg* (6298'); then across the dreary *Karren-Alp* between the *Kirchberg* and *Faulen* (p. 62), and down the *Braunwald-Alp* to (41/2 hrs.) *Stachelberg*. Another and more interesting route is the following (10-11 hrs., with guide). From Schwarzenbach through wood and meadows (path generally well discernible) to the (11/4 hr.) *Waldibach Fall*, the finest waterfall of Central Switzerland; ascend thence to the left to the (2 hrs.) *Glatt Alp*, with the pretty blue *Glatten-See* (6090'), surrounded by lofty cliffs, and to the (3 hrs.) top of the *Ortstock* or *Silberstock* (8908'; p. 62); descend via the *Bräch-Alp* to (3-31/2 hrs.) *Stachelberg*. — Or from the Waldibach Fall we may ascend to the right over the *Waldi-Alp* and *Russ-Alp* to the (3 hrs.) *Russalper Kulm* (7125'), descend to the *Käsern-Alp*, turn to the left, and reach the (11/4 hr.) *Balmalp* on the Klausen route (see p. 72).

TO SISIKON THROUGH THE RIEMENSTALDEN-THAL and across the *Katzenzagel* (4888'), a footpath, 7 hrs. (unattractive; comp. p. 92).

From Muotathal the path leads to the (1/2 hr.) foot of the **Stalden**, and then ascends a toilsome and stony slope to (1 hr.) a group of houses (fine retrospect); 1/4 hr. farther on, it crosses the *Starzlenbach* by the *Klosterberg Bridge*, to the left, and ascends rapidly to the right to two houses; 40 min., by a gate, we descend to the right, and cross the brook; 10 min., a cross; 5 min., a cattle-shed in a picturesque valley; 1/4 hr., the *Sennebrunnen*, with excellent water; 5 min., refuge-hut; 5 min., a cross. Lastly, almost level, to the (25 min.) chalets on the marshy **Pragel** (5060'; no view).

The path, at first steep and stony, now descends to the (3/4 hr.) chalets of the *Schwellaui* (4367'), and then leads through wood; 1/4 hr., the *Neuhüttli* (4193'); here we turn to the right towards a large pine, where the pretty Klönthal and its lake become visible; 1/2 hr. **Richisau** (3590'; *Curhaus*, moderate, pens. 5-7 fr.), a rich green pasture with fine groups of trees, to the N. of which tower the *Wannenstock* (6495') and *Ochsenkopf* (7155'), and to the S. the furrowed slopes of the *Silbern* (7570').

The *Schwannhöhe*, an old moraine, 1/2 M. to the E. of the Curhaus, affords a beautiful view of the Klönsee, Schild, Glärnisch, and (to the S.) the

Faulen. Attractive excursions may be made to the W. to the (2½ hrs.) Cross on the *Saasberg* (6226'); pass to the Sihlthal and Einsiedeln) and to (5 min.) the *Sihlseeli* (5985'); to the S. to (3 hrs.) the top of the *Silbern* (7570'), with fossils and interesting furrowed slopes; to the *Glärnisch* (see below; to the club-hut 4 hrs., thence to the top 3 hrs.); to the top of the *Faulen* (*Griesst*, 8953') viâ the *Drackloch-Alp* in 6 hrs. (with guide), descending to (4 hrs.) *Stachelberg* (p. 69); to the N., viâ (1 hr.) the *Schweinalp* to (3½ hrs.) *Hinterägyithal* (comp. p. 43); to the top of the *Ochsenkopf* (7155'; 3½ hrs.; with guide); to the top of the *Scheye* (5 hrs.; see p. 67) viâ *Längenegg*, etc.

From Richisau a road descends, across a fine open pasture, in full view of the imposing Glärnisch, to (1 hr.) **Vorauen** (2640'; *Hôtel-Pension Klönthal*, pens. 6½-7½ fr.; *Vorauen Inn*, at the lower end of the village, plain), beautifully situated in the Klönthal.

The *Glärnisch, the huge rocks of which bound the Klönthal on the S. side, one of the most picturesque mountains in Switzerland, culminates in the *Vorder-Glärnisch* (7648'), the *Vrenelisgärtli* or *Mittler-Glärnisch* (9055'), the *Ruchen-Glärnisch* (9557'), and the *Bächistock* or *Hinter-Glärnisch* (9583'). The ascent of the Ruchen-Glärnisch is laborious, but not difficult for mountaineers (7½ hrs.; guide 20 fr.; see p. 69). We cross the Richisauer and Rossmatter Klön, to the W. of Vorauen, to the huts on (40 min.) the *Klönstalden* (3450'; direct path hither from Richisau in 25 min.), then enter the narrow *Rossmatter-Thal* (red marks), pass the chalets of *Käsern* (3968') and *Werben* (4562'), and reach the (3½ hrs.) *Club-Hut* in the *Steinthäli* (6613'; Inn in summer). We next ascend steep stony slopes and cross the *Glärnischfirn*, regain the rock, and reach the top in 3 hrs. from the hut. Grand view (panorama by Heim). — Ascent of the *Vorder-Glärnisch* from Glarus, 5½-6 hrs. (comp. p. 68).

The *Klönthal is a picturesque dale, with meadows of freshest green, carpeted with wild-flowers until late in the autumn, and thinly peopled. To the S. rise the almost perpendicular precipices of the *Glärnisch* (see above). The pale-green *Klönthaler See* (2640'), 1½ M. from Vorauen, a lake 2 M. long and ⅓ M. broad, enhances the beauty of the valley, reflecting in calm weather the minutest furrows on the side of the Glärnisch. The rocks on the S. bank, near a waterfall, bear an inscription to the poet *Salomon Gessner* (d. 1788), who often spent the summer in a neighbouring chalet. The road skirts the N. bank. Boat across the lake in 50 min., 1½ fr. At the (3½ M.) *Seerüti*, at the lower end of the lake (fine views), is a small *Inn*.

Below the lake the valley narrows to a gorge, through which dashes the *Löntsch*, the discharge of the lake, forming a series of cascades amid grand rocky scenery down to its confluence with the Lint, below Netstall. To the left rise the huge perpendicular cliffs of the *Wiggis Chain* (p. 67). We obtain a pretty view of the deep ravine from the iron foot-bridge, reached by a footpath opposite the Staldengarten Inn.

The road divides at the (¾ M.) *Staldengarten Inn*. The left branch leads to (2 M.) *Netstall* (p. 67), the right crosses the Löntsch to (1 M.) *Riedern* and (1¼ M.) Glarus (p. 67). In descending we enjoy a fine view of the *Fronalpstock*, the *Schild*, and the *Freiberge* (between the Lint and Sernf valleys).

24. From Glarus to Coire through the Sernf-Thal.

Comp. Map, p. 66.

16-18 hrs. RAILWAY from Glarus to Schwanden, 17 min.; DILIGENCE (2 fr. 55 c.) from Schwanden to (9½ M.) Elm twice daily in 2¾ hrs. (descent, 1¾ hr.). — From Elm to Flims over the Segnes Pass, 8-9 hrs., guide 20 fr. (p. 68); to Ilanz over the Panixer Pass, 9 hrs., guide 18 fr. — From Flims to Coire DILIGENCE twice daily in 2¼ hrs.; from Flims to Reichenau a pleasant walk; thence to Coire driving is preferable (diligence 4 times daily).

At *Schwanden* (p. 68), 3 M. to the S. of Glarus, the deep *Sernf-Thal*, or *Klein-Thal*, diverges to the left from the Lintthal. The high-road gradually ascends the N. slope. Beyond (1½ M.) *Wart* is a pretty waterfall on the left; fine retrospective view of the Glärnisch. 3 M. *Engi* (2540'; pop. 1164; *Sonne), with cotton-mills, at the mouth of the narrow *Mühlebach-Thal*. (Passage of the *Widerstein-Furkel* to the *Murgthal*, see p. 47.) The slate-quarries *(Plattenberge)* on the left bank of the Sernf are noted for their fossil fish. From (2 M.) *Matt* (2710') a path to the N. E. leads in 6 hrs. through the *Krauchthal* and over the *Rieseten Pass* (6644') to *Weisstannen* (p. 47).

3 M. (9½ M. from Schwanden) **Elm** (3215'; *J. Elmer*; *Zentner*), the highest village in the valley, in a fine basin encircled by snow-mountains, was partly destroyed on 11th Sept., 1881, by a landslip from the *Tschingelberg* (S.E.), through which 114 persons lost their lives (memorial tablet at the church).

ASCENTS (for experts only; guides, *Heinrich* and *Peter Elmer*, see p. 69). *Kärpf*, or *Kärpfstock* (9180'), by the *Wichlen-Alp*, 6 hrs. (laborious; but, with good guides, free from danger). — *Vorab* (9925'), by the *Sether Furka* (see below), 7-8 hrs. — *Hausstock* (10,340'), by the *Richetli Pass* and the *Ruch Wichlenberg*, or by the *Panixer Pass* (see below) in 7-7½ hrs., laborious. — *Piz Segnes* (10,180'), from the *Segnes Pass* (see below) in 1½-2 hrs., or from the *Segneslücke* (see below) by the S. arête in 1 hr. (6½-7½ hrs. from Elm), not difficult for experts.

PASSES. TO FLIMS OVER THE SEGNES PASS (pron. 'Senyes'), 8 hrs., fatiguing, but interesting (guide, 18 fr., necessary). We cross the Sernf, amidst the remains of the landslip, and the *Raminbach*, and ascend the wild gorge of the *Tschingelbach*, which forms several picturesque falls, to the *Tschingeln-Alp*. We then mount steep grassy and stony slopes to the (5 hrs.) Segnes Pass (8615'), lying to the S.W. of the *Piz Segnes* (10,230'). To the right rise the jagged *Tschingelhörner* or *Mannen* (9351'), perforated by the *Martinsloch* (8648'), a hole through which the sun shines on the church of Elm twice a year. We descend over the short but steep *Segnes Glacier* (easy except in the absence of snow; when rope and ice-axe are useful), then by a steep path, which afterwards improves, to the *Flimser Alpen*, and thence past a pretty waterfall (to the left the huge *Flimser Stein*, p. 368) to (3 hrs.) *Flims* (p. 365).

TO ILANZ OVER THE PANIXER PASS, 9 hrs. (guide 18 fr.), fatiguing and unattractive, but historically famous for Suvoroff's retreat of 5th-10th Oct., 1799 (comp. p. 73). A road ascends on the left bank of the Sernf from Elm by *Hinter-Steinibach* to the (40 min.) *Erbser-Brücke*; 25 min. farther up, at *Wallenbrugg*, we cross the Sernf and ascend by a steep, rugged path to the chalets of the *Jätzalp* (*Im Loch*, 4822'; *Ober-Staffel*, 5587'). We next cross the *Walenboden*, pass the *Rinkenkopf*, traverse a patch of snow (with a small tarn on the left), and reach the (3½ hrs.) Panixer Pass (*Cuolm da Pignieu*; 7897'), with its refuge-hut. On the right rises the *Hausstock* (ascent from the pass in 3½-4 hrs., see above), with the *Meer*

Glacier. Descent over the *Meer-Alp* and the wild *Ranasca-Alp* to (2½ hrs.) *Panix* (1332'; Panixer Pass Inn), and viâ *Ruis* to (2 hrs.) *Ilanz* (p. 369). — Another route, fatiguing **and** uninteresting, crosses the **Sether Furka** (8565'). It diverges from the Panix route to the left, by the tarn above mentioned, and ascends steeply to the pass, between the *Rothhorn* and the *Voralb* (ascent of the latter from the pass in 2 hrs., see p. 75). Descent by the *Ruscheiner Alp* and the *Sether Tobel* to (9 hrs.) *Ilanz* (p. 361).

To WEISSTANNEN BY THE FOO PASS, 7 hrs., rather rough (guide 15 fr.). We ascend the right bank of the Raminbach, chiefly through wood, to the *Ramin-Alp*, and past the chalets of *Matt* (6179'), to the (4 hrs.) **Foo Pass** or Ramin Pass (7290'); then descend by the *Foo-Alp* and the *Unter-Sies-Alp* (4377') to the *Sees Valley* and (3 hrs.) *Weisstannen* (p. 47; 3 hrs. from *Mels*).

To VÄTTIS OVER THE SARDONA PASS, 10-11 hrs., difficult, **and** rarely traversed (guide 30 fr.). From the Segnes Pass (see above) we cross the glacier to the E., climb up the steep rocky S. arête of the Piz Segnes to the *Segneslücke* (9351m), just to the S. of the Piz Segnes, descend by a very steep snow slope to the *Segnes Glacier*, and cross it to the **Sardona Pass** (about 9680'), between the Piz Segnes and the *Trinserhorn* (*Piz Dolf*, 9235'). We then cross the *Sardona Glacier* to a rocky ridge between two arms of this glacier, whence a rugged descent leads to the left to the *Sardona-Alp* (5735'), in the *Kalfeisen-Thal*, 3 hrs. above *Vättis* (p. 66). — Another difficult and laborious pass from Elm to Vättis (9-10 hrs.) is the Sauren Pass **or** Scheibe **Pass** (9680'), to the S. of the *Saurenstock* (10,020'; easily ascended from the pass in 20 min.). — OVER THE MUTTENTHALER GRAT, 10-11 hrs. to Vättis, less difficult, but rough and fatiguing (guide 25 fr.). From the (4 hrs.) **Foo** *Pass* (see above) we first descend to the *Obere Foo-Alp*, then ascend to the right through the *Muttenthal* to the basin of the *Haibützli*, with **a** small tarn (7693'), and thence to the (3 hrs.) **Muttenthaler Grat** (8104'). Rough descent over the *Malanser Alp* to **(2 hrs.)** *St. Martin* (4433') in the *Kalfeisen-Thal* and (2 hrs.) *Vättis* (p. 66).

To LINTTHAL, by the **Richetli Pass** (7425'), 8 hrs., not difficult; °View of the Hausstock, Vorab, and Glärnisch. Descent by the *Durnach-Thal*.

II. LAKE OF LUCERNE AND ENVIRONS.
THE ST. GOTTHARD.

25. From Zürich to Zug and Lucerne 78
    i. Railway Journey 78
        Hausen, 79. — Excursions from Zug: Felsenegg and Schönfels. Stalactite Caverns in the Hölle, 79. — Schönbrunn. Menzingen. Ageri-Thal, 80.
    ii. From Zürich to Zug viâ Horgen 81
26. Lucerne 81
        From Lucerne to Kriens and Herrgottswald, 86.
27. Lake of Lucerne 86
        Weissenfluh. From Beckenried to Seelisberg. Niederbauen (Seelisberger Kulm), 88. — Buochser Horn. Hochfluh. Curhaus Seelisberg, 89. — Schwendifluh. Gütsch, 90. Morschach. Axenfels. Axenstein. Stoos. Fronalpstock, 91. — Riemenstalden-Thal. Rophaien. Rossstock. Kaiserstock, 92. — Isenthal. Schönegg Pass. Rothgrütli. Uri-Rothstock, 93.
28. The Rigi 94
29. From Lucerne to Alpnach-Stad. Pilatus 100
        Bürgenstock. From Stansstad to Sarnen, 101.
30. From Zug and Lucerne to Arth 103
    i. From Zug to Arth. Lake of Zug 103
    ii. From Lucerne to Küssnacht and Arth 104
31. From Zürich viâ Wädensweil to Arth-Goldau. From Biberbrücke to Einsiedeln 105
        Feusisberg. Hütten, 105. — Gottschalkenberg. From Pfäffikon to Einsiedeln; the Etzel, 106. — From Einsiedeln to Schwyz over the Hacken or the Iberger Egg, 107. — The Schlagstrasse. Rossberg, 108.
32. From Lucerne to Bellinzona. St. Gotthard Railway . 108
        Goldau Landslip, 109. — The Myten, 110. — Bürglen; Schächen-Thal; Rossstock; Belmistock, 111. — Erstfelder-Thal. Bristenstock; Hohe Faulen. The St. Gotthard Road from Amsteg to Göschenen, 112. — Pizzo Rotondo; Passo del Sassi; Val Piora; Taneda, etc., 114, 115.
33. From Göschenen to Airolo over the St. Gotthard . . 117
        The Göschenen Valley. Passes to Realp, the Trift Glacier, and the Steinalp. The Fleckistock, 118. — The Stock. The Badus or Six-Madun. Gurschenstock and Gamsstock. Lucendro Lake, 120. — The Pizzo Centrale; Prosa; Fibbia; Piz Lucendro; Pizzo Rotondo; Sorescia. From the St. Gotthard over the Orsino Pass to Realp, and over the Lecki Pass to the Furka, 121.
34. The Maderaner-Thal 122
        Hüfi Glacier. Düssistock; Oberalpstock, etc., 123. — Clariden Pass; Hüfi Pass; Kammlilücke; Ruchkehlen Pass; Scheerhorn-Griggeli Pass; Brunni Pass, 124.
35. From Göschenen to the Rhone Glacier. The Furka . 124
        From Realp over the Cavanna Pass to the Val Bedretto. Tiefen Glacier; Tiefen-Sattel; Winterlücke, 125. — Furkahorn; Blauberg; Muttenhorn; Galenstock. From the Furka over the Nägeli's Grätli to the Grimsel Hospice, 126.

36. From Lucerne to Altdorf viâ Stans and Engelberg.
The Surénen Pass 127
  Stanser Horn, 127. — Nieder-Rickenbach, 128. — Excursions from Engelberg: Schwand; Tätschbach Fall; Arnitobel; Fürrenalp; Rigithalstock; Engelberger Rothstock; Uri-Rothstock; Spannort; Titlis, 129, 130. — From Engelberg to Erstfeld over the Schlossberg-Lücke or the Spannort-Joch; to Wasen over the Grassen Pass; to the Steinalp over the Wenden-Joch, 130.
37. From Lucerne over the Brünig to Meiringen and Brienz (Interlaken) 131
  The Melchthal; over the Storegg or the Juchli to Engelberg; Nünalphorn; Hutstock. Excursions from Melchsee-Frutt, 132. — The Schwendi-Kaltbad. Flühli. Giswiler Stock, 133. — Footpath from the Brünig to Meiringen, 134.
38. From Meiringen to Engelberg. Engstlen-Alp. Joch Pass 134
  From the Engstlen-Alp to Melchsee-Frutt. Schafberg, Graustock, etc. Ascent of the Titlis from the Engstlen-Alp, 135. — From the Engstlen-Alp over the Sätteli to the Gadmen-Thal, 136.
39. From Meiringen to Wasen. Susten Pass 136
  Triftthal. Excursions from the Trift Hut (Dammastock, etc.); over the Trift-Limmi to the Rhone Glacier; Furtwang-Sattel and Stein-Limmi, 137. — From the Stein Inn over the Susten-Limmi or the Thierberg-Limmi to the Göschener-Alp. Brunnenstock, 137.
40. From Lucerne to Bern. Entlebuch. Emmen-Thal . 138
  Schwarzenberg; Farnbühl-Bad, 138. — Schimberg Bad. From Schüpfheim to Flühli and Sörenberg. From Flühli viâ the Seewenegg to Sarnen. The Napf, 139. — Rüttihubelbad, 140.
41. From Lucerne to Lenzburg (Aarau). The Seethal Railway 140
  Excursions from Hochdorf: Hohenrain; Horben; Oberreinach, etc., 141. — From Hitzkirch to Wohlen by Fahrwangen, 141. — From Beinwyl to Reinach and Menzikon; Homberg, 141. — From Boniswyl to Fahrwangen; Brestenberg, 141.

25. From Zürich to Zug and Lucerne.

Comp. Maps, pp. 40, 86.

i. Railway Journey.

41½ M. RAILWAY to Zug in 1½ hr. (4 fr. 5, 2 fr. 85, 2 fr. 5 c.); to Lucerne in 2½ hrs. (7 fr., 4 fr. 90, 3 fr. 50 c.

Zürich, see p. 33. On leaving the station the train crosses the *Sihl*, and at (2½ M.) *Altstetten* diverges from the Bâle line (p. 21). To the left rises the long *Uetliberg* (p. 39), which the line skirts in a wide curve. To the right the pretty valley of the Limmat. 5½ M. *Urdorf;* 8 M. *Birmensdorf.* We now ascend the pleasant *Reppisch-Thal* and pass through a tunnel under the *Ettenberg* to (12 M.) *Bonstetten-Wettschwyl* (1805'). To the right the Bernese Alps and Pilatus, and to the left, farther on, the Uri-Rothstock and the Titlis become visible. 14 M. *Hedingen* (1712'); 15½ M. *Affoltern* (Löwe, pens. 4½-6 fr.), with an institute for the 'Kneipp Cure'. To the left rises the *Aeugster Berg* (2723'), at the foot of which lie *Aeugst* and the *Baths of Wengi*. — 18 M. *Mettmenstetten* (1550').

Diligence thrice daily in 55 min. to Hausen (1980'; *Löwe*), at the W base of the *Albis* (p. 40); near it is the excellent *Albisbrunn Hydropathic* (Dr. Paravicini). Near *Kappel*, 1½ M. to the S., Zwingli was slain on 11th Oct., 1531, in a battle against the Roman Catholic cantons (comp. p. 33).

20 M. *Knonau* (Adler). Near Zug we cross the *Lorze*, which descends from the *Ägeri-See* (p. 80).

24½ M. Zug. — Hotels: *HIRSCH, R., L., & A. 2-3, B. 1¼, lunch 2-3, D. 2¼-3, pens. 6-7 fr.; *OCHS; *LÖWE, on the lake, R., L., & A. 2½-3½, B. 1¼, lunch 2½-3, D. 3, pens. 5½-7½ fr., good beer in the restaurant; *HÔTEL BAHNHOF, with garden-restaurant, R., L., & A. 2½-3½, B. 1¼, lunch 3, D. 3½, pens. 6-8 fr.; HÔTEL RIGI, near the station, R. from 1½, B. 1, D. 2½-3 fr.; FALKEN; BELLEVUE; WIDDER; *PENS. GUGGITHAL, on the road to Felsenegg, 4-4½ fr.; RESTAURENT AKLIN, near the Zeitthurm.

Zug (1385'), the capital of the smallest Swiss canton, with 5161 inhab., lies on the lake of that name. The lower town, part of which was undermined by the lake on July 5th, 1887, has fine *Quays*, commanding beautiful views of the lake, the Rigi, Pilatus, and the Bernese Alps. The upper and old towns still retain a quaint and mediæval appearance, with their walls, towers, and substantial mansions. In the *Old Rathhaus* (now a restaurant) is a handsome late-Gothic apartment containing a museum of wood-carvings and other antiquities of Zug (adm. 50 c.). The Gothic *Church of St. Oswald* (15th cent.) contains a Last Judgment by P. Deschwanden, and the *Church of the Capuchins* an Entombment by Calvaert. In the *Arsenal* are preserved ancient captured weapons and flags, and a scarf stained with the blood of its bearer Peter Collin, who fell at Arbedo in 1422. Handsome new *Government Buildings* in the Italian style. Well-equipped *Fish-breeding Establishment*. Above the town are the educational institutions of *Minerva* and *St. Michael*, and the nunnery of *Maria Opferung*. On the (¾ M.) *Rosenberg* (Restaurant) is the interesting *Swiss Museum of Bee-Culture*.

On the W. slope of the *Zuger Berg*, 1½ hr. from Zug (good road; omnibus from the station at 11 and 6; fare 2½ fr.), are the *Hotel Felsenegg (3085'; pens. 7-8 fr.; *English Church Service* in summer); with a fine view towards the W., and (5 min. to the N.) the *Curhaus Schönfels (R. 2-3, B. 1¼, board 6 fr.), with hydropathic establishment and pleasant grounds, also commanding a beautiful view. This spot is recommended for a prolonged stay; pleasant wood-walks. The (¼ hr.) *Hochwacht* (3250'), ¼ M. to the N.E., commands a complete survey of the Alpine chain; below us, to the E., lies the Lake of Ägeri (p. 80). — Pretty walks also to the (20 min.) *Hüngigütsch* (2400'; view interrupted by trees) and the (½ hr.) *Horbachgütsch* (3070'), which affords a charming view of the lakes of Zug and Lucerne and the Rigi. — The ascent of the (2½ hrs.) *Wildspitz* (*Rossberg*, p. 108) is an attractive expedition, over mountain-pastures with rich flora.

In the wild valley of the *Lorze*, to the N.W. of Zug, are the interesting *Stalactite Caverns in the Hölle*, to which a road leads viâ *Baar* (p. 81) in 1½ hr. (carriage with one horse from Zug and back, 5-7 fr. and fee), and a footpath (1 hr.) viâ *Thalacker* (road to Ägeri, see below) and the *Tobel-Brücke*. The caverns, at one time full of water, were made accessible in 1887 and are open from Easter Monday to Oct. 15th. They

contain magnificent stalactite formations of various shapes, besides
stalagmites. Admission, 1 fr.; guide and key at the (1/4 M.) *Restaurant
Höll* (trout). From the caverns a route leads viâ the Tobel-Brücke to
(2 M.) *Schönbrunn* (see below).

On the Menzingen hills above the Lorze, 4½ M. to the E. of **Zug**
(diligence twice daily, 1 fr. 35, coupé 1 fr. 60 c.) and ½ M. from the diligence
station of *Edlibach*, is Dr. Hegglin's well-managed *Schönbrunn Hydropathic (2215'; board 6, R. 1½-4 fr.), with sunny terrace and forest-walks, much frequented by French visitors. The view from the chapel
(2230') extends as far as the Jura. — About 6 M. to the E. of Zug (diligence twice daily in 1¾ hr.) is the prettily situated village of **Menzingen** (2635'; *Löwe; Hirsch*) with a large convent-school for girls; and
1 M. farther on, beyond the *Edlibach*, is the *Pens. Schwandegg* (2770'; pens.
4½-5 fr.), with pine-cone and other baths. The summit of the *Schwandegg-Gütsch* commands a view of the Lake of Zürich and of the Sentis range.

Ägeri-Thal. A road (diligence to Ober-Ägeri twice daily in 2 hrs.) ascends
through a fruitful district viâ *Thalacker* (route at the head to the left to
Schönbrunn, the Hölle caverns, and Menzingen, see above) and *Inkenberg* to
(3 M.) *Allenwinden* (2320'). Thence it descends into the valley of the winding
Lorze (on a hill on the other side of the stream is the nunnery of *Gubel*)
to (1½ M.) *Neu-Ägeri*, and past *Mühlebach*, with its large cotton factories, to
(1½ M.) **Unter-Ägeri** (*Ägerihof; Brücke; Post*), a handsome industrial village with a new Gothic church, on the *Ägerisee* (see below). The road
skirts the lake, flanked by pretty villas, to (1½ M.) the pleasant mountain
village of **Ober-Ägeri** (*Löwe*, pens. 4½ fr.; Hirsch; Ochs). In a picturesque
situation on the lake, between Unter-Ägeri and Ober-Ägeri, is Dr. Hürlimann's private *Hospital* for children; and on the hill, farther back, is a
Sanitarium for scrofulous children, erected by the Zürich Benevolent
Society. — EXCURSIONS from Unter-Ägeri through the *Hürithal* and viâ the
Rossberg-Alps to the (2½ hrs.) summit of the *Wildspitz* (Rossberg, see
p. 108); from Ober-Ägeri to the (1½ hr.) *Gottschalkenberg* (p. 106), etc.

On the pretty **Ägerisee** (2380'; 3½ M. in length) a steamboat plies 4
times daily from Unter-Ägeri in ¾ hr., past the stations of *Ober-Ägeri, Ländli,*
and *Eierhals*, to **Morgarten**, at the E. extremity; omnibus thence to the
rail. station of *Sattel-Ägeri* (p. 107; 50 c.). Stat. *Eierhals* (Pension) commands a picturesque *View, comprising the Uri-Rothstock, Krönte, etc.
Between Eierhals and Morgarten are the houses of *Haselmatt*, where on
16th Nov., 1315, the Confederates in the *Battle of Morgarten* won their
first victory over their Hapsburg oppressors commanded by Duke Leopold of
Austria. A memorial chapel, containing a representation of the battle,
was erected at *St. Jakob*, 1 M. from the S.E. end of the lake and ¾ M.
from Sattel, where a commemoration service is held annually on the day
of the battle.

The train backs out of the station and skirts the flat N. bank of
the *Lake of Zug* (p. 103), crosses the *Lorze* near its influx into the
lake, and recrosses it at its efflux near (27½ M.) **Cham** (*Rabe*), a village with a slender zinc-covered church-tower and a large manufactory
of condensed milk. Pretty view of the lake to the **left**; on the hill
above Zug are the summer-resorts just mentioned; in the middle
rises the Rigi; and to the **right** are the Stanser Horn, the Engelberg
Alps, and Pilatus. Beyond (31 M.) **Rothkreuz** (1410'; *Rail. Restaurant*), the junction of the St. Gotthard (p. 109) and the Muri and
Aarau (p. 22) lines, we enter the valley of the *Reuss*. 33 M. *Gisikon*. Through an opening to the left we survey the Rigi, from the
Kulm to the Rothstock. 37 M. *Ebikon*. To the right rises the wooded
Hundsrücken. The train skirts the *Rothsee*, 1½ M. long, and crosses
the Reuss by a bridge 178 yds. long. The line now unites with the

Swiss Central (p. 21) and the Lucerne and Bern lines (p. 138), and finally passes through a tunnel under the *Gütsch* (p. 85).

41½ M. Lucerne, see below

ii. From Zürich to Zug viâ Horgen.

RAILWAY from *Zürich* to (11 M.) *Horgen* in ½ hr. (steamer in 1¼ hr., see p. 39). POST OMNIBUS daily (8.45 a. m.) from *Horgen* to (12½ M.) *Zug* in 2 hrs. 35 min. (2 fr. 80 c.); one-horse carr. in 2 hrs., 12 fr.

To *Horgen* (1394′), see p. 42. The road ascends in windings, passing the *Curhaus Bocken*, to (3 M.) *Haurüthi*, where, by the finger-post, it joins the road from Wädensweil. Several fine views of the lake, the Sentis, Speer, Curfirsten, and the Glarus Mts. At (1½ M.) *Hirzel* (2245′), on the saddle of the hill, is the *Inn Zum Morgenthal*. We then descend gradually into the valley of the *Sihl*, which here separates the cantons of Zürich and Zug, to the (2 M.) covered Sihl-Brücke (1745′; *Krone, good wine).

Pedestrians should take the road from Horgen over the HORGER EGG to the Sihl-Brücke (4½ M.), which shortens the route by 2 M., and affords far finer views. Near (2 M.) *Wydenbach* rises the *Zimmerberg (2586′), ¼ hr. to the right, with a beautiful view of the Lake of Zürich, the sombre valley of the Sihl, the Lake of Zug, the Alps, and particularly the Myten, the Rigi, and Pilatus. About ¾ M. beyond Wydenbach the road reaches the *Hirzelhöhe* (2415′; Inn), its highest point, with another fine prospect. We join the high-road near the Sihl-Brücke.

The Zug road leads through an undulating tract, passing on the left the wooded hill of the *Baarburg* (2180′). Beyond the wood (2 M.) we obtain a view of Baar, the Lake of Zug, the Rigi, and Pilatus. To the left, ¼ M. farther on, on the *Lorze*, which we cross, is a large cotton-factory. Near (1¼ M.) Baar (1465′; *Lindenhof*, moderate; *Krone; Schwert; Rössli*), a straggling village with 4065 inhab., is the hamlet of *Blickenstorf*, with the house in which Hans Waldmann, burgomaster of Zürich and conqueror of Charles the Bold at Morat, was born. — In the prettily wooded valley of the *Lorze*, 2 M. to the E., are the curious *Stalactite Caverns* in the *Hölle* (p. 79).

From Baar we continue straight on to (2½ M.) Zug, see p. 79.

26. Lucerne.

RAILWAY STATION (Pl. D, E, 4) on the left bank of the lake (new building in progress); BRÜNIG STATION (Pl. E, 4) ¼ M. farther to the E. (Restaurants at both). — The STEAMBOATS to Flüelen and Alpnach generally touch on the left bank after leaving the Schweizerhof Quay; those from Flüelen touch first here, and then at the quay. — In the busy season travellers arriving by steamer or railway with luggage cannot be sure of getting on by the corresponding train or boat unless they and their luggage are booked through to some station beyond Lucerne. If luggage is booked to Lucerne only, it is often impossible to reclaim it and get it rebooked in time.

Hotels. *SCHWEIZERHOF (Pl. a; D, E, 2), a spacious hotel with two 'dépendances', and *LUZERNER HOF (Pl. b; E, 2), both on the Schweizerhof Quay, R., L., & A. from 5, B. 1½, déj. 3½, D. 5, pension 10-12 fr.; *GRAND HÔTEL NATIONAL (Pl. c; E, F, 2), on the Quai National, R., L., & A. from 5½, B. 1½, déj. 4, D. 5, pens. from 10 fr.; *HÔTEL BEAURIVAGE (Pl. d; F, 2), in the Halden-Strasse, R., L., & A. 3½-6, B. 1½, déj. 3, D, 4½, pens.

BAEDEKER, Switzerland. 16th Edition. 6

8-12 fr.; *Hôtel de l'Europe, Halden-Strasse, R., L., & A. 8-6, B. 1½, déj. 2½-3, D. 4, pens. 7-12 fr.; *Englischer Hof (Pl. e); *Hôtel du Cygne (Pl. f), R., L., & A. 4-6, D. 4½, pens. 10-12 fr.; *Hôtel du Rigi (Pl. g), R., L., & A 3, R. 1½, déj. 3, D. 4, pens. from 8 fr. (these all on the lake, on the right bank); *Hôtel du Lac (Pl. h, D, 4), on the left bank of the Reuss, with bath-house, R., L., & A. 4-6, B. 1½, D. 4, S. 3, pens. 7-12 fr.; °Hôtel St. Gotthard (Pl. i), with restaurant, near the station, R., L., & A. from 3, B. 1½, D. 4 fr. (no gratuities); *Hôtel Victoria (Pl. u; C, 4), R., L., & A. from 3, B. 1½, D. 3½, pens. from 8 fr.; *Wage (Balances; Pl. k, C 3), near the third bridge over the Reuss, R., L., & A. 3-4½, B. 1½, D. 3½, pens. from 7½ fr. — Less expensive: *Rössli (Pl. n; C, 3), R. & A. 2½, B. 1¼, D. 3 fr.; *Engel (Pl. l; B, 3), R., L., & A. from 2, D. 3 fr.; *Adler (Pl. m; C, 3), R. 2-3, B. 1½ fr.; *Hôtel de la Poste (Pl. o; C, 4), R. from 2, D. 3 fr.; Hôtel des Alpes (Pl. p; D, 2), R., L., & A. 2½-4 fr.; °Goldner Löwe, Kappelgasse 22, R., L., & A. 2-2½, B. 1, déj. with wine 2½, D. 2½, pens. 6 fr.; *Storchen (Cigogne), Kornmarkt (Pl. C, 3), R., L., & A. 1½-2, B. 1, D. 2, S. 1½ fr., good wine; Union, Löwen-Str. (Pl. E, 1), with café-restaurant; *Bären (Ours), R., L., & A. 2-3, D. 2½-3 fr.; *Hôtel Rütli; *Rebstock (Pl. v; E, 2), beside the Hofkirche; Mohr (Pl. u; D, 3); *Hirsch (Pl. q; C, 3); *Krone (Pl. r; C. 3), R., L., & A. from 1½, B. 1, pens. 4½ fr.; *Weisses Kreuz (Pl. s; D, 3); *Sauvage (Pl. t; C, 4), R. & A. 2-2½, D. 3 fr.; *Raben, R., L., & A. 1½-2½, pens. 5½-6½ fr.; Pfistern, Metzgern, °Schlüssel, *Schiff, *Sonne, all on the Reuss.

Pensions. *Eden House (R. from 2½, pens. from 8 fr.); *Tivoli (6-10 fr.); *Kaufmann; Kost-Häfliger; °G'segnet-Matt; Belrédère (7-8 fr.). All these are on the Küssnacht road, close to the lake. *Bienz, above the Cursaal; Faller, above Beaurivage (from 6 fr.); *Neu-Schweizerhaus (Kost), Gyger, *Felsberg (Pietzker), all three loftily situated (Pl. E, F, 1); *Alt-Schweizerhaus; *Pens. Villa Maria, well situated near the Hofkirche; Mme. Trüb's Pension Anglaise, on the Drei Linden hill; *Hôt.-Pens. Gütsch (D. 3½, pens. 8-10 fr.) and °Hôt.-Pens. Wallis, on the Gütsch (p. 85), with charming view; *Suter (pens. 5-6 fr.), on the hill of Gibraltar (Pl. A, 4). Still higher, to the S. of Lucerne (from the Gütsch in ¾ hr.; brake from Lucerne thrice daily; one-horse carr. 8, two-horse 12 fr.; comp. p. 85), *Curhaus Sonnenberg (2500'), with pleasant grounds and a fine view (7 fr. per day). Pens. Stutz, see p. 101.

Restaurants. °Stadthof (Pl. E, 2; Pilsener beer, music in the evening); *St. Gotthard, near the station, see above; Café-Restaurant Flora, Chalet, both at the station; Café du Théâtre, Café Alpenclub, on the Reuss; Café du Lac; °Cigogne (see above; good wine on draught); °Hungaria (Hungarian wines). — Beer. Stadthof, St. Gotthard, Union, see above; Löwengarten, near the Lion Monument, with garden and a large concert-hall; Rosengarten, Grendel-Strasse; Muth, Zürcher-Str. 3; Kreuz (see above); Seidenhof, on the left bank of the Reuss, etc. — Confectioners. Huguenin, near the Stadthof; Gnandt, next door to the Hôtel du Rigi.

Cursaal on the Quai National (Pl. F, 2), with reading, concert, and ball-rooms, restaurant, theatre, and garden. Band every afternoon and evening. Admission to the garden free; theatre (French operettas) 2 ½ fr

Panorama of the French army entering Switzerland in Feb., 1871, by E. Castres, in the Löwen-Platz (p. 84; adm. 1 fr.).

Baths in the lake by the Quai National; swimming 20, separate bath 40 c. (towels extra). — Lake-baths also near the Tivoli (see above). Baths in the Reuss below the town, at the Nollethor, with swimming-basin. Warm baths at the Hôtel du Lac and at Felder-Lehmann's, Spreuer-Brücke.

Post and Telegraph Office (Pl. D, 4), near the railway-station.

Cabs. For ¼ hr., 1-2 pers. 80 c., 3-4 pers. 1 fr. 20 c. (to or from the station 1 or 2 fr.); for ½ hr. 1 fr. 50 or 2 fr. 20 c.; for 1 hr., 2 fr. 50 or 3 fr. 60 c.; each box 50 c. To Seeburg 1½ or 2 fr.; Dreilinden-Stiege 2½ or 4, Dreilinden-Plateau 3½ or 5, Meggen 3½ or 5, Küssnacht 6½ or 9 fr. — Double fares at night (10-6).

Rowing Boats and Naphtha Launches at the Quai National (Rud. Herzog), Schweizerhof Quay, and Schwanen-Platz. Fare without boatman 50 c. per hr., with canopy 1 fr., gondolas 1 or 1½ fr.; boatman 1 fr. per hr. Launch from 5 fr. per hour and 45 fr. per day.

United States Vice-Consul, *E. Williams*, Villa Geissenstein (in summer, Seehofstrasse). — **British Consular Agent**. *L. Falck*, banker, Schwanen-Platz.

English Church Service in the Protestant Church (Pl. D, 2) in summer (7.45, 11, & 5). *Presbyterian Service* in the Boys' School, Museggstrasse, at 11 and 4. *American Service* at Christ Church (Old Catholic), Museggstrasse (Pl. D, 2), at 7.45, 11, and 5.

Physicians. *Dr. Otto Stocker-Freiss*, Kapell-Platz 9; *Dr. Rob. Steiger*, Hertenstein-Str. 56. — Dentist: *Dr. Alfred Steiger*, Hertenstein-Str. 56. — *Anglo-American Pharmacy* (C. Kopp), Schwanen-Platz, opposite the Hôt. du Rigi.

Official Enquiry Office, Schwanen-Platz 7, opposite the Hôtel du Rigi.

Lucerne (1437'; pop. 24,236), the capital of the canton of that name which joined the Forest Cantons in **1332**, lies picturesquely on the *Lake of Lucerne* or *Vierwaldstätter See*, at the efflux of the *Reuss*. It is enclosed by well-preserved walls with nine watch-towers, erected in 1385, while its amphitheatrical situation surrounded by low hills, facing the Rigi and Pilatus and the snow-clad Alps of Uri and Engelberg, is of surpassing beauty.

The clear, emerald-green *Reuss* issues from the lake with the swiftness of a torrent. Its banks are connected by five bridges. The highest, the iron **Seebrücke** (Pl. D, 3), erected in 1869-70, 500' long and 50' wide, crosses from the town to the railway-station and the post-office, and affords an excellent view of the town and the lake. The two interesting mediæval bridges, the **Kapellbrücke** (Pl. D, 3) and the **Spreuerbrücke** or **Mühlenbrücke** (Pl. B, C, 3), are both carried obliquely across the stream. Each is covered with a roof, which, in the case of the former, is painted with 154 scenes from the lives of St. Leodegar and St. Mauritius, the patronsaints of Lucerne, and from Swiss history; and in the case of the latter, with a Dance of Death. The paintings all date from the 18th century. Adjoining the Kapellbrücke, in the middle of the river, rises the old **Wasserthurm** (Pl. D, 3), containing the *Municipal Archives*. According to tradition, this building was once a lighthouse *(lucerna)*, and gave its name to the town. *St. Peter's Chapel*, on the N. bank, has four modern altar-pieces by Deschwanden (p. 127). — The Reuss and the lake are enlivened with swans and flocks of half-tame waterfowl (Fulica atra; black, with white heads).

The *Schweizerhof Quay and the *Quai National (Pl. D, E, F, 2), with their umbrageous avenue of chestnuts, extend in front of the large hotels along the N. bank of the lake and afford a delightful view. The stone indicators or 'toposcopes', on a projecting platform about the middle of the quays, point out the chief places in the environs.

VIEW. To the left the *Rigi Group*; to the left is the *Kulm* with the hotels; on the saddle between the Kulm and the *Rothstock* is the Staffel Inn; more to the right the *Schild*; the *Dossen*, and the isolated *Vitznauer Stock*. To the left of the Rigi, above the hills by the lake, rises the *Rossberg*; to the right of the Vitznauer Stock, in the distance, are the singularly indented peaks of the *Liedernen Chain*, the *Clariden*, the *Tödi*, and the *Kammlistock*; then the *Nieder-Bauen* or *Seelisberger Kulm* and the *Ober-Bauen*; nearer are the dark *Bürgenstock*, with its hotel, and the *Buochser Horn*; to the left and right of the latter tower the *Engelberg Alps*, the last to the right being the *Titlis*; farther to the right the *Stanserhorn*, the mountains of *Kerns* and *Sachseln*, and to the extreme right *Pilatus*.

6*

At the E. end of the Schweizerhof Quay is the handsome office of the administration of the St. Gotthard Railway. Farther on, on the Quai National, is the *Cursaal* (p. 82).

On rising ground overlooking the quays is the *Hofkirche, or *Stiftskirche* (*St. Leodegar*; Pl. E, F, 2), said to have been founded in the 7th cent., restored in the 17th cent., with two slender towers erected about 1506. It contains a carved pulpit and stalls of the 16th cent., two altars with gilded reliefs in carved wood, that on the N. side representing the death of the Virgin (15th cent.), a fine crucifix by the Engelberg wood-carver Custer, and stained-glass windows. Organ-concert daily 6.30-7.30 p. m. (1 fr.). In the arcades enclosing the old *Churchyard* are several frescoes by Deschwanden.

We next follow the Alpen-Strasse and Zürcher-Strasse, passing *Meyer's Diorama of the Rigi and Pilatus* (Pl. D, E, 2; adm. 1 fr., interesting), the *Panorama* (p. 82), and *Stauffer's Museum* of stuffed Alpine animals (Pl. E, 1; adm. 1 fr.), and in 5 min. reach the famous *Lion of Lucerne (Pl. E, 1), a most impressive work, executed in 1821 to the memory of 26 officers and about 760 soldiers of the Swiss guard, who fell in the defence of the Tuileries on 10th Aug., 1792. The dying lion (28' in length), reclining in a grotto, transfixed by a broken lance, and sheltering the Bourbon lily with its paw, is hewn out of the natural sandstone rock after a model (exhibited in the adjoining building) by the celebrated Danish sculptor *Thorwaldsen*. Inscription: *Helvetiorum fidei ac virtuti. Die X Aug., II et III Sept. 1792. Haec sunt nomina eorum, qui ne sacramenti fidem fallerent, fortissime pugnantes ceciderunt. Duces XXVI. Solerti amicorum cura cladi superfuerunt Duces XVI.* The rock which bears the inscription and names of the officers is overhung with trees and creepers. A spring at the top flows down on one side and forms a dark pool at the base, surrounded by trees and shrubs. — The neighbouring *Chapel* (inscription: *Invictis Pax*) contains the escutcheons of the deceased officers, and the '*Museum*', opposite the Lion, contains a painting of the last struggle of the Swiss guard in the Tuileries, a diorama of the Jungfrau from the Männlichen, by Ernest Hodel, and a view of the Arth Rigi-Railway (adm. 1 fr.).

To the N. of the monument is the entrance to the *Glacier Garden (adm. 1 fr.), an interesting relic of the ice-period, with 32 'pot-holes' or 'giants' cauldrons', of different sizes (the largest being 26' wide and 30' deep), well-preserved 'Gletscherschliffe', or rocks worn by the action of the ice, etc., discovered in 1872, and connected by means of steps and bridges. Small park with deer and chamois. The house contains a reconstruction of a lacustrine village, with some genuine relics; a relief of a glacier, with erratic blocks, by Prof. Heim (1:10,000); a *Relief of the St. Gotthard Railway (1:25,000), by Imfeld and Becker; and a relief of the Muota valley (1:2500), with a representation of the battle between the French and the Russians in 1799. A kiosque contains *Pfyffer's Relief* of Central Switzerland, on a scale of 5⅛ inches to the mile, 25' long, and 14' wide. On fine evenings a concert (Alpine horns) is given here by electric light.

Many quaint and picturesque houses of the 16-17th cent. are still to be seen in the crooked streets of the older parts of the town (Pl. C, D, 3). — The ancient **Rathhaus** (Pl. C, D, 3), in the corn-market, dates from 1519-1605. A fresco on the tower represents the death of the Lucerne burgomaster Gundoldingen at the Battle of Sempach.

On the groundfloor is the *Historical Museum* (adm. 9-6, 1 fr.). ROOM I. contains the armoury from the Arsenal, embracing weapons, flags, and trophies of the battles of the 14th cent. and of the Burgundian and Milanese wars; in the glass-case on the right are the coat-of-mail of Duke Leopold of Austria, and several banners captured by the townsmen at the battle of Sempach. A chased sword-handle ('Tellenschwert', i.e. 'Tell's sword') of the 16th cent., and the uniforms of different Swiss guards (in the middle of the large glass-case) should also be noticed. At the windows is exhibited a *Collection of Stained Glass* of the 14-18th cent., including a series of armorial bearings of the 17th century. — ROOM II. contains the collections of the Historical Society, comprising relics of the prehistoric, Celtic-Roman, Germanic, and mediæval periods; in glass-cases in the centre are Roman objects (bronze statue of Mercury; tripod) and the blue and white banner presented to Lucerne by Pope Julius II. — On the first floor is the *Council Chamber*, with beautiful 16th cent. carving on the ceiling and walls. In the ante-chamber are a number of portraits of magistrates, most of which are by Reinhart.

The late-Gothic *Fountain* in the Weinmarkt (Pl. C, 3) is by Conrad Lux (1481).

On the left bank of the Reuss are the *Jesuit Church* (Pl. C, 4), built in 1667 in the rococo style, and the former Jesuit College, now the *Government Building*, with a picturesque court, the state archives, and a collection of coins. Opposite are the *Museum* (Pl. C, 4), with the cantonal library of 80,000 vols. (including many rare books; adm. 10-12), and the *Civic Library*, on the Reuss, containing a valuable collection of works on Swiss history and copies of Holbein's frescoes on the Harter house, pulled down in 1824.

The *Gütsch (1722'), an eminence on the left bank of the Reuss, at the W. end of the town, affords a splendid survey of the town, the lake, the Rigi, and the Alps of Uri, Unterwalden, and Engelberg, best from the view-tower (1920'; lift 30 c.). It is reached from the Schweizerhof Quay or the railway-station by a walk of 10-12 min. and then by *Cable Tramway* in 3 min. more (188 yds. long; gradient 53 : 100; train every 10 min.; fare 30, return-ticket 50 c.). *Hotel and Restaurant, with wooded grounds, at the top.

A pretty walk by wood and field leads from the Gütsch to the (½ hr.) *Curhaus Sonnenberg* (p. 82) and thence to (10 min.) the *Kreuzhöhe* (2560'), a charming point of view. From the Curhaus a road descends to (½ hr.) Kriens (p. 86).

Another beautiful point in the neighbourhood of the town is the *Drei Linden (1810'), to which a good road leads in about 20 min. from the Hofkirche. We ascend to the right behind the church, turn to the left at the café ('Terrassenstieg'), and soon reach the top, where a number of villas are rising. The 'Drei Linden' stand in private grounds (no admission). In front is a terrace, with benches, commanding the finest view of the environs of Lucerne and the Alps, with the Titlis and Stanserhorn in the middle and the Finsteraarhorn and Schreckhorn in the distance to the right.

86   II. Route 27.   LAKE OF LUCERNE.   Lake of

The return may be made to the N.W., past the Capuchin Convent on the *Wesemlin*, to the (20 min.) Lion Monument (p. 84).

From Lucerne to Kriens, 2½ M., steam-tramway in 12 min., skirting the *Krienbach*. — Kriens (1670'; "*Pilatus*; *Linde*), a considerable manufacturing village, is situated in a fertile valley at the N. foot of Mt. Pilatus. To the S. a road ascends to (1 M.) the château of *Schauensee* (1950') and to the (1¼ M.) *Hôtel-Pension Himmelreich* (2264'; pens. 4-5 fr.), a health-resort amid woods, with a fine view. To the N. is the *Sonnenberg* (2560', to the Curhaus, 1 hr.; see p. 85]. The road ascends the valley beyond Kriens to the *Renggbach*, whence a footpath leads through wood to (1 hr.) *Herrgottswald* (2800'; *Hôt.-Pens. Haas, pens. 5-7 fr.), an inexpensive health-resort in a picturesque situation, and to (1 br.) *Eigenthal* (3375'; "Pens. Burri, 5-5½ fr.), another cheap health-resort (hence to *Schwarzenberg*, ¾ hr.; see p. 138). — From Eigenthal a path ascends by the *Rümligbach* past the huts of *Buchsteg* and *Rothstock*, and finally mounts steeply to the left to (1½-2 hrs.) the *Eründlen-Alp* (4985'), with the little *Pilatus Lake* (generally dry in summer), where, according to an old tradition, Pontius Pilate drowned himself in the bitterness of his remorse. From this point the *Widderfeld* (6825') may be ascended in 1¾ hr.; and a rough and not always distinct path leads round the slopes of the Widderfeld and Gemsmättli and past the *Kastelen-Alp* to the (1½ hr.) *Hôtel Klimsenhorn* (p. 103). Neither expedition should be attempted without a guide.

27. Lake of Lucerne.
Comp. also Map, p. 94.

Steamboat 6-7 times daily between Lucerne and Flüelen in 2¾ hrs., express in 2¼ hrs. (to Hertenstein 35 min., Weggis 45 min., Vitznau 1, Buochs 1¼, Beckenried 1½, Gersau 1¾, Treib 2, Brunnen 2 hrs. 5 min., Rütli 2 hrs. 12 min.; Sisikon 2 hrs. 20 min., Isleten 2 hrs. 20 min., Bauen 2 hrs. 25 min., Tells-Platte 2½, Flüelen 2¾ hrs.; the steamers do not all touch at Hertenstein, Buochs, Treib, Rütli, Sisikon, and Tells-Platte, while Bauen and Isleten are called at twice a day only). Fare to Flüelen 3 fr. 65 or 2 fr. 60 c.; return-tickets available for two days at a fare and a half. Trunk 40-80 c., including embarcation and landing. Sunday excursion trips from Lucerne to Flüelen and back, first class 1½ fr. All the steamers, except the express-boat at 5.45 a.m., touch at the railway-station of Lucerne after leaving the quay (comp. p. 81). Good restaurants on board. Time-tables and maps of the lake to be had at the steamboat-offices gratis.

The **Lake of Lucerne** (1435'; *Vierwaldstätter See*, or 'Lake of the Four Forest Cantons'), which is bounded by the 'forest cantons' of *Uri*, *Schwyz*, *Unterwalden*, and *Lucerne*, is unsurpassed in Switzerland, and even in Europe, in magnificence of scenery. Its beautiful banks are also intimately associated with those historical events and traditions which are so graphically depicted by *Schiller* in his *William Tell*. The lake is nearly cruciform in shape, the bay of Lucerne forming the head, the bays of Küssnacht and Alpnach the arms, and those of Buochs and Uri the foot. Length from Lucerne to Flüelen 23 M.; width ½-2 M.; greatest depth 700'.

The **wind** on the lake is apt to change with extraordinary rapidity, and the boatmen declare that it blows from a different quarter as each promontory is rounded. The most violent is the *Föhn* (S. wind), which sometimes renders the S. bay of the lake impracticable for sailing or rowing-boats, and dangerous even for steamers. In fine weather the *Bise* (N. wind) usually prevails the whole day.

Soon after leaving Lucerne the steamer affords a strikingly picturesque view of the town, with its towers and battlements. To

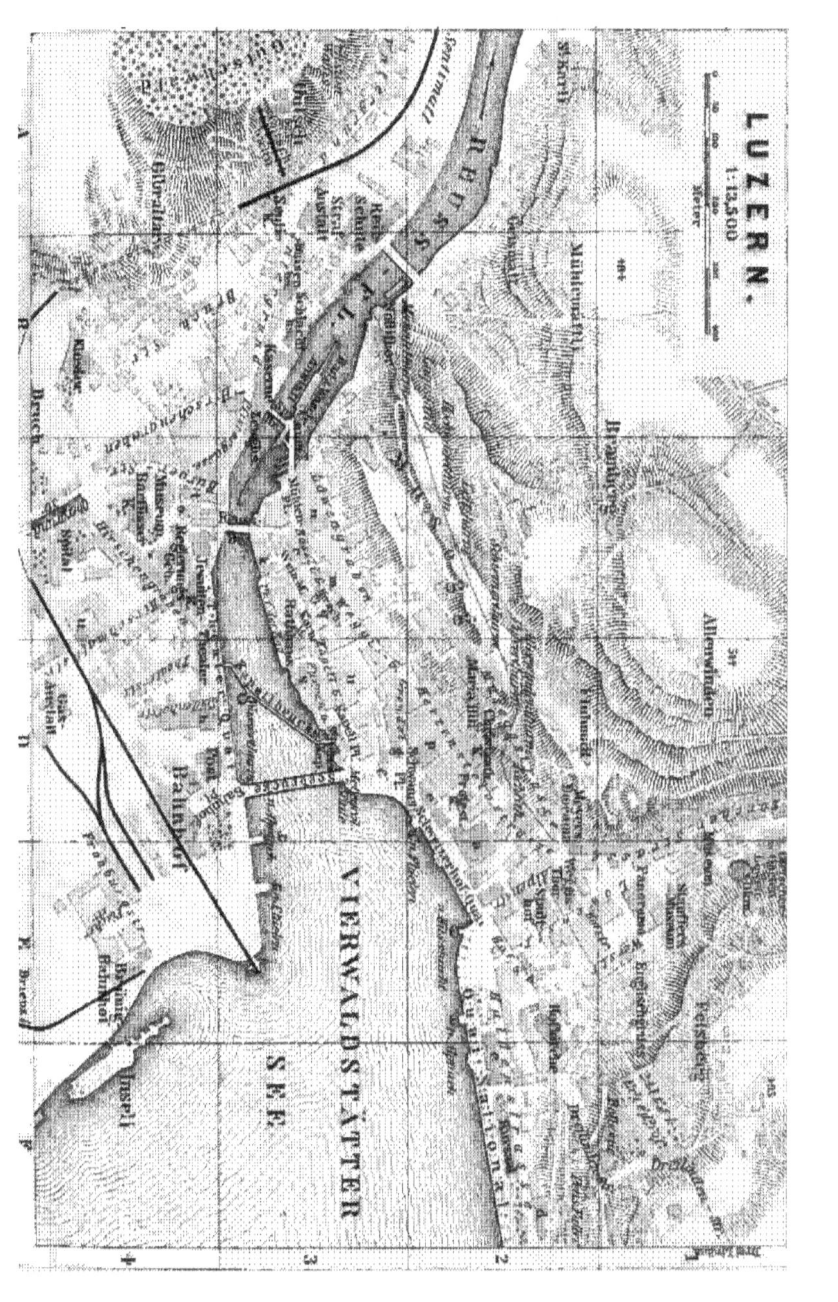

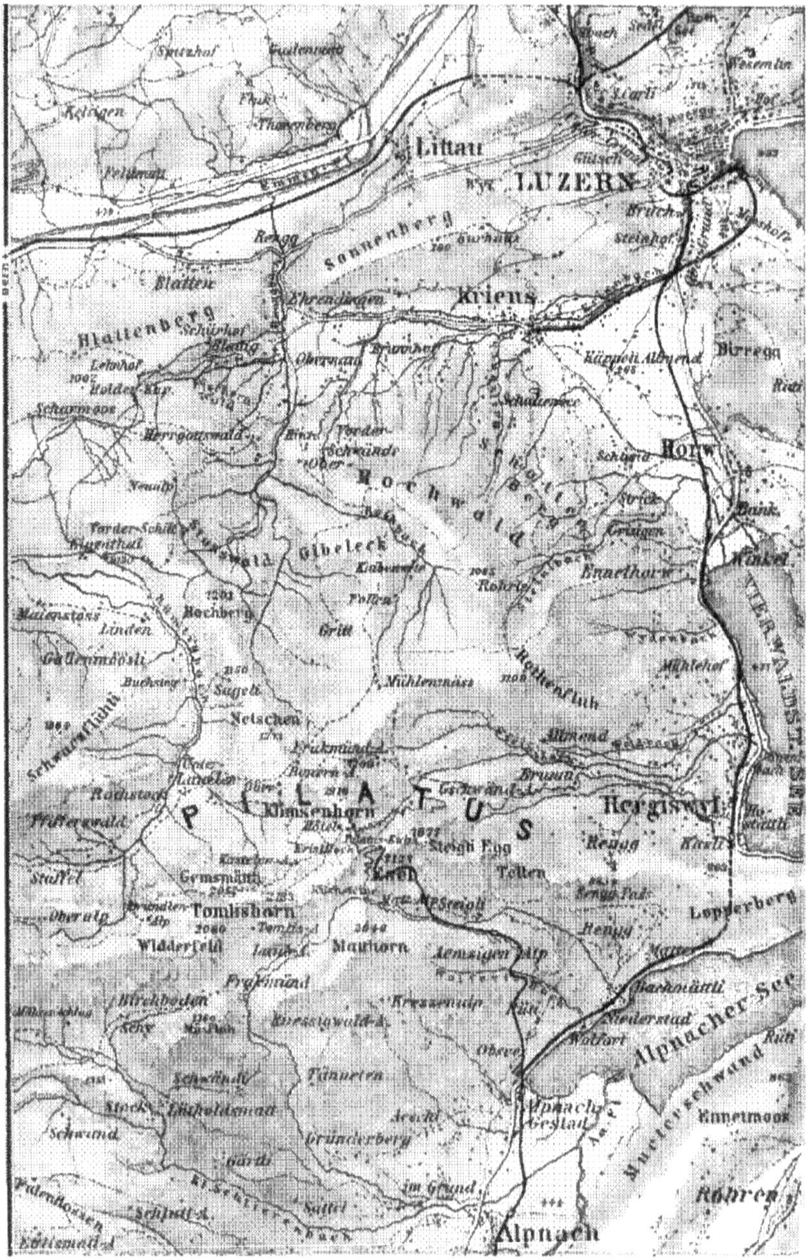
Littau
LUZERN
Sonnenberg
Kriens
Blattenberg
Blatten
Horw
Hochwald
Gibeleck
Hochberg
Netschen
Ennethorw
Winkel
Rothenfluh
PILATUS
Klimsenhorn
Hergiswyl
Steigli Egg
Gemsmättli
Tomlishorn
Widderfeld
Mauhorn
Lopperberg
Alpnacher See
Alpnach
Gründerberg
Tännenen
Ennetmoos
Röhren

the left rises the Rigi, to the right Pilatus, and facing us the Bürgenstock, the Buochser Horn, and Stanser Horn; to the left of Pilatus, above the hills of Sachseln, the Wetterhörner (Rosenhorn, Mittelhorn, Wetterhorn), Schreckhorn, Mönch, Eiger, and Jungfrau gradually become visible. The small promontory to the left, with a pinnacled villa, is the *Meggenhorn*. In front of it lies *Altstad* ('old shore'), an islet planted with poplars, on which fragments of an old custom-house are still to be seen.

Beyond the Meggenhorn the lake of Küssnacht opens to the left, and the bay of Stansstad to the right, and we have now reached the central part (*'Kreuztrichter'*) of the cross formed by the lake. In the distance to the left, *Küssnacht* (p. 104) is visible; in the foreground, *Neu-Habsburg* (p. 104). To the right the forest-clad *Bürgenstock*, with its hotel and railway, rises abruptly from the water (see p. 101). From this part of the lake the *Pilatus* (p. 102) is very striking. Its barren, rugged peaks, seldom free from cloud or mist, form a marked contrast to the *Rigi* on the opposite bank, the lower slopes of which are covered with gardens, fruit-trees, and houses, and the upper with woods and green pastures.

Beyond the promontory of *Tanzenberg*, in a small bay to the left, is the *Hôtel Schloss Hertenstein* (pens. 7-10 fr.; reached either on foot through the park in 10 min., or by boat in 5 min.). Straight on, in the distance, appears the double-peaked Scheerhorn (p. 123). Stat. *Hertenstein* (Pens. Hertenstein, dépendance of the above); then —

Weggis. — *HÔT.-PENS. DU LAC, R. 2½, D. 3, S. 2, pens. 6-8 fr.; *LION D'OR, R. 1½-2, B. 1, D. 2½, pens. 5-6 fr.; *HÔT.-PENS. SCHÖNAU, from 5 fr.; *HÔT.-PENS. DE LA POSTE, at the pier, D. 2½, pens. 5 fr.; *HÔT.-PENS. PARADIES, 6-8 fr.; *PENS. BELVEDERE, with pleasant grounds, lake-baths, etc., pens. 8-10 fr.; PENS. ZIMMERMANN-SCHÜRCH, with garden; *HÔT.-PENS. BELLEVUE, ³/₄ M. to the W., with extensive grounds, baths, etc., R., L., & A. 3-7, B. 1¼, D. 4½, pens. 7-11, omnibus at the pier; PENS. BAUMEN, ¼ M. farther up (4 fr.). On the lake are several furnished villas which are let to families.

Weggis, a thriving village in a very sheltered situation, is frequented as a health-resort. — Bridle-path to the Rigi, see p. 96.

Immediately to the N. of the pier of Weggis rises the *Rigiblick*, a hill affording a fine survey of the lake (permission from proprietor necessary). — From Weggis to *Greppen* (p. 104) a nearly level road leads in ³/₄ hr. The pleasant path across the hill, between the Rigi and the Rigiblick, is preferable (1 hr.). It starts from the back of the school-house, a little to the E. of the church (ascend in ¼ hr. to a farm; for 4 min. level; by a second farm ascend again; by the third farm descend to the left). — Beautiful walk to the E., by the road skirting the lake, to the *Hôt.-Pens. Lützelau* (pens. 5-6 fr.) and (3 M.) *Vitznau*. A new road leads on from Vitznau by the *Obere Nase* (fine view of the lake) to (1¼ hr.) *Gersau* and past the *Kindlimord Chapel* (p. 89) to (1½ hr.) *Brunnen*.

Nearing Vitznau, we observe on the hillside to the left the railway-bridge across the Schnurtobel (p. 95); high above it the Hôtel Rigi-First (p. 99) and, farther to the right, the Hôtel Unterstetten (p. 99).

Vitznau. — *HÔT. & RESTAURANT RIGIBAHN & PENSION KOHLER, R., L., & A. 2-3½, B. 1¼, D. 2-3, pens. 6½-8 fr.; *HÔTEL-PENSION RIGI, R. 2-3, D. 2½-3, pens. 5½-7 fr.; *HÔTEL-PENS. DU PARC, ⅓ M. to the W., with

88 *II. Route 27.*   BECKENRIED.   *Lake of*

baths and extensive grounds, pens. 7-10 fr.; *Pension Zimmermann zum Kreuz; Pens.-Restaurant Bellevue*. Furnished Rooms at *Zimmermann's* at Unterwylen, 1 M. from the village, with fine view. Beer at the hotels and at the *Restaurant zur Alpenrose*, 3 min. from the Rigi station, on the Gersau road; *Flora Alpina Restaurant*, on the Gersau road, 1 M. from Vitznau, with a charming view.

Vitznau, prettily situated at the base of the *Vitznauer Stock* (see below), is the terminus of the *Rigi Railway* (p. 95). High above the village rises the precipitous *Rothfluh*, with the *Waldisbalm*, a stalactite grotto 200 yds. long (difficult of access).

On the S.W. slope of the Vitznauer Stock (bridle-path in 1¼ hr. from Vitznau, shady in the early morning) is the charmingly situated *Hotel-Pens. Weissenfluh (3100'; pens. from 5½ fr.), frequented as a health-resort, with beautiful view (finest from the *Blümlismatt*, 5 min. to the S.). Pretty walks to *Aeusser-Urmi* (3525'; ¼ hr.); *Ober-Urmi* (3740'; ½ hr.); to the top of the *Vitznauer Stock* (4775'; 1¼ hr., the last ½ hr. steep); *Dossen* (5510'; 2 hrs.), etc. Descent from Weissenfluh to Gersau 50 min. (ascent 1½ hr.; path rough in places).

Beyond Vitznau two rocky promontories, called the *Nasen* (noses), project far into the lake, apparently terminating it, the one being a spur of the Rigi, the other of the Bürgenstock (p. 101). To the left of the E. Nase, above the Pragel, the Glärnisch (p. 74) becomes visible. Beyond this strait the lake is called the *Buochser See*, from Buochs (*Krone*, R., L., & A. 1-2, B. 1, D. 2-3, pens. 4½-5 fr.; *Hirsch; *Restaurant Kreuzgarten*), a village to the right, above which rise the *Buochser Horn* (p. 89) and the E. slopes of the Bürgenstock. Diligence to *Stans* (p. 127) thrice daily in ¾ hr. Between Buochs and Beckenried (a pretty walk of ¾ hr.) extensive operations have been carried out to regulate the torrents descending from the Buochser Horn and the Schwalmis. — Farther on, on the S. bank, is —

Beckenried, or *Beggenried* (*Sonne*, R. from 1½, L. ½, B. 1¼, D. 3, pens. from 6 fr.; *Mond*, R. & B. 3, D. 3, pens. 6-8 fr.; *Nidwaldner Hof*, R., L., & A. 2-3½, D. 3, pens. 5-7½ fr.), where the delegates from the Four Forest Cantons used to assemble. In front of the church rises a fine old walnut-tree. In the neighbourhood are several cement-factories and the picturesque *Riseten Waterfall*.

One-horse carriage to Engelberg (p. 128) 18 fr., two-horse 30 fr. (from Buochs 17 or 28 fr.); to Stans 6 or 12, Stansstad 8 or 16, Alpnach 11 or 18, Grafenort 12 or 20, Saelisberg 13 or 25, Schönegg 6 or 12 fr., and fee.

From Beckenried to Seelisberg (2¾ hrs.). The road leads by the (1 hr.) charmingly situated *Hôtel & Curhaus Schöneck* (water and whey-cure, board 6 fr.; S. B. C. H.) to (1¼ hr.) the village of Emmetten (2590'; *Post*, well spoken of; *Engel*, pens. from 4½ fr.); then through a somewhat monotonous dale between the *Stutzberg* and *Niederbauen*, past the picturesque little *Seelisberger-See*, to the (1½ hr.) *Curhaus Seelisberg* (p. 89).

The *Niederbauen or Seelisburger* Kulm (6315'), a very attractive ascent, is best made from **Emmetten** (3 hrs.; guide not indispensable). The path (fine views) starts at the E. end of the village and for a short way follows the right bank of the brook. Beyond a group of three houses it ascends through wood, then in numerous windings through young pine-trees, and farther on over mountain-pastures, leaving the chalet to the right. In 1½ hr. we reach the middle of the rocky ridge on the W. side of the mountain, which is visible from the village. Here we proceed in a straight direction towards the E. and ascend broad grassy slopes to (1½ hr.) the spacious summit. — Another route (somewhat easier but

¹/₂ hr. longer) diverges from the road to the S., near the Hôt.-Pens. Engel, and ascends the narrow *Kohlthal*. At (1 hr.) a group of huts we cross a bridge to the left and ascend a steep path to (1 hr.) the W. arête, where it joins (1¹/₂ hr.) the route first described. — The routes from *Beroldingen* (p. 90) and the *Seelisberger Seeli* (p. 88; each 3¹/₂-4 hrs.) are rough and should not be attempted by novices. — The summit commands an imposing and highly picturesque view of the Lake of Lucerne, in its whole length from Lucerne to Flüelen, of the Uri-Rothstock, Bristenstock, Tödi, Scheerhorn, and Windgällen, and of the Reuss valley as far as Amsteg. The distant view is more limited than that from the Rigi, as we are much nearer the lofty mountains.

The **Buocheer Horn** (6260') may be ascended in 3¹/₂ hrs. from Deckenried or Buochs (guide desirable; fine view). Descent to (1¹/₄ hr.) *Nieder-Rickenbach* (p. 128) and viâ *Büren* to (2 hrs.) *Stans* (p. 127).

On the opposite bank, on a fertile strip of land between the *Vitznauer Stock* and the *Hochfluh*, lies the pretty village of —

Gersau. — Hotels. *Hôt.-Pens. Müller, R., L., & A. 3-5, D. 3¹/₂, pens. 7-10 fr. (depot of the S.B.G.H.); Seehof, on the lake, ¹/₄ M. to the E.; *Hof Gersau, R., L., & A. 1¹/₂-2, B. 1, D. 2, pens. 5-6 fr.; *Hôtel-Pens. Beau-Séjour, R., L., & A. 2, B. 1, D. 2, pens. 4¹/₂-5¹/₂ fr.; Bellevue; Hirsch; Sonne; Zur Ilge, plain. Furnished Rooms at *Müller's zur Säge* and at *Waad's*. — *English Church Service*.

Gersau, in **a sheltered situation in** the midst of orchards, with its broad-eaved cottages scattered over the hillside, is a pleasant resort of invalids and others. In the ravine behind it is a silk-factory. and on the mountain above is the *Rigi-Scheidegg Hotel* (p. 100).

The ascent of the *Rigi-Hochfluh* (5564') from Gersau, viâ the *Zihlistock-Alp* in 3-3¹/₂ hrs., is attractive. The last part of the route has been improved (see p. 100). From the Hochfluh to the Scheidegg, 1¹/₂-2 hrs. — The *Vitznauer Stock* (4775') may be ascended in 2¹/₂ hrs. from Gersau or Vitznau viâ *Ober-Urmi;* the last ¹/₂ hr.'s climb is toilsome (comp. p. 88). — From Gersau to (4¹/₂ M.) *Brunnen* (p. 90) a pleasant walk by the road skirting the lake (Axenstrasse).

The chapel on the bank to the E. of Gersau is called *Kindlimord* ('infanticide'). To the E. rise **the bare peaks of the** *Myten*, at the base of which, 3 M. inland, lies *Schwyz* (p. 109); nearer is the church of *Ingenbohl*, and to the right the broad *Fronalpstock*.

The steamer now crosses **to Treib (***Inn***,** rustic), in Canton Uri, at the foot of the precipitous *Sonnenberg*, the landing-place (telephone) for the village of **Seelisberg** (2628'; *Hôt.-Pens. Bellevue*, 5 fr.; *Pens. Aschwanden*, behind the church, 5 fr., unpretending; *Pens. Löwen*) on the hill above, to which a road leads in 1¹/₂ hr. through the orchards of *Folligen* (omnibus four times daily in 1 hr., up 2, down 1¹/₂ fr.; one-horse carr. 5, two-horse 10, to the Curhaus 6 or 12 fr., with fee of 2 fr.). The more direct footpath ascends to the left behind the Inn (1 hr.; stony but shady part of the way). By the *Chapel of Maria-Sonnenberg* (2770'), 12 min. from the church of Seelisberg, is the *Pension Grütli* (5-7 fr.), and 100 paces farther on is the little *Hôtel Mythenstein*, beyond which is the *Curhaus **Sonnenberg-Seelisberg** (three houses, with 350 beds; R. from 2, board 7, A. ¹/₂ fr.; Engl. Ch. Serv. in summer), a sheltered spot with pure mountain air, and a favourite health-resort. The terrace in front of the Curhaus commands a beautiful *View of the lake of

Uri lying far below and of the surrounding mountains **from the Myten to the Uri-Rothstock.**
An attractive walk may be taken to (25 min.) the *Schwendifluh (2723'), by a route diverging to the left from the Beroldingen road (guide-post) about 1 M. to the S. of the Curhaus. The view from the top of the perpendicular rocks, the *Teufelsmünster* of Schiller ('Wilhelm Tell', Act IV., Sc. 1), is highly picturesque. — Beautiful view from the *Känzeli* (3303'; in the wood to the right at the S. end of the Curhaus, ¹/₂ hr.), over the lake and the plain as far as the Weissenstein. — About 20 min. to the S.W. of the Curhaus lies the picturesque little *Seelisberger See*, or '*Seeli*' ('little lake', 2470'; with bath-house, 50 c.) on the precipitous N. side of the *Niederbauen* or *Seelisburger Kulm* (ascent, see p. 88).
Those who desire to walk from Seelisberg to *Bauen*, on the Lake of Uri, and thence to cross the lake to Tell's Platte or Flüelen, go straight on from Sonnenberg (finger-post; the path to the Schwendifluh leads to the left) to (³/₄ hr.) the little château of *Beroldingen* (beautiful view) and thence by a safe, though steep and rather uncomfortable path to (1/₂ hr.) *Bauen* (Tell, plain). Boat from Bauen to Tellsplatte 2, Rütli 3, Flüelen 4 fr. (higher charges at the 'Tell'). — Path to the (¹/₂ hr.) *Rütli*, see p. 92.

Opposite Treib, on the E. bank, lies the large village of —

Brunnen. — °WALDSTÄTTER HOF, on the lake, with baths, R., L., & A. from 2, déj. 3, D. 4, S. 2¹/₂, pens. 8-11, in spring 7-9 fr. (concerts in the large entrance-hall); *HÔT.-PENS. AUFDERMAUER AU PARC, ¹/₄ M. from the lake, R., L., & A. from 2, B. 1¹/₂, déj. 2¹/₂, D. 4, pens. from 7 fr.; °HÔT.-PENS. ADLER, R., L., & A. from 2, B. 1¹/₄, D. 3¹/₂. S. 2¹/₂ fr.; °HÔT.-PENS. HIRSCH, at the quay, R. 2-3, pens. 7-10 fr., both by the pier; *HÔT.-PENS. BELLEVUE (R., L., & A. from 2, D. 3, pens. 5-7 fr.) and PENS. MYTHENSTEIN (6 fr.), same prices, both on the Axenstrasse, close to the lake; °HÔT.-PENS. SCHWEIZERHOF, with restaurant, R., L., & A. from 2, B. 1¹/₄, D. 3, pens. 5-6 fr.; RÖSSL, R., L., & A. from 1¹/₂, B. 1, D. 3, pens. 5-6 fr.; BRUNNER-HOF, all three near the quay; *HÔT.-PENS. RIGI, on the Gersau road, R. L., & A. 2, D. 3, pens. 5 fr.; °HÔT.-PENS. GÜTSCH, with fine view, R., L., & A. 2, D. 2¹/₂, pens. 5-6 fr.; °PENS. DU LAC, ¹/₄ M. to the W. of the village, with lake-baths, R. 1³/₄, board 5-5¹/₂ fr.; HÔTEL-PENSION ST. GOTTHARD, near the rail. station, pens. 5-7 fr.; HÔT. BAHNHOF, EGW, °FREIHOF, °SONNE, °RÖTLI, and others, plain (pens. 5-6 fr.). Furnished roms at *Villa Schoeck*, above the Gütsch, etc. — *Zur Drossel Beer-Garden*, on the quay; beer also at *Kleis's*. — Confectioner, *Jos. Nigg*, Bahnhof-Str. Preserves, chocolate, etc., at *Fassbind's*, near the Adler (telephone to Lucerne).
ROWING BOATS: to Treib and back with one boatman 1 fr., with two 2 fr.; Rütli (and back) 2¹/₂ or 4, Tellsplatte 3 or 6, Rütli and Tellsplatte 5 or 8 fr.
BATHS (warm and lake-baths) at the **Waldstätter Hof** (lake-bath and towel, 50 c.). — Wood-carvings, photographs, books, newspapers, etc., at *Leuthold's*, by the steamboat-pier.
BOOK DEPÔTS of the *Bibliothèque des Grands Hotels* (p. xviii) at the Waldstätter Hof and the Hôtels Adler, Axenfels, Axenstein, Frohnalp, Stoos, Sonnenberg-Seelisberg, etc.
ENGLISH CHURCH SERVICE at the Waldstätter Hof.

Brunnen, the port of Canton Schwyz, a station on the St. Gotthard Railway (p. 110), and one of the most beautiful places on the lake, **is** partly situated in a **flat** valley near the mouth of the Muota. In **the background rise** the two Myten. The old Susthaus, or goods-magazine, is decorated with quaint frescoes. New Protestant Church on the Schwyz road, opposite the railway-station.

The Gütsch (1700'; hotel, see above), a hill behind Brunnen, overlooks the two arms of the lake and the pretty valley of Schwyz. Shady walks in the neighbouring woods. — FROM BRUNNEN TO MORSCHACH a good road (in shade in the morning) ascends in 1 hr. from the Axenstrasse,

LAKE OF URI. II. Route 27. 91

The shady footpath which diverges at the (³/₄ M.) guide-post to the left cuts off a long curve. 50 min. *Hôtel Axenfels (2055'; R. from 2¹/₂, D. 4, pens. from 7 fr.), with gardens, park, and a fine view. A few min. farther on is the charmingly situated hamlet of **Morschach** (2155'; *Hôt.- Pens. Frohnalp & Curhaus Morschach*, with garden and view, pens. 5¹/₂-8 fr.; *Pens. Bettschardt*, 5 fr.; *Pens. Degenbalm*, beautifully situated on an eminence 230' above the village, pens. 5-8 fr.). The road forks immediately behind the Pens. Hettschardt, the right branch leading viâ *Ober-Schönenbuch* to (4¹/₂ M.) *Schwyz*, while the left branch ascends past the *Pens. Rütliblick* (fine view) to (10 min.) the °**Grand Hôtel Axenstein** (2460'; R. 9-12, D. 4-6, board 7 fr.; less in June and Sept.), splendidly situated on the *Brändli*, with a magnificent **Survey of both arms of the lake. Large covered promenade and beautiful shady grounds close to the hotel, containing numerous erratic blocks and interesting traces of glacier-action. Strangers are admitted to the park, but if residing at the Hôtel Axenfels or at Morschach only by special permission. Besides the road, there is a path from the Gütsch to the hotel, for the most part in shade (³/₄ hr.). Adjacent is an *English Church* (*All Saints*). Omnibuses run between the Axenstein Hotel and Brunnen (50 min., 2 fr.; one-horse carr. 5, two-horse 10 fr.).

The **Stoos** (4230'), the N. spur of the Fronalp (°*Curhaus*, R., L., & A. 3¹/₂-4¹/₂, B. 1¹/₄; pens. 8-12, in June and Sept. 7-10 fr.; *Pens. Balmberg*, 5-6 fr.), another good point of view (best from the *Stooshorn*, 5 min. to the N.), with varied walks, is reached by a road (in shade in the morning for most of the way) from Morschach in 1³/₄ hr. (carr. and pair from Brunnen in 2¹/₄ hrs., 20 fr., there and back 25-30 fr.; with one horse 15 fr.; riding-horse 10, porter 5 fr.). — The **Fronalpstock** (6305'; small *Inn*, ten beds), 1¹/₂ hr. to the S.W. of the Stoos, reached by a rough path (milk at a chalet halfway), affords a magnificent view, hardly inferior to that from the Rigi, of the Alps and of the entire Lake of Lucerne. — A footpath leads from the Stoos to (1¹/₂ hr.) *Ried* (p. 73) in the *Muota-Thal*, at first traversing meadows, but beyond the *Stoosbach* descending in steep zigzags through wood to the bridge over the Muota.

Other excursions from Brunnen: by the St. Gotthard Railway **to** (12 min.) Schwyz-Seewen, and then by boat (in 25 min. from Seewen) to the island of Schwanau in the Lake of Lowerz (p. 109); to the Muota- Thal as far as the (1³/₄ hr.) Suvoroff Bridge (p. 72), viâ Ingenbohl, Unter- and Ober-Schönenbuch, and back on the right bank viâ Ibach or Schwyz in 2¹/₄ hrs.; by the Axenstrasse (see below) to Tellsplatte and Flüelen (9 M.; best by carr., the road being shadeless after 10 a.m.; to Flüelen with one horse 8 fr.); to the Kindlimord Chapel (p. 89) and Gersau (1¹/₂ M.; p. 89); to the Rütli (see below; boats, see p. 90), and thence, or viâ Treib, to Seelisberg (p. 89); ascent of the Rigi (p. 94; 1 day); by the St. Gotthard Railway to Göschenen-Andermatt and back (R. 32; 1 day).

At Brunnen begins the S. arm of the lake, called the *Urner See* or *Lake of Uri*. The **mountains now rise** very abruptly, and the lake **narrows**. Lofty peaks, often snow-clad, peep through the gorges which **open** at intervals; conspicuous among these is the mighty Uri-**Rothstock** with its glacier. By the sharp angle which juts into the lake from the W. bank rises the *Mytenstein*, a pyramid of rock, 80' **high**, bearing an inscription in huge gilded letters to the memory **of Schiller**, the 'Bard of Tell'. A little farther on, below Seelisberg (p. 89), and 8 min. above the lake, are the three springs of the **Rütli**, or *Grütli*, trickling from an artificial wall of stone, in the midst of an open space planted with trees. This spot, with the adjacent timber-built guard-house in the old Swiss style (refreshments) and pretty grounds, belongs to the Confederation. At a fine point of view, 5 min. to the W., is a block of granite, 10 ft. high, with bronze medallions, commemorating the author (J. G. Kraus,

1792-1845) and the composer (Jos. Greith, 1798-1869) of the Song of Rütli.

On this spot, on the night of 7th Nov., 1307, thirty-three men, from Uri, Schwyz, and Unterwalden, assembled and entered into a solemn league for the purpose of driving their oppressors from the soil. Tradition relates that these three fountains sprang up on the spot where the three confederates, *Werner Stauffacher* of Steinen in Schwyz, *Erny* (Arnold) *an der Halden* of Melchthal in Unterwalden, and *Walter Fürst* of Attinghausen in Uri, stood when the oath was taken. — A shaded path ascends in 1¼ hr. from the Rütli to the *Curhaus Seelisberg* (p. 89). Small boat from Brunnen to Rütli, see p. 90; an excursion by boat (3-4 fr.) from *Treib* is also attractive.

On the E. bank of the lake runs the *Axenstrasse, leading from Brunnen to (9 M.) Flüelen, and remarkable for the boldness of its construction, being to a great extent hewn in the rock. It is the joint creation of the cantons of Uri and Schwyz (1863-65). Below, alongside, or above the road, runs the *St. Gotthard Railway* (p. 110), skirting the lake in a succession of tunnels and cuttings. About ¼ hr after leaving Brunnen the steamer touches at *Sisikon* (Pens. Urirothstock), at the entrance to the narrow *Riemenstalden-Thal* (p. 73).

From the hamlet of (1½ hr.) *Riemenstalden* (3410'; *Inn*) the *Rophaien* (6830'), commanding a fine view of the Lake of Lucerne, especially good by morning light, may be easily ascended in 2½ hrs. The descent may be made by a path, obvious beyond the *Buggisgrat*, to (2¼ hrs.) Tell's Chapel or to (3 hrs.) Flüelen. The *Rossstock* (8080'; 3½-4 hrs.), also with a charming view, is another easy ascent from Riemenstalden (comp. p. 111). The *Liedernen* or *Kaiserstock* (8255'; 4-4½ hrs., with guide) should be attempted only by experienced mountaineers not subject to dizziness. — Over the *Katzenzagel* to the *Muota-Thal*, see p. 73.

We next reach stat. Tell's Platte (*Restaurant*, with baths, at the landing-place), 3 min. above which, on the Axenstrasse, is the *Hôt.-Pens. Tellsplatte* (1680'; pens. 6 fr.), with pleasure-grounds and a charming view. A little to the S. of the landing-place (path in 1 min.) is a ledge of rock at the base of the *Axenberg*, where, shaded by overhanging trees and washed by the lake, stands the romantic **Tell's Chapel**, rebuilt in 1880, and adorned with four frescoes by Stückelberg of Bâle. It is said to have been originally erected by Canton Uri in 1388 on the spot where the Swiss liberator sprang out of Gessler's boat. On Friday after Ascension Day mass is performed here at 7 a.m., and a sermon preached, the service being attended by the inhabitants of the neighbourhood in gaily decorated boats. Near the chapel the lake is upwards of 700' deep.

The grandest part of the Axenstrasse is between Tell's Platte Inn and Flüelen (2½ M.), where it pierces the curiously contorted limestone strata of the *Axenfluh*, 360' above the lake, by means of a tunnel. Beyond the chapel, Flüelen (which the steamer reaches in ¼ hr. more) becomes visible. The scenery of this part of the lake is very striking. Opposite the chapel, on the W. bank, lies the hamlet of *Bauen* (Tell; p. 90), and, farther on, the dynamite-factory of *Isléten* (now abandoned), at the mouth of the *Isenthal* (p. 93).

Flüelen. — Hotels. *Tell & Post, R. 2, B. 1, D. 3 fr. ; Adler, R., L., & A. 2-3, B. 1¼, D. 3, pens. 7-9 fr.; *St. Gotthard, R., L., & A. 1½-2, B. 1, D. 2, pens. 4½-5 fr.; *Kreuz, R., L., & A. 2½-3, B. 1¼, D.

3-4 fr.; *HIRSCH, R. 1-2, B. 1, D. 1½-2½, pens. 4-6 fr., all on the lake; FLÜELERHOF; STERN. — *Rail. Restaurant* (beer-garden). — *Baths* in the lake, ½ M. to the N. of Flüelen.

Flüelen is the port of Uri and a station (close to the pier) on the St. Gotthard Railway (p. 110). Beyond the church is the small château of *Rudenz* which once belonged to the Attinghausen family. The *Reuss*, which falls into the lake between Flüelen and *Seedorf*, has been 'canalized' here to prevent inundations (½ hr.'s walk, or ¼ hr. by boat to its influx).

The Isenthal (see Map, p. 126) may be reached from Flüelen or Altdorf on foot in 3 hrs. viâ *Seedorf* (see above), by a path skirting the lake and ascending to the *Kreuzhöhe* (1860'), with a picturesque view, where the path turns to the left into the valley; or by the steamer from Flüelen, which touches at Isleten twice daily. These two routes unite at the Kreuzhöhe. The pleasantest and shortest route is by row-boat or sail-boat (1½ fr.) from the baths of Flüelen to the path from Altdorf along the W. bank (½ hr.). From *Bauen* (p. 92) a pleasant path, affording splendid views of the lake, ascends round the slope of the *Furkelen* direct to Isenthal in 1½ hr. — About 1 hr. from Isleten we reach the prettily situated village of Isenthal (2152'; *Gasser's Inn*, three beds, rustic but clean; guides, *Albin Imfanger, Mich.* and *Joh. Gasser, Andreas, Josef*, and *Jost Aschwanden*), at the S. base of the precipitous *Oberbauen* or *Schynsgrat* (6955'), which may be ascended viâ the *Bauberg* in 3½-4 hrs. (recommended to adepts; guide necessary). The valley divides here into the *Grossthal* to the right and the *Kleinthal* to the left. Through the GROSSTHAL, in which lies the Alpine hamlet of (¾ hr.) *St. Jakob* (3215'), we may either proceed to the W., passing over the Schönegg Pass (6315'), between the *Hohe Brisen* (7690') and the *Kaiserstuhl* (7877'), to *Ober-Rickenbach* and (5½ hrs.) *Wolfenschiessen* (p. 128); or to the S.W., over the Rothgrätli (8420'), between the Engelberger-Rothstock and the *Hasenstock*, to (10 hrs.) *Engelberg* (p. 128). The *Engelberger-Rothstock* (9250') may be ascended without difficulty from the Rothgrätli in ¾ hr. (comp. p. 129). Viâ the *Jochli* and the *Bühlalp* to (4½-5 hrs.) *Nieder-Rickenbach*, see p. 128.

Through the KLEINTHAL leads the shortest route to the summit of the Uri-Rothstock (6-8½ hrs.; not easy; guide 15, or with descent to Engelberg 25 fr.). A fatiguing path leads to the *Nefen-Alp* and (2 hrs.) *Musen-Alp* (4885'; night-quarters in the chalet); then a toilsome ascent across two torrents and along precipices of slate-rock to the upper snow-fields of the *Kleinthal Glacier*, to the E. of the *Kesselstock* (8455'); next an ascent in sweeping curves over the névé to the (4½ hrs.) arête separating it from the Blümlisalp Glacier (striking view of the Bernese Alps); lastly by an obvious path over slopes of rubble to the (¼ hr.) summit of the *Uri-Rothstock* (9620'). An easier, but longer route through the Grossthal, passing *St. Jakob* (see above) and the *Schlossfelsen*, ascends by a steep and rough path to the (3 hrs.) *Hangbaum-Alp* (5660'), grandly situated (fine cascades), where the night is spent (hay-beds); thence (starting early in the morning) over pastures, loose stones, and along the N. edge of the *Blümlisalpfirn* to the ridge between the Grossthal and Kleinthal; and lastly up the arête towards the W. to the summit (3½-4 hrs. from Hangbaum), which is usually free from snow in summer. The mountain-group which culminates in the Uri-Rothstock and the Brunnistock (9683'), like the Titlis, is almost perpendicular on the E. and S.E. sides (towards the Gitschen-Thal and Surenen), and is composed of gigantic and fantastically contorted limestone rocks. The *View from the summit is exceedingly grand: to the S. the chain of the Alps, from the Sentis, Rhätikon, and Bernina on the E. to the Diablerets on the W.; at our feet, 8000' below, the Lake of Lucerne and the entire Schächen-Thal; to the N.E., N., and N.W. the Myten, Rossberg, Rigi, Pilatus, and the Entlebuch Mts., the lower hills of N. Switzerland, and the plains of S. Germany. — Easy descent by the Blümlisalp Glacier, the *Schlossstock-Lücke*, and the *Rothstock-Lücke* to the (3hrs.) *Plankenalp Club-Hut*, and to (3hrs.) *Engelberg* (p. 129).

28. The Rigi.

The **Mountain** Railways which ascend the Rigi from Vitznau and from Arth are now used by the vast majority of travellers who visit this admirable and justly famous point of view. The journey is further facilitated by the numerous trains and steamboats which connect Arth and Vitznau with places both near and distant, so that a visit to the Rigi and back may now be accomplished easily from Lucerne or Zürich in one day. The ascent from *Vitznau*, which is more convenient for many travellers, affords beautiful views all the **way**, while that from *Arth* offers the advantage that the view bursts **upon the** spectator far more strikingly as he approaches the top.

Both lines are constructed on the rack-and-pinion system. The gauge is of the usual width. Between the rails runs the toothed rail, which consists of two rails placed side by side and connected with cross-bars at regular intervals. Into the spaces thus formed works a cog-wheel under the locomotive, which is always placed below the passenger-car. The maximum gradient of the Vitznau line is 1:4, and of the Arth line 1:5. Each train on the Vitznau line consists of one carriage only, with 54 **seats**, not divided into classes, and, on the Arth line, of two carriages holding 40 persons each. The average speed is 4-6 M. per hour.

The Footpaths to the top of the Rigi are now very little used, but the *Descent to Weggis* on foot (2-2½ hrs.; see p. 96) is recommended.

Hotels. On the Kulm (p. 97), *SCHREIBER'S RIGI-KULM HOTELS (three houses, the two higher and older being now dépendances of the lower; Restaurant on the groundfloor of the last); high charges, R., L., & A. 4-7, déj. 4, D. 5, pens. 12-14 fr. — On the Rigi-Staffel (p. 95), where all the **routes converge,** ½ hr. below the Kulm, 'HÔT.-PENS. RIGI-STAFFEL, R., L., & A. 3-3½, D. 4, S. 3, pens. 7½-9 fr., adapted for a stay of some time; HÔTEL STAFFEL-KULM and HÔTEL RIGIBAHN, both immediately above the station, **moderate.** — The *CURHAUS RIGI-KALTBAD (p. 95), 1½ hr. below the Staffel, **to the W.,** is a large, first-class establishment (pens. from 9 fr., cheaper in June and September; hot and cold baths; Engl. Church Service); *BELLEVUE, below stat. Kaltbad, D. 3½, **pens.** from 7 fr. — *HÔTEL RIGI-FIRST, on the Scheidegg railway **(p. 99),** ¼ hr. from the Kaltbad, pleasant for some stay, R. 2½-6½, L. ¾, **A.** ½, **D.** 4½, pens. from July to Sept. 8½-13½ fr., earlier or later in the **season** 8-11½ fr. (depot of the S.B.G.H.). — *SONNE and *SCHWERT, by **the** *Klösterli* (p. 96), R., L., & A. 1½-2½, D. 2½-3, pens. 5-6½ fr.; KRONE. — **Pens.** RIEDBODEN, between the Klösterli and the Staffel, 4 fr. — *HÔT.-PENS. RIGI-FELSENTHOR (p. 96), 10 min. from stat. *Romiti-Felsenthor* (p. 95), pens. 5-5½ **fr.** HÔTEL-PENS. GRUBISBALM, ¼ hr. from stat. *Freibergen* (p. 95), unpretending. — *HÔTEL-PENS. RIGI-UNTERSTETTEN, near stat. Unterstetten (p. 100), plain, pens. 5-6 fr. — *CURHAUS RIGI-SCHEIDEGG **(p.** 100), R. 3-5, D. 4, B. 1½, S. 2½, pens. **in** July and August 7-12, in **June** and Sept. 7-10 fr. (Engl. Ch. Serv.).

The ****Rigi** (5905', or **4470'** above the **Lake** of Lucerne; originally 'die Rigi', *i.e.* the strata), a group of mountains about 25 M. in circumference, lying between the lakes of Lucerne, Zug, and Lowerz, is chiefly composed of conglomerate (p. 109), while the N. and W. sides belong to the meiocene formation. The N. side is precipitous, but the S. side **consists of** broad terraces and gentle slopes, covered **with fresh green** pastures which **support** upwards of 4000 head of **cattle,** and planted towards the **base with** fig, chestnut, and almond **trees.** Owing to its isolated situation, **the** Rigi commands a most extensive view, 300 M. in circumference, and unsurpassed for beauty **in** Switzerland. The mountain was known to a few travellers during the latter part of the 18th cent., but it was not till after the peace of 1815 that it became a resort of tourists. In 1816 a very modest inn was erected on the Kulm by public subscription, and in 1848 this was superseded by the oldest of the three houses on the summit. Since then the number of inns has been steadily increasing, and the Rigi is now one of the most popular of Swiss resorts.

FROM VITZNAU TO THE RIGI-KULM, 4½ M., MOUNTAIN RAILWAY in 1 hr. 20 min., fare 7 fr. (to Kaltbad 4½, Staffel 6 fr.); descent also in 1 hr. 20 min., fare 3½ fr.; 10 lbs. of luggage free. First-class return-tickets from Lucerne to the Rigi via Vitznau 13½ fr.; Sunday tickets 7 fr. Return-tickets do not permit of an alternative return-route; *e.g.* holders of tickets from Vitznau may not return to Arth, or vice versâ.

Vitznau (1443'), see p. 87. The station is close to the quay. The train (views to the left) ascends gradually through the village (1 : 15), and afterwards more rapidly (1 : 4), skirting the precipitous slopes of the *Dossen*. A *View of the lake is soon disclosed, becoming grander as we ascend. Opposite us first appears the dark Bürgenstock, then the Stanser Horn, Pilatus, and Lucerne. Farther up, the Alps of Uri, Engelberg, and Bern come in sight above the lower mountains. The train (20 min. after starting) penetrates a tunnel 82 yds. long, crosses the *Schnurtobel*, a ravine 75' deep in which the *Grubisbach* flows, by a bridge borne by five iron pillars, and soon reaches the watering and passing station of *Freibergen* (3333'), beyond which the line is double. Stat. *Romiti-Felsenthor* (3890'; comp. p. 96) and (54 min. from Vitznau) —

2¾ M. Rigi-Kaltbad (4700'); to the left is the large *Curhaus* (p. 94), with its covered promenade, a health-resort on a plateau sheltered from the N. and E. winds.

A path leads through a narrow opening in the rock, to the left of the hotel, to (5 min.) **St. Michael's Chapel**, the walls of which are hung with numerous votive tablets. One of these on the left side records that two pious sisters sought refuge here from the persecutions of a governor of the district in the time of King Albert, and built the chapel. The spring (42° Fahr.) which bubbles forth from the rock adjoining the chapel was formerly called the 'Schwesternborn'.

A level path among the blocks of conglomerate near the chapel, and afterwards traversing park-like grounds, leads to the (¼ hr.) *Känzeli (4773')*, a pavilion on a projecting rock, commanding an admirable view of the snow-mountains, and of the plain towards the N. with its numerous lakes, similar to that from the Staffel, but with a more picturesque foreground. — A path leads hence to the Staffel in the same time as from the Kaltbad (50 min.), ascending to the right as far as the point where the S. part of the Lake of Lucerne becomes visible, and following the crest of the mountain until it joins the path from the Kaltbad, at the (½ hr.) Staffelhöhe.

Railway from the Kaltbad to the *Scheidegg*, see p. 99.

In 5 min. more the train reaches stat. *Staffelhöhe* (5090'), where the view towards the W. and N. is suddenly disclosed. It then ascends to the left, round the *Rigi-Rothstock*, in 9 min. to (4 M.) **Rigi-Staffel** (5270'), the junction of the Arth line (see p. 96).

The *Rigi-Rothstock (5455'), ¼ hr. to the S.W. (direct path from the Kaltbad in 35 min.), affords a very picturesque survey of the central part of the Lake of Lucerne, which is not visible from the Kulm. A clear view is often enjoyed from this point while the Kulm is enveloped in dense fog. The sunset is said to be sometimes seen in greater perfection from the Rothstock than from the Kulm, but the sunrise should certainly be witnessed from the latter.

The railway (here parallel with the Arth line) now ascends steeply
to the Kulm (in 7 min.; a walk of ½ hr.), skirting the precipices on
the N. side of the hill. 4½ M. Station *Rigi-Kulm* (5740'), see p. 97.

FROM ARTH TO THE RIGI-KULM, 7 M., MOUNTAIN RAILWAY in 1½ hr.,
fare 8 fr. 30 (to the Klösterli 5 fr. 50, Staffel 7 fr. 40 c.; from *Arth-Goldau*, on
the St. Gotthard Railway, to the Kulm in 1¼ hr., 8 fr.); descent in 1½ hr.,
4 fr. 80 c.; return-tickets from Arth 11½, from Arth-Goldau 11 fr.; 10 lbs.
of luggage free.

Arth (1345'; Rail. Restaurant), see p. 104. As far as Arth-
Goldau the line is of the ordinary kind. The train ascends gradually
to *Ober-Arth* (1490'), passes through the *Mühlefluh Tunnel* and
under the St. Gotthard Railway, and reaches (1½ M.) **Arth-Goldau**
(1683'), on the St. Gotthard line (p. 109), where the toothed-
wheel system begins, and where we change our direction (Seats
should if possible be secured at Arth on the left side, that
farthest from the waiting-room.) The Rigi line traverses part of
the scene of the Goldau landslip (p. 109), crosses the Schwyz road,
and describes a wide curve to the W.; then, ascending more rapidly,
it skirts the slope at the foot of the Scheidegg and reaches (2¾ M.)
stat. *Kräbel* (2513'), where the engine is 'watered'. Farther on,
ascending 1' in 5', we skirt the precipitous *Kräbelwand*, and obtain
a fine view of the valley and lake of Lowerz, with the island of
Schwanau, the Myten near Schwyz, **the Rossberg and scene of the
great landslip**, and the Lake of Zug. Beyond the *Rothenfluh Tunnel*
we are carried through a picturesque wooded valley, and across the
Rothenfluhbach, to the passing-station *Fruttli* (3730'). Still ascending
rapidly, the train traverses the *Pfedernwald*, crosses the *Dossen-
bach* and (beyond the *Pfedernwald Tunnel*) the *Schildbach*, and
reaches (5 M.; 1¼ hr. from Arth) —

Stat. **Rigi-Klösterli** (4320'), lying in a basin enclosed by the
Rigi-Kulm, the Rothstock, and the First. The 'Klösterli' is a small
Capuchin monastery and hospice, with the pilgrimage-chapel of *Maria
zum Schnee*, founded in 1689 and rebuilt in 1712, and the inns already mentioned (p. 94). The chapel is much visited by pilgrims,
especially on 5th Aug. and 6th Sept.; and on Sundays there is mass
with a sermon for the herdsmen. This spot has no view, but is
sheltered, and the air is often quite clear while the Kulm, Staffel, and
Scheidegg are shrouded in mist. Walk from the Klösterli to the Rigi-
First 20 min., Unterstetten ½ hr., to the Staffel, the Rothstock, or the
Schild ¾, to the Dossen or Kulm 1¼ hr., to the Scheidegg 1½ hr.

At (6¼ M.) stat. **Rigi-Staffel** (p. 95) a strikingly beautiful
view is suddenly disclosed towards the W. and N. (comp. p. 94).
From this point to the (7 M.) *Rigi-Kulm*, see above.

Foot and Bridle Paths to the Rigi (comp. p. 94). FROM WEGGIS (p. 87) a
bridle-path (3¼ hrs.), which cannot be missed (finger-post 5 min. from
the landing-place), winds at first through productive orchards. It crosses
the track of a mud-stream which descended from the mountain in 1795,
taking a fortnight to reach the lake. 50 min. *Sentiberg Restaurant* (2643');
25 min. *Heiligkreuz-Capelle* (3150'); ½ hr. *Hôtel-Pension Felsenthor* (p. 94),

near the *Hochstein* or *Käsbissen*, an arch formed of two huge masses of conglomerate, on which rests a third block. (Stat. *Römiti*, a little higher up, see p. 95.) The path runs parallel to the railway part of the way. (³/₄ hr.) *Kaltbad*, see p. 95. This route commands beautiful views and is especially recommended for the descent (comp. p. 94).

From Küssnacht (p. 104) a bridle-path (3¹/₄ hrs.). From the Tell Fountain, in the middle of the village, we follow a lane to the E. and reach a finger-post indicating the good and easily followed path to the (1¹/₂ hr.) *Vordere Seeboden-Alp* (3372'; *Hôtel-Pension Seebodenalp, 5-7 fr.), a splendid point of view. About 5 min. farther on our path unites with those from Immensee and Tell's Chapel; 18 min., *Hintere Seeboden-Alp*. Then a steep zigzag ascent of 1¹/₄ hr. to *Rigi-Staffel* (p. 95).

From Goldau (p. 109), 3¹/₂ hrs., an excellent bridle-path, and not to be mistaken. To the W. of the railway-station we cross the *Aa*, and proceed to the left of the brook through meadows, pine-wood, and rocky débris, ascending by steps at places. To the left the precipitous slopes of the *Rothfluh* (5233'). 1 hr. Untere Dächli (3083'; *Inn*), where the path from Arth comes up on the right; good view of the valley of Goldau, the Lake of Lowerz, and the Myten of Schwyz. By the cross adjoining the tavern begin the thirteen stations or oratories which lead to the chapel of Our Lady of the Snow. At (20 min.) the *Obere Dächli* (rfmts.), with its fresh spring, the wood is quitted; on the opposite side of the valley runs the railway. This point is about halfway to the top; the second half (1³/₄ hr.) is easier. 10 min. *Malchus-Kapelle*, the 8th station; then (¹/₂ hr.) *Klösterli* (p. 96); thence to the *Rigi-Staffel* (p. 95) 40 min., to the *First* 20 min. (p. 99).

The **Rigi-Kulm** (5905'), the highest and northernmost point of the Rigi, descends abruptly on the N. to the Lake of Zug, while on the S.W. side it joins that part of the mountain which encloses the basin of the Klösterli and extends to the Scheidegg. At the top rises a wooden belvedere. The hotels (p. 94) stand about 130 paces below the summit, sheltered from the W. and N. winds.

The Kulm almost always presents a busy scene, but is most thronged in the morning and evening. The sunset is always the chief attraction. A performer on the Alpine horn blows the 'retreat' of the orb of day, after which the belvedere is soon deserted.

Half-an-hour before sunrise, the Alpine horn sounds the reveille. All is again noise and bustle; the crowded hotels are for the nonce without a tenant; and the summit is thronged with an eager multitude, enveloped in all manner of cloaks and mantles. Unfortunately a perfectly cloudless sunrise is a rare event.

A faint streak in the E., which gradually pales the brightness of the stars, heralds the birth of day. This insensibly changes to a band of gold on the horizon; each lofty peak becomes tinged with a roseate blush; the shadows between the Rigi and the horizon gradually melt away; forests, lakes, hills, towns, and villages reveal themselves; all is at first grey and cold, until at length the sun bursts from behind the mountains in all his majesty, flooding the superb landscape with light and warmth.

View. The first object which absorbs our attention is the stupendous range of the snow-clad Alps, 120 M. in length (comp. the Panorama). The chain begins in the far E. with the *Sentis* in Canton Appenzell, over or near which the first rays of the rising sun appear in summer. Nearer the Rigi rises the huge snowy crest of the *Glärnisch;* then the *Tödi*, in front of which are the *Clariden*, and to the right the double peak of the *Scheerhorn;* next, the broad *Windgälle*, immediately opposite, and the sharp pyramid of the *Bristenstock*, at the foot of which lies Amsteg on the St. Gotthard road; then the *Brunnistock* and the *Uri-Rothstock*, side by side, both so near that the ice of their glaciers can be distinguished; next, the broad *Schlossberg* and the serrated *Spannörter*, and more to the right the *Titlis*, the highest of the Unterwalden range, easily distinguished by its vast mantle of snow. The eye next travels to the Bernese Alps, crowning the landscape with their magnificent peaks clad with perpetual snow. To the extreme left is the *Finsteraarhorn*, the loftiest of all (14,025′); adjacent to it the *Lauteraarhorn* and the *Schreckhorn*, the three white peaks of the *Wetterhorn* (*Rosenhorn*, *Mittelhorn*, and *Wetterhorn*), the *Mönch*, the *Eiger* with its perpendicular walls of dark rock on the N. side, and the *Jungfrau* with the *Silberhorn*. To the W. tower the jagged peaks of the *Pilatus*, forming the extreme outpost of the Alps in this direction. — Towards the NORTH the entire *Lake of Zug* is visible, with the roads leadings to Arth, and the villages of *Zug*, *Cham*, *Risch*, and *Walchwyl*. To the left of the Lake of Zug, at the foot of the Rigi, stands *Tell's Chapel*, midway between Immensee and Küssnacht, a little to the left of a white house; then, separated from the Lake of Zug by a narrow strip of land, the Küssnacht arm of the Lake of Lucerne; more to the W. *Lucerne* with its crown of battlements and towers, at the head of its bay. Beyond Lucerne is seen almost the entire canton of that name and farther to the N. the canton of *Aargau*, with the *Emme* meandering through the open landscape like a silver thread; the *Reuss* is also visible at places. More distant are the *Lake of Sempach*, the W. side of which is skirted by the railway to Bâle, and the lakes of *Baldegg* and *Hallwyl*. — Towards the WEST and NORTH-WEST the horizon is bounded by the *Jura Mts.*, above which peep some of the crests of the Vosges. To the N., but to the left of the Lake of Zug, in the distance, rises the castle of *Habsburg;* still farther off is visible the *Black Forest*, with its highest peaks, the *Feldberg* (to the right) and the *Belchen* (to the left). Beyond the Lake of Zug is seen the crest of the *Albis* with the *Uetliberg*, which nearly conceals the Lake of Zürich; the long cantonal hospital and the cathedral in the town of *Zürich* are, however, visible. In the extreme distance rise the basaltic cones of *Hohenhöwen* and *Hohenstoffeln* (close together) and the *Hohentwiel* in Swabia. — Towards the EAST, behind the N. slope of the Rossberg, a glimpse is obtained of the *Lake of Ägeri*, on the S. bank of which was fought the famous battle of Morgarten (p. 80). Beyond Arth, opposite the Kulm, is the *Rossberg*, the S. slope of which was the scene of the disastrous Goldau landslip (p. 109). Between the Rossberg and the E. ramifications of the Rigi lies the

Lake of Lowerz, with its two little islands; beyond it, the town of *Schwyz*, at the foot of the bald heights of the *Myten*, overtopped by the imposing *Glärnisch*. To the right opens the *Muota-Thal*, celebrated in military annals. — To the SOUTH-EAST and SOUTH the different heights of the Rigi form the foreground: viz. the *Hochfluh* (below it the *Rothfluh*), *Scheidegg*, *Dossen*, and *Schild*, at the foot of which lies the Klösterli. To the left of the Schild part of the *Lake of Lucerne* is seen near Beckenried, and to the right the bay called the *Lake of Buochs*, with the *Buochser Horn* above it; a little more to the right the *Stanser Horn* with *Stans* at its base; nearer, the less lofty *Bürgenstock* and the *Rigi-Rothstock*. Beyond these, to the left, is the *Lake of Sarnen*, embosomed in forest, to the right, the *Bay of Alpnach*, connected with the Lake of Lucerne by a narrow strait formed by the *Lopperberg*, a spur of Pilatus. — Good panorama by *Keller*, upon which that annexed is based.

For a quarter of an hour before and after sunrise the view is clearest; at a later hour the mists rise and condense into clouds, frequently concealing a great part of the landscape. To quote the chamois-hunter in Schiller's Tell:

> 'Through the parting clouds only
> The earth can be seen,
> Far down 'neath the vapour
> The meadows of green.'

But the mists themselves possess a certain charm, surging in the depths of the valleys, or veiling the Kulm, and struggling against the powerful rays of the sun. The effects of light and shade, varying so often in the course of the day, are also a source of constant interest. In the morning the Bernese Alps are seen to the best advantage, and in the evening those to the E. of the Bristenstock. One whole day at least should be devoted to the Rigi. A visit may also be paid (on foot or by rail) to the Staffel (p. 95) and the Rothstock (p. 95), the Kaltbad (p. 95), the Klösterli (p. 96), or the Scheidegg (p. 100).

As the temperature often varies 40-50° within 24 hours, overcoats and shawls should not be forgotten. During the prevalence of the Föhn, or S. wind, the Alps seem to draw nearer, their jagged outlines become more definite, their tints warmer; and during a W. wind the Jura Mts. present a similar appearance. These phenomena generally portend rain.

FROM THE KALTBAD TO THE RIGI-SCHEIDEGG. — 4¼ M. RAILWAY (ordinary cars, without toothed rail) in 25 min.; fare 2 fr. 50, there and back 3 fr. 60 c.

Rigi-Kaltbad (4700'), see p. 95. The railway skirts the S. slope of the Rothstock, being hewn in the rock the greater part of the way, and ascends gradually to stat. **Rigi-First** (4747'; *Hotel*, see p. 94), which commands a beautiful view of the Lake of Lucerne, the Uri and Unterwalden Mts., and the Bernese Alps. The train now describes a wide curve round the N. slopes of the *Schilt* (5062'; 20 min. from the Hôtel Rigi-First), affording a pleasant view, towards the E., of the Myten, the Glärnisch, and the Alps of **Appenzell**. Beyond stat. *Unterstetten* (Hotel, see p. 94) we traverse the saddle of the hill and cross a bridge 55 yds. long, with a view to the N. and S. We pass through the *Weissenegg Tunnel*, 55 yds. long, cross the *Dossentobel* by a viaduct 84' high, and beyond the ridge which connects the Dossen with the Scheidegg, where a view towards the S. is again disclosed, reach *Unter-Dossen*.

Stat. **Rigi-Scheidegg**, 190' below the *Hotel & Curhaus* (5462') mentioned at p. 94. The view hence is less extensive than that from the Kulm, but it also embraces the principal mountains, and some points not visible from the Kulm (view-tower 70' high; panorama at the hotel). The plateau of the Scheidegg, about 1 M in length, affords a pleasant promenade which may be prolonged by the 'Seeweg' along the slope of the Dossen as far as Unterstetten. The *Dossen* (see below), commanding a splendid view, is $^3/_4$ hr. distant.

The *Rigi-Hochfluh* (5584') may be ascended in $1^1/_2$-2 hrs. from the Scheidegg, by a new path constructed by Dr. Stierlin-Hauser, which steadily follows the ridge, passing the *Gätterli* (pass from Gersau to Lowerz; 3720') and *Scharteggli* (4625'). In the couloir, on the N.W. side of the summit, an almost perpendicular iron ladder, 80' high, must be ascended (wire-rope railing, but steady head indispensable). This interesting ascent affords a most picturesque view of the Lake of Uri, the Alps of Uri and Schwyz, and the Glarner Alps. The older route ($2^1/_2$-3 hrs.), crossing the saddle towards the *Zihlistock-Hütte*, and then ascending among the rocks on the S. side, has also been improved and is preferable to the above-mentioned route on the N. side (see p. 89).

Paths to the Scheidegg. FROM GERSAU (p. 89) a bridle-path ($3-3^1/_2$ hrs.), steep at places. Beyond the village we cross the brook and ascend by a paved path between orchards and farm-houses; 40 min., the *Brand*; $^1/_2$ hr., a saw-mill, where we again cross the brook; 10 min., *Unter-Gschwend* (3200'; tavern); 10 min., *Ober-Gschwend* (3330'; halfway). To the right, the precipitous slopes of the *Hochfluh* (see above); below lies the little chapel of *St. Joseph*. We now turn to the left (to the right is the path to Lowerz viâ the *Gätterli*, see above) and ascend by the *Hasenbühl-Alp* and the *Krüselboden* to the sharp crest of the hill, where a view is suddenly disclosed of the Rossberg, the lakes of Lowerz and Zug, and the Curhaus of Rigi-Scheidegg.

FROM LOWERZ (p. 109) a bridle-path (3 hrs.), ascending towards the S. to the *Gätterli* (see above) and thence to the right over the ridge to the hotel.

FROM THE KLÖSTERLI (p. 96) a bridle-path ($1^1/_2$ hr.), ascending from the Schwert Inn to the ($^1/_2$ hr.) *Hôtel Rigi-Unterstetten* (see above), situated on the saddle between the Schild and *Dossen* (5510'), 40 min. below the summit, which commands the whole of the Lake of Lucerne and Canton Unterwalden. Descent viâ Unterdossen to Scheidegg in 40 minutes. Refreshments may be obtained at a chalet, halfway between Unterstetten and Scheidegg.

29. From Lucerne to Alpnach-Stad. Pilatus.
Comp. Map, p. 87.

BRÜNIG RAILWAY from Lucerne to ($8^1/_2$ M.) Alpnach-Stad in 27-32 min., (1 fr. 40, 1 fr., 70 c.; return-tickets 2 fr. 25, 1 fr. 60, 1 fr. 15 c.), see p. 131. — STEAMBOAT, 9 times daily in $^3/_4$-$1^1/_2$ hr. (6 times viâ Kehrsiten, thrice viâ Hergiswyl, twice direct viâ Stansstad), connecting at Alpnach-Stad with the Brünig and Pilatus Railways. Passengers with through-tickets may use as far as Alpnach either the Brünig Railway or the steamboat. — The ascent or descent by the PILATUS RAILWAY (p. 102) takes 1 hr. 25 min.; fares, up 10, down 6 fr.; return-fare for the first and the last train 12 fr.; combined tickets for railway and hotel, including R., D., and B. 25 fr.; Sunday tickets, valid in May and Oct. for the first, in June-Sept. for the first and second trains (return by any train) 9 fr. (from Lucerne 10 fr.).

The BRÜNIG RAILWAY to Alpnach-Stad, viâ Hergiswyl, see p. 131. — The STEAMBOAT steers towards the 'Kreuztrichter' (p. 87), keeping near the W. bank and passing the country-seat of *Tribschen*, the *Pension Stutz* (p. 82), the *St. Niklauscapelle*, and the station of *Kastanienbaum*, and enters the bay of Stansstad. To the left rises the *Bürgenstock*, with its precipitous N. slopes, at the N. E. angle of which lies the station of *Kehrsiten* (Restaurant).

A RACK-AND-PINION AND WIRE-ROPE RAILWAY ascends the *Bürgenstock from Kehrsiten in ¼ hr. (fares, up 1½, 1 fr., down 1 fr., 50 c.), traversing a distance of 1025 yds., with an average gradient of 45 : 100. The motive power is electricity, which is also utilized for pumping water and for purposes of lighting. At the top of the railway (2855'; 1420' above the level of the lake) is a *Restaurant* (high charges), with view-terrace, beside which is the *Park Hotel*; 3 min. farther to the S. the large *Hôtel Bürgenstock* (R. 2½-6½, B. 1½, D. 5, board 7 fr.; resident physician; Engl. Ch. Service; S. B. G. H.), a favourite health-resort, with extensive and shady grounds. The hotel and several points near it command beautiful views of the lakes of Lucerne, Zug, Sempach, and Baldegg, the Rigi, etc. A good path leads to the S.E. to (½ hr.) *Honegg* (2906'); another (lately improved) through wood to the N.E. to the (¾ hr.) *Hammetschwand* (3720'), the summit of the Bürgenstock, which descends abruptly to the Lake of Lucerne: striking view of the greater part of the lake, and of the lakes of Sarnen, Sempach, Baldegg, Hallwyl, and Zug, of the Rigi, Pilatus, Myten, Weissenstein, and of the Alps of Glarus and Unterwalden, and part of the Bernese Alps (Panorama 50 c.).

To the right the promontory of *Spissenegg* extends far into the lake, forming a bay which extends to the N. to *Winkel*. The steamer steers (except on the direct voyages, see p. 100) to the S.W. to Hergiswyl (*Hôt.-Pens. Rössli*, *Hôt.-Pens. Schweizerheim*, both moderate), at the foot of Pilatus (see p. 103), and then to the E. to **Stansstad** (1445'; *Hôtel Winkelried*, R. 3-4, B. 1¼ fr.; *Freienhof*, pens. 4-6 fr., well spoken of; *Rössli*; *Schlüssel*), the 'harbour of Stans'. The square pinnacled *Schnitz-Thurm* was erected by the Swiss in 1308 to vindicate their new-won independence.

Steam-tramway from Stansstad to *Stans* and cable-line thence to the top of the *Stanserhorn*, see p. 127. — From Stans to *Engelberg*, see R. 36.

WALK FROM STANSSTAD TO SARNEN, 3 hrs. The path skirts the lake for a short way, enters the Rotzloch, and at *Attweg* (*Inn), 2 M. from Stansstad, where there is a chapel in memory of Winkelried (pp. 19, 127), joins the *Stans and Sarnen Road* (no diligence). This road leads past the W base of the *Stanserhorn* (p. 127), and by *Röhren* to (2 M.) *St. Jakob*, a village with an old church, then across the *Mehlbach*, and through the *Kernwald* to (3 M.) *Kerns* and (1½ M.) *Sarnen* (p. 132).

The *Lopper*, the E. spur of Pilatus, extends far into the lake. At its base runs the Lucerne and Alpnach road, while the Brünig railway (p. 131) penetrates the hill by a tunnel. The brook opposite, which falls into the lake at Stansstad, has further narrowed the channel between the Lake of Lucerne and the **Lake of Alpnach** with its alluvial deposits, and the strait is now crossed by an embankment and a swing-bridge *(Acheregg-Brücke)*, which is opened for the passage of steamers. Within the bay of Alpnach rises the *Rotzberg* (2214'), crowned by a ruined castle of the same name (ascent from the Rotzloch ¾ hr.; view). The hill is separated from the *Plattiberg* by the **Rotzloch**, a narrow ravine. Portland cement factory (the dust sometimes very unpleasant). On the lake is situated *Hôtel-Pension Rotzloch*, with a sulphur-spring and grounds (pens. 4-5 fr.).

At the S.W. angle of the Lake of Alpnach lies **Alpnach-Stad** (1443'; *Hôt.-Pens. Pilatus*, R., L., & A. 1½-3½, D. 3½, B. 1¼, pens. 5-6 fr., with veranda and garden; *Rössli*, moderate; *Stern*), a station of the Brünig Railway and the **starting-point of the Pilatus Railway**.

*Pilatus (6995'), the lofty mountain to the S.W. of Lucerne, rises boldly in a rugged and imposing mass, almost isolated from the surrounding heights. The W. and N. portions belong to the canton of Lucerne, the E. and S. to Unterwalden. The lower slopes are clothed with beautiful pastures and forests, while the upper part consists of wild and serrated cliffs, from which its ancient name *Fractus Mons* (broken mountain) is derived. The names 'Fracmont', 'Frakmund', have in later times been occasionally applied to it, but the name Pilatus (probably from the tradition mentioned at p. 86) came into general use about the close of last century. The mountain is the popular barometer of the district; if the summit is free from clouds and fog in the morning, the weather cannot be depended on; but if shrouded in fog till midday, a clear evening may be expected. The flora of Pilatus is very rich, including nearly 500 species.

The names of the different peaks from W. to E. are the *Mittaggüpfi* or *Gnepfstein* (6300'), the *Rothendossen* (5833'), the *Widderfeld* (6817', the wildest), the *Tomlishorn* (6998', the highest), the *Gemsmättli* (6792'); to the S. the *Matthorn* (6693'); to the N. the *Klimsenhorn* (6205', which, seen from Lucerne, is the farthest W.); in the centre the *Oberhaupt*, then the *Esel* (6962', the most frequently ascended), and lastly the *Steigli-Egg* (6485').

The PILATUS RAILWAY (duration of journey and fares, see p. 100; best views to the right), constructed in 1886-88 by Col. Locher of Zürich, is nearly 3 M. long, with an average gradient of 42 : 100 and a maximum gradient of 48 : 100. The line rests throughout on a substructure of massive granite blocks and slabs, to which an upper framework of iron and steel is securely fastened with huge screws. The toothed rail has vertical teeth on both sides, into which two pairs of toothed wheels attached to the train work horizontally. The engine and the passenger-carriage (32 seats) form a single car with two axles.

The railway begins near the Hôtel Pilatus (1443'; see above), and immediately **ascends**, traversing orchards and afterwards wood. 21 min. **Wolfort** (2985'), a watering-station, immediately beyond which the train crosses a stone bridge, with a span of 82', across the gorge of the *Wolfort*; fine view of the Lake of Alpnach to the right. We then enter the *Wolfort Tunnel* (48 yds.), beyond which the line is carried along the stony slope of the *Risleten*, the most difficult portion of the railway to construct (gradient 48 : 100), and then traverse the *Lower* (56 yds.) and *Upper Spycher Tunnel* (106 yds. long; 3773' above the sea-level) to the (43 min.) **Aemsigen-Alp** (4130'), a passing-station with pumping-works which force water to the Pilatus-Kulm, 2355' above. The railway now ascends through wood on the edge of a gorge, crosses the *Mattalp* (to the right the *Steigli-Egg*, in front the Esel), turns to the N., and is next carried up the precipitous rocky slope of the *Esel* through four tunnels (48, 60, 50, and 12 yds. long). The terminus **Pilatuskulm** (6790') adjoins the former Hôtel Bellevue, now a dépendance of the large *Hôtel Pilatuskulm* (R., L., & A. 6-8, B. 2, luncheon 4, D. 5 fr.; restaurant on the groundfloor cheaper). The terrace commands a fine mountain view. — An easy path leads from the station to (6 min.) the summit of the *Esel, or *Etzel* (6962'), the chief point of view, with a spacious summit-plateau, surrounded by a parapet. The view surpasses that from the Rigi in grandeur and variety, the Bernese Alps in particular looming nearer and more massive (comp. the panorama). — A similar but less picturesque view may be enjoyed from the *Tomlishorn (6995'), the highest peak of Pilatus, to which a good path (varying views), skirting the slopes of the Oberhaupt and Tomlishorn and crossing the Tomlishorngrat (railings; no danger even for novices), leads from the Hôtel Pilatuskulm in 1/2 hr. (panorama by Imfeld). — Another new path, cut in the rocks, leads to the top of the *Matthorn* (6693'; from the Hôtel Pilatuskulm 2 hrs., there and back).

Pedestrians will find the ascent of Pilatus best made from *Hergiswyl* (p. 101), at the N.W. foot of the mountain. There is a bridle-path as far as the (3 1/2 hrs.) Hôtel Klimsenhorn, whence a footpath ascends to (40 min.) the Pilatuskulm. In front of the church we take the broader path to the left, and after 3 min. turn to the right, traversing orchards and meadows, and afterwards wood. At (1 hr.) the *Curhaus Brunni* (pens. 6 fr.), a health-resort, there is a terrace affording a fine view. After 1/2 hr. the path leads through a gate to the *Gschwänd-Alp*; 20 min. farther up, near a chalet (Inn, with beds), we pass through another gate and ascend in steep zigzags to the left, at first through beautiful pine-wood, and then across slopes of grass and debris, to (1 1/4 hr.) the *Hôtel Klimsenhorn*, situated on the saddle (5940'; 35' higher than the Rigi-Kulm) connecting the Oberhaupt with the (15 min.) "Klimsenhorn (6265'), which affords an extensive and picturesque prospect to the E., N., and W., from the Uri Mts. to the Lake of Neuchâtel. The view to the S. is hidden by the loftier peaks of Pilatus.

From the Hôtel Klimsenhorn a well-constructed zigzag path (iron railing higher up) ascends the steep slope of the *Oberhaupt*, to the (40 min.) *Kriesiloch*, an aperture in the rock resembling a chimney, 20' high, through which 52 easy steps ascend to the arête between the Oberhaupt and the Esel. The *View of the Bernese Alps is suddenly disclosed here. The path then leads in 4 min. to the Hôtel Pilatuskulm (p. 102).

The Pilatuskulm may also be reached by bridle-paths from *Alpnach-Stad* (4 1/2-5 hrs.; viâ the *Aemsigen-Alp* and *Mattalp*) and from *Alpnach* (p. 131; 4 1/2-5 hrs.; viâ the Alps of *Lüthoidsmatt*, *Schwändi*, and *Hinter-Frakmünd*). — From *Kriens* (p. 86) a path leads to (3 1/2-4 hrs.) the Hôtel Klimsenhorn, passing the château of *Schauensee*, and traversing the *Hochwald* and marshy pastures viâ the *Mühlenmäss-Alp* and *Frakmünd-Alp* (guide indispensable). Viâ the *Brändlen-Alp* (last part of the route very rough), see p. 86.

30. From Zug and Lucerne to Arth.
Comp. Maps, pp. 86, 94.
i. From Zug to Arth. Lake of Zug.

STEAMBOAT (in connection with the Zürich and Lucerne and the Rigi railways) in 50 minutes. (Quick train from Zug by Rothkreuz to Arth-Goldau in 48 min., ordinary in 1 hr. 40 min.

The **Lake of Zug** (1368'), 8 3/4 M. long, 2 1/2 M. wide, and 650' deep, is very picturesque. Its richly wooded banks rise gently to

104   *II. Route 30.*   KÜSSNACHT.

a moderate height, while to the S., above its azure waters, towers the Rigi, visible from base to summit. On the flat N. bank of the lake many remains of lake-dwellings have been discovered. *Zug*, see p. 79. Soon after the steamer has left the pier, Pilatus appears to the S.W., and then the Bernese Alps and the Stanserhorn to the left. On a promontory on the W. bank is the handsome new château of *Buonas*; on the E. bank lie the village of *Oberwyl* and the houses of *Ottersuyl* and *Eielenegg*. Looking back, we observe the church-tower of *Cham* (p. 80), rising above the plain. On the W. bank, farther on, the wooded promontory of **Kiemen** projects far into the lake. To the left of the Rigi-Scheidegg are the Frohnalpstock and the Rossstöcke. The steamer touches at *Lothenbach* on the E. bank, and then crosses to **Immensee** *(Hôt. Rigi)*, charmingly situated at the foot of the Rigi. (Rail. stat., see p. 109; omnibus to Küssnacht in 1/2 hr.) The steamer then steers diagonally across the lake to **Walchwyl** *(*Pens. Hürlimcnn*, with hydropathic, well situated, pens. 4 1/2 - 6 fr.; **Stern)*, on the E. bank. The mildness of the climate is indicated by chestnut-trees and vines. To the left lies *St. Adrian*, at the foot of the *Rossberg* (see p. 108), which on this side is clothed with wood and pasture. As Arth is approached, one of the Myten **of Schwyz (p. 110)** peeps from behind the Rossberg.

Arth (1345'; **Adler*, with garden on the lake; **Hôt. Rigi*) lies at the S. end of the lake, between the Rigi and the Rossberg, but not exposed to the landslips of the latter, the strata of which dip in another direction.

Arth-Rigi Railway, see p. 96. — From Arth to *Küssnacht* and *Lucerne*, see below.

ii. From Lucerne to Küssnacht and Arth.

STEAMBOAT from Lucerne to (8 M.) Küssnacht in 45-55 min.; POST-OMNIBUS from Küssnacht to (2 M.) stat. Immensee thrice daily in 25 min.; RAILWAY from Immensee to (5 M.) Arth-Goldau in 19 minutes. (From Lucerne by Rothkreuz to Arth-Goldau 55-75 min.; see pp. 80, 109.)

Departure from Lucerne, see p. 87. The steamer **touches at** *Pens. Seeburg*, rounds the promontory **of** *Meggenhorn* **(p. 87)**, and enters the *Bay of Küssnacht*. To the left, near stat. *VorderMeggen*, rises the picturesque château of *Neu-Habsburg*, **behind which peeps the ancient tower of** the castle of that name, once a frequent resort of the Emp. **Rudolph** when **Count of Hapsburg**, and **destroyed by the Lucerners in 1352.** The incident which induced **Rudolph** to present his horse to the priest is said to have occurred here (see Schiller's ballad, 'The Count of Hapsburg').

Stat. **Hinter-Meggen** (**Curhaus & Pens. Gottlieben*, suitable for some stay, prettily situated 1/4 M. from the lake, pens. 5 1/2 - 7 1/2 fr.). The steamer now crosses to *Greppen*, skirts the well-wooded slopes of the Rigi, and soon reaches —

8 M. **Küssnacht** (1395'; pop. 2940; **Hôtel-Curhaus Mon-Séjour*, with hydropathic, garden, and sea-baths, R. 1 1/2-2, déj. 2, D. 2 1/2,

pens. 5-6 fr.; *Hôt.-Pens. du Lac; *Schwarzer Adler; Rössli; Tell), a village prettily situated at the N. end of this bay of the lake, with a fine mountain-view. Omnibus to *Immensee* (p. 104) from the landing-place; one-horse carr. 3 fr. — Ascent of the *Rigi*, see p. 97

The *St. Michaelskreuz* (2615'), locally known as the 'Kleine Rigi', 1½ hr. to the N.W. of Küsnacht (easily reached viâ *Altikon*), commands a beautiful view of the lakes of Zug and Lucerne, the Alps and the hilly landscapes of N. Switzerland. Unpretending *Inn and chapel on the top. A more extensive view is enjoyed from the *Ochsencaldhöhe* (2685'), 5 min. from the inn. The St. Michaelskreuz may also be ascended by good roads from Rothkreuz (viâ Meierskappel in 1½ hr.), from Gisikon (in 1 hr.) and from Lucerne (viâ Adligenschwyl and Udligenschwyl in 3 hrs.).

The road ascends through the **'Hohle Gasse'** or 'hollow lane'; see Schiller's Tell), now half filled up, but still deserving the name at one point where it is shaded by lofty beeches. At the upper end of it, 1¼ M. from Küssnacht, to the left, is Tell's **Chapel** (1585'), rebuilt in 1834, marking the spot where the tyrant Gessler is said to have been shot by Tell. Over the door is a painting of the event, with an inscription. By the (½ M.) inn *Zur Eiche* the road divides. A few paces to the right is stat. *Immensee-Küssnacht* (p. 109). The road to the left descends to (¼ M.) the village of *Immensee* (p. 104).

31. From Zürich viâ Wädensweil to Arth-Goldau.
From Biberbrücke to Einsiedeln.

Comp. Maps, pp. 40, 86.

36 M. RAILWAY in 3-3½ hrs. This is the shortest route from the Lake of Zürich to the Rigi and the St. Gotthard Railway, as well as to Einsiedeln. — Railway from Rapperswil viâ *Pfäffikon* to Einsiedeln in 1 hr. 6 min. (see p. 42).

From Zürich to (15½ M.) *Wädensweil* (1348'), see p. 42. The line ascends the fertile slopes on the S. bank of the Lake of Zürich, commanding beautiful views of the lake, with the Curfirsten and Sentis in the background. 17½ M. *Burghalden* (1741'); 19½ M. *Samstagern* (2080'; Restaurant), junction of the line (to the left) to Rapperswil-Pfäffikon viâ *Wollerau* (p. 43). — Beyond (21 M.) **Schindellegi** (2483'; *Freihof*; *Hirsch*) we cross the brawling *Sihl*.

Diligence twice daily in ½ hr. to Feusisberg (2233'; *Curhaus Feusisgarten*), a health-resort, pleasantly situated, with fine view of the lake of Zürich and the Alps of Appenzell. — 1½ M. to the S.W. of Schindellegi (diligence twice daily in ½ hr.) is the whey-cure resort of Hütten (2428'; *Bär*; *Kreuz*), charmingly situated on the idyllic *Hüttensee*, at the foot of the wooded *Hohe Rhonen* (see below). — The *Dreiländerstein* (4127'), the highest point of the Hohe Rhonen, marking the boundaries of cantons Zürich, Zug, and Schwyz, may be reached from Schindellegi in 1 hr.; and the walk may be continued along the crest of the hill to the *Gottschalkenberg* (p. 106).

The line rounds the E. slopes of the *Hohe Rhonen* (see above), and approaches the *Alp*, which falls into the Sihl here. Towards the S. appear the Myten (p. 110). — Beyond (23 M.) Biberbrücke (2730'; *Post*), where the *Biber* falls into the Alp, the Glarus Mts., bounded on the left by the pyramidal Köpfenstock (6240'), form the background.

106   *II. Route 31.*   EINSIEDELN.   *From Zürich*

Pleasant excursion from Biberbrücke (by road 1½ hr.; damp footpath, to the right, about halfway, 1¼ hr.) to the top of the °**Gottschalkenberg** (3780'; *Hotel*, pens. 6-7 fr.), the W. prolongation of the *Hohe Rhonen* (p. 105), commanding a fine view of the Alps (finest from the *Belvedere*, 20 min. to the S.). The descent may be made to (2½ M.) *Ober-Ägeri* (p. 80), to (1½ hr.) *Richterswell* (p.42), or by *Menzingen* to (6 M.) *Zug* (p. 79).

FROM BIBERBRÜCKE TO EINSIEDELN, 3 M., branch-railway in ¼ hr. The train follows the narrow *Alpthal* (several cuttings and embankments, and a short tunnel).

FROM PFÄFFIKON (p. 43) BY THE ETZEL TO EINSIEDELN, 3½ hrs. A narrow road commanding fine views of the lake ascends in windings, past the *Pens. Lugeten*, to the (3 M.) pass of the **Etzel** (3145'; *Inn*), with the *Chapel of St. Meinrad*. The *Hohe-Etzel* (3610'; steep ascent of ½ hr. from the inn) is wooded, and commands no view, but the °**Schönboden** (3518'), ¾ hr. to the E., affords a splendid view of the lake, the Limmatthal as far as Baden, the Alps of Appenzell and Glarus, the Sihlthal and Alpthal, with Einsiedeln, the Myten of Schwyz, the Rossberg, and the Rigi; to the W. rises the Hohe Rhonen (p. 105). Travellers bound for Einsiedeln may descend from the Schönboden towards the S.W. direct to *Egg*, visible below, cross the *Sihl*, and join the road from the Etzel. — From the Etzel Inn the road descends to the (¾ M.) *Teufelsbrücke* (2755') over the Sihl. Thence 3¼ M. to Einsiedeln.

Einsiedeln (2770'; pop. 8512; *Pfau*, R. & A. from 2½, B. 1-1¼, D. with wine 3 fr.; *Sonne*; *Drei Könige*; *St. Catharina*; *Schwan*; *Restaurant Oechslin*, with rooms), or *Notre-Dame-des-Ermites (Monasterium Eremitarum)*, in a green valley, watered by the *Alpbach*, vies with Rome and Loreto in Italy, St. Jago de Compostella in Spain, and Mariazell in Styria as one of the most famous pilgrim-resorts in the world.

Its foundation is attributed to Count Meinrad of Sulgen, who built a chapel here in honour of a wonder-working image of the Virgin presented to him by the Abbess Hildegard of Zürich. After the death af Meinrad, who was assassinated in 861, a monastery of Benedictine Hermits ('Einsiedler') sprang up here. In 1274 it was created an independent principality by Emp. Rudolph of Hapsburg, and owing to the constantly increasing throng of pilgrims which it attracted soon vied with St. Gallen as one of the richest monasteries in Switzerland.

In the large open space between the houses (a great many of which are inns for the entertainment of the pilgrims) and the conspicuous buildings of the monastery rises a black marble *Fountain* with fourteen jets, surmounted by an image of the Virgin, from which the pilgrims are wont to drink. Under the *Arcades*, which form a semicircular approach to the church on the right and left, as well as in the Platz itself, there are numerous stalls for the sale of prayer-books, images of saints, rosaries, medals, crucifixes, and other 'devotional' objects. So great is the demand for engravings, religious works, and other souvenirs of the place, that at *Benziger & Co.*'s establishment no fewer than 900 workmen are employed in printing and stereotyping, engraving on wood and zinc, chromolithographing, book-binding, etc. The pilgrims, chiefly from Switzerland, Bavaria, Swabia, Baden, and Alsace, number about 150,000 annually. The chief festival takes place on 14th September.

The extensive *Abbey Buildings*, in the Italian style, which were

re-erected for the sixth or seventh time in 1704-19, are 148 yds. long, 41 yds. of which are occupied by the *Church* and its two slender towers. On the right and left of the entrance are *Statues* of the Emperors Otho I. and Henry II., two benefactors of the Abbey.

The INTERIOR of the church is gaudily decorated with gilding, marble, and pictures of little value. In the nave stands the CHAPEL OF THE VIRGIN, of black marble, the 'Sanctum Sanctorum', with a grating, through which, illuminated by a solitary lamp, a small image of the Virgin and Child is visible, richly attired, and decked with crowns of gold and precious stones. In the chapel to the right a Crucifix by J. Kraus; in the choir an Assumption by the same artist, skilfully restored by Deschwanden in 1858. — The Abbey contains a well-arranged LIBRARY of 50,000 volumes, chiefly historical, a number of MSS., and a small natural history collection. The FÜRSTENSAAL is hung with good lifesize portraits, including those of Pius IX. and the emperors William I., Francis Joseph, and Napoleon III. The PRIVATE CHAPEL of the abbot is adorned with paintings of ecclesiastical events. — Connected with the Abbey are a SEMINARY and a LYCEUM.

The *Herrenberg* (3050'; 1/2 hr.), a hill above the Abbey to the S.E., commands a beautiful view of the neighbourhood. Similar views are obtained from the Kreuz or from the *Meinradsberg*, 3/4 M. to the S. of the town.

About 1/4 M. to the W. of the town, near the rail. station, is an interesting *Panorama *of the Crucifixion*, by Leigh, Frosch, and Krieger (adm. 1 fr.).

FROM EINSIEDELN TO SCHWYZ OVER THE HACKEN (3 1/2 hrs.), destitute of shade, and very disagreeable in bad weather. We ascend the monotonous *Alpthal* (with the nunnery of *Au* on the right) to the (1 1/2 hr.) village of *Alpthal* (3258'; *Stern), where the somewhat rough and steep log-path ascending the Hacken begins. In 1/2 hr. we gain a point where the space between the two Myten (p. 110), shaped like the letter V, is distinctly observed, and in 1/2 hr. more reach the *Inn* on the Hacken Pass (4588'), which commands a splendid view of the lakes of Lucerne and Lowerz, etc. (The view is still finer from the *Hochstuckli*, 5105', 1/2 hr. higher up, to the N., and embraces the N. part of the lake and the town of Zürich.) Descent to (1 hr.) Schwyz steep and stony.

FROM EINSIEDELN TO SCHWYZ OVER THE IBERGER EGG, **13 M.** Good road through the *Sihlthal* or *Euthal* by *Steinbach* and *Euthal* to (8 M.) *Iberg* (3488'); thence to the **Iberger Egg** (4823') or *Heilighäuschen*, affording a fine survey of the Lake of Lucerne and the Alps, and by *Bülisberg* and *Rickenbach* to (5 M.) Schwyz.

Beyond Biberbrücke (p. 105) the railway crosses the *Biber*, and ascends across a monotonous plateau. From (25 1/2 M.) *Altmatt* (3035'; Rössli), a poor hamlet on a large moor, a carriage-road leads in 1 3/4 hr. to the *Gottschalkenberg* (p. 106).

28 M. **Rothenthurm** (3050'; *Ochs, R. 1 1/2, B. 1, D. 2, pens. 3 3/4-4 fr.; *Schlüssel*), with a new Romanesque church, where to the left the Myten, to the right the long back of the Rigi **and the hotels** on the Kulm become visible, is named after a red tower belonging to fortifications *(Letze)* once erected by the Schwyzers to protect their N.W. boundary. In the vicinity, on the E. slope of the *Morgarten* (p. 80), on 2nd May, 1798, the Schwyzers under Reding defeated the French, who lost 2000 men. The railway then descends in the wooded valley of the *Steinen-Aa* to (31 M.) *Sattel-Ägeri*; to the left

is the pleasantly situated village of **Sattel** (2345'; *Neue Krone*, at the station, pens. 4½-6 fr.; *Alte Krone*, in the village).

The *Schlagstrasse, as the picturesque road from Sattel to Schwyz is called (6 M.; a fine walk), crosses the Steinen-Aa and ascends on the W. slope of the *Hacken* (see above), affording beautiful views of the fertile valley of Steinen, the Lake of Lowerz with the Schwanau, the scene of the Goldau landslip, and the Rigi. At (3¾ M.) the *Hirsch Inn* (a little father on the *Burg Inn*), Schwyz and the Myten become visible. Thence to stat. *Seewen* 1¼ M., to *Schwyz* (p. 109) 2 M.

From Sattel-Ägeri to *Morgarten*, 2 M., omnibus in ½ hr. (50 c.); steamboat on the *Ägeri Lake*, see p. 80.

The railway descends the slopes of the *Rossberg*, by several viaducts and a short tunnel to (34 M.) **Steinerberg** (1950'; *Rössli*; *Löwe*), a mountain-village with a fine view of the valley of Lowerz, framed by the slopes of the Rigi, the Frohnalpstock (with the Liedernenstöcke and Möhrenberge in the distance), and the two Myten. The *Rossberg (highest peak, the *Wildspitz*, 5190') may be ascended from Steinerberg by a new bridle-path in 2½ hrs. At the top, which commands a fine view (panorama by Imfeld), is the *Hôtel Rossberg-Kulm*. From the *Grippen* (5127'), or W. summit of the Rossberg, reached from the hotel by a level path in 20 min., we obtain a good survey of the scene of the landslip of 1806 (comp. p. 109). — The descent may be made to *Ägeri* (p. 80).

The railway traverses the scene of the **Goldau Landslip**, and joins the St. Gotthard Railway (p. 109) at (36 M.) **Arth-Goldau** (Hôt. Hof Goldau, etc.). — *Rigi Railway*, see p. 96.

32. From Lucerne to Bellinzona. St. Gotthard Railway.

Comp. *Maps, pp. 86, 94, 126, 112, 118, 382*.

109 M. RAILWAY. Express ('Blitzzug'; first class only) in 4½, fast trains in 5¼, ordinary trains in 7 hrs.; fares 24 fr. 60, 17 fr. 20, 12 fr. 30 c. (To Lugano 127½ M., express in 5½-6½ hrs.; 29 fr. 30, 20 fr. 50, 14 fr. 65 c.; to Milan 176 M., in 7½-9 hrs.; 35 fr. 70, 25 fr., 17 fr. 65 c.). — *Rothkreuz* (p. 80), a station between Zug and Lucerne, the starting-point of the St. Gotthard line, is reached by express from *Zürich* in 1-1½ hr.; from *Bâle* by *Lucerne* in 2¾ hrs., or by *Aarau* and *Muri* in 3½ hrs. — For the day express there is a table d'hôte at Göschenen, where the traveller should be careful to avoid an involuntary change of carriages, or even of trains. Finest views from Lucerne to Flüelen to the right, from Flüelen to Göschenen to the left, and from Airolo to Bellinzona to the right.

The **St. Gotthard Railway**, constructed in 1872-82 at a **cost of 238 million francs**, is one of the grandest achievements of **modern engineering**. The highest point of the line, in the middle of the great tunnel, is 3787' above the sea-level, and the maximum gradient is about 1' in 4'. At places the ascent is rendered more gradual by means of curved tunnels, piercing the sides of the valley; there are three such tunnels on the N. side, and four on the S. side of the mountain (comp. Map, p. 113). Altogether the line has 56 tunnels (of an aggregate length of 25½ M.), 32 bridges, 10 viaducts, and 24 minor bridges. In order to examine the most interesting structure of the line itself, the traveller may drive in an open carriage or walk from Amsteg to Göschenen (12 M.) and from Airolo to Giornico (15 M.). Those who are not pressed for time should take the steamboat from Lucerne to Flüelen, in preference to the train (holders of through tickets and circular tickets have the choice of either route); or, if they have not yet visited the Rigi, they may take the railway to Rothkreuz, Arth-Goldau, the **Rigi-Kulm**, and Vitznau, and the steamer thence to Flüelen.

From *Lucerne* to (11 M.) **Rothkreuz** (1410′), see p. 80. Our line diverges to the right, traversing a hilly and wooded tract. To the right the Rigi, the Uri and Engelberg Alps, and Pilatus. Before reaching *Immensee* (p. 104), which lies below us, on the left, we obtain a survey of the E. part of the *Lake of Zug* (p. 104). On the N. bank lies *Walchwyl*; then *St. Adrian* (p. 104).

16 M. **Immensee-Küssnacht** (1585′; omnibus to *Küssnacht* in 25 min., see p. 105). To the right are the wooded slopes of the *Rigi*, with the Kulm Hotel far above us (p. 97). The train runs high above the Lake of Zug, passing through several cuttings. At the E. end of the lake, on the left, lies the thriving village of *Arth* (p. 104), at the foot of the wooded Rossberg, behind which rise the Myten (see below). Threading the *Rindelfluh Tunnel* (220 yds.) and several rock-cuttings, we reach —

21 M. **Arth-Goldau** (1845′; *Hôtel Central; Hôtel Hof Goldau*, R. 2, B. 1 fr.; *Restaurant Bellevue*, all three near the station; **Rössli*, unpretending, R. 1½, B. 1, D. 2 fr.), also a station on the *Arth-Rigi Railway* (p. 96), and the junction for *Einsiedeln* and *Wädensweil* (p. 108). The station is situated on the scene of the great *Goldau Landslip*, which occurred on Sept. 2nd, 1806. This landslip, which descended from the *Gnippen* (5127′), the W. summit of the *Rossberg* (p. 108), buried four villages with 457 of their inhabitants. The railway traverses part of this scene of desolation, which extends a considerable way up the Rigi. Time has covered the fragments of rock with moss and other vegetation, and picturesque pools of stagnant water have been formed between them at places. The track of the landslip may be distinctly traced on the side of the Rossberg, which is still entirely barren.

On the slope to the left lie the houses of *Steinerberg* (p. 108); on the right, high above, is the *Curhaus Rigi-Scheidegg* (p. 100). The train skirts the pretty **Lowerzer See** (1475′; 3 M. long). To the right lies the village of *Lowerz* (Pens. Bücheler-Peter, 4½-5 fr.), and in the middle of the lake the island of *Schwanau* with its ruined castle, a chapel, and a fisherman's house (Inn; boat from Lowerz or Seewen in 25 min.). — 24½ M. **Steinen** (1540′; **Rössli*), a considerable village in a fertile situation, the traditional birthplace of *Werner Stauffacher* (p. 92). On the supposed site of his house stands the *Chapel of the Holy Rood* with new frescoes by Ferd. Wagner of Munich. The train crosses the *Steinen-Aa* to —

26 M. **Schwyz-Seewen** (1500′; **Hôt.-Pens. Schwyzerhof; Railway Inn*, both at the station). The village of *Seewen* (**Rössli, R., L., & A. 1½-2, B. 1, D. 3, pens. 6-8 fr.; *Stern*, R., L., & A. 1½, D. 2, pens. 4½-5 fr.; *Pens. Seehof*), to the W. of the line, at the foot of the E. spur of the Rigi, has a chalybeate bath which attracts visitors. About 1 M. to the E. lies **Schwyz** (1685′; pop. 6668; **Rössli*

R., L., & A. 2-2½, D. with wine 3, pens. 5½-6½ fr.; *Hôtel Hediger*, same charges; *Café Central*, near the church, with garden), a straggling town, lying picturesquely at the base and on the slopes of the *Little Myten* (5955') with its two peaks, and the *Great Myten* (6245'). The *Parish Church* (1774) is considered one of the handsomest in Switzerland. The *Town Hall*, restored in 1891 and embellished on the exterior with frescoes from Swiss history by Ferd. Wagner of Munich, contains portraits of 43 'landammanns' (magistrates) from 1534 downwards, and an old carved ceiling. The large *Jesuit Monastery*, above the town, is now a grammar-school.

The *Great Myten (6245'; 3½ hrs.; guide 6 fr., unnecessary for the experienced; horse to the Holzegg 8-10 fr.), ascended without difficulty by a new path, is a magnificent point of view, hardly inferior to the Rigi and Pilatus. Road from Schwyz to (1 M.) *Rickenbach* (Bellevue; Stern, pens. 4 fr.); bridle-path thence to the (2 hrs.) *Holzegg* (4642'; small Inn), which may also be reached by a direct path from Schwyz viâ the *Hölle* and the pastures of *Hasli* and *Holz* (guide desirable). — From Brunnen (p. 90; diligence to Schwyz five times daily, 80 c.) by *Ibach* and (3 M.) *Rickenbach* to the Holzegg in 3 hrs., Schwyz remaining on the left. — Good path from Einsiedeln by *Alpthal* to the Holzegg in 2¾ hrs. — From the Holzegg the new Myten path (railings at the steepest parts) ascends in 49 zigzags on the E. side of the mountain, and then follows the narrow arête to the (1¼ hr.) summit (*Inn, plain, 10 beds). Good panorama by A. Heim. — The ascent of the *Little Myten* (5955') is difficult; view inferior to that from the Great Myten.

An interesting walk may be taken from Schwyz to the *Suvoroff Bridge* in the *Muota-Thal*, returning viâ *Ober-Schönenbuch* (2 hrs. in all); comp. p. 72.

We now turn to the S. (on the left the *Fronalpstock* with the *Curhaus Stoos* far above us, p. 91), cross the *Muota* near *Ingenbohl*, passing the large nunnery of *Mariahilf*, and reach —

28½ M. **Brunnen** (1445'; p. 90), the most frequented spot (after Lucerne) on the *Lake of Lucerne*. (Station ½ M. from the lake.)

Passing through a tunnel under the *Gütsch* and the *Axenstrasse* (p. 92), the train now reaches the *Lake of Uri, or S.E. bay of the Lake of Lucerne (p. 91), and is carried along its bank by a series of tunnels and rock-cuttings. Splendid views of the lake to the right. High above it, on the opposite bank, lie the houses of *Seelisberg*, at the foot of which are the *Mytenstein* and *Rütli* (p. 91); and farther to the left towers the *Uri-Rothstock* with its glacier (p. 93). We pass through the *Hochfluh Tunnel*, the *St. Franciscus Tunnel*, and the *Oelberg* or *Schiefernegg Tunnel* (2160 yds.), the longest but one on the line. — 32½ M. **Sisikon**, at the mouth of the narrow *Riemenstalden-Thal* (p. 92). Crossing the Axenstrasse, we traverse the *Stutzeck Tunnel* (1082 yds.) and others, passing *Tell's Platte* (chapel not visible; p. 92), the *Axenberg* (3670' long), and the *Sulzeck*.

36 M. **Flüelen** (1435'; *Rail. Restaurant*); see p. 92. Omnibus to Altdorf 50 c.

We now ascend the lower Reussthal, with the *Bristenstock* (p. 112) in the background, and the two *Windgällen* (p. 123) to the left of it.

38 M. **Altdorf**, or *Altorf* (1466'; pop. 2553; *Schlüssel*, R., L.,

& A. 1¹/₂-3, B. 1¹/₄, D. 2¹/₂-3, pens. 5-7, omn. ¹/₂-1 fr.; *Löwe*, moderate; *Krone*; *Bär*; *Tell*, with garden, pens. 4-5 fr.; *Hôtel de la Gare*, at the station, R. 1-2 fr.), the capital of Canton Uri, 1 M. from the station, lies in a fertile valley surrounded by mountains. This pleasant little town is the traditional scene of the exploits of William Tell, the liberator of Switzerland from the Austrian yoke. A bronze statue of the intrepid archer, with the child by his side, from Kissling's model, was erected in 1895 to the N.W. of the tower (dating from the 13th cent.) in the principal 'Platz' of the village. Opposite is a fountain with the statue of a village magistrate. The *Church* contains a Madonna in relief, by *Imhof*. The *Capuchin Monastery*, above the church, and the neighbouring *Pavillon Waldeck* command beautiful views. (Ascent near the tower, or from below Tell's statue.) Above the monastery lies the *Bannwald*, a 'sacred grove', in which the woodman's axe is proscribed, as it protects Altdorf from falling rocks (see Schiller's Tell, Act iii, Scene 3).

To the right, beyond the town, is a *Nunnery*, to the left the *Arsenal*; then, about 1 M. to the left, the village of Bürglen (1805'; *Tell*, pens. 4¹/₂-5 fr.), prettily situated on a height at the entrance to the *Schächen-Thal* (p. 72), the traditional birthplace of Tell. The supposed site of his house is marked by a *Chapel*, erected in 1522, and adorned with paintings of his exploits.

Through the *Schächen-Thal* and over the *Klausen* to (28 M.) *Stachelberg*, see R. 20. A glimpse at the Schächen-Thal is best obtained by ascending from *Weiterschwanden* or *Spiringen* (p. 72) in about 1¹/₂ hr. to one of the farmhouses in the *Kessel* (4505'), which afford a most picturesque survey of the grand head of the valley (Scheerhorn, Griesgletscher, Kammlistock, and Claridenstock), with beautiful fresh pastures and dark pine-forest in the foreground. — The *Rossstock* (8080'; 5 hrs.; guide 12 fr.), a splendid point of view, is ascended without difficulty by experts from Bürglen, viâ the *Mettenthal-Alp*. Descent, if preferred, through the *Riemenstalden-Thal* to *Sisikon* (p. 92). — *Belmistock* or *Belmeten* (7933'), from Altdorf in 5¹/₂ hrs. (guide 8 fr.), also interesting.

The train now crosses the wild *Schächenbach* in its artificial bed, near its confluence with the Reuss. From among fruit-trees to the left peeps the church of *Schattorf*. To the right, beyond the Reuss, we observe the church-tower and the ruined castle of *Attinghausen*, in which the Baron Werner of Attinghausen, one of the characters in Schiller's Tell, is said to have died in 1307 (*Inn at the foot of the castle-hill). The background of the valley towards the S. is formed by the pyramidal *Bristenstock* (p. 112); to the right rise the bold precipices of the *Gitschen* (8335') and the *Bockli* (6810'); to the left the *Schwarzgrat* (6636'), *Belmistock* (7933'), *Hohe Faulen* (8260'), and lastly the two *Windgällen* (*Grosse*, or *Kalkstock*, 10,463'; *Kleine*, or *Sewelistock*, 9800').

41¹/₂ M. **Erstfeld** (1503'; *Hof Erstfeld*, *Hôt. Bahnhof*, both at the station, unpretending), a large railway-depôt, where the ascent begins and a heavier locomotive is attached to the train. The village lies on the opposite bank of the Reuss, at the mouth of the *Erstfelder-Thal*, above which peep the jagged *Spannörter* and the *Schlossberg* (10,280'; p. 131), with its strangely contorted glacier.

112 *II. Route 32.* **AMSTEG.** *From Lucerne*

The Erstfelder-Thal (comp. Map, p. 128) extends to the S.W to the *Glattenfirn*. At the head of the valley are two Alpine lakes, the gloomy *Faulensee*, 1/2 hr. from the glacier, and the *Obersee* (6463'), 1/2 hr. farther to the S. Above the Faulensee, 3 1/2 hrs. from Erstfeld, is the *Krönte-Hütte* of the Swiss Alpine Club, whence the *Krönte* or *Krönlet* (10,197') may be ascended by the *Weissen Platten* and the *Glattenfirn* in 4 1/2 hrs. (guide from Erstfeld 20 fr.), and the *Great Spannort* (10,515') in 5 hrs. (guide 25 fr.). The *Faulenbach*, which flows out of the Obersee, forms a beautiful fall. Fatiguing passes lead hence to the W. over the *Schlossberg-Lücke* (8635'; guide 25 fr.) and over the *Spannort-Joch* (9610'; guide 35 fr.) to (6 1/2 hrs.) *Engelberg* (comp. p. 130); also to the S. over the *Leidensee Pass* (7695') to the *Leutschach-Thal* and (7-8 hrs.) *Inschi* (see below). Guide, *Gebhard Püntener* of Erstfeld.

From Erstfeld or Altdorf over the *Surenen Pass* to (8 1/2 hrs.) *Engelberg* (guide 20 fr.), see p. 131.

The Reussthal narrows, and the train begins to ascend on the right bank. 45 M. Stat. *Amsteg* (1795'), above *Silenen*, a village in the midst of fruit-trees. Near the station, on a rocky hill to the right, are the ruins of *Zwing-Uri*, the traditional castle of Gessler (rooms in the adjoining house). About 1 M. farther on lies the village of **Amsteg** (1760'; **Stern* or *Post*, R. 1 1/2-2 1/2, D. 3, pens. 5-7 fr.; **Hirsch; *Weisses Kreuz; *Engel; Freihof*, R. 1 1/2-2, B. 1, pens. 4-6 fr.), prettily situated at the mouth of the *Maderaner-Thal*, from which the *Kärstelenbach* descends to the Reuss.

°MADERANER-THAL (bridle-path in 3 1/4 hrs. to the Hôtel Alpenclub), see R. 34. — Over the *Kreuzli Pass* or the *Brunni Pass* to *Disentis* and over the *Clariden Pass* to *Stachelberg*, see p. 124.
The Bristenstock (10,090'), ascended from Amsteg in 7-8 hrs. by the *Bristenstäfeli* (5000') and the *Blacki-Alp* (6133') and past the small *Bristen-Seeli* (7000'), affords a grand panorama, but is very fatiguing (guide 25 fr.). Descent to the *Etzlithal* or *Pellithal* difficult. — *Oberalpstock* (10,925'), *Kleine* and *Grosse Windgälle* (9800' and 10,463'), etc., see p. 123. — The **Hohe Faulen** (8260'), ascended from Silenen in 5 hrs. (guide 10 fr.) through the *Evithal* and over the *Strengmatt, Rhonen*, and *Balmeten Alps*, is attractive and not difficult.
The St. Gotthard Road from Amsteg to Göschenen (comp. Map, p. 126) should be traversed on foot (or in an open carriage), both for the sake of the scenery and for the opportunity it affords of examining the interesting railway. It crosses the Kärstelenbach and then the Reuss by a bridge of two arches. To the left runs the railway; below us the Reuss dashes through its deep ravine, forming a succession of waterfalls. In the early summer huge masses of avalanche-snow, looking like earth or detritus, are seen in the gorges. Beyond (1 3/4 M.) **Inschi** (2168'; *Lamm*) we pass a fall of the *Inschi-Alpbach*. From Inschi a visit may be paid to the picturesque *Leutschach-Thal* (to the *Obersee*, at the foot of the *Männtliser*, 3 1/2 hrs.; hence over the *Leidensee Pass* to the *Erstfelder-Thal*, see above). — A second bridge carries the road back to the right bank of the Reuss (the railway remaining on the left bank), on which lies (1 1/2 M.) *Meitschlingen*, with a chapel. About 1/2 M. farther on we cross the *Fellibach*. (Through the narrow *Felli-Thal* or *Fellenen-Thal*, which abounds in crystals, the *Oberalp-See* may be reached by the *Felli-Lücke* in 6 hrs.; p. 375.) On the hill opposite stands the hamlet of *Gurtnellen* (3015'). Beyond the village of *Wyler* is (3 M.) a third bridge (2660'), called the *Pfaffensprung*, by which the road recrosses to the left bank. The first of the curved tunnels of the railway begins here (p. 113). Far below, the river dashes through a narrow gorge. View beautiful in both directions. The road crosses the turbulent *Meien-Reuss* (p. 133) shortly before reaching (1 1/2 M.) Wasen (p. 113). To the right are the three railway-bridges. A path to the right, 50 yds. beyond the bridge, cuts off the windings of the road which ascends to the loftily situated church.

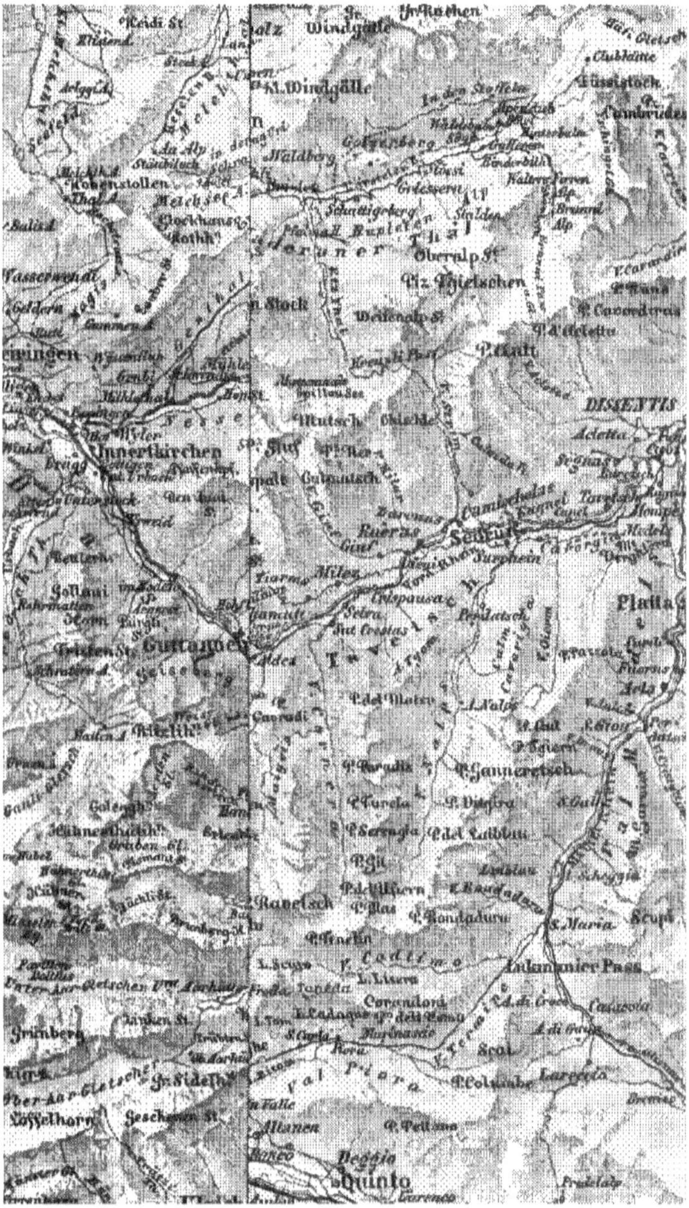
Kl. Windgälle
Oberalp St.
Piz. Tgietschen
DISENTIS
Innertkirchen
Guttannen
Sedrun
Rueras
Platta
P. Ganneretsch
S. Maria
Lukmanier Pass
Scai
Ober-Aar-Gletscher
Unter-Aar-Gletscher
Altanca
Deggio
Quinto

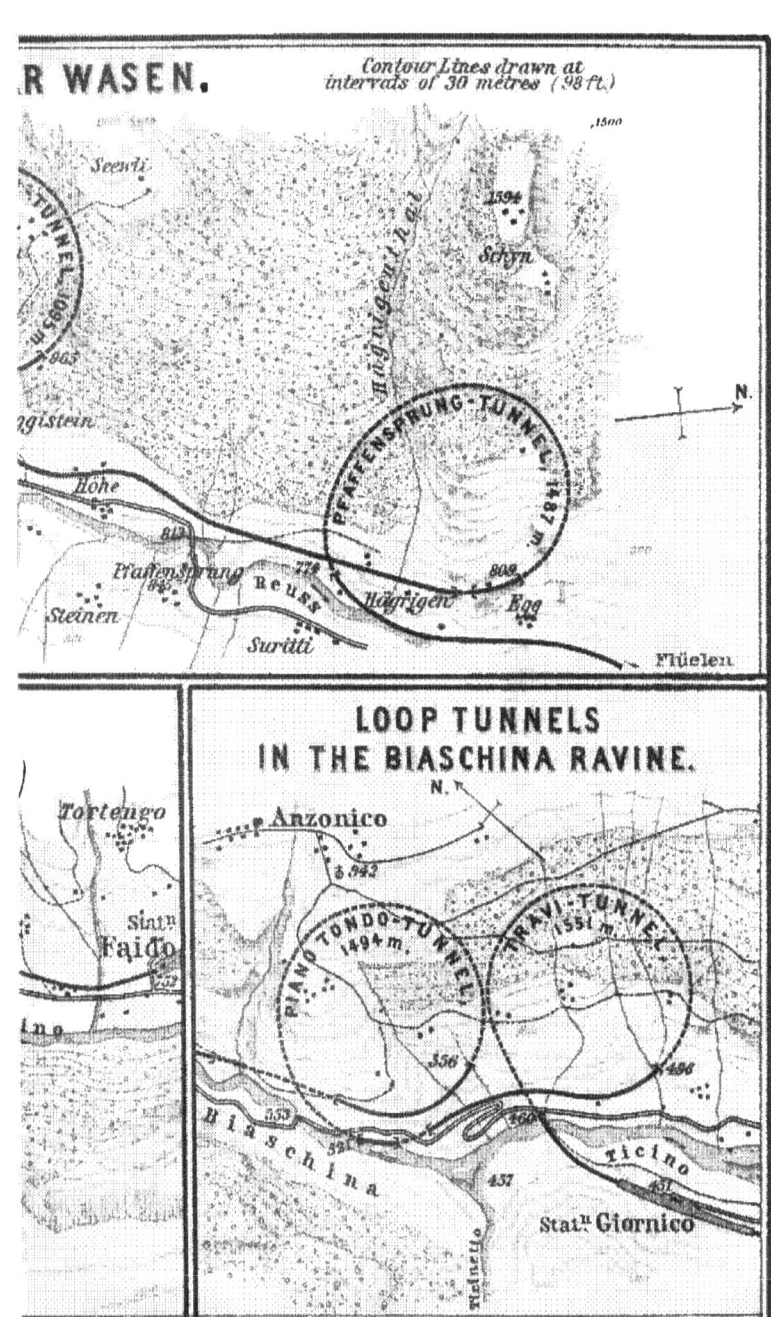

to Bellinzona. **WASEN.** *II. Route 32.* 113

Near (³/₄ M.) **Wattingen** (2908') is the fourth bridge over the Reuss, above which, to the right, is a fall of the *Rohrbach* (see below). The (1 M.) fifth bridge (*Schönibrück*, 3212') crosses to the left bank of the Reuss. To the left rises the *Teufelsstein*, a huge mass of rock. The next place (1¹/₂ M.) is *Göschenen* (3640'; p. 114).

Above the village of Amsteg the line pierces a projecting rock by means of the *Windgälle Tunnel* (1828'; 189 yds. long), crosses the *Kärstelenbach* by an imposing iron bridge (147 yds. long, 177' high), affording a fine view of the deeply-cut *Maderaner-Thal*, with the *Grosse Windgälle*, to the left, and of the Reussthal to the right, and is then carried through the slope of the *Bristenstock*, which is much exposed to avalanches, by means of the two *Bristenlaui Tunnels* (436 yds. and 234 yds. long), and across the brawling Reuss by an iron bridge 256' high. We now follow the left bank of the picturesque Reussthal (views to the left), traverse the *Inschi Tunnel*, cross the *Inschi-Alpbach* and the *Zraggen-Thal* (viaduct about 100 yds. long), thread the *Zgraggen, Breiten*, and *Meitschlingen* tunnels and a long cutting, and skirt the hillside by a viaduct to (50 M.) **Gurtnellen** (2427').

Above Gurtnellen we come to one of the most remarkable parts of the line, which in order to facilitate the ascent to Göschenen (p. 114) passes through three curved tunnels and describes a wide double bend. It crosses the *Gornerenbach* and the *Hägrigenbach* (fine waterfall on the right), enters, near the *Pfaffensprung-Brücke*, the *Pfaffensprung* **Loop Tunnel** (1635 yds., 3 min.), in which it mounts 115', traverses the short *Mühle Tunnel*, recrosses the Hägrigenbach (overlooking the Pfaffensprung bridge on the left), and then traverses the *Mühren Tunnel* (2822'; 93 yds. long). Next follow a handsome bridge over the deep ravine of the *Meienreuss* (p. 138), the *Kirchberg Tunnel* under the 'church-hill' of Wasen (330 yds.), a bridge across the Reuss to the left, the *Wattinger* **Loop Tunnel** (1199 yds.; ascent of 76'), another bridge over the Reuss, and the *Rohrbach Tunnel* (242 yds.). — 55 M. **Wasen** or *Wassen* (3055'), a considerable village (**Hôt. des Alpes; *Ochs*, plain; *Krone*, R., L., & A. 2, B. 1, D. 2, pens. 5 fr.; *Walker's Restaurant*) with a loftily situated church commanding an admirable survey of the bold structure of the railway. — Over the *Susten* to *Meiringen*, see R. 39.

The imposing *Mittlere Meienreuss Brücke* (69 yds. long, 260' high) and the *Leggistein Loop Tunnel* (1204 yds.; ascent of 82') carry us to the *Upper Meienreuss Bridge* (59 yds. long; 148' high), beautifully situated, the third bridge over the deep, wild gorge of the Meienreuss. We then pass through the short *Meienkreuz Tunnel* (3250'; 84 yds. long), skirt the hillside, and obtain a view of Wasen and the windings just traversed. Opposite rises the *Rienzer Stock* (9785'). Crossing the *Kellerbach* and the *Rohrbach*, the train passes through the *Naxberg Tunnel* (1719 yds.; ascent of 118'), crosses the deep gorge of the *Göschenen-Reuss* (bridge 69 yds. long, 161' high;

BAEDEKER, Switzerland. 10th Edition. 8

view of the *Göschenen-Thal* to the right, with the beautiful *Dammafirn*, p. 118), and reaches —

69½ M. **Göschenen**, or *Geschenen* (3640'; *Rail. Restaurant*, D. with wine 3½ fr., cheaper and also good, in the third-class waitingroom; *Hôt. Göschenen*, opposite the station, R., L., & A. 3, B. 1½, déj. 3, D. 4, pens. from 7 fr.; *Rössli*, R. & A. 2½, B. 1¼, D. 3 fr.; *Hôt. St. Gotthard*, R. 1½-2, D. 2½, pens. 6 fr.; *Löwen*; *Krone*). In the little cemetery is a tasteful monument (1889), by Andreoletti, to L. Favre, the engineer of the St. Gotthard Tunnel, who died of apoplexy in the tunnel on July 19th, 1879. — From Göschenen to **Airolo** by the *St. Gotthard Road*, 22 M., see R. 33.

Immediately beyond the station the train crosses the Gotthard-Reuss (p. 118) by a bridge 105' high, and enters the great *St. Gotthard Tunnel*, which is 16,309 yds. (9¼ M.) in length, being 2930 yds. (1⅔ M.) longer than the Mont Cenis Tunnel. The central point is 3786' above the sea-level, from which it descends on both sides, about 6' in 1000' towards Göschenen and 2' in 1000' towards Airolo. The work was begun in June, 1872, at Göschenen, and a month later at Airolo, and the boring was completed on 29th Feb., 1880. During seven years and a half no fewer than 2500 workmen were on an average employed here daily, and the number sometimes rose to 3400. The cost was 56¾ million fr. (2,270,000*l*.). The tunnel, 28' broad and 21' high, is lined with masonry throughout, and is laid with a double line of rails. In the interior there is always a strong current of air; temperature 70° Fahr. The tunnel runs at a depth of 1083' below Andermatt, 6076' below the Kastelhorn (which rises above the centre of the tunnel), and 3350' below the Sella Lake. Express trains take 16 min. to pass through the tunnel, slow trains 25 min.; lanterns are placed on each side of the tunnel at intervals of 1000 mètres (even numbers on the right, uneven on the left). To the right, above the exit from the tunnel, fortifications have recently been erected.

69½ M. **Airolo** (3755'; pop. 1800; *Posta*, R., L., & A. 3-3½, B. 1½, D. 4, pens. 7-8 fr.; *Hôt. Lombardi*, *Hôt. Airolo*, R. & A. 2½ fr., *Hôt. des Alpes*, *Hôt. Rossi*, R. from 2, B. 1, D. 2½, pens. 7 fr., all these near the station; *Rail. Restaurant*), in the upper valley of the Ticino (*Valle Leventina*, p. 116), the first Italian-Swiss village, rebuilt since a fire in 1877. The scenery retains its Alpine character until near Faido. To the W. is the imposing Pizzo Rotondo group.

EXCURSIONS (guides, *Clem. Dotta* and *Basil Jori* of Airolo). From Airolo to the picturesque *Stalvedro Gorge* (p. 116), 20 min.; to the *Lombard Tower*, 35 minutes. — **Pizzo** Rotondo (10,490'), the highest peak of the St. Gotthard, may be ascended from Airolo in 8-9 hrs. (difficult, for experts only; guide 40 fr.). Drive or walk in the afternoon to (3 hrs.) *All'Acqua* in Val Bedretto (p. 315; Inn), where the night is spent; steep ascent thence over grassy slopes, debris, and snow-fields to the (3½ hrs.) *Passo Rotondo* (9515'); whence the rocky summit is reached in 1½-2 hrs. by a difficult climb up a steep snowy couloir (foot-irons desirable) and over loose stones. The *View is extremely grand and picturesque (comp. p. 121).

PASSES. To the St. Gotthard, see p. 118 (rich Alpine flora as far as the Tremola gorge). — Through the *Val Bedretto* and over the *Nufenen Pass* to the *Valais*, see p. 315; over the *S. Giacomo Pass* (7572') to the *Falls of the Tosa*, see p. 820. Through the *Val Canaria* and over the *Unteralp Pass* (8303') to *Andermatt* (8 hrs.), fatiguing; the ascent very steep. Over the *Bocca di Cadlimo* (8357') to *S. Maria* (p. 376) in 8 hrs., attractive. — By the *Passo Bornengo* to *Val Maigels*, see p. 374. Over the *Sassello Pass* to *Val Maggia*, see p. 443. — To Val Maggia over the *Passo dei Sassi* (ca. 8200'), interesting, but fit for steady climbers only (to Fusio 8 hrs.). From Airolo past the hamlet of *Nante* and the (2 hrs.) *Alp Piscium* (5630') to (³/₄ hr.) *Comaschne* (6734') and along precipitous rocky slopes, where the path entirely disappears, to the (2¹/₂ hrs.) pass, between the *Poncione di Vespero* and *Poncione di Mezzodi*, with fine retrospective view of the St. Gotthard mountains. Descent across steep grassy slopes (plenty of edelweiss) into the Val Maggia, to (2 hrs.) *Corte* and (³/₄ hr.) *Fusio* (p. 443).

FROM AIROLO TO DISENTIS THROUGH THE VAL PIORA (10 hrs., guide, unnecessary, to Piora 6, to 8. Maria 10 fr.; porter, at the Hôtel Lombardi at Airolo, 15 c. per kilogramme up to Piora, 10 c. down; horse to Piora, 3 hrs., 15 fr.). Descending the St. Gotthard road for ³/₄ M., we cross the *Canaria* to the left, and ascend to (20 min.) *Madrano* (4110'). After ¹/₄ hr. more the path ascends to the left to (20 min.) *Brugnasco* (4548'). It then runs at nearly the same level, overlooking the picturesque valley of the Ticino, and afterwards through wood. From (³/₄ hr.) *Altanca* (4567'; Inn) we ascend to the left in zigzags past a small chapel to the (40 min.) *Alp in Valle* (a spring by the wayside). The rock below it bears a very ancient inscription. In the gorge to the right the *Fossbach* forms several picturesque waterfalls. Fine retrospect of the mountains of Ticino. We next cross a rocky saddle to the (¹/₂ hr.) sequestered *Lake Ritom* (6000'), on a hill to the left of which is the "Hôtel Piora (R. 2, B. 1, D. 4, pens. 7-9 fr.), a health-resort, suitable for a stay. Pine-woods close to the hotel; great variety of geological formations and of plants. Pleasant walks in the vicinity; in secluded basins lie six small lakes, and there are four others just beyond the ridges in the direction of the *Val Cadlimo*. Delightful view of the lake, the Ticino valley, etc., from the *Bella Vista* (¹/₄ hr.); a more extensive prospect is enjoyed from *Fongio* (7257'), 1 hr. farther on (skirt the hillside to the W.), and from the *Cima di Camoghè* (7740'; 1³/₄ hr.). — *Taneda* (8760'), an easy ascent of 2¹/₂ hrs., past *Lake Tom* to the ridge separating Val Piora from Val Cadlimo, between Taneda and Punta Nera, where we keep to the right to the broad summit. The splendid view commands the Val Piora, the Val Bedretto, and the Alps of Valais, Bern, Uri, Ticino, and the Grisons. — Other interesting points near Val Piora are the *Punta Nera* (8925'; 2³/₄ hrs.), *Corandoni* (8788'; 3 hrs.), *Piz dell' Uomo* (9020'; 3¹/₂ hrs.), *Pizzo Lucomagno* (9115'; 5 hrs.), and *Piz Blas* (9920'; 5¹/₂ hrs.). — The path to *S. Maria* (3³/₄ hrs.; porter 7 fr.) leads round the lake, to the left. By the (20 min.) *Ritom Chalets* we ascend the slope to the left by a good path to the (20 min.) chapel of *S. Carlo*. Crossing the brook, and passing a cross on the right (leaving the small lake of *Cadagno*, with its chalets, to the left), we reach (¹/₂ hr.) the *Alp Piora* and (¹/₄ hr.) *Marinascio*, a group of huts. The path, indicated by crosses, leads straight on for ¹/₄ hr., and then ascends to the left. Farther on it always bears to the left. [The last huts of *Piano de' Porci* lie to the right, below us. Persons bound for Olivone may from this point cross direct by the *Passo Columbe* (7792'), between the *Scai* and *Piz Columbe*, to the Casaccia hospice; p. 376.] We ascend the secluded *Val Termine*, with the *Piz dell' Uomo* (9020') on the left, to the (³/₄ hr.) summit of the Uomo Pass (7257'; 10 min. before reaching which we pass a good spring by a heap of stones), with its deserted hut. Descent on the other side marshy at places. To the left, the *Medelser Rhine* descends from the Val Cadlimo in a copious waterfall. Before us, to the right, rises the Scopi, to the left in the distance the Tödi chain. The (1¹/₂ hr.) *Hospice of St. Maria*, see p. 376. Thence to *Disentis*, or across the *Lukmanier* to *Olivone*, see R. 94.

Below Airolo the train crosses the *Ticino*, which descends from

8*

the *Val Bedretto* (p. 315), passes through the *Stalvedro Tunnel* (209 yds.), and enters the *Stretto di Stalvedro*. On the left bank of the Ticino the high-road runs through four rock-cuttings. The valley expands. 73 M. **Ambri-Piotta** (3250'; *Restaurant Soldini; Brasserie Piotta*). To the left lies *Quinto*. Beyond (76 M.) **Rodi-Fiesso** (3100'; *Hôtel Monte Piottino*) we come to one of the most curious parts of the line (comp. the map, p. 112). The *Platifer (Monte Piottino)* here projects into the valley from the N.; the Ticino has forced its passage through the barrier, descending in a series of falls through a wild rocky gorge to a lower region of the valley, while the railway accomplishes the descent by means of two circular tunnels. At **Dazio Grande** it crosses the Ticino (striking view down the valley), is carried through the *Dazio Tunnel* and the short *Artoito Tunnel*, and enters the *Freggio Loop Tunnel* (1712 yds.), from which it emerges into the *Piottino Ravine*, 118' lower down. It then recrosses the Ticino (fine scenery), passes through the *Monte Piottino* and *Pardorea* tunnels, and descends 118' more by means of the *Prato Loop Tunnel* (1711 yds.), beyond which opens the beautiful valley of Faido. Crossing the Ticino by the *Polmengo Bridge*, and beyond another tunnel, we reach —

81 M. **Faido** (2485'; pop. 991; **Hôtel Faido, Hôt.-Pens. Suisse*, both at the station; **Angelo*, R., L., & A. 3; B. 1¼, lunch 2½, D. 3½, pens. 6-7 fr.; **Hôt. - Pens. Fransioli*, R., L., & A. 2, B. 1, lunch 2, D. 3, pens., incl. wine, 7 fr.; *Hôt. Vella; Restaurant Belgeri; Birraria Rosian*), the capital of the *Leventina*, very picturesquely situated, and frequented as a summer-resort. On the right the *Piumogna* descends to the Ticino in a fine waterfall.

The **Valle Leventina**, or Valley of the Ticino, formerly belonged in common to the thirteen confederated cantons (with the exception of Appenzell), and was governed in the most despotic manner by bailiffs, who purchased their appointment at auction. A revolt broke out in 1755, but was suppressed with the aid of the Swiss troops. The French put an end to this mode of government in 1798, and in 1814 the Congress of Vienna formed the Leventina and other Italian districts into the **new** canton of Tessin or Ticino.

From Faido over the *Predelp Pass* to the *Lukmanier*, see p. 376; over the *Campolungo Pass* to the *Val Maggia*, see p. 443.

The train now carries us through beautiful scenery, richly wooded with walnut and chestnut trees, on the left bank of the Ticino; the numerous campanili in the Italian style, crowning the hills, have a very picturesque effect. To the right lies *Chiggiogna*, with an old church. From the cliffs on both sides fall several cascades, the veil-like fall of the *Cribiasca* on the right, near (85½ M.) **Lavorgo** (2025'), being the finest. Huge masses of rock lie scattered about, interspersed with fine chestnut-trees. Below Lavorgo the Ticino forces its way through the picturesque **Biaschina Ravine** to a lower region of the valley, and forms a fine waterfall, while the railway descends about 300' on the left bank by means of two loop-tunnels, one below the other in corkscrew fashion. We pass through the *La Lume* *to Bellinzona.* BIASCA. *II. Route 32.* 117

Tunnel, cross the *Pianotondo Viaduct* (114 yds. long), and then enter the *Pianotondo Loop Tunnel* (1643 yds.; descent of 115'). Next follow the short *Tourniquet Tunnel,* the *Travi Viaduct,* and the *Travi Loop Tunnel* (1706 yds.; descent of 118'), from which we emerge upon the floor of the lower Valle Leventina. Crossing the Ticino, we next reach —

90 M. **Giornico** (1480'). The large village (1295'; *Posta, Cervo,* both well spoken of), picturesquely situated among vineyards on the left bank, 1¼ M. to the S., has an old Lombard tower and remains of fortifications near the church of *S. Maria di Castello.* The early Romanesque church of *S. Niccolò da Mira* is said to occupy the site of a heathen temple. Below Giornico the train crosses the Ticino by a bridge 132 yds. long. On the right is the pretty fall of the *Cramosina.* — 94 M. **Bodio** (1090'). Beyond *Polleggio* (Corona) the *Brenno* descends from the *Val Blenio* (p. 377) on the left, and is crossed by two bridges. The valley of the Ticino now expands and takes the name of *Riviera* down to the mouth of the Moësa. Luxuriant vines, chestnuts, walnuts, mulberries, and fig-trees now remind the traveller of his proximity to 'the garden of the earth, fair Italy'. The vines extend their dense foliage over wooden trellis-work supported by stone pillars, 6-10' in height.

98 M. **Biasca** (970'; *Rail. Restaurant;* in the village, ½ M. to the N., *Union & Poste,* unpretending), with an old Romanesque church on a hill (1112'). A series of oratories near the station ascends to the *Petronilla Chapel,* near which is a pretty waterfall. — To *Olivone,* and over the *Lukmanier* to *Disentis,* see R. 94.

The train skirts the base of the richly clothed E. slopes of the valley, which is very hot and dusty in summer, and traverses two tunnels. 101½ M. **Osogna** (870'; *Posta*) lies at the foot of an abrupt rock with a rounded summit. 105 M. **Claro** (830') lies at the base of the *Pizzo di Claro* (8920'), a beautiful mountain with luxuriant pastures, on the slope of which, on a projecting eminence to the left, stands the monastery of *S. Maria* (2074'). Beyond (107½ M.) **Castione** (800') we pass the mouth of the *Val Mesocco* (p. 387) and cross the *Moësa.* To the left lies *Arbedo* (p. 387). We pass through a short tunnel and approach Bellinzona, with its three old castles.

109 M. **Bellinzona** (760'), see p. 433.

From Bellinzona to *Lugano* and *Como,* see p. 434; to *Locarno,* see p. 440; to *Laveno,* see p. 443.

33. From Göschenen to Airolo over the St. Gotthard.

22 M. DILIGENCE from Göschenen to *Andermatt* twice daily in 1 hr. (fare 1 fr. 40, coupé 1 fr. 70 c.); to *Hospenthal* twice in 1½-1½ hr. (2 fr. 10 or 2 fr. 55 c.). No diligence from Hospenthal over the St. Gotthard. OMNIBUSES from the Göschenen station to the Andermatt (1-1½ fr.) and Hospenthal hotels (2 fr.). CARRIAGE and pair from Göschenen to Andermatt or Hospenthal 10, to the Hospice 35, to Airolo 60 fr.; from Andermatt to the Hospice 30, to Airolo 50 fr.; from Hospenthal to the Hospice 25 (there

118 *II. Route 33.*   SCHÖLLENEN.   *From Göschenen*

and back 30 fr.), to Airolo 45 fr. Carriage with one horse from Göschenen to Andermatt or Hospenthal 6 fr.; from Hospenthal to the Hospice 15 (there and back 25 fr.), to Airolo 25 fr. Driver's fee, 10 per cent of the fare. The St. Gotthard was probably the most frequented of the Alpine passes down to the beginning of this century, but being crossed by a bridle-path only it was gradually deserted for the new roads over the Simplon, the Splügen, and the Bernardino. In 1820-82 the cantons of Uri and Ticino constructed the carriage-road, which for half-a-century was the scene of busy traffic; but since the completion of the railway it has again become deserted. Travellers will, however, be repaid by a drive in an open carriage or a walk over the pass. On foot from Göschenen to Andermatt 4 hr. 10 min.; thence to Hospenthal, 3/4 hr.; thence to the Hospice, 2¼ hrs.; and thence to Airolo, 2¾ hrs. or by footpaths, 1¾ hr. Those whose chief object is to make excursions from the Hospice will **reach** it more quickly from Airolo than from Göschenen (telephone).

Göschenen or *Geschenen* (3640'), on the St. Gotthard Railway, see p. 114.

The Göschenen-Thal (3 hrs. to the Göschener-Alp; guide unnecessary) deserves a visit. A good path leads by *Abfrutt* to (1¼ hr.) *Wichi* (4350'), where the *Voralp-Thal* opens to the right (see below); then by *St. Niklaus* and the *Brindlistaffel* (5043') to the (1¾ hr.) Göschener-Alp (6040'; *Hôtel Dammagletscher*), grandly situated. To the W. descends the beautiful *Dammafirn* from the *Winterberg* range (which culminates in the *Dammastock* and *Rhonestock*); and 1 hr. farther up the valley the Göschenen-Reuss issues from the *Kehle Glacier*, imbedded between the Winterberg and Steinberg. — A toilsome but very interesting path (6½-7 hrs., guide 15 fr.) leads from the Göschenen-Alp over the *Alpligen Glacier* and the Alpligen-Lücke (9110'), between the *Lochberg* and *Spitzberg* (p. 125), to *Realp* (p. 125). The *°Lochberg* (10,130'), which affords a splendid view of the Galenstock and St. Gotthard groups, etc., is easily ascended in 3/4 hr. from the pass. — Several difficult passes, fit for experts only, cross from the Göschener-Alp to the Rhone and Trift Glaciers (*Winterjoch, Damma Pass, Maasplank-Joch*; comp. p. 137). Over the *Susten-Limmi* (10,180') or the *Thierberg-Limmi* (about 10,500') to the *Steinalp*, 9 hrs., laborious (see p. 137). — Ascent of the Fleckistock (*Spitzliberg*, 11,215'; 7-8 hrs., guide 85 fr.) for experts only, difficult. We ascend from *Wichi* (see above) through the *Voralp-Thal*, past the chalets of *Hornfell, Bodmen*, and *Flochenstein* to the (2½ hrs.) *Voralp-Thal Hut* of the Swiss Alpine Club (6390'), finely situated at the foot of the *Wallenbühlhirn*; thence we mount to the right to the *Flühen* (7874'), and over loose stones and steep rocks to the summit (5 hrs. from the club-hut). Over the *Wallenbühlhirn* and the *Susten-Joch* (8717') to the *Meien-Thal*, with descent to the *Kalchthal* (p. 138), steep and difficult; fine view from the Joch.

Above the Göschenen station the **⁕St. Gotthard** Road crosses the Reuss by the *Vordere*, or *Häderli-Brücke* (3720'). On the left are the railway-bridge and the N. end of the great tunnel. Here, ¼ M. beyond Göschenen, begins the sombre rocky defile of the ⁕Schöllenen (2½ M. long), bounded by lofty and almost perpendicular granite rocks, at the base of which dashes the Reuss. The road ascends by numerous windings, most of which may be cut off by footpaths or the old bridle-path passing the dilapidated *Lange Brücke* (a little above are the Göschenen water-works, with a considerable waterfall), and crossing the (1 M.) *Sprengibrück* (4048'). The road in the Schöllenen is much exposed to avalanches, and at one of the most dangerous points is protected by a gallery, 60 yds. long, at the farther end of which is the bull's head of Uri.

The road next crosses (3 M. from Göschenen) the (1½ M.) ⁕**Devil's**

Bridge (*Teufelsbrücke*, 4593'), amidst wild and grand rocky scenery. The Reuss here falls into an abyss 100' below, bedewing the bridge with its spray. The wind (aptly called 'Hutschelm', or 'hat-rogue', by the natives) sometimes comes down the gorge in violent gusts, and endangers the hats of the unwary. The new bridge, built of granite in 1830, has a single arch of 26' span. The old bridge, 20' below, was carried away by a flood in 1888. Bloody contests took place here in Aug. and Sept., 1799, between the French on the one side and the Austrians and Russians under Suvoroff on the other, the former being compelled to retreat to the Lake of Lucerne.

Beyond the Devil's Bridge (cabaret; good collection of St. Gotthard minerals) the road winds upwards, passing a chapel and a new fort (see below), to the (1/4 M.) **Urner Loch** (4642'), a tunnel 70 yds. long cut through the rock in 1707, originally broad enough for a bridle-path only. Both above and below the Urner Loch, as well as at Andermatt and Hospenthal, strong fortifications have recently been erected; while new roads have been made from the Devil's Bridge to the *Bäzberg* and from the Oberalp to the top of the *Musch*, two points commanding fine views.

The **Valley of Urseren**, upon which the road emerges from the dark Urner Loch, presents a striking contrast to the wild region just traversed. This peaceful valley (p. 125), with its green pastures watered by the Reuss, is about 8 M. in length and 1/2-1 M. in breadth, and is surrounded by lofty and barren mountains partially covered with snow. Corn grows here but scantily, and trees are scarce. Winter lasts nearly eight months, and during the short summer fires are often necessary. — 3/4 M. —

4 M. **Andermatt**. — Hotels: *Hôt.-Pens. Bellevue, a large house, in an open situation, 1/4 M. from the village, R., L., & A. from 5, B. 1½, lunch 3½, D. 5 fr.; adjacent, Hôtel-Restaurant du Touriste, moderate; opposite, Hôt.-Pens. Nager, small; *Grand Hôtel Andermatt, at the upper end of the village, R., L., & A. 3½, B. 1½, lunch 3, D. 4 fr.; *Hôt.-Pens. Oberalp, R. from 1½, pens. from 6 fr.; *St. Gotthard, R., L., & A. 2½-3½, D. 4, pens. 6-9 fr.; *Hôt. des Trois Rois, R., L., & A. 2, B. 1½, D. 2½, pens. 7 fr.; *Couronne, R., L., & A. 2-2½, B. 1, lunch 2½, D. 3, pens. 6-7 fr.; Sonne. — *English Church*.

Andermatt (4738'; pop. 711), or *Urseren*, Ital. *Orsĕra*, 1¼ M. from the Devil's Bridge, the principal village in the valley, is a winter-resort of invalids. Adjoining the church is a charnel-house adorned with skulls bearing inscriptions. At the exit of the Urner Loch, beside the cliffs to the left, is a much older church said to date from the time of the Lombards (recently restored and embellished with ceiling-frescoes representing the spread of Christianity in the Urseren valley). The *Mariahilf Chapel* affords a good survey: to the W. rises the barren grey Bäzberg, in the background the Furka with its inn, to the left the Muttenhorn; a few paces beyond the chapel, the Six-Madun or Badus (p. 120) is visible; to the E., in long zigzags, ascends the road over the Oberalp (p. 375). St. Gotthard minerals sold by *Frau Meyer-Müller*.

EXCURSIONS. To the Stock, or *Stöckle* (8070′), 3 hrs. (guide unnecessary for experts), by the Oberalp road and the *Grossboden-Alp*, easy and interesting (descent past the *Lautersee* to the *Oberalpsee Hotel*, p. 375). — The **Badus** or **Six-Madun** (9615′), the huge outpost of the Alps of the Grisons, is ascended from Andermatt in 4½–5 hrs. (toilsome; guide 15 fr.; better from Tschamut, p. 374). The summit, which consists of blocks of gneiss, commands numberless peaks of the Alps of the Grisons, Bern, and the Valais, and the whole of the Vorder-Rheinthal. The descent may be made to the *Toma See* in the valley of the Rhine (to Sedrun, 4 hrs., comp. p. 374). — The *Ourschenstock* (9428′; 4 hrs.; guide 15 fr.) and *Gamsstock* (9728′; 4½ hrs.; guide 12 fr.) are also fine points of view (guide necessary).

From Andermatt over the *Oberalp* to *Coire*, see R. 93; over the **Furka** to the *Rhone Glacier*, see R. 35; over the *Unteralp Pass* to *Airolo* (8 hrs.), see p. 115.

Between Andermatt and Hospenthal we observe the *Glacier of St. Anna*, high above the brow of the mountain to the left.

5½ M. **Hospenthal** (4800′; *Meyerhof*, R., L., & A. 2-4, B. 1½, lunch 3, D. 4, pens. 7–10 fr.; *Goldner Löwe*, with restaurant, R., L., & A. 2½, B. 1¼, D. 3-4, pens. from 6 fr.; *Post; Schäfli*, unpretending) was formerly the seat of the barons of Hospenthal, of whose castle the ancient tower on the hill is a relic. Eng. Ch. Service in summer. The *Furka Road* (R. 35) diverges to the right beyond the village.

The St. Gotthard road ascends in numerous windings through a bleak valley, on the left bank of that branch of the Reuss which descends from the Lake of Lucendro (see below). A short-cut diverges to the left by the second house beyond the Reuss bridge. Pleasant retrospects of the Urseren-Thal and the jagged peaks of the Spitzberge (p. 125), as far as the Galenstock to the W. To the left of the bleak (3 M.) *Gamsboden* opens the abrupt *Guspis-Thal*, at the head of which are the *Guspis Glacier* and the *Pizzo Centrale* (p. 121). At a bend in the road (¾ M.) is the first *Cantoniera* (5876′; closed), at the foot of the *Winterhorn* or *Piz Orsino* (8747′). The road enters Canton Ticino, passes the dilapidated second Cantoniera, and crosses the Reuss for the last time, near its outflow from the *Lake of Lucendro* (to the right; not visible), by the (3 M.) *Rodont Bridge* (6620′).

To the *Lake of **Lucendro** (6835′) a digression of ½ hr. only. The path diverges below the Rodont Bridge (on the left bank), leads over masses of rock to the (¾ hr.) beautiful green lake, environed with snow-peaks and glaciers, and skirts its N. bank. To the S. rises *Piz Lucendro* (9708′), to the W. the *Twerberhörner* (9265′), *Piz dell' Uomo* (8820′), etc. — The path crosses the Reuss at its exit from the lake, and rejoins the St. Gotthard road near the top of the pass.

On the (1 M.) **Pass of St. Gotthard** (6935′) the road passes between several small lakes.

The ST. GOTTHARD is a mountain-group, 160 sq. M. in area, with a number of different peaks, extensive glaciers, and about thirty small lakes. The pass is a barren depression, destitute of view, bounded on the E. by the precipitous *Sasso di S. Gottardo* (8235′), and **on the** W. by the rocks of the *Fibbia* (8995′) and the *Pizzo la Valletta* (8334′). The chief peaks of the St. Gotthard are: E., the *Prosa* (8983′) and *P. Centrale* (9850′; p. 121); W., the *Piz Lucendro* (9708′), *Twerberhorn* (9265′), *Piz dell' Uomo* (8820′), and *Winterhorn* or *Piz Orsino* (8747′); then, more to the W., the *Leckihorn* (10,070′),

Muttenhorn (10,184'), *Pizzo Pesciora* (10,250'), *Pizzo Rotondo* (10,490'), *Kühbodenhorn* (10,080'), etc.

13³/₄ M. **Albergo del S. Gottardo** (6867'), ¹/₄ M. to the S. of the culminating point, is a 'dépendance' of the *Hôtel du Mont Prosa* which stands opposite (telephone to Airolo). The latter is adjoined by the former *Hospice*. On a rock a little to the S. is the old *Mortuary Chapel*.

EXCURSIONS (guides for the shorter ascents at the hotel). To the **Sorescia** or *Scara Orell* (7350'), a pleasant excursion (1 hr.). We descend the road to the S. to the Ticino bridge, and beyond it ascend a narrow path to the left. Fine view, especially of the Ticino Alps, the Cristallina, Campo Tencia, Basodino, etc. Descent to the Sella valley unadvisable, there being no bridge over the Ticino.

*°**Pizzo Centrale** (9850'; 3¹/₂ hrs.; guide 10 fr.), somewhat laborious but highly interesting. Beyond the hospice we cross the brook to the left, and ascend the slope of the Sasso San Gottardo over detritus to the entrance of the *Sella Valley*, through which the route leads. To the left rises *Mte. Prosa* (see below). We skirt the slope high above the *Sella Lake* (7320') and ascend snow-fields to the base of the peak, which consists of crumbling hornblende. The °View is one of striking magnificence, embracing almost all the highest mountains in Switzerland (panorama by A. Heim). The ascent may also be made from Hospenthal in about 5 hrs., viâ the *Gamsboden* and the *Guspis-Thal* (see p. 120). — **Monte Prosa** (8983'; 2¹/₂ hrs.; guide 7 fr.), less interesting. By the hut above the Sella Lake (1¹/₄ hr.) we diverge to the left from the Pizzo Centrale path, and ascend across poor pastures and patches of snow to the (³/₄ hr.) saddle (8520') between the Prosa and Blauberg. Thence to the left, up the arête, and lastly over sharp rocks to (¹/₂ hr.) the summit. The W. peak, 41' higher than the E., is separated from it by a chasm 20' deep.

The **Fibbia** (8995'; 2¹/₂ hrs.; guide 7 fr.), a gigantic rock which commands the St. Gotthard road on the W. and descends suddenly to the Val Tremola, is fatiguing. Excellent survey of the St. Gotthard group, the valley of the Ticino, and the Ticino Alps. — *Piz Lucendro** (9708'; 3¹/₂-4 hrs.; guide, 10 fr., unnecessary for the experienced), a fine point, free from difficulty. From the *Lucendro Lake* (p. 120) we ascend by the *Lucendro Alp* and the depression between the Ywerberhörner and the Pizzo la Valletta to the *Lucendro Glacier* and gradually mount to the rocky summit. — *Leckihorn* (10,070'), see below. — °**Pizzo Rotondo** (10,490'), the highest peak of the St. Gotthard, from the Hôtel Prosa 7-8 hrs. (guide 30 fr.), difficult. We follow the Lecki Pass route (see below) past the Piz Lucendro to the *Wyttenwasser Glacier*, ascend to the left to the *Wyttenwasser Pass* (9365') and skirt the precipitous slopes of the Pizzo Rotondo to the *Passo Rotondo* (9692'), whence we climb to the left to the summit (p. 114).

PASSES. OVER THE ORSINO PASS TO REALP, not difficult (4¹/₂ hrs.; adepts need no guide). We ascend either from the *Rodont Bridge* (p. 120) across the stony *Rodont Alp* and past the *Orsino Lake* (7515'), or from the Lucendro Lake to the N.W. over grassy slopes, past the *Orsirora Lake* (8058'; to the left), to the **Orsino Pass** (about 8530'), S.W. of Piz Orsino (p. 120); striking view (S.) of the St. Gotthard group from the Furka to the Fibbia, (N.W.) of the Finsteraarhorn and Agassizhorn, and (N.) of the Galenstock and Dammastock range as far as the Sustenhörner and Titlis. Descent over the pastures of the *Eisenmanns-Alp* and through brushwood to Realp (p. 125).

OVER THE LECKI PASS TO THE FURKA (10 hrs., guide 30 fr.), fatiguing, but repaying at places. From the Lucendro Lake to the *Lucendro Glaciers* see above; thence across the depression to the N. of *Piz Lucendro* (ascent highly recommended, see above) to the *Wyttenwasser-Thal* and the *Caranna Pass* (p. 125). We then traverse the *Wyttenwasser Glacier*, pass the *Hühnerstock*, and reach (5¹/₂-6 hrs.) the **Lecki Pass** (9555'), lying to the N. of the *Leckihorn* (10,070'; easily ascended from the pass in ¹/₂ hr.). Descent across the *Mutten Glacier*, past the *Muttenhörner*; then an ascent between the *Thierberg* and *Blauberg* to the small *Schwärze Glacier*, and down to the (3¹/₂ hrs.) *Furka Hotel* (p. 126). — Or we may proceed from the Wyttenwasser Glacier to the *Wyttenwasser Pass* (9365') and the *Passo Rotondo* (p. 121) and thence descend to *All' Acqua* in Val Bedretto (p. 315; 10 hrs. from the Hôtel Prosa, an interesting expedition for experienced mountaineers).

From the Hospice to Airolo is a walk or drive of 2-2½ hrs.; in the reverse direction 3 hours. In winter and spring the snow-drifts on the roadside are often 30-40' high, and sometimes remain unmelted throughout the summer. Snow-storms and avalanches are most prevalent on the S. side. About ½ M. to the S.E., below the hospice, the road crosses that branch of the *Ticino* which issues from the *Sella Lake* (see p. 121), and enters the *Val Tremöla*, a dismal valley into which avalanches often fall; it then descends past the *Cantoniera S. Giuseppe* (6010') in numerous windings, avoided by the old bridle-path. Rich Alpine flora. At the *Cantoniera di Val Tremola* (5504') the Val Tremola ends and the *Valle Leventina* (p. 116) begins. *View down to Quinto. To the right opens the *Val Bedretto* (p. 315), from which the main branch of the *Ticino* descends.

22 M. Airolo (3755'), 8½ M. from the St. Gotthard Pass, see p. 114. Travellers going from the St. Gotthard to the *Val Bedretto* need not descend to Airolo, but save an hour by leaving the road below the *Cantoniera di Val Tremola* (see above), at the angle of the first great bend in the direction of the Val Bedretto. The path descends to the right, and at *Fontana* (p. 315) joins the road leading from Airolo to All'Acqua.

34. The Maderaner-Thal.

Comp. Map, p. 68.

The Maderaner-Thal, a picturesque valley about 12 M. in length, enclosed by lofty mountains (N., the *Great* and *Little Windgälle*, the *Great* and *Little Rüchen*, and the *Scheerhorn*; S., the *Bristenstock*, *Weitenalpstock*, *Oberalpstock*, and *Düssistock*), and watered by the turbulent *Kärstelenbach*, is worthy of a visit. Bridle-path (shaded in the early morning) from Amsteg to the (3¼ hrs.) *Hôtel Alpenclub* (3030' above Amsteg; porter 6, horse 12 fr., there and back within two days 24 fr.). Beautiful return-route viâ the *Stafeln* (p. 123), 6-7 hrs., practicable even for ladies.

Amsteg (1760'), see p. 112. We diverge from the St. Gotthard road on the left bank of the *Kärstelenbach* and ascend by a good zigzag path, passing under the huge railway-bridge (p. 113; 178' high), to the *St. Antons-Kapelle;* then over gently sloping pastures, shaded with fruit-trees, to (50 min.) the hamlet of *Bristen* (2615'; Café Fedier, with garden, beyond the chapel, to the right). The path descends a little, crosses by (5 min.) an iron bridge to the right bank of the foaming Kärstelenbach, and again ascends. After 7 min. we avoid a bridge to the right, leading to the narrow *Etzlithal* (see p. 124), in which, ¼ hr. farther up, is a fine waterfall. After 20 min. the path recrosses to the left bank and leads to the (5 min.) houses *Am Schattigen Berg*. It then ascends rapidly to (40 min.) the top of the *Lungenstutz* (3600'; two small inns), and (8 min.) a cross commanding a fine view. Passing through wood at places, we next cross the *Griessenbach* and the *Staldenbach* to (1½ hr.) the chalets of *Stössi* (3904'). Crossing the Kärstelenbach at a (5 min.) *Saw-Mill*, and passing the houses of *Balmwald* on the left, in 25 min. more we reach the *Balmenegg* (4790'; *Hôtel zum Schweizer Alpenclub*, R., L., & A. 3, D. 4, pens. 8-10 fr.; *Engl. Church Service*). Fine view from the terrace on the W. side of the house. Pleasant woodwalks in the vicinity. About 1/2 M. from the hotel is the small *Butzli-See*.

To the Hüfi Glacier, an interesting walk (1 hr., guide unnecessary). From the inn a path, at first through wood, ascends the grassy slopes on the N. side of the valley (passing opposite the falls of the *Brunnibach*, the *Stäuberbach*, and the *Lämmerbach*), crosses the *Schleierbach*, the *Seidenbach*, and the *Milchbäche*, and ascends to (1 hr.) a rocky height (5230'), overlooking the glacier (which has greatly receded), from which the Kärstelenbach issues. We may now descend to the end of the glacier (guide necessary, 3-4 fr.) and return to the hotel on the left bank of the Kärstelenbach, passing the waterfalls above mentioned, and crossing the *Alp Oufern* (3-4 hrs. in all).

Beautiful return-route to Amsteg by the *Stafeln (6-7 hrs.; guide 8 fr.), the lofty pastures on the N. side of the valley. The path first leads to the above-mentioned rock overlooking the Hüfi Glacier (1 hr.), and then ascends to the (1 hr.) *Alp Gnof* (6235'), the (3/4 hr.) *Stäfel-Alpen* (6290'), and the (1/4 hr.) *Alp Bernetsmatt* (6553'; Alpine fare and accommodation), commanding a magnificent view of the Hüfi Glacier, Clariden Pass, Düssistock, Tschingel Glacier, Oberalpstock, Weitenalpstock, Crispalt, Bristenstock, Galenstock, Spitzliberg, the Windgällen, and Ruchen. [A still finer view, especially of the conspicuous Windgällen, is commanded by the *Widderegg* (7840'), 1 1/4 hr. from Bernetsmatt, with guide.] We then descend rapidly to the pretty *Golzern-See* (4636') and the (1 hr.) *Golzern-Alpen* (4583'; excellent drinking-water), and lastly in zigzags through underwood to the hamlet of (1 1/2 hr.) *Bristen* and (1/2 hr.) *Amsteg* (to the station 1/4 hr. more).

EXCURSIONS FROM THE HÔTEL ALPENCLUB. (Guides: *Ambros, Carl Ambros*, and *Josef Zgraggen; Josef, Josef Maria, Melch.*, and *Joh. Jos. Tresch; Joh., Jos.*, and *Melchior Gnos; David* and *Jos. Furger; Alpin Walker*, and others; ordinary excursions, 6 fr. per day.) The ascent of the Düssistock (*Piz Git*, 10,702'; 6-7 hrs.; guide 25 fr.) is difficult and requires experience. The path leads up the *Brunni-Thal* to the (2 hrs.) *Wallersfirren Alp* (6330'), ascends to the left to the (2 hrs.) *Restli-Tschingel Glacier*, and crosses it; we then clamber over the precipitous rocks of the *Kleine Düssi* (10,280') and ascend the arête to the (2 hrs.) summit. Splendid view. — The *Oberalpstock (*Piz Tgietschen*, 10,925'; guide 20 fr.) presents no serious difficulty to adepts. We either proceed from the Alpenclub Hotel by the Brunni Pass route (p. 124) to the upper part of the (4 1/2-5 hrs.) *Brunni Glacier* (p. 124), and mount the snowy slopes, to the right, to the summit in 2-2 1/2 hrs.; or ascend from the *Kreuzli Pass* (p. 124) across the *Strim Glacier* (7-8 hrs. to the top). Ascent from Sedrun (5 1/2-6 hrs.), see p. 373. — Weitenalpstock (9870'), from the *Alp Culma*, on the Kreuzli Pass route (4 hrs. from Amsteg), over the *Weiten-Alp* in 4 1/2 hrs., very toilsome. — Bristenstock (10,090', see p. 112. — Piz Cambriales (10,590'), 4-5 hrs. from the Hüfi Club-Hut (see p. 124), and Claridenstock (10,720'; 25 fr.), 5 hrs. from the club-hut; not very difficult for practised climbers. Kammlistock (10,624'; 25 fr.), 5 hrs. from the club-hut, by the *Kammlilücke*, laborious. — The Grosse Windgälle or Kalkstock (10,463'), from the Alp Bernetsmatt (see above) by the *Stäfel Glacier* in 5 hrs. (guide 30 fr.), and the Grosse Scheerhorn (10,815'), from the Hüfi Club-Hut by the *Kammlilücke* in 6 hrs. (guide 25 fr.), both very difficult. — Grosse Ruchen (10,295'), less difficult, but extremely fatiguing from the *Alp Gnof*, 4-5 hrs.; guide 20 fr.). — The **Kleine Windgälle** (9800'), rom the *Ober-Käsern* huts (6390'; 3½ hrs. from Amsteg, 1½ hr. from Bernetsmatt) in 3½ hrs. (guide 20 fr.), not very difficult.

PASSES. To STACHELBERG over the *Clariden Pass* (9843'), 11-12 hrs. from the Alpenclub Hotel, a grand and most interesting expedition, presents no serious difficulty to experts with able guides (35 fr.). The route ascends the slopes of the Düssistock (p. 123), on the left bank of the *Hüfi Glacier*, to the (2½-3 hrs.) *Club-Hut* on the finely situated *Hüfi Alp* (5905'; spend night). Then a steep ascent for a short distance, over the moraine to the (40 min.) *Hüfi Glacier*, and gradually up the *Hüfifirn* and *Claridenfirn* to the (3-8½ hrs.) *Pass*, between the *Hinter Spitzalpeli-Stock* (9853') and the *Claridenhorn* (10,184'), commanding a fine view of the Tödi, the Rheinwaldgebirge, etc. We then descend the Claridenfirn, passing the *Bockschingel*, a rock with a hole through its middle, and the *Gemsfayrenstock* (p. 69), and traverse the difficult *Wallenbach Gorge* to the *Altenoren-Alp* and (5 hrs.) *Stachelberg*. Or from the Claridenfirn (keeping to the right before reaching the Clariden Pass) we may cross the **Hüfi Pass** or **Planura Pass** (9645'), between the *Hinter Spitzalpelistock* and the *Catscharauls* (10,045'), to the *Sandfirn*, and then either descend to the left to the *Upper Sandalp* (p. 70) or to the right by the *Sandgrat* to *Disentis* (p. 372; guide 30 fr.). — Another pass to Stachelberg (12-13 hrs. from the Alpenclub Hotel; guide 30 fr.) is the **Kammlilücke** (*Scheerjoch*; 9268'), lying between the *Scheerhorn* and the *Kammlistock* (p. 123). Descent over precipitous ice-slopes to the crevassed *Gries Glacier*, the *Kammli Alp*, and the *Klausen Pass* (p. 72).

To UNTERSCHÄCHEN over the **Ruchkehlen Pass** (8790'), 8-9 hrs., laborious (guide 25 fr.). From the *Alp Gnof* (p. 123) we ascend precipitous grass-slopes, rock, and glacier to the pass, between the *Sattelhörner* and the *Kleine Ruchen*, and descend steeply through the ice-clad *Ruchkehle* into the *Brunni-Thal* and *Schächen-Thal* (p. 72). — The **Scheerhorn-Griggeli Pass** (9180') is also toilsome. From the Hüfi Club-Hut we mount the Hüfi Glacier and the *Bockischingelfirn* to the pass, between the Scheerhorn and the Kleine Ruchen, and descend to the *Upper Lammerbach-Alp* and *Unterschächen*.

To DISENTIS over the Brunni Pass (8875'), 8 hrs., interesting but fatiguing (guide necessary, 25 fr.). We ascend the *Brunni-Thal* by *Rinderbiel* and *Waltersfirren* (p. 123) to the (2½-3 hrs.) *Brunni-Alp* (6990'), cross the E. lateral moraine and the upper snow-fields of the *Brunni Glacier* to the (2½ hrs.) pass between the *Piz Cavardiras* (9505') on the left and the *Piz d'Acletta* (9570') on the right, and descend through the *Val Acletta*, past the small *Lac Serein*, to *Acletta* and (3½ hrs.) *Disentis* (p. 372).

FROM AMSTEG over the KREUZLI PASS (7645') TO SEDRUN, 8 hrs., fatiguing. Through the *Etzlithal* to the pass, 5½ hrs.; thence down the *Strimthal* to *Sedrun* (p. 373), 2½ hrs.

35. From Göschenen to the Rhone Glacier. The Furka.

Comp. Map, p. 118.

25 M. DILIGENCE in summer twice daily in 6½ hrs. (9 fr. 85, coupé 11 fr. 85 c.); from Göschenen to Brig daily in 12 (Brig to Göschenen 14) hrs., with ½ hr.'s halt at Tiefenbach, and dining at the Rhone Glacier (20 fr. 85 c., coupé 25 fr. 15 c.); from Göschenen over the Furka and Grimsel to Meiringen in 11½ hrs. (19 fr. 15 c., coupé 23 fr. 5 c.). — PEDESTRIANS should allow the following times from Göschenen: to Andermatt 1¼, Realp 2, the Furka 3½ (return 2½), Rhone Glacier 2 (return 2½) hrs. — HORSE from Realp to Tiefenbach 5, Furka 8 fr. — CARRIAGES with one horse from Göschenen to Realp 10 fr., with two horses 15 fr.; to the Rhone Glacier ('Gletsch') 35 and 65, Fiesch 55 and 100, Brig 75 and 140, Meiringen 72 and 135 fr.; carr. and pair from Andermatt to Realp 15, the Furka 40, Rhone Glacier 60, Fiesch 90, Brig 125 fr.; from Hospenthal to Realp, with one horse 6, two horses 10, to Furka 20 (there and back 25) and 35, Rhone

Glacier 30 and 50, Fiesch 50 and 90, Brig 70 and 120 fr.; from Realp to the Furka, with one horse 12, two horses 20 fr., Rhone Glacier 18 and 25 fr.; from the Rhone Glacier to the Furka 15 fr.

The *Furka Road, constructed chiefly for military purposes, and forming a convenient route to or from the Grimsel and the Bernese Oberland, commands striking views of the Rhone Glacier and the neighbouring mountains, and from Realp onwards should be traversed in an open carriage or on foot. Rich flora.

To (5 1/2 M.) *Hospenthal* (4800'), see pp. 119, 120. At the upper end of the village the road diverges to the right from the St. Gotthard route, ascends a little, and skirts the level bank of the *Realper Reuss* in the bleak *Urseren-Thal* (p. 119). On each side rise steep grassy slopes, furrowed by numerous brooks, and overshadowed on the N. by the jagged pinnacles of the *Spitzberge* (10,053'). 2 1/4 M. *Zumdorf* (4965'), a group of huts with a chapel. Farther on we cross the Reuss and the *Lochbach*, which descends from the Tiefen Glacier (see below), and soon reach (1 3/4 M.) —

9 1/2 M. Realp (5060'; *Hôt. des Alpes, Post*, both plain), a poor hamlet at the W. end of the Urseren Valley.

Over the *Alpligen-Lücke* to (6 hrs.) the *Göschener-Alp*, see p. 118; over the *Orsino Pass* to the *St. Gotthard*, see p. 121. — From Realp to' *Villa* in the *Val Bedretto* (p. 315) by the Cavanna Pass (8565'), between the *Piz Lucendro* and *Hühnerstock*, 5 hrs., uninteresting.

Beyond Realp the road begins to ascend in long windings, which the old road to the right, 50 paces beyond the second bridge, 1/2 M. from Realp, avoids. (In descending from the Furka we quit the new road a few hundred paces beyond the 50th kilomètre stone, and descend by a few steps to the left.) We soon obtain a fine retrospective view of the broad Urseren-Thal, with the zigzags of the Oberalpstrasse in the background (p. 375); on the left are the Wyttenwasser-Thal with the glacier of that name, the Ywerberhörner, and the Piz Lucendro. At the last winding of the road (*Fuchsenegg*, 6595'), 3 1/2 M. from Realp, stands the small *Hôt.-Pens. Galenstock* (R. 2, D. 3 1/2, pens. 6 fr.). About 1 1/2 M. farther on, beyond the *Ebneten-Alp*, is Tiefenbach (6790'; *Hôtel Tiefengletscher*, well spoken of, R., L., & A. 2 1/2, lunch 2 1/2, D. 3 1/2, pens. 5-7 fr.), where the diligence halts some time.

By following the slope from this point and crossing the moraine, we reach (1 1/4 hr.; guide) the Tiefen Glacier, imbedded between the Galenstock and the *Gletschhorn* (10,850'), where beautiful crystals (more than 12 1/2 tons) were found in 1868 (p. 150). — Over the *Tiefen-Sattel* to the *Rhone Glacier* (*Grimsel, Trift-Hütte*), see p. 137. — Over the *Winterlücke* (9450') to the *Göschener-Alp* (p. 118), 6 hrs., with guide; descent to the Winter Glacier steep.

The road crosses the *Tiefentobel* and ascends, running high up on the N. slope. The old bridle-path (not recommended) follows the *Garschen-Thal* on the left, far below. On the right lies the *Siedeln Glacier*, the discharge of which forms a fine waterfall; above it rise the pinnacles of the *Bielenstock* (9670'). Before us rises the *Furkahorn* (p. 126). The (3 1/2 M.) —

17 1/2 M. Furka (7990') is a saddle between the Muttenhörner on the left and the Furkahörner on the right, descending abruptly

on both sides. We first reach the barracks for the garrison of the fortifications (see below) and the new *Hôtel-Restaurant Furkablick* (R. 2, B. 1¹/₄, lunch 2¹/₂, D. 3¹/₂ fr.). A little farther on, to the left of the road, is the *Hôtel-Pension Furka* (R., L., & A. 3-5, déj. 4, D. 5 fr.; post and telegraph office). Magnificent view of the Bernese Alps with the imposing Finsteraarhorn; to the left of it the Oberaarhorn, Walliser Fiescherhörner, Siedelhorn, and Wannehorn, and to the right the Agassizhorn and Schreckhörner. From the *Signal*, about ¹/₂ M. from the hotel, we obtain a view of the Upper Valais and its Alps (Mischabelhörner, Matterhorn, Weisshorn, etc.), while the *Känzli*, ¹/₂ M. farther on, also commands the upper part of the Rhone Glacier.

EXCURSIONS. °Furkahorn (9935'; 2¹/₂ hrs.; guide 7 fr., not necessary for adepts), to the N. of the pass, by an easy bridle-path; very interesting. Admirable panorama of the Alps of Bern and the Valais, the Galenstock, St. Gotthard group, etc. The nearer summit (9248') may be scaled in 1¹/₄ hr. It is not advisable to descend direct to the Rhone Glacier. — The Blauberg (9110'), to the S. of the Furka road, is easily ascended by a new path in 2 hrs. (attractive). — *Muttenhorn (10,184'; 3 hrs.; guide 10 fr.), to the S. of the Furka, a very fine point, not difficult.

Galenstock (11,805'; 5 hrs.; guide 15 fr.), not difficult for adepts under favorable conditions of the snow (axe and rope required). From the Furka we ascend to the (³/₄ hr.) *Rhone Glacier* (see below), skirt its left margin, climb a steep snowy slope to the right, follow a rocky arête, and lastly mount very steep névé to the overhanging snowy summit (caution required). View exceedingly grand.

From the Furka over the *Lecki Pass* to the *St. Gotthard Hospice* (10 hrs., with guide), see p. 121; over the *Trift-Limmi* to the *Trift-Hütte*, see p. 137.

TO THE GRIMSEL HOSPICE (p. 188), 5 hrs. (guide 10 fr.; alpenstock and nailed boots requisite). Walkers may descend from the Furka by a good path, diverging to the right from the road ¹/₂ M. from the inn, to the (³/₄ hr.) upper part of the *Rhone Glacier*, cross it above the ice-fall in 1¹/₂ hr., ascend the (³/₄ hr.) *Nägeli's Grätli* (8470'), affording a splendid view of the Bernese and Valaisian Alps, and descend to the (2 hrs.) Hospice. The path issues at the N. extremity of the small Grimsel Lake (p. 189).

The road follows the slope to the right, passing the new fortifications of the Furka, to the (1¹/₄ M.) *Galen-Hütten* (7900') and descends to the left in long zigzags, high above the huge *Rhone Glacier (p. 314), affording admirable views of its fantastic icemasses. [A shorter footpath descends to the left from the Hôtel Furka, skirting the Gratschlucht Glacier.] At the second bend of the road is the small *Hôtel Belvedere*. A path leads hence in ¹/₄ hr., over the moraine, keeping to the left, to a point commanding the upper part of the glacier. Adjacent is an artificial glacier-cave. In the valley we cross the *Muttbach* (the discharge of the *Gratschlucht Glacier*). The road is joined here on the left by the steep old bridle-path from the Furka. It then gradually descends the slope of the *Längisgrat*, and again describes several long bends, which the old bridle-path, to the right, cuts off. Crossing the infant *Rhone*, we reach the (6¹/₄ M.) —

25 M. *Rhone Glacier Hotel*, in the '*Gletsch*' (5750'; p. 314).

From the Rhone Glacier to *Brig*, see p. 315; over the *Grimsel* to *Meiringen*, see R. 52.

36. From Lucerne to Altdorf viâ Stans and Engelberg. The Surënen Pass.

Comp. Map, p. 86.

STEAMBOAT from Lucerne to Stansstad 8 times daily in 40 min., fare 1 fr. 40 or 80 c. (see p. 100). — DILIGENCE from Stansstad to (14 M.) Engelberg twice daily in summer in 3½ hrs. (fare 4 fr. 60, coupé 6 fr. 40 c.; from Lucerne, incl. steamboat, 6 fr. 60, 7 fr. 75 c.); one-horse carriage 15, two-horse 25 fr. — Walkers may dismiss their vehicle at Grafenort (7 M. from Stans, a drive of 1¼ hr.), one-horse carr. 7-8, two-horse 12 fr.), beyond which the road is so steep that travellers usually alight and walk. (One-horse carr. from *Beckenried* to Engelberg, the route for travellers from the St. Gotthard, 15-18, two-horse 25-30 fr.; see p. 88.) — From Engelberg to Altdorf over the Surënen Pass, rather fatiguing (bridle-path, 9 hrs.; guide, 14 fr., unnecessary in fine weather; travellers from Altdorf need a guide to the top of the pass only, 8 fr.).

To *Stansstad*, see p. 101. The road (electric tramway in ¼ hr.) leads between the *Bürgenstock* (p. 101) on the left and the *Stanser Horn* (see below) on the right, through orchards and pastures.

2 M. **Stans** or Stanz (1510'; pop. 2458; *Engel*, R., L., & A. 2-3, B. 1, D. 2½-3 fr.; *Krone*, R. 1-1½, B. 1, D. 1½, pens. 3-4 fr.; *Winkelried; Rössli*), the capital of *Nidwalden*, the E. half of Canton Unterwalden, lies in the midst of a vast orchard, on which, however, from 11th Nov. to 2nd Feb. the sun shines for one hour only in the morning, between the *Hohe Brisen* (7890') and the *Stanser Horn* (see below). Adjoining the handsome *Parish Church* is the *Monument of Arnold von Winkelried* (p. 19), a fine group in marble by *Schlöth*. A tablet by the *Burial Chapel* in the churchyard, on the N. side of the church, commemorates the massacre perpetrated here in 1798 by the French, who were exasperated by the obstinate resistance they met with. The **Town Hall** contains portraits of all the mayors from the year 1521; below them is a collection of Unterwalden flags; a picture by the artist Würsch, who afterwards became blind, and perished in 1798; another by Volmar, representing Brother Klaus taking leave of his family (p. 132). In the studio of the late painter *Deschwanden* a number of his paintings are exhibited gratis. The *Historical Museum*, in the Bahnhof-Platz, contains objects of historical and antiquarian interest, weapons, coins, minerals, a library, and an interesting relief of Stans on the scale of 1:500 (key kept by *Jac. Christen*, behind the Hôt. Winkelried; 30 c.). Fine view from the *Knieri*, above the *Capuchin Monastery*.

The *Stanser Horn* (6230') is a splendid point of view, scarcely inferior to Rigi and Pilatus. Cable-railway (opened in August, 1893) in 50 min.; 5 fr., return-ticket 8 fr. or, including S., R., and B. at the hotel, 15½ fr. The line (4265 yards in length; maximum gradient 60:100) is divided into three sections, and carriages are changed twice. Each section has its own power house; the electric motors are supplied from the central station at Buochs. In the middle of each section is a crossing, where the ascending and descending cars pass each other; there is no toothed rail, but safety is guaranteed by strong automatic brakes. — The line ascends gradually (12:100) from the entrance of the village through luxuriant meadows, and farther on more rapidly (27 : 100) to the (13 min.) station of *Kälti* (2343'), where carriages are changed. The second section has a gradient at first of 40 : 100, afterwards of 60 : 100; the line ascends a wooded **ravine**, crosses a torrent, and intersects a deep cutting to the (13 min.) second station of *Blumatt* (4606'), whence it proceeds (third section) with the same gradient (3 : 5) through a tunnel (150 yds.) to the terminal station (6076'), at the *Hotel Stanserhorn* (pens. 8 fr.). A good path leads hence to the top (60' higher), which commands a highly picturesque *View of the Bernese Apps (with the Titlis rising in the foreground), the Lake of Lucerne, and the hills of N.W Switzerland.

The road to (12 M.) Engelberg **traverses the valley of the** *Engelberger Aa*, **between the Stanser Horn on the right and the Buochser Horn on the left. In the background rises the snow-clad Titlis.** Near (2 M.) *Dallenwyl* we cross the Aa. On a mound of detritus at the mouth of the *Steinbach*, to the right, stands the church of the village.

A good bridle-path, diverging to **the left, ascends to** (4½ M.; 6 M. from Stans viâ *Nieder-Büren*; one-horse **carr. from Stansstad to Büren in** 1 br., 4 fr.; from Buochs 5 fr.) the finely-situated health-resort of Nieder-Rickenbach (3830'; *Curhaus zum Engel*, pens. 5-6 fr.). Hence to the *Buochser Horn* (5395'), 1¾ hr., repaying; to the *Steinalp-Brisen* (7890'), 3 hrs., viâ the *Ahorn-Alp* and the *Steinalp*, interesting (guide not indispensable for adepts). Another attractive ascent is that of the *Schwalmis* (7373'; 3 hrs.; guide unnecessary), by the *Ahorn-Alp*, the *Bärfalle* (with a cross), and the *Bähl-Alp*, and thence up the E. arête. The descent may be made to (3 hrs.) Isenthal viâ the *Jochli* (see below). — An interesting pass (4½+5 hrs., with guide) leads from Nieder-Rickenbach by the *Bähl-Alp* (see above) and the Jochli (6925') between the Schwalmis and the Reissendstock, descending by the *Boljen-Alp* and the *Laueli* to *St. Jakob* in the Isenthal (p. 93).

1¾ M. **Wolfenschiessen** (1710'; *Eintracht*, unpretending; *Kreuz*). Beside the church is the hermit-hut (brought hither from Altzellen) of *Conrad Scheuber*, **grandson of Nikolaus von der Flüe** (p. 132), whose worship he shares.

From Wolfenschiessen viâ *Ober-Rickenbach* and the *Schönegg Pass* (6315') to (5½-6 hrs.) *Isenthal*, see p. 93. Guide advisable, the descent from the pass to the *Sulzthal-Alp* being steep and pathless.

Beyond (2½ M.) *Grafenort* (1885'; Inn, good wine) the road ascends through beautiful wood. To the right, far below, flows the brawling Aa. We next pass (4 M.) the Inn 'Im Grünen Wald', below which, in the valley to the right, **the brook** descending from the Trübsee (p. 136) falls into the Aa. After another slight ascent, we turn to the left, and suddenly obtain a view of the *Engelberger-Thal*, a green Alpine valley, 5 M. long and 1 M. broad, bounded on three sides by lofty, snow-clad mountains The *Titlis* with its ice-mantle stands forth majestically, and to the left rise the **rocky pinnacles of the Great** and *Little Spannort* (p. 130); in the foreground is the **Hahnenberg** or *Engelberg* (8566'). Then (2 M.)—

12 M. **Engelberg.** — **Hotels.** '**HÔT.-PENS.** SONNENBERG, finely situated, R., L., & A. 3½-6½, B. 1½, D. 4½, S. 3, pens. 8-11 fr.; °HÔT. CURHAUS TITLIS, R., L., & A. 2½-6, B. 1½, lunch 3, D. 4½, pens. from 7½ fr.; HÔT. NATIONAL, R., L., & A. 2½-6, B. 1½, D. 4, pens. from 8 fr.; 'HÔT.-PENS. ENGEL, R., L., & A. 1½-3, B. 1¼, D. 3½, S. 2½, pens. 6½-8 fr.; *CURHAUS-PENS. MÖLLER, R., L., & A. from 1, B. 1¼, D. 3, pens. 6-8 fr.; °HÔT.-PENS. ENGELBERG, R., L., & A. from 2, D. 3, pens. 6-8 fr.; °HÔT.-PENS. HESS, R., L., & A. from 2, D. 3, pens. 6-8 fr.; °HÔT. DES ALPES, R., L., & A. from 1½, D. 3, pens. 6-8 fr. Rooms at several other

to Altdorf. **ENGELBERG.** *II. Route 36.* 129

houses; usual charges, R. 1½, B. 1. Beer at *Waser's*. — *English Church* in the grounds of the Hôtel Titlis. — Guides: *Karl, Eugen*, and *Jos. Hess; Leodegar Feierabend; Jos. Kuster*, father and son; *Placidus Hess; Jos. Amrhein; Jos. Infanger; N. Hurschler; C.* and *Joh. Waser.*

Engelberg (3315'; pop. 1973), loftily and prettily situated, and sheltered from the N., is a favourite health and summer resort. At the upper end of the village rises the handsome Benedictine Abbey of the name, founded in 1121, named *Mons Angelorum* by Pope Calixtus XI., and rebuilt after a fire in 1729.

The *CHURCH contains modern pictures by *Deschwanden, Kaiser*, and *Würsch* (p. 127). High-altar-piece, an Assumption by *Spiegler*, 1734. In the chapter-house, two transparencies by Kaiser, the Conception and the Nativity. The LIBRARY (20,000 vols., 210 MSS.), which was pillaged by the French in 1798, contains a good relief of the Engelberg Valley. Permission to visit the monastery is now not very often granted. — The SCHOOL connected with the abbey is well attended. The FARM BUILDINGS, with the labourers' dwellings, are very extensive, and in the cheese-magazine several thousand cheeses are frequently stored at one time. The revenues of the abbey, which formerly exercised sovereign rights over the surrounding district, were considerably reduced by the French in **1798**.

Opposite the abbey, to the S., on the left bank of the Aa, are pleasant shady walks, which are **reached** in 10 min. (Café Bänklialp).

EXCURSIONS. A favourite promenade, with pretty views, leads to the Schwand (3970'; *Inn*), in 1¼ hr. — The Bergli (4300'; *Inn*) and the Flühmatt (4355'), each 1 hr., command an excellent view of the valley and the Titlis. — Pleasant walk (brake several times daily, 60 c.; one-horse carr. there and back 5-6, two-horse 9 fr.) to the (¾ hr.) *Tätschbach Fall. We may either follow the road to the right of the abbey, passing (1¼ M.) the coffee-garden of *Eienwäldchen*, or we may take the shorter footpath, to the left of the abbey, which passes (10 min.) the *Neue Heimat Inn*, at the mouth of the *Horbis-Thal*, and the (¼ hr.) *Schweizerhaus Inn*. [The rocky basin at the head of the Horbisthal, reached in ½ hr., is known as the *End der Welt*.] The road ends at the Tätschbach Fall (3575'; Inn), which descends from the Hahnenberg or Engelberg. The bridle-path (way to the Surenen Pass, p. 130) goes on through the wood and crosses the *Fürrenbach*, which also forms several falls. It then traverses pastures, passing the dairy-farm of *Herrenrütti* (left bank), to (¾ hr.) the Nieder-Surenen Alp (4133'; rfmts.), which affords a fine view of the pyramidal Schlossberg, the serrated Spannörter, the Firnalpeli and Grassen glaciers, and the huge precipices of the Titlis. — The Arnitobel, a gorge with a waterfall, ¾ hr. to the W., a pleasant and shady walk; thence to the right to the (1 hr.) *Lower Arnialp* (4850'; Inn), with a good view of the Engelberger Rothstock, and to (1 hr. farther) the *Upper Arnialp* (5800'), commanding a beautiful survey of the Engelberg valley. — *Fürrenalp (6073'; 2½ hrs.); the path ascends to the left before reaching the Tätschbach Fall, and then skirts the slope above (beautiful view of the Titlis).

ASCENTS. Rigidalstock (8515'; 5 hrs.; guide 9 fr.), the last part difficult, fine panorama. *Widderfeld (7723'; 4 hrs.; guide 8 fr.), vià the *Arni-Alp*, less fatiguing; preferable vià the *Zingel-Alp* and *Hohlicht* (5 hrs.; guide 8 fr.). — Hutstock (8790'; 6-7 hrs.; guide 12 fr.), by the *Juchli* (p. 132), not difficult for mountaineers. — The *Hanghorn (8790'), an attractive point, is reached in 6-7 hrs (guide 12 fr.) by crossing the slope of the Schattband, in front of the Hutstock. — *Engelberger Rothstock (9250'; 6 hrs.; guide 9, with a night out 12 fr.), interesting and not difficult. We ascend by the *Alp Obhag* to the (4 hrs.) *Club-Hut* above the *Planken-Alp* (7560'), on the *Ruckhubel*, not far from the *Griessen Glacier*; thence below the *Rothgrätli* (p. 96) to the top in 2 hrs. more.

*Uri-Rothstock (9620'; 8½ hrs.; guide 17, with descent to Isenthal 22 fr.), very interesting. From the club-hut above the Plankenalp to the

BAEDEKER, Switzerland. 16th Edition. 9

130 II. Route 36. SURENEN PASS.

(1¼ hr.) gap (8878') on the S. of the Engelberger-Rothstock; thence across snow to the (1 hr.) *Porta* or *Schlossstock-Lücke*, adjoining the *Schlossstock* (9055'); then a rather steep descent to the *Blümlisalpfirn*; again an ascent to the arête separating it from the Kleinthal, and lastly up the *Kleinthalfirn* to the (2½ hrs.) top (comp. p. 93).

The *Great Spannort (10,515') is ascended from the *Spannort Club-Hut* (6500'), 3½ hrs. from Engelberg, by the *Schlossberg-Lücke* and the *Glattenfirn*, or direct by the *Spannort-Joch* (see below) in 4½ hrs.; interesting, though toilsome (comp. p. 112; guide 25 fr.). — The Little Spannort (10,380') is climbed from the Spannort Hut by the Schlossberg-Lücke or the Spannort-Joch in 6-7 hrs. (guide 35 fr.); difficult, for expert climbers only. — Schlossberg (10,280'), from the *Blacken-Alp* (see below) in 4½ hrs., laborious (guide 25 fr.). The admirable view is scarcely inferior to that from the Titlis. Edelweiss abundant.

The *Titlis (10,627'; 6-7 hrs.; guide 12, to Engstlen-Alp 17 fr.) is very interesting, though for novices somewhat trying. It is advisable to go on the previous evening to the *Hôtel Hess* (p. 136; 2¼ hrs.; horse 10 fr.), in order not to have the steep *Pfaffenwand* (p. 136) to ascend at starting. From this point the guides like to start at 2 a.m., in order that on the return-route the snow may be traversed before the heat of the day; but the ascent by lantern-light is very disagreeable and toilsome, and it is better to wait till daybreak. From the Hôtel Hess the path ascends over the *Laubersgrat* to the (1½-2 hrs.) *Stand* (8033'); it then mounts a steep slaty incline in zigzags, over rock and detritus, to the (¾ hr.) *Rothegg* (9030'), where the glacier is reached and a short rest is taken. We ascend the glacier, at first gradually, then more rapidly (step-cutting sometimes necessary), and if the snow is in good condition we reach the (1½-2 hrs.) summit, called the *Nollen*, without material difficulty. The view, highly picturesque and imposing, embraces the entire Alpine chain from Savoy to Tyrol, N. Switzerland, and S. Germany (panorama by Imfeld). The ascent of the Titlis, though requiring perseverance, is perhaps the least difficult of glacier-excursions. Descent to the Joch Pass (Engstlen-Alp), see p. 135.

PASSES. From Engelberg over the *Joch Pass* to *Meiringen* (9½-10 hrs.; guide, unnecessary, to Engstlen 6 fr.), see R. 38; over the *Storegg* (5 hrs.; guide 12 fr.) or the *Juchli* (6½ hrs.; guide 12 fr.) to the *Melchthal*, see p. 132; over the *Rothgrätli* to the *Isenthal* (10 hrs.; guide 17 fr.), see p. 93.

FROM ENGELBERG TO ERSTFELD (p. 111) viâ the Schlossberg-Lücke (8635') and the *Glattenfirn* (10 hrs.; guide 25 fr.), a fine route, but fatiguing. By spending a night in the *Spannort Hut* (see above; 2 hrs. below the pass) mountaineers may combine the ascent of the *Great Spannort* (see above) with this pass. — To Erstfeld across the Spannort-Joch (9610'; 10-11 hrs.; guide 25 fr.), between the Great and the Little Spannort, toilsome.

To WASEN over the Grassen Pass (*Bärengrube*, 8917'), 10 hrs., difficult (guide to Meien 25 fr.). — To THE STEINALP over the Wenden-Joch (8695'), 10-11 hrs., fatiguing, but interesting (guide 25 fr.).

The route to the Surenen Pass leads past the Tätschbach Fall and the dairy-farm of Herrenrüti (p. 129), follows the right bank of the Aa to (1¾ hr.) the frontier of Canton Uri by the *Nieder-Surenen Alp* (4133'), and ascends to the (½ hr.) *Stäffeli* (4652'). After a steep ascent to the (50 min.) *Stierenbach Fall* (best viewed from below), we cross (5 min.) the brook, and in 40 min. more recross it to the *Blacken-Alp* (5833'), with its chapel. The path then ascends gradually over snow, which melts in July, to the (1½ hr.) Surênen Pass (7560'), on the S. side of the *Blackenstock* (9587').

The Titlis becomes grander as we ascend, and we observe a long range of peaks and glaciers, particularly the Klein- and Gross-Spannort and the Schlossberg, extending as far as the Surênen. On the other side we survey the mountains enclosing the Schächen-Thal, on the opposite side of the Reuss, the Windgälle being most conspicuous. On the E. side of the Surenen the snow, which never entirely melts, is crossed in 1/4 hr. in the height of summer. Then a steep descent to the (11/4 hr.) *Waldnacht-Alp* (4754'), which is visible in the long valley below. At a stone bridge (3/4 hr.) the road divides. The very steep path in a straight direction leads to (13/4 hr.) *Altdorf* (p. 110); that to the right, crossing the bridge, to (2 hrs.) *Erstfeld* (p. 111). By the latter we reach the (5 min.) *Bockitobel*, with the picturesque falls of the *Waldnachtbach* (beyond which the guide may be dismissed), descend through wood into the valley, traverse the pastures to the village of *Erstfeld*, and cross the Reuss to the station on the St. Gotthard line (p. 111).

37. From Lucerne over the Brünig to Meiringen and Brienz (Interlaken).

Comp. Maps, pp. 86, 87, 152.

RAILWAY from Lucerne to (281/2 M.) *Meiringen* in 3 hrs. (fares 7 fr. 90, 5 fr. 45, 3 fr. 55 c.); to (36 M.) *Brienz* in 33/4 hrs. (fares 10 fr. 30, 7 fr. 25, 4 fr. 25 c.). From Brienz to *Interlaken*, railway and steamboat in 11/3-2 hrs. — STEAMBOAT (preferable if time permit) from Lucerne to Alpnach-Stad (3/4-11/2 hr.; p. 100); the direct trips are timed to connect with the Brünig Railway at Alpnach-Stad. From Alpnach-Stad to *Vitznau (Rigi)* direct steamer thrice daily in 1-11/2 hr.

The "Brünig Railway, opened in 1888-89, is, as far as (10 M.) Giswyl, *i. e.* about halfway, an ordinary narrow-gauge line, but from that point it surmounts the pass (3295') alternately by means of the 'rack-and-pinion' system and the ordinary system, with a maximum gradient of 18:100. Best views to the *right*. In picturesque beauty, however, the old Brünig Road is superior, and those who visit the Bernese Oberland for the first time may still cross the Brünig to Meiringen on foot, from Giswyl or Lungern.

Lucerne (Brünig Railway Station, Pl. E, 4; restaurant), see p. 81. The BRÜNIG RAILWAY runs to the S.W. in a wide curve into the broad valley of the *Allmend*, and leaving *Kriens* (p. 86), at the foot of the Sonnenberg, to the right, passes (3 M.) *Horw* (the village with its pretty church lies to the left), and approaches the S.W. arm of the *Lake* of *Lucerne* (p. 101). 51/2 M. Hergiswyl (*Rössli*), at the foot of *Pilatus* (bridle-path to the *Hôtel Klimsenhorn*, p. 103). The railway now pierces the rocky *Lopperberg* by means of a tunnel, 3/4 M. in length, and skirts the *Lake of Alpnach* to —

8 M. Alpnach-Stad (*Hôt.-Pens. Pilatus; Rössli; Stern*), the starting-point of the *Pilatus Railway*; see p. 102.

Thence the line proceeds through the partly marshy valley of the *Aa* and across the *Kleine Schlierenbach* to (91/2 M.) Alpnach or *Alpnachdorf* (1530'; *Krone; Sonne; Schlüssel*). The church of Alpnach with its slender spire was erected with the proceeds of the sale of timber from the forests of Pilatus, which were rendered accessible by a wooden slide, 8 M. long, and were cut down in 1811-19.

Beyond Alpnach the train crosses the broad stony bed of the *Grosse Schlieren* and the *Saarner Aa*, the right bank of which it follows past *Kägiswyl* (on the right), with its large parquetry-factory, to (1 M.) *Kerns-Kägiswyl* (1620'), the station for the *Melchthal*. The Grosse Melchthal, an idyllic valley, 15 M. long, studded with numerous chalets and watered by the *Melch-Aa*, well repays a visit. From the station a diligence plies daily in 2¾ hrs. to the village of Melchthal, viâ (1/2 hr.) Kerns (1865'; *Krone; Hirsch; Rössli*), a considerable village with a pretty church, finely situated at the foot of the *Arvigrat* (6920'). At the entrance of the Melchthal, 3 M. from Kerns and 3¾ M. from Sarnen, is *St. Niklaus* (2752'), or *St. Klaus*, the first Christian church erected in this district. The ancient tower adjoining it is locally known as the *Heidenthurm* (heathens' tower). In the ravine of the Melchaa, opposite, below *Flühli* (p. 139), is the *Ranft*, formerly a barren wilderness, with the hermitage of ST. NIKOLAUS VON DER FLÜE, who is said to have lived here for twenty years without other food than the sacramental elements, of which he partook monthly. After their victory over Charles the Bold of Burgundy in 1482, the confederates assembled at Stans disagreed about the division of the spoil, but through the intervention of the venerable hermit the dispute was soon amicably settled. After his death (1487) he was canonised. His memory is still revered by the people, and there is scarcely a hut in the Forest Cantons that does not possess a portrait of Brother Klaus.

From the hermitage the road proceeds to the (3 M.) village of Melchthal (2933'; *Höt.-Pens. Alpenhof*, pens. 5 fr.; accommodation also at the curé's) and the (3 M.) *Alp Stöck*, at the foot of the precipitous *Rumisfluh* (6115'), whence a new road, practicable for light vehicles, leads to (6 M.) Melchsee-Frutt (see below). At the *Ohr-Alp* (3075'), 3 M. to the E. of Melchthal, is one of the largest maple-trees in Switzerland, with a girth of 37½ ft. at about 5 ft. from the ground. From Melchthal (guide Jos. Imdorf) a rough path crosses the *Storegg* (5710') to Engelberg (p. 123) in 4½ hrs.; another, more interesting but more difficult (steep descent; guide 12 fr.), leads thither in 6 hrs. over the *Juchli* (7120'). The *Nünalphorn (Juchlistock*, 7830'; fine view of the Titlis and the Bernese Alps) may be ascended in 1 hr. from the Juchli. View still finer from the *Hutstock* (8190'), reached by good climbers from the Juchli in 2 hrs. (comp. p. 129). — The basin of the Melchsee (6115'; *Höt.-Pens. Frutt, Pens. Reinhard*, both unpretending) affords an attractive picture of Alpine life. Rich flora. Interesting excursions may be made to *Bonі*, 1 hr.; *Spicherfluh* (6690'), 1½ hr.; *Hohmatt*, 2-2½ hrs.; *Erzegg (7138'), 1¼ hr.; *Balmereggborn (7280'), 1½ hr.; *Abgschütz*, 1¾ hr.; *Hohenstollen (8150'), 2¼ hrs., with fine view (comp. p. 182); *Glockhaus* (8320'), 2 hrs., toilsome; *Wildgeissberg* (8710'), 3 hrs. viâ the Tannen-Alp (comp. p. 135), etc. To the E. an easy pass crosses the *Tannen-Alp* (6500') in 2 hrs. to the *Engstlen-Alp* (p. 135); to the W. an interesting pass leads viâ the *Weit Riss* (about 7700'), to the S. of the Hohenstollen, in 4 hrs. (guide 10 fr.) to *Meiringen* (p. 181).

13 M. Sarnen (1545'; pop. 3928; *Obwaldner Hof; *Seiler, R., L., & A. 1½-2, B. 1, D. 2½, pens. 5 fr.; *Adler; Metzgern*, moderate; *Pens. Landenberg*, see below; *Pens. Niederberger*, on the 'Boll', ¾ M. to the E.; *Wylerbad*, on the S.W. bank of the lake, 1½ M. from Sarnen), the capital of *Obwalden*, the W. part of Canton Unterwalden, with its nunnery and Capuchin monastery. The *Rathhaus* contains portraits of all the magistrates of Obwalden from the year 1381 to 1824, and one of St. Nikolaus von der Flüe (see above), and a relief model of Unterwalden and Hasli. The large church, on a hill, the cantonal hospital, the poor house, the *Niklaus von Flüe* **Pensionat** (for students), and the arsenal on the *Landenberg* (1667'; fine view; pension, see above), are conspicuous buildings. The castle of *Landenberg*, destroyed by the Confederates on New Year's Day, 1308, formerly stood on the last-mentioned hill.

At the head of the *Schlieren-Thal*, 3½ hrs. to the W. of Sarnen, lies the sequestered *Schwendi-Kaltbad* (4737'), **with a chalybeate spring and whey-cure**. The road ascends the W. slope **of the** *Schwendiberg* to (1 hr.) *Stalden* (2614'; refreshments at the cure's), whence a bridle-path crosses the meadows of *Schwendi* and goes on, often through wood, to the (2½ hrs.) Kaltbad. Thence to the top of the *Feuerstein* (6760') 2½ hrs.; to the *Schimberg Bad*, 2 hrs., see p. 139. Viâ *Seewenegg* and *Seewenalp* to (3½ hrs.) *Flühli*, in the Entlebuch (p. 139), an attractive route.
To the *Melchthal* (3½ M. to St. Niklaus), see p. 132.

The railway (views to the right) crosses the *Melchaa*, which has been conducted into the **Sarner See** (1552'), a lake 4 M. long and 1-1¼ M. broad, well stocked with fish, which it continues to skirt. The valley of Sarnen is pleasing, though without pretension to Alpine grandeur. — At (15 M.) **Sachseln** (1598'; **Kreuz*, pens. 4½-5 fr.; *Engel*; pop. 1556), a thriving village near the E. bank of the lake, is **a large church, erected in** 1663, containing the bones of St. Nikolaus **and other relics**.

From Sachseln a pleasant shady route leads into the Grosse Melchthal (p. 132) viâ (2¼ M.) Flühli (2454'; **Pens. Anderhalden*, 4½ fr.; *Pens. Stolzenfels*, 4 fr.), the birthplace of Brother Klaus (p. 132), above the Ranft, with a chapel commanding a fine view. Hence to the village of Melchthal, 1¼ hr.

Ascending a short distance, from the S. end of the lake, and passing (on the left) the entrance of the *Kleine Melchthal*, the train next halts at (18 M.) **Giswil** (1665'; *Hôtel de la Gare*; *Krone*), **partly destroyed in** 1629 by inundations of the *Laulbach*. A lake was thus formed, and 130 years later was drained into the Lake of Sarnen. Fine view from the churchyard, beside the high-lying church; to the S.W. rise the *Giswiler Stock* (6605') and the *Brienzer Rothhorn* (7713'). Above the station are the relics of a château of the *Rudenz* family.

The Giswiler Stock (6605'), affording a beautiful view, may be ascended in 4 hrs. from Giswil, viâ *Kleintheil* and *Ivi*. The descent may be made to the *Marien-Thal* (*Entlebuch*, p. 139). — The Brienzer Rothhorn (p. 182) may be ascended from Giswil in 8 hrs.; path for the first 3 hrs. good, afterwards steep and toilsome. — Pedestrians are recommended to walk by the old **Brünig Road* from Giswil to (3 hrs.) the Brünig Pass (3395'; **Curhaus Brünig*, p. 134), whence they may descend to (1¾ hr.) *Meiringen* or (3 hrs.) *Brienz* (p. 182).

At Giswil, where the railway meets its first serious obstacle, the 'rack-and-pinion' system begins. The line ascends the side of the valley **at a considerable** gradient (10 : 100), traversing wood and crossing two torrents and traversing two rock-cuttings, and at *Bürgeln* reaches the **summit of the** *Kaiserstuhl* (2305'). From the top the triple peak of the Wetterhorn is visible to the S. over the depression of the Brünig. The railway proceeds, high above the picturesque **Lake of Lungern** (2162'; 1½ M. long) and through a short tunnel, to —

22½ M. **Lungern** (2475'). The large village (pop. 1756'; **Löwe & Hôt. Brünig*, pens. 5-6 fr.; *Hôt.-Pens. Alpenhof*; *Bär*) is, with the adjoining *Ober-Seewies*, the last village in the valley and lies ½ M. from the S. end of the lake, half of which was drained into the Lake of Sarnen in 1836, by means of a channel ¾ M. long. —

134   II. Route 37.   BRÜNIG.

The *Dundelsbach* forms a picturesque fall on the hillside to the W. The *Giebel* (6680'; fine view), to the S.E., may be easily ascended from Lungern in 3½ hrs.

The second steep gradient begins beyond Lungern; picturesque retrospect. The train passes through the *Käppeli Tunnel* (2970'; 150 yds. in length) and ascends the wooded *Brünigmatt-Thal* (above us, to the right, is the road), at a moderate gradient, which becomes steeper before (25½ M.) Brünig (3295'; *Rail. Restaurant*, D. incl. wine 3½ fr.; *Hôt.-Pens. & Curhaus Brünig*, 3 min. from the station, pens. 9-12 fr.), situated on the crest of the saddle, not far from the old Brünig Pass. Fine view; opposite us tower the Engelhörner (p. 185) and the Faulhorn chain (p. 179); to the left we overlook the valley of Meiringen as far as the Kirchet (p. 187); at the foot of the mountains to the S. is the lower fall of the Reichenbach (p. 185); opposite is the fall of the Oltschibach (p. 182); below us flows the Aare, and to the right is part of the Lake of Brienz.

Fine prospect from the *Wyler Alp* (4856'), 1½ hr. to the N.W. of the Brünig; more extensive from the *Wylerhorn* (6580'), 3 hrs. from the pass.

FROM THE BRÜNIG TO MEIRINGEN, on foot in 2 hrs., attractive. From the road, about ¼ M. below the station, a footpath diverges to the right, and crossing the railway, runs chiefly through wood to (3 M.) *Hohfluh* (p. 182). Before reaching the inn we turn to the left, take the first turning to the right, and cross the pastures to the right again viâ *Wasserwendi* and *Goldern* to the Hôtel Alpbach and (3 M.) *Meiringen* (p. 181). After Hohfluh we have a continuous and picturesque view of the Wetterhörner and Oberhasli.

The railway has been carried down the steep rocky wall at a considerable gradient (maximum 12 : 100) by means of blasting, retaining-walls under overhanging cliffs, and cuttings. We cross the brawling *Grossbach*, *Kehlbach*, and *Hausenbach* (charming view at the *Brunnenfluh*), enter the Aarethal, and beyond *Hausen* reach —

28½ M. *Meiringen* (p. 181). Thence to *Brienz* and *Interlaken*, see R. 50.

38. From Meiringen to Engelberg.   Engstlen-Alp. Joch Pass.

Comp. Maps, pp. 112, 126.

9¾ hrs. Im-Hof 1¼, Engstlen-Alp 4½-5 (Lauenen direct from Meiringen 2½, Engstlen-Alp 2½ hrs.), Joch 1½, Trübsee ½, Engelberg 1½ hr. — Horse from Im-Hof to Engstlen-Alp 15, to Engelberg 50, for two days 45 fr.; guide (unnecessary) 16; horse from Engstlen-Adt to Engelberg 15, guide 8 fr. — If the traveller can devote two days to this interesting journey (still more attractive in the reverse direction), he should sleep on the Engstlen-Alp, where an afternoon may be pleasantly spent.

From Meiringen to (1¼ hr.) *Im-Hof* (2054'), see p. 187. Two routes lead thence to the Genthal. We follow the Susten route (p. 136) to the (¾ hr.) foundry in the *Mühlethal*; then, beyond the (¾ hr.) bridge over the *Genthalwasser*, ascend to the left through wood to the (1 hr.) **Genthal-Alp** (3900'; *Inn zur Wagenkehr*, plain, good wine). Or we may diverge to the left from the Susten route at *Wyler*, 20 min. from 1m-Hof, cross the *Gadmenbach*, turn to the left again **after 5 min.**, and ascend rapidly through pastures and wood. Near the (1 hr.) chalets of *Lauenen* (3800') begins the Genthal-Alp.

A path called the '*Hundschäpf*', shorter by ½ hr., but very narrow at places, and somewhat dizzy (guide advisable), leads from Meiringen straight on for ½ M. beyond the bridge over the brook and then, ascending to the left, skirts the brow of the *Hasliberg*, affording a striking view of the valleys which unite at Im-Hof far below, to the (2¼ hrs.) chalets of *Lauenen* (see above).

The path soon approaches the *Genthalbach*, and follows its right bank. On the (¼ hr.) *Leimboden* (3920') our path is joined on the right by that from Mühlethal above mentioned (small auberge on the left bank). We now gradually ascend the monotonous Genthal. Behind us rise the Wetterhörner and the Hangend-Gletscherhorn at the head of the Urbach-Thal (p. 187). In 20 min. we pass the *Genthal-Hütten* (3993'), on the left bank of the brook, and after a slight ascent reach (1 hr.) the *Schwarzenthal-Hütten* (4596'; rfmts.).

The valley now becomes more interesting. From the precipices of the *Gadmer Flühe* (9750') on the right, which become grander as we proceed, falls a series of cascades, varying in volume according to the state of the melting snow, and we at last come to eight of these close together (*Achtelsassbäche*). The *Engstlenbach*, as the brook is named above this point, also forms several considerable falls. The path crosses the stream and ascends, often steeply, through venerable wood, to the (1½ hr.) *Engstlen-Alp (6033'; *Immer's Hotel, with dépendances, R., L., & A. 3-5, D. 4, S. 3, pens. 7-9 fr.), a beautiful pasture, with fine old pines and Alpine cedars'. (Excellent water, temperature 40-42° Fahr.) *View, to the S.W., of the majestic Wetterhorn; to the left the Schreckhörner; to the right the Blümlisalp; to the E. the Wendenstöcke and the Titlis. — The *Wunderbrunnen* ('miraculous spring'), near the inn, is an intermittent spring which only flows in wet weather and in spring during the melting of the snow, usually about noon.

Excursions. To MELCHSEE-FRUTT (2 hrs.; guide, 4 fr., unnecessary; horse 10 fr.). From the inn we walk to the N.W. to the waterfall and ascend rapidly on the right side, soon obtaining a splendid view of the Bernese Alps (among which the Finsteraarhorn comes in view to the left of the Schreckhörner). At the top we round the grassy *Spicherfluh* (6690'), pass a small lake, and reach the (1 hr.) *Tannen-Alp* (6500'), with its numerous chalets. We next traverse beautiful level pastures, pass two other small lakes, and reach (1 hr.) *Melchsee-Frutt* (6210'; Hôt.-Pens. Frutt, Pens. Reinhard; see p. 132).

Ascents. *Schafberg* (*Gwärtler*; 7950'; 2 hrs.), not difficult; *Grauslock* (8737'; 2½-3 hrs., with guide), fatiguing; *Wildgeissberg* (8710'; 3 hrs.; with guide, 5 fr.), an admirable point, but rather laborious (comp. p. 132). — *Wendenstock* (9990'; 4 hrs.; with guide), difficult, for experts only; imposing view.

The ascent of the *Titlis (10,627') is shorter from the Engstlen-Alp than from Engelberg (p. 128). From the (1½ hr.) Joch Pass we ascend to the right over rocks, débris, and snow, and reach the (3½-4 hrs.) top after a steep and fatiguing climb. Guide from the hotel 15 fr. (charged in the bill) and gratuity (with descent to Engelberg 20 fr.).

OVER THE SÄTTELI TO GADMEN, 3½-4 hrs. (guide to the **Sätteli** 4, Gadmen 10, Steinalp 14, Wasen 21 fr.), a fine route. At the W. end of the Engstlen-See (see below) we cross the Engstlenbach to the *Alp Scharmadläger*, and ascend a narrow path on the slope of the Gadmer Fluh to the (2 hrs.) **Sätteli** (splendid view of the Gadmen-Thal, Trift Glacier, and Bernese Alps). Then a long and steep descent to (1½-2 hrs.) *Gadmen* (p. 137). A still finer view is obtained from the °*Achtelsassgrat* (*'Grätli*'), ½ hr. beyond the Sätteli and a few hundred feet lower.

For ½ hr the bridle-path to (3½ hrs.) Engelberg skirts the *Engstlen-See* (6075′), a lake 1¼ M. long, and then ascends, in view of the *Wendenstöcke*, with the *Pfaffen* and *Joch Glaciers*, to the (1 hr.) **Joch Pass** (7245′; view limited). A tolerable path now descends over rock and detritus to the (½ hr.) *Obere Trübsee-Alp* (Inn), on the S.E. side of the turbid *Trübsee* (5795′), and then leads to the N.E. through the flat and marshy valley (with the Trübsee on the left), and across the brook which descends from the glaciers of the Titlis, to the (¾ M.) *Hôt.-Pens. Hess* (R., L., & A. 2½-3½, B. 1½, D. 3½, pens. from 7 fr.), on the margin of the *Pfaffenwand* (5870′). The view hence of the Titlis and the Engelberg Valley is surpassed by that from the *Bitzistock* (6225′; easily ascended in 20 min. from the hotel), which includes also the Schlossberg, Spannörter, and other mountains. Ascent of the *Titlis*, see p. 130.

The path now descends the steep *Pfaffenwand* in zigzags, leads over the *Gerschni Alp* (4125′) towards a clump of pines, enters a wood, crosses the *Engelberger Aa* at the foot of the hill, and reaches — 1½ hr. *Engelberg* (p. 128).

39. From Meiringen to Wasen. Susten Pass.

Comp. Maps, pp. 112, 126, 118.

12 hrs.: Im-Hof 1½, Gadmen 3, Am Stein 2¾, Susten-Scheidegg 1¼, Meien 2¾, Wasen 1 hr. Horse 35 (or, for two days, 40), guide 18 fr. (unnecessary).

From Meiringen to *Im-Hof* (2055′), 1¼ hr., see p. 187. The SUSTEN ROAD, constructed by Bern and Uri in 1811, and still tolerably well kept on the Bernese side (practicable for driving as far as the Stein Inn), diverges here to the E. from the Grimsel route. It traverses pleasant meadows and wooded slopes, and skirts the winding *Gadmenbach*. At one time the Wetterhorn, Wellhorn, and Engelbörner, at another the Schwarzhorn group form the background towards the W.

The lower valley is called the *Mühle-Thal*, above which is the *Nessen-Thal*. Beyond (20 min.) *Wyler* the path to the *Engstlen-Alp* (p. 134) diverges to the left. The road crosses (10 min.) the Gadmenbach, and at (¼ hr.) an old iron-foundry the *Genthalbach*, on the left bank of which a second path (see p. 134) to the Engstlen-Alp diverges. At (¾ hr.) *Mühlestalden* (3117′) the narrow *Triftthal* opens towards the S.E., with the *Trift Glacier* in the background.

GADMEN. *II. Route 39.* 137

Triftthal (comp. Map, p. 118; 4½ hrs. to the club-hut; guide necessary; *Andreas von Weissenfluh* of Mühlestalden, *Joh. Moor* and *Joh. Luchs* of Gadmen). The path ascends on the left bank of the *Triftbach* and on the left side of the ice-fall to the (3 hrs.) simple *Windegg-Hütte* (6237'). We now cross the glacier, here tolerably level, and mount the steep rocks of the *Thäitistock* to the (1½ hr.) *Trift Hut* of the Swiss Alpine Club (8250'), affording a good survey of the upper basin of the Trift Glacier. From the club-hut over the **Trift-Limmi** (10,170') and the *Rhone Glacier* to the *Furka* (p. 125) or to the *Grimsel Hospice* (p. 188), 9 hrs., fatiguing. — The *****Dammastock** (11,910'; splendid view) is ascended without very serious difficulty from the club-hut in 4½-5 hrs. (guide from Meiringen, 40 fr.; descent by the Rhone Glacier and Nägeli's Grätli to the Grimsel, 7 hrs.). — The *Schneestock* (11,667'; 5 hrs.), *Thieralplistock* (11,140'; 5 hrs.), and *Diechterhorn* (11,120'; 4 hrs.) may also be ascended from the Trift Hut without difficulty. — Passes to the *Göschener-Alp* over the *Winterberg Range* (*Maasplank-Joch, Damma Pass, Winterjoch*), 8 hrs., difficult (comp. p. 118). — Over the *Tiefen-Sattel* (about 10,820') and the *Tiefen Glacier* (p. 125) to the Furka, 9 hrs., interesting, and in certain states of the snow not difficult. — Interesting passes also cross the **Furtwang-Sattel** (8302') to *Guttannen* (a steep ascent of 3 hrs. from the Windegg; descent by the *Steinhaus-Alp* to Guttannen in 2 hrs.), and the **Stein-Limmi** (8970') to the *Stein-Alp*. The latter route leads from the *Graggi-Hütte*, opposite the Windegg on the right side of the glacier, in 3 hrs. to the col, between the *Giglistock* and *Vorder-Thierberg*, and descends over the *Stein-Limmi Glacier* and round the slopes of the *Thaleggli* to the (2 hrs.) *Stein Inn* (see below). By combining the two last-named passes, a good walker may reach the Stein Inn from Guttannen in a single day (11-12 hrs.).

The road crosses the Gadmenbach and ascends by *Schuftelen* to (1 hr.) *Unterfüren* (3848'), where the beautiful *Gadmen-Thal* begins, and (20 min.) the village of **Gadmen** (3945'; *Bär*, moderate), consisting of the hamlets of *An der Egg, Bühl*, and *Obermatt*. (Path over the *Sätteli* to the *Engstlen-Alp*, see p. 135.) The green valley with its fine old maple-trees contrasts strikingly with the barren *Gadmer Fluh* (see p. 135). To the E., on the slope of the *Uratstöcke* (9545'), lies the *Wenden Glacier.*

After a level stretch, the road ascends through wood in numerous windings to the chalets of *Feldmoos* (4935'), and then traverses a wild rocky region ('Hölle') to the (2½ hrs.) **Stein Inn** (6122'), at the foot of the huge Stein Glacier.

OVER THE SUSTEN-LIMMI TO THE GÖSCHENER-ALP, 9 hrs., laborious (guide from Meiringen 35 fr.). We ascend the slopes of the *Thaleggli* (on the W. side of the Stein Glacier), cross the *Stein-Limmi Glacier* to the *Thierbergli*, and traverse the névé of the *Stein Glacier* to the Susten-Limmi (10,180'), lying to the S.W. of the *Brunnenstock* (see below). Descent **over** the *Susten Glacier* to the *Kehlen-Alp* (7562') and **across** the *Kehle Glacier* to the *Hintere Röthe* and *Göschener-Alp* (p. 118). — A similar pass is the Thierberg-Limmi (about 10,500'): we cross the Stein **Glacier** to the Juch between the *Steinberg* and the *Hinter-Thierberg*, and descend the Kehle Glacier to the (9 hrs.) Göschener-Alp. — Ascent of the *****Brunnenstock** (11,520'), the highest of the *Sustenhörner*, viâ the *Stein-Limmi* and *Steinen Glaciers*, toilsome but interesting (7-8 hrs. from the Stein Inn; guide 35 fr.). The descent may be made viâ the *Susten-Limmi* to the *Göschener-Alp* (p. 118).

Over the *Stein-Limmi* to the *Trift Glacier* (5 hrs. to the Graggi Hut), see above. Another route crosses the snow-saddle of Zwischen-Thierbergen (about 9780'), between the *Vorder-* and the *Hinter-Thierberg*, to the (5-6 hrs.) *Trift-Hütte* (see above). — To *Engelberg* over the *Wenden-Joch*, see p. 130.

The bridle-path now ascends above the moraine, describing a long circuit to the right (which a footpath cuts off), and overlooking the grand Stein Glacier, environed by the Sustenhörner, Susten-Limmi, Gwächtenhorn, Vorder- and Hinter-Thierberg, and Giglistock, to the (1¼ hr.) **Susten Pass** (7420'), which affords an admirable survey of the imposing mountains bounding the Meien-Thal on the N. and culminating in the Spannörter (p. 130).

The path, now uninteresting, winds down to the *Meienbach*, a brook issuing from the *Kalchthal*, a wild gorge on the right, into which avalanches frequently fall from the *Stücklistock* (10,855') and the *Sustenhörner* (p. 137). Below us lie the *Susten-Alp* (5767'), on the right, and the (1 hr.) *Guferplatten-Alp* (5725'), on the left. The path traverses the stony valley of the Meien-Reuss, which consists here of several branches, and crosses the brook twice. It next crosses the deep ravine of the (³/₄ hr.) *Gorezmettlenbach* (5137'), and passes the *Gorezmettlen-Alp*. Several brooks issue from the *Rüttifirn* on the right.

The first group of houses (20 min.) is *Färnigen* (4787'; Inn, poor); then (40 min.) Meien (4330'; *Kreuz*, *Stern*, both unpretending) and (20 min.) the hamlet of *Husen* (3865'). Above Wasen we pass the *Meienschanz* (3600'), an intrenchment erected in 1712 during the Religious War (p. 62), and destroyed by the French in 1799. Descending rapidly for a short distance, and crossing the St. Gotthard Railway, we at length reach (40 min.) *Wasen* (p. 113).

40. From Lucerne to Bern. Entlebuch. Emmen-Thal.

59 M. RAILWAY in 2¼-4 hrs. (9 fr. 90, 6 fr. 95, 4 fr. 95 c.).

Lucerne, see p. 81. — Near the Reuss bridge the train diverges to the left from the Zürich line (p. 81), and passes through a tunnel under the *Zimmeregg*, 1248 yds. long, into the broad dale of the *Kleine Emme*. 3 M. *Littau*, at the base of the wooded *Sonnenberg* (p. 86); 7½ M. *Malters* (1693'; *Kreuz*), with a handsome church.

Road hence to (3¼ M.) Schwarzenberg (2760'; *Weisses Kreuz; Pfisterhaus; Curhaus Matt*, very unpretending), on the hill to the S., a pleasant summer-resort. About 2 M. above it is the rustic *Curhaus Eigenthal* (3475'), in a sheltered situation. (Fine view of Lucerne and its lake from the *Würzenegg*.) Hence to (6 M.) *Kriens*, viâ *Herrgottswald*, see p. 86.

From *Schachen* (see below) the old BRAMEGG ROAD leads to the (2 M.) prettily-situated Bad Farnbühl (2460'; *Curhaus*, pens. 5-6 fr.), with chalybeate and mineral springs, and thence over the *Bramegg* (3366') to (5 M.) *Entlebuch*.

Above **Schachen** (1½ M. from Malters) the valley contracts. The train approaches the Emme, and crosses it near *Werthenstein* (on the left), with its handsome old monastery, now a deaf-and-dumb asylum. Beyond a short tunnel we reach (12½ M.) **Wohlhausen** (1873'; pop. 1661; *Rössli; Kreuz*), a large village, divided by the Emme into *Wohlhausen-Wiggern* on the left bank, and *Wohlhausen-Markt* opposite. — About 6 M. to the W., at the foot of the *Napf* (see p. 139), lies the *Curhaus Mensberg* (3314'), a health-resort.

We here enter the **Entlebuch**, a valley 15 M. long, with rich pastures. The train recrosses the Emme and ascends the E. side of the valley (several embankments and four tunnels).

17½ M. **Entlebuch** (2225'; *Hôtel du Port; Drei Könige; *Pension Jenni; pop. 2720), a well-built village, picturesquely situated. — Ascent of the *Napf*, see below.

In the *Entlen-Thal*, on the W. side of the *Schimberg* (see below), 8 M. to the S., is the Schimberg Bad (4677'), with an alkaline sulphur-spring. Road from Entlebuch to the (4½ M.) Engstleamatt Inn, whence a new road descends to the *Entlen-Brücke* (carr. to the bridge, 1-2 pers. 5 fr., to the Baths, 1 pers. 10, 2 pers. 14, 3 pers. 18, 4 pers. 22 fr.). The road then ascends in windings to the (3½ M.) well-equipped *Curhaus* (pens. from 6 fr.). Fine mountain-view to the N. and N.W. A good path ascends in 1 hr. to the top of the *Schimberg* (5975'), which affords an admirable panorama. Still more extensive and imposing are the views from the (2½ hrs.) *Feuerstein* (6700') and from the (2½ hrs.) *Schafmatt* (6505'). Footpaths lead also to (1½ hr.) *Heiligkreuz* (see below), to the (2½ hrs.) *Schwendi-Kaltbad* (p. 133), etc.

The train crosses the rapid *Entlenbach*, which here falls into the Emme. On the left lies the village of *Hasle*, prettily situated.

22 M. **Schüpfheim** (2388'; pop. 2808; *Adler; Rössli*), the capital of the valley. About ½ M. from the station is the **Bad** *and Curhaus Schüpfheim* (chalybeate spring containing iodine). To the E. (1½ hr.) is *Heiligkreuz* (3700'; rustic Inn), a summer-resort, with fine view.

A road (diligence twice daily in 1¾ hr.; carriage for one pers. 5, two pers. 7 fr.) gradually ascends hence to the S. through the romantic valley of the Waldemme or Kleine Emme, to the (5 M.) pretty mountain-village of **Flühli** (2930'; *Hôt.-Pens. Kreuzbach*, pens. 4½-5 fr.), with a sulphur-spring. Fine woods; rich flora. Pleasant excursions may be made to (1 hr.) the *Kessiloch*, a rocky gorge with a high waterfall; to (3 hrs.) the *Bäuchlen* (5810'); to the (3½ hrs.) *Hayleren* (6400'); and to the (4 hrs.) *Schrattenflühe* (6310'), with interesting slopes of debris and a splendid view.

From Flühli a road leads to (4½ M.) *Sörenberg* (3812'; *Inn, pens. 4-4½ fr.), a health-resort in the upper Emmen-Thal or *Marien-Thal*. The *Brienzer Rothhorn* (p. 132) may be ascended hence by a good path in 3 hrs.

From Flühli to Sarnen via the Seewenegg, 6¾ hrs., an attractive route. About ¾ M. to the S. of Flühli the path diverges to the left, passes the hamlet of *Kragen* and the Alps of *Holzhack*, *Städtelt*, and *Blüttli*, traverses wood, passing a saw-mill, and reaches (3 hrs.) the *Seewen-Alp* (5640'), a health-resort on the *Seewenalp-See*, with a chalet and a small Curhaus (pens. 3½ fr.). Splendid view of the Bernese Alps. The *Feuerstein* (6700'), which affords an imposing survey of the Alps, from the Sentis to Mt. Blanc, is easily ascended hence in 1 hr. — From the chalet the footpath ascends the (¼ hr.) *Seewenegg*, another fine point of view (still better from a height 250 yds. to the left). It then descends to the right into the valley, passing a saw-mill and leaving the Schwendi-Kaltbad (p. 133) to the left, to *Stalden* and *Sarnen* (p. 132).

We now cross the Kleine Emme, which rises on the Brienzer Rothhorn, and ascend the valley of the **Weisse Emme** to —

26 M. **Escholzmatt** (2815'; *Löwe; Krone*), a scattered village (3086 inhab.), on the watershed between the Entlebuch and Emmen-Thal; then descend to (29 M.) *Wiggen* (2600'; Rössli), follow the right bank of the *Ilfis*, and reach (32½ M.) *Trubschachen* (2396'), at the confluence of the *Trubbach* and Ilfis, the first village in Canton Bern.

The *Napf* (4620'; 3½-4 hrs., guide unnecessary; */Inn* at the top, frequented as a health-resort, pens. 5-6 fr.), to the N. of Trubschachen, deserves a visit. A carriage-road leads viâ (2¼ M.) *Trub* (2675'; Inn) to (6 M.) *Mettlen* (3454'; carriage for 1 pers. to this point, 6 fr.), and a bridle-path thence to the (¾ hr.) top of the Napf, whence there is a fine panorama from the Sentis to the Dôle, and a beautiful view of the Bernese Alps. — From Entlebuch (p. 138) a road crosses the Grosse and the Kleine Emme, to the W.; we then either follow the road by *Dopleschwand* to (6 M.) *Romoos* (2592'; Inn), or reach it by a direct path in 1 hr.; from Romoos a good bridle-path leads to the top in 2½ hrs. more. — From the Napf a footpath, with an almost continuously fine view, leads viâ the (2 hrs.) *Lusshütte* (rustic inn), the *Lüderen-Gässli* (Hotel zu den Alpen, moderate), and the *Rafrüti* (see below) to (4 hrs.) *Langnau* (guide convenient, 5-6 fr.).

35½ M. **Langnau** (2245'; pop. 7644; *Hirsch*, R., L., & A. 2, B. 1, D. 2, pens. 5 fr.; *Löwe; Bär; Hôt. Bahnhof; Hôt. Emmenthal*), a large and wealthy village, the capital of the **Emmen-Thal**, a valley about 25 M. long, 10-12 M. wide, watered by the *Ilfis* and the *Grosse Emme*, and one of the most fertile in Switzerland. The cheese of the Emmen-Thal is much esteemed; the carefully kept pastures, the fine breed of cattle, and the neat dwellings with their pretty gardens bear witness to the prosperity of the natives.

Railway to *Burgdorf*, see p. 17. — The *Bageschwand Höhe*, 1 hr. to the N.W., commands a fine view of the Emmen-Thal and **the Alps;** the view from the *Rafrüti* (3950'), 2¼ hrs. to the N., is still more extensive (panorama by G. Studer).

Beyond Langnau the train crosses the Ilfis and the Emme. 38 M. *Emmenmatt*, 40 M. *Signau* (2090'; Thurm; Bär), 44 M. *Zäziwyl* (Krone), thriving villages. It then skirts the *Hürnberg* in a wide curve to (46 M.) *Konolfingen*, 3 M. to the S.E. of which is the frequented *Schwendlenbad* (2830'), surrounded by fine woods. 48½ M. *Tägertschi*. — 51 M. *Worb* (Löwe; Stern), a large village with an old Schloss. Pleasing view of the Stockhorn chain to the left.

From Worb a carriage-road runs to the E. to (2 M.) the frequented watering-place of *Enggistein* (2264'; Inn), situated in a pleasant mountain-valley, and (1 M. farther) *Rüttihubelbad* (2414'; pens. 3½-4½ fr., unpretending), situated among woods, with a saline chalybeate spring, pleasant walks, and a good view, especially fine from the *Knörikubel* (3027'; 55 min.). Magnificent views are also afforded by the *Gummegg* (3208'), reached viâ *Walkringen* in 1½ hr., and by the *Ballenbühl*, the W. summit of the Hürnberg, reached viâ *Schlosswyl* in 1¼ hr. (descent to the station at *Tägertschi* in 20 min.).

54 M. *Gümlingen*, junction of the **Bern** and **Thun** line (change carriages for Thun, p. 151). Thence to (59 M.) *Bern*, see p. 151.

41. From Lucerne to Lenzburg *(Aarau)*. The Seethal Railway.

29½ M. STEAM TRAMWAY in 2¾-4 hrs.; 2nd cl. 4 fr. 85, 3rd cl. 3 fr. 30 c. — This 'Seethal Railway' from Emmenbrücke to Lenzburg offers a pleasant tour, though dusty in summer. The gauge is that of the ordinary railways, the carriages of which can run on this line.

From Lucerne to (2½ M.) *Emmenbrücke*, see p. 19; here we change carriages for the 'Seethalbahn' which diverges to the right.

4 M. *Emmen* (1410'; Stern), near the *Reuss*, on the right bank of which, ½ M. to the E., is the old nunnery of *Rathhausen*, now

an asylum for poor children. We traverse the fertile *Emmenboden* to (6 M.) *Waldibruck*. The line quits the road, here unsuitable for a tramway, and ascends, affording a fine view of the Rigi to the right, to (8 M.) *Eschenbach* (1560'; Rössli; Löwe), with its large Cistercian Abbey and valuable gravel-pits in the vicinity. (Diligence twice daily in 40 min. to Gisikon, p. 80.)

At (9½ M.) *Ballwyl* (1693') the line crosses the watershed between the Reuss and the *Aa*, and descends into the Seethal, belonging partly to Lucerne and partly to Aargau, one of the most fertile and attractive valleys in Central Switzerland. This 'lake-valley', 18½ M. long, is bounded on the E. by the long *Lindenberg* (2953') and on the W. by the *Ehrlose* (2670') and the *Homberg* (2595'), and in the middle of it lie the pretty **Baldegg Lake** (or *Obere See*) and the larger *Hallwyl Lake* (or *Untere See*), amidst pastures sprinkled with fruit-trees.

11 M. **Hochdorf** (1653'; *Hirsch*), a picturesque and prosperous village, with beautiful pine-woods in the vicinity.

EXCURSIONS. On a hill to the E. (½ hr.) is the cantonal deaf-and-dumb asylum of *Hohenrain* (2014'), formerly a commandery of the knights of St. John, with a fine view of the Alps. Thence in 1½ hr. to *Schloss Horben* (2625'; p. 22), a health-resort, affording a superb view to the N. and E.; then to the (½ hr.) ruined castle of *Lieli*, another fine point of view, to (½ hr.) *Augstholz* (Hydropathic Establishment), and back to (½ hr.) Hochdorf. The whole excursion may be made by carriage.

To the W. of Hochdorf roads lead by *Römerswyl* to (4 M.) *Oberrainach*, a ruined castle, with an admirable view of the Seethal and the Jura; by the pilgrimage-shrine of *Hildisrieden* to the (5 M.) chapel commemorative of the battle of *Sempach* (p. 19); and by *Urswyl* to (3½ M.) *Rain*, near which is *Oberbuchen* (2133'), where we obtain a picturesque survey of Pilatus and the Entlebuch Mts.

12½ M. **Baldegg** (Löwe) a pretty village with an old castle, now a nunnery and girls' school, lies at the S.E. end of the **Baldegger See** (1532'), a lake 3 M. long. Skirting the E. bank of the lake, we next reach (15 M.) *Gelfingen* (Stern), where the culture of the vine begins. On the right is the castle of *Heidegg*, and ¾ M. to the N. is the pretty village of *Hitzkirch* (Kranz; Engel), once a Teutonic commandery, with a seminary for teachers.

To the N. of Hitzkirch a road leads by *Altwis* and *Aesch* to (5 M.) *Fahrwangen* (Bär) and *Meisterschwanden* (Löwe; *Pens. Seerose*), two large villages, where straw-plaiting is the chief industry (see below); thence by *Sarmensdorf*, past *Schloss Hilfikon*, to *Villmergen* and (5 M.) *Wohlen* (p. 22).

16¼ M. *Richensee*, with the ruins of the *Grünenburg*, which was destroyed in 1386, standing upon an enormous erratic block. 17 M. *Ermensee*, a well-to-do village on the Aa. At (18 M.) *Mosen* the tramway reaches the **Hallwyler See** (1383'), a lake 5½ M. long and 1¼ M. broad (small steamer), and ascends on its W. bank to —

20 M. **Beinwyl** (1700'; 1679 inhab.; *Löwe*), a busy, thriving village with considerable **cigar-manufactories**, commanding a charming view of the lake.

RAILWAY in 5 min. to (1¼ M.) *Reinach* (Bär) and in 9 min. to (2½ M.) *Menziken* (Stern), two industrial villages in the upper *Winen-Thal*. — A pleasant excursion from Beinwyl is the ascent of the *Homberg* (2595'), ³/₄ hr. to the N.W.; beautiful view of the Alps and the Jura Mts.

The cars now run high above the lake to (21¼ M.) *Birrwyl*, with its large factories, and descend thence to (23½ M.) *Boniswyl* (Rail. Restaurant), a busy wine-trading place.

To FAHRWANGEN diligence twice daily in 1 hour. The road leads past the handsome old château of *Hallwyl*, the ancestral seat of the distinguished family of that name, to (1½ M.) *Seengen* (Bär), a large village, with the burial-vaults of the Hallwyl family. About ½ M. to the S. E. is the Brestenberg Hydropathic, formerly a château of Hans Rudolf v. Hallwyl, built in 1625, prettily situated among vineyards at the N. end of the Lake of Hallwyl. From Brestenberg we follow the E. bank to *Tennwyl*, *Meisterschwanden*, and (2 M.) *Fahrwangen* (p. 141).

24½ M. *Niederhallwyl-Dürrenäsch*; 25½ M. *Seon* (Stern), a large manufacturing village (1794 inhab.).

29½ M. **Lenzburg** (1300'; 2501 inhab.; *Krone; Löwe*), a busy little town on the *Aa*, with the large cantonal prison. On a hill above the town, to the E., stands the old *Schloss Lenzburg* (1663'), the property of Mr. Jessup, an American, at whose expense it is being restored. Opposite, to the W., rises the *Staufberg* (1710').

From Lenzburg to *Aarau* and *Baden*, see p. 22.

III. BERNESE OBERLAND.

42. Bern 144
    Enge, 150. — Gurten; Zimmerwald, 151.
43. From Bern to Thun 151
    Environs of Thun; the Gurnigelbad, 152, 153.
44. The Niesen 153
45. From Thun to Interlaken. Lake of Thun. 155
    a. Thunersee Railway 155
    b. Steamboat Journey 155
       Sigriswyl, 156. — From Spiez to Aeschi, 156. — Tanzbödeli
       Pass; Morgenberghorn; Schwalmern, 157. — St. Beaten-
       berg; Amisbühel; Gemmenalphorn; Niederhorn; Burg-
       feldstand, 158.
46. Interlaken and Environs 158
    Heimwehfluh; Abendberg; Saxeten-Thal; Sulegg;
    Harder; Habkern-Thal; Hohgant; Augstmatthorn;
    Schynige Platte, 163-164.
47. The Lauterbrunnen Valley and Mürren 164
    Isenfluh 165. — Schmadribach Fall. Upper Steinberg.
    Oberhornsee 167. — Allmendhubel. Schilthorn, 168. —
    The Sefinen-Thal, 169. — From Lauterbrunnen over the
    Sefinenfurgge to the Kienthal; over the Hohthürli to
    Kandersteg; and over the Tschingel Pass to Kandersteg,
    170. — From Lauterbrunnen over the Petersgrat to the
    Lötschen-Thal, 171. — Wetterlücke, Schmadri-Joch,
    Lauinenthor, Roththal-Sattel, and Ebnefluh-Joch, 171.
48. From Interlaken to Grindelwald 171
    a. Direct Line 171
    b. Wengernalp Railway 172
       Mettlen-Alp, 172. — Jungfrau; Silberhorn, 173. — Guggi-
       Hütte; Lauberhorn; Tschuggen; Männlichen, 174. —
       From Grindelwald over the Eismeer to Zäsenberg.
       Mettenberg; Schreckhorn, 177. — Mönch; Eiger. From
       Grindelwald over the Strahlegg and the Finsteraar-Joch
       or Lauteraar-Sattel to the Grimsel Hospice. From Grindel-
       wald over the Jungfrau-Joch, Mönchjoch, Eiger-Joch, and
       Fiescher-Joch to the Eggishorn, 178.
49. The Faulhorn 178
    From Grindelwald to the Faulhorn, 179. — From the
    Schynige Platte to the Faulhorn, 180. From the Faul-
    horn to the Great Scheidegg. Röthihorn, 180. — Schwarz-
    horn, 181.
50. From Meiringen to Interlaken. Lake of Brienz. . . 181
    Gorge of the Aare. Hasleberg. Hohenstollen. Brienzer
    Rothhorn, 182. — Giessbach, 183. — Rauft. Enge. Axalp.
    Hinterburg-See. Ascent of the Faulhorn from the Giess-
    bach. From the Giessbach to Interlaken, 184.
51. From Meiringen to Grindelwald 185
    Falls of the Reichenbach. Baths of Rosenlaui, 185. —
    Rosenlaui Glacier; Dossen-Hütte; Wetterlimmi, 186.

52. **From Meiringen to the Rhone Glacier. Grimsel** . . **187**
Urbach-Thal; Gauli Pass; Berglijoch; Dossen-Hütte, 187. — Kleine Siedelhorn; Unteraar Glacier; Dollfus Pavilion; Ewigschneehorn; Finsteraarhorn, 189. — From the Grimsel over the Oberaar-Joch or the Studer-Joch to Fiesch, 190.

53. **From Spiez to Leuk over the Gemmi** **191**
Kienthal; Gamchilücke; Büttlassen; Gspaltenhorn; Wilde Frau, 191. — The Blaue See, 192. — Oeschinen-See; Blümlisalp, 192. — Doldenhorn; Fründenhorn; Dündenhorn; Gastern-Thal; Alpschelenhubel; Tschingel Pass; Petersgrat; Balmhorn; Altels, 193. — Excursions from Bad Leuk; Torrenthorn, etc., 195.

54. **The Adelboden Valley** **196**
Excursions and Ascents from Adelboden. Bonderspitz. Elsighorn. Albrist. Gsür. Gross-Lohner. Wildstrubel. etc. From Adelboden to Lenk viâ the Hahnenmoos; to Kandersteg viâ the Bonderkrinden; to Schwarenbach viâ the Engstligen-Grat, 196, 197.

55. **From Gampel to Kandersteg. Lötschen Pass** . . . **197**
Hohgleifen; Bietschhorn. From Ried to Leuk over the Ferden Pass, the Gitzi-Furgge, the Resti Pass, the Faldum Pass, or the Niven Pass, 198.

56. **From Thun to Sion over the Rawyl** **199**
Source of the Simme; Oberlaubhorn; Mülkerblatt; Iffigensee; Wildhorn; Rohrbachstein; Wildstrubel. From Lenk to Gsteig, 200. — From Lenk to Saanen; to Adelboden, 201.

57. **From Thun to Saanen through the Simmen-Thal** . . **202**
From Latterbach to Matten through the Diemtig-Thal. Stockhorn; Bad Weissenburg, 202. — Over the Gantrist Pass to the Gurnigelbad. From Reidenbach to Bulle, 203. — From Saanen to Château d'Oex, 204.

42. Bern.

Hotels. *BERNERHOF (Pl. a; D, 4), Bundesgasse 3, R., L., & A. 4-8, B. 1½, luncheon 4, D. 5, pens. 10-14 fr.; *BELLEVUE (Pl. b; E, 4), Inselgasse 3, R., L., & A. from 3½, B. 1½, luncheon 3, D. 4½, pens. from 8 fr.; both these command a view of the Alps. — *SCHWEIZERHOF (Pl. c; C, 3), R., L., & A. from 2½, B. 1½, luncheon 3, D. 4, pens 7-10 fr.; *HÔTEL DE FRANCE (Pl. e; C, 3), R., L., & A. from 2½, B. 1, D. 3, pens. 7-10 fr.; *HÔTEL DU JURA (Pl. d; C, 4), R., L., & A. 2½-3½, B. 1½, D. incl. wine 3½, S. incl. wine 3 fr.; these three near the station. — In the town: *FAUCON (Pl. f; E, 4), Marktgasse, R., L., & A. from 2½, B. 1½, D. 3 fr.; *PFISTERN (*Hôtel des Boulangers*; Pl. g, E 3), near the clock-tower, R., L., & A. from 2½, D. 3½, pens. 7-10 fr.; STORCH (Pl. h; D, 3, 4); *GOLDNER LÖWE (Pl. i; C, D, 4), Spitalgasse, R., L., & A. 2½, B. 1½, D. 2½, pens. 8 fr.; *SCHMIEDEN (*Maréchaux*; Pl. k, E 3), unpretending; HÔTEL-PENSION RUOF (Pl. l; D, 3), Waisenhaus-Platz, R. 2, B. 1 fr.; *STERNEN (Pl. m; D, 3), Aarbergergasse, plain, R. 1½-2½, D. 2½ fr.; HÔTEL ZU ZIMMERLEUTEN (Pl. n; E, 3), Marktgasse; *HIRSCH (Pl. o; D, 3), *BÄR, both near the station, R. 2½-3, D. 3 fr.; HÔTEL DU SAUVAGE (WILDER MANN; Pl. p, D 3), Aarbergergasse, R. 2-2½, B. 1½, D. 3 fr.; EMMENTHALER HOF, Neue Gasse; these last all moderate.

Pensions. °HERTER (Pl. q; F, 4), well situated, near the cathedral; *VILLA FREY, Schwarzthor-Str. 71 (Pl. r; A, 4), pens. 4-7 fr.; BEAU-SITE, Niesenweg 3; JOLIMONT, Äussere Enge (1½ M.; p. 150), with fine view

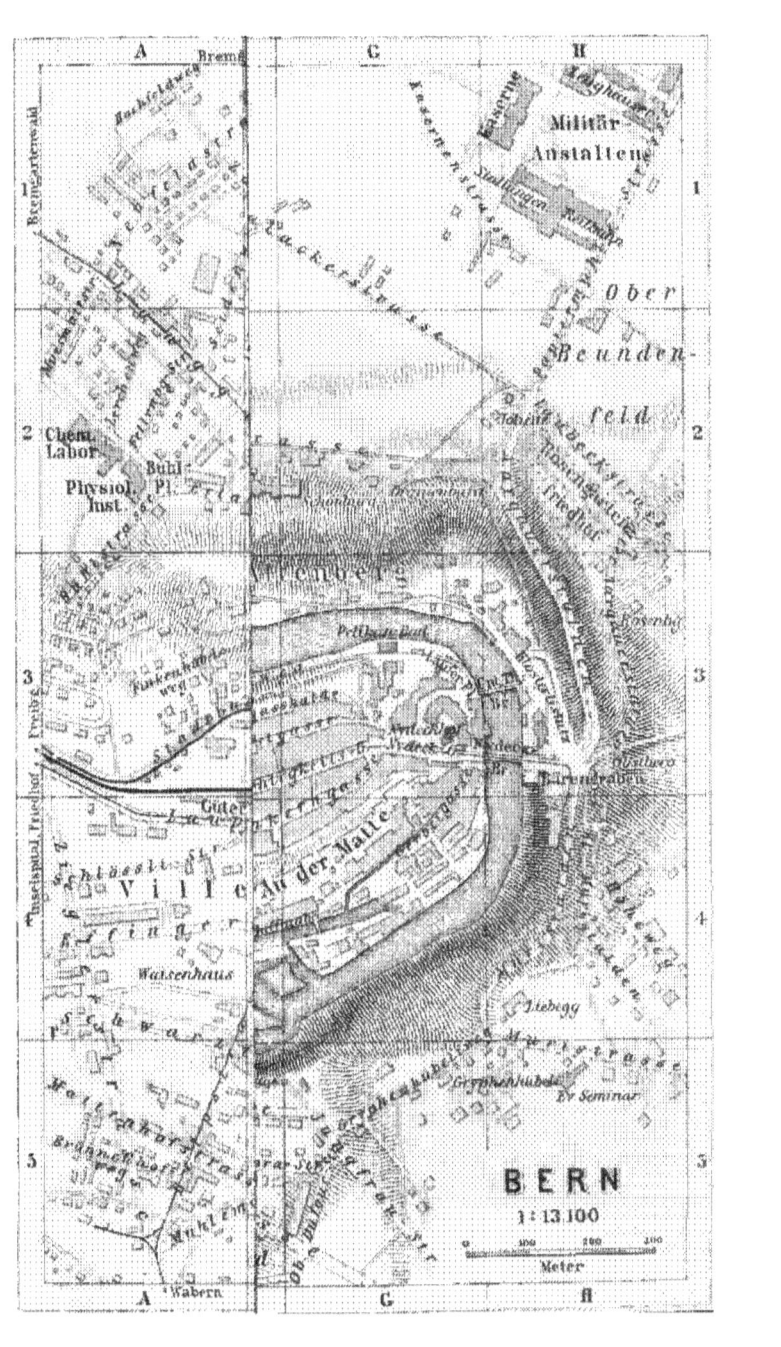

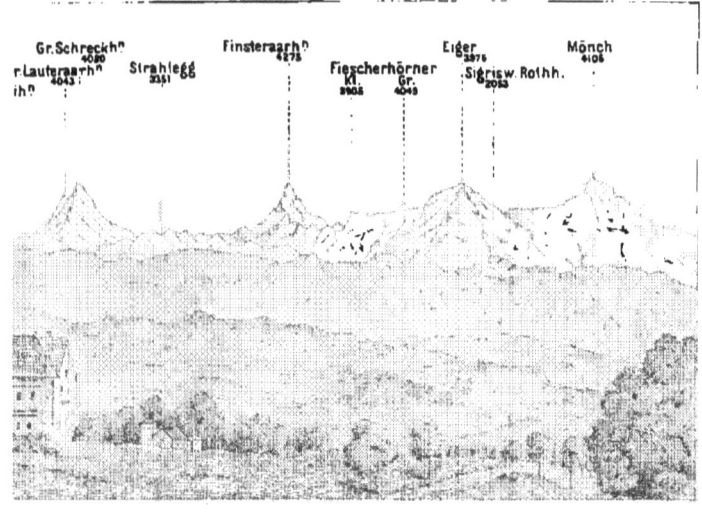

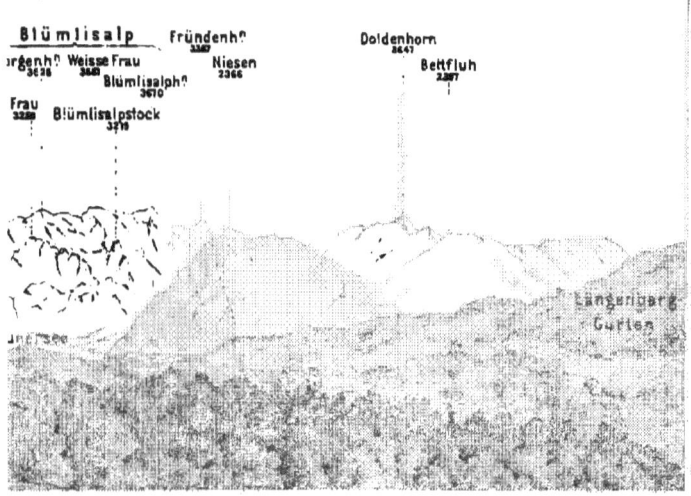

JS BERN
·ldbrücke (538m).

History. BERN. *III. Route 42.* 145

(5-6 fr.); SCHLOSS BREMGARTEN, prettily situated on a peninsula in the Aare, 2¼ M. to the N. (road viâ the Neubrück).

Cafés and Restaurants. *Rail. Restaurant*, D. 2½ fr.; *Café du Jura*, *Café de France*, at the hotels of these names, near the station; *Gesellschaftshaus Museum*, Bären-Platz, corner of the Bundesgasse, déj. or D. 1-2½ fr.; *Café National*, Schauplatzgasse 3 (mural paintings of old Bern); *Café du Pont*, beyond the Kirchenfeld bridge, to the right, with a fine view, déj. 2 fr.; *Schwellenmätteli*, adjacent, below, to the left (Pl. E, 4; fish). — Good *Wine* on draught at the *Café Bär*, Schauplatzgasse 4 (quaint wall-paintings); *Weibel*, Zeughausgasse. — BEER GARDENS. *Café Sternwarte*, on the Grosse Schanze (p. 146); *Café Schänzli* (p. 150; daily concert or theatrical performance in summer, 2 fr., 1 fr., 50 c.); *Café in the Innere Enge* (p. 140), 1 M. from the Aarberg Gate; *Restaurant Schloss Bremgarten*, 2¼ M. to the N. (see above). — Confectioner. *G. Stroebel-Durheim*, Bahnhofs-Platz.

Baths. *River Baths* in the Aare (58-68° Fahr.), at the Marzili (Pl. D, 6; p. 149); *Warm Baths* in the Sommerlustbad, Laupen-Str. (Pl. B, 4, 5; also Turkish and Swimming Baths).

Cabs. One-horse, for ¼ hr. 1-2 pers. 80 c., 3-4 pers. 1 fr. 20 c.; each additional ¼ hr. 40 or 60 c. Two-horse: same fares as for 3-4 pers. with one horse. Box 20 c., small articles free. From 10 p.m. to 6 a.m., double fares. Whole day, *i.e.* over 8 hrs., 1-2 pers. 15 fr., 3-4 pers. 20 fr.

Tramway (moved by compressed air) from the Bears' Den through the chief street to the railway-station, and thence on to the 'Linde' (Bremgarten Cemetery; fares 10-20 c.) and back. *Steam Tramway* from the railway-station to Wabern (p. 151; 25 c.) and to the Länggasse (Bremgartenwald, 10 c.).

Post and Telegraph Office (Pl. C, 3), near the station. Branch-office in the Kramgasse.

British Minister, *F. R. St. John*, Effinger-Strasse 49, office-hours 10-12. — **American Minister**, *James O. Broadhead*; Vice-Consul, *J. E. Hinnen*, Hirschengraben 7 (9-12 and 2-4). — **English Church Service** in the Hall of the Lerber Schule, Nœgeligasse 2, at 10.30 a.m. and 5 p.m. (4 p.m. in winter).

The **Official Enquiry Office** (*Verkehrsbureau*), at the E. corner of the railway-station, opposite the Church of the Holy Ghost, furnishes information gratis as to sights, excursions, etc. — *Money-Changer*, opposite the chief entrance of the railway-station.

ATTRACTIONS. Visit the 'Kleine Schanze' and walk past the Federal Buildings to the Kirchenfeld-Brücke and the Historical Museum; then to the Cathedral (Münster-Terrasse); follow the Kreuzgasse to the Rathhaus; across the Nydeck-Brücke to the Bears' Den; return past the Zeitglockenthurm to the Kornhaus-Platz and cross the Waisenhaus-Platz to the Art and Natural History Museum; lastly cross the railway-bridge to the Schänzli and then return to the station.

Bern (1765'), the capital of Canton Bern, with 50,000 inhab. (including its extensive suburbs), has been the seat of the Swiss government since 1848. It is also the seat of a university (500-600 students), founded in 1834, and of the Central Office of the International Postal Union. Founded by Duke Berthold V. of Zähringen in 1191, the town became independent of the Empire in 1218. By 1288 its powers had so increased that it warded off two sieges by Rudolph of Hapsburg, and in 1339 the Bernese overthrew the Burgundian nobles at the battle of Laupen (p. 215). In 1353 Bern joined the Confederation, and in 1528 the citizens embraced the reformed faith. In 1415 they conquered part of Aargau, and in 1536 they wrested the Pays de Vaud from the princes of Savoy; but in 1798 they were deprived of these territories.

The city, in a striking situation, is built on a peninsula of sand-

BAEDEKER, Switzerland. 16th Edition. 10

stone-rock, formed by the *Aare*, which flows 100' below. Most of the broad principal streets run from E. to W. Those in the old part of the town are flanked with arcades *(Lauben)*, which form a covered way for foot-passengers. One of the chief characteristics of Bern consists in its numerous fountains, most of them dating from the 16th cent., and recently restored. In other respects also Bern still retains more mediæval features than any other large town in Switzerland. Bern is celebrated for its splendid views of the Alps (comp. pp. 147-150), and the phenomenon of the 'Alpine glow' (p. xvii) is seen here to great advantage.

The chief artery of traffic is a series of broad streets, called the Spitalgasse, the Marktgasse, the Kramgasse, and the Gerechtigkeitsgasse, which extend from the Obere Thor (Pl. C, 4) to the Nydeck Bridge (p. 147), a distance of nearly a mile. In the SPITALGASSE is the pretty *Bagpiper Fountain*, dating from early in the 16th century. At the beginning of the MARKTGASSE, where the Bären-Platz and the Waisenhaus-Platz mark the W. limit of the town down to 1346, stands the *Käfigthurm* (Pl. D, 3), restored in the 17th century. The Marktgasse contains the fine *Schützenbrunnen (Archer Fountain)* of 1527 and the *Seilerbrunnen*, the latter with a statue of the foundress of the Insel Hospital (p. 150) on an ancient marble column. Farther on, beyond the interesting old guild-houses *(Webern, Schmieden, Zimmerleuten)*, is the Zeitglockenthurm (Pl. E, 3), the W. gate of the town in its earliest phase, but now its central point, rebuilt in the 15-17th cent., and recently decorated with frescoes. On the E. side is a curious clock, which announces the approach of each hour by the crowing of a cock, while just before the hour a troop of bears marches in procession round a sitting figure. Being the heraldic emblem of Bern, the bear frequently recurs. Thus, on the neighbouring *Zähringer-Brunnen* (Pl. E, 3, 4), in the KRAMGASSE, Bruin appears with shield, sword, banner, and helmet. The *Samson Fountain* and the *Gerechtigkeitsbrunnen*, the latter in the GERECHTIGKEITSGASSE, also deserve notice.

The KORNHAUS-PLATZ (Pl. E, 3) is embellished with the grotesque *Kindlifresser-Brunnen (Ogre Fountain)*, with a procession of armed bears on the shaft of the column. The Kornhaus (Pl. E, 3), built in 1711-16, contains a large open hall on the groundfloor, with 34 columns, in which the weekly corn-market takes place (winecellar below, much frequented; the largest cask contains about 8800 gal.). On the first floor is the cantonal *Industrial Museum* (collection of samples and models, open gratis, 10-12 and 2-5, Sun. 10-12, and on Frid. evening, 7-9, with electric light). — The handsome new *Kornhaus-Brücke* leads hence to the Spitalacker, to the E. of the Schänzli (p. 150).

Opposite, at the E. end of the METZGERGASSE, are the modern Old Catholic Church (Pl. F, 3), in the Romanesque-Gothic style, designed by Deperthes of Rheims, and the cantonal Rathhaus or

Town Hall (Pl. F, 3), erected in 1406-16 in the Burgundian late-Gothic style and restored in 1862, approached by a handsome flight of steps, and adorned with the arms of the Bernese districts. The interior contains the rooms of the Great Council and of the Government Council. — Adjacent is the *State Chancellery*, a late-Gothic building of 1520-41.

On the E. side of Bern, where the old castle of *Nydeck* stood, the Aare is crossed by the handsome *Nydeck Bridge* (Pl. H, 3), in three arches, built in 1844 by K. E. Müller (tramway, see p. 145). The central arch has a span of 165', and is 100' high. On the right bank of the Aare is the Bears' Den (*Bärengraben*), where Bruin is maintained, according to immemorial usage, at the cost of the municipality. Bread and fruit are the only offerings permitted. — From this point the *Muri-Stalden*, a handsome avenue of plane-trees, affording a fine view of the town, ascends to the right, whence we may return to the (20 min.) centre of the town by the Marien-Strasse and the Kirchenfeld-Brücke (p. 146).

The *Cathedral or *Münster* (Pl. F, 4), a fine late-Gothic structure, 93 yds. long, 37 yds. broad, and 76' high, was begun in 1421, completed in 1598, and restored in 1850. Round the whole of the roof runs a beautiful open *Balustrade*, the design of which is different between each pair of buttresses. The W. *Portal* is remarkably fine; the sculptures **represent** the Last Judgment; in the outer arches are Christ, above, with the Virgin and John the Baptist on the left and right, and the Twelve Apostles; in the inner (smaller) arches are the Prophets and the Wise and Foolish Virgins. The *Tower*, 328' high, was finished in 1890-94 from the plans of the German architect Beyer.

INTERIOR (adm. 20 c.; Sun., 2-6, free). The *Stained Glass* on the N. side of the Choir (one window representing the dogma of Transubstantiation) dates from 1496; that on the S. side is modern (1867). The *Choir Stalls* (1522) are adorned on one side with Christ and the Apostles, on the other with Moses and the Prophets. A monument with the armorial bearings of *Berthold von Zähringen*, the founder of Bern (see p. 145), was erected by the city in 1600. Another in memory of the magistrate *Friedrich von Steiger*, bears the names of the 702 Bernese who fell on 5th March, 1798, at the Graubolz and at Neuenegg, in an engagement with the French. In front of this is a Pietà in marble, by *Tscharner* (1870). The great organ dates from 1849 (performance four times weekly in summer at 8½ p.m.; adm. 1 fr.). — The octagonal gallery of the TOWER (223 steps; 20 c.) commands a magnificent view.

The Platz in front of the cathedral is adorned with an *Equestrian Statue of Rudolph von Erlach*, the victor at Laupen (p. 215), in bronze, designed by *Volmar* of Bern, and erected in 1848, with bears at the corners and inscriptions and trophies on the pedestal.

The *Cathedral Terrace (*Münster-Terrasse*; Pl. F, 4), rising abruptly 110' above the Aare, formerly the churchyard, is now a shady promenade with seats, adorned with a bronze statue of *Berthold von Zähringen* (p. 145), designed by *Tscharner*, with Bruin as a helmet-bearer. The view from this terrace, as indeed from every open space in Bern, is justly celebrated. In clear weather the panorama of the Bernese Alps witnessed here is more extensive than from any other spot in the Oberland.

10*

Views. The most important mountains are marked in the annexed Panorama. From other points (the Klosterhof, Bundes-Terrasse, Kleine Schanze, Café Schanzli, and the Enge outside the Aarberger Thor) the following mountains are also visible: — To the right of the Doldenhorn, the *Balmhorn* (12,180') with the *Altels* (11,930'; 37 M. distant), and over the *Gurten*, the bell-shaped summit of the *Stockhorn* (7195'; 18 M.); also, to the extreme left, the peaks of the *Spannörter* (10,515'; 53 M.) and the *Schlossberg* (10,280'; 54 M.), both in the canton of Uri; the crest of the *Bäuchlen* near Escholzmatt (5810'; 24 M.), and the *Feuerstein* above the Entlebuch (6700'; 30 M.). Comp. p. 145.

From the Cathedral Square we follow the Herrengasse to the *Municipal Library* (Pl. E, 4; adm. on week-days, 2-4), containing numerous works on Swiss history, and to the *University* (Pl. E, 4). We then turn to the left and cross the *Klosterhof* (the point whence our Panorama was taken) to the *Kirchenfeld - Brücke (Pl. E, 4; splendid view), a bold iron bridge built in 1882-83, 115' above the Aare, which crosses the Aare Valley in two graceful spans of 260' each and connects the old town with the new quarter in the *Kirchenfeld*.

Here, in the Helvetia-Platz, rises the imposing new *Bern Historical Museum (Pl. E, 5), a picturesque building in the mediæval style, designed by Lambert (adm. in summer daily 8-12 and 4-6, 50 c.; Sun. 10½-12 and Tues. and Sat. 3-5, free).

MIDDLE FLOOR (first entered). The vestibule contains a number of models for a monument to Adrian von Bubenberg, the leader of the Bernese in the battle of Morat. — To the left (E.) is the *Ethnographical Collection*, consisting chiefly of objects from Greenland, Canada, China, Japan, Persia, Central Africa, Borneo, and Java. — To the right (W.) is the *Archaeological Collection*, including antiquities from lake-dwellings, implements of the flint, bronze, and iron periods, chiefly from the Jura, and Roman remains (bust from Aventicum, fragments of a mosaic floor from Herzogenbuchsee, bronze vase from Grächwyl).

UPPER FLOOR. In the handsome staircase are *Armour* of the 15-16th cent. and modern *Weapons* and *Banners*, all from the Bern Arsenal. — To the right (E.). ROOM I. *Tapestry* from Burgundy and the Netherlands, including embroidered *Antependia* from Lausanne and the Convent of Königsfelden (p. 20), of the 13-15th cent.; table from the Bern Town Hall, 1576; *Ecclesiastical Vestments* of the 14-16th cent. (by the windows); *Stained Glass* of the 16th century. ROOM II. *Articles in Wrought Iron*, including some well-preserved swords of the 13-14th cent. (in a case). ROOM III (Silver Chamber). More than 200 silver *Guild, Family, and Church Cups; Diptych* of 1537, formerly supposed to be the field-altar of Charles the Bold, known to have been given to the Convent of Königsfelden by Queen Agnes, and in Bern since the Reformation; *Bernese Coins*. — To the left (W.) of the staircase. ROOM I. *Tapestry* from Lausanne and Burgundy (with the Burgundian and other arms); Bernese *Magistrate's Chair* of the 18th cent.; views of Bern in the 17-18th centuries.

GROUND FLOOR. Reproductions of *Early Swiss Rooms*.

To the N.W. of the Kirchenfeld-Brücke, conspicuously situated on the edge of the town-hill, are the *Federal Buildings*, or Bundes-häuser (Pl. D, 4), two handsome edifices in the Florentine palatial style, which are to be connected by a domed building not yet completed. The *Bundeshaus-Ost*, erected from Auer's designs in 1888-92, accommodates the departments of war and agriculture; the *Bundeshaus-West*, built by Stadler and Studer in 1852-57, contains the chambers of the two legislative assemblies (the 'National-rath' and the 'Ständerath'). Both buildings are shown (free) by the porter, 9.30-11.30 and 2-4. In front of the Bundeshaus-West is a fountain-figure of *Berna*, in bronze, on a pedestal adorned with figures of the four Seasons (1863). The *Bundes-Terrasse*, adjoining the S. façade, commands a splendid view. — Near the Bundeshaus-Ost is the *Federal Statistical Office*, in the former house of A. von Haller (d. 1777), the well-known physician and author; adjacent, at the corner of the Inselgasse, stands the *Mint* (1790-93). — Between the two federal buildings, to the N., extends the Bären-Platz, in which, to the left, is the *Museum* (now a restaurant, p. 145), adorned with statues of celebrated Bernese.

A *Cable Tramway*, 360' long (gradient 3:10) descends on the W. side of the Bundes-Terrasse to the bathing establishments in the *Marzili* (p. 145). Trains every 5 min.; fare 10 c.

To the W. of this point, passing the Bernerhof, a few paces bring us to the promenades on the *Kleine Schanze (Pl. C, 4), which affords a superb survey of the Bernese Alps (comp. p. 148; panorama by Imfeld on a round stone in the upper promenade): in the foreground the Aare Valley and the Kirchenfeld-Brücke, with the cathedral-tower to the left and the Historical Museum to the right. In the grounds is a bust of *Niggeler* (d. 1887), the 'Turnvater' (promoter of gymnastics).

The **Kunst-Museum** (Pl. D, 2) in the Waisenhaus-Str., built by Stettler in 1879, is open on week-days, 9-12 and 1-5 (adm. 50 c.; free on Tues. and on Sun., 10.30-12 and 1-4; catalogue 50 c.).

On the GROUND FLOOR are two rooms to the left containing sculptures and casts.

The vestibule of the UPPER FLOOR contains statues of Rebecca, Miriam, Ruth, and David, by *Imhof*; busts of Bianca Capello and of an Arab sheikh, after *Marcello* (p. 204); *Burnand*, Herd leaving the mountain-pasture. On the left, three cabinets with early German, Italian, and Netherlandish pictures, including several, by *Nic. Manuel* (1484-1520) and others, from Bern Cathedral. — ROOM I (left). 97. *K. Girardet*, Battle of Morat; above, 115. *Ch. Humbert*, Cattle at a ford; 187. *Rüdisühli*, Deserted castle; 228. *A. Veillon*, Spring on the Lake of Brienz; 39. *Arth. Calame*, Lake of Geneva at Hermance; 6. *Anker*, Bernese village-school; 210. *A. Stäbli*, After the storm; above, 214. *Carl Stauffer*, The sister of the artist; 142. *Annie Hopf*, Prayer-meeting in the house of G. Monod at Paris; 47. *Fr. Diday*, View of the Lauterbrunnen-Thal from Wengen; 163. *A. Potter*, Evening in South Italy; 137. *T. Massarani*, Oriental scene; above, 243. *R. Weiss*, Street in Cairo. — In the adjoining CABINET: 61-68. *Joh. Dünz* (1645-1736, Bern), Bernese portraits. — ROOM II. On the right wall: 212-219. Pictures and studies by *Carl Stauffer* (1857-91); to the left, 218. Portrait of Gustav Freytag, by the same; to the left, farther on, 121. *R. Koller*, Strayed cow and calf; 104. *C. Grob*, Family prayers; 226. *B. Vautier*, Saying grace; *Anker*, Luncheon hour at the school; 143. *A. de Meuron*, Chamois-hunter; *Anker*, Boy reading to his sick grandfather; 37. *Al. Calame*, Waterfall at Meiringen. — ROOM III. 201. *Jul. Schrader*, Abdication of Emp. Henry IV.; 229. *A. Veillon*, Tombs of the Califs near Cairo; 17. *A. Böcklin*, Idyll of the Sea; above, 169. **A. de Regny**, Arch of Titus at Rome; 240. *Fr. Walthard*, Last battle between the Bernese and the French at Grauholz (1798); above the door, 215. *Carl Stauffer*, Sister of the artist; 160. *Th. Pixis*, Huss taking leave of his friends before his execution; 8. *Bachmann*, Going to baptism in winter; 256. *Zünd*, Forest-landscape in autumn, with figures by *Koller*; 147. *D. Meyer*, Woman of the Simmen-Thal; 92, 93. *E. Girardet*

150   III. Route 42.   BERN.

Going to school, Alms-giving; 42. *G. Castan*, Lake of Oeschinen. — CABINET. 38. *K. Gehri*, Golden wedding; 38. *A. Calame*, Handeck. This and the three following cabinets chiefly contain works by Swiss masters of the end of the 18th and beginning of the 19th centuries.

Opposite is the **Natural History Museum** (Pl. D, 3), built by A. Jahn in 1879-81 (open in summer, Tues. and Sat., 2-5, and Sun., 10.30-12.30, free; on other days, 8-6, adm. 1 fr.; for 2-5 pers., 50 c. each; larger parties 3 fr.).

GROUND FLOOR. In the entrance-hall are busts of *A. von Haller* (p. 149) and *E. L. Grunner* (d. 1883), the geologist; also a geological map of Switzerland. By the staircase is a group of chamois. The room to the right contains the *Collection of Minerals*, which includes two cases of magnificent crystals from the St. Gotthard. Bust of *B. Studer* (d. 1887). To the left is the *Palaeontological Collection*, rich in Alpine fossils. Perfect skeletons of the Irish elk and the cave-bear. Relief of the Bernese Oberland by Ed. Beck. — On the first and second floors is the *Zoological Collection*. In the central saloon (1st floor), with ceiling-frescoes by Baldancoli, are large ruminants. In the **room on the left**, birds and eggs. In the room on the right, mammalia. **Adjacent**, a small room devoted to the Swiss fauna; Barry, the celebrated **St. Bernard dog**. — On the 2nd floor, to the left, reptiles, amphibia, fish, corals, and sponges; to the right, mollusks, crabs, insects, echinodermata, and worms.

Adjoining the Museum on the E. is the large new *School Building* (Pl. D, 3), accommodating the Gymnasium and the Commercial and Elementary Schools. — The old *Cavalry Barracks* (Pl. D, 3), near the post-office, contain the *Industrial Exchange* (groundfloor), the *Pharmaceutical Institute* (1st floor), and the *Permanent Educational Exhibition* (2nd floor), the last of great interest for teachers (open free on week-days, 8-11 and 1-4).

The promenades on the Grosse Schanze, above the station to the W. (Pl. B, C, 3), afford an extensive panorama, but the view of the city is less picturesque than from other points. At the top are the *Observatory* (1880'), the *Head Offices of the Jura-Simplon Railway*, the *Women's Hospital*, and a bust of *President Stämpfli* (d. 1879).

To the W. of the town, in the continuation of the Laupen-Strasse (Pl. A, 3, 4), is the *Inselspital*, a large hospital on the pavilion system (1880-84), originally founded in 1354 (in the Inselgasse, p. 146). Adjacent are the *University Clinical Institutes*.

Crossing the *Railway Bridge* (p. 17), at the N.W. end of the town, we pass the *Botanic Garden* (Pl. D, 2) and reach (¹/₂ M.) the *Schänzli (Pl. E, 2; *Café-Restaurant*, see p. 135; adm. for noncustomers 50 c.), with a terrace and grounds commanding the finest view near Bern. In the foreground lies the picturesque city; above it rises the wooded Gurten; to the left are the Bernese Alps, and to the right the Stockhorn chain, adjoined by the Freiburg Mts.; to the extreme W. is the Moléson. — Adjacent is the *Pension Victoria*, for invalids.

The large **Military Depôt** of Canton Bern, in the *Beundenfeld* beyond the Schänzli, erected in 1874-78 at a cost of 4¹/₂ million francs, comprises an arsenal, offices, stables with riding-schools, and barracks. Adjacent is the large *Drill Ground*.

To the N., 1 M. from the Aarberg Gate, on the left bank of the Aare past the *Deer and Chamois Park* (comp. Pl. C, 1), is the *Enge (Café see p. 145), rising high above the Aare, with promenades and view of the town and the Alps. A monument commemorates *Gottlieb Studer* (1804-90), the Alpine authority. Adjacent is the beautiful *Bremgarten Forest*, with marked paths.

The view from the *Gurten* (2825'; *Inn*), a long hill to the S. of Bern, embraces, besides the Bernese Alps (p. 187), the Stockhorn chain, the Freiburg Alps, the Jura for a distance of 100 M., with parts of the Lake of Neuchâtel; and, to the left, the Unterwald and Lucerne Mts. as far as Pilatus. A steam-tramway runs half-hourly from Bern railway-station to (10 min.) *Wabern*, whence several paths ascend to the top. On the hillside are the *Bächtelen* asylum for deserted boys and the girls' institute *Victoria*.

Above *Belp* (p. 153), 7 M. to the S. of Bern, lies *Zimmerwald* (2815'; Hôt.-Pens. Beau-Séjour), charmingly situated, and (4 M. farther) *Bütschelegg* (3470'; Inn), with an extensive view. — During a longer stay, excursions may be undertaken to the *Frieswylhubel* (2385'; 4 hrs.), to the *Belpberg* (3592'; 4¾ hrs.), and to the *Falkenfluh* (3410'; 4 hrs., see below).

43. From Bern to Thun.
Comp. Map, p. 152.

19½ M. RAILWAY *(Centralbahn)* in 1 hr. (3 fr. 35, 2 fr. 35, 1 fr. 70 c.). View to the *right* as far as Münsingen; thence to Uttigen on the *left*. — Through-trains from Bern to *Interlaken* (*Thunersee Railway*, p. 155).

Bern, see p. 144. On the *Wylerfeld* (p. 18) the train turns to the right, affording an admirable survey of the Alps to the S.; to the left is the lunatic asylum of *Waldau*. 3 M. *Ostermundingen*. — 5 M. *Gümlingen* (Hôt. Mattenhof), junction for Lucerne (p. 140). About 2¼ M. to the E. is the *Pension Dentenberg* (2325'). The *Giebel* (1¼ hr.) commands a fine view. — 8 M. *Rubigen*; 10 M. *Münsingen*. On the right rise the Stockhorn and Niesen, on the left the Mönch, Jungfrau, Blümlisalp, and (farther on) Eiger. 12½ M. *Wichtrach*; 14½ M. *Kiesen*. From this point a road ascends viâ *Diesbach* in 2½ hrs. to the *Falkenfluh* (3410'), a health-resort with an unpretending inn and a fine view. Near (15½ M.) *Uttigen* we cross the Aare.

19½ M. Thun. — Railway Stations. *Thun*, the chief station, on the N.W. side of the town; *Scherzligen* (formerly *Thun-See*), to the S. (for Interlaken), where passengers alight for the steamer. — The Steamer (p. 155) calls at *Thun-Stadt*, near the Hôtel Freienhof, at *Thun-Hofstetten*, above the large hotels, and at *Scherzligen*, close to the railway-station (see above).

Hotels. *THUNER HOF or GRAND HÔTEL*, a large first-class house, with a garden on the Aare, R. 3½-7, L. ¾, A. 1, B. 1½, déj. 3, D. 5, pens. 8-15, omn. 1 fr.; *BELLEVUE*, with grounds, R. 2-5, L. & A. 1, B. 1½, déj. 3, D. 4½, pens. 7½-12, omnibus 1 fr.; *HÔT.-PENS. BAUMGARTEN*, with grounds, and dépendance (CHOISY) on the Aare, R., L., & A. 3-5, D. 4, pens. 6-10 fr. — *FREIENHOF* (Pl. c), in the town, with café-restaurant and garden on the Aare, R., L., & A. 2½, B. 1½, D. 3, S. 2, pens. 6-7 fr.; *FALKEN* (Pl. a), near the station, with terrace on the Aare, same prices; *WEISSES KREUZ* (Pl. d), next the post-office, D. 3 fr.; *KRONE*, Rathaus-Platz (Pl. R P), R., L., & A. 1½-2½, B. 1, D. 2, pens. 4-6 fr.; *SCHWEIZERHOF* (Pl. b), at the end of the street leading straight from the station, R. 1½-3, D. 2½, pens. 5-7 fr.; *BÄR*, farther on, beyond the bridge, unpretending. — *PENS. ITTEN*, on the Amsoldingen road, 6½ fr.; *PENS. EICHBÜHL*, on the lake, near Hilterfingen, 2 M. to the S.E.

Munich beer at the *Falkenhalle* (in the Hôtel Falken; Pl. a). Native beer at the *Steinbock* and in several beer-gardens. — *CURGARTEN*. Concerts daily 3.30-5 and 8-10 p.m. Adm. 50 c.; weekly ticket 2, monthly 5 fr.

152   *III. Route 43.*   THUN.

BATHS in the very rapid and cold Aare, to the N. of the town, 50 c. Warm Baths at the *Bällix Baths*. — BOAT on the lake, according to tariff, 3 fr. per hour, 2 hrs. 5 fr., 3 hrs. 7, ½ day 8, whole day 10 fr.; but better terms may sometimes be made. — MONEY-CHANGER, *A. Knechtenhofer*, opposite the Thuner-Hof.
POST & TELEGRAPH OFFICE (Pl. P), on the Aare island.
CAB to or from the station 1 fr. Carriage with one horse the first hour 5, with two horses 10 fr., each addit. hour 3 and 5 fr. To Wimmis 8 or 15, to Kandersteg 22 or 40, to Weissenburg 13 or 24, to Zweisimmen 28 or 50, Gessenay 35 or 60, Gsteig 40 or 70, Château d'Oex 40 or 70, Aigle 80 or 150, Gurnigel 25 or 45 fr.
ENGLISH CHAPEL in the grounds of the Bellevue.

Thun (1844′), a quaint old town with 5500 inhab., charmingly situated on the rapid green *Aare*, ³/₄ M. below its efflux from the lake, forms a fitting portal to the beauties of the Oberland. All the open spaces in the town command splendid views to the S.E. of the snowy peaks of the Blümlisalp and the Doldenhorn (comp. lower row of the Niesen panorama, to the left), with the Niesen in the foreground and the Stockhorn chain to the left of it. Thun is the seat of the *Federal Military School*, for commissioned and non-commissioned officers. The *Artistic Pottery* of Thun has a considerable reputation; one of the chief manufactories is that of *Wanzenried* at *Schwäbis*, 1 M. to the N.W. (depôt in Thun-Hofstetten).

Above the town rises the large and conspicuous square tower of the old *Castle of Zähringen-Kyburg* (Pl. S) with a turret at each corner, erected in 1182, and within the walls of the castle is the *Amts-Schloss*, or residence of the Bernese bailiffs, erected in 1429. It may be reached from the N. gate (¼ M. from the station viâ the bridges), by a covered flight of steps from the market-place (Pl. RP), and on the S.E. by another flight of steps or by an easy path from the Hôtel-Pension Baumgarten. The tower contains a small historical museum (50 c.). A walk round the castle discloses a series of beautiful views. Still more picturesque are the views from the *Church* (Pl. K; 1738), to the S.E. of the castle, and from the pavilion in the corner of the churchyard.

Walks. On the right bank of the Aare, about 110 yds. above the Thun-Hofstetten landing-place (p. 151), is a finger-post (left) indicating a flight of steps, which ascends, at first between houses, to the (¼ hr.) *Pavillon St. Jacques* (*Jakobshübeli*; 2100′), commanding the lake, the Alps from the Finsteraarhorn to the Doldenhorn, Thun, and the valley of the Aare. [Guests of the Hôtel Bellevue can reach this point by a prettier route through the hotel-grounds.] A sign-post here shows the way to (10 min.) the *Pavilion* (fine view of Thun); to the (³/₄ hr.) Rabenfluh; to the (25 min.) Kohleren Waterfalls; and to (1 hr.) the Haltenegg (see below). Close to the Pavillon St. Jacques is the *Pension-Curhaus Obere Wacht* (pens. 5 fr.). — Another walk is by the road on the right (N.) bank of the Aare and of the lake across the *Bächimatt*, with its pretty grounds and Alpine view, to the (20 min.) *Chartreuse* (the property of the Parpart family). Here (or by a shorter path 8 min. farther back) we turn to the left, passing the *Bächihölzli*, cross (10 min.) the *Hünibach*, and follow a path through the picturesque *Kohleren Ravine*, where the brook forms several small falls. This path ascends to the Grüsisbergwald (see below) and the Goldiwyl road (½ hr.; see below).

The *Goldiwyl Road*, which diverges to the right from the Steffisburg road, at the '*Hübeli*', a few hundred yards to the N. of the town (shorter

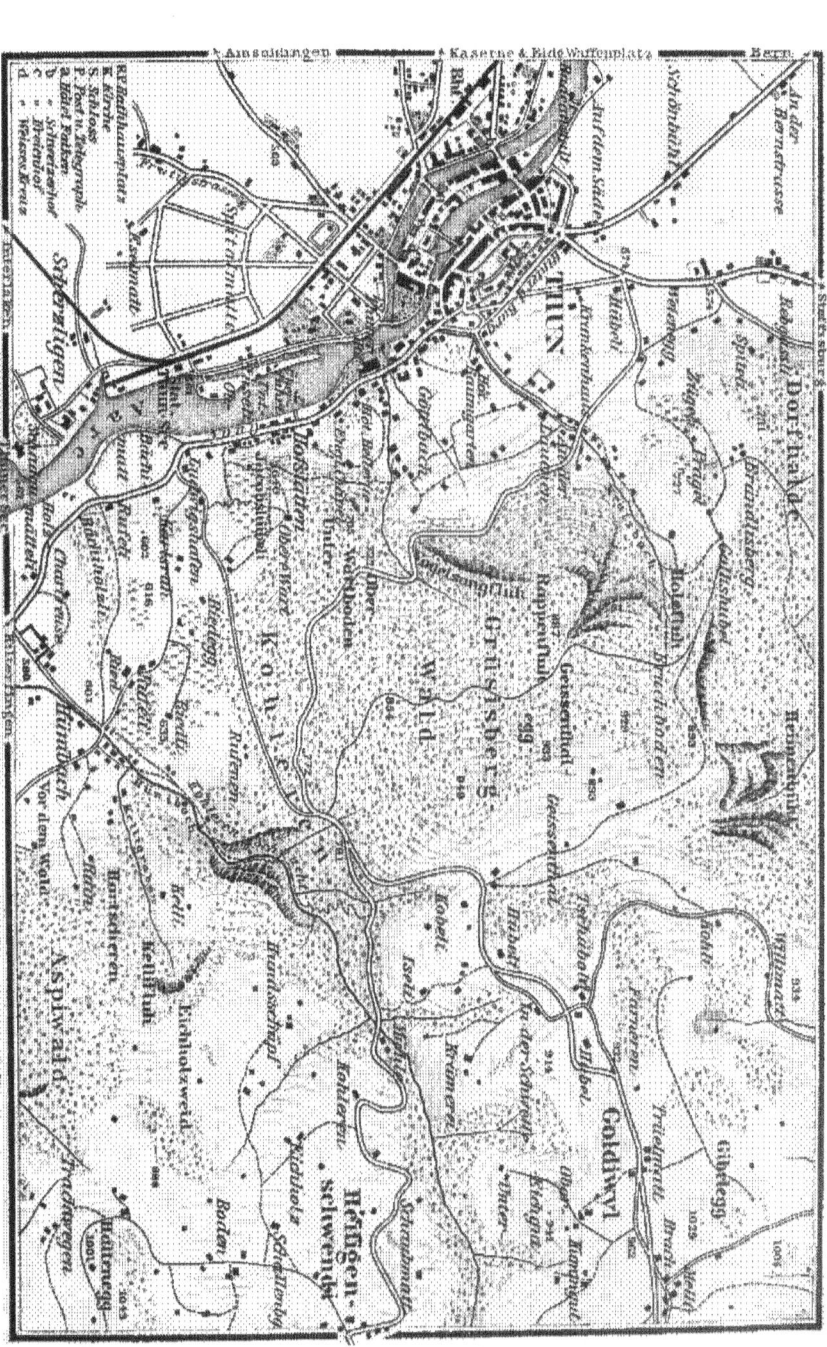

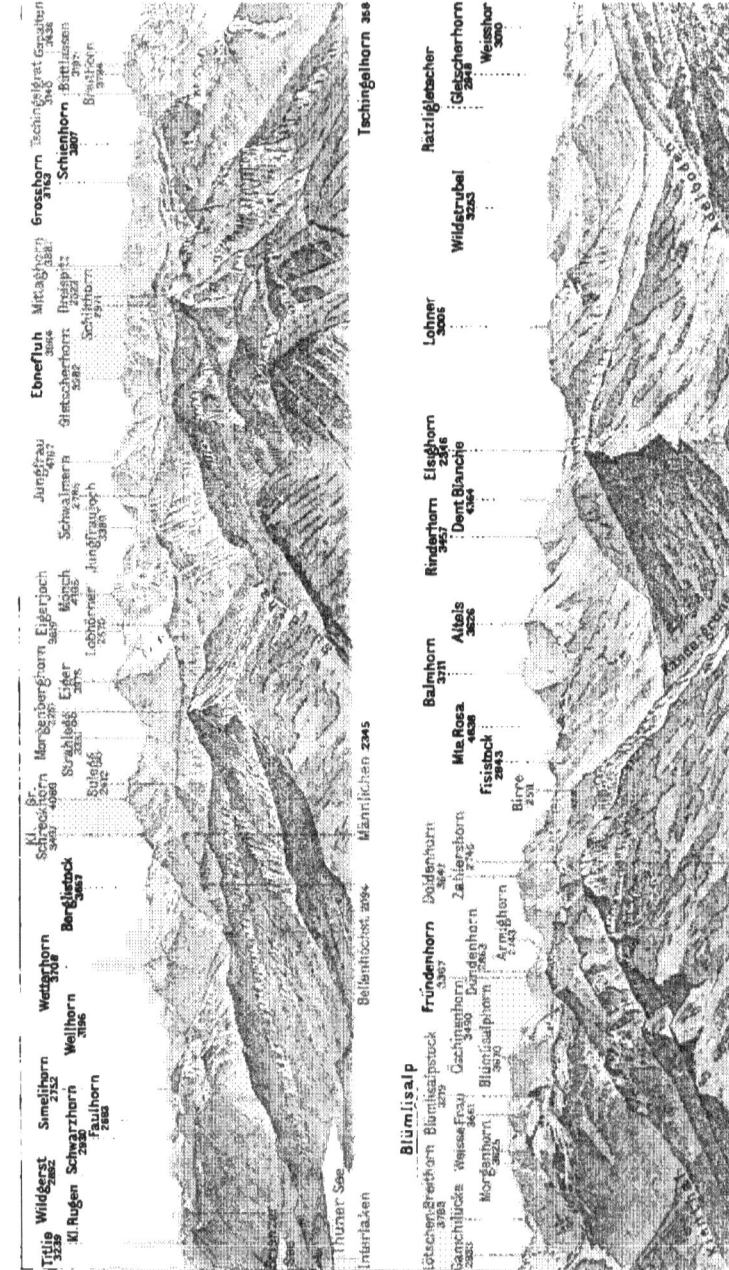

GURNIGELBAD.   *III. Route 43.* 153

path to the right at the Hôt.-Pens. Baumgarten, with numerous guideposts), leads along the slope of the *Grüsisberg*, the fine woods of which are intersected by numerous walks. Fine view of the town, the valley of the Aare, and the Stockhorn chain from the *Rappenfluh* or *Rabenfluh* (2844'; 1 hr.). Hence we may return to the town, in a curve towards the N., viâ the *Brändlisberg* (2397'; 20 min.), another view-point, and the Pavillon St. Jacques (¹/₂ hr.), or we may go to the S. direct to the Pavillon St. Jacques. After about 2¹/₄ M. the Goldiwyl road joins a road connecting it with the above-mentioned road on the right bank of the Aare, and divides. The left branch leads to (1¹/₂ M.) *Goldiwyl* (3155'; Zysset's Inn), the right to (2³/₄ M.) *Heiligenschwendi* (3324'), ³/₄ M. to the S. of which is the *°Hattenegg* (3287'), affording a magnificent view of the lake and the Alps.

Schloss Schadau, about ¹/₄ M. to the S. of Scherzligen, see p. 155 (park open on Sun.). — *Schloss Hünegg*, 1¹/₂ M. to the S. of Thun-Hofstetten, see p. 156.

Longer Excursions. To the N. of Thun (1¹/₂ M.; diligence 5 times daily in 20 min.; carr. with one horse 3 fr.) is the considerable village of *Steffisburg* (brewery), on the *Zulg*, whence we may ascend in ¹/₂ hr. to the small *Schmittweyer-Bad* (2625'; pens. 5 fr.), with its mineral spring and pretty walks. *Thierachern* (1867'; Löwe), with fine view, 3 M. to the W.; 3 M. farther to the W., *Bad Blumenstein* and the *Fallbach*; thence through wood in 1¹/₂ hr. to the *Gurnigel-Bad* (see below). *Baths of Schwefelberg* (3¹/₂ hrs. to the W. of Blumenstein, beyond the *Gantrist Pass*), see p. 203. — *Burgistein* (2690'), a village and castle with fine view, 8 M. to the N.W. of Thun. *Amsoldingen* (Roman tombstones), 3¹/₂ M to the S.W., and the ancient tower of *Strättligen* (p. 145), 3¹/₂ M. to the S. of Thun, a splendid point of view. The undulating district between the Stocken-Thal and Thun abounds in beautiful walks and mountain-views. — The *Stockhorn* (from Blumenstein or Amsoldingen 4-4¹/₂ hrs.), see p. 202.

To the Gurnigel-Bad, from Thun a walk of 3¹/₂ hrs. (guide desirable), or a drive of 4 hrs. (carr. with one horse 25, with two horses 45 fr.); or from Bern direct (17 M.; diligence twice daily in 4¹/₂ hrs.; fare 7 fr. 15 c., coupé 8 fr. 60 c.). The road from Bern leads by *Wabern* and *Kehrsatz*, and (leaving *Belp* on the left) follows the W. side of the Gürbethal, soon affording a fine view of the Bernese Alps; to (7¹/₂ M.) *Toffen*. At (12¹/₂ M.) *Kirchenthurnen* (1995') it ascends to the right to the large village of *Riggisberg* (2500'; Sonne), beyond which we follow a road to the left to (15 M.) *Rüthi* and ascend steeply through the *Laaswald* to the (17 M.) *Gurnigelbad* (3783'), a favourite health-resort, with a spring impregnated with lime and sulphur, situated on a broad plateau (600 beds; R. 2¹/₂-6, board 6-8 fr.; S. B. G. H.). Extensive wood-walks in the environs: to (¹/₄ hr.) *Seftigschwend* (Inn); past the Laashöfe to the (1 hr.) *Längnel-Bad*; to the (1 hr.) *Obere Gurnigel* (5070'), an admirable point of view; to the (1¹/₂ hr.) *Seelibühl* (5750'). — Over the *Seelibühlgrat* to (2¹/₂ hrs.) *Bad Schwefelberg*, see p. 203; over the *Gantrist* to *Bad Weissenburg* (5-6 hrs.), see p. 205. — From *Wattenwyl*, 5 M. to the W. of Thun and 3 M. to the S.W. of stat. *Uttigen* (p. 151), a pleasant path, which cannot be mistaken, ascends to Bad Gurnigel in 2¹/₂ hrs.

To *Saanen* through the valley of the *Simme*, see R. 57.

44. The Niesen.

Two bridle-paths lead to the top: on the N. side from *Wimmis* (see below), and on the E. from the *Heustrich-Bad* (p. 154), each in 4¹/₂ hrs. The former has the more shade in the morning, the latter in the afternoon. Horse to the top and back 15 fr., or, if the start is later than 10 a.m., 20 fr.; to Heustrich over the Niesen (or in the reverse direction from Heustrich over the Niesen to Wimmis), 22 and 28 fr. — Guide (unnecessary) 10 fr. — Chair-porters 12 fr. (four porters are required for one chair).

Spiez (p. 156) is the station both for Wimmis (2³/₄ M.; diligence thrice daily in 40 min., 85 c.) and for the Heustrichbad (4 M.;

diligence twice daily in 55 min., 1 fr. 10 c.). — The diligence starts at the railway-station; carriages may also be obtained at the steamboat-wharf. — The Kander-Thal road (comp. p. 191) crosses the railway near *Spiezmoos*, at its junction with the Thun road, and leads to the left to (1¼ M.) *Spiezwyler* (Bär). It then divides, the left branch leading to Heustrichbad and Frütigen (see below), while the right branch descends in a wide curve (to the left a direct footpath through wood) to the *Kanderbrücke*, with a fine view of the Blümlisalp, and thence proceeds in a straight direction to (2¼ M.) —

Wimmis (2080'; pop. 1242; *Löwe*), a pretty village in a very fertile district, at the E. base of the *Burgfluh* (3248'), overlooked by a castle which is now occupied by a school and the local authorities. The church is mentioned in ancient documents as early as 533.

ASCENT OF THE NIESEN FROM WIMMIS. The route (at first a narrow cart-track) ascends on the S. side of the Burgfluh. After 35 min. it crosses the *Staldenbach*; 3 min. later, by a gate, is a finger-post indicating the path to the left ('Niesen 3¾ hrs.'), which ascends in zigzags through pastures and wood, passing the chalet on the Bergli. By the (2 hrs.) chalets of *Unterstalden* (4940') the path crosses to the right bank of the Staldenbach, and winds up the slopes of the Niesen, past the chalets of *Oberstalden* (5888'). The prospect first reveals itself beyond the (1¼ hr.) *Staldenegg* (6345'), a sharp ridge connecting the *Bettfluh* (7924') or *Fromberghorn* with the Niesen. Thence to the top 1 hr. more.

Beyond Spiezwyler (see above) the road to Heustrichbad and Frutigen continues to follow the ridge, affording views of the Blümlisalp, the Niesen (r.), and, beyond the lake (l.), the Sigriswyl Rothorn and the Ralligstöcke. After ¾ M. a branch diverges to the left to Aeschi (p. 156), while the main road descends gradually to the Kander-Thal. At the *Casino Inn* in *Emdthal*, 4 M. from Spiez, our road diverges to the right from that to Frutigen (p. 191) and crosses the *Kander*. On the left bank, at the foot of the Niesen, lies the much-frequented —

*Heustrichbad (2300'; board 3½-6 fr.; S. B. G. H.), with an alkaline-saline sulphur-spring, pleasure-paths, and a view of the Blümlisalp.

A good bridle-path ascends the grassy slopes behind the baths in windings (whenever it divides, the steeper branch must be selected), as far as an ancient lime-tree, with a bench (1½ hr.); then through wood (1 hr.) and over pastures, past the chalets of *Schlechtenwaldegg* and the *Hegern-Alp* (6308'), in numerous windings, to the (2½-3 hrs.) summit, with the extreme top to the right and the Niesen Inn to the left. This route affords beautiful and diversified views; milk at the two upper chalets.

The *Niesen (7763'; *Weissmüller's Inn*, 5 min. below the top, R. 4, B. 2 fr.), the conspicuous N. outpost of a branch of the Wildstrubel, and like Pilatus regarded as an infallible barometer (see p. 102), rises in the form of a gently sloping pyramid. The rocks at the base are clay-slate, those of the upper part sandstone-conglomerate. The view vies with that from the Faulhorn (comp. the Panorama, p. 153). The beautiful snow-clad Blümlisalp is seen to great advantage. Best light towards sunset or in the morning before 10 o'clock.

45. From Thun to Interlaken. Lake of Thun.

Comp. Map, p. 152.

a. Thunersee Railway.

16³/₄ M. Railway in 1-1¹/₄ hr. (3 fr. 25, 2 fr. 20, 1 fr. 50 c.); from Bern to Interlaken in 2-2¹/₂ hrs. (6 fr. 60, 4 fr. 50, 3 fr. 20 c.). — Through railway tickets may be also used for the steamboat (see below), but allow no break in the journey.

Thun, see p. 151. — ¹/₂ M. *Scherzligen* (see p. 151), on the left bank of the Aare, which here emerges from the lake; the station is opposite the steamboat-pier (see below). The train skirts the W. bank of the lake, with a view of the Stockhorn chain to the right, and the Bernese Alps from the Wetterhorn to the Blümlisalp to the left. 3 M. *Gwatt* (Schäfle; Post). Beyond *Strättligen*, with its old tower (p. 143), we cross the gorge of the *Kander* (p. 191) by a handsome bridge, 65 yds. long and 98' high.

6 M. **Spiez** (2090'). The station is high above the village (p. 156); splendid view of the Lake of Thun and the mountains on its N. bank (Ralligstöcke, Sigriswyler Rothhorn, etc.); in the foreground Spiez with its château, and to the S.E. and S. the Bernese Alps.

Beyond Spiez the line descends past the village of *Faulensee* (p. 156); it then skirts the precipitous slopes of the S. bank, passing through three tunnels near *Krattigen*. 11 M. *Leissigen* (*Steinbock; *Weisses Kreuz), pleasantly situated at the foot of the *Morgenberghorn* (p. 162; road to Aeschi, see p. 157). Beatenberg (p. 158) is visible high above the N. bank of the lake. — 13¹/₂ M. *Därligen* (Pens. Seiler, Schärz, Schwalbenheim). To the left, near the influx of the Aare, is the ruin of *Weissenau*. The train skirts the new *Aare Channel* and reaches the station of *Interlaken* (p. 158).

b. Steamboat Journey.

Steamboat (Restaurant on board, D. 2¹/₂ fr.), 8-9 times daily in 1-1¹/₂ hr. from *Thun-Stadt* (p. 151) to *Interlaken* (fare 2 fr. 95 c.). — Stations *Hofstetten, Scherzligen, Oberhofen, Gunten, Spiez, Merligen, Beatenbucht, Leissigen, Därligen* (the last two not always touched at).

The steamboat starts from the Freienhof Hotel (p. 151), ascends the *Aare*, stops at *Hofstetten* on the right bank (p. 151), and then at the rail. station of *Scherzligen* (see above). To the left, among trees, is the *Chartreuse* (p. 152); to the right, on the peninsula where the Aare emerges from the lake, *Schloss Schadau*, a building in the English Gothic style, with numerous turrets and a large park.

The **Lake of Thun** (1850'), which the steamer now enters, is 11 M. long and nearly 2 M. wide; its greatest depth is 1130'. The *View from the steamer is magnificent. The Stockhorn (7195'), with its conical summit, and the pyramidal Niesen (7763') rise on the right and left of the entrance to the valleys of the Kander and Simme (p. 202). To the left of the Niesen are the glittering snow-fields of the Blümlisalp; on the right, at the head of the Kander-Thal, the Fründenhorn, Doldenhorn, Balmhorn, Altels, and

Rinderhorn gradually become visible (from left to right). In the direction of Interlaken appear successively (from left to right) the Mittaghorn, Jungfrau, Mönch, Eiger in the foreground, and farther off the Schreckhorn and Wetterhorn. The steamer skirts the N.E. bank, which is clothed below with villas and gardens and higher up with woods, and passes the pretty village of *Hilterfingen*. To the left is the château of *Hünegg*, in the French Renaissance style (adm. to the park on application to the gardener, who lives on the road, close by; no fee). The boat touches at Oberhofen *(Pensions Moy; *Oberhofen, *Blau; Restaurant Zimmermann)*, which has a picturesque château of Countess Pourtalès, and at Gunten (*Weisses Kreuz; *Pens. du Lac*, 5 fr., *Hirsch*, both on the lake; *Pens. Schönberg*, on the hill). In the vicinity (1½ M. from the lake) the water of the Guntenbach has worn a curious gorge for itself, with waterfalls (path and bridges at present much damaged).

A road ascends from Gunten to (¾ hr.) **Sigriswyl** (2620'; *Pens. Bär*, from 5 fr.), a prettily situated village. The *Blume* (4577'; fine view) is ascended hence in 2 hrs. viâ *Schwanden*; the *Sigriswyl-Grat* (*Vorder-Bergli*, 5508'; *Hinter-Bergli*, 6056') by the *Alpiglen Alp* in 2½-3 hrs.; the °*Sigriswyler Rothhorn* (6737'), the highest point of the Sigriswyl-Grat, in 4 hrs. (with guide). — On the steep slope of the Sigriswyl-Grat towards the *Justisthal* (p. 158) is the *Schafloch* (6840'), a grand ice-cavern, reached from the Obere Bergli by a giddy path in ¾ hr. (guide, ice-axe, and torches necessary).

The steamer now crosses the lake, at its broadest part, to —

Spiez. — Hotels. °SPIEZER HOF, by the pier, with garden and lakebaths, R. 2½-5, L. ½, A. ¾, B. 1½, déj. 3, D. 4, pens. 6-12 fr., Eng. Church Serv. in summer; °HÔT.-PENS. SCHONEGG, ½ M. from the lake, near the rail. station, R., L., & A. 2½-4, B. 1½, déj. 2½, D. 3½, pens. 7-10 fr.; PENS. ITTEN, 3 min. to the W. of the station; RAILWAY RESTAURANT, with rooms and fine view of Spiez. — POST & TELEGRAPH OFFICE, at the rail. station. — CARRIAGE from the rail. station or pier to Wimmis 4, with two horses 7 fr.; to Heustrichbad 5 or 10 fr. (see p. 158); to Aeschi 6 or 12 fr.

The village of *Spiez*, the starting-point for an ascent of the **Niesen** (p. 153) and for excursions to the Kander and Simme valleys (pp. 191, 202), is the most attractively situated place on the Lake of Thun. The picturesque old château, which formerly belonged to the Erlach family, is now the property of a Berlin gentleman, who has restored it and surrounded it with pretty grounds. The road ascends among the houses and orchards of the village and divides into three branches at the (½ M.) *Pension Itten*. That in a straight direction leads to the *Railway Station* (235' above the lake; ¾ M. from the pier), that to the left to Faulensee (p. 157), and that to the right to Wimmis and the Kander-Thal (pp. 154, 191).

FROM SPIEZ TO AESCHI, 2¼ M. (carr., see above). The road diverges to the left from the Kander-Thal road, about ¾ M. to the S. of Spiezwyler (p. 154). Walkers may follow the Faulensee road from the rail. station and then (20 min.) ascend the path to the right (finger-post; ½ hr.). The village of **Aeschi** (2818'; **Hôt.-Pens. Blümlisalp*, pension 5-7 fr.; *°Hôt.-Pens. Niesen*) lies on the height between the Lake of Thun and the Kander-Thal, with a charming view of the lake, and is visited as a health-resort. A pleasant road also leads in 2 hrs. from *Leissigen* (p. 155) to Aeschi viâ *Krattigen* (Stern). From Aeschi to the *Heustrichbad* (p. 154), footpath in 40 min.; to the Mülinen road, ¾ hr. (The *Faulenseebad*, see p. 157, is 1 M. to the S.E.)

Descent to Emdthal or Mülinen, 1½ M. — FROM AESCHI TO THE SAXETEN-
THAL, a pleasant route (7½ hrs.; guide unnecessary). Road by *Aeschi-Ried*
in the *Suldthal* to the (6 M.) *Untere Suldalp* (3418'); then a bridle-path,
past a fine waterfall of the Suldbach, to the (1¼ hr.) *Schliuren-Alp* (4675');
ascent to the left to the (1½ hr.) Tanzbödeli Pass (6168'), between the
Morgenberghorn and the *Schwalmern*; then descent by the *Hinter-Bergli-
Alp* to (1½ hr.) *Saxeten* (p. 162). The Morgenberghorn (7883') may be
ascended from the pass in 1½ hr. (guide desirable for the inexperienced),
or direct from Aeschi viâ *Aeschi-Allmend*, the *Sonnenberg*, and the *Hutmad
Alp* in 5 hrs. The ascent of the Schwalmern (9137') from the Suldthal is
more interesting, but fit for experts only, with guide; descent past the
Sulegg (p. 162) to Saxeten or Isenfluh.

From Spiez two black peaks are visible for a short time towards
the E., above the S. bank of the Lake of Brienz; that to the right
is the Faulhorn, the broader (left) the Schwarzhorn. The next station
on the S. bank is *Faulensee*, above which (3 M. from Spiez) is the
Faulensee-Bad (2625'; *Hôtel Victoria, pens. 6-12 fr.), with a mineral
spring, pleasant grounds, and beautiful view (Engl. Ch. Serv. in
summer).

On the N. bank we next observe the abrupt *Sigriswyl-Grat*, with
the bold *Ralligstöcke* (6066') and the *Sigriswyler Rothhorn* (6737').
On the lake is *Schloss Ralligen*. Beyond stat. **Merligen** (**Hôt.
Beatus*, with garden on the lake, pens. 5-6 fr.; *Löwe)*, at the mouth
of the *Justisthal*, the steamer proceeds to the (¼ hr.) *Beatenbucht*
(Restaurant), the station for *St. Beatenberg* (see below).

The *Nase*, a rocky headland, here projects into the lake. High
up on the steep bank runs the boldly constructed road (Merligen to
Interlaken 6 M.), hewn in the rock and passing through two tunnels. On the lake is the château of *Lerow*, near the *Beatenbach*,
which issues from the *Beatushöhle*, ¾ M. above the road, making
a noise like thunder in spring and after heavy rain. Farther on the
road threads three more tunnels and then runs above the ravine of
the *Sundgraben* (p. 158), in which lie the houses of *Sundlauenen*.
It next passes the *Küblibad* or *St. Beatusbad*, the *Neuhaus*, and the
Pension Simpkin, and reaches Unterseen.

The steamer, which sometimes calls at *Leissigen* (p. 155) on the
S. bank, next enters the *Aare Channel* (1¾ M. long; to the left, the
ruin of Weissenau, p. 155) and stops at the landing-place near the
W. or principal station of *Interlaken* (p. 158).

FROM BEATENBUCHT TO ST. BEATENBERG, *Cable Tramway* in
16 min. (ascent 2½ fr., descent 1 fr., return-fare 3 fr.). The line,
opened in 1889, is 1 M. long and has an average gradient of 1 : 3.
The station at the top is 5 min. from the Curhaus.

FROM INTERLAKEN TO ST. BEATENBERG, by road, 7 M. This
diverges to the left from the Habkern road (p. 163), about 1 M. from
Unterseen, crosses the *Lombach*, and winds upwards through the
wood (one-horse carr. 13, two-horse 24, to the Curhaus 14 or 25 fr.).
Walkers, with the aid of short-cuts, take 1 hr. from the Lombach
bridge to a roadside inn, and ¾ hr. thence to the Hôtel des Alpes.

158 III. Route 45. ST. BEATENBERG.

St. Beatenberg. — **Hotels.** *CURHAUS, at the W. end of the village, near a wood, with 130 beds and 2 'dépendances', R. 3-5, D. 4½, S. 3, pens. 7½-12 fr. (S.B.G.H.). The following are named in their order from W. to E.: PENSION EDELWEISS; *PENSION BEATRICE, 4½-6, in July and Aug. 5-7 fr.; *HÔT.-PENS. BLÜMLISALP, ¾ M. from the Curhaus, R., L., & A. from 2, B. 1, D. 3, pens. from 5 fr.; *HÔT.-PENS. WALDRAND, similar prices; *HÔT.-PENS. SCHÖNEGG (an Evangelical resort), in the middle of the village, 4½-6½, in July and Aug. 5½-7½ fr.; FEUZ, village inn; *GRAND HÔTEL VICTORIA, a first-class house 1¼ M. from the Curhaus, rebuilt after a fire in 1894; *HÔT.-PENS. ZUR POST, R., L., & A. 2½-4½, B. 1½, D. 3, pens. 6-10, omn. 1 fr.; *HÔT.-PENS. BELLEVUE, frequented by the English, 7½-9 fr.; *PENS. SILBERHORN, 2¼ M. from the Curhaus, 6-7½ fr.; PENS. BALMER; on the other side of the Sundgraben: *HÔT.-PENS. ALPENROSE, 6-8 fr.; *HÔT.-PENS. DES ALPES, 3 M. from the Curhaus, R., L., & A. 2½, B. 1½, M. 3, pens. from 5 fr. — *Private Lodgings.* — *English Church.* — Good woodcarvings at moderate prices.

The village of *St. Beatenberg* (3775'), a favourite health-resort, stretches along the flank of the Beatenberg for 2½ M., overhung by the rocky ridge of the *Güggisgrat* and occupying both sides of the *Sundgraben*, **the deep-sunken bed of a mountain-torrent. Admirable view of the Alps, from the Schreckhorn to the Niesen, including the Eiger, Mönch, Jungfrau, Blümlisalp, Doldenhorn, and Wildstrubel. Pleasant paths, with benches in commanding situations, have been laid out above and below the road.**

At the Pens. Edelweiss is a finger-post indicating the way to the *Waldbrand* (25 min.; green marks), the *Vorsass*, and the *Niederhorn;* one at the Hôt.-Pens. Blümlisalp indicates the *Parallel Promenade* (blue marks); another between the church and the Victoria shows the way to the (¾ M.) *Beatushöhle* (p. 157; red marks); a fourth, at the Bellevue, points upwards towards the (½ hr.) *Känzli* (white and blue marks).

The finest point of view is the *Amisbühel (4383'; Inn* at the top), 25 min. to the E. of the Hôtel Alpenrose. Walkers from Interlaken diverge from the road to the right by a finger-post, 2 M. beyond the Alpenrose and ½ M. from the Hôtel des Alpes, at a point where a sign-post indicates the route down to the Beatushöhle; thence to the top ½ hr.

Ascent of the *Gemmenalphorn (6770'), the highest point of the *Güggisgrat*, from the Amisbühel over the *Waldegg-Altmend, Leimern*, and *Gemmenalp*, or from St. Beatenberg through the *Kieschnen Valley* in 2½ hrs., not difficult (path marked red and white; guide 4 fr., unnecessary). Superb view, ranging from Pilatus to the Stockhorn chain and the Diablerets; at our feet lies the Justis-Thal (p. 156), beyond it are the Aare valley, Bern, and the Jura Mts. The Lake of Thun is not visible. — The *Niederhorn (6445') and Burgfeldstand (6780'), each 2½-3 hrs. from Beatenberg, are also fine points of view. The route to the former is indicated by white and yellow marks, that to the latter, passing the Känzli, by white and blue marks. By following the arête, all three points may be visited in one excursion.

46. Interlaken and Environs.
Comp. Map, p. 160.

Railway Stations. THUNERSEE RAILWAY or PRINCIPAL STATION (p. 155), at the W. end of the town; BERNESE OBERLAND RAILWAY (station *Interlaken-Ost*, pp. 160, 165), at the E. end, 1 M. from the first-named. They are connected by the BÖDELIBAHN (change carriages; 1¼ M., in 7 min.; fares 60, 35, 25 c.), on which 15 trains run daily in each direction, four going on to Bönigen (p. 181). Hotel-omnibuses and other vehicles at both stations. — **Steamboat Piers** for the Lake of Thun near the Principal Station (p. 160); for the Lake of Brienz opposite the station Interlaken-Ost (p. 184).

Hotels and Pensions (omnibus 1 fr.). On the *Höheweg*, from W. to E.: *Hôt. Métropole (Pl. 1), R., L., & A. from 3, déj. 3, D. 5 fr., pens. from 7 fr.; *Victoria (Pl. 2), R., L., & A. from $4^{1}/_{2}$, B. $1^{1}/_{2}$, déj. $3^{1}/_{2}$, D. 5, pens. 9-12, in July and Aug. 10-15 fr.; beyond it, *Hôt. Horn (Pl. 30), unpretending; *Jungfrau (Pl. 3), R., L., & A. from $4^{1}/_{2}$, déj. 3, D. 5 fr.; *Schweizerhof (Pl. 4), R., L., & A. from $3^{1}/_{2}$, déj. $3^{1}/_{2}$, D. $4^{1}/_{2}$-5 fr., good cuisine; *Belvedere (Pl. 5), R., L., & A. from $3^{1}/_{2}$, B. $1^{1}/_{2}$, D. $4^{1}/_{2}$, pens. from 8 fr.; *Hôt. des Alpes (Pl. 6), R., L., & A. from 4, déj. 3, D. 4, pens. from 9 fr.; *Hôtel Beaurivage (Pl. 9), R., L., & A. from $4^{1}/_{2}$, déj. $3^{1}/_{2}$, D. 5 fr; *Hôt. du Nord (Pl. 7), R., L., & A. from 3, D. 4, pens. 7-8 fr.; Hôt. St. Georges (Pl. 22), R. $2^{1}/_{2}$-3 fr., well spoken of; *Hôt.-Pens. Interlaken (Pl. 8), R., L., & A. from 3, D. 4 fr.; *Hôt. du Lac (Pl. 10), near the pier and the E. station, R. $2^{1}/_{2}$, B. $1^{1}/_{4}$, D. 3, pens. from 6 fr., unpretending.

To the W. of the Höheweg, in the direction of the railway-station (all second-class): *Hôt. Oberland (Pl. 12), R., L., & A. 3, D. 3, pens. 6-7 fr.; opposite to it, *Post (formerly *Cheval Blanc;* Pl. 26), moderate; Cerf, pens. 5-6 fr., well spoken of; *Croix Blanche (Pl. 11), R. $1^{1}/_{2}$-2, D. 3, B. $1^{1}/_{4}$ fr.; *Hôt. Berger (Pl. 28), R., L., & A. $2^{1}/_{2}$-3, D. $2^{1}/_{2}$, pens. 6-7 fr.; *Hôt.-Pens. Krebs (Pl. 27), R. $2^{1}/_{2}$-3, B. $1^{1}/_{4}$, D. 3 fr.; *Hôt. Terminus & de la Gare (Pl. 29), R., L., & A. from 3, B. $1^{1}/_{4}$, D. 4, pens. from 7 fr., the last three near the station; Schwan, R. 1-2 fr. — Near the lower bridge over the Aare: *Bellevue (Pl. 15), R., L., & A. 2-3, B. $1^{1}/_{4}$, D 3, pens. from 6 fr.

On the small island of *Spielmatten*: *Hôt. du Pont (Pl. 16), with garden, R., L., & A. from 3, B. $1^{1}/_{2}$, D. $3^{1}/_{2}$, pens. from 7 fr.; *Krone, unpretending. — At *Unterseen*: *Hôt. Unterseen (Pl. 17), R. 2, B. 1, D. $2^{1}/_{2}$, pens. 6 fr.; *Beau-Site (Pl. 18), pens. from 6 fr.; *Hôt.-Pens. Eiger, on the Neuhaus road, pens. $5^{1}/_{2}$-7 fr; Pens. Alpenruhe, on the Beatenberg road, 5-6 fr.; Pens. Simpkin, near the Lake of Thun. — Furnished apartments in the *Villa Alpina*, Jungfrau-Str.

To the S. of the Höheweg, on the road to the Kleine Rugen: Deutscher Hof (Pl. 20), R., L., & A. 3-$4^{1}/_{2}$, B. $1^{1}/_{4}$, D. 3, pens. from 6 fr.; *Hôt. National & Pension Wyder (Pl. 19), R., L. & A. from $2^{1}/_{2}$, déj. $2^{1}/_{2}$, D. $3^{1}/_{2}$, pens. 8-9 fr.; Union Hôtel & Pens. Reber (Pl. 21), pens. 6 fr.; *Hôtel-Pens. Ober & Villa Silvana (Pl. 23), 6-9 fr.; *Hôt. St. Gotthard, 6-7 fr.; Hôt.-Pens. Eden, *Pens. Schönthal, 5 fr. — *Hôt. Jungfraublick (Pl. 22), a first-class house, in an elevated position close to the Rugen Park (p. 161), commanding a splendid view; R., L., & A. from 6, B. $1^{1}/_{2}$, déj. 4, D. 6, omn. $1^{1}/_{2}$ fr.; pens. in July and August 15-18, at other times 12-15 fr. — *Hôt.-Pens. Mattenhof (Pl. 24), at the foot of the Kleine Rugen, pens. in July and August 6-8, at other times 5 fr.; Pens. Zwahlen-Spycher, 4-5 fr.

In the Environs of Interlaken good and inexpensive quarters may be obtained. At *Wilderswyl* (p. 168), $1^{1}/_{2}$ M. to the S. *Pens. Schönbühl, *Hôt.-Pens. Wilderswyl, both in a fine lofty situation, pens. 5-6 fr.; *Bär, in the village, pens. 5-6 fr.; Pens. Jungfrau, 4 fr.; Stern; Kreuz, modest. — At *Gsteigwyler*, $^{1}/_{2}$ M. from the railway-station of Wilderswyl-Gsteig; Pens. Schönfels. — On the Brienz road, on this side of the church-hill of Goldswyl, ($^{3}/_{4}$ M.) Pens. Schönegg, $5^{1}/_{2}$ fr. — At *Bönigen* (p. 181), on the S. bank of the Lake of Brienz, terminus of the Bödelibahn (p. 158): *Pens. Bellerive, *Hôt.-Pens. Bönigen, *Chalet du Lac, and *Hôt.-Pens. de la Gare (near the steamboat-pier), R. from $1^{1}/_{2}$, B. 1, D. $1^{1}/_{2}$-3, pens. 5 fr.

Restaurants in the hotels *Métropole*, *Victoria*, etc. — Beer. *Cursaal*, see below; *Café Oberland*, in the hotel of that name (see above); *Baierische Bierbrauerei*, with garden, next to Hôt. Beaurivage (concert in the evening); *St. Georges*, see above; *Adlerhalle*, to the W. of the Métropole; *Hôt. du Pont*, on the Aare, with garden and view; *Berger*, *Krebs*, by the Thunersee rail. station. — **Confectioners**: *Weber*, on the Höheweg, at the entrance to the Cursaal; *Schuh*, opposite the Métropole; *Seitz*, Bahnhof-Str.

Cursaal on the Höheweg, with café-restaurant, reading, concert, and billiard rooms, garden, etc.; music in the morning, afternoon, and evening; admission 50 c., per day 1 fr., per week 4 fr., month 12 fr.; for extra entertainments higher charges. At the back of the Casino is a whey-cure establishment (open 7-8 a.m.).

160   *III. Route 46.*   INTERLAKEN.   *Bernese*

Chemists. *Seewer*, opposite the Hôt. Oberland; *Pulver*, Postgasse. — Money Changers: *Volksbank* (Pl. 26), *Betschen*, both Bahnhof-Str.
Carriage from the station to Interlaken, Unterseen, and Matten 1 fr. each person, to Bönigen, Gsteig, Wilderswyl, and Ringgenberg 2 fr.; per hour with one horse 4, with two horses 6, each additional hour 3 or 5 fr.; to Lauterbrunnen and Grindelwald, see p. 165. — Post and Telegraph Office (Pl. P) adjoining the Oberländer Hof. — The Official Enquiry Office (*Verkehrsbureau*), on the Höheweg, adjoining the Cursaal, supplies information gratis, sells railway-tickets, etc.

English Church Service in the old Convent Church. Presbyterian Service (Scottish Free Church) in the Sacristy of the Schloss at 11 and 4. American Services (in summer) at the Hôtels Victoria and Métropole.

The low land between the lakes of Thun and Brienz, which are 2 M. apart, is called the '*Bödeli*'. These lakes probably once formed a single sheet of water, but were gradually separated by the deposits of the *Lütschine*, flowing into the Lake of Brienz, and the *Lombach*, which falls into the Lake of Thun. These accumulations, first descending from the S., out of the valley of Lauterbrunnen, and then from the N. out of the Habkěren valley, account for the curve which the *Aare* has been compelled to describe. On this piece of land, 'between the lakes', lies **Interlaken** (1863'), consisting of the villages of *Aarmühle, Matten*, and *Unterseen*, and extending nearly as far as the Lake of Brienz (total pop. 5385). Interlaken is a favourite summer-resort, and is noted for its mild and equable temperature. The purity of the air, the whey-cure, and the beauty of the situation attract many visitors, while others make it their headquarters for excursions to the Oberland.

The principal resort of visitors is the *HÖHEWEG, an avenue of old walnuts, now past its best, extending from the village of Aarmühle to 'the upper bridge over the Aare, and flanked with large hotels and tempting shops. It commands a beautiful view of the Lauterbrunnen-Thal and the Jungfrau (finest by evening-light). To the right, near the upper or N.E. end of the Höheweg, rises the old monastery and nunnery of *Interlaken*, founded in 1130, and suppressed in 1528, surrounded by beautiful walnut-trees. The E. wing of the monastery has been used as a hospital since 1836; the rest of the building, with the *Schloss* added in 1750, is occupied by government-offices. Different parts of the monastery-church are now used for Anglican, Scottish Presbyterian, French Protestant, and Roman Catholic services. The prolongation of the Höheweg leads to the rail. station *Interlaken-Ost* (p. 158; also a landing-place of the Brienz steamer) and on to Bönigen (2 M. ; p. 184). The Brienz road, diverging to the left at the Hôtel Beaurivage (to Ringgenberg 2 M., to Brienz 10 M.), crosses the Aare (Brückwald, see p. 162).

At the S.W. end of the Höheweg, opposite the Hôtel Oberland, the road to the Kleine Rugen (p. 164) diverges to the S.E., while that in a straight direction leads past the new *Post Office* (Pl. P) and the *Volksbank* (Pl. 26) to the *Thunersee Station* (p. 158). — The road diverging to the N.W. at the Volksbank crosses the two

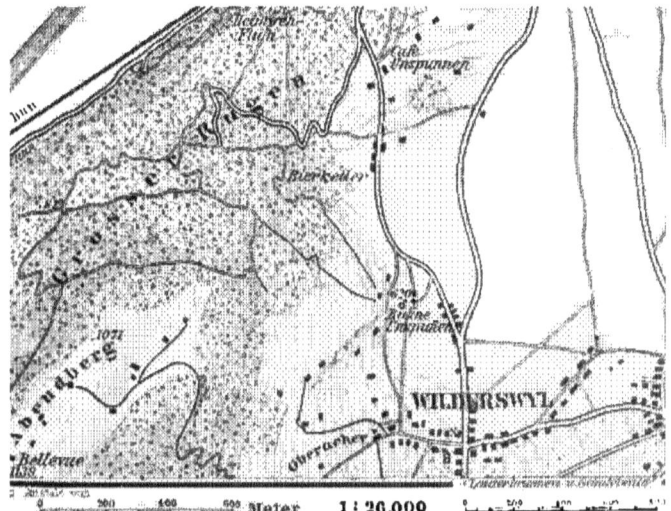

Geogr. Anst. v. Wagner & Debes, Leipzig

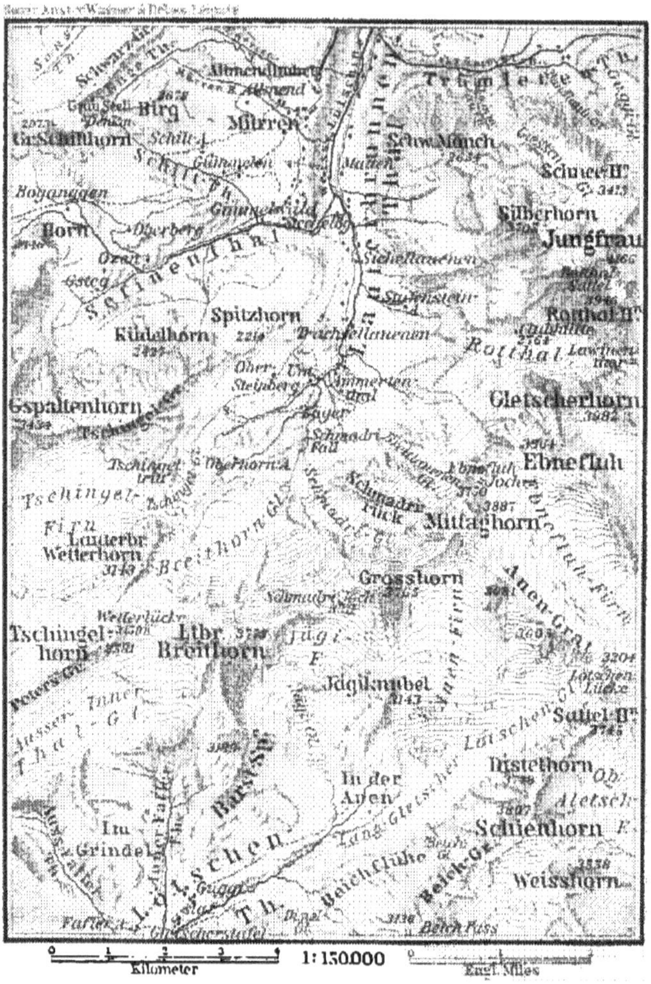

Oberland. INTERLAKEN. *III. Route 46.* 161

islands of *Spielmatten* (fine view from the middle bridge of the Jungfrau, rising to the S. between the two Rugen) and leads to the small town of *Unterseen*, which consists chiefly of wooden houses, with a large square and a modern church. Large manufactory of parquetry. The road to Merligen and Thun (p. 157) begins here to the left, at the hotels mentioned on p. 159. The road to the Habkern valley and to St. Beatenberg leads to the N.W. (pp. 163, 157).

The *Kleine Rugen, the beautiful wooded spur of the Grosse Rugen, offers the most attractive walks, with frequent benches and ever-changing views. The principal path ascends by the Hôtel Jungfraublick in a straight direction and then leads round the whole hill on its lower slopes. Turning to the left we pass the 'Humboldtsruhe' (view of the Jungfrau and Lake of Brienz) and reach the ($^1/_2$ hr.) *Trinkhalle* (Café-Restaurant), commanding the Jungfrau, Mönch, and Schwalmern. Farther on, beyond the 'Scheffel Pavilion' (with a view of the Lake of Thun), is the *Kasthoferstein*, erected in memory of the chief forester Kasthofer, by whom, about the beginning of the century, the hill was planted with specimens of the principal trees of Switzerland. Thence the path proceeds past a reservoir and a chamois-preserve back to the Hôtel Jungfraublick. Other paths, with benches and points of view, ramify in every direction; *e.g.* to the (25 min.) *Rugenhöhe* (2425'), where three artificial openings in the foliage permit views of the Jungfrau and the lakes of Thun and Brienz.

View from the Heimwehfluh. (Heights in mètres.)

162 *III. Route 46.* INTERLAKEN. *Bernese*

Just beyond the Trinkhalle a path diverges to the left, and by a (1 min.) bench (where the path straight on leads in 10 min. to the Café Unspunnen) descends to the right to the *Wagnerenschlucht*, which separates the Kleine and Grosse Rugen. Near the Studer memorial (see below) our footpath joins a road, which traverses the ravine and leads past the *Café *Unspunnen* and the *Bairische Bierkeller*, and below the ruin of *Unspunnen*, to Wilderswyl (p. 163), with continuous views of the Lauterbrunnen valley and the Jungfrau, and of the Lake of Brienz to the left.

In the middle of the Wagnerenschlucht, about 300 paces from the fork at its W. end, is a block of rock with an inscription in honour of Bernh. Studer (d. 1887), the geologist. Here we diverge by a path to the right (W.), which ascends rapidly, passing a fine point of view on the right, to the (20 min.) *Heimwehfluh (2218′). The terrace in front of the restaurant commands a charming view (best in the afternoon) of the Bödeli and the lakes of Thun and Brienz; the Jungfrau, Mönch, and Eiger are seen from the belvedere. A more comprehensive view (extending to the Schreckhorn on the left) is commanded by the *Abendberg, ascended viâ the Grosse Rugen in 1½-2 hrs. on foot (horse 8, mule 6 fr.). We follow the Heimwehfluh route in the Wagnerenschlucht (see above), and after ¼ hr. turn to the left and farther on (guide-board) to the left again, without leaving the wood. On the top is the *Hôtel Bellevue (3735′; pens. 5½-7 fr.). — A path ascends above the hotel, across grass and past some chalets, to (20 min.) a tall dead fir-tree, known as the *Siebenuhrtanne* (2125′), whence there is a charming *View of the Lake of Thun, lying far below.

A footpath leads past the different peaks of the Abendberg to the (3 hrs.) *Rothenegg* (6230′; shortest way from the hotel, 2 hrs.). The next peaks of the range are the *Fachsegg* (6346′), the *Grosse Schifti* (6674′), the *Kleine Schifti* (6585′), and finally the *Morgenberghorn* (7333′). The last is very difficult from this side (better from Saxeten, by the Tanzbödeli Pass, see p. 157). — A footpath leads from the Hôtel Bellevue to Saxeten in 1 hr. (the upper path to the right in the meadow, behind the second chalet).

The Saxeten-Thal, between the *Abendberg* and the *Bellenhöchst* (6870′), is reached by a pleasant bridle-path (mule 7 fr.) to *Mülinen* and the (7 M.) village of *Saxeten* (3600′; Kreuz). About 1¼ M. higher up are the falls of the *Gürben* and *Weissbach*, and the valley is picturesquely closed by the *Schwalmern* (9137′).

The Sulegg (7915′; 3½-4 hrs.), an excellent point of view, is ascended from Saxeten. We ascend by the (35 min.) *Gürben Fall* to the *Untere Nesslern-Alp* (4805′), cross the Gurbenbach to the left, and several other brooks descending from the Sulegg. Beyond the (1¼ hr.) *Bellen-Alp* (6205′), we turn to the right between the *Bellenhöchst* (6870′) and the Sulegg, skirt the E. slope of the latter, nearly as far as the *Suls-Alp*, for ¾ hr., and reach the top in 1 hr. more. The ascent is easier from *Isenfluh* (p. 165), viâ the *Gummen-Alp* and *Suls-Alp* (3½ hrs.; guide 10 fr.). — From Saxeten over the *Tanzbödeli Pass* and through the *Suldthal* to (6 hrs.) *Aeschi*, see p. 157 (interesting; guide not indispensable).

About 100 paces beyond the bridge over the Aare on the Brienz road (p. 160) a guide-board at the Brückwald, which stretches to the left up the slopes of the *Harder*, indicates a number of walks,

A fine view of the Bödeli, the lakes, and the mountains is commanded by the (20 min.) *Hohbühl* (2070'), on which a pavilion commemorates the sojourn at Interlaken of Weber, Mendelssohn, and Wagner. The Jungfrau is better seen from the pavilion at the *Lustbühl*, 1/4 hr. farther along the slope, whence we may return viâ the middle Aare bridge or viâ Unterseen (a walk of 1-1 1/4 hr. in all).

The view from the *Obere Bleiki* is opener but not finer than that from the Hohbühl, 1 hr. above which it lies. About 1/2 hr. higher is the *Pavillon Falkenfluh* (see below). These two points may also be reached by a bridle-path diverging to the left (guide-post) from the Brienz road, 350 paces from the Aare bridge.

The Thurnberg of Goldswyl (1/2 hr.), beyond *Schönegg* on the Brienz road (p. 184), overlooks the Lake of Brienz and the small, sombre Faulensee or Lake of Goldswyl. — A walk may be taken by the same road (or by the new picturesque path crossing the hills between the road and the Lake of Brienz) to (1/2 hr.) Ringgenberg (*Pens. & Rest. Seeburg*, with garden, at the pier; *Bär, Chalet zur Post*, in the village, pens. 4-5 fr.), with a picturesque church built among the ruins of the castle (view), and to the *Schadburg* (2888'; 1 1/2 M. farther on), on a spur of the Graggen, an unfinished castle of the ancient barons of Ringgenberg.

A guide-post at the upper bridge over the Aare in Unterseen (p. 161) indicates the road to Beatenberg, which skirts the S.W slope of the *Harder* and enters the (1/4 hr.) *Habkern-Thal*, watered by the *Lombach*. Here it diverges (3/4 M. from Unterseen) from the road going on to the (3 1/2 M.) village of *Halkern* (3500'). One-horse carr. from Interlaken to Habkern and back 15 fr.

Immediately before the above-mentioned parting of the roads, a bridle-path diverges to the right and ascends through woods to the (1 1/2 hr.) *Hardermatte* (3990'), which commands a fine view of the Bernese Alps. We proceed a little farther on the same level before descending to (1/4 hr.) the pavilion on the *Falkenfluh* and return to the Brienz road (p. 157) viâ the Obere Bleiki and to Interlaken (3-3 1/2 hrs. in all).

Three fine points of view may be visited from Habkern. The *Gemmenalphorn (6770') is reached by crossing the *Brändlisegg*, or by following the *Bühlbach*, in 4 hrs. (comp. p. 158). The Hohgant (7215') is ascended in 4 hrs. viâ *Bohl* (5902') and the *Hagletsch-Alp*, or by the *Alp Bösälgäu* and through the *Karrholen*. To the S.W. of the Hohgant is the *Grünenberg* (5095'), over which a pass leads from *Habkern* to *Schangnau* in the Emmen-Thal (6 hrs.). The Augstmatthorn (*Suggithurm*, 6844'; 3 1/2 hrs.) is ascended viâ the *Bodmi-Alp*.

A pleasant morning-walk may be taken viâ *Gsteig* (see below; 1 1/2 M. from the Höheweg), where the cemetery for the surrounding districts lies, down the right bank of the Lütschine either to (1 1/4 M.) Bönigen (p. 184), or to the bridge over the Lütschine halfway and back to Matten. Another walk leads from Gsteig up the right bank of the Lütschine to (1/4 hr.) *Gsteigwyler*.

From Bönigen to the *Giessbach* viâ Iseltwald, see p. 184.

The *SCHYNIGE PLATTE, one of the finest points of view in the Bernese Oberland, is reached by a RACK- AND -PINION RAILWAY (opened in 1893) from station *Wilderswyl-Gsteig* (1870'; p. 154) in 1 1/4 hr. (fare 8 fr., down 4 fr., return-ticket 10 fr.), or from *Interlaken-Ost* (change carriages at Wilderswyl-Gsteig) in 1 1/2 hr. (fares 9, 5, 11 fr. 60 c.; 3rd cl. 8 fr. 60 c., 4 fr. 60 c., 11 fr.). — The line (maximum gradient 1 : 4) crosses the Lütschine and ascends in curves to the *Rothenegg Tunnel*, beyond which it enters a wood of beeches and pines, affording pretty glimpses to the left of Interlaken, the Lake of Brienz, etc. 3 M. Stat. *Breitlauenen* (5068'; Curhaus, pens. from 6 fr.), with fine view of the lakes of Brienz and Thun and the hilly country towards the N.W. Describing a wide curve, the line then ascends to the mountain-crest and passes through the *Grätli Tunnel* to the S. side of the ridge, where we obtain a view of the Lauterbrunnen Valley, and then of the Lütschine Valley; to the left towers the majestic Jungfrau. Following the S. slope of the crest, overlooking the Grindelwald Valley with the Schreckhörner and Wetterhörner, and threading a short tunnel, we reach the (4½ M.) **Schynige Platte** (6463'), the terminus. A broad path leads from the station along the Platte, a slope of crumbling and 'shining' slate rock, in a few minutes to the *Hôtel-Restaurant* (R., L., & A. from 4½, lunch 4, D. 5, 'plat du jour' 1½ fr. Engl. Ch. Serv. in July).

To the S. we enjoy a magnificent *VIEW of the Bernese Alps; from left to right, the Wellhorn, Wetterhörner, Berglistock, Upper Grindelwald Glacier, Schreckhörner, Lauteraarhörner, Lower Grindelwald Glacier, the Finsteraarhorn peeping over the Eigergrat, the Fiescherhörner, Eiger, Mönch, Jungfrau, Ebnefluh, Mittaghorn, Grosshorn, Breithorn, Tschingelhorn, Tschingelgrat, Gspaltenhorn, Weisse Frau, Doldenhorn, and numerous nearer peaks. The ridge concealing the base of the Jungfrau group is the Männlichen (p. 174).

An easy winding path ascends from the hotel past the *Geisshorn* (view like that from the hotel) and the precipitous *Gummihorn* (6893'; recently made accessible for experts) to the (20 min.) *Daube* (6772'), whence the survey of the lakes and of the peaks towards the N. is particularly fine; to the N.E. is the Brienzer Rothhorn, with Pilatus to the right in the distance. Towards evening the lakes of Neuchâtel and Bienne are seen glittering in the distance.

From the Schynige Platte to the *Faulhorn*, see p. 180. — Descent from the Platte by *Gündlischwand* to *Zweilütschinen*, 2½-3 hrs., steep at places. At the small pond near the Platte to the right we descend across meadows to the (³/₄ hr.) lower chalets of the *Iselten-Alp* (5116'; guide advisable to this point, 2 fr.); thence through wood (unmistakable).

FOOTPATH FROM GSTEIG TO THE SCHYNIGE PLATTE (3½ hrs.). We may either cross the bridge by the church of Gsteig and follow the road to the right (³/₄ M.) *Gsteigwyler*; in the middle of the village take the bridle-path to the left, and very soon to the left again; after 17 min. ascend to the right, through wood; or, shorter, we may ascend from Gsteig to the left, by a path between the church and the inn (Steinbock), turning to the right where the path divides, and in 20 min. reach the bridle-path at the point where it enters the wood. We now ascend by numerous zigzags, crossing the railway twice, to the (1½ hr.) *Schönegg* (4754') and the (¼ hr.) *Curhaus Breitlauenen* (see above). Thence to the top, 1½ hr.

47. The Lauterbrunnen Valley and Mürren.

From Interlaken to *Lauterbrunnen*, 8 M., BERNESE OBERLAND RAILWAY in 42 min. (fares 3 fr. 25, 1 fr. 95 c., return 5 fr. 20, 3 fr. 15 c.); circular tour from Interlaken to Lauterbrunnen, the Kleine Scheidegg, Grindelwald, and back to Interlaken, 20 fr., 13 fr. 95 c. (tickets valid for 6 days). The railway is on the ordinary system (maximum gradient 35:1000), with short sections on the rack-and-pinion system (maximum gradient 120:1000). — CARRIAGE from Interlaken to Lauterbrunnen (in fine weather preferable to the railway) and back, including 2 hrs. stay, with one horse 9, two horses

15 fr.; to Trümmelbach 12 and 22, to Stechelberg 14 and 27 fr. — The following *EXCURSION (one day) is highly recommended: by railway to Mürren (p. 168), walk to the Obere Steinberg (p. 167; 2³/₄-3 hrs.), descend to (1 hr.) *Trachsellauenen* (p. 166), and return by the valley, past the falls of the *Trümmelbach* and *Staubbach* (p. 166) to *Lauterbrunnen* (2³/₄ hrs. to the railway-station). The views from Mürren and the Obere Steinberg are among the finest in Switzerland.

The line begins at the *Interlaken-Ost* station (1865'; p. 158) and describes a wide curve through the fertile plain to (1¹/₂ M.) *Wilderswyl-Gsteig* (1925'; change carriages for the Schynige Platte, p. 164). To the right is the village of *Wilderswyl*, at the foot of the *Abendberg*; to the left is the church of *Gsteig* (see p. 163). — The railway crosses the *Lütschine* and ascends its right bank through wood. On the left bank is the high-road. To the right rises **the precipitous** *Rothenfluh*, overtopped by the Sulegg; in the foreground, to the left, is the Männlichen, with the Mönch and the Jungfrau adjacent. The railway next crosses the *Black Lütschine*, which descends from Grindelwald. To the left, in the background of the Lütschen-Thal, rises the finely-shaped **Wetterhorn**.

5 M. **Zweilütschinen** (2150'; *Hôt.-Pens. Zweilütschenen*, formerly *Bär*), junction for the railway to *Grindelwald* (p. 171; passengers not in through-carriages change for Lauterbrunnen).

Interesting excursion to (1¹/₄ hr.) Isenfluh (3600'; *Pens. Isenfluh*, 5 fr.). About ¹/₂ M. from Zweilütschinen the bridle-path diverges to the right from the Lauterbrunnen road and ascends the steep W. slope of the valley (shade after 3 p.m.; a second path ascends by the *Sausbach* opposite the *Hunnenfluh*, see above; a third ascends from Lauterbrunnen, opposite the Hôtel Steinbock). Isenfluh commands a splendid *View of the Jungfrau. A still finer view is obtained from the PATH FROM ISENFLUH TO MÜRREN (3¹/₄ hrs.; guide desirable for novices; from Zweilütschinen to Mürren 7 fr.). At the upper end of the village (¹/₄ hr.) this path turns to the left and ascends to the (³/₄ hr.) *Sausbach* (5050'), and then more steeply for 25 min. to the *Flöschwaldweid* (b608'). Here we turn to the left and proceed to the chalets of *Alpligen* (5792'), where we descend. The path, which commands a fine view of the Jungfrau and its neighbours, next traverses the *Pletschen-Alps*, crosses the *Pletschbach* and the *Spissbach* (1¹/₄ hr.) the station of *Crüttsch-Alp* and (35 min.) *Mürren* (p. 168). — Ascent of the *Sulegg* (7815'), 3¹/₂ hrs., see above.

The train crosses to the left bank of the *White Lütschine*, and ascends (two rack-and-pinion sections) the wooded *Valley of Lauterbrunnen*, which begins at the **Hunnenfluh**, a rock resembling a gigantic round tower, and is bounded by precipitous limestone cliffs, 1000-1500' in height. The railway crosses first the *Sausbach*, which descends on the right, and then the road several times.

8 M. **Lauterbrunnen**. — The RAILWAY STATION lies 2620' above the sea-level; 3 min. higher up, to the right, is the station for the cable-railway to Mürren (p. 167); change carriages for Wengen, Scheidegg, and Grindelwald (p. 172). — Hotels: *STEINBOCK, at the station, with the railway restaurant, R., L., & A. 3-5, B. 1¹/₂, lunch 3, D. 4, pens. 7-9 fr.; *Hôt. STAUBBACH, with view of the Staubbach, R., L., & A. 3-4, lunch 3, D. 4 fr.; *ADLER, near the station, plain, R., L., & A. 2-3, B. 1¹/₄, D. 3, pens. 5-6 fr. — *Lauener's Beer and Wine Saloon*, to the S. of the Adler. — Guides: *Christ., Joh., Ulrich,* and *Peter Lauener, Heinr.* and *Fritz v. Almen, Fritz Graf, Friedr. Fuchs, Ulrich Brunner, Fritz Schlunegger, Karl Schlunegger* (at Wengen), etc.

Lauterbrunnen (2640'), a pretty, scattered village, lies on both banks of the Lütschine, in a rocky valley ½ M. broad, into which in July the sun's rays do not penetrate before 7 a.m., and in winter not till 11 a.m. It derives its name (*lauter Brunnen*, 'nothing but springs') from the numerous streams which descend from the rocks, or from the springs which rise at their bases in summer. The snow-mountain to the left, rising above the lower mountains, is the Jungfrau; to the right is the Breithorn.

About 8 min. from the station, at the Hôtel Staubbach, the village-street forks. The left branch descends past the church to the Trümmelbach (see below); the right branch leads straight on to the (5 min.) *Staubbach ('dust-brook'), the best-known of the falls at Lauterbrunnen. This brook, which is never of great volume, and in dry summers is disappointing, descends from a projecting rock in a single fall of 980', most of it, before it reaches the ground, being converted into spray, which bedews the meadows and trees far and near. In the morning sunshine it resembles a transparent, silvery veil, wafted to and fro by the breeze, and by moonlight also it is very beautiful. The best point of view is in a meadow in front of the fall, to the left of a seat marked by a flag (20 c.).

The road to the left at the fork (see above) crosses the White Lütschine near the church, and ascends its right bank, with a view of the snowy Breithorn and the Schmadribach Fall (to the left a bridle-path diverges to Wengen, p. 172). In ½ hr. we reach the *Hôtel-Pension Trümmelbach (R., L., & A. 3-4, déj. 3, D. 4; omn. at Lauterbrunnen station; carr. there and back, including stay, 4 fr.). A path (adm. 50 c.) here diverges to the left to the (7 min.) *Trümmelbach Fall. The narrow gorge, with the copious *Trümmelbach*, fed by the glaciers of the Jungfrau, is rendered accessible by steps and railings. During sunshine three rainbows are formed in the spray.

Through the *Trümleten-Thal to the *Wengern-Alp* (p. 172), 3 hrs., with guide, somewhat trying but highly interesting. — To the *Roththal Hut*, see p. 171.

The road continues to ascend the valley, in view of several waterfalls, and passes the (18 min.) *Dörnigen-Brücke*, where we join the old route passing near the Staubbach. Beyond *Stechelberg* we reach (40 min. from the Trümmelbach) the *Café-Restaurant & Pension Stechelberg* (3020'; pens. 5 fr.), where the road degenerates into a bridle-path. The main path (to the left; that to the right leads to the Sefinen Valley and Mürren, p. 169) skirts the right bank of the brawling Lütschine, and near the (¼ hr.) chalets of *Sichellauenen* crosses the stream. Thence we traverse finely wooded meadows to (50 min.) Trachsellauënen (4145'; *Hôt. Schmadribach*, R., L., & A. 2-2½, B. 1½, pens. 5 fr.), a picturesque cluster of chalets on the left bank of the Lütschine, 1¾-2 hrs. from the Trümmelbach.

The path hence to the (1 hr.) Schmadribach Fall ascends on the left bank of the Lütschine to the (12 min.) 'Bergwerk', a ruined house with a chimney. Here it diverges to the left from the main path (which goes on to the Upper Steinberg, p. 167), and ascends (guide-boards) round projecting rocks (from the top, a view of the waterfall) and past the chalets of the (1/2 hr.) *Lower Steinberg Alp* (4480'), where it crosses (to the left) the *Thalbach* (two bridges). Ascending the pastures on the right bank, we pass a waterfall, mount the *Holdri*, and reach (1/2 hr.) the *Läger Chalet*, in sight of the *Schmadribach Fall. There is nothing to be gained by approaching closer to the fall. — From the 'Bergwerk' it is preferable to follow the main path, to the right, which zigzags up a gorge, clad with firs and ferns, to the chalets of the *Ammerten-Alp*, and thence to the Upper Steinberg. Here (1½ hr. from Trachsellauenen) are the small *Hôtel Tschingelhorn* (well spoken of; R. 3 fr., B. 1 fr. 60 c., pens. 4½-5 fr.), and (20 min. farther up) the *Hôtel Ober-Steinberg* (unpretending; pens. from 5 fr.). The *View of the mountains and glaciers surrounding the upper valley of Lauterbrunnen is very fine; from right to left are seen the Lauterbrunner Wetterhorn, with the Tschingelhorn behind it, the Breithorn, the beautiful Tschingel Glacier between these, then the Grosshorn, the Mittaghorn, the Ebnefluh, the Gletscherhorn, and the Jungfrau, while directly opposite is the Schmadribach Fall.

A pleasant walk (boy as guide 1½-2 fr.) may be taken from the Upper Steinberg along the *Tschingel Glacier*, at the end of which is an interesting ice-grotto, and viâ the *Oberhorn-Alp* to the (1½ hr.) *Oberhornsee (6822'), a beautiful little blue lake, magnificently situated in the rocky hollow between the Tschingel and Breithorn glaciers.

From Lauterbrunnen to Mürren. — *Cable and Electric Railway* in 55 min. (fares 3 fr. 75 c., return-ticket, valid for 3 days, 6 fr.). The station of the cable-railway in Lauterbrunnen lies 3 min. from that of the Bernese Oberland Railway (see p. 165). On the arrival of the trains from Interlaken the number of passengers is often so great that the traveller has to wait until the despatch of more than one train before finding a seat. This crowding for the best places is repeated in changing from the cable to the electric trains, so that some will doubtless prefer to book only to *Grütsch-Alp* (2 fr. 75 c.) and to walk thence to (1 hr.) Mürren by the picturesque footpath skirting the railway.

The Cable Railway mounts straight uphill (maximum gradient 60 : 100), through meadows and wood, to the *Grütsch - Alp* (4975'). Here we change carriages (comp. above) for the Electric Railway, which continues to follow the slope, crossing several streams, to (2½ M.) *Mürren*. To the left (even below the Grütsch-Alp) a magnificent **View of an amphitheatre of mountains and glaciers unfolds: the Eiger and the Mönch, the Jungfrau with its dazzling Schneehorn and Silberhorn, the huge precipices of the Schwarze Mönch rising abruptly from the valley, the wall of the Ebne-Fluh with its conical peak to the left and its mantle of spotless snow; then as we approach Mürren (near which the Jungfrau disappears behind the Schwarze Mönch), the Mittaghorn, the Grosshorn, the Breithorn (from which the Schmadribach descends), the Tschingelhorn, the Tschingelgrat, and the Gspaltenhorn.

The Bridle-Path from Lauterbrunnen to Mürren, 2½ hrs., which is very muddy after rain, ascends rapidly to the right about 3 min. from the station, beyond the Adler Hotel, at the guide-post ('Mürren 5.7 Kil.', *i.e.* 3½ M.), and crosses the *Greifenbach* twice. Beyond the second bridge (20 min.) it ascends through wood, crosses the *Fluhbächli*, the (20 min.) *Laubach* (fine waterfall), and the *Herrenbächli*, and reaches (25 min.) the bridge over the scanty *Pletschbach* or *Staubbach* (4037'; rfmts.). In 5 min. more, where the wood has been much thinned, we obtain a beautiful view of the Jungfrau, Mönch, and Eiger, which remain in sight for the rest of the way. Farther up, by (½ hr.) a saw-mill (4925'), we cross two branches of the *Spissbach*, in 25 min. more reach the top of the hill (°View see above) and then walk alongside the railway to (½ hr.) *Mürren*.

Mürren. — Hotels. *GRAND HÔTEL & CURHAUS MÜRREN,* 5 min. from the station, with restaurant, **Cursaal,** and several dépendances (*Bellevue, Fontana, Victoria*), R., L., & A. 5-8, B. 1½, déj. 3, D. 5, pens. from July 15th to Sept. 10th 10-16, at other times 9-13 fr.; *GRAND HÔTEL DES ALPES,* nearer the station, with restaurant (Munich beer on draught 60 c.), R., L., & A. 4-6, B. 1½, déj. 3½, D. 5, pens. 9-15 fr; 8. B. G. H. at both. — *HÔT. JUNGFRAU,* R. 3½-4, D. 3½, S. 3, pens. 8-10 fr., near the English Church, above the Curhaus; *HÔT. EIGER,* close to the station, R., L., & A. 2½-3, B. 1½, déj. 2½, D. 3½, pens. 6-8 fr. — *English Church.*

Mürren (5350'), situated on a terrace high above the Lauterbrunnen Valley, is one of the most frequented points in the Bernese Oberland. It commands a famous view, including **not only the above-mentioned peaks, but also the Wetterhorn to the left, and the Sefinen-Furgge to the extreme right** (p. 159). We ascend between the restaurant and the beer-saloon of the Hôtel des Alpes to the walks that lead along the slope of the *Allmendhubel,* a height to the W., on which firs grow higher up.

The top of the *Allmendhubel* (6358') is reached in ¾ hr., by following the above-mentioned path to the left to (¼ hr.) the first chalets of *Allmend,* then the path to the Schilthorn to (20 min.) a solitary chalet, and finally to the right for 6 or 8 min. more. The view includes the snowy Jungfrau in addition to the peaks seen from Mürren. — Another good view-point is the *Obere Winteregg* (5738'), ½ hr. to the N.W. of Mürren, by a path diverging to the left above the electric railway, 10 min. from Mürren.

The *Schilthorn (9747'; 3½-4 hrs., guide 8 fr. and fee) is an admirable point of view. The path ascends past the chalets of *Allmend* (on the right is the Allmendhubel, see above), and farther up enters the dreary *Engethal,* which ends in a rocky basin at the foot of the Schilthorn (to this point, 2½ hrs. from Mürren, riding is practicable; horse 12 fr.). Then a steep ascent over snow, loose stones, and rock, past the monument to Mrs. Arbuthnot, who was killed here by lightning in 1865, to the arête between the *Kleine* and *Grosse Schilthorn*, and without difficulty to the (1 hr.) flattened summit. Magnificent survey of the Jungfrau, the queen of the Bernese Alps, and of the whole chain (including the Blumlisalp, to the S.W., quite near), and of N. Switzerland (the Rigi, Pilatus, etc.); panorama by Impfeld. Mont Blanc is not visible hence, but is seen from the arête, about 250 yds. to the W., a little below the summit. — The descent through the imposing *Sefinen-Thal* (p. 169), by the *Sefinen-Alp* and the *Teufelsbrücke* (a fine point above Gimmelwald), is longer by 1½ hr. than the direct path, but far more interesting (unsuitable for ladies). A shorter way back leads past the *Graue Seeli* and down the steep *Schiltflühe* (guide advisable), and afterwards through the beautiful pastures of the *Schiltalp,* with **views of** the Jungfrau, etc. — Another route (interesting; guide advisable) crosses the *Rothe Herd* and the *Telti* (a saddle between the Grosse Hundshorn and the Wild-Andrist) to the *Dürrenberg Chalets* in the *Kienthal* (see p. 170).

At the Chalet Bellevue, beyond the Curhaus, a guide-post indicates the way (to the left) to Stechelberg, and 100 paces farther on another points the way to Gimmelwald and Stechelberg. In 5 min. more we cross a bridge over a fall of the *Mürrenbach*, and at (20 min.) the beginning of **Gimmelwald** (4545') the road forks. The branch to the right leads straight on to the (8 min.) *Hôt.-Pens. Schilthorn* (5-6 fr.; Engl. Church Service in summer), on the brink of the grand *Sefinen-Thal*, which is enclosed by the Büttlassen, the Gspaltenhorn, and the Tschingelgrat. The branch to the left descends in 4 min. to the *Hôt.-Pens. Gimmelwald* (4$^1/_2$-5 fr.).

To the **Sefinen-Thal**, an interesting walk (as far as the Gspaltenhorn Glacier and back 3 hrs.; guide unnecessary). To the W. of the Pension Schilthorn we cross the (5 min.) *Schiltbach*, and ascend on the left side of the Sefinen-Thal (with the superb Jungfrau behind us); then (³/₄ hr.) cross a bridge and enter a pine-wood, and lastly, in a grand basin, with numerous waterfalls, traverse stony débris to the (³/₄ hr.) *Gspaltenhorn* (or *Kirchspalt*) *Glacier*, at the foot of the Gspaltenhorn.

The road to Stechelberg descends to the left past the Hôtel Gimmelwald and (¹/₄ hr.) crosses the *Sefinen-Lütschine*. After a short ascent we again descend through wood, and cross a brook descending from the right, enjoying a view, to the left, of the beautiful *Fall of the Sefine*. About 12 min. farther on the path divides: the branch to the left descends steeply to (¹/₄ hr.) *Stechelberg* (p. 166); that to the right goes on at the same level to *Trachsellauenen* ('Hôt. Schmadribach 40 min.'; p. 166). A footpath diverges to the right from the latter after 6 min., passes a deserted shaft, and, after affording

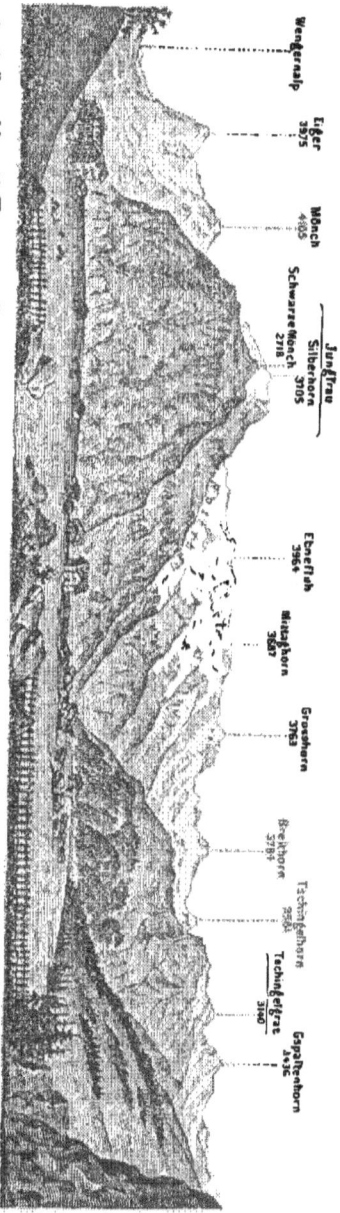

170   *III. Route 47.*   TSCHINGEL PASS.   *Bernese*

a view of the Schmadribach, reaches (1½ hr.; not 1 hr.) the *Hôtel Tschingelhorn* on the Upper Steinberg (p. 167; in all about 3 hrs. from Mürren; guide, 7 fr., not necessary in good weather).

Passes. FROM LAUTERBRUNNEN OVER THE SEFINEN-FURGGE TO THE KIEN-THAL, not difficult, and on the whole attractive (10-11 hrs. to Reichenbach; guide 25 fr.). From (2½ hrs.) *Mürren* (p. 156) the path ascends viâ the *Alp Bogangen* to the (3 hrs.) Sefinen-Furgge (8583'), between the *Hundsfluh* (9620') and the *Büttlassen* (10,490'; p. 191. (The path by Gimmelwald and through the Sefinen-Thal is easier, but 1 hr. longer.) Descent (fine view of the Wilde Frau and Blümlisalp) past the chalets of *Dürrenberg* (6545'), and of *Steinenberg* (4856'; night-quarters) to the huts of *Gorneren*, by the *Bärenpfad* to the (2 hrs.) *Tschingel-Alp* (3783') and down the *Kienthal* to (2½ hrs.) *Reichenbach* (p. 191). — From the Steinenberg-Alp over the *Gamchilücke* to the *Tschingelfirn*, see below and p. 191.

FROM LAUTERBRUNNEN TO KANDERSTEG OVER THE SEFINEN-FURGGE AND THE HOHTHÜRLI, a long and fatiguing walk (14 hrs.; guide necessary, 30 fr.). The night may, if necessary, be passed at the Dürrenberg chalets or in the Dünden Hut. Over the *Sefinen-Furgge* to the *Kienthal*, see above. Before the path reaches the *Steinenberg-Alp* we descend to the left, cross the *Pochtenbach* (the discharge of the *Gamchi Glacier*, p. 180), ascend to the *Lower* and *Upper Bund-Alp*, and traverse pastures, stony slopes, and snow to (4½ hrs. from the Furgge) the Hohthürli or Dünden Pass (8875'), a depression of the *Oeschinengrat* between the *Schwarzhorn* (9150') and the *Wilde Frau* (10,693'), affording a superb view of the Blümlisalp, Doldenhorn, etc. (To the left of the pass is the *Dünden Club-Hut*, p. 193.) We now descend over loose stones and the rocky ledges of the *Schafberg* (with the *Blümlisalp Glacier* quite near us on the left) to the *Upper Oeschinen-Alp* (6470'), and by steep steps cut in the rock, to the *Lower Oeschinen-Alp*, pass round the N.W. side of the *Oeschinen-See* (5223'), and reach (4 hrs.) *Kandersteg* (p. 192).

*FROM LAUTERBRUNNEN TO KANDERSTEG OVER THE TSCHINGEL PASS (14 hrs.; 6-7 hrs. on snow and ice; guide 30, porter 25 fr.), a grand route, fatiguing, but for tolerable mountaineers free from difficulty. A night had better be spent at the *Upper Steinberg* (see p. 167). We now follow the W. slope of the valley to the (¾ hr.) *Lower Tschingel Glacier*, cross it, and toil up the left lateral moraine to the (½ hr.) base of the W. rocks, the ascent of which is very steep at first (a nearly perpendicular part, called the *Tschingeltritt*, about 13' high, is now avoided by means of a narrow path). Farther up (40 min.) we come to turf (pleasanter; a halt usually made here; superb view). Then again across débris in ½ hr. to the upper *Tschingelfirn*, an immense expanse of snow; for 20 min. we follow the left moraine, and then take to the glacier, where the rope becomes necessary. A gradual ascent of 1¾ hr. brings us to the top of the Tschingel Pass (9267'), where a view of the mountains of the Gastern-Thal is disclosed; behind us towers the most majestic Jungfrau with her S. neighbours, and to the left is the Eiger. On the right are the furrowed *Gspaltenhorn* (p. 191) and the *Gamchilücke* (9235'; pass to the Kienthal, p. 191). An additional hour may be devoted to visiting the Gamchilücke, which affords a striking survey of the Kienthal, the Niesen, and the Bernese plain. To the left of the Tschingel Pass rises the *Mutthorn* (9978'). The descent across the snow is easy. (The W. arm of the glacier, bounded on the right by the rocky walls of the Blümlisalp and the Fründenhorn, and on the left by the Petersgrat, is called the *Kanderfirn*.) After 1½ hr. we quit the snow for the left lateral moraine. The route descends steeply, over loose stones and then over grass, to the *Gastern-Thal*, passing a spur which overlooks the magnificent ice-fall of the Kander Glacier. We then for a considerable time follow the narrow crest of a huge old moraine, which descends precipitously on the right to the former bed of the glacier, 170-200' below; 1½ hr., bridge over the *Kander*; 6 min., the first chalet (coffee, milk, and two beds); ¼ hr., *Selden*; 2 hrs., *Kandersteg* (p. 192).

*FROM LAUTERBRUNNEN TO THE LÖTSCHEN-THAL OVER THE PETERSGRAT (from the Steinberg to Ried 9-10 hrs.), trying and recommended only to experienced mountaineers, but **very** grand (guide 50 fr.; for one tourist 2 guides or a guide and a porter are required). From the Upper Steinberg to the (2½ hrs.) upper *Tschingelfirn*, see above. On the glacier we ascend to the left, between the *Mutthorn* and the *Tschingelhorn*, to the (3 hrs.) **Petersgrat** (10,515'), a lofty snow-arête commanding a superb view of the Alps of the Valais. Then a steep descent over snow, rocky slopes, and turf, either through the *Ausser Faster-Thal* to the *Faster Alp* (good quarters at the Chalet Seiler), or through the *Tellithal* to *Blatten* and (8½ hrs.) *Ried* (p. 193). — The **Wetterlücke** (10,365'), between the Tschingelhorn and Breithorn; the **Schmadri-Joch** (10,863'), between the Breithorn and Grosshorn; and the **Mittagjoch** (12,150'), between the Grosshorn and Mittaghorn, are difficult (guides 45-50 fr.).

FROM LAUTERBRUNNEN TO THE EGGISHORN over the *Lauinenthor* (12,000'), a difficult and hazardous expedition (18 hrs.; the night being spent in the Rothhal Hut; guide 100 fr.), through the wild *Roththal*, across the huge rock-arête connecting the *Jungfrau* (13,670') and *Gletscherhorn* (13,064'), and down the *Kransberg-Firn* and the *Great Aletsch Glacier* to the *Concordia Hut* and the *Eggishorn Hotel* (p. 316). — Over the **Roththal-Sattel** (12,330'), close to the Jungfrau (p. 173), also very difficult and dangerous (19-20 hrs. to the Eggishorn). — Over the **Ebnefluh-Joch** (12,300'), between the *Ebnefluh* and *Mittaghorn*, very laborious, but without danger to experts (15-16 hrs.; guide 80 fr.). — It will repay a good walker to go as far as the *Roththal Hut* (8860'; 6 hrs. from Lauterbrunnen, crossing the *Stufenstein-Alp*), and to return the same way (a good day's **walk**; for experts only; guide 15 fr.). Ascent of the *Jungfrau* from this hut, see p. 173.

48. From Interlaken to Grindelwald.

BERNESE OBERLAND RAILWAY: *a.* Direct (12 M.) in 1 hr. 12 min. (fares 5, 3 fr., return 8 fr.; 4 fr. 80 c.). *b.* Viâ Lauterbrunnen and Wengern-Alp (18½ M.) in 4½ hrs.; from Lauterbrunnen, 11 M. in 2¼ hrs. (fares 14 fr. 40 c., 9 fr.; circular tickets for both lines, valid for three days, 20 fr., 13 fr. 95 c.). — A CARRIAGE from Interlaken to Grindelwald and back in one day (one-horse 13, two-horse 25 fr.) is pleasanter and not much dearer for a party. — PEDESTRIANS still often prefer the beautiful WALK over the Wengern-Alp to Grindelwald: bridle-path to the Wengern-Alp 3 (descent 2), Little Scheidegg ¾ (descent ½), Grindelwald 2½ hrs. (ascent 3½ hrs.); in all 6¼ hrs. from Lauterbrunnen. Small trunks may be sent by train unaccompanied by passengers, but not open handbags.

a. DIRECT LINE. From **Interlaken** to (5 M.) *Zweilütschinen* (2150'), see p. 165; carriages are usually changed here. The railway to Grindelwald ascends the left bank of the *Black Lütschine*, traversing a tunnel and a snow-shed, in the finely wooded and populous Lütschen-Thal. The road runs on the other bank, beneath the slopes of the Schynige Platte (p. 164). Beyond (7½ M.) *Lütschenthal* (2355') the railway also crosses to the right bank and ascends the *Stalden* by means of a rack-and-pinion section (1935 yds. long; gradient 12 : 100) to (9 M.) *Burglauenen* (2915'). In front appear the **Wetterhorn** and the Berglistock. Farther on we pass through the defile of the *Ortweid*, after which a view of the beautiful valley of Grindelwald is suddenly disclosed: to the right is the massive **Eiger**, adjoined by the Jungfrau with the Schneehorn and the Silberhorn; in the middle are the Mettenberg and the Schreckhörner; farther off the Finsteraarhorn and the Grosse Fiescherhorn; and to the left the graceful Wetterhorn. The railway finally ascends another toothed rail section (1420 yds.) to (12 M.) *Grindelwald* (p. 175).

b. By the Wengern-Alp Line (rack- and-pinion railway on Riggenbach's system). There is only one car on this line, but when passengers are numerous extra trains are despatched (duration of journey and fares, see p. 171). — *Lauterbrunnen* (2640'), see p. 165. The railway describes a curve, crosses the Lütschine, and rapidly ascends the steep slopes below the village of Wengen, where it passes over several viaducts and bridges. Hence we enjoy a fine retrospect of **Lauterbrunnen** and its valley **and** of the Schmadribach Fall in the background, with the Breithorn and Grosshorn above it. Higher up, to the right of the former, is the Tschingelhorn, and to the left of the precipitous Schwarze Mönch are the Silberhorn and Jungfrau. On the opposite side of the valley ascends the cable-railway to Mürren, above which (right) rises the Sulegg-Grat, with the serrated rocks of the Lobhörner, resembling the fingers of a huge hand. Beyond a wide curve we reach —

1½ M. **Wengen.** — Hotels. *Hôt.-Pens. Blümlisalp; *Hôt.-Pens. Victoria (R. 2½-3, pens. 6-9 fr.); Hôt.-Pens. Silberhorn (R. 2, B. 1, D. 2½, pens. 5-6 fr.); these three near the station; *Hôt.-Pens. Alpenrose, 7 min. from the station (pass under the line near the Hôt. Blümlisalp), R., **L.,** & A. 2-2½, B. 1¼, lunch 2½, D. 3, pens. 5½-6 fr.; Hôt.-Pens. Mittaghorn, farther on and lower down, similar charges, well spoken of; *Pens. Wengen, ¾ M. beyond the **Alpenrose,** 6-7 fr. — *English Church Service* in summer.

Wengen (4190'), situated amidst meadows interspersed with trees, below the precipitous *Tschuggen* (p. 174), with a view of the Lauterbrunnen Valley and of the Jungfrau to the S., is much visited as a summer-resort. Attractive walks may be taken hence to the *Leiterhorn,* 1 hr. from the station, beneath the Männlichen (p. 174); to the *Mettlen-Alp* and *Wengern-Alp* (see below), etc.

Bridle-Path from Lauterbrunnen to the Wengern-Alp (3 hrs.). From the station, we descend to the left, cross the Lütschine, and ascend straight on, soon joining the path mentioned at p. 166. ¾ hr. *Restaurant Linder*, with a pavilion which affords a beautiful view. Farther up a (20 min.) finger-post shows the way to the left viâ the *Hôt. Mittaghorn* and the *Hôt. Alpenrose* to the (20 min.) *Wengen* station; and to the right to the (10 min.) *Pens. Wengen*, and thence uphill (after 10 min. to the left) to a point below the watering-station (see below). — The first part of this steep ascent may be avoided by taking the railway to *Wengen*. From the station we cross the terrace in front of the Hôt. Blümlisalp, turn to the left and a little farther on to the right, crossing the line and following the fenced path amidst houses and fields; ½ hr. a chalet (rfmts.); 10 min. we join the above-mentioned path from the Pens. Wengen; 8 min. pass through a gate into the pine-wood, from which we emerge 20 min. farther on, and turn to the left. In ¾ hr. more, passing beneath the line, **we** reach the station of *Wengern-Alp* (p. 173). — If we go straight on after emerging from the wood, we reach the (¾ hr.) ***Mettlen-Alp** (5680'), **on the** N. side of the *Trümleten-Thal*, directly facing the Jungfrau. Hence we **may** either ascend to the Wengern-Alp in ¾ hr., or walk round the head **of the** Trümleten-Thal to the (1 hr.) *Biglen-Alp*, with the *Kühlauenen Glacier*, and thence to the (¾ hr.) Wengern-Alp.

Beyond **Wengen** the railway curves towards the Tschuggen, affording a continuous view of the snow-mountains and glaciers from the Grosshorn to beyond the Gspaltenhorn, with the Breithorn in the centre. After a short halt at a *Watering Station* below the Lauberhorn (p. 174), we skirt the *Galtbachhorn* (7610') and reach —

Oberland. WENGERN-ALP. *III. Route 48.* 173

4½ M. **Wengern-Alp** (6158'; *Hôt. Jungfrau*, R., L., & A. 4-5, B. 1¾, déj. 3½, D. 4-5, pens. 8-10 fr.). Hence we command a celebrated view across the *Trümleten-Thal* of the ***Jungfrau** (13,670'), with her dazzling shroud of eternal snow, flanked by the *Silberhorn* (12,155') on the right and the *Schneehorn* (11,205') on the left. The proportions of the mountain are so gigantic, that the eye in vain attempts to estimate them, and distance seems annihilated by their vastness. To the left of the Jungfrau, the highest summit of which is not visible, rise the *Mönch* (13,468') and the *Eiger* (13,040'); while to the right are the summits of the upper valley of Lauterbrunnen as far as the Gspaltenhorn. — The view from the (20 min.) top of the *Hundsschopf* (bench and signal) is little superior to that from the Hôtel Jungfrau. A fine view of the Lauterbrunnen valley is obtained from the *Gürmschbühl* (6223'), reached by diverging to the left from the way to Wengen, ¼ hr. below the station, and turning 8 min. farther on to the right (the path to the left here leads to the Mettlen-Alp, p. 172).

On the Wengern-Alp, at Grindelwald, and elsewhere the traveller may have an opportunity of witnessing *Snow Avalanches* or *Ice Avalanches*, which, on warm, sunny days, generally occur several times an hour. Except that the solemn stillness which reigns in these desolate regions is interrupted by the echoing thunders of the falling masses, the spectacle can hardly be called imposing. The avalanche, as it descends from rock to rock on the mountain-side to disappear finally at its foot, resembles a huge white cascade. The more destructive avalanches, bearing with them rocks, earth, and gravel, occur only in spring and winter.

ASCENT OF THE JUNGFRAU. Between 1811, when the Jungfrau was scaled for the first time by the two *Meyers* of Aarau, and 1856 the ascent was only accomplished five times; but it has since been undertaken frequently, and is now made several times almost every year. Though extremely fatiguing, it is unattended with danger to experts (guides 80 fr. each; with descent to the Eggishorn 100 fr.; porter 60 and 80 fr.). The easiest ascent is that by the S. side, the night being spent in the *Concordia-Hütte* (p. 317), 5 hrs. from the Eggishorn Hotel; thence to the summit 6-7 hrs. The ascent from Grindelwald is facilitated by spending a night in the *Bergli-Hütte* (p. 178), 8-9 hrs. from Grindelwald; thence over the *Mönchjoch* and the *Jungfraufirn* to the *Rothtal-Sattel* (p. 171) 4-4½ hrs., and to the top in 1½ hr. more. — The ascents from the Little Scheidegg and from Lauterbrunnen by the *Rothtal-Sattel* are difficult and hazardous. From the *Rothtal Hut* (p. 171), leaving the Rothtal to the right, the ascent requires 7½ hrs. (trying, but safe). — The Silberhorn (12,155') was ascended for the first time, in 1863, by *Ed. v. Fellenberg* and *Karl Baedeker* (from the Wengern-Scheidegg by the *Eiger*, *Guggi*, and Giessen *Glaciers*, in 12½ hrs.; difficult and trying; guide 50 fr.). The ascent by the W. arête was first performed in 1887 by *Mr. Seymour King*.

The projected JUNGFRAU RAILWAY (electric line; maximum gradient 26 : 100), for which a concession has been granted by the Federal authorities, ascends from the Kleine Scheidegg to the station *Eigergletscher* (7480'), on the margin of the glacier, beyond which it is carried through a constant succession of tunnels. Stations *Eiger* (10,567'), on the S. side of the Eiger, near the Bergli Hut (p. 178), and *Mönch* (11,886'), on the S. side of the Mönch, near the Jungfrau-Joch (p. 178). The terminus *Jungfrau* (13,450') will be connected with the summit by a lift 210' high, with a winding staircase on the outside.

From the Wengern-Alp the railway ascends gradually, with continuous fine views to the right. Pedestrians follow the bridle-path,

174 *III. Route 48.* LITTLE SCHEIDEGG. *Bernese*

which crosses the line near the Hôtel Jungfrau and then **skirts** it to the ($3/4$ hr.) station of Scheidegg; this walk is especially recommended for the descent.

$5^3/4$ M. **Scheidegg** (carriages changed in both directions; detention frequent), on the summit of the Little or **Lauterbrunnen Scheidegg** (6788'; *Hôtel Bellevue*, R., L., & A. 4-5, B. $1^3/4$, D. 4 fr.; S. B. G. H.; Engl. Ch. Service in July and August). This ridge affords a striking view of the valley of Grindelwald to the N., as far as the Great Scheidegg, dominated on the right by the broad summit of the Wetterhorn, with its rocky peaks and snow-fields, bounded on the N. by the **Faulhorn** range (to the extreme left is the blunt cone of the Faulhorn with its inn). On the S. opens a splendid **view** of the Mönch, Eiger, and Jungfrau, with the Silberhorn and Schneehorn (but more in profile than from the Wengern-Alp).

A tolerable path leads from the Hôtel Bellevue to (1 hr.) an *Ice Grotto* in the *Eiger Glacier* (fee); but the chief attraction of this expedition is the view, especially from the °*Fallbodenhubel* (7136'; about halfway; viewbench), of the Mönch, from which descend the Eiger Glacier (on the left), and the Guggi Glacier (on the right). — At the lower end of the Guggi Glacier we can descry, with a telescope, the *Guggi Club-Hut* (7972'; $1^3/4$-2 hrs. from the Little Scheidegg; guide 5 fr.), seldom used, as the Mönch is now ascended from the Bergli-Hut, and the Eiger direct from the Hôtel **Bellevue** (see above).

The ascent of the **Lauberhorn** (8120'), in about 1 hr. by a path (guideposts) between the station and the Bellevue Hotel, is especially recommended for the magnificent view it commands and the ease by which it is accomplished. The entire chain of the Bernese Alps is in sight. To the right of the imposing Wetterhorn **are** the broad Berglistock and the **Upper** Grindelwald Glacier, the Mettenberg, Great and Little Schreckhorn, Lauteraarhorn, Eiger, Mönch, and Jungfrau; still farther to the right, a piece of the Mittaghorn, the Grosshorn, Breithorn, Tschingelhorn, with the dark Lauterbrunner Wetterhorn in front of it, between the Tschingel Glacier (on the left) and the **Petersgrat** (on the right); then follows the range as far as the Gspaltenhorn; farther **back**, the Blümlisalp; in front, the plateau of Mürren, with Lauterbrunnen and the Staubbach below; above are the Schilthorn, the Sulegg-Grat with the **Lobhörner** (p. 172), **and** projecting above it, to the right, the Niesen; then the Abendberg, Wilderswyl, Unterseen with St. Beatenberg above it; above the Grindelwald valley appears the Faulhorn range, with the Schwarzhorn (usually snowclad); and in the distance beyond the Great Scheidegg, **the Sustenhörner** and the Titlis.

On the N. the Lauberhorn is adjoined by the precipitous *Tschuggen* (8278'; 2 hrs. from the Scheidegg) and, farther **on**, by the °**Männlichen** (7695'), another good point of view, ascended in 2-$2^1/2$ hrs. from the Little **Scheidegg**. A path is projected to skirt the slope of the Lauberhorn, but pending its completion, we descend by the **stables of the** Hôtel Bellevue to the chalets of *Bustigein* (6250'), which are visible from the hotel, to the **left**. Above these ($1/2$ hr.) a guide-board on an old stone-pine indicates the **path**, ascending to the left. In a short time we come in sight of the white **inn** on the top of the Männlichen. The distinct path skirts the Tschuggen; 40 min. crosses the *Mehlbaumen-Graben* streamlet; 50 min. *Hôt. Grindelwald-Rigi* (about 7220'; R., L., & A. $3^1/2$-4, B. $1^1/2$, déj. $3^1/2$, D. $4^1/2$ fr.). The top is reached in $1/2$ hr. more. The view of the Eiger, Mönch, and Jungfrau is inferior to that from the Lauberhorn, owing to the intervening Tschuggen, but the more distant peaks to the right and left are better seen (panorama by G. Studer). — The Männlichen is ascended from Grindelwald without difficulty in 4 hrs. (descent $2^1/2$-3 hrs.; horse 18 fr.; guide, unnecessary, 10 fr.). Near the station of Grund (see below), beyond the bridge over the Lütschine, we turn to the right and follow the road and afterwards the bridle-path (finger-posts) crossing the *Mehlbaumen-Graben* at the *Steinenweid* (about 3935') and ascending the *Raufte* (5085', view-hut at the top), whence we see the Hôt. Grindelwald-Rigi in front of us. Thence to the top by the *Itramen-Alp*, nearly 2 hrs.

Both railway and bridle-path (2½ hrs). walk to Grindelwald) follow the slope to the right immediately behind the Hôtel Bellevue. To the right is a retrospect of the Jungfrau. Then over the stony *Wergisthal-Alp*, at the foot of the Eiger, to (8 M.) Alpiglen (5287'; Hôt. des Alpes, ¼ M. from the station, plain but not cheap), on a commanding terrace. [The direct path hence to the 'Eismeer' (p. 177), fatiguing but repaying, should be attempted only with guide, ice-axe, and rope.] The Wetterhorn becomes more and more conspicuous as we advance, with the Mettenberg in front of it; farther on the Schreckhorn is seen through the gap between the Mettenberg and the Eiger. The line makes a steep descent into the valley of the *Black Lütschine* and crosses the stream. — 10½ M. *Grund* (3100'), the lower station for Grindelwald, whence the train backs out to ascend to the (11 M.) principal station of *Grindelwald* (see below). — Walkers from Grindelwald to the Little Scheidegg cross the bridge over the Lütschine near the station of Grund, and thence follow the bridle-path to the left, which crosses the line farther on; to Alpligen 2 hrs., thence to the top 1½ hr.

Grindelwald. — Hotels (all of which have restaurants and usually also seats in the open air). °BEAR (*Messrs. Boss*), 3 min. from the station, a large new house of five stories but without a lift, R., L., & A. 4-6, B. 1½, lunch 3½, D. 5, pens. from 10 fr. (S. B. G. H.); °EAGLE, ½ M. from the station, with pleasant garden and several dépendances (same proprietors and similar charges), D. 5, pens. 9-12 fr.); °EIGER, R. from 3, B. 1½, lunch 3, D. 4, pens. 7-8 fr. — °HÔT.-PENS. BURGENER, R., L., & A. from 3, B. 1¼, lunch 2½, D. 3, pens. 8 fr.; HÔT.-PENS. GRINDELWALD, somewhat plainer, pens. from 5-6 fr. — At the station: HÔT. ALPENRUHE, R. 2, B. 1¼, D. 3, pens. from 6 fr.; HÔT.-PENS. OBERLAND; HÔT. DE LA GARE, close to the station (wine on draught). — °HÔT. DU GLACIER, 7-8 min. below the principal station and as far from Grund, R., L., & A. from 2, B. 1¼, lunch 2½, D. 3, pens. from 5 fr. — VICTORIA, in an open situation on the Dürrenberg, 1 M. above the station, new. — °HÔT.-PENS. SCHÖNEGG, in a quiet situation, 8 min. from the station, above the Hôt. Eiger, to the left, with garden, pens. from 5 fr. — *Restaurant Bellevue*, beside the Hôt. Eiger, beer on draught, with rooms. — Confectioner, *J. Zbären*, beside the Bear.

Post and Telegraph Office, between the Eiger and Eagle Hotels.
English Church Service in the Protestant Church.
Guides. *Rud. Kaufmann* (Obmann), *Peter Baumann* ('am Guggen'), *Peter Baumann-Tuftbach*, *Christ. Almer*, father and son, *Ulrich* and *Hans Almer*, *Chr. Bohren-Trychelegg*, *Peter Kaufmann* (two of this name), *Hans Kaufmann*, *Hans Baumann*, *Hans Bernet*, *Ul. Rubi*, *Christ.*, *Franz*, and *Sam. Jossi*, *Joh. Heimann*, *Peter*, *Sam.*, and *Hans Brawand*, *Joh.* and *Christ. Burgener*, and many others. — Good ice-axes (18 fr.) from *Ch. Schenk*.

The authorities at Grindelwald recommend tourists not to yield to any of the attempts made to obtain money from them by songs, performances on the Alpine horn, exhibitions of Alpine animals, etc. — all of which are merely forms of begging in disguise. It is of course impossible to escape payment of the numerous tolls and pontages, but in these cases one payment is supposed to frank the visitor for the entire season, no matter how many visits he may make. The paths are mostly in very poor repair.

Grindelwald (3468' at the church; 3415' at the station), properly *Gydisdorf*, a large village (3087 inhab.) almost entirely rebuilt since the disastrous fire of 18th Aug., 1892, is an excellent starting-point for mountain-excursions, and also a favourite summer-resort, **the situation being** sheltered and healthful. Three gigantic mountains bound **the** valley on the S., the *Eiger* (13,040'), the *Mettenberg* (10,197'), which forms the base of the Schreckhorn, and the beautiful three-peaked *Wetterhorn* (12,150'), which impresses its character on the entire landscape. Between the two former lies the *Upper Grindelwald Glacier*, and between the two latter the *Lower Grindelwald Glacier*. These glaciers are **the feeders of the** *Black Lütschine*.

Most visitors **content themselves with a visit to the** *Upper Glacier (horse there and back 8 fr.)*. From the station we follow the principal village-street, passing the hotels and the (10 min.) *Church*, and beyond the new school-house, decorated with mottoes, take the footpath to the right (straight on is the bridle-path). The undulating path leads past some refreshment-huts to the (3/4 hr.) *Hôtel Wetterhorn* (4040'; R. 1½, pens. 4½-5 fr.), just before which we pass a memorial to *Dr. A. Haller* of Burgdorf, who perished on the Lauteraar Glacier in 1880. Here we diverge to **the right from the main path** (which goes on to the Great Scheidegg, see p. 186), cross the Lütschine (3935'), and in 10 min. reach the glacier. The artificially hewn *Ice Grotto* (adm. 50 c.; a small fee is also usually **given)** is the finest near Grindelwald **and the only one worth a visit.**

Another way back to Grindelwald (**guide**, 6 fr., not indispensable) is by a path ascending the left moraine to the **Chalet Milchbach** (4130'; rfmts.; visible from below), which affords a good view of the ice-fall. The path (finger-posts) then enters the wood to the right, passing between the Mettenberg and the *Halsegg*, and descends on the left bank of the Lütschine and across the *Sulz* to the bridge near the saw-mill mentioned below, and back to (1¼ hr.) Grindelwald. — From the Chalet Milchbach we may, by means of numerous ladders (not recommended to novices; guide necessary; 1 fr.), ascend to the Wetterhorn path (comp. p. 177), and pass through the *Milchbach-Schlucht* to the (3/4 hr.) edge of the glacier above the ice-fall (about 5250'; fine **survey** of the glacier).

A narrow, and in wet weather very muddy, path leads **to the E. from** the Hôtel Wetterhorn, past the 'Camera Obscura' and the small pavilion, and through shrubs and pines, to (20 min.) **the *Eisboden ('Ischbode';** 4400'), a beautiful pasture close to the base of the Wetterhorn, affording a noble survey of the glacier, the Mettenberg, Schreckhörner, Eiger, and the Grindelwald Valley.

The **Lower Glacier has so** retrograded **that the ascent to the** *Bäregg* will alone repay the visitor (p. 177; guide, 7 fr., unnecessary for moderately experienced walkers; horse to the Weissenfluh, ½ hr. below the Bäregg, 10 fr., not recommended), while **the only other** interesting point is the imposing *Gorge of the Lütschine*. Bridle-paths, above the Hôtel Eiger and between the Eagle Hotel and the church, descend **to the** right to the bridge spanning **the branch of the** Lütschine that issues from the upper glacier. On the opposite bank, on which is a saw-mill, the **path** straight on ascends to **the** Bäregg, while we continue to the right at the same level, and finally cross a wooden bridge over the discharge of the glacier to the entrance of the Gorge of the Lütschine, which has been rendered accessible by means of wooden galleries and steps (50 c.). The ascent of the left lateral moraine to the ($^1/_2$ hr.) upper part of the glacier, where there is an artificial *Ice Grotto* (50 c.), is not worth the trouble. It is preferable either to return to Grindelwald by the pretty wooded path **on the left bank** of the Lütschine, finally crossing the foot-bridge below the Hôtel du Glacier; or to follow the right bank for about 70 paces from the wooden bridge and then to ascend the right lateral moraine to the Bäregg path. On this latter ascent we pass ($^1/_4$ hr.) a refreshment-hut beside a wooden bridge, affording an interesting view of the gorge (50 c.), and in $^1/_4$ hr. more a second refreshment-hut whence another artificial *Ice Grotto* is accessible (50 c.). — From the bridge and saw-mill mentioned above a path ascends straight uphill to ($1^1/_2$-$1^3/_4$ hr.) the *Chalet Bäregg* (5410'; dear), which commands a good survey of the ***Lower Eismeer*** ('sea of ice'), the large basin of névé in which the glacier accumulates before it descends to the valley. Above it rise the Zäsenberghorn, Grindelwalder Grünhorn, Great and Little Grindelwalder Fiescherhorn, and Eiger. A projecting rocky knoll, 20-25 min. farther on, affords a still opener view.

A flight of wooden steps (1 fr.), about **5 min.** from the chalet, descends to the edge of the 'Eismeer'. The glacier may be crossed, with guide (from Grindelwald, 9 fr.), to (1 hr.) the *Zäsenberg* (6050'), surrounded by pastures, and occupied by shepherds in summer. — The ascent of the *Zäsenberghorn* (7687'; magnificent survey of the glaciers) takes $1^1/_2$ hr. from the Zäsenberg (guide 12 fr.). On every side tower huge and wild **masses** of ice, and the view is bounded by the imposing summits of the Eiger, Schreckhörner, Fiescherhörner, etc. An interesting and comparatively easy trip may be made from the Bäregg to the *Zäsenberghorn, Fiescherfirn*, and back by the *Kalli* (7-8 hrs.; guide 20 fr.).

The **Mettenberg** (*Mittelberg*, 10,197'; 10 hrs. from Grindelwald viâ the Bäregg; guide 30 fr.) commands an imposing view of the Schreckhorn, the Finsteraarhorn, and the Eismeer, but is comparatively seldom ascended.

The favourite ascent is that of the **Wetterhorn** (12,150'; guide 60, porter 45 fr.), which was first scaled in 1844. The ascent, now made almost daily in summer when the weather is good, is free from serious difficulty, though requiring perseverance and a steady head. From the Chalet Milchbach by the ladders to the upper glacier, see p. 176. We cross the glacier to the *Schlupf* and traverse the precipitous *Zybachsplatten*, with numerous brooks in wet weather, to the *Gleckstein Club-Hut* (7695'; $5^1/_2$-6 hrs. from Grindelwald), where the night is spent. Thence over the *Krinnen-Firn* and by a steep ascent to the snow-covered *Saddle* between the *Mittelhorn* (12,165') and the *Vordere Wetterhorn* or *Hasli-Jungfrau* (12,150'), and thence to the top of the latter, 5-6 hrs. The *Rosenhorn* (12,110'), the third peak, is seldom ascended. — Descent to the *Dossen Hut* (and Rosenlaui or Innertkirchen), see pp. 186, 187 (guide from Grindelwald, 70 or 80 fr.). — From the Gleckstein Hut over the *Bergli-Joch* to the *Urbach-Thal*, see p. 187. — The *Berglistock* (12,000'), to the right of the Bergli-Joch ($4^1/_2$-5 hrs. from the club-hut; guide 70 fr.), commands a superb view of the Schreckhörner, Wetterhörner, etc.

Ascent of the *Jungfrau*, p. 173; *Finsteraarhorn* (from Grindelwald viâ the *Agassiz-Joch*, dangerous as a descent on account of falling stones), p. 189. — Gross-Schreckhorn (13,386'; from the *Schwarzegg Club-Hut* 6-7 hrs.; guide 80 fr.), ascended for the first time by Mr. Leslie Stephen in 1861, very difficult. — Klein-Schreckhorn (11,475'), from the Schwarzegg Club-Hut 4-6 hrs., or from the Gleckstein Hut (see above) 5-6 hrs., interesting and for experts not difficult (guide 60 fr.). — Mönch (13,465'; first scaled by Dr. Porges of Vienna in 1857), ascended either from the *Bergli-Hütte* by the *Mönchjoch* (see below) in 5-6 hrs., or from the *Guggi-Hütte* (p. 162) by the N. side in 8-9 hrs. (not without danger; guide 70-80 fr.). — Eiger (13,040'; first ascended by Mr. Chas. Barrington in 1858), from the Little Scheidegg by the Eiger Glacier and up the W arête, 5½-7 hrs. (guide 80 fr.). All these are for thorough adepts only.

Passes. To THE GRIMSEL HOSPICE over the *Strahlegg (10,995'; 14 hrs.; two guides, 40 fr. each), a grand but toilsome route. The night is passed in the *Schwarzegg Club-Hut* (8200') on the upper Eismeer, 5 hrs. from Grindelwald. Thence a steep ascent over ice and rock to the (3 hrs.) pass, lying between the Gross-Lauteraarhorn and the Strahlegghörner; descent over the *Strahleggfirn* and the *Finsteraar* and *Unteraar Glaciers* to the (3-4 hrs.) *Pavillon Dollfus* (p. 189) and the (3 hrs.) *Grimsel Hospice* (p. 188). In the reverse direction (especially if a night be spent in the Pav. Dollfus) the route is less trying and more interesting. — Finsteraar-Joch (11,025'; 15-16 hrs.; guides 40 fr. each), between the Strahlegghörner and the Finsteraarhorn, very trying, with splendid views of the Finsteraarhorn, etc. — Lauteraar-Sattel (10,355'; 16-17 hrs.; guides 50 fr. each), between the Schreckhörner and the Berglistock, a fatiguing pass, but without serious difficulty to proficients. The night is spent in the *Gleckstein-Hütte* (p. 171); thence we ascend the *Upper Grindelwald-Firn* in 5-6 hrs. to the pass, which affords a grand survey of the Gross-Schreckhorn, Lauteraarhorn, etc. We then descend a steep snow-slope to the *Lauteraarfirn* (sometimes guarded by a wide 'Bergschrund' or chasm) and the (3 hrs.) Pav. Dollfus (p. 189). — Over the *Bergli-Joch* to the *Urbach-Thal*, see p. 187.

PASSES FROM GRINDELWALD TO THE EGGISHORN (p. 305), for experts only, with able guides. The Jungfrau-Joch (11,090'; two guides, 100 fr. each), between the Jungfrau and Mönch, from the Little Scheidegg to the Eggishorn Hotel in 16 hrs., viâ the *Guggi Glacier*, is very difficult and trying. — The passage of the Mönchjoch (11,910'; guides 60 fr. each), 15 hrs. from Grindelwald to the hotel, less difficult, is facilitated by spending a night in the *Bergli-Hütte* (see below), or when the journey is made in the reverse direction, in the *Concordia-Hütte* (p. 305). This is comparatively the easiest and finest of these glacier expeditions. From the Bäregg we cross the lower Eismeer to the opposite moraine, and ascend the precipitous *Kalli* for 2½ hrs.; then cross the much crevassed *Grindelwald-Fiescher Glacier* to the (3 hrs.; 7-9 hrs. from Grindelwald) *Bergli Club-Hut* (10,825'), commanding a grand though not extensive view of the Fiescherwand, Schreckhörner, Eiger, etc. From the hut a steep climb of ³/₄ hr. over rock and glacier to the *Lower Mönchjoch* (11,810'), between the Mönch and Fieschergrat; thence either to the right over the *Upper Mönchjoch* (11,930'), between the Mönch and Trugberg, to the *Jungfraufirn* (p. 162) and down to the *Great Aletsch Glacier* and the (5-6 hrs.) *Eggishorn Hotel*; or to the left, over the vast *Ewigschneefeld*, to the Aletsch Glacier (the two routes unite at the *Concordia Hut*). — The Eiger-Joch (11,875'; guides 100 fr.), between the Eiger and Mönch, 22 hrs. from the Wengern-Alp to the Eggishorn, a night being spent in the *Guggi-Hütte* (see p. 174), whence the Eiger Glacier is ascended, is very difficult. — The Fiescher-Joch or Ochsen-Joch (about 11,700'), to the E. of the *Kleine Fiescherhorn* or *Ochs* (12,812'), 22 hrs. from Grindelwald to the Eggishorn, is very toilsome and lacks interest.

49. The Faulhorn.

Guide (unnecessary): from Grindelwald and back 10, if a night be spent at the top 13 fr.; from the Schynige Platte 6, with descent to Grindelwald 12, or viâ the Great Scheidegg to Meiringen or Imhof 25 fr. — *Chair-Porters* 6 fr. each; if they pass the night on the top, 12 fr. (three generally suffice; a bargain should be made beforehand). — *Horse* from Grindelwald and back 20 (or with one night out, 25) fr.; to the top and back by the Great Scheidegg 30, with descent to Meiringen or Im Hof 40 fr.; from the Schynige Platte to the top 20 fr.; from Interlaken by the Schynige Platte and the

Faulhorn to Grindelwald 40, or to Meiringen or Imhof (two days) 50 fr.; from Meiringen to the Faulhorn in 1 day 30 fr., to the Faulhorn and Grindelwald 36 fr. — °INN on the summit (not cheap, R. 5, L. & A. 1½, B. 2½, D. 5 fr., cup of coffee 75 c.).

The *Faulhorn (8803'), rising between the Lake of Brienz and the valley of Grindelwald, and composed of friable, calcareous schist (the name being probably derived from *faul*, 'rotten'), commands a closer survey of the giants of the Bernese Oberland (see Panorama) than is obtained from the Rigi. To the N., at our feet, lies the Lake of Brienz, with its surrounding mountains, from the Augstmatthorn to the Rothhorn; part of the Lake of Thun, with the Niesen and Stockhorn, is also visible; to the N.E. are parts of the Lakes of Lucerne and Zug, with Pilatus and the Rigi; then the Lakes of Morat and Neuchâtel.

FROM GRINDELWALD TO THE FAULHORN (4¾ hrs.; descent 3 hrs.). From the *Station* the route leads past the Hôtel Oberland by the narrower road to the left, while from the *Bear Hotel* we cross the road and proceed in the direction of the large hotel-stables, where we take the footpath ascending to the left between the stables and the little wash-house. After 3 min., to the right (in the direction of the lower edge of the pine-forest on the slope of the Dürrenberg); 10 min., at a cross-way, straight on; 5 min., to the right; 2 min., to the left past a cottage. The footpath soon unites with the bridle-path that begins opposite the *Eagle Hotel*, diverging from the road at the Pension Schlössli, then leading to the left past the stables to this point (½ hr.). We now follow the main path, partly through wood. After 40 min., in the middle of the *Hertenbühl* (5157'), a large pasture with several chalets, the path turns sharply to the left, ascending past a little cabaret into (10 min.) wood; 10 min., to the right, past a small pond; 20 min., a gate (persons descending here keep to the left, also passing through the gate); 25 min., *Waldspitz* (6200'; *Hôt.-Pens. Alpenrose, unpretending, R. 2½ fr.), with a splendid view. This point is nearly halfway. Farther on (20 min.), to the left, is a fall of the *Mühlibach*, which we cross near the upper chalets of the *Bach-Alp* (6496'). Farther on the path holds to the left and crosses a brook. Then a moderate ascent of ¾ hr. to the *Bach-See* (7428'), in a stony basin, bounded on the left by the *Röthihorn* (9052') and *Simelihorn* (9030'), and on the right by the *Ritzengrätli* (8282'). (By the stone hut the path for travellers descending to the Scheidegg diverges to the left, see p. 180.) The top of the Faulhorn is now in view. The path, indicated by stakes for guidance in fog or snow, ascends rapidly for nearly 1 hr. over crumbling slate and limestone. We pass another stone hut (Alpine horn performer), cross the nearly level pastures at the foot of the peak, and reach the top by a zigzag path in ¼ hr. more. The path from the Schynige Platte joins our path on the left, below the Inn.

For the RETURN TO GRINDELWALD (3 hrs.) the path viâ the *Buss-Alp* is recommended, which diverges to the right at the last-mentioned stone

12*

180 *III. Route 49.*   FAULHORN.   *Bernese*

hut. To the W. of the upper chalets rises the *Burg* (7247'), which is sometimes ascended from Grindelwald direct in 2½ hrs. for the sake of the view.

FROM THE SCHYNIGE PLATTE TO THE FAULHORN (4 hrs.; descent 3 hrs.). The picturesque bridle-path, the beginning of which is indicated by a finger-post below the station (p. 175), first crosses the *Iselten-Alp*, below the steep *Oberberghorn* (6791'). Beyond (20 min.) a gate we keep to the right, skirting the S.W. and S.E. slopes of the *Laucherhorn* (8333'), and traverse the rocky debris of the *Bütschi*, beyond which, at the foot of the *Sägishörner*, a footpath descends along the brook to the right. (In descending, therefore, we here keep to the right, with the brownish inn on the Schynige Platte in sight, and the Geisshorn and Gummihorn above it.) We turn to the left, cross the (8 min.) watershed of the *Egg* (6985'), and descend slightly into the *Sägisthal*. The signal on the top of the Faulhorn soon comes in sight; to the left is the *Rothhorn* (7535'), also with a signal. In 35 min. more we reach the chalet on the *Sägisthal-See* (6258'), follow the marshy path round the left side of the lake (right side also practicable, but no path), and ascend (to the right) the barren slope of the *Schwabhorn* (7795'), at the top of which our path is joined by that from the Giessbach (p. 184). Beyond the (1½ hr.) saddle between the Schwabhorn and the *Winteregg* (to the right), the way is indicated by piles of stones. We reach the foot of the peak in 12 min. and ascend to the right to the inn in ¼ hr. more.

FROM THE FAULHORN TO THE GREAT SCHEIDEGG (3 hrs.; ascent 4 hrs.). The path diverges to the left from the Grindelwald path, near the (¾ hr.) hut on the *Bach-See*, traverses the stony slopes of the *Ritzengrätli*, and keeps nearly the same level for some distance; ½ hr., a gate between the *Bach-Alp* and the *Widderfeld-Alp;* we follow the main path to the left, not down the bed of the brook; 12 min., we cross the ridge of the *Langenbalm-Egg* (7106'), with a magnificent view; 8 min., we keep to the left and cross the brook; 7 min., we descend to the left over black, crumbling slate. Beyond a gate the path becomes indistinct at places, the general direction being to the left along the slope above the upper *Grindel-Alp*, towards the conspicuous Scheidegg Hotel (in misty weather a course may be steered slightly to the left of the Wetterhorn); ¼ hr., a small brook is crossed; 5 min., another brook; 10 min., a rough bridge over the *Bergelbach;* 5 min., the *Oberläger* (upper chalets) *of the Grindel-Alp* (6410'), with a spring. At (¼ hr.) a gate we ascend to the right on this side of the fence, pass through the next gate (12 min.), and make for the top of a hill; 8 min., Scheidegg Inn.

In ascending from the Scheidegg, be careful not to turn to the left at the bridge over the *Bergelbach;* farther on, where the path is lost on the pastures, again avoid turning to the left, but follow a direction parallel with a long enclosure lying a little to the left, and make for the slope of the mountain, at the foot of which the path is regained.

The view from the Faulhorn is partially intercepted by the neighbouring group of the *Simelihorn* (9030') and the **Röthihorn** (9052'), rising between the Finsteraarhorn and the Schreckhorn, which conceals part of the Alpine chain and the valley of Grindelwald. The Röthihorn, from which the magnificent view is uninterrupted, is easily ascended from the Bach-See in 1½ hr. (guide advisable; from the Faulhorn 5 fr.).

The view is still grander and more extensive from the *Schwarzhorn (9613'), which, with the *Wildgerst* (9488'), intercepts the view from the Faulhorn on the E. side. (The lakes of Lungern, Sarnen, Alpnach, and Küssnacht are visible hence, all lying in the same line.) The ascent is made from the Great Scheidegg by the *Grindel-Alp* and the *Krinnenboden* in 3½-4 hrs.; or from Rosenlaui by the upper *Breitenboden-Alp* (6560'), to which there is a bridle-path, and the little *Blue Glacier*, in 5 hrs.; or from *Axalp* (p. 184) in 4 hrs. (guide 12 fr.).

50. From Meiringen to Interlaken. Lake of Brienz.

From Meiringen to Brienz (8 M.) RAILWAY in 25 min. (fares 2 fr. 60, 1 fr. 95, 80 c.). — From Brienz (station) to *Interlaken* STEAMBOAT 7 times daily in 1 hr., fare 2 or 1 fr.; luggage additional, 50 c. for each box.

Meiringen. — Hotels. *HÔTEL DU SAUVAGE (Zum Wildenmann)*, 3 min. from the station (omnibus), with garden, R., L., & A. 5-6½; D. 5 fr.; *HÔT. DE L'OURS, *HÔT. BRÜNIG, both near the station, R., L., & A. 2-3, B. 1¼, D. 1½-3, pens. from 5 fr.; HÔT. DE LA GARE, opposite the station, unpretending; *CROIX BLANCHE, in a side-street in the direction of the church, R., L., & A. 2½, B. 1, D. 2½, pens. 6 fr.; POST, in the main street, R., L., & A. 2-2½, B. 1 fr. 20 c., D. 2-3, pens. 5-7 fr.; *HIRSCH, ½ M. from the station (omnibus), near the Willigen-Brücke, R., L., & A. 2½, B. 1¼ fr. — *HÔTEL-PENSION REICHENBACH, beyond the Aare, on the way to the Reichenbach waterfalls, 1 M. from the station (omnibus), with dépendance *(Pens. des Alpes)*, R. 2-4, B. 1½, lunch 3, D. 4, pens. 6-9 fr.

Restaurants at the station and in the hotels; *Café-Restaurant Victoria*, near the station, D. 1½-2 fr.; *Brauerei Stein*, with garden, beyond the Hirsch.

English Church, in the garden of the Hôtel du Sauvage.

Guides. *Melchior, Jakob, Joh.,* and *Peter Anderegg, Joh.* and *Kaspar v. Bergen, Heinrich Führer jr., Joh.* and *Andr. Jaun, Kaspar Moor, Kaspar Maurer, Andreas Stähli, Melchior Zenger*, etc.

Meiringen (1968'), the principal station on the *Brünig Railway* (R. 37), is the chief village of the *Haslithal*, the inhabitants of which, according to tradition, immigrated with the Schwyzers from Scandinavia. The village, almost entirely burned down in Oct., 1891, but since rebuilt in an improved style, lies on the right bank of the *Aare*, in a level valley 3 M. in width, surrounded by wooded mountains, above which rise several snowy peaks. To the S. appear the *Reichenbach Falls* (p. 185), with the snow-fields of the Wellhorn and the Rosenlaui Glacier above them. The *Mühlebach, Alpbach*, and *Dorfbach*, descending from the *Hasleberg* to the N. of the village, form considerable waterfalls. They often overflow their banks, and cover the whole district with rocks and mud. Several finger-posts in the village indicate the way to the 'Alpbach-Schlucht' (20 min.; adm. 80 c.; at the head is a small restaurant, with view). The massive detached church-tower of Meiringen originally belonged to a castle. Both tower and church have had repeatedly to be freed from the deposits of debris, which have raised the ground around them. — To the E. of the village rises the ruined tower of *Resti*.

182 *III. Route 50.* BRIENZ. *Bernese*

The *Gorge of the Aare (*Aareklamm;* carriage there and back with stay of 1 hr., 4 fr., with 1½ hr.'s stay and back from the Lammi Inn, 6 fr.; two-horse carr. 7 and 10 fr.) is the chief point of interest near Meiringen, next to the Falls of the Reichenbach (p. 185). We follow the main road to beyond the Hirsch, then diverge to the right, cross the (½ M.) *Willigen-Brücke* (see p. 187), and take the road to the left (that to the right goes on to the Hôt. Reichenbach, p. 181). At the entrance to the gorge is a small *Restaurant*, where tickets of admission (1 fr.) are obtained. The wild and romantic rocky gorge, which affords passage to the Aare through the Kirchet (p. 174), is about 1500 yds. long, and has been made accessible by means of an iron gallery. After about 10 min. we pass a pretty waterfall on the left, and farther on we have a glimpse of the Ritzlihorn through the opening at the top of the ravine. We now either return or ascend by a path diverging by a side-gorge to the right to (15-20 min.) the *Lammi Inn*, on the road over the Kirchet, by which we may return to the Willigen-Brücke in 25 minutes. — A finger-post, about 2 min. from the Lammi Inn, indicates the way to the upper Reichenbach Fall (½ hr.; comp. p. 185).

On the **Hasleberg**, ¾ hr. to the N. of Meiringen and about 750' above it, is the **Hôt.-Pens. Alpbach* (R. 2½, D. 3, S. 2, pens. 5½-8 fr.), with a view of the Wellhorn and Wetterhorn group. About 1½ hr. farther on (good path by *Goldern* and *Wasserwendi*) lies the village of *Hohfluh* (3443'; *Pens. Willi- von Bergen, 4½-6 fr.), another fine point of view. (Hohfluh may also be reached direct from Meiringen by *Unterfluh* in 1½ hr.) From this point the ***Hohenstollen** (8150'; splendid view) may be ascended by the *Balisalp* in 4 hrs. (with guide; from Meiringen 12, from the Hôt. Alpbach 7 fr.), or from Meiringen direct, by the *Mägisalp* and the *Schwarzenfluh* in 5 hrs. Descent to the *Melchthal* viâ *Frutt*, see p. 132. — In *Reuti* or *Rüti* (3150'), to the E. of the head of the gorge of the Alpbach (p. 181), is the *Pens. & Restaurant Kohler*.

The railway skirts **the right bank of the Aare**. The beautiful *Oltschibach* and other cascades fall from the precipices on the left. Beyond (5 M.) *Brienzwyler* (Hôtel Balmhof), where it crosses the Brünig road, the line skirts the geologically interesting *Ballenberg* (2385'), then bends to the right and follows the shore of the Lake of Brienz, viâ *Kienholz*, to —

8 M. **Brienz.** — The *Station* is situated at Tracht, to the E. of Brienz, beside the station of the *Rothhornbahn* and a few yards from the *Steamboat Pier*. Most of the steamers also touch at the pier near the Bär in Brienz. — Hotels. HÔTEL DE L'OURS, ½ M. from the stations, with a terrace on the lake, well spoken of, R., L., & A. 3, B. 1¼ fr.; WEISSES KREUZ, in Tracht, near the stations, R. L. & A. 2½-3, B. 1½ fr.; SCHÜTZEN, farther to the E., plain. — *English Church Service* in summer (at the Hôt. de l'Ours).

The village of *Brienz* (2531 inhab.), adjoined on the E. by *Tracht*, stretches for 1½ M. on the bank of the Lake of Brienz, backed by green **pastures** dotted with fruit-trees, above which rises the *Brienzer Grat*, whence descend the falls of the *Trachtbach* and the *Mühlbach*. The latter, to the W. of Brienz, is often dry in summer. Brienz is the chief centre of the Oberland wood-carving industry, which employs about 600 persons; a large selection of carvings may be inspected in the *Industrie-Halle*, next the Bear Hotel. On a hill about ¼ M. farther to the W. is the pleasantly-situated *Church*, commanding a view of the valley of Meiringen, with the Sustenhörner in the background. The **view** from the pavilion on the *Fluhberg*, ¼ hr. above the station, is now somewhat impeded by foliage.

The ***Brienzer Rothhorn** (7713'), the highest peak of the Brienzer Grat, is a famous point of view. RACK-AND-PINION RAILWAY (opened in 1892) in 1 hr. 20 min. (up 8 fr., down 4 fr., there and back 10 fr., on Sun. 8 fr.). This line (4³/₄ M. in length; maximum gradient 25:100) ascends through luxuriant meadows, soon affording a view of the Lake of Brienz and the Schwarzhorn range. Beyond the bridge across the *Trachtbach* the ascent becomes steeper; the line approaches the *Mühlbach*, turns to the right by means of the short *Schwarzfluh Tunnel* and mounts to the (1¹/₃ M.) station of *Geldried* (3360'). To the right, we overlook the valley of Meiringen and the Sustenhörner. Describing a large loop, we pass through the *Stockisgraben Tunnel* and the five tunnels of the *Planalpfluh* to the (2 M.) station *Hausstadt* (4415'; rfmts.), commanding a view of the Blümlisalp, Doldenhorn, and Wildstrubel. We now proceed on the left bank, and farther up on the right bank of the *Mühlbach* over the pastures of the *Planalp*, past the chalets of *Mittelstaffel* (5023'), and beyond the *Kühmatt Tunnel* (100 yds.) attain the (3¹/₂ M.) watering station of *Oberstaffel* (5980'). Finally the line sweeps in a wide curve round the uppermost valley, bends back by means of the two *Schönegg Tunnels*, and reaches its terminus at (4³/₄ M.) station *Rothhorn-Kulm* (7288'), on the *Breitengrat*, 3 min. below the *Hôtel Rothhorn-Kulm* and 12 min. below the summit, on which a triangular stone marks the meeting-place of the cantons of Bern, Lucerne, and Unterwalden. The *View (panorama at the hotel; best in the morning and evening) vies in extent and picturesque charm with that from the Rigi, especially as the great peaks are nearer, while the glimpses of the surrounding valleys excel those obtained from the Faulhorn. The prospect embraces the entire chain of the Urner, Engelberg, **and** Bernese Alps, with the Lake of Brienz in the foreground; a glimpse of the Lake of Thun beyond Interlaken; the Haslithal from Meiringen **nearly to the Grimsel**; on the other side the small Ey-See, **the Lake of Sarnen**, a considerable part of the Lake of Lucerne with the **Rigi, part of the Lake of** Zug, and a long strip of the Lake of Neuchâtel.

The Lake of Brienz (1857'), 8³/₄ M. long, and 1¹/₄-1¹/₂ M. wide, 500' deep near the Giessbach and 859' near Oberried, lies 20' higher than the Lake of Thun. It is enclosed by lofty wooded rocks and mountains. A beautiful road skirts its N. bank (from Brienz to Interlaken, 10¹/₂ M.; one-horse carr. 8-10 fr.). To the S.E. in the background are the snow-clad Sustenhörner, to the right the Thierberge. Farther on we lose sight of the snow-mountains. The steamboat crosses the lake to the (10 min.) —

Giessbach. — From the landing-place (small restaurant) we may walk to the terrace opposite the falls by a broad road in 20 min., or ascend by the *Cable Tramway* (380' long; gradient 28:100) in 6 min. (there and back 1 fr.; luggage ¹/₂-1 fr.).

Hotels. *Hôtel Giessbach, a large new building, with a restaurant on the terrace and a pension (the old hotel), R., L., & A. from 4, B. 1¹/₂, lunch 3¹/₂, D. 4¹/₂-5, S. 5, pens. 7¹/₂-10 fr. (less before July 1st and after Sept. 5th); illumination of the falls 1 fr. (for the first evening only), music 2 fr. per week; also well-equipped hydropathic, with electric baths; post, telegraph, and railway ticket office, etc. *English Church Service* at the hotel. — *Hôtel Beau-Site, ¹/₄ M. higher, less pretentious, R., S., & B. 3, D. 3, pens. 6 fr. Both hotels belong to the *Messrs. Hauser*.

The *Giessbach is one of the prettiest and most popular spots in the Bernese Oberland. The stream, which is copious at all seasons, rises on the N. slope of the Schwarzhorn (p. 181), and on its way to the Lake of Brienz, 980' below, forms a series of seven cascades falling from rock to rock, and framed in dark green foliage. Only the lowest fall is seen from the steamer; the terrace in front of the hotel affords a complete view. The falls are crossed by three bridges. Paths ascend on both banks to the (¹/₄ hr.) second bridge,

184   III. Route 50.   ISELTWALD.

from which to the third (½ hr.) there is a path on the right bank only. A wooden gallery enables visitors to pass behind the second fall. Those who have time should ascend to the *Highest Fall*, where the Giessbach, issuing from a sombre ravine, is precipitated **under the bridge** into an abyss, 190′ in depth. (Best view from **a projecting rock to the right of the bridge**.) **Above** the highest bridge there is no attraction. About noon rainbows are formed in the falls. —The falls are illuminated with Bengal lights every evening at 9.30 from May 15th to the end of Sept. (spectators not living in one of the hotels, 1½ fr.).

A guide-post behind the 'Etablissement Hydrothérapique' indicates the way, to the left, to the (20 min.) Rauft, a wooded rock on the N. side of the valley, rising abruptly 600′ above the lake, commanding a view **of the Lake of Brienz**. — The path to the right from the guide-post leads **to the** Alpine hamlet of Enge, situated among beautiful pastures. Pretty view at the point (½ hr.) where the path reaches the lake. We then descend past the *Nüseli* to the *Aare Bridge* and the Meiringen and **Brienz** road (p. 172). — About 3 hrs. above the Giessbach (porter 5 fr.) **lies Axalp** (5580′), a health-resort with an unpretending 'Inn (pens. 4½-5 **fr.**), whence the *Axalphorn* (7635′; 2 hrs.), the *Faulhorn* (p. 179; 5 hrs.), **and** the *Schwarzhorn* (9610′; 4 hrs.; guide 10 fr.; comp. p. 181) may be ascended. — About 1 hr. from Pens. Axalp (2½ hrs. from the Giessbach) is the *Hinterburg-See* (5000′), charmingly situated **in wood at the base** of the *Oltschikopf*.

ASCENT OF THE FAULHORN (p. 179) FROM THE GIESSBACH, 7 hrs. (guide 12 fr.), fatiguing at places, especially on the *Bätten-Alp*, which is exposed to the morning sun. To the S. of the Schwabhorn this path joins **the** bridle-path from the Schynige Platte to the Faulhorn (p. 180).

FROM THE GIESSBACH TO INTERLAKEN (3½ hrs.). A good, well-shaded path, crossing the first bridge over the falls, and bearing to the right (see finger-posts), leads to the (½ hr.) *Hochfluh*, a charming point of view. It then runs high above the lake and descends to (1 hr.) *Iseltwald* (see below), from which a road (steep ascent at first) leads to (1½ M.) **Sengg**, (3 M.) *Bönigen*, and (1½ M.) *Interlaken*.

From the Giessbach the ordinary steamers proceed to *Oberried*, on the N. bank of the lake, but the express-boats steer along the precipitous S. bank, past the **small** wooded *Schnecken-Insel*, with its little chapel, direct to the **pretty** village of Iseltwald (*Pens. Iseltwald, 1¼ M. to the W., 5-6 fr., unpretending; Zum Strand), on the S. bank. The village (telegraph office) is united with Interlaken by a picturesque road (6 M.; see above). — Then to *Niederried*, charmingly situated on the N. bank among **fruit-trees** at the foot of the *Augstmatthorn* (p. 163). Farther on, beyond a wooded promontory, is *Ringgenberg* (p. 163), beside the old castle and **church of that name**, surrounded by underwood and orchards. On the opposite bank is the influx of the *Lütschine*, which descends from the valley of Lauterbrunnen. The steamer stops at *Bönigen* (p. 159) and enters the canalized Aare. On a hill to the right is the ruined **tower** of the *Church of Goldswyl* (p. 163). The steamboat-station at *Interlaken* is opposite the railway-station *Interlaken-Ost* (p. 158).

51. From Meiringen to Grindelwald.

7½-8 hrs. Bridle-path. From Meiringen, past the Reichenbach Falls, to the *Zwirgi Inn* 1½ hr.; thence to *Rosenlaui* 1¾ hr. (descent from Rosenlaui to Meiringen 2 hrs.); from Rosenlaui to the *Great Scheidegg* 2¾ (descent 1¾) hrs.; from the Scheidegg to *Grindelwald* 2 (ascent 3) hours. — *Guide* (unnecessary) 12 fr., including the Faulhorn, 20 fr. — *Horse* from Meiringen to Rosenlaui 10, Scheidegg 15, Grindelwald 25 fr.

Meiringen, see p. 181. Crossing the *Willigen-Brücke* (p. 182), we turn to the right at the road to the Gorge of the Aare, and reach the (5 min.) *Hôtel Reichenbach*, situated at the foot of the hill from which the celebrated *Falls of the Reichenbach descend. The *Lower Fall* is 5 min. to the W. of the hotel by the road; beside it is a saw-mill. — Returning to the hotel we follow the broad bridle-path to the left between the barn and the fountain. After 10 min. a footpath diverges to the right to the falls and to Rosenlaui; 5 min., hut (30 c.; not worth it) commanding a view of the *Central* or *Kessel Fall*. Thence we keep to the left, soon coming in sight of the spray of the upper fall; 18 min., several huts with a guide-post. In 8 min. more we reach the *Upper Fall*, with its beautiful jets; beside it is a hut (50 c.; rfmts.), whence a narrow footpath, passing a gallery (view of the fall from above), leads back to the bridle-path in 25 min. The latter (guide-posts for the descent) brings us in 5 min. to the little inn *Zur Zwirgi* (3202'), commanding a retrospect of the Hasli-Thal and the mountains surrounding the Brünig and Susten. In a gorge to the right the Reichenbach forms a picturesque fall (30 c.).

Travellers from Rosenlaui to In-Hof (the Grimsel, Engstlen-Alp, etc.), may, omitting the Falls of the Reichenbach and Meiringen, save nearly an hour by following the bridle-path for 5 min. beyond the path to the falls, and then turning to the right by a footpath to the village of (25 min.) *Geisshotz* (2623'), hidden among fruit-trees. Here we ascend the pastures, and then rapidly descend the *Kirchet* (p. 174) to (40 min.) *Im-Hof* (p. 175).

Our path now ascends the Reichenbach, at a considerable height above the right bank. In front of us soon appears the Wellhorn, with the Wetterhorn to the right of it, and the Rosenhorn behind it, to the left; farther on the Rosenlaui Glacier also comes in sight. Beyond a *Saw Mill* (3986'; Inn) we cross a bridge (4238') to the left bank, and reach the (1⅓ hr.) *Gschwandenmad-Alp*, commanding a celebrated **View the bare *Engelhorn* (9130'), the beautiful *Rosenlaui Glacier* between the *Dossenhorn* (10,303') and the *Wellhorn* (p. 186), and the snow-clad cone of the *Wetterhorn* (p. 177) to the right, together with the beautiful foreground, present a picture unsurpassed in Switzerland. Immediately beyond the bridge the path forks; the main branch, to the left, leads to (20-25 min.) Rosenlaui; the right branch is a shorter route to the Schwarzwald-gletscher Hotel (see below). Both routes offer the same views.

The **Baths of Rosenlaui** (4363'; *Hôt.-Pens. Curhaus*, R., L., & A. from 3, B. 1½; board 6-7 fr.; Engl. Ch. Serv.) occupy a secluded situation in the well-watered, fir-clad valley of the Reichenbach, which forms a pretty waterfall in the gorge behind the Curhaus.

186   III. Route 51.   GREAT SCHEIDEGG.

From the other side of the bridge opposite the Curhaus a path to the left leads to the **Rosenlaui Glacier**. One of the several guide-posts on this path shows the way (wooden steps) to the glacier stream. The glacier, famed for the beauty and purity of its ice, has receded of late years so much that an ascent of 1½-2 hrs., very rough towards the end, up the left lateral moraine to the height of about 5740', must be made in order to obtain a survey of it. The **Dossen-Hütte** (8860'), grandly situated about 5 hrs. above Rosenlaui, affords a highly interesting expedition for good mountaineers (reached also from Im-Hof through the *Urbach-Thal* in 8 hrs.; see p. 187). This is the starting-point for the *Dossenhorn* (10,303'; 1 br.), the *Renfenhorn* (10,777'; 2½ hrs.), the *Hangend-Gletscherhorn* (10,810'; 4 hrs.), and above all for the *Wetterhorn* (12,150'; 4 hrs.). Descent from the Wetterhorn to the (3½ hrs.) *Gleckstein Hut* and (3½ hrs.) Grindelwald, see p. 177. — From the Dossen Hut we may cross the *Wetterlimmi* (10,443'), the *Gauli Glacier*, and the *Gauli Pass* (10,260') to the *Grimsel*, 10 hrs., fatiguing; with this route the ascent of the *Ewigschneehorn* is easily combined (p. 189).

The path to Grindelwald now ascends the right bank of the Reichenbach, at first on the wooded N. slope of the *Welligrat*, and then continues level for some time. After 20 min. it crosses the stream, on the left bank of which debouches the above-mentioned direct route from the Gschwandenmad-Alp. We ascend the left bank viâ the *Breitenboden-Alp* (4650'), cross the *Gemsbach*, and traverse the *Schwarzwald-Alp* (4810'; passing through a gate) to the (1-1¼ hr.) **Hôtel-Pension Schwarzwaldgletscher* (5020'; R. 2-2½ fr., unpretending), prettily situated amidst wood. To the left are the precipitous cliffs of the *Wellhorn* (10,486') and the *Schwarzwald Glacier*. We pass a *Saw Mill*, and at the exit from the wood a (25 min.) bridge (5315'), beyond which we ascend the mountain-slope.

The (1 hr.) **Great Scheidegg or Hasli-Scheidegg** (6430'; *Hôt. Grosse Scheidegg*, R. 2½, B. 1½, S. 3½ fr., fairly good) commands a striking view towards the W. The smiling valley of Grindelwald, bounded on the S.W. by the pastures and woods of the Little Scheidegg, contrasts picturesquely with the bare precipices of the Wetterhorn, which tower above us to a giddy height. To the S.W. of the Wetterhorn are the Mettenberg, Fieschergrat, Mönch, Eiger, and lastly the Tschingelgrat, Gspaltenhorn, and Blümlisalp. Towards the N. the view is intercepted by the sombre Schwarzhorn and other peaks of the Faulhorn chain.

The ROUTE TO THE FAULHORN (4 hrs.; see p. 180) diverges to the right close beside the hotel and cannot be mistaken in clear weather. The (⁸/₄ hr.) chalets of the *Oberläger of the Grindel-Alp*, where the Faulhorn view begins to open, are visible from the Great Scheidegg. Thence to the top, see p. 180.

As we descend from the Scheidegg, the church of Grindelwald is in sight below. Passing the *Obere Lauchbühl-Hütte* (5900'; ascent to the Scheidegg, ½ hr.), and at various other points of the way, we are greeted with a blast of the Alpine horn, an instrument of bark or wood, 6-8' long, the not unpleasing notes of which are echoed by the Wetterhorn. To the left of the Mettenberg, the Little and Great Schreckhorn and the Lauteraarhorn gradually become visible. In not less than 1 hr. we reach the *Hôtel Wetterhorn*, near the Upper Grindelwald Glacier. Thence to Grindelwald, ³/₄-1 hr., see p. 176.

52. From Meiringen to the Rhone Glacier. Grimsel.

23 M. DILIGENCE in summer twice daily in 7 hrs. (from the Rhone Glacier to Meiringen in 5½ hrs.), fare 9 fr. 30 c. (coupé 11 fr. 20 c.); to Göschenen in 13½ (Göschenen-Meiringen 11½) hrs., fare 19 fr. 15 c. (coupé 23 fr. 5 c.). Not more than 20 passengers are booked for each trip; no extra-post supplied on the Grimsel road. — One-horse carriage from Meiringen to the Rhone Glacier ('Gletsch') 35, two-horse 65, three-horse 90 fr. (to Guttannen 12, 22, 30 fr.; Handegg 17, 32, 40; Grimsel Hospice 27, 50, 65 fr.); from Meiringen to Andermatt 65, 120, 165, Göschenen 72, 135, 175, Fiesch 55, 100, 135, Brig 75, 140, 185 fr. — ON FOOT (10-11 hrs.): Im-Hof 1¼ hr., Guttannen 2¾ hrs., Handegg 1¾ hr., Grimsel Hospice 2½ hrs., summit of the Grimsel 1, Rhone Glacier ¾ hr. (in the reverse direction about 8½ hrs. in all).

Meiringen, see p. 181. The road crosses the *Aare* by the (½ M.) *Willigen-Brücke* (on the left the road to the *Gorge of the Aare*, p. 182) and ascends the **Kirchet** (2313′), a wooded hill, sprinkled with blocks of granite, which divides the valley into the *Lower* and *Upper* **Haslithal**. Near the top (1¼ M.) is the auberge '*Zur Lammi*', where the path from the Gorge of the Aare (p. 182) debouches.

The road descends the Kirchet in long windings (short-cuts), with views of the Gelmerhörner at the head of the valley and of the Ritzlihorn to the right, traverses the fertile basin of *Hasli im Grund*, and, at the inn *Zur Alpenrose (unpretending), crosses the Aare to (2¼ M.) **Im-Hof** (2054′; *Hôt. Hof*, with the dépendance *Alpenhof*, R. & L. 2-2½, pen s. 5-6 fr.), the principal village in the parish of *Innertkirchen*, where the Susten (p. 136) and Joch Pass (p. 134) routes diverge to the left.

Travellers from the Grimsel on their way to Rosenlaui and Grindelwald may go from Im-Hof direct, by *Geissholz*, to the Upper Reichenbach Fall (comp. p. 185; enquire for the beginning of the path).

The **Urbach-Thal** (comp. Map, p. 160), opening here towards the S.W., deserves a visit. The path ascends to the (½ hr.) narrow mouth of the valley, is then nearly level for 1 hr., and afterwards mounts steeply to the (2 hrs.) *Alp Schrättern* (4910′; beds), where the path to the *Dossen-Hütte* diverges to the right (see below), and to the (1 hr.) *Matten-Alp* (6102′), at the foot of the huge *Gauli Glacier*. In 1 hr. more we reach the new *Gauli Club-Hut* on the *Urnen-Alp* (7218′). Thence over the *Gauli Pass* (10,260′) to the Grimsel, combined with the ascent of the *Ewigschneehorn*, 8-9 hrs., fatiguing, but very grand (guide 35 fr.; see p. 189). — Over the *Bergli-Joch* (11,290′) to Grindelwald, 16-17 hrs. from Im-Hof, very toilsome and hardly repaying (guide 35 fr.). From the *Urnen-Alp* (where we pass the night) we ascend the *Gauli Glacier* to the pass, lying between the *Berglistock* (p. 177) and the *Rosenhorn*, and descend the *Grindelwaldfirn* to the *Gleckstein Hut* (comp. p. 177). — The *Dossen Hut* (p. 186) is reached in 4½-5 hrs. from the *Alp Schrättern* (see above), by the Alps *Illmenstein*, *Enzen*, and *Fläschen* (guide from Meiringen or Im-Hof 20 fr.). Thence to Rosenlaui, ascent of the Wetterhorn, and to Grindelwald, see p. 186. All these expeditions are for adepts only, with good guides. (At Innertkirchen, *Joh. Tännler, Heinr. & Ulrich Fuhrer, Joh. Moor, Joh. & Melch. Thöni*, etc.)

Beyond Im-Hof the road is at first level, and then gradually ascends on the right side of the fir-clad valley, running high above the rapid **Aare**. Beyond a short tunnel, over which a waterfall descends, it reaches the (3¼ M.) *Innere Urweid* (2464′; small inn), and beyond another tunnel arrives at (1¼ M.) *Im-Boden* (2933′), where it crosses the Aare by a new bridge.

3/4 hr. **Guttannen** (3480'; *Bär*, R., L., & A. 3, B. 1½ fr.) is the last village in the Oberhasli-Thal. The meadows are covered with stones, brought down by torrents and avalanches, and collected into heaps in summer to permit the grass to grow. (Over the *Furtwang Sattel* to the *Trift Glacier*, see p. 137; guide, *Andr. Sulzer*.)

About 1½ M. beyond Guttannen the road crosses the wild and foaming Aare by the *Tschingel-Brücke* (3733'). The valley becomes wilder, and barren black rocks rise on the right. Huge masses of débris testify to the power of avalanche and torrent. In another ½ hr. we recross to the left bank of the Aare by the (1½ M.) *Schwarzbrunnen-Brücke* (3976'). The stream becomes wilder and descends in noisy rapids. The road skirts the cliffs of the *Stäuben-den* and ascends the Handegg Saddle in long windings, often hewn in the rock. From the (2 M.) *Restaurant zum Handeggfall* we may reach (on the left) a view-point below the ****Handegg Fall**, and about 100 yds. from it. This cascade of the Aare, which descends amidst a cloud of spray into an abyss, 250' in depth, falls unbroken halfway to the bottom, and in its rebound forms a dense cloud of spray, in which rainbows are formed by the sunshine between 10 and 1 o'clock. The silvery water of the *Ærlenbach* falls from a height to the left into the same gulf, mingling halfway down with the grey glacier-water of the Aare. Passengers by the diligence may alight at the restaurant and rejoin the vehicle at the Hôtel Handegg. The road leads through a tunnel and, above the fall, crosses the Ærlenbach, near which is a view-terrace, and in 6 min. reaches the *Hôtel Handegg* (4570'), situated above the road, to the right.

The road now traverses the boulder-strewn Handegg-Alp, with a view of a fall of the *Gelmerbach*, which descends from the *Gelmersee* (5968'), a lake on the mountain to the left, between the *Gelmerhorn* and *Schaubhorn* (1½ hr. from the Handegg; rough path viâ the Hellenmad-Brücke). The old bridle-path diverges to the right at the Handegg-Alp and leads over rounded slabs of rock, called the *Helle* or *Hehle* ('slippery') *Platte*, worn by glacier-friction. The road crosses to the right bank of the Aare below a waterfall by means of the *Hellenmad-Brücke* and ascends in a wide curve. To the left, above us, is the Ærlen Glacier, with the rocky ridge of the Ærlengrätli appearing over it. The last dwarf-pines now disappear, and the road gradually ascends, with the brawling Aare below it. On the opposite bank appear the chalets in the *Räterichsboden* (5595'), and high up, to the left, is the *Gersten Glacier*. Beyond a wild defile traversed by the Aare, with interesting marks of glacier striation, the bridle-path joins the road on the right. The Zinkenstöcke with their glacier come into sight on the right; behind them, to the right, rise the Finsteraarhorn and the Agassizhorn; and farther on in the distance appears the Great Grindelwalder Fiescherhorn. In 2¼ hrs. from the Handegg the road reaches the —

Grimsel Hospice (6160'; *Inn, R., L., & A. 4-5, B. 1½, D.

5 fr.), situated at the W. end of the sombre little *Grimsel Lake*, in a desolate basin, enclosed by bare rocks with occasional patches of scanty herbage or moss.

EXCURSIONS from the Grimsel Hospice (comp. Maps, pp. 118, 160; guide, *Caspar Roth*). The *Kleine Siedelhorn* (9075'; 3 hrs.; guide 6 fr., not indispensable). We follow the old bridle-track (p. 190) for about 3/4 hr., then turn to the right, beyond the brook descending to the Grimsel Lake, at the point where the bridle-track cuts off the highest great curve of the carriage-road, and make for the height marked by a signal-cross (the Siedelhorn is not in sight), over pasture, debris, and rocks (no path at first). We keep somewhat to the right, as the signal-cross should later remain to the left. A distinct path leads over the ridge to the Siedelhorn. The last part of the mountain is covered with fragments of granite. The view is imposing. Gigantic peaks surround us on every side: to the W. the Schreckhorn, the Finsteraarhorn, and the Fiescherhörner; to the N.E. the Galenstock, from which the Rhone Glacier descends; to the S. the Upper Valais chain with its numerous ice-streams, particularly the Gries Glacier; to the S.W., in the distance, the Alphubel, Mischabel, Matterhorn, Weisshorn, etc. (comp. Dill's Panorama). — Travellers bound for *Obergestelen*, farther on, descend on the S.E. side of the mountain and there regain the bridle-path (guide advisable; comp. p. 190).

To THE PAVILLON DOLLFUS, 3 1/2–4 hrs. (there and back 7 hrs. guide 10 fr.), an easy and attractive ascent. The *Aare* is formed, to the W. of the hospice, by the discharge of two vast glaciers, the Unteraar and the Oberaar Glacier, which are separated by the *Zinkenstöcke*. The Unteraar Glacier is formed by the confluence of the *Finsteraar* and *Lauteraar Glaciers*, which unite at the foot (8286') of the rock-arête named '*Im Abschwung*', beyond a huge medial moraine, 100' high at places. At the foot of this arête the Swiss naturalist *Hugi* erected a hut in 1827. In 1841 and several following years the eminent naturalist Agassiz, with Desor, Vogt, Wild, and other savants, spent a considerable time here, dating their interesting observations from the 'Hôtel des Neuchâtelois', a stone hut on the medial moraine. These huts have long since disappeared. M. Dollfus-Ausset next erected the Pavillon Dollfus (7676') lower down, on the N side of the Lauteraar Glacier, now used as a club-hut (comp. p. 178). A bridle-path leads from the hospice across the stony *Aareboden* to (1 1/4 hr.) the foot of the Unteraar Glacier (6160'). Here we ascend the rocky slope to the right by a narrow path and then traverse the rocks and débris of the terminal moraine. After about 40 min. we take to the glacier, which affords good walking, pass several fine 'glacier-tables', and cross the medial moraine and the Lauteraar Glacier, which is here often considerably crevassed. Lastly we ascend a steep slope to the (1 hr.) Club-Hut, admirably situated on a rocky height overlooking the Unteraar Glacier. Opposite rise the Zinkenstöcke, Thierberg, Scheuchzerhorn, and Escherhorn; in the background, above the Finsteraar Glacier, the Finsteraarhorn; and to the right of the Abschwung the huge Lauteraarhörner and Schreckhörner. — We may continue our walk on the glacier as far as (3/4 hr.) the foot of the Abschwung (see above), where we enjoy a full view of the majestic Finsteraarhorn. In the medial moraine adjoining the Lauteraar Glacier, nearly opposite the Pav. Dollfus, is a fragment of rock bearing the names of 'Stengel 1844; Otz, Ch. Martins 1845', inscribed there during the scientific observations above referred to. The rock, re-discovered in 1884, was then about 2650 yds. from its original site.

The ascent of the *Ewigschneehorn* (10,930'; 4 1/4 hrs.) presents little difficulty to adepts. From the Pav. Dollfus across the Lauteraar Glacier to the foot of the mountain (8390') 1 1/2 hr., to the *Gauligrat* (10,260') 2 hrs., to the top 3/4 hr. (comp. p. 187).

The **Finsteraarhorn** (14,025'; guide from Hof or Meiringen 70, from Grindelwald 90, from the Concordia Hut 60 fr.), the highest of the Bernese Alps, was scaled for the first time in 1812, then in 1829 and twice in 1842, and has pretty often been ascended since. Travellers from the Grimsel spend the night in the (7 hrs.) *Oberaarjoch Hut* (see p. 190). The route then ascends to

III. Route 52. GRIMSEL PASS.

the *Gamslücke* (c. 11,150') between the Rothhorn and Finsteraarhorn, and skirts the W. flank of the latter to the *Hugi-Sattel* (13,205') and the top (7-9 hrs.). This is the most advisable route. On the ascent from Grindelwald, the *Schwarzegg Hut* (p. 178) affords night-quarters; thence to the top in 9-10 hrs., over the *Finsteraar-Joch* (11.122') the *Agassiz-Joch* (12,630'; beside which rises the steep *Agassizhorn*, 12,960'), and the *Hugi-Sattel*. It is by no means advisable to descend by this route, which is dangerous from falling stones. If the Eggishorn be the starting-point, the night is spent in the (5 hrs.) *Concordia Hut* (p. 305), from which we ascend to the summit in 8 hrs. over the *Grünhornlücke* (10,843'), the *Walliser Fiescherfirn*, and the *Hugi-Sattel*. The expedition is for experts only, with first-rate guides. Even when the ice is favourable the ascent is difficult and very trying.

From the Grimsel to the Furka direct over the **Nägeli's Grätli** (8470'), 5½ hrs. (guide 10 fr.), a fine walk, though fatiguing, for good walkers preferable to the Grimsel, see p. 126.

From the Grimsel to Fiesch, or to the Eggishorn (p. 316), over the *Oberaar-Joch*, 13 hrs., fatiguing, but interesting (two guides, 40 fr. each, including the Oberaarhorn 50 fr. each). We ascend the *Oberaar Glacier* in 7 hrs. to the finely situated and well-appointed *Oberaarjoch Hut* of the S. A. C. (10,430') on the Oberaar-Joch (10,625'), lying to the S. of the *Oberaarhorn* (11,953'; which experts may scale from the hut in 1½ hr.). We next descend the *Studerfirn*, passing the *Rothhorn* (11,345'), and then either cross the difficult and sometimes dangerously crevassed *Fiesch Glacier* to the *Stock-Alp* (p. 310) and to the *Hôtel Jungfrau-Eggishorn* (p. 316; 7 hrs. from the club-hut), or, preferably, descend by the *Grünhornlücke* (see above) to the *Concordia Hut* (p. 305), and thence cross the *Great Aletsch Glacier* to the Hôtel Eggishorn. — Over the Oberaar-Rothjoch (10,906'), to the S. of the Oberaar-Joch, not difficult. — Over the Studer-Joch to Fiesch, 14-15 hrs., difficult. The route ascends the *Unteraar* and *Finsteraar Glaciers* to the Studer-Joch (11,550'), between the *Oberaarhorn* (see above) and the *Studerhorn* (11,935'; a splendid point of view, easily attained from the pass in ¾ hr.). Descent over the *Studerfirn* and the *Fiesch Glacier*, as above.

From the Grimsel over the *Strahlegg* and the *Finsteraar-Joch* or *Lauteraar-Joch* to *Grindelwald*, p. 178; over the *Triftlimmi* to the *Trift-Hütte*, p. 137.

From the Hospice the bridle-path, which pedestrians follow, ascends **direct**. The carriage-road skirts the Grimsel Lake, and, with a retrospect of the Schreckhorn, winds up the (1 hr.) **Grimsel Pass** (7103'), which connects the Haslithal with the Upper Valais and marks the boundary between the Canton of Bern and the Valais. The small *Todtensee* ('lake of the **dead**'), on the Valais side, recalls the struggle in 1799 between the Austrians and the French advancing from the Haslithal.

A footpath to the right, at the uppermost bend on the top of the pass, ascends over a stony tract to the height of 7230' and then descends to (2 hrs.) *Obergestelen* (p. 315; in the opposite direction 2½-3 hrs.; guide, 4 fr., advisable in unsettled weather). — Those who have seen the Rhone Glacier and intend to climb the *Kleine Siedelhorn* (p. 189) do not ascend direct from the summit of the pass, but follow the road for some distance beyond the curve on the Bern side before diverging to the left.

From the pass the road descends the *Maienwang*, a steep grassy slope carpeted with rhododendrons and other Alpine plants, in view of the imposing Rhone Glacier, the Dammastock, and the Galenstock. The bridle-path (shorter) is in bad preservation. The (1 hr., up 1½ hr.) *Rhone Glacier Hotel* (5750'), see p. 314. Thence to *Brig*, see R. 81; over the **Furka** to *Andermatt*, R. 35.

53. From Spiez to Leuk over the Gemmi.

14 hrs. DILIGENCE daily from Spiez to (17½ M.) Kandersteg in 6 hrs. (5 fr. 65, coupé 7 fr. 75 c.). One-horse carriage to Frutigen 10, two-horse 18 fr., to Adelboden 18 and 32, to the Blaue See 12 and 22, to Kandersteg 18 and 32, with use of the horse for riding to the Gemmi 30 and 55 fr. — From Kandersteg an admirably kept bridle-path leads over the *Gemmi*, one of the grandest and most frequented of the Alpine passes, to the Baths of Leuk (5½ hrs.; walk; guide unnecessary). — Carriage-road from Leuk to the Rhone Valley (2½ hrs.; walk down, 3½ up).

From *Spiez* to (4 M.) *Emdthal*, where the road to the Heustrich-bad diverges to the right, see p. 154. Our road descends to the *Kander*, with a fine view of the Blümlisalp at the head of the Kienthal. We cross the *Suldbach* before reaching (25 min.) **Mülenen** (2260'; *Dr. Luginbühl's Pension*, 4-5 fr., unpretending; *Bär*).

The road forks, the right branch being the shorter. The diligence passes through (5 M.) **Reichenbach** (2335'; *Bär*, plain), at the mouth of the *Kienthal*.

A narrow road ascends the attractive Kienthal, affording fine views of the Büttlassen, Gspaltenhorn, and Blümlisalp, to the (4 M.) village of *Kienthal* (rustic inn); cart-road thence to (8½ M.) the extensive *Tschingel-Alp* (5783'), 10 min. from which is the *Pochtenbach Fall* with the interesting *Hexenkessel*, a kind of 'glacier mill' (guide advisable). Thence over the *Sefinen-Furgge* to *Mürren* (8-9 hrs.), and over the *Hohthürli* to *Kandersteg*, see p. 170. To the E. the valley is closed by the crevassed *Gamchi Glacier*, the source of the *Pochtenbach*. Experts with able guides will find it interesting to cross the Gamchilücke (9295'), between the Blümlisalp and the Gspaltenhorn, to the *Tschingelfirn* (p. 170). We may then either cross the *Petersgrat* to Ried in the Lötschen-Thal (p. 171), or the *Tschingel Pass* to Kandersteg (p. 170), or the *Tschingeltritt* to Lauterbrunnen (p. 170). Distances: from the Tschingel-Alp to Steinenberg 1 hr., end of the Gamchi Glacier 1½ hr., Gamchilücke 2½, Ried 6-7, Kandersteg 6, Lauterbrunnen 4 hrs. — Ascents from the Kienthal: Büttlassen (10,490'; guide 25 fr.), from the *Dürrenberg-Hütte* (2½ hrs. above the Tschingel-Alp, see p. 170), 3½-4 hrs., toilsome, but repaying. — Gspaltenhorn (11,275'; guide 70 fr.), reached by the *Leitergrat* between the Büttlassen and the Gspaltenhorn, very difficult (first scaled by Mr. Foster in 1869). — Wilde Frau (10,693'), from the *Dünden Hut* (p. 193) and up the *Blümlisalp Glacier*, 3 hrs., laborious.

The road crosses the Kander (fine view up the Kienthal to the left to the Blümlisalp), and beyond (8 M.) *Wengi* reaches —

9½ M. **Frutigen** (2717'; pop. 4021; *Bellevue*, with view, R., L., & A. 2½, B. 1½, D. 3-3½, pens. from 5 fr.; *Adler*; *Helvetia*, R. 1½-2½, D., incl. wine, 3½-4, pens. 5 fr.), a village situated in a fertile valley on the *Engstligenbach* (p. 196), which falls into the Kander lower down. Matches are largely manufactured here. From the church and other points we obtain beautiful views of the Kander-Thal, the Balmhorn, the Altels, etc., and of the Ralligstöcke (p. 157).

A still more extensive view is commanded by the *Ueblenberg* (4780'), to the N.W., 1¼ hr. above the village. — The *Gerihorn* (6695'; 3½-4 hrs.; guide not indispensable) is an easy and attractive ascent. — The road to Adelboden (p. 196) diverges to the right at Frutigen and ascends the valley of the Engstligenbach.

Our road crosses the Engstligenbach and turns into the Kander-

Thal on the left, between the Gerihorn on the left, and the **Elsighorn** on the right. In front appear the Balmhorn and Altels. At the (1 M.) ruins of the *Tellenburg* we cross the Kander (walkers may follow the left bank), **traverse** the pleasant *Kandergrund*, and finally ascend the new road, leaving the church of *Bunderbach* (2880') on the left, to the (3½ M.) *Hôtel-Restaurant Blauseehöhe*.

About ½ M. to the right is the *Blaue See, picturesquely embosomed in wood, and remarkable for its brilliant colour (best by morning-light; adm. 1 fr.). *Pension* on the bank of the lake for resident guests only.

Near (4½ M.) *Mittholz* (3154') we pass the ruined *Felsenburg*. We then ascend the *Bühlstutz* in windings (old road shorter; fine view of the Blümlisalp at the top) to the district of Kandersteg, pass the (3 M.) *Bühlbad* (3885'; *Inn, R. 1½, D. 2½, pens. 4½–5 fr.), and reach (¾ M.) —

17½ M. **Kandersteg** (3840'). — Hotels. *HÔT. VICTORIA, R., L., & A. 2½-5, B. 1½, lunch 2½, D 4, board 6 fr.; *HÔT. GEMMI, *BEAR, similar charges, both in *Eggenschwand*, 1¼ M. farther on (see p. 193); ALPENROSE, to the N. of the Hôt. Gemmi, unpretending; PENS. J. REICHEN. — **Guides** (*Abraham Müller; Hans Ogi-Müller; Fritz Ogi; Christian Hari; Joh. Künzi*): to Schwarenbach (unnecessary; 2 hrs., descent 2 hrs.) 5 fr.; to the Gemmi (summit of the pass 1, descent ¾ hr.) 7 fr.; to the Baths of Leuk (1½, descent 2½ hrs.) 10 fr. — Horse to Schwarenbach 10, to the Gemmi 15 fr. (the descent on horseback to the Baths of Leuk is prohibited). — **Carriages** (return-vehicles cheaper): one-horse to Frutigen 10, two-horse 18 fr.; Spiez, 18 or 32; Thun, 22 or 40; Interlaken, 25 or 45 fr. — *English Church* near the Hôtel Victoria.

A grand panorama is disclosed between Bühlbad and the Hôtel Victoria : to the N.E. is the jagged Birrenhorn; to the E. the glistening snow-mantle of the Blümlisalp or Frau, the beautiful Doldenhorn, and the barren Fisistöcke. Farther on the snow-peaks disappear, leaving only the Gellihorn and some other rocks at the end of the valley in sight. On the W. side of the valley is an old moraine. The road comes to an end at the Bear Hotel in *Eggenschwand*, 1½ M. beyond the Hôtel Victoria.

To the E. lies the interesting OESCHINEN-THAL, containing the beautiful *Oeschinen-See (5223'), 1 M. in length. The path to it (1½ hr.; guide 4 fr., unnecessary; horse 8 fr.) diverges to the left by the Hôtel Victoria, ascends for 50 min. on the left bank of the *Oeschinenbach*, partly through wood, then **crosses** to the right bank (pretty waterfall to the right), and descends to the lake (Hôt.-Pens. Oeschinensee, well spoken of, 4-5 fr.). Above the lake **tower** the huge, snow-clad *Blümlisalp*, *Fründenhorn*, and *Doldenhorn*, from the precipices of which fall several cascades. A row on the lake is very enjoyable (to the gorge at the S.E. angle and back 1 hr.). Walkers may proceed round the lake to the left as far as the *Berglibach*, opposite the glaciers. Thence to the *Oeschinen-Alp* and over the *Dündengrat* into the *Kienthal* (guide to Reichenbach, 20 fr.), see p. 170.

The **Blümlisalp** or **Frau**, a huge mountain-group, covered on the N. side with a dazzling mantle of snow, and on the S. side descending in bold precipices to the Kandergletscher, culminates in three principal peaks. To the W. is the *Blümlisalphorn* (12,042'), the highest; in the centre is the snowy peak of the *Weisse Frau* (12,012'); and to the E. is the *Morgenhorn* (11,894') with the lower *Wilde Frau* (10,693'; p. 191), *Blümlisalpstock* (10,562'), *Blümlisalp-Rothhorn* (10,828'), and *Oeschinenhorn* (11,450'). The Blümlisalphorn was first ascended by Mr. Leslie Stephen in 1860, the Weisse Frau by Dr. Roth and Hr. E. v. Fellenberg in 1862, and both have frequently been ascended since. (Both toilsome, but very interesting; guide, 50 fr. for each. The night is spent in the *Dünden Club Hut* on the Dünden Pass; thence up the *Blümlisalp Glacier*, 4-5 hrs. to the summit.) — The **Doldenhorn** (11,965; guide, 40 fr.), first ascended by Messrs. Roth and Fellenberg in 1862 (from Kandersteg by the *Biberg Alp* in 8 hrs.), is difficult. — The **Fründenhorn** (11,030'; guide 40 fr.), first ascended in 1871 by Messrs. Ober and Corradi (from Kandersteg by the *Alp In den Fründen*, 10½ hrs.), is also difficult. — Interesting but toilsome passes lead from the Oeschinen-Thal to the Kander Glacier, across the *Oeschinen-Joch* (about 10,430'), between the Oeschinenhorn and the Fründenhorn, and across the *Fründen-Joch* (about 10,030'), between the Fründenhorn and the Doldenhorn.

The *Dündenhorn or *Wittwe* (9410'; guide 20 fr.), ascended from Kandersteg by the *Obere Oeschinen-Alp* in 6 hrs., rather difficult, for experts only, affords a splendid survey of the Blümlisalp group. We may then follow the arête to the *Dünden Hut* (see above), and descend thence to Kandersteg (13-14 hrs. in all).

The wild **Gastern-Thal**, from which the Kander descends in picturesque falls, deserves a visit (3/4-1 hr.). A good path, diverging between the Bear and Gemmi hotels, skirts the left bank and ascends steeply through the *Klus* (p. 199) to the upper part of the valley, bounded on the S. by the precipices of the Tatlishorn and **Altels**. Splendid fall of the *Geltenbach*.

The *Alpschelenhubel (7385'; 3 hrs.; guide advisable, 8 fr.), to the W of Kandersteg, presents an easy and attractive ascent. We diverge to the right from the Gemmi road, 7 min. from the Bear Hotel, ascend via the *Alpbach* and the *Oeschinen-Thal* to the (1 hr.) *Ueschinen-Alp* (p. 197), and thence to the right by the Bonderkrinden route (p. 197; steep at places but perfectly safe) to the *Alpschelen-Alp* (6870'). Thence to the (2 hrs.) summit, over pasture to the N.E. (fine view).

From Kandersteg over the *Bonderkrinden* to *Adelboden*, see p. 197 (guide 10 fr.); **over** the *Lötschen Pass* to *Gampel* (in the Valais), see R. 55 (guide 20 fr.); over the *Tschingel Pass* to *Lauterbrunnen*, see p. 170 (guide 30 fr.; preferable in the reverse direction, as there are no inns in the Gastern-Thal, and the ascent thence is very long and fatiguing). — Over the *Petersgrat to the *Lötschen-Thal* (11-12 hrs. from Kandersteg to Ried; guide 40 fr.), a very fine route. We follow the Tschingel Pass route to the top of the Kanderfirn; then turn to the right and ascend snow-slopes to the pass on the *Petersgrat* (10,515'; splendid view). Descent through the *Faflerthal* or *Tellithal* to *Ried* (comp. p. 171).

Beyond the *Bear Hotel* (p. 192) the road contracts to a well-kept bridle-path, and ascends straight in the direction of the *Gellihorn* ('Mittaghorn'; 7530'), which closes the Kander-Thal. On the right is the *Alpbach*, issuing from the *Ueschinen-Thal*, with several small falls. The path ascends in windings on a slope at the base of the Gellihorn for about 1¾ hr., and then leads through pine-forest high above the Gastern-Thal (p. 199) and then above the *Schwarzbach Valley*, affording fine views of the Fisistock, Doldenhorn, etc. About ¾ hr. farther on, we reach the *Spitalmatte* (6250'), a pasture which was entirely devastated in Sept., 1895, by a rupture of the glacier covering the slopes of the *Altels* (11,930'), to the left. Between the Altels and the black rocky peak of the *Kleine Rinderhorn* (9865'; adjoining which is the snow-clad *Grosse Rinderhorn*, 11,372'), lies imbedded the **Schwarz Glacier**, drained by the *Schwarzbach*. We next traverse a stony wilderness to the (½ hr.) **Inn** on the **Schwarenbach** (6775'; R., L., & A. 3¼, B. 1½ fr.).

The *Balmhorn (12,180'), ascended in 5-6 hrs., over the *Schwarz Glacier* and the *Zagengrat* (toilsome, but free from danger; guide 30 fr.), affords a magnificent panorama of the Alps of Bern and the Valais, extending to N. Switzerland. — The **Altels** (11,930') is also interesting (5-6 hrs.; guide

25 fr.; much step-cutting necessary when there is little snow). Those who are not subject to dizziness may combine the Balmhorn with the Altels (guide 50 fr.). — The Wildstrubel (10,670'; guide 25, with descent to Leuk 35 fr.), ascended from the Gemmi over the *Lämmern Glacier* in 4-4½ hrs., is fatiguing, but repaying (comp. p. 200).

We next reach the (½ hr.) shallow *Daubensee* (7265'), a lake 1¼ M. long, fed by the Lämmern Glacier (see below), with no visible outlet, and generally frozen over for seven months in the year. The path skirts the E. bank of the lake for nearly ½ hr., and, 10 min. beyond it, reaches the summit of the pass, the **Gemmi** or **Daube** (7553'; *Hôtel Wildstrubel*, R., L., & A. 3-3½, B. 1½, lunch 3, D. 3½, pens. 9 fr.), at the base of the *Daubenhorn* (9685'), commanding a magnificent *View of the Alps of the Valais (panorama by Imfeld). The mountains to the extreme left are the Mischabelhörner (Balfrinhorn, Ulrichshorn, Nadelhorn, Dom, and Täschhorn); more to the right and farther off rise Monte Rosa, the Barrhorn, the Brunegghorn; in the centre, the huge Weisshorn, the Zinal-Rothhorn, the Ober-Gabelhorn, the blunt pyramid of the Matterhorn, the Pointe de Zinal, the Dent Blanche, the Bouquetins, and the Dents de Veisivi. To the right of the Daubenhorn is the range of the Wildstrubel, with the Lämmern Glacier, and far below lie the Baths of Leuk. Rich flora.

About 4 min. below the pass we reach the brink of an almost perpendicular rock, 1660' high, down which, in 1736-41, the Cantons of Bern and Valais constructed one of the most curious of Alpine routes, nowhere less than 5' in width. The windings are hewn in the rock, often resembling a spiral staircase, the upper parts actually projecting at places beyond the lower. The steepest parts and most sudden corners are protected by parapets. Distant voices reverberating in the gorge sometimes sound as if they issued from its own recesses. The descent on horseback is now prohibited; a marble cross, ¼ hr. from the top, commemorates an accident to a rider. At the foot of the cliff succeeds a slope of debris, the lower part of which is covered with firs. The descent from the pass to the Baths takes 1½ hr. (ascent 2½ hrs., of which 1½ hr. represents the ascent of the cliff).

Baths of Leuk. — Hotels. ⁕HÔTEL DES ALPES, R. & A. 3½, B. 1½, lunch 3½, D. 4½, pens. 7-10 fr.; ⁕MAISON BLANCHE, with its dépendance GRAND BAIN; ⁕HÔTEL DE FRANCE, R., L., & A. 4, B. 1½, D. 4 fr.; ⁕UNION, R., L., & A. 3, D. 4 fr.; ⁕HÔT. DES FRÈRES BRUNNER, R., L., & A. 3, D. 3½, pens. 5-6½ fr.; ⁕BELLEVUE, R., L., & A. 2-3, B. 1 fr. 30, lunch 2½, D. 3. pens. 5 fr., recommended to passing tourists; ⁕GUILL. TELL, similar charges; RÖSSLI, unpretending, R., L., & A. 1½, B. 1, D. 2, pens. 4-5 fr. — BEER at the *Maison Blanche*, *Bellevue* (Cursaal), and *Restaurant des Touristes* (opposite the Hôt. Tell). — **Horse** to Kandersteg 20, Schwarenbach 12, Gemmi 8 fr.; *Porter* to Kandersteg 10, Schwarenbach 6, Gemmi 4 fr. — **Diligence** (from the Hôtel de France) to the Leuk station every forenoon in summer in 2 hrs. (3 fr. 95 c.); one-horse carr. 12-15, two-horse 25 fr. — *English Church*.

Bad Leuk (4630'), Fr. *Loëche-les-Bains*, locally known as *Baden*, a village (620 inhab.) consisting chiefly of wooden houses and the large hotels and bath-houses, lies on green pastures in a valley opening to the S., and watered by the *Dala*. In July and August the baths are much frequented by French, Swiss, and Italian visitors. In the height of summer the sun disappears about 5 p.m. The huge, perpendicular wall of the Gemmi presents a weird appearance by moonlight. The *Thermal Springs* (93-123° Fahr.), impregnated with lime, about 22 in number, are chiefly beneficial in cases of cutaneous disease and rheumatism. The bath-houses (*Grosses Bad*, *Neues Bad*, *St. Lorenz-Bad*, and three others) are connected with the hotels, and contain both private and common basins, in which the patients under full treatment spend several hours daily. Spectators are admitted to the galleries of the common basins, where they are expected to contribute a small sum 'pour les pauvres'. The loud and animated conversation of the patients, who appear to enjoy excellent spirits, is chiefly in French. Small tables or trays float upon the water, bearing cups of coffee, newspapers, books, and other means of passing the time. The baths are open from 5 to 10 a.m. and from 2 to 5 p.m. — The *Cur-Promenade*, an avenue $1/2$ M. in length, leading from the Neue Bad past the Hôtel Bellevue, is frequented in the morning by the patients drinking the waters and in the afternoon by promenaders (music).

Excursions. A walk leads from the end of the Cur-Promenade to the (20 min.) foot of a lofty precipice on the left bank of the Dala. Here we ascend by eight rude *Ladders* (échelles), attached to the face of the rock, to a good path at the top, which leads in 1 hr. to the village of *Albinen*, or *Arbignon* (4252'). The fine view obtained from a projecting rock above the second ladder will repay the climber; but persons liable to dizziness should not attempt the ascent. The descent is more difficult. — Excursions may also be made to the *Fall of the Dala*, $1/2$ hr. to the N.E., above Leuk; to the *Foljeret* or *Feuillerette Alp* (5850'), $3/4$ hr. to the E., with fine view of the Altels, Balmhorn, and Gemmiwand; to the *Fluh Alp* (6710'), $2^{1}/_{2}$ hrs. and to the *Torrent Alp* (6345') via the *Wolfstritt*, $1^{1}/_{2}$ hr.

The *Torrenthorn* (9852'; $4^{1}/_{2}$ hrs.) commands a magnificent view of the Bernese and Valaisian Alps; bridle-path nearly to the summit (horse 15 fr.; guide desirable, 10 fr.). About $1^{1}/_{2}$ hr from the summit and $2^{1}/_{2}$ hrs. from the Baths is a new *Hotel*, commanding fine views. The route may be varied by descending across the *Majing Glacier* (guide indispensable). Travellers from the Rhone Valley save considerably by going direct from the town of Leuk (p. 198) to Albinen, and thence with a guide by *Chermignon* (6284') to the Torrenthorn, whence they may descend to the Baths of Leuk. The descent by the above-mentioned ladders, which is usually chosen by the guides, should be avoided, especially in wet weather. The *Galmhorn* (8080'), near Chermignon, is also frequently ascended ($2^{1}/_{2}$ hrs. from the Baths, by the Torrent Alp). Those who do not care to ascend higher will be repaid by a visit to Chermignon, which affords a capital survey of the Rhone Valley and the Valaisian Alps. — Passes: To the LÖTSCHEN-THAL over the *Gitzifurgge*, or to KANDERSTEG over the Gitzifurgge and the *Lötschen Pass*, laborious (comp. p. 198). To the Lötschen-Thal over the *Ferden Pass*, interesting and not difficult (comp. p. 198). To ADELBODEN over the *Engstligengrat* (7-8 hrs.), repaying (p. 197).

The road to Leuk crosses the *Dala* immediately below the Baths (retrospect of the Rinderhorn and Balmhorn) and descends on the right bank to (3 M.) Inden (3730'; *Restaurant des Alpes*), whence pedestrians should follow the shorter bridle-path to the left. The road, after following the slope a little farther, descends in windings, and recrosses the (1½ M.) torrent by a handsome bridge affording fine views of the ravine.

Pedestrians bound for SIERRE (p. 309) take the old road, which diverges to the right from the above road, below the last curve and about 500 yds. before the bridge, passes through several tunnels, and gradually descends the slope by *Varen* and *Salgesch* (to Sierre 2 hrs.).

The road quits the Dala ravine about 1¼ M. farther on at a point high above the Rhone Valley, of which a beautiful view is disclosed. From the angle (2998′) pedestrians follow the fingerposts direct to **Leuk**, or *Loêche-Ville* (2470′; p. 309), while the carriage-road describes a curve of nearly 2½ M. From the town to **Leuk Station** (2044′; p. 309), about 1 M.; from the Baths of Leuk to the station is a walk of 2-2½ hrs.

54. The Adelboden Valley.
Comp. Map, p. 192.

From Spiez to (19½ M.) *Adelboden*, DILIGENCE daily in 6½ hrs. (5 fr. 40 c.), at 7 a. m. (from Frutigen at 10.30). Carriage with one horse 18, with two horses 32 fr., from Frutigen 10 and 18 fr. — The verdant Adelboden Valley, watered by the *Engstligenbach*, is one of the most attractive upland valleys in the Oberland. The upper end of the valley, shut in by the Lohner and the Wildstrubel, presents imposing scenery, while the village of Adelboden is a convenient centre for numerous shorter and longer excursions, and is much frequented as a summer-resort.

Frutigen (2717′), see p. 191. The new road gradually ascends on the left bank of the *Engstligen*, crossing several impetuous tributary brooks descending from the wooded mountain-slopes on the right, and passes beneath the *Linterfluh* (slate quarries). At (5½ M.) *Rinderwald* it crosses to the right bank by means of a bold bridge, and passes the inn of *Steg* and the *Pochtenkessel* (2 min. below the road, see below) to *Hirzboden*, where it returns to the left bank near the *Hospital for the Poor*. It continues to ascend to (4½ M.; 10 M. from Frutigen) Adelboden (4450′; *Hôt.-Pens. Wildstrubel*, R., L., & A. 3, D. 3½, S. 2½, pens. 7½-8½ fr.; *Curhaus*, prettily situated above the village, R., L., & A. 2-4, B. 1, déj. 2, D. 3½, pens. 6-9 fr.; *Pens. Edelweiss*, pens. 5-6 fr.; *Adler*, *Pens. Hari*, both unpretending; English Church Service in summer), beautifully situated on a sunny terrace, 400′ above the Engstligenbach, with interesting old timber buildings and an old church containing mediæval frescoes. Huge maple-tree in the churchyard. Pine-forests in the vicinity.

EXCURSIONS (guides; *G. Fühndrich*, schoolmaster; *Joh. Pieren*, *David Spori*, *Sam. Zryd*). SHORT WALKS: To the N., through the *Aeusser-Schwand* to the (¾ hr.) Bütscheggen (4480′), at the mouth of the *Tschenten-Thal*, commanding a view of the Frutig valley and the Niesen chain. The *Hörnli* (4910′), 1½ hr. farther up towards the Tschenten-Alp, commands a still more extensive view. — To the (1 hr.) Köleren Gorge, in the Tschenten-Graben, with a curious grotto excavated by the Tschenten-Bach (entrance from below). — To the (1½ hr.) Pochtenkessel, a deep gorge of the Engstligenbach near the inn of Steg (see above), 2 min. below the road to Frutigen. — To the (1 hr.) **Wettertanne** or **Schermtanne** in the *Allenbach-Thal*, viâ *Stiegelschwand*, at the foot of the tremendous precipices of the Albrist and Gsür. — To the **Bonderlen-Thal** and the *Lohner Waterfalls* (2 hrs. to the foot of the cliffs of the Lohner), a charming Alpine glade and a beautiful cascade. Farther up towards the Bonder-Alp are abundant rhododendrons. — To the (2 hrs.) ***Engstlig Falls**, a copious waterfall, 490' high, in two leaps (the ascent to the imposing upper fall not advisable for novices). To the *Engstlig-Alp*, see below. — SHORT ASCENTS: To the **Kunisbergli** and **Höchst** (5380'), 2½ hrs., viâ the farm of *Boden*, a picturesque Alp, with rhododendrons; the Höchst commands a **view of the Adelboden valley** (guide 3 fr., not indispensable). — To the **(2 hrs.) Schwandfeldspitze** (6660'; good view), above the village to the W. (guide 4 fr., not indispensable). — To the **Regenbolshorn** (7200'; **3 hrs.**; guide 6 fr.), to the left of the Hahnenmoos (see below), attractive. — **To the** (3½-4 hrs.) ***Laveigrat** (7952'; guide 6 fr.), viâ the *Alp Sillern* and **along the** *Sillern Grat;* fine view of the Bernese Alps and the Vaud and Freiburg mountains. At the W. foot of the mountain are the Baths of **Lenk**.

LONGER MOUNTAIN TOURS: ***Bonderspitz** (8360'; 4 hrs.; guide 8 fr.) and **Elsighorn** (7695'; 5 hrs.; guide 8 fr., not indispensable), two easy and interesting ascents. On the *Elsigalp* is a small lake, with stone-pines in the vicinity. — ***Albrist** (9065'; 5-6 hrs.; guide 12 fr.), not difficult; fine view of the Bernese and Valaisian Alps. The ascent leads viâ the elevated *Furggi-Alp* (6885'), and an attractive descent may be made viâ the *Hahnenmoos* (guide 15 fr.). — **Gsür** (8895'; 5 hrs.; guide 12 fr.), viâ *Schwandfeld*, difficult, for experts only; fine view of the Bernese Alps. — **Gross-Lohner** (10,020'; 7-8 hrs.; guide 30 fr.), a fatiguing ascent, adapted only for experts; fine view. — **Wildstrubel** (*Gross-Strubel*, or E. summit, 10,670'; 8-10 hrs.; guide 30 fr.), an interesting glacier expedition, not difficult for adepts, viâ the *Engstlig-Alp* (see below), and the *Strubelegg* (9610'). The summit commands an imposing view of the entire chain of the Valaisian Alps, the Mont Blanc group, the Lämmern Glacier, the Plaine Morte, etc. The descent may be made over the crevassed *Lämmern Glacier* to the *Gemmi* (p. 194; guide 40 fr.), or over the *Plaine Morte* to *Lenk* (p. 199). — **Felsenhorn** (9175'; 7 hrs.; guide 15 fr.), viâ the *Engstligen-Grat* (see below), a very interesting expedition, with a fine view of the neighbourhood of the Gemmi, and of the Bernese and Valaisian Alps. — ***Männlifluh** (8705'), viâ *Rinderwald* and *Otterngrat* (pass to Diemtigen, 7220'), 5½ hrs., also interesting.

PASSES. To LENK a path, marshy at places, leads over the **Hahnenmoos** (6410'), passing a large dairy establishment near the top, in **4-5** hrs. (guide 8, horse 15 fr.). Beautiful view, during the descent, of the upper Simmen-Thal, the Wildstrubel, the Weisshorn, and the Räzli Glacier. In the reverse direction 1-1½ hr. longer.

FROM ADELBODEN TO KANDERSTEG, an interesting route over the **Bonderkrinden** (8300'; 6 hrs.; guide 10 fr.), with which the ascent of the *Bonderspitz* (see above) may be conveniently combined. — To SCHWARENBACH, somewhat fatiguing (8-9 hrs.; guide 15 fr.), viâ the *Bonderkrinden*, *Ueschinen-Thal*, and *Schwarzgrätli* (see below). — To SCHWARENBACH OVER THE ENGSTLIGENGRAT, 7-9 hrs., with guide (15 fr.), a fine route. From Adelboden we ascend to the S., passing the *Engstlig Falls* (see above) or by the steep and stony 'Fahrweg' (a digression of 1 hr.), to the (3 hrs.) *Engstlig Alp* (6360'; small inn), a wide Alpine basin at the base of the *Wildstrubel* (see above). We then cross the (2 hrs.) **Engstligen-Grat**, passing the serrated *Tschingelochtighorn* (8990'), and descend into the *Ueschinen-Thäll*, with its little lake (far below to the left lies the *Ueschinen-Thal*). Then to the left, over the *Schwarzgrätli* (see above), to (2 hrs.) *Schwarenbach* (p. 193); or we may traverse the *Ueschinen-Thäli Glacier*, on the W. side of the *Felsenhorn* (9175'), and descend through the *Rothe Kumme* to the *Daubensee* and (4 hrs.) *Gemmi Pass*. The route passes through a rich Alpine flora, with abundant edelweiss.

55. From Gampel to Kandersteg. Lötschen Pass.

Comp. Map, p. 192.

This excursion (12 hrs.) is suited for good walkers only, in fine weather Guide from Ferden or Ried to Kandersteg necessary (15, from Gampel 20 fr.). The *Lötschen-Thal* itself deserves a visit. A rough and steep cart-road leads to Goppenstein; thence to Ried and Gletscherstaffel a bridle-path.

From **Gampel** (2100'; *Hôtel Lötschenthal*), on the right bank of the Rhone, 1 M. to the N. of the station of that name (p. 309), the road ascends the *Lötschen-Thal*, or gorge of the *Lonza*, which is much exposed to avalanches. Mounting rapidly at first, it passes the chapels of (1 hr.) *Mitthal* (3425') and (½ hr.) *Goppenstein* (4035'). Beyond Goppenstein the bridle-path crosses the (¼ hr.) *Lonza*, where the valley expands, and leads to (1 hr.) *Ferden* (4557') and (¼ hr.) *Kippel* (4514'; bed at the curé's). It then ascends gradually by *Wiler* to (40 min.) **Ried** (4950'; *Hôt. Nesthorn*, unpretending), finely situated at the N.W. base of the *Bietschhorn* (12,965').

EXCURSIONS. (Guides, *Jos. Rubin, Jos. Kalbermatten*, etc.) The **Hohgleifen** (*Adlerspitze*, 10,828'; 6-7 hrs., with guide) is not difficult. Superb view of the Valaisian Alps from the Simplon to Mont Blanc, the W. Bernese Alps, the Lötschen-Thal and Rhone Valley, and to the E. in the foreground the huge Bietschhorn. — The **Bietschhorn** (12,965'; 9 hrs., guide 60 fr.), first ascended by Mr. Leslie Stephen in 1859, is very fatiguing and difficult, and fit for experts only. The night is spent in the *Club-Hut* on the *Schafberg* (8440'), 3 hrs. from Ried.

The following ascents may also be made from Ried: °*Lauterbrunnen-Breithorn* (12,400'; 7-8 hrs., guide 3 ¹ fr.), not difficult for experts; ° *Hockenhorn* (11,817'; 5½-6½ hrs.; guide 8 fr.), not difficult (see below); *Tschingelhorn* (11,748'; over the Petersgrat in 6 hrs., guide 20 fr.), not difficult; and *Grosshorn* (12,352'; 8 hrs.; guide 35 fr.), not difficult for experts.

PASSES. Over the *Petersgrat* (10,515') to *Lauterbrunnen* (12 hrs.; 25 fr.), fatiguing but highly interesting, see **p. 171**. — *Wetterlücke* (10,305') and *Schmadri-Joch* (10 863'), both difficult, see p. 171. — Over the *Lötschenlücke* to the *Eggishorn*, see p. 318; over the *Beichgrat* to the *Belalp*, p. 311 (accommodation in the Chalet Seiler, on the *Fafler Alp*, see p. 171).

Over the **Baitschieder-Joch** (about 11,150') to the Rhone Valley (from Ried to Visp 12 hrs., guide 20 fr.), interesting but fatiguing. — The **Bietschjoch** (10,633'), 8 hrs. from Ried to Raron, is a fine route, free from difficulty (guide 12 fr.).

FROM RIED TO BAD LEUK OVER THE FERDEN PASS, 8-9 hrs., with guide, a very fine route, and not difficult. At the *Kummen-Alp* (see below) the path diverges to the left from the Lötschen Pass route and ascends the *Ferden-Thal* to the Ferden Pass (8593'), between the *Majinghorn* and the *Ferden-Rothhorn*. Descent over long stony slopes to the *Fluh-Alp* and through the *Dalathal* to *Bad Leuk* (p. 194). — Over the **Gitzifurgge** (9613'), 9-10 hrs. to Bad Leuk, an interesting but laborious route. The pass lies to the S.W. of the Lötschen Pass, between the *Ferden-Rothhorn* and the *Balmhorn*. Descent over the *Dala Glacier* to the *Fluh-Alp* (see above). — OVER THE RESTI PASS, 7-8 hrs., also interesting (guide 12 fr.). From Ferden we ascend over the *Resti-Alp* (6926'; two beds) in 4 hrs. to the **Resti Pass** (8658'), between the *Resti-Rothhorn* and the *Laucherspitze* (9400'; easily ascended from the pass in ¾ hr.; admirable view) and descend over the *Bach-Alp* to the town of Leuk in 3-4 hrs. more. — To Leuk-Susten over the **Faldum Pass** (8675'), between the Laucherspitze and the *Faldum-Rothhorn* (9810'), or over the **Niven Pass** (8563'), between the Faldum-Rothhorn and the *Niven* (9110'; a fine point of view, ½ hr. from the pass), both easy.

The Lötschen Pass is reached from Ried in 3½ hrs. by *Weissenried*, the *Lauchern-Alp*, and *Sattlegi*. Another route ascends from

Ferden (p. 198) to the N.W., through beautiful larch-wood and over pastures, to the (2 hrs.) *Kummen-Alp* (6808'); then over rock, débris, and patches of snow to the (2 hrs.) **Lötschen Pass** (8840'), commanded on the W. by the steep slopes of the *Balmhorn* (p. 193), and on the E. by the *Schilthorn* or *Hockenhorn* (10,817'; ascended from the pass in 2½ hrs.; splendid view). We obtain the finest view on the route a little before reaching the pass itself: to the S.E. rises the Bietschhorn, to the S. the magnificent group of the Mischabel, Weisshorn, and Monte Rosa; to the N. are the rocky buttresses of the Doldenhorn and Blümlisalp; to the N.E. the Kander Glacier, overshadowed by the Mutthorn (9978').

The path descends on the right side of the *Lötschenberg Glacier*; near the end of the glacier it crosses to the left side and leads over the *Schönbühl* to the (1¼ hr.) *Gfäll-Alp* (6036'; milk), overlooking the upper Gastern-Thal. At the bottom of the valley we cross the *Kander* to (½ hr.) *Gasterndorf* or *Selden* (5315'), a group of hovels (the first, a small cabaret). The Gastern-Thal was more thickly peopled at the beginning of the century than now; but indiscriminate felling of timber has so exposed it to avalanches that the inhabitants have to leave it from February to the hay-harvest. Beyond a beautiful forest, which for centuries has resisted the avalanches of the *Doldenhorn*, we next reach (1 hr.) *Gasternholz* (4462'), amidst a chaos of rocks. The valley bends here and soon expands, being bounded on the S. by the snow-clad *Altels* (11,930') and the *Tatlishorn* (8220'), and on the N. by the *Fisistöcke* (9200'). Of the various waterfalls that descend the abrupt cliffs to the S., the finest is that of the *Geltenbach*.

At the end of the valley the road enters the (1 hr.) *Klus*, a defile ¾ M. long, through which the Kander forces its way in a series of cascades. In the centre of the gorge we cross to the left bank of the river, and beyond its outlet we reach the Gemmi route, and (½ hr.) *Kandersteg* (see p. 192).

56. From Thun to Sion over the Rawyl.
Comp. Map, p. 192.

22 hrs. DILIGENCE from Thun to Lenk (33 M.) daily in 8 hrs. (9 fr., coupé 11 fr. 80 c.; one-horse carr. 35, two-horse 60 fr.). From Lenk to Sion (10½ hrs.) a BRIDLE PATH, good on the Bern side, but rough on the other. Guide desirable (to Sion 16 fr.; horse 30 fr.). The Gemmi is far preferable to the Rawyl as a route to the Valais.

To (25½ M.) *Zweisimmen*, see pp. 202, 203. The Lenk road crosses the *Simme* near *Gwatt*, and ascends the *Upper Simmen-Thal* by *Bettelried*, passing *Schloss Blankenburg* on the right (p. 203), to the prettily situated (3 M.) *St. Stephan* (3297'; Adler); then to *Grodei*, *Matten* (Inn), at the mouth of the *Fermel-Thal* (p. 202), and (5 M.) —

33½ M. **Lenk** (3527'; *Hirsch*, pens. 5 fr.; *Krone*, R. & A. 2½, B. 1 fr. 20 c., pens. 6 fr.; *Stern*, pens. 5 fr.; *Kreuz*), a village rebuilt since a fire in 1878, situated in a flat and somewhat marshy part of the valley of the Simme. About 1/2 M. to the S.W. (path in 9 min.), lies the *Curanstalt Lenk* (3624'; R., L., & A. from 2 1/2, pens. 9 1/2 fr.), with well fitted-up sulphur-baths and grounds. The *Wildstrubel* (10,670'), with its huge precipices and glaciers, whence several streams descend, forms a grand termination to the valley.

EXCURSIONS. (Guides, *Chr.* and *Joh. Jac. Jaggi; Gottlieb Ludi.*) The *Simme* rises, 4 M. to the S. of Lenk, in the so-called *Siebenbrunnen*, to which an interesting walk may be taken (4 hrs. there and back). Road (passing on the left the *Burgfluh*, an isolated nummulite rock with a 'glacier mill', and view of the Wildhorn) by *Oberried* (Inn) to (1 1/4 hr.) *Stalden* (1232'), at the foot of the falls of the Simme. A path now ascends in front of the saw-mill, between alders, describing a curve on the right bank of the stream, and skirting a deep gorge with fine waterfalls. It passes two chalets, traverses pastures, and crosses the brook to (3/4 hr.) the chalets of the Räzliberg (4583'; Fridig's Inn, small). To the S., the '*Seven Fountains*' (4744'), now united into a single stream, issue from the perpendicular rocks. Farther on, to the left, is the *Upper Fall of the Simme*, which is conspicuous from a long distance. To the right rise the *Gletscherhorn* (9672') and *Laufbodenhorn* (8878'), to the left the *Ammertenhorn* (8740').

The Oberlaubhorn (6570'), rising to the W. of the Räzliberg, is frequently ascended from Lenk either by *Trogegg* in 3 1/2 hrs., or by *Pöschenried* and the *Ritzberg Alp* (5710') in 4 hrs., with guide; back by the Räzliberg, Stalden, and Oberried. — The *Mülkerblatt* (6855') is well worth ascending for the fine view of the Wildstrubel, etc. (2 1/2 hrs.). Beyond the Curhaus we ascend on the left bank of the *Krummbach*, (10 min.) cross it, traverse pastures and wood, passing several chalets, and mount the *Bettelberg* to the top.

The Iffigensee (6326'), 3 1/2–4 hrs., is also worth seeing. By the (2 hrs.) *Iffigen Inn* (see below) we turn to the right to the (1/2 hr.) *Stieren-Iffigenalp* (5512'; refreshmts.). The path, steep and stony at places, then ascends to the (1 hr.) saddle which bounds the lake, and leads round its bank to the right (where edelweiss abounds) to the (1/4 hr.) humble chalet at the W. end. — At the base of the Niesenhorn (8113'), 3/4 hr. higher up, is the *Wildhorn Club-Hut* (about 7880'), from which the Wildhorn (10,705') is ascended in 2 1/2–3 hrs. (laborious and fit for experts only; guide from Lenk 25, porter 18 fr.). The route ascends the moraine of the *Dungel Glacier*, and the steep and toilsome E. slope of the *Kirchli* (9157') to the top of the glacier, whence a gentle incline leads to the summit. Splendid view of the Jura, the Tödi, Mte. Leone, Mte. Rosa, Mt. Blanc, Mte. Viso, and particularly of the Plaine Morte on the Wildstrubel, and of the Diablerets. Descent, if preferred, to the S., by the *Glacier du Brozet*, to the *Hôtel Sanetsch* at *Zanfleuron* (2 1/2–3 hrs.; see p. 250).

The '**Rohrbachstein** (9690'; 6 1/2 hrs., guide 15 fr.) is a capital point of view, free from difficulty. From the (4 hrs.) Rawyl Pass (p. 201) we turn to the left and mount to the (1 1/2 hr.) saddle between the Rohrbachstein and the *Wetzsteinhorn*, and to the summit in 1 hr. more. Fossils are found here.

The **Wildstrubel** (W. peak 10,666'; central peak 10,656'; E. peak or *Gross-Strubel*, 10,670') is best ascended from the Rawyl Pass. From the Iffigen Inn, where the night is spent, to the Rawyl 2 hrs.; we then ascend to the left to the snow-arête between the Weisshorn and the Rohrbachstein (2 1/2 hrs.), cross the *Glacier de la Plaine Morte*, and mount the slopes of a snow-arête to the W summit in 2 1/2 hrs., and the central peak in 1/2 hr. more (from Iffigen 7 1/2 hrs. in all). Guide from Lenk 27, down to the Gemmi 30 fr. — From the Räzliberg (see above) a steep path ascends the *Fluhwände* above the Siebenbrunnen to the (2 hrs.) *Fluhseeli* (6710'); thence over débris, moraine, and the *Räzli Glacier* to the W. peak (4 hrs.). — A third route (toilsome) ascends steeply from the (2 1/2 hrs.) *Ritzberg Alp* (see above; bed of hay) past the *Laufbodenhorn* (8878'), viâ the

Thierberg and the *Thierberg Glacier*, and past the *Gletscherhorn* (9672') to the *Räzli Glacier* and to the W. peak (8 hrs. from Ritzberg). Descent to the N.W. by the *Ammerten Glacier*, difficult; to the E. over the crevassed *Lämmern Glacier* to the *Gemmi* (p. 191); to the N.E. over the *Strubelegg* to the *Engstlig-Alp* and *Adelboden* (p. 197).

FROM LENK TO GSTEIG (7 hrs.): over the *Trüttlisberg* (6713') to (4½ hrs.) *Lauenen* (p. 250), and thence over the *Krinnen* (5463') to (2½ hrs.) *Gsteig* (p. 250). Path bad at places (guide 12, horse 5 fr.), see R. 67.

FROM LENK TO SAANEN (p. 204), 6 hrs., path over the *Reulissenberg* or *Zwitzer Egg* (5636'), and down the *Turbach-Thal* (guide 8 fr.). — To ADELBODEN over the *Hahnenmoos* (guide 8, horse 15 fr.), see p. 197. Over the *Ammerten Pass* (8032'), to the S.E. of the *Ammertengrat* (8580'), interesting (7 hrs., with guide).

The RAWYL ROUTE (at first a carriage-road) gradually ascends on the W. side of the valley to (1¼ M.) the left bank of the *Iffigenbach* and the pleasant *Pöschenried-Thal*. The road ends 2 M. farther on. By the (5 min.) **Iffigen Fall** (4483'), 400' high, the bridle-path ascends to the right. After 20 min. we turn, above the fall, into a wooded valley, through which the Iffigenbach dashes over its narrow rocky bed, and traverse a level dale (with the precipices of the Rawyl on the left) to the (½ hr.) **Iffigen-Alp** (5253'; rustic *Inn*). Here we turn sharply to the left (fingerpost), ascend through a small wood on a stony slope, skirt the face of a cliff, cross (10 min.) a brook, and reach (50 min.) a stone hut on a height overlooking the Simmen-Thal. We skirt the W. side of the small (¾ hr.) *Rawyl-See* (7743') and reach (¼ hr.) a cross *(la Grande Croix)* which marks the boundary of Bern and Valais and the summit of **the Rawyl** (7943'; 4¼ hrs. from Lenk), with a refuge-hut. The pass consists of a desolate stony plateau *(Plan des Roses)*, enclosed by lofty and partially snow-clad mountains: to the W. the long *Mittaghorn* (8842'); S.W., the *Schneidehorn* (9640') and the snow-clad *Wildhorn* (10,705'; p. 200); S., the broad *Rawylhorn* (9540') and the *Wetzsteinhorn* (9114'); E., the *Rohrbachstein* (9690'; p. 200); N.E., the extremities of the glaciers of the *Weisshorn* (9882').

Beyond the pass **the path is bad**. It passes a second small lake, and (¾ hr.) reaches the margin of the S. slope, which affords a limited, but striking view of the mountains of the Valais. It descends a steep rocky slope (leaving the dirty chalets of *Armillon*, 6926', to the left), and (½ hr.) crosses a bridge in the valley (5970'; a good spring here). Instead of descending to the left to the chalets of (¼ hr.) *Nieder-Rawyl* (Fr. *Les Ravins*, 5768'), we ascend slightly by a narrow path to the right, and skirt the hillside. Then (25 min.) a steep ascent, to avoid the *Kändle* (see below); 20 min., a cross on the top of the hill (6330'), whence we again descend to (½ hr.) *Praz Combeira* (5344'), a group of huts; and lastly a long, fatiguing descent by a rough, stony path, ascending at places, to (1½ hr.) **Ayent** (3400'; 3¾ hrs. from the pass; accommodation at the curé's, good wine, or at the merchant *Mosoni's*).

The footpath from Nieder-Rawyl to Ayent, shorter by 1 hr., leads by the so-called 'Kändel' (i.e. channel), Fr. *Sentier du Bisse*, along the edge of a water-conduit skirting a steep slope 1800' in height. Being little more than 1' in breadth, the path is practicable only for persons with steady heads.

The path, which now improves, next leads by *Grimisuat* (2594'; Ger. *Grimseln*) and *Champlan* to (2 hrs.) Sion (p. 308; 10½ hrs. from Leuk) and to (1¼ hr.) *St. Léonard* (p. 305).

57. From Thun through the Simmen-Thal to Saanen.

31½ M. DILIGENCE twice daily (7 a. m. and 12.30 p.m.) direct to Saanen in 8½ hrs. (fare 9 fr. 30, coupé 12 fr. 5 c.); another to Zweisimmen daily at 3.30 p.m. in 5 hrs. 40 minutes. — One-horse carr. to Weissenburg 13, two-horse 24 fr.; to Zweisimmen 28 or 50, to Saanen 35 or 50, to Château d'Oex 40 or 70, to Aigle 80 or 130, to Bulle 70 or 120 fr. — From Spiez (p. 156) a diligence plies thrice daily viâ *Wimmis* to (1 hr.) *Brodhäsi* (see below), in connection with the Thun diligence. One-horse carriage from Spiez to Weissenburg 10, two-horse 18, to Zweisimmen 22 or 40, to Lenk 32 or 55, to Saanen 35 or 60, to Château d'Oex 40 or 70, to Aigle 75 or 135 fr.

The road skirts the Lake of Thun as far as (3 M.) *Gwatt* (p. 155) and gradually ascends towards the *Niesen* (p. 154). On a hill to the right rises the slender tower of *Strättligen* (p. 153). At the bottom of the valley flows the *Kander*, in an artificial channel made in 1714. The road follows its left bank, and then the left bank of the *Simme*, which falls into the Kander near *Reutigen*, a prettily situated place.

6 M. **Brodhäsi** *("Hirsch")*. About 1 M. to the E. lies the substantial village of *Wimmis*, with its picturesque old castle (see p. 154). The road passes through a defile *(Porte)* between the *Simmenfluh* and the *Burgfluh* into the **Simmen-Thal** (locally called the *Sieben-Thal*), a fertile valley with numerous villages.

8½ M. **Latterbach** (2308'; *Bär*). To the S. is the *Diemtig-Thal*.

FROM LATTERBACH TO MATTEN a shorter, but uninteresting route (7 hrs.) leads through the Diemtig-Thal. At Latterbach it crosses the *Simme* and follows the right bank of the *Kirel* (passing the village of Diemtigen on the hill to the right) and then the left bank to *Wampflen* and (2½ hrs.) *Tschnepis* (3163'), where the valley divides into the *Mänigrund* to the right and the *Schwenden-Thal* to the left. We follow the latter, which after ¾ hr. again divides at *Waritannen* (8370'). The path now diverges from the road, ascends to the W. through the *Grimbach-Thal* to the (2 hrs.) **Grimmi** (6644'), a little-frequented pass, and descends through the fertile *Fermel-Thal* to (2 hrs.) *Matten* (p. 199).

10 M. **Erlenbach** (2320'; *Krone*, *Löwe*, both unpretending), with well-built wooden houses.

The **Stockhorn** (7195') is sometimes ascended hence by experts in 4½ hrs.; better from *Thun*, by *Amsoldingen* and *Ober-Stocken* ("Bär, rustic) in 5½ hrs., or from *Blumenstein* (p. 153) by the *Wahl-Alp* (new chalet, dear) in 4 hrs.; descent, if preferred, by the Wahl-Alp to *Bad Weissenburg*, which is reached by means of ladders. Splendid flora and grand view.

14½ M. **Weissenburg** (2418'; *Hôtel Weissenburg*, R. & A. 2½ fr.), a group of neat houses.

In a steep defile, so narrow at places as almost to exclude the sun, about 1¼ M. to the N.W., lies the favourite **Weissenburg-Bad** (2770'; a drive of 20 min., for which 4 fr. are demanded).

The mineral water, impregnated with sulphate of lime (70°, at its source 81°) and beneficial for bronchial affections, is used exclusively for drinking. The *Neue Bad*, situated in a sheltered basin, consists of two large houses (reading and billiard rooms; post and telegraph office; board 8, R. 2½-5, D. 3½, warm bath 1½ fr.); the *Alte Bad*, buried in the ravine ½ M. higher up, is inferior (pension 5-7 fr.). The baths, with the extensive pine-forests round them, belong to *Messrs. Hauser*.

FROM WEISSENBURG TO THE GURNIGELBAD (6 hrs.) Attractive path through the *Klus*, passing the *Morgetenbach Fall*, 300' high, and the *Morgeten-Alp* to the (3½ hrs.) *Bürglen Sattel* (6434'); then down (passing *Bad Schwefelberg*, 1¼ M. to the left) to the *Gantrisi Pass* (5247'), with a charming view, and over the *Obere Gurnigel* to the (1½ hr.) *Gurnigelbad* (p. 153).

20½ M. **Boltigen** (2726'; *Hôt. Imoberstag, Bär*, both moderate), a thriving village with handsome houses, is reached beyond the *Simmenegg*, or *Enge*, a defile formed by two rocks between which the road passes. Above the village rise the two peaks of the *Mittagfluh* (6198'). To the left peep the snow-fields to the E. of the Rawyl (p. 201). The coal-mines in a side-valley near *Reidenbach* (2756'; ¾ M. from Boltigen) account for the sign of the inn (a miner).

FROM REIDENBACH TO BULLE, 24 M. A little above Reidenbach the road diverges to the right and ascends in numerous windings (which footpaths cut off) to the (6 M.) pass of the *Bruchberg* (4940'). It then descends gradually (preferable to the bad footpath) to (3 M.) *Jaun*, Fr. *Bellegarde* (3336'; *Hôt. de la Cascade*, poor), a pretty village with a waterfall 86' high. (Path to the *Schwarzsee-Bad* by *Neuschels*, 3 hrs., see below.) [A cart-track to the S. ascends on the left bank of the Jaunbach to (1½ hr.) *Ablantschen* (4280'; Inn) at the foot of the bare rocky chain of the *Gastlose* (6542'). Easy passes thence over the *Grubenberg* (5118), to the S. of the *Dent de Ruth* (7074'), to (3 hrs.) *Saanen*, and over the *Schländi* to (2½ hrs.) *Reichenstein* (see below).] We next traverse the beautiful pastures of the *Jaunthal* or *Bellegarde Valley*, which yield excellent Gruyère cheese (see p. 204), and the picturesque *Défilé de la Tzintre* to (7½ M.) *Charmey*, Ger. *Galmis* (2957'; *Hôt. du Sapin*; "*Maréchal Ferrant*, pens. 5 fr.), a well-to-do village and summer-resort, charmingly situated (diligence to Bulle twice daily in 1¾ hr.). Fine view from the church. The road next passes *Crésus*, *Châtel*, and the ruin of *Montsalvens* (rare flora), crosses the *Jaun*, and beyond *Broc* (Pens. de la Grue), the *Sarine*, and leads through wood to *La Tour-de-Trème* (p. 253) and (7½ M.) *Bulle* (p. 252). — From Crésus (see above) a pleasant route leads by *Cerniat* and the old monastery of *Valsainte*, and over the *Chésalette* (4659') to the (3½ hrs.) *Schwarzsee-Bad* (p. 217). On the *Kalte Sense*, 4 hrs. to the N.E. of the Schwarzsee (diligence daily in summer from Freiburg viâ Plaffeyen), are the sequestered but well-kept *Baths of Schwefelberg* (4578), with springs impregnated with lime, at the foot of the *Ochsen* (7180'), 2½ hrs.; fine views). Hence a route leads over the *Scilbühlgrat* to the (2½ hrs.) *Gurnigelbad* (p. 153); and a bridle-path crosses the *Gantrisi Pass* (see above) to (3 hrs.) *Bad Blumenstein* (p. 153).

The road crosses the Simme at (2 M.) *Garstatt* and turns suddenly round the *Laubeggstalden* rock, passing a fine waterfall. We recross the stream and pass the ruined castle of *Mannenberg* to (3 M.) —

25½ M. **Zweisimmen** (3215'; pop. 1910; *Krone*, R., L., & A. 3, B. 1½, D. 3 fr.; *Hôt. Simmenthal*; *Bär*), the chief village in the valley, with an old church, situated in a broad basin on the

SAANEN.

Kleine Simme. Pleasant views from the churchyard, and from *Schloss Blankenburg*, now containing public offices and a prison, 1/2 hr. to the S.E. (p. 199).

The road ascends gradually for 5 M., crossing the *Schlündibach* at (3 1/2 M.) *Reichenstein.* (To *Ablantschen,* see above.) In a pine-clad valley on the left flows the *Kleine Simme*; the road crosses four deep lateral ravines and finally the Kleine Simme itself. At the top of the hill (4227'; Inn) **begin** the *Saanen-Möser,* a broad Alpine valley, sprinkled with chalets and cottages. A striking view is gradually disclosed of the frowning *Rüblihorn* (7570'), the barometer of the surrounding country (comp. p. 102), the serrated *Gumfluh* (8068'), the snow-fields of the *Sanetsch* beyond it, and lastly the huge *Gelten Glacier* (p. 250) to the left. Lower down we obtain a fine survey of the *Turbach, Lauenen,* and *Gsteig* valleys (p. 250).

34 1/2 M. **Saanen,** Fr. *Gessenay* (3382'; pop. 3733; *Grand Logis,* or *Gross-Landhaus;* **Ours,* unpretending), is the capital of the upper valley of the *Saane (Sarine).* The inhabitants rear cattle and **manufacture** the famous *Gruyère* and *Vacherin* cheese.

To *Gsteig,* and over the *Col de Pillon* to **Aigle, see p. 250;** over the *Sanetsch* to *Sion,* see p. 250.

FROM SAANEN TO CHÂTEAU D'OEX (p. 254) 7 M.; diligence twice daily in 1 1/2 hr., by *Rougemont,* or *Rothenberg* (Pens. du Rubli), the frontier between cantons Bern and Vaud, where the language changes from German to French, and *Flendruz.*

IV. WESTERN SWITZERLAND. LAKE OF GENEVA. LOWER VALLEY OF THE RHONE.

58. From Bern to Neuchâtel 206
    Twannberg; Isle of St. Peter; Chasseral; Cerlier, 202. — Chaumont, 209.
59. From Neuchâtel to Chaux-de-Fonds and Locle . . 209
    Tête de Rang; Col des Loges. Côtes du Doubs. From Chaux-de-Fonds to Bienne through the Val St. Imier, 210. — From Locle to Morteau and to Brenets; Saut du Doubs, 211.
60. From Neuchâtel to Pontarlier through the Val de Travers 211
    Creux du Van. Ravine of the Raisse, 212.
61. From Neuchâtel to Lausanne 213
    Gorges de l'Areuse, 213. — Chasseron, 214.
62. From Bern to Lausanne *(Vevey)* 215
    From Flamatt to Laupen, 215. — From Freiburg to Payerne and Yverdon. Schwarzsee-Bad; Berra, 217. — From Romont to Bulle. Signal de Chexbres. From Chexbres to Vevey, 218.
63. From Lausanne to Payerne and Lyss 219
    From Morat to Neuchâtel, 220.
64. From Lausanne to Vallorbes and Pontarlier . . . 220
    From Vallorbes to Le Pont. Dent de Vaulion; Lac de Joux, 221.
65. Geneva and its Environs 221
    Pregny; Ferney; Bois de la Bâtie; Salève; Voirons, etc., 230-233.
66. From Geneva to Martigny viâ Lausanne and Villeneuve. Lake of Geneva *(North Bank)* 233
    Divonne, 234. — The Dôle, 235. — Signal de Bougy; Gimel; Col de Marchairuz, 236. — From Lausanne to Bercher, 239. — Hauteville and Blonay; the Pléiades, 241. — Excursions from Montreux: Glion; Rochers de Naye; Gorge du Chaudron; Les Avants, etc., 243, 244. — From Aigle to Villars; Chamossaire; Corbeyrier, 246, 247. — From Bex to Les Plans, 247. — Baths of Lavey; Morcles, 248. — Pissevache; Gorge du Trient, 249. — Arpille; Pierre-à-Voir, 250.
67. From Saanen to Aigle over the Col de Pillon. . 250
    The Lauenen-Thal. From Gsteig to Sion over the Sanetsch, 250. — Excursions from Ormont-Dessus: Creux-de-Champ, Palette, Oldenhorn, Diablerets, etc. From Ormont-Dessus to Villars or Gryon over the Col de la Croix, 251. — Pic de Chaussy; Leysin, 252.
68. From Bulle to Château d'Oex and Aigle 252
    Montbarry. Ascent of the Moléson from Bulle or Albeuve. Châtel St. Denis, 252, 253. — From Montbovon over the Jaman to Montreux, 253. — Mont Cray, 254.
69. From **Bex** to Sion. Pas de Cheville 255
70. From **Geneva** to St. Maurice viâ **Bouveret**. Lake of Geneva *(South Bank)*. Val d'Illiez. 256
    From Thonon to Samoëns. Valley of the Drance, 257. — The Blanchard. Dent d'Oche. Grammont. Cornettes de Bise, 258. — Excursions from Champéry: Culet; Dent du Midi; Tour Sallières; Dents Blanches. From Champéry to Samoëns, Sixt, or Vernayaz (Cols de Coux, de la Golèse, de Sageron, de Clusanfe), 259, 260.

14

58. From Bern to Neuchâtel.

41 M. Railway in 1³/₄-2³/₄ hrs. (fares 6 fr. 90, 5 fr., 3 fr. 65 c.).

Bern, see p. 144; from Bern to (21 M.) Bienne, see p. 12. (Münster-Thal Railway to Bâle, see R. 2; by St. Imier to Chaux-de-Fonds, see p. 210.) Near the beautiful avenues to the S.W. of Bienne the train reaches the **Lake** of Bienne (1425'; 9¹/₂ M. long, 2¹/₂ M. broad). As the train skirts the W. bank, we obtain a very pleasing view of the lake, enhanced in clear weather by the magnificent chain of the Bernese Alps. — Beyond (27¹/₂ M.) *Douanne*, Ger. *Twann* (*Ours), we pass a fall of the *Twannbach*.

Interesting excursion through the gorge of the *Twannbach* to the (1¹/₂ hr.) Curhaus Twannberg (2887'; pens. 4¹/₂-5 fr., well spoken of), with view of the lakes of Bienne and **Morat** and the High Alps. Hence to (1¹/₂ hr.) *Macolin* (p. 12).

29 M. *Gléresse*, Ger. *Ligerz.*

To the left, in the lake, lies the **Isle of St. Peter**, clothed with beautiful old oaks, vineyards, and fruit trees, where Rousseau spent two months in 1765. (His room is shown in the *Hôtel.*) Boat from Douanne or from Gléresse, there and back, 4, from Neuveville 6 fr. A steamboat also plies from Neuveville to Cerlier and the Isle of St. Peter. — The island of St. Peter is now connected on the S. side with the mainland near Cerlier (see below).

30¹/₂ M. **Neuveville**, Ger. *Neuenstadt* (*Faucon; *Trois Poissons*; *Pens. Zur Guten Quelle*, with bath-estab.), a pleasant little town (2368 inhab.), the last in Canton Bern, is the first place where French is spoken. The *Museum*, near the station (adm. 50c.), and the house of *Dr. Gross* contain interesting antiquities from the lake-dwellings and the Burgundian wars. In the latter also is Beck's collection of nephritoides. On the *Schlossberg* (1752'), 20 min. from the station, stands a ruined castle of the Bishops of Bâle (fine view from the top and on the way up), near which the *Béon* forms a waterfall (often dry in summer).

To the N. of Neuveville rises the *Chasseral*, or *Gestler* (5280'), studded on the S. side with numerous villages amid green meadows. Road from Neuveville via *Lignières* (2654'; *Hôtel-Pension Beau-Séjour, 4·5 fr.) to the (4 hrs.) top (*Chalet-Hôtel du Chasseral*, with 20 beds, fair). The view from the (10 min.) signal embraces W. Switzerland, the Black Forest, **the Jura**, and the Alps. — The ascent may be made from Macolin (p. 12) in 4 hrs.; from St. Imier in 2¹/₂-3 hrs. (see p. 198).

The old town of **Cerlier**, or *Erlach* (*Ours*), lies opposite Neuveville, at the N. foot of the wooded *Jolimont* (1988'; ³/₄ hr.), a charming point of view (*Curhaus*, with view-tower). The '*Teufelsbürde*' is a group of large erratic blocks on the summit. — On the E. bank of the lake, at *Lüscherz*, and at *Mörigen*, farther to the N., numerous remains of lake-dwellings have been discovered.

Near (33 M.) *Landeron* we quit the Lake of Bienne; the little town lies on the left; farther to the E. rises the *Jolimont* (see above). 34¹/₂ M. *Cressier*, with its church on a lofty rock; 35¹/₂ M. *Cornaux*. Beyond a tunnel the train reaches (38 M.) *St. Blaise* (tramway hence to Neuchâtel), near which is the lunatic asylum of *Préfargier*, built in 1844. At *Marin* (*Pens. Nusslé) are the celebrated lake-dwellings of *La Tène*, the name of which is sometimes used as

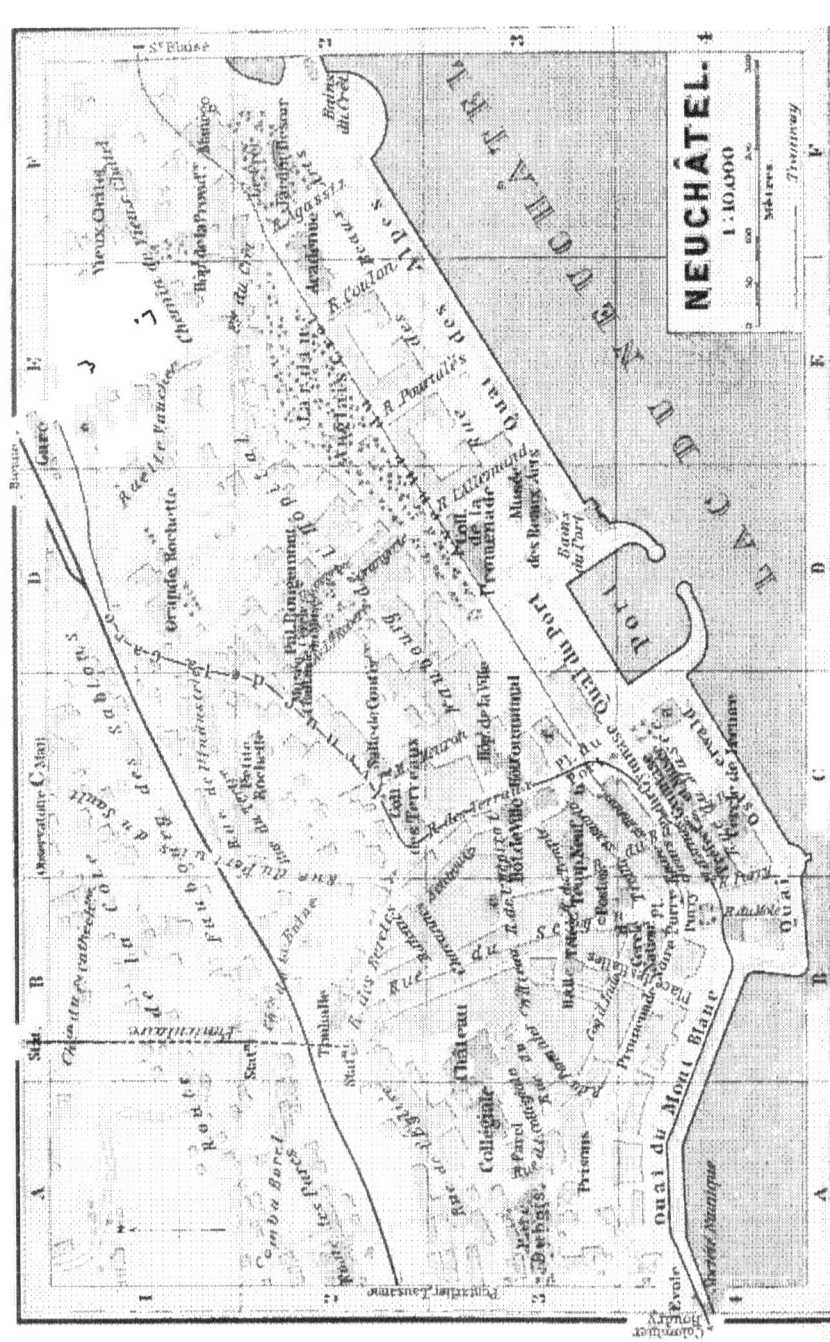

descriptive of the civilization of the peoples to the N. of the Alps during the last centuries before the Roman period. — The train next reaches the **Lake of Neuchâtel** (1427'), the Roman *Lacus Eburodunensis*, which is 25 M. long and 4-6 M. broad (greatest depth 500'). Near the N.E. end the *Thièle* or *Zihl* issues from the lake, the level of which has been lowered 6' by the enlargement of this outlet. The vine-clad W. bank, above which rise the abrupt Jura Mts., affords a view of the Alps from the Bernese Oberland to Mont Blanc.

44 M. **Neuchâtel**. — RAILWAY STATION (Buffet) on the hillside above the town, 1 M. from the principal hotels, which send omnibuses to meet the trains. A *Cable Railway* (fares 20, 10 c.) also descends hence in 9 min. to the harbour (*Port*; Pl. C, 3), and thence continues to the W. as an ordinary steam-tramway, past the station of *Evole* (Pl. A, 4), to Colombier and Boudry (p. 213). — Another tramway plies to *St. Blaise* (p. 206). — STEAMBOAT on the Lake of Neuchâtel, see pp. 213, 220.

Hotels. *GRAND-HÔTEL BELLEVUE (Pl. a; C, 4), in an open situation on the lake, R., L., & A. 4½-5½, B. 1½, lunch 4, D. 5, pens. 8-10, omnibus ½-1 fr. — *GRAND-HÔTEL DU LAC (Pl. b; C, 3), R., L., & A. 3-4, D. 3½, pens. from 8, omnibus ½-¾ fr.; *FAUCON (Pl. c; B, 3), R., L., & A. 2-4, B. 1¼, lunch 3, D. 3, pens. 8-9, omn. ½-¾ fr.; *HÔT. DU SOLEIL (Pl. d; B, 3, 4), R. 2, D. incl. wine 2½ fr.; HÔT. DU PORT (Pl. f; C, 3). — PENS. BOREL (*Villa Surville*), well situated above the town, board 4-5 fr.

Cafés. *Chalet du Jardin Anglais* (Pl. E, 2); *Brasserie Gambrinus*, on the harbour, etc. — BATH-ESTABLISHMENT, at the harbour.

English Church Service, at the Divinity Library (*Rev. J. H. K. Best*).

Neuchâtel (1433'; 18,000 inhab.), Ger. *Neuenburg*, the capital of the canton of that name (formerly a principality of the Orange family, under Prussian sway from 1707 to 1815, when it joined the Confederation, and finally given up by Prussia in 1857), is charmingly situated on the Lake of Neuchâtel, to the E. of the mouth of the *Seyon* (p. 211), and at the base and on the slopes of the Jura. The banks of the lake are skirted for about 1½ M. by a tree-shaded *Quay, known at different parts of its length as the *Quai du Mont Blanc*, *Quai Osterwald*, and *Quai des Alpes*, and commanding a beautiful Alpine view. Near the middle of this avenue is the little *Harbour* (Pl. D, 3, 4), beside which is the handsome *Post Office* (Pl. C, 3), built in 1893-95.

To the W. of the harbour is the COLLÈGE LATIN (Pl. C, 4), containing a valuable natural history collection, founded by Agassiz (p. 189) and Coulon (open Thurs., 10-12 & 2-4, and Sun., 2-4), and a public library (100,000 vols.; open daily, except Sun. & Mon., 10-12 & 2-4). — In the adjoining Place Purry (Pl. B, 4) rises a bronze statue of *David de Purry* (1709-1786), a native of Neuchâtel, who bequeathed 4½ million francs to the town. The *Halles* (Pl. B, 4; now a club), a picturesque little Renaissance edifice of 1570, stand in the Place des Halles.

To the E. of the harbour is the *MUSÉE DES BEAUX-ARTS (Pl. D, 3), a handsome Renaissance building, containing the interesting municipal *Collection of Antiquities* and *Picture Gallery* (adm. to each collection 50 c., free on Sun. and Thurs. 10-12 and 1-5).

208 *IV. Route 58.*   NEUCHÂTEL.

GROUND FLOOR. The rooms to the right **and** left of the entrance contain the valuable *Historical & Archaeological Collections*, including numerous reminiscences of the period of the Prussian rule. — On the STAIRCASE is a bronze bust of *M. de Meuron* (d. 1868), the founder of the museum. At the top are three *Paintings by *Paul Robert*, executed in 1886-84. The central picture represents the intellectual life as mirrored in the Christian dispensation: among clouds at the top appears the Saviour, with the Gospel below him, to the left rises a procession of female forms symbolizing Art, Science, and Morality, in blessed harmony, to the right the Archangel Michael stands upon the defeated dragon, and in the background is a view of Neuchâtel. In **the** composition to the left Celestial Grace is shown enriching the earth **with** flowers **and** fruits, while the evil spirits are driven off. The picture to **the** right depicts industrial life: in the foreground are workmen and workwomen, a manufacturer, employers of labour, and merchants; in the centre of the background rises the golden statue of Industry, round which throngs an eager multitude; a beam of divine light falls upon the group on the right; at the top are the Angel of Justice, to the right, and the Recording Angel, to the left. — The balcony offers a beautiful view of the lake and the Alps. — To the right is the —
PICTURE GALLERY. ROOM I. (right) *Dubois*, Autumn evening, Summer morning; *P. Robert*, Evening air; *Jacquand*, Arrest of Voltaire at Frankfort; *Al. Calame*, Monte Rosa; *Berthoud*, The Jungfrau; *Jeanmaire*, Street at Sion; *E. Tschaggeny*, Draught-horses. — ROOM II. Engravings and Drawings. — ROOM III. *K. Girardet*, Old Franciscan monastery at Alexandria; *Isabey*, Sea-piece; *E. de Pury*, Lucifer; *Robert-Fleury*, Massacre of St. Bartholomew; *L. Robert*, Italian street-scene; *K. Girardet*, Cromwell reproached by his daughter Mrs. Claypole for the condemnation of Charles I. — ROOM IV. Small landscapes, cattle-pieces, etc. — ROOM V. Sketches by *Léopold Robert*, and copies of all his works by his brother *Aurèle*. L. Robert, born in 1794 at Chaux-de-Fonds (d. in Italy, 1835) is famous for his representations of popular life in Southern Europe. — ROOM VI. *E. de Pury*, Venetian fishermen; *Guillarmod*, Watering horses; *A. de Meuron*, Betten-Alp; *Coleman*, Campagna di Roma; *Imer*, Evening on the lake-shore, Ruins of Crozant; *E. de Pourtalès*, Valley of Meiringen; *Guillarmod*, Freight-waggon; *Bocion*, Canal Grande; *A. de Meuron*, Pasture near Iseltwald; *Bocion*, On the Riviera; *Schuler*, Floating timber. — ROOM VII. (left) *E. Girardet*, Maternal love, El Kantara (Algiers), The father's blessing, The little culprit; *K. Girardet*, Huguenots surprised by Roman Catholic soldiers; *Léopold Robert*, *Basilica of S. Paolo Fuori le Mura near Rome after the fire of 1823, *Fishermen of the Adriatic, Brigands pursued by soldiers, Improvisatore; *Anker*, The French army under Bourbaki entering Switzerland in 1871. — ROOM VIII. (left) *Gaud*, Harvest fire; *A. Calame*, The Wetterhorn; *Anker*, Sunday afternoon; landscapes by *M. de Meuron*, *Alb. de Meuron*, *A. Veillon*, *Berthoud*, and others. — ROOM IX. (left) *Grosclaude*, Desdemona; *Jeanmaire*, Midday rest on the Alp; on the end-wall a number of good ancient paintings of different schools, from Count Pourtalès's collection, then *Bachelin*, Entry of the French army into Switzerland in 1871 (p. 213); *Dan. Jean Richard* (p. 211) promising to repair a traveller's watch (1697); *E. Burnand*, The village engine; *Anker*, Pilgrimage to Gleyresse; *E. de Pury*, The fencing-master; *Tschaggeny*, Mother and child pursued by a bull.
Next the museum is an interesting '*Sépulcre Préhistorique*', discovered at Auvernier in 1876.

A little to the N.E. **is** the new *Academy* (Pl. E, **F**, 2; 40 teachers, 150 students), **between** the *Jardin Anglais* and the *Jardin Desoir*; and to the N. and N.E. of the latter are the *Hôpital de la Providence* and the *Pourtalès Hospital*. — Near the *Palais Rougemont* (Pl. D, 2), on the groundfloor **of which is the** *Cercle du Musée*, is the *Musée Alpestre*, a collection of stuffed Alpine animals (1 fr.).

The CHÂTEAU (Pl. B, 3), on the hill above the town, dating in its present form partly from the 12th cent., but mainly from the

15-17th cent., was restored in 1866, and is now the seat of the cantonal government. Near it is the *ABBEY-CHURCH (*Collégiale*; Pl. A, 3; key at 6 Rue du Château), built in 1149-1190 and restored in the 13th cent.; the two pointed Gothic towers date from the 15th century. The choir contains a handsome Gothic monument with 15 lifesize figures, erected in 1372 to the Counts of Neuchâtel, Freiburg, and Hochberg, and restored in 1840. There are also memorial-stones to two Prussian governors. — The *Place* in front of the church is adorned with a *Statue of Farel*, the Reformer (d. 1565), erected in 1875. The graceful cloisters on the N. side, rebuilt after a fire in 1450, were restored in 1860-70. — A bridge crosses the old castle-moat to the *Public Park*.

The *Observatory (Observatoire Cantonal)*, 25 min. above the town, erected for the benefit of the watch-manufacturers, is in telegraphic communication with Chaux-de-Fonds, etc. (p. 210). The adjoining *Mail*, a grass-plot planted with trees, commands a charming view of the lake and the Alps. Another good view is enjoyed from the new *Parc du Plan*, to which a cable-tramway ascends (Pl. B, 2, 1).

Near the town there are pleasant wood-walks to the *Roche de l'Ermitage*, *Pierre à Bot*, *Gorges du Seyon*, *Chanélaz* (p. 213), etc.

The *Chaumont (3845'; *Hôtel de Chaumont*, a large house near the top, 3700', pens. 6-9 fr.; *Hôtel du Château*, lower down, 3 min. to the S.E.; *Eng. Church Service* in summer), a spur of the Jura, rising to the N., is the finest point of view near Neuchâtel. The road to it diverges from the Chaux-de-Fonds road, 1¼ M. from Neuchâtel, and leads to the top in 1½ hr. (diligence twice a day in summer in 2½ hrs., 2 fr., down in 1 hr., 1½ fr.; carr. with one horse 10, with two horses 20 fr.). Near the hotels at the top are a chapel and a school-house. The view from the *Signal*, ¼ hr. above the hotels (at the top indicator of the Swiss Alpine Club, by Imfeld), embraces the lakes of Neuchâtel and Morat, and the Alpine chain from the Sentis to Mont Blanc in the background. The afternoon light is best, but a perfectly clear horizon is rare. A charming view of the Val de Ruz and the Jura, to the W., is obtained from the (¼ hr.) *Pré Louiset*. — An attractive route, following the mountain-ridge the whole way, viâ *La Dame* and *Chuffort* (guide advisable), leads in 4 hrs. from the Chaumont to the *Chasseral* (p. 206). — *Gorges de l'Areuse*, see p. 213; *Tête de Rang*, see 210.

59. From Neuchâtel to Chaux-de-Fonds and Locle.

RAILWAY from Neuchâtel viâ Chaux-de-Fonds to (23½ M.) Locle in 2¼ hrs. (fares 5 fr. 25, 3 fr. 80, 2 fr. 80 c.). This route, as far as Les Hauts-Geneveys, is very attractive; views to the left.

Neuchâtel, see p. 207. The train skirts the slopes behind the town and crosses the *Seyon*, which descends from the Chasseral and since 1839 has entered the lake through a tunnel above the town. Beyond a tunnel of 748 yds. the line affords a superb *View of the lake and the Bernese Alps (to the S., the Mont Blanc) 3 M. *Corcelles* (1880'). The train ascends through wood; two short tunnels.

7 M. *Chambrelien* (2300'), beautifully situated high above the valley of the *Reuse* (p. 212). The train backs out from the station towards the N.E. and skirts a wooded chain of hills. To the right

IV. Route 59. LA CHAUX-DE-FONDS.

is the fertile *Val de Ruz*, with its numerous villages, above which rises the *Chaumont* (p. 209).

10½ M. *Les Geneveys-sur-Coffrane* (2870'; Hôt.-Brasserie du Jura); then (12½ M.) *Les Hauts-Geneveys* (3135'; Buffet; Hôt. du **Jura**, Hôt. du Nord, both plain), the highest point of view **on the line**, where Mont Blanc becomes very conspicuous.

The "**Tête de Rang** (4668'; *Inn*), ascended in 1¼ hr. from Hauts-Geneveys (by a lane to the left, 10 min. beyond the village), commands a magnificent and extensive view of the Jura westwards to the plateau of Langres, of the Vosges, and of the Alps from the Sentis to Mont Blanc and the mountains of Geneva. — A path leads hence along the hill to the "**Col des Loges** (4220; *°Hôtel à la Vue des Alpes*), on the road from Neuchâtel to Chaux-de-Fonds. View similar, but less extensive. Descent either to (1½ M.) Hauts-Geneveys or to (3 M.) Chaux-de-Fonds.

The train passes through a tunnel, 2 M. long (9 min.), **under the** *Col des Loges* to (16 M.) *Les Convers*, a solitary station in a rock-girt valley. Beyond a tunnel, ¾ M. long (3 min.), under *Mont Sagne*, and a shorter one, we reach —

18½ M. **La Chaux-de-Fonds** (3255'; *°Grand Hôtel Central*, R. from 2 fr.; *°Fleur de Lys*, R. & A. 3, B. 1¼ fr.; *Lion d'Or; Croix d'Or; Balance;* U. S. Consular Agent), an important watch-making town (30,000 inhab.), with handsome streets and public buildings. If time permit, the traveller may visit the **Church** with its skilfully vaulted roof, and the *Collège*, containing the municipal picture-gallery (good pictures by Swiss masters), the library, etc.

A pleasant walk may be **taken by** a new path to the N. to (1 hr.) the hill of *Pouillerel* (4200), commanding a view over Franche-Comté to the Vosges and of the Bernese Alps to the Wildstrubel and Mont Blanc. — *Aqueduct*, see p. 212.

From Chaux-de-Fonds to the picturesque "**Côtes du Doubs**, a pleasant excursion of one day. The road leads past the *°Restaurant Bel-Air* to a *Restaurant and Hotel*, **near** the *Combe de la Greffière* (view of the Doubs below); then descends through wood (short-cuts for walkers) towards the *Doubs*, reaching the river at (5¼ M.) the charmingly-situated *Maison Monsieur*, and skirting its bank via the *°Pavillon des Sonneurs* (Restaurant) to (2¼ M.) *Biaufond*. Then by boat to (½ hr.) *Le Refrain*, and on foot through grand and wild scenery to the (2¼ M.) picturesque *Moulin de la Mort*. Opposite are the curious *Echelles de la Mort*, used by the inhabitants as means of communication. Here, and for several leagues farther **to the N.**, the Doubs (lower course also attractive) forms the boundary between France and Switzerland. Visitors may take a boat to (50 min.) the *Verrières du Bief d'Etoz*, **then** below the Fall of the Doubs continue either by boat or on foot along the French bank past (right) *La Goule* to (¾ hr.) *Bief d'Etoz*. Thence we proceed on the Swiss bank to the (¾ hr.) mill of *Theusseret*, ascend to the right to *Belfond*, and again descend to (1 hr.) *Goumois* (*Couronne*, good trout), a village charmingly situated on both banks of the river. A road ascends hence to the E. in wide curves to (3 M.) *Seignelégier* (Cheval Blanc), whence a railway (Chemin de fer régional) runs to (1½ hr.) Chaux-de-Fonds.

A pleasant road leads to the W. of La Chaux-de-Fonds to (1¼ hr.) *Les Planchettes* (Restaurant) and the (1½ hr.) *Saut du Doubs* (p. 211).

From Chaux-de-Fonds to Bienne, 28 M., railway in 1½-2 hrs. (fares 4 fr. 75, 3 fr. 35, 2 fr. 40 c.). The line passes the station of (2½ M.) *Halte du Creux*, and enters the industrious *Val St. Imier*, watered by the *Suze* or *Schüss*. 5½ M. *Renan*; 8 M. *Sonvillier*, with the picturesque ruins of the castle of *Erguel* on a pine-clad rock. 10 M. St. Imier (2670'; 7114 inhab.;

Hôt. de la Ville; Hôt. des Treize-Cantons; Couronne), the capital of the valley, with considerable watch-manufactories. (Ascent of the *Chasseral*, p. 206, by a bridle-path, 2½-3 hrs.) — 11M. *Villeret*; 13½ M. *Cormoret*; 15½ M. *Courtelary*; 17 M. *Cortebert*; 18½ M. *Corgémont*. — 20 M. *Sonceboz*. and thence to (28 M.) *Bienne*, see p. 11.

The railway bends suddenly to the S.W. — 21 M. *Eplatures*.

23½ M. **Le Locle** (3020'; 11,312 inhab.; *Hôt. des Trois Rois*, *Hôt. du Jura*; *Hôt. National*), famed for its watches and jewellery. In front of the Watchmakers' School a bronze statue was erected in 1888 of *D. J. Richard* (d. 1741), founder of the watch-making industry in Le Locle and La Chaux-de-Fonds. The top of the *Sommartel* (4350'), 1 hr. to the S., affords a wide view of a great part of the Jura.

FROM LOCLE TO MORTEAU (Besançon), 8 M., railway in 35 min. viâ *Col des Roches* (where an interesting road diverges to the right to *Les Brenets*, 2 M., see below) and *Villers-le-Lac*, 1 M. to the S.W. of the *Lac des Brenets* (see below). From Morteau to *Besançon* 42 M. (see *Baedeker's Northern France*).

FROM LOCLE TO BRENETS, 2½ M., railway in ¼ hr. This narrow-gauge line ascends to the right, passing through a tunnel, to the station of *Les Frêtes*, whence it proceeds through wooded valleys and meadows. Farther on the train skirts the deep gorge of the *Bied* (beyond which runs the line to Morteau, see above) and passes through two tunnels into the valley of the Doubs, with the large village of *Les Brenets* (*Couronne*; *Lion d'Or*; *Bellevue*). From the station we descend through the village to the (15, ascent 20 min.) *Pré du Lac*, on the *Lac des Brenets*, a lake 3 M. in length, which the Doubs forms above the waterfall. A boat (3 fr. there and back, more than 3 pers. 1 fr. each), or the small steamboat which plies on Sundays (for large parties also on week-days) now conveys us down the dark-green lake, gradually narrowing between wooded sandstone rocks, and presenting a series of picturesque scenes. In ½ hr. we reach the *Saut du Doubs* (*Hôt. du Saut du Doubs*, with garden, on the Swiss side; *Hôt. de la Chute*, on the French side, both unpretending). In about 6 min. from the French inn we obtain a fine view from a point high above the picturesque waterfall, which is 80' in height. A road through woods, affording charming glimpses of the basin of the Doubs, leads back to (3 M.) *Les Brenets*.

60. From Neuchâtel to Pontarlier through the Val de Travers.

33½ M. RAILWAY in 1¾-2¾ hrs.; fares 6 fr. 75, 4 fr., 2 fr. 80 c. (From Pontarlier to Paris by Dijon, express in 10½ hrs.; from Bern to Paris 14¼ hrs.) This Jura Railway (comp. p. 209) also traverses a most picturesque country. The most striking points are between Neuchâtel and Noiraigue, between Boveresse and the last tunnel above St. Sulpice, and between St. Pierre de la Cluse and Pontarlier. Finest views to the left.

Neuchâtel, see p. 207. The line, running parallel with that to Yverdon (p. 213) as far as Auvernier, crosses the *Seyon* (p. 207). Beyond a short tunnel under the Val de Travers road we enjoy a beautiful *View of the lake and the Alps (comp. p. 207). The train skirts lofty vine-clad slopes, and crosses the *Gorge of Serrières* by a bold viaduct. In the valley is *Suchard's* large chocolate factory, and above it rises the small château of *Beauregard*.

4 M. **Auvernier**; the little town lies below, to the left (1480'; *Hôtel du Lac*, moderate). The train diverges to the right from the

Yverdon line (p. 213), and as it ascends we enjoy an admirable view of the lake and the Alps. On entering the rocky and wooded ravine of the *Reuse* or *Areuse* we observe the lofty viaduct of the Lausanne line (p. 214) far below us to the left. The last glimpse of the lake down this romantic valley is particularly picturesque. We soon enter a tunnel, high on the N. slope of the valley, almost under the station of Chambrelien (p. 209). Seven more tunnels, beyond the fourth of which is the (8½ M.) station of *Champ du Moulin* (2020'; Hôt. des Gorges, trout) in a picturesque situation (hence to the *Gorges de l'Areuse*, see p. 214).

Artificial conduits supply Neuchâtel and Chaux-de-Fonds (13 M. distant) with spring water from this point; the engine-house (2067'), ¼ hr. up the Reuse to the left, is interesting. The neighbouring house of Lieutenant-Colonel Perrier was, according to the inscription, once occupied for some time by J. J. Rousseau. A footpath, behind the water-wheels, leads along the left bank of the Reuse to the (1½ hr.) *Saut de Brot*.

12 M. **Noiraigue** (2360'; *Croix Blanche*), at the N. base of the *Creux du Van*. The valley, called the *Val de Travers* from this point to St. Sulpice, changes its character here, and the Reuse now flows calmly through a grassy dale.

From Noiraigue a path ascends the **Creux du Van** (4807') in 2 hrs., a better route than from *Boudry* (p. 213) or *St. Aubin* (p. 202), as the striking view, extending from Pilatus to Mont Blanc, is suddenly revealed. At the top is a basin, 500' deep, shaped like a horseshoe, and nearly 3 M. in circumference. Within this is an excellent spring, to which the descent is steep and fatiguing but without danger. When the weather is about to change, this 'hollow of the wind' is filled with surging white vapour, which rises and falls like the steam in a boiling cauldron, but does not quit the basin. Rare plants and minerals are found here. Simple refreshments may be obtained at the *Ferme Robert*, at the top.

From (14½ M.) *Travers* (2392'; Ours) a branch-line runs along the bottom of the valley viâ *Couvet*, *Môtiers*, and *Fleurier*, to *Buttes* and *St. Sulpice* (see below). Farther on, on the opposite side of the valley, are asphalt-mines. — 17 M. **Couvet** (2418'; *Ecu de France*) a pretty town. Here, and at Môtiers and Fleurier, excellent absinth is manufactured.

The line again ascends the N. slope of the valley. Opposite, far below, lies *Môtiers-Travers* (2415'; Maison de Ville), where, by permission of the Prussian governor Lord Keith, Rousseau lived in 1762 after his expulsion from Yverdon by the government of Bern, and wrote his 'Lettres écrites de la Montagne'.

The Ravine of the Raisse (affluent of the Reuse), with its picturesque rocks and waterfalls, deserves a visit. About ½ M. from Môtiers we pass a bridge and follow the brook to the right, ascending a pretty wooded gorge. In 1 hr. we reach a new path, leading to the top (35 min.). From this point, with the aid of a guide or a good map, we may ascend the Chasseron (p. 214). — Behind Môtiers is the *Grotte de Môtiers*, a limestone cavern, one arm of which is 3½ M. long. It may be safely explored for about ½ M. (rough walking; swarms of bats). At the entrance is a waterfall.

19 M. **Boveresse**, above the village of the name. In the valley, farther on, is **Fleurier** (2455'; *Poste; Couronne*), with extensive watch and absinth-factories. **Hence** to the top of the *Chasseron* in 2½ hrs., see p. 215. Beyond a long tunnel, we observe *St. Sulpice* (2557')

below us, on the left, with a large Portland cement factory. Scenery again very picturesque. Two bridges and two tunnels. In the valley, 1½ M. to the W., of Fleurier, the Reuse, which probably flows underground from the *Lac des Taillères*, rises in the form of a considerable stream, soon capable of working a number of mills. Road and railway pass through the defile of *La Chaîne*.

The line attains its highest point, and then enters a monotonous green valley with beds of peat. At (25 M.) Verrières Suisse (3060'; *Balance*), the last Swiss village, the French 'Army of the East' under Bourbaki crossed the frontier in Feb., 1871. The train enters France (luggage examined at Pontarlier, see below) before reaching (26 M.) *Verrières-France* (3015'). Near *St. Pierre de la Cluse* the scenery again becomes interesting. The defile of *La Cluse*, which railway and road both traverse, is fortified; on the left rises the ancient *Fort de Joux*, which was blown up with dynamite in 1877, overtopped by a new fort on a bold rock to the right. Mirabeau was imprisoned here in 1775 at the instance of his father; and in 1803 Toussaint Louverture, the negro chieftain of St. Domingo, died in the fort, where he had been confined by Napoleon. We cross the *Doubs*, which drains the *Lac de St. Point*, 3½ M. to the S.W., and follow its left bank to Pontarlier.

33½ M. **Pontarlier** (2854'; *Hôtel de la Poste*, Grande Rue, R. 2 fr.; *Hôt. de Paris; Hôtel National*; *Rail. Restaurant*, D. incl. wine 3-4 fr.), a small town on the *Doubs*. Luggage examined here. See *Baedeker's Northern France*.

From Pontarlier to *Cossonay* and *Vallorbes*, see R. 64.

61. From Neuchâtel to Lausanne.

47 M. Railway in 2-2½ hrs.; fares 7 fr. 80, 5 fr. 50, 3 fr. 90 c. (to Geneva in 2¾-5 hrs.; fares 12 fr. 70, 8 fr. 90, 6 fr. 30 c.). — Steamboat on the *Lake of Neuchâtel* between Neuchâtel and *Morat* (p. 219), and between Neuchâtel and *Estavayer* only (twice daily in 1½ hr., corresponding with the train to Freiburg, p. 217).

Neuchâtel, see p. 207. Route to (4 M.) *Auvernier*, see p. 211. The Lausanne train quits the lake, to which it returns beyond Bevaix (p. 214). — 5 M. **Colombier** *(Cheval Blanc)*, with an old château converted into a barrack, and beautiful avenues, yields excellent white wine. (On the lake, 1½ M. to the E., is the *Chanélaz Hydropathic*, with pleasure-grounds and views; pens. 6-8 fr.) — 6 M. **Boudry** (1693'); the little town (1542'; *Maison de Ville*), the birthplace of Marat (1744-1793), lies below the line, on the right bank of the Areuse, 1 M. from the station. Steam-tramway to Neuchâtel, see p. 207.

The *Gorges de l'Areuse* are interesting. Leaving stat. Boudry, we cross the line (passing the viaduct on the left) and pass through the village of *Trois-rods*. Before the last house we turn to the left, between walls, and descend in 20 min. to the entrance to the ravine. A path, hewn in the rock at places, affords striking views of the narrow, wooded gorge. In 5 min. we come to a path to the left, leading to the *Chalet aux Clées* (donation for the use of the path expected). In 20 min. more we observe the *Grotte aux Fours*, above us, on the right, with a large entrance (easily accessible). Farther on the Pontarlier railway runs high above the gorge, on the right, and still higher is the carriage-road. We next reach (55 min.; 1 hr. 40 min. from Boudry station) the *Champ du Moulin*, picturesquely situated (station for several trains, p. 212). — Perhaps a more convenient way of making this excursion is to take the train to Champ du Moulin and then to walk down through the Gorges to Boudry. Another path descends to the Gorges from *Chambrelien* (p. 209). Circular ticket from Neuchâtel and back viâ Chambrelien and Boudry, 2nd cl. 1 fr. 40 c., 3rd cl. 1 fr.

From Boudry to the *Creux du Van* (p. 212), 3 hrs.

Beyond Boudry the train is carried by a great viaduct over the deep valley of the *Areuse* or *Reuse*. The stream falls into the lake near *Cortaillod*, where the best red wine in the canton is produced. 9 M. *Bevaix* (1568'). The line returns to the bank of the lake, which it follows to Yverdon. 11 M. *Gorgier-St-Aubin*; 14 M. *Vaumarcus*, with the fine well-preserved castle of that name. At (16 M.) **Concise** (1453'; *Ecu de France*) many traces of ancient lake-villages have been found. To the right, above, lies *Corcelles*, near which are three blocks of granite, 5' to 8' in height, placed in the form of a triangle, but not visible from the line. They are said to commemorate the battle of Grandson, but are more probably of Celtic origin. — 18 M. *Onnens-Bonvillars*.

21 M. **Grandson** *(Lion d'Or; Croix Rouge; Hôtel de la Gare)*, a picturesque little town (1708 inhab.) probably of Roman origin, has a handsome old *Château* of Baron de Blonay, now restored. (*View from the terrace.) The old *Church*, Romanesque with a Gothic choir, once belonged to a Benedictine abbey.

The château of Grandson, originally the seat of a family of that name and said to have been built about the year 1000, was taken by the Bernese in 1475, and in Feb., 1476, was captured by Charles the Bold, Duke of Burgundy. A few weeks later, on 3rd March, 1476, the Duke was surprised by the advancing Confederates near Grandson, and notwithstanding his numerical superiority (50,000 Burgundians, it is said, against 20,000 Swiss) was utterly defeated. Part of the enormous booty captured on the occasion is still preserved in the Swiss arsenals.

The train skirts the S.W. end of the lake, and crosses the *Thièle* near its influx into the lake.

24 M. **Yverdon** (1433'; 6330 inhab.; *Hôt. de Londres*, R. & A. 2½, D. 3 fr.; *Paon*), the Roman *Eburodunum*, is a thriving little town on the Thièle, with pleasant promenades and fine views. The *Château*, erected by Duke Conrad of Zähringen in 1135, and the seat of Pestalozzi's famous school in 1805-25, is now occupied by the town-schools, a library, and a museum of Celtic, Roman, and other antiquities. Near the churchyard are some mural fragments of a Roman fort. To the S.E. (³/₄ M.) are the *Bains d'Yverdon*, with a sulphur spring and a Curhaus (pens. 7 fr.), halfway to which are the *Pension La Prairie* (5-6 fr.) and the *Maison Blanche* (pens. 4-4½ fr.), both with gardens.

The **Chasseron** (5285'), a height of the Jura, to the N.W. of Yverdon, commands a fine view. Diligence twice daily in 3¼ hrs. to *Ste. Croix* (3635'; Pens. Jacques; 1½-2 hrs. from the top), noted for its musical boxes. The descent may be made, if desired, by a good road to (1½ hr.) *Fleurier* (p. 212). — The *Aiguille de Beaulmes* (5128') and *Mont Suchet* (5238') are also fine points (3½-4 hrs.; comp. p. 221).

From Yverdon to *Payerne* and *Freiburg*, see p. 217.

The train quits the lake, and enters the broad valley of the *Thièle*, a stream formed by the confluence of the *Orbe* (p. 220) and the *Talent* near stat. *Ependes*. To the W. rises the long chain of the Jura: the Aiguille de Beaulmes and Mont Suchet (see above), between which are the Mont d'Or, the Dent de Vaulion (p. 221), and Mont Tendre.

30 M. *Chavornay-Orbe* (the town of *Orbe* lies 1½ M. to the N. W.; omnibus at the station; p. 220). **Two tunnels under the *Mauremont*.** Then (33½ M.) *Eclépens*. The train enters the wooded valley of the *Vénoge*, which is connected with the Thièle by the *Canal d'Entreroches*, passes *La Sarraz* (p. 220), and stops at —

38 M. **Penthalaz-Cossonay** (1850'; *Hôt. des Grands Moulins*); the little town of Cossonay lies on a wooded hill to the right. — To *Vallorbes* and *Pontarlier*, see R. 64.

Beyond (43 M.) *Bussigny*, to the S., appear the mountains of Savoy. 44½ M. *Renens*.

47 M. *Lausanne*, see p. 236.

62. From Bern to Lausanne *(Vevey)*.

61 M. RAILWAY to Freiburg in ¾-1¼ hr. (3 fr. 35, 2 fr. 35 c., 1 fr. 70 c.; to Chexbres in 3-3½ hrs. (8 fr. 95, 6 fr. 30, 4 fr. 50 c.); to Lausanne in 3¼-4 hrs. (10 fr. 20, 7 fr. 15, 5 fr. 10 c.); to Geneva in 5½-6½ hrs. (16 fr. 55, 11 fr. 60, 8 fr. 30 c.). — Travellers to Vevey had better alight at Chexbres (comp. p. 218). Best views on the left.

Bern, see p. 144. To the left we obtain a glimpse of the Bernese Alps, and the mountains of the Simme and Sarine valleys, among which the serrated Brenleire (7743') and Foliéront (7690') are conspicuous; more to the right is the Moléson. This view is soon hidden by wood. 3 M. *Bümplitz*; 6 M. *Thörishaus*. The train descends and crosses the *Sense*, the boundary between the cantons of Bern and Freiburg. — 9 M. *Flamatt*.

To the W. (5½ M.; diligence thrice daily in 50 min., viâ *Neuenegg*) lies Laupen (*Bär*), a small town with an ancient château, at the confluence of the *Sense* and the *Sarine*, famed in the annals of Switzerland for a victory gained in 1339 by the Bernese under *Rudolph von Erlach* (p. 147) over the army of Freiburg and the allied nobility of the Uechtland, Aargau, Savoy, and Burgundy. The *Bramberg*, ½ M. to the N. of the road to Neuenegg, is marked by a monument, erected in 1829.

Beyond the next tunnel we enter the green valley of the *Taferna-Bach*. 12½ M. *Schmitten;* 16 M. *Düdingen* (Fr. *Guin*), where we cross a viaduct, 100' high. Beyond *Balliswyl*, which lies to the left, the **train crosses the profound gorge of the *Saane* or *Sarine*** by means of the huge iron **Viaduc de Granfey*, 360 yds. in length and 250' in height.

20 M. **Freiburg**. — *HÔTEL SUISSE, R., L., & A. 2-3, B. 1¼, D. 3, pens. 6-8 fr.; FAUCON; TÊTE NOIRE, R., L., & A. 1½-2, B. 1, D. 2, pens. 6½ fr.; CROIX BLANCHE, unpretending; HÔT.-PENS. BELLEVUE, ½ M. from the town, beyond the suspension-bridge, well spoken of.

Freiburg (2100'; pop. 12,239), Fr. *Fribourg*, the capital of Canton Freiburg, the ancient *Uechtland*, founded in 1178 by Berthold IV. of Zähringen, stands like Bern on a rocky height nearly surrounded by the *Sarine (Saane)*. The town retains some of the ancient walls and towers, and is the seat of a Roman Catholic university opened in 1889. Most of the inhabitants speak French. The town lies on the boundary between the two tongues, and German is still spoken in the lower quarters. As the picturesque situation of the town and its bridges is not seen from the railway-station, the following walk of 1 1/2 hr. is recommended.

From the station, to the left, past the little Protestant church and through the suburb to (7 min.) an open space (where the **Rue du Musée** ascends to the left, see p. 217), and thence by the Rue de Lausanne to the PLACE DE L'HÔTEL-DE-VILLE. Here stands a venerable lime-tree, 14' in circumference, supported by stone pillars.

According to tradition, this tree was originally a twig, borne by a young native of Freiburg when he arrived in the town, breathless and exhausted from loss of blood, to announce to his fellow-citizens the victory of Morat (1476). 'Victory' was the only word he could utter, and having thus fulfilled his mission, he expired.

To the right rises the old *Hôtel de Ville*, on the site of the palace of the Dukes of Zähringen. The octagonal clock-tower dates from 1511. — To the left of the lime-tree the Rue du Tilleul leads past a bronze *Statue of Father Grégoire Girard* (d. 1850) to the Gothic —
*CHURCH OF ST. NICHOLAS, founded in 1283, rebuilt in the 15th cent., and restored in 1860. The handsome tower, 280' high, was erected in 1470-92; the portal is adorned with curious reliefs of the Last Judgment.

The large *Organ*, with 67 stops and 7800 pipes, some of them 32' in length, was built by *Al. Mooser* (d. 1839), whose bust has been placed to the left of the entrance. Performances in summer at 1.30 and (except Sat. and the eves of festivals) 8 p.m. daily. Adm. 1 fr. — The late-Gothic carved *Stalls* deserve notice. The second chapel on the S. side contains a picture by *Deschwanden*, St. Anne and St. Mary. The choir has three modern stained-glass windows (St. Nicholas and other saints). A tablet on the S. pillar at the entrance to the choir is to the memory of *Canisius* (d. 1597), a famous Jesuit, who is buried in St. Michael's Church (p. 217).

Behind the choir of St. Nicholas is the *Post Office*, and a little to the left the great *SUSPENSION BRIDGE, or *Grand Pont Suspendu*, constructed by Chaley in 1834, 270 yds. long and 168' above the Sarine. It is supported by six wire-ropes, 410 yds. in length, the extremities of which are secured by 128 anchors attached to blocks of stone far below the surface of the earth.

Ascending the right bank for about 1/3 M., we reach the PONT DE GOTTERON (250 yds. long, 245' high), a similar bridge, constructed in 1840 over the *Vallée de Gotteron*, a deep ravine descending to the Sarine. — We cross this bridge and follow the road on the other side. After 5 min. we take a short-cut to the right, regain the road, and descend to the right, through the old *Porte de Bourguillon*, to the (12 min. from the Pont de Gotteron) picturesquely situated *Loretto Chapel*, built in 1648, restored in 1888 (fine view of the town). Farther on, we obtain to the left a view of the valley of the Sarine, which has been converted into a reservoir to supply the town. A path with steps descends 5 min. from the chapel to the lower town, turning to the left at the fountain and passing the church of *St. John* (founded by the Knights of Malta), beyond which we cross the Sarine by a stone bridge, and either ascend by the steps to the (5 min.) Hôtel de Ville, or follow the road to the left leading to the (¹/₄ hr.) station.

Those who have sufficient time may follow the Rue du Musée, mentioned at p. 216, to the old Jesuits' COLLÈGE ST. MICHEL, founded in 1580 by Father Canisius. The *Lycée*, next the Collège, contains the valuable CANTONAL MUSEUM.

Two rooms on the groundfloor contain the *MARCELLO MUSEUM, bequeathed to the town by the sculptress Duchess Adèle Colonna (d. 1879), a native of Freiburg, who assumed the name of *Marcello*: busts and statues (Abyssinian sheikh; Pythia, from the Opera House at Paris) by Marcello; pictures by her, and by Regnault, Hébert, Delacroix, Fortuny, Courbet, and others; furniture, etc.; also the *Cantonal Picture Gallery* of ancient and modern works. — On the first floor (five rooms) is a valuable collection of antiquities from lake-dwellings, Roman and Swiss relics, ethnographical objects, weapons and armour, coins, etc. — The second floor (two rooms) contains zoological and physical, the third floor mineralogical and botanical collections.

FROM FREIBURG TO YVERDON, 31¹/₂ M., railway in 2 hrs. (3 fr. 75 c. or 2 fr. 65 c.). Near (3¹/₂ M.) *Belfaux* is a huge embankment, forming an aqueduct for the *Sornas*, 150 yds. in length. Stat. *Grolley*, *Léchelles*, *Cousset*, *Corcelles*, and (14¹/₂ M.) *Payerne* (p. 219), the junction of the 'Ligne de Broye'. We cross the *Broye* and the *Glane*. 16¹/₂ M. *Cugy*. — 20 M. Estavayer (*Maison de Ville*; *Cerf*), a little town with the picturesque château of *Chênaux*, on the Lake of Neuchâtel. (Steamer twice daily by *Cortaillod* and *Auvernier* to *Neuchâtel*, p. 207.) — 23¹/₂ M. *Cheyres*; 26 M. *Yvonand*, on a tongue of land projecting far into the lake, at the mouth of the *Mentue*, where Roman relics have been found. 31¹/₂ M. *Yverdon* (p. 214).

To the S.E. of Freiburg (15 M.; road by *Rechthalden* and *Plaffeyen*; diligence in summer daily in 4 hrs.), in the valley of the *Sense*, is the Schwarze See (*Lac Noir*, 3365'), amidst lofty mountains, and well stocked with fish. On its bank lies the *Schwarzsee-Bad*, or *Bains Domène* (R. 1-3, board 4-6 fr. per day), with sulphur-springs. The *Kaiseregggschloss* (7188'), to the S.E. (3 hrs., with guide), commands the Bernese and Valaisian Alps. — From the Schwarze See over the *Chésalette* to (10¹/₂ M.) *Charmey*, see p. 208; over the *Gantrisch Pass* to *Thun*, p. 203.

Berra (*Birrenberg*, 5655'), 4¹/₂-5 hrs. from Freiburg, interesting. Road by *Marly*, a village prettily situated on the *Gérine* (*Aergerenbach*), to (6 M.) *Le Mouret*; thence a bridle-path up the *Käsenberg* to the (2¹/₂ hrs.) top. Extensive view of the Jura, the lakes of Neuchâtel, Morat, and Bienne, and the Alps. Descent to *Valsainte* (p. 208) ³/₄ hr., to the *Schwarze See* 1¹/₂ hr.

As the train proceeds we enjoy a view of the Simmen-Thal and Freiburg Mts. to the left, the Moléson being conspicuous. The *Glane*, with its precipitous banks, and a bridge of four arches which carries the road across it, are also seen to the left. 24 M. *Matran*; 25¹/₂ M. *Rosé*; 27 M. *Neyruz*; 28¹/₂ M. *Cottens*; 30 M. *Chénens*. Near (33 M.) *Villaz-St-Pierre* the train enters the valley of the *Glane*; on the left are the fertile slopes of the *Gibloux* (3947'). Near Romont, to the left, is the nunnery of *La Fille Dieu*.

36 M. **Romont** (2325'; pop. 1885; *Cerf; Couronne; *Croix Blanche*), a little town on the Glane, with ancient walls and watchtowers, is picturesquely situated on a hill. The *Castle* on the S. side, founded by the Burgundian kings in the 10th cent., is now occupied by the local authorities. The old Gothic *Church* contains choir-stalls with grotesque carving. At the S. end of the hill rises a massive round tower; the adjoining grounds afford a pleasing view.

From Romont to Bulle (p. 262), 12 M., branch-line in 53 minutes. Stations *Vuisternens, Sales, Vaulruz* (p. 262).

39½ M. *Siviriez*. A tunnel pierces the watershed between the Glane and the Broye. 42 M. *Vauderens*. To the right lies the valley of the *Broye*, with the Payerne railway and the town of *Rue* (see below). At (46 M.) *Oron-le-Châtel* (2378') we pass through a cutting in the castle-hill to the station on the S. side; *Oron-la-Ville* lies below, to the right (see below). The train now descends and crosses the *Mionnaz* and the Broye. 48 M. *Palézieux* (see p. 219). We again ascend slightly, traversing a smiling tract, to (53½ M.) *Chexbres* (2043').

The *Signal de Chexbres* (2145'; '*Hôt. du Signal*, with extensive grounds), 25 min. from the station, affords a superb view. At our feet lies the greater part of the Lake of Geneva; to the left Vevey; above it, from left to right, are the saddle of the Col de Jaman, the tooth-like **Dent** de Jaman, the broad back of the Rochers de Naye, and the Tour d'Aï and Tour de Mayen; farther back, the Grand-Moveran and the Dent de Morcles. In the centre of the background is the pyramid of Mont Catogne; on its left rise the snowy cones of Mont Velan and Grand Combin; to the right the Savoy Mts., with the Dent d'Oche. — Travellers bound for Vevey may descend direct from the Signal to the (25 min.) village of Chexbres.

From Chexbres to Vevey, 4½ M., diligence **thrice** daily in 50 min. (ascent from Vevey to Chexbres 1½ hr.). On a cool morning or evening travellers will find the walk from Chexbres to **Vevey** (1½ hr.) very pleasant, but in the reverse direction it is apt to be hot and tiring. Luggage may be forwarded by railway. — The road leads through (1 M.) the large village of *Chexbres* (1940'; "**Hôt.** Victoria, with garden and fine view, pens. from 5 fr.; '**Lion d'Or**), with its old castle (whence a path descends direct to *Rivaz-St-Saphorin*, a station on the W. Railway, p. 245), and then descends, in view of the beautiful lake and the Savoy Mts., to the Lausanne and Vevey road and (3 M.) *Vevey* (p. 239).

Beyond the next tunnel (506 yds.) a ****View of singular beauty**, embracing the greater part of the Lake of Geneva and the surrounding mountains, is suddenly disclosed. In the direction of Vevey, which is not itself visible, are the Pléiades, the Dent de Jaman, the valley of the Rhone, and the Savoy Mts.; in the foreground lie numerous villages amidst vineyards. Beyond a tunnel (through which the setting sun shines in summer) and stat. *Grandvaux (Cully)* we observe the villages of Lutry, Pully, and Ouchy on the lake, and Lausanne on the hill above them. Beyond another tunnel and a viaduct we reach (58½ M.) *La Conversion (Lutry)*, and cross the valley of the *Paudèze* (p. 239) by a viaduct of nine arches. After another short tunnel our train reaches the Lausanne and Vevey line.

61 M. *Lausanne*, see p. 236.

63. From Lausanne to Payerne and Lyss.

63 M. RAILWAY in 4½ hrs.; fares 7 fr. 45, 5 fr. 35 c. (no 1st class).

To *Palézieux* (13 M.), see p. 218. We follow the pleasant valley of the *Broye*. 15 M. *Palézieux-halte*; 17½ M. *Châtillens* (½ M. to the N.E. is *Oron-la-Ville*, p. 218). — 20 M. *Ecublens-Rue*. The little town of **Rue** (2323'; *Maison de Ville*; *Fleur de Lis*) lies on a hill to the right, commanded by an old château. — 23 M. *Bressonaz*.

24½ M. **Moudon** (1690'; pop. 2647; *Hôt. du Pont*; *Couronne*; *Hôt. de la Ville*), with the châteaux of *Carouge* and *Rochefort*, an old town, the Roman *Minodunum*, and long the capital of the Pays de Vaud. Handsome Gothic church. — Farther on we cross the Broye twice. 27½ M. *Lucens*, with a picturesque old château; 30 M. *Henniez*, to the left of which are the old château and church of *Surpierre*, on a lofty crag; 32 M. *Granges-Marnand*.

37 M. **Payerne** (1480'; pop. 3673; *Ours*; *Croix Blanche*), an old town, the Roman *Paterniacum* (?), was in the 10th cent. a frequent residence of the Kings of Burgundy, whose rule then extended over the modern Franche-Comté, Switzerland as far as the Reuss on the E., and part of Savoy.

Bertha of Swabia, wife of Rudolph II. (912-937), erected a church and Benedictine abbey here, the former now a granary, the latter a school. Her bones, with those of her husband and her son Conrad, were discovered in 1864, and were buried in the Parish Church, where the queen's saddle with a hole for her distaff is shown. To this day the expression, 'Ce n'est plus le temps où Berthe filait', is a regretful allusion to the 'good old times'.

From Payerne to *Freiburg* and *Yverdon*, see p. 217.

The valley of the Broye becomes broad and marshy. 38½ M. *Corcelles*; 40½ M. *Dompierre*; 42 M. *Domdidier*.

43½ M. **Avenches** (1519'; pop. 1864; *Couronne*; *Hôtel de Ville*), now a small town, was the ancient capital of the Helvetii, the Rom. *Aventicum*.

Remains of an *Amphitheatre* and other buildings, and of the old town-walls, testify to its former prosperity. The mediæval *Castle*, at the entrance to the town, occupies the site of the Roman capitol. To the N.W. rises a solitary Corinthian column 39' high, the remnant of a temple of Apollo, now called *Le Cigognier*, from the stork's nest which has occupied it for centuries. The *Museum* (custodian lives near the church; small fee) contains mosaics, inscriptions, and other relics recently found here; in its garden is the above-mentioned amphitheatre.

In his *Childe Harold* (iii. 65) Lord Byron alludes to the 'Cigognier': —
'By a lone wall a lonelier column rears
A grey and grief-worn aspect of old days.'

At (45½ M.) *Faoug* (Soleil; Hôt.-Pens. Wicky) we approach the **Lake of Morat** (1428'), the Roman *Lacus Aventicensis* and the *Uecht-See* of the middle ages, 5½ M. long. It is separated from the Lake of Neuchâtel by the narrow *Mont Vully* towards the N. and the *Charmontel* to the S., but connected with it by the *Broye*.

47½ M. **Morat**, Ger. *Murten* (1522'; pop. 2360; *Couronne*; *Croix, R. 1½-2, D. incl. wine 2½, pens. 4½ fr.; *Lion*; *Pens. Kauer*, on the lake, moderate; *Rail. Restaurant*), an ancient little town with well-preserved gates and walls, which in 1476, with a garrison of 1500 Bernese under Adrian von Bubenberg, resisted the artillery of Charles the Bold for ten days before the battle of Morat. Its narrow arcaded streets are overshadowed by an old *Castle*. The *School* contains a collection of Burgundian weapons. *Lake Baths* next the Pension Kauer, at the S. end of the town.

About 1½ M. to the S. of Morat, near the lake, rises a marble *Obelisk*, erected in 1822 in memory of the Battle of Morat, which was fought on 22nd June, 1476. This was the bloodiest of those three disastrous contests (Grandson, Morat, and Nancy), in which the puissant Duke of Burgundy successively lost his treasure, his courage, and his life ('Gut, Muth, und Blut'). The Burgundians lost 15,000 men and all their military stores.

The STEAMBOAT FROM MORAT TO NEUCHÂTEL (twice daily in 2½ hrs.) crosses the lake to *Motier* and *Praz*, at the E. base of the vine-clad *Mont Vully* (2267'); at *Sugiez* it passes under a wooden bridge and enters the *Broye*. To the W. stretches the Jura, from the Weissenstein to the Chasseron. Near *La Sauge* we enter the *Lake of Neuchâtel* (p. 207), steering first S.W. to *Cudrefin*, and afterwards N.W. to *St. Blaise* and *Neuchâtel* (p. 207).

Near (50½ M.) *Galmitz*, Fr. *Charmey*, we leave the lake. To the left is the *Grosse Moos*, an extensive marshy tract, partly reclaimed of late. 52½ M. *Kerzers*, Fr. *Chiètres* (*Pens. Mösching, 4-4½ fr.); 54½ M. *Fräschels*, Fr. *Frasse*; 57 M. *Kallnach*.

59½ M. **Aarberg** (1470'; pop. 1249; *Krone*), an old town on an island in the *Aare*. Adjoining the church is the old castle of the Counts of Aarberg, who sold their dominions to Bern in 1351.

The train crosses the Aare to (63 M.) *Lyss*, on the Bienne and Bern line (p. 12).

64. From Lausanne to Vallorbes and Pontarlier.

45 M. RAILWAY in 2½-3 hrs. (7 fr. 70, 5 fr. 35, 3 fr. 70 c.). Express from Lausanne to Paris by this route (327 M.) in 10½ hrs. (58 fr. 50, 39 fr. 65, 26 fr. 5 c.).

To (9 M.) *Penthalaz-Cossonay*, see p. 215. The train diverges to the left from the Yverdon line at *Villars-Lussery*. 15 M. **La Sarraz** (1647'; *Maison de Ville*), a small town with an old château. Two short tunnels. We then ascend to (18 M.) *Arnex* (1791'); ¾ M. to the N. lies the picturesque little town of **Orbe** (1460'; 1947 inhab.; *Deux Poissons*), on the *Orbe*, which is crossed here by two bridges. In the 10th cent. Orbe was one of the capitals of Little Burgundy, to which period belong the two towers of the château (view from the terrace).

The line then leads in long windings, affording a splendid view, at first to the right, then to the left, of the entire Alpine chain from the Mont Blanc to the Jungfrau, to *Bofflens* and (22 M.) *Croy-Romainmotier*, 1½ M. from the small and ancient town of *Romainmôtier* (2295'; Maison de Ville). Farther on the train skirts wooded hills; on the right, in the deep valley of the *Orbe*, lies the village of *Les Clées* with its castle, and high on the left bank are the villages of *Lignerolles*, whence *Mont Suchet* (5235') is easily ascended in

2 hrs., and *Ballaigues* (*Hôt.-Pens. la Sapinière; *Pens. Maillefer, ¹/₂ M. to the E.), visited as a summer-resort (Engl. Church service). Two short tunnels; then (26 M.) *Le Day*, the junction for Le Pont (see below). Near Vallorbes we cross the Orbe by a handsome iron bridge above the influx of the *Jougnenaz*.

28¹/₂ M. **Vallorbes** (2520'; 2147 inhab.; *Hôtel de Genève*, at the station; *Maison de Ville*, *Croix Blanche*, both moderate), a watchmaking place, at the base of the *Mont d'Or* (4818'), almost totally rebuilt since the fire of 1883.

FROM VALLORBES TO LE PONT, 7¹/₂ M., railway in 40 minutes. To (2¹/₂ M.) *Le Day*, see p. 208. The line to Le Pont diverges here to the right and, skirting the wooded slopes of the Dent de Vaulion, gradually ascends to the tunnel (500 yds. long) under the *Mont d'Orzeires* (3395'), whence it descends along the *Lac Brenet* (see below) to —

7¹/₂ M. **Le Pont** (*Truite*), a hamlet at the N. end of the *Lac de Joux* (3310'; 5 M. long, 1¹/₄ M. broad), which is separated from the little *Lac Brenet* by an embankment with a bridge. On the N. side of the Lac Brenet are a number of apertures (*entonnoirs*) in the rocks, serving to drain the lake, the waters of which, after a subterranean course of 3 M., re-appear as the so-called *Source of the Orbe*, 750' lower.

Le Pont lies at the S. foot of the *Dent de Vaulion (4880'), the W. side of which presents a barren and rugged precipice, 1600' high, while the E. side is a gentle, grassy slope. The top is reached in 1¹/₂ hr. from Le Pont (guide convenient). View of the Lac de Joux, the Lac des Rousses, the Noirmont, and the Dôle; to the S.E., part of the Lake of Geneva, and beyond it Mont Blanc and the Alps of the Valais; lastly the Bernese Oberland.

A small steamboat plies on the idyllic Lac de Joux (3010'; 5 M. long, 1¹/₄ M. broad) to Rocheray (50 min.; 60 c.). It crosses from Le Pont to *L'Abbaye*, a prettily situated hamlet on the E. bank, whence the *Mont Tendre* (5512') may be ascended in 2 hrs. (fine view). The following stations are *Le Lieu*, on the W. bank; *Grosjean* and *Bioux*, on the E. bank; and *Le Rocheray* (Hôt. Bellevue), at the S. extremity of the lake. Omnibus hence to (³/₄ M.) *Le Sentier* (*Pens. Guignard; Union; Hôt. de Ville; Lion d'Or). About 2 M. higher up the Orbe is the village of *Le Brassus* (3412'; Hôt. de la Lande; Hôt. de France); thence over the *Col de Marchairuz* to (16¹/₂ M.) *Rolle*, see p. 236.

The train backs out from the station, describes a wide curve and ascends the pretty, wooded valley of the *Jougnenaz*, where it soon enters French territory. A short and a long tunnel are passed through before (35 M.) *Hôpitaux-Jougne*. We then cross the highest ridge of the Jura and descend through wooded and rocky valleys to (42 M.) *Frambourg*. Near the *Fort de Joux*, before the defile of *La Cluse* (p. 213), we join the Neuchâtel line.

45 M. *Pontarlier*, see p. 213.

65. Geneva and its Environs.

Arrival. PRINCIPAL STATION (*Gare de Cornavin*; Pl. D, 2), for the Swiss Jura-Simplon and the French Paris, Lyons, & Mediterranean lines, on the right bank, at the upper end of the Rue du Montblanc. *Omnibus* from the station to all the hotels (and from the hotels to the station) 50 c., luggage 30 c. — STATION OF EAUX-VIVES (*Gare des Vollandes*), for Annemasse, Cluses, Annecy, Bouveret, and Bellegarde, on the left bank (Pl. F, 8; tramway to the Place du Molard and the Cornavin Station). French railwaytime is 55 minutes behind Central European time. — STEAMBOAT PIERS on the S. (left) bank by the Jardin Anglais, and on the N. (right) bank by the Quai des Pâquis and (for the express boats only) the Quai du Montblanc.

Hotels. *On the Right Bank*, with view of the lake and the Alps *Hôtel National (Pl. f.; F, 2), a large house on the Quai du Léman; *Hôt. des Bergues (Pl. a; D, 4), Quai des Bergues; *Hôt. de Russie (Pl. b; D, 4) and *Hôt. de la Paix (Pl. c; D, 4), on the Quai du Montblanc, R., L., & A. from 4, B. 1½, lunch 3-4, D. 5, pens. 10, omn. with luggage 1¼ fr.; *Hôt. Beau-Rivage (Pl. d; E, 4), on the Quai des Pâquis, R., L., & A. from 5, lunch 3-4, D. 5, pens. 10 fr.; *Hôt. d'Angleterre (Pl. e; E, 4), Quai du Montblanc, R., L., & A. from 3, B. 1½, lunch 3, D. 4, pens. from 8 fr. — *Hôt. Richemond (Pl. r; E, 3, 4), Rue Adhémar Fabri, with view of the Pont du Montblanc, R., L., & A. from 3, B. 1½-1½, lunch 3, D. 5½, pens. from 7 fr. — Also on the right bank, near the station, without view, and rather of the second class: Schweizerhof (Pl. p; D, 3), Rue du Montblanc, R., L., & A. from 3, B. 1½, lunch 3, D. 4 fr.; *Hôt. de Genève (Pl. q; D, 3), Rue du Montblanc, E., L., & A. 2½, D. incl. wine 3½ fr.; Hôt. Terminus-Baur (Pl. u; D 3); Hôt. de la Gare; Hôt. de la Monnaie. — *On the Left Bank:* *Hôt. Métropole (Pl. g; D, 5), by the Jardin Anglais, frequented by Americans, R., L., & A. from 4, B. 1½, lunch 4, D. 5, pens. from 10 fr.; *Hôt. de l'Ecu (Pl. h; C, 4), R., L. & A. from 3, lunch 3, D. 4, pens. from 8 fr., both with view of the lake. *Hôt. du Lac (Pl. k; D, 5), R., L., & A. 3-5, D. incl. wine 4, S. incl. wine 3½, pens. from 11 fr.; *Hôt. de la Poste (Pl. i; B, 4), frequented by Germans, R., L., & A. 3-4, D. incl. wine 3½, S. incl. wine 3 fr.; Hôt. de Paris (Pl. l; D, 5), with view of the lake, R. & A. 2-2½ fr.; *Hôt. Victoria (Pl. m; E, 6), Rue Pierre-Fatio 1, R., L., & A. 2-4, B. 1½, déj. 2½, D. 3, pens. 7-10 fr.; Hôtel du Mont Blanc, Balance (Pl. n; C, 4), Grand Aigle (Pl. o; D, 5), and *Hôt. du Nord (R., L., & A. from 2, D. 2½ fr.), **all in the Rue du Rhône.**

Pensions (*Pensions alimentaires*: most of them good). On the Right Bank (Pl. B-F; 1-4): *Hôt.-Pens. Roth* (Pl. s; **D**, 4), Rue du Montblanc 10 (6-9 fr.); *Mme. Richardet*, Rue du Montblanc 6-8 (6 fr.); *Jackson-Fromont*, Rue Pradier 1 (5-6 fr.); *Mmes. Cossen*, Rue **des** Alpes 5 (6 fr.); *Maitre*, Rue Gevray 2, Place des Alpes; *Morhardt*, Boul. James-Fazy 2 (5-6 fr.); *Hôt.-Pens. Bellevue*, Route de Lyon 29-33, with garden (5-7 fr.). — On the Left Bank, at Eaux-Vives, the S.E. quarter of the old town (Pl. D-F; 5-8): *Picard*, Place de la Métropole 2, Jardin Anglais (42-45 fr. per week); *Vultier*, Quai Pierre-Fatio 12 (6 fr.); *Mmes. Livet & Grobet*, Quai des **Eaux-Vives 2** (6 fr.); *Mme. Bovet*, Quai des Eaux-Vives 2 (5-6 fr.); *Fischer*, Quai **des** Eaux-Vives 20 (5-6 fr.); *Bérard*, Rue du Rhône 59 (6 fr.). — On the Left Bank, at Plainpalais, the S.W. quarter of the old town (Pl. A-C, 4-8): *Faure-Matthey*, Maison des Trois-Rois, Place Bel-Air 2 (5 fr.); *Beau-Site*, Rue Général Dufour 20 (from 5 fr.); *Breuleux*, Boul. de Plainpalais 4-6 (6-8 fr.); *Pens. du Rhône*, Boul. de Plainpalais 26 (5-6 fr.); *Mmes. Labarthe*, Rond-Point de Plainpalais 5 (5-7 fr.); *Fleischmann*, Rond-Point de Plainpalais 6 (6-8 fr.); *Mme. Duraffourd*, Boul. des Philosophes 3 (1½-5 fr.); *L. Monard*, Boul. des Philosophes 7 (5½-6 fr.); *Mme. Chappuis*, Boul. des Philosophes 15 (4½-5 fr.); *Durand*, Chemin Dancet 3 (4-5 fr.); *Mlle. Tallon*, Chemin des Minoteries 7 (from 4 fr.). — Between Plainpalais and Eaux-Vives, to the S.: *Wetten-Amberny*, Place Töpffer 5 (5-6 fr.); *Rererchon*, Petit-Florissant 12 (150 fr. per month). — At Champel-sur-Arve: *Hôt.-Pens. Beau-Séjour* (board 6, B. from 1½ fr.); *Hôt.-Pens. de la Roseraie*.

Cafés-Restaurants. *Café du Nord*, *de la Couronne*, and *de Genève*, all on the Grand Quai du Lac (Pl. D, 6); *du Théâtre*, in the Theatre, D. incl. wine, at 12.15 and 7 p.m., 2½ fr.; *Kiosque des Bastions*, with large garden, on the Promenade des Bastions (p. 228), open in summer only, with frequent concerts, lunch 2½ fr. — Beer. Left Bank. *Ackermann's Successor*, Rue du Rhône 92, near the Jardin Anglais (much frequented); *Berger*, Rue du Rhône 48; *L. Müller*, Rue du Rhône 50, near the Place du Lac; *Landolt*, opposite the University and the Jardin des Bastions; *Brasserie Bâle*, *Café-Brasserie de l'Opéra*, near the theatre. — Right Bank.

Tavernes Anglaise, Rue des Alpes 4, D. incl. wine 2½ fr., from 11 to 2; *Brasserie du Jardin des Alpes*, Place des Alpes. Geneva beer at the breweries outside the gates. *Treiber*, Route de Chêne, with a pleasant shady terrace; *Brasserie St. Jean* (Pl. B, 3), with fine view, etc.

Baths. *Bains de la Poste*, Place de la Poste, well fitted up, hot, cold, shower, and vapour baths; *Bains des Alpes*, Rue Lévrier 5, etc. — Lake Baths. *Swimming* and other baths by the Quai des Eaux-Vives (left bank); also by the pier on the opposite bank (Pl. 10; F, 4); both open for ladies 8-11 o'clock. — *Baths in the Rhone* above the *Pont de la Machine* (Pl. C, 4; p. 225), well fitted up; swimming-bath 30, plunge-bath with towels 60 c.

General Post Office, Rue du Montblanc (Pl. D, 3), a handsome new edifice with a colonnaded façade adorned with statues, open 7 a.m. to 8 p.m.; on Sun. 8-10 and 11-1. — **Central Telegraph Office** (open day and night), Rue du Stand (Pl. B, 4).

Tramway from the Gare de Cornavin (Pl. D, 2) by the Pont du Montblanc, Place du Molard (Pl. D, 5), Place Neuve, and Rond Point de Plainpalais to *Carouge* (p. 282), and by the Place du Molard and Cours de Rive to the *Eaux-Vives Station* (p. 221) and to *Chêne* (p. 271) and *Annemasse* (p. 271). — **Electric Tramway** from *Petit Saconnex* viâ Gare de Cornavin and Place Bel-Air to *Champel* (10-30 c.). — **Steam Tramways** (*Chemins de Fer à voie étroite*) to *Veyrier*, *St. Julien*, *Chancy*, *Vernier*, *Ferney*, etc.; see p. 230.

Cabs. Drive in the town and suburbs, 1-4 pers. 1½ fr., trunk ½ fr.; per hr., 1-4 pers. 2½ fr., each additional ¼ hr. 05 c. At night (April 1 to Sept. 30, 10-5; other seasons 8-8) per drive, 1-4 pers. 2¼/, per hr. 3³/₄, each additional ¼ hr. 1 fr. Over-charges are not uncommon; it is advisable to arrange the fare beforehand and to note the number of the cab.

Steamboats to the N. bank of the Lake of Geneva, see p. 233; to the S. bank, see p. 256. — Piers in Geneva, see p 222. — The *Tour du Petit Lac* (3 hrs.; without disembarking) is made by steamers several times daily, viâ Bellevue, Versoix, Coppet, Céligny, Nyon, Tougues, Anières, Corsier, Bellerive, La Belotte, Cologny, and back to Geneva. The tour of the entire lake is also frequently made (9.30 a.m. to 7.30 p.m.).

Rowing-Boats, 60 c. - 1 fr. 20 c. per hr.; each ½ hr. more 30-60 c.; boatman 1 fr. 20 c. per hr. extra, each ½ hr. more 60 c. The best boats are those at the Jetée des Pâquis and the Jardin Anglais. — **Sailing-Boats**, small 1½, large 2½ fr. per hr., each ½ hr. more ⅜ or 1¼ fr. extra. Sailing-boats are not let without a boatman (see above). — A printed tariff is handed to the hirer on embarking; after 6 or 7 p.m. the charges are about 50 per cent higher. Rowers are prohibited from approaching the Pont du Montblanc on account of the dangerous rapids.

Shops. Geneva is noted for its watches and jewellery. About 110,000 watches are annually manufactured here; those that have been officially tested have an official stamp on the movement. — Among the watch-makers of repute may be mentioned *Vacheron & Constantin*, Rue des Moulins 1; *Golay, Leresche, & Fils*, Quai des Bergues 31; *Bachmann, Koehn, Patek, Philippe & Co.*, all on the Grand-Quai; *Plojoux*, *J. Rossel*, *Henry Capt*, Rue du Rhône 4, 12, and 17; *Wirth*, Place Molard 11. — Engraver, *M. H Bovy*, chiefly for medals, Rue Chantepoulet. — Musical boxes: *F. Conchon*, Place des Alpes 9 & Rue des Pâquis 2; *G. Baker-Troll & Co.*, Rue Bonivard 6. — Optician, *Th. Stichling*, Quai des Bergues 29. — Jewellery, etc., *Kleinafeldt*, Rue du Commerce 5. — Photographic materials. *Fabre & Borrey*, Rue du Marché 14. — Alpine plants (living), *Jardin Alpin*, Chemin Dancet 2.

Booksellers. *Georg & Co.*, Corraterie 10; *Burkhardt*, Molard 2; *Stapelmohr*, Corraterie 24. — **Reading Room** (free) with English and American newspapers at the office of the 'Geneva Telegraph', Rue Lévrier 3.

Theatre (p. 229). Performances daily in winter (adm. 2-5 fr.; seats secured in advance, or 'en location', higher). — **Cursaal** on the Quai des Pâquis (Pl. E, 3); variety performance every evening at 8 p.m., adm. 1-3 fr.

Music. *Organ Concert* in the Cathedral (p. 227) on Mon., Wed., and Sat., at 7. 30 p. m.; tickets (1 fr.) obtainable from the concierge and at the hotels. — Concerts in the *Bâtiment Electoral* (Pl. B, 5) every Sunday afternoon in winter; also fortnightly in the *Theatre* (p. 229). — Concerts frequently on Thurs. in summer at the *Jardin Anglais*, with illumination of the fountain on the quay ('fontaines lumineuses'), at the *Place des Alpes* (Pl. D, E, 3), and in the *Kiosque des Bastions* (p. 223).

Exhibition of Art, belonging to the *Société des Amis des Beaux-Arts*, in the Athénée (p. 223), open daily 10-6, Sun. 11-4; adm. 1 fr. — *Exposition Municipale des Beaux-Arts* in Aug. and Sept. annually, in the Bâtiment Electoral (p. 223). — **Panorama** (Pl. B, 4; 1 fr.) at present containing the *Siege of Belfort in 1870-71*, by Berne-Bellecour; in an adjoining building is a *Relief Map of Geneva in 1850* (adm. 50 c.). — **Public Lectures** (*Cours publics et gratuits*) in the University Hall, in winter daily at 8 p.m.

Physicians. Prof. *D'Espine*, Rue Beauregard 6; *Dr. Cordès*, Rue Bellot 12; *Dr. Tucker-Wise*, Pens. Sütterlin (Oct.-May); *Dr. Batault* (homeopathist), Rue de l'Université 6; *Dr. Collardon* (aurist), Rue de Candolle 17; *Dr Wyss* (aurist), Rue Calvin 7. — **Chemists.** *Baker*, Place des Bergues 3; *Finck*, Rue du Montblanc 26; *Goegg*, Corraterie 18; *Ackermann*, Rue des Allemands 13, etc.

Hydropathic Establishment (physician, *Dr. Glatz*) at Champel-sur-Arve (p.222; tramway-station *La Cluse*), with a view-tower (*Tour de Champel*; ¹/₂ fr.).

Official Enquiry Office of the Association des Intérêts de Genève, Quai du Montblanc 5 (daily 10-12 and 2-4, except Sun. and holidays). — **Cook & Son's office**, Rue du Rhône 90.

British Consul (for the French-speaking cantons), *D. P. F. Barton, Esq.*, Rue Bonivard 10 (10-12 a.m.). — **American Consul**, *Benj. H. Ridgely, Esq.*, Rue Pécolat 3 (9-2); vice-consul, *Peter Naylor, Esq.* — *Union Bank*, Rue Petitot 10.

English Church (*Holy Trinity*; Pl. D, 3, 4) on the right bank, in the Rue du Montblanc; chaplain, *Rev. A. S. Douglas*. — **American Church**, Rue des Voirons (Pl. E, 3), not far from the Brunswick Monument and the Cursaal. — *Presbyterian Services* (8-11 a.m.), Place de la Fusterie 7.

Geneva (1243'; pop. 78,000, including the suburbs), Fr. *Genève*, Ital. *Ginevra*, the capital of the smallest canton next to Zug (total pop. 106,738), is the richest town in Switzerland. It lies at the S. end of the lake, at the point where the blue waters of the *Rhone* emerge from it with the swiftness of an arrow, and a little above the confluence of the Rhone and the *Arve* (p. 231). The Rhone divides the town into two parts: on the left bank lies the *Old Town*, the seat of government and centre of traffic, with the suburbs of *Plainpalais*, to the S.W., and *Eaux Vives*, to the S.E.; on the right bank is the *Quartier St. Gervais*, with the suburb of *Pâquis*, to the E. Since the removal of the old fortifications (after 1850) both parts of the town have extended with extraordinary rapidity.

History. Geneva makes its appearance in the 1st cent. B. C. as *Genava*, a town of the Allobroges (Caes. de Bell. Gall., i. 6-8), whose territory became a Roman province. In 433 it became the capital of the Burgundian kingdom, with which it came into the possession of the Franks in 533, was annexed to the new Burgundian kingdom at the end of the 9th cent., and fell to the German Empire in 1033. In 1034 Emp. Conrad II. caused himself to be crowned here as King of Burgundy. In the course of the protracted conflicts for supremacy between the Bishops of Geneva, the imperial Counts of Geneva, and the Counts (afterwards Dukes) of Savoy, the citizens succeeded in obtaining various privileges. In 1518 they entered into an alliance with Freiburg, and in 1526 with **Bern**. Two parties were now formed in the town, the Confederates ('Eidgenossen', pronounced by the French 'Higuenots', whence the term '*Huguenots*'), and the *Mamelukes*, partisans of the House of Savoy.

In the midst of these discords dawned the REFORMATION, which Geneva zealously embraced. In 1535 the Bishop transferred his seat to Gex, and the following year the theologian *Jean Calvin* (properly *Caulvin* or *Chauvin*), who was born at Noyon in Picardy in 1509, a refugee from Paris, sought an asylum at Geneva. He attached himself to *Farel*, the chief promoter of he new doctrines at Geneva, and soon obtained great influence in all affairs of church and state. In 1538 he was banished, but on his return three years later he obtained almost sovereign power and succeeded in establishing a rigid ecclesiastical discipline. His rhetorical powers were of the highest order, and the austerity which he so eloquently preached he no less faithfully practised. In accordance with the spirit of the age, however, his sway was tyrannical and intolerant. *Castellio*, who rejected the doctrine of predestination, was banished in 1540; and *Michael Servetus*, a Spanish physician who had fled from Vienne in Dauphiné in consequence of having written a treatise against the doctrine of the Trinity (*de Trinitatis erroribus*), and was only a visitor at Geneva, was arrested in 1553 by Calvin's order and condemned to the stake and executed by order of the Great Council. In 1559 Calvin founded the Geneva Academy, which soon became the leading Protestant school of theology, so that the hitherto commercial city now acquired repute as a seat of learning also. Calvin died on 27th May, 1564, but his doctrine has been firmly rooted in Geneva ever since. — The attempts made by the Dukes of Savoy at the beginning of the 17th cent. to recover possession of Geneva were abortive, and Protestant princes, who recognised the town as the bulwark of the Reformed church, contributed considerable sums towards its fortification.

In the 18th cent. Geneva was greatly weakened by dissensions, often leading to bloodshed, between the privileged classes, consisting of the old families (*citoyens*), who enjoyed a monopoly both of power and of trade, and the unprivileged and poorer classes (*bourgeois*, *habitants*, and *sujets*). To these differences the writings of *Jean Jacques Rousseau*, the son of a watchmaker, born here in 1712, materially contributed. At the instigation of Voltaire and the University of Paris, his '*Émile*' and '*Contrat Social*' were burnt in 1763 by the hangman, by order of the magistrates, as being 'téméraires, scandaleux, impies, et tendants à détruire la religion chrétienne et tous les gouvernements'. — In 1798 Geneva became the capital of the French *Département du Léman*, and in 1814 it joined the Swiss Confederation, of which it became the 22nd Canton. In 1846, under the leadership of *James Fazy*, the Conservative government was overthrown, and in May, 1847, a democratic constitution was adopted, which is still essentially in force.

The two halves of the city separated by the Rhone are connected by six bridges. The highest of these, the handsome *Pont du Montblanc (Pl. D, 4, 5), 280 yds. long, leads from the *Rue du Montblanc*, a broad street descending from the railway-station, to the Jardin Anglais (see p. 226), and with this garden forms the centre of attraction to visitors in summer. In the Rue du Montblanc is the Gothic *English Church* (Pl. D, 3, 4), erected by Monod in 1853. Between the Pont du Montblanc and the *Pont des Bergues* is *Rousseau's Island* (Pl. D, 4), united to the latter by a chain-bridge, and planted with trees. In the centre rises the bronze statue of the 'wild self-torturing sophist', by Pradier (1834). At the third bridge, the *Pont de la Machine* (Pl. C, 4, above which are the Rhone baths, p. 223), is the *Central Station of the Geneva Electricity Works*. The *Island*, on which lies one of the oldest quarters of the town (recently partly pulled down), divides the Rhone into two branches (p. 230).

Handsome quays with tempting shops flank the river near these bridges, the principal being the *Grand-Quai* on the left bank, and the *Quai des Bergues* on the right. Adjacent to the latter is the **Quai du Montblanc** (Pl. D, E, 4), extending from the Pont du Montblanc towards the N.E., and affording a beautiful survey of the *Mont Blanc group, which presents a majestic appearance on clear evenings (mountain indicator on the railing).

BAEDEKER, Switzerland. 16th Edition.

An idea of the relative heights of the different peaks is better obtained from this point than at Chamonix. Thus Mont Blanc is 15,730' in height, whilst the Aiguille du Midi on the left is 12,605' only. Farther to the left are the Grandes Jorasses and the Dent du Géant; in front of the Mont Blanc group are the Aiguilles Rouges; then, more in the foreground, the Môle, an isolated pyramid rising from the plain; near it the snowy summit of the Aiguille d'Argentière; then the broad Buet; lastly the long crest of the Voirons, which terminate the panorama on the left, while the opposite extremity is formed by the Salève.

On the left side of the Quai du Montblanc rises the large and sumptuous **Monument Brunswick** (Pl. E, 4), erected to *Duke Charles II. of Brunswick* (d. 1873), who bequeathed his property (about 20 million fr.) to the town of Geneva.

The monument (in all 66' in height) is a modified and slightly enlarged copy of that of Can Signorio della Scala at Verona. It was designed by *Franel*, and consists of a hexagonal structure in the form of a pyramid, in three stories, standing upon a platform, 220 ft. long and 78 ft. broad, the approach to which is guarded by two colossal lions in yellow marble by *Cain*. In the central story is a sarcophagus, on which is a recumbent figure of the duke, and reliefs (scenes from the history of Brunswick), all by *Iguel*. At the projecting corners are marble statues of six celebrated Guelphs (Augustus; Otho the Child; Charles William Ferdinand; Frederick William; Henry the Lion; Ernest the Confessor), by *Schoenewerk*, *Thomas*, *A. Millet*, and *Kissling*. On the roof are the Christian virtues, the Twelve Apostles, etc. The bronze equestrian statue of the duke (by *Cain*), which crowned the monument, proved too heavy and has been taken down.

The Quai du Montblanc is continued by the *Quai des Pâquis*, on which, to the left, is the *Cursaal* (Pl. E, 3; p. 223). Behind it is the *American Church* (p. 224). Beyond the *Jetée*, or pier, at the end of which is a lighthouse (flash-light), the *Quai du Léman* extends to the villas of *Sécheron*.

On the S. (left) bank of the lake, to the left of the Pont du Montblanc, rises the *National Monument* (Pl. D, 5), a bronze group of Helvetia and Geneva by *Dorer*, commemorating the union of Geneva with the Confederation in 1814. — Adjacent, on the lake, are **the pleasant grounds of the Jardin Anglais** *(Promenade du Lac)*, with a café-restaurant, where a band often plays on summer-evenings. To the left of the entrance is a 'barometer column', and in the centre of the garden are a pretty fountain and bronze busts of *Al. Calame* (p. 229) by Iguel, and *Fr. Diday* by Bovy. A pavilion here contains an interesting *Relief of Mont Blanc* (adm. 50 c.; Sun., 9-3, gratis), on a scale of 1 : 6000 (Mont Blanc 31 in. in height; proportion of vertical to horizontal dimensions, 2 : 1).

In the lake, off the *Quai des Eaux-Vives* (Pl. E, F, 5), rise two **granite** rocks, the larger of which, the *Pierre du Niton*, is traditionally said to have been a Roman altar to Neptune. At the end of **the Quai is** a pier, on which a *Fountain*, with a jet 115 ft. in height (illumination, see p. 224), plays on Sun. and holidays in summer. — Near the Quai is the *Salle de la Réformation* (Pl. E, 6), containing a large concert-hall, the *Calvinium*, with memorials of Calvin, articles brought home by missionaries, etc. (adm. 50 c.), and an interesting *Relief Model of Jerusalem* by Illès.

Ascending the **Rue d'Italie**, to the right near the Hôtel Métropole, for a few paces, we reach the *Promenade de St. Antoine* (Pl. C, D, 6), a terrace planted with trees. On the right is the *Collège de St. Antoine*, founded by Calvin in 1559; to the left (E.) is the *Observatory*, and on a height farther off (S.E.) rises the *Russian Church*, with its gilded domes. Adjacent is a bronze bust of *R. Töpffer* (d. 1846), the author.

The highest point on the left bank is crowned by the **Cathedral** (*St. Pierre*; Pl. C, 6), completed in 1024 by Emp. Conrad II. in the Romanesque style, altered in the 12th and 13th cent., and disfigured in the 18th by the addition of a Corinthian portico. The building is at present under restoration.

The verger lives at the back of the church, **Rue Farel** 8. Adm. by the side-door, next the choir; week-days 1-3, free; at other hours, except Sun. 10-12, each pers. 20 c., parties of more than five, 1 fr.; ascent of the tower, 1-5 pers. 1 fr., each additional pers. 20 c.

INTERIOR. To the right of the entrance, Monument of *Duke Henri de Rohan* (leader of the French Protestants under Louis XIII.), who fell at Rheinfelden (p. 19) in 1638, of his wife *Marg. de Sully*, and of his son *Tancrède*; the black marble sarcophagus rests on two lions; the "Statue of the duke, in a sitting posture, by Iguel, is modern (original destroyed in 1798). Beneath a black tombstone in the nave lies *Cardinal Jean de Brogny* (d. 1426), president of the Council of Constance. A black stone in the S. aisle is to the memory of *Agrippa d'Aubigné* (d. 1630 at Geneva, in exile), the confidant of Henry IV. of France, erected to him, in gratitude for his services, by the Republic of Geneva. Under the pulpit is a chair once used by Calvin. — Adjoining is the beautiful Gothic *Chapelle des Macchabées* (1406; restored 1878-83), with modern stained-glass windows. — Admirable *Organ* (concerts, see p. 223).

Near the cathedral is the **Hôtel de Ville** (Pl. C, 5, 6), a clumsy building in the Florentine style, which is entered by an inclined plane, once enabling the councillors to ride, or be conveyed in litters, to or from the council-chambers. — Opposite is the **Arsenal** (Pl. C, 5; Sun. and Thurs., 1-4), containing the *Musée Historique Genevois*, a collection of old armour and weapons, the ladders used at the 'Escalade' (see below), etc.

In the vicinity, Grand' Rue No. 40, is the house in which *Jean Jacques Rousseau*, the son of a watchmaker, was born (1712, d. 1778 at Ermenonville near Paris). His grandfather lived at that time at the back of Rue Rousseau 27, on the right bank of the Rhone, which bears an erroneous inscription that Rousseau was born there.

The *Musée Fol* (Pl. C, 5; Sun. and Thurs., 1-4), Grand' Rue 11, founded by *Mr. W. Fol*, contains (in the court to the right) a valuable collection of Greek, Roman, and Etruscan antiquities, the yield of recent excavations, and mediæval and Renaissance curiosities.

The Rue de la Cité, the lower prolongation of the Grand' Rue, leads to the Rue des Allemands, where a tasteful *Fountain Monument* (Pl. C, 4) commemorates the last and nearly successful attempt of the Savoyards to gain possession of the town (comp. p. 225). The day on which the 'Escalade' was repulsed (early on 12th Dec., 1602) is still kept with public rejoicings.

A gateway adjoining the Hôtel de Ville leads to the promenade of *La Treille*, which is planted with chestnut-trees. Below this terrace is the BOTANIC GARDEN (Pl. B, C, 5, 6), laid out in 1816 by the celebrated *Aug. De Candolle* (d. 1841). The hot-house is adorned

15*

with marble busts of famous Genevese scientists, **and there** are others in the vicinity.

The PROMENADE DES BASTIONS, a favourite resort, on which is the *Kiosque des Bastions* (p. 222), separates the Botanic Garden from the University buildings. In the grounds are a statue of David by *Chaponnière* and the '*Pierre aux fées*', or '*aux dames*', with four figures, said to be a Druidical stone. To the E. is a monument of *Gosse*, the geologist.

The **University Buildings** (Pl. B, 6), erected in 1868-72, consist of three different parts connected by glass galleries. The *Central Part* contains the lecture-rooms and laboratories (except the medical school, which lies **on the Arve, to the S.E.**), the *E. Wing* the collections of antiquities **and coins, and the Library**, and the *W. Wing* the Nat. Hist. Museum. In the square in front of it is a bronze bust of *Ant. Carteret* (d. 1889), the statesman and educationist, by Charmot (1891). In the vestibule of the central building **is a bronze bust** of the Swiss author *Marc Monnier* (d. 1885), by Dufaux; behind is a model of the Saussure Monument at Chamonix (p. 277). The university has 70 professors and about 700 students. **Ladies are** admitted to the lectures.

The Library, founded about the middle of the 16th cent., contains about 190,000 vols. and 10,000 MSS. The SALLE LULLIN on the groundfloor, to the right of the entrance (open Sun. and Thurs., 1-4; at other times apply to the concierge; fee) contains 250 ancient and modern portraits of reformers, statesmen, and scholars, either of Genevese origin or of importance in the history of Geneva. This room also contains a collection of MSS., including autographs of Calvin and Rousseau. The most valuable MSS. are exhibited in glass-cases: homilies of St. Augustine on papyrus (6th cent.); house-keeping accounts of Philip le Bel (1308) on wax tablets, many with miniatures, some of them captured from Charles the Bold at Grandson (p. 202). On an old reading-desk is a French Bible (printed at Geneva in 1588), richly bound in red morocco, and bearing the arms of France and Navarre, which was destined by the Council of Geneva as a gift to Henry IV., but never presented owing to his abjuration of Protestantism. — On the groundfloor is also the *Cabinet of Coins*; and on **the** sunk-floor is the *Archaeological Museum*, containing prehistoric and other antiquities, chiefly of local interest (Sun. and Thurs., 1-4). The first floor contains the reading-room (open on week-days, 9-12 and 1-6; closed in the afternoon during the university vacations). — In the court is the *Musée Epigraphique*, a collection of Roman and mediæval inscriptions found at Geneva.

The Natural History Museum, admirably arranged by F. J. Pictet, contains the famous collection of conchylia of B. Delessert, which has been described by Lamarck; Pictet's collection of fossils; De Saussure's geological collection; Melly's collection of about 35,000 coleoptera; a complete collection of the fauna of the environs of Geneva; valuable rock-crystals from the Tiefengletscher (p. 117), etc. — Admission to the Museum on week-days (except Tues. and Sat.), 1-4, and Sun., 11-4, gratis; at other times apply to the concierge (fee).

The ATHÉNÉE (Pl. C, 6), to the **S.E. of** the Botanic Garden, contains lecture-rooms, a library **of works on the history of art,** and an exhibition of art (p. 224). Near it is the *Ecole de chimie* (Pl. C, 6).

To the N.W. of the Botanic Garden and the Bastion Promenade extends the round *Place Neuve* (Pl. B, 5), in which **is an equestrian** statue of *Gen. Dufour* (d. 1875), in bronze from a model by Lanz.

On the S.W. side of the Place is the *Conservatory of Music*. On the N.W. side rises the *Theatre, designed by **Gosse**, and erected in 1872-79, a handsome Renaissance building, with a façade enriched with columns and figures. The interior (with 1300 seats and a handsome foyer), deserves a visit (adm. on week-days, 1-4). — To the N.E., at the beginning of the Corraterie, is the —

*Musée Rath, a collection of pictures, casts, etc., founded by the Russian general *Rath* (1766-1819), a native of Geneva, and presented to the city by his sisters. It has since been much extended. The building was erected in 1825. Adm. in summer, Mon., Wed., Thurs., and Frid. 1-4, and Sun. 11-4, gratis; at other times, ½ fr. (catalogue ½ fr.).

VESTIBULE. Antique statue of Trajan as Mars; busts, chiefly of distinguished natives of Geneva, many of which are by *J. Pradier* (b. at Geneva 1790; d. at Paris 1852); Molière, Necker, by *Houdon*; Dumont, Jeremy Bentham, by *David*. Also plaster-casts, and a few paintings crowded out of the picture-gallery.

Picture Gallery (three rooms). CENTRAL ROOM. In the middle, busts of General Rath, by *Pradier*, and L. Favre (p. 114), by *Ch. Töpffer*. Entrance-wall: *39-41. *Al. Calame* of Vevey (1810-64), The Seasons; on the side-walls to the right and left, 229-232. Four pictures by *Léopold Robert* (p. 208). Farther on, to the left: 1. *Agasse*, At the smithy; several paintings by *A. W. Töpffer* (d. 1847) and his son *R. Töpffer*, better known as an author (d. 1846); 104. *Feyen-Perrin*, 'Vanneuse' (girl winnowing corn); 147. *Humbert*, The ford; 55-59. *Corot*, Landscapes; 102. *Favas*, General *Dufour*; 148. *Humbert*, Landscape with cattle. — 29. *Bocion*, Lake of Geneva; 137. *Hornung*, The Eve of St. Bartholomew; 76. *Fr. Diday* of Geneva (1802-77), The Giessbach. — 179. *J. L. Lugardon*, Arnold of Melchthal; *Diday*, *77. Lake of Lucerne; *78. Thunder-storm on the Handegg; 136. *Hornung*, Calvin's farewell to the councillors of Geneva; 288. *Veillon*, Lake of Tiberias; 152. *Jacot-Guillarmod*, Cattle fighting; 269, 266, 268. by *A. W. Töpffer*; above, *Carl Stauffer*, Study.

ROOM TO THE LEFT (older paintings, chiefly Dutch). 299. *P. Wouverman*, Naval battle; 45. *Caravaggio*, Four singers; 204. *G. Netscher*, Portrait; 122. *Greuze*, Study of a child's head; 297. *Weenix*, Dead game; 262. *D. Teniers*, The five senses; 159. *Largillière*, Portrait; *Velazquez*, 289, 290. Philip IV. of Spain and his consort Maria Anna of Austria, 291. Spanish singers; 275. *Van der Helst*, Portrait; 261. *Teniers*, The smoker; 274, 273, and farther on 272. *Van Goyen*, Landscapes; 14. *Jac. Bassano the Elder*, Adoration of the Shepherds. — 241. *Ryckaert* and *Molenaer*, Flemish tavern; 52. *Phil. de Champaigne*, Dead nun; 178. *J. L. Lugardon*, Liberation of Bonivard (p. 214); 61. *A. Cuyp*, Pasture; 197. *Mirevelt*, Portrait. — In the adjoining CABINET: Portraits, the majority by *Liotard* (141, 142, 143); 198. by *Mirevelt*.

ROOM TO THE RIGHT (chiefly modern paintings). At the main entrance are three busts by *Carriès*, *Ch. Töpffer*, and *Dufaux* (41, the painter Diday), in the rear a bust by *Bovy*. — To the left of the entrance: 296. *Veillermet*, Portrait. — 49. *Castres*, Swiss field-hospital, 1871; 184. *A. Lugardon*, Wengern-Alp; 95-98. by *S. Durand* of Geneva; 287. *B. Vautier*, The sick mother; 117. *J. Girardet*, Flight of the Vendéans after the battle of Cholet; 47. *Castres*, The tale of the prisoner-of-war (1871); 87. *Dufaux*, Marketboat to Vevey; 219. *Ravel*, Drawing-lesson; 286. *Vautier*, Peasants carrying on a lawsuit; 7. *Anker*, Communal meeting in the Canton of Bern; 119. *Giron*, Education of Bacchus; 64. *Durier*, Choristers; 150. *Ihly*, Child's funeral; 99. *Dunal*, On the Upper Nile; 217. *E. de Pury*, Venetian beadstringers; 80, 81, 79. *Fr. Diday*, Landscapes; 118. *E. Girardet*, Arab at prayer; 43. *Ant. Calame*, Vevey; 95. *Durand*, After the review; 208. *Palérieux*, Return from the market.

Below the *Pont de la Coulouvrenière* (Pl. B, 3, 4), the lowest of the Rhone bridges, are the new **Waterworks** *(Forces Motrices du Rhône)*, constructed in 1883-86, which not only supply the houses of Geneva but afford motive power equal to 4200 horses for the use of manufactories (at a charge of 60 fr. annually per litre and hr.). The entire left branch of the Rhone (p. 225) is dammed up for the purposes of these works, the right branch being left open to accommodate itself to the variations in the level of the lake. A visit to the large hall, reached from the Quai de la Poste, will be found highly interesting even by the unscientific tourist; the huge waterwheels here each represent 210 horse-power and describe 26 revolutions per minute. — Similar works, to supply motive power equal to 12,000 horses, are being constructed $3^1/_2$ M. downstream.

On the RIGHT BANK, to the left of the Pont de la Coulouvrenière, is the *Promenade St. Jean* (Pl. B, 3), with a bronze bust of *James Fazy* (d. 1878; p. 225), the Genevese statesman, by Rolland. We next pass the *Ecole d'Horlogerie* (built in 1874-78), with the *Musée des Arts Décoratifs* (on the first floor; adm. daily, except. Sat., 11-4, Sun. 9-12), containing an important collection of engravings, and the *Musée Industriel*, in which are the machines used by L. Favre in boring the St. Gotthard tunnel. Thence we proceed past the *Ecole des Arts-Industriels* (built in 1877), and the Place des vingt-deux Cantons (see below), with the old-Catholic church of *Notre-Dame*, and soon reach the railway-station.

Environs of Geneva (see Map, p. 223). An extensive system of STEAM TRAMWAYS *(Chemins de fer à voie étroite)* much facilitates a visit to the charming environs of Geneva, which are studded with villas and country-houses with beautiful gardens. The termini of the cars at Geneva are, on the right bank of the Rhône, the *Place des vingt-deux Cantons* (Pl. C, 3), and on the left bank, the *Quai de la Poste* (Pl. B, 4) and the *Cours de Rive* (Pl. D, 6). Return-tickets are obtained only at the ticket-offices in the waiting-rooms; single tickets only on the cars. The time-tables give Central Europe time even for the sections of the lines on French territory.

TO PREGNY AND FERNEY. From the Place des 22 Cantons, 14 times daily, to Pregny in 10 min. (fare 20 c.), to Ferney in 35 min. (60 c.). Comp. Pl. C, 4, 3 and D, 2, 1. The first station is *Voie-Creuse*, the second *Ariana*, for the Musée Ariana (5 min.) and for Baroness Rothschild's Château.

The *Musée Ariana, founded and bequeathed to the town by M. Gust. Revilliod (d. 1890), a handsome Renaissance building erected in 1880, is situated in an extensive park (adm. daily, 9-7), commanding a magnificent view of the lake and the Alps. The most diverse branches of art are represented in the museum, corresponding to the catholic interests of its collector. (Adm. on Thurs. and Sun. 10-6, gratis; Mon., Tues., Frid., and Sat., 1 fr.)

Genthod
Bellevue
FERNEY-VOLTAIRE
Chambésy
Sacconnex
Pt Saconnex
Corsier
Collonge
St Maurice
Meinier
Vésenaz
Vandœuvres
Choulex
Cologny
LAC DE GENÈVE
les Pâquis
GENÈVE
Eaux Vives
Puplinge
Chêne Bougeries
Chêne-Bourg
Annemasse
Thônex
Villette
Gaillard
Etrembières
Plainpalais
Lancy
CAROUGE
Plan les Ouates
Troinex
Veyrier
Bossey
Collonges
Monnetier
les Esserts
Salève

The imposing CENTRAL HALL, with a double tier of marble columns, contains a group of Sleep and Death (in the centre) by *Guglielmi*, marble busts, vases, etc. The CENTRAL CORRIDOR (right and left) is hung with valuable tapestry of the 17th cent.; the ceiling-paintings (the seasons, etc.) are by *Dufour*. To the left of the hall are Asiatic porcelain, inlaid work, European faïence, ivory carvings, and bronzes (statuette representing a contest with a serpent, by the mirror in the Japanese room); to the right are the collections of European porcelain, antique vases, articles from Alemannic graves, etc. — FIRST FLOOR. On the staircase is a Chinese boudoir, and at the top, antique furniture, weapons, and stained glass. The PICTURE GALLERY occupies four rooms on this floor. *Room I*: Portraits by *Bronzino, Giorgione, Guercino, Holbein, Rigaud*, and others; in the centre, a small antique head of Venus. — *Room II*: *Seb. del Piombo*, Bearing of the Cross; *Ribera*, John the Baptist; *Lucas van Leyden*, Madonna; *Fyt*, Boarhunt; *Raphael*, Madonna of Vallombrosa (replica of the Madonna del cardellino; not original); Madonnas by *L. Credi, Van Dyck*; and others. — *Room III* contains chiefly flower-pieces, studies of still-life, and other small examples of the Netherlands school; marble busts of M. Revilliod and his mother Ariana (née De la Rive) by *Duphot*. — *Room IV*: Landscapes by *Diday, Calame, Duval, Veillon, Loppé*, and *Lugardon*; Cattle-pieces by *Humbert, Agasse*, and *Delarive*; Genre-scenes by *Vautier, S. Durand, Rubio, Töpffer*, and others. — On the other side of the large hall are paintings by *Horace Revilliod*; portraits, pastels, and drawings by early Genevese masters; engravings (10,000 plates); a library, with glass-cases containing interesting autographs; glass, ivory carvings, antique Genevese tinware; and the Silver Chamber, containing ornaments, coins, medals, enamels, etc. Fine view from the balcony. — In the grounds close to the museum is the sumptuous *Tomb of Revilliod* (see above).

The *Château of Baroness Adolphe Rothschild* ('Pavillon de Pregny'), built in 1860 by Gindroz, lies 1/4 hr. from the tramway-station of Ariana. The fine park is open on Tues. and Frid., 3-6 in July and Aug., 2-5 in Sept. and Oct. (tickets obtained gratis at the hotels in Geneva).

The steam-tramway next passes the pretty villages of *Petit-Sacconnex* (to the left) and *Grand-Sacconnex*, crosses the French frontier before the *Tuilerie*, and reaches (4 M.) Ferney, officially *Ferney-Voltaire* (*Truite*; *Hôtel de France*), a place of some size, founded by Voltaire in 1758. Opposite the station is a bronze *Statue of Voltaire* ('au patriarche de Ferney, 1694-1758-1778'), by E. Lambert, presented by the artist (1890). Following the street leading straight from the station, then turning to the left, we reach the (1/2 M.) *Château* erected by Voltaire, now containing various memorials of the founder (adm. in summer on Mon., Wed., and Frid., 2-5; fee to the concierge). Over the former chapel is the well-known inscription: 'Deo erexit Voltaire'. The garden-terrace commands a beautiful view. — From Ferney an omnibus plies four times daily in 1 hr. to (6 M.) *Gex* (p. 235).

To VERNIER, ten times daily in 25 min. (from the Place des XXII Cantons, p. 230; fare 40 c.). The line (comp. Pl. C, B, 2; A, 1) runs viâ *Les Délices*, and *Les Charmilles*. Beyond the hamlet of *Châtelaine*, with the 'Théâtre Voltaire' (now a store), we pass the much-frequented *Bois des Frères* (on the left) and reach the prettily situated village of *Vernier*.

To THE BOIS DE LA BÂTIE. Starting from the Quai de la Poste (comp. Pl. A, B, 4), the line runs past the Abattoirs to the *Pont de St. Georges* over the Arve. On the other side of the river a path ascends to the right to the Bois de la Bâtie (1 1/4 M. from Geneva), a plateau covered with woods and meadows (several cafés), affording a fine survey of the town and environs. The blue water of the Rhone and the gray water of the Arve flow side by side without mingling for several hundred yards below their confluence. — From the bridge the tramway goes on viâ *Rampe Quidort, Petit Lancy*, and *Onex* to (3 1/2 M.) Bernex (several small restaurants); a considerable village whence the *Signal de Bernex* (1655'; fine view) may be ascended in 1/4 hr.; and thence viâ *Laconnex* to (9 M.) the railway-station of *Chancy* (p. 264).

To St. Julien, 5 1/2 M., twelve times daily in 3/4 hr. (to Carouge in 13 min.), starting from the Quai de la Poste (p. 231). Beyond the Pont d'Arve our line diverges to the left from that to Lancy, and reaches (1 3/4 M.) **Carouge** (1260'; *Balance*; *Ecu de Savoie*), a suburb (5700 inhab.) of Geneva, founded in 1780 by Victor Amadeus III. of Savoy, who attracted a number of Genevese artisans hither by the offer of special advantages. There are two stations: *Grand-Bureau*, at the N. end, and *Carouge-Rondeau*, at the S. end, near the terminus of the tramway to Geneva and Annemasse (p. 271). — The tramway next passes *Bachet-Pesay*, *Plan-les-Ouates*, with the drill-ground and rifle range of the Geneva troops, *Arare*, and *Perly*, and reaches (5 1/2 M.) St. Julien, a little French town, with 900 inhab., on the *Aire*, a station on the railway from Bellegarde to Bouveret (p. 264). About 1 M. to the W. are the picturesque ruins of the château of *Ternier*. — The *Pitons* (1505'), adjoining the Salève on the S.W., may be ascended from St. Julien viâ *Beaumont* in 3 hrs.

To the Salève. — Steam Tramway (50 c.), fifteen times daily, in 25 min., starting from the *Cours de Rive* (comp. Pl. D, 6-8) and running, viâ *Florissant* and across the Arve between the hamlets of *Villette* and *Sierne*, to (3 1/2 M.) Veyrier (°*Hôt. Beau-Séjour*), a village prettily situated at the foot of the Salève. — Thence the tramway goes on viâ *Bossey* (p. 323) to *Collonges*.

The *Salève, a long hill of limestone rock to the S.E. of Geneva, is a favourite object of excursions. It consists of two portions, separated by the valley of Monnetier; to the N.E. the *Petit-Salève* (2800'), and to the S.W. the *Grand-Salève* (4290'), adjoined by the *Petit* and *Grand Piton* (4505'). Electric Railways, starting respectively from *Etrembières* and from *Veyrier*, ascend to *Monnetier-Mairie*, where they unite, and continue as one line to the terminus *Treize-Arbres* on the Grand-Salève. From Etrembières to Monnetier 27 min., to Treize-Arbres 60-67 min.; from Veyrier 1/2 hr. and 1 hr.; fare from either terminus to Monnetier 90 c., return 1 1/2 fr.; to Treize-Arbres 3 fr. 20 c. and 5 fr. First-class circular ticket from Geneva (Molard) viâ Etrembières, Treize-Arbres, Veyrier, and back to Geneva (Cours de Rive), 8 fr. — From Etrembières (p. 271) reached from Geneva-Molard by tramway viâ Annemasse, 10 times daily in 50 min.) the electric railway runs past the old ivy-mantled château (beneath which are the *Trous de Tarabara*, two large rock-caves said to date from Celtic times) and ascends the slope of the *Petit Salève*, viâ the stations of *Bas-Mornex* (1394') and *Haut-Mornex* (2230'), to the junction at *Monnetier-Mairie*. Mornex (*Hôt.-Pens. Bellevue*, at Haut-Mornex station, with a full view of the Alps; °*Hôt. Beau-Site*; *Hôt. de Savoie*; °*Pension Bain*, in the old château; *Pens. Chevalier*, etc.), a charming village on the S. slope of the Petit-Salève, is visited as a health-resort. — From Veyrier (see above) the electric railway crosses the Annemasse and Bellegarde line (p. 257), skirts the extensive limestone quarries of Veyrier, ascends above the *Pas de l'Echelle* (see below), passes through a tunnel (120 yds.), and reaches (2 M.) Monnetier-Eglise (2335'; °*Hôt.-Pens. de la Reconnaissance*; *Hôt. du Château de Monnetier*; *Chalet de Monnetier*; *Hôt.-Pens. Trottet*, R., **L., &** A. 3 1/2, B. 1 1/4, D. 2 1/2, pens. 6 1/2 fr.; *Hôt. Belvedere*), situated in a depression between the Petit and Grand-Salève. From this point the *Petit Salève* is easily ascended in 1/2 hr.; the *Grand-Salève* in 1 1/2 hr. (see below). — The line then goes on to the (3 M.) junction of *Monnetier-Mairie* (see above), and thence ascends the partly wooded slopes of the Grand-Salève to the (3 3/4 M.) terminus at *Treize-Arbres* (3746'; Buffet; Auberge des Treize Arbres, 5 min. farther up). We now ascend the ridge, passing the Grande Gorge (see below), and in 1/4 hr. reach the *Crêt de Grange Tournier* (4524'), the highest point of the *Grand-Salève, whence we survey the Mont Blanc chain, the Lake of Geneva, the Jura, the cantons of Geneva and Vaud, and a part of France. The walk may be extended to the (1 1/4 hr.) *Pitons* (see above). — *Veyrier* (see above) is the best starting-point for the ascent of the Salève on foot. We follow the *Pas de l'Echelle*, running below the electric railway, then ascend a flight of 101 steps in the rock to (1/2 hr.) *Monnetier-Eglise* (see above), whence a good bridle-track, to the right, winds up to the

(1½ hr.) Treize **Arbres**. — The ascent from *Etrembières* (p. 271) is longer but easier. We cross the Arve, and after 5 min. turn to the left and follow the road to (1½ hr.) *Mornex*, and thence take the upper road viâ the (20 min.) *Hôtel Bellevue*, at the station of Monnetier-Mairie, to (¼ hr. more) *Monnetier-Eglise* (see p. 232). — A third, but somewhat more fatiguing route ascends from *Bossey* (steam-tramway station, p. 232) viâ *Crevin* and through the *Grande Gorge*, by a steep but well-made path, to the (2 hrs.) plateau (p. 232). The route through the *Petite Gorge*, to the left of the Grande Gorge, is dangerous.

On the left or E. BANK of the lake a picturesque walk (tramway from the Cours de Rive to Vésenaz, 50 c.) may be taken along the Quai **des Eaux-Vives**, planted with plane-trees to (3 M.) Vésenaz (garden-restaurants by the lake, in *La Belotte*); return to (3½ M.) Geneva viâ Cologny (*Chalet Suisse; Café-Restaurant des Alpes*), with a charming view of the lake, or farther to the E. viâ *Vandoeuvres* and *Chougny* (see below), with a fine survey of Mont Blanc. — The steam-tramway goes on from Vésenaz to (10 M.) the little French town of *Douvaine*.

The long range of the *Voirons, to the N.E. of Geneva, commanding a superb view of the Alps of Savoy, the Jura Mts., etc., is another favourite point. Railway (Geneva and Eaux-Vives Station, p. 221) viâ *Annemasse* (p. 271) to (50 min.) *Bons-St-Didier*; thence a drive of 3 hrs., or a walk of 2½ hrs. to the summit. In summer omnibus from Bons-St-Didier to the top on three afternoons weekly (Mon., Wed., Sat.) in 3 hrs. (4 fr.; one-horse carr. 10 fr.). On the E. slope, 100' below the summit, is the *Hôtel de l'Ermitage* (pens. 6-8 fr.; frequented by the French), in the midst of pine-wood, visited as a health-resort; and 10 min. below it is the *Hôt. Chalets des Voirons* (pens. 8-12 fr.). Charming walks to the (10 min.) pavilion on the *Calvaire* or *Grand Signal*, the highest point (4375'); to the (20 min.) old monastery (4590') on the N.W. slope; to the *Crête d'Audox*, an eminence ½ hr. to the S.W.; and to the (1 hr.) *Pralaire* (4630'), the S. **peak**.

66. From Geneva to Martigny viâ Lausanne and Villeneuve. Lake of Geneva (North Bank).

81 M. **Railway** in 4¾-6 hrs. (to Lausanne 1¼-2, to Vevey 2½-3¼ hrs.); fares 13 fr. 35, 9 fr. 35, 7 fr. 70 c. (to Lausanne 6 fr. 35, 4 fr. 45, 3 fr. 20 c.; to Vevey 8 fr. 35, 5 fr. 85, 4 fr. 20 c.). Return-tickets from Geneva to St. Maurice, and from Bouveret to Brig, are available for two days, **and may be used for the steamers, and *vice versâ*.

Steamboats along the NORTHERN BANK, far preferable to the railway: to Morges (4 fr., 1 fr. 70 c.) in 2-2½ hrs.; to Ouchy (for Lausanne, 5 fr., 2 fr.) in 2½-3 hrs.; to Vevey (6 fr. 50, 2 fr. 70 c.) in 3½-4 hrs.; to Villeneuve (7½ fr., 3 fr.) in 4-4¾ hrs.; to Bouveret (7½ fr., 3 fr.) in 4¾-5 hours. Return-tickets for three days at a fare and a half, available also for returning by railway, but not unless specially asked for. The cabin-tickets are available for the second class only; if the holder desires to travel first class he may obtain a supplementary ticket from the guard of the train. Steamboat-stations on the N. bank (all with piers): *Bellevue, Versoix, Coppet, Céligny, Nyon, Rolle, St. Prex, Morges, St. Sulpice, Ouchy* (Lausanne), *Pully, Lutry, Cully, Rivaz-St-Saphorin, Corsier* (near the Grand Hôtel de Vevey), *Vevey-Marché, Veney-La-Tour, Clarens, Montreux-Vernex, Territet-Chillon, Villeneuve, Bouveret*. The express-steamers leaving Geneva (Quai du Montblanc) at 9.30 a.m. and 1.50 p.m. touch at the following stations only: Nyon, *Thonon*, and *Evian* on the S. bank, Ouchy, Vevey, Clarens, Montreux, Territet, Villeneuve, and *Bouveret*. — Several steamboats also ply daily between the N. and S. banks (Nyon-Nernier, Nyon-Thonon, Evian-Ouchy). — Good restaurants on board the larger steamers (D. 2½-3 fr.); those on the smaller boats are mediocre.

The *Lake of Geneva (1230'), Fr. *Lac Léman*, Ger. *Genfer See*, the *Lacus Lemanus* of the Romans, is 45 M. in length, upwards of

8 M. broad between Morges and Amphion, and 1½ M. between the Pointe de Genthod and Bellerive; 250' deep near Chillon, 940' near Meillerie, 1100' between Ouchy and Evian (deepest part), and 240' between Nyon and Geneva. The area is about 225 sq. M., being 15 sq. M. more than that of the Lake of Constance. This lake differs in the deep blue colour from the other Swiss lakes, which are all more or less of a greenish hue. The Lake of Geneva has for centuries been a favourite theme with writers of all countries — Byron, Voltaire, Rousseau, Alex. Dumas, and many others. On the N. side it is bounded by gently sloping hills, richly clothed with vineyards and orchards, and enlivened with numerous smiling villages. To the E. and S. a noble background is formed by the long chain of the mountains of Valais and Savoy.

The BIRDS which haunt the lake are wild swans *(Cycnus olor)*, the descendants of tame birds introduced at Geneva in 1838, gulls *(Larus ridibundus)*, sea-swallows *(Sterna hirundo)*, and numerous birds of passage, such as ducks and divers. There are twenty-one different kinds of FISH, the most esteemed of which are the trout, the 'Ritter', the 'Féra' *(Coregonus;* the 'Felchen' of the Lake of Constance), and the perch.

A phenomenon frequently observed on the Lake of Geneva, and sometimes on other lakes also, consists in the so-called 'SEICHES', or fluctuations in the level of the water, caused by sudden alteration in the atmospheric pressure. The *seiches longitudinales* are those running from one end of the lake to the other; the *seiches transversales* cross from the Swiss to the Savoy side in 10 minutes. The highest longitudinal swell on record was over 6 ft. in height. — The TEMPERATURE of the lake varies from 45° in winter to 75° or even 85° in summer, while in the deeper parts it never rises above 42-44°. The lake has never been known to freeze over entirely.

STEAMBOAT JOURNEY (piers by the Jardin Anglais and the Quai des Pâquis; express-steamers at the Quai du Montblanc; comp. p. 221). The banks of the lake are clothed with rich vegetation and studded with charming villas. On the left, the Musée Ariana, and the château of Pregny (p. 231); farther on, Genthod, once the residence of the famous naturalists Cb. Bonnet (d. 1793), H. B. De Saussure (d. 1799), Theod. De Saussure (d. 1845), Pictet de la Rive (d. 1872), and Ed. Boissier (d. 1885). The steamer stops at *Bellevue*.

Versoix, a considerable village (1379 inhabitants), once belonged to France.

Coppet *(Hôt.-Pens. du Lac; Garden-Restaurant*, near the pier). The château, now the property of M. d'Haussonville, was inhabited from 1790 till 1804 by *Necker*, a native of Geneva, who became minister of finance to Louis XVI. His daughter, the celebrated *Mme. de Staël* (d. 1817), also resided at the château for some years. Her portrait as Sappho by David, several paintings by Gérard, and a bust of Necker are shown to visitors (Thurs. only, 2–6).

From Coppet (carr. at the station) a road leads by *Commugny* and *Chavannes de Bogis* to (3½ M.) Divonne (1543'; excellently fitted up hydropathic estab.), charmingly situated beyond the French frontier in the *Pays de Gex* (from Nyon 5 M., diligence in connection with the express trains in 55 min.; from Geneva 12 M., carr. in 1½ hr., with one horse 15-18, with two horses 25 fr.). Ascent of the *Dôle* from Divonne, see p. 235.

English miles

Céligny is prettily situated on a hill a little inland. Farther on is the handsome château of *Crans*, belonging to Herr Van Berchem.

Nyon (**Hôt. du Lac*, small; *Beaurivage*, with terrace on the lake; *Ange*, pens. 5–6 fr.) was the *Colonia Julia Equestris*, or *Noviodunum*, of the Romans (4225 inhab.). The ancient castle, with walls 10′ thick, and five towers, built in the 16th cent., and now the property of the town, was occupied towards the end of last cent. by Victor von Bonstetten (d. 1832), the author, who was district governor, and was visited here by J. v. Müller, Salis, Matthisson, etc. The terrace and the pleasant promenades of the upper part of the town afford a beautiful view of the lake, the Jura, and the Alps, with Mont Blanc. Several relics of the Roman period still exist here.

Ascent of the Dôle, very interesting. A high-road (diligence) leads from Nyon through the Jura by (1 hr.) *Trélex*, (2 hrs.) **St. Cergue**, and (2 hrs.) *Les Rousses*, a small French frontier fort, to (1 hr.) *Morez*, a little town in the French department of Jura. Walkers ascend from Nyon in 2¼ hrs. to **St. Cergue** (3432′; **Hôt.-Pens. Capt; *Pens. Auberson; Observatoire Amal*, a hotel and pension on a height, 5 min. to the E., with a splendid view of the Lake of Geneva and Mont Blanc; Engl. Ch. Serv.), a village and summer-resort in a green valley at the N.E. base of the Dôle. The traveller should drive from Nyon as far as the beginning of the well-shaded old road, 1½ M. beyond Trélex, which follows the telegraph-wires, and ascends straight to St. Cergue (3 M.). From St. Cergue (guide 5 fr.; not indispensable) we ascend to the (1 hr.) *Chalet de Vuarne*, and through the depression (*Sur Porta*, 5127′) between the Vuarne and the Dôle, to the (1 hr.) top of the ***Dôle** (5505′), the highest summit of the Swiss Jura. The view (best in the afternoon) is picturesque and extensive, and Mont Blanc is seen in all its majesty. — From *Gingins*, 1½ M. to the W. of Trélex, a good road leads to the (7½ M.) *Chalets de la Divonne*, ½ hr. from the top of the Dôle. — Another route leads by *La Rippe*, 3¾ M. from Céligny (see above), and 1½ M. from Divonne (p. 234), and before reaching (¾ M.) *Vendôme*, enters the broad path (to the right) through the wood, which after 3 M. joins the road from Gingins. — The best route for pedestrians from Geneva (7½ hrs. to the summit of the Dôle) is by the *Col de la Faucille*, a deep depression in the Jura chain, to the N.W. of Geneva. Steam-tramway to *Ferney*, see p. 231; omnibus thence in 1 hr. to *Gex* (2120′; Hôt. de la Poste; Hôt. du Commerce), a small French town, at the foot of the Jura; whence we proceed (shorter by the old road) to the (2 hrs.) *Col de la Faucille* (4355′; Hôt.-Pens. de la Faucille, unpretending; Couronne, still smaller). We keep to the road to Morez, see above) for 1¼ hr. more, finally diverging to the right beyond the *La Vasserode Inn*, whence we ascend to the summit in 1½ hr.

Diligence from *Les Rousses* (see above) to **Le Brassus**, to the *Lac de Joux*, and *Le Pont*, a pleasant route (comp. p. 221).

Farther on, among trees, is the château of **Prangins**, formerly occupied by Joseph Bonaparte, and now a Moravian school for boys. A great part of the estate of *La Bergerie*, or **Chalet de Prangins**, was afterwards the property of Prince Jérôme Napoléon (d. 1891).

On a promontory lies *Promenthoux*, and on the opposite (Savoyard) bank, 3 M. distant, *Yvoire* (p. 257). The Jura Mts. gradually recede; the most conspicuous peaks are the Dôle, and to the right of it the *Noir-Mont* (5118′). The lake forms a bay between the mouth of the *Promenthouse* and the *Aubonne* (p. 246) beyond Rolle, and here attains its greatest width. The banks of this bay, called *La Côte*, yield one of the best Swiss white wines.

Rolle (*Tête Noire*, plain, with garden), the birthplace of the Russian general *F. C. Laharpe*, tutor of Emp. Alexander I., and one of the most zealous advocates for the separation of Canton Vaud from Bern (1798). An artificial islet in the lake contains an *Obelisk* to his memory.

On a vine-clad hill, 1 hr. to the N. of Rolle, above the village of *Bougy*, is the *Signal de Bougy (2385'; pavilion, with rfmts.), a famous point of view, which commands the lake, the Savoy Mts., and Mont Blanc. The best way to it is from stat. *Aubonne-Allaman* (p. 245) by omnibus (twice or thrice daily) or on foot to (2 M.) Aubonne (*Couronne), a very old and picturesque little town, with numerous gardens, a beautiful avenue, and pleasant public grounds, and thence on foot to the top in less than an hour. Carriage from the station to Aubonne 2, there and back 3, to the Signal and back, with 1 hr.'s stay, 7 fr. — About ½ M. to the S.W. of Aubonne, and 1½ M. from stat. Aubonne-Allaman, lies the finely-situated château of *Tréselin* in a large park (1645'; Hôt.-Pens., 5-7 fr.); hence to the Signal in 40 minutes. — About 5 M. to the W. of Aubonne, and 5½ M. to the N. of Rolle, is **Gimel** (2395'; *Union*, pens. from 5 fr.), with wood-walks, a favourite summer-resort of the Genevese.

A road (diligence to St. Georges daily) leads from Rolle to the N.W. by *Gilly*, **Burtigny**, and *Longirod* to (9 M.) *St. Georges* (3067'; Inn) and over the (4 M.) **Col de Marchairuz** (4767'; *Inn*) to (4½ M.) *Le Brassus* (p. 221). On the way from St. Georges to the col, we enjoy charming views of the Lake of Geneva and the Rhone Valley down to the Fort de l'Ecluse, and beyond the col we overlook the Lac de Joux and the Dent de Vaulion.

The bank of the lake between Rolle and Lausanne is somewhat flat. On a promontory lies the village of *St. Prex*; then, in a wide bay, **Morges** (*Hôt. du Montblanc*, pens. from 5 fr., adapted for a stay of some time; *Hôt. du Port; Couronne*), a busy little town (pop. 4088), with a harbour and an old château now used as an arsenal. Good lake-baths. From Morges we obtain a fine view of *Mont Blanc in clear weather through a valley on the S. bank. The mediæval château of *Vufflens*, on a height at some distance to the N., is said to have been erected by Queen Bertha (p. 219). The steamer next reaches the station of *St. Sulpice*, and then —

Ouchy (1230'), formerly called *Rive*, the port of Lausanne.

Hotels. *HÔTEL BEAURIVAGE, with pleasant garden, baths, etc., R., L., & A. 5-7, déj. 3½, D. 4-3, pens. 10-16 fr.; *HÔT. DU CHATEAU, near the steamboat-pier, a castellated building with view-tower (lift), R., L., & A. from 3, B. 1½, déj. 3½, D. 5, pens. from 8 fr.; *HÔT. D'ANGLETERRE, R., L., & A. 3-4, B. 1¼, D. 4, pens. 6-9 fr.; *HÔT. DU PORT, plain, all on the lake. PENS. DU CHALET, Avenue Roseneck; PENS. LA PRINTANNIERE. — *Lake Baths*, two establishments, one ½ M. to the W., the other ¼ M. to the E. of the landing-place; bath 80 c., including towels, etc. — *Boat* 60 c. per hour, or with boatman 1½ fr.

The RAILWAY STATION of the Jura-Simplon line (p. 245) is ¾ M. from Ouchy, and Lausanne lies fully ½ M. higher. CABLE RAILWAY (commonly called *Ficelle*) from Ouchy to Lausanne in 9 min. (station at Ouchy 3 min. from the steamboat-quay; station at Lausanne, called 'Gare du Flon', under the Grand-Pont; 46 trains daily; fare 50 or 25 c., return-ticket 80 or 40 c.; intermediate stations *Jordils* and *Ste. Luce* ('Gare'), the latter near the Jura-Simplon station. — *Porterage* of small articles to or from the steamer 10 c., trunk 20 c., if over 100 lbs. 30 c.

Lausanne. — **Hotels.** *HÔTEL GIBBON (Pl. a; E, 4), opposite the post-office, R., L., & A. 4-6, B. 1½, déj. 3, D. 4-5, pens. 6-10 fr.; in the garden behind the dining-room the historian Gibbon wrote the concluding portion of his great work in 1787. *HÔT. RICHE-MONT (Pl. b; D, E, 5), with pleasant grounds, R., L., & A. 4-6, D. 4-5 fr.; *FAUCON (Pl. c; F, 3), R., L., & A. 3-5, B. 1½, D. 4, pens. 6-10 fr.; *HÔT. TERMINUS, at the Jura-Simplon station (p. 245), R., L., & A. 3-5, B. 1¼, D 3½-4, pens. 7-10 fr.; HÔT. DU GRAND-PONT (Pl. d; E, 4), near the bridge, R., L., & A. 3½, B. 1¼, D, 3½, pens. 11 fr.; *HÔT.-PENS. BEAU-SITE (Pl. e; D, 4), E. L., & A. 2-4, B. 1¼, D. 3½, S. 2½ fr.; *HÔT.-PENS. VICTORIA, Avenue de Rumine, R., L., & A. 2½-4, D. 3, pens. 6-8 fr.; *HÔT. DU NORD (Pl. f; F, 3, 4), Rue St. Pierre, R., L., & A. 2½-3, B. 1¼, D. 3, pens. 6-9 fr.; HÔTEL BELLEVUE, R. from 2 fr.; HÔT. DES MESSAGERIES, Place St. François 4; HÔT DE LA POSTE, Petit-Chêne 4. — Pensions: *Beauséjour, Avenue de la Gare; *Campart, Route d'Ouchy, opposite the English church; Pittet, at Ste. Luce (see above; 5 fr. per day), and many others. — Restaurants: Hôtel du Nord, Hôtel du Grand-Pont, see above; Café du Banque; Restaurant du Théâtre (see below), with garden; Rail. Restaurant, D. 2½ fr.; Café Vaudois, Place de la Riponne 3; Gambrinus (beer), Rue Haldimand, near the Place de la Riponne; Bavaria, Rue du Petit-Chêne 3; Brasserie des Alpes, near the station.

Theatre (Pl. G, 4; open in winter only), Avenue du Théâtre (with café).

OMNIBUS from the station into the town 1 fr.; to the steamboat at Ouchy, only if ordered. — CABS: with one horse ½ hr. 1½, with two horses 3 fr.; 1 hr., 3 and 5; 1½ hr., 4 and 7; 2 hrs., 5 and 9 fr.; from Lausanne to Ouchy 2 and 4, to the rail. station 1½ and 3, from Ouchy to the rail. station 2 and 4 fr. — Booksellers, with lending library, etc.: Benda, Rue Centrale 3; Th. Roussy, F. Payot, both Rue de Bourg. — Pianos, music: E. R. Spies, Place St. François 2.

ENGLISH CHURCH, Avenue de Grancy. Scottish Free Church, Rue Rumine. Wesleyan Church, Rue du Valentin, Place de la Riponne. — ENGLISH PHYSICIAN, Prof. A. Ganges, Avenue de la Gare 8.

Lausanne (1690′; pop. 34,049), the *Lausonium* of the Romans, now the capital of the Canton de Vaud, occupies a beautiful and commanding situation on the terraced slopes of *Mont Jorat*, overshadowed by its cathedral on one side, and its castle on the other. The interior of the town is less prepossessing. The streets are hilly and irregular, and the houses in the older part are poor; but the new quarters contain a number of handsome houses. The two quarters are connected by the handsome *Grand-Pont* (135 yds. long), also named *Pont Pichard* after its builder (1839-44). The valley of the *Flon*, spanned by the bridge, has been largely filled up and built over. A nearly level street, passing the castle and cathedral, skirts the town and leads under the castle to the N. by a tunnel, 50 paces long. Lausanne possesses many excellent schools; the College, founded in 1806, was raised to the dignity of a University in 1891.

The *Cathedral (Pl. E, 2; Prot.), erected in 1235-75, is a simple but massive Gothic edifice. In 1875-87 it was judiciously restored from plans by *Viollet-le-Duc* (d. 1879). The terrace on which it stands is approached from the market-place (Place de la Palud) by a flight of 160 steps. The church is open in summer on week-days, 9-12 and 1-4; at other hours, adm. 30 c. each person. Bell for the sacristan by the entrance.

The *INTERIOR (352′ long, 150′ wide) is remarkable for its symmetry of proportion. The vaulting of the nave, 82′ in height, is supported by 20 clustered columns of different designs. Above the graceful triforium runs another arcade, which serves as a framework for the windows. The choir contains a semicircular colonnade. In the arcades of the choir-ambulatory appears an ancient form of pilaster, a relic of the Burgundian-

Romanesque style. The beautiful but sadly damaged rose-window, the sculptured portals, and the carved choir-stalls (completed in 1509) at the S. wall also merit inspection. (The W. portal is being restored; the S. portal was restored in 1884.) Above the centre of the church rises a slender tower (213'), erected in 1874. The finest MONUMENTS are those of *Otho of Grandson*, who fell in 1398 in a judicial duel with Gerard von Estavayer (hands on the cushion, a symbol of the ban; statue accidentally deprived of its hands); *Bishop Guillaume de Menthonex* (d. 1406); the Russian *Princess Catherine Orloff* (d. 1782); the *Duchess Caroline of Courland* (d. 1783); *Henrietta Stratford-Canning* (d. 1818), first wife of Lord Stratford de Redcliffe, then ambassador in Switzerland (by Bartolini); *Countess Wallmoden Gimborn* (d. 1783), mother of the Baroness of Stein, the wife of the celebrated Prussian minister. A tablet on the wall of the N. transept commemorates *Major Davel*, executed in 1723 for attempting to free the Vaudois from the dominion of Bern. — In 1536 a famous Disputation took place in this church, in which *Calvin*, *Farel*, and *Viret* participated, and which resulted in the separation of Vaud from the Romish Church, and the overthrow of the supremacy of Savoy.

The *Terrace* (1735'), formerly the churchyard, commands a view of the town, the lake, and the Alps of Savoy, which is, of course, still more extensive from the church-tower (137'); and the prospect from the terrace of the old episcopal PALACE *(Evêché;* now occupied by the cantonal authorities), higher up, is also very fine. The *Bishop's Hall* contains old carved furniture and stained-glass windows.

The CANTONAL MUSEUM (Pl. E, 2; Wed. and Sat. 10-4, Sun. 11-2 o'clock), in the *Collège* near the cathedral, contains natural history collections, a valuable collection of freshwater conchylia, presented by M. de Charpentier (d. 1855), relics from Aventicum (p. 207) and Vidy, the ancient Lausanne, interesting Celtic antiquities from lake-dwellings, coins, medals, etc. The same building contains the *Cantonal Library* (120,000 vols.).

The MUSÉE ARLAUD (Pl. D, 3; Sun. 11-2, Wed. and Sat. 10-4; at other times, 50 c., each person more 30 c.), founded by an artist of that name in 1846, in a building in the *Riponne* opposite the corn-hall *(Grenette)*, contains a small picture-gallery.

On the groundfloor is a room with paintings by *Bocion*. On the staircase: *Koller*, Cattle-pond. — First Floor. In the room to the left: *Domenichino*, Joseph's Dream; *Corracci*, Joseph cast into the pit; *Jouvenet*, Healing of the man with the palsy; *Gleyre*, Execution of Major Davel (see above), Adam and Eve, Divico's victory over the Romans, etc. In the room to the right: *Anker*, New-born child; *Calame*, Lake of Brienz; *Diday*, Wellhorn, Fall of the Reichenbach; *Girardet*, Return from the mountain-pasture; *Muyden*, Hide-and-seek; *Vautier*, Sabbath morning; *Burnand*, Bull, etc.

On the MONTBENON, a hill immediately to the W. of the town, planted with fine avenues, and affording a charming view of the lake, is situated the handsome new *Palais de Justice Fédéral*, or supreme court of appeal for the whole of Switzerland.

The BLIND ASYLUM *(Asile des Aveugles)*, to the W. of the town (Pl. A, 3), was founded by Mr. Haldimand (d. 1862), who amassed a fortune in England, and Miss Cerjat. — In the *Champ de l'Air*, to the N.E., the highest point in the town, are the HÔPITAL CANTONAL (250 beds), a *Station Viticole et Météorologique* (wine-growing and meteorology) and an *Ecole d'Agriculture*.

The *Signal (2125'), ¼ hr. above the castle, is a famous point of view. We cross the Place de la Barre (Pl. E, 1) and follow the road straight on for about 100 paces; then ascend to the right by a paved path and flights of steps to the carriage-road and follow this to the right till the hut with the trigonometrical pyramid and grounds are seen on the right. (This point may also be reached by a broad path diverging from the road to the right.) The view embraces a great part of the lake, the Diablerets, Grand Mœveran, etc.; Mont Blanc is not visible from this point, but is seen from the *Grandes Roches* (½ hr. from the town, to the right of the Yverdon road). — A pleasant way back from the Signal is through the wooded valley of the *Flon*, on the E. side of the hill, and then by the Rue des Eaux to the Place de la Barre. Cab from the town to the Signal, and thence to the station, 5 fr.

FROM LAUSANNE TO BERCHER, 12½ M., a local narrow-gauge railway (1 hr. 27 min.). Near (2 M.) *Jouxtens-Cery*, the second station, is a large lunatic asylum (Asile des Aliénés). 8¾ M. *Echallens* (2054'; 1089 inhab.; *Balances) is a thriving little town, with an old castle now used as a boys' school. — 12½ M. *Bercher*.

The slopes rising to the E. of Lausanne are named *La Vaux*, and yield good wine. Above the station of *Pully*, on the hillside, is the lofty viaduct crossing the *Paudèze* (p. 218), below which is the bridge of the Martigny Railway (p. 245); above *Lutry* is the viaduct near La Conversion, mentioned at p. 218. The amphitheatre of mountains becomes grander as the steamboat advances: the Rochers de Verraux, Dent de Jaman, Rochers de Naye, Tour d'Aï, Tour de Mayen, Dent de Morcles, and Dent du Midi; between these, to the S., Mont Catogne, and in the background the snowy pyramid of the Grand Combin. Stations: *Cully* and *Rivaz-St-Saphorin*.

Vevey. — Steamboat Piers: (1) *Corsier*, to the W., near the Grand Hôtel de Vevey; (2) *Vevey-Marché*, at the town itself; (3) *Vevey-la-Tour*, to the E. near the Grand Hôtel du Lac. — Railway Station (*Buffet*) on the N. side of the town. For excursions to the E. (Montreux, etc.) the station of *La Tour de Peilz* (p. 246) is more convenient.

Hotels. *GRAND HÔTEL DE VEVEY, to the W. of the town, on the right bank of the Veveyse, with lift, large grounds, swimming and other baths (closed in winter) R., L., & A. 3-10, B. 1½, déj. 3½, D. 5, board 8 fr.; *GRAND HÔTEL DES TROIS COURONNES, on the Quai Perdonnet; *GRAND HÔTEL DU LAC, on the Quai Sina, R., L., & A. 3½-6½, déj. 3, D. 4½, pens. 7-12 fr.; these three hotels, all on the lake, are large and comfortable; pension from 15th Oct. to 1st May. — To the E. of the town, *HÔT. MOOSER (p. 241). — *HÔT.-PENS. D'ANGLETERRE, R., L., & A. 2½-3½, D. 3, pens. 5½-8 fr., *HÔT.-PENS. DU CHÂTEAU, pens. 6-12 fr., both on the lake, with gardens and lake-views; *HÔTEL DU PONT, at the station, with garden, R. 2½, B. 1¼, D. 2½ fr.; *TROIS ROIS, not far from the station, R., L., & A. 2-2½, B. 1-1¼, D. 2½-3 fr.; *HÔT.-PENS. DE FAMILLE, opposite the station, R. 1½-2 fr., D. 80 c., pens. 3½-4 fr.; HÔTEL DE LA GARE, plain but good. — Pensions, see p. 241.

Cafés. *Café du Lac* (Munich beer), *Bellevue*, both on the quay; *Café du Théâtre*; *Cercle du Léman*, with reading-room and a large garden on the lake (open to strangers). — *Casino Restaurant*, at Vevey-La-Tour.

Lake Baths at the E. end of the town, beyond the Hôtel du Lac.

Post and Telegraph Office, Place de l'Ancien Port. — Bankers: *Crédit du Léman*, Rue du Lac; *A. Cuénod-Churchill*, Place du Marché 21.

Omnibus from the station to the hotels 20, box 10 c.; to La Tour de Peilz 30, box 15 c.; to Chexbres from the post-office 1 fr. (see p. 218). — Cab with one horse, per drive in the town 1½, with two horses 2 fr.; ½ hr. 1½ or 2 fr., 1 hr. 3 or 4 fr., for every ½ hr. more 1 or 1½ fr. From the station to Montreux 7 fr.

Electric Tramway from Vevey to Chillon every 10 min. from 6.30 a.m.,

in 1 hr. (fares 10-60 c.). Stations: Grand-Hôtel, Vevey-Gare, Hôtel du Lac, Villa Thamine, Maladaire, Clarens, Vernex, Cursaal, Territet, and Chillon.
Rowing-boats 1 fr. per hr.; with one rower 2, with two rowers 3 fr.; to Chillon 6 or 10 fr.; to St. Gingolph (p. 258) same charges; to Meillerie (p. 258) with two rowers 12, with three rowers 15 fr.
Bookseller. *Benda*, Hôtel Monnet (also music, etc.). Pianos at *Ratzenberger's* (also at Montreux and Bex). — Theatre, Rue du Théâtre, behind the Grande Place, to the right.
United States Consular Agent, *Mr.* **William Cuénod.**
English Church at the E. end of the town.

Vevey (1263'), Ger. *Vivis*, the *Vibiscus* of the Romans, with **8144** inhab., situated mainly on the left bank of the sometimes turbulent *Veveyse*, near its influx into the lake, is the second town in the Canton de Vaud, and has considerable manufactories of tobacco, infants' food, etc. It is the scene of Rousseau's famous romance, the **'Nouvelle Heloise'** (1761). Vevey commands a beautiful view of the head of the lake, with the mouth of the Rhone and, in the background, the Alps of the Valais, the jagged, snow-covered Dent du Midi, Mont Velan, and Mont Catogne (the 'Sugar Loaf'); on the S. bank of the lake, the rocks of Meillerie, overshadowed by the Dent d'Oche; and to the left, at the foot of the Grammont, St. Gingolph (p. 258). Beside the pier of Vevey-Marché are the turretted *Château of M. Couvreu* (beautiful garden with exotic plants, fee 1 fr.) and the large Grande Place or Marché. The *Quais Sina* and *Perdonnet*, to the E., with the pier of Vevey-la-Tour, afford a beautiful **walk, sheltered** from the N. wind.

Ascending **across the market-place,** with the theatre to the right, and then the Rue de Lausanne, we reach the *Railway Station*, to the E. of which are the *Russian Chapel* with its gilded dome and the handsome new *Musée Jenisch* (not yet opened). The road passing in front of the Russian chapel and **crossing the railway** leads to the —

Church of St. Martin, erected in 1498, on a hill *('Terrasse du Panorama')* outside the town, surrounded by lime and chestnut-trees, and commanding a charming view (see the *'Indicateur des Montagnes'*). Service in summer only (organ-concerts).

In this church repose the remains of the regicides Ludlow (*'potestatis arbitrariae oppugnator acerrimus'*, as the marble tablet records) and Broughton. The latter read the sentence to King Charles (*'dignatus fuit sententiam regis regum profari, quam ob causam expulsus patria sua'* is the inscription on his monument). Charles II. on his restoration demanded the extradition of the refugees, a request with which the Swiss government firmly refused to comply. Ludlow's House, which stood at the E. end of the town, has been removed to make way for an addition to the Hôtel du Lac. The original inscription chosen by himself, *'Omne solum forti patria'*, was purchased and removed by one of his descendants. A new memorial tablet was erected in 1887 on the Quai Sina.

At the E. end of the town are the pretty *Roman Catholic Church* and the *English Church*. The tower among the trees on the lake farther on, the *Tour de Peilz **(Turris Peliana),*** said to have been built by Peter of Savoy in the 13th cent., **was** once the seat of a court of justice, and was afterwards used as a prison. The neighbouring château of *M. Sarasin* contains a collection of ancient weapons.

The château of **Hauteville** (1650'), 2 M. to the N.E. of Vevey, with an admirably kept park, commands a beautiful view from the terrace and the temple (fee to the gardener). In the same direction, 2 M. higher, is the mediæval château of **Blonay** (2118'), which has belonged the family of that name for centuries. The road from Hauteville to Blonay passes through the villages of *St. Légier* and *La Chiésaz*, several houses in which are adorned with clever sketches by A. Béguin, a native of the place, now an artist in Paris. In returning, we may descend by a path to the right beyond the bridge (finger-post: 'Montreux 5 Kil'.) to the carriage-road below, which leads to (1 M.) *Chailly* (see p. 230), the bridge of (1 M.) *Tavel*, below the *Château des Crêtes* (see below), and (¹/₄ M.) the Clarens station. — About 1 hr. to the N.E. of Blonay are the **Pléiades** (4483'), a famous point of view (auberge near the top), at the E. base of which, ³/₄ hr. from the top, are the small sulphur-baths of *L'Alliaz* (3428'; pens. 4-5 fr.).

From Vevey to Freiburg, see R. 62. — Pleasant excursion to *St. Gingolph* (p. 258; 1¹/₂ hr. by boat), on foot to *Novel*, in the valley of the Morge, and thence to the top of the *Blanchard* (p. 258). Inn at Novel now very fair; unnecessary to bring provisions from Vevey.

On the lake, 3¹/₂ M. from Vevey, lies the beautiful village of **Clarens** *(English Church Service)*, immortalised by Rousseau. On a height to the N.W. rises the *Château des Crêtes* (1493'; 'crêtes' = edge or ridge), with its pleasant grounds, and a beautiful view from the terrace (visitors admitted). Adjoining it is a chestnut copse, called the '*Bosquet de Julie*'; but Rousseau's 'Bosquet' has long since disappeared. Splendid view from above Clarens, near the churchyard; at *Tavel*, ¹/₄ hr. to the N., is the old château of *Châtelard* (1645'). Between Clarens and *Vernex* is the *German Protestant Church*.

Pensions (p. xviii) abound on this favourite S.E. bay of the Lake of Geneva. The best-known are here mentioned in their order from Vevey. Charges often raised in the busy season.

At **Vevey**: *Hôt.-Pens. du Château*, see p. 239; *Pens. du Panorama*, at the back of the town, recommended to ladies; *Hôtel-Pens. Mooser*, at Chemenin, 10 min. above Vevey, charming view (6-10 fr.); *Pens. Florentina*. At St. Légier: *Pens. Béguin*. — At La Tour de Peilz, near Vevey: *Pens. Comte*; *Pens. des Alpes*. — At St. Légier, 3 M. above Vevey (see above): *Pens. Richemond* (English landlady; 5 fr.).

Near Clarens, 'au Basset': *Hôt.-Pens. Ketterer*, sheltered (6-8 fr.); lake-baths adjacent. This is the beginning of the region which, being sheltered from the 'Bise' or bitter N. wind, is often recommended to persons with delicate lungs as a winter-residence. The gay cluster of 22 villas near Clarens was built and fitted up by M. Dubochet of Paris (d. 1877), at a cost of 2¹/₂ million francs. They now belong to Mr. J. Guichard, and are let furnished for 3 months or upwards at rents varying from 4000 to 8000 fr. per annum (apply to the 'régisseur', at Villa No. 6). — At Clarens: on the left, *Beausite (Moser)*; on the right, *Hôt.-Pens. Verte-Rive* (5-7 fr.); on the left, *Hôt.-Pens. Sanssouci* (5 fr.); on the right, *Hôtel Roth*, with a garden on the lake. At the station: *Hôt.-Pens. des Crêtes* (5-6 fr.); *Hôt.-Pens. du Châtelard* (5-7 fr.; good cuisine). — At Baugy (1545'), 10 min. above Clarens, *Pens. Baugy*. At Chailly (1600'), 10 min. farther on, *Pens. Mury*, with garden; *Pens. la Colline*. At Charnex (1925'), 1¹/₂ M. above Clarens, *Hôt.-Pens. Dufour*. — Between Clarens and Vernex (all on the lake): to the left, *Grand Hôt. Roy*, with pleasant garden; *Pens. Germann*; to the right, *Pens. Clarenzia*; *Hôt.-Pens. Continental*, with garden on the lake; *Lorius* (three houses; 6 fr. and upwards), with fine garden.

At **Montreux-Vernex**: To the left of the pier: *Grand-Hôtel Monney & Beau-Séjour au Lac*, R., L., & A. from 3, B. 1¹/₂, lunch 3, D. 4, pens. from 6-7 fr.; *Cygne*, with three dépendances and a garden on the lake. R., L., & A. 4, déj. 2¹/₂, D. 4³/₂, pens. from 7 fr.; *Pens. Pilivet*, 6¹/₂ fr.,

BAEDEKER, Switzerland. 16th Edition. 16

with garden on the lake; *Hôt.-Pens. Suisse*, on the left side of the road, with a garden on the lake. R., L., & A. 2-3, B. 1¼, D. 3, pens. from 5½ fr. At the station, *Hôt.-Pens. Bellevue*, 5½-8 fr.; °*Hôtel de la Gare*, R., L., & A. 2-3, D. 3, pens. from 5 fr.; *Hôt. Victoria & Pens. Barbier*, R. 2, B. 1, D. 2½, pens. 6 fr.; *Hôt. de Montreux*, R. from 1½, B. 1, S. incl. wine 2 fr.; *Hôt. Central*, moderate. — In the Avenue de Belmont, 12 min. to the N. of the station, *Hôt. Belmont*, with open view. By the pier, *Hôt.-Restaurant Tonhalle*, for single gentlemen, moderate; *Hôt. du Parc & Restaurant Nicodet* (see below). — Beer at the *Tonhalle*. *Café des Alpes*, and at *Nicodet's* (all near the pier). — Strangers' Enquiry Office at the *Collège* (passports). — English Doctor: *Dr. Tucker Wise*, Villa Champod, Bon Port. — American Dentist: *J. J. Patterson*, Grand' Rue 74. — Chemists: *Buhrer* at Clarens; *Engelmann* at Territet; *Schopfer, Rouge, Rapin & Schmidt* at Montreux. — Bookseller: *Benda*. Reading-Rooms at the Cursaal; lending libraries at *Benda's* and *Faist's*. — Boarding and Day School for Girls: *Mlle. Hélène Guenther*, Ave. du Kursaal 17. — *Visitors' Tax* (after a week's residence): one pers. 1, 1½, or 2 fr. per week, two pers. 1½, 2½, or 3, three pers. 2, 3½, or 4 fr. The visitor receives an 'estampille' admitting him to the Cursaal; but a special ticket is required for balls, concerts, theatrical performances, etc.

In BONPORT, on the Territet road (where the *Cursaal* is on the right, music daily at 3 and 8 p.m.; adm. 1 fr.; weekly subscription, see above), on the lake, farther to the S.E.: on the right, *Hôt. du Léman*, °*Hôt.-Pens. des Palmiers*, from 6 fr.; *Hôt. Richemond & Pens. des Fougères*, from 6 fr.; on the left, °*Hôt. de Paris*, 7-10 fr.; *Maison Blanche*; °*Hôtel National*, with a terrace high above the lake, 7-10 fr. On the right, °*Hôt.-Pens. Beaurivage*, °*Hôt.-Pens. Brewer* (R., L., & A. 4-5, D. 4, pens. 7-12 fr.), both with gardens on the lake; °*Hôt.-Pens. Bonport*. The last four, ½ M. from the station, command a fine view. — In the village of LES PLANCHES, ½ M. from the lake and the station: °*Hôt.-Pens. Vautier*, 6-10 fr.; °*Pens. Visnand*, the oldest in Montreux; °*Pens. Mooser*, 5-6 fr.; °*Pens. Biensis*, 5-7 fr., all with view.

At Territet (just to the E. of stat. Territet-Glion): °*Grand Hôtel & Hôt. des Alpes*, pens. 7½-15 fr., an extensive establishment with handsome rooms, cold-water cure, and terraced grounds on the lake, with a fine view. °*Hôtel Mont-Fleuri* (1886), finely situated higher up, 6-8 fr. — To the left, *Hôtel du Lac*, very plain; *Hôtel d' Angleterre*, 6-8 fr.; to the right, *Hôt. Bristol & Pens. Mounoud*, 5-8 fr.; *Hôt.-Pens. Béatrice*; °*Hôt.-Pens. Richelieu*, 6-7 fr.

At Veytaux: °*Hôtel Bonivard*, R., L., & A. from 3 fr.; °*Hôt.-Pens. Masson*, higher up, 5-7 fr.; *Hôt.-Pens. Chillon*, near the castle, 5½-6 fr. — Between Chillon and VILLENEUVE, the handsome °*Hôtel Byron*, finely situated, 6-9 fr. (omnibus from the Villeneuve station, p. 245).

At Glion (2270'; cable-tramway, see p. 243): *Hôtel du Righi-Vaudois*, 8-12 fr.; °*Hôtel Victoria*, 8-14 fr.; °*Hôtel de Glion* (6-8fr.); *Hôt. Bellevue*, all with beautiful gardens; °*Hôt. du Midi* (4-5 fr.); *Pens. Champ-Fleuri* (from 5 fr.); *Hôt.-Restaurant Nicodet* (5 fr.); these usually closed in winter. — Above Glion, °*Grand-Hôt. de Caux*; *Grand-Hôt. de Naye* (p. 243).

Most of these pensions receive passing travellers at hotel-charges, but in autumn they are generally full. At many other houses rooms with or without board may also be obtained. The GRAPE CURE begins towards the end of September and lasts about a month. — AIGLE (p. 246) and BEX (p. 247) are also pleasant resorts in early summer and in autumn. In the height of summer, when the heat on the lake and in the valley of the Rhone becomes overpowering, the pensions at *Château d'Oex* (p. 254), *Ormont-Dessus* (p. 251), *Villars* (p. 246), etc., are much frequented.

ENGLISH CHURCH at Territet, daily services from Oct. to June, three services on Sun. during the whole year. Subscription library in the Parish Room ('St. John's Institute') next the church. — PRESBYTERIAN CHURCH at Montreux-Vernex, Rue de la Gare (serv. Sun. 10.30 a.m. and 4 p.m.).

Clarens, Charnex, Vernex, Glion, Colonges, Veytaux, and the other villages which lie scattered about, partly on the lake and partly on the hillside, are collectively called **Montreux** (pop. 10,696). The parish of Montreux, which extends to the Dent de Jaman, is divided into three parts, *Le Châtelard*, *Les Planches*, and *Veytaux*, by the brook *(Baye)* of Montreux and the Verraye. The central point of the district is the village of *Montreux-Vernex*, on the lake, with a railway-station and pier, quays with gardens, and a large market-hall on the lake. About 1/4 M. from the S. end of it is the *Cursaal*, with pleasant grounds (see p. 242); opposite is the *Roman Catholic Church*, in the Romanesque style. About 1/2 M. higher up, at the foot of the mountain, lies the village *Les Planches*, separated from *Sâles*, to the W., by the *Baye de Montreux*, which descends from the Gorge du Chauderon (see below) and is spanned by the handsome *Pont de Montreux*, 100' high. Above Les Planches rises the quaint old *Church of Montreux*, the shady terrace in front of which commands a superb and far-famed *View of the lake.

EXCURSIONS FROM MONTREUX (electric tramway from Chillon to Vevey, see p. 240). Chief excursion to *GLION AND THE *ROCHERS DE NAYE. To Glion (2270'; *Hotels*, see p. 242) a cable-tramway ('Chemin de fer funiculaire') ascends in 9 min., starting from the Territet-Glion station on the Jura-Simplon Railway (21 trains daily: fare 1, return-ticket 1 1/2 fr.). The line, constructed by Hr. Riggenbach on the same system as the Giessbach tramway, but much steeper, is about 750 yds. long, the maximum gradient being 1 : 1 3/4. At the top is a *Buffet-Restaurant*, which commands a delightful survey of the upper end of the Lake of Geneva and the mountains enclosing it, with the snow-clad Dent du Midi in the centre. Pleasant way back through the *Gorge du Chauderon* (see below) to the village of Montreux in 1 hr. (enquire for beginning of path).

*FROM GLION TO NAYE, 4 1/2 M., rack-and-pinion railway in 1 1/4 hr. (return-fare 10 1/2 fr.; from Territet to Naye and back 12 fr.). The station adjoins that of the 'Funiculaire' to the right. The line is carried beneath the houses of Glion by means of a tunnel, beyond which to the left we look down into the gorge of the *Baye de Montreux* (see above); on the opposite bank are the village of *Sonzier* and the reservoir of the Montreux electric works. We ascend gradually through meadows and pass over a viaduct, enjoying a fine view to the left of Montreux and the Lake of Geneva and of the large Hôtel des Avants below us (p. 244). Ascending more rapidly, we pass through a cutting and the curved tunnel of *Tremblex* (147 yds. long) to the E. side of the ridge and the (1 1/4 M.) station of *Caux* (3457'; Buffet). Above is the *Grand Hôtel de Caux (3580'; R. & L. 4 1/2, B. 1 1/2, lunch 3, D. 5 fr.; Engl. Ch. Serv.), commanding a splendid view of the lake and the Alps. — We now skirt the head of the valley of the *Veraye* (to the right, the Rochers de Naye) and beyond the chalets of *Myoux* pass again to the N. side of the ridge, where the conical *Dent de Jaman* (6493') suddenly appears. The line ascends rapidly to the ridge (5593') between Jaman and Naye and passes through a tunnel (82 yds. long) to the (3 3/4 M.) station of *Jaman* (5708') in the sequestered *Combe d'Amont*; to the left below us is the small *Lac de Jaman* (5144'). [The *Dent de Jaman*, a fine point of view, may be climbed hence in 1 1/2 hr.; see p. 254.] Farther on we are carried over a narrow arête, commanding a view of the Lake of Geneva to the right and of the mountains of the Gruyère to the left. We then pass through the rocky wall of the Rochers de Naye by a tunnel (6055') 267 yds. in length and ascend round the uppermost valley to the (4 1/2 M.) station of *Naye* (6485'; *Grand Hôtel, R. from 3, B. 1 1/2, D. 4-5, pens. from 8 fr.), 230' (10 min.) below the summit of the *Rochers de Naye (6708'). The splendid view (Panorama 1 fr. 80 c.) commands the Bernese Alps (Wetterhorn, Eiger, Mönch, Jungfrau, Finsteraarhorn), the Alps of the Canton de Vaud (Diablerets, Grand-Mœveran, Tour de Mayen, and Tour d'Aï), part of the Valaisian (Grand Combin, Dent du Midi) and Savoyan Alps (Aiguille d'Argentière, Aig. Verte), and the whole of the Lake of Geneva. Close to the station is a *Jardin Favrat* of the Montreux Botanical Society (adm. 30 c.).

16*

To the **Gorge du Chauderon**, a wooded ravine between *Glion* and *Sonzier*, watered by the *Baye de Montreux* (p. 243). From the bridge of Montreux to the head of the gorge, and back, 1 hr., or returning by Glion 2 hours. — **Les Avants** (3188'; *Hôtel des Avants*, pens. 6-12 fr., Engl. Ch. Serv. in summer), a charmingly situated health-resort for both summer and winter, lies 1³/₄ hr's. drive from Montreux viâ *Charnex* and *Chaulin* (omnibus from April 15th to Oct. 15th., from Montreux railway-station at 9 a.m., in 1³/₄ hr., returning at 4 p.m. in ³/₄ hr.; fares, up 3, down 2, return-ticket 4 fr.; carriage with one horse 12, with two horses 18 fr.). Les Avants may be reached on foot from Montreux viâ *Sonzier* (Maison Blanche, moderate) in 1½ hr., or from Glion viâ the Gorge du Chauderon in 1³/₄ hr. The fields of narcissus at Les Avants are a lovely sight in early summer. From Les Avants to the top of *Mont Cubly* (3505'), with charming view, 1 hr.; *Dent de Jaman* (6165'), viâ the *Col de Jaman* (p. 254), 2½ hrs., etc. — By *Charnex* and *Chaulin* to the *Bains de l'Alliaz* and the *Pléiades* (4488'), returning by *Blonay* (p. 241), 8 hrs. — By Aigle to the *Ormonts*, see R. 67. — To *Villars*, see p. 246. — To the *Pissevache* **and** *Gorges du Trient* (p. 249) by railway, and back, in one day.

Stat. *Territet-Chillon* (*Hôt. des Alpes, etc.; see p 242), opposite the railway-station of *Territet-Glion* (p. 246). The *Castle of Chillon, with its massive walls and towers, ³/₄ M. from the pier (¹/₄ M. from stat. Veytaux-Chillon), stands on an isolated rock 22 yds. from the bank, with which it is connected by a bridge. Above the entrance (fee) are the arms of the Canton de Vaud.

'Chillon! thy prison is a holy place,
And thy sad floor an altar, — for 'twas trod,
Until his very steps have left a trace,
Worn, as if the cold pavement were a sod,
By Bonivard! — may none those marks efface,
For they appeal from tyranny to God.'

The author of these beautiful lines has invested this spot with much of the interest which attaches to it, but it is an error to identify Bonivard, the victim to the tyranny of the Duke of Savoy, and confined by him in these gloomy dungeons for six years, with Byron's 'Prisoner of Chillon' (composed by him in the Anchor Inn at Ouchy in 1817). The author calls his poem a fable, and when he composed it he was not aware of the history of Bonivard, or he would, as he himself states, have attempted to dignify the subject by an endeavour to celebrate his courage and virtue. Francis Bonivard was born in 1496. He was the son of Louis Bonivard, Lord of Lune, and at the age of sixteen inherited from his uncle the rich priory of St. Victor, close to the walls of Geneva. The Duke of Savoy having attacked the republic of Geneva, Bonivard warmly espoused its cause, and thereby incurred the relentless hostility of the Duke, who caused him to be seized and imprisoned in the castle of Grolée, where he remained two years. On regaining his liberty he returned to his priory, but in 1528 he was again in arms against those who had seized his ecclesiastical revenues. The city of Geneva supplied him with munitions of war, in return for which Bonivard parted with his birthright, the revenues of which were applied by the Genevese to the support of the city hospital. He was afterwards employed in the service of the republic, but in 1530 when travelling between Moudon and Lausanne fell into the power of his old enemy, the Duke of Savoy, who confined him in the castle of Chillon. In 1536 he was liberated by the Bernese and Genevese forces under Nögelin, and returning to the republic, he spent the rest of his life as a highly respected citizen. He died in 1570 at the age of 74 years.

It is an historical fact that in 830 Louis le Débonnaire imprisoned the Abbot Wala of Corvey, who had instigated his sons to rebellion, in a castle from which only the sky, the Alps, and Lake Leman were visible (*Pertz, Monum. ii. p. 556*); this could have been no other than the Castle of Chillon. Count Peter of Savoy improved and fortified the castle in the 13th cent., and it now stands much as he left it. The strong pillars in the vaults are in the early-Romanesque style, and belonged to the original edifice. The Counts of Savoy often resided in the castle, and it was afterwards converted into a state-prison. It is now used for the cantonal archives. — A fine effect is produced by the beams of the setting sun streaming through the narrow loopholes into these sombre precincts, which are also lighted by means of two small electric lamps. Among the names on the pillars are those of Byron, Eugène Sue, George Sand, and Victor Hugo.

Between Chillon and Villeneuve is the handsome *Hôtel Byron* (p. 242). The Ile de Peilz, an islet $1/3$ M. to the W. of Villeneuve, commanding a fine view, was laid out and planted with three elms a century ago, and recalls Byron's lines: —

'And then there was a little isle,
Which in my very face did smile,
The only one in view.'

In the E. bay of the lake, $1^1/_2$ M. from Chillon, lies **Villeneuve** (*Hôt. du Port*, at the pier; *Hôt. de Ville*), a small walled town, the *Pennilucus* or *Penneloci* of the Romans. The 'Clos des Moines' is a good wine grown here. (Railway-station, see p. 246.)

Footpath to Montbovon (p. 253) over the *Col de la Tinière* (5340') in $4^1/_2$ hrs., to Château d'Œx (p. 254) in 6 hrs.

RAILWAY JOURNEY. *Geneva*, see p. 221. The train runs high above the lake, overlooking the hills on the E. bank with their numerous villas, above which rises the long ridge of the Voirons and in clear weather Mont Blanc. $2^1/_2$ M. *Chambésy* (station for Pregny, p. 231); 4 M. *Genthod-Bellevue*; $5^1/_2$ M. *Versoix* (p. 234); $8^1/_2$ M. *Coppet* (p. 234). At (11 M.) *Céligny* the *Dôle* (p. 235) becomes visible to the left. Beyond ($14^1/_2$ M.) *Nyon* (p. 235) the line skirts *Prangins* with its château, and then quits the bank of the lake.

The tract of country between the *Promenthouse*, which the train crosses near ($17^1/_2$ M.) *Gland*, and the Aubonne (see below) is called *La Côte* and is noted for its wine. 20 M. *Gilly-Bursinel*; $21^1/_2$ M. **Rolle** (p. 236). The height to the left is the *Signal de Bougy* (2910'; p. 236), a splendid point of view, easily reached from Rolle or from the next station (25 M.) *Aubonne-Allaman*.

The train crosses the *Aubonne* and returns to the lake. 28 M. *St. Prex*; the village lies on a promontory below, on the right. From ($30^1/_2$ M.) **Morges** (p. 236; station 8 min. from pier) Mont Blanc is seen in all its majesty in clear weather, but soon disappears. In the distance to the N.W., above the valley of the *Morges*, which the train crosses here, is the château of *Vufflens* (p. 236).

The line again leaves the lake, crosses the *Venoge*, and joins the Neuchâtel railway (p. 215). $35^1/_2$ *Renens*.

38 M. **Lausanne** (*Hôt. Terminus & Rail. Restaurant*), see p. 236.

The train (views on the right) skirts the lake the greater part of the way to Villeneuve. We cross the *Paudèze* by a handsome bridge (above which, to the left, is the lofty nine-arched viaduct of the Freiburg line, p. 218), pass through a short tunnel, and skirt the vine-clad slopes of *La Vaux* (p. 239). 42 M. *Lutry*.

From (44 M.) *Cully* (p. 239) to (47 M.) *Rivaz-St-Saphorin* the train runs close to the lake, then quits it, and crosses the *Veveyse*. 50 M. **Vevey** (*Buffet*; p. 239); 50¹/₂ M. *La Tour de Peilz* (p. 240); 52 M. *Burier*. Beyond a tunnel we obtain a fine view of Montreux, Chillon, and the E. bay of the lake. 53 M. *Clarens* (p. 241).

54 M. **Montreux-Vernex** (p. 242), beyond which we again approach the lake. 55 M. *Territet-Glion*(Café-Restaurant, and small bazaar), immediately **above** the pier of **Territet-Chillon** (p. 242), and the **starting-point** of the **cable-tramway to** *Glion* (p. 243). 55¹/₂ M. *Veytaux-Chillon* (p. 242) is ¹/₄ M. from the castle.

57 M. **Villeneuve**, see p. 245. The train now enters the broad and somewhat marshy *Rhone Valley*, bounded by high mountains. The *Rhone* flows into the lake 3 M. to the W., near Bouveret. Its grey waters, the deposits of which have formed an extensive alluvial tract, present a marked contrast to the crystalline azure of the same river where it rushes through the bridges at Geneva.

The first station in the Rhone Valley is (59¹/₂ M.) *Roche*. Part of the mountain near *Yvorne* (1560'), to the left, was precipitated on the village by an earthquake in 1584. Excellent wine is grown in the gorge ('Crosex-Grillé' and 'Maison Blanche' or 'Clos du Rocher'). To the right towers the jagged *Dent du Midi* (p. 259).

63 M. **Aigle.** — °Grand Hôtel, 1 M. above Aigle in the valley of the Grande-Eau, with extensive grounds, suitable for a prolonged stay, R. 2-6, L. 1, A. 1, B. 1¹/₂, lunch 3, D. 5, pens. 6-12, omn. 1-1¹/₂ fr. — *Hôt.-Pens. Beau-Site, at the station, with grounds, R. 2-3, B. 1¹/₄, lunch 3, D. 3¹/₂, pens. 6-7 fr.; *Victoria, next the post-office, 3 min. from the station, with garden, R. 2, D. 3, S. 2 fr.; Hôt. du Midi and Hôt. du Nord, both unpretending. — *English Church* (St. John the Evangelist).

Aigle (1375'; pop. 3555), a **small** town with a large château, is prettily situated on the turbulent *Grande-Eau*.

The *Plantour* (1601'), a wooded hill ¹/₂ hr. to the S., with grounds, affords charming views of the Rhone Valley.

Villars, 3¹/₄ hrs. to the E. of Aigle, 2¹/₂ hrs. above Ollon (see below), a very favourite summer-resort, lies on the hillside, high above the right bank of the Rhone. It is best reached from Aigle (carr. 15, with two horses 30 fr., down 25 fr., and fee; a drive of 3 hrs.; diligence daily at 3.30 p.m. in 4¹/₂ hrs., returning from Villars at 8.20 a.m. in 4¹/₄ hrs.; fare 3 fr. 75 c.), as the hotel and other accommodation at Ollon is poor. High-road to (2 M.) *Ollon* (Hôtel de Ville, poor); thence a good road in numerous windings, with fine views. Pedestrians follow the Panex road, which diverges to the left immediately above Ollon. After 1 min., where the path divides, we follow that to the extreme right. At (40 min.) *La Pousaz* we **take** the path to the left, by the second fountain, in the middle of the village; 35 min. *Huemoz* (3307'; pron. *Wems* by the natives), charmingly situated; 40 min. *Chesières* (3970'; °Hôtel du Chamossaire, pens. 5¹/₂-9 fr.; Pens. Mon Repos), with beautiful view; 20 min. Villars (4166'; °*Grand Hôtel Muveran*, patronized by the French, pens. 6-10 fr.; *Bellevue, a little higher up, R., L., & A. 2¹/₂-3, pens. 6-9 fr.; *Pens. Victoria*, 5-6 fr.; *Engl. Church*). Pleasant park-like environs, affording a variety of walks, with benches at all the best points of view and shady spots. The air is bracing but mild, and there is no N. or E. wind. Magnificent view of part of the Diablerets, the Grand and Petit Mœveran, the Dent aux Favres, Tête Noire, Dent de Morcles, the N. spurs of the Mont Blanc group with the Glacier de Trient, the Dent du Midi, Rhone valley, etc. The finest excursion is the ascent (2½ hrs.; guide unnecessary) of the *Chamossaire (6950'), which commands a most picturesque view of the Bernese Alps, the Weisshorn, the Diablerets, Grand Mœveran, Dent de Morcles, Mont Blanc, Dent du Midi, Valley of the Rhone, and Sepey. The route is by a cart-track nearly to *Bretayes* (5365'; Inn), 1 hr. from the top, a little below which we ascend by a path to the left to the stone signal on the summit. — From Bretayes a tolerable path leads past the small lakes *des Chalets*, *Noir*, and *des Chavonnes* (Inn), to (2 hrs.) *La Forclaz* (4144'), and, crossing the *Grande-Eau*, to (½ hr.) *Le Sepey* (p. 252). We may return to Villars the same day by carriage, viâ Aigle; or the next day on foot by *Au Pont*, *Plambuit*, and *Chesières* (see above). — Shorter excursions may be made from Villars to (¼ hr.) *Les Closalets*, a point commanding a fine view of the Rhone valley and of Mont Blanc; to (2 hrs.; horse 10 fr.) *Panex* or *Plambuit* viâ *Chesières* and *Les Ecovets*; to the (1¼ hr.) *Montagne de la Truche* (fine view) viâ Chesières, etc. — From Villars to *Ormont-Dessus* over the *Col de la Croix* (5687'), 4 hrs.; guide (6 fr.) unnecessary, if the traveller is shown the beginning of the route (comp. p. 251). — From Villars by *Arveye* to *Gryon* (p. 255), 1 hr.; to *Les Plans* (see below), 2½ hrs.

From Aigle a road leads by *Feorne* (p. 246) to (2 hrs.; one-horse carr. 8, two-horse 15 fr.) Corbeyrier (3235'; *Hôt.-Pens. Dubuis*, 5 fr.), a village in a sheltered situation, with fine views. The *Signal* (¼ hr.) overlooks the Rhone Valley from St. Maurice to the Lake of Geneva; more extensive view, particularly of the Tour Sallières and Dent du Midi, from the plateau of the *Agittes* (4997'; road, 1¾-2 hrs.). The *Tour de Mayen* (7628'), from Corbeyrier by the *Alp Luan* and *Alp Ai* in 3½-4 hrs., and the *Tour d'Ai* (7657'; 4 hrs.) are attractive ascents (not difficult).

FROM AIGLE TO LEYSIN (Grand Hôtel, p. 252), road by Le Sepey in 3½ hrs. (carriage in 3 hrs., with one horse 15 fr., two horses 25 fr.), direct footpath in 2½-3 hrs. — FROM AIGLE TO THE ORMONTS see (p. 251), one-horse carr. to Le Sepey 10, to Ormont-Dessus 15 fr, and fee of 1 fr.; diligence to Le Sepey daily in 2¼ hrs., to Ormont-Dessus in 5½ hrs.

Between Aigle and (65 M.) *Ollon-St-Triphon*, on the left, rises a wooded hill with an ancient tower. The village of *St. Triphon* lies on the S. slope of a hill, 1 M. from the railway; *Ollon* is on another hill, to the N.E. (Road to Villars 2½ hrs., see p. 246.) To the left tower the Grand Mœveran and the **Dent de Morcles**.

68 M. **Bex**. — *GRAND HÔTEL DES SALINES*, with salt and other baths, hydropathic establishment, etc., in a fine sheltered situation, 2 M. from the station, R., L., & A. 2½-6, D. 4, pens. 7-13 fr. (in August the visitors are almost exclusively French); adjacent, *HÔT.-PENS. VILLA DES BAINS*; in the village, *GRAND HÔTEL DES BAINS*, R., L., & A. 2-3, B. 1¼, déj. 2, D. 3, pens. 6-7 fr.; *HÔT.-PENS. DES ALPES*, R., L., & A. 2½, B. 1, D. 2½, pens. 5-6 fr.; *HÔT.-PENS. DU CROCHET*, 5-6 fr.; *UNION*, pens. from 5 fr.; PENS. DES MÛRIERS; MONDE. At *Chiètre* near Bex; *PENS. MŒSCHING*, 4-4½ fr. — *English Church*, opposite the Gr. Hôt. des Bains.

Bex (1427'; pop. 4420; pronounced *Bay*), prettily situated on the *Avançon*, and affording many beautiful walks, lies ¾ M. from the station (omnibus 30 c.). Bex is a favourite resort in spring; and in autumn it is frequented by patients undergoing the 'grape-cure'.

Fine view from *Le Montet*, a hill to the N. (½ hr.), from the *Buet*, and from the *Tour de Duin*, a ruin on a wooded hill (¾ hr. to the S.E.). — The extensive Salt Works of *Dévens* and *Bévieux*, 3 M. to the N.E., reached by a shady road of gradual ascent, may be visited in half-a-day (guide 5 fr.). Visitors usually drive to Dévens, see the salt-works, and then visit the mines, where the salt is obtained from the saline argillaceous slate by a process of soaking. Salt is also obtained from the salt-springs by evaporation. In the wood at the back of the salt-works are two huge erratic blocks.

A road leads to the E. of Bex, on the left bank of the Avançon, to (3½ M.) *Frenières* (2850') and (2 M.) Les Plans (3674'; *Pens Tanrer*,

248 *IV. Route 66.*   ST. MAURICE.   *From Geneva*

Marletaz, 5-7 fr., both unpretending; guides, *Felix Cherix*, *Philippe Marletaz*, *Charles*, *Jules*, and *Vincent Veillon*), in the sequestered *Vallée des Plans*, a good starting-point for excursions. Thus, to the *Pont de Nant* (4110'; rfmts.), with view of the glaciers of the Dent de Morcles, 1/2 hr.; to the *Croix de Javernaz* (6910') 3 hrs.; to the *Glacier de Plan-Névé* 3 hrs.; ascent of the *Argentine* (7985') 4 hrs.; *Dent de Morcles* (9775'), with an imposing view of the Mont Blanc chain and the Alps of the Valais, 7 hrs. viâ *Nant* and the *Glacier de Martinet* (descent to Morcles, see below, 31/2 hrs.); *Tête à Pierre-Grept* (9515') 7 hrs.; *Grand-Mœveran* (10,043'), by the *Frête de Sailles* (8527'; a pass to the Rhone Valley between the Grand and the Petit Mœveran), 7 hrs.; to *Anzeindaz* (p. 255) over the *Col des Essets* (6090') 4 hrs., etc.

From Bex to *Gryon*, and over the *Pas de Cheville* to *Sion*, **see R. 69.**
To *Chesières* and *Villars* (**by** *Dévens*, **3 hrs.**), see p. 246.

The train crosses the Avançon and the Rhone, joins the line on the S. bank (p. 260), and passes through a curved tunnel.

71 M. **St. Maurice** (1377'; pop. 1666; *Hôtel-Pens. Grisogono*, in connection with the *Rail. Restaurant*, R., L., & A. 31/2, D. incl. wine 41/2, pens. 8-10 fr.; *Hôt. des Alpes*, moderate; *Hôt. des Bains*; *Union*; *Ecu du Valais*; *Dent du Midi*, mediocre), a picturesque old town with narrow streets, on a delta between the river and the cliffs, the Roman *Agaunum*, is said to derive its name from St. Maurice, the commander of the Theban legion, who, according to the legend, suffered martyrdom here with his companions in 302 (near the Chapelle de Véroilley, see below). The abbey, probably the most ancient on this side of the Alps, supposed to have been founded at the end of the 4th cent. by St. Theodore, is now occupied by Augustinian monks, and contains some interesting old works of art (shown by special permission only): a vase of Saracenic workmanship, a crozier in gold, a chalice of agate, Queen Bertha's chalice, and a rich MS. of the Gospels, said to have been presented to the abbey by Charlemagne. On the walls of the churchyard and on the tower of the venerable abbey-church are Roman inscriptions. — To the W. of the station, halfway up an apparently inaccessible precipice, is perched the hermitage of *Notre-Dame-du-Sex* (sax, *i.e.* rock), to which a narrow path has been hewn in the rock. Farther to the N., above the mouth of the tunnel, halfway up the hill, is the *Grotte aux Fées*, an interesting stalactite cavern with a lake and a waterfall (1/4 hr. from the station; tickets and guides at the old château).

Travellers descending the valley change carriages at St. Maurice for Bouveret, where steamers (far preferable in fine **weather) correspond** with the trains. Comp. pp. 233, 256.

The **Baths of Lavey** (1377'; *Hôtel*, **D.** 31/2, S. 23/4, omnibus 3/4 fr.), 11/2 M. above St. Maurice, are much frequented. The warm spring (100° Fahr.), first discovered in 1831, impregnated with sulphur and common salt, rises in a pump-room on the bank of the Rhone, 8 min. from the hotel. — A narrow road (one-horse carr. 11 fr.) ascends through wood in zigzags, to the E. of the baths, to (21/2 hrs.) Morcles (3822'; *Pens. Cheseaux*; guides, *Ch. Guillot* and *Jul. Cheseaux*), prettily situated at the foot of the Dent de Morcles. Ascent of the *Croix de Javernaz* (6910'; fine view from the top) from Morcles viâ *Planhaut* in 23/4 hrs. (descent to Les Plans, see above); of the *Dent de Morcles* (9775'), 51/2 hrs. (see above); bed of hay if required on the *Haut de Morcles* (5740'), 11/2 hr. from Morcles.

Beyond St. Maurice, on the right, is the *Chapelle de Véroilley*, with rude frescoes. Opposite, on the right bank, are the *Baths of Lavey* (p. 248). The line approaches the Rhone, and passes the spot where huge mud-streams from the Dent du Midi inundated the valley in 1835, covering it with rocks and débris.

75 M. **Evionnaz** occupies the site of *Epaunum*, a town which was destroyed by a similar mud-stream in 563. Before us rises the broad snow-clad *Grand Combin* (p. 300). Near the hamlet of *La Balmaz* railway and road skirt a projecting rock close to the Rhone. On the right is the *Pissevache, a beautiful cascade of the *Salanfe* (p. 260), which here falls into the Rhone Valley from a height of 230' (3/4 M. from Vernayaz; best light in the forenoon) A path ascends on the right side, and passes behind the waterfall (adm. 1 fr.).

77 M. **Vernayaz** (1535'; *Grand-Hôtel des Gorges du Trient*, 1/2 M. from the station, finely situated at the entrance of the Gorge, first-class, R., L., & A. 2-5, B. 1 1/2, lunch 3, D. 4 1/2 fr.; *Hôt. des Alpes*, R. 2 1/2 fr.; *Hôt. de la Poste*, plain; *Hôt. de la Gare* at the station, with restaurant, moderate), the starting-point of the routes to Chamonix viâ Salvan (p. 285) and viâ Gueuroz (p. 286; guide to the Tête-Noire or Châtelard 6, **Chamonix 12**, Cascade du Dailey 4 fr.).

On the right, beyond Vernayaz, we observe the bare rocks at the mouth of the *Gorges du Trient, a ravine worn by the Trient Glacier, which at one time extended into the valley of the Rhone. The Gorges may be ascended for 1/2 M. by means of a wooden gallery attached to the rocks above the torrent. Tickets (1 fr.) at the Grand Hôtel des Gorges du Trient.

The view at the entrance to the gorge is imposing. The rocks, here about 420' high, approach each other so closely at every turn, that the gorge almost resembles a huge vaulted cavern. Where the path crosses the Trient for the second time, the stream is said to be 40' deep; at the end of the gallery it forms a waterfall, 30' high. The gorge (inaccessible farther up) is 7 1/2 M. long, extending almost to the Hôtel de la Tête-Noire (p. 284), from which its entrance is visible. — The interval between two trains suffices for a visit from Vernayaz to the Pissevache and the Gorges du Trient.

To the left of the entrance to the Gorges a path ascends to (25 min.) *Gueuroz* (2205'), commanding a beautiful view of the Rhone valley, the Grand Combin, Dent de Morcles, etc. (Hence to the Tête-Noire, see p. 286.)

Near Martigny, at the right angle which the Rhone valley here forms, on a hill to the right, stands **La Batiaz** (1985'), a castle of the bishops of Sion, erected in 1260, and dismantled in 1518. The tower (ascent from the Drance bridge in 1/4 hr., adm. 30 c.) commands a splendid view of the Rhone Valley and its environs. — The train crosses the *Drance* (p 298).

81 M. **Martigny**. — Hôtel Clerc, R., L., & A. 5 1/2, D. **5** fr.; *Hôtel du Montblanc, R., L., & A. 3-5, D. 4, pens. 7-12 fr.; Aigle, second class, R., L., & A. 2 fr., B. 1 1/4, D. 3 1/2, pens. 6 fr.; National, next the post-office, R. from 1 1/2, D. 2 1/2 fr., unpretending; *Grand St. Bernard, R., L., & A. 2-3, B. 1 1/4, D. 3 1/2, pens. 5 fr.; *Hôtel-Restaurant de la Gare, the last two at the station, 1/2 M. from the town.

Martigny-Ville (1560'; pop. 1552), the Roman *Octodurus*, is a busy little town in summer, being the starting-point of the routes over the Great St. Bernard to Aosta (R. 78), over the Tête-Noire and Col de Balme (RR. 74, 75) to Chamonix, and for the Val de Bagnes (R. 79). In the market-place, which is planted with trees, is a bronze bust of Liberty by Courbet. A large Roman building has recently been excavated at Martigny. — Above Martigny, on the road to the Great St. Bernard, is situated (1 M.) *Martigny-Bourg* (Trois Couronnes, good 'Coquempey' wine), the vineyards of which yield excellent wine (*Coquempey* and *Lamarque*, both known to the Romans).

EXCURSIONS. Near *Branson*, on the right bank of the Rhone, 3 M. to the N.E. of Martigny, is the rocky hill of *Les Follaterres*, famed for its flora. Ascent of the Arpille (6830'; 4-5 hrs., with guide). The bridle-path ascends beyond *La Battaz* (p. 249) through vineyards to the hamlet of *Sommet des Vignes*; then past the hamlets of *Ravoire*, through wood, to the chalets of *Arpille* (5965') and the summit. Superb view. Descent to the S., through wood, in 1 hr. to the *Col de la Forclaz* (p. 286).
The *Pierre-à-Voir* (8123'), a peak of the limestone range which separates the Rhone Valley from the valley of the Drance, is ascended from Martigny, the Baths of Saxon (p. 306), Sembrancher (p. 299), or Chable (p. 304). From Martigny a bridle-path, 6 hrs. (guide 8, mule 10 fr.). From the *Col*, 1/4 hr. below the summit, the descent to Saxon may be made rapidly, but not very pleasantly, on a sledge in 1-1 1/2 hr., or on foot in 3 hours. Beautiful view of the Valaisian and Bernese Alps, of the Rhone, Entremont, and Bagnes valleys, and the glacier of Giétroz (p. 305).
Gorges du Durnant (3-4 hrs. from Martigny, there and back), see p. 299.

67. From Saanen to Aigle over the Col de Pillon.

32 M. DILIGENCE from Saanen to Aigle daily in 9 1/3 hrs. (from Aigle to Saanen 8 1/2 hrs.); 11 fr. 15, banquette 14 fr. 95 c. From Saanen to Gsteig 8 M.; Ormont-Dessus 9 M.; Sepey 7 1/2 M.; Aigle 7 1/2 M. One-horse carr. from Saanen to Gsteig 8, two-horse 15 fr., to Ormont-Dessus 20 and 38, to Aigle 40 and 70 fr. and fee; from Thun, see p. 152.

Saanen (3382'), see p. 204. The road leads to the S. through the broad and smiling *Gsteigthal* to *Ebnit* and (1 3/4 M.) **Gstad** (3455'; *Bär*), at the mouth of the *Lauënen-Thal*.

A road ascends on the right bank of the *Louibach*, crossing the *Turbach* after 1/2 M., to (4 M.) **Lauënen** (4130'; *Bär*, rustic), the chief place in the valley, beautifully situated. The picturesque *Lauenen-See* (4557'), 1 hr. higher up, is best surveyed from the *Bühl*, a hill on the E. side. To the S. the brooks descending from the *Gelten* and *Dungel* glaciers form fine waterfalls on both sides of the *Hahnenschritthorn* (9304'). — From Lauenen to Lenk over the *Trüttlisberg*, and to Gsteig by the *Krinnen*, see p. 200. Over the *Gelten Pass (Col du Brozet,* 9270') to *Sion* (to *Zanfleuron*, see below, 8 hrs., with guide), toilsome. — The *Wildhorn Club-Hut* (p. 200) is reached in 5 hrs. from Lauenen.

Gsteig, Fr. *Châtelet* (3937'; *Ours*, pens. 5-6 fr.), 6 1/4 M. from **Gstad**, is finely situated. To the S. rise the *Sanetschhorn* (9665') and the *Oldenhorn* (10,250').

TO SION OVER THE SANETSCH, 8 1/2 hrs., attractive on the whole (guide 13, horse 25 fr.; experts may dispense with a guide in fine weather). The path crosses the *Sarine*, and ascends steeply over pastures, and afterwards through the *Rothengraben*, in windings partly hewn in the rock, to the (2 1/2 hrs.) dreary *Kreuzboden* (6565'); thence 1 hr. to the pass of the **Sanetsch** (7287'), on this side of which there is a cross (*La Grande Croix*). Descent (passing the large *Zanfleuron Glacier* on the right) to the (1/2 hr.) *Alp Zanfleuron* (6775'; Hôt. Sanetsch, plain), with fine view of the Alps of the Valais. From this point the *Oldenhorn* (p. 251) may be ascended in 4 hrs., the *Wildhorn* (p. 200) in 4½ hrs., the *Sanetschhorn*, or *Montbrun* (9665') in 5 hrs., and the *Diableret* (see below) in 6 hrs. (ascent of the latter easiest from this side). The *Sublage* (8973'), 2½ hrs. from the hotel, affords a magnificent view of the valleys and mountains of the S. Valais as far as Mont Blanc. Then by a winding path down to the *Alp Glary* (4920') and through the wild ravine of the *Morge* to the bold *Pont Neuf*, whence a road leads to (3 hrs.) *Chandolin*, and by *Granois* and *Ormona* to (1½ hr.) *Sion* (p. 307). Ascent from Sion to the pass 6, descent thence to Gsteig 3 hrs.

The new road here turns to the S.W., and ascends the valley of the *Reuschbach* through woods and pastures, in view of the precipices of the *Oldenhorn* (see below) and the *Sex Rouge* (9767'), to (5 M.) the **Col de Pillon** (5085'), at the S. foot of the *Palette* (see below). In descending (passing the *Cascade du Dard*, above us on the left) we soon obtain a view of a valley bounded by fine wooded mountains, and thickly studded with houses and chalets known collectively as **Ormont-Dessus**. To the left is the rocky *Creux de Champ*, the base of the Diablerets, the numerous brooks falling from which form the *Grande-Eau*. We first reach (3 M. from the Col) **Le Plan** (3815'; *Hôtel des Diablerets*, with baths, R., L., & A. 3½, D. 4, pens. 6-8 fr., opposite the post-station for Ormont-Dessus; *Pens. Bellevue*, moderate; *Pens. du Moulin*; *Pens. du Chamois*; English Church); and about 1½ M. farther on, beyond the prettily-situated *Hôtel Pillon*, lies **Vers l'Eglise** (3650'; *Pens. Mon Séjour*; *Pens. Busset*; *Hôtel de l'Ours*, all unpretending), with the church of the upper part of the valley.

EXCURSIONS from Le Plan. (Guides: *Mollien*, *V. Gottraut*, *Fr. Bernet*, *Fr.* and *Moïse Pichard*.) To the Creux de Champ (4275'), a grand rocky basin at the N. base of the Diablerets, with waterfalls on every side, 1½ hr. (to the foot of the largest fall). A good survey of the Creux de Champ, the Oldenhorn, etc., is obtained from *La Layaz* (5340'), 1½ hr. to the S. of Plan. — Ascent of the *Palette* (7133'; guide 5, horse 12 fr.), easy as far as the (2¼ hrs.) chalets of *Isenaux*; thence, without path, and rather rough, ¾ hr. more to the top; view of the Bernese Alps from the Diablerets to the Jungfrau and of the Dent du Midi to the S.W.; at the N. base of the mountain lies the pretty *Arnen-See*. Or we may ascend from the *Col de Pillon* in 1½-2 hrs., past the small *Rettau-See*. — **Pointe de Meilleret** (6404'), 2½ hrs. from Vers l'Eglise, not difficult; view extending to Mont Blanc. — Good walkers need no guide for any of these.

The **Oldenhorn** (10,250'), Fr. *Becca d'Audon*, a superb point of view, is ascended from Gsteig (7 hrs.), or from Le Plan (8 hrs.; guide 15 fr.). A steady head and sure foot necessary. Travellers from Ormont spend the night in the chalet of *Pillon*; those from Gsteig on the *Upper Oldenalp*.

The **Diableret** (10,650'; 7 hrs.; guide 18 fr.), from the Hôtel des Diablerets, difficult. Descent over the *Zanfleuron Glacier* to the *Hôt. Sanetsch* (comp. above).

To VILLARS (4 hrs.), OR GRYON (4½ hrs.) BY THE COL DE LA CROIX, a fine route (or over the Col de la Croix and the Chamossaire to Villars 6½ hrs.); guide, 6 fr., not indispensable. From the Hôtel des Diablerets we ascend the valley of the Grande-Eau for 1¼ M., and then enter a lateral valley by a bridle-path to the right (S.W.). After a somewhat steep ascent of 1¾ hr., with almost uninterrupted views of the Diablerets, we reach the **Col de la Croix** (5687'), 5 min. to the N. of the hamlet of *La Croix*. View limited. (Travellers who do not ascend the Chamossaire should at least mount the pastures to the right of the Col de la Croix for ½ hr. in order to obtain a fine view of Mont Blanc.) The path descends on the right bank of the *Gryonne*, and after 1¼ hr. divides to the left to *Arveye* 10 min.;

to the right to *Villars* 20 min. (p. 246). — The path to *Gryon* descends to the left a little above Arveye, crosses the brook, and reaches Gryon in 40 min. (p. 243). This route is preferable to a path to Gryon which crosses the Gryonne ½ hr. from the pass and follows the left bank.

Adjoining Ormont-Dessus are the houses of the lower part of the valley, known as **Ormont-Dessous**. About 4½ M. from Vers l'Eglise the road joins that from Château d'Œx (p. 242); to the S. appears the Dent du Midi. 1½ M. Le **Sepey** (3704'; *Hôt. des Alpes*; **Mont d'Or*; *Cerf*; Engl. Ch. Serv. at the Mont d'Or in summer; one-horse carr. to Plan 8 fr., fee 2 fr.), the chief village in the lower part of the valley.

Excursions. *Pic de Chaussy* (7798'), 4½ hrs., not difficult (comp. p. 255). — Ascent of the *Chamossaire viâ *Bretaye* (3½-4 hrs.), and descent to *Villars* (1½ hr.), see p. 246. — A road, with fine views, leads from Le Sepey by *Les Crêtes* to the lofty village of (2½ M.) **Leysin** (4150'; **Grand Hôtel de Leysin*, 650' above the village, in a sheltered situation, with splendid view towards the S., pens. 8-15 fr.; **Hôtel du Mont-Blanc*, pens. from 6 fr.; *Pens. Cullaz*, *Pens. de l'Espérance*, in the village, well spoken of; good Yvorne wine at the 'Capitaine Tauxe'). Pretty new walks near the hotel; excursions to (¾ hr.) *Prafondaz*, with view of the Lake of Geneva, and to the *Lac d'Aï*, on the Tour d'Aï (2½ hrs., fatiguing). From Leysin to *Aigle* a good path, mostly through wood (1½ br., ascent 2½-3 hrs.). — Footpath to (1½ hr.) *Corbeyrier* (p. 247).

The road turns suddenly to the S.W. in a fine wooded valley. Far below, the *Grande-Eau* forms several falls; to the left rises the *Chamossaire* (p. 246). Near Aigle we cross the Grande-Eau.

Aigle, 7 M. from Le Sepey, see p. 246.

68. From Bulle to Château d'Œx and Aigle.
Comp. Maps, pp. 231, 250.

41 M. Diligence thrice daily to (18 M.) Château d'Œx in 3½ hrs. (5 fr. 70 c.); thence to (23 M.) Aigle daily in 5⅓ hrs. (8 fr. 90 c.). — Carriage and pair from Bulle to Aigle in 7 hrs., 75-80 fr.

Bulle (2487'; pop. 2797; **Hôt. des Alpes*, near the station, R. 2, B. 1, D. 2½-3, pens. 6-7 fr.; *Union; *Cheval Blanc*; **Hôtel de la Ville* or *Poste*, R. 1½-2, D. 2½, pens. 5 fr.), a busy little town, the chief place of the *Gruyère* and the centre of the Freiburg dairy-farming district, is the terminus of the Romont and Bulle railway (p. 218). The environs consist of rich pasture-land, famed for Gruyère cheese and the melodious 'ranz des vaches'. The natives speak a Romanic dialect, known as 'Gruérien'.

On the slopes of the Moléson, 2 M. to the S. (carriage in 20 min.), lie the sulphur-baths of **Montbarry** (2712'; **Hôt. Montbarry*, pens. 5-6 fr.; **Hôt.-Pens. du Moléson*), commanding a charming view. Ascent of the Moléson hence, 3-3½ hrs.

Ascent of the Moléson from Bulle, 4 hrs.; guide (8 fr.) unnecessary for the experienced. We follow the Châtel St. Denis road (see below) for ¾ M., and diverge to the left by a saw-mill. The path gradually ascends by the brook *La Trême*, which it crosses by a (20 min.) mill, to the (½ hr.) red-roofed buildings of *Part-Dieu*, formerly a Carthusian monastery (3133'), and leads along the W. slope (guide-posts) of the mountain, crossing several brooks. We pass (½ hr.) the *Gros-Chalet-Neuf*; (1 hr.) *Gros-Planay* (4855'; a rustic inn in a large pasture); (¾ hr.) the chalet of *Bonne Fontaine* (5945'). Thence by a steep path to the summit in ½ hr. more.

MONTBOVON. *IV. Route 68.* 253

The *Moléson (6578'), the Rigi of W. Switzerland, is a bold rock, precipitous on every side, surrounded with meadows and forests, which afford an excellent field for the botanist. The view embraces the Lake of Geneva, the Mts. of Savoy, the Dent d'Oche, and the Dent du Midi, and stretches to the Mont Blanc chain, of which the summit and the Aiguille Verte and Aiguille d'Argentière are visible. To the left of the latter, nearer the foreground, rises the Dent de Morcles, the first peak of a chain which culminates in the Diablerets in the centre, and extends to the heights of Gruyère at our feet. The only visible peak of the Valaisian Alps is the Grand Combin, to the left of the Mont Blanc group. Most of the Bernese Alps are also concealed. To the extreme left, the Titlis. To the W., the Jura.

Ascent of the Moléson from Albeuve (see below; 3½-4 hrs.). On the outskirts of the village the path, marked with red, white, and red, crosses the left bank of the brook, traverses pastures, enters a picturesque ravine, and follows a well-shaded slope to a small chapel. Here we cross the stream, recross it ½ hr. farther on, and reach (5 min.) the first chalet. Towards the N.N.E. the ridge separating the Moléson from the Little Moléson is now visible. The path continues traceable to the vicinity of the highest chalet, which we leave on the left. Thence a somewhat fatiguing climb of 1¼ hr. to the arête, which is easily found, though there is no path, and to the summit, which rises before us, in 10 min. more.

From Bulle through the *Jaunthal* to *Boltigen* in the Simmen-Thal, see p. 205. (Diligence in summer daily in 6½ hrs.) — From Bulle diligence every afternoon, by *Vuadens*, *Vaulruz* (Hôt. de la Ville), and *Semsales*, to (2½ hrs.) Châtel St. Denis (2670'; *Hôt. de la Ville*), a small town prettily situated on the *Veveyse*. (The Moléson may be ascended hence, by the *Alp Tremeltaz*, in 4 hrs.) From Châtel St. Denis a diligence plies thrice a day in 50 min. to the railway-station of *Palézieux* (p. 219); another runs every morning in 1 hr. 40 min. to *Vevey* (p. 239).

The road from Bulle to Château-d'Œx leads past (¾ M.) *La Tour-de-Trême*, with its picturesque old tower, to (1½ M.) **Epagny** (2390'; Croix Blanche; one-horse carr. to Montbovon 7 fr.). On a steep rocky hill to the right lies the old town of **Gruyères** (2723'; *Fleur de Lys*, plain), with a well-preserved old castle of the once powerful Counts of Gruyères, who became extinct in the 16th cent., flanked with massive towers and walls, and now containing frescoes, a collection of old weapons, etc. (fee to attendant).

We enter the pretty valley of the *Sarine* or *Saane*. At (1½ M.) *Enney* (2410') we observe the tooth-like **Dent de Corjeon** (6460') in the background; on the right are **Les Vudalles** (5207'), spurs of the Moléson. At the mouth of a ravine opposite (2¼ M.) *Villard-sous-Mont* lies the large village of *Grand-Villard* (Hôt.-Pens.). Passing *Neirivue*, we next reach (1 M.) **Albeuve** (2487'; *Ange*, moderate; ascent of the Moléson, see above), cross the *Hongrin* (below, to the left, is a picturesque old bridge), and arrive at (3 M.) **Montbovon** (2608'; *Hôt.-Pens. du Jaman*, moderate; horses and guides).

From Montbovon over the Jaman to Montreux (6 hrs.; to Vevey 7½ hrs.), guide unnecessary (8 fr.); horse to the top of the pass 15, to Les Avants 20, to Montreux or Vevey 25 fr. A most attractive walk; but the pass should be reached as early as possible, as the midday mists are apt to conceal the lake from view. — From the hotel we follow the road for 30 paces, and then ascend to the right; 25 min., we turn to the right by a house; 35 min., bridge over the *Hongrin*; ¼ hr., church of the scattered village of **Allières**; ¼ hr., *Croix de Fer Inn*. (A direct route from Albeuve to this point follows the Montbovon road for ½ M., and diverges to the right by a path to *Sciernes* and Allières, 1¾ hr.; beyond Sciernes we take the path descending a little to the left.)

CHÂTEAU-D'ŒX.

The path now ascends gradually to the foot of the pass, then more rapidly over green pastures (not too much to the left), to the chalets of the *Plan de Jaman*, a little beyond the boundary between cantons Freiburg and Vaud, and the (1½ hr.) *Col de la Dent de Jaman* (4974'). A most beautiful prospect is suddenly disclosed here, embracing the Rochers de Naye and the entire range to the S. as far as the Tour d'Aï, and to the N. as far as the Dent de Lys and the Moléson; also the rich Canton de Vaud, the S. part of the Jura chain, the long range of the Savoy Alps, the E. angle of the Lake of Geneva, and the huge Valaisian Mts. to the S. From the Dent de **Jaman** (6165'; fatiguing ascent of 1¼ hr. from the Col) the view is still **more** extensive (descent to station Jaman of the Glion and Naye railway, **see** p. 243). — The *Rochers de Naye* (p. 248) may be reached from the col **in** 2 hrs.

From the pass to Montreux the path cannot be mistaken; 12 min. **from** the chalets it turns to the right (the path to the left, skirting the E. slope of the Baye, or brook of Montreux, being shorter but rough); 25 min., a bridge over the brook; then a slight descent by easy paths to the left at the division of the roads, to (½ hr.) **Les Avants** (3230'; p. 232). A road descends the W. slope of the valley. Where it trends to the W., 2 M. from Les Avants, at the beginning of the region of fruit-trees, we descend by a paved path to the left to (10 min.) *Sonzier*, and then rapidly to the left again to (½ hr.) *Montreux-Vernex* (p. 242).

The valley of the Sarine now turns to the E., **and** we enter **a** wooded ravine, the stream flowing far below in a deep rocky channel. In a wider part of the valley lies (2¼ M.) *La Tine* (Inn), with beautiful meadows. Farther on (2½ M.), on the opposite bank, is the pretty village of Rossinière (**Hôt.-Pens. Grand-Chalet*, 5-6 fr.; *Pens. de la Tour;* Eng. Ch. Serv. in summer). At (1½ M.) *Les Moulins* the road to Aigle diverges to the right (see below). We cross the Sarine by the (¾ M.) bridge of *Le Pré*, and ascend to (1 M.) —

18 M. **Château-d'Œx.** — **Hôt. Berthod*, in an open situation, R., L., & A. 2½-3, B. 1½, lunch 3, D. 4, pens. from 5 fr., patronized by English visitors; *Ours, in the village, R., L., & A. 2½-3, B. 1½, D. 2½, pens. 6-8 fr.; H. de Ville; **Pens. Rosat, Bricod, de la Cheneau, Martin, du Midi, Morier-Rosat,* etc., pens. from 5 fr. — *Turrian*, confectioner, ices, also a few rooms, opposite the Berthod. — *Engl. Church Service* in summer.

Château-d'Œx, Ger. *Oesch* (3498'; pop. 2691), is a scattered village **and summer-resort in a** green valley. The church, situated on a hill, commands a **good** view. To the E. rise the jagged *Rüblihorn* (7570') and the *Gumfluh* (8068').

***Mont Cray** (6795') may be ascended from Château-d'Œx in 3 hrs. (guide desirable). The view embraces the Bernese and Valaisian Alps as far as Mont Blanc, and the lakes of Bienne and Neuchâtel to the N.

From Château-d'Œx to (2½ hrs.) *Saanen*, see p. 201.

From Château-d'Œx **to** Aigle (23 M.; diligence daily in 5⅓ hrs.). The road diverges **from** the Bulle road at (1¾ M.) *Les Moulins* (see above) to **the left,** and ascends the valley of the *Tourneresse (Vallée de l'Etivaz)* in long windings. (Walkers follow the old road, diverging at *Le Pré*, just beyond the Sarine bridge.) The road runs high above the valley, affording picturesque views of the profound rocky bed of the brook. At (3¼ M.) *Au-Devant* the road enters a more open tract, and its continuation is seen on the mountain to the right, but it remains in the valley as far as (2 M.) *L'Etivaz* (3865'), where it turns and quits the ravine. (Pedestrians avoid this long bend by a rough, stony path ascending to the right by a saw-mill in the valley, and rejoining the road considerably higher up.) From Etivaz (5 min. farther up, the *Hôt. des Bains, with sulphureous springs) to the top of the hill (5070') 2 M.; then a slight descent to (3/4 M.) *La Lécherette* (4520'; Inn). From (1 1/4 M.) *Les Mosses* (Inn) we have a splendid view of the Dent du Midi. The road now descends the valley of the *Raverette* to (2 1/4 M.) **La Comballaz** (4476'; *Couronne, pens. 6 fr., Engl. Ch. Serv. in summer), charmingly situated, and much frequented for its mineral spring and its pure air. (*Pic de Chaussy*, 7798', an easy ascent of 3 hrs.; see p. 252.) Beyond this the road overlooks a picturesque basin, with the Diablerets and Oldenhorn in the background, and winds down to (3 M.) *Le Sepey* (p. 252) and (7 M.) *Aigle* (p. 246).

69. From Bex to Sion. Pas de Cheville.
Comp. Map, p. 250.

12 hrs. From Bex to Gryon 7 M. (diligence daily in 3 1/2 hrs., 2 fr. 90 c.; one-horse carr. 12 fr., descent 8 fr.); then a bridle-path. Guide to Aven desirable (*P. L. Amiguet, P. F. Broyon,* and *O. F.* and *Henri Aulet* at Gryon; a guide may generally be found at Anzeindaz also; from Gryon to Sion 12 fr.). Horse 20 fr. — This route, cutting off the right angle formed by the Rhone Valley at Martigny, presents an almost continuous series of wild rocky landscapes, especially on the Valais (S.) side, and commands the Rhone Valley towards the end of the journey.

Bex, see p. 247. The road leads to the N. to *Bévieux* (p. 236), crosses the *Avançon,* and ascends in zigzags (which the old path cuts off), passing the villages of *La Chêne, Fenalet,* and *Aux Posses.* Fine view of the Dent du Midi (p. 269). Near Gryon we obtain to the right a pleasing glimpse of the village of Frenières and the falls of a branch of the Avançon, descending from the Vallée des Plans (p. 247).

7 M. **Gryon** (3632'; *Pens. Morel, 4 1/2 - 5 fr.; Pens. Ouendet) is a considerable village in a picturesque situation (to *Villars* and *Ormont-Dessus,* see p. 251).

BRIDLE PATH. By the (10 min.) last house of Gryon we follow the path to the right, in view of the four peaks of the *Diablerets,* and skirt their steep S. slopes in the valley of the Avançon. On the right rise the *Argentine* (7985') and the *Grand Mœveran* (10,043'). Above the (1 hr.) chalets of *Sergnement* (4245') we cross the Avançon, and for a short distance traverse a pine-forest on the abrupt limestone slopes of the Argentine, which glitter like silver in the sunshine. Crossing the Avançon again, and passing the (3/4 hr.) chalets of *Solalex* (4810'), we ascend a stony slope in a long curve, and next reach the chalets of (1 1/2 hr.) **Anzeindaz** (6220'; *Inn* with 9 beds, open from the middle of July to Sept. only). To the S. lies the *Glacier de Paneyrossaz,* descending from the *Tête à Pierre Grept* (9545'), adjoined on the E. by the *Tête du Gros-Jean* (8567'). To the N. rise the rugged and riven limestone cliffs and peaks of the *Diablerets* (highest peak 10,650'; ascent from this side difficult and dizzy; experts take 4 hrs. from Anzeindaz; comp. p. 251). Our path now ascends gradually to (3/4 hr.) the **Pas de Cheville** (6720'). In the distance to the E. are the Alps of Valais, over which towers the Weisshorn. The path now descends to the left, round the mountain, where a wall and gate mark the frontier of Valais, and over steep and stony slopes, past a waterfall, to the (1/2 hr.) *Chalets de Cheville* (5710'). Here we cross the brook, follow the slope to the right, and then descend in zigzags, passing the chalets of *Derborence* (5213'), to (1/2 hr.) the *Lac de Derborence* (4698'), in a gloomy basin formed by a fall of rocks from the Diablerets in 1749. To the left, high above us, lies the large *Zanfleuron Glacier* (p. 250).

We skirt the S. side of the lake, then cross (3/4 hr.) the *Lizerne*, follow the left bank, and, passing the chalets of *Besson* (4370'), descend into the *Val de Triquent* and skirt a wooded slope descending steeply from the E. into the profound gorge of the Lizerne. The path, for the most part protected by a low stone wall, and quite safe, except that at certain times it is exposed to showers of stones, gradually descends to (1 3/4 hr.) the *Chapelle St. Bernard* (3530'), at the end of the Lizerne gorge, where an extensive view of the Rhone Valley is suddenly disclosed. We now descend to the left to (20 min.) *Aven*, surrounded by fruit-trees, follow the slope to (20 min.) *Erde* and (25 min.) *St. Séverin*, a thriving village belonging to *Conthey*, one of the chief wine-growing villages in the Rhone Valley, which extends to the (1 1/2 M.) bridge over the *Morge*. From this point by the high-road to (2 1/4 M.) *Sion*, see p. 307. Instead of following the dusty road, we may cross the vine-clad hill of *Muraz* from St. Séverin by a path commanding a fine view.

A shorter route (shady in the afternoon) on the right bank of the Lizerne diverges to the right 5 min. before the Lizerne bridge (see above). It crosses débris at first, and is not easy to trace. Beyond the (10 min.) chalets of *Mottelon*, we ascend to the right and pass above the chalets of *Servaplana* (4070'; milk) to (1 hr.) those of *L'Airette*. Then nearly level, with fine views of the Rhone Valley; lastly a zigzag descent to (1 1/2 hr.) *Ardon* (Hôtel du Pont), 1/2 M. from the station of that name (p. 307).

70. From Geneva to St. Maurice viâ Bouveret. Lake of Geneva (South Bank). Val d'Illiez.

Comp. Maps, pp. 234, 272.

STEAMBOAT to Bouveret along the S. Bank four times daily, in 4 3/4-5 hrs. (fare 6 or 3 fr.). Stations: *Cologny, La Belotte, Bellerive, Corsier, Anières, Hermance, Tougues-Douvaine, Nernier, Yvoire, Sciez, Anthy-Séchex, Thonon, Amphion, Evian, Tourronde, Meillerie, St. Gingolph*, and *Bouveret*. See p. 233. — RAILWAY viâ *Annemasse* to (42 M.) *Bouveret* in 2 1/2 hrs. (fares 8 fr. 30, 6 fr. 25, 4 fr. 55 c.; comp. p. 264).

Geneva, see p. 221. On leaving the quay the steamer affords a fine retrospect of the town with its numerous villas. It touches at *Cologny* (the village lying on the hill above, p. 233), *La Belotte* (for *Vésenaz*, p. 233), *Bellerive* (for *Collonge*, a little inland), *Corsier*, and *Anières*. At *Hermance* (*Pens. Gillet, 5 fr.) the brook of that

THONON-LES-BAINS. *IV. Route 70.* 257

name falls into the lake, forming the boundary between the Canton of Geneva and Savoy (France). Then *Tougues* and *Nernier*, opposite which Nyon (p. 235) is conspicuous on the N. bank.

Beyond *Yvoire* with its ancient castle, situated on a promontory, the lake suddenly expands to its greatest width (8 1/4 M.). The N. bank is now so distant that its villages are only distinguished in clear weather. A large bay opens to the S., in which lies *Excenevex*. The next stations are *Sciez* and *Anthy-Séchex*.

Thonon-les-Bains (1400'; pop. 5780; **Grand Hôtel des Bains*, at the W. end of the town, with open lake-view; *Hôtel de l'Europe*, on the terrace; *Hôtel du Léman*, unpretending), rising picturesquely from the lake, the ancient capital of the province of *Chablais*, possesses handsome buildings and a lofty terrace in the upper town; the site of a palace of the Dukes of Savoy which was destroyed by the Bernese in 1536. (Cable-tramway from *Rive*, the lower part of the town, in 1 1/2 min.; fare 10 c.). Near the railway-station is a new bath-establishment, with mineral springs.

Railway to *Bellegarde*, see p. 265. — To the S. of Thonon (8 M.) is the village of *Les Allinges*, with a ruined castle (2335'; ascent 1/2 hr.; fine view). At the top are a convent and chapel of St. Francis de Sales (rfmts.).

From Thonon a road ascends the pretty Valley of the Drance by *La Baume*, *Le Biot*, and (16 M.) *St. Jean d'Aulph* (Hotel), with ruins of a monastery, to (18 1/2 M.) a bridge which crosses the Drance opposite *Montriond*, beyond which the road divides. The road to the right leads by *Les Gets* (3645') to (10 M.) *Taninges* (p. 275); that to the left to (21 M.) *Morzine* (Hôtel des Alpes). From Morzine over the *Col de Jouplane* or the *Col de la Golèse* to (4 hrs.) *Samoëns*, see p. 275; over the *Col de Coux* to (5 1/2 hrs.) *Champéry*, see p. 260; to the *Baths of Morgin*, see p. 259.

The **steamer** next passes the ancient château of *Ripaille*, a little to the N. of Thonon, once the seat of Duke Victor Amadeus VIII. of Savoy. The long promontory round which the vessel now steers has been formed by the deposits of the *Drance*, which falls into the lake here. To the E. in the bay lie the baths of *Amphion* (Grand Hôtel; Hôt. des Bains), with a chalybeate spring, in a chestnut-grove.

Evian-les-Bains. — Hotels. *GR. HÔT. DES BAINS, above the town, R. 3-8, L. & A. 2, B. 1 1/2, lunch 2 1/2, D. 5, pens. 12-15, omn. 1 fr.; *GRAND HÔT. D'EVIAN, with garden, R., L., & A. from 4 1/2, D. 5 fr.; DE FONBONNE, on the lake; DE PARIS, all these of the first class with corresponding charges. — HÔT. DE FRANCE, R., L., & A. 2-3, B. 1, lunch 2 1/2, D. 3 1/2, pens. 8-10 fr.; DE LA PAIX; DES ÉTRANGERS, pens. 8 fr.; DES ALPES; DU NORD, etc. — *Restaurant at the Casino, lunch 3, D. incl. wine 3 1/2 fr.

Evian-les-Bains, a small town picturesquely situated (2777 inhab.), with a conspicuous church-tower, is, like Amphion (omnibus 50 c.), frequented almost exclusively by French visitors. In the centre of the town is the *Bath-House* (water containing bicarbonate of soda), the terraced garden of the Hôtel des Bains (see above) behind which affords a beautiful view. On the pleasant lake-promenade are the pretty theatre and the *Casino*, with a garden.

Railway to *Bouveret* and *Bellegarde*, see p. 265. Evian has two stations: *Evian-les-Bains* and *Les Bains-d'Evian*, 1/2 M. to the W., 3 min. from the Grand Hôtel des Bains.

BAEDEKER, Switzerland. **16th Edition.** 17

258   *IV. Route 70.*   BOUVERET.   *From Geneva*

On the lake, near station *Tourronde-Lugrin*, is the old château of
Blonay. Opposite lies Lausanne (p. 236), picturesquely situated on
the hillside; more to the right is visible the lofty Paudèze viaduct,
on the Freiburg Railway (p. 218). The hills of the S. bank, which
the boat now skirts, become steeper and higher. In a romantic
situation close to the lake is **Meillerie**, where the railway is carried through a tunnel. Beautiful view near *Les Vallettes*.

St. Gingolph (**Hôtel Suisse; Lion d'Or)*, on a promontory
opposite Vevey (p. 239), belongs half to Savoy and half to Valais,
the boundary being the *Morge*, which flows through a deep ravine.
The grotto of *Viviers*, with its springs, may be visited by boat.

Interesting excursion, with fine views, up the ravine of the *Morge* and
across the mountain to *Port Valais* (see below). We may extend our walk
on the left bank of the Morge to (1¼ hr.) *Novel* (*Inn), ascend the Blanchard (4642'; with guide), 1¾ hr.; milk, etc., to be had in a chalet near the
top), and return by the right bank of the Morge through beautiful forest to
St. Gingolph. — Ascent of the **Dent d'Oche** (7300') from Novel, interesting,
5 hrs. (with guide), viâ (1½ hr.) *Les Granges* and the (2½ hrs.) *Chalets
d'Oche*. **Fine** view. — The "**Grammont** (7145'; see below) is an easy and
attractive ascent of 4 **hrs.** from St. Gingolph, viâ the chalets of *Fritaz*
and *La Chaumeny*, then over grassy slopes, and finally over rough rocks.
The ascent from Novel (4 hrs. with guide) is more fatiguing. From Vouvry, see below. — To the E. of Novel a tolerable bridle-path leads round
the W. and S. sides of the Grammont, and past the lakes of *Lovenex* and
Taney, in 4½ hrs. to *Vouvry* (see below).

Bouveret (*Tour;* **Hôt.-Restaurant Chalet de la Forêt*, with extensive grounds) lies at the S.E. end of the Lake of Geneva, ⅜ M.
to the S.W of the mouth of the *Rhone*, which has converted the
adjoining land into a marsh. Its impetuous current, called *La
Battaglière*, may be traced for upwards of 1 M. in the lake. — Railway to *Annemasse* and *Geneva*, see p. 265.

The RAILWAY enters the Rhone Valley to the S.E. and follows
the left bank. At the foot of a rocky hill to the right lies *Port
Valais*, the *Portus Vallesiae* of the Romans, once on the lake, but
now 1½ M. inland. Near the defile of *La Porte du Sex* (1290'),
which was once **fortified**, the rock approaches so near the river as
scarcely to leave room for the road. A wooden bridge crosses to *Chessel*
on the right bank. To the right rises the Dent du Midi (p. 259).

4 M. **Vouvry** (*Poste)*, on the right, is the first station; beautiful view by the church (3 M. from the station of Roche, see p. 246).
The Rhone is joined here by the *Stockalper Canal*, begun a century
ago by a family of that name, but never finished.

The ascent of the *Grammont (7145'; 5 hrs.; guide not necessary for
adepts) from Vouvry is very attractive and not difficult. A bridle-path
(horses at Vouvry) ascends viâ *Miex* (Inn) to (3½ hrs.) *Taney* (rustic inn), at
the W. end of *Lac Taney*; thence in 1½ hr. to the summit, which commands
a magnificent view, ranging from Mont Blanc to the Matterhorn and the
Jungfrau and over the Lake of Geneva. Descent to *St. Gingolph*, see above.

The *Cornettes de Bise (8005'; 6 hrs.; guide not indispensable) may
also be ascended without difficulty from Vouvry. The route ascends viâ
Miex (see above) to the (3½-4 hrs.) *Col de Vernaz*, then crosses the ridge
to the (¾ hr.) chalet of *La Challaz* (hay-bed), about ½ hr. below the top,
which commands a magnificent view. Descent (with guide) to *Lovenex* or

Taney (p 258), or (without guide) to *La Chapelle* in the *Vallée d'Abondance*, whence we may descend by a good road to the right to (5 hrs.) *Evian*, or ascend to the left viâ *Châtel* (*Hôt.-Pens. Villa Chatel, pens. 5-6 fr.) and the *Pas de Morgin* to (2½-3 hrs.) *Morgin* (see below).

To the right are the villages of *Vionnaz* and *Muraz*, at the foot of the hills. Opposite the former lies Yvorne (p. 246), to the right of which rise the Diablerets and the Oldenhorn. We next pass *Colombey*, with its nunnery (fine view). A suspension-bridge, 70 yds. long, crosses the Rhone here to Ollon-St-Triphon (p. 247).

10 M. **Monthey** (1380'; **Cerf*; **Hôt. des Postes*, both **moderate**), with an old château and glass-works. In a chestnut-grove 20 min. above it, among a number of boulders, is the huge *Pierre-a-dzo*, balanced on a point not exceeding a few square inches in area.

To the S.W. of Monthey opens the **Val d'Illiez*, about 15 M. in length, remarkable for its fresh green pastures, picturesque scenery, and stalwart inhabitants. (Diligence from Monthey in summer daily in 3¼ hrs., 2 fr. 90 c.; one-horse carr. from Monthey to Troistorrents 6, two-horse 10, to Champéry 10 & 20, to Morgins 12 & 24 fr. and fee.) Near Monthey the new road ascends on the left bank of the *Vièze* through vineyards, and afterwards for 2 M. through a chestnut-wood, in numerous windings (cut off by the old paved bridle-path, following the telegraph posts, the beginning of which had better be asked for at Monthey). Beautiful retrospect of the valley of the Rhone, Bex and Aigle, the Diablerets, and the Grand Mœveran. About ¾ M. above Monthey the old path joins the road, which we now follow to the left where the telegraph-wires turn in that direction, and do not again quit. (The path to the right ascends to Morgin.) We next reach (1½ M.) the prettily situated village of *Troistorrents* (2600'; Hôtel-Pens. Troistorrents), with a good fountain near the church. (Here to the W. opens the VAL DE MORGIN, in which lie the *Baths of Morgin*, 4405', 3 hrs. from Monthey; the chalybeate water is chiefly used for drinking; *Grand Hôtel, pens. 6-8 fr.; *Hôt.-Pens. du Chalet, 8-10 fr.). The road in the Val d'Illiez gradually ascends, in view of the Dent du Midi all the way, to (2½ M.) *Val d'Illiez* (3145'; Hôt.-Pens. du Repos) and (2 M.) Champéry (3450'; **Hôtel de la Dent du Midi*, R. 2, lunch 2½, D. 3½, pens. 6-9 fr.; *Hôtel des Alpes*; **Hôt.-Pens. Berra*, R., L., & A. 2, B. 1 fr. 20 c., lunch 2½, D. 3½, pens. 5 fr.; *Hôt.-Pens. de Champéry*, pens. from 5 fr.; **Croix Fédérale*, R. 2-3, D. 3 fr.; **Pens. du Chalet*, 5 fr.; *Pens. du Nord*), the highest village in the valley, beautifully situated. Engl. Church Service at the Dent du Midi Hotel in summer.

EXCURSIONS FROM CHAMPÉRY. (Guides, *Maur. Caillet*, the brothers *Grenon*, *Ant. Clement*, *E. Joris*, etc.) To the (20 min.) **Galleries*: we descend to the Vièze and cross it, passing a saw-mill, to the passage constructed along the sheer cliffs opposite the village, which commands a charming survey of the valley as far as Troistorrents (adm. 50 c.). — The *Roc d'Ayerne* (1 hr.) affords a good survey of the environs. — The **Culet* (6448'; 3 hrs.; guide 4 fr.) commands a splendid view, especially of the Dent du Midi. We follow the path to the Col de Coux (p. 260) for ¾ hr., turn to the right by a small shrine where the path divides, pass a large chalet on the left, and another on the right, farther up; then through pine-wood, and by a narrow path to the cross on the top. Frequent opportunities of asking the way.

*Dent du Midi (10,450'; 7-8 hrs.; guide 18, with a night at Bonaveau 20, with descent to Vernayaz 24 or 26 fr.). The previous night is spent in the chalets of (2 hrs.) *Bonaveau* (5103'; good quarters); thence by the *Pas d'Encel*, the *Col de Clusanfe*, and the *Col des Paresseux* to the summit 5-6 hrs., the last 3 hrs. very fatiguing, but without danger to the sure-footed. Late in summer the path is almost free from snow, and there is no glacier to cross. The view of Mont Blanc and the Alps of the Valais and Bern is imposing; the background to the S. is formed by the Alps of

Dauphiné and Piedmont; the Lake of Geneva is visible from Villeneuve to Vevey. We may descend to Salvan (5³/₄ hrs.); at first a toilsome descent over débris to (3¹/₄ hrs.) the meagre pastures of the upper *Salanfe Alp* (8278'; occupied in August only); then across the Alp and past the picturesque falls of the *Salanfe* by a steep and stony path to (1¹/₂ hr.) *Van d'en Haut* (milk), where we cross the Salanfe. A better path now skirts the S. side of the valley (affording a view of Mont Blanc as the corner of the *Col de la Matze* is turned), and then descends to (1 hr.) *Salvan*.
  Tour Sallières (10,587'; 9-10 hrs., guide 30 fr.; spend night at Bonaveau, see p. 259), a difficult and fatiguing ascent, crossing the *Glacier du Mont-Ruan*. Superb view of Mont Blanc. Descent to *Salvan*, see p. 274. — Similar view from the Dents Blanches (9100'), ascended by the *Barmaz Alp* in 6 hrs., without danger for proficients (guide 15 fr.).
  PASSES. FROM CHAMPÉRY TO SAMOËNS OVER THE COLS DE COUX AND DE LA GOLÈSE, 7 hrs.; guide (13 fr.) unnecessary. At the (³/₄ hr.) small shrine mentioned at p. 259, we keep to the left, and, passing several chalets and looking back on the imposing Dent du Midi, reach (2 hrs.) the Col de Coux (6310'; *Inn*), the frontier of Switzerland and Savoy, which towards the W. overlooks the valley of the Drance. The saddle to the left is the Col de la Golèse. In descending, partly through wood, we avoid the paths leading to the right to Morzine (p. 257). On leaving the wood we see the continuation of the path bearing to the left to the (1¹/₂ hr.) Col de la Golèse (5480'; fine view). We descend past the chalets of *Les Chavannes*, leaving the hamlet of *Les Allamans* to the left, then by the valley of the Giffre, to (1³/₄ hr.) *Samoëns* (p. 275). A good road thence to (5 M.) *Sixt* (p. 275).
  FROM CHAMPÉRY TO SIXT OVER THE COL DE SAGEROU, 8-9 hrs., arduous, for adepts only (guide necessary, 18 fr.). From the Hôtel de la Dent du Midi we descend by a narrow road leading towards the head of the valley to a (20 min.) bridge, and beyond it, at (3 min.) the point where two brooks unite to form the *Vièze*, we cross another bridge, and avoid the path to the left. After 10 min. more we take the path to the left, ascending rapidly for 1 hr., and 10 min. from the top of the ascent reach the *Chalets de Bonaveau* (p. 259); thence we ascend gradually, skirting precipitous rocks, to the (40 min.) *Pas d'Encel*, where a little climbing is necessary (caution required). In ¹/₄ hr. more the path to the Col de Clusanfe diverges to the left (see below). Our route now ascends slowly over the pastures of the *Clusanfe Alp*, on the left bank of the brook, crosses the brook (¹/₂ hr.), and then mounts a very steep path to the (1 hr.) Col de Sagerou (7917'), a sharp arête descending abruptly on both sides, between the (r.) *Dents Blanches* (see above) and (l.) *Mt. Ruan* (9995'; 3 hrs. from the pass; attractive). We descend thence to the (³/₄ hr.) chalets of *Vogealle* (6115') and (¹/₂ hr.) *Boray*, and along a sheer rocky slope into the (¹/₂ hr.) valley of the *Giffre*. In 1¹/₄ hr. we reach *Nant-Bride*, and in 1¹/₄ hr. more *Sixt* (p. 275).
  FROM CHAMPÉRY TO VERNAYAZ over the COL DE CLUSANFE or SEZANFE (7940'; 10-11 hrs.; with guide), fatiguing. Beyond the *Pas d'Encel* (see above) we ascend to the left to the col, between the Dent du Midi and the Tour Sallières, and descend through the *Salanfe Valley* (see above) to *Salvan* and *Vernayaz*. — Or we may ascend to the right from the chalets of *Salanfe*, 1 hr. beyond the Col de Clusanfe, and cross the *Col* or *Chieu d'Emaney* (7960'), lying between the Tour Sallières and the Luisin (p. 286), to the valley of the *Trièqe*, *Emaney*, and (5-6 hrs.) *Triquent* (p. 286), or the Col d'Emaney and *Col de Barberine* (8136') to the valley of the *Eau Noire*, *Barberine*, and (7 hrs.) *Vatorcine* (p. 284), or finally to the E. by the *Col de Salanfe* (7290) to (3¹/₂ hrs.) *Evionnaz* (p. 249).

  The train crosses the *Vièze*, which descends from the **Val d'Illiez**, and at *Massongex* approaches the Rhone. At (14¹/₂ M.) *St. Maurice* (p. 248) our line is joined by that of **the right bank**.

V. SAVOY, THE VALAIS, AND THE ADJACENT ITALIAN ALPS.

71. From Geneva viâ Culoz and Aix-les-Bains to Chambéry and back viâ Annecy 264
    Perte du Rhône. From Bellegarde to Bouveret, 264. — Excursions from Aix-les-Bains: Lac du Bourget; Haute-Combe, Revard, etc. From Aix-les-Bains to Annecy, 266. — Excursions from Chambéry. Dent du Nivolet, 267. — From Albertville to Moûtiers and to Beaufort; to Contamines viâ the Col Joli; to Chamonix viâ Flumet, 268. — Excursions from Annecy: Semnoz; Parmelan; Tournette, 269, 270. — From Annecy to Chamonix, to Cluses viâ Grand Bornand, and to Sallanches over the Col des Aravis, 270.
72. From Geneva to Chamonix 271
    i. Viâ Cluses 271
    Môle; Pointe d'Andey, 272. — Pointe Percée. St. Gervais-les-Bains, and over the Col de la Forclaz to Les Houches. Gorges de la Diosaz, 273.
    ii. Viâ Sixt. 274
    Pralaire; Môle. Pointe de Marcelly, 274. — Excursions from Sixt: Vallée du Fer à Cheval; Fond de la Combe; Pic de Tanneverge; Pointe Pelouse. From Sixt to Chamonix over the Buet, 275.
73. Chamonix and Environs 276
    Mont Blanc, 281. — From Chamonix over the Col du Géant to Courmayeur; Cols de Triolet, de Talèfre, de Pierre-Joseph, des Hirondelles, de Miage, 282.
74. From Chamonix to Martigny over the Tête-Noire, or to Vernayaz viâ Triquent and Salvan 282
    Glacier d'Argentière; Col du Chardonnet; Col d'Argentière; Col du Mont Dolent; Col des Grands Montets, etc., 283. — Gouffre de la Tête-Noire, 284. — Col de la Gueula; Cascade du Dalley; Luisin; Dent du Midi; Tour Sallières, 285, 286. — From Vernayaz to Chamonix viâ Gueuroz, 286.
75. From Martigny to Chamonix. Col de Balme 286
    Glacier de Trient, 286. — From the Col de Balme to the Tête-Noire. To Orsières over the Col du Tour, 287.
76. From Chamonix to Courmayeur over the Col du Bonhomme and the Col de la Seigne. Tour du Mont Blanc. 288
    Col de Voza, 288. — Mont Joli; Cols du Mont Tondu and de Trelatête, 289. — From Chapieux to Pré-St-Didier over the Little St. Bernard. Col d'Enclaves, 290, 291. — Excursions from Courmayeur: Col de Chécouri; Mont de Saxe; Grandes Jorasses. From Courmayeur to Martigny over the Col Ferret, 292.
77. From Courmayeur to Aosta and Ivrea 293
    Crammont. From Pré-St-Didier to Bourg-St-Maurice over the Little St. Bernard. Mt. Valaisan, Belvédère, Lancebranlette. From Bourg-St-Maurice to Tignes, 293, 294. — Becca di Nona; Mont Emilius; Mt. Fallère, 295, 296. — From Aosta to Zermatt over the Col de Valpelline, Mont Lusency. Passes from the Val Pellina to the Val St. Barthélemy, 296.

SAVOY AND VALAIS.

78. From Martigny to Aosta. Great St. Bernard . . . **298**
Gorges du Durnant. Mont Chemin. Pas du Lens. From Martigny to Oraières viâ Champex. Excursions from the Lac de Champex and the Cabane d'Orny, 299. — Mont Brûlé. Tête de Bois. Valsorey; Grand Combin, 300. — Mont Velan, 301. — Chenaletta; Pointe des Lacerandes; Mont Mort. From St. Bernard's Hospice over the Col de Fenêtre to Martigny, and over the Col Ferret to Courmayeur, 302, 303. — Col de la Seréna, 303.

79. From Martigny to Aosta over the Col de Fenêtre. Val de Bagnes **304**
Col de Sexblanc. Cabane de Panossière; Grand Combin; Cols du Crêt, de Sevreu, de Cleuson, and de Louvie, 304, 305. — Excursions from Mauvoisin and Chanrion. Mont Avril; Tour de Boussine; Grand Combin; Mont Blanc de Seilon; Mont Pleureur, etc., 305. — Cols du Sonadon, des Maisons Blanches, de Crête-Sèche, de Seilon, de Breney, and de Vasevay. From Chanrion to the Val Pelitna over the Col d'Otemma or Col de la Reuse d'Arolla, 305, 306.

80. From Martigny to Domodossola over **the** Simplon . . **306**
Col des Etablons, 307. — Montana, 308. — Belalp; Upper Aletsch Glacier; Sparrhorn; over the Beichgrat to the Lötschen-Thal, 309, 310. — Excursions from Berisal; Wasenhorn, Beitlihorn, and Bortelhorn; to Iselle viâ Alp Veglia; Passo di Valtendra. Schönhorn; Monte Leone, 311. — From Simplon to Saas; Rossböden Pass; Laquin-Joch; Sirvolten Pass; Simell Pass; Gamser Joch; Fletschhorn, 312. — From Gondo to Saas over the Zwischbergen Pass, 313. — Valle di Bognanco. From Domodossola over the Antrona Pass to Saas, and over the Antigine Pass to Mattmark, 314.

81. From the Rhone Glacier to Brig. The Eggishorn . . **314**
Geren-Thal. From Ulrichen to Airolo over the Nufenen Pass. Löffelhorn, 315. — Blindenhorn. Eggishorn; Märjelen-See; Concordia Hut, 316. — Great Aletschhorn. Viâ the Lötschenlücke to Ried; to the Riederalp and Belalp. Rieder Furka. From the Riederalp to Mörel, 317. — From Fiesch over the Albrun Pass to Baceno, or to the Tosa Falls; Binnen-Thal; Ofenhorn. From Fiesch to Baceno over the Geisspfad Pass or the Kriegalp Pass, and to Iselle over the Ritter Pass, 317, 318.

82. From Ulrichen to Domodossola. Gries Pass. Falls of the Tosa. Val Formazza **319**
Basodino. From the Tosa Falls to Airolo over the S. Giacomo Pass; to Bignasco over the Bocchetta di Val Maggia; to Binn over the Hohsand Pass or the Albrun Pass. — From Andermatten to Cevio over the Criner Furka, 320.

83. The S. Valleys of the Valais, between Sion and Turtmann (Val d'Hérens, Val d'Anniviers, Turtmann Valley) . **321**
i. From Sion through the Val d'Hérens to Evolena, and over the Col de Torrent to the Val d'Anniviers **321**
Mayens de Sion, 321. — Val d'Hérémence, 322. — Excursions from Evolena; Pic d'Arzinol; Mt. de l'Etoile; Pointe de Vouasson, 322, 323. — Excursions from Arolla: Lac Bleu de Lucel; Pigno d'Arolla; Aig. de la Za; Dents de Veisivi; Mont Collon; Evêque; Dent Perroc; Dent des

Bouquetins, 323, 324. — Cols de Collon, de Za-de-Zan, and de Riedmatten; Pas de Chèvres; Cols de Chermontane, de l'Évêque, de Bertol, du Mont Brûlé, and de Valpelline, 324, 325. — Ferpècle; Alp Bricolla. Dent Blanche; Grand Cornier. Cols du Grand Cornier, de la Pointe de Bricolla, d'Hérens, and des Bouquetins, 325, 326. — Sasseneire; Pas de Lona; Becs de Bosson. Col and Corne de Sorebois, 326.

ii. From Sierre through the Val d'Anniviers to Zinal. ... **327**
From Sierre to St. Luc viâ Chandolin; Illhorn. From Vissoye to the Hôtel Weisshorn, 327. — Alp de l'Allée; Alp d'Arpitetta; Roc de la Vache; Constantia Club-Hut; Roc Noir; Corne de Sorebois; Garde de Bordon; Pointe d'Arpitetta; Besso; Pigno de l'Allée; Bouquetin; Diablons; Grand Cornier; Zinal Rothhorn; Ober-Gabelhorn, 328, 329. — Col de l'Allée; Col de Couronne; Triftjoch; Col Durand; Morning Pass; Schalli-Joch. From Zinal to St. Luc, 329.

iii. St. Luc. Bella **Tola.** Over the Pass du Bœuf (or the Meiden Pass) into the Turtmann Valley, and over the Augstbord Pass to the Valley of the Visp. . **329**
Turtmann Glacier; Col des Diablons, 330. — Pas de la Forcletta. From Gruben to Turtmann. The Schwarzhorn. Jung Pass; Barrjoch; Brunnegg-Joch; Biesjoch, 331.

84. From Visp to Zermatt **331**
From Stalden to the Simplon over the Bistenen Pass, 332. — Excursions from Zermatt: Gorner Gorge; Riffelberg and Gorner-Grat, 334. — Findelen Glacier, 335. — Schwarzsee Hotel; Hörnli; Théodule Pass; Staffel-Alp, 336, 337. — Mountain Excursions from Zermatt and the Riffelhaus: Breithorn; Cima di Jazzi; Riffelhorn; Mettelhorn; Unter-Gabelhorn; Wellenkuppe; Ober-Rothhorn; Strahlhorn; Rimpfischhorn; Dom; Lyskamm; Monte Rosa; Matterhorn; Ober-Gabelhorn; Zinal-Rothhorn; Weisshorn; Dent Blanche; Dent d'Hérens, 337-339. — Glacier Passes from the Riffel: Théodule Pass; Furggjoch; Col de Tournanche; Schwarzthor; Zwillinga-Joch; Lysjoch; Felik-Joch; Sesia-Joch; Piode-Joch; New and Old Weissthor, 339. — Glacier Passes from Zermatt to Zinal, Evolena, Chermontane, Val Pellina, Châtillon, and the Saas Valley, 340.

85. From Visp to Saas and Mattmark **340**
Excursions from Saas im Grund: Triftalp; Weissmies; Sonnighorn; Latelhorn, 341. — Excursions from Saas-Fee. Gletscher-Alp. Mittaghorn. Egginerhorn. Allalinhorn; Alphubel; Nadelhorn; Südlenzspitze; Ulrichshorn; Balfrinhorn, 341. — Alphubel-Joch; Fee Pass; Mischabel-Joch; Domjoch; Nadel-Joch; Lenzjoch; Ried Pass; Windjoch, 342. — Excursions from Mattmark. Stellihorn; Schwarzberg-Weissthor; Adler Pass; Allalin Pass, 343.

86. From Piedimulera to Macugnaga, and over the Monte Moro Pass to Mattmark **343**
Excursions from Macugnaga: Belvédère; Petriolo Alp; Pizzo Bianco; Monte Rosa. From Macugnaga over the Weissthor to Zermatt, 345.

87. From Macugnaga to Zermatt round Monte Rosa . . . **346**
Turlo Pass; Col delle Loccie, 346. — Pile Alp; Corno Bianco. Colle Mond and Bocchetta Moanda. Col d' Olen;

Gemsstein. Col delle Pisse; Col di Valdobbia, 347. — Excursions from Gressoney: Cortlys; Linty Hut; Hohe Licht; Gnifetti Hut; Sella Hut; Lyskamm; Monte Rosa; Castor; Colle Ranzola; Col de Joux; Mont Taille; Becca di Frudiera, 347, 348. — Bettaforca; Bettliner Pass; Pinter Joch; Val d'Ayas or Challant, 348. — Col des Cimes Blanches; Grand' Semetta, 349.

88. From Châtillon to Valtournanche and over the Théodule Pass to Zermatt 349
Grand Tournalin. Col du Val Cournère. Château des Dames, 350.

71. From Geneva viâ Culoz and Aix-les-Bains to Chambéry, and back viâ Annecy.

RAILWAY to Aix-les-Bains (55½ M.) in 3½ hrs. (11 fr. 30, 8 fr. 5, 6 fr. 10 c.), to Chambéry (64 M.) in 4 hrs. (12 fr. 75, 9 fr. 60, 7 fr. 5 c.), to Albertville (93½ M.) in 7 hrs. (18 fr. 70, 14 fr. 10, 10 fr. 35 c.); from Aix-les-Bains to Annecy (25 M.) in 1½-2 hrs. (4 fr. 95, 3 fr. 65, 2 fr. 65 c.); from Annecy to Geneva (37½ M.) in 2½ hrs. (7 fr. 30, 5 fr. 50 c., 4 fr.). DILIGENCE between Albertville and (28 M.) Annecy daily in 4 hrs. — See also *Baedeker's South-Eastern France.*

Geneva, see p. 221. 3 M. *Vernier-Meyrin;* 5½ M. *Satigny;* on the left flows the *Rhone*. Near (8½ M.) *La Plaine* we cross the valley of the *London*. 12½ M. *Chancy-Pougny;* 14½ M. *Collonges*. The Rhone here separates the steep slopes of the *Mont Vuache* (3444') from the Jura chain. The lofty **Fort de l'Ecluse** (1387'), to the right, guarding the entrance to France, was founded by the Dukes of Savoy, rebuilt by Vauban, destroyed by the Austrians in 1815, and enlarged by the French in 1824. Beyond the short tunnel (200 yds.) under the fort we pass through the *Tunnel du Crédo*, 2½ M. long, and cross the deep valley of the *Valserine* by an imposing viaduct, 275 yds. long and 170' high.

21 M. **Bellegarde** (*Buffet; Hôt. des Touristes; Hôt. de la Poste*, at the station), with the French 'douane'.

Above the confluence of the Valserine and the Rhone, about ½ M. from the station, is what was once the so-called **Perte du Rhône**. Formerly, when the river was low (Nov. to Feb.), it disappeared entirely in a cleft in the rock for about 100 paces. Although this attraction has now ceased to exist, the valley here is very picturesque. The street to the left of the hotels leads down to a bridge over the deep bed of the Valserine, 430 yds. to the right of which is another bridge over the Rhone, at the point where that river used to plunge beneath the rocks, now blasted away. Higher up, to the left, is the entrance to a conduit 820 yds. long, 600 yds. being underground, at the other end of which, below the bridge, are 3 turbines (water-wheels on vertical axes) giving motive power to two factories. To see the turbines, apply at the first of the factories; they cannot be seen from the opposite bank. — We may also visit the *Valserine Viaduct* (near the station), mentioned above, and the *Gorge*, 85 ft. deep, which the river has hollowed out of the limestone rock, forming a 'Perte', or subterranean passage, more than 400 yds. in length, about 1½ M. from the viaduct. — Very numerous fossils have been found in the cretaceous formations on the Savoy side near the Perte du Rhône.

FROM BELLEGARDE TO BOUVERET (62½ M.), railway in 3¼ hrs. Stations: *Valleiry; Viry;* 15 M. *St. Julien* (steam-tramway to Geneva, see p. 232);

to Chambéry. AIX-LES-BAINS. *V. Route 71.* 265

20 M. *Bossey-Veyrier*, at the N.W. base of Mt. Salève (p. 282). The Arve is then crossed to (24 M.) *Annemasse* (p. 271), the junction for Annecy and Geneva (p. 271), and Cluses (Chamonix, p. 271). 28 M. *St. Cergues;* 33 M. *Bons-St-Didier* (ascent of the Voirons, see p. 263); 37 M. *Perrignier;* 43 M. *Thonon* (p. 257); 49 M. *Evian* (p. 257); 52¹/₂ M. *Lugrin;* 56 M. *Meillerie;* 59¹/₂ M. *St. Gingolph;* 62¹/₂ M. *Bouveret* (p. 258).

Four tunnels (1121, 917, 493, and 165 yds. in length respectively). Beyond (28 M.) *Pyrimont* (with asphalt-mines near it) a handsome viaduct crosses the *Vezeronce.* 32¹/₂ M. *Seyssel* (Hôt. du Rhône; etc.), an old town, on both banks of the Rhone, here crossed by a double suspension-bridge. The river, now navigable, flows through a broad channel with numerous islands, and the valley expands.

41¹/₂ M. Culoz (774'; *Hôt. Folliet;* *Rail. Restaurant*), at the base of the *Colombier* (5033'), is the junction for Lyons, Mâcon (Paris), and Turin. Carriages generally changed, and a long halt.

The Mont-Cenis train crosses the Rhone, and at (46 M.) *Chindrieux* reaches the N. end of the Lac du Bourget (757'), which is 10 M. long and 3 M. broad. To the right, on a wooded hill projecting into the lake, is the old château of *Châtillon.* The train skirts the rocky E. bank, passing through four tunnels. To the right a pleasing view of the lake, the monastery of Haute-Combe, the château of Bourdeau, and the Dent du Chat (p. 266).

55¹/₂ M. **Aix-les-Bains.** — Hotels. *Grand Hôtel d'Aix*, Avenue de la Gare; *Grand Hôtel de l'Europe*, *Grand Hôtel du Nord*, Métropole, and *Hôt. Venat et Bristol*, all in the Rue du Casino; Grand Hôtel du Louvre, Avenue de la Gare; Splendide Hôtel, finely situated above the Jardin Public. All these are of the first class, with corresponding charges: R., L., & A. from 4, B. 1¹/₂, lunch 3, D. 5, pens. from 9 fr. — Hôt. des Bergues, *Hôt. International*, Savoy Hotel, Avenue de la Gare; Hôtel des Bains, Rue du Casino; Beau-Site, above the Jardin Public; Hôt. Gaillard, Hôt. de Paris, Rue Despine; *Hôt. de la Poste*, Hôt. du Grand-Café, Place Centrale; Hôt. Britannique & Thermal, by the Baths; Hôt. Damesin & Continental, Grand-Hôtel du Parc, Rue de Chambéry; Hôt. Germain. Bossut, Gabin, Laplace, de Genève, etc. — *Pensions* and *Maisons Meublées* also abound.

Cafés-Restaurants. *Grand Café*, Place Centrale; *Café-Restaurant de la Gare*, etc.

Cab, per drive, 1-2 pers., 1 fr., 3-4 pers. 2 fr.; per hour with one horse 3, with two horses 4 fr. — Voitures Publiques for excursions (to Marlioz, Port Puer, etc.), Place du Revard and Place Centrale.

Casinos. *Cercle*, Rue du Casino, adm. 3 fr.; season-ticket 40, for 2 pers. 60 fr. — *Villa des Fleurs*, Avenue de la Gare, similar.

English Church Service during the season.

Aix-les-Bains (850'; pop. 6296), the Roman *Aquae Gratianae*, a famous watering-place, picturesquely situated, is visited annually by upwards of 12,000 patients. It possesses warm (113°) sulphur-springs, used for drinking and for baths. The large *Etablissement Thermal*, erected in 1854, is well fitted up. In front of it rises the *Arch of Campanus*, a monument erected in the 3rd or 4th cent. A. D., in the form of a triumphal arch, in memory of L. Pomp. Campanus and his family. The eight niches contain the urns of the persons whose names are recorded on the monument. The well-preserved *Château* (16th cent.), now the *Hôtel de Ville*, contains a *Museum* of antiquities, chiefly from the lake-dwellings of the Lac du Bourget, and other curiosities (open daily 9-12 and 2-50; 5 c.). The rallying-points of visitors are the sumptuous *Casino*, with its handsome saloons, and the *Villa des Fleurs* (see above), with its pleasant garden, where concerts are frequently given. Queen Victoria resided at the *Villa Mottet* during her visit to Aix in April, 1885. — Omnibuses run from the Place Centrale every 20 min. to (1 M.) *Marlioz* (in 10 min.; 40 c., there and back 60 c.), which possesses cold sulphur-springs (chiefly used for drinking and inhaling), a château, and a large and beautiful park (restaurant).

EXCURSIONS. Pleasant shady walks in the *Parc*, the *Promenade du Gigot*, and the *Avenue Marie*. — The Lac du Bourget (p. 265) may be reached by the 'Route du Lac', leading to the (2 M.; omnibus 50 c.) *Port de Puer* (steamboat-pier; rowing-boats for hire). On the bank of the lake extends the beautiful wooded hill of *Tresserve*, 3 M. in length, with shady walks and fine views.

*Hautecombe, a Cistercian monastery on the N.W. bank of the lake, at the foot of the *Mont du Chat*, is another interesting point. (Steamboat thither several times a week; trip round the lake on Sundays, allowing an hour at Hautecombe. Boat with two rowers to Hautecombe and back, with one hour's stay, 4 fr.; each hour more 1½ fr.; to Bourdeau 5 fr.; a bargain should be made beforehand.) The abbey, which was the burial-place of the Princes of Savoy until 1731, when the Superga near Turin was chosen for that purpose, was destroyed during the French Revolution, and handsomely rebuilt in 1824 by Charles Felix, King of Sardinia. The church (open 7.30-9, 10-11.30 a.m., 2-3, 3.45-6 p.m.) is very richly decorated and contains upwards of 300 statues, besides basreliefs, paintings, etc. The statue of Charles Felix, by Cacciatore, and Albertoni's group of Maria Christina protecting the Arts should be noticed. Not far from the church is a restaurant. — Farther to the S., at the influx of the *Leisse*, lies the village of Le Bourget (*Hôt. Ginet*) with a ruined castle and a church in the transitional style, the choir of which contains fine basreliefs of the 13th century. — A favourite drive from Aix is viâ Le Bourget and *Bourdeau* to the Col du Chat (2093'), to the N. of the Dent du Chat, with beautiful view (carriage there and back in 5 hrs., 15-20 fr.; brake 3 fr. 50 c. each person); another to La Chambotte (3034'), a hotel and restaurant on the top of the *Mont Gigot* or *de Corsuel*, to the N. of Aix (same time and fares). — Ascent of the *Dent du Chat* (4595'), 3 hrs. from Le Bourget, by a good bridle-path; splendid view of the Alps, including Mont Blanc.

From Aix a good road leads to the N E. viâ *St. Simon*, with chalybeate springs, to the (1½ M.; omnibus 60 c., there and back 1 fr.) *Gorges du Sierroz, a romantic defile ¾ M. in length. A steam-launch (there and back 1 fr. 50 c.) plies to the upper end, whence a footpath ascends to the *Cascade de Grésy* (Restaurant; see below).

To the *Revard (5070'), a summit of the *Montagne de la Cluse*, to the S.E. of Aix, a mountain-railway 5¾ M. in length, leads from Aix in about 1½ hr. Ordinary line to (1¼ M.) *Mouxy* (1353') and rack- and-pinion line (on Abt's system) thence viâ *Pugny* (1968') and *Pré-Japert* (3280') to the top, with Chalet-Hotel and splendid view (Mont Blanc, etc.). — The mountain-group terminates in the *Dent du Nivolet* (p. 267), to the S.

FROM AIX-LES-BAINS TO ANNECY, 25 M., a branch-line (1½ hr.). The train runs at first to the N. through the valley of the *Sierroz*, which has worn a deep channel for itself, passing near the entrance of the Gorges (see above). 2½ M. *Grésy-sur-Aix*, with a ruined castle and a pretty waterfall (see above). 7½ M. *Albens*. Through an opening to the right appear the Semnoz and the Tournette (p. 270). 10½ M. *Bloye*. At (13 M.) Rumilly (1095'; *Hôt. de la Poste*; *Restaurant Ducret*), a little town of Roman origin, we cross the *Chéran*. The train turns to the E. and enters the pretty

to Chambéry. **CHAMBÉRY.** *V. Route 71.* 267

valley of the Fier. 17 M. *Marcellaz-Hauteville*. We now traverse the wild and romantic *Défilé du Fier* (twelve bridges and two short tunnels). On the left, near the end of the gorge, rises the château of *Montrottier*, of the 14-16th centuries. — 20½ M. *Lovagny* (restaurant at the station and at the entrance to the gorge). About ½ M. to the E. are the *Gorges du Fier*, a grand ravine 275 yds. long, enclosed by limestone rocks nearly 300′ high, rendered accessible by a wooden gallery (1 fr.). Beyond Lovagny we obtain a fine view, to the right, of the Parmelan, the Semnoz, and the Tournette. Tunnel of 1270 yds.; then a bridge across the Fier. 25 M. *Annecy*, see p. 269.

As the train proceeds, the lake is concealed by the wooded hill of Tresserve (p. 266). Fine view to the right. — 58 M. *Viviers*.

64 M. **Chambéry.** — Hotels. *Gr. Hôt. de France*, Quai Nézin 5, near the Boulevards, R., L., & A. 3½, B. 1½, lunch incl. wine 3, D. incl. wine 4, pens. 9-12 fr.; *Hôt. du Commerce*, Rue Vieille-Monnaie; *Des Princes*, Rue de Boigne 4; *Gr. Hôt. de la Poste & Métropole*, Rue d'Italie, R., L., & A. from 2½, B. 1, lunch 2½, D. 3, pens. 8 fr.; *de la Paix*, opposite the station.

Chambéry (885′), the capital of the department of Savoy, a handsome looking town with 20,922 inhab., lies on the rapid *Leisse*. On the promenade between the railway and the town rises a *Monument* commemorating the first union of Savoy with France in 1792, with a bronze figure of a Savoyard woman, by Falguière; and farther on is a large *Fountain-Monument*, by Sappey, adorned with lifesize elephants, in memory of *General de Boigne* (d. 1830), who bequeathed to Chambéry, his native town, a fortune of 15 million francs amassed in the East Indies. Of the ancient and loftily situated *Château* of the counts and dukes of Savoy, erected in 1232, **now restored and occupied** by the Préfecture, two towers and the chapel, in the Gothic and Renaissance styles, belong to the original building. At the back of the château (reached by going to the left round the building, through the gate, and up the avenue) is a small *Natural-History Museum* with a botanic garden. The *Theatre* is richly decorated in the interior. Near it is the archiepiscopal *Cathedral*, a Gothic edifice (14th and 15th cent.). In front of the Palais de Justice rises a bronze statue of **Ant. Favre** (d. 1624), a famous jurist, erected in 1864. Opposite is the *Museum*, containing archæological collections, sculptures, a library, and a picture-gallery (adm. 50 c.; Sun. & Thurs. 1-5, free).

Walks. To the N., above the town (10 min.) rise the *Rochers de Lemenc*, with a church in which Gen. de Boigne and Mme. de Warens, Rousseau's friend, are interred. Charming view. — To *Buisson-Rond* (20 min.), a pleasant park; the *Cascades de Jacob* (1/4 hr.); the chapel of *St. Saturnin* (1¼ hr.). — *Bout du Monde* (1 hr.), a rocky gorge at the base of the Dent du Nivolet, with a fine waterfall of the *Doria*. — *Les Charmettes* (½ hr.; adm. ½ fr.), a country-house once occupied by Rousseau and Mme. de Warens (1736). — The sulphur-baths of **Challes** (*Hôt. du Château; de France; du Centre; du Pavillon*), 3½ M. to the E. of Chambéry (omnibus and tramway), possess a bath-establishment, a casino, and a large park. — The ascent of the Dent du Nivolet (5115′, 4½-5 hrs.) is attractive and free from difficulty. Road for about 8 M.; then a bridlepath nearly to the top. Magnificent view.

Beyond Chambéry we traverse a picturesque district, passing the ruins of *Bâtie* and *Chignin*. The precipitous *Mont Granier* (6358′) on the right owes its peculiar form to a landslip in 1248, which buried sixteen villages. 70 M. *Chignin-les-Marches*. 72 M. *Montmélian* (921'; Rail. Restaurant), junction for *Grenoble*. The castle, on a hill, of which a few fragments only are left, long served as a bulwark of Savoy against the French, but was destroyed by Louis XIV. in 1705. Pleasing survey of the valley of the *Isère*, which the train now ascends. 74½ M. *Cruet; 79 M. St. Pierre d'Albigny*, junction of the Mt. Cenis Railway; the small town, 1½ M. to the N., is dominated by the ruined castle of *Miolans*, a state-prison of Savoy in the 16-18th centuries.

The MONT-CENIS RAILWAY quits the Isère here and ascends to the right in the *Maurienne Valley*, watered by the *Arc*. Stations *Chamousset, Aiguebelle, Epierre, La Chambre, St. Jean-de-Maurienne, St. Michel, La Praz*, and (46 M.) *Modane*. Then through the great *Mont-Cenis Tunnel* (7½ M. long) to *Bardonnecchia* and *Turin* (see *Baedeker's Northern Italy*).

The line to Albertville ascends the right bank of the Isère. 85 M. *Grésy-sur-Isère*, with Roman antiquities. On the left is *Montailleur*, with an old castle. 89 M. *Frontenex*, whence a road leads to the N. over the *Col de Tamié* (2980') to (11 M.) *Faverges* (p. 269).

93½ M. **Albertville** (1180'; pop. 5854; *Hôt. Million*, in the market-place; *Hôt. des Balances*, Grande Rue; *Hôt. de la Gare*), a pleasant town of 5854 inhab., received its present name in 1835 in honour of King Charles Albert of Sardinia. It consists of two parts separated by the *Arly*: on the right bank *L'Hôpital*, on the left the picturesque little old town of *Conflans*.

FROM ALBERTVILLE TO MOÛTIERS-EN-TARENTAISE, 17 M., railway in 1½-2 hrs. The line leads through the Isère Valley, which gradually narrows and becomes grander as we ascend, by *Tours, La Bathie*, and (8 M.) *Cevins*, at the N.E. base of the *Tournette* (8060'), and then below the ruined castle of *Briançon* and viâ (12 M.) *Notre Dame de Briançon* and (15½ M.) *Aigueblanche* to (17 M.) *Moûtiers* (1575'; 2397 inhab.; *Hôt. Vizioz, Hôt. Bertoli*, both very fair), the ancient capital of the *Tarentaise*, the seat of a bishop, and named after a monastery founded here in the 5th century. The treasury of the cathedral is worth seeing. A little to the S., in the pretty valley of the Doron, are the baths of (¾ M.) *Salins* and (3½ M.) *Brides-les-Bains*. — A road leads to the E. of Moûtiers (diligence twice daily) through the picturesque valley of the Isère viâ *Aime*, with Roman remains, and *Bellentre* to (17 M.) *Bourg-St-Maurice* (p. 294).

FROM ALBERTVILLE TO BEAUFORT, 12 M. (diligence daily in 3 hrs., 2 fr.), by a road through the picturesque *Doron Valley*. The little town of Beaufort (2620'; *Montblanc*), prettily situated, is commanded by the château of *La Salle*. Thence through the *Gitte Valley* to the *Col du Bonhomme* and over the *Col des Fours* to *Mottets*, 9-10 hrs., with guide (16 fr.; comp. 290). — FROM BEAUFORT OVER THE COL JOLI TO CONTAMINES, 8-9 hrs., with guide, interesting on the whole. Carriage-road and bridle-track through the *Dorine Valley* viâ *Haute-Luce* to (3 hrs.) *Belleville*, thence footpath over the Col Joli (6558'), lying to the S. of *Mont Joli* (p. 289), with a view of Mont Blanc, to (5 hrs.) *Contamines* (p. 289).

FROM ALBERTVILLE TO CHAMONIX, 46 M., diligence daily in 10 hrs. (16 fr.; two-horse carriage for 4 pers. 90 fr.), by a good new road, viâ (5 M.) *Fontaines d'Ugines*, at the junction of the road to Annecy (see below), and through the picturesque valley of the Arly to (18½ M.) **Flumet** (3008'; *Cheval Blanc*), a village at the influx of the *Arondine* into the Arly. (Over the *Col des Aravis* to *St. Jean-de-Sixt*, see p. 270.) (Travellers in the reverse direction have to undergo custom-house formalities here.) Then (25 M.) **Mégève** (3690'; *Hôt. du Soleil*, lunch incl. wine 3½ fr.; *Tissot*), on the watershed between the Isère and the Arve. The road forks 1¾ M.

farther on, the left arm leading viâ (30 M.) *Combloux* to (31½ M.) *Sallanches*. We descend to the right, enjoying a superb view; opposite us towers the Aignille de Varens (8831'), to the left lies the valley of the Arve as far as Magland (p. 272); to the right rises the entire Mont Blanc chain, with its glaciers and the highest summit. Beyond (30 M.) *Le Freney* we pass by an imposing bridge over the gorge of the Bon-Nant to (31½ M.) *St. Gervais-le-Village* (p. 273), and thence down to (34 M.) *Le Fayet*, on the road from Cluses to (46 M.) Chamonix.

The ROAD TO ANNECY (28 M.) ascends to the N., on the right bank of the *Arly*. To the left, on a steep hill, stands the church of *Pallud;* on the right the *Doron* issues from the *Vallée de Beaufort* (see p. 268). Near (5 M.) *Fontaines d'Ugines* (1350'; Hôt. de Chamonix, Hôt. Carvin) the road quits the valley of the Arly, and enters that of the *Chaise* to the left. To the right, on a hill, stands the small town of *Ugines* (1510'; 3000 inhab.). Here the culture of the vine begins on the lower slopes facing the S. Beyond *Marlens* the road quits the valley of the Chaise and crosses the hardly perceptible watershed of the *Eau Morte*, which we now follow. 7½ M. *Faverges* (1700'; Hôt. de Genève; Poste), with its old castle, now a silk factory. (To *Frontenex* over the *Col de Tamié*, see p. 256.) We next reach (6 M.) *Bout du Lac*, near the hamlet of *Doussard*, at the S. end of the *Lac d'Annecy (1463'; 9 M. long), on which a steamer plies five times daily to Annecy in 1½ hr. To the right rise the rocky pinnacles of the *Tournette* (p. 270). On a promontory extending far into the lake, to the left, is the prettily situated (3 M.) *Château Duingt* (1476'). On the opposite bank lie *Talloires* (Hôt. de l'Abbaye), the birthplace of Berthollet (see below), and *Menthon*, with sulphur-baths and an old château in which St. Bernard was born (p. 301). To the left lies *Sévrier*, at the foot of the long *Semnoz* (see below).

28 M. **Annecy** (1465'; pop. 11,947; *Gr.-Hôt. d'Angleterre*, Rue Royale, R., L., & A. 3-10, B. 1½, déj. 3½, D. 4, pens. 10-12 fr.; *Gr.-Hôt. Verdun*, Promenade du Pâquier; *Aigle*, Rue Royale, R., L., & A. 3, B. 1, D. 3½, pens. 7½ fr.), a picturesque, old-fashioned town, the capital of the department of Haute-Savoie, with linen-manufactories. The lofty old *Château* is now a barrack. In the chapel of the monastery *De la Visitation* repose St. Francis de Sales (d. 1622) and St. Johanna of Chantal (d. 1641). The *Promenade du Pâquier* on the lake affords a pleasant walk and fine view. In the middle of it rises the *Préfecture*, in front of which stands a bronze statue, by Becquet, of *Sommeiller* (1815-71), one of the engineers of the Mont-Cenis Tunnel. On the other side of the canal issuing from the lake lies the *Jardin Public*, with a bronze statue of the famous chemist *Berthollet* (1748-1822), by Marochetti. The *Hôtel de Ville* contains a *Museum*, chiefly of natural history and industrial objects (open on Sun., Tues., Thurs., 9-12 and 1.30-4). — Annecy, with its beautiful environs, is recommended as a pleasant resting-place.

EXCURSIONS. The *Semnoz* (5590'), to the S. of Annecy, a fine point, is easy (5 hrs.). We take the Albertville road on the S. bank of the lake to (3 M.) *Sévrier*, and ascend by a road to the right to the (7½ M.) *Col du Leschaux* (3028'; Inn); bridle-path thence to the top or *Crêt du Châtillon* in 1½-2 hrs. A little below the top is a hotel. Omnibus from Annecy to the Col on Wed., Thurs., & Sat. at 12 p.m. (3 fr.; mountain-railway projected). Beautiful view. — The *Parmelan (6018'), to the N.E. of Annecy, is chiefly interesting on account of its grotesque rock-formations. Road by *Sur-les-Bois* and *Dingy St. Clair* to (9 M.; carr. in 2½ hrs., 15 fr.) *La Blonnière;* thence (guide not necessary for experts) by the *Chalet Chapuis* and the *Grand Montoir* to the top in 2½-3 hrs. (admirable panorama). — Ascent of the *Tournette (7738'), the fine mountain to the S.E. of Annecy, attractive but laborious (only for experts; guide 10 fr.). Road to (9 M.) *Thônes* (see below), thence with guide, via *Belchamp* and the *Chalets du Rosairy*, to the top in 5½ hrs. Superb view, especially of the Mont Blanc group.

Railway to *Aix-les-Bains*, see p. 267. Near *Lovagny*, the first station (11 min.), are the interesting *Gorges du Fier (p. 267).

From Annecy to Chamonix, diligence daily in 12 hrs. (by steamer to *Doussard*, thence by carriage via *Fontaines d'Ugines*, *Flumet*, *Megève*, and *St. Gervais;* comp. p. 269); fare to St. Gervais 18, to Chamonix 21 fr.; return fare 32 fr.

From Annecy viâ Grand Bornand to Cluses, 12½ hrs., attractive. A carriage-road runs by *Veyrier* and *Alex* to (4 hrs.) Thônes (2054'; *Hôt. de Plain-Palais*), a little town prettily situated at the confluence of the *Nom* and the *Fier* (ascent of the Tournette, see above). Thence it ascends the valley of the Nom to the E., passing *Les Villards*, to (1¾ hr.) *St. Jean-de-Sixt* (3319; to Sallanches, see below), beyond which it divides. The left branch runs by *Petit-Bornand* to (4½ hrs.) *Bonneville* (p. 271); the right leads through (½ hr.) Grand-Bornand (3053'; *Hôt. Milhomme*), a considerable village on the *Borne*, seriously damaged by fire in 1894, to (1½ hr.) *Venay*. From Venay a bridle-path ascends over the *Col des Anaes* (5608') to (2 hrs.) *Reposoir* or *Pralong* (Inn), where it joins the carriage-road leading through the picturesque *Valley of Reposoir* to (2 hrs.) *Scionzier* and (½ hr.) *Cluses* (p. 272). — From Annecy over the Col des Aravis to Sallanches, 14 hrs., attractive. To (5¾ hrs.) *St. Jean-de-Sixt*, see above. Thence a carriage-road leads to the S.E. through the valley of the Nom to *La Clusaz* and to the (2½ hrs.) Col des Aravis (4913'; Inn), which commands a fine view of Mont Blanc. From the Col a bridle-path descends to (¾ hr.) *La Giettaz* (3640'; Hôt. du Col des Aravis), whence another carriage-road leads to (2 hrs.) *Flumet*, on the road from Ugines (p. 269) to (4¾ hrs.) *Sallanches* or *St. Gervais*. A shorter route is offered by a footpath leading from *La Giettaz* over the *Col Jaillet* direct to (4 hrs.) Sallanches.

The Railway from Annecy to Geneva traverses a tunnel, crosses the *Fier*, and turns to the N. into the valley of the *Fillière*. On the right rises the *Parmelan* (see above). 3 M. *Pringy-la-Caille;* 6 M. *St. Martin-Charvonnex;* 10 M. *Groisy-le-Plot* (3½ M. to the W. are the sulphur-baths of *La Caille*, in a picturesque gorge). At (14½ M.) *Evires* (2516'; Buffet), beyond another tunnel and a lofty viaduct, the line reaches its highest point. Travellers in the opposite direction are subjected to the formalities of the customhouse here, as that part of the Department of Haute-Savoie which adjoins Switzerland is exempt from French duties (see p. 271). Two tunnels, the first 1320 yds. long.

The train now descends, making a long bend to the E., and enters the valley of the *Arve*, of which it affords a beautiful survey. Beyond (20 M.) *St. Laurent* is a viaduct 157' high. — 23½ M. *La Roche-sur-Foron*, junction of the railway from Cluses to Annemasse. Hence to (37½ M.) *Geneva*, see p. 271.

72. From Geneva to Chamonix.

i. Viâ Cluses.

54½ M. RAILWAY from Geneva (*Eaux-Vives* station) to (27½ M.) *Cluses* in 1½ hr. (fares 5 fr. 70, 4 fr. 25, 2 fr. 35 c.). Thence to (27½ M.) *Chamonix* OMNIBUS in 4¾ hrs. (8 fr., there and back 14 fr.). Through-fares from Geneva to Chamonix, 13 fr. 95, 11 fr. 65, 10 fr. 15 c.; return-tickets, 22 fr. 5, 19 fr. 80 c. Tickets may be obtained in Geneva at Grand Quai 10 as well as at the station, and in Chamonix at the Bureau de Messageries, near the Hôtel Impérial. Cook's tickets also are offered at the hotels in Geneva (to Chamonix and back, or viâ Chamonix and the Tête Noire or Salvan to Martigny and Vernayaz, etc.). — At Eaux-Vives (tramway from the Place Molard; cab 2 fr.) the railway-time is about 55 min. behind that of Geneva. Tickets are changed at the omnibus-office at Cluses (to the right of the station), after which no time should be lost in securing a seat in the omnibus (front seats preferable). A seat in one of the supplementary carriages, which are provided when the main vehicle is full, is preferable to an inside seat in the latter. A carriage and pair (4 pers.), from Cluses to Chamonix in 4 4½ hrs., costs 40-50 fr. and fee. Luggage is not examined at the French frontier, as the department of Haute-Savoie is free of customs.

Geneva (*Eaux-Vives* station; Pl. F, 8), see p. 221. The train ascends at first through a tunnel, then traverses a plateau, with the Salève on the right and the Jura chain on the left. At (2½ M.) *Chêne* (1385'), a thriving village belonging to Geneva, Mont Blanc appears on the right, between the pyramidal Môle (p. 272) and the double peak of the Pointe d'Andey (6165'). We now cross the *Foron*, the French boundary, and reach (3¾ M.) **Annemasse** (1420'; *Rail. Restaurant; Hôt. de la Gare*, at the station; *National*, in the town), the junction for the Bellegarde and Bouveret line (p. 264) and the steam-tramway to Samoëns (p. 274). The train backs out from the station, describes a wide curve through the straggling little town, and crosses the *Arve* at *Etremblères* (electric railway to the *Salève*, see p. 232). Fine view of the Mont Blanc chain in the distance to the left. At (5½ M.) *Monnetier-Mornex* the charmingly situated village of *Mornex* (p. 232) lies above us to the right, and the deep gorge of the Arve to the left. Then, beyond a handsome viaduct over the *Vaison*, the railway ascends through orchards, with a continuous view of the Arve valley, and crosses the *Foron* to (9½ M.) *Reignier* (Hôt. du Mont Blanc). Beyond (11 M.) *Pers-Jussy-Chevrier* is (12 M.) **La Roche-sur-Foron** (1905'; *Hôt. de la Croix Blanche*), the junction of the line to Annecy (p. 270; change carriages for Chamonix). To the left is the village with its ruined castle, picturesquely situated high above the Arve valley.

The railway crosses the *Foron* and one of its tributaries, and traverses a short tunnel. To the right tower the cliffs of the Pointe d'Andey, to the left the Môle, and farther back the Voirons. We now descend through brushwood into the Arve valley. Beyond (16 M.) *St. Pierre-de-Rumilly* we cross the *Borne* and soon afterwards the Arve, and skirt the town to the station of (18½ M.) **Bonneville** (1457'; pop. 2271; *Couronne*, dear; *Balance*), a place of some importance, picturesquely situated among vine-clad hills. A handsome bridge crosses the *Arve*, on the N. side of which, to the right, stands a monument to the Savoyards who fell in the campaign of 1870-71. On the opposite bank rises a monument, 73' high, to King Charles Felix of Sardinia. To the right, we obtain a superb *View of Mont Blanc, whose dazzling peaks towering majestically at the head of the valley seem to annihilate the intervening distance of nearly 30 M. The Aiguille du Goûter appears first; then, from right to left, the Dôme du Goûter, Mont Blanc itself, Mont Maudit, Mont Blanc du Tacul, Aiguille du Midi, and Aiguille Verte. — Steam-tramway to the N. to *Bonne*, see p. 274.

The **Môle** (6130'; fine view), to the N.E. of Bonneville, is ascended in $3^1/_2$-4 hrs. viâ (20 min.) *Lépagny*, *Gallinous*, and the couloir of *Pertuis*; or viâ *Reyret*, the *Col de Reyret* (3040'), the *Grange à Béroud* ($1^3/_4$-2 hrs.), and ($^3/_4$ hr.) the *Lardère* (4980'), on which there is a refuge-hut, $^3/_4$ hr. below the summit. — Ascent from St. Jeoire, see p. 274.

The **Pointe d'Andey** (6165'; good view), to the S. of Bonneville, is ascended in 3 hrs. viâ ($^1/_4$ hr.) *Pontchy* and ($^3/_1$ hr.) *Andey*; or in $3^1/_2$ hrs. viâ ($^3/_4$ hr.) *Thuet*, (1 hr.) *Brison* (Inn), and (1 hr.) *Solaizon*, $^3/_4$ hr. from the top. Carriages may proceed as far as Brison viâ *Vougy*. The fine view is partly obstructed to the S. by the *Rochers de Leschaux*.

Beyond Bonneville the railway traverses, in an almost straight line, the broad valley of the Arve, bounded by lofty mountains. To the right winds the road ascending to *Brizon*. Crossing the *Giffre* we reach (23 M.) *Marignier* (1530'; Inn; steam-tramway to *Pont du Risse*, see p. 274). To the right rise lofty limestone hills. We continue to skirt the right bank of the Arve (on the hill to the left is *Châtillon*, see below) to ($27^1/_3$ M.) **Cluses** (1605'; *Hôtel-Buffet de la Gare*, lunch incl. wine $3^1/_2$, D. incl. wine 4 fr.; *Hôt. Revuz*), a small town, chiefly inhabited by watchmakers, at the beginning of the narrower part of the Arve valley, and at present the terminus of the railway.

A winding road ascends hence to the N. viâ *Châtillon* to (6 M.) *Taninges*, on the line from Geneva to Samoens (p. 274); and another new road, formed by blasting the rock, leads to the right to (2 M.) *Nancy-sur-Cluses*. — To Annecy viâ *Grand-Bornand*, see p. 276.

The Road to Chamonix leads from Cluses through the narrowing gorge of the Arve, on the right bank of the stream. Beyond (28 M.) *Balme* (1824'), in the bluish-yellow limestone precipice to the left, 750' above the road, is seen the entrance to the *Grotte de Balme*, a stalactite-grotto hardly worth visiting (2 hrs. there and back; 3 fr. each pers.). $31^1/_2$ M. *Magland*. On the right, farther on, rise the *Pointe d'Arreu* (8097') and the *Pointe Percée* (9025'; see p. 273), and on the left, the bold precipices of the *Aiguilles de Varens* (8165'). The conspicuous *Cascade d'Arpenaz* is imposing after rain.

The valley expands. The road crosses the Arve, and leads straight on, affording a continuous view of the Mont Blanc group.

38 M. **Sallanches** (1788'; *Hôt. des Messageries*), a straggling little town, with a fountain commemorating the Revolution, adorned with a statue of Peace, by Gambos (1890).

to Chamonix.   ST. GERVAIS.   *V. Route 72.*   273

The **Pointe Percée** (9025'), commanding a fine view of Mont Blanc, may be ascended from this point in 5-5½ hrs. (with guide) viâ the (2½ hrs.) *Praz-ès-Ros* and the (2 hrs.) *Col des Verts* (no difficulty for experts). The **Pointe d'Arreu** (8095'; 6 hrs.) and the *Aiguille de Varens* (8165'; 6½ hrs.; with guide) are more difficult. — Route from Sallanches by *Flumet* to *Albertville*, see p. 268; to *Annecy* over the *Col des Aravis*, see p. 270.

The road next leads by *Domancy* to (44 M.) **Le Fayet** (1860'; *Hôtel du Pont*), by the bridge over the *Bon-Nant*.

In a wooded ravine of the *Vallée de Montjoie*, ½ M. from Le Fayet, lies **St. Gervais-les-Bains** (2075'; *°Curhaus*), a frequented sulphur-bath, which was totally destroyed by an outburst of the Glacier de Tête-Rouge (p. 281) in July, 1892, but has since been rebuilt in a higher and safer position. A shady path leads in 20 min. from the baths to the **Village of St. Gervais** (2680'; *°Hôt. du Mont Joli;* *°Hôt. du Montblanc;* *°Hôt. des Etrangers*; several pensions), on the road to Contamines (p. 289), a health-resort, finely situated. About ¾ M. below the village (4 min. from the foot-path to the Baths) is the *°Cascade de Crépin*, a waterfall of the Bon-Nant. From St. Gervais a road leads viâ *Bionnay* to (6 M.) *Contamines* (p. 289). To *Albertville* or *Annecy* viâ *Mégève* and *Flumet*, see p. 268. — The *Mont Joli* (8238') may be ascended without difficulty from St. Gervais in 5 hrs. (comp. p. 289).

Pedestrians may quit the diligence at Le Fayet and walk over the **Col de la Forclaz** (5105'), between the *Tête-Noire* (5800'; not to be confounded with the Tête-Noire between Chamonix and Martigny) and the *Prarion* (6460'), direct to *Le Fouilly* and *Les Houches* in 5-6 hrs. (guide desirable, 6 fr.). A longer but more interesting route (6-7 hrs.) is over the *Col de Voza* (p. 288).

From Le Fayet a road crosses the Arve to Chède and Servoz (see below). The road to Chamonix on the left bank of the Arve ascends gradually, with the torrent almost immediately below it, passes through a cutting and enters the wooded valley of (46½ M.) *Le Châtelard* (tavern). Through the opening of the valley appear the *Dôme du Goûter* (14,210') and the jagged *Aiguille du Midi* (12,610'). Beyond the inn is a short tunnel; the road returns to the Arve for a short distance.

A road diverges here to the left and crosses the Arve to (½ M.) *Servoz* (Hôt.-Pens. Diosaz; A la Fougère, well spoken of), whence we may visit (in 1 hr., there and back) the *°Gorges de la Diosaz* (adm. 1 fr.), a grand ravine, through which the *Diosaz*, a torrent rising on the Buet, dashes in fine cascades. Easy access to the gorge is afforded by a gallery, ½ M. long, attached to the rocks. Visitors should penetrate as far as the *Cascade de Soufflet*, the most imposing part.

48½ M. *Les Montées* is an inn by the *Pont Pélissier*, over which the old road from Servoz comes to join ours. About ½ M. farther on the old road ascends to the right to *Le Fouilly* and *Les Houches* (p. 288), while the new road traverses the wild ravine of the Arve, crossing the stream by the *°Pont Ste. Marie* (fine view of the gorge) and again higher up. The glaciers now gradually become visible, but owing to the vastness of the mountains in which they are framed it is impossible at first to realise their extent. The first are the *Glaciers de Griaz* and *de Taconnay*; then the *Glacier des Bossons* (p. 280) near the village of that name, which, as it extends farthest into the valley, is apparently the largest. A little above it the road crosses the Arve by the *Pont de Perralotaz*, and 1¾ M. beyond it reaches —

54½ M. **Chamonix** (p. 276).

BAEDEKER, Switzerland. 16th Edition.   18

ii. Viâ Sixt.

RAILWAY from Geneva to (3³/₄ M.) *Annemasse* in 18 minutes. STEAM TRAMWAY from Annemasse to *Samoëns*, 27¹/₂ M., in 3 hrs. (3 fr. 55, 2 fr. 20 c.). From Samoëns to *Sixt*, 5 M., OMNIBUS in 1 hr. From Sixt to *Chamonix*, BRIDLE-PATH (10-11 hrs.) over the Col d'Anterne and Col du Brévent (guide, 18 fr., unnecessary in good weather), a somewhat fatiguing expedition, as both passes and the ascent and descent between them must usually be accomplished in the hot midday hours, but affording splendid views of Mont Blanc. Provisions should be carried, as nothing except milk is to be obtained on the way.

From Geneva to (3³/₄ M.) *Annemasse*, see p. 271. At the N. end of Annemasse the line turns to the E. (right), leaving the hill of *Monthoux* to the left (stations *Malbrande, Bas-Monthoux*) and skirts the foot of the *Voirons* viâ *Borly* to (8 M.) *La Bergue* (1680′).

The Pralaire (4630′), the S. peak of the *Voirons* (p. 233), may be ascended hence in 2 hrs. viâ (³/₄ hr.) *Lucinges* and *Les Gets*.

9¹/₂ M. *Bonne*, on the *Menoge*. Branch-line to (8 M.) Bonneville, see p. 271. — 10¹/₂ M. *Pont de Fillinges*, at the confluence of the Menoge and de Foron. — 13¹/₂ M. *Viuz-en-Sallaz*. To the left is the *Pointe des Brasses* (4940′), to the right the *Môle* (see below).

16 M. St. Jeoire (1925′; *Hôt. de Savoie;* *Pens. des Alpes*, unpretending), a market-village of 1750 inhab., with the château of *Fléchère* and a statue of *Sommeiller*, one of the engineers of the Mont Cenis Tunnel (see p. 268).

The Môle (6130′), which commands a fine view of the valley of the Arve and of Mont Blanc, may be ascended in 3¹/₂ hrs. from St. Jeoire, viâ the hamlet of *Montrenaz* and the chalets of *Pinget, Char d'Amont, Char d'Aval*, and *L'Ecutieux*. Riding is practicable to within ¼ hr. of the summit. The club-hut lies farther to the S. (p. 272).

The road now ascends a narrow gorge, which it quits for the valley of the *Giffre*, to the left. From (17¹/₂ M.) *Pont du Risse* a branch-line leads in ¹/₂ hr. viâ *Le Breuillet* to *Marignier* (p. 272). 20 M. *Mieussy* (2225′; inns), at the W. base of the *Pointe de Marcelly* (see below). In front rise the Buet and Mont Blanc. The line rounds the conical *Roc de Suets* (3002′) and reaches —

24¹/₄ M. Taninges (2100′; *Balances*), a busy little town with the old abbey of *Mélan*, now a seminary. Route to Cluses, see p. 272.

The Pointe de Marcelly (7105′) may be ascended hence in 4¹/₂ hrs. by a steep path viâ *Les Pontets* and the chalets of *Grand Planay*, or from Mieussy (see above) in 5 hrs., with guide. — A road leads N.E. from Taninges, viâ *Les Gets* (3865′), to (13¹/₂ M.) *St. Jean d'Aulph* (p. 257), in the valley of the Drance. A diligence plies daily in 3¹/₄ hrs. from Taninges to Morzine.

We proceed through the valley of the *Giffre*, viâ *La Palud*, *Jutteninge*, *Verchaix-Morillon*, *Les Chenets*, and *Le Bérouze*, to —

31 M. Samoëns (2490′; *Croix d'Or*, moderate; *Commerce*, unpretending), a small town of 2540 inhab., on the Giffre. Fine view from the little chapel above the church (10 min.).

From Samoëns to (7 hrs.) *Champéry* in the Val d'Illiez, over the *Col de la Golèse* and the *Col de Coux*, see p. 260. — From Samoëns two passes, to the left the *Col de Jouplane* (5635′), to the right the *Col de la Golèse* (5480′), lead to the N. to (4 hrs.) *Morzine* (p. 257).

Beyond Samoëns the road enters a defile in which the Giffre **forms a fall, 160′ in height.** As the valley expands we see in front

of us the precipices of the Buet, to our right the **Pointe de Salles**
and the Pointe des Places, and to our left the Pic de **Tanneverge**
(see below).

36 M. **Sixt** or *L'Abbaye de Sixt* (2480'; *Hôt.-Pens. du Fer à
Cheval*, in an old monastery, R. & L. 3, B. 1½. D. 3 fr.).

ENVIRONS. In spring, when the brooks are swollen by the melting
snow, the neighbourhood of Sixt abounds in fine waterfalls, there being
no fewer than thirty in the upper part of the valley alone, called from
its shape **Vallée du Fer à Cheval**. In summer and autumn, however, the
number dwindles to five or six. An attractive excursion may be made
through the debris of a landslip of 1602, to the (3 hrs.) *Fond de la Combe*
(3274'), at the head of the valley, with a waterfall.

From Sixt over the *Col de Sageron* (7917') to Champéry (with ascent
of the *Mont Ruan*), see p. 260. — The Pic de Tanneverge (*Pointe des Rosses*,
9780'; 9 hrs., with guide), by the Col de Sagerou or the *Col de Tanneverge*
(7745'), is a difficult ascent, but commands a splendid view. The descent
from the Col may be made into the valley of the *Barberine* to *Emosson* and
thence over the *Col de la Gueula* to *Finhaut* (p. 285). — The **Pointe Pelouse**
(8118'), ascended past the *Lac de Gers* in 6 hrs., presents no difficulty;
fine view of Mont Blanc. The descent may be made by the *Désert de
Platé* and the *Escaliers de Platé* to *Chède* (p. 273; dizzy path, recalling
the Gemmi).

FROM SIXT TO CHAMONIX OVER THE BUET, 12-13 hrs., fatiguing but interesting (guide necessary, 23 fr. incl. return). To the *Chalets des Fonds*, see
below. Thence the route leads to the left to the (2½ hrs.) *Col Léchaud*
or *des Fonds* (7325'), and ascends over loose stones and snow to the top
of the *Buet (10,200'), which commands a magnificent view of the Mont
Blanc range, Monte Rosa, the Matterhorn, the Bernese Alps with the Jungfrau and the Finsteraarhorn, the Dent du Midi, and the Jura as far as
the mountains of the Dauphiny. A toilsome descent leads down to (2 hrs.)
the *Chalet de la Pierre à Bérard* (6380; Inn), and through the *Vallée de
Bérard* (p. 284) to *Argentière* and (4 hrs.) *Chamonix*.

The bridle-path from Sixt to the Col d'Anterne (to **Chamonix**
11 hrs., fatiguing) ascends the *Vallée des Fonds* to the S., past
a picturesque waterfall on the right, to (½ hr.) *Salvagny* (in front
rises the beautiful Pointe de Sales; 8182'), beyond which it zigzags
up a grassy hill to the left, past the *Cascade du Rouget* (right), to
the (1½ hr.) *Chalets des Fonds* (4530'; Alpine fare). Near this point
is 'Eagle's Nest', the summer-residence of Sir Alfred Wills, at the
foot of the *Buet* (see above). About 5 min. **farther up, beyond the
bridge, we ascend to the right** (the path to the left leads to the
Col Léchaud, see above), passing the *Chalets de Grasse-Chèvre* in
a wide curve, to (1 hr.) the saddle **of the Plateau du Bas du Col.**
Then, leaving the *Chalets d'Anterne* below us to the right, we cross
the pastures of that name and skirt the *Lac d'Anterne* to (1½ hr.)
the *Col d'Anterne (7425'), where a magnificent survey of Mont
Blanc suddenly breaks upon our sight. We descend to the left
(the path to the right leads in 2½ hrs. to Servoz), in view of the
Aiguilles Rouges, into the valley of the *Diosaz*, which we cross after
1½ hr. by a wooden bridge (5592'). We once more ascend to the
(1½ hr.) **Col du Brévent** (8076'), which also commands a fine view
of Mont Blanc. Thence the descent leads chiefly through wood,
viâ *Planpraz* and *Les Chablettes* (p. 280) to (2 hrs.) *Chamonix*.

18*

73. Chamonix and Environs.

Hotels. *Hôt. de Londres et d'Angleterre, *Gr. Hôt. Royal et de Saussure, *Gr. Hôt. Impérial; at these, R., L., & A. 4-5 fr. and upwards, B. 1½, D. 5 fr. *Hôt. du Montblanc, R., L., & A. 2½-5, D. 5, pens. from 9 fr.; *Hôt.-Pens. Couttet, frequented by the English, R., L., & A. from 3, D. 4, pens. from 8 fr.; *Hôt. des Alpes, R., L., & A. from 3, D. 4, pens. from 8 fr. — *Hôtel Beau-Site, in an open situation at the S. end of the village, R. from 2, D. 3½, pens. from 6 fr.; *Hôtel de France et de l'Union, R. from 2, B. 1¼, D. 3, pens. 6-7 fr.; *Hôt. Suisse, R. 2, B. 1¼, D. 3, pens. 7 fr.; *Hôt.-Pens. de la Poste, R., L., & A. 2-3, D. 3½, pens. from 6 fr.; *Hôt. de la Paix, R., L., & A. from 1½, B. 1¼, D. 3, pens. from 5 fr.; *Hôt.-Pens. de la Mer de Glace, on the Martigny road, pens. 6 fr.; Hôtel-Garni Beauséjour; *Croix Blanche, R., L., & A. from 1½, B. 1¼, D. 3, pens. from 5 fr.; Hôt.-Pens. du Lac, prettily situated 1 M. to the W. (p. 280).

Guides. A guide is unnecessary for the *Montanvert*, the *Flégère*, the *Brévent*, and the *Pierre Pointue*. The paths are so minutely described in the following pages that they can hardly be mistaken, while opportunities of asking the way are frequent. Visitors to the *Chapeau* need only engage a guide for the passage of the Mer de Glace to or from the Chapeau (p. 278). — The guides at Chamonix have formed themselves into a society under a *Guide-Chef*, who assigns them in regular order to travellers applying for them; but travellers may also choose guides for themselves. An ordinary tour is paid for as completed, if more than half the distance has been traversed and the expedition is given up at the traveller's desire; when less than half the distance has been completed, two-thirds of the tariff must be paid (special tariff for Mont Blanc, see p. 281). The guides are bound on the 'courses ordinaires' to carry baggage not exceeding 26 lbs.; on the 'courses extraordinaires', 15 lbs. only. — The following are recommended for difficult expeditions: *Ed.* and *Aug. Cupelin; Henri Devouassoud; Jules Bossoney; Michel* and *Adolphe Folliguet; Alph., Michel* and *Fréd. Payot; Ben. Simon*, surnamed *Benoni; A. Tournier; Jules Simond* of Les Praz; *Franç., Alfred,* **and** *Joseph Simond* of Lavancher; *Gasp.* and *Joseph Simond* of Les Mossons; *Michel Savioz; Franç. Meugnier,* etc.

Horses and Mules. With the exception of the excursion to the Montanvert and Chapeau (9 fr.), and to the Montanvert for the purpose of visiting the Jardin, and back to Chamonix in the evening (9 fr.), the same charges are made as for the 'courses ordinaires' of the guides, and as much more is charged for the attendant.

The **Collection of Pictures** of *M. Loppé*, the well-known painter of Alpine scenery, situated behind the Hôtel Royal, on the way to the Montanvert, is worth seeing. Admission gratis (small fee to the attendant).

English Church Service during the season.

Points of Interest. The traveller should devote three or four days at least to Chamonix, but those who have one day only at command should ascend the Montanvert (p. 277) in the morning (2½ hrs.), cross the Mer de Glace (p. 278) to the (1½ hr.) Chapeau (p. 278), descend to (1 hr.) Les Tines (p. 278), ascend the Flégère (p. 279; 2½ hrs.), and descend thence in 1¾ hr. to Chamonix. Early in the morning the path to the Montanvert is in shade, in the afternoon that to the Flégère at least partly so; and by this arrangement we reach the Flégère at the time when the light is most favourable for the view of Mont Blanc. For this excursion a guide (to be found on the Montanvert) is necessary for the Mer de Glace only. Riders send their mules round from Montanvert to Les Tines or the Chapeau to meet them. The excursion to the Flégère alone takes 5 hrs., and that to the Montanvert or the Chapeau about the same time. — Those who come from the E., and have spent the night at *Argentière*, should leave the road **near** *Lavancher* (p. 283), take a guide there, and proceed by the Chapeau, the Mer de Glace, and Montanvert to Chamonix. The Flégère may also be reached from *La Joux* (p. 283), on the right bank of the Arve; but the path is bad and unsuitable for riding, and cannot be found without a guide (boy 1-1½ fr.).

On a cloudy afternoon, when the views from the heights are concealed, the GLACIER DES BOSSONS (p. 280) is the best object for a walk (there and back 3 hrs.). — To the CASCADE DE BLAITIÈRE, on the hillside to the E. of Chamonix, 1/2 hr. (adm. 1/2 fr.). — To the PAVILLON DE LA PIERRE-POINTUE (p. 280) and back, 5-6 hrs. ; or, including the Aiguille de la Tour and Pierre à l'Echelle, a whole day. — Ascent of the BRÉVENT (p. 279) and back, 7 hrs. ; ascent or descent by the Flégère 2 hrs. more.

The *Valley of Chamonix (3445'; pop. about 4000), or *Chamouny*, 12 M. long, 1/2 M. wide, watered by the *Arve*, runs from N.E. to S.W., from the Col de Balme to Les Houches. It is bounded on the S.E. by the *Mont Blanc* chain, with its huge ice-cataracts, the *Glaciers du Tour, d'Argentière, des Bois (Mer de Glace)*, and *des Bossons;* and on the N.W. by the *Aiguilles Rouges* and the *Brévent*. A Benedictine priory first brought the valley into cultivation at the beginning of the 12th cent., but it remained practically unknown until the 18th cent. when it was visited by the English travellers Pococke and Windham (1748) and the Genevese naturalists De Saussure (d. 1799), Pictet (d. 1825), and Deluc (d. 1817). It is inferior to the Bernese Oberland in picturesqueness of scenery, but superior in the grandeur of its glaciers, in which respect it has no rival but Zermatt.

In front of the Hôtel Royal rises the *Saussure Monument, unveiled in 1887, on the centenary of the first ascent of Mont Blanc, and consisting of a bronze group (by J. Salmson) on a granite pedestal, representing Saussure conducted by Balmat (p. 281); inscription: 'à H. B. de Saussure Chamonix reconnaissant'. Another small monument to Balmat stands in front of the church.

The *Montanvert or *Montenvers* (6303'; 2 1/2 hrs. ; guide unnecessary), an eminence on the E. side of the valley, is visited for the sake of the view it affords of the vast 'sea of ice'which fills the highest gorges of the Mont Blanc chain in three branches *(Glacier du Géant* or *du Tacul, Glacier de Leschaux*, and *Glacier de Talèfre)*, and which descends into the valley in a huge stream of ice, about 4 1/2 M. long and 1/2-1 1/4 M. broad, called the *Mer de Glace* above the Montanvert, and the *Glacier des Bois* below it. The bridle-path leads to the left by the Hôtel Royal, passes the little English church, and crosses the meadows (to the left of the cemetery-wall) to the (1/4 hr.) houses of *Les Mouilles*. We then ascend through pinewood to the right (again turning to the right after 1/4 hr.), past the (10 min.) *Chalets des Planards*, to (40 min.) *Le Caillet* (4880'; rfmts.), a spring by the wayside. Farther on (12 min.) a bridle-path to the left descends to Les Bois (p. 278). Our path ascends gradually, at first through wood, to the (1 hr.) **Hôtel du Montanvert* (R., L., & A. 4, déj. 4, D. 5, pens. 9 fr.), at the top of the hill, commanding the **Mer de Glace* and the mountains around it: opposite us rises the huge *Aiguille du Dru* (12,517'); behind it, to the left, is the snow-clad *Aiguille Verte* (13,540') and lower down the *Aig. du Bochard* (8765'), to the right the *Aig. du Moine* (11,214'); farther distant are the *Grandes Jorasses* (13,800'), the *Mont Mallet* (13,085'), and the *Aig. du Géant* (13,160'); and immediately to our right tower the *Aiguilles de Charmoz* (11,295') and *de Blaitière* (11,595').

From the Montanvert travellers usually cross the **Mer de Glace** to the (1¼-1½ hr.) *Chapeau*, opposite. A path descends the left lateral moraine to (¼ hr.) the glacier. The passage of the glacier (15-20 min.; guide from the Montanvert Hotel, unnecessary for the experienced, 3 fr., or to the Chapeau 5 fr.; woollen socks to prevent slipping, 1 fr.) presents no difficulty. On the opposite side we ascend over débris to the (5 min.) top of the right lateral moraine, skirting which we then descend to the '*Mauvais Pas*', where the path is hewn in steps and flanked with iron rods attached to the rocks, and the (40 min.) Chapeau.

Elderly travellers and those subject to giddiness are to be dissuaded from attempting the Mauvais Pas. — Guides for travellers making this excursion in the reverse direction are not always to be found at the Chapeau; if required, they should be brought from Lavancher (6 fr., see below).

The *Chapeau (5082'; Inn), a projecting rock on the N.E. side of the Glacier des Bois, at the base of the *Aiguille du Bochard*, is considerably lower than the Montanvert, but commands an excellent survey of the ice-fall of the Glacier des Bois and the Chamonix Valley. In the background *Mont Mallet* (13,085') and the *Aiguille du Géant* (13,160'); to the right the *Aiguilles de Charmoz* (11,295'), *de Blaitière* (11,595'), *du Plan* (12,050'), and *du Midi* (12,610'), the *Bosses du Dromadaire* (14,950'), the *Dôme du Goûter* (14,210'), and the *Aig. du Goûter* (12,710').

A bridle-path descends the moraine from the Chapeau, in view of the precipices of the Glacier des Bois and the Aiguille du Dru, and then through pine-wood to (40 min.) the *Hôt.-Pens. Beau-Séjour* (p. 283). Here it divides: to the right to (10 min.) *Lavancher*, to the left to (20 min.) *Les Tines* (p. 283). The route hence to the Flégère crosses the Arve at the inn 'à la Mer de Glace', then leads to the left through wood and pastures to (20 min.) the beginning of the zigzag path (p. 279). A shorter path, rough and unfit for riding, diverges from the path to Les Tines (20 min. from the Chapeau) to the left, and descends the moraine (passing the *Source of the Arveyron* below on the left) to *Les Bois* and (40 min.) *Les Praz* (p. 279).

The °*Jardin* (9145'; guide necessary, 14 fr.) is a triangular rock rising from the midst of the *Glacier de Talèfre*, and walled in by moraines. Around a spring in the midst of this oasis Alpine flowers bloom in August. From the Montanvert, where the night is passed, we skirt the somewhat dizzy rocks of *Les Ponts* to the right and traverse the moraine to the *Angle*; here we take to the crevassed Mer de Glace, and ascend it for 2½-3 hrs. to the foot of the *Séracs de Talèfre*. We now turn to the right, ascend past the *Pierre à Béranger*, on the S. side of the Séracs (¾-1 hr.; a wooden hut halfway up), and cross the Talèfre Glacier to the (25 min.) Jardin. This excursion makes us acquainted with the grand icy wilds of the Mont Blanc group, though somewhat fatiguing, it presents no difficulty to good walkers, and is even undertaken by ladies. Provisions necessary.

The **Aiguille de Charmoz** (11,295') is scaled (with guide) from the Montanvert in 5½ hrs. or more, according as one or more of its five peaks are climbed. We first reach (3 hrs.) a rocky platform at the foot of a couloir above the *Glacier des Nantillons*, to the S. of the Aiguille, and thence ascend to the (2½ hrs.) N. peak by the E. side of the mountain. About 2-3 hrs. are required to reach the fifth peak. The fourth appears to be the highest. — The **Aiguille du Dru** (12,517'), a difficult peak, adapted only to expert climbers, is ascended from the Montanvert in about 12 hrs. We climb a couloir exposed to falling stones; ascend a vertical chimney 160' high; traverse the couloir to the col by means of an insecure ladder; cross another col with the precarious aid of a rope; and, beyond a narrow cornice and several difficult 'chimneys', finally reach the top by passing astride along rocks and a snow-arête, with precipices of 3000' on either side.

The ***Flégère** (5925'; 2½-3 hrs., descent 2 hrs.), to the N. of Chamonix, is a buttress of the *Aiguille de la Floria* (9690'), one of the highest peaks of the *Aiguilles Rouges*. We follow the Argentière road to (1½ M.) *Les Chables*. The direct footpath diverges to the left immediately on this side of the **Arve** bridge, leading in 12 min. through pastures (very marshy at places) to the foot of the mountain, where the ascent begins. [The bridle-route, a few minutes longer, crosses the Arve to *Les Praz* (p. 283), diverges to the left at the last house (guide-post), crosses the Arve, and is joined by the path mentioned above.] We now ascend the stony slope in long zigzags. After 35 min. we enter the wood to the right, pass (35 min.) the *Chalet des Praz* (rfmts.), and in 1 hr. more reach the *Croix de la Flégère* (*Inn, déj. 3½, D. 4, pens. 6 fr.). The *View (comp. panorama) embraces the entire chain of Mont Blanc, from the Col de Balme to beyond the Glacier des Bossons. Opposite us lies the basin of the *Glacier des Bois (Mer de Glace)*, enclosed by the sharply defined Aiguilles: to the left the *Aig. du Dru* and the huge snow-clad *Aig. Verte;* to the right the *Aiguilles de Charmoz, de Blaitière, du Plan*, and *du Midi*. The summit of Mont Blanc is also distinctly seen, but is less striking than the lower peaks owing to its greater distance. The jagged pinnacles of the *Aiguilles Rouges* also present a singular appearance. Evening light most favourable.

From the Flégère the bridle-path goes on to (1 hr.) the *Pavillon de la Floria*, from which the Aiguille de la Floria (9085'), affording a magnificent view to the W. as far as the Lake of Geneva, may be ascended, with guide, in 3 hrs. The ascent of the **Belvédère** (9780'; 3½ hrs. from the pavilion), the highest peak of the Aiguilles Rouges, is also interesting but difficult. Splendid view. — Those bound from the Flégère to Argentière or to the Chapeau may descend direct to **La Joux** (comp. p. 283; path hardly to be mistaken on the descent).

The ***Brévent** (8285'), the S.W. prolongation of the *Aiguilles Rouges*, affords a similar but finer view. While from the Flégère the Mer de Glace and the Aiguille Verte are the chief features, Mont Blanc is here revealed in all its grandeur; to the right of the Buet we also see the Bernese Alps, and to the S.W. the Alps of Dauphiny. The bridle-path (4½ hrs.; guide, 10 fr., unnecessary) leads from Chamonix to the W., passing the hamlets of *La Mola* and *Les Mossons*, ascends through wood to (1½ hr.) *Plan Nachat* (4833'; rfmts.), an admirable point of view, and then in numerous zigzags to the (1¾ hr.) *Plan Bel-Achat* (6975'; *Restaurant, bed 2, D. 4 fr.), on a saddle to the S.W. of the summit. Thence to the top, passing the sombre little *Lac du Brévent*, 1¼ hr. more.

Or we may ascend the 'Chemin Muletier de Chamonix à Sixt' past the *Restaurant des Chablettes* to (3 hrs.) *Planpraz* (6770'; Inn, well spoken of); then mount rather steeply to the left, and lastly through a rocky gully (*la Cheminée*) to the (1¼ hr.) summit (guide 10 fr.). Iron bars are fixed in the chimney to assist climbers and steps are cut in the rock, so that the expedition is quite safe. — The Brévent may also be combined with the Flégère. The 'Route de Planpraz', a well-defined path, diverges to the right from the Flégère path, about 20 min. below the Croix de la Flégère, and undulates along the slope of the mountain, in full view of the Mont Blanc chain, passing the *Chalets de Charlanos* halfway, to the (2 hrs.) inn of *Planpraz* (p. 278), which is visible from the Flégère.

To the *Glacier des Bossons an interesting walk (3 hrs. there and back). We follow the Geneva road (p. 273) past the *Hôtel-Pension du Lac*, cross by the (1½ M.) *Pont de Perralotaz* to the left bank of the Arve, diverge to the left at the hamlet of *Les Bossons* by a good path, and ascend to the *Pavillon Foncière* on the left moraine. Fine view of the huge glacier, which has begun to advance of late, overshadowed by the *Mont Blanc du Tacul* (13,940'). On the left rise the *Aiguilles du Midi* (12,610') and *de Blaitière* (11,595'). We descend to the grotto hewn in the glacier (85 yds. long, interesting; adm. and lights 1 fr.) and cross the glacier (guide necessary, 2, from Chamonix 6 fr.; woollen socks to prevent slipping 1 fr.) to the (½ hr.) top of the right lateral moraine. Descending over debris, and farther on through wood, we join the path to the Pierre Pointue at the Nant des Pèlerins (see below; to Chamonix 1 hr.).

The *Pavillon de la Pierre-Pointue (6722') is another favourite point (bridle-path, 3 hrs.; horse 8 fr.; guide 8 fr., unnecessary). On the left bank of the Arve we pass the hamlets of *Le Praz Conduit*, *Les Barats*, and (by the upper path, to the left) *Les Tsours*; here we turn to the left, ascend through wood on the right bank of the brook to the (25 min.) *Cascade du Dard* (cantine), a fine double fall, and then cross the broad stony bed of the *Nant des Pèlerins*. (After 10 min. the path to the Glacier des Bossons diverges to the right; see above.) We then ascend to the left in zigzags on the side of a wild valley, through which the *Nant Blanc* dashes over rocks, to the (¾ hr.) *Chalet de la Para* (5265') and the (1½ hr.) *Pavillon de la Pierre-Pointue* (Restaurant, déj. 3½ fr.), on the brink of the huge Glacier des Bossons, with its beautiful ice-fall. Opposite, apparently quite near, rise Mont Blanc, the Dôme du Goûter, the Aiguille du Goûter, etc.; also a superb view to the N. and W.

An interesting point is the Aiguille de la Tour (7650'), which commands the best survey of the Glacier des Bossons (1 hr., guide desirable; ascend to the left by the pavilion). — The Pierre à l'Echelle (7910') is another fine point (1¼ hr.; guide advisable). The narrow path (route to Mont Blanc, see below) leads by the pavilion to the right, round an angle of rock, and ascends to the brink of the Glacier des Bossons (where falling stones are sometimes dangerous). Admirable view of the riven ice-masses of the glacier; above them the Aiguille du Goûter, the Dôme du Goûter, the Bosses du Dromadaire, and the highest peak of Mont Blanc; in the foreground are the *Grands-Mulets* (see p. 281), 2½ hrs. distant (guide necessary). — The Aiguille du Midi (12,610') may be ascended from the Pierre-Pointue viâ the Pierre à l'Echelle and the *Col du Midi* (11,810') in about 8½ hrs. (guide 60 fr.); difficult. The *View is very fine. The descent may be made viâ the *Vallée Blanche* and the *Glacier du Géant* to the Montanvert. — A pleasant way back from the Pierre-Pointue is by the **Plan de l'Aiguille** (1½ hr.; no defined path, guide advisable), over grassy slopes and the moraine of the *Glacier des Pèlerins*. We then ascend a little to the *Plan de l'Aiguille* or *La Tapias* (7487'), lying at the foot of the pinnacles of the *Aiguille du Plan* (12,050') and the *Aiguille du Midi* (12,610'). Superb view of the valley of Chamonix, with the Bernese Oberland and Dauphiny Mts. in the distance. We descend by the *Chalets sur le Rocher* to *Les Tsours* (p. 280) and (2 hrs.) *Chamonix*.

Mont Blanc (15,730'), the monarch of European **mountains** (Monte Rosa 15,217', Finsteraarhorn 14,025', Ortler 12,800'; the Pic de Néthou, the highest of the Pyrenees, 11,170'), which since 1860 has formed the boundary between France and Italy, is composed chiefly of Alpine granite or protogine. It was ascended for the first time in 1786 by the guide Jacques Balmat, and by Dr. Paccard **the same year**. In 1787 the ascent was made by the naturalist H. B. de Saussure, with eighteen guides, and described by him with his valuable scientific observations; in 1825 it was accomplished by Dr. E. Clarke and Captain Sherwill, and in 1827 by Mr. Auldjo. In summer the ascent is now made almost daily, but travellers are cautioned against **attempting it in foggy** or stormy weather, as fatal accidents have not unfrequently occurred on the mountain. The view from the summit is unsatisfactory in the common sense. Owing to their great distance, all objects appear indistinct; even in the clearest **weather the outlines** only of the great chains, the Swiss Alps, the Jura, and the Apennines are distinguishable.

According to the regulations of the guides at Chamonix, one traveller ascending Mont Blanc requires two guides (100 fr. each) and one porter (50 fr.), each additional member of the party one guide more; but for experienced mountaineers one guide and one porter suffice. When the 'hotel bill' on the Grands-Mulets and other items are added, the minimum cost of the ascent usually comes to 220-250 fr. for one person. On the first day travellers usually ascend by the *Pavillon de la Pierre-Pointue* (p. 280) to the (7 hrs.) **Grands-Mulets** (10,007'; *Inn* with four rooms; bed, L., & A. 12, B. 3, D. 6, vin ordinaire 4½ fr.); on the second they proceed by the *Petit-Plateau* to the (3 hrs.) *Grand-Plateau* (12,900'), and, bearing to the right (the usual route), ascend by the *Col du Dôme* to the left of the *Dôme du Goûter* to the (1½ hr.) *Cabane des Bosses* (14,327'; erected in 1890-91 by Mr. Vallot; 9 beds), and thence by the *Bosses du Dromadaire* (14,950') and the snowy arête to the (1½ hr.) summit. Another route leads to the left from the Grand-Plateau by the *Corridor*, the *Mur de la Côte*, the *Rochers Rouges*, and *Petits Mulets* in 3-4 hrs. On the top of Mont Blanc is an *Observatory*, built in 1893, which rests entirely upon the snow, as borings failed to find the rock even at a depth of 40 ft. — From St. Gervais (p. 278), the ascent is made by the *Col de Voza* (p. 288) and the *Glacier de Tête-Rouge*, on which was the water-filled cavity the bursting of which caused the catastrophe at St. Gervais in 1892 (p. 289). The night is spent in the (8-10 hrs.) *Cabane* (12,530'), on the S. side of the *Aiguille du Goûter* (12,710'); thence by the *Dôme du Goûter* and the *Cabane des Bosses* (see above) in 5-6 hrs. to the top. — From Courmayeur (p. 292) about 14 hrs.: from the *Combal Lake* (p. 291) across the *Glacier de Miage* to the (7½ hrs. from Courmayeur) *Cabane du Dôme* of the Italian Alpine Club (10,885'), at the foot of the *Aiguille Grise*; thence across the *Glacier du Dôme* and the S.W. arête of the *Dôme du Goûter* to the (5 hrs.) *Cabane des Bosses* and the (1½ hr.) summit. — Another route leads from the Combal Lake across the *Glacier de Miage* and *Glacier du Mont Blanc* to the (8½ hrs. from Courmayeur) *Rifugio Quintino Sella* (11,155'), on the *Rocher du Mont Blanc*, whence the top is attained in 6-7 hrs.; but in the middle of the day this route is exposed to stone avalanches and should be avoided as a descent. — The ascent by the *Glaciers du Brouillard* and *du Fresnay* is very difficult and dangerous. — A most interesting excursion, free from danger, is the ascent of the **Dôme du Goûter** (14,210'; see p. 281), 4-4½ hrs. from the Grands-Mulets; guide from Chamonix 60 fr.

Tour du Mont Blanc, see R. 76.

From Chamonix to Courmayeur over the Col du Géant, 15-16 hrs., a trying glacier-pass, but most interesting, and for adepts not difficult (guide 50, porter 30 fr.). After a night at the *Hôtel du Montanvert* (p. 277) we traverse the upper part of the *Mer de Glace* and the *Glacier du Tacul* or *du Géant*, the jagged 'séracs' of which must be crossed with the necessary precaution. On the right we pass the *Mont Blanc du Tacul* (13,940'), and on the left the *Dent du Géant* (13,160'; p. 292), and in about 6 hrs. reach the **Col du Géant** (11,030'), between the *Aiguille de Saussure* (11,570') on the right and the *Aiguilles Marbrées* (11,605') on the left, with two refuge-huts and splendid view. We then descend almost perpendicular rocks on the S. side to the *Pavillon du Mont Fréty* (p. 292) and Courmayeur. — Other passes over the Mont Blanc Range from Chamonix to Courmayeur (all very difficult, and for thorough adepts only): the **Col de Triolet** (11,455') and the **Col de Taléfre** (11,790'), both at the head (E. end) of the *Glacier de Talèfre*, between the *Aig. de Triolet* and the *Aig. de Talèfre* (guide 50 fr.); the **Col de Pierre-Joseph** (11,415'), to the S. of the *Aig. de Talèfre* (60 fr.); the **Col des Hirondelles** (11,420'), between the *Petites* and the *Grandes Jorasses* (60 fr.); and the **Col de Miage** (11,165'), between the *Aig. de Bionnassay* and the *Dôme de Miage* (60 fr.).

From Chamonix to Sixt over the *Col du Brévent* and the *Col d'Anterne*, see p. 275. Over the *Buet*, see p. 275.

74. From Chamonix to Martigny over the Tête-Noire, or to Vernayaz viâ Triquent and Salvan.

Comp. Maps, pp. 272, 276.

Two Roads and a Bridle-Path connect the valley of Chamonix with the Valais. A road leads from Chamonix by Argentière and Valorcine to (4¼ hrs.) Châtelard, whence one road to the right leads by the Tête-Noire, Trient, and the Col de la Forclaz to (4¼ hrs.) Martigny, and the other to the left to Finhaut, Salvan, and (4 hrs.) Vernayaz. The bridle-path diverges to the right from the road at Argentière, crosses the Col de Balme, and rejoins the road at the Col de la Forclaz. Of these routes the road over the Tête-Noire to Martigny is the most frequented, but is less interesting than that to Salvan and Vernayaz, which affords finer and more varied views. The path over the Col de Balme, on the other hand, though less interesting on the whole, commands a superb view of the valley of Chamonix and Mont Blanc, which are not seen to advantage from the other routes. Travellers from Martigny, approaching Mont Blanc for the first time, should therefore choose the Col de Balme in clear weather.

a. From Chamonix to Martigny viâ the Tête-Noire.

8½ hrs. Diligence from Chamonix to Martigny, or vice versâ, 16 fr. (tickets at Chamonix in the office of the 'Messageries' near the Hôtel Impérial, at Martigny in the Gr. Hôt. du Montblanc). Carriage and pair from Chamonix to Martigny 35-40 fr. (bargain with the driver; return-carriages may sometimes be had); from Martigny to Chamonix 40, 50, 60 or 70 fr. for 1, 2, 3, or 4 persons. — Walkers (guide, of course, superfluous) may send their luggage by the diligence on arrangement at the office.

The road ascends the valley and crosses the *Arve* between *Les Chables* (ascent of the Flégère, see p. 279) and (1½ M.) *Les*

Praz (Hôt.-Pens. du Chalet des Praz; Hôt.-Pens. National, R. 1½, pens. 4½-5 fr.; both good and moderate). The village of *Les Bois* and the *Glacier des Bois* remain on the right. At (1½ M.) *Les Tines* (*À la Mer de Glace; Au Touriste) a path to the Chapeau diverges to the right (p. 278). The road ascends through a wooded defile to (¾ M.) *Lavancher* (3848'; *Hôt.-Pens. Beau-Séjour, 10 min. above the road. R. 2, pens. from 5 fr.); to the Chapeau, see p. 278. About ½ M. farther on a bridge crosses the Arve to *La Jouz*, situated to the left, behind a hill. (Ascent of the Flégère, see p. 279.) We next pass the hamlets of *Les Iles*, *Grasonet*, and (1 M.) *Les Chosalets*, cross the Arve, and reach (¾ M.) —

6 M. **Argentière** (3963'; *Couronne, R., L., & A. 3, déj. 3, D. 3½, pens. 5-7 fr.; *Bellevue*, mediocre), a considerable village, where the huge glacier of that name descends into the valley between the *Aiguille Verte* (13,540') and the *Aiguille du Chardonnet* (12,540').

Glacier d'Argentière. Bridle-path (guide 5, mule 6 fr.) from Argentière to the (2 hrs.) *Pavillon de Lognan* or *du Chardonnet* (6563'; Devouassoud's Inn); ¼ hr. higher we obtain a splendid survey of the grand 'séracs' of the glacier (where ice-avalanches are frequent). In ½ hr. more (guide necessary, usually to be found at the inn) we reach the flat upper part of the glacier, almost free from crevasses (*Mer de Glace d'Argentière*). The middle of it affords a striking view of the surrounding Aiguilles (du Chardonnet, d'Argentière, Tour Noire, Mt. Dolent, Les Courtes, Les Droites, Aig. Verte). We may then ascend the glacier to (3 hrs.) the '*Jardin*' (c805'), a rocky 'islet' at the base of the Aiguille d'Argentière, with fine flora in summer. — EXCURSIONS from the Pavillon de Lognan. *Aiguille du Chardonnet* (12,540'; 7 hrs., with guide) and *Aiguille d'Argentière* (12,833'; 8 hrs., with guide), two difficult ascents. — To ORSIÈRES over the Col du Chardonnet (10,978'; 11 hrs., guide 50 fr.), difficult but very interesting. We ascend the steep Glacier du Chardonnet to (4½ hrs.) the Col, between the Aiguille du Chardonnet and the Aiguille d'Argentière, then cross the *Glacier de Saleinaz* to the *Cabane de Saleinaz* and descend (steep and fatiguing) along the right side of the imposing glacier-fall to *Praz de Fort* and (6 hrs.) *Orsières* (p. 298). — To Orsières over the Col d'Argentière (11,548'; 12 hrs., guide 60 fr.), very difficult. The summit of the pass, which commands a fine view, lies between the *Tour Noire* (12,545') and the *Aiguilles Rouges* (12,026'). The dangerous descent leads across the *Glacier de la Neuva* to the chalets of *La Folly* in the *Val Ferret* (p. 293). — To COURMAYEUR over the Col du Mont Dolent (11,960'; 14 hrs. with guide), between *Mont Dolent* (12,565') and the *Aiguille de Triolet* (12,725), another difficult expedition. The descent leads by the *Glacier du Pré de Bar* to the chalets of that name or to the *Cabane de Triolet* and into the *Val Ferret* (p. 292). — TO THE MONTANVERT over the Col des Grands-Montets (10,680'; 8 hrs., with guide), difficult. The summit of the pass lies between the Aiguille Verte and the Aiguille du Bochard, at the top of the steep *Glacier de la Pendant*. — From the Pavillon de Lognan we may return to the chalets of *Lognan* and *Pendant*, and follow the Chapeau route to (2½ hrs.) *Les Tines* (see above).

Beyond **the** village the new Tête-Noire road ascends to the left in bold windings. Beyond (25 min.) *Tréléchamp* (4593'; Restaurant du Col des Montets) we obtain a fine retrospect of the Glacier du Tour and the magnificent Aiguille Verte. The (¼ hr.) *Col des Montets* (4740') is on the watershed between the Rhone and the Arve.

The road now turns to the W. side of the valley and gradually descends, passing (20 min.) a finger-post which indicates the way to the left to the (25 min.) picturesque *Cascade à Bérard or à Poyaz, in a wild ravine, a digression to which adds $1/2$ hr. to the walk (adm. 50 c.). Through this ravine, the *Vallée de Bérard*, runs the route to the *Buet* (10,200′), the top of which is visible in the background (see p. 275). Our road crosses the ($1/4$ hr.) **Eau-Noire** (Cantine; to the waterfall from this point, $1/4$ hr.).

We next traverse a lonely valley bounded by lofty, pine-clad mountains. Before us rises the *Bel-Oiseau* (8655′). In 10 min. more we reach the first houses of the scattered village of **Valorcine** (pop. 640), the church of which lies to the left farther on. At a (20 min.) Cantine, we have a final retrospect of the summit of Mont Blanc. The valley contracts. The road descends to the Eau-Noire, which dashes over the rocks, and (5 min.) crosses it. The ($1/4$ hr.) *Hôtel de Barberine* (closed) stands at the confluence of the Eau-Noire and the *Barberine*, which forms a waterfall here, and a finer one $1/2$ hr. higher up. We cross (5 min.) the Eau-Noire by a bridge (3684′), the boundary between France and Switzerland, pass the *Hôt. Suisse au Châtelard*, and reach (6 min.) the *Hôtel Royal du Châtelard* (burned down in 1886), where the two routes to the Rhone Valley separate: to the right the road over the Tête-Noire to Martigny; to the left the road viâ Salvan to Vernayaz (see below).

The Martigny road crosses the (5 min.) Eau-Noire. The once dangerous *Mapas (mauvais pas)* descends to the left, while the **new road** leads high above the deep and sombre valley, being hewn in the rocks of the (40 min.) **Tête-Noire** or *La Roche-Percée*. We next reach (10 min.; from Argentière 3 hrs.) the *Hôtel de la Tête-Noire* (4003′). A wooden belvedere, which we pass 2 min. before the inn, affords a fine survey of the grand gorge of the Eau-Noire.

A steep path descends by the inn to the left to the (20 min.) Gouffre de la Tête-Noire, a ravine of the *Trient*, with a waterfall and a natural bridge ('*Pont Mystérieux*'). Tickets at the inn (1 fr., incl. guide). The steep ascent back to the hotel requires 25-30 minutes. — A path leads direct from the ravine to Finhaut (see below).

The road **here** turns to the right into the sadly thinned forest of Trient, skirting the base of the Tête-Noire. In the valley, far below, is the brawling *Trient*, which joins the Eau-Noire a little farther down. In $1/2$ hr. we reach the village of **Trient** (4250′; *Hôt.-Pens. des Alpes*; *Hôt. du Glacier de Trient*, mediocre), a little beyond which the road is joined by the path from Chamonix over the Col de Balme (p. 286). At the end of the valley rises the *Aiguille du Tour* (11,585′) with the fine *Glacier de Trient* (p. 286).

From Trient the road ascends somewhat steeply to the (40 min.) *Col de Trient*, better known as **Col de la Forclaz** (4997′; two inns, see p. 286). The view hence is limited, but $1/2$ hr. lower down we enjoy a noble survey of the Rhone Valley as far as Sion. At our feet lies *Martigny*, reached in $2^1/4$ hrs. by the road (p. 286), or in $1^1/2$ hr. by the steep old path. — 6 M. *Martigny*, see p. 249.

to Martigny. FINHAUT. *V. Route 74.* 285

b. From Chamonix to Vernayaz viâ Finhaut and Salvan.

7³/₄ hrs. Road to Châtelard, 3³/₄ hrs. Thence to Vernayaz in 4 hrs. by a route, practicable only for light vehicles, but more picturesque than the preceding (see p. 282). Carr. for 1 or 2 pers. 50 fr. — In 1894 a service of diligences like the Tête-Noire diligences plied on this route (fare 16 fr.).

To *Le Châtelard*, see p. 284. The narrow road ascends from the ruins of the Hôtel Royal (p. 284) to the left, partly by zigzags, for 40 min., turns to the right at a cross, and continues at nearly the same level. — ³/₄ hr. (1 hr. 25 min. from Le Châtelard) **Finhaut** or *Fins-Hauts* (4060'; *Hôt.-Pens. Beauséjour*, 4¹/₂-5¹/₂ fr.; *Hôt.-Pens. du Bel-Oiseau*, 5-7 fr.; *Pens. de la Croix*; *Hôt.-Pens. du Montblanc*; *Pens. de la Croix Fédérale*, unpretending), beautifully situated. Engl. Church Service in summer.

A path (the beginning of which should be asked for) leads hence direct to the (1 hr.) Tête-Noire Inn. It descends steeply to a wooden bridge over the Eau-Noire, crosses it, ascends to the right, and passes several houses, where, if necessary, a boy may be found to show the way, to the Hôtel de la Tête-Noire (p. 284). — From Finhaut we may ascend to the W. by a good path to the (2 hrs.) *Col de la Gueula* (6380'), to the S. of the Bel-Oiseau (see below), where we enjoy a splendid view, across the Barberine Valley, of Mont Blanc, the Glacier de Trient, etc., and to the E. of the Bernese Alps. We may descend into the Barberine Valley to *Emosson* and skirt the shoulder of the *Perron* (8890'), passing the picturesque falls of the *Barberine* (p. 284). to *Châtelard*; or we may re-ascend from Emosson to the *Col de Tanneverge* (8133') and descend to *Sixt* (p. 275). — The ascents of the *Bel-Oiseau* (8665'; from Finhaut 4 hrs., with guide); *La Rionda* (7800'; 3 hrs., with guide), and *La Rebarmaz* (8115'; 3¹/₂ hrs., with guide) are interesting and not difficult.

Ascending a little, then level again, the road passes (¹/₄ hr.) a Cantine (continuous fine view), descends through wood in many windings, and leads along the slope of the hill, past the hamlet of *Triquent* (3260'; Hôt.-Pens. du Mont-Rose; Hôt.-Pens. de la Dent-du-Midi, pens. 3¹/₂-5 fr.), to the (1 hr.) *Gorges du Triège* (restaurant at the bridge), with its picturesque waterfalls framed with rocks and dark pines (rendered accessible by wooden pathways; 1 fr.). The road gradually ascends for the next 20 min., and then descends, between interesting marks of glacier striation and past the *Hôt.-Pens. de la Creusaz* (4 fr.), to (¹/₂ hr.) **Salvan** (3035'; *Hôt.-Pens. de la Dent du Midi*, R. 2-2¹/₂, pens. 4-5 fr.; *Grand-Hôt. de Salvan*, R. 2¹/₂, B. 1¹/₂, D. 3, pens. 6-7 fr.; *Hôt. Bellevue*; *Union*, moderate; Engl. Church Service in summer). In the village is a large erratic boulder; interesting prehistoric sculptures.

To the *Cascade du Dailley*, a fine fall of the *Salanfe*, a good path leads in 40 min. by the hamlet of *Les Granges*, on the slope facing the Rhone Valley. The finest point of view is opposite the fall. Lower down the Salanfe forms the Pissevache Fall (p. 249). — A fine view of Mont Blanc, the Grand Combin, etc. may be obtained from the *Mayens de la Creuse* (5790'; 2¹/₂ hrs., with guide), to which an attractive path (suited for riding) leads through wood. — The Luisin (9140'; 6 hrs. from Salvan; with guide), ascended by the *Alp* and *Col* or *Chieu d'Emaney* (7960'), affords a superb view of the Alps of Savoy, the Valais, and Bern. Descent in 5 hrs., by *Salanfe* and *Van* (p. 260). — The ascent of the Dent du Midi (10,695'; 8 hrs., with guide), a difficult but attractive expedition, leads viâ *Les Granges* and *Van d'en Haut* to the (3 hrs.) *Alp Salanfe* (6215'; night-quarters), whence the W. summit is reached in 5 hrs. viâ the *Col de Clusanfe*, where our route is joined by that from Champéry p. 259). The E. summit is more difficult (recommended to experts). — Tour (Sallières (10,587'; 7-8 hrs., with guide), laborious, for experts only; the night is spent at the (2½ hrs.) *Emaney Alp* (6072'), whence the summit is reached by the (2 hrs.) *Col de Barberine* (8136') in 3-4 hrs. Descent to *Champéry* (p. 260).

From Salvan a good road, shaded by chestnut and walnut-trees and crossing the stream about 50 times, descends the steep slopes in thirty windings to (3/4 hr.; up 1½ hr.) *Vernayaz* (rail. stat., p. 249).

Pedestrians have an agreeable alternative to the road viâ Salvan in the so-called 'NOUVEAU CHEMIN', which leads on the right bank of the Trient from Vernayaz to the Tête-Noire (3 hrs.; guide advisable for novices). The path, beginning at the exit of the *Gorges du Trient* (p. 249), ascends the cliffs to the left to the hamlet of (½ hr.) *Gueuroz* (2200'), and continues through beech-woods to (1/4 hr.) *La Taillat*, whence a footpath diverges through the Gorge de Trient to (½ hr.) Salvan, joining the Finhaut and Vernayaz road beside the Maison de la Commune at Salvan. Thence the path mounts steeply to (3/4 hr.) the prettily situated *La Crête* (3385'; simple fare), and then keeps along the level viâ *Plan à Jeur* to (1 hr.) *L'Itroz* (3880'), lying high above the junction of the Trient and Eau-Noire. We descend to the left into the valley of the Trient, cross the stream by a timber-bridge, and ascend to the road, which we strike a little above the (3/4 hr.) *Tête-Noire Hotel* (p. 281).

75. From Martigny to Chamonix. Col de Balme.

Comp. Maps, pp. 272, 276.

10 hrs. From Martigny to the Col de Balme 6, thence to Chamonix 4 hours. Road from Martigny to Trient, and from Le Tour to Chamonix. Guide (12 fr.) unnecessary, if the following directions be observed. Luggage may be sent on by arrangement at the diligence-office (comp. p. 282). Horse or mule and attendant 24 fr.; but from the Col to Le Tour the path is unfit for riding.

Martigny, see p. 249. We follow the Great St. Bernard road through the long village of *Martigny-Bourg* (p. 250) to the (1½ M.) *Drance Bridge* (1640'), and (4 min.) reach the hamlet of *La Croix*. A notice on a house here indicates the road to Chamonix, ascending to the right, through vineyards, orchards, and meadows, in numerous windings, which the rugged old path cuts off: 20 min. *Les Rappes*; 25 min. *La Fontaine*; 35 min. *Sergnieux* (3820'); 1/4 hr. *Le Fay*. The road here takes a wide bend to the right, which the old path cuts off. By the (3/4 hr.) *Chalet de Bellevue* we enjoy a fine retrospective survey of the Rhone Valley. Then (20 min.) *Les Chavans* (auberge), and an ascent of 40 min. more to the Col de la Forclaz (4997'; *Hôtel Gay-Descombes*, R., L., & A. 2, D. 2½-3 fr.; *Restaurant Fougère*, 2 min. farther on, plain), 3½ hrs. from Martigny.

From the pass a **nearly** level path, lately damaged in several places, leads to the (1½ hr.) "Glacier de Trient (lower end 5560'), the northernmost glacier of the Mont Blanc range (good view from a point about ½ hr.'s climb up the left side). Over the *Fenêtre d'Arpette* (8800') to *Champex*, 5½-6 hrs., with guide, see p. 299. — *Mont d'Arpille* (6830'), ascended in 1½ hr. from the Col de la Forclaz, see p. 250.

After a descent of 1/4 **hr.** the bridle-path to the Col de Balme diverges to the left from the Tête-Noire road (p. 284), and in 10 min. crosses a bridge opposite the upper houses of *Trient* (p. 284). We now ascend the meadows to the left (with the *Glacier de Trient* to the left, see above) and (20 min.) cross the *Nant-Noir* ('nant', probably from *natura*, being the Savoyard word for a torrent), which descends from the *Mont des Herbagères*. We follow the right bank for about 200 paces, and then mount to the left in steep zigzags through the *Forest of Magnin*, which has been thinned by avalanches. After 1 hr. the path becomes more level, passes (¼ hr.) a cantine and (¼ hr.) the chalets of *Zerbazière* (6660′), and (½ hr.) reaches the *Col de Balme (7225′; *Hôtel Suisse*), 6 hrs. from Martigny, the boundary between Switzerland and France. This point commands a superb view of the whole of **the Mont Blanc** range: the Aiguilles du Tour, d'Argentière, **Verte, du Dru, de Charmoz, and du** Midi, Mont Blanc itself, and the Dôme du Goûter; and also of the valley of Chamonix as far as the Col de Voza. On the right are the Aiguilles Rouges, to the left of them the Brévent, and to the right the snow-clad Buet. In the opposite direction, over the Forclaz, we survey the Valais and the mountains which separate it from the Bernese Oberland, the Gemmi with its two peaks, the Finsteraarhorn, Grimsel, and Furka.

A still finer *View is obtained from *La Balme* (7590′), the second eminence to the right, with a wooden cross, about ¼ hr. to the N.W. of the inn, at the foot of the *Croix de Fer* or *Aiguille de Balme* (7677′), the last spur of the hills which rise abruptly above the Col de Balme. From this point Mont Blanc looks still grander; to the N.E. we see the entire chain of the Bernese Alps, and to the E., at our feet, lies the Tête-Noire ravine, with the Dent du Midi rising beyond it. The descent may be begun immediately from this point. The ascent of the Aiguille itself is recommended to good climbers (1 hr., with guide).

From the Col de Balme to the Tête-Noire (2½ hrs.; no guide required in fine weather), fatiguing but interesting, and recommended to the traveller who desires to visit both these points in one day either from the Rhone Valley or from Chamonix. The views are less striking in the reverse direction. To the W. of the Col, behind the above-mentioned eminence with the cross, a narrow path leads nearly to the (10 min.) brink of the Tête-Noire Valley, and then becomes indistinct. We turn to the right (N.) and follow a slight depression for a few minutes until a number of heaps of stone become visible, to the right of which the path re-appears. The chalets of *Catogne* (6570′) are left to the right as we descend. The path next crosses the stream, and descends abruptly along the right bank to a lower plateau of the mountain, then bends to the N.E., and reaches (40 min.) the chalets of *Grangettes*. Beside the most northerly chalet, beyond the stream, are two boulders, conspicuous by their light colour, between which the path descends to the N., steep and stony at places, but henceforward easily traced, to the scattered chalets of *Les Jeurs* and (1¼ hr.) the *Tête-Noire* (p. 284).

From the Col de Balme to Orsières over the Col du Tour (10,990′; 11-12 hrs., with guide), a fatiguing route, suited for adepts only. The route skirts the cliffs of the *Grands Autannes* to the *Glacier du Tour*, over which it leads to the pass, between the *Aiguille du Tour* (11,605′) and the *Petite Fourche* (11,605′; both ascended from the pass; fine views). The descent is made viâ the *Glacier de Trient* and the *Glacier d'Orny* to the *Cabane d'Orny*, and thence to *Som la Proz* and *Orsières* (p. 299).

The path, now rough and steep, descends over pastures carpeted with rhododendrons and other Alpine flowers. On the right flows the *Arve* (p. 277), which rises on the **Col de Balme**. We cross several small brooks, pass (¾ hr.) a heap of stones and (¼ hr.) a second heap, resembling a hut without a roof, and reach (1/4 hr.) *Le Tour* (4695'), to the left of which is the fine *Glacier du Tour*. Carriage-road hence to Chamonix (7 3/4 M.; carr. with one horse 6, with two 9-10 fr.; those who intend to drive should take a carriage here if possible). The fragments of slate brought down by the Arve are carefully collected by the peasants, who cover their fields with them in spring, thus causing the snow under them to melt several weeks earlier than would otherwise be the case. About 1/2 M. beyond Le Tour we cross the *Buisme*, which drains the Glacier du Tour, and (1 M.) the Arve, and soon reach (1/4 M.) *Argentière* (p. 283).

76. From Chamonix to Courmayeur over the Col du Bonhomme and the Col de la Seigne.

Comp. Maps, pp. 272, 276.

BRIDLE-PATH. Three days: 1st, to Contamines 5 3/4 hrs. (or to Nant-Borant, best night-quarters, 7 1/2 hrs.); 2nd, to Mottets from Nant-Borant, 5 1/2 hrs. viâ the Col des Fours, or 6 1/2 hrs. viâ Chapieux; 3rd, to Courmayeur 8 1/2 hrs. — Good walkers may reach Courmayeur from Nant-Borant in one day. Or, omitting the Col de Voza, we may drive from Chamonix viâ St. Gervais to Contamines or to Notre Dame de la Gorge, in which case Mottets is easily reached on the first day and Courmayeur on the second. — Guide (not needed by good walkers in fine weather, but advisable for others, especially over the Col des Fours) from Chamonix to Courmayeur in two days 20, in three days 24 fr; return-fee 16 fr. extra.

The *Tour of Mont Blanc*, as this route is called, is easy and interesting. To complete our circuit of Mont Blanc, we may return to Martigny over the Great St. Bernard or over the Col Ferret; good walkers proceed from Aosta to Châtillon, and cross the Théodule Pass to Zermatt (in the opposite direction, beginning from Zermatt, the route is less interesting). A passport will be found convenient in satisfying the enquiries of the Italian and French custom-house officers.

We follow the Geneva road (p. 273) from Chamonix to (3 1/2 M.) the hamlet of *La Griaz*, turn to the left at a large iron cross, and cross the deep bed of the *Nant de la Griaz* to (3/4 M.) *Les Houches* (Hôt. du Glacier, poor), with a picturesquely situated church. A few paces beyond the church, and on the other side of the brook (guide-post), a tolerable footpath (hardly to be mistaken) diverges to the left, enters the (1/2 hr.) wooded ravine to the right, and ascends in 1 1/2 hr. to the **Pavillon de Bellevue** (5947'), a rustic inn on a saddle of *Mont Lachat* (see below), affording a superb *View (best by evening-light) of the Chamonix Valley as far as the Col de Balme, the Mont Blanc range (summit hidden by the Dôme du Goûter), and the valley of the Arve.

Another path (easier at first, but disagreeable after rain) diverges by a cross 18 min. beyond Les Houches, and ascends in 1 1/2 hr. to the Col de Voza (5495'; Inn closed; simple refreshments in the chalet), a depression between *Mont Lachat* (6926') and the *Prarion* (p. 273), 20 min. to the W. of the Pavillon de Bellevue, with a fine view, but inferior to that from the Bellevue. We may descend either on the right bank of the stream by *Bionnassay* to Contamines, or by a better and shorter route on the slopes to the left to the under-mentioned bridge over the Bionnassay, where we join the route from the Pavillon de Bellevue, and thence along the left bank.

From the Pavillon de Bellevue the path descends to the S. over pastures (the *Aiguille de Bionnassay*, 13,360', rising on the left) and crosses the stream issuing from the *Glacier de Bionnassay* below the chalets **near** the end of the glacier. The flood from the Glacier de Tête Rouge which destroyed St. Gervais-les-Bains in 1892 (p.273) descended by the course of this stream. Our route, now a tolerable bridle-path, descends on the left side of the valley to (1¼ hr.) *Champel* and turns to the left by the fountain. We descend rapidly, enjoying a fine view of the wooded and well-cultivated *Montjoie Valley*, bounded on the W. by the slopes of *Mont Joli* (see below), with the *Mont Roselette* (8825') in the background, while to the E., above the green lower hills, peep several of the W. snow-peaks of the Mont Blanc group (*Aig. de Tricot, de Trélatête*, etc.). Beyond (18 min.) *La Villette* the path joins (6 min.) the carriage-road from St. Gervais (p. 273), which we follow to the left. The road crosses the brook descending from the *Glacier de Miage* just before the hamlet of *Tresse* (to the right, on the slope of Mont Joli, stands the church of *St. Nicolas de Véroce*). The road then ascends high on the right bank of the *Bon-Nant* to *La Chapelle* and (1 hr.) —

Les Contamines (3927'; *Union*, R., L., & A. 3½, B. 2 fr.; *Hôt. du Bonhomme*, well spoken of), a large village with a handsome church.

The *Mont Joli (8290') is ascended from *St. Nicolas* (see above) without difficulty in 3 hrs. (guide 6 fr.; auberge ¾ hr. from the top). Splendid view of Mont Blanc. — The *Pavillon de Trélatête* (see below) is more easily reached from Contamines than from Nant-Borant (path ascending to the left, 20 min. above Contamines). From Contamines by the Pavillon de Trélatête to Nant-Borant, 3 hrs., interesting. — From Contamines over the *Col Joli* to *Beaufort*, see p. 268.

Beyond Contamines the road descends to the hamlet of *Pontet*, and overlooks the valley as far as the peaks of the Bonhomme. The valley contracts. At (1 hr.) the bridge which crosses to the pilgrim-age-chapel of *Notre-Dame de la Gorge* the road ends.

The bridle-path now ascends to the left, passing a bridge and frequent traces of glacier-friction. Then through wood, past two waterfalls, and (½ hr.) across the deep gorge of the Bon-Nant; 10 min. **Chalets of Nant-Borant** (4780'; *Inn*, R. & A. 3-4, D. 3 fr.). We cross the wooden bridge to the left, and traverse the pastures by a somewhat stony path. On the left the *Glacier de Trélatête* and the *Col de Béranger* are visible; looking back, we survey the valley as far as the Aiguilles de Varens (p. 273).

From Nant-Borant, or better from Contamines (see above), we may reach Mottets or the Col de la Seigne in 7 hrs. by the Col **du Mont Tondu** (10,130'); trying, but without danger (guide 30 fr.). From Nant-Borant we ascend to the left (fine waterfalls) to the (1½ hr.) *Pavillon de Trélatête* (6483'; Inn, well spoken of), which overlooks the *Trélatête Glacier*, and mount the glacier towards the S.E. to the pass, to the left of *Mt. Tondu* (beautiful view, especially from a height on the left). We may either descend to the right to *Mottets* (p. 290), or to the left over shelving rocks and across the *Glacier des Lancettes* or *des Glaciers* to the *Col de la Seigne* (p. 291). — Over the Col de Trélatête (11,424'), immediately to the S. of the Aiguille de Trélatête, to the *Glacier de l'Allée Blanche* and *Combal Lake* (p. 291), very difficult (2 guides, 60 fr. each).

BAEDEKER, Switzerland. 16th Edition. 19

290   V. Route 76.   COL DU BONHOMME.   *From Chamonix*

We next reach (50 min.) the **Chalet à la Balme** (5627'), a plain inn, situated at the head of the Montjoie Valley.
In doubtful weather, or if evening is approaching, a guide should be taken from this point to the summit of the pass (8 fr.); but, as guides are not always to be had here, it is safer to engage one at Contamines (to the Col du Bonhomme 6-8, Col des Fours 6-8, Chapieux 8-10, Mottets 10-12 fr., the higher fees being charged when the guide cannot return the same day). If the guide be taken to the Col du Bonhomme only, his attendance should be required as far as the highest point (Croix du Bonhomme, see below). Mule from Nant-Borant to the Croix 8 fr.

The path, indicated by stakes, ascends wild, stony slopes, passing a waterfall on the left, to the ($1/2$ hr.) *Plan Jovet* (6437') with a few chalets. (To Mottets over the Col d'Enclaves, see below.) On the ($1/2$ hr.) *Plan des Dames* (6543') rises a conical heap of stones, where a lady is said to have perished in a snow-storm. At the end of the valley (20 min.) the path ascends the **slope to the** right, and (25 min.) reaches the **Col du Bonhomme (7680')**, whence we look down into the desolate valley of the *Gitte*.
A path, at first ill-defined, descends into this basin, passes the lonely *Chalet de la Sauce*, turns to the left and crosses the brook, and leads to (2 hrs.) the chalets of *La Gittaz* and to *Beaufort* (p. 268) in $3^1/_2$ hrs. more. Guide to La Gittaz advisable.

Two curious rocks, the *Rochers du Bonhomme* and *de la Bonnefemme*, here tower aloft, like two ruined castles. Beyond these we follow the rocky slope to the left (path indicated by stakes), passing an excellent spring (good resting-place), and next reach (40 min.) the **Croix du Bonhomme** (8153'), with a fine view of the mountains of the Tarentaise, in the centre of which rises the beautiful snow-peak of *Mont Pourri* (12,425'). Here the path divides. In a straight direction the path descends, partly over loose stones, to ($1^3/_4$ hr.) —

Les Chapieux or *Chapiu* (4950'; *Soleil; Hôt. des Voyageurs*), an Alpine hamlet in the *Val des Glaciers*, $1^3/_4$ hr. below Mottets.
FROM CHAPIEUX TO PRÉ-ST-DIDIER over the *Little St. Bernard* (11 hrs.; preferable to the Col de la Seigne in doubtful weather). The path to (3 hrs.) *Bourg-St-Maurice* (p. 294), at first very stony, but afterwards better, passes the chalets of *Le Crey* and *Bonneval*, commanding a beautiful view of the upper Isère Valley (Tarentaise), and at length unites with the high-road. From *Bourg-St-Maurice* to *Pré-St-Didier*, see p. 294.

The direct route to Mottets ($2^1/_2$ hrs.) ascends **from the Croix du Bonhomme to the** left, rarely free from snow (guide advisable for less experienced travellers) to the (35 min.) **Col des Fours** (8695'), to the right of which is the *Pointe des Fours* (8920'; 10 min.), a splendid point of view. Then a steep descent over slate-detritus, and over pastures to ($1^1/_4$ hr.) a group of chalets (6573') and the (20 min.) *Hameau du Glacier*, where the path from Chapieux comes up from the right. We descend to the left, cross the bridge (5840'), and ascend the left bank to (20 min.) the two houses of —

Mottets (6227'; *Veuve Fort's Inn*; mule to the Col de la Seigne, 6 fr.), at the head of the *Val des Glaciers*. To the N. rises the *Aiguille du Glacier* (12,520'), with the extensive *Glacier des Glaciers*.
Over the *Col du Mont Tondu* to *Contamines*, see p. 289. Another route to Mottets (4 hrs. from Nant-Borant; shorter, but trying) is from the *Plan*

to Courmayeur. COL DE LA SEIGNE. *V. Route 76.* 291

Jovet (p. 290), past the small lake of that name, and over the Col d'Enclaves (8810'), between Mt. Tondu and the Tête d'Enclaves.

A bridle-path ascends hence in zigzags to the (1³/₄ hr.) **Col de la Seigne** (8240'), where a cross marks the frontier between France and Italy. Magnificent *View of the **Allée Blanche**, an Alpine valley several miles long, bounded on the N.W. by the tremendous precipices of the Mont Blanc chain.

To the left of the pass rise the *Aig. du Glacier* (12,523') and *Aig. de Trélatête* (12,900'); then the imposing snowy dome of *Mont Blanc*, borne by the huge rocky buttresses of the *Rocher du Montblanc*, near which is the *Mont Maudit*; farther on, to the left of the *Aig. d'Estelette*, towers the bold and isolated *Aig. Blanche de Peuteret* (13,490'), ascended for the first time in 1885 by Mr. Seymour King. Farther to the right, in the distance, rise the *Mt. Velan*, *Grand Combin*, etc. In the valley lies the green Lac de Combal. The retrospective view of the Tarentaise Mts. is also fine, but it cannot compete with the imposing scene just described.

Beyond the pass we descend over snow and débris, keeping to the left, then across pastures, to the (¹/₂ hr.) upper *Chalets de l'Allée Blanche* (7230'; occupied for a few weeks in the height of summer only), and the (25 min.) lower chalets (7135'), at the end of a level plateau. We round the hill to the right, cross the brook, with a splendid view of the imposing *Glacier de l'Allée Blanche*, and descend to a second level reach of the valley, at the end of which (³/₄ hr.) lies the green **Lac de Combal** (6365'), bounded on the N. by the huge moraine of the *Glacier de Miage*. Near a sluice at the lower end of the lake (10 min.) we cross the *Doire*, which issues from the lake, and descend the side of the moraine through a wild ravine, filled with fragments of rock. (The Miage Glacier is not visible.) After 40 min. the Doire is again crossed. The valley, now called *Val di Veni*, expands. The *Cantine de la Visaille* (5420'), about 5 min. farther on, commands a splendid survey of the valley, with the Jorasses and the Dent du Géant towering on the left.

The path descends through wood and pastures, passing (³/₄ hr.) the *Chalet de Pertud* (4945', on the left bank). On the left is the fine *Glacier de la Brenva*, which once filled the whole valley, but has receded greatly within the last few decades. Beyond the (20 min.) *Chalet de Notre-Dame de Guérison*, a little below the exit from the wood, we have a comprehensive view of the Brenva glacier; on the left is the Aiguille de Peuteret with the snowy summit of Mont Blanc towering above it; on the right the pavilion on the Mont Fréty (p. 292) and the tooth-shaped Dent du Géant (p. 292). By the chapel of *Notre-Dame de Guérison* or *de Berrier* (4710'), a few minutes farther on, the path rounds an angle of rock, overlooking the village of *Entrèves* to the left, at the mouth of the *Val Ferret*, and then descends to the Doire, which unites here with the Doire du Val Ferret and takes the name of *Dora Baltea*. Opposite the little sulphur-baths of *La Saxe* (¹/₂ hr.) we cross the Dora, pass the (¹/₄ hr.) *Hôtel du Montblanc*, and in 10 min. more reach —

19*

292 V. Route 76.   COURMAYEUR.

Courmayeur. — *Hôtel Royal*, *Angelo, in both R., L., & A. 5-6, B. 1½, déj. 3½, D. 5 fr.; *Union, R., L., & A. 3, B. 1½, déj. 2½, D. 4½, pens. 8-10 fr.; *Mont Blanc, ½ M. to the N. of the village, R. & A. 2½, D. incl. wine, 4 fr. — *Restaurant Verney* (also rooms); *Café du Montblanc*. — Diligence to Aosta, see p. 293; one-horse carriage to Aosta 15, two-horse 25 fr. (return vehicles cheaper). As at Chamonix, there is a society of guides here with similar regulations (see p. 276). *L.* and *Julien Proment, G. Petigax, J. M. Lanier, J. Gardin, Al. Berthod, P.* and *A. Puchoz, J.* and *L. Croux*, and *P. Revel* are recommended.

Courmayeur (3963'; 600 inhab.), a considerable village, with mineral springs, beautifully situated at the head of the Aosta Valley, is much frequented by **Italians in summer.** Though higher than Chamonix, the climate is warmer and the vegetation far richer. The highest peak of Mont Blanc is concealed from Courmayeur by the *Mont Chétif* (7685'), but is seen from the Pré-St-Didier road, ½ M. to the S.

Excursions. From the hamlet of **Dollone**, opposite Courmayeur, at the base of Mont Chétif, we obtain an excellent survey of the enormous precipices of the Jorasses and the glacier of that name. Pleasant **walk** thither, crossing the *Dora Bridge* (10 min.); then through the village, down to the Dora by a shady path at its N. end, and back by the left bank (1½ hr.). — A bridle-path (guide unnecessary) leads from Dollone to the W. to the (2 hrs.) Col de Chécouri (6337'), on the S.W. side of the Mont Chétif (see above), commanding a fine view of Mont Blanc. We may return by the Allée Blanche, see p. 291.

The *Mont de Saxe (7735'; 2½-3 hrs.; guide, 6 fr., unnecessary) affords a complete view of the S.E. side of Mont Blanc with its numerous glaciers, the Col du Géant and the Jorasses being close to us. A good bridle-path ascends from Courmayeur, by *La Saxe* (p. 291) and *Le Villair*, to the (2 hrs.) *Chalets du Pré* (6670') and the (1 hr.) nearer peak. The descent may be made by the *Chalets de Leuchi* into the Val Ferret.

The *Crammont* (9080'), commanding a grand view of Mont Blanc, is more conveniently ascended from Pré-St-Didier (see p. 293).

Ascent of *Mont Blanc* (14 hrs. from Courmayeur), see p. 281. — The **Grandes Jorasses** (13,800'), 14 hrs., with 2 guides, are difficult, and dangerous after fresh snow. Diverging at (1¼ hr.) *Planpansière* from the Ferret route (see below), we ascend through wood, over grassy slopes, snow, and rocks (extremely steep and difficult at last), to the (5½ hrs.) *Cabane des Grandes Jorasses* of the Italian Alpine Club (9515'), whence we reach the summit by the *Rocher du Reposoir* in 7-8 hrs.

To Chamonix over the Col du Géant (comp. p. 282), 14-15 hrs. (guide 50, porter 30 fr.; two guides, or a guide and a porter required). Interesting excursion (bridle-path, 3 hrs.) to the **Pavillon du Mont Fréty** (7130'; small inn; fine view); thence to the *Col du Géant* (11,030'; **two** refuge-huts), with most magnificent view, a steep ascent of 3½ hrs. (guide to the Pavillon 6 fr., unnecessary; to the pass and back 12, in two days 15 fr.). The ascent of the *Aiguille* or *Dent du Géant* (13,160'), from the Col du Géant in 7-8 hrs., is very difficult (first ascended by the brothers Sella in 1882).

From Courmayeur to Martigny over the Col Ferret (14 hrs.), bridle-path (guide to the Chalets de Ferret advisable, 15 fr.). From *La Saxe* (p. 291) **we** follow the left bank of the *Dora* (leaving the village of Entrèves on the left) to the chalets of (1¼ hr.) *Planpansière*; **we** then **cross** the *Doire du Val Ferret*, and ascend on its right bank. By the (1 hr.) chalets of *Prax-Sec* (5336') we again cross the stream. (The path on the right bank is soon lost among the huge debris of a moraine.) We now ascend the steep and narrow Val Ferret, passing the poor huts of *La Vachey* (5382'), *Péraché* (5795'), *Gruetta* (5782), and *Sagivan* (6370'); to the left are the moraines of the *Glacier de Triolet*, and high up on the rocks of the *Mont Rouge* is the *Cabane de Triolet* of the C. A. I. (8475'), the starting-point for the difficult *Aig. de Triolet* (12,725'; 8-9 hrs.), the *Col du Mt. Dolent* (p. 283), etc. The last chalets are **those** of (2½ hrs.) *Pré de Bar* (6756';

rĩnts.), at the base of the glacier of that name (p. 283), which descends from *Mont Grapillon* or *Mont Dolent* (12,565'). The bridle-path ascends to the right in numerous windings to the (1½ hr.) Col Ferret, or *Col de la Peulaz* (8323'), the frontier of Switzerland and Italy, with a superb view of the Val Ferret and the S. side of the Mont Blanc group with its huge glaciers (de Triolet, etc.), of the Jorasses, the Aiguille du Géant, and the Allée Blanche as far as the Col de la Seigne. [Another pass, called the *Pas de Grapillon* or *Col du Petit Ferret* (8179'), farther to the N., close to the foot of the precipices of Mont Dolent, is shorter, but more fatiguing and devoid of view.] We descend to the (1 hr.) *Chalets de la Peulaz* (6843'), below which we cross the *Drance* and (½ hr.) reach the Col de Fenêtre route. (From this point to the St. Bernard Hospice 4-4½ hrs.; comp. p. 303.) The path then descends to the left to the (½ hr.) *Chalets de Ferret* (5566'; cabaret, with a few beds, clean and moderate), and through the N. (Swiss) *Val Ferret* or *Ferrex* to (½ hr.) *La Folly* (5240'), with the *Glacier de la Neuva* above it, on the left (p. 283). Then (½ hr.) *La Seiloz* (4920'; small Inn), (1¼ hr.) *Praz de Fort* (where we reach the road), *Ville d'Issert*, *Som la Proz*, and (1¼ hr) *Orsières* (p. 299). Good walkers starting from Courmayeur at 3 or 4 a. m. may catch the afternoon-diligence for Martigny at Orsières.

77. From Courmayeur to Aosta and Ivrea.

68 M. From Courmayeur to (21 M.) *Aosta*, an OMNIBUS (6 fr.) plies thrice a day in July & Aug. in 4 hrs. (In the reverse direction 5 hrs.), starting (1895) at 6 a.m. and 1 and 5 p.m., returning from Aosta at 6 and 11.30 a.m. and 8.30 p.m. (fare 3 fr., banquette 3½ fr.); one-horse carr. 18, two-horse 30 fr. From Aosta to (42 M.) *Ivrea*, RAILWAY in 2½ hrs. (fares 7 fr. 60, 5 fr. 30, 3 fr. 45 c.). The railway, a fine example of engineering enterprise, traverses a highly picturesque district.

Courmayeur, see p. 292. — The road to Aosta (21 M.; walking not recommended) winds down to the Doire and follows its left bank through a wooded ravine. (Walkers will prefer the old road, with fine views, on the hillside to the left, descending to the new road below Pré-St-Didier.) Passing (2¼ M.) *Palesieux*, we cross the *Doire* to (¾ M.) **Pré-St-Didier** (3280'; *Hôt. de l'Univers*; *Restaurant de Londres*), a picturesquely situated village with baths, where the road to the *Little St. Bernard* diverges to the right. Near the hot springs (¼ M. lower) the *Thuile* forces its way between perpendicular rocks towards the Dora valley.

EXCURSIONS. (Guides: *Jos. Barmaz, F. Brunod*, and others). The ascent of the *Crammont* (9080'; 3½ hrs.) is highly interesting. Following the St. Bernard road to the first tunnel (½ hr.; shorter footpath in 20 min.), we thence ascend to the right to the (½ hr.) hamlet of *Chanton* (5970'), whence we reach the summit in 2½ hrs. more. Splendid view of Mont Blanc and the Graian Alps. About 5 min. below the top is the *Pavillon De Saussure*, a refuge-hut of the C. A. I. Another route (bridle-path) diverges to the right from the St. Bernard road at *Elevaz*, 3 M. from Pré-St-Didier, joining the above route before the final ascent. Experts may dispense with a guide.

To BOURG-ST-MAURICE OVER THE LITTLE ST. BERNARD, 8 hrs., a route preferred by some to that over the Col de la Seigne (p. 291). The fine new road ascends the valley of the *Thuile* viâ *La Balme* to (2 hrs.) *La Thuile* (4726'; two small Inns), where we have a view of the great glacier of the *Rutor* (11,435'), which may be ascended hence (2 hrs. to the S. are the beautiful *Rutor Waterfalls*). Thence the road ascends, passing (1¼ hr.) *Pont Serrand* (5416') and the *Cantine des Eaux-Rousses* (6740'), to the (¾ hr.) pass of the Little St. Bernard (7176'). The boundary between France and Italy is on the S. side, about ¼ hr. beyond the summit, and near a *Hospice* (7060') affording good accommodation. [The *Mt. Valaisan* (9453'), 3½ hrs. to the S.E., the *Mt. Belvedère*

(8665'), 1½ hr. to the E., and the *Lancebranlette* (9605'), 3 hrs. to the W., all afford admirable views of the Mont Blanc chain.] We now descend gradually, overlooking the beautiful upper valley of the Isère (*La Tarentaise*) and the Savoy Mts. the whole way, to *St. Germain*, *Séez*, and (12 M.) **Bourg-St-Maurice** (2805'; *Hôt. des Voyageurs* or *Mayet*, R. & A. 3½, D. 3 fr.), a small town on the Isère, whence a diligence runs twice daily in 4½ hrs. to (16 M.) *Moûtiers-en-Tarentaise* (p. 268).

From Bourg-St-Maurice to *Chapieux*, see p. 290. — A road, practicable for carriages nearly all the way, leads to the E. from Bourg-St-Maurice through the romantic Isère valley, viâ *Ste. Foy*, at the W. base of the *Ormelune* (10,770'), and *La Thuille* (with the beautiful Mont Pourri, 12,430', to the right), to (6½ hrs.) Tignes (5405'; *Hôt. du Club Alpin; Hôt. des Touristes*), at the junction of the Isère with the *Sassière*, which here forms a fine waterfall. Excursions from Tignes to the *Aig. de la Grande Sassière*, etc., see *Baedeker's South-Eastern France*. — Over the *Col de Rhèmes* to the *Val de Rhèmes*, and over the *Col de la Galise* to *Ceresole-Reale*, see *Baedeker's Northern Italy*.

Below Pré-St-Didier we again cross the Doire (grand retrospective view of Mont Blanc, which continues visible as far as Avise), follow the lofty slope for some distance, and then descend through vineyards into a broad and rich valley. To the S. appears the beautiful pyramid of the *Grivola* (13,018'). On the hill to the left of (2¼ M.) **Morgex** (3017'; *Angelo*) is the picturesque ruined château of *Châtelar* (3840'); farther on is *La Salle* with the ruins of a castle. On the right bank is the pretty waterfall of *Derby* in several leaps. The valley contracts. The road crosses to the right bank by the (4½ M.) *Pont d'Equilive* (2570') and leads through a wild defile (*Pierre Taillée*) to *Ruinaz* (2580'; Croix, poor). Opposite lies *Avise*, with a ruined castle and an ancient church. Mont Blanc is now lost to view. The road traverses another rocky gorge, where the pyramidal Mt. Emilius comes in sight. Near the beautifully situated, but dirty village of (2¼ M.) Liverogne (2390'; *Hôt. du Col du Mont*) we cross the deep gorge of the *Dora di Valgrisanche* and soon see the first chestnut trees. Behind us is the snowy *Rutor* (11,435'); to the left is the church of *St. Nicolas* (3922'), on a precipitous rock. Beyond (¾ M.) **Arvier** we descend rapidly and cross the *Savaranche*; to the right, on the hill, are the château and church of *Introd*. Then (2½ M.) **Villeneuve** (2295'; *Cervo*, poor), beautifully situated, and commanded by the ruined castle of *Argent* on a lofty rock.

Excursions from Liverogne and Villeneuve, see *Baedeker's Northern Italy*.

We next ascend a little on the left bank of the Doire, passing a massive old tower. Beautiful retrospective view of the three-peaked Rutor, the Grivola with the Trajo Glacier, etc. Opposite *St. Pierre* (2168'), with its church and old castle on a rocky hill, opens the *Val de Cogne* on the S.; on the right bank lies *Aymavilles*, with iron-foundries and the château of Count Castiglione with its four towers. The road passes the handsome château of *Sarre* (2154'), and traverses a broad shadeless valley to (6 M.) —

21 M. **Aosta.** — *Hôtel Royal Victoria, at the station, R. from 2, L. & A. 1¾, B. 1½, déj. 3, D. 5, pens. 9-12 fr.; *Hôt. du Montblanc, to the W. of the town, R., L., & A. 3-3½, B. 1½, D. 5 fr. These two are

closed in winter. — Hôt. PAUL LANIER, in the Hôtel de Ville in the principal piazza, good cuisine; CORONA, opposite the last; Hôt.-PENS. CENTOZ, Piazza Carlo Alberto, well spoken of. — *Caffè Nazionale, in the Hôtel de Ville; beer at *Zimmermann's*, near the Hôtel de Ville; *Rail. Restaurant*, poor. — One-horse carriage to Courmayeur 18 or 30 fr.; to St. Rémy 15 or 25 fr. Omnibus to Courmayeur, see p. 293 (office in the market-place, with rooms, 3 fr. incl. L. & A.); to St. Rémy, see p. 298.

Aosta (1913'; pop. 5700), the *Augusta Praetoria Salassorum* of the Romans, and now the capital of the Italian province of Aosta, is beautifully situated at the confluence of the *Buthier* and the *Doire* or *Dora Baltea*. The still existing antiquities testify to the importance of the place during the Roman period. The *Town Walls*, flanked with strong towers, enclosing a rectangle, 790 yds. long by 650 yds. broad, still exist throughout their entire circuit; while on the S.W. side the flagged top and cornice are still intact. The walls of the ancient *Theatre* and the arcades of the *Amphitheatre* may be seen from the market-place, rising above the modern houses.

The principal street, running eastwards, passes through the ancient *PORTA PRÆTORIA* to the ($1/4$ M.) *TRIUMPHAL ARCH OF AUGUSTUS*, with ten Corinthian pilasters, and then crosses the Buthier, which has deserted its ancient channel, to the imposing arch of a Roman *Bridge*, half sunk in the ground.

The church of ST. OURS, the ancient crypt of which is supported by Roman columns, is situated in the suburb; in the choir are the tombstone of Bishop Gallus (d. 546) and finely-carved stalls. Adjacent are cloisters with interesting early-Romanesque columns (12th cent); and immediately beside the church is a 12th cent. *Tower*, built of Roman hewn stones. Opposite are portions of two antique columns in front of a chapel. — The *Priory of St. Ours*, in the same square, is a picturesque building of the 15th cent., with terracotta ornamentation, and an octagonal tower. The wood-carvings and frescoes in the interior are interesting.

The CATHEDRAL (14th cent.) has a gaudily-painted relief above its main portal, and in the interior two mosaics of the 10th cent., and some early Renaissance carved stalls. The treasury contains two reliquaries of the 13th and 15th cent. respectively, a cameo of a Roman empress in a setting dating from the 13th cent., and a diptych of Probus (consul in 406) and the Emperor Honorius.

Beside the S. gate is the tower known as *Bramafam* (12th cent.) in which a count of Challant is said to have starved his wife to death; and on the W. wall is the *Tour du Lépreux* (described in one of Xavier de Maistre's tales), in which a leper named Guasco (d. 1803) and his sister Angelica (d. 1791) suffered.

Near the railway-station stands a bronze *Statue of Victor Emmanuel II*, by Tortone, in hunting dress, on a lofty rock pedestal. — The natives of the town are sadly afflicted with cretinism.

EXCURSIONS. The *Becca di Nona (Pic Carrel*, 10,805'; 6-7 hrs.; guide, 12 fr.; provisions necessary; tolerable night-quarters at the Comboé Alp, see below) is an admirable point of view. The bridle-path, dusty at first, crosses the Doire and ascends somewhat rapidly to the village of *Charvensod*

(2446'; guides Grégoire and Grat. Jos. Comè), traverses a wood, and passes the hermitage of *St. Grat* (5815') and the chalets of *Chamolé* to the (4¹/₄ hrs.) *Col de Plan Fenêtre* (7298'). [The *Signal Sismonda* (7698'), to the S., 20 min. above the Col Pian Fenêtre, commands an excellent view of the Rutor and the Pennine Alps.] From the col we reach in ¹/₄ hr. **the** *Alp Comboé* (6930'), in a basin at the foot of the Becca di Nona, and in 2¹/₂ hrs. more the summit. A few yards below **the** top is the *Capanna Budden* of the I. A. C. Superb *View (panorama by Carrel), embracing the whole of the Mont Blanc and the Monte Rosa chains, and the Graian Alps to the S. — We may, for variety, in descending from **the** Becca, leave Comboé to the left, and go straight through the valley of the *Comboé*. Below the basin of Comboé there is a fine waterfall, at the foot of which we cross the brook and then descend to the left to *Chareensod*.

Mont Emilius (11,673') may be ascended by **experts** from Comboé in 4¹/₂ hrs. (guide 30 fr.). We follow the Col d'Arbole route (pass to *Cogne*, see *Baedeker's N. Italy*) as far as the (1 hr.) *Chalets d'Arbole* (8200'), and then turn to the left, passing a small glacier lake. View still more extensive than from the Becca di Nona.

Mont Fallère (10,045'), easily ascended from **Aosta in 7 hrs.**, **by a new bridle-path**, viâ *Ville-sur-Sarre* (guide, unnecessary, 10 fr.), commands a splendid view of the entire Pennine and Graian chains. On the arête, ¹/₄ hr. below the top, is the *Capanna Regina Margherita* of the C. A. I.

FROM AOSTA TO ZERMATT (p. 333) an interesting but fatiguing route leads through the *Val Pellina*, and over the **Col de Valpelline** (11,685'), in two days: to the chalets of *Pra-Rayé* (p. 324) 9 hrs.; thence a difficult ascent over the *Glacier de Za-de-Zan* to the pass, to the S. of the *Tête Blanche* (12,300'), and down the *Stock* and *Zmutt* glaciers to (10-12 hrs.) *Zermatt* (comp. p. 340). — From Bionaz (p. 324), 3 hrs. above Valpelline and 5 hrs. from Aosta, the **Mont Luseney** (11,500'), which commands a grand view, may be ascended in 7 hrs. (difficult, for experts only). — Several passes lead from the Val Pellina to the *Val St. Barthélemy* (see below) from Oyace (p. 324) or Bionaz over the *Col de Vessona* (about 8950'), easy and attractive; from Bionaz over the *Colle Montagnaia* (9643'), easy; from Pra-Rayé over the *Col de Livournea* (9613'), laborious.

From Aosta to *Evolena* over the *Col de Collon*, see p. 324; over the *Col de Fenêtre* to the *Val de Bagnes*, see p. 306; over the *Great St. Bernard* to *Martigny*, see R. 78.

From Aosta to *Cogne (Graian Alps)*, see *Baedeker's Northern Italy*.

Leaving Aosta the RAILWAY crosses the *Buthier* and the *Baynère*, and approaches the *Dora*, the course of which is here interrupted by numerous islands. As we look back we enjoy a splendid **view** of the valley of Aosta, surrounded by lofty mountains: to the S. rise the Becca di Nona and Mt. Emilius, to the N. the Grand Combin **and Mt. Velan**, and to the W. the Rutor (see p. 293). Shortly before reaching the station of (5 M.) *Quart-Villefranche* (1755') we see the château of *Quart* (2486') on a hill to the left. The train now crosses the Dora, but beyond (7 M.) *St. Marcel* it returns to the left bank. On the slope above St. Marcel is the pilgrim-resort of *Plou*. Near (8 M.) *Nus* (1755'), with its ruined castle, the *Val St. Barthélemy* (see above) opens on the N. The line once more crosses and recrosses the Dora. To the right appears the picturesque château of *Fénis*, at the mouth of the *Clavalité Valley*, above which towers the snowy pyramid of the *Tersiva*. We **now** intersect, near *Diemoz*, a large deposit of débris and traverse a tunnel to (12¹/₂ M.) *Chambave* (1623'), noted for its wine, where we command for the last time **a** retrospect as far as the Rutor.

The valley now contracts; the railway runs between the river and the cliffs, traversing two tunnels and a deep cutting through a deposit of débris, and crosses the Matmoire or Marmore descending from the Valtournanche. — 15½ M. **Châtillon** (1805'; *Hôt. de Londres*, R., L., & A. 3 fr.; *Hôt.-Pens. Suisse*, both in the village, near the bridge; *Hôt. des Alpes*, at the station), the district capital, with 900 inhab. and a castle of the ancient counts of Challant, is beautifully situated, 1 M. above the station (1480'), at the mouth of the *Valtournanche*. The deep wooded gorge of the Matmoire, which is picturesquely studded with houses, is spanned in the centre of the village by an imposing single-arched bridge. — To *Valtournanche* and over the *Théodule Pass* to *Zermatt*, see R. 88.

From Châtillon the railway continues along the left bank of the Dora. On a steep hill to the right is the old château of *Ussel*, also once belonging to the Challant family. Beyond two short tunnels is (16½ M.) St. Vincent (1415'), the station for the baths of the same name (1885'; *Lion d'Or*; *Corona*), situated 1 M. to the left, at the foot of *Mt. Zerbion* (8924'). We next enter the *Montjovet Defile*, the most striking part of the entire journey; a series of tunnels, separated by massive retaining and sheltering walls, follow each other in the narrow rocky gorge, while far below the foaming Dora descends in cascades. The exit of the pass is commanded by the ruined castle of *Montjovet* or *St. Germain*, high up on the left. An imposing viaduct here spans the Dora, and the train enters a tunnel. Beyond (20 M.) the station of *Montjovet*, the valley again expands. Extensive vineyards begin to appear; on the right rise lofty cliffs. Farther on we see on the slopes the village of *Champ de Praz*, at the mouth of the *Val Chalame*, watered by a stream, which has scattered stones far and wide over the valley of the Dora. The train crosses the Dora and the *Evançon* and reaches —

23½ M. **Verrés** (1280'). The village of that name (1100 inhab.; *Italia*; *Ecu de France*), with an old château of the former counts of Challant, is picturesquely situated at the mouth of the *Val de Challant* (p. 348). *Issogne*, on the opposite bank of the Dora, has another old château of the same family. To the N.E. appears the rocky pyramid of the *Becca di Viou* (9370').

25½ M. *Arnaz*, with a ruined castle. The line traverses an extensive alluvial deposit, and at *Campagnola* crosses to the right bank of the Dora. 28 M. **Hône-Bard**, in a superb situation. To the right the *Val Champorcher* or *Camporciero*, with its picturesque rocky summits; to the N.W., in the background of the Dora valley, the *Mont Luseney* (p. 297). On a steep crag on the left bank of the Dora rises *Fort Bard (1282')*, captured in 1242 after a long siege by Count Amadeus IV. of Savoy, and in 1800, before the battle of Marengo, gallantly defended by 400 Austrians against the French army. Beyond this point Italian only is spoken.

The railway crosses the river and passes under the fortress by means of a tunnel, 650 yds. long. Then through a narrow rocky ravine to (30 M.) *Donnas* (Rosa), prettily situated, and over the wild *Lys* torrent in a broad valley surrounded by imposing mountains to (31 M.) **Pont St. Martin**, the station for the village of the same name (1005'; *Rosa Rossa; Cavallo Bianco*), in a highly picturesque situation, at the mouth of the deep and narrow *Lys Valley*, with a ruined castle, foundries, and an ancient Roman bridge across the *Lys* (new road to *Gressoney-la-Trinité*, $20^1/_2$ M. ; see p. 336).

The railway again crosses and recrosses the Dora, which here forms a large island. On the slope to the left is the village of *Carema*, surrounded by vineyards and fruit-gardens. On the right bank is (33 M.) *Quincinetto*, at the foot of the *Becco delle Steje* (9184'); on the left bank is the ruin of *Cesnola*. 35 M. *Tavagnasco*, the village lies to the right. Opposite, at the foot of the *Colma di Monbaron* (7773'), is the larger village of *Settimo Vittone*. The lower terraces of the hills enclosing the picturesque and highly cultivated valley are covered with vines, higher up are woods of walnut and chestnut trees, above which rise bare rocky peaks. We cross the Dora again at *Montestrutto*, pass (on the left) *Terrassa* and *S. Germano*, with ruined castles, and reach ($37^1/_2$ M.) **Borgofranco** (924'), with an arsenical spring, prettily situated $1^1/_4$ M from the station.

The mountains now recede. 39 M. *Montallo-Dora*, with a pinnacled ruined castle on a rocky hill. The train enters a tunnel ($1^1/_4$ M. long) under the hill of Ivrea, crosses the Dora, and stops at ($41^1/_2$ M.) **Ivrea** (768'; *Scudo di Francia; Universo; Corona d'Italia*), a town with 5400 inhab., picturesquely situated on the left bank of the Dora, with an ancient castle, several lofty round towers, and numerous churches. Comp. *Baedeker's Northern Italy*.

78. From Martigny to Aosta. Great St. Bernard.

17 hrs. From Martigny to the Hospice 11, thence to Aosta 6 hrs. (from Aosta to the Hospice 8, thence to Martigny 9 hrs.); new road to the hospice (32 M.), thence to (2 hrs.) St. Rémy bridle-path; road again to Aosta ($13^1/_2$ M.). *Diligence* daily from Martigny (station) to (13 M.) Orsières in $3^1/_2$ hrs. (back, in the afternoon, in 2 hrs.; 3 fr. 25 c.). *Carriage* to Orsières 15, with two horses 20, Bourg-St-Pierre 25 or 40, Great St. Bernard 50 or 60 fr.; one-horse carr. from St. Rémy to Aosta, 1 pers. 10, 2 pers. 12, 3 pers. 15 fr. (from Aosta to St. Rémy, 1·2 pers., 15 fr.). Omnibus from Aosta to St. Rémy daily in 4 hrs. (6 fr.), returning in 3 hrs.

The **Great St. Bernard Route**, though less attractive than most of the other Alpine passes, presents some very fine scenery, and is a direct and convenient approach to Italy (Aosta, Courmayeur) from the Rhone Valley. A visit to the Hospice is also interesting. Those who do not intend going farther should not omit the ascent of the Chenaletta, and may return over the Col de Fenêtre (p. 308) and through the Val Ferret.

Martigny, see p. 249. Beyond *Martigny-Bourg* (p. 250) we cross the ($1^1/_2$ M.) *Drance*, 4 min. beyond which the road to Chamonix diverges to the right (p. 286). The St. Bernard road leads through the deep ravine of the Drance, **by Le Brocard** and *Le Borgeau*, to (3 M.) *Les Valettes* (1978'; Restaurant des Gorges du Durnant).

*Gorges du Durnant (from Martigny and back 4 hrs., one-horse carr. 7, two-horse 10 fr.). A road leads from Les Valettes to the right to the (1 M.) entrance of a rocky gorge, through which the *Durnant* is precipitated in 14 falls (made accessible by a wooden gallery 1/2 M. in length; (adm. 1 fr.; Inn by the entrance). From the upper end of the gorge the path ascends to the bridle-path to Champex (see below). — Fine view from the hill of *Lombard* (2888', see below), ascended to the left from the lower end of the gorge in 1/2 hr., by a shady path.

Beyond (3/4 M.) *Bovernier* (2037') **the Drance traverses a wooded gorge**, where its course is impeded by huge masses of rock, especially near the (1 1/2 M.) *Galerie de la Monnaie* (2362'), a tunnel 70 yds. long. In 1818 a great fall of rock was caused here by the bursting of a lake in the *Val de Bagnes* (p. 305). At (1 1/2 M.) **Sembrancher** (2330'; Inn) the *Drance d'Entremont*, descending from the St. Bernard, unites with the *Drance de Bagnes* (p. 304). On a hill **stands a ruined castle.** To the right rises the abrupt *Catogne* (8460').

FROM MARTIGNY TO SEMBRANCHER, over the Mont Chemin, 4 hrs., interesting, especially in the reverse direction (fine views of the Rhone Valley). From Martigny-Bourg the path ascends to the left, through wood, by *Chemin d'en Bas* to *Chemin* (3786'), leads to the right past iron-mines to *Vence* (3701'), and descends in windings to Sembrancher.

FROM SEMBRANCHER TO SAXON over the *Pas du Lens* (5446'), 5 hrs., a bridle-path. — The *Pierre-à-Voir* (8123'; guide 7 fr.) may be ascended from Sembrancher in 5-6 hrs. (comp. p. 250).

The road enters the *Val d'Entremont* to the S., crosses the Drance twice, and leads by the left bank via *La Donay*.

12 M. **Orsières** (2894'; *Hôt. des Alpes*), at the **mouth of the** *Ferret Valley* (p. 293), has a curious old tower.

FROM MARTIGNY TO ORSIÈRES BY THE VAL CHAMPEX (5 1/2 hrs.), bridle-path, more interesting and not much longer than the high-road. Road to (1 1/2 M.) *Les Valettes*, see above. Here we diverge to the right (or we may go through the Gorges du Durnant), and ascend gradually through pastures and wood, by *Lombard* (p. 287), *Crettet*, and *Les Grangettes* to (2 1/2 hrs.) the village of *Champex* (4496'). Thence across the pass (4900') to the (1/2 hr.) pretty **Lac de Champex** (4807'; *Hôt. du Lac*, pens. 5-7 fr.; Engl. Ch. Service in summer; Pens. Crettet; Pens. Biselx, 4 fr.), visited as a summer-resort, whence we descend either to the left by *Biollay* to (1 hr.) *Orsières*, or to the right by a direct path to (1 hr.) *Som la Proz*. — Excursions from Lac Champex: to the *Grand Plan* (6560'; 2 hrs.); *La Breya* (7800'; 3 hrs.); *Catogne* (8160'; 3 1/2 hrs.). More difficult is the *Pointe des Ecandies* (9170'; 4 1/2 hrs.). A bridle-path ascends through the monotonous *Val d'Arpette* to the (3 1/2 hrs.) *Fenêtre d'Arpette* (8800'), to the N. of the Pointe des Ecandies, affording a survey of the beautiful *Glacier de Trient* (p. 286), whence we may descend to the *Col de la Forclaz* (p. 286; 3 hrs.). — From Lac Champex to the *Cabane d'Orny* (see below) the shortest route is by the *Col de la Breya* (8200'); 5-6 hrs., with guide (not difficult for experts).

From Orsières to *Courmayeur* over the *Col Ferret*, see p. 292. — **Passes to** *Chamonix* (*Cols du Tour*, *du Chardonnet*, *d'Argentière*, etc.), see pp. 287, 283. The Cabane d'Orny (8835') may be reached from Orsières through the uninteresting *Combe d'Orny* in 6 hrs. (with guide). Excursions hence (guides see above): to the *Pointe d'Orny* (10,755'; 2 hrs., not difficult); *Portalet* (10,990'), by the *Col des Plines* in 3 hrs.; *Aiguille du Tour* (11,585'), by the *Glacier du Tour* and *Glacier de Trient* in 3-4 hrs., not difficult for adepts and highly interesting; *Le Darrei* (11,605'; 5 hrs.); *Grande Fourche* (11,877'; 5 hrs.), etc. Fine view from the (3 hrs.) *Fenêtre de Saleinaz*

(10,860'), between the Grande Fourche and the Aiguilles Dorées, whence we may descend across the *Saleinaz Glacier* to the *Cab. de Saleinaz* (p. 284). — Over the *Col du Tour* to the *Col de Balme*, see p. 287.

*Mont Brûlé (8450'; 4½ hrs.; guide 6 fr.), an easy and attractive expedition from Orsières. The view from the top embraces the entire chain of the Bernese and Valaisian Alps, with the Lake of Geneva and the Jura in the background; in the foreground are the Dent du Midi, the Orny and Trient chain, the Grand Combin, etc. The ascent may also be conveniently made from Liddes (p. 289; 4 hrs.) or Chable (p. 304; 5 hrs.). — To Chable viâ the *Col de Sexblanc* (7 hrs.), interesting, see p. 304.

The road crosses the Drance, which is seldom visible in its deep bed, and ascends in a long bend (which the old bridle-path cuts off). On entering the upper part of the valley we obtain an admirable view of *Mont Velan* (p. 301), which with its snow and ice fills the background. The slopes of the broad valley are covered with pastures and corn-fields. Between *Fontaine-Dessous* (3800') and *Rive Haute* (4010') the road again describes a long curve which walkers may cut off. It passes the chapel of *St. Laurent*, and reaches (4³/₄ M.) —

16³/₄ M. Liddes (4390'; *Hôt. du Grand St. Bernard; Union; Angleterre*), a considerable village. On the left rise the finely shaped *Merignier* (10,403') and the *Maisons-Blanches* (12,137'). Above Liddes is the chapel of *St. Etienne*. At *Allèves* we cross the brook of that name, coming from the *Glacier de Boveyre*, pass the chapel of *Notre-Dame de Lorette* on the left, and reach (3¹/₄ M.) —

20 M. Bourg-St-Pierre, or *St. Pierre-Mont-Joux* (5358'; *Au Déjeuner de Napoléon*), a large village at the mouth of the *Valsorey*, with a church of the 11th century. (On the wall by the tower is a Roman milestone.) Some traces of old fortifications, with an ancient gateway, are to be seen on the S. side of the village. On a hill to the left of the road, on which formerly stood the château of *Quart*, is the 'Jardin Alpin' **'Linnaea'** (key at the 'Déjeuner de Napoléon'; keeper, the guide Jules Balley).

Excursions. (Good guides, *Dan., Eman.*, and *Jules Balley*, and *Michel Genoud.*) The *Tête de Bois* (2¹/₂ hrs.; guide 6 fr.; mules also) commands the Mont Blanc and Combin group, and the Val d'Entremont below.

A good path leads through the interesting Valsorey, on the right bank of the *Valsorey*, to a fine waterfall and to the (2¹/₂ hrs.) *Chalets d'Amont* (7190'), in a grand situation. The background is formed by the *Glacier du Valsorey*, and others uniting with it, (l.) that of *Sonadon*, descending from the Grand Combin, and (r.) that of *Tzeudet*. Beautiful view of the dazzling snows of Mont Velan and the jagged rocks of the *Luisettes*. — The night is passed at these chalets by travellers about to cross the *Col des Maisons-Blanches* (11,240') or the *Col du Sonadon* (11,447') to the Val de Bagnes (p. 305), or the *Col du Valsorey* or *des Chamois* (10,213') to the Val Ollomont (p. 306). — The Grand Combin (14,163') may be ascended from the Chalets d'Amont by the *Col des Maisons-Blanches*, or better by the *Glacier du Sonadon*, in 8-9 hrs. (grand, but difficult; for experts only; guide 40 fr.). Ascent easier from the *Cabane de Panossière* (comp. p. 304).

Beyond St. Pierre the road crosses the deep gorge of the *Valsorey*, which forms a waterfall above the bridge. It was here that Napoleon, during his famous passage of the Alps with 30,000 men on 15th-21st May, 1800, encountered the greatest difficulties. The road, hewn in the rock, and avoiding the steep parts of the old route, traverses the forest of St. Pierre and the *Défilé de Charreire*. 4 M. **Cantine de Proz** (5982'), a lonely inn, at the beginning of the *Plan de Proz*. To the E. rises the snow-clad *Mont Velan*, from which descends the *Glacier de Proz*, with its extensive moraines.

For the ascent of *Mont Velan (12,355'; 6-7 hrs.; difficult; for experts only; guide 25 fr.), the starting-point is either the Cantine de Proz (6 hrs. to the top, crossing the *Glacier de Proz*, very steep at places), or the *Chalets d'Amont* (see p. 300; ascent rather longer, but less difficult). Above the chalets we ascend a 'cheminée' to the E. moraine of the *Glacier du Valsorey*, cross the glacier to the E. rocky slope of *Mt. de la Gouille*, and mount (an interesting climb) to the upper and grandest part of the glacier; cross it, ascend another cheminée, traverse masses of rock, and reach the summit in 6-7 hrs. in all. Magnificent view: N., as far as the Lake of Geneva; S., to the Val d'Aosta. Immediately to the W. towers Mont Blanc; to the N.E. the Grand Combin.

The road ascends the boulder-strewn pastures of the Plan de Proz to the *Cantine d'en Haut*, traverses the *Pas de Marengo*, a rocky defile, and reaches the (4 M.) *Hospitalet* (6890'), two stone chalets and an Alpine dairy in a broader part of the valley, across the stream, to the right. It next (1 M.) crosses the Drance by the *Pont Nudrit* (7336'), recrosses it farther on by the (1 M.) *Pont Tronchet* (7457'), and leads through the dreary *Grande Combe* to the (2 M.) —

32 M. **Hospice of St. Bernard** (8120'), on the pass, consisting of two buildings. One contains the church, the dwellings of the brethren, and the rooms for travellers; the other *(Hôtel de St. Louis)* is a refuge in case of fire, containing the storehouse and lodging for poor wayfarers. On arriving, strangers are welcomed by one of the brethren, who conducts them to a room and presides over the meals (at 12 and 6 or 7; Frid. and Sat. are fast-days). Travellers are boarded and lodged gratuitously, but few will deposit in the alms-box ('offrandes pour l'Hospice', in the church, on the first pillar on the left), *less* than they would have paid at a hotel. Adjacent is a small *Restaurant*.

In 962 St. Bernard de Menthon (p. 209) founded the monastery here. The inmates now consist of 10-15 Augustinian monks and 7 attendants *(maroniers)*, whose office it is to receive and lodge strangers gratuitously, and to render assistance to travellers **in danger** during the snowy season, which here lasts nearly nine months. In this work of benevolence they are aided by the famous St. Bernard dogs, whose kennels are worth visiting. Their keen **sense** of smell enables them to track and discover travellers buried in the snow, numbers of whom have been rescued by these noble and sagacious animals. The stock is said to have come originally from the Spanish Pyrenees, but the genuine old breed is extinct.

The brotherhood of St. Bernard consists of about 40 members. Some of the monks minister in the Hospice on the Simplon (p. 311); others perform ecclesiastical functions. The sick and aged have an asylum at Martigny. Next to the fourth Cantoniera S. Maria on the Stelvio Pass (p. 436) St. Bernard is the highest winter habitation in the Alps. *Humboldt* in his 'Kosmos' mentions that the mean temperature at the Hospice of St. Bernard (45° N. latitude) is 30° Fahr. (in winter 15°, spring 25°, summer 43°, autumn 32°), and that such a low temperature would only be found on the sea-level at a latitude of 75° (the S. Cape of Spitzbergen).

The monastery was very wealthy in the middle ages. The beneficence of its object was widely recognised by extensive grants, chiefly by

302   *V. Route 78.*   ST. BERNARD HOSPICE.   *From Martigny*

the emperors of Germany, and gifts from various parts of Christendom; but it was afterwards impoverished by various vicissitudes. The 30-40,000 fr. required for its annual support are in part derived from the revenues of the monastery, and in part from annual collections made in Switzerland; the gifts of travellers, it must be said with regret, form a very insignificant portion of the sum. Of late years 16-20,000 travellers have been annually accommodated, while the sum they have contributed barely amounts to what would be a moderate hotel-charge for 1000 guests. The expenses of the establishment are increasing. Provisions are generally brought from Aosta, and in July, August, and September about twenty horses are employed daily in the transport of fuel from the Val Ferret (see below), 4 hrs. distant.

The traveller will hardly quit the hospice without a feeling of veneration and compassion for this devoted fraternity. They generally begin their career at the age of 18 or 19. After about fifteen years' service the severity of the climate has undermined their constitutions, and they are compelled to descend with broken health to the milder climate of Martigny or some other dependency. Amid the pleasure and novelty of the scene, the traveller is too apt to forget the dreariness of the eight or nine months of winter, when all the wayfarers are poor, when the cold is intense, the snow of great depth, and the dangers from storms frequent and imminent. It is then that the privations of these heroic men are most severe, and their services to their fellow-creatures most invaluable.

During the Italian campaigns of 1798, 1799, and 1800, the pass was crossed by several hundred thousand soldiers, French and Austrian. In 1799 the Austrians endeavoured to pass the hospice, but after several fierce engagements the French remained masters of the pass, and kept a garrison of 180 men in the hospice for a whole year. Napoleon's famous passage has already been mentioned (p. 300). The Romans used this route in B.C. 100. After the foundation of *Augusta Praetoria Salassorum* (Aosta, B.C. 26) it became more frequented. Constantine caused the road to be improved in 339. The Lombards made the passage about 547; Bernard, an uncle of Charlemagne, marched an army by this route into Italy in 773, and, according to some, gave his name to the pass.

The present substantial edifice dates from the middle of the 16th cent., the church from 1680. The walls of the dining-room are hung with engravings and pictures, the gifts of grateful travellers. In the library on the upper floor is a collection of ancient and modern coins, relics found in the environs (fragments of votive brass tablets offered to Jupiter Pœninus. p. 303, after escape from danger, statuettes, etc.), and a small natural history collection. The visitors' books contain many well-known names. A chapel to the left of the entrance to the church contains the monument of General Desaix, who fell at the battle of Marengo in 1800. Relief by Moitte.

Near the hospice is the *Morgue*, a receptacle for bodies found in the snow. The small lake to the W. of the monastery is sometimes coated with ice even on summer mornings. On the hillside to the right is a small botanic garden with Alpine plants. — Towards the E. of the hospice we observe the snow-capped *Mont Velan*, adjoined on the left by the *Combin de Corbassière* (12,210').

The "**Chenaletta** (9475'; 1½ hr., steep at places; guide necessary), to the N. of the Hospice; the Pointe des Lacerandes (*Pic de Dronaz*; 9675'; 2½- 3 hrs., with guide; trying), to the N.W.; and the **Mont Mort** (9405'), 1½ hr. to the S.E., all command magnificent views of Mont Blanc, the Graian Alps, Monte Rosa, and (N.) the Bernese Alps, while the Mont Velan and Grand-Combin are quite near

FROM THE HOSPICE TO MARTIGNY OVER THE COL DE FENÊTRE (9 hrs.; guide necessary for the inexperienced), recommended as a return-route from the

Hospice to Martigny. From the path to the Vacherie (see below) the (20 min.) bridle-path ascends rather steeply to the right, to the (1 hr.) **Col de Fenêtre** (8855'; fine view). It descends over débris and sometimes snow, past the three small *Lacs de Fenêtre*, to the chalets of (1¼ hr.) *Plan la Chaud* and (1 hr.) *Ferret* (5565'), where it unites with the route from the Col Ferret (p. 293). — FROM THE HOSPICE TO COURMAYEUR (9-10 hrs.) the direct route is across the *Col de Fenêtre* and the *Col Ferret*. In order to reach the Col Ferret we need not descend from the Col de Fenêtre to Ferret, but (guide advisable) beyond the third lake we turn to the left, descend steep grassy slopes to a bridge over the Drance, follow its left bank for a time, and then ascend on the right bank of the brook coming from the Col Ferret, until after about 50 min.) we can cross it. A steep ascent of ½ hr. more brings us o the Col Ferret route (p. 293; from the Hospice to the Col, 5 hrs.).

On the N.W. side of the lake on the St. Bernard Pass, near a small brook, are stones marking the Italian frontier. On the adjacent *Plan de Jupiter* once rose a temple to *Jupiter Poeninus*. The mountain has thence derived its Italian name of *Monte Jove*, locally *Mont Joux*, and the range is called the *Pennine Alps*. The path rounds an angle of rock and descends in a wide bend to *La Vacherie*, a green pasture, where the cattle of the hospice graze, with several chalets, and the *Cantine* (7270'), or road-menders' house. To the W. rises the conical *Pain de Sucre* (9515'). A shorter footpath, diverging to the left at a cross, before the above-mentioned angle of rock, rejoins the bridle-path here. The path zigzags down the left side of the valley, and then descends gradually to (1 hr.) **St. Rémy** (5353'; *Hôt. des Alpes Pennines*, well spoken of; *Croix Blanche*, at the other end of the village), the first Italian village, where the road begins. The first house on the right is the customhouse. Carr., see p. 298. Mule and attendant to the Hospice 4½ fr.

FROM ST. RÉMY TO COURMAYEUR over the Col de la Seréna (7580'), 9-10 hrs., fatiguing and not very interesting. (From the hospice over the Col de Fenêtre and Col de Ferret, preferable, see above.)

The deep and narrow *Val des Bosses* diverges from the valley of St. Bernard beyond St. Rémy. Cultivation on both sides of the valley begins at (2¼ M.) *St. Oyen* (4515'), and becomes richer at (1½ M.) **Etroubles** (4200'; *Croix Blanche; National*). The road crosses the *Buthier* here, and skirts the right side of the valley, soon running high above the river. Opposite, on the slope, is the church of *Allein*. 2 M. *Les Echevenoz* (4050'), a hamlet; (1½ M.) *La Cluse* (3940'), a solitary house. By (1 M.) the village of *Condemine* a view is disclosed of the long *Val Pellina*, with the snow-clad *Dent d'Hérens* in the background. To the N. tower the rounded summit of *Mont Velan* and the imposing pyramid of the *Grand Combin*. The road descends in long windings to (1½ M.) **Gignod** (3260'; *Osterie*), with a square tower of the 14th cent., most picturesquely situated opposite the entrance to the Val Pellina, from which the main arm of the Buthier descends. Far below is the church-tower of *Roysan*, and farther up the village of *Valpelline*.

The scenery now assumes a softer character; walnuts, chestnuts, vines, and maize thrive luxuriantly. The road, running high up on the right side of the valley, gradually descends. Before us the fine pyra-

mid of the *Grivola* is visible for a time. To the left is the blunted cone of *Mt. Mary* (9230'). Beyond (2¹/₄ M.) **Signayes**, where the extensive vineyards of Aosta begin, the three-peaked *Rutor* appears on the right. Before us rise the *Becca di Nona* and *Mt. Emilius*; to the left, the S. spurs of Mte. Rosa. — 1¹/₂ M. —
13¹/₂ M. *Aosta*, see p. 294.

79. From Martigny to Aosta over the Col de Fenêtre. Val de Bagnes.
Comp. Map, p. 298.

From Martigny to Mauvoisin 8¹/₄ hrs. (Sembrancher 2³/₄, Chable 1¹/₂, Champsec 1, Lourtier ¹/₂, Mauvoisin 2¹/₂ hrs.). To Lourtier a good road (diligence from Martigny to Chable daily in 3¹/₂ hrs.; one-horse carr. 18 fr.); thence a bridle-path. — Travellers going to Aosta over the COL DE FENÊTRE (guide 18 fr.; *Séraphin* and *Justin Bessard, F. Basse, Maur. Ant. Troillet*, and others at Chable) should pass the night at Mauvoisin, or at **Chermontane**, 2³/₄ hrs. farther up. From Chermontane to the pass 1¹/₂, Val Pellina 4, Aosta 2 hrs. Good carriage-road from Val Pellina to Aosta.

To (8 M.) *Sembrancher*, see p. 299. We diverge here to the left from the St. Bernard road, cross the Drance, and follow the right bank of the *Drance de Bagnes* to (4¹/₂ M.) **Chable** (2743'; *Hôt. du Giétroz*, moderate), the capital of the *Val de Bagnes*, picturesquely situated. In the background to the S.E. is the snow-clad *Ruinette* (12,727'); to the left *Mont Pleureur* (12,155') and the *Glacier de Giétroz*.

The *Pierre-à-Voir* (8123') may be ascended hence in 5 hrs. (guide 6 fr.; comp. p. 250). — *Mont Brûlé* (8450') in 5 hrs., viâ *Zeppelet* and *Mille* (comp. p. 300). — To *Orsières* or *Liddes* (pp. 299, 300), over the **Col de Sexblanc** (about 7380') in 7 hrs., attractive and not difficult (guide convenient). Fine view of Mont Blanc from the top of the pass. — Over the *Col des Etablons* to *Riddes*, see p. 307.

We now follow the left bank of the Drance, pass *Montagnier* on the right bank, and reach *Versegère* and (2¹/₂ M.) *Champsec* (2965'). Here we cross the Drance and ascend to (¹/₂ hr.) *Lourtier* (3655'; rustic inn), where the road ends. Between Lourtier and Mauvoisin the Drance forms several falls; at (1 hr.) *Granges Neuves* it receives a large contribution from the *Glacier de Corbassière*. Then (20 min.) **Fionney** (4910'; *Hôt. du Grand-Combin; *Hôt.-Pens. Carron, 5 fr., Engl. Ch. Serv. in Aug.).

To the Cabane de Panossière (8900') a most interesting excursion (from Fionney, by the *Corbassière Alp*, in 4¹/₂ hrs., with guide; from Mauvoisin, over the *Col de Plangolin* or *Col des Otanes*, 9350', in 3¹/₂–4 hrs.). This club-hut, finely situated on the margin of the huge *Corbassière Glacier*, is the starting-point for the *Combin de Corbassière* (12,210'), the *Tournelon Blanc* (12,180'), the *Col des Maisons-Blanches* (p. 305), etc. The Grand Combin (14,163'; 7-8 hrs.) is best ascended from this point, but requires experience and a steady head (comp. pp. 300, 305).

PASSES. To the E. of Fionney a fatiguing route crosses the *Alp Le Crêt* (7575') to the Col du Crêt (10,380'; splendid view), on the S. side of the *Parrain* (10,702'); descent over the *Glacier des Ecoulaies* to the (6-7 hrs.) Alp *La Barma* in the *Val des Dix* (1 hr. below *Liappey*, p. 322). A similar pass is the Col de Sevreu (10,500'), between the *Parrain* and the *Rosa Blanche;* ascent by *Alp Sevreu* and the small glacier of that name to the (4¹/₂ hrs.) col, with fine view; descent to (2 hrs.) *La Barma* (guide

VAL DE BAGNES. *V. Route 79.* 305

over the Col du Crêt, or Col du Sevreu, and the Col de la Meina to Evolena 18 fr.). — Two other passes (trying, for mountaineers only), one the **Col de Cleuson** (9565'), to the W. of the *Rosa Blanche* (10,985'; an admirable point, easily ascended from the pass in 1¼ hr.), the other the **Col de Louvie** (9510'), to the S.E. of the *Mont Fort* (10,925'), lead to the N.E. to the glacier of the *Grand Désert*. Descent thence to the (8-9 hrs.) *Alp Cleuson* (6975') in the *Val de Nendaz*, whence a good bridle-path leads to (3 hrs.) *Nendaz* (3340') and (2½ hrs.) *Sion* (p. 307). From the Col de Cleuson the traveller may prefer to cross the Grand Désert towards the N.E. and the *Col de Prazfleuri* (9705') to the Val des Dix (p. 322).

Above Fionney the valley becomes narrower and wilder. The bridle-path leads on the right bank of the Drance by *Bonatchesse* to the (1½ hr.) bridge of **Mauvoisin** (5570'), spanning the Drance, which flows 100' below. On the opposite bank, 20 min. higher, is the *Hôtel du Giétroz* (5847'; 24 beds.)

About 1 M. to the S. of the hotel, on the right side of the valley, is the *Cascade du Giétroz*, the discharge of the **Glacier de Giétroz**, which has receded much of late. A good view of it may be obtained from the *Pierre à Vire* (7823'), ascended by the chapel behind the inn in 1¼ hr. In the winter of 1817-18 the fallen masses of ice and snow so impeded the Drance that a considerable lake was formed above Mauvoisin. In June, 1818, this sheet of water burst its barriers and caused terrible devastation throughout the entire Val de Bagnes as far as Sembrancher and Martigny.

The path again descends to the Drance and intersects the former bed of the lake. It next leads through the ravine of *Torrembey* and passes the chalets of (1½ hr.) *Petite Chermontane* (6290'), where it crosses to the left bank, and *Vingthuit*. Beyond the (½ hr.) chalets of **Boussine** (6570') the path divides: the right branch crosses the moraine and the flat tongue of the *Glacier du Mont-Durand* to the alp *Grande Chermontane* (7315'); the left branch crosses the Drance to the chalets of *Lancey* (6716') and ascends to the (1½ hr.) **Cabane de Chanrion** (8660'), a well fitted-up club-hut, beautifully situated on the W. slope of the *Pointe d'Otemma* (10,985'). The head of the valley is encircled from W. to E. by the *Grand Combin* (14,163'), *Tour de Boussine* (12,590'), *Amianthe* (11,810'), *Tête de Buy* (11,225'), *Mont Avril* (10,960'), and *Mont Gelé* (11,540').

EXCURSIONS. *Mont Avril* (10,960'), from Chermontane (½ hr. from Chanrion) by the Col de Fenêtre, 3 hrs., easy (see below; guide 10 fr.). — Tour de Boussine (12,590'), by the *Glacier du Mont-Durand*, 7-8 hrs., laborious (guide 25 fr.). — Grand Combin (14,163'), by the *Col du Sonadon* (see below) in 10-12 hrs., difficult (guide 40 fr.; comp. p. 304). — Mont Blanc de Seilon (12,700'), from Mauvoisin by the *Glacier de Giétroz*, 10 hrs. (guide 30 fr.); or better from Chanrion over the *Glacier de Breney*, 6-7 hrs.; magnificent view. — Mont Pleureur (12,155'), from Mauvoisin, by the *Alp Giétroz*, 8 hrs. (guide 15 fr.), not very difficult. — The *Pointe d'Otemma* (10,985'), from Chanrion 2½-3 hrs. (guide 12 fr.); *Pigne d'Arolla* (12,470'), from Chanrion 5-6 hrs. (20 fr.; not difficult, comp. p. 323); *Tournelon Blanc* (12,130'), from Mauvoisin 8 hrs. (15 fr.); *La Luette* (11,625'), from Mauvoisin 7-8 hrs. (15 fr.); *Serpentine* (12,110'), from Chanrion 5-6 hrs. (20 fr.); and *Ruinette* (12,725'), from Chanrion 6-7 hrs. (30 fr.), may also be ascended by mountaineers (tariff from Mauvoisin).

PASSES. Over the **Col du Sonadon** (11,445') to Bourg-St-Pierre, a difficult glacier-pass (11-12 hrs.; guide 30 fr.). From Chermontane to the W., up the *Glacier du Mont-Durand* to the pass, on the S. side of the Grand Combin; descent over the *Glacier du Sonadon* to the *Valsorey* (p. 300) and Bourg-St-Pierre (p. 300). — Over the **Col des Maisons-Blanches**

(11,240'), 12-13 hrs. from Mauvoisin or Fionney to Bourg-St-Pierre, grand, but difficult (guide 25 fr.; spend night in the *Cabane de Panossière*, p. 304). — To the S., besides the Col de Fenêtre (see below), another route crosses the Col de Crête-Sèche (9475'), traversing the lower end of the *Glacier d'Otemma* and the *Glacier de Crête-Sèche*, to the *Val Pellina* (from Chanrion to Valpelline 8 hrs.; guide 18 fr.). — To the Val d'Hérémence over the Col de Seilon (10,665'; 6³/₄ hrs. from Mauvoisin to Liappey; 6¹/₄ hrs. to Arolla), by the *Glacier du Giétroz* and the crevassed *Glacier de Durand* or *Seilon*, fatiguing (better from Chanrion over the *Glacier de Lyrerose* and the *Col du Mont Rouge*, comp. p. 324). Over the Col de Breney (11,975'; 7-8 hrs. from Chanrion to the *Alp Seilon*, p. 322), difficult. From the Col de Breney the "*Pigne d'Arolla* (12,470'), a superb point of view, may be ascended in 1/2 hr. (comp. pp. 305, 323). From the Glacier Durand or Seilon we may cross the *Col de Riedmatten* or the *Pas de Chèvres* to the E. to *Arolla* (see p. 324). — Over the Col de Vasevay (10,705'; 6-7 hrs. from Mauvoisin to Liappey), interesting, and not very difficult. — To Arolla over the *Glacier d'Otemma* and *Col de Chermontane* (10 hrs. from Chanrion), see p. 324; *Col de l'Evêque* (13 hrs.), see p. 324. — From the upper Glacier d'Otemma over the *Col d'Otemma* (about 11,025') or the *Col de la Reuse d'Arolla* or *Col d'Oren* (10,635') to Valpelline, difficult (8-9 hrs. from Chanrion to Pra-Rayé; guide 20 fr.).

The route from Chermontane to the (1¹/₂ hr.) **Col de Fenêtre** (9140') ascends at first over pastures and then over loose stones and moraine-deposits, skirting the *Glacier de Fenêtre*. To the left rises the *Mt. Gelé* (11,540'), to the right the *Mont Avril* (10,960'), a splendid point of view (1¹/₂-2 hrs. from the pass; no difficulty). The col commands a fine view of the *Val d'Ollomont* and the Graian Alps. A bridle-path descends past the chalets of *Balme* and *Vaux* to (3 hrs.) **Ollomont** (4385'; small inn) and (³/₄ hr.) *Valpelline* (3130'; two small inns), whence a good road leads to (9 M.) *Aosta* (p. 294).

80. From Martigny to Domodossola over the Simplon.

Comp. *Maps*, pp. *298*, *306*, *310*, *316*, *322*.

88¹/₂ M. RAILWAY from Martigny to (47¹/₂ M.) Brig in 2¹/₂-3 hrs., fares 8 fr. 20, 5 fr. 80, 4 fr. 10 c. (from Lausanne to Brig in 5-6 hrs., fares 15 fr. 20, 10 fr. 70, 7 fr. 60 c.). — DILIGENCE from Brig to Domodossola (41 M.) twice daily in summer, in 8³/₄ hrs. (fares 16 fr. 5, coupé 19 fr. 30 c.). — Luggage to be sent by post over the Simplon must be booked the previous night. It cannot be conveyed beyond Iselle (Italian frontier, p. 314) unless the keys are sent with it to the custom-house there. — Extra-post with two horses from Brig to Domodossola 88 fr. 40 c.; carriage with one horse 35 fr., with two horses 70-80 fr. (apply at the hotels).

A kind of gnat, with black gauzy wings, is a source of great annoyance in the marshy parts of the lower Rhone valley, especially in the evening; bedroom-windows should therefore be closed early.

Martigny (1560'), see p. 249. — The wide *Rhone Valley* is enclosed by lofty mountain-chains, whose lower slopes, as far as Leuk, are covered with vineyards. Extensive improvements in the river-channel have reclaimed for cultivation much of the valley which used formerly to be covered with gravel and debris. — 3 M. *Charrat-Fully*.

5¹/₂ M. **Saxon** (1570'; *Gr.-Hôt. des Bains*) has springs impregnated with iodine. The Etablissement des Bains lies ¹/₄ M. to the right of the station. The village, commanded by a ruined castle, is picturesquely situated, 1 M. from the station, in a gorge at the foot of the Pierre-à-Voir.

SION. V. Route 80. 307

Ascent of the *Pierre-à-Voir* (8125') from Saxon in 5-6 hrs., by a bridle-path (guide 6, horse and man 12 fr.); see p. 260. — To CHABLE in the Val de Bagnes (p. 304) from Saxon (or from Riddes, see below), a bridle-path over the Col des Etablons (7130'; fine view) in 7 hrs. (guide unnecessary). — To *Sembrancher* over the *Pas du Lens*, see p. 299.

On a hill on the right bank of the Rhone is *Saillon*, with a ruined castle. The train crosses the Rhone (1570') beyond (8½M.) *Riddes*, and the *Liserne* at (12 M.) *Ardon* (Hôt. du Pont). Ardon, *Vétroz*, and *Conthey*, all yielding excellent wine (see p. 256), lie at the foot of the hills to the left. The train crosses the *Morge*.

16 M. **Sion**, Ger. *Sitten* (1710'; pop. 5513; *Hôt. de la Poste*, R., L., & A. 3½, D. 3 fr.; *Hôt. du Midi*, moderate, good wine; *Pens. Beerli-Peter*), the capital of Canton Valais, which formed the French *Département du Simplon* in 1810-15, lies on the *Sionne*, which flows through it in an artificial channel covered with logs (*Rue du Grandpont*, forming the principal street). From a distance the town, with its castles on isolated hills, has a handsome appearance. On the height to the N. are the ruins of the episcopal castle of *Tourbillon* (2150'), erected in 1294, and burned down in 1788 (reached in 20 min. by the Rue du Château, to the left by the town-hall); extensive view, down to Martigny, and up to Leuk. On the lower hill to the right, on the site of a Roman fort, stands the old castle of *Valeria* (2040'), surrounded by towers and other buildings, among which is the *Church of Notre Dame de Valère* (9-13th cent.), with remarkable capitals, pictures, carved choir-stalls, etc. The newly-founded cantonal *Antiquarian Museum* occupies an adjacent room. — Close to the town, near Tourbillon, is the castle of *Majoria*, also burned down in 1788; part of it is now a barrack.

In the town itself the Gothic *Cathedral* (end of the 15th cent., with a tower of the 9th cent.) and the elegant church of *St. Théodule* adjoining it are objects of interest. In the old mansion of the *Supersaxo* family, in the Gundisgasse, is a fine hall with an artistically carved Renaissance ceiling of 1505 (visitors admitted).

From Sion over the *Rawyl* to Thun, see R. 56; over the *Pas de Cheville* to *Bex*, see R. 69; over the *Sanetsch* to *Gsteig*, see p. 259 (the *Hôtel Sanetsch* at *Zanfleuron* may be reached from Sion in 5 hrs.). — To the *Mayens de Sion* and *Evolena*, see R. 83. — In the deep ravine of the Borgne, about 1 M. from *Bramois* (p. 324; 3 M. to the E. of Sion), is the hermitage of *Longeborgne*, hewn out of the rock, and much frequented by pilgrims.

Above Sion the *Borgne* descends from the *Val d'Hérens* (p. 321), at the head of which we obtain a glimpse of the *Dents de Veisivi*. Near (19½ M.) *St. Léonard* we cross the *Rière*, which rises on the Rawyl. 21 M. *Granges*, the village, with a ruined castle and a church on the hill, lies on the left bank of the Rhone, ½ M. to the S.

25½ M. **Sierre**, Ger. *Siders* (1765'; pop. 1342; *Bellevue*, with garden, R. & L. 2-3, B. 1¼, D. 4, pens. 6 fr.; *Poste*, R., L., & A. 3 fr., D. 3 fr.; *Terminus Hotel*, R. 2 fr.; *Eng. Church Service*), with a number of interesting, but mostly dilapidated mediæval houses, lies picturesquely on a hill, amidst luxuriant vegetation.

20*

308 V. Route 80. LEUK-SUSTEN. From Martigny

On the side next the Rhone is the *Tour de Goubin*, or *Schinderthurm*, with a fine view of the Val d'Anniviers. On a rocky hill above the Rhone, 1/2 M. to the S., is the *Géronde* (2043'), formerly a Carthusian monastery, now a deaf and dumb asylum, with two little lakes (baths).
Above Sierre, to the N.W. (bridle-path viâ *Cortin* in 2 1/2-3 hrs.; mule 10 fr.; footpath viâ *Loc* and *Vogne* in 2 hrs.), is Montana (4048'; *Hôt. du Parc*, pens. 7-12 fr.; Eng. Ch. Serv. in summer), pleasantly situated near extensive pine-woods and several lakes, and commanding a magnificent view of the Valaisian Alps. Excursions: to the *Pointe de Vermala* (1/2 hr.), *Pointe de Mentahry* (1988'; 1 hr.), *Pépinet* (6500'; 2 hrs.), *Mont Lachaud* (7294'; 3 hrs.), *Col de Pochet* (8195'; 3 1/2 hrs.), *Mort Tubang* (9356'; 4 1/2 hrs.), *Glacier de la Plaine morte*, *Wildstrubel*, etc. (Letters should be addressed: Hôt. du Parc, Crans sur Sierre).
From *Sierre* to *St. Luc* in the Val d'Anniviers (*Bella Tola*) and *Zinal*, and passes to the *Turtmann Valley* and the *Val d'Hérens*, see R. 88. — To the *Baths of Leuk* viâ *Salgesch* and *Varen*, see p. 196.

Beyond Sierre a short tunnel and a deep cutting. Opposite, on the left bank of the Rhone, is the *Forest of Pfin*, a range of pine-clad hills. The village of *Pfin*, Fr. *Finge (ad fines)* is the boundary between the French and German languages. — 27 1/2 M. *Salgesch*, Fr. *Salquenen*, a wine-growing village. The line, hewn in the rock at places, approaches the Rhone, the valley of which is strewn with débris. We cross the deep gorge of the *Dala* (view to the left), **pass through** another tunnel, and cross the Rhone to —

30 1/2 M. **Leuk-Susten**, Fr. *Loëche-Souste* (2045'; *Hôtel de la Souste*, R., L., & A. 2 1/2-3 1/2, D. 4 fr.; *Rail. Restaurant*). The little old town of Leuk, Fr. *Loëche-Ville* (2470'; *Krone*, R. 2, D. 3 fr.), with its castle and towers, lies 1 M. distant, on the right bank, high above the Rhone (see p. 196).

One-horse carr. from the station to the *Baths of Leuk*, 12, two-horse 25 fr. Walkers reach the Baths (p. 194) in 3-3 1/2 hrs., by turning to the left (finger-post) beyond the church in the town, crossing the bridge over the Dala (p. 196), and following the old bridle-path to the right.

As the train leaves Leuk-Susten **we have** a retrospect, to the right, of the *Illgraben* or *Höllengraben*, a vast semicircular basin with bleak, yellowish slopes. The line passes the château of Baron Werra (on the right), and is carried **by a** stone embankment along the artificial channel of the river. We cross the *Turtmannbach* to (34 M.) **Turtmann** (2080'), Fr. *Tourtemagne*; the village (*Poste* or *Lion; Soleil*, both plain), lies 1/2 M. to the right, at the mouth of the *Turtmann Valley* (p. 331). The torrent forms a fine waterfall, 85' high, 8 min. from the Post Inn.

35 1/2 M. *Gampel*. The village, with deserted smelting-works, lies on the right bank, 1 M. distant, at the narrow mouth of the *Lötschen-Thal* (p. 197), through which peeps the snowy *Petersgrat* (p. 171). **Near** *Niedergestelen* are the scanty ruins of the *Gesteinburg*. 39 M. **Raron**, Fr. *Rarogne*; on the opposite bank, at the mouth of the *Bietschthal*, **lies** the village, with its old church on a rocky hill On a wooded height on the left bank, above the hamlet of *Turtig*, **is the little** pilgrimage-church of *Wandfluh*, reached by a **winding path** flanked with oratories. — We now cross the turbid *Visp*, which **has covered the** Rhone Valley here with its debris.

to Domodossola. **BRIG.** *V. Route 80.* 309

42$^1/_2$ M. **Visp** or **Vispach**, Fr. *Viège* (2160'; pop. 858; *Post*, R., L., & A. 2$^1/_2$, B. 1$^1/_4$, D. 4 fr.; *Sonne*, R. 2-2$^1/_2$, B. 1$^1/_4$, D. 3, pens. from 5 fr.; *Hôt. des Alpes*, near the station, R. 2$^1/_2$-3, D. 3$^1/_2$ fr.; *Rail. Restaurant*, mediocre), a picturesque village at the mouth of the *Visp Valley* (p. 331), has several old mansion-houses and handsome churches. The beautiful snow-mountain at the head of the Visp Valley is the *Balfrinhorn* (12,475'; p. 341), the first peak of the *Saasgrat*, which separates the valleys of Saas and Nicolai. — Railway to *Zermatt*, see p. 331.

Above Visp we traverse the stony tract at the influx of the *Gamsen*, which descends from the *Nanzer-Thal*. To the right is the pilgrim-resort of *Glis*, with a large church, at the base of the *Glishorn* (8290'); to the E. rises the fine pyramid of the *Bortelhorn* (p. 311). — We then cross the artificial channel of the *Saltine* to —

47$^1/_2$ M. **Brig** or **Brieg**, Fr. *Brigue* (2245'; pop. 1172; *Hôt. Couronne & Poste*, R., L., & A. 3-4, D. 4 fr.; *Angleterre*, R., L., & A. 3-4, lunch 3, D. 4$^1/_2$ fr.; *Hôt.-Pens. Suisse*; *Hôt.-Pens. Müller*, R., L., & A. 1$^1/_2$-2, B. 1, D. 2$^1/_2$, pens. 4$^1/_2$-5 fr.; *Hôt. de Londres*, opposite the post-office; *Rail. Restaurant*, with beds, lunch 2$^1/_2$ fr.), a small town, where the railway terminates. The turreted *Stockalper Château*, containing an interesting interior court, a large hall, etc., is the largest private residence in Switzerland. Kasper Stockalper (d. 1691), who built it, dominated the trade over the Simplon, which he protected by a guard of 70 men. The terrace in front of the former Jesuits' monastery commands a fine view. The fine snow-mountain to the S.E. is the Wasenhorn; to the N. the Sparrhorn, Belalp, and Eggishorn are visible.

To **Belalp**, a beautiful excursion (bridle-path, 4$^1/_2$-5 hrs.; porter 6, horse 15 fr.). Just before (1M.) *Naters* (p. 318), on the right bank of the Rhone, we ascend to the left (finger-post) by an almost shadeless bridle-path, steep at places, viâ *Geimen* (3440'), to (2 hrs.) the village of *Platten* (4330'; rustic Inn); then through wood and over the *Rischenen* and *Eggen-Alps*.

The conspicuous °*Hôtel Belalp* (7155'; R., L., & A. 4, B. 1$^1/_2$, D. 5, pens 9-11 fr.; English Church), situated on the *Lüsgen-Alp* at the base of the Sparrhorn, and high above the Aletsch Glacier, is a good centre for excursions. (Splendid view of the Valaisian Alps. Sunrise particularly fine.) The little *Villa Lüsgen*, 5 min. above the hotel, belonged to Prof. Tyndall (d. 1893). A pleasant walk may be taken on the hillside, past the hamlet of *Belalp* (6735'), to that of (1$^1/_2$-2 hrs.) *Nessel* (6675'; milk, etc.), high above the Rhone Valley, with beautiful view.

To the °**Upper Aletsch Glacier**, very attractive. Bridle-path from the hotel to the (1$^1/_2$ hr.) W. moraine; then across this and a second moraine to the almost uncrevassed glacier, with its numerous 'ice-tables', 'glacier-mills', etc., as far as the (1$^1/_4$ hr.) *Oberaletsch Hut* of the S. A. C. on the E. side (8685'), at the foot of the *Fusshörner* (p. 310). We may walk up the glacier to the right to the foot of the *Great Aletschhorn* (p. 310), or traverse the *Beichfirn* to the left to the snow-slopes of the *Beichgrat* (p. 310; 8-10 hrs. in all; guide, 5 fr., and provisions necessary).

*°**Sparrhorn** (Belalphorn*, 9890'), 2$^1/_2$-3 hrs. from the inn, bridle-path most of the way (guide 4 fr., unnecessary for the experienced). Beautiful view, finer on the S. side than from the Eggishorn, but inferior to it on the N. side. (Panorama at the inn.) To the N., above the Aletsch Glacier, and to the left of the Fusshörner, the Great Aletschhorn is most prominent;

adjoining it are the Sattelhorn, Ebnefluh, Distelhorn, Breithorn, and the Tschingelhörner, and to the left, adjacent to the Hochstock, is the Nesthorn. Towards the S. rises the broad mass of the Monte Leone; more to the right are the Fletschhorn, Monte Rosa, Mischabel, Matterhorn, Weisshorn, Brunnegghorn, Dent Blanche, Grand Combin, and Mont Blanc. To the left of Monte Leone are the Bortelhorn, Hüllehorn, Helsenhorn, Punta d'Arbola, Güschihorn, Ofenhorn, the peaks of the St. Gotthard group, and lastly the Walliser Fiescherhörner.
The *Great Aletschhorn* (13,775'; 7-8 hrs., guide 40 fr.); *Great Nesthorn* (12,530'; 6-7 hrs., 40 fr.; grand view); *Lötschenthaler Breithorn* (12,410'; 5-6 hrs., 40 fr.); *Fusshörner* (11,900'; 4 hrs.), an interesting climb; *Sattelhorn* (12,290'; 4½ hrs.); and *Schienhorn* (12,490'; 6-7 hrs., very difficult) may be ascended from the Oberaletsch Hut (by experts only).
From the Belalp to the *Eggishorn Hotel* (5½ hrs.), see p. 317. Guide 8 fr.; necessary only for the passage of the Great Aletsch Glacier (3 fr.).
FROM BELALP TO RIED OVER THE BEICHGRAT, toilsome, but very interesting (8-9 hrs.; guide 25 fr.). We ascend the *Upper Aletsch Glacier* and the *Beichfirn* to the (4½-5 hrs.) Beichgrat (10,235'), between the *Schienhorn* and the *Lötschenthaler Breithorn* (see above); then descend rapidly over the *Distel Glacier* to the beautiful *Gletscherstaffel Alp*, the *Fafler Alp* (Inn, see p. 171), and (3½-4 hrs.) *Ried* (p. 198).

The **Upper Valais**, and the *Grimsel*, **Furka**, and **Gries** passes, see RR. 81, 52, 35, 82.

The SIMPLON ROAD, the first great Alpine route after the Brenner, constructed by order of Napoleon I. in 1800-6, quits the Rhone Valley here. The scenery is far finer than on the Splügen route, but the engineering of the road is less striking. The road is kept open for carriages in winter also (diligences, etc., see p. 306).

WALKERS should allow the following times: from Brig to Berisal, 3¼ hrs. by the road, 2¾ hrs. by the short-cuts; from Berisal to the Hospice 2¼ hrs.; to Simplon, 1¾ hr.; to Algaby, 35 min. (footpath in 20 min.); to Gondo, 1¼ hr.; to Iselle ¾ hr.; Domodossola 3¼ hrs. In the reverse direction: from Domodossola to Iselle 4 hrs.; Gondo 1 hr.; Algaby 1¾ hr.; Simplon ¾ hr. (by the footpath); Hospice 2¼ hrs.; Berisal 2¼ hrs.; Brig 2½ hrs. (or 1¾ hr. by the footpath).

The ascent begins at the post-office (2320') at Brig. (A somewhat steep footpath diverging to the left outside the town, then following the telegraph-wires and part of the old bridle-path, and joining the road about ½ M. before the second refuge, is a considerable short-cut.) The road is soon joined (½ M.) by the old road from *Glis* (p. 309), which crosses the gorge of the Saltine by the lofty *Pont Napoléon* (2485'). Opposite rises the *Glishorn* (8290'). The road winds over green pastures to the E., in the direction of the *Klenenhorn* (8840'). Fine retrospective view of the Rhone Valley; high above the right bank of the Rhone is the Hôtel Belalp, commanded by the Sparrhorn, with the Nesthorn on the left; to the right, farther up, the cone of the Eggishorn; above us, to the S., is the Kaltwasser Glacier, which the road afterwards passes, and the Schönhorn. Beyond the hamlet of *Schlucht*, by the (2¾ M.) **First Refuge** (3200'), the road turns back and ascends the wooded slope in many windings, affording splendid views of the Rhone Valley and the mountains of the Aletsch region. Beyond the *Bleiche Kapelle* (4110') it again approaches the deep ravine of the Saltine. By the (2¾ M.) *Second*, or *Schallbery*, *Refuge* (4380'; auberge), where we come in sight of the pass, two brooks from the *Staldhorn*, one on each side, unite far below with the Saltine, the valley of which *(Ganter-Thal)* now turns to the E. Fine view of the picturesquely grouped valleys, and of the Wasenhorn, Furggenbaumhorn, and Bortelhorn. The road, now nearly level, traverses the Ganter-Thal to the (2 M.) *Ganter Bridge* (4820') and ascends in a wide curve (steep short-cut to the left) to (1 M.) —

9 M. **Berisal**, the *Third Refuge* (5005'; *Hôt.-Pens. Berisal* or *de la Poste*, R., L., & A. 3½, B. 1½, D. 4 fr., finely situated; *Engl. Ch. Service* in summer; opposite, a *Restaurant*).

EXCURSIONS. *Wasenhorn (Punta di Terrarossa*, 10,680'; 5-6 hrs.; guide 8 fr.), interesting, and not difficult. — *Bettlihorn* (9720'), over the *Saflisch Pass* (8650') in 5 hrs. with guide, not difficult (comp. p. 317). — *Bortelhorn (Punta del Rebbio*, 10,512'), by the *Bortel-Alp* and the *Bortel Glacier* in 5 hrs., laborious (guide 10 fr.).

FROM BERISAL TO ISELLE VIÂ VEGLIA, 8-9 hrs., with guide, an attractive route, but fatiguing. We either ascend viâ the *Bortel-Alp* and the glacier on the N. side of the *Furggenbaumhorn (Punta d'Aurona*, 9820') to the *Forca del Rebbio* (9040'), and descend over rocks, débris, and grassy slopes to the *Alp Veglia* (p. 318); or we may proceed viâ the *Laub-Alp* (6265') and the Furggenbaum Pass (*Passo di Forchetta* or *Forca d'Aurona*; 8826'), between the Furggenbaumhorn and the Wasenhorn. From Veglia we descend to *Trasquera* and (6 hrs.) *Iselle*, see p. 313. — From the Alp Veglia over the *Passo di Vallendra* (7995') and the *Passo di Buscagna* (7743') to *Devero* (p. 318), 6-7 hrs., with guide, a fine route, not difficult. — From Veglia over the *Kaltwasser Pass (Bocchetta d'Aurona*; 9250') and the *Kaltwasser Glacier* to the Simplon, 6-7 hrs., with guide, for adepts only.

In 1 hr. more the road reaches the *Fourth Refuge* (5645'). To the right the top of the pass is again visible; above it rise the Rauthorn with the Raut Glacier and the finely shaped Fletschhorn with the Rossboden Glacier; beautiful retrospective view of the Aletschhorn, Schienhorn, etc. Beyond the (1¼ M.) *Kapfloch*, hewn in the rock for 33 yds., is the (¾ M.) *Fifth*, or *Schallbett, Refuge* (6345'). Between this point and the top of the pass is the most dangerous part of the road during the period of avalanches and storms. Over the (¾ M.) *Wasser Gallery* (6460') is precipitated the stream which issues from the *Kaltwasser Glacier*. To the left is *Monte Leone* (see below). The road then passes through the *Old Gallery* and the long *Joseph Gallery*, beyond which, to the left, is a third gallery used in winter. The (1 M.) *Sixth Refuge* (6540') commands a splendid final view of the Bernese Alps.

About 5 min. farther on we reach the highest point of the Simplon (6590'; 6¼ M. from Berisal), ½ M. beyond which is the **Hospice** (6570'), at the base of the *Schönhorn* (10,505'), a large building with a lofty flight of steps, founded by Napoleon for the reception of travellers, and subject to the same rules as that of the Great St. Bernard. It remained unfinished from want of means till 1825, when the St. Bernard Hospice purchased the buildings.

EXCURSIONS. *Schönhorn* (10,505'; 3½ hrs., with guide), laborious but interesting. — **Monte Leone** (11,684'; 6-7 hrs.; guide 12 fr.), by the *Bretthorn Pass* (10,990') and the *Alpien Glacier*, difficult and unfit for novices. A preferable route descends the Simplon road to *Algaby* (p. 313) and mounts thence viâ *Alpien* to the (3½-4 hrs. from the village of Simplon) huts of the *Schwartze Balmen*, or *Upper Fraxinado Alp* (6890'), where the night is spent. A somewhat steep ascent, passing two pretty little lakes, leads thence up to the S.E. corner of the *Alpien Glacier*, from which the summit may be gained by either the S. or the S.W. arête (4½ hrs., guide 12 fr.). — From the hospice to *Stalden* by the *Bistenen Pass* (guide 12 fr.; mule-track), see p. 382; to *Saas*, see below.

A broad open valley resembling a dried-up lake, **bounded by snow-capped peaks**, forms the highest part of the Pass. The hardy rhododendron alone thrives here. The (¾ M.) *Old Hospice* (5700'), a high square building with a **tower**, on the right, below the new road, is now occupied by herdsmen. ¾ M. **Seventh Refuge**, by the *Engeloch* (5855'). Farther down we cross the (1¼ M.) *Krummbach* (5305'), pass the chalets of (½ M.) **Eggen (5250'**; to the right is the *Rossboden Glacier* with its huge **moraine, see below**), and cross the (¼ M.) *Sengbach* (5115') to (½ M.) —

21 M. **Simplon** (4855'), Ital. *Sempione*, Ger. *Simpeln* (*Poste*, R., L., & A. 3½, D. 4, pens. from 6 fr.; *Hôt. Fletschhorn*, at the lower end of the village, R., L., & A. 2½, D. 4 fr.), **among pastures, at the N. E. base of the *Fletschhorn* (see below).**

FROM SIMPLON TO SAAS several routes. The finest is across the *Rossboden Pass* (10-11 hrs.; difficult, suitable for adepts only; guide 20 fr., *Jos. Dorsaz* of Simplon). At the (20 min.) chalets of *Eggen* (see above) we diverge to the left from the Simplon road and ascend on the left side of the Sengbach to the (1 hr.) *Rossboden Alp* (6360'), with fine view of the séracs of the Rossboden Glacier. Farther on we mount over grassy slopes and débris of moraine to the *Griesseren Glacier*, beyond which we climb a steep rocky wall to the (4 hrs.) pass (about 10,500'), to the S. of the *Rauthorn* (10,725'), commanding a splendid view (to the right below us is the Gamsen Glacier). We descend across the *Mattwald Glacier* to the *Hofers Alp* (see below) and *Saas im Grund* (p. 340). — Another grand, but still more difficult pass, not without danger, is the **Laquin-Joch** (11,473'), between the Laquinhorn and the Weissmies (10-11 hrs.; guide 30 fr.).

TO SAAS OVER THE SIRVOLTEN AND SIMELI PASSES (or the GAMSER JOCH), 10-11 hrs., fairly interesting (guide 20 fr.). By the Seventh Refuge (see above) we descend to the left, cross the Krummbach to the *Klusmatten Alp*, and ascend by a narrow path towards **a waterfall** visible from below. On the left side of it we mount a 'couloir', steep at first (leaving **the *Sirvolten Lake*** to the left), to the (4 hrs.) Sirvolten Pass (8740'), **to the N.** of the *Sirvoltenhorn* (9344'); view limited. Descent **over** rock and débris (keeping well to the left) to the head of the *Gamser-Thal*, **into** which the *Gamsen Glacier* descends. We ascend the glacier gradually to the S.W. towards an arête coming down from the Magenhorn on the E., at the foot of which the route divides: to the right to **the Simeli Pass** (9935'); to **the left to the Gamser Joch** (about 9190'; each 2-2½ hrs. from the Sirvolten Pass). These passes, between which rises the pointed *Magenhorn* (10,243'), command beautiful views of the majestic *Mischabel* group; immediately to the left is the Fletschhorn with the Mattwald Glacier; to the E. are the Monte Leone and the St. Gotthard group; and to the N. are the Bernese Alps from the Furka to the Diablerets. A still grander point is the *Mattwaldhorn* (10,672'), easily ascended from the Simeli Pass in 1 hour. Toilsome descent from the Gamser Joch over the moraine of the *Mattwald Glacier*. Rounding the lower part of the valley to the left, we next come to the *Sattel* (9025'), on the E. side of the *Aeusser-Rothhorn* (10,854'), and to the *Hofers-Alp* (6854'). The path now improves and leads by *Bodmen* to (3½ hrs.) *Saas im Grund* (p. 340). Descent by the *Alp Sevenen* to *Balen* (p. 340) much longer and not advisable.

The **Fletschhorn** (*Rossbodenhorn*; 13,125'), 9-10 hrs. from Simplon (guide

to Domodossola. GONDO. *V. Route 80.* 313

25 fr.), fatiguing, but safe for proficients. A night is spent in the *Hohsaas-Hütte* (about 8000'), 3 hrs. from Simplon, above the Laquin-Thal (see below). Thence up the S.E. arête to the top in 6-7 hrs.

Beyond the (¹/₄ M.) *Löwenbach*, the road forms a wide bend and enters the *Laquin-Thal*, and at the (1³/₄ M.) hamlet of *Algaby* or *Gstein* (4042'; Inn, primitive) it crosses the Krummbach, into which the *Laquinbach* falls. Below this the brook is named the *Doveria*. Beyond the (¹/₄ M.) *Gallery of Algaby* begins the *Ravine of Gondo, watered by the brawling Doveria, one of the wildest and grandest gorges in the Alps, which becomes narrower and deeper at every step, till its smooth walls of mica-slate quite overhang the road. The road passes the (1¹/₄ M.) *Eighth Refuge* (3841'), beyond which the path to *Alpien* (p. 312) diverges on the left, and crosses the Doveria by (¹/₂ M.) the *Ponte Alto* (3747'), and by another bridge near the (¹/₂ M.) *Ninth Refuge* (3514'). A huge mass of rock, which seems to terminate the road here, is pierced by the **Gallery of Gondo**, a tunnel 245 yds. long, with the inscription, '*Aere Italo 1805 Nap. Imp.*' At the farther end of the gallery the *Fressinone*, or *Alpienbach*, forms a waterfall, which is crossed by a slender bridge. On both sides the rocks tower to a dizzy height (about 2000'). The sombre entrance to the tunnel contrasts strikingly with the white spray of the cascade, while in the rear the beautiful *Bodmer Glacier* is visible beyond the ravine. Traces of the old road are still visible opposite the waterfall. Farther on are several smaller falls. The hamlet of (2¹/₂ M.) **Gondo** (2815') is the last Swiss village (custom-house). The tall square tower here (now an inn, uninviting) was erected by the Stockalper family as a refuge for travellers, long before the new road was made. Opposite is a more attractive looking 'Osteria'.

To the S. opens the narrow *Val Vaira* or *Zwischbergen-Thal*, from which we may without difficulty cross the Zwischbergen Pass (10,735'), between the *Weissmies* (p. 341) and the *Portjengrat* (*Pizzo d'Andolla*, 12,010'), to *Saas im Grund* (p. 340; 12 hrs.; guide 20 fr.).

A column of granite on the left, ¹/₂ M. from Gondo, marks the boundary of Italy (2610'). The first Italian village is (¹/₄ M.) *S. Marco*. Below this the valley is called *Val di Vedro*. We next pass through a new tunnel, and reach (1³/₄ M.) —

30 M. **Iselle** (2155'; *Posta*, well spoken of, R., L., & A. 3¹/₂-4, B. 1¹/₂ fr.), where luggage is examined. Below the church of *Trasquera*, which stands on the hill to the left, the road crosses the (2 M.) *Cairasca*. (To the Rhone Valley by the *Alp Veglia*, see pp. 311, 318.) Near *Varzo* (1865'; Inn, on the road), a large village on the left, the vegetation becomes more luxuriant (chestnuts, figs, mulberries, maize, vineyards). Passing through a picturesque ravine and the (3¹/₂ M.) *Gallery of Crevola* (1286'), we descend by a curve past the village of **Crevola** (1100') to the *Osteria della Stella*, where for the last time we cross the Doveria by a bridge 100' high, near its confluence with the *Tosa*, which here emerges from the *Val Antigorio* (p. 321). The fertile valley, now called *Valle d'Ossola*,

though frequently ravaged by inundations, is strikingly picturesque and thoroughly Italian in character. We next reach (21/4 M.) —
41 M. **Domodossŏla** (905'; *Gr. Hôt. de la Ville et Poste*, R., L., & A. 3½, lunch 3, D. 4-5 fr., with a good café; *Hôt. d'Espagne*, well spoken of, with a restaurant on the ground floor), a small town with 2200 inhab., charmingly situated on the Tosa, which becomes navigable here. The *Palazzo Silva* (16th cent.) contains a few antiquities. In the Via Garibaldi is a bust of Garibaldi (1890). The *Mount Calvary, 20 min. to the S., commands a superb view.

To *Locarno* through the *Val Vigezzo*, see p. 441. Diligence to (10½ M.; fare 3 fr.) *S. Maria Maggiore*, daily, at 5 p. m. — On the W. opens the **Valle di Bognanco**, with mineral springs, from which several passes lead to the *Zwischbergen* and *Antrona Valleys* (see p. 313 and below).

RAILWAY to *Gravellona* (Pallanza, Stresa), *Orta, Novara*, see R. 110.

From the first station (3½ M.) *Villadossola* an interesting route leads OVER THE ANTRONA PASS TO SAAS (12-13 hrs. guide unnecessary). A carriage-road ascends the left, and afterwards the right bank of the *Ovesca*, viâ *Viganella* and *Schieramco*, to (2½ hrs.) *Antronapiana* (2955'; bed at the syndic's). Footpath thence, past the charming little *Antrona Lake* (3650'), formed by a landslip from the *Pizzo Pozzolo* (8300') in 1632, to the (3½ hrs.) *Cingino Alps* (6860') and along the slopes of the *Pizzo Cingino* (10,570'), far above the little *Lago Cingino* (7190'), to the (2½ hrs.) **Saas** or **Antrona Pass** (9320'), between the *Jäzzihorn* (*Pizzo di Cingino*, 10,570') on the left and the *Latelhorn* (10,525'; easily ascended from the pass in 1½ hr.; see p. 341) on the right. Descent on the right side of the *Furggen Glacier* to the *Furggthal*, *Almagell*, and (4 hrs.) *Saas* (p. 340). — To MATTMARK from Antrona a direct but rough route crosses the Antigine or Ofenthal Pass (9300'). From the ascent to the Cingino Alp (see above) we diverge to the left to the *Lombraoro Alp*, whence a steep ascent leads viâ the *Laugera di Sopra Alp* to the pass, between the *Pizzo Cingino* and the *Pizzo d'Antigine* (10,465'; a fine point, 1¼ hr. from the pass); descent through the wild *Ofenthal* to the *Mattmark Alp* (8-9 hrs. from Antronapiana; p. 343).

At the third station (6¾ M.) *Piedimulera* (p. 451) the picturesque *Val d'Anzasca* opens to the right (to *Macugnaga* and over the *Moro Pass* to *Saas*, see p. 341).

81. From the Rhone Glacier to Brig. The Eggishorn.

Comp. *Maps, pp. 118, 316, and 310*.

31 M. DILIGENCE to Brig twice daily (7.30 a.m. and 2.10 p.m.) in 4¾ hrs. (10 fr. 80, coupé 13 fr. 30 c.; to Fiesch in 2¾ hrs.; 7 fr. 5, coupé 8 fr. 60 c.). Walking is preferable from Münster on. In the reverse direction the diligence takes 7¼ hrs. — One-horse carr. from the Rhone Glacier to Münster 10, two-horse 20 fr.; to **Fiesch** 18 or 35, to Brig 30 or 60 fr.; from Brig to Fiesch 12 or 25, to Ulrichen 20 or 40, the Rhone Glacier 30 or 60 fr.; from the Rhone Glacier to Andermatt and Göschenen 30 or 60, Grimsel Hospice 10 or 15, Handegg 15 or 25, Meiringen 30 or 60 fr.

Hotel at the Rhone Glacier. HÔT. DU GLACIER DU RHÔNE, first class but not quite satisfactory in some respects, R., L., & A. 2½-5 and upwards, B. 1½, lunch 3½, D. 5 fr., beer on draught at the café, 50 c. *Engl. Ch. Serv.* in summer. — The *Hôtel Belvedere*, seen high up on the Furka road, belongs to the same proprietor.

The ***Rhone Glacier**, imbedded between the *Gerstenhörner* (10,450') and *Gelmerhörner* (10,500') on the W., and the *Galenstock* (11,805'), *Rhonestock* (11,825'), and *Dammastock* (11,910') on the E., ascends in terraces for about 6 M. A few centuries ago, this glacier, from which issues the *Rhone*, filled more than half of the *Gletsch* (5750'), the valley covered partly with debris, partly with grass, lying at the junction of the three great roads from the *Furka* (Andermatt, p. 126), the *Grimsel* (Meiringen, p. 190), and the *Rhone Valley* (see below). From the hotel the glacier is reached in ½ hr. by a road crossing the bridge and ascending the left bank of the *Rhone*. That river issues from a beautiful vault of blue ice; and an artificial ice-grotto, hewn in the glacier, is worth seeing (½ fr.). — The natives give the name of *Rotten*, or *Rhodan*, to three partially warm springs rising at the back of the hotel, to the W., which they regard as the source of the river.

A short distance from the hotel, and again a little farther on, the road crosses the infant Rhone, which dashes through its rocky ravine far below, and descends in long windings through pine-woods on the right bank to (3¾ M.) Oberwald (4455'; *Hôtel Furca*, very plain), at the bottom of the valley of the *Upper Valais*, a broad expanse of pasture, enclosed by monotonous chains of mountains. In front rises the majestic Weisshorn, and behind us the Galenstock. The valley consists of three regions, the highest extending a little beyond Fiesch, the second to the bridge of Grengiols, and the third lying below this bridge. The inhabitants (Rom. Cath.) speak German; French begins near Sion (p. 307).

From the wild Geren-Thal, a ravine opening to the E. of Oberwald, a fatiguing pass crosses the *Kühboden Glacier* and the *Geren Pass* (9052'), to the S. of the *Kühbodenhorn* (10,080'), to the *Alp Nuova* and *All' Acqua* in the Val Bedretto (see below; 8 hrs., guide 18 fr.).

2¼ M. Obergestelen (4450'), a village rebuilt since a fire in 1868 (footpath to the Grimsel, see p. 190). Opposite (1¾ M.) Ulrichen or *Urlichen* (4380'; *Hôt. zum Griesgletscher*, plain), a village with a new church, is the mouth of the *Eginen-Thal*. (Over the *Gries Pass* to the Tosa Falls, see p. 319.)

To AIROLO OVER THE NUFENEN PASS (8½ hrs.), a rough, uninteresting bridle-path (guide necessary, 12, horse 25 fr.). Beginning of route, see p. 319. At (2¼ hrs.) *Attstaffel* (p. 319) the path leads to the left, ascends in zigzags, and crosses the (1¼ hr.) Nufenen Pass (*Passo di Novena*, 8005'), between the *Pizzo Gallina* (10,066') on the left and the *Nufenenstock* (9400') on the right, to the Val Bedretto. Immediately to the N. of the pass rises the *Ticino*, which the path follows, first on the right, and below the *Alp Cruina* on the left bank, to the (1¾ hr.) Hospice all' Acqua (5265'; poor inn; route over the *S. Giacomo Pass* to the *Tosa Falls*, see p. 320). The lofty Val Bedretto is bleak and barren. The wooded slopes are overtopped by bare pinnacles of rock. Our path frequently crosses the tracks of avalanches. 1 hr. the hamlet of *Bedretto* (4610'; Inn, rustic); 20 min. *Villa* (very poor inn; route over the *Cavanna Pass* to *Realp*, see p. 125). Near (20 min.) Ossasco (4365'; Albergo delle Alpi, rustic, but dear) the road crosses the Ticino. Beyond (25 min.) *Fontana* is the picturesque *Val Ruvino* to the right, with waterfalls. Then (1 hr.) *Airolo* (p. 114).

The following villages are *Geschenen* (4395') and (2¼ M.) —

10 M. Münster (4530'; *Goldnes Kreuz*; one-horse carr. to Brig 18 fr. and fee). Fine view from the chapel-hill.

The *Löffelhorn (10,140'; 4½ hrs., fatiguing; guide 6 fr.) is ascended from Münster, partly over snow and granite-rocks. View like that from the Eggishorn (see below), with the addition of the Finsteraarhorn in the foreground. — The ascent of the *Blindenhorn (11,095'; 6-7 hrs.; guide 12 fr.) is very attractive. From *Reckingen* (see below) we follow a good path through the *Blinden-Thal* to the end of the *Blinden Glacier*. Thence we ascend on the left bank of the *Hohstellbach* and across the *Sulz Glacier*, to the *Griesgletscher Pass* (10,585'), between the Merzenbachschien and the Blindenhorn, and reach the summit by a steep ascent to the right. Magnificent view

To the left, over the *Blinden-Thal* (see above), appears the *Rappenhorn* or *Mittaghorn* (10,374'), adjoined on the left by the *Blindenhorn* (see above). The next villages are *Reckingen* (with the finest church in the valley), *Gluringen*, *Ritzingen*, *Biel*, *Selkingen*, and *Blitzingen* (*Pens. Seiler, 4-5 fr.). Beyond (5 M.) *Niederwald* (4050'), the Rhone forces its passage to a lower region of the valley. The road leads on the right bank, high above the river, and finally descends through wood in two great curves.

19½ M. **Fiesch** (3460'; *Hôt. du Glacier et Poste*, R., L., & A. 3, D. 4, pens. 6 fr.; *Hôt. des Alpes*, R., L., & A. 2½-3, D. 3 fr.), prettily situated at the influx of the *Fieschbach* into the Rhone.

*ASCENT OF THE EGGISHORN, very interesting (5 hrs.; guide unnecessary; to the inn 3 hrs.; porter 5, horse 10 fr.). From the bridge (or below the Hôt. des Alpes) the good bridle-path ascends to the right, somewhat steeply, chiefly through wood, past (1½ hr.) a little *Inn* (night-quarters) and several earth pyramids lying a little to the left, to (40 min.) the *Fiescher Alp* (6210'); then over pastures (where we may take the direct path following the telegraph-posts, ½ hr. shorter, but steep) to the (50 min.) *Hôt.-Pens. Jungfrau* (7195'; R. & A. 3½, lunch 3, D. 5, pens. 8-11 fr.), a favourite resort of English tourists, and suitable for some stay *(English Church)*. From the hotel to the top 2 hrs. more (guide 4 fr., but not needed; horse 7 fr.). The bridle-path ascends in zigzags, then turns to the right, and after ¾ hr. to the left (the path to the right leads to the Märjelen-See; see below). After ¾ hr. more the bridle-path ends, and we ascend by a good footpath and lastly mount steps of rock to the (½ hr.) summit of the *Eggishorn (9625'), the highest peak of the ridge which separates the *Great Aletsch Glacier* from the Rhone Valley, with a superb view of the Bernese and the **Valaisian** Alps (compare the annexed Panorama by *Imfeld*). We also overlook a great part of the Simplon Route and of the Nicolai-Thal to the S. (p. 332).

EXCURSIONS. From the Hôtel Jungfrau a good path leads to the N., at first nearly level and afterwards undulating, skirting the slope above the *Fiesch Valley* and affording an excellent survey of the beautiful ice-fall of the *Fiesch Glacier*, and then crosses the *Thaelligrat* to the left to the Märjelen-See (7710'), on which floating ice is frequently seen. On the left bank of the *Seebach* emerging from the lake is the (1¼ hr.) *Märjelen-Alp*. [The Fiesch Glacier may be visited hence by a path descending to the right to the *Stock-Alp*.] On the N. side of the Märjelen-See, a path leads in 25 min. to the margin of the *Great Aletsch Glacier*. Hence to the Concordia Hut of the S. A. C. (9415') a beautiful glacier-walk of 3 hrs. (5 hrs. from the Jungfrau Hotel; guide 15 fr.). The hut is grandly situated at the foot of the *Trugberg*. From the hut to the (3 hrs.) *Jungfrau-Joch* (p. 178), with splendid view, not difficult for experts (2 guides, 20 fr. each). Ascent of the *Jungfrau*, see p. 173; *Finsteraarhorn*, p. 189. — The **Great Aletschhorn** (13,775'; guide 40 fr.), the second-highest of the

Bernese peaks, is ascended either from the Concordia Hut (in 7 hrs.) or from the Oberaletsch Hut (p. 309; in 7-8 hrs.); difficult, for experts only.

From the Eggishorn Hotel to *Grindelwald* over the *Mönchjoch*, 15-16 hrs., see p. 178; to the *Grimsel Hospice* over the *Oberaar-Joch* or *Studer-Joch*, 14 hrs., see p. 190. From the Concordia Hut to the Grimsel Hospice, viâ the *Grünhornlücke* (p. 190), *Gamslücke* (p. 190), and *Oberaar-Joch* (p. 190), a fine glacier-tour of 10-12 hrs., not difficult for adepts with good guides. — From Lauterbrunnen to the *Eggishorn* by the *Lauinenthor*, *Roththal-Sattel*, and *Ebnefluh-Joch*, see p. 171.

FROM THE EGGISHORN HOTEL VIÂ THE LÖTSCHENLÜCKE TO RIED 13-14 hrs., a grand glacier-route (guide 30 fr.). We ascend the *Great Aletsch Glacier* to the Lötschenlücke (10,515'), a depression of the *Anengrat*, to the N. of the *Sattelhorn* (12,290'), and descend the crevassed *Lötschen Glacier* to the *Fafler Alp* (Chalet Seiler, p. 171) and *Ried* in the Lötschen-Thal (p. 198).

FROM THE EGGISHORN HOTEL TO THE RIEDERALP AND BELALP (guide 8, horse 20 fr.; but riding not possible on the glacier), 5½ hrs., a beautiful walk. The bridle-path from the hotel, past the little English Church, running nearly at the same level, high above the Rhone Valley, leads over the *Bettmer-Alp*, with its little lake (6530'; abounding in fish), and the *Goppisberg-Alp*, and turns to the right at the cross to the (2½ hrs.) Riederalp (6315'; *Hôt.-Pens. Riederalp*, R., L., & A. 3, D. 4½, pens. 7½ fr.; Engl. Ch. Serv. in summer). The beautiful situation and mild climate of this alp adapt it for a stay of some time. Here we ascend to the right to the (25 min.) *Rieder Furka* (6820'; *Pens. Rieder Furka*), whence we may scale the *Riederhorn* (7348'; ½ hr.), a very fine point of view. (We command practically the same view by following for 6 min. in the direction of the glacier the small path ascending from the Furka on the slope of the Riederhorn.) The *Bettmerhorn* (9400', 2½-3 hrs. from the Rieder Furka) is not difficult; experts may go on by the arête towards the Egishorn and descend by the *Eislücke* (8950') to the Hôtel Jungfrau (a grand but rough route; guide desirable). Descent, with splendid views of the *Upper Aletsch* or *Jägi Glacier*, lying between the Sparrhorn and the Fusshörner and overshadowed by the Schienhorn, to the (½ hr.) *Great Aletsch Glacier* (5185'), which is safely crossed here in ½ hr. (with guide) to *Aletschbord*; then a steep ascent, past the chalets of *Unter-Aletsch*, to the (1¼ hr.) *Hôtel Belalp* (p. 309). — FROM THE RIEDERALP TO MÖREL, 2-2½ hrs. (guide, 5 fr., unnecessary). A bridle-path, leading at first across pastures, then through wood, makes a long sweep to the right (steep paths to the left to be avoided) and crosses pastures again (very hot about midday), with splendid views of the Rhone Valley, the Simplon Mts., etc. It then descends to *Ried* (3890'), a finely situated village, and *Mörel* (p. 318; from Mörel t the Riederalp 3-3½ hrs.; porter 5, horse 10 fr.).

FROM FIESCH OVER THE ALBRUN PASS TO BACENO OR TO THE TOSA FALLS, 12-13 hrs. (guide from Im Feld desirable; to Baceno 12 fr.). A good bridle-path leads by *Aernen* and the *Binnegg*, with fine view of the Binnen-Thal and Valais, to (1¾ hr.) *Ausser-Binn* and (1½ hr.) *Schmidhäuser* or *Binn* (4720' *Hôt. Ofenhorn*, finely situated, Engl. Ch. Serv. in summer), a village, with an attractive church, in the Binnen-Thal, a valley interesting to mineralogists. (Guides, Jos. Welschen and J. J. Gorsat, of Binn; Ad. and Elias Walpen of Im Feld.) The *Bettlihorn* (9720'; 4½-5 hrs.; guide 8 fr.) is easily ascended from Binn by the *Furggen-Alp* (comp. p. 311). Another easy ascent is that of the *Mittaghorn* or *Rappenhorn* (10,374'; 5½ hrs., with guide), viâ *Feldbach* and the *Rappen Glacier*. — *Ofenhorn* (*Punta d'Arbola*; 10,637'), 6-7 hrs. (guide 10 fr.), not difficult for experts. We may either diverge beyond *Auf dem Platt* (p. 318) from the Albrun Pass route and ascend by the *Eggerofen Valley* to the *Passo del Ghiacciaio d'Arbola*, whence the summit is gained by the S.W. arête; or (preferable) we may ascend from the *Hohsand Pass* (9608'; easy glacier-pass from Binn to the Tosa Falls, 9 hrs.) by the N. arête. — *Helsenhorn* (10,712'), by the *Ritter Pass* in 6½ hrs., not difficult (see p. 318). — *Hüllehorn* (10,450'), by the *Mätt thal* and the *Rämi Glacier* in 6 hrs., difficult, for experts only. — We now follow the left bank of the *Binna*, by *Giessen*, to (¾ hr.) *Im*

BINNEN-THAL.

Feld (5145'), where the path, now indifferent, crosses to the right bank (guide advisable). We enter (¼ hr.) a pine-wood, pass a number of chalets, and reach (1¾ hr.) the last huts *Auf dem Platt* (6925'; chalybeate spring). We now ascend steeply to the (1 hr.) Albrun Pass (*Bocchetta d'Arbola*, 7910'), between the *Ofenhorn* (p. 317) on the left and the *Albrunhorn* (9450') on the right. We descend to the (1 hr.) *Beuli Alp*, past the (1 hr.) *Lago di Codelago* (5055'), and by *Crampiolo* to (1 hr.) *Devero* (5380'; poor inn at the hamlet of *Ai Ponti*) and (2½ hrs.) *Baceno* (p. 321). From Devero over the *Buscagna Pass* and the *Valtendra Pass* to the *Alp Veglia*, see p. 311. — To the TOSA FALLS. From the Albrun Pass we descend to the left to the *Forno Alp*; then over the *Scatta-Minojo* (8520') by a bad path to the *Lago di Lebendun* (*Lago Vannino*, 7065'), whence we descend on the left bank of the brook to *Zum Steg* in the *Val Formazza* and re-ascend to *Auf der Frutt* (10 hrs. from Binn). In fine weather the route over the *Hohsand Pass* is preferable (see pp. 317, 320).

FROM FIESCH TO BACENO OVER THE GEISSPFAD PASS, 11-12 hrs., a fine route (guide from Im Feld 12 fr.). At (4 hrs.) *Im Feld* (see above) we diverge to the right to the *Messern-Alp* (6175') and ascend past the *Geisspfad Lake* (7975') to the (4 hrs.) Geisspfad Pass (8365'); then proceed over the rock-strewn plateau to the (½ hr.) *Bocca Rossa* and descend a steep rockwall into the *Val Rossa*, to (1½-2 hrs.) *Devero* (see above).

FROM FIESCH TO BACENO OVER THE KRIEGALP PASS, 13-14 hrs., toilsome and of little interest (guide from Binn 12 fr.). From Binn we ascend to the S. through the *Längthal* to (1 hr.) *Heiligkreuz* (4862') and then to the left through the *Kriegalp-Thal* to the (3½ hrs.) Kriegalp Pass (*Passo di Cornera*; 8465'), between the (l.) *Gilschihorn* (*Pizzo Cornera*; 10115') and the (r.) *Helsenhorn* (10,742'; ascent from here difficult, see below). Descent to the *Val Buscagna* and to (2 hrs.) *Devero* (see above).

FROM FIESCH TO ISELLE OVER THE RITTER PASS, 14-15 hrs., trying, but very interesting (guide from Binn 12 fr.). From Binn (p. 317) we ascend the *Läng-Thal* to the S. to the (5 hrs.) Ritter Pass (*Passo Boccareccio*, 8832'), between the (r.) *Hüllehorn* (10,450') and the (l.) *Helsenhorn* (10,742'; easily ascended, with guide, in 1½ hr. from the pass; magnificent panorama). Descent to the (1½ hr.) beautifully situated *Alp Veglia* (5800'; *Alb. del Monte Leone*, unpretending) in the *Val Cairasca*, and by *Trasquera* to (3 hrs.) *Iselle* (p. 313).

Beyond Fiesch the road descends the fertile valley (numerous trees), passing *Aernen* on the hill opposite (see p. 317), to (1½ M.) **Lax** (3425'; *Kreuz*), with a new church, whence the Eggishorn Inn may be reached in 4 hrs. It then winds down to the bridge of Grengiols (2905'; Inn), by which we cross the deep bed of the Rhone. (Good path hence by *Grengiols*, which lies 390' higher, in 5 hrs. to **Binn**, p. 317.) We recross to the right bank by the *Kästenbaum Bridge* (2670'), pass through a short rocky ravine, and, after the valley again expands, reach (6 M.) **Mörel** (2525'; *Hôt. Eggishorn*, R. & B. 2½ fr.; *Hôt. des Alpes*, moderate). — To the *Riederalp*, **3 hrs.**, see p. 317.

The river now dashes wildly over sharp slate rocks. On a bold rock below Mörel, which the road has difficulty in passing, rises the picturesque (1½ M.) *Hochfluhkirche*. We next cross the *Massa*, the discharge of the Great Aletsch Glacier, which issues from a fine gorge about ¾ M. from the road. 3 M. **Naters** (2236'), a large village amidst fruit-trees, is commanded by the ruined castles of *Weingarten* and *Supersax*. At the other end of the village, to the right, a finger-post indicates the route to the Belalp (p. 317). We then cross the Rhone by an iron bridge to (1 M.) the station of —

31 M. *Brig*, see p. 309.

82. From Ulrichen to Domodossola.
Gries Pass. Falls of the Tosa. Val Formazza.
Comp. Maps, pp. 118, 312.

15-16 hrs. Two days, spending the night at the Tosa Falls. Bridle-path from Ulrichen to the Tosa Falls (6½ hrs.); thence to Foppiano a rough cart-track (3 hrs.). Guide (to Frutwald 12, porter 10, horse 20 fr.) unnecessary in fine weather; otherwise advisable as far as the other side of the glacier (6 fr.); indispensable in the reverse direction. — Road from Foppiano to Domodossola, 21 M.; diligence from Crodo to Domodossola daily. One-horse carriage from Foppiano to Domo 20, from Premia 15 fr. (not always to be had). Porter from the Falls to Domodossola 6-8 fr.; horse (for which a carriage and harness may generally be had at Foppiano) 30 fr.

At *Ulrichen* (4880'; p. 315) a bridge crosses the *Rhone* to (10 min.) *Zum Loch*, a group of deserted huts at the entrance to the **Eginen-Thal**. The path crosses the *Eginenbach* above a fine waterfall, and leads through larch-wood and a rock-strewn valley to (1¼ hr.) the *Alp Hohsand* (5720'). It then ascends a steeper part of the valley, overgrown with alders and rhododendrons, where the brook forms several falls on the left. In front of us rises the *Nufenenstock* (9387'). In ½ hr. we cross the brook by the *Ladtsteg* (6340'), beyond which are the chalets of *Im Ladt*. (The more obvious path leading straight on, before the bridge, must be carefully avoided.) To the right, above us, is the *Gries Glacier* (see below). Following the right bank of the brook, we then traverse the highest reach of the valley and ascend to (20 min.) *Altstaffel* (6585'), the last chalet, where the path to the Nufenen Pass diverges to the left (p. 315). A steep ascent of 1¼ hr. more brings us to the level *Gries Glacier*, which we cross in 20 min., towards the S.W., passing a small glacier-lake on the left and a still smaller one on the right. The **Gries Pass** (8025'), between the *Bettelmattenhorn* (9800') on the right and the *Grieshorn* (9600') on the left, is the boundary between Switzerland and Italy, and in clear weather commands a fine retrospect of the Bernese Alps. (A path, little frequented, leads hence to the N.E. through the *Val Corno* to *All' Acqua* in the Val Bedretto, p. 315.)

The S. side of the pass, as is usually the case among the Alps, is steeper than the N. side. The narrow path at first keeps to the left (rich vegetation). The *Griesbach* rises here, and unites at Kehrbächi (see below) with the *Tosa* or *Toce*, descending from the Val Toggia. The upper part of the Formazza valley consists of three distinct reaches, each with its chalets: *Bettelmatt* (6900'; two chalets, generally empty) in the highest (the slope below which is called *Wallisbächlen*), *Morast* (or *Morasco*, 5840') in the second, and *Kehrbächi* (or *Riale*, 5640') and *Auf der Frut* (*Sopra la Frua*) in the third, with a small chapel and the unpretending *Hôt. de la Cascade (5490'; R., L., & A. 3, B. 1½ fr.). This inn (2½ hrs. from the Gries Pass) stands on the brink of a precipice over which the *Tosa* falls in three cascades, widening as it descends. The **Tosa Falls**, or *Cascata della Frua*, 470' high and 85' broad, are perhaps the grandest among the Alps, especially when the river is high. (We descend by the bridle-path to the left for 1/4 hr., to a mass of rock by the wayside, which affords the best survey. A still finer point is beyond the bridge.)

The *Basödino (10,748') may be ascended by good climbers without difficulty from the inn in 4-5 hrs. (the landlord, Ant. Zertanna, acts as guide). Spendid view. Descent to the Val Bavona, see p. 443.

FROM THE TOSA FALLS TO AIROLO, 8 hrs. (guide desirable to All' Acqua, and necessary in the reverse direction). The bridle-path diverges by the chapel above the falls to the right from the path to the Gries Pass, and after 20 min. crosses the brook descending from the Basodino. It then ascends to the right by the wall (leaving Kehrbächi below to the left), and mounts in steep zigzags to the (3/4 hr.) upper reach of the sequestered *Val Toggia;* 1/2 hr., a bridge; 20 min., chalets *Im Moos.* (To the right the *Bocchetta di Val Maggia,* see below.) The small *Fisch-See,* well stocked with trout, lies on the right. By the *Alp Königin,* 1/2 hr. farther on, we recross the brook. In the highest part of the valley we pass another small lake on the right, and reach (1/2 hr.) the S. Giacomo Pass (7570'), the boundary between Italy and Switzerland (Canton Ticino). Below the pass on the N.E. side, stands the (20 min.) chapel of *S. Giacomo* (7370'). In descending, we enjoy a beautiful view of the southern St. Gotthard Mts., the Kühbodenhorn, Pizzo Rotondo, Pesciora, Lucendro, etc., and also, for a short time, of the Finsteraarhorn and Fiescherhörner. Farther on (keeping to the left at the first chalet) we descend through a growth of rhododendrons and larch-wood into the valley, where we cross two brooks, and then the Ticino, and reach the (11/2 hr.) *Hospice all' Acqua* (p. 315). Thence to *Airolo,* see p. 315.

FROM THE TOSA FALLS TO BIGNASCO, 9 hrs., with guide, a fine route. By the *Fisch-See* (see above) we diverge to the right from the S. Giacomo path and ascend over débris and rock to the Bocchetta di Val Maggia (8710'), between the (r.) *Kastelhorn* and the (l.) *Marchhorn;* then descend through the *Val Fiorina* (with the snowy *Basodino* on the right, see above) to the *Alp Robiei,* and through the picturesque *Val Bavona* to *Bignasco* (p. 442).

From the Tosa Falls to *Binn* over the Hohsand Pass (a fine glacier expedition, 9 hrs., with guide), or over the Albrun Pass (10 hrs., with guide), see pp. 317, 318.

Below the Tosa Falls begins the Val Formazza, or *Pommat Valley,* the upper part of which, as far as Foppiano, is German-speaking In this valley are the villages of (1/2 hr.) *Frutwald* (*Canza,* 4755'), (10 min.) *Gurf* (*Grovella,* 4475'), (1/4 hr.) *Zum Steg* (*Al Ponte,* 4200'), with the town-hall and archives of the valley, (1/4 hr.) *Pommat* (*San Michele,* 4210'), where we cross the bridge (not straight on), and (1/2 hr.) Andermatten (*Alla Chiesa,* 4050'), with the church of the valley. Below (1/4 hr.) *Staffelwald (Fracchie)* the path enters a grand *Defile, in which it crosses the Tosa twice. At (3/4 hr.) *Foppiano* (*Unterwald;* 3075'; Valduga's Inn, well spoken of), the first village where Italian is spoken, the carriage-road begins (vehicles not always to be had; see p 319).

TO THE VAL MAGGIA (p. 442), a toilsome route and deficient in attraction (from Andermatten to Cevio 8 hrs., not without guide): from *Staffelwald* a steep ascent of 3 hrs. over the *Staffel-Alp* to the Criner Furka (7925', fine view); descent of 11/2 hr. to *Bosco* and (31/2 hrs.) *Cevio* (p. 442).

The ROAD follows the right bank to (1 M.) *Rivasco* (2790'; Inn) and (1 M.) *Passo* (2680'). The valley of the Tosa, called *Val Antigorio below this point, is one of the most beautiful on the S. side of the Alps, and enlivened with waterfalls 11/2 M.

S. Rocco (Alb. del Sole, good **Asti** wine); 3¾ M. *Premia* (2620'; Agnello; Restaurant Antigorio, modest). At (1½ M.) **Baceno** (2245'; *Alb. Devero*, moderate), at the mouth of the *Val Devero*, a bold bridge spans the deep gorge of the *Devero*. (From Baceno to *Fiesch* over the *Albrun Pass* or the *Kriegalp Pass*, see p. 318.) To the W. rises *Monte Cistella* (9450').

The Italian custom-house is at (3 M.) *Crodo* (1650'; Inn), below which is (1 M.) a 'stabilimento di bagni'. Then by *Rencio* and the finely situated *Oira* ('il Giardino dell' Ossola') to (6 M.) *Crevŏla* on the Simplon route, and (2¼ M.) —

21 M. *Domodossola*, see p. 314.

83. The S. Valleys of the Valais between Sion and Turtmann.
(Val d'Hérens, Val d'Anniviers, Turtmann Valley).
Comp. Maps, pp. 308, 310, and 332.

Good walkers on their way from the Lake of Geneva to *Zermatt* (R. 84) may avoid the Rhone Valley and reach their destination by an interesting mountain-route in 4-5 days. 1st day. By rail to Sion, and walk through the Val d'Hérens to Evolena, 16 M. — 2nd day. Over the Col de Torrent to St. Luc in the Val d'Anniviers, 8-9 hrs. — 3rd day. Ascend the Bella Tola, and cross the Pas du Bœuf or the Meiden Pass to Gruben in the Turtmann Valley, 8½ hrs. — 4th day. Over the Augstbord Pass to St. Niklaus in the Visp Valley, 7 hrs. (or, including the Schwarzhorn, 8½ hrs.), and thence to Zermatt by rail.

i. From Sion through the Val d'Hérens to Evolena, and over the Col de Torrent to the Val d'Anniviers.

To Evolena (15½ M.) a post-vehicle with 2-3 seats runs daily at 6.30 a.m. in 5¾ hrs. (6 fr. 40 c.; surplus passengers are sent on in open carriages), returning at 1.50 p.m. in 3⅓ hrs. One-horse carr. from Sion to Evolena, 20-25 fr. (carriages from the hotel at Evolena are frequently waiting at the station). — From Evolena over the Col de Torrent to Vissoye a bridle-path in 8-9 hrs. (guide 12 fr., **unnecessary** for experts). Horse to Vissoye 24, to St. Luc 26 fr.

Sion, see p. 307. The road to Evolena leads from the Rhone bridge (1625') straight to the (½ M.) foot of the mountain, which it ascends in long windings. (Short-cut by the old bridle-path.) To the left, below, lie *Bramois* and *St. Léonard* (p. 307), the latter at the mouth of the gorges descending from the Rawyl. Near the old cemetery chapel of (4½ M.) **Vex** (3140'; *Inn*, rustic) we obtain a view of the head of the valley, first of the *Dents de Veisivi* and the *Pic d'Arzinol*, and then of the great *Ferpècle Glacier*, commanded by the round summit of the *Tête Blanche*, to the left of which are the *Dent Blanche* and the *Dent d'Hérens*. The cultivation of maize, vines, chestnuts, and walnuts extends as far as Vex.

A bridle-path ascends from Vex to the right, by *Presse* and *Les Agettes*, to the (1 hr.) Mayens de Sion or *Mayenberg* (4267'; *Pens. des Mayens*, 6 fr.), a summer-resort of the Sionese, in a beautiful and healthy situation, commanding a magnificent view of the entire chain of the Bernese Alps. Hence to Hérémence, ¾ hr.

BAEDEKER, Switzerland. 16th Edition. 21

The road, nearly level, skirts the W. slope, high above the
Borgne. The valley divides farther up. The W. branch is the *Val
d'Hérémence* (see below), and the E. the **Val d'Hérens** *(Eringer
Thal)*. The road passes the large village of *Hérémence* on the hill
to the right, and near (1¼ hr.) *Sauterot* (3050′) crosses the *Dixense*,
which **descends from the** Val d'Hérémence. It then penetrates the
remains of the terminal moraine of that valley by means of two
tunnels. Near the second, where the road re-enters the Val d'Hérens, are a number of *Pyramids of earth, each covered with a stone,
which belong to the old moraine and are not unlike 'glacier-tables'.

Val d'Hérémence (the upper part *Vallée des Dix*). A cart-track leads
from Vex (p. 310) to (1 hr.) *Hérémence* (4055′; accom. at the curé's); thence
a bridle-path by the hamlets of *Ayer*, *Prolin*, *Cerise*, and *Mars* to the
(3 hrs.) *Mayens de Prazlong* (5275′), at the W. base of the *Pic d'Arzinol*
(see below; over the *Col de la Meina* to Evolena, 4 hrs.). Farther on
we pass the *Méribé Alp* (l.) and ascend a ravine to the upper part of the
valley, called *La Barma*, with the Alp of that name on the right (8095′;
thence over the *Col du Crêt* to Fionney, see p. 293). Passing the chalets
of *Lautaret*, we next reach (3 hrs.) the *Seilon Alp* (7455′), opposite which,
on the left bank of the Dixenze, is the *Liappey Alp* (7630′; good quarters).
From Liappey over the *Col de Riedmatten* or the *Pas de Chèvres* to *Arolla*
Evolena), see p. 313; *Cols de Vaseray*, *de Seilon*, *du Mont Rouge*, and *de
Breney* to the *Val de Bagnes*, see p. 294. The **Pigne d'Arolla* (12,470′)
is best ascended from this point **over** the *Glacier de Durand* and the *Col
de Breney* (comp. pp. 306, 323).

We next reach (1 M.) *Euseigne* (3182′; **wine at the post-station**),
prettily situated. High above, **on the opposite bank, is the church of**
St. Martin. Beyond (1½ M.) the hamlet of *La Luette* (**3345′**) **the road**
crosses the Borgne. (Near the *Chalets de Praz-Jean*, higher up, is
the old bridge of the bridle-path.) We ascend on the right bank,
below the small chapel of *La Garde*, to (6 M.) —

Evolēna (4520′; **Hôt.-Pens. de la Dent Blanche*, R., L., & A.
from 3, B. 1½, lunch 3, D. 4, pens. 6-8 fr.; *Gr.-Hôt. d'Evolène*,
connected with the Arolla Hotel, R. 2-4, B. 1½, lunch 3, D. 4,
pens. 6-9 fr.; Engl. Ch. Service in summer), the capital of the
valley, lying picturesquely in a broad green dale flanked with
pine-clad rocks. On the E. rises the *Sasseneire*, on the W. **the**
Mont de l'Etoile and *Pic d'Arzinol*. Looking up the valley we see
the *Dents de Veisivi*. On the left, high above, are the snow-fields
of the *Ferpècle Glacier* and the huge *Dent Blanche*; to the N., beyond
the Rhone valley, the large *Zanfleuron Glacier* with the *Oldenhorn*
(p. 251) behind it. The picturesque native costumes, especially
those worn by the women on Sunday, are interesting.

EXCURSIONS. (Guides: *Jean* and *Pierre Maitre*, *Jos. Quinodoz*, *Pierre*
and *Jean Beytrison*, *M.* and *Jos. Métrailler*, *M. Gaspoz*, *J.* and *M. Vuigner*,
M. Pralong, *M. Cherrier*, and *Ant. Bovier*.) The guides at Evolena,
Arolla, etc., raised their charges very considerably (especially for short
excursions) in 1894; but as this was done without any authorisation we
adhere below to the former official tariff. — Pleasant walk (shade early
in the morning) to Villa, returning viâ La Sage (2¼-2½ hrs.). About
12 min. to the S. of Evolena we diverge to the left from the road to
Haudères (p. 323) and ascend a steep footpath to (¾ hr.) *Villa* (5655′;
fine view); before reaching the (2 min.) village-fountain we turn to the
right (the path **to the left** leads **to the** Col de Torrent, see p. 327) and in

5 min. to the right again; 20 min. *La Sage* (5482'); descend to the right at the church (to the left, to Forclaz, p. 325); 5 min. to the right again; 20 min.;carriage-road; 22 min. Evolena. — *Sasseneire* (10,690'; guide 6 fr.), 5 hrs., see p. 326. The *Couronne de Bréonna* (10,380'; 5½ hrs.; guide 7 fr.), viâ *La Sage* and *Alp Bréonna*, is also interesting. — *Becs de Bosson* (10,348'; guide 7 fr.), 6 hrs., see p. 326.

W. side: The *Alpe de Niva* (6625'), 2 hrs., affords an admirable survey of Ferpècle and Arolla. — The *Pic d'Arzinol* (9845'; guide 12 fr., unnecessary for adepts), ascended by the *Col de la Meina* (bridle-path thus far) in 4½ hrs., is very interesting and not difficult. Below Evolena we cross the Borgne, ascend to the left (avoiding the path to the right to *Lanna*, ½ hr.), through wood, and cross (1½ hr.) the *Merdesson*, the discharge of the Glacier de Vouasson, to the (½ hr.) *Alpe de Vouasson* (6850'). Thence we ascend pastures (keeping to the right) to the (2 hrs.) *Col de la Meina* or *Col de Méribé* (8878'; to the *Val d'Hérémence*, see p. 322). Leaving the col on the left, we mount a rocky arête to the right to the top (1 hr.). Magnificent panorama, especially towards the S. (Mont Blanc, Aiguille Verte, Grand Combin, Mont Velan, Matterhorn, Weisshorn), and of the Bernese Alps to the N. Descent 2½ hrs. — *Mont de l'Etoile* (11,066'; guide 6 fr.), by the alps *Niva* and *Creta* in 6 hrs., repaying, but for adepts only; so also the *Pointe de Vouasson* (11,470'; guide 10 fr.), 6-7 hrs., whence we may descend the *Glacier des Aiguilles Rouges* to the *Alp Lucel* (see below) and Arolla.

The Val d'Hérens divides at **Haudères** (4747'; *Hôt.-Pens. Haudères*, plain), ¾ hr. to the S. of Evolena. To the W. is the *Combe d'Arolla*; the E. branch, terminated by the *Glacier de Ferpècle*, retains the name of the main valley.

(a.) *Combe d'Arolla. The bridle-path (from Evolena to Arolla 3½ hrs.; horse or mule 8, there and back 10 fr.) crosses the Ferpècle at Haudères, turns to the right, and crosses the Borgne to *Pralovin*. It then ascends the W. slope of the valley, overlooking the wild ravine, and through wood to the (1¼ hr.) *Chapel of St. Barthélemi* (5960'), by a huge rock. We next ascend gradually, following the telegraph-poles, past the (10 min.) chalets of *Gouille, Satarma, Praz Mousse*, and *La Montaz*, to the (1½ hr.) **Mayens d'Arolla** (6570'; *Hôt. du Mont Collon*, pens. 7-8 fr.; Eng. Ch. Service in summer), amid stone-pines ('Arolla' or 'Alpine cedar'), splendidly situated opposite the grand pyramid of *Mont Collon* (11,857'), at the base of which the *Glaciers d'Arolla* (r.) and *de Vuibez* (l.) unite. To the right rise the rocks of the *Serra de Vuibez* (10,150') and the snow-clad *Pigne d'Arolla* (12,470'), and close to the inn is the old moraine of the *Glacier de Zigiorenove*.

EXCURSIONS (guides, see p. 322). To the Lac Bleu de Lucel, a pleasant walk, 1½ hr. (or from Evolena 3 hrs.; without guide). At *Satarma*, 2½ M. from Arolla (see above), a steep path ascends to the left (N.W.) to the chalets of *Lucel* (6820'), a little beyond which is the little lake, fed by a brook falling from the rocks. Beautiful view of Mt. Collon; to the W. tower the *Aiguilles Rouges* (11,975'); to the left is the *Cascade des Ignes*, descending from the Glacier des Ignes.

The *Pigne d'Arolla* (12,470'; guide 25 fr.) is ascended by the *Glacier de Pièce* in 6-7 hrs.; very grand and not difficult; comp. pp. 305, 306, 322. — The Aiguille de la Za (12,050'; 4½-5½ hrs.), an interesting ascent for experienced climbers, with good guide (30 fr.), is accomplished either from the Arolla valley direct viâ the *Glacier de la Za* (step-cutting for 2-3 hrs.), or viâ the *Glacier de Bertol*. Either route brings us in 4-5 hrs. to the foot of the last peak, which is surmounted after ½ hr.'s steep rock-climbing. Descent only by the Glacier de Bertol. — The *Petite Dent* (10,465'; guide 15 fr.), one of the **Dents de Veisivi**, is ascended without difficulty viâ the *Alp Zarmine*. The *Grande Dent* (11,240'; 20 fr.) is more difficult. Between the Petite and Grande Dent the *Col de Zarmine* (10,045'), not easy, leads from Arolla to Ferpècle. — The **Mont Collon** (11,857'; guide 50 fr.), best ascended from the W side (Col de Chermontane), is only fit for adepts with steady heads; so also the **Evêque** (12,265'; guide 50 fr.), rising to the S. of Mt. Collon. — The **Dent Perroc** (12,073'; 35 fr.) **and the Dent des** Bouquetins (12,625'; 40 fr.) involve difficult climbing.

PASSES. TO THE VAL PELLINA OVER THE COL DE COLLON, a grand route and not difficult (7-8 hrs. from Arolla to **Pra-Rayé**, two guides, 30 fr. each). We ascend the *Glacier d'Arolla*, skirting **the E. base** of the almost perpendicular rocks of *Mt. Collon*, remarkable for their echoes, **to the** snow-basin of *Za-de-Zan* and the (4 hrs.) summit of the **Col de Collon (10,270')**, to the S.E. of the Evêque (see above). View grand, but not extensive. Descent over the *Glacier de Collon* to the profound *Combe d'Oren* **and (3** hrs.) *Pra-Rayé* (6540'; small Inn, 6 beds), and in 3 hrs. more (bridle-path) **to** *Bionaz* (5248'), and thence viâ (1 hr.) *Oyace* (4490') to (1½ hr.) *Valpelline* (p. 306). (Passes from the Val Pellina to the *Val St. Barthélemy*, see p. 296.) Those who cross in the reverse direction should bring provisions with them; good guides not easily found at Aosta, but a peasant who knows the pass may be found at Bionaz. From Pra-Rayé to the Col 3½-4, descent to Arolla 2½-3 hrs. — From the basin of *Za-de-Zan* (see above) we may ascend to the left **to the Col de Za-de-Zan** (11,660'), between *Mont Brûlé* (11,880') **and the Col du Mont Brûlé** (see below); descent, steep and difficult, to the *Glacier de Za-de-Zan* (p. 296) and *Pra-Rayé*.

TO **THE** VAL D'HÉRÉMENCE from Arolla there are two passes close together: **the Col de Riedmatten** (9567'; 4 hrs. to Liappey), and to the S. of it **the Pas de Chèvres** (9355'; 3¼ hrs. from Arolla; rather more difficult). From the latter we descend steep rocks and over the *Glacier de Durand* or *Seilon* (beware of numerous concealed crevasses) to the (4½ hrs.) chalets of *Seilon* (7455'), opposite *Liappey* (p. 322). (The Riedmatten route descends the rocks and grass-slopes on the right side of the glacier.) Then down the *Vallée des Dix* to (4½ hr.) *Hérémence*, see p. 322. — Or, from the Durand or Seilon Glacier (see below) we may ascend to the *Col de Seilon* (10,665'; 4½-5 hrs. from Arolla; p. 306) and thence either descend the *Glacier de Giétroz* to (2½ hrs.) *Mauvoisin* (p. 305), or cross the *Col du Mont Rouge* (10,960') and descend the *Glacier de Lyrérose* to (3½ hrs.) *Chanrion* (p. 305; guide 25 fr.).

TO THE VAL DE BAGNES **OVER THE** COL DE CHERMONTANE, 11 hrs., a long and fatiguing glacier-route (guide 25 fr.). We ascend over the moraine, **the lower** end of the *Glacier de Ziglorenove*, and the *Glacier de Pièce* or *Torgnon* to a snowy saddle (10,235') on the W. side of the *Serra de Vuibez*, and thence by the *Glacier de Vuibez* to the **Col de Chermontane** (10,120'), between the *Petit Mt. Collon* (11,630') and the *Pigno d'Arolla* (p. 323). Striking view of the Mont Collon, the Dents with the Aiguille de Za, the **Dent** Blanche, and to the N. the Bernese Alps. **Descent** across the vast snow-fields of the *Glacier d'Otemma* to Chanrion **(p. 305). —** Longer, but far more striking, is the route to **Chermontane** over the **Col de l'Evêque** (11,485'; 13 hrs.; guide 30 fr.). Route over the *Glacier d'Arolla* to the *Col de Collon*, see above; here we ascend to the right to the *Col de l'Evêque*, (11,130'), lying to the S.W of the *Evêque* (p. 324), cross a snow-arête between the (l.) *Sengla* (12,155') and the (r.) *Petit Mont Collon* (see above) to the *Glacier d'Otemma*, and descend as above to Chanrion.

TO ZERMATT OVER **THE** COL DE BERTOL, 11-12 hrs., fatiguing but repaying (guide 30 fr.). We follow **a** narrow footpath along the moraine of the Glacier d'Arolla to the *Plan de Bertol*, and ascend rocks and the steep *Glacier de Bertol* to the **Col de Bertol** (10,925'), between two of the *Dents de Bertol* (11,505' and 11,145'). We then cross the vast snow-fields of the *Glaciers du Mont Miné* and *de Ferpècle*, past the *Tête Blanche* (which takes 1¼ hr. more to ascend; see below), to the *Col d'Hérens* and the *Stockje* (p. 326); thence to *Zermatt*, 3½-4 hrs. — OVER THE COL DU MONT BRÛLÉ

AND THE COL DE VALPELLINE, another grand route, 12-13 hrs. (guide 30 fr.). We follow the Col de Collon route to the basin of *Za-de-Zan*, ascend steeply to the left to the **Col du Mont Brûlé** (10,900'), cross the crevassed upper *Za-de-Zan Glacier* (passing on the left the *Dents* and *Col des Bouquetins*, p. 326), and mount laboriously to the Col de Valpelline (11,685'), on the S. side of the *Tête Blanche* (12,300'; ascended from the col in ¾ hr.; splendid view; see below). Then down the *Stock Glacier* to the *Stockje* (see p. 326).

(b.) *Ferpècle.* (Bridle-path, 2¼ hrs. from **Evolena**; horse or mule 8, there and back 10 fr.) At (¾ hr.) Haudères, by the third house before the bridge (p. 323), we diverge to the left, ascend gradually, and then more rapidly, passing four, and shortly beyond them six chalets. Beyond the next ridge we ascend to the left to (¾ hr.) *Sepey* (5580'), where the bridle-path from Evolena viâ *La Sage* and *Forclaz* (see p. 323; ¼ hr. longer, but finer) joins ours on the left. The imposing head of the valley (Glacier de Ferpècle and Dent Blanche) is now revealed, the view being finest from the chalets of *Prazfleuri*, the second group beyond Sepey. Then through wood to (¾ hr.) the chalets of *Salay* or *Ferpècle* (5910'; *Hôt. du Col d'Hérens, plain, R., L., & A. 2½-3, B. 1½, lunch 2½, D. 3, pens. 6½-7 fr.; post and telegraph office), splendidly situated opposite the *Mont Miné* and *Ferpècle Glaciers*, both of which have much receded.

Just beyond the hotel a narrow path ascends to the left through larch-wood and over débris and pastures to the (1½ hr.) *Alp Bricolla* (7960'; milk to be had), a strikingly grand point of view. At our feet lies the huge Ferpècle Glacier, overtopped by the snow-clad Wandfluh, while to the left rise the huge Dent Blanche and the Grand-Cornier. To the right, separated from the Ferpècle Glacier by the Mont Miné, is the Glacier du Mont Miné, with the Dents de Bertol, Aiguille de la Za, and Dents de Veisivi.

ASCENTS. **Dent Blanche** (14,320'), very difficult and in some years impossible (10-14 hrs. from Ferpècle; guide 70 fr.). The night is usually spent below the rocks on the right side of the *Glacier de la Dent Blanche*; hence to the top 6-8 hrs. — **Grand Cornier** (13,022'), from Ferpècle by the *Col de la Pointe de Bricolla* (see below) in 7-8 hrs., toilsome, but without danger (guide 60 fr.).

PASSES. To ZINAL OVER THE COL DU GRAND-CORNIER, 10-11 hrs., not very difficult (guide 50 fr.). Beyond (1½ hr.) Bricolla (see above) we turn to the E. to the *Glacier de la Dent Blanche*, and ascend it rapidly to the (3½ hrs.) Col du Grand-Cornier or de la Dent Blanche (11,625'), between the Dent Blanche and the Grand Cornier. We descend an arête to the right and snow-slopes, passing the *Roc Noir*, to the (2½ hrs.) *Mountet Club-Hut* (p. 328), and over the *Durand* (*Zinal*) *Glacier* to (3 hrs.) *Zinal* (p. 328). — OVER THE COL DE LA POINTE DE BRICOLLA, 10 hrs. to Zinal (guide 35 fr.), rather fatiguing. From Bricolla (see above) we ascend to the N.E. across the *Glacier de Bricolla* and over steep rocks, partly covered with ice, to the (3½ hrs.) Col de la Pointe de Bricolla (about 10,170'; splendid view), immediately to the E. of the *Pointe de Bricolla* (12,015'). We descend across the *Glacier de Moiry*, and by the *Col de l'Allée* and the *Alp de l'Allée* to (5 hrs.) Zinal. With this excursion may be easily combined the ascents of the *Pte. de Bricolla*, the *Bouquetin* (11,430'), and the *Pigne de l'Allée* (11,170). — Over the *Col de Couronne* (*Col du Zaté* or *Col de Bréonna*) and the *Col de l'Allée*, see p. 329.

To ZERMATT OVER THE COL D'HÉRENS, 11 hrs., fatiguing (guide 80 fr.). From Bricolla in ¾ hr. to the *Ferpècle Glacier*, which we ascend, at first steeply, to the (3 hrs.) Col d'Hérens (11,415'), between the *Wandfluh* and the *Tête Blanche* (12,300'; easily ascended from the pass in ¾ hr.; grand view; we may descend to the *Col de Valpelline*, and regain the Zermatt route at the Stockje; this adds 1¼-1½ hr. to the route; see p. 325, Col de Valpelline). To the E. towers the overwhelming Matterhorn. From the pass we descend steep rocks and the crevassed *Stock Glacier* to the (1 hr.) *Stockje* (9052′), a rocky island at the head of the *Zmutt Glacier*, between the *Stock Glacier* (left) and the *Tiefenmatten Glacier* (right). The club-hut was destroyed in 1891. We descend the stone-covered Zmutt Glacier, and at length regain a firm footing at the (3 hrs.) *Staffel-Alp* (p. 337). Thence to *Zermatt* 1½ hr.

To PRA-RAYÉ OVER THE COL DES BOUQUETINS (10-11 hrs.; guide 30 fr.), also fatiguing. We either follow the Col d'Hérens route (see above), or ascend the left moraine, past *Mont Miné*, to the upper Ferpècle Glacier, and mount to the right to the Col des Bouquetins (11,215′), to the E. of the *Dent des Bouquetins* (12,625′). Descent over the *Glacier de Za-de-Zan* to *Pra-Rayé* (p. 324).

FROM EVOLENA TO VISSOYE OVER THE COL DE TORRENT, bridle-path, 8-9 hrs. (guide 12 fr., convenient; horse 24 fr.). Walkers should ascend direct to (1 hr.) *Villa* (p. 322). Riders follow the road for 10 min. more to (22 min. from Evolena) a tall wooden cross, at which the bridle-path diverges to the left. 40 min. *La Sage* (5482′), where we turn to the left immediately above the church; 25 min. *Villa* (5645′), where we turn to the right a few yards beyond the fountain. We next ascend the *Alp Cotter* in long zigzags, and then across slate-débris, to the (4 hrs.) *Col de Torrent* (9595′), on the S. side of the *Sasseneire* (see below), commanding a splendid view of the Val d'Hérens and the grand mountains encircling it (from right to left: Pointe de Vouasson, Aiguilles Rouges, Mt. Pleureur, Mt. Blanc de Seilon, Serpentine, Pigne d'Arolla, Petites and Grandes Dents, Dents de Bertol, Mont Miné, Tête Blanche, etc.).

The *Sasseneire (10,690′), 1 hr. from the col (guide desirable), affords a magnificent panorama: to the N., the Jura appears like a blue line beyond the Col de Cheville (p. 256); to the S. the attention is chiefly arrested by the Dent Blanche (p. 325).

To the N. of the Sasseneire another bridle-path (easy and attractive) crosses the *Pas de Lona* (9075′) to the Val d'Anniviers (from the Chalets de Praz Jean to Grimence 8 hrs.; guide 12 fr.). The *Becs de Bosson* (10,348′; superb view) may be ascended from the pass in 2 hrs.

The path descends in long windings, passing the N. side of the little *Lac de Zozanne* (8870′), in full view of the lofty range between the Anniviers and Zermatt valleys (Gabelhorn, Trifthorn, Rothhorn, Weisshorn, Brunnegghorn, Diablons), to the *Torrent-Alp* (7940′) and the (1½ hr.) *Alp Zatelet-Praz* (7085′), in the *Val de Moiry* or *de Torrent*, the W. branch of the Val d'Anniviers, watered by the *Navigenze*. The valley is grandly terminated by the *Glacier de Moiry*, overshadowed by the (r.) Couronne de Bréonna, Za de l'Ano, Pointe de Bricolla, Grand Cornier, Dent Blanche, (l.) Pigne de l'Allée, and the black slaty cone of the Garde de Bordon

ZINAL (p. 328) may be reached from this point in 3½ hrs. by remounting the E. slope of the valley and crossing the Col de Sorebois (8970′). From the *Corne de Sorebois* (9210′), 20 min. to the N. of the pass, we obtain a splendid view of the Weisshorn, Rothhorn, Gabelhorn, Grand Cornier, Dent Blanche, etc. Descent by an easy path, or (shorter, with guide) direct through wood to *Zinal*. — To Zinal over the *Col de l'Allée* and to Evolena over the *Col de Couronne* or the *Col de Bréonna*, see p. 329.

Beyond the Alp we traverse a level and monotonous valley and descend a rocky defile to (1½ hr.) *Grimentz* (5015'; Hôt. du Bec des Bossons), a large village with copper-mines. In front appear St. Luc and Bella Tola. Thence viâ *St. Jean* to (1 hr.) a bridge over the *Navigenze* (3743') and to (¼ hr.) *Vissoye* (see below).

ii. From Sierre through the Val d'Anniviers to Zinal.

Road to (11 M.) Vissoye (diligence daily at 6 a m. in 3½ hrs., 6 fr.); horse 10, one-horse carr. 12 fr.; thence mule-path to (6¼ M.) Zinal.

Sierre, see p. 307. We follow the road to the E. to the (1¼ M.) *Rhone Bridge* (1775'), ½ M. beyond which the road to the Val Anniviers diverges to the right and ascends rapidly through wood. Below, to the right, lies *Chippis*, at the influx of the *Navigenze* into the Rhone. After an ascent of 2½ M. we enter the **Val d'Anniviers** (Ger. *Einfisch-Thal* or *Eivisch-Thal*; 3050'), with the deep and inaccessible gorge of the Navigenze to the right. To the S. the beautiful snow-mountains enclosing the valley, the Rothhorn, Trifthorn, Besso, Dent Blanche, etc., are gradually revealed. Beyond (1¼ M.) *Niouc* (Cantine des Alpes, dear), the road is carried by means of galleries across a wild ravine, descending from the left; and immediately before the hamlet of *Barmes* we cross a similar ravine.

A direct route to Niouc for walkers diverges to the right beyond the station of Sierre, passes under the railway-embankment, and crosses a hill to the new Rhone bridge and (20 min.) *Chippis*. Beyond the second house we turn to the left and cross the *Navigenze*; then, leaving the church to the left, a narrow path following the telegraph-wires, frequently crossing the road, and at length uniting with it, leads us to (1¼ hr.) *Niouc*.

A footpath, diverging to the left beyond the first gorge (see above), leads from Niouc by *Sussillon* (4545') in 3 hrs. to the lofty village of *Chandolin* (6340'; accommodation at the curé's, 4 beds). The eminence beside the church affords a beautiful view of the Val d'Anniviers, the Rhone Valley, and the Bernese Alps. From Chandolin a good path through pine-forest, with beautiful views, descends to St. Luc in 1¼ hr. — The **Illhorn** (8935'), which overlooks the Illgraben (p. 308), the Rhone Valley, and the Bernese and Valaisian Alps, is ascended without difficulty from Chandolin in 2½ hrs.

To the right below the road (3½ M.) is the prettily situated village of *Fang*. (Travellers bound for St. Luc must take the bridle-path to the left, beyond the saw-mill, about ½ M. before Fang, which ascends gradually to St. Luc in 1½ hr.; see p. 329.) The road follows the valley, passing several small ravines. On the opposite slope lies *Painsec*. Then (3 M.) **Vissoye** (4006'; *Hôt.-Pens. d'Anniviers*, R. 2½, D. 3, pens. 5-6 fr.), the capital of the valley, on a hill on the right bank of the Navigenze, with a handsome church.

A picturesque walk leads hence over the Navigenze (see above) and then to the right viâ *Mayoux* and *Frasse* to (¾ hr.) *Painsec* (4297'), a prettily situated village, with a fine view of the snow-mountains above Zinal. A good footpath leads hence to the N. to (2 hrs.) *Vercorins* (4400'), with a view of the mountains to the N. of the Rhone Valley, and thence **down** to *Chippis* (p. 327) and (2 hrs.) *Sierre* (p. 307).

About 2½-3 hrs. above Vissoye (bridle-path, diverging to the left at the upper end of the village from the Zinal route beyond the bridge over the Bella Tola brook, and ascending generally through wood) is the *Hôtel

Weisshorn (7690'; R., L., & A. 4, lunch 3½/2, D. 4, pens. 8-10 fr.; Engl. Ch. Serv. in Aug.), in an open situation on the *Tête à Payas*, a spur of the *Rochers de Nava*, with 'view and rich flora. Excursions may be made hence to the *Pointe de Nava* (9090'; 1¼ hr.); to the *Tounot* (9915'; 2½ hrs.); to the *Lac de Tounot* (8726'; 1½ hr.); to the *Pas la Forcletta* (p. 331; 2 hrs.); to the *Meiden Pass* (p. 330; 2 hrs.); to the *Bella Tola* (p. 330; 3 hrs.), etc.

St. Luc (steep ascent of 1 hr. from Vissoye), see p. 329; thence to **the Hôtel** *Weisshorn* 1¾-2 hrs. To *Evolena* over the *Col de Torrent*, see pp. 327, **326.**

Beyond Vissoye the bridle-path leads towards a saw-mill (not to the left) to (1¾ M.) *Mission* (4288'), with an interesting chapel, at the mouth of the *Val de Moiry* (see p. 326), and (1 M.) *Ayer* (4870'), with deserted nickel-mines. (To St. Luc, see p. 329.) The path ascends a little, crosses a torrent, and traverses a stony wilderness, the scene of a landslip. It then (1½ M.) crosses the Navigenze, passes a chapel on the left bank, recrosses by the second bridge to the right bank, and reaches (2 M) —

6¼ M. **Zinal** (5505'; *Gr. Hôt.-Pens. des Diablons*, R. & L. 2½, B. 1½, D. 4, pens. 5-6 fr., good spring-water; *Hôt.-Pens. Durand*, similar charges; *Hôt. du Besso*; English Ch. **Service** in summer). The valley ends towards the S., 1 hr. from Zinal, in the *Durand* or *Zinal Glacier*.

EXCURSIONS (guides, *Elie* and *Joachim Peter, Elie Cottar, Joseph Monnet*, etc.). The *Alpe de l'Allée* (7187'), to the W., above the lower end of the glacier, 2 hrs. from Zinal, commands a noble survey of the head of the valley, from the Dent Blanche to the Weisshorn, and of the glaciers of Durand and Moming, separated by the beautiful double-peaked pyramid of the *Besso* (12,055'). A guide (5 fr.) is not necessary. After ¼ hr. we cross to the left bank, and traverse pastures; ½ hr., fragments of rock, where we ascend gradually; 20 min., a ravine with a waterfall above; beyond the ravine we turn to the right, and 10 min. farther, right again (the more level path leads to the glacier), ascending in zigzags; 35 min., a stone chalet on the first mountain terrace. Then rather a steep ascent; ¾ hr. to the left; 10 **min.**, chalet. Descent in 1½ hr.

The *Alpe d'Arpitetta* (7430'), opposite the last-mentioned alp, to the E., affords an even finer view, particularly of the Weisshorn, the Moming Glacier, and the Rothhorn. A still more imposing and complete view is commanded from the *Roc de la Vache* (8485'), ascended from the alp in 1 hr., or from Zinal direct vià **the** *Alp Tracuit* in 2½ hrs. (guide **6** fr.). By crossing the terminal moraine of the Durand Glacier (with guide), the Alp de l'Allée and Arpitetta may be combined. — Good walkers, however, should not fail to extend the excursion up the Durand Glacier **to the Constantia Club-Hut** or *Cabane de Mountet* (9495'; *Inn*, well spoken of), at the S. base of the Besso (4½ hrs. from Zinal; guide **10 fr.**), overlooking the grand amphitheatre of the glacier, encircled by **the Rothhorn, Triftnorn**, Gabelhorn, Dent Blanche, Grand Cornier, and Bouquetin. **The view is** still grander from the Roc Noir (10,360'), rising from the ice opposite **the** Mountet, reached in 1 hr. from the club-hut (guide from Zinal 14 fr.).

ASCENTS. The *Pointe de Sorebois* (9210'), 3½ hrs., with guide (6 fr.), easy and attractive, see p. 326. A much finer mountain-view is obtained from **the Garde de Bordon** (10,880'), reached in 2½ hrs. from the Corne de Sorebois vià the arête, for adepts only (guide 12 fr.). The ascent direct from Zinal is very steep. — The **Pointe d'Arpitetta** (10,300'), from the Alp Arpitetta 3-4 hrs. (guide 10 fr.; laborious). — Besso (12,055'), rather steep and toilsome, for experts only (8-4 hrs. from the Mountet Club-Hut; guide 30 fr.); view exceedingly grand. — Pigno de l'Allée (11,170'), from Zinal by the *Alp de l'Allée* in 6-7 hrs. (guide 15 fr.), and Bouquetin (11,430'), from Zinal over the *Col de l'Allée* and the *Glacier de Moiry* in 6-7 hrs. (20 fr.), neither very difficult. — Diablons (11,825'; 12-15 fr.), by the *Alp Tracuit* in 6 hrs., laborious. — The **Grand Cornier** (13,020'; 50 fr.) is best ascended from the plateau of névé below the *Col du Grand-Cornier* (p. 325), the last part difficult. — **Zinal-Rothhorn** or *Moming* (13,855'; 80 fr.), a difficult and hazardous climb (from the Mountet Club-Hut over the W. flank in 6-7 hrs.). — **Ober-Gabelhorn** (13,365'; 70 fr.), by the W. arête, also very difficult. Comp. p. 339.

Passes. To Evolena over the *Col de Sorebois* and *Col de Torrent*, see p. 326; by the *Col du Grand-Cornier* and the *Col de la Pointe de Bricolla*, see p. 325; by the *Pas de Lona*, see p. 326. — Over the Col de l'Allée and the Col de Couronne, 10-11 hrs. (guide 20 fr.), trying, and for adepts only. From the Alp de l'Allée we ascend steep grassy and rocky slopes to the Col de l'Allée (10,485'). Descent to the *Glacier de Moiry*, and another steep ascent to the Col de Couronne (9895'), between the *Couronne de Bréonna* and the *Za de l'Ano*. Then a steep descent to *Ferpècle* (p. 325). — Instead of the Col de Couronne we may cross the *Col de Bréonna* (9575'), lying to the N., between the Couronne de Bréonna and the *Serra Neire*, or the *Col du Zaté* (9485'), between the Serra Neire and the *Pointe du Zaté* (both toilsome).

To Gruben in the Turtmann Valley over the *Pas de Forcletta* or the *Col de Tracuit (des Diablons)*, see p. 330.

To Zermatt over the Triftjoch, 11-12 hrs., trying and difficult, for steady-headed climbers only (guide 30 fr.). From the (4½ hrs.) *Constantia Hut* (p. 328) we traverse the *Durand Glacier* towards the E. to the (1¼ hr.) foot of the precipitous rocks of the *Trifthorn* (12,260'), and clamber up at first by a ladder, with the aid of a rope, and then along narrow ledges of rock and through perpendicular couloirs. The (1½ hr.) Triftjoch (11,615'), between the Trifthorn and the *Ober-Gabelhorn* (13,365'), affords a striking view of Monte Rosa and the Mischabel. Then down the *Trift Glacier* and its huge moraine to (4 hrs.) Zermatt (p. 333).

To Zermatt over the Col Durand, 13-14 hrs. (guide 35 fr.). From the Constantia Club-Hut we ascend towards the S., passing the **Roc Noir** (see above), at first gradually, but soon rapidly, and in some years with difficulty, to the (4 hrs.) Col Durand (11,400'), between the *Mont Durand* (*Arbenhorn*, 12,234') and the *Pointe de Zinal* (12,437'), where we obtain a most striking view of the Matterhorn towering opposite. Descent not direct over the *Hohwäng Glacier* to the *Zmutt Glacier*, as the lower part of tho former is full of crevasses, but to the left, along the rocks of the *Ebihorn* (11,968'), to (3½-4 hrs.) *Zmutt* (p. 337) and (1 hr.) *Zermatt* (p. 338).

To Zermatt over the **Moming Pass** (12,445'), between the Rothhorn and Schallihorn (14 hrs.; guide 50 fr.), and to Randa over the Schalli-Joch (12,805'), between the Schallihorn and Weisshorn (14 hrs.; 40 fr.), both difficult and toilsome.

From Zinal to St. Luc (3½ hrs.). We return to (5 M.) *Ayer* (p. 328), ascend to the right, and skirt the hillside, traversing pastures and wood (guide desirable, 5 fr.; or enquiry may be made at Ayer). — From Zinal to the *Hôtel Weisshorn* (p. 323) direct in 4 hrs. (guide desirable, 8 fr.).

iii. St. Luc; Bella Tola; over the **Pas du Bœuf** (or the **Meiden Pass**) into the Turtmann Valley, and over the **Augstbord Pass** to the Valley of the Visp.

Road from Sierre to Vissoye (11 M.); ascent thence to St. Luc, 1 hr. (from Sierre direct to St. Luc 5 hrs.; horse 10 fr.; comp. p. 316). Luggage under 10 lbs. may be sent by post. Ascent of the Bella Tola from St. Luc 3½ hrs.; from the Bella Tola to Gruben over the Pas du Bœuf in 4½, or the Meiden Pass in 3½-4 hrs. (guide 10, horse 16 fr.). From Gruben over the Augstbord Pass to St. Niklaus 7 (or including the Schwarzhorn 8½) hrs. (guide 12, horse 30 fr.).

St. Luc (5390'; *Hôtel-Pension Bella Tola*, R. & L. 2-3, lunch 3, D. 4, pens. 5-6 fr.; *Gr.-Hôt. Mont Cervin*, new, well spoken of; Engl. Ch. Service in summer), lying on a steep slope, amid pastures and fields, commands a superb view of the Val d'Anniviers and the snow-mountains at its head (Schallhorn, Lo Besso, Ober-Gabelhorn, Mont Durand, Matterhorn, and Pointe de Zinal). The *Pierre des Sauvages* (5623'), 1 M. above the village, is an ancient altar-stone, ascribed to the 'Druids'.

The *Bella Tola (9845'; 3 hrs.; guide 6 fr., not **necessary**; horse 8 fr.), an admirable and favourite point of view, is the N.W. peak of a group of mountains enclosing the large crater-like basin of the *Bella Tola Glacier* on the S. side. From the Hôtel Bella Tola we proceed to the church, immediately beyond which we turn to the left; 1 min. more, again to the left (the path straight on leads to the Hotel Weisshorn, see p. 328); 40 min., ascend to the left; 4 min., turn to the right; 10 min., ascend in zigzags, then through wood and across two brooks to (40 min.) a white hut known as the 'Chalet Blanc'. Then to the left over an old moraine; 5 min., to the right in a straight line for the centre of the Bella Tola; 1¼ hr., we reach its base and ascend in steep zigzags to (50 min.) a refuge-hut, and (left) to (¼ hr.) the summit. The N.W. peak, that usually ascended, is marked by a metal vane, but a path ascends the S.E. peak (9934') also. The *View embraces the whole of the Bernese and Valaisian Alps; opposite, to the N., the whole gorge of the Dala is visible, up to the Gemmi. The mountains to the S., from Monte Leone (p. 311) to Mont Blanc, are particularly grand.

In order to reach the Turtmann Valley we descend from the Bella Tola to the S., and ascend to the left to the (1 hr.) **Pas du Bœuf** (9155'). In descending into the *Borter-Thal* we keep to the left, and in some seasons cross a patch of snow. At (1½ hr.) the chalets of *Pletschen* the track divides: to the left to (2½ hrs.) *Turtmann* (p. 308), to the right to (2 hrs.) *Gruben* (see below).

The direct bridle-path from St. Luc to Gruben crosses the **Meiden Pass** (9095'; 5 hrs.; guide hardly needed). After 50 min. we cross the brook descending from the Bella Tola, then proceed straight on (passing in 6 min. a path diverging to the left to the Bella Tola) to the (1 hr.) *Alp Tounot* (7225'). The path ascends pastures and then over rocky débris to (1½ hr.) the pass, with a fine view of the Weisshorn, Brunnegghorn, etc. We descend past a small lake (with the *Meidenhorn*, 9426', on the right) to the *Upper* (7670') and the *Lower Alp Meiden* (7352'; fine view of the great Turtmann Glacier, see below). Lastly a zigzag descent, through larches and stone-pines, to the Alpine hamlet of (2 hrs.) **Gruben** or *Meiden* (5960'; *Hôt. Schwarzhorn*, plain, but dear, closed till the end of June), in the *Turtmann Valley*.

The **Turtmann Valley** ends to the S. in the magnificent *Turtmann Glacier*, imbedded between the *Diablons* (11,825'), *Weisshorn* (14,805'), *Brunnegghorn* (12,630'), and *Barrhorn* (11,920'). At its base lie the chalets of *Sennthum* (1½ hr. from Gruben). A difficult but interesting route crosses this glacier and the **Col des Diablons** or **de Tracuit** (10,675'), between the Diablons and the Weisshorn, to Zinal (9-10 hrs. from Gruben; guide 16 fr.).

FROM GRUBEN TO VISSOYE OVER THE PAS DE LA FORCLETTA, 8 hrs., not *of the Valais.* AUGSTBORD PASS. *V. Route 83.* 331

difficult, and fairly interesting (guide 12 fr.). By the *Lower Plumatt-Alp*, 1/2 hr. above Gruben, we ascend to the right through wood to the (1 hr.) *Upper Plumatt* (7355'), with a fine view of the Turtmann Glacier, Weisshorn, etc. Then past the *Kaltenberg Alp* (8152') and through a dreary valley to the (2 1/2 hrs.) Pas de Forcletta (9475'), between the (r.) *Roc de Budri* and the (l.) *Crête de Barneusa* (9997'). Fine view of the Valaisian and Bernese Alps. Descent to the *Hôtel Weisshorn* and (4 hrs.) *Vissoye* (p. 327).

FROM GRUBEN TO TURTMANN (3 1/2 hrs.). The bridle-path follows the right bank of the *Turtmannbach*, vià *Tschafel* and *Niggeling*, to the (1 1/2 hr.) *Vollensteg*, which carries the path to the left bank. Thence we proceed through the *Taubwald* or *Dubenwald*, a pine-forest now much thinned. In the middle is a little white chapel with numerous votive tablets. At (1 1/2 hr.) *Tummenen* (3200') we recross the stream, by the second bridge, then descend the steep left bank of the brook, with fine views of the Rhone Valley, to (1/2 hr.) *Turtmann* (p. 308).

From Gruben to St. Niklaus (7 hrs., with the Schwarzhorn 8 1/2 hrs.; guide desirable, 12 or 15 fr.), a bridle-path ascends the steep E. slope of the valley and the *Gruben-Alp* to the (3 hrs.) **Augstbord Pass** (9490'), between the *Steinthalhorn* (10,213') on the S. and the *Schwarzhorn* (10,512') on the N., affording a fine view of the Fletschhorn, Simplon group, and Mischabel.

The *Schwarzhorn* (10,512') is easily ascended in 3/4-1 hr. from the pass. Superb view, finer than from the Bella Tola (p. 318): N., the Bernese Alps, from the Doldenhorn to the Finsteraarhorn; E., the St. Gotthard group, Alps of Ticino, Mte. Leone, Fletschhorn, Weissmies, and the imposing Mischabel; S., Monte Rosa, the Lyskamm, Brunnegghorn, Weisshorn, Dent Blanche, Diablons, etc.

The path descends over débris (bad for riding) into the *Augstbord Valley*. We may then either go to the right, skirting the Steinthalgrat, to *Jungen* (6490'; splendid view of the Vispthal from the church; to the left the Gassenried Glacier, Dom, and Grabenhorn, to the right the Brunnegghorn and Weisshorn; in the centre the Breithorn and Zwillinge), and descend to (3 hrs.) *St. Niklaus* (p. 332).

From Gruben to St. Niklaus by the Jung Pass (9822'), farther to the S., interesting on the whole (6 hrs.; guide 12 fr.). — The Barrjoch (11,990'), Brunnegg-Joch (11,100'), and Biesjoch (11,644') are glacier-passes, fit for experts only, with able guides. The last is very difficult, but by keeping up a nearly perpendicular rock-gully on the Randa side, access is obtained to the great snow-basin of the Bies Glacier by a route safe from falling stones.

84. From Visp to Zermatt.
Comp. Maps, pp. 332, 308, 330.

22 M. RAILWAY in 2 hrs. 40 min. (2nd cl. 16, 3rd cl. 10 fr.), an attractive journey. — The railway from Visp to Zermatt, opened in 1891, combines the ordinary and rack-and-pinion systems, and is worked by engines on Abt's system. The maximum gradient on the ordinary sections is 45:1000; on the rack-and-pinion sections 125:1000.

DISTANCES ON FOOT: from Visp to Zermatt 9 hrs. (Stalden 1 3/4 hr., St. Niklaus 2 1/2 hrs.; Randa 5 1/2 M., Täsch 2 1/2 M., Zermatt 3 1/2 M.). Bridle-path to St. Niklaus, carriage-road thence to Zermatt. — The route from Visp to Zermatt is easy and attractive, being varied with picturesque rock-scenery and waterfalls.

Visp (2150'), see p. 309. The railway makes a wide curve to the S. towards the rapid and turbid *Visp*, and gradually ascends on the right bank of the stream, which fills the entire breadth of the valley. The train passes under the *Neubrücke* (2280'), by which the bridle-path crosses to the left bank, then crosses the river by an iron bridge, 40 yds. long, and ascends (rack-and-pinion section, 1050 yds. in length; gradient 120-125 : 1000) to the (5 M.) station of **Stalden** (2630'; *Buffet*), 2 min. to the S. of the village of the same name (2735', **Hôt. Stalden*, R., L., & A. 3 fr.), situated in a very fertile region, on a mountain-spur, at the foot of which the *Saaser Visp* and the *Matter Visp* unite. The valley divides here. The vast group of the *Saasgrat*, the N.E. spur of the Monte Rosa mass, separates the Nicolai Valley from the Saas Valley. The culture of the vine extends about 2 M. beyond Stalden.

From Stalden to (4½ hrs.) *Saas-Fee*, see p. 340.

To THE SIMPLON HOSPICE OVER THE BISTENEN PASS, 11 hrs., fatiguing (guide 15 fr.; *Joh. Furrer* of Stalden or *J. Dorsaz* of Simplon; horse 30 fr.). From Stalden the route leads by *Staldenried* and *Gspon* (6230') to a pass (about 7200') lying to the N. of the *Ochsenhorn* (9547'), also reached in 4 hrs. from Visp by *Visperterbinen* (Pens. Zimmermann). Descent to the chalets of *Bististaffel* (6170') in the upper *Nanzer-Thal*, ascent again to the *Bistenen Pass* (about 7870'), and descent thence to the Simplon Hospice (p. 311).

Immediately beyond Stalden is another rack-and-pinion section, 1030 yds. long. The train ascends through a rock-cutting and a tunnel to the height of 2940', and for a considerable time follows a level course, high up on the left side of the deep valley of the Visp; ahead of us the Weisshorn, Brunnegghorn, and Barrhorn are visible. Three short tunnels are traversed, and, beyond the imposing viaduct (177' long, 165' high) over the *Mühlbach*, two more tunnels and two other viaducts in the gorge of the *Faulkinn*. At (6½ M.) the station of *Kalpetran*, the bottom of the valley is once more reached. Above, to the right, are the little church and hamlet of *Emd*, situated on so shelving a pasture that, according to the local wits, the very fowls must be shod with iron to enable them to keep their footing. Another level section follows, after which the line crosses to the right bank and ascends by a rack-and-pinion section, 1½ M. long, through the gorges of *Kipfen* and *Seeli*, keeping close to the brawling Visp, which here forms a series of waterfalls amidst huge blocks of gneiss. We return to the left bank, and reach —

10 M. **St. Niklaus** (3705'; pop. 806; *Buffet*; *Gr. Hôtel St. Nicolas*, R., L., & **A.** 3½, B. 1½, lunch 3½, D. 5 fr.; *Hôt.-Pens. Lochmatter*), the capital of the valley. English Church Service in summer. (To Gruben over the *Augstbord Pass*, see p. 331.)

Beyond a short rack-and-pinion section the railway crosses the *Blattbach*, descending on the right from the Brunnegghorn, and then returns by a skew-bridge to the right bank of the Visp. To the right a lofty waterfall in several leaps. 13½ M. *Herbrigen* (4120'). Another steep gradient (1835 yds. long) begins at the chalets of *Breitenmatt*. High up on the left is the *Festi Glacier*, descending from the *Dom* (p. 338); to the right is the *Weisshorn* (14,805') with the fissured *Bies Glacier*; and to the S. rise the *Little Matterhorn* and the superb

ZERMATT. V. Route 84. **333**

Breithorn. Between (16 M.) **Randa** (4620'; *Hôt. Weisshorn*) and (18 M.) **Täsch** (4770') the traces of a landslip which is said to have buried a whole village are still visible. The line is now carried along the right bank of the Visp on a massive embankment. To the right opens the *Schalli-Thal*, with the **Hohlicht** *Glacier*, commanded by the *Rothhorn;* to the E. is the *Täschthal* (p. 342). At the chalets of *Zermettje* the line crosses the Visp for the last time, and then ascends a gradient, 970 yds. long, on the *Bühl*, high above the Visp, which foams in its narrow ravine below. We then enter a gorge, scarcely broad enough for both road and railway. At its end, to the right, the stupendous *Matterhorn* suddenly comes in sight; in the middle distance lies the **Gorner Glacier**; and above it stretches the vast *Upper Théodule Glacier*, with the *Little Matterhorn* and the *Breithorn* on the left. The line then passes through a short tunnel and reaches (22 M.) *Zermatt* (Railway Restaurant, with rooms).

Zermatt. — Hotels. HÔTELS DU MONT-CERVIN, *DU MONT-ROSE*, and *ZERMATT*, all belonging to the *Seiler* family; R., L., & A. 3½-5, B. 1½, lunch 3, D. 5, pens. 7-16 fr.; *HÔT. TERMINUS*, near the rail. station, R., L., & A. 3½-4, B. 1½, lunch 3½, D. 5, pens. 7-14 fr. — Outside the village, near the rail. station: *HÔT. D'ANGLETERRE*, R., L., & A. 4, B. 1½, lunch 2½, D. 4, pens. 6-9 fr.; HÔT. GORNERGRAT, R., L., & A. 3-4, B. 1½, lunch 3, D. 4, pens. 6-8 fr.; HÔT.-PENS. BELLEVUE, R., L., & A. 2-3, B. 1½, D. 3½-4, pens. 6-7 fr. — In the village: *POST, R. 2-3, B. 1½, D. 4, pens. from 6 fr. — *HÔT.-PENS. EIFFELALP (*Seiler's*), admirably situated 2 hrs. above Zermatt on the way to the Riffelberg, excellently managed, R., L., & A. 5, D. 5, lunch 10-16 fr. (patronised by the English). — *HÔT.-PENS. RIFFELHAUS (*Seiler's*), on the Riffelberg, 2½-3 hrs. from Zermatt, R., L., & A. 5, lunch 3½, D. 5, pens. 10-16 fr.; Engl. Ch. Serv. in summer. — SCHWARZSEE HOTEL (*Seiler's*), 2½ hrs. from Zermatt (p. 336), R., L., & A. 3½, lunch 3, D. 4, pens. 9-12 fr. The Seiler hotels issue coupons for lunch, etc. to their guests, which may be used at the Riffelalp, Riffelhaus, and Schwarzsee. — *Beer* at the Hôt. du Mont Cervin and Hôt. Gornergrat.
Post and Telegraph Office near the Mont-Cervin Hotel.
Guides abound, and several are first-rate (*Alex. Burgener; Weisshorn Biner; Jos., Raphael, Joh.* (two of this name), *Peter Anton Biner; Peter* and *Caesar Knubel; Fridolin* and *Alois Kronig; Jos. Maria Chanton; Roman Imboden;* the brothers *Gentinetta; Alois Pollinger; Joh., Jos. Maria, Clemens, Franz,* and *Fridolin Perren; Jos. Moser; Jos.* and *Ambros. Imboden, Quirin Schwarzen,* etc.). An agreement should be made with the guide as to the luggage he is to carry. — Horse to the Riffelalp 8, Riffel 10, Gorner Grat 12, Schwarzsee 10, Upper Théodule Glacier 15 fr. — Horses for the Gorner Grat are rarely to be had at the Riffel.
English Church (*St. Peter's*) opposite the Mont-Cervin Hotel.
BOOKSELLER: *B. Benda.* — DRIED PLANTS from the neighbourhood (an excellent field for botanists), insects, **and minerals are sold by** *Biner*.

Zermätt (5315'; pop. 525), called by the Piedmontese *Praborgne*, lies in a green valley with pine-clad slopes, above which, to the S., rises the snowy Théodule Glacier, commanded by the Breithorn on the left and the huge rock-pyramid of the Matterhorn on the right.

Zermatt surpasses the Bernese Oberland in the magnificence of its glaciers, although inferior in variety. In no other locality is the traveller so completely admitted into the heart of the Alpine world, the very sanctuary of the 'Spirit of the Alps'. The panorama from the Gorner Grat, in particular, though destitute of the common attributes of the picturesque, cannot fail to impress the spectator with its unparalleled grandeur.

The Hôt. Zermatt contains an admirable *Relief of the Environs of Zermatt*, from the Weisshorn to Macugnaga, by *Imfeld* (admission gratis).

The *Churchyard* contains the tombstones of *M. v. Grote* (p. 343), *Ch. Hudson* and *R. Hadow* (who perished on the Matterhorn in 1865), *W. K. Wilson* (Riffelhorn, 1865), *H. Chester* (Lyskamm, 1869), and to the right of the church *Michel Croz* (p. 338). Beside the *English Church* (p. 333) repose several other travellers who have perished among the mountains.

EXCURSIONS FROM ZERMATT. To the **Gorner Gorge** (1½ hr. there and back). We diverge from the Théodule route (p. 325) after ¼ hr., cross the Zmuttbach by the (3 min.) lower bridge, and turning to the left reach (7 min.) the entrance (adm. 1 fr.) to the picturesque gorge, through which the Matter-Visp dashes in brawling cascades. Bridges and paths lead to (5 min.) the upper end of the Lower Gorge, whence a stair ascends to the top of the right bank. A shady path leads thence to the (10 min.) *Upper Gorge* (worth seeing; 50 c.), where we may either cross the bridge to the left bank of the Visp and ascend to the hamlet of *Zum See* and the Schwarzsee path, or ascend to the right to the *Upper Moos* and return to (¾ hr.) Zermatt by the Riffel route.

Deservedly foremost among the attractions are the *Riffelberg and **Gorner Grat, easily visited in a single day (electric mountain-railway under construction). The bridle-path (to the Riffelhaus 2½-3, descent 1½-2 hrs.; guide unnecessary) cannot be mistaken. From the Hôtel du Mont-Rose we follow the road, leaving the church to the left, for 8 min., and cross the Visp; on the right bank we ascend through pastures, 8 min., at the church of *Winkelmatten* (5500'), we turn to the right, 2 min., bridge over the *Findelenbach*, descending from the left (p. 325); here we turn to the right, cross the pasture to the right, and ascend more rapidly, **passing between** (8 min.) **four huts**, to the *Obere Moos* (rfmts.; sign-post to the Gorner Gorge, see above). The path now ascends the *Fällistuts*, traversing a wood of larches and stone-pines; 25 min., a hut (rfmts.) above the *Schwegmatt*, whence we observe the lower end of the Gorner Glacier, the Furggbach issuing from the Furgg Glacier, and in the Zmutt valley, to the right, the Hohwäng Glacier (p. 329); 25 min., chalets **on the** *Augstkummen-Matt* (7110'; rfmts.). The steep old path to the Riffel now ascends straight on, while the new and easier path to the left describes a wide bend through a wood of stone-pines, and passes the (¼ hr.) *Hôtel Riffelalp* (7305'; p. 333), commanding a **superb view** of the colossal Matterhorn, of the Zmutt Valley with the Dent Blanche, and of the Ober-Gabelhorn, Trifthorn, Rothhorn, Weisshorn, etc. Adjacent are an English chapel and a Roman Catholic chapel. Above the hotel the two paths unite. At the foot of the *Riffelberg* (12 min.) we cross the brook, and then ascend in easy zigzags to the (40 min.) *Riffel Hotel* (8430'; p. 333), which enjoys a fine view of the Breithorn and Matterhorn and other peaks. The *Gugel* (8680'), the height to the N.E., commands also the Findelen and Adler glaciers and the Adler Pass.

The **Gorner Grat** (10,290'; 4975' above Zermatt), a rocky ridge rising from the plateau of the Riffelberg, and reached from the Riffelhaus in 1½ hr. (bridle-path, guide unnecessary; refreshmts. on the way; new hotel at the top, which sadly mars the view), commands a most imposing scene (see Panorama). The spectator is entirely surrounded by snow-peaks and glaciers. The *Mischabelhörner* (*Täschhorn*, 14,757'; *Dom*, 14,940'), huge spurs of the Monte Rosa extending towards the N., between the twin-valleys of Zermatt and Saas, as well as the mountains opposite them (*Ober-Gabelhorn*, 13,365'; *Zinal-Rothhorn*, 13,855'; *Weisshorn*, 14,805'), contest the palm with the giants of the central chain themselves. Of the peaks of *Monte Rosa* itself, the highest (15,215') and two others are a lone visible, and its appearance is less imposing than from the Italian side. The most striking object in the panorama, and incontestably the lion of Zermatt, is the *Matterhorn* (14,705'; p. 338). Around the base of the Riffelberg, from E. to W., winds the huge *Gorner Glacier*, which is joined in its course by six other glaciers; from its base issues the *Visp (Matter-Visp)*.

From the *Hohthäli-Grat* (10,796'), the E. prolongation of the Gorner Grat, 1¼ hr. more (laborious, for tolerably expert climbers only; guide convenient), the view is still finer and embraces the Findelen Glacier also.

From the Riffelalp (p. 334) there is another path to the Riffel Inn, ½ hr. longer than the above, but more interesting from its frequent proximity to the Gorner Glacier. At the Hôtel Riffelalp it diverges to the right from the bridle-path and skirts a stony slope (*Riffelbord*), the haunt of the marmot, at first in the direction of the Matterhorn, then towards the beautiful and dazzling snows of the Breithorn, beside which, farther on, appear the Zwillinge ('Twins'), the Castor (13,880') on the E. and the Pollux (13,430') on the W. After ½ hr. a path diverges to the right to the *Lower Gorner* or *Boden Glacier*, which at this point, below the icefall, may be crossed in safety (with guide). The path to the Riffelhaus continues to ascend the slope and now mounts to the left; 12 min. turn to the left; 20 min. *Gagenhaupt* (8480'), a huge mass of rock to the W. of the *Riffelhorn* (p. 337); then, to the N., to the (20 min.) Riffelhaus. — The following extension of this walk is recommended. From the Gagenhaupt we ascend to the E. towards the col, passing close to the N. side of the Riffelhorn; ½ hr., a small pool. The path leads towards an opening through which Monte Rosa is visible, passes the little *Riffelhorn-See*, and in ¼ hr. reaches the Rothe Boden (9128'), a rocky ridge to the E. of the Riffelhorn, commanding a splendid view of the Gorner Glacier and Monte Rosa. The Riffel Inn, ½ hr. to the N.W., is visible as soon as the brow of the mountain is reached. The rock-arête to the E., near the Rothe Boden, is the *Gorner Grat* (p. 334), the ascent of which from this point takes 1 hr.

The return-route from the Riffel to Zermatt viâ FINDELEN is strongly recommended (3 hrs.; comp. p. 337). At the Hôtel Riffelalp (p. 323) a new bridle-path diverges to the right, skirting the slopes of the Riffelberg, into the *Findelen Valley*. Keeping at first at nearly the same level, it finally ascends through pine-wood to the (1 hr.) *Inn* near the small *Grünsee* (7580'), at the foot of the huge moraine of the *Findelen Glacier. The top of the moraine commands a splendid view of the ice-fall of the glacier, with the Rimpfischhorn, Strahlhorn, and Stockhorn, and to the S.W. and W. of the Matterhorn, Dent Blanche, Gabelhorn, Rothhorn, Weisshorn, etc. — We return from the inn by the same path, descend at a (10 min.) finger-post to the (16 min.) bridge across the *Findelenbach*, and re-ascend to the village of (6 min.) *Findelen*, whence a bridle-path, affording a grand view of the Matterhorn and the Zmutt Valley, descends to (10 min.) *Winkelmatten* and (¼ hr.) *Zermatt*.

Mountain ascents and passes from the Riffelhaus, see p. 337.

To the *Schwarzsee Hotel (8490'), a favourite excursion (2½ hrs.; mule-path; guide, 6 fr., unnecessary; horse 10 fr.). From the Mont-Cervin Hotel the path, which as far as Hermättje is also the way to the Théodule Pass, ascends the left bank of the Visp to (10 min.) the Visp Bridge (across which leads the Riffel path) and to (8 min.) the confluence of the Visp and the Zmuttbach. Here the path divides, the left branch leading to the Gorner Gorge, while our route ascends to the right. In 20 min. more we reach the hamlet of *Zum See* (about 5575'), in the middle of which the path again forks, the bridle-path leading to the left and a shorter footpath over meadows to the right. After 7 min. the paths re-unite; and in 8 min. more the bridle-path to the Staffel-Alp (p. 337) diverges to the right. Our route ascends to the left to (25 min.) the chalets of *Hermättje* (6790'; rfmts.), where we obtain a splendid survey of the Gorner Glacier (p. 335), the Breithorn, and Zwillinge. We here turn to the right (to the left the way to the Théodule Pass, see below), and follow the bridle-path, which ascends steeply over scanty and stony pastures, partly through wood. In 40 min. the wood ceases and the path becomes less steep; and in ¾ hr. more we reach the *Schwarzsee Hotel* (p. 333), which stands on a detached hill, high above the Furgg Glacier. Below, 5 min. to the W., lies the little *Schwarzsee* (8385'). The view from the hotel is splendid, though inferior to that from the *Hörnli (9490'; 1 hr. from the hotel; guide advisable, 2-3 fr.), whence the stupendous Matterhorn is seen to great advantage.

The ascent from the Hörnli to the lower *Matterhorn Hut* (10,820'; 2½ hrs., with guide) is attractive for adepts, but somewhat laborious (comp. p. 338). — An easy return-route from the Schwarzsee to Zermatt leads over the *Staffel-Alp* (p. 337); a more difficult route (guide, including the Hörnli, 10 fr.) leads over the rock-strewn *Furgg Glacier* and the crevassed *Gorner Glacier* to (4 hrs.) the Riffelhaus.

To the THÉODULE PASS, 5-5½ hrs. (guide 10 fr.), usually combined with the passage to *Valtournanche* (p. 350) or the ascent of the *Breithorn* (see below). To (1¾ hr.) *Hermättje*, see above (route to the Schwarzsee). The Théodule route crosses the *Furggbach* (fine waterfall a little farther up), and ascends the stony slopes in many windings, very dusty in dry weather. On the right is the dirty *Furgg Glacier;* above it towers the Matterhorn, which faces us the whole way with varying outline. An ascent of 2 hrs. more brings us to the moraine of the *Upper Théodule Glacier* (about 8855'), where the bridle-path ceases. We may then either ascend the glacier (a good deal crevassed, but presenting no difficulty; rope necessary), to the (1¾-2 hrs.) *Théodule Pass*; or (preferable) we may follow the path to the left, over rocks and débris, to the (¾ hr.) *Gandegg Hut* (9800'; Inn), finely situated on the rocks of the *Leichenbretter*, between the *Lower* and *Upper Théodule Glaciers*, and thence ascend the upper glacier to the (1¼ hr.) **Théodule Pass** or *Matterjoch* (10,900'; small *Inn* with fourteen beds, plain; 'vin brûlé, 2½ fr.), to the S. of the *Theodulhorn* (11,395'), on the frontier between

Breithorn. ZERMATT. *V. Route 84.* 337

Switzerland and Italy. View limited. Descent to *Breuil* or **Fiery**, see pp. 349, 350; ascent of the **Breithorn*, see below.

To the **Staffel-Alp** (3½ hrs. from Zermatt and back; without guide). Above (¾ hr.) *Zum See* the path diverges to the right from the Théodule route (p. 336) and follows the right side of the deep *Zmutt Valley*, through beautiful stone-pine and larch wood, to the (2 hrs.) *Staffel-Alp* (7045'), commanding an admirable view of the huge Matterhorn with its glacier, the rock-strewn Zmutt Glacier with the Stockje, the Stock Glacier and Tête Blanche, and (r.) the Hohwäng Glacier; behind us, the Rimpfischhorn, Strahlhorn, and Stockhorn. From the Staffel-Alp to the *Schwarzsee* (p. 336), 1½ hr. A shorter way back (stony) leads by the hamlet of *Zmutt* on the left side of the Zmuttbach, to which we cross by a bold bridge.

To the **Findelen Glacier**, 3 hrs. (guide 6 fr., unnecessary). We follow the Riffel path to the (¼ hr.) church of *Winkelmatten* (p. 334) and ascend to the left through wood to (1¼ hr.) *Findelen* (6810') and the (½ hr.) *Eggen-Alp* (7180'), where the path divides; both paths lead past the *Stellisee* (8343') to the (1¼ hr.) *Fluh-Alp* (8570'; small Inn), whence the glacier is well surveyed. — From Findelen to the *Gränsee* (Inn) and the *Hôtel Riffelalp* (1½-2 hrs.), see p. 335. — Ascent of the *Ober-Rothhorn*, *Strahlhorn*, and *Rimpfischhorn*, see p. 333; *Adler Pass*, see p. 343.

MOUNTAIN ASCENTS from Zermatt or the Riffelhaus (guide-tariff from Zermatt).

The **Breithorn* (13,685') is ascended from Zermatt without difficulty in 7½-8 hrs. (guide 25, if the night be spent 30 fr.). We follow the Théodule route to the (3½ hrs.) *Gandegg Hut* (p. 336), where the night is spent (or at the inn on the Théodule Pass). From the Gandegg Hut we gradually ascend across the *Upper Théodule Glacier*, leaving the Théodule Pass on the right and (farther on) the rocky peak of the *Kleine Matterhorn* (*Petit Mont-Cervin*, 12,752') to the left, over frozen snow to the Breithorn plateau, and lastly mount a steeper slope of ice, where step-cutting is sometimes necessary, to the top (4-4½ hrs., from the Théodule Pass 2½-3 hrs.). Imposing **View: towards the W., towers the gigantic Matterhorn; to the left of it Mont Blanc; to the right of it the Dent Blanche, Grand Cornier, Gabelhorn, Trifthorn, Rothhorn, Schallihorn, Weisshorn; N., the Bernese Alps, the Saasgrat (Balfrinhorn, Nadelgrat, Dom, Täschhorn, Alphubel, Allalinhorn, Rimpfischhorn, Strahlhorn; E., Monte Rosa, Lyskamm, the Zwillinge; S., the Graian Alps (Gran Paradiso and Grivola) and Mte. Viso. Descent to the Théodule Pass 1½-2 hrs.

The **Cima di Jazzi* (12,525'), also easy (5-5½ hrs. from the Riffelhaus; guide 15 fr.). From the Riffel Hotel we follow the Gorner Grat route to the (½ hr.) *Rothe Boden* (9120), then turn to the right and skirt a steep slope as far as the (1¼ hr.) *Gorner Glacier*, reaching it at the '*Gadmen*' (6620'). A gradual ascent on the ice brings us to the (1 hr.) *Stockknubel* (9055'), a resting-place at the rocky base of the *Stockhorn* (11,595'); thence 2½ hrs. to the summit. Superb view, but often hazy on the Italian side. Care must be taken not to approach the overhanging snow on the E. (Macugnaga) side; were it to give way, the traveller would be precipitated to a depth of 3-4000'. If strength permits, we go on to the (1 hr.) *New Weissthor Pass* (p. 339), whence the view of Macugnaga below, apparently only a stone's-throw distant, is very striking. Back to the Riffel 3-4 hrs. — Descent to Zermatt across the *Findelen Glacier* (see above) not recommended.

The **Riffelhorn** (9615'), from the Riffelhaus 1¼ hr. (an interesting climb; guide with rope 6 fr.), affords a fine survey of the Visptbal.

*Mettelhorn (11,190'; 5 hrs. from Zermatt; guide 10 fr.), an **admirable point** (panorama by Imfeld). Bridle-path for 3½ hrs.; then **over débris and snow**, not difficult.

Unter-Gabelhorn (11,150'; 5-6 hrs. from Zermatt; guide 20 fr.), only for experts. We ascend the *Triftthal* to the (2 hrs.) *Restaurant Bellevue* at *Trift* (R. 3½, B. 2, D. 4-5 fr.); thence for 3 hrs. over grass and débris, then through a steep couloir generally filled with hard snow (step-cutting necessary), lastly a climb over rocks. Beautiful view of the Matterhorn (quite near), the Dent Blanche, Ober-Gabelhorn, Rothhorn, Weisshorn,

BAEDEKER, Switzerland. 16th Edition. 22

Mischabel, Monte Rosa, Lyskamm, Breithorn, and Mont Blanc in the distance. The Trift Inn is a starting-point also for the Ober-Gabelhorn, Zinal-Rothhorn, Triftnorn, Triftjoch. etc. (comp. p. 389). — *Wellenkuppe (12,830'; 4-5 hrs. from the Trift Inn), an interesting climb, not difficult for adepts (guide 40 fr.).

Ober-Rothhorn (11.215'; 5 hrs. from Zermatt; guide 10 fr.), attractive and not difficult. Ascent via *Findelen* (p. 337), the *Rothe Boden*, and the *Furggje*, to the E. of the *Unter-Rothhorn* (10,190'; another easy ascent). — Strahlhorn (13,760'; 8 hrs.; 80 fr.) up the *Findelen* and *Adler Glaciers*, and Rimpfischhorn (13,790'; 8-9 hrs.; 85 fr.), by the *Langenfluh Glacier*, both not very difficult. For the last four, the inn at the Fluh Alp (p. 337) is a convenient starting-point. — Dom (14,940'; 10-11 hrs. from Randa; 50 fr.), very toilsome, but without serious difficulty for adepts. From Randa to the (4½-5 hrs.) *Dom Hut* of the S. A. C. on the *Festi* (9630'); then cross the *Festi Glacier* and the arête which separates it from the *Hohberg Glacier*; and lastly ascend over steep snow and ice to the (5-6 hrs.) summit. *View one of the grandest among the Alps.

The Lyskamm or *Silberbast* (14,890'; guide 80 fr.), ascended by the *Lysjoch* (p. 339) in 9-10 hrs. from the Riffelhaus, is difficult, and dangerous on account of the snow-cornice on the E. arête. (The ascent from the *Sella Hut* by the S. arête is without danger. pp. 348 and 339. There is also a fine, but difficult route from the Gnifetti Hut.)

°**Monte Rosa**, *Höchste*, or *Dufour-Spitze* (15,215'; 8-9 hrs. from the Riffelhaus, there and back 14 hrs.; two guides, 50 fr. each; porter 35 fr.) was first ascended by Messrs. Smith, Birkbeck, and Stephenson in 1855 (comp. p. 345). For experts the ascent is free from danger or serious difficulty, but it is attended with much fatigue, and requires a perfectly steady head. The route descends near the Riffelhorn to the *Gadmen* rock (p. 337), crosses the *Gorner Glacier* and the *Monte Rosa Glacier*, and then ascends over rocks to the (2½ hrs.) *Monte Rosa* or *Bétemps Hut* of the S. A. C. (inn in summer), on the *Untere Plattje* (9810'); then over snow to (1 hr.) *Auf'm Felsen* (*Obere Plattje*, 10.970'). Again an ascent of 3 hrs. over snow, very steep at places, to the *Sattel* (13,285'), where the S. peaks of Monte Rosa are revealed. We now (the most difficult part) ascend precipitous snow-arêtes, and at last gain the top (1-3 hrs., according to the state of the snow), by clambering over perpendicularly piled slabs of rock. *View exceedingly grand (panorama by Imfeld). The Dufour-Spitze may also be ascended from the *Grenz Glacier*, by the S.W. face (for adepts only; guide 60 fr.). — Besides the Dufour peak, the following also belong to the Monte Rosa group: *Nord-End* (15,130'), *Zumstein-Spitze* (15,000'), *Signal-Kuppe* (*Punta Gnifetti*; 14,965'; club-hut, see p. 348), *Parrot-Spitze* (14,575'), *Ludwigshöhe* (14,350'), *Balmenhorn* (14,185'), *Schwarzhorn* (13,895'), and *Vincent-Pyramide* (13,830').

The Matterhorn. Fr. *Mont Cervin* (14,705'), was ascended for the first time on 14th July, 1865, by *Messrs. Whymper, Hudson, Hadow*, and *Lord Francis Douglas*, with the guides *Michael Croz* and the two *Taugwalders*. In descending Mr. Hadow lost his footing not far from the summit, and was precipitated along with Mr. Hudson, Lord **Francis Douglas**, and Croz, to a depth of 4000' towards the Matterhorn Glacier. Mr. Whymper and the two other guides escaped by the breaking of the rope. — Three days later the ascent was again made by four guides from *Breuil* (p. 350), and it is now frequently undertaken both from Zermatt and Breuil. The rock has been blasted at the most difficult points, and a rope attached to it, so that the most formidable difficulties have been removed; but even now the ascent should not be attempted by any but proficients, accompanied by guides of the first class (100 fr. with descent to Breuil 150 fr.; porter 70, to the upper hut 15 fr.). The ascent takes 7-10 hrs., including halts, from the *Schwarzsee* Hotel, where the preceding night is usually spent: to the *Matterhorn Hut* at the beginning of the N.E. arête (10,745'), 2½ hrs; thence to the unserviceable upper hut (12,610') 3 hrs., and over the *Schulter* to the summit 2 hrs. more (excl. of halts). — The ascent from *Breuil* (p. 350) is more difficult: over the *Col du Lion* (11,845') to the *Cabane de la Tour* (12,760') of the Italian Alpine Club in 5-6 hrs., and thence by the *Mauvais Pas*, the *Col Tyndall*, the *Cravate*, with the old Italian refuge-hut, and the *Pic Tyndall* to the top in 6-7 hrs. more.

Very difficult (for thorough experts only, with first-rate guides): Ober-Gabelhorn (13,365'; 8-9 hrs.; guide 70 fr.), from the *Trift Inn* (p. 337) on the E. side straight up (solid firm rock), finally crossing the narrow snow-arête in the 'Gabel' (no danger when the snow is in good condition). The descent to Zinal is very difficult (comp. p. 329). — Zinal-Rothhorn (*Moming*, 13,855'; 5½-7 hrs.; 80 fr.; ascent from Zinal, p. 329). — Weisshorn (14,805'; 80 fr.), from Randa 10-11 hrs.; by the *Schalliberg-Alp* to the *Weisshorn Hut* on the *Hohlicht* (9380'), where the night is passed, 4 hrs.; thence up the E. arête to the summit, 6-8 hrs. — Dent Blanche (14,320'; 80 fr.), from the *Stockje* (p. 326; club-hut destroyed but being rebuilt), up the *Wandfluhgrat* in 8-10 hrs.; better from Ferpècle (comp. p. 326). — Dent d'Hérens (*Mont Tabor*, 13,713'; 80 fr.), 7-8 hrs. from the Stockje, by the *Tiefenmatten Joch* (11,788').

PASSES. To BREUIL in the Val Tournanche over the *Théodule Pass* (10,900'), 9-10 hrs., not difficult (guide 20 fr.; see pp. 336, 350). The Théodule Pass may also be reached from the Riffelhaus (p. 334) viâ the *Gorner* and *Lower Théodule Glaciers*, or (easiest route) from the Schwarzsee Hotel (p. 336) viâ the *Furgg Glacier* and the *Upper Théodule Glacier*. Horse from Zermatt to the *Gandegg Hut* (p. 336) 10 fr. Descent from the pass to *Fiery* viâ the Cimes Blanches (guide 25 fr.), see p. 349. — To Breuil over the *Furggjoch* (10,990'), to the E. of the Matterhorn, shorter but more difficult than the Théodule Pass (the Schwarzsee Hotel is the best starting-point, see p. 336); over the Col de **Tournanche** (11,380'), to the W. of the Matterhorn, difficult (guide 40 fr.).

To FIERY over the Schwarzthor (12,777'), 10-11 hrs. from the Riffel (guide 40 fr.), difficult. The track ascends the *Gorner Glacier* and the crevassed *Schwärze Glacier* to the summit of the pass, between the Breithorn and the Pollux, and descends the *Verra Glacier* and *Klein-Verra Glacier* to the Val d'Ayas. Over the Zwillings-Joch (*Verra Pass*; about 13,100'), between the Castor and Pollux, also difficult (guide 40 fr.).

To GRESSONEY OVER THE LYSJOCH, 12-14 hrs. from the Riffel, laborious (guide 45 fr.). The Monte Rosa route is followed to the *Plattje* (p. 338), and the right side of the crevassed *Grenz Glacier* ascended, skirting the slopes of the *Dufour-Spitze* (beware of ice-avalanches), to the upper snow-basin of the glacier, enclosed by a majestic amphitheatre of the peaks of Monte Rosa, and to the (6-7 hrs.) Lysjoch (14,040'), between the *Lyskamm* (14,890') and the *Ludwigshöhe* (14,250'), affording to the S. a superb °View of the plain of Piedmont enclosed by the Apennines and the Maritime Alps. Descent across the *Lys Glacier* (with the *Vincent Pyramide*, 13,820', rising on the left; ascent 1 hr.), to the (1½ hr.) *Capanna Gnifetti* (11,965'; see p. 347) of the C. A. I.; thence either to the left across the *Garstelet* and *Indren Glaciers* to the (1½ hr.) *Col delle Pisse* (p. 347); or to the right by the **Garstelet Glacier** to the (1½-2 hrs.) *Capanna Linty* (10,300') and (3½ hrs.) *Gressoney-la-Trinité* (p. 347). — From the Riffel to Gressoney over the Felik-Joch (13,515'), to the E. of the Castor, difficult, and dangerous owing to frequent ice-avalanches; 12 hrs. to Gressoney-la-Trinité (guide 40 fr.). On the S. side of the pass, 2 hrs. below it, is the *Capanna Quintino Sella* of the C.A.I. (11,910'); comp. p. 348.

To ALAGNA over the Sesia-Joch (13,858'), between the Signalkuppe and the Parrot-Spitze, and the *Vigne Glacier*, very difficult and dangerous (guide 60 fr.). Over the Piode-Joch (*Ippolita Pass*, 14,185'), between the Parrot-Spitze and the Ludwigshöhe, also dangerous (feasible in the reverse direction only, from the *Bors Alp*, p. 347, and up the *Piode Glacier*). — All these passes are for experts only, with first-rate guides.

To MACUGNAGA over the New Weissthor (12,010'; 9-10 hrs.; guide 35 fr.). The route as far as the pass (5 hrs.; including the Cima di Jazzi, a digression of ¾-1 hr.; see p. 326) is an easy glacier-excursion. Beyond the pass a farther ascent is made for a short distance over abrupt rocks; then a giddy descent, along perpendicular cliffs and over precipitous snow-fields. The *Capanna Eugenio Sella* (p. 345) is reached in 1¼-1½ hr. from the pass, and *Macugnaga* (p. 344) in 3½ hrs. more. — The Old Weissthor (11,730'), between the Cima di Jazzi and the *Fillarkuppe* (12,070'), one of the most difficult of Alpine passes, has of late years been crossed by Messrs. Schlagintweit, Tyndall, Tuckett, and other mountaineers. Several different routes: to the N. is the *Jazzi Pass*, close by the Cima di Jazzi; to the S. of it, on the Weissgrat, is the *Jazzikopf*, with the couloirs descending from it; then the *Old Weissthor* proper, immediately to the N. of the Fillarkuppe. Between the Fillarkuppe and the *Jägerhorn* (13,042') is the *Fillar-Joch* (about 11,800'), and between the Jägerhorn and the *Nordend* is the *Jäger-Joch* (about 12,800'). Descent from all these to the *Jazzi* (or *Castelfranco*) *Glacier* exceedingly steep, and dangerous owing to falling stones (guide 40 fr.). — To SAAS over the *Schwarzberg-Weissthor*, see p. 343.

To ZINAL over the *Triftjoch* (11,615'; guide 35 fr.), difficult, see p. 329; over the *Col Durand* (11,400'; 35 fr.), less difficult, but longer, see p. 329; over the *Moming Pass* (12,445') and the *Schalli-Joch* (12,805'), both very difficult (guide 50 fr.), see p. 329. — To EVOLENA in the Val d'Hérens over the *Col d'Hérens* (11,415'; 30 fr.), see p. 326. To AROLLA over the *Col de Bertol* (10,925'; 30 fr.), laborious, see p. 324; over the *Col de Valpelline* and *Col du Mont Brûlé* (10,900'; 30 fr.), see p. 325. — To CHERMONTANE over the *Col de Valpelline*, *Col du Mont Brûlé*, *Col de l'Evêque*, and *Col de Chermontane* (the 'High-level Route'; 60 fr.), a long day's journey. To VAL PELLINA over the *Col de Valpelline* (11,685'; guide 35 fr.), see p. 296. — To CHÂTILLON in the Aosta Valley over the *Théodule Pass* (10,900'), easy; guide to Breuil 15 fr.; see p. 350. — The *Schwarzthor*, *Lysjoch*, and *Weissthor*, see p. 339. — To THE SAAS VALLEY six glacier-passes: the *Schwarzberg-Weissthor* (11,850'; guide 30 fr.), *Adler Pass* (12,460'; 30 fr.), *Allalin Pass* (11,715'; 30 fr.), *Fee Pass* (12,500'; 30 fr.), *Alphubel-Joch* (12,475'; 35 fr.), and *Mischabel-Joch* (12,650'; 35 fr.); comp. pp. 342, 343.

85. From Visp to Saas and Mattmark.

From Visp to *Stalden*, 5 M., railway in 26 min. (2nd cl. 3 fr. 55, 3rd cl. 2 fr. 25 c.); from Stalden to *Mattmark*, bridle-path in 7½ hrs. (to Balen 2¾, Saas-Grund 3½, **Saas-Fee** 3¾ hr., Almagell 50 min., Mattmark 2½ hrs.). Horse from Stalden **to Saas** 15, to Saas-Fee 18, from Saas to Mattmark 10 fr. Luggage may be sent by post as far as Saas-Fee.

To (5 M.) *Stalden* (2680'), see p. 332. The bridle-path descends to the left from the station, and crosses the *Kinnbrücke* (2570'), a bridge, 160' high, over the *Matter-Visp*, a little above its junction with the *Saaser-Visp*. On the hill to the left is the small church of *Staldenried*. Where the path divides, beyond two chalets on the other side of the bridge, we follow the left branch into the deep and narrow **Saasthal**, skirting the Saaser-Visp, which descends in foaming waterfalls. Beyond the chalets of *Resti* (3045') we reach (1¼ hr.) *Zen Schmieden* or *Eisten* (3555'), and thence ascend more steeply to (40 min.) *Huteggen* (4088'; Inn), with a retrospect of the Bietschhorn and its glaciers. Farther on we pass the chalets of *Im Boden*, cross the (10 min.) *Bodenbrücke* (4300') to the right bank, near a fine waterfall of the *Schweibbach* (on the right) descending from the Balenflen Glacier, and (20 min.) return to the left bank, on which lies (20 min.) the village of *Balen* (4985') in a fertile expansion of the valley, at the base of the *Balfrinhorn* (12,475'). Above the village the path once more crosses to the right bank, passes through a wild rocky defile, in which lies the chapel of *St. Anton*, and leads straight on in the open valley to —

¾ hr. **Saas im Grund** (5125'; *Hôt. Monte Moro*, R., L., & A.

SAAS-FEE. *V. Route 85.* 341

3-4, D. 4, pens. 6 fr.; wine at the *Restaurant du Dôme*), the principal place in the valley. Eng. Ch. Service in summer in St. Augustine's Church, adjoining the hotel.

EXCURSIONS. (Guides, *Clemens Zurbriggen, Xaver Andenmatten, Alois* and *Abraham Imseng, Alois* and *Peter Supersaxo, Joh. Jos.* and *Alois Anthamatten, Emanuel Burgener, J. M. Blumenthal.*) On the *Triftalp*, 3 hrs. above Saas on the E. side of the valley, is the little Hôtel Weissmies (ca. 7875'), commanding an admirable survey of the Saasgrat from Monte Rosa to the Balfrinhorn. This hotel is a good centre for the ascents of the *Triftgrätli* (9100'; 1 hr.; guide 10 fr.), *Trifthorn* (11,155'; 3 hrs.; guide 20 fr.), *Jagihorn* (10,540'; 2½ hrs.; guide 15 fr.), and *Inner Rothhorn* (11,290'; 3 hrs.; guide 15 fr.). The *Weissmies* (13,166'; 4 hrs.; guide 25 fr.), ascended via the *Trift Glacier*, is laborious but very attractive. The *Laquinhorn* (13,140'; 4 hrs.; guide 40 fr.) and *Fletschhorn* (13,125'; 4 hrs.; guide 40 fr.) are both difficult. Difficult glacier-passes lead to the Simplon via the *Laquin-Joch* (11,473'; guide 30 fr.) in 7-8 hrs. or via the *Trift Pass* (12,060'; guide 30 fr.) in 8 hrs. — Other excursions from Saas-Grund: Sonnighorn or Bottarello (11,155'), by Almagell and the *Furgg Alp* (7 hrs.; guide 30 fr.), toilsome. — Latelhorn (10,525'; 5½-6 hrs.; guide 10 fr.), not difficult. Bridle-path via Almagell and the *Furgg-Alp* to the (4½ hrs.) *Antrona Pass* (p. 314); thence to the left by the S.W. arête to the (1¼ hr.) summit.

From Saas to the *Simplon* over the *Rossboden Pass*, the *Simeli Pass*, or the *Gamser-Joch*, see p. 312; to *Gondo* over the *Zwischbergen Pass*, see p. 313; to *Domodossola* over the *Antrona Pass*, see p. 314.

A bridle-path leads from Saas to the W., crossing the Visp and ascending through wood, past the chapel of *St. Joseph*, to (¾ hr.) **Saas-Fee** (5900'; *Grand-Hôt. du Dom; *Grand-Hôt. Bellevue; *Grand-Hôt. Saas-Fee*, R., L., & A. 4, lunch 3, D. 5, pens. from 7 fr.; all belonging to the same proprietors and well adapted for a stay of some time; *Hôt.-Pens. Saas-Fee*, R., L., & A. 2¼-4, pens. 6-8 fr., at the entrance to the village), charmingly situated amidst **pastures**, with a magnificent view of the *Fee Glacier*, environed by **the Mittaghorn, Egginerhorn, Allalinhorn, Alphubel, Täschhorn, Dom, Süd-Lenzspitze, and Ulrichshorn** in a wide amphitheatre. To the E. rise the Weissmies with the Triftgrat, the Laquinhorn, and the Fletschhorn. *English Church*, **with services in summer**.

EXCURSIONS AND ASCENTS. (Guides, see above.) Pleasant walks on the pastures and in the wood near Fee, and in the romantic gorge of the *Feekinn*. — On a moraine between the two arms of the Fee Glacier, 1 hr. from Fee, lies the *Gletscher-Alp* (7008'; small restaurant), a pasture once surrounded by the glacier (interesting). The *Plattje* (8460'), by the *Gaden-Alp*, 2 hrs., and the *Mellig* (8812'), by the *Hannig-Alp*, 2 hrs., are interesting and not difficult (guide unnecessary). — Mittaghorn (10,380'; 4 hrs.; guide 10 fr.), and Egginerhorn (11,060'; 5 hrs.; 20 fr.), both very interesting and not difficult. — Allalinhorn (13,235'; 7-8 hrs.; 30 fr.), trying, but without pifficulty for experts. Above the (3 hrs.) *Lange Fluh* we diverge to the left from the Alphubel route, ascend to the (4-5 hrs.) *Fee Pass* (12,500'), and to the left to the (¾ hr.) summit (magnificent view). — The Alphubel (13,800'; guide over the *Alphubel-Joch* 35 fr., over the *Mischabel-Joch* 40 fr.); the Nadelhorn (*West-Lenzspitze*, 14,220'; guide 40 fr.), and the Süd-Lenzspitze (14,100'; guide over the *Lenzjoch* 80 fr., over the *Eggfluh* 100 fr.) may be ascended from Fee by experts (difficult). The ascents of the *Täschhorn* (14,460') and the *Dom* (14,040') on this side are dangerous from falling stones and are not recommended. — Ulrichshorn (12,890'; from the *Ried Pass* (p. 312) 1 hr. (or from Fee 7-8 hrs.; guide 30 fr.), and Balfrinhorn (12,475'), from Saas up the *Bider Glacier* and *Balenfirn* 6-7 hrs., or from the *Ried Pass* ¾ hr. (guide 30 fr.), both without difficulty.

PASSES. FROM FEE TO ZERMATT OVER THE ALPHUBEL-JOCH, 11-12 hrs., **very** attractive and without difficulty for experts (guide 30 fr.). From Fee 1 hr. to the *Gletscher-Alp* (p. 341); then a steep ascent to the *Lange Fluh*, at the (2 hrs.) upper end of which (9945') we reach the magnificent *Fee Glacier*. We gradually ascend this glacier, which is seamed at places with numerous crevasses, and finally cross snow-fields to the (6 hrs.) °**Alphubel-Joch** (12,475'), between the *Alphubel* (13,800') and the *Mellichenhorn* (12,834'), commanding a splendid view of the Matterhorn, Weisshorn, etc. Descent over the *Wand Glacier*, and then over rock, moraine, and grassy slopes to the *Obere* and (3 hrs.) *Untere Täsch Alp* (7270; small Inn, dear) in the *Mellichen Valley*. A direct but disagreeable forest-path leads hence to the left, round the slope, to Zermatt in 1½ hr.; better to descend to (½ hr.) *Täsch* (p. 333) and follow the road (or by train) thence to (4 M.) *Zermatt*. — A similar pass is the **Fee Pass** (12,505), between the Mellichenhorn and **Allalinhorn** (12 hrs. from Saas to Zermatt; guide 30 fr.).

FROM FEE TO ZERMATT over **the Mischabel-Joch** (12,650'; 13 hrs.; guide 35 fr.), between the Täschhorn **and** Alphubel, fatiguing, but not very difficult for adepts. Over the **Domjoch** (14,060'; 13 hrs.; 100 fr.), between the Täschhorn and Dom, and over the Nadel-Joch (13,670'; 14 hrs.; 50 fr.), between the Dom and the Süd-Lenzspitze, both very difficult and dangerous from falling stones. Over the **Lenzjoch** (about 12,200'), between the Süd-Lenzspitze and Nadelhorn, grand but difficult.

FROM FEE TO ST. NIKLAUS over the **Ried Pass** (12,050'; 10-11 hrs.; guide 20 fr.), difficult. The route leads from Fee to the (1 hr.) *Alp Hannig* (7065'; p. 341) and ascends steeply, to the left of the *Mellig*, over the *Hochbalen Glacier* and the rocks of the *Gemshorn* to the pass, between (r.) the *Balfrinhorn* (12,475') and (l.) the *Ulrichshorn* (12,890'), both of which may be ascended from the pass (see p. 341). Descent over the *Ried Glacier* to the *Schalbett Alp* (6915') and by *Hellenen* to *St. Niklaus* (p. 332). — A similar **pass** is the **Windjoch** (10,660'), between the Ulrichshorn and Nadelhorn.

FROM FEE TO MATTMARK over the **Egginer Pass** (about 9340') between the Mittaghorn and Egginerhorn, 7-8 hrs., with guide, not difficult for experts.

Beyond Saas-Grund the bridle-path degenerates. It ascends gradually passing the chalets of *Zerbrüggen* and *Moos*. The *Almagell-Bach* forms a fine waterfall, on the left, just before we reach (50 min.) *Almagell* (5505'), where the path from the Antrona Pass descends (p. 314). [A direct path leads from Fee to Almagell in 35 min., so that the excursion to Fee forms but a short digression from the direct route to Mattmark.] The path continues hence along the right bank of the Visp, occasionally leading through wood, and crosses the *Furggbach* near the (20 min.) chalets of *Zermeiggern* (5630'; on the left bank). To the right rise the precipices of the *Mittaghorn* and *Egginerhorn* (p. 341), with the glittering snow-fields of the *Allalinhorn* (p. 341) above them. We next cross the stony *Eienalp* to the (1 hr.) ruined chapel of *Im Lerch* (6375'). On the right lie the huge moraines of the *Allalin Glacier*, which descends from the Allalinhorn, filling up the entire valley and forming the Mattmark Lake (see below). The moraine contains blocks of 'gabbro', mingled with smaragdite, like those common in W. Switzerland, but hitherto found nowhere as ingredients of the soil except on the Saasgrat; whence geologists infer that the glaciers of this region once extended to the Jura.

The path ascends in zigzags over the débris of the moraine, past the light-green little *Mattmark Lake*, to the (1 hr.) *Hôtel Mattmark* (6965'; homely, R. & A. 3½, D. 4-5 fr.; not open before July) on the **Mattmark Alp**. Down to 1818 the *Schwarzberg Glacier* extended across the bed of the lake, but afterwards receded, leaving behind it its moraines and a huge block of serpentine called the *Blaue Stein* to mark its former extent. It is now visible only high up above the cliffs.

EXCURSIONS. (Guides should be brought from Saas, as they are rarely to be found at Mattmark.) — The Stellihorn (11,393'), ascended from the Mattmark Inn by the *Ofenthal* in 3½ hrs. (guide 10 fr.), affords an imposing view of the Eastern Alps.

GLACIER PASSES TO ZERMATT, for mountaineers, with good guides:

The Schwarzberg-Weissthor (11,850'; 10 hrs.; guide 30 fr.). The route skirts the left side of the *Schwarzberg Glacier*, ascending rock and moraine, and crossing the crevassed glacier to the (5 hrs.) pass, lying to the S. of the *Strahlhorn*. (The *New Weissthor*, leading from Zermatt to Macugnaga lies farther S.; comp. p. 339.) From this point to the *Riffelhaus*, see p. 337.

The **Adler Pass** (12,460'; 11-12 hrs.; guide 30 fr.). From the inn we cross the Thällibach to the chalets of the *Mattmark Alp*, and ascend rapidly below the *Schwarzberg Glacier* (see above) and past the *Schwarzenberg Chalets* (78°0'). In 2 hrs. we reach the *Allalin Glacier* at a height of 9435', and ascend on its E. margin to the (½ hr.) *Aeussere Thurm* (9945') and (¾ hr.) the *Innere Thurm* (10,880'). We now turn to the W., to the middle of the glacier, where the route divides. To the right, crossing in the direction of the *Allalinhorn* (13,235'), is the route to the *Allalin Pass* (see below), while we ascend very steeply in a straight direction to the (2-3 hrs.) *Adler Pass*, between (l.) the *Strahlhorn* (13,750'; from the pass in 1½ hr.) and (r.) the *Rimpfischhorn* (13,790'). The view of Monte Rosa and the Matterhorn is very striking, but the view to the N. and N.W. is shut out by the Rimpfischhorn. Descent across the *Adler Glacier* to the foot of the *Rimpfischwänge*, difficult in certain states of the snow; we then skirt the latter, crossing rock and moraine, and next traverse the *Findelen Glacier* to the (3 hrs.) *Fluh-Alp* (8370'; Inn), 2½ hrs. from Zermatt (p. 337). — HR. v. Grote (p. 334), a Russian traveller, lost his life in 1859 by falling into a crevasse of the Findelen Glacier.

The **Allalin** or **Täsch Pass** (11,715'; 10-12 hrs.; guide 30 fr.) is sometimes impracticable owing to the crevasses of the upper Allalin Glacier. From the Innere Thurm (see above) to the top 2 hrs.; descent over the *Mellichen Glacier*, and along the N. base of a ridge separating the latter from the *Wand Glacier*, to the *Mellichen Valley*. Thence to Zermatt, p. 342.

FROM MATTMARK TO MACUGNAGA via the **Monte Moro Pass** (5½ hrs.; guide from Saas 15 fr., incl. night spent at Mattmark; see below). The previous night should be spent **at Mattmark**, as in that case the summit of the pass may be reached before the noonday mists rise from the S. valleys to conceal the view.

From Mattmark to *Antrona* (and Domodossola) over the *Antigine* or *Ofenthal Pass* (guide 15 fr.), see p. 314.

86. From Piedimulera to Macugnaga, and over the Monte Moro Pass to Mattmark.

Comp. Maps, pp. 320, 332, 310.

From Piedimulera (p. 443) to Macugnaga 6¾ hrs. (to Pontegrande 2¼ hrs., Vanzone ¾, Ceppomorelli 1, Pestarena 1½, Macugnaga 1¼ hr.). Carriage-road as far as Ceppomorelli (one-horse carr. from Piedimulera 10-12 fr.) From Macugnaga to the Moro Pass 4, Mattmark 2, Saas 2½, Stalden 3 hrs. — Guides necessary only from Macugnaga to the Thälliboden (10 fr.; to the Mattmark Alp 12 fr.).

The **Moro Pass** was the usual route from the Valais to Italy before the construction of the Simplon road, but is now frequented by pedestrians only. Its great attraction consists in the immediate proximity of Monte Rosa, and the views will **compare** with the finest in Switzerland.

Piedimulera (795'), see p. 443. The road ascends the *Val d'Anzasca, passes through two tunnels, and skirts fertile and vine-clad slopes high above the left bank of the Anza. Charming and varied views. 1¹/₂ M. *Gozzi di Sotto* (1280') belongs to *Cimamulera*, which **lies above, to the right. We obtain a temporary glimpse of the Monte Rosa** group shortly before **we** reach (4¹/₂ M.) **the considerable** village of *Castiglione d'Ossola* (1685'). The road proceeds at the same level; **above, to the** right, is (1¹/₂ M.) *Calasca*. Near (2¹/₄ M.) **Pontegrande (*Hôt. du* *Grand Pont*, clean), where** Monte Rosa again becomes visible, **the** stream descending **from the** *Val Blanca* forms a waterfall.

On the hill opposite, on the **right bank of the Anza, lies** *Bannio* (2237'; Osteria del Pino, very plain). (Over the *Col di Baranca* **to** *Fobello*, **and** over the *Col d'Egua* to *Carcoforo*, see p. 454.)

The road ascends **past** *S. Carlo* (1890'), **with its large church, near which are some** gold-mines worked **by an English company,** to (2¹/₄ M.) **Vanzone** (2220'; pop. 470; **Alb. dei Caccilatori*, plain), **the** chief village **in the** valley. Immediately beyond **the village we enjoy a** superb **view of** Monte Rosa. The road ends at (3 M.) **Ceppomorelli** (2427'; *Hôt. des Alpes*, R. & A. 2, B. 1-1¹/₂ fr.; *Mondo d'Oro*), **where the** bridle-path begins (mule to Macugnaga 10 fr.; **road under** construction) Near (20 min.) *Prequartero* a path **diverging to the** right crosses the *Mondelli Pass* (9320') **to the** Saas Valley **(p. 346), but commands no view of** Monte Rosa. Our path then **crosses the** *Anza*, **ascends the rather steep** hill to (35 min.) **the hamlet of** *Morghen*, **and again descends to the** stream.

At (40 min.) **Pestarēna** (*Albergo delle Alpi*, well spoken of; *Alb. deil Minieri*, plain) are gold-mines. Near (40 min.) *Borca* (3945'; **A bergo del Passo del Turlo*, R. 1 fr.), **the first village where German is spoken, a fine waterfall descends from the** *Val Quarazza* **on the left (p. 346), and 20 min. farther on Monte Rosa is fully revealed for the first time.**

The parish of **Macugnāga** consists of **six** different villages *Borca*, *In der Stapf* (or *Staffa*), *Zum Strich* (or *Pratti*), *Auf der Rive* (or *Rippa*), *Das Dorf* (or *La Villa*), and *Zertannen* (or *Pecetto*). Staffa lies 1³/₄ M. from Borca; the other villages are only a few minutes' walk apart. The hamlet *Zum Strich* is generally named *Macugnaga* (4125'; **Hôt. Monte Rosa*, kept by *Lochmatter*, R. & A. 3, B. 1¹/₂, D. 4¹/₂ fr.; **Hôt. Monte Moro*, kept by *Oberto*, same **charges;** *Hôtel Belvedere*, **at the lower end** of **the** village, well **spoken of).** **The village** is situated **in a** pleasant grassy dale, **enclosed by a** majestic ampitheatre of snow-clad mountains: (1.) **the four peaks of Monte Rosa:** *Signalkuppe* (*Punta Gnifetti*; 14,965'), *Zumsteinspitze* (15,005'), *Höchste* (or *Dufour*) *Spitze* (15,215'), and *Nord-End* (15,130'); then the *Jägerhorn* (13,040'), *Fillarkuppe* (12,070'), *Old Weissthor* (11,730'), *Cima di Jazzi* (12,525'), *New Weissthor* (12,010'), *Roffelhörner* (11,690'), *Rothhorn* (10,620'), and *Faderhorn* (10,550'). The church of the old 'village' (the greater part of which was buried by a landslip), **built in the 16th cent., with the old communal linden-tree**, is worth a visit (10 min. from the Hôt. Monte Rosa).

Excursions. (Guides, *L. Burgener, Clemens Imseng, Aless. Corsi, G. Oberto, L. Zurbriggen*, etc.) From the *Belvedere (6640'), 2 hrs. above Macugnaga, to the W., the above-mentioned amphitheatre is surveyed at a glance from summit to base; and the view embraces the parish of Macugnaga with its **pastures** and fields, the larch-forest on the right side, and the grassy slopes **above them**. Guide (5 fr.) convenient for novices. From the hotels we pass the old church of Macugnaga (see above), and proceed in the direction of the church of the uppermost hamlet of *Zertannen* or *Pecetto*, where a guide-post directs us to the right to the Weissthor and to the left to the *Belvedere*. We cross the Anza in about ¼ hr., and then again after 10 min. walking over loose stones. We next follow a good path through bushes and pastures to the wood-clad hill, which separates the two tongues of the *Macugnaga Glacier* (last ³/₄ hr. steep). — Over the Macugnaga Glacier to the Petriolo Alp (there and back 6 hrs.; guide 6 fr.), repaying. About ¼ hr. above Zertannen we ascend to the right (leaving the Belvedere path on the left) and traverse the *Roffelstafel Alp* (where the route to the New Weissthor diverges to the right) to the *Jazzi-Alp*; then past the *Fillar Alp* (above which to the right is the *Castelfranco Glacier*, crossed on the way to the *Old Weissthor*) to the *Macugnaga Glacier*, and across the latter (superb view) to the (8 hrs.) *Petriolo Alp* (6730'; milk). We return either by the high-lying *Croza Alp*, or by a shorter route across the glacier, the S. arm of which is called the *Petriolo Glacier*, passing the *Belvedere* (see above).

Pizzo Bianco (10,190'; 5-6 hrs.; guide 10 fr.), a splendid point of view, fatiguing but without danger; last hour over steep snow.

Monte Rosa, *Höchste* or *Dufourspitze* (15,215'; guide 150, porter 100 fr.), very difficult and hazardous from Macugnaga (first time, 1872). The night is spent in the (7 hrs.) *Capanna Marinelli* of the I. A. C. (10,500'), on the *Jägerrücken*. Thence to the Dufourspitze 9-10 hrs. (p. 338).

To Zermatt over the New Weissthor (12,010'; guide 30, porter 25 fr.; 10-12 hrs. from Macugnaga to the Riffel Inn, p. 334), a grand route for adepts with good guides, without danger or serious difficulty. About 5 hrs. from Macugnaga and 1½-2 hrs. below the pass is the *Capanna Eugenio Sella* of the Ital. Alpine Club (about 10,500'), grandly situated at the margin of the large *Roffel Glacier*. — The Old Weissthor (11,730'), very difficult (guide 35 fr.), is better from this side than from Zermatt; see p. 339.

From Macugnaga to Alagna over the *Col del Turlo* or the *Col della Loccie*, see p. 346; to Carcoforo over the *Passo della Moriana* or the *Col della Bottiglia*, see p. 451; to Rima by the *Col del Piccolo Altare*, see p. 454.

The path to the **Moro Pass (guide necessary,** see p. 343) leads to the old church (p. 344), and then ascends steeply to the right through larch-wood, over stony pastures, past the *Galkerne Alp* (6890'; milk), and lastly over rock and a shelving patch of snow. The (4 hrs.) ***Monte Moro Pass** (9390'), between (l.) *Monte Moro* (9803') and (r.) the *St. Joderhorn* (9970'), affords an admirable survey of the grand Monte Rosa group to the S.W., flanked by (l.) the Punta delle Loccie, Pizzo Bianco, and Fallerhorn, and (r.) the Fillarkuppe, Old Weissthor, Cima di Jazzi, **and** Roffelhörner; to the N. are the valley of Saas and the Mischabel, **with** the Bietschhorn in the background.

The St. Joderhorn (9970'), to the E. of the pass, ascended without difficulty in ³/₄ hr., commands a still finer view, though seldom clear towards the Italian side.

We descend by the side of the *Thälliboden Glacier* by rude steps **of rock,** the remains **of** the old bridle-path, to the (³/₄ hr.) *Thälli-*

boden (8190'), a small moss-grown plain at the foot of the glacier, where the route from the *Mondelli Pass* (p. 344) comes down on the right. Towards the N.W. the Mischabelhörner (Dom and Täschhorn) are revealed; nearer are the Allalinhorn, Innere Thurm, and Strahlhorn. Crossing the *Thälibach* (above, to the left, the *Seewinen Glacier*), we next reach ($^3/_4$ hr.) the chalets of the *Distel-Alp* (7190') and the ($^1/_2$ hr.) *Hôtel Mattmark* (p. 342).

87. From Macugnaga to Zermatt round Monte Rosa.

Four Days: 1st. Over the *Turlo Pass* to *Alagna*. 2nd. Over the *Col d'Olen* to *Gressoney-la-Trinité*. 3rd. Over the *Bettaforca* to *Fiery*, and over the *Col des Cimes Blanches* to the *Théodule Pass*. 4th. Ascent of the *Breithorn*, and descent to *Zermatt*. (Or: 1st day, to *Riva*; 2nd, over the *Col di Valdobbia* to *Gressoney-St-Jean*; 3rd, over the *Pinter-Joch* to *Fiery*; 4th, over the *Théodule Pass* to *Zermatt*.) Guide 8-10 fr. per day. — Less robust walkers who wish to avoid the Turlo Pass may cross the *Col di Baranca* from *Pontegrande* (p. 344) to *Fobello* and reach *Alagna* thence through the *Val Sesia* in 2-3 days, an easy route (comp. p. 445). The Col di Valdobbia, Bettaforca, and Col des Cimes Blanches are also practicable for **mules**.

FROM MACUGNAGA TO ALAGNA OVER THE TURLO PASS, 9-10 hrs., fatiguing and not very interesting (guide 14 fr.). Below Macugnaga ($^1/_4$ hr.) we quit the path to Borca (p. 344), cross the *Anza* to the hamlet of *Isella*, and ascend a wooded hill to the (1 hr.) chalets of *Spissa*, at the entrance to the rock-strewn *Val Quarazza*, which we enter to the right. The slopes are wooded, and several waterfalls are passed on each side. The path, at first level, afterwards ascends a rocky barrier, and (1 hr.) crosses to ($^1/_2$ hr.) *La Piana*, the highest Alp (5978'), on the right bank of the stream. Opposite, on the W. side of the valley, the discharge of the *Loccie Glacier* forms a fine waterfall *(La Pissa)*. Ascending more rapidly, the path describes a wide bend round the desolate head of the valley, passes ($^3/_4$ hr.) a ruined hut (6560'), and comes to an end. We next climb steep grass-slopes, and lastly rocks and snow-slopes, to the (2 hrs.) **Turlo Pass** (9090'), a sharp ridge with a cross, between (r.) the *Fallerhorn* (10,300') and (l.) the *Piglimohorn* (9470'). Descending over an expanse of snow and poor stony pastures, we enjoy a fine view of the Sesia Glacier, the Signalkuppe, and the Parrot-Spitze. We pass the small *Turlo Lakes* and the *Alp Faller*, and descend to the *Alp Iazza* and the *Val Sesia* ($2^1/_2$-3 hrs. to the Sesia bridge). A good path now leads on the right bank of the stream, past the deserted gold-mine of *S. Maria Maddalena* to ($^3/_4$ hr.) **Alagna** (3955'; *Hôt. Monte Rosa*, R. $2^1/_2$, B. $1^1/_2$, D. $3^1/_2$ fr.; *Gr. Hôt. Alagna*, well spoken of), frequented by Italians as a summer-resort.

FROM MACUGNAGA TO ALAGNA over the **Colle delle Loccie** (11,965'), 14-15 hrs., difficult; for proficients only, with good guides (40 fr.). A toilsome and **even** hazardous climb of 8-10 hrs., over the *Petriolo Alp* (p. 345) and the crevassed *Macugnaga Glacier*, leads to the pass, between the *Punta delle Loccie* and the *Cima della Pissa* (12,475'). Descent over the *Vigne Glacier* to the *Vigne-Alp* and *Pile Alp* (p. 347).

EXCURSIONS (guides, *G. Barone*, *Franc.* and *Giov. Bottoni*, *M. Cerini*, *G. Gilardi*.) Up the Val Sesia to the (2 hrs.) *Pile Alp* (5300'; superb survey of the S.E. peaks of Monte Rosa); then to the (3/4 hr.) *Alp Bors* and (1/2 hr.) *Alp Decco*. (Over the *Col delle Pisse* to Gressoney, see below.) — The *Corno Bianco* (10,945'; 5-6 hrs.; 12 fr.). a difficult peak, with fine view of Monte Rosa and the Graian Alps, is ascended either from Alagna or Gressoney. — Towards the E., two passes lead from Alagna to (5½-6½ hr.) *Rima* in the *Val Piccola* (p. 454): the *Colle Mond* (7447') to the N. of the *Taglia ferro* (9730'), and the *Bocchetta Moanda* (7835') on its S. side (preferable). — To *Zermatt* over the *Lysjoch*, the *Sesia-Joch*, and the *Piode-Joch*, see p. 339. — From Alagna to *Mollia* and *Varallo*, see p. 454.

FROM ALAGNA TO GRESSONEY-LA-TRINITÉ OVER THE COL D'OLEN, 6½-7 hrs., attractive and easy (bridle-path; guide, 14 fr., unnecessary, but enquire for the beginning of the path). We ascend to the W. through meadows and wood, passing several groups of houses, to the (2 hrs.) *Alp Seon* or *Laglietto*, cross the brook, and mount pastures and afterwards over débris to the (2½ hrs.) Col d'Olen (9420'; *Guglielmina's Inn*). View towards the N.W. very fine. The *Gemsstein* or *Corno del Camoscio* (9928'), to the N., easily ascended from the pass in 25 min., affords a striking view of Monte Rosa, Mont Blanc, the Grand Combin, the Graian Alps, and Monte Viso. — We descend by a good path to the *Gabiet-Alp* with its little lake, and through the *Val Gressoney* or *Lysthal* to (2 hrs.) *Orsia* (5740') and (20 min.) Gressoney-la-Trinité, Ger. *Oberteil* (5370'; *Hôt.-Pens. Thedy*, R. 2, D. incl. wine 4½, pens. 7½ fr.). A new road descends the picturesque valley past (1 M.) the *Hôt. Miravalle*, a large new house opened in 1895, by *Castel*, *Perletoa*, and *Chemonal* to (3 M.) Gressoney-St-Jean (4495'; *Hôt. Delapierre*, R. & A. 2½, pens. 8½ fr.; *Hôt.-Pens. du Mont-Rose*), the capital of the valley, the upper part of which is German.

FROM ALAGNA TO GRESSONEY over the Colle delle Pisse (10,374'), 9-10 hrs., rather fatiguing. The route leads viâ the *Stoffel-Alp* and the *Bocchetta della Pisse* (7877'), round the N. side of the *Gemsstein* (see above), into the *Bors Valley* (to the right the *Bors Glacier*, with a fine waterfall), and thence in 5-6 hrs. to the pass, with the ruined *Vincent Hütte*. (Hence to the Colle d'Olen, 1 hr.; to the Gnifetti Club-Hut over the *Indren* and *Garstelet Glaciers* 2 hrs., see p. 339.) Descent by a good path to the left through the *Mos Valley* to the *Gabiet-Alp* and (3½ hrs.) *Gressoney-la-Trinité* (see above).

An easier route is across the Col di Valdobbia (8800'), from *Riva Valdobbia* (2 M. below Alagna, p. 454) to Gressoney-St-Jean (7 hrs.; guide 14 fr.). A road ascends the *Val Vogna* to the (2½ hr.) *Casa Janzo* (4598'; *Alb. & Pens. Alpina*), whence the bridle-path viâ (3 M.) *Peccia* (5023'), mounts steeply to the right to the (2 hrs.) *Ospizio Sottile* on the col. The view is limited, but we enjoy a charming survey of the Val Gressoney with its rich pastures, pine-clad slopes, and waterfalls. Steep descent over snow and stones, then through pine-forest, to (1½ hr.) *Gressoney-St-Jean*.

EXCURSIONS from Gressoney (guides, *G. Cugnod*, *Val. Laurent*, *G. Monterin*, *S. G. Vicquery*, *Al. Welf*). Beautiful view from the (1 hr.) *Boden Alp* at the foot of the Grauhorn and from the promontory of *Castel*, halfway to St. Jean. — Interesting excursion viâ (2 hrs.) *Cortlys* (*Cour de Lys*, 6570') to the (1 hr.) plateau of the *Alps Salza inferiore* and *superiore* (7667'), affording a splendid view of the Lys Glacier. — The Hohe Licht (11,635'), ascended from the (3 hrs.) *Linty Hut* (10,300'; very small) in 1 hr., is another fine point. — Two club-huts of the C. A. I. are useful for glacier expeditions. From the Gnifetti Hut (11,365'), at the W. side

348   V. Route 87.   COLLE DI BETTAFORCA.

of the *Garstelet Glacier*, 5 hrs. from Cortlys and 3 hrs. from the Col d'Olen, the *Vincent Pyramid* (13,830') may be ascended in 2 hrs., the *Parrot-Spitze* (14,575'; guide 30 fr.) in 3 hrs., the *Signalkuppe* (*Punta Gnifetti*; 14,965'; guide 35 fr.), with the *Capanna Osservatorio Regina Margherita* of the C. A. I., in 4½ hrs., and the *Zumstein-Spitze* (15,005'; guide 35 fr.) in 4½ hrs. The last two ascents may be combined in one tour. The *Dufourspitze* (15,215') was ascended hence for the first time in 1886 (7 hrs. from the Gnifetti Hut). — The **Quintino Sella Hut** (11,910), on the rocks on the W. side of the *Felik Glacier*, 3 hrs. below **the** Felik-Joch (p. 330) and 5 hrs. from Cortlys, is the starting-point for the ascent of the *Lyskamm* (14,890'; 5-6 hrs.; 50 fr.) and the *Castor* (13,850'; 4½ hrs.; 30 fr.). The descent from the latter may be made to *Breuil* (guide 40 fr.) or *Zermatt* (50 fr.).

From Gressoney to Zermatt over the *Lysjoch*, *Felik-Joch*, *Zwillings-Joch*, or *Schwarzthor*, see p. 339. Guide in each case 50 fr.

An excellent new road (diligence daily in 3 hrs. 10 min., 2 fr. 50 c.; from Pont-St-Martin to Gressoney in 5 hrs., 4 fr.) leads from Gressoney-St-Jean through the beautiful *Lysthal* viâ *Gaby* to (8 M.) *Issime* ('Posta) and thence through beautiful chestnut-woods viâ *Fontainemore* and *Lillianes* to (8¼ M.) *Pont-St-Martin* (p. 293). — To **the W.** an easy bridle-path leads from St. Jean in 3½ hrs. over the *Colle Ranzola* (7182') to *Brusson* (see below) in the *Challant Valley*, and in 2½ hrs. more over the *Col de Joux* to *St. Vincent* and *Châtillon* (p. 297). — Two very attractive ascents are those of the *Mont Taille* (7935'), in 1 hr. from the Ranzola Pass, and the *Becca di Frudiera* (*Marienhorn*, 10,790'), rising farther to the S., between the Gressoney and Challant valleys (6-7 hrs. from Gressoney-St-Jean; guide 12 fr.).

From Gressoney - la - Trinité to Fiery over the **Bettaforca**, 4½ hrs., pleasant and easy (with guide). At (20 min.) *Orsia* (p. 347) we diverge to the left, (5 min.) cross the *Lys*, and mount rapidly past the houses of *Betta* to (1 hr.) the chapel of *St. Anna* (7120'; below it, a fine waterfall), where we have a beautiful view of the **Lyskamm** and Monte Rosa. Then up a monotonous valley **(keeping to the right)** past the *Sitten Alp*, to the (1 ¾ hr.) Colle **di Bettaforca** (8640'), where we see the Graian Alps peeping above the **Val d'Ayas**, and the Grand Combin to the right. We descend (still to the right) to (1 hr.) the hamlet of *Résy* (6780'; auberge), turn to the right, and cross the **Verra** to (½ hr.) Fiery or *Fière* (6160'; *Hôt. des Cimes Blanches*, plain), on the slope 20 min. above *S. Giacomo* (5500'), overlooking the wooded Val d'Ayas.

A longer but more interesting route leads over the **Bettliner Pass** (*Passo Bettolina*; 8500') from Trinité to Fiery in 6¼ hrs. From (1¾ hr.) *Cortlys* we ascend to the left viâ the *Bettolina Alp* to the (2½ hrs.) pass, which commands a fine view of Monte Rosa. The descent skirts the W. slopes of *Monte Bettolina* (8330') to the Bettaforca route, which it follows to (2 hrs.) *Fiery* (see above).

From Gressoney-St-Jean to Fiery over the **Pinter-Joch** (8200'), 6 hrs., easy and repaying. From the pass (extensive view) experts may ascend the *Grauhaupt* (10,702'; toilsome) in 2 hrs.; view strikingly grand.

A new road descends the picturesque Val **d'Ayas** (called **Val Challant** in its lower part), watered by the *Évançon*, to **Champlan**, (3 hrs.) *Brusson* (4520'; Lion d'Or), and (3 hrs.) *Verrès*, in the Dora Valley (p. 297).

From Fiery to Breuil, or to the Théodule Pass, over the Col des Cimes Blanches. The rough mule-track to Breuil (5 hrs., guide advisable) at first ascends rapidly through wood, passes the *Alp Aventina*, and then traverses poor pastures and a dreary valley, with the *Aventina Glacier* on the right. Beyond (2 hrs.) *Varda*, the last alp, it ascends steeply, crossing (½ hr.) a brook descending from the right, and in ½ hr. more the *Cortoz*, which flows out of the Grand Lac (where the path to the Théodule diverges to the right, see below). We next traverse a rocky chaos to the right of the small *Lacs de Vent* and reach the (½ hr.) **Col des Cimes Blanches** (9910'), with a fine view of the Matterhorn and Dent d'Hérens. The *Gran Semetta* (10,595'), to the N.E., a splendid point of view, is easily scaled from the pass in ³/₄ hr. Then a descent over snow, stones, and pastures, past the little *Lacs de la Barmas* and the chalets of *Goillet* and *La Barmaz*, to (1½ hr.) *Breuil* (p. 350).

The route to VALTOURNANCHE diverges to the left from the above route about 10 min. above the bridge over the Cortoz (see above), and reaches the (12 min.) pass (9500') to the S. of the Gran Semet'a (see above; still farther to the S. is a third pass, 9298'). The route then descends, with a fine view to the W., to the beautifully situated *Alp Cleva Grossa* (7352'), and to the left to (2 hrs.) *Valtournanche* (p. 338). After fresh snow, this direct path to Valtournanche is not practicable and the descent viâ Breuil must be taken.

Travellers bound for the **Théodule Pass and Zermatt** need not descend to Breuil, but (with guide) ascend to the right from the Col des Cimes Blanches (see above), traverse rocks and stony slopes, skirt the little *Grand Lac* (9135'), and reach the (³/₄ hr.) S. edge of the *Valtournanche Glacier* (10,125'). The crevassed glacier is then crossed (rope advisable), and lastly a steep snow-slope ascended to the (1½-2 hrs.) *Théodule Pass* (p. 336). Ascent of the *Breithorn*, see p. 337; route to *Zermatt*, see p. 336.

88. From Châtillon to Valtournanche and over the Théodule Pass to Zermatt.

Comp. Map, p. 332.

Carriage-road to (11½ M.) *Valtournanche* (diligence twice daily in 3½ hrs., 3 fr.; one-horse carriage 15-20 fr., two-horse, 25-30 fr.). From Valtournanche to Breuil 2½ hrs., Théodule Pass 3½-4 hrs., Zermatt 3 hrs. Guide from Châtillon to Zermatt 25, from Valtournanche 20 fr., incl. the Breithorn 40 fr.; mule and attendant from Châtillon to Valtournanche 15 fr. — This is a very attractive expedition, often undertaken by ladies. It is the most picturesque route back into Switzerland for those who have made the Tour of Mont Blanc (R. 76). The guide should be taken all the way to Zermatt, as the path beyond the Gandegg-Hütte, after the glacier is quitted, is poorly kept and easy to miss.

Châtillon (1805'), see p. 297. The road ascends the right bank of the deep gorge of the *Marmoire*, among fine walnut and chestnut trees. On the hillside to the right appear occasionally the dilapidated arches of Roman aqueducts. Beyond (1½ M.) *Champlong* we cross to the left bank, but we return to the right bank at (3 M.) *Grand-Moulin*, where the imposing *Matterhorn suddenly appears in the opening of the valley. On the slope to the right lies the church of *Antey - St - André*; to the left the remains of an aqueduct of the 12th century. The last walnut-trees are seen at (1½ M.) *Fiernaz* (Cantine de la Rose). High up to the right is the hamlet of *Chamois* (5950'), where oats are grown notwithstanding the elevation. At (2½ M.) Ussin (4180') we cross again to the left bank (to the left the pretty *Cascade du Moulin*), and ascend in windings to (3 M.) the village of **Valtournanche** (5000'; *Hôt. du Mont-Rose*, plain), with the church of the upper valley (adjoining the church door is a tablet in memory of Chanoine Carrel, d. 1870). To the E. rises the finely shaped *Mont Roisetta* (10,895').

To the *Col des Cimes Blanches*, see p. 319; *Col de Val Cournère*, see below. Guides: *Louis Carrel*, P. *Maquignaz*, J. B. *Bich*, J. *Barmasse*, C. and *Max. Gorret*, A. and E. *Pession*, and others. — The Grand Tournalin (11,065'), reached viâ *Cheneil* in 5 hrs, with guide (12 fr.), is not difficult for experts. On the top is the *Capanna Carrel* of the C. A. I. Splendid view.

The bridle-path crosses to the left bank of the Matmoire, but returns to the right bank ¾ hr. farther on. Near the second bridge (5715') is a fine waterfall in a sombre gorge (*Gouffre de Busserailles* or *Grotte du Géant*), approached by a wooden gallery (1 fr.); adjacent is the unpretending *Hôt. des Alpes*. The path now ascends steeply through a wild and romantic defile to the (1 hr.) Chalets d'*Aouil* (6495'), and then traverses an open valley surrounded by imposing mountains: to the left the Jumeaux du Vallon, the Pointe des Cors, Dent d'Hérens, Tête du Lion, and Matterhorn, and to the right the Cimes Blanches. In ½ hr. more we reach the chalets of **Breuil** or Breil (6560'; *Hôt. des Jumeaux*, new), ¼ hr. above which lies the solitary *Albergo del Monte Cervino* at Jomein (6830'; R., L., & A. 3½-4, D. 4 fr.), amidst imposing scenery.

Ascent of the *Matterhorn* from Breuil, see p. 327. — Over the *Col des Cimes Blanches* to *Fiery*, and thence to *Macugnaga*, see R. 87. — Guides are not always to be found at Breuil, but always at Valtournanche, a fact to be noted by travellers coming from Châtillon.

To Pra-Rayé over the Col du Val Cournère, 6 hrs. with guide, rough but repaying. We cross the Matmoire ¾ hr. below Breuil and ascend to the right to the (1½ hr.) *Col de Dra* (8010'), enjoying a superb survey of the Matterhorn; descend a little, then ascend over grass, rocks, and snow, round the S.E. side of the Château des Dames (see below) and past some small lakes, to the (2 hrs.) Col de Val Cournère (10,325'), to the S. of the *Pointe de Fontanelle* (11,100'), with a fine view of Mt. Velan, the Grand Combin, etc. (From Valtournanche a bridle-path leads to the chalets of *Cignana*; thence a steep and laborious climb to the pass, 4½-5 hrs.) Descent through the *Val Cournère* to (1½ hr.) Pra-Rayé in the Val Pellina (p. 324). — The Château des Dames (11,435') may be ascended from the pass in 2½ hrs. (not very difficult, guide 15-18 fr.).

The route to the Théodule Pass (riding practicable as far as the glacier) ascends over stones and turf, past the *Chalet des Cors*, to the (2½ hrs.) spot known as *Les Fourneaux*, at the end of the *Valtournanche Glacier*, where the rope should be brought into use. We then cross the tolerably easy glacier to (1¼-1½ hr.) the **Théodule Pass** or *Matterjoch* (p. 336). Ascent of the *Breithorn*, see p. 337; to (3½ hrs.) Zermatt, see p. 336.

VI. S.E. SWITZERLAND. THE GRISONS.

89. Coire 354
    Excursions from Coire: Mittenberg; Schönegg; Spontisköpfe; Bad Passugg; Calanda, 355, 356.
90. From Landquart to Davos through the Prätigau and to Schuls over the Flüela Pass 356
    Valzeina; Seewis; Scesaplana; Fideris, 357. — From Kühlis to the Montavon. Serneus, 358. — Excursions from Klosters: Vereina Pass; Fless and Jöri-Fless Passes; Fuorcla Zadrell; Silvretta Pass, 358, 359. — Schwarzhorn, 360.
91. From Davos-Dorf to Coire viâ Lenz 360
    Excursions from Davos, 362. — From Davos to Scanfs over the Scaletta Pass; to Bergün over the Sertig Pass, 362. — Excursions from Wiesen, 363.
92. From Coire to Davos through the Schanfigg-Thal. Arosa 364
    From Langwies to Kühlis over the Duranna Pass. Weissfluh, 365. — Excursions from Arosa: Aroser Rothhorn. From Arosa to Davos over the Maienfelder Furka; to Coire viâ Tschiertschen; to Parpan over the Urden Fürkli; to Alvaneu over the Furcletta; to Davos over the Strela Pass, 365, 366.
93. From Coire to Göschenen. Oberalp. 366
    From Reichenau to Ilanz viâ Versam and Kästris, The Safier-Thal; over the Löchliberg to Splügen, and over the Glas Pass to Thusis, 367. — Flimserstein; Vorab, 368. — Piz Mundaun, 369. — Lugnetz Valley; passes to Hinter-Rhein and the Val Blenio, 369, 370. — Brigels; Val Frisal. Val Puntaiglas. Val Somvix; over the Lavaz-Joch to Curaglia, 371. — Piz Muraun; Sandalp Pass, 372. — Piz Pazzola; Val Nalps; Kreuzli Pass; Oberalpstock; Pass da Tiarms, 373. — Lake Toma; Badus; Piz Nurschallas; Val Cornera, 374. — Stock, 375.
94. From Disentis to Biasca. The Lukmanier 375
    Val Cristallina; Piz Cristallina, 375. — Piz Medel; Scopi; Piz Rondadura. From Casaccia to Faido over the Predelp Pass; Passo Columbe, 376.
95. From Coire viâ Thusis to Tiefenkasten (Schyn Road) or Splügen (Via Mala) 377
    Muttnerhorn, 379. — Piz Beverin; Piz Curvèr, 380. — Piz Vizan; Piz la Tschera. From Andeer to Stalla through the Ferrera and Averser Valleys. Stallerberg, Forcellina, and Duana Passes, 381. — From Canicül to Pianazzo over the Madesimo Pass. Guggernüll; Einshorn; Piz Tambo, 375.
96. From Splügen to the Lake of Como 383
    Madesimo, 383.
97. From Splügen to Bellinzona. Bernardino 385
    Source of the Hinter-Rhein. Rheinwaldhorn, 385, 386. — From Cama to Chiavenna; Val Calanca, 387.
98. From Coire to the Engadine over the Albula Pass . 388
    Piz Michel; Tinzenhorn; Piz d'Aela, 388. — Fuorcla Pischa; Piz Kesch; Aela Pass, 389.

99. From Coire to the Engadine over the Julier . . . 390
Churer Joch; Stätzer Horn, 390. — Aroser Rothhorn. Lenzerhorn, 391. — Piz Curvèr, Fianell Pass; Tinzenthor Pass; Errjoch; Val da Faller; Piz Platta. From Stalla over the Septimer to Casaccia in the Val Bregaglia; to Sils by Gravasalvas, 391, 392.

100. The Upper Engadine, from the Maloja to Samaden 394
Lake Cavloccio; Orlegna Fall; Forno Glacier; Piz Lunghino. From Maloja to the Averser-Thal over the Forcellina Pass; to Chiesa over the Muretto Pass, 395. — Casnile and Cacciabella Passes, 396. — Excursions from Sils. Fex Valley. Piz Margna; Piz Fora, etc. Tremoggia Pass; Fuorcla Fex-Scerscen, 396, 397. — Fuorcla Surlej; Piz Julier, 398. — Piz Nair; Val Suvretta, 401. — Muottas Muraigl; Piz **Padella**; **Piz** Ot, 402.

101. Pontresina and Environs 402
Schlucht-Promenade, 404. — Morteratsch and Roseg Glaciers, 404, 405. — Schafberg; Sruors; Muottas Muraigl, 406. — Piz Languard, Diavolezza Tour, 407. — Piz Rosatsch; Chalchagn; Surlej; Corvatsch; Morteratsch; Chapütschin; Tschierva; Sella; Glüschaint; Palü; Zupò; Crastagüzza; Bernina; Roseg; Scerscen, 408, 409. — From Pontresina to Sils over the Fuorcla da Fex, the Chapütschin Pass, or the Fuorcla Glöachaint, 409. — From Pontresina to Malenco over the Sella Pass or the Fuorcla Bellavista; from Fellaria to the Bernina Hospice over the Cambrena Pass; to Poschiavo over the Confinale or the Canciano Pass; from Pontresina to Malenco by the Chapütschin Pass and the Fuorcla Fex-Scerscen, 409, 410. — From Pontresina to the Bernina Hospice. Val del Fain; over **the** Fieno Pass to Livigno, 410. — **Piz** Campascio; Piz Lagalb; Sassal Masone; Alp Grüm; by Cavaglia to Poschlavo, 411. — Val Lagone; over the Forcola to Livigno. From the Bernina through the Val Viola to Bormio. Capanna di Dosde, Passo di Verva, 411, 412.

102. From Samaden to Nauders. Lower Engadine . . . 412
Munt Müsella, 412. — Piz Uertsch. From Ponte to Livigno over the Lavirum Pass. Piz Griatschouls; Piz Mezaun; Piz Kesch. From Scanfs to Livigno through the Val Casana. Munt Baseglia; Piz d'Arpiglia; Piz Nüna, 413. — Piz **Sursura**. From Zernetz over the Ofen Pass to Münster; to Livigno and Bormio, 414. — Piz Mezdi; Piz Linard. Guarda; Fetan; Val Tasna; Futschöl Pass, 415. — Excursions from Tarasp: Castle of Tarasp; Val d'Uina; Muotta Naluns; Piz Glüna; Piz Champatsch; Piz Lischanna, 417. — From Schuls through the Scarl Valley to S. Maria, and to the Ofenberg through the **Val** Plavna. Piz Arina; Flimber Pass, 418.

103. From Samaden-Pontresina over the Bernina to Tirano, and through the Valtellina to Colico 419
Sassalbo, 420. — Corno Stella; Val Malenco; Monte della Disgrazia, 421. — Val Masino; Piz Badile, 421, 422.

104. From the Maloja to Chiavenna. Val Bregaglia . . 422
Albigna Valley; Forcella di S. Martino; Piazzo della **Duana**, 423. — Val Bondasca; over the Forcella di Bondo to Masino, 423, 424. — Soglio; Piz Gallegione, 424.

105. From Tirano to Nauders over the Stelvio 425
From Bormio over the Foscagno Pass to Livigno, 425.

— Wormser Joch; Piz Umbrail, 426. — Three Holy Springs, 427. — From Prad to S. Maria in the Münster-Thal via Taufers and Münster, 428.
106. From Nauders to Bregenz over the Arlberg 429
Lünersee; Scesaplana, 430. — The Montafon; over the Vermunt Pass to Guarda, 431. — From Feldkirch to Buchs. Gebhardsberg; Pfänder, 432.

THE GRISONS.

The region which now forms the Canton of the Grisons *(Graubünden)* was inhabited at the beginning of the Christian era by the Rhaetians, who were subjugated by the Romans in A. D. 15. After the fall of the Roman Empire, Rhaetia came into the possession of the Ostrogoths and afterwards into that of the Franks. In the middle ages the country became the residence of many noble families, including the Bishops of Coire, the Abbots of Disentis and **Pfäfers, the** Counts of Montfort, Werdenberg, and Mätsch, and the **Barons of Vatz,** Rhäzüns, Belmont, and Aspermont, whose ruined castles are still seen on the heights. The inhabitants were grievously oppressed by these magnates, and banded together on several occasions they met and entered into associations with a view to obtain redress. Thus **in** 1396 they formed the *'League of the House of God' (Lia da Ca Dè,* **or** *Casa Dè)*; in 1424 the '*Upper*' or '*Gray League*' *(Lia Grischa)*; and between 1428 and 1436 the *'League of the Ten Jurisdictions' (Lia dellas desch dretturas)*. These coalitions gave rise in 1471 to the establishment of the *'Three Perpetual Leagues of Rhætia'*. In 1512 the confederates conquered the Valtellina, which they governed by means of bailiffs down to 1797. By **the year** 1521 more than half the population had embraced the Reformation, **but** a powerful minority remained steadfast adherents of the Roman Catholic faith. The dissensions of these two parties gave rise to the invasion **of their** country during the Thirty Years' War by Austro-Spanish and French armies, but owing to the energy of *George Jenatsch* (d. 1639) the land **at length** succeeded in recovering its independence. From the 15th cent. onwards the 'Gray Confederates' were **on** friendly terms or in alliance **with the** Swiss, and in 1803 their territory was incorporated with Switzerland **as the** 15th Canton.

Down to 1848 the canton was divided into 26 small and almost entirely independent republics, called *Hoch-Gerichte* (jurisdictions), but these were abolished by the new constitution. It is now the largest, though not the most populous canton in Switzerland (2800 sq. M.; 96.291 inhab.), embracing more than one-sixth of the area of the whole country; and it is remarkable for the variety of its scenery, climate, productions, and languages, as well as for its national peculiarities and political constitution. The country consists of an immense network of mountains, **furrowed by** about 150 valleys. Barren rocks are surrounded by luxuriant cultivation; wild deserts, where winter reigns during three-fourths of the **year,** lie amid forests of chestnuts, under the deep blue sky of Italy.

Not less varied are the inhabitants themselves in origin, language, religion, and customs. The population includes 52,842 Protestants, and 43,521 Roman Catholics, of whom 37,708 are of Romanic and 44,271 of Teutonic race. Of the curious Romanic language there are two distinct dialects: the *Ladin* of the Engadine, the Albula, and Münster valleys, and the *Romance* of the valleys of Disentis and Ilanz, the Oberhalbstein, Schams, etc. This dialect is spoken generally amongst the people, but German is gaining ground, and is taught in the schools so successfully that the younger natives speak it better than the inhabitants of German Switzerland. Several small Romanic newspapers appear at Coire, Disentis, etc. — *Italian* is spoken to the S. of the Alps, in the valleys of Poschiavo, Bregaglia, Mesocco, and Calanca.

BAEDEKER, Switzerland. 16th Edition. 23

89. Coire.

Germ. **Chur**, Ital. **Coira**, Roman. **Cuera**.

Hotels. *STEINBOCK (Pl. a; C, 4), on the Churwalden road, outside the town, R., L., & A. 3½-6, B. 1½, D. 4-4½, pens. from 8 fr.; *LUKMANIER (Pl. b; D, 2), opposite the post-office, R., L., & A. from 4, D. 4, omnibus ³/₄ fr. — Second-class: °WEISSES KREUZ (Pl. c; D, 2), R., L., & A. from 2, B. 1¹/₄, D. 2¹/₂. pens. 7 fr.; °STERN (Pl. d; E, 1), R. & A. 2¹/₂, B. 1¹/₄ fr.; *ROTHER LÖWE (Pl. e; D, 3), R. 1¹/₂-2, B. 1 fr.; HÔT.-PENS. HOFKELLER (see below); DREI KÖNIGE, moderate. — PENSION RUÄTIA.

Restaurants. °*Caianda* (Pl. g; D, 2); *Chalet Restaurant*, with garden, opposite; °*Rhätia*; *Zanolari*, at the station (Valtellina wine); *Rail. Restaurant*. — Beer at the *Casino*, in the Rothe Löwe; *Franziskaner Leistbräu*; *Löwenhof*, near the market; *Rohrer*, with garden, at the Steinbock.

Baths (swimming and other) at *Willi's*, on the Plessur (Pl. F, 4; 50 c.).

Wines. *Valtellina* (red, see p. 421), abundant and not dear. *Kompleter*, grown near Malans (p. 356) in the valley of the Rhine, near the lower Zollbrücke, good but expensive. The '*Landwein*', or ordinary wine of the country, of which the best is the *Herrschäftler*, is a good red wine, similar to Valtellina. Good wine at the *Hôtel Hofkeller*, to the left in the Episcopal Court (see below), and at the auberges '*Zu den Rebleuten*', by the church of St. Martin, **and** '*Zum Süssen Winkel*'.

English Church Service at the Steinbock Hotel.

Coire (1935'; pop. 9381; ²/₃ Prot.), the capital of the Canton of the Grisons, the *Curia Rhaetorum* of the Romans, and since the 4th cent. the seat of a bishop, is picturesquely situated on the banks of the *Plessur*, which falls into the Rhine 1¹/₂ M. from the town. Most of the Roman Catholic inhabitants dwell in the *Bischöfliche Hof*, or '*Episcopal Court*' (Pl. E, F, 3), the upper and most interesting quarter of the town, surrounded with walls. Here is the episcopal *Cathedral of St. Lucius (Pl. F, 3), part of which dates from the 8th cent. (adm. to the treasury 1 fr., tickets in the court, Sun. and holidays 3-7, other days 8.30-2 and 3-7).

The very ancient PORTAL OF THE ENTRANCE COURT is borne by columns resting on lions; above is another lion, and on the columns are Apostles. The PORTAL OF THE CATHEDRAL, with its projecting slender columns with graceful capitals, is Romanesque.

The INTERIOR is interesting owing to the succession of different styles it presents. The aisles are only about half the height of the nave. The pillars of the latter, strengthened by semi-columns, have bases adorned as was usual in the 12th cent., with leaves at the corners and heads of animals, and have curious capitals of Corinthian tendency. The vaulting is effected by Gothic arches, which, in the aisles, are stilted. SOUTH AISLE: Sarcophagus of *Bishop Ortlieb de Brandis* (d. 1491). Altar-piece, a Madonna by *Stumm*, a pupil of Rubens. Tombstone of *Count de Buol-Schauenstein* (d. 1797), and opposite, that **of** his son (d. 1838). SOUTH TRANSEPT. 1st Altar: above it, Herodias by *Cranach*; in the centre a Madonna of *Rubens's School*; the side-pictures by the elder *Holbein* and his school. The finely ornamented altar itself dates from **the fifth** century. 2nd Altar: altar-piece, a Crucifixion and Saints, a work **of the** German School of the 15th cent.; reliquary of the 16th century. Choir: °High-altar **gilded** and richly carved by *Jacob Russ* (1491). Stalls and a °Tabernacle of 1484 (the latter attributed to Adam Krafft). The CRYPT is a low chamber **with** flat ceiling of the 5th century. NORTH AISLE: 1st Altar; St. Aloysius by *Angelica Kauffmann*. Over the central altar, °Christ bearing the Cross, by *Dürer*. Adjacent, the tomb of *Jürg Jenatsch* (p. 358). In the Sacristy is the rich °TREASURY: reliquaries, crucifixes, candelabra, vestments, etc.; reliquary in embossed copper (8th cent.); reliquary in the form **of** a Gothic church, with Christ and

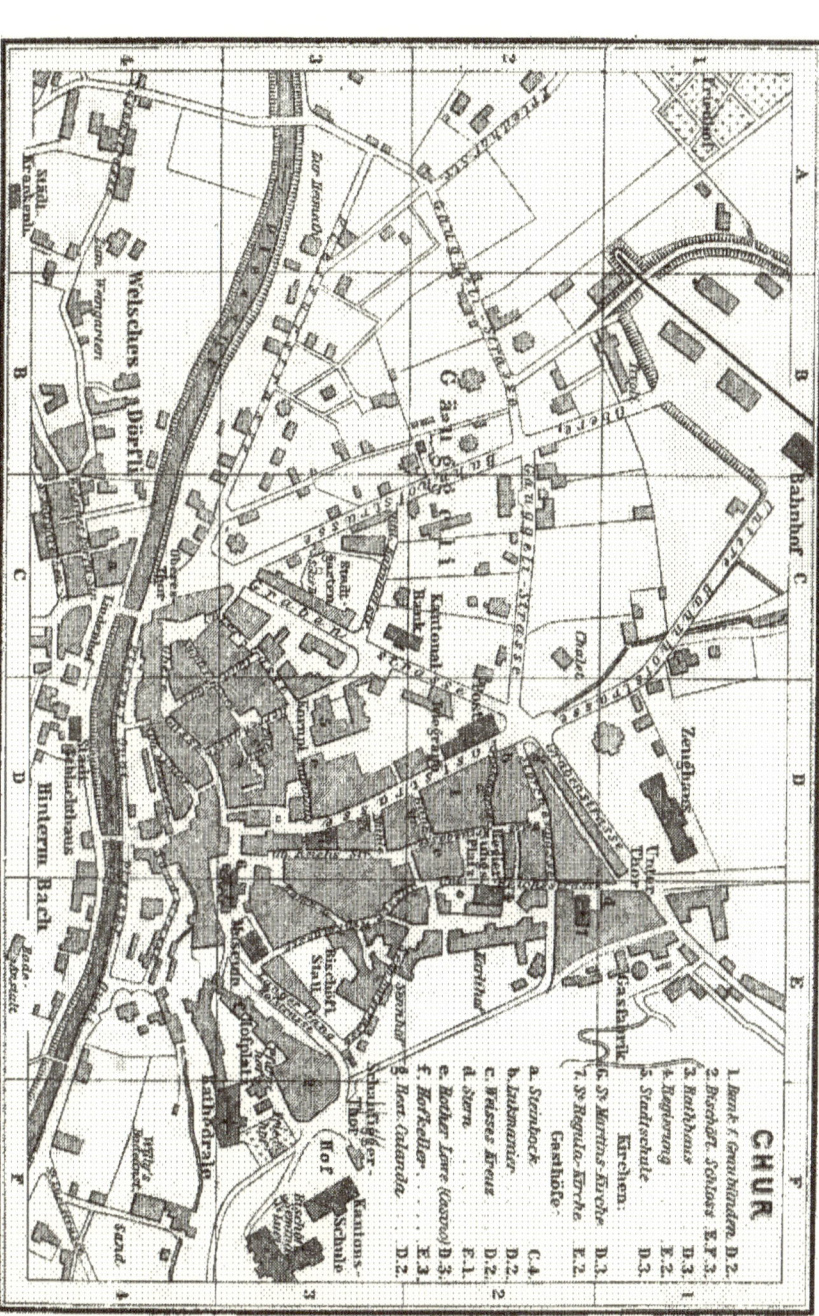

the Apostles in the arches (13th cent.); embroidered stuffs of the Saracenic period; fragments of silk dating from the time of Justinian; Christ and Peter on the sea, a miniature-painting on lapis-lazuli by C. Dolci. The glass-cabinets contain charters granted by Charlemagne, Louis le Débonnaire, Lothaire, etc.

Adjoining the church is the venerable **Episcopal Palace** (Pl. 2; E, F, 3). The *Chapel*, one of the earliest of Christian edifices, lies to the N., within the walls of the ancient Roman tower of *Marsoel* (*'Mars in oculis'*), which is connected with the palace. This tower and another named *Spinoel* (*'Spina in oculis'*, containing the 'Hofkeller', see p. 354; fine view from the windows) form the N. angles of the 'Hof' An ancient tower to the N.W., with the adjacent wall, appears also to be Roman. The names of these towers imply that the Rhætians were kept in subjection by the threats of their conquerors.

In the Hof-Platz rises the *Hofbrunnen*, a tasteful Gothic fountain (1860). Behind the cathedral are the *Priests' Seminary of St. Lucius* and the *Cantonal School* (Pl. F, 3; for both creeds).

The town itself contains few objects of interest. The Protestant *Church of St. Martin* (Pl. 5; D, 3), the *Government Buildings* (Pl. 4; E, 2), and the *Hospital* (Pl. A, 4) founded by the Capuchin Father Theodosius (d. 1865) are the chief buildings. The *Vazerols Monument*, an obelisk in the Regierungs-Platz (Pl D, 2), commemorates the leagues of Truns (1424), Davos (1436), and Vazerols (1471).

Opposite the Martinskirche, to the left of the approach to the cathedral court, is the *Rhaetian Museum* (Pl. E, 3; Sun. 10-12, gratis; at other times 1 fr.), containing antiquities, old mural paintings from the episcopal palace (Death-dance after Holbein), the cantonal library, a natural history collection, etc. In front of the museum is a bust of *Dr. E. W. Killias* (d. 1891), the naturalist. — Three windows in the hall of the *Rathhaus* (Pl. 3; D, 3) contain stained glass of the 16th century. — The old cemetery, now a public garden (Pl. C, 3), in the Graben-Strasse, opposite the Cantonal Bank, contains numerous gravestones of the 16-18th cent., in good preservation, and a monument to the poet *Gaudenz von Salis-Seewis* (d. 1834), by Kayser of Zürich.

ENVIRONS. Fine view of the town and the Rhine Valley from the *Rosenhügel* (Restaurant) on the Churwalden road, 1/2 M. from the Plessur bridge, with pleasant grounds and a monument to Moritzi, the botanist. The "'Haldenanlagen" on the *Mittenberg* also afford a good view. From the 'Hof' (p. 354) we follow the *Schanfigg Road* (p. 361) to the pavilion at the first bend of the road, then ascend to the left by the avenue and through wood, with charming views of the town and its environs. Forest-paths ascend to the (3/4 hr.) *St. Luciuskapelle*, situated under an overhanging rock in the middle of the wood; to the (1 3/4 hr.) *Mittelbergweide* (3610'), a fine point of view; to the *Kaltbrunner Tobel*, and to other points. — About 1 M. to the N.E. of the town (pleasant path from the Untere Thor through the 'Steinbruch') is the Lürlibad (*Hôtel-Restaurant Montalin*), with a fine view toward Reichenau. Hence we may follow the new Loe road to the (3/4 M.) lunatic asylum of *Waldhaus*, the *Fürstenwald*, the romantic *Scalüra Tobel*, etc.

On the *Pisokel*, a wooded hill to the S. of Coire, on the E. side of which the Churwalden road ascends (p. 390), a pleasant forest-path leads to the chalet ('Maiensäss') of (1 1/2 hr.) Schönegg (rfmts. in May and June). It

356　*VI. Route 89.*　　COIRE.

diverges by the Rosenhügel to the W. from the first bend in the road, leading to a finger-post 'nach Schönegg'. Fine view of the Vorder-Rhein Valley. Another pleasant path diverges from the same road 2 M. from Coire (finger-post), to the right, turning back, to the ($3/4$ M.) Känzeli (about 3930'). Thence to the '*Maiensässe*', the ($2^1/2$ hrs.) Spontisköpfe (6360'), and the (1 hr.) *Dreibündenstein* (7066'), affording a view of the Schanfigg Valley as far as Peist, of the Vorder-Rhein Valley, and of the Domleschg. — The *Stätzer Horn* (8408'), farther to the S., see p. 390.

Bad Passugg, with a chalybeate spring containing soda and carbonic acid, lies 3 M. from Coire in the wild valley of the *Rabiusa* (p. 39.). A path leads to it in $1^1/4$ hr. from the *Todtenguat* on the *Sand*. Or we may follow the Churwalden road to the end of the fourth great bend, diverge to the left to the *Hof Bruck* (Inn. good wine), and then ascend (left) to ($1^1/2$ hr.) *Bad Passugg* (2720'; "Hotel). The springs are 20 min. higher up, to the right. Thence a path to ($1^3/4$ hr.) Churwalden (p. 390), on the right bank of the Rabiusa, lastly crossing it and turning to the left.

The **Calanda** (9215') may be ascended from *Haldenstein*, 3 M. to the N. of Coire (p. 62), in 6-7 hrs. (fatiguing). Guides: Joh. Peter Lütscher, G. Batänjer, schoolmaster, and Andr. Cyger, of Haldenstein. The night is spent in the Calanda Hut of the S. A. C., $4^1/2$ hrs. from Haldenstein and 2 hrs. below the top. Magnificent view; more striking when the ascent is made from Vattis (p. 66; 7-8 hrs.; more fatiguing).

The following excursion of $2^1/2$-3 days is recommended: in the afternoon by Malix to Parpan 3 hrs.; next morning ascend the Stätzer Horn in 3 hrs. (p. 390); descend to Lenz; go by Alvaschein, and the Schyn road to Thusis and the Via Mala; drive to Reichenau and Coire.

From Coire to the *Schanfigg Valley* and to *Arosa*, see R. 92.

90. From Landquart to Davos through the Prätigau and to Schuls over the Flüela Pass.

Comp. Maps, pp. 356, 360, 412.

NARROW-GAUGE RAILWAY from Landquart to ($31^1/2$ M.) *Davos-Platz* in $3^1/2$-4 hrs. (fares 15 fr. 30, 10 fr., 4 fr. 70 c.); to *Klosters* in 2-$2^1/2$ hrs. (9 fr. 80, 6 fr. 60, 3 fr.). Diligence from Davos-Platz to ($31^1/2$ M.) *Schuls* twice daily in summer in $7-7^1/2$ hrs. (12 fr. 85, coupé 15 fr. 45 c.). Passengers arriving by railway from Landquart make direct connection with the Flüela diligence in Davos-Dorf, where the diligence stops at the Post Hotel, opposite the railway-station. One-horse carriage from Davos to Schuls-Tarasp 32, two-horse 60 fr. This is the direct route from Rorschach and Coire to the Lower Engadine.

The **Prätigau** ('meadow-valley'; Roman. *Val Partenz*), a somewhat narrow valley, richly sprinkled with fruit-trees, is noted for its fertility, its excellent pasturage, and its fine breed of cattle. At its mouth and in other places it is covered with the deposits of the Landquart. Among the surrounding mountains are several snow-peaks. Population (Prot.) about 10,000. German is spoken, but, as in Tyrol, most of the villages have Romanic names, that language having once been spoken here. The *Rhaetikon* chain, to the N., culminating in the *Scesaplana* (p. 367), separates the Prätigau from the Montafon (p. 431).

Landquart (1730'), see p. 62. The railway crosses the *Landquart* and describes a wide curve to the E. to (2 M.) *Malans* (1865'; Krone; Kreuz), charmingly situated $1/2$ M. from the railway, with the château of *Bodmer*. 'Kompleter', the best wine in the Rhine valley, is grown here. We again approach the Landquart, and enter the **Klus**, a narrow gorge, $3/4$ M. long, the entrance to the *Prätigau*. On a projecting rock are a few fragments of the castle of *Fragstein*, which once commanded the mouth of the gorge. In 1799 the

French had to make a détour in order to capture this defile, which was bravely defended by the peasants. — 3½ M. *Felsenbach-Valzeina* (1870'), the second station, is situated in the gorge.

A narrow road ascends the left bank of the Landquart to the (4½ M.) hamlet of **Valzeina** (4137'; *Curhaus*, unpretending, R. from 2, pens. 4-5 fr.), frequented as a summer-resort. Thence to the top of the *Valzeinerspitz* or *Hampt* (4598'; fine view), 1 hr., easy; to the *Cipriansspitz* (5883'), 2½ hrs., viâ *Hinter-Valzeina*. A bridle-path leads from Valzeina over the *Sternaboden* (4505') and through the *Schlundtobel* to (2½ hrs.) *Zizers* (p. 62).

Beyond the Klus the valley expands. 4½ M. *Seewis-Pardisla*.

A road leads hence to the N. (diligence twice daily in 1¼ hr.) to (2½ M.) **Seewis** (2985'; *Curhaus*, pens. 5½-8 fr.; "*Hôt.-Pens. Scesaplana*, at the E. end of the village, pens. 5-7 fr.), a summer-resort, charmingly situated on the hillside amidst rich pastures. Pleasant walks to the *Tantboden*, above the school, and to the *Markusplatz* (¾ hr.); to the *Emilienbrücke* (¼ hr.); to *Marnein* (3660'; ¾ hr.); to the *Malensäss* or chalet of *Malan* (4282'; 1 hr.); to *Stutz* (4280'; 1¼ hr.); to *Fadera* (3177'; 1 hr.); and to the *Mannas* (3812'; 1 hr.). — Ascents (guides, *Joh.* and *Martin Sprecher*): The Vilan (7802'; 3½-4 hrs.; guide 8 fr.) affords a splendid view. — **Scesaplana** (9740'; 6-7 hrs.; guide 14 fr.), by the *Alp Palus* and the (4½ hrs.) *Schamella Club Hut* (7800'; defective); thence to the top by a steep path in 2-2½ hrs. more (comp. p. 431). — Passage of the *Casell-Joch* (7563') to the Douglas-Hütte, 6 hrs. (guide 8 fr.), rather toilsome (comp. p. 431).

On the slope to the left is the ruin of *Solavers*. Farther on, on the hill, rises the church-tower of *Fanas*. — 5 M. **Grüsch** (2113'; *Krone*; *Rosengarten*), on the *Taschinesbach*. Large embankments were constructed across the valley in 1847-48 with a view to reclaim the land devastated by the Landquart.

7½ M. **Schiers** (2155'; *Post*; *Stern*; *Löwe*), a pretty village. On 24th April, 1622, the villagers defeated the Austrians in the churchyard. The women chiefly contributed to the victory, and they have since enjoyed the privilege of first receiving the sacrament.

Over the *Schweizerthor* (7055') or the *Drusenthor* (7710') to (8-9 hrs.) *Schruns*, see p. 481 (both toilsome, and rarely traversed). — Ascent of the **Kreuz** (7218') by *Faiauna* and *Stelserberg*, in 4 hrs., interesting.

The railway crosses the wild *Schraubach* and skirts the left bank of the Landquart, passing through a tunnel in the *Fuchsenwinkel*, 250 yds. long. 10 M. *Furna* (2360'; *Sommerfeld*). We then cross the *Fornezabach* to (10½ M.) **Jenatz** (2400'; *Sonne*; *Krone*), a large village to the right. — 11 M. **Fideris** (2445'; *Niggli*, plain).

A road (diligence to Bad Fideris thrice daily in 1¼ hr.) ascends here to the right to (1 M.) the village of *Fideris* (2962'; Inn, belonging to the owner of the baths; several pensions), where a monument to the judge *Schneider*, the 'Hofer' of the Vorarlberg, was erected by Archduke John. To the S. of the village (¾ M.) is the prettily situated *Hôtel Aquasana* (3330'; R. & A. 2-3½, board 4½ fr.); ¾ M. farther on are the **Baths of Fideris** (3463'), situated in a gorge. The chalybeate water, containing carbonate of soda and carbonic acid gas, is beneficial in pulmonary complaints, like that of St. Moritz, but it is less powerful (R. 2-3½, pens. 5-6 fr.).

The railway follows the Landquart through a magnificent rocky wooded gorge. To the left, high above, lies the hamlet of *Putz*, with the ruined stronghold of *Castels*, destroyed by the 'Gray Confederates' in 1622. From a pine-clad hill to the right peeps the ruin of *Strahlegg*. We cross the Landquart to the hamlet of *Dalvazza*, belonging to the parish of *Luzein* higher up, and then the wild *Schanielenbach* to (13½ M.) **Küblis** (2690'; *Krone; Steinbock*), a pleasant village, 1½ M. to the E. of the railway.

FROM KÜBLIS TO THE MONTAVON (p. 431), over the *St. Antönier-Joch* (7665'), 8 hrs. to Gallenkirch, easy. From the village of (8 hrs.) *St. Antönien* (4660'; Lötscher; guide, And. Flütsch) the *Sulzfluh* (9265'; superb view) may be ascended in 4-5 hrs. (trying; with guide). — To SCHEUNS over the *Partnun* or *Gruben Pass* (7330'), 7-8 hrs.; over the *Plasseggen Pass* (7694'). 8 hrs.; both without difficulty. On the *Partnun-Staffel*, 1½ hr. above St. Antönien, is the finely-situated *Hôt.-Pens. Sulzfluh* (5866'; modest, pens. 5 fr.). — To LANGWIES by *Conters* and the *Duranna Pass*, 5 hrs., see p. 365.

The railway begins to ascend (above, to the right, is *Conters*, p. 365). It skirts the N. slope, affording fine views, crosses several valleys with waterfalls, and passes a tunnel to (15 M.) **Saas** (3260'; Post); then high above the Landquart to (16½ M.) **Serneus-Mezzaselva** (3400'; *Hôt. Mezzaselva, at the station, moderate).

A carriage-road, descending to the right and crossing the Landquart, leads hence to the (1 M.) considerable village of *Serneus*. Up the valley to the left. on the left bank of the Landquart, are (1 M.) the **Baths of Serneus** (3303'; *Curhaus*, pens. from 5 fr.), noted for their sulphur-spring. The route hence to (3 M.) *Klosters* crosses both arms of the Landquart, and then keeps to the right, traversing pastures, and ascending the stream.

The line continues to ascend, high above the Landquart, and crosses the *Schlappinbach* to (19 M.) *Klosters-Dörfli* (4190'; *Curhaus Klosters-Dörfli, R. 1-2, B. 1, D. 2½, pens. 5-6 fr.; Pens. Schweizerhaus). From the height we survey the Prätigau, with the finely vaulted *Silvretta Glacier* closing the valley to the E.; to the right rise the *Canardhorn* (8566') and the *Gatschieferspitz* (8770').

20½ M. **Klosters** is prettily situated among pastures and woods in a broad valley, shut in on all sides by lofty mountains, and is much frequented in summer. It consists of the three hamlets of *Klosters-Dörfli* (see above), *Platz* (3937'), 1 M. farther on, with the church, and *Bei der Brücke* (3374'), adjoining, with the station (Restaurant). In the last two are the hotels: *Hôt.-Pens. Silvretta*, or *Curanstalt Mattli*, R., L.. & A. 1½-4, B. 1½, D. 3½, S. 2½, pens. 7-10 fr.; *Hôt.-Pens. Vereina; *Hôt.-Pens. Brosi*, R., L., & A. 1½-5, B. 1½, D. 3½, pens. 7-9½ fr.; *Alpenrose; *Hôt.-Pens. Florin; Pens. Belvedere, 6 fr., well spoken of. — The *Rütiwald*, ¼ M. from the 'Brücke', is well provided with benches.

EXCURSIONS. (Guides: *C. C. Hew, Chr.* and *W. Jann,* and *L. Guler.*) Attractive short walks to *Selfranga* (¼ hr.), *Marienhöhe* (20 min.), *Fluhstein* (25 min.), the *Fischweier* (½ hr.), *Auje* (½ hr.), *Monbiel* (1 hr.), the *Schwarzsee* (1¼ hr.), *Obere Rüti* (1½ hr.), etc. — To the **Silvretta Club-Hut** (5 hrs.; guide 7 fr., to the glacier 10 fr.), see below. From the hut to the séracs of the *Silvretta Glacier*, 1½ hr. there and back; to the top of the glacier, 3 hrs. — **Gotschna** (7435'), 3½ hrs. with guide, past the Schwarzsee (p. 359) and crossing the meadows of *Parsenn*; Canardhorn (8566'; 5 hrs., viâ *Novai*, see p 359; guide 9 fr.); **Aelpelispitz** (8825'; 5 hrs.; 7 fr.), ascended through the *Schlappin-Thal*; **Weissfluh** (9343'; viâ Ober-Laret in 5 hrs.; guide 9 fr.): these four fine points, free from difficulty. **Casanna** (8405'; 3½-4 hrs.; guide 8 fr.); the last part requires a steady head. **Pischahorn** (9790') viâ *Vereina* in 6 hrs. (guide 10 fr.), or through the *Mönchalp-Thal* in 7 hrs. (guide 12 fr.), not difficult (comp. p. 361). More laborious are the **Ungeheuerhorn** (9843'; 5 hrs. from the Vereina Hut,

to Davos. KLOSTERS. *VI. Route 90.* 359

through the Süser-Thal; 25 fr.) and the **Plattenhörner** (highest peak 10,587'; 6 hrs. from Vereina; 25 fr.). — The **Silvrettahorn** (10,655'), 4 hrs. from the Silvretta Hut (see below; guide 17, from the hut 10 fr.), the **Signalhorn** (10,538'; from the hut in 4 hrs. (guide 16 or 9 fr.), and the "**Great Piz Buin** (10,870'), 6 hrs. from the hut (guide 20 fr.), present no danger to experts. More difficult are the *Klein-Buin* (10,710'), *Verstanklahorn* (10,835'), and *Seehörner* (*Gross-Litzner*, 10,200'; *Gross-Seehorn*, 10,250').

FROM KLOSTERS TO SÜS, 9-10 hrs., with guide. A narrow road ascends the right bank of the Landquart, which is formed by the confluence of the *Sardasca* and *Vereina*, 1½ hr. above Klosters, and leads by *Monbiel* to (1½ hr.) the *Novai Alp* (1770'), on the left bank of the Sardasca. We now follow a bridle-path to the right, and ascend the *Vereina Valley*, passing the *Stutzalp* (6158'), to the (1½ hr.) *Vereina Hut* (6395'), at the mouth of the *Vernela Valley* (see below), and to the (½ hr.) *Alp Fremdeareina* (6437'), where the valley divides into the *Jörithal* to the right and the *Süser-Thal* to the left. We ascend the latter to the (2½ hrs.) pass of *Val Torta*, or **Vereina Pass** (8725'), traverse the snow to the left of the *Hörnli*, and descend rapidly by a rough path through the *Val Sagliains* to (3 hrs.) **Süs** (p. 414). Or, at the upper end of the Süser-Thal, we may turn to the right to the **Fless Pass** (8133') and descend thence through the *Val Fless* to the *Susasca Valley* and the Flüela road (p. 390), 3 M. above Süs. A third route, the finest of all (guide 18 fr.), leads through the *Jörithal* (see above), with the seven *Jöri Lakes* and the extensive *Jöri Glacier* overshadowed by the *Weisshorn* (10,130'), and across the **Jöri-Fless Pass** (8422') to the Val Fless and the Flüela road.

FROM KLOSTERS TO LAVIN BY THE FUORCLA ZADRELL, 10-11 hrs. (guide 18 fr.), suited for adepts only. From the Vereina Hut (see above) the path ascends the *Vernela Valley* (see above), passing the cavern of *Baretta-Balma*, to the *Pitter Glacier*; then a toilsome ascent on the ice to the (6-7 hrs.) **Fuorcla Zadrell** (*Vernela Pass*, or *Laviner Joch*; 9130'). Steep descent into the *Val Lavinuoz*, to *Marangun*, and below the precipices of the *Piz Linard* by the *Alp da Mest* and *Alp da Dowra* to *Lavin* (p. 414).

FROM KLOSTERS TO GUARDA BY THE SILVRETTA PASS, 10-11 hrs. (guide 20 fr.), fatiguing, but presenting no difficulty to adepts. Road to *Novai* (see above; shorter path on the right bank of the Sardasca by *Pardenn* and *Garfiun*) and through the *Sardasca Valley* to the (3 hrs.) *Sardasca Alp* (5364'); then a new bridle-path to the (2 hrs.) *Silvretta Club-Hut* (about 7480'; Inn in summer) on the *Medje-Kopf* (8225'), close to the crevassed *Silvretta Glacier*. We then ascend the crevassed glacier to the (3 hrs.) **Silvretta Pass** to the W. of the *Signalhorn* (10,520'), skirt the *Kleine Piz Buin* (10,710'), and finally descend the steep and troublesome *Plan-Rai Glacier* and the *Val Tuoi* to (3 hrs.) *Guarda* (p. 415). — From the Silvretta Hut to Lavin over the *Verstankla-Thor* or the *Tiatscha Pass* (*Fuorcla del Confin*), 7 hrs., two trying routes, for adepts only (guide 22 fr.).

To the Montafon over the *Schlappina-Joch* (3 hrs. to Gallenkirch), see p. 481. — Over the *Kloster Pass* (9180') to the *Madlener-Haus* and (11-12 hrs.) *Patenen* (p. 431), fatiguing but interesting (guide necessary).

At Klosters the locomotive is transferred to the other end of the train. The railway crosses the Landquart and ascends through the *Rüttwald*, with a pretty view to the right, as far as the *Drostobel*, where it reverses its direction by means of the Cavadürli spiral tunnel, ¼ M. long. Thence it ascends the steep *Klostersche Stütz*, a wooded hill, with fine views of the Silvretta Glacier on the left, to (25½ M.) *Laret* (4740'; Buffet). Beyond the little *Schwarzsee* (4945), with the village of *Unter-Laret* to the left, we cross the *Stützbach* to (27 M.) **Wolfgang**, at the top of the pass (5357'). The line descends through wood, skirts the E. side of the *Davoser See* (5125'; 1 M. long), a lake abounding in fish and drained by the *Davoser Landwasser*, and passes the mouth of the *Flüela Valley*

360   VI. Route 90.   FLÜELA PASS.

(see below) to (30 M.) *Davos-Dorf* (p. 361). Thence it follows the right bank of the Landwasser to (3¼1 M.) **Davos-Platz** (p. 361).

The FLÜELA ROAD crosses the Landwasser, at the station of Davos-Dorf (diligence, see p. 356; halt of ½ hr.). To the right, at the head of the *Dischma Valley*, rises the beautiful *Piz Vadret* (10,565'). We ascend the sequestered *Flüela Valley*, on the right bank of the stream, traversing wood, and passing the (4 M.) *Inn Zur Alpenrose* (6005') and (4½ M.) the *Tschuggen Inn* (6370'), to the bleak upper part of the valley, bounded by barren slopes. (The old bridle-path cuts off the windings of the road.) On the (4 M.) — 38 M. **Flüela Pass** (7835'; *Flüela Hospice, R. 2, D. 2½ fr.) the road passes between the *Schottensee* (right), with greenish-white glacier-water, and the *Schwarzsee* (left), with clear spring-water. To the N. rises the *Weisshorn* (10,130'), to the S. the *Schwarzhorn*.

The *Schwarzhorn (10,310'; 3-3½ hrs.; guide from the Flüela Hospice 8 fr., not indispensable for adepts), an admirable point, is not difficult. We descend the road to the E. (or 1 M. and then ascend the *Radön-Thal* by a good path to the right, over stony and grassy slopes, to the (1½ hr.) glacier. This we cross to the (20 min.) base of the peak, and ascend its steep S. arête to the (¾ hr.) top. Imposing panorama: most conspicuous from S. to W. are the *Piz Vadret*, and beyond it the Bernina, Piz Dosdè, etc.; the Piz Kesch, Piz d'Aela, Tinzenhorn, Piz Michel (and. farther off, the Valaisian and Bernese Alps); Lenzerhorn, Tödi, Glärnisch, Sentis, Scesaplana, in the foreground the Silvretta, the Oetzthaler Ferner, Piz Lischanna, Pisoc, Ortler; then the valleys of Flüela, Dischma, Davos, and the Lower Engadine with Ardetz and the castle of Tarasp. The descent to the *Dürrboden* in the Dischma-Thal (p. 362) leads over debris and steep slopes and should not be attempted except by adepts with a guide.

The rich flora of the Flüela Pass affords constant entertainment to the pedestrian. In the season the S. slopes are covered with the brilliant hues of masses of rhododendrons. The Primula villosa, Primula farinosa, Alpine anemones, Empetrum nigrum, the Saxifraga Seguieri, and the Saxifraga androsacea (near the hospice) also grow here.

The road descends the rock-strewn valley in windings, and crosses the *Susasca* at (2¼ M.) *Chant Sura*, by a road-menders' hut (7143'). To the right opens the dreary *Val Grialetsch*, at the head of which rises the jagged *Piz Vadret* (10,565'), with the great *Grialetsch Glacier*. The road crosses a torrent from the *Val Fless* (p. 359) on the left. Fine retrospect of the Schwarzhorn. Farther down, we cross to the right side of the valley and pass through a gallery, beyond which Süs, with its ruined castle, becomes visible in the valley, with the three-peaked *Piz Mezdi* (p. 415) above it. Then a descent in windings (old road to the left shorter) to (3¾ M.)—
44 M. **Süs** (p. 414); thence to (57 M) *Schuls*, see R. 102.

91.   From Davos-Dorf to Coire viâ Lenz.

36½ M. DILIGENCE daily in 8 hrs. (from Coire to Davos-Platz twice daily in 8-10 hrs.); 14 fr. 65, coupé 17 fr. 60 c. — EXTRA-POST, with two horses, from Coire to Davos-Platz 93 fr. 80 c.; through the Schyn Pass 118 fr. — Two-horse carr. from Coire to Wiesen 77, to Davos 110 fr., incl. fee. — The *Landwasser Road, constructed in 1870-73, vies in boldness of structure with the Schyn-Strasse and the Via Mala.

The district of **Davos** (Rom. *Tavau*), a lofty Alpine valley, about 8 M. long and ½ M. broad, with 3800 Prot. inhab., consists of pastures and a few corn-fields, sprinkled with cottages and chalets. It is enclosed by wooded mountains, and watered by the *Landwasser*. Around the five churches of the valley are grouped the hamlets of *Dörfli, Am Platz* (or *St. Johann am Platz*), *Frauenkirch, Glaris*, and, in a lateral valley, *Monstein*. Down to 1848 the district formed one of the 26 sovereign jurisdictions of the Grisons (p. 353). The inhabitants are said to have been originally German immigrants from the Valais, who settled here in the 13th century.

Railway from Landquart to Davos, see R. 90.

Davos-Dorf (5160'; *Curhaus Davos Dörfli*, well sheltered, R., L., & A. 2-6, B. 1¼, D. 3½, pens. 7½-10 fr.; *Hôt. Flüela & Post*, R., L., & A. 2-4, B. 1¼, D. 3½, S. 2½, pens. 7½-11 fr., in winter open for transient guests only; *Pens. Gredig; Mühlehof*, pens. 6-7 fr.; *Pens. Bellevue*, 4½-5 fr.; *Pens. Paul; Villa Windsor*, etc.) is prettily situated at the base of the *Schiahorn* (8900'). Opposite, at the head of the Dischma Valley, to the S.E., is the Scaletta Glacier with the Piz Vadret (p. 362); and to the left rises the Schwarzhorn (p. 360).

Pleasant walk to the (¾ hr.) *Davoser See* (p. 359). The *Weissfluh* (9345'; viâ *Meierhof* in 4½ hrs.; guide advisable) is a fine point of view (alternative descent to *Langwies*, p. 395, or *Klosters*, p. 358). — The *Pischahorn* (9790'; 5½ hrs.; guide 10 fr.) is ascended without difficulty viâ *Tschuggen*, see p. 358.

1¾ M. **Davos-Platz**. — *CURANSTALT HOLSBOER*, including the *Curhaus Davos* and several villas, R., L., & A. from 2½, B. 1½, lunch 2½, D. 4, pens. from 8 fr.; *Hôt. Pens. d'Angleterre*, R., L., & A. from 2, D. 3½, pens. from 8 fr.; *Hôt. Pens. Buol*, similar prices; *Grand Hôtel Belvedere*, with a large terrace ('Solarium'), R., L., & A. from 3½, B. 1½, lunch 3, D 4, pens. 7-10 fr.; *Hôt. Victoria; *Hôt.-Pens. Gabre; *Hôt.-Pens. Strela*, 5-7½ fr.; *Schweizerhof*, pens. 8-15 fr.; *Hôt.-Pens. Christiana*, pens. from 6½ fr.; *Hôt.-Pens. Charlotte*, pens. from 5 fr.; *Post*, reasonable charges; Hôt. Rhætia, R., L., & A. 3-5, B. 1¼, pens. 6-7½ fr.; *Davoserhof*, near the station; *Hôt. Bahnhof*, opposite the station, pens. 5-6 fr.; Hôt.-Pens. Löwe; Hôt.-Pens. Bergadler; Rathhaus, moderate; Hôt.-Pens. Geltra; Hôt.-Pens. Villa Eisenlohr; Villa Collina; Villa Freitag; Villa Frei; Pens. van Ryn; Centralhof; Tobelmühle Hotel. — Café in the *Curhaus Holsboer; Schweizerhof*, see above; *Café-Restaurant Frenziscaner; Restaurant Alpina; Gemiona Luncheon Rooms*. — *Visitors' Tax* 75 c. per week. — *Dr. Turban's Sanatorium* for consumptive patients, at the S.W. extremity of the village, in an elevated position, R. from 3½, board 11, children 8 fr. — *Mr. F. Faris-Barlow's* school for delicate boys (130-140*l*. per annum). Similar establishments are the *Fridericianum*, for boys and *Frl. Dickens's* school, for girls.

Conversations-Haus and Cur-Garten at the Curanstalt Holsboer (tickets at the hotels); concerts in the afternoon and evening, theatrical performances twice a week in winter.

English Church (*St. Luke's*); chaplain. *Rev. J. Wagstaff*. — English Physician, *Dr. W. R. Huggard*. — Information of every kind at the *Curverein Davos-Platz*.

Carriages. One-horse, to Davos-Dorf 3 fr., two-horse 5 fr.; to Spinabad and Glaris 6 or 12. Tschuggen 10 or 18, Hoffnungsau 10 or 18, Flüela Hospice 14 or 26. Wiesen 13 or 24, Tiefenkasten 25 or 45. Thusis 35 or 65, Coire 55 or 90, Tarasp 38 or 70. Samaden 50 or 90, Pontresina 55 or 100, Nauders 80 or 105, Meran 130 or 240 fr. — An Omnibus plies between Davos-Platz and Davos-Dörfli hourly; 30 c., there and back 50 c.

Davos-Platz, or *St. Johann am Platz* (5115'; pop. 4780), the capital of the district and of the ancient league of the ten jurisdictions, with picturesque houses scattered among the pastures, is a

DAVOS.

favourite summer and winter resort of consumptive patients. It is sheltered by lofty mountains from the N. and E. winds, and the air is remarkably pure and dry. The hall of the handsome *Rathhaus* contains old weapons, stained glass, and other curiosities. — *John Addington Symonds* (d. 1893) lived **for many years at Davos and wrote** most of his books here.

WALKS. Fine view above the **Hôtel Buol**, 25 min. from the rail. station. — To the *Waldhaus* (Hôt.-Pens.) at the entrance to the Dischma Valley, 1/4 hr. — To *Davos-Dörfli* and the *Davoser See* (p. 350), 1 hr. — *Gemsjäger*, 1/2 hr. — *Schatzberg* (6150'; **rfmts.**), 1 hr.; *Strela Alp* (6195'), 1 1/4 hr.; *Grüne Alp* and *Ischa Alp*, each **1 hr.** — To *Frauenkirch*, 3/4 hr.; baths of *Claradel*, 3/4 hr., etc.

ASCENTS (guides. *A. Mettier, J. Engi.* and *Chr. Clavadetscher*). — "*Schiahorn* (8900'), by a new path in **4 hrs. (guide 7 fr.)**; easy and interesting. — *Alteingrat* (7810'), by *Glaris*, 4 1/2 hrs., not difficult (guide 8 fr.); easily ascended from Wiesen also, viâ the *Alvascheiner Alp*. — "*Schwarzhorn* (10,340'), from the Flüela Pass in **3 hrs.** (10 fr.), **see p. 360**. — *Piz Vadret* (10,565'), by the Scaletta Pass in 6 hrs. (guide 20 fr.), **an interesting glacier-expedition for experts**. — *Hoch-Ducan* (10,060'), from *Sertig-Dörfli* (see below) 6 hrs. (20 fr.), difficult and fatiguing.

FROM DAVOS TO SCANFS OVER THE SCALETTA PASS, 8 1/2 hrs., attractive (direct route from Davos to the Upper Engadine; bridle-path, guide not indispensable). From Davos-Dorf we follow the high-road to Davos-Platz for a few hundred paces, turn to the left into the *Dischma Valley*, and reach (2 3/4 hrs.) the *Dürrboden* (6538'; 'Inn, rustic), with a fine view of the *Scaletta Glacier*. To the left rises the *Schwarzhorn* (10,340'), ascended hence in 4 hrs. (better from the *Flüela Pass*, p. 360). The steep and stony path ascends in 2 hrs. more to the **Scaletta Pass** (8590'), lying between the *Kühalphorn* (10,110') and the *Scalettahorn* (10,065'), on which is a ruined hut. View limited. Descent, steep at places, but enlivened by waterfalls and views of the lateral valleys with their glaciers, to the *Alp Fontauna* (7210'), and through the *Val Sulsanna* to (2 1/2 hrs.) *Sulsanna* (two poor inns) and (1/2 hr.) *Capella* in the Innthal; then to (1 1/2 M.) Scanfs (p. 413).

From Davos to Coire by the *Strela Pass* (*Schanfigg, Arosa*), see R. 92; to Arosa, by the *Maienfelder Furka*, see p. 366.

TO BERGÜN OVER THE SERTIG PASS, 8 hrs., interesting (road to Sertig-Dörfli; guide not indispensable for adepts with good maps). About 1 M. to the S. of Davos-Platz the road diverges **from the Frauenkirch road to** the left, crosses the Landwasser, enters the pretty, wooded *Sertig Valley*, and leads past (2 M.) the small sulphur-bath of *Claradel* (5400'; Curhaus, **pens.** 5 1/2–7 1/2 fr.) and many scattered chalets to (4 M.) *Sertig-Dörfli* (6102'; **Gadmer**, rustic), with the church of the valley. Above the village, 'Hinter den Ecken', the valley divides into the *Ducan-Thal* to the right, from which a fatiguing route leads over the *Ducan Pass* (8768') to **Filisur**, and the *Kühalp-Thal* to the left, through which our path now **ascends. At the** head of the valley, where the path loses itself (1 1/4 hr.), we **cross the stream to the** right and proceed to the S.W. over turf, debris, and **screes**, where the path re-appears, to the (1 1/4 hr.) Sertig Pass (9062'), **between the** *Kühalphorn* (see above) and the *Hoch-Ducan* (10,060'). **Fine view of the** *Porchabella Glacier* and *Piz Kesch* (see below) towards the S. We then descend to the right past the *Ravéisch Lakes*, where the bridle-path begins again, and through the *Val Tuors* to the chalets of *Chiaclavuot* (6106') and (3 hrs.) Bergün (p. 389); or we may descend from the pass to the S. to the (1 1/2 hr.) *Kesch-Hütte* (8630'), finely situated at the foot of the Porchabella Glacier. A fine route for adepts leads across this glacier and over the *Fuorcla d'Eschia* (9868') to (6 hrs.) *Madulein* (p. 413); splendid view of the Bernina, Ortler, Inn valley, etc. The *Piz Kesch* (11,290') may be ascended from the Kesch Hut in 2 1/2–3 hrs. by adepts with guide (comp. p. 413).

The beautiful *Landwasser Road* crosses several torrents with **their broad stony** deposits and follows **the right side** of the valley,

WIESEN. *VI. Route 19.* 363

which is sprinkled with houses and chalets. In front rises the tooth-like *Tinzenhorn* (p. 388). 2½ M. *Frauenkirch* (4793'; *Post, R. 1½-3, pens. 5-7 fr.), protected from avalanches by a bulwark, with a picturesquely situated old church. To the left opens the *Sertig Valley*, in which lies *Clavadel* (p. 362). The valley contracts. We cross the Landwasser near the (1½ M.) *Spinabad* (4816'), a sulphur-bath (good, though plain; pens. 4½ fr.) prettily situated amidst pines, and pass (¾ M.) *Glaris* (4785'; Post), scattered on the pastures of the right bank. The road then leads through the picturesque, wooded valley, on the left bank of the stream, in the direction of the *Piz Michèl* (p. 388), to the (2½ M.) *Schmelzboden Hoffnungsau* (4362'; Inn), an abandoned foundry. To the right rise the precipitous pine-clad and stony slopes of the *Züge*.

Below the foundry the valley contracts to a wild gorge. The new road (*Zügenstrasse*) follows the left bank for ¾ M. more, leads through a tunnel and an avalanche-gallery, and crosses to the right bank, where it soon begins to ascend. Three more tunnels and another avalanche-gallery. The *Bärentritt*, a projecting platform, 250' above the Landwasser, affords a striking view of the grand and wild valley, into which the *Sägentobel Fall*, 105' high, is precipitated on the right. The road crosses the *Sägentobel*, the *Mühlentobel*, and the *Brückentobel*, and ascends in long windings to (2¾ M.) —

12½ M. **Wiesen** (4720'; *Hôt.-Pens.Bellevue*, pens. 6-8 fr.; Engl. Ch. Serv.), on the sunny slope high above the Landwasser, sheltered from the N. and N.E. winds, and frequented as a health-resort. To the S., beyond the deep gorge of the Landwasser, on the green slopes of the *Stulsergrat* (8790'), lies *Jenisberg*. Farther distant are the huge *Tinzenhorn* (10,430') and the *Piz Michèl* (10,375').

WALKS. Viâ *Süsswinkel* to the upper *Brückentobel* and the *Mühlentobel*, with their pretty waterfalls (½ hr.). — To the *Tiefentobel* (see below), 20 min.; the road commands a beautiful view of the Tinzenhorn, Piz Michel, and Piz d'Aela; farther on is (40 min.) *Schmitten*. Beyond the Tiefentobel we may descend to (¼ hr.) *Bodmen* (4162'), with its ruinous houses; pleasant forest-path thence into the gorge of the Landwasser, to the *Theerhütte*, and to the *Leidboden* (20 min.); then either return to (¾ hr.) Wiesen, or cross the stream and traverse the larch-forest interspersed with pleasant glades to (1 hr.) *Filisur* (p. 388). — To the (35 min.) *Jenisberg Bridge* (3900'), 273' above the Landquart. A few paces to the left, before reaching the bridge, we have a fine view of the *Känzeli Waterfall*. From the bridge a steep ascent to (1¼ hr.) *Jenisberg* (5010'); then by a path, very rough at places, high above the Zügenstrasse, with fine views of the Davos valley, to the (1¼ hr.) *Schmelzboden Hoffnungsau* (see above). — To the (¾ hr.) *Bärentritt*, and by the romantic *Zügenstrasse* to *Hoffnungsau* and to *Davos* (see above). — The *Wiesener Alp* (6310'; good forest-path, 1½ hr.) is a good point of view; a finer is the *Sandhubel* (9080'), ascended from the alp in 2½ hrs. (riding practicable; comp. p. 386).

Beyond Wiesen (1 M.) the road crosses the profound *Tiefentobel* (with a large avalanche-bulwark above it), and passes through a tunnel. The church of (2 M.) **Schmitten**, Roman. *Farrēra* (4150'; *Adler*; *Kreuz*; *Krone*), on a grassy hill, now becomes visible. Below the village the *Albula* unites with the Landwasser.

To Filisur (p. 388) a footpath, which diverges from the road to the left near the church, descends in a wide curve, crosses the Schmitterbach (impassable after rain), and joins the Albula road to the W. of the (1/2 hr.) bridge across the Landwasser between Bad Alvaneu and Filisur. — From Wiesen across the *Leidboden* to (11/2 hr.) Filisur (boy as guide), see p. 363. The road crosses the *Schmittertobel* to (11/2 M.) *Alvaneu*, Rom. *Alvagne* (3887'). To the S.E. we obtain a pleasant view of the Bergün Valley, separated from the valley of the Landwasser by the *Stulsergrat* (8790'); in the background rises the *Piz Uertsch* (10,740'). The road soon descends by a long curve into the large *Crapanaira Tobel*, where it divides. The road to Tiefenkasten descends to *Surava* (Bad Alvaneu lies to the left, see p. 388), in the Albula valley, and leads to (51/2 M.) *Tiefenkasten* (p. 391), 10 M. from Wiesen. — The road to Coire follows the hillside, and crosses a covered wooden bridge at the base of the castle of *Belfort* (3575'), destroyed in 1499, a picturesque ruin on an almost inaccessible rock. Then (3 M.) *Brienz* (3713'), and (2 M.) —

22 M. **Lenz** (p. 391); thence to *Churwalden* and (141/2 M.) *Coire*, 361/2 M. from Davos-Dorf, see R. 99.

Those bound from Davos to *Thusis* do not descend to Tiefenkasten but follow the road to *Brienz*, where a path to the left, at the end of the village, leads to (50 min.) *Alveschein* (comp. p. 379).

92. From Coire to Davos through the Schanfigg-Thal. Arosa.

Comp. Map, p. 360.

From Coire to *Arosa*, 20 M., diligence in summer twice daily in 6 hrs. (descent in 31/2 hrs.; fare 6 fr. 35 c.); carriage with one horse 30, two horses 50 fr. From Langwies to *Davos*, by the *Strela Pass*, bridle-path in 31/2-4 hrs. (guide or horse 10 fr.).

Coire, see p. 354. The new Schanfigg road ascends the steep slope of the *Mittenberg* (p. 355) in long windings, and commands a fine retrospect of Coire and the valley of the Vorder-Rhein. At (3 M.) the *Strela Inn*, below *Maladers* (3320'), which is not within sight at first, it enters the picturesque **Schanfigg-Thal**, with its woods and meadows. The *Plessur*, far below in its wooded gorge, is fed by many affluents from both sides. On the left bank are the Baths of Passugg (p. 356); above on the Churwalden road is Malix (p. 390). Beside the bridge which spans the deep ravine of the *Calfreiser Tobel* is a pretty waterfall. The road passes through a short **tunnel** below (3 M.) *Calfreisen* (4095'); to the left, above the road, rises the ruin of *Bernegg*. Crossing the *Castieler Tobel*, we pass through another tunnel **and reach** (1 M.) **Castiel** (3960'; Hemmi, good wine), a charmingly situated village with a mineral spring. The road now winds along the mountain-slopes, maintaining a tolerably uniform level and crossing the *Glasaurer-Tobel* and the *Gross-Tobel* (earth-pyramids), to (31/2 M.) **St. Peter** (4125'; Löwe; Pens. Badrutt, 4 fr. daily), and goes on by *Peist* (4382'; Inn) and over the *Peister Tobel*, the *Frauen-Tobel*, and the *Gründje-Tobel*, to (31/2 M. —

14 M. **Langwies** (4285'; *Hôt.-Pens. Strela*, R. 2, B. 1, pens. 4-5 fr.; *Bär*), the largest parish of the Schanfigg, in a sheltered position. To the S. opens the *Arosa-Thal* (see below).

FROM LANGWIES TO KÜBLIS OVER THE DURANNA PASS, 5 hrs., an easy and attractive route. A carriage-road ascends to (1½ hr.) *Fondei* or *Strassberg* (6275'), whence a bridle-path leads to the (1 hr.) mareny summit of the **Duranna Pass** (6970'), between the *Weissfluh* (see below) on the right, and the *Kistenstein* (8135') on the left. View of the Rhætikon chain, etc. We descend by the *Fideriser Alps* to (2 hrs.) **Conters** (3715'), whence a carriage-road leads to (1½ M.) *Küblis* (p. 354). — The **Weissfluh** (9345') may be ascended in 3½ hrs. from Langwies, either viâ *Fondei* or viâ *Sapün* and the *Haupter Alp* near the Strela Pass (easy and attractive; descent if desired to Klosters or Davos, comp. pp. 363, 364).

Arosa (ca. 5900'), which has lately come into favour as a health and summer resort, may be reached in 2 hrs. from Langwies by a new road (diligence twice daily, see p. 364), which descends to the E. to the *Sapüner Bach*, flowing from the Strela Pass (p. 366). Crossing the brook the road ascends through wood on the left bank, passing the (½ hr.) gorge of the *Bühlenbach* with its waterfalls. It then gradually descends to the bridge over the *Plessur*, whence it again ascends to the (½ hr.) *Rüti* (4810'; Hôt. Alpenhof; Pens. Rütihof). About 1 M. farther on the road divides. The new road winds up to the right in wide curves and continues at a high level, past the small *Schwarzsee* and the *Obere See* (see below), to the (3 M.) *Post Office* (see below). The old road ('Waldweg'; preferable for walkers) leads through wood to the (50 min.) **Hôt.-Pens. Seehof* (5625') and the **Pension Belvedere*, both prettily situated on the little *Untersee*, in the *Seegrube*, or lower part of Arosa. The other hotels (pens. daily about 6-8 fr.) are situated round the partly wooded valley: to the right, ¼ M. above the Untere See, are the *Hôt. Rhätia*, *Villa Germania*, *Hôt. zur Post*, **Hôt.-Pens. Rothhorn*, with the Post and Telegraph Office, the **Hôt.-Pens. Victoria*, and the **Hôt.-Pens. Hof-Arosa*; to the left, a little below the Hof Arosa, lies the **Pens. Waldhaus*; beyond it, in the wood, the *Grand-Hôtel (5692'; 100 beds; R., L., & A. 4-6, B. 1½, lunch 3½, D. 5, pens. 11-14 fr.). About ¾ M. above the Hôt. Rothhorn, in *Inner-Arosa* (6070'), at the head of the wooded region of the valley, are the *Villa Zürrer* (pens. from 7 fr.), *Hôtel Bellevue* (5-6 fr.), the **Pens. Brunold* (6003'), with the Post Office for Inner-Arosa, the *Curhaus Arosa (6½-8 fr.), and to the right on the slope of the Tschuggen the *Villa Dr. Herwig*, *Villa Dr. Janssen*, and, on the hill, the *Sanatorium Arosa* (6090'), in a sunny situation. Most of these houses are closed in winter. Visitors' Tax 1 fr. per week.

EXCURSIONS (guides, *Jakob Janett*, *Joh.* and *Lucius Brüsch*, *Heinr. Henani*, *Jacob Juon*, *Alb. Scheller*). From the Villa Herwig by a shady path, or from the Seegrube past the *Obere See* (5705'), to the (½ hr.) hamlet of Maran (6102'; *Pension-Restaurant Hof Maran*) and to the (½ hr.) *Alp Pretsch* (fine view). — From the Seehof to the (1 hr.) pretty waterfall in the *Welschtobel*. — From the Sanatorium to the top of the *Tschuggen* (6725'; ½ hr.; easy). From the Curhaus to the (1 hr.) blue *Schwellisee* (6295') and the (¾ hr.) *Aelplisee*

(7055'), at the foot of the *Rothhorn* (see below). — The Aroser Weisshorn (8710'; 2½-3 hrs. from the Sanatorium, with guide, 5 fr.) is an easy and attractive ascent viâ *Tschuggen* and the *Mittlere Hütte*. — Schiesshorn (8533'), 3 hrs., with guide, viâ *Furka-Obersäss*, not difficult. — The *Aroser Rothhorn (9790'; splendid view) is most conveniently ascended through the *Welschtobel* (5 hrs.; guide 15 fr.); the descent past the Aelplisee and the Schwellisee takes 3-4 hrs. — Tiejerfluh (9135'; 4 hrs.; guide 12 fr.), viâ the Maienfelder Furka (see below), attractive, and not difficult for experts. — Sandhubel (9080'; 4½ hrs.; guide 12, with descent to Wiesen 15 fr.), through the Welschtobel, also not difficult (comp. p. 368).

PASSES. FROM AROSA TO DAVOS by **the Maienfelder Furka** (8020') between the *Furkahorn* (8960') and the *Amselfluh* (9135'), 5 hrs. to *Frauenkirch* (p. 363; guide 10, to Davos 15 fr.). — To COIRE by the *Ochsen-Alp* (6890'), an interesting walk (5-6 hrs., guide **not** indispensable) commanding a succession of beautiful views, viâ *Maran* to *Tschiertschen* (4430'; Bruesch, good wine), whence a new road leads viâ **Prada** to *Passugg* (p. 356). A more fatiguing route leads over the Carmenna **Pass** (7800'), between the Weisshorn and the Plattenhorn, with a steep descent to the *Urden-Thal* and to Tschiertschen (guide to Coire, 15 fr.). — To PARPAN, 4½-5 hrs. with guide (10 fr.), repaying; we pass to the S. of the *Hörnli* (8190') to the *Urder Augstbery* (7380') with its small lake, and **cross the Urden Fürkli** (8510'), between the *Parpaner Weisshorn* and the *Parpaner Schwarzhorn*, to Parpan (p. 390). — To ALVANEU through the *Welschtobel* and across the Furcletta (8455') to the E. of the *Piz Naira* (9420'), descending by the *Alp dil Guert* and the *Alvaneuer Maiensässe*, 5-6 hrs. with guide (15 fr., to the Furcletta only 10 fr.), toilsome but interesting.

FROM LANGWIES TO DAVOS, 3½-4 hrs. The bridle-path (guide unnecessary) ascends through woods on the right bank of the *Sapüner Bach*, then (10 min.) crosses the *Fondeier Bach*, and (20 min.) the *Sapüner Bach*, and ascends more steeply, at one place high up on the brink of the cliffs. It returns once more to the right bank of the stream, and leads through meadows past *Dörfli*, *Schmitten*, and *Küpfen* (all belonging to the parish of *Sapün*), through a treeless upland valley, finally ascending in zigzags to the (2 hrs. from Langwies) **Strela Pass** (7800'; fine view), between the *Küpfenfluh* (8650') on the right, and the *Schiahorn* (8900'; easily ascended from the pass in 1 hr.; see p. 362) on the left. We descend to the (¾ hr.) *Schatz-Alp* and thence either to the right to (¾ hr.) **Davos-Platz** (p. 361) or to the left to (1 hr.) *Davos-Dorf* (p. 361).

93. From Coire to Göschenen. Oberalp.

See Maps, pp. 368, 112.

63 M. DILIGENCE twice daily in 14½ hrs. (24 fr. 15, coupé 29 fr. 20 c.), once direct viâ Flims, **and** once viâ *Bonaduz*, a night in this case being spent at Disentis. — EXTRA-POST with two horses from Coire to Andermatt 157 fr., with three horses **215 fr.**; to Göschenen 155 fr. 40 c. and 227 fr. — CARRIAGE with one horse from Coire to Reichenau 6 fr.; with two horses to Reichenau 12, Flims 30, Ilanz 45, Disentis 80, Andermatt 135, Göschenen 145 fr.; from Göschenen to Disentis 70, to Coire 150 fr.; from Andermatt to Disentis 50-60, to Coire or Thusis 130-135, to St. Moritz or Samaden 270 fr.; fee 10 per cent of the fare. — Walkers should allow 2 hrs from Coire to Reichenau, thence to Flims 2½, Flims to Ilanz 2¼, Ilanz to Truns 4, Truns to Disentis 3¾, Disentis to Oberalp 4¼, and Oberalp to Andermatt 2 hrs.

REICHENAU. *VI. Route 93.*

Coire, see p. 354. Beyond the Rhine, at the foot of the *Calanda* (p. 356), lies the village of *Felsberg*, which is menaced with a fate similar to that of Goldau (p. 109). Part of the rock fell in 1850. The road passes through the large village of (4 M.) **Ems**, Rom. *Domat* (1880'), with the scanty ruins of the castle of *Oberems*. The mounds of earth here and near Reichenau are probably remains of an old moraine. Near Reichenau the road crosses the Rhine by a new iron bridge.

6 M. **Reichenau** (1935'; *Adler*), a hamlet at the confluence of the *Vorder-Rhein* and the *Hinter-Rhein*. The best view of the rivers is obtained from a pavilion in the garden of *Dr. von Planta*, adjoining the Adler. At their junction, the Vorder-Rhein, in spite of its superior volume, is driven back by the boisterous Hinter-Rhein, which descends from the Bernardino. To the W. towers the Brigelser Horn. The pleasant garden is open to visitors, and may be seen during the halt of the diligence; curious old inscription on the gardener's house. The *Château*, opposite the entrance to the garden, erected by the Bishops of Coire, and named by them after the Abbey of Reichenau on the Lake of Constance (p. 25), now belongs to Dr. A. v. Planta. In 1793 Louis Philippe sought refuge here under the name of Chabot, and his room and other memorials still exist (fee 1 fr.).

From Reichenau to *Thusis (Via Mala)* and over the *Splügen* to *Cotico*, see RR. 95, 96; by the *Bernardino* to *Bellinzona*, see R. 97. — *Schyn Road* from *Thusis* to *Tiefenkasten*, see p. 379; *Kunkels Pass* to *Ragatz*, see p. 66.

ROAD FROM REICHENAU TO ILANZ, 18½ M., on the right bank of the Rhine (diligence daily; see above). From Reichenau to (1 M.) *Bonaduz*, see p. 378. The road here diverges to the right from the Splügen road. For 1½ M. it is perfectly straight and level; then, gradually ascending, it traverses wood for ¾ M., beyond which it leads high above the picturesque Vorder-Rheinthal, hewn in the rock at places, and commanding a fine view. The bold construction of the road is itself interesting. We next (½ M.) turn sharply to the left into the picturesque valley of the *Rabiusa* (see below), and descend gradually, passing through a short tunnel, to (¾ M.) a covered wooden bridge over the *Versamer Tobel* (2390'; 260' above the stream). We now ascend through pine-woods by numerous windings (which walkers may cut off) to (2 M.) *Versam* (2980'; *Hôt. Siguina*), a charmingly situated village, with a fine view. After a level stretch of 1½ M., the road descends towards the valley of the Vorder-Rhein, of which we have a striking view. Opposite, on the left bank, high above the river, lies *Laax* (p. 369). Farther off, on the same bank, rises the *Brigelser Horn* (p. 371). We next reach (1 M.) *Carrera*. Still descending, we cross a picturesque ravine, pass through a tunnel, and reach (1 M.) *Valendas* (2700'; Krone, rustic). Again descending, with a fine view before us, we next pass (2½ M.) *Kästris*, and cross the (1¼ M.) broad stony bed of the *Glenner* to (¼ M.) *Ilanz* (p. 369).

Through the Safier-Thal, watered by the *Rabiusa*, a new road leads from Versam to the S. to (12 M.) *Safien-Platz* (4255'; *Gredig's Inn), with a fine fall of the *Carnusa* on the left. Bridle-path thence over the large *Camana Alp* to *Thalkirch* (5545') and the (2½ hrs.) *Curtnätscher-Hof* (5307') at the head of the valley, with a splendid waterfall. Then a steep ascent to the (2 hrs.) pass of the *Safterberg* or *Löchliberg* (8170'), from which the path descends by the *Stutzalp* to (1½ hr.) *Splügen* (p. 382). — To the E. of Safien-Platz an easy route crosses the *Heinzenberg* by the *Glas Pass* (6056'; small inn, cheap), and leads through the villages of *Tschappina* and *Urmein* to (5 hrs.) *Thusis* (p. 378). Above Tschappina lies the *Lake of Lüsch* (6309'), which has no visible outlet. Its water softens the porous slate of its banks to the consistency of mud, and large masses of the strata adjoining it periodically slide down to the Nolla (p. 379). Tschappina itself is built in part on a shifting foundation.

The ROAD ON THE LEFT BANK ascends from Reichenau to ($^3/_4$ M.) **Tamins**, Rom. *Tumein* (2245'; *Post Restaurant*), where **we obtain an admirable** survey of the *Vorder-Rheinthal* with the *Unterhorn* (9180') and the *Piz Riein* (9030'). The *Lavoi*, descending on the right beyond Tamins, forms a fine waterfall after rain. At (2 M.) **Trins** (2820'; good wine at *Caflisch's*) rises the ruined castle of *Hohentrins*. At ($^1/_2$ M.) *Digg* the road turns suddenly to the N., passes through a cutting *(Porclas)*, and at the base of the precipitous *Flimser Stein* (see below) **sweeps round the** *Seeboden*, a nearly circular basin enclosed by wooded hills. Near ($1^1/_2$ M.) *Trinser Mühle*, Rom. *Mulins* (2720'; Inn), **are several** waterfalls **on the** right. To the left, farther on, is the small *Cresta Lake*, surrounded by pines. About 2 M. farther on is ——

$13^1/_2$ M. **Flims** (3615'; *Hôt.-Pens. Bellevue*, R. $1^1/_2$-$2^1/_2$, pens. 6-$7^1/_2$ fr.; *Post*), Rom. *Flem*, an ancient little town **(pop. 797)** with several mansions of the Capaul family.

The road leads through the valley of the *Flembach* to the (1 M.) *Hôt.-Pens. Segnes* (3445'; R. $2^1/_2$, B. $1^1/_4$, pens. 8-9 fr.), opposite the **Waldhäuser**. About $^1/_2$ M. farther on, on a hill (3707') a few min. to the right of the road, is the large and well-situated *Curanstalt Waldhaus-Flims*, with five 'dépendances' (R., L., & A. from 5, board $7^1/_2$ fr.), a pleasant **summer-resort**, with beautiful pine and **beech woods**. Near it is the *Flimser See* or *Cauma Lake* (3280'), embosomed in wood, without visible outlet. Pleasant swimming baths ($^1/_2$ fr.), to which a path descends in 10 minutes.

EXCURSIONS (**Guides**, **Rich.** and **Conr.** *Joos*, *Pankraz Koch*). A picturesque walk may be taken from **Flims** to the ($^1/_2$ hr.) *Segnes Waterfall* and the ($^3/_4$ hr.) *Runca* **Bridge**. — For the *Buchen* we proceed to the E. from the Hôtel Segnes, passing between the Waldhäuser, **and** then take the direction indicated by the finger-post. This walk may be extended to the **Cresta** Lake (see above). — For *Mutta* we follow the Laax road for $1^1/_4$ M from the Waldhäuser **and** then ascend to the left through wood (finger-post; **1 hr.**). We may return by the *Cauma Lake*. — Flimserstein (*Crap da Flem*, 8635'; 5 hrs.; guide 6 fr., unnecessary), repaying. **The** path ascends gradually by *Fidaz*; then through wood, round the S.E. angle of the mountain, to the ($1^3/_4$ hr.) pastures of *Bargis*. **Here** we ascend to the left by a good path to the hilly plateau of the *Alp Sura* (6896'; milk and bread; $^1/_4$ br. to the S. of which is a rock affording a good survey of **the** Bündner Oberland Mts. and the Tödi). In 2 hrs. more we ascend to the arête and the summit, where we obtain a splendid view, especially towards the N., of the Ringelspitz and Piz Dolf. We may descend to the N.W towards Segnes, and return to Flims by the *Cassons* and *Foppa Alps*. — **Vorab** (9925'; $6^1/_2$-7 hrs.; 20 fr.), a very fine point, also easy (comp. p. 75). From Flims to the brink of the *Bündnerbergfirn*, which has receded greatly, $4^1/_2$ hrs.; then up the easy glacier to the (2 hrs.) summit, consisting of fragments of slate. Superb view, particularly of the neighbouring Tödi group; also of the Sernf-Thal and the Bernese Alps from the (20 min.) N. peak, the *Elmer Vorab* (9910').

Over the *Segnes Pass* to *Elm* (8 hrs.; guide 20 fr.), see p. 75. The *Martinsloch* (p. 75) may be reached in 4-5 hrs. from Flims (guide 18 fr.). — A visit to the *Segnes Glacier* (guide 10 fr.) hardly repays the fatigue.

Traversing sequestered dales and skirting the deep *Laaxer Tobel* on the left, we next reach (2¼ M.) *Laax* (3324'; *Hôt.-Pens. Seehof*, close to the Laaxer See, with baths, pens. 7-8 fr.). (A road to the right ascends in ½ hr. to the village of *Fellers*, Rom. *Fallera*, 3997'; splendid *View.) We now descend into the Rhine Valley (passing *Sagens* far below, to the left) and reach *Schleuis*, Rom. *Schluein* (2507'), with the old château of *Löwenberg*, now an orphan-asylum. Opposite lies the large village of *Kästris* (p. 368); before us, above Ilanz, rises the Piz Mundaun. — 3 M. —

20½ M. **Ilanz**, Rom. *Gliōn* (2345'; pop. 802; *Hôt. Oberalp*, R., L., & A. 3, B. 1¼, D. 3 fr.; *Lukmanier, Krone*, moderate, both on the left bank; one-horse carr. to Disentis 20 fr. and fee), mentioned in a charter of the 8th cent. as the 'first town on the Rhine', built on both sides of the river, was the capital of the 'Gray League' (p. 353). The upper part has narrow streets and old-fashioned houses adorned with armorial bearings. The population is partly Romanic, partly German; Romanic alone is spoken higher up the valley. Ilanz is beautifully situated, overlooking the Rhine Valley in both directions, and the broad Lugnetz Valley to the S.

The views are still finer from the old *Church of St. Martin* (2570'), ¼ hr. to the S., on the left slope of the Lugnetz Valley, and from the chapel of the pretty village of *Luvis* (3280'), ½ hr. higher. A most superb prospect of the Grisons Oberland, and especially of the Tödi chain to the N., immediately opposite, and of the Rhine Valley down to Zizers (p. 343), is commanded by the *Piz Mundaun* or *Piz Grond* (6765'), to the S.W. of Ilanz. The path (4 hrs.; guide, not indispensable, 7 fr.) leads by Luvis (see above), ascends on the S.E. side of the wood, crosses a flat basin obliquely towards the left, and mounts the pastures to the conspicuous (2½ hrs.) *Ina* (closed and falling to decay). Then in the same direction, through a depression in the mountain, to the crest, which we ascend to the W. to the top in 1 hr. more. The mediæval chapel of *S. Carlo* remains to the left. Those who intend visiting the Lugnetz Valley (see below) may descend direct to *Villa* (p. 370; thence to the top 2 hrs., best way to reach it, guide 3 fr.), or by *Mortissen* (4420'; Hôt. Piz Mundaun) to (2 hrs.) *Cumbels* (p. 370). — Travellers bound for Disentis, instead of returning to Ilanz, may follow a beautiful path through the district of *Obersaxen*, the chief village of which is (1¼ hr.) *Maierhof* (4270'; *Casanova, rustic), whence *Tavanasa* (p. 371) may be reached by a pretty forest-path in ¾ hr. (guide advisable), or Ilanz by a new road in 2 hrs. — Those who ascend the Piz Mundaun from Truns diverge from the road about 3 M. below the village, by the telegraph-post No. 222, to the right, and ascend by a good path, at first through wood. Farther on it overlooks the Rhine Valley and passes the ruin of *Axenstein*. After 2 hrs., beyond the chapel of *St. Valentin*, by a crucifix on this side of a ravine, we descend to the left into the valley and reach (½ hr.) *Maierhof* (see above). Then up sunny pastures to the top of the Piz Mundaun in 2½ hrs. more.

The Lugnetz Valley, watered by the *Glenner*, 18 M. in length (pop. Rom. Cath. and Romanic), is one of the finest in the Grisons. Road to Vals-Platz (14 M.; diligence from Ilanz daily in 4½ hrs., fare 3 fr. 35 c.; carr. from Coire to Vals 40, with two horses 70 fr., and fee of 10 per cent) on the left bank, past the ruin of *Kastelberg* and through the (3 M.) *Frauenthor*, Rom. *Porclas* (3336'), once the key to the upper valley. On the opposite bank of the Glenner, high above the *Riemer Tobel*, lies the village of *Riein*, and beyond it are *Pitasch* and *Duvin*. Beyond (3¼ M.) the chapel of *St. Moritz* (3504') the road divides: the right branch ascends to Vrin (p. 370); that to the left descends to the village of *Peiden* and the

(1½ M.) *Peidner Bad* (2690'), on the right bank of the Glenner, at the mouth of the *Duvner Tobel* (a haunt of the chamois), with three chalybeate springs. Then (1½ M.) *Furth* (2980'; *Schmid's Inn; Piz Mundaun*), at the confluence of the *Vriner* and *Valser Rhein*, which are separated by the *Piz Aul* (10,250'). Opposite lies the picturesque *Oberkastels* (3274'). We now ascend the wild *Valser-Thal*, or *St. Petersthal*, by *St. Martin* and *Lunschania*. At *Campo*, where the valley expands, we recross to the left bank. 7½ M. **Vals-Platz**, or *St. Peter* (4094'; *Hôt. Albin*, °*Hôt.-Pens. Piz Aul*, both plain, 5 fr.), is splendidly situated and possesses a chalybeate thermal spring (*Hôt.-Pens. Therme in Vals*, with baths and postal telegraph office, pens. from 7 fr.). From Vals-Platz (guides: Andr. Furger, Ben. Schnyder) a well-trodden bridle-path leads through the *Peiler-Thal*, a side-valley to the **S.E.**, to the *Vallatsch Alp* (6178'), the *Valser Berg* (8225'), and (5 hrs.) *Nufenen* or *Hinterrhein* (p. 385). The °*Weissensteinhorn* (9675'; 4½ hrs.; guide 7 fr.), ascended from Vals-Platz, is an admirable point of view; another is the *Bärenhorn* (9620'; 5 hrs., with guide); from both we may descend into the Safier-Thal (p. 367). The *Piz Aul* (10,250'; 6 hrs., with guide), also a fine point, is laborious (via the Sattelelücke, see below). To *Vrin* over the *Fuorcla de Patnaul* (9113'), to the S., between the Piz Aul and the Faltschonhorn, or over the *Sattelelücke* (9082'), between Piz Aul and Piz Seranastga, both laborious (6-7 hrs., with guide).

The **S.W.** branch of the valley (*Val Zervreila*), watered by the **Valser Rhein, divides** at the hamlet of Zervreila (5840'; *Tönz's Inn*), 8¼ hrs. above Vals-Platz, **into** the *Lenta-Thal* to the **S.W.** and the *Kanal-Thal* **to** the **S.** — A toilsome route, requiring a guide, leads through the latter, across the *Kanal Glacier* and the *Plattenschlucht* (*Zapportgrat*, 9314'), and down to the *Zapport-Thal* and (9 hrs.) *Hinterrhein* (p. 385). — In the grand and interesting Lenta-Thal, 1 hr. above Zervreila, is the beautiful *Lampertsch-Alp* or *Sorreda-Alp* (6580'; bed of hay). Thence over the *Vernok* or *Vanescha Pass* (9350') to *Vrin* (see below) in 6-7 hrs., or over the *Sorreda* or *Scaradra Pass* (9088') to *Olivone* (p. 377), 8 hrs., both routes toilsome; over the *Lentalücke* (9692') to *Hinterrhein* (p. 385; 9-10 hrs.), difficult, for experts only, with good guides.

The road ascending to the right by the chapel of St. Moritz (p. 369) leads to *Cumbels*, *Villa* (4080'; Post, rustic), *Vigens*, *Lumbrein*, and (4 hrs.) **Vrin** (4770'; *°Post*, plain; *Casanova*, poor), the principal village in the *Vrinthal* or *Upper Lugnetz Valley* (from Ilanz to Vrin, 13½ M., diligence daily in 4¼ hrs.). From Vrin we may easily ascend the *Piz Regina* (8294'; 4 hrs.; guide advisable), a fine point. *Piz Cavel* (9660'; 5-6 hrs.), ascended by the *Ramosa Alp* and the *Fuorcla de Ramosa* (8694'), also easy; descent to the N. to the *Cavel-Joch* (p. 371), if preferred. *Piz Aul* (10,250'; 6-7 hrs.; with guide; superb view), by *Val Seranastga* (route to the Sattelelücke, see above), laborious. *Piz Terri* (9996') is ascended from *Vanescha*, 1¾ hr. from Vrin, in 5 hrs., by the *Blengias Alp* and the *Güda Glacier* (no serious difficulty). Route over the *Vanescha Pass* to *Zervreila*, see above. Over the *Cavel-Joch* to *Somvix*, see p. 371. — From Vrin, with a guide (to Olivone 18 fr.), we ascend past **the mouth of the** *Val Vanescha* (see above), to *St. Giusepp*, *Puzatsch*, the **Alp Diesrut**, and the (3 hrs.) **Pass Diesrut** (7958'), on the S. side of the *Piz Tgietschen* (9877'). Descent to the *Camona Alp* (7339'), at the head of the **Val** *Somvix* (p. 371), and again a gradual ascent, passing the *Piz Vial* (10,387') **and** the *Piz Gaglianera* (10,243') on the right, and the *Piz Coroi* (9130') **on** the left, to the **Greina Pass** (*Passo Crap*, 7743'). We next descend through the wild *Val Camadra* or upper part of the *Val Blenio*, with the *Piz Medel* (10,510') on the right, by *Daigra*, *Cozzera*, and *Ghirone*, to (3½ hrs.) *Olivone* (p. 377). Or, half-way between the Camona Alp and the Greina Pass, we may cross the low *Monterascio Pass* (7415'), to the left, to the *Monterascio Alp*, and descend the picturesque *Val Luzzone* to *Lorcioto*, *Cavallo*, *Davresco*, and *Olivone* (shorter than the Greina route).

Road from Ilanz by *Versam* to *Bonaduz* and *Reichenau*, see p. 367. — From Ilanz to *Elm* over the *Panixer Pass* or the *Sether Furka*, see p. 75. To *Linttthal* over the *Kisten Pass*, **see p. 71.**

The road follows the N. side of the narrow Rhine Valley, here called *Pardella;* beyond (1 M.) *Schnaus* it crosses the *Sether-Bach*, and beyond (1½ M.) *Ruis*, beautifully situated on the hill to the right, the *Panixer-Bach*. On a rocky hill to the right rise the picturesque ruins of the robbers' stronghold of *Jörgenberg* (3100′).

To the right, 1 M. above the bridge of Ruis, a road (diligence from Ilanz daily in 3 hrs.), commanding fine views, ascends by the village of *Waltensburg* (3800′) to (4 M.) Brigels (4260′; *Hôt.-Pens. Fausta - Capaul; Hôt. Kistenpass*), prettily situated amid pastures. Above it the *Val Frisal*, with the glacier of that name, ascends to the *Bifertenstock* (11,240′), which, as well as the *Piz Frisal* (10,810′) and the *Brigelser Horn* (10,663′), may be ascended from the Val Frisal (all difficult; see below).

Farther on, the scenery is inferior. To the right rises the *Brigelser Horn* (see above). The Rhine is crossed near (4½ M.) **Tavanasa** (2620′; *Kreuz*), and again near (3 M.) *Zignau* or *Rinkenberg*. High up on the N. slope lies *Brigels* (see above); then *Dardin* and *Schlans*. Before reaching Rinkenberg we observe on the left the stony chaos formed by the inundations of the *Zignauer Bach* descending from the *Zavragia Ravine*. By the bridge we enjoy a delightful view, embracing numerous villages, chapels, and ruined castles on the richly clothed slopes.

We next pass (1½ M.) the *Chapel of St. Anna*, on the right, erected in 1778 on the spot where the 'Upper' or 'Gray League' (p. 353) was founded in March, 1424, and adorned with old frescoes and verses. A few paces farther on is —

32 M. **Truns** (2820′; *Krone; *Zum Tödi*, plain). The hall of the old Statthalterei of the abbey of Disentis is adorned with the arms of the members of the Gray League, and of the magistrates since 1424.

The **Val Puntaiglas**, ascending rapidly to the N., ends in the *Puntaiglas Glacier*. Ascent of 2 hrs. from Truns to the *Alp Puntaiglas* (about 5050′), with a fine view of the Brigelser Horn, Piz Mut, Piz Ner, etc. The S. peaks of the Tödi group, *Piz Urlaun* (11,060′), *Bündner Tödi* (10,226′), and *Brigelser Horn* (or *Kavestrau Grond*, 10,663′; very difficult), may be ascended hence. Ascent of the *Tödi-Russein* by the *Gliemspforte*, see p. 71.

Beyond (1¾ M.) *Rabiüs* (3133′) we obtain a glimpse, to the left, of the grand *Piz Gaglianera* (10,243′), with its glaciers, at the head of the Val Somvix. Then (1½ M.) **Somvix** (3458′; *Weisses Kreuz*, poor), conspicuously situated on a height, as its name ('*summus vicus*') intimates.

The **Val Somvix**, which here opens to the S., deserves a visit. We cross the Rhine to (¼ hr.) *Surrhein*, and ascend by a good bridle-path on the left side of the valley, through wood and pastures, to *Val* and the (1½ hr.) *Somvixer* or *Teniger Bad* (4176′; good and moderate, pens. 4 fr.). Farther up (½ hr.), we pass the *Alp Vallenigia*, where the glaciers of the *Piz Vial* (10,337′) are revealed, and the mouth of the *Val Lavaz*, and reach (1½ hr.) the rock-girt head of the valley, where the *Greina* forms a fine waterfall on the left. The path ascends steeply on the E. side of the valley to the rocky defile of *La Fronscha*, and divides higher up: to the left to the *Diesrut Pass* (p. 370), and to the right to the *Greina Pass* (p. 370). — PASSES. From the Teniger Bad (see above) over the *Cavel-Joch* (8320′) to Villa, 7 hrs., not difficult. From the pass the *Piz Cavel* (9060′; fine view) may be ascended in 1¼ hr. — Over the *Valgronda-Joch* (9120′) to *Tavanasa* or *Materhof*, 7-8 hrs.; with guide. — OVER THE LAVAZ-JOCH TO

24*

372   *VI. Route 93.*   DISENTIS.   *From Coire*

Curaglia, 7-8 hrs., with guide, an attractive route. From the Toniger Bad (p. 371) we ascend on the left side of the valley, through wood and rhododendrons, to the *Alp Rentiert*, where from the cairn (6640') we get a splendid view of the Tödi. We may now either cross the *Fuorcla de Stavelatsch* (8376') to the right, or turn to the left and skirt the E. slopes of *Piz Rentiert* (keeping to the right on the hill, by the chalet of *Rentiert-Dadans*), to the (2 hrs.) chalet of *Stavelatsch* (7632') in the *Val Lavaz*. Opposite are the two glaciers descending from the *Piz Vial* and *Piz Gaglianera* (10,243') and the *Lavaz Glacier*. Thence to the Lavaz-Joch (8232') an easy ascent of ³/₄ hr.; the ridge to the N. of the pass commands a fine survey of the *Medelser Glacier* and of the Bernese Alps to the W. Steep descent over grassy slopes to the *Alp Sura* (6526'), and through *Val Plattas* to (2 hrs.) *Curaglia* (p. 375).

Beyond Somvix the road is very boldly constructed. A lofty wooden bridge (2¹/₄ M.) carries it over the profound *Russeiner Tobel*. (Below, to the right, a finger-post indicates the path to the Sandalp Pass; see below.) Above the (³/₄ M.) *Stalusa Bridge* is a small waterfall. 1¹/₄ M. *Curhaus Disentiser Hof* (see below), built on the site of the château of *Castelberg*, which was burned down in 1830.

39¹/₂ M. Disentis (3773'; *Desertinum, Disiert, i.e.* desert), Rom. Mustèr (*Disentiser Hof*, with fine view, R. L., & A. 4-6, D. 4¹/₂, S. 2¹/₂, pens. 9 fr., whey and chalybeate water; *Krone & Post*, R. 2-3, B. 1¹/₄, D. 3, S. 2¹/₂, pens. 5-6 fr.; Eng. Ch. Serv.), a small town of 1329 inhab., is protected against avalanches by a forest. The foundation of the Benedictine Abbey in the 7th cent. soon brought Christianity into the remote valleys of the Grisons; and the abbots, enriched by liberal endowments, afterwards acquired great power in Rhætia. The large abbey-buildings, on a height, now contain a school.

Near Disentis the *Medelser-Rhein* or *Mittel-Rhein* (p. 375) unites with the *Vorder-Rhein*. From the *Chapel of Acletta*, with an old altar-piece, at the entrance to the Acletta Valley (4236'), ¹/₂ hr. to the W. of Disentis, a fine view (especially by evening-light) is obtained of the Medelser Glacier, and far down the valley.

Excursions. (Guides: *J. Peischen*, the schoolmaster; *J. M. Schnoler*, hunter; *P. Tenner* and *Jos. Huonder*.) Walk on the *Lukmanier Road* to (4¹/₂ M.) *Curaglia* (p. 375), interesting. Also by the chapel of *St. Gada*, with old frescoes, to *Mompé-Medel* (1 hr.), on the right bank of the Rhine, with fine view. Viâ *Catardiras* to the (2¹/₂ hrs.) *Alp Lans* (5260'), with the charmingly situated little lake of that name. To *Crest-Muntatsch* (1¹/₂ hr.); *Alp Lumpegnia* (6520'; 2¹/₂ hrs.), etc.

The fine pyramid of *Piz Muraun* (9510'; 5¹/₂ hrs. from Disentis) is best ascended from *Curaglia* (4 hrs.; guide 8 fr.; p. 375). Superb view from Monte Rosa to the Ortler, especially of the neighbouring Tödi group, grander than from Piz Mundaun (p. 369). — *Piz Pazzola*, see p. 373; *Piz Medel, Piz Cristallina*, see pp. 375, 376. — *Crap Alv (Piz Glendusas; 9784')* and *Piz Ault* (9957'), from the *Val Acletta* (each 5 hrs.; guide; not difficult for experts).

From Disentis **over the** *Lukmanier* (6290') to *Olivone*, see p. 375; through the *Val Piora* to *Airolo*, **see** p. 115. — Over the Sandalp Pass to Stachelberg, 11-12 hrs., with guide (26 fr.), trying. We ascend the *Val Russein* (see above) to the Sandalp Pass (*Sandgrat;* 9120') between the *Lesser Tödi* or *Crap Glarun* (10,072') on the E., and the *Catscharauis* (10,050') on the W., and descend the *Sand-Firn* to the *Upper Sandalp*. Thence to *Linthhal*, see p. 70. — Ascent of the *Tödi* by the *Porta da Spescha*, and descent to Linthhal, 18-19 hrs., for thorough adepts only, with able guides (see p. 71).

From Disentis over the Brunni Pass (8875') to the Maderaner-Thal (to the Hôt. Alpenclub 8-9 hrs.; guide 20 fr.), see **p. 124.**

The road to (19½ M.; a walk of 7-8 hrs.) Andermatt, which lies lower than the old route, ascends the valley of *Tavetsch*, leaving the hamlets *Acletta*, *Segnas*, and *Mompè Tavëtsch* (4584') to the right. From the height, where the road enters a wood, we obtain a *View of the Disentis district, particularly striking when approached from Andermatt. The valley contracts. The road traverses woods and pastures, commanding a view of the infant Rhine in its deep valley, and of the snow-clad mountains we are approaching. — 5¼ M. —

45 M. Sedrun (4587'; *Krone*, plain, pens. 5 fr.), locally known as *Tavëtsch*, is the principal village in the Val Tavetsch. The church contains an old altar in carved wood.

'Piz Pazzola (8170'; 4 hrs.; guide unnecessary), to the S., between the *Val Medel* (p. 375) and the *Val Gierm*, is worth visiting. We cross the Rhine to *Surrhein*, and the gorge of the *Val Nalps* (see below) to the (½ hr.) hamlet of *Cavorgia* (4426'); then cross the *Gierm* and ascend to the right, over pastures and through wood, to the (1½ hr.) *Pazzola Alp* (6150'), with a fine view, and thence to (2 hrs.) the top without difficulty. Magnificent view, particularly of the Tödi and the Medel Mts.

In the lonely Val Nalps, the head of which is enclosed by lofty mountains and glaciers, 3 hrs. from Sedrun, lies the *Alp Nalps* (5991'), and 2 hrs. higher is the *Ufiern Hut* (7550'), the starting-point for the *Piz del Laiblau* (9720'), *Piz Rondadura* (9905'; comp. p. 376), *Piz Blas* (9920'), *Piz dell' Ufiern* (9900'), *Piz Gli* (9744'), *Piz Serengia* (9803'), etc. (each about 6 hrs.). A tolerably easy route (with steep descent) leads hence across the *Nalps Pass* (9035') to the *Val Cadlimo* and the *Uomo Pass* (p. 115). Another (trying) leads to the E. over the *Rondadura Pass* (8904') to the *Hospice of S. Maria* (p. 376). A third crosses the *Fuorcla da Paradis* (8556'), between the *Piz Furcla* and the *Piz Paradis*, to the *Val Cornera* (p. 374).

FROM SEDRUN TO AMSTEG over the Kreuzli Pass (7645'), 8 hrs., rather trying (guide 15 fr.). The steep path ascends the bleak rocky *Strimthal*, at the head of which the pass lies to the left (W.), at the S. base of the *Wellenalpstock* (p. 123). Guide necessary only to the point beyond the pass where the *Etzlibach*, descending from the *Spitallaui-See* to the W., becomes visible. We cross the stream to *Culma* (6322'), the highest Alp, and descend the *Etzlithal*, past the chalets of the *Hintere* and *Vordere Etzlialp* to *Bristen* and *Amsteg* (comp. p. 124). — The Oberalpstock (*Piz Tgietschen*, 10,925') may be ascended from Sedrun in 5-5½ hrs. (guide 15 fr.). We follow the Kreuzli Pass route to the head of the Strimthal, at the foot of the *Calmut* (2 hrs.), where we ascend to the right by grass slopes and the moraine to (1½ hr.) the small glacier lying on the S. flank of the Oberalpstock. We cross this glacier to the S.E. arête (1 hr.) and, mounting the snow-slopes on the S.E. side and finally over stones, reach the summit in ½ hr. Comp. p. 115.

From Sedrun the road leads through *Camischolas*, **Zarcuns**, and (1½ M.) **Rueras** or *S. Giacòmo* (4597'), crosses the brook descending from the *Val Milar*, and soon afterwards, near the hamlet of *Dieni*, that which issues from the *Val Gluf* (both N. lateral valleys). To the left, on a rock above the ravine of the infant Rhine, stands part of the ancient tower of *Pultmenga*, once the ancestral seat of the Pontaningen or Pultingen family.

Walkers will prefer the so-called 'SUMMER ROUTE' to the high-road, for the sake of the views (guide desirable). It diverges to the right by a finger-post (to 'Pass Tiarms'), ascends a spur of the *Crispalt* (10,105'), above the hamlet of *Crispausa* which lies to the left, and leads past the chalets of *Milez* and *Scharinas* amidst the richest pastures in this district. It now skirts the brink of the slope, overlooking the Rheinthal, turns to the right into the bleak *Val Terms* or *Tiarms*, crosses the *Gämmer-Rhein* (Rom. *Vala*) by the *Alp Culm de Val* (6420'), and ascends to the **Pass da Tiarms** (7067'), between (r.) the *Piz Tiarms* or *Berglistock* (9564') and (l.) the *Calmot* (7598'), where we get a fine view of the Vorder-Rheinthal as far as the Vorarlberg and Rhætikon Mts. Descending to the *Oberalpsee* (p. 375), we keep to the left in order to avoid a marsh, and regain the high-road 2½ hrs. from Sedrun. The high-road follows the direction of the old 'Winter Route' on the left bank of the Vorder-Rhein and passes the *Chapel of St. Brida*, below the hamlet of *Crispausa*, and the poor villages of *Selva* (5046') and (2 M.) **Chiamūt**, or *Tschamut* (5380'; *Zur Rheinquelle, plain; minerals), which consist of a few wooden huts and a chapel. In front of us rises the *Six-Madun* or *Badus*, behind the second terrace of which lies the **Toma Lake** (see below). Chiamut is probably the highest village in Europe where rye is grown. The road crosses (½ M.) the *Gämmer-Rhein* near its influx into the Vorder-Rhein, and (1 M.), opposite the *Alp Milez*, turns to the right (N.W.) into the *Val Surpalix*, between the *Piz Nurschallas* on the left and the *Calmot* on the right. The *Vorder-Rhein (Aua da Toma,* or *Darvun)* descends in a series of falls from the slope to the left.

Source of the Vorder-Rhein. The Vorder-Rhein rises in the Toma Lake (7690'), on the N.E. slope of the *Six-Madun* or *Badus* (p. 120). The path to the lake (guide advisable) diverges from the road to the left, 1¼ M. above Chiamut (see above); near the *Alp Milez* it crosses the brook emerging from the Val Surpalix, and ascends to the (½ hr.) *Alp Tgiettems*. Above this alp (avoid path to the left, crossing the brook) we ascend the pastures to the right, on the left bank of the *Fil Toma*, the brook descending from Piz Nurschallas. After about 1 hr. we ascend steeply to the left and soon reach the rocky barrier behind which the lake lies. The *Toma Lake* (2½ hrs. from Chiamut), a green lake, very deep, and destitute of fish, about 270 yds. long and 130 yds. broad, is bounded on the S. and S.W. sides by precipitous rocks and stony slopes, and on the N. and N.W by pastures. The Badus (9615'; comp. p. 120) ascends vertically from the lake, but good cragsmen may reach the top in 2 hrs. by keeping to the N. side of the rocks (guide 10 fr.).

The Piz Nurschallas (9063'), running out from the Badus to the N. (from the Oberalp Pass 2, from Chiamut 3½ hrs.; guide desirable for novices), is easier and interesting. We follow the Toma Lake route, diverge to the right where it turns to the left, ascend steep pastures, and lastly mount toilsomely over rocky debris to the summit. Superb survey of the Reuss and Vorder-Rhein valleys and the mountains enclosing them. Easy descent to the Oberalp Pass, 1¼ hr.

To the S. of Chiamut the Val **Cornera**, the mouth of which is a pathless ravine, ascends to the frontier-chain **of Ticino**, and from it the *Val Maigels* diverges to the W., 1½ hr. from Chiamut. Toilsome routes lead from the Val Cornera over the *Passo Vacchio* (8908') to the *Val Cadlimo* and *Piora* (p. 115); from the Val Maigels, to the S., over the *Passo Pian Bornengo* (8650') to the *Val Canaria* and *Airolo* (p. 114); and to the W., over the *Maigels Pass* (8078') or the *Lohlen Pass* (7835'), to the *Unteralp-Thal* and *Andermatt* (p. 119).

The road ascends the sequestered Val Surpalix in long windings (which paths cut off; one ascending to the left by the first bend, and bearing to the right, leads to the pass in ¾ hr.). It affords views of the Crispalt and Berglistock, and of the Piz Cavradi, Piz dell' Uffern, and Piz Ravetsch behind us. The (52 M.) **Oberalp Pass** (6710'), 3½ M. from Chiamut, forms the boundary between the Grisons and Uri. Extensive turf-diggings and new fortifications. (The diligence ascends to the pass from Chiamut in 70 min.; descent 40 min.; descent to Andermatt 1 hr. 10 min., ascent 2 hrs.)

The road rounds the E. end of the sombre *Oberalpsee* (6654'; 1 M. long), abounding in trout (to the right the route to the Pass da Tiarms, p. 374), and skirts its N. bank to the (1/2 hr.) *Hôt.-Pens. Oberalpsee*, at the W. end (trout).

A pleasant excursion may be made hence uphill to the N. to the beautiful clear *Lautersee* (7743') and thence viâ the *Strahlboden-Alp* to the summit of the *Stock, or Stückli (8070'), commanding a splendid panorama. We may descend viâ the *Grossboden-Alp* to the Oberalp road and (2 hrs.) Andermatt (comp. p. 120). — Over the *Felli-Lücke* to *Amsteg*, see p. 112.

The road traverses the nearly level *Oberalp* (6443'). About 2 M. from the hotel we obtain a view of the Ursern-Thal, with the Furka towards the W. (p. 125). The old path descending here to the left direct to (1/2 hr.) Andermatt is steep and stony, and affords little view. The road remains on the hill a little longer, and then descends by nine long windings to (6 M. from the lake) —

59 M. **Andermatt** (4738'); thence to (4 M.) —
63 M. **Göschenen**, see pp. 118, 119.

94. From Disentis to Biasca. The Lukmanier.
Comp. Maps, pp. 368, 112, 382.

39½ M. DILIGENCE in summer daily in 9 hrs.; fare 13 fr. 40, coupé 17 fr. 60 c. Carriage and pair from Coire to Olivone 140, to Biasca 180 fr. Except the lower part of the road, as far as Curaglia, the scenery is not very striking. Inns unpretending. — Walkers take 5 hrs. from Disentis to Sta. Maria, 4½ hrs. thence to Olivone, and 4½ hrs. from Olivone to Biasca.

Disentis, see p. 372. — The road crosses the *Vorder-Rhein* by a handsome bridge (3488') and enters the **Val Medel**, the wild ravine of the *Mittel-Rhein*, along the left bank of which it is carried by means of cuttings and tunnels (eleven as far as Curaglia). At the end of the gorge, of which we obtain several striking views, we cross (2¾ M.) to the right bank of the Rhine and ascend in long windings (cut off by paths) to (3/4 M.) —

3½ M. Curaglia (4370'; *Hôt. Lukmanier*), a village at the entrance to the *Val Plattus* (over the *Lavaz-Joch* to *Somvix*, p. 372). To the S. appears the *Piz Cristallina* (10,265'), with its glacier. — *Piz Muraun* (9510'; 4 hrs.), see p. 372.

Following the right side of the pleasant Val Medel, the road passes the (5 M.) straggling village of **Platta** (4528'; *Post*, well spoken of), a picturesque waterfall on the Rhine (to the right of the road), the hamlets of *Pardi*, **Fuorns**, and *Acla* (beautiful *Fumatsch* waterfall of the Rhine), and (2¼ M.) *Perdatsch* (5093'), at the mouth of the **Val Cristallina**.

The wild **Val Cristallina**, noted for its cheese, contains several fine waterfalls, particularly in the *Höllenschlund* (*Val Ufiern*). From the head of the valley two easy passes, the *Passo Cristallina* (7887'), passing the *Lago Retico* (*Redig-See*; 7802'), and the *Passo d'Ufiern* (8727'), between the *Cima Camadra* and the *Cima Garina*, lead to *Olivone* (p. 377). — The Piz Cristallina (10,265'; 4½ hrs.; good guide necessary) is ascended from

Perdatsch by the *Forcella Cristallina* (9862'; not to be confounded with the Passo Cristallina) without difficulty. Grand survey of the Medel and Rheinwald Mts. *Piz Uffern* (10,346'; 5½ hrs.) is more difficult. — The *Piz Medel* (10,510'; 5-6 hrs.), a splendid point of view, presents no difficulty to experts. The route leads from Fuorns (p. 375) to the E. up the *Buora Glen* nearly to the pass of that name, then ascends (right) the stony slopes to the W. of the *Miez Glatsché* ridge till it reaches the upper snowfields of the *Buora* and *Medel Glaciers*, and, passing the rock island called *Rifugl Camolsch*, gains the top by the N.E. arête. The descent may be made either over the *Camadra Glacier* to the *Passo d' Uffern* (see p. 375) or to *Ghirone* in the *Val Camadra* (p. 370).

Above Perdatsch the road ascends by a long bend to *St. Gion* (5298'), a group of hovels with a hospice, and traverses a wild, rock-strewn valley, scantily overgrown with grass, willows, and rhododendrons. The hospice of *St. Gall* (5514') is passed on the opposite bank. By the *Alp Scheggia* we cross to the left bank and reach (5 M. from Perdatsch) the hospice of —

12¼ M. **Sta. Maria** (6043'; *Inn*), anciently called Sancta **Maria** '*in loco magno*', whence perhaps the name of the pass.

To the E. rises the Scopi or *Skupil* (10,500'; '*Tschupe*', summit, or crown); ascent from the hospice in 3½-4 hrs. (guide 12 fr.), not difficult, at first over steep grassy slopes, finally over debris of weathered slate and the rocky arête. Extensive view. The descent may be made to the E., to the *Boarina Alp* (6140') in the *Val di Campo* (3 hrs.) and viâ *Campo* (beyond which there is a carriage-road) to (3 hrs.) *Olivone* (p. 377). — Piz Rondadura (9905'), to the W. of S. Maria (3½ hrs.), also easy.

From S. Maria to the *Hôtel Piora* (3 hrs.; guide 10, horse 25 fr.) and *Airolo*, see p. 115. — Over the *Rondadura Pass* to *Val Nalps*, p. 373.

The road now crosses for the last time the Mittel-Rhein, which rises in several little lakes in the *Val Cadlimo*, opening on the right, and ascends gradually to the (1¼ M.) **Lukmanier Pass** (6290'), the second-lowest between Switzerland and Italy (p. 394). To the left rises the black, slaty summit of the *Scopì*; on the right are the *Piz dell' Uomo*, *Piz Blas*, *Piz dell' Uffern*, and *Piz Rondadura*. We now descend, over beds of avalanches and mud-streams which have been precipitated from the yellowish slopes of the *Piz Corvo* (9340') on the left, and which frequently endanger the road in wet weather, to the (2 M.) former hospice of *Casaccia* (5975'), prettily situated. To the E. towers the huge *Rheinwaldhorn* (p. 386).

A path leads hence over the *Predelp Pass* (8058') to (5 hrs.) *Faido* (p. 107). Another crosses the *Passo Columbe* (7792'), between the *Piz Scai* and the *Piz Columbe*, to the (3½ hrs.) *Hôtel Piora* (p. 115).

The road is level as far as the (1½ M.) *Lukmanier Inn*, at the beginning of the *Piano di Segno* (5415'), and then descends, high above the *Brenno*, on the steep N. side of the *Val S. Maria*, being hewn at places in the perpendicular rock. Below lie the chalets of *Campra*. We descend by a long curve to the left to (4¼ M.) the hospice of *Camperio* (4028'), cross the Brenno, and skirt the wooded S. side of the valley, soon obtaining fine views of the **Val Blenio**. Far below, among walnut-trees, lie the villages of *Somascona*, *Scona*, and *Olivone*, commanded by the conical *Sosto* (7280'). Descending another long bend (footpath shorter), we reach (3 M.) —

24½ M. **Olivone**, Rom. *Luorscha*, locally *Rivōi* (2925'; *Hôt. Olivone*, R. 2, D. 3 fr.; pop. 711), the highest village in the *Val Blenio*, or *Pollenzer-Thal*, picturesquely situated. To the E. tower the abrupt spurs of the Rheinwald range. To *Vrin* by *Ghirone*, see p. 370. No guides to be had at Olivone.

The road crosses the *Brenno* by a stone bridge, and descends on its left bank to (2 M.) *Aquila* and to (³/₄ M.) *Dangio* (2645'), beautifully situated at the entrance to the *Val Soja*. Vines and mulberries now appear, and the slopes are clothed with walnuts and chestnuts. Next villages (½ M.) *Torre* and (1½ M.) *Lattigna*. [Opposite, above *Prugiasco*, stands the little church of *S. Carlo* with some frescoes of interest to students of art.] Then (30½ M.) **Acquarossa** (1740'; *Albergo delle Terme*), with a chalybeate spring, at the foot of the pyramidal *Simano* (8475'), which may be ascended without difficulty in 6 hrs., with guide (fine view; rich flora).

The valley contracts. Then (1½ M.) *Dongio*, a long village (Inn, carriages), and (1 M.) *Motto* (1445'), where the road divides. The road to the left (on the left bank of the Brenno) passes *Malvaglia*; that to the right (shorter, and shady in the afternoon) leads by *Ludiano* and (2 M.) *Semione* (1320'), beside the ruined château of *Serravalle*. The two roads re-unite at (2½ M.) the bridge of *Loderio* (1190'), a village destroyed by a flood in 1868. The lower part of the valley is monotonous; its broad floor is covered with stony deposits, and the slopes are furrowed by torrents. After crossing a mound of débris, the road descends to (1¼ M.) —

39½ M. **Biasca** (p. 117), where the Val Blenio unites with the Riviera (Val Ticino). The **station** of the St. Gotthard Railway is ³/₄ M. to the S. of the village. Post-office at the station.

95. From Coire viâ Thusis to Tiefenkasten (Schyn Road) or to Splügen (Via Mala).

Comp. Map, p. 368.

FROM COIRE VIÂ THUSIS TO TIEFENKASTEN, 25 M., Diligence (Julier and Landwasser Routes) once daily in 5 hrs. (9 fr. 10 c., coupé 11 fr. 15 c.); to Thusis 4 times daily in 3-3¼ hrs. (5 fr. 50, coupé 6 fr. 80 c.). — One-horse carriage to Thusis 15 fr., two-horse 30 fr.; to Tiefenkasten 45 fr.

FROM COIRE TO SPLÜGEN, 32½ M., Diligence twice daily in 7 hrs. 10 min. (12 fr., coupé 14 fr. 60 c.); to Chiavenna in 13 hrs. (21 fr. 95, coupé 26 fr. 60 c.). — EXTRA-POST with two horses from Coire to Splügen 77 fr. 90 c., to Chiavenna 130 fr. 40 c., with three horses 181 fr. — CARRIAGE with two horses from Coire to Splügen 65, with three 100 fr.; to Chiavenna 135 or 185 fr. (fee 10 per cent of the fare).

The following times should be allowed by walkers: Coire to Reichenau 2, Reichenau-Thusis 3½, Thusis-Tiefenkasten 3, Thusis-Andeer 2¾, Andeer-Splügen 3 hrs.

From Coire to (6 M.) **Reichenau** (1935'; **Adler*), see p. 367. The road through the *Vorder-Rheinthal* to *Disentis* and *Andermatt* diverges here to the right (see R. 93). A new iron bridge crosses the *Vorder-Rhein* above Reichenau, immediately before its confluence with the **Hinter-Rhein**. In the vicinity are a large saw-mill and several workshops for cutting and polishing marble.

The fertile valley, called **Domleschg**, *Domliaschga*, or *Tomiliasca* (the W. side *Heinzenberg*, Romanic *Montagna*), through which the road to Thusis leads on the left bank of the Hinter-Rhein, is 7 M. long and 2 M. wide. The Rhine, which formerly occupied nearly the whole valley, is now confined within due limits by large embankments. The sides of the valley are remarkable for their fertility, while on the right bank numerous castles peep down from almost every hill and rock.

The road ascends slightly to (1 M.) **Bonadūz** (2145'; *Post; Degiacomi; Simones*). To the left, on the Rhine, is the *Chapel of St. George*, adorned with ancient frescoes (road to Ilanz, see p. 360). Then (3/4 M.) **Rhäzüns** (2125'), on a rock rising from the Rhine, with a handsome château of the Vieli family. Fine view of the mountains to the S. (see below); behind us rises the Calanda.

On the EAST SLOPE is the ruin of *Nieder-Juvalta*; farther on are the baths of *Rothenbrunnen*, containing iron, iodine, and phosphorus, and specially adapted for childish ailments (°Curhaus, pens. 8 fr.); above them the ruins of *Ober-Juvalta*; then the châteaux of *Ortenstein* (recently rebuilt, in a picturesque situation) and *Paspels*. We next observe the ruined church of *St. Lorenz* and the châteaux of *Canova*, *Rietberg*, *Fürstenau*, *Baldenstein* (on the Albula), and *Ehrenfels*, below *Hohen-Rhätien* (see below).

We next reach (3¼ M.) *Realta* (2058'; Gasthaus zur Rheincorrection), with the ruin of *Nieder-Realta* (not visible from the road), and pass (1¼ M., on the left) the large cantonal *Prison* and *Lunatic Asylum*. Beyond (1¼ M.) **Katzis** (2185'; *Kreuz*) we pass a nunnery and school on the right, and the venerable little church of *St. Martin* on the left. Beautiful scenery. To the S. rises the snowy *Piz Curvèr* (9760'); beyond it, to the left, is the Schyn Pass, with the majestic *Piz Michel* (10,375') in the background; to the N. the *Ringelspitz* (10,660') and the *Trinserhorn* (9935'). About 3/4 M. to the E. the *Albula* falls into the Rhine; beyond it lies the pretty village of *Scharans*. Near (2½ M.) Thusis, above the pleasant village of *Masein*, stands the château of *Nieder-Tagstein*.

16 M. **Thusis**. — Hotels. °HÔT.-PENS. VIA MALA, at the beginning of the Via Mala, with garden, R., L., & A. 4-5½, B. 1½, D. 4, pens. 9-11 fr.; °POST & CURHAUS, with baths, R., L., & A. 2-3, B. 1¼, D. 3½, pens. 6-8 fr.; *RHAETIA, R., L., & A. 3, D. 3½, B. 1¼, pens. 7½ fr.; *WEISSES KREUZ, R. 2-2½, D. 2½-3, pens. 6-7 fr.; GEMSLI, plain. — Beer at the ' *Felsenkeller*' on the *Rosenbühel*, to the right of the entrance to the Via Mala, fine view (not always open). — *One-horse carr.* to the third bridge o' the Via Mala and back, 2 pers. 6, 3 pers. 8 fr., *two-horse carr.* 12 fr.; to Andeer 11½, 14, or 22½ fr.; to Splügen 22½ or 39 fr.; to Schyn (Solis Bridge) 7, 9, or 14 fr.; to Tiefenkasten 13½, 16½, or 24½ fr.; to Reichenau 11, 15, or 22 fr.; to Coire 17 or 33 fr. Fees included in each case. — *English Church Service* in the Swiss Church.

Thusis (2450'; 1098 inhab.), Rom. *Tuseun*, beautifully situated at the foot of the Heinzenberg, is well adapted for some stay and as a starting-point for excursions. — Immediately above Thusis the turbid *Nolla*, a torrent which has frequently devastated this district (p. 368), falls into the Rhine, the valley of which appears to be entirely shut up by lofty mountains. The right bank of the gorge from which the Rhine issues is guarded by the ruined castle of *Hohen-Rhaetien*, or *Hoch-Realta (Hoch-Ryalt)*, 807' above the river, the most ancient in Switzerland, having been founded, according to tradition, in B.C. 589, by the legendary hero *Rhaetus*, leader of the Etruscans when retreating before the Gauls. The ruin is on the S. side of the hill, which overlooks the whole of the Domleschg (ascent, see below); on the N. side is the dilapidated *Church of St. John*, the oldest Christian church in the valley.

WALKS AND EXCURSIONS (guides, *Daniel Pappa, Peter Beeli*). To the (5 min.) *Rosenbühel* (see p. 378); to the *Belvedere* (20 min.); to the *Boval* wood (1/4 hr.); to the first weir in the *Nolla Valley* (20 min.); to the (1 hr.) *Crapteig*, to the right above the Via Mala. To *Hohen-Rhätien*, a zigzag path ascends beyond the Rhine bridge in 3/4 hr.; the descent may be made by the *Alp Carschenna*, through wood, past the church of *St. Cassian* and the château of *Baldenstein*, to *Sils* (see below). — Through the *Schlosswald* to the *Taubenstein* and (40 min.) the château of *Tagstein* (see above), with pleasure-grounds. Past (3/4 hr.) *Rongellen* (p. 380) to the (1 1/2 hr.) *Matensäss Aclasut* (4095'), situated high above the second bridge in the Via Mala. — By the Schyn Road to the *Solis Bridge* (1 3/4 hr.; one-horse carr. there and back 6 fr. and 1 fr. fee). — On the Heinzenberg rises the *Präzerhöhe* (6065'), a fine point of view, ascended in 4 1/2 hrs. by *Masein, Portein*, and *Sarn* (3363'; Inn). — The *Slätzerhorn* (8450'), 5-6 hrs., is toilsome from this side (comp. p. 390). — Viâ *Tschappina* and the *Glas Pass* to the *Safier-Thal* (4 hrs. to Platz), see p. 367.

The *Schyn Road, constructed in 1868-69 and vying in grand and picturesque scenes with the Via Mala, crosses the Rhine at the foot of Hohen-Rhætien, immediately above the Nolla bridge, passes the ruin of *Ehrenfels* on the right, and beyond (3/4 M.) Sils (2283'; *Post), a village almost entirely burnt down in 1887, the small château of *Baldenstein* on the left. It next ascends the left bank of the *Albula* to *Campi (Campo Bello*, ruin of the ancestral seat of the Campell family; Ulrich Campell was a Rhætian reformer and historian), picturesquely situated to the left, and the farm of (2 M.) *Runplanas*. Pretty view hence of the church of Solis. Then through the ancient forest of *Versasca*. By a ravine we observe above us, to the right, a bridge of the old Mutten road, and we pass the Freihof, an auberge on the left. The road is next carried through the *Pass Mal*, which begins here, by means of galleries of masonry and extensive cuttings and tunnels. 1 1/2 M. Inn 'Zum Pass Mal' (plain).

About 1/4 M. farther on, by the chalets of *Calabrien*, a narrow road to the right ascends to (3 1/2 M.) *Unter-Mutten* (4833'; °Inn, plain; closed in summer when all the inhabitants migrate to Ober-Mutten). Thence to (1 1/4 br.) *Ober-Mutten* (6148'; Hosang's Inn), whence the Muttnerhorn (8070'; °View) may be ascended in 1 1/2 hr.; good path at first, then up grassy slopes. Descent from Ober-Mutten to (2 1/2 hrs.) Zillis or to Thusis interesting, but rather rough.

The bridge across the *Muttner Tobel* affords a fine view of the gorge. 1 1/4 M. *Unter-Solis*, a hamlet with a spring containing iodine. High above, to the left, lies *Obervatz* (p. 391). Looking back near the last tunnel, we obtain a fine survey of the Heinzenberg, and before us a view of Alvaschein and the peaks of the Albula group. The road now crosses the gorge of the Albula by the *Solis Bridge*, 250' above the foaming stream, and ascends in a curve (cut off by a path to the right beyond the bridge) to the village of (2 M.) Alvaschein *(Augustin)*. Opposite, below the loftily situated *Stürvis*,

380   VI. Route 95.   VIA MALA.   From Coire

is a waterfall. Farther on, to the right, below the road, is the **church**
of *Müstail*, the oldest in the Albula valley, formerly a burial-place.
At *Unter-Müstail* there is an alkaline spring. The road unites with
the Julier route near (1½ M.) *Tiefenkasten* (p. 391).

The famous ***Via Mala**, forming the first part of the **Splügen**
Road, which ascends the valley of the Hinter-Rhein, was constructed
in 1822. Formerly the route ascended the bank of the Nolla through
wood, and entered the deep gorge of the Rhine, then known as the
'*Verlorne Loch*', and traversed by a path only 4' wide, at a point
above Rongellen. The limestone-rocks rise almost perpendicularly
on both sides to a height of 1600'. At the (½ hr.) *Känzeli* the
retrospective view is very fine. A little farther on, 1½ M. from
Thusis, the road passes through a tunnel (2685'), 55 yds. long.
From a point beyond the tunnel, where the side-wall ceases and
the wooden railings recommence, the boisterous river is visible at
the bottom of the profound gorge. Below the (¾ M.) hamlet of
Rongellen (Hôt.-Pens. Via Mala; Hôt.-Pens. Alte Post, both moderate) the gorge expands into a small basin, and soon contracts
again. The road crosses the river three times at short intervals:
1 M., first bridge, built in 1738 (refreshments at a pavilion above);
¼ M. *Second Bridge* (2844'), built in 1739, the grandest point.
The Rhine, 160' below the road, winds through so narrow a ravine
that the precipices above almost meet. At the (¾ M.) third bridge
(2903'; built in 1834) the Via Mala ends.

We now enter the **Schamser-Thal**, the green meadows of
which contrast pleasantly with the gloomy Via Mala. In the background to the S. rises the pointed *Hirli* (9373'). The first village
in the valley of Schams is (1 M.) —

2 M. **Zillis**, Rom. *Ciraun* (3060'; *Rathhaus* or *Post*, plain), with
the oldest church in the valley (nave and tower Romanesque; interesting ceiling-paintings of the 12th century).

Ascents. *Piz Beverin* (9843'; 6-7 hrs.; guide 7, horse to the Obrist
Alp 12 fr.), a superb point of view, but trying. Bridle-path by *Donath*
and *Mathon* to the (3½-4 hrs.) *Obrist Alp* (7172'); thence by the S.E. arête
to the top 2½ hrs. more. The ascent from Thusis direct, by *Glas* (simple
entertainment) in 7-8 hrs. (guide 10 fr.), is more interesting but should
not be attempted except by experts. — Piz Curvèr (9760'; 6 hrs.; 6 fr.), from
Zillis or Pignieu, also interesting and for experts not difficult. The descent
may be made to the chapel of *Zilteil* and *Savognin* (p. 391).

On the hill to the right, on the left bank of the Rhine, above
the village of *Donath*, and overshadowed by the *Piz Beverin*, stands
the ruined castle of **Fardün** or **La Turr** (3820'), once the seat of
the governors of the valley. About the middle of the 15th cent.
the brutality of one of these officials, like that of Gessler 150 years
earlier, is said to have given rise to the emancipation of this
district from their sway. Entering the cottage of a peasant whom
he disliked, the tyrant spat into the boiling broth prepared for
dinner. The peasant, Johann Caldar, seized him by the throat,

to Splügen.   ANDEER.   *VI. Route 95.* 381

plunged his head into the scalding liquid, exclaiming, '*Malgia sez il pult cha ti has condüt*' ('Eat the soup thou hast seasoned'), and strangled him. This was the signal for a general rising.

Near **the ruined *Baths of Pignieu*** (the waters of which, containing iron **and alkali**, are conducted to Andeer, **and** there **used for baths**), the *Pignieuer Bach* is crossed by a bridge, the last completed on this route, and bearing the inscription on the E. parapet: '*Jam via patet hostibus et amicis. Cavete, Rhaetii simplicitas morum et unio servabunt avitam libertatem*'. To the left is the village of *Pignieu;* opposite, on the left bank of the Rhine, are *Clugin* and the square tower of the ruin of *Cagliatscha*. Then (2½ M.) —

25½ M. **Andeer** (3210'; pop. 581; *Krone, or *Hôtel-Pens. Fravi*, with chalybeate baths, R. & L. 2½, B. 1¼ fr.; **Hôt.-Pens. Beverin*, well situated; *Sonne, rustic), the principal village in the valley. Fine view from the loftily situated church (built in 1673).

ASCENTS. **Piz Vizan** (8110'; 4½ hrs., with guide), by the *Burgias Alp;* splendid view. — Piz La Tschera (8015'; 5 hrs., with guide), by *Alp Albin*, also interesting. — *Piz Beverin* and *Piz Curver*, see above.

FROM ANDEER TO STALLA (11 hrs.; guide unnecessary), an attractive walk. The new road to Canicül (2¼ M.) quits the Splügen road 2 M. above Andeer and enters the wild *Ferrera Valley to the left, leading first on the left, and then on the right bank of the *Averser-Rhein*, which forms several fine waterfalls. On the left is *Piz Grisch* (10,000'), on the right the *Surettahorn* (9926'). We pass (40 min.) a deserted silver-foundry and reach (½ hr.) *Ausser-Ferrera* (4334'; two modest inns, where the valley expands slightly. (Over the *Fianell Pass* to *Savognin*, see p. 392.) We then follow the right bank to (1½ hr.) *Inner-Ferrera* or *Canicül* (4856'; rustic inn), at the mouth of the *Val d'Emet* (see below). Descending by a bridle-path (road under construction), we cross the Rhine and ascend its steep left bank (20 min.). The path skirts the slope, passing through wood; after 25 min. it rounds a projecting rock (view of the Surettahorn, etc., behind us), and then again descends to the river, which is augmented here by the torrents from the *Val Startera* on the left and the *Valle di Lei* on the right. The narrow path crosses (25 min.) the latter. (By the bridge is the frontier-stone of Italy, to which the Valle di Lei belongs.) The path ascends rapidly, and then immediately descends. Near (1 hr.) *Campsut* (5500'; Inn) it crosses the Rhine, and beyond (¼ hr.) *Crot*, another poor village, recrosses it. Beyond the bridge (view, to the right, of the *Madris Valley*, with the *Piz Gallegione* and the *Cima di Lago* at its head) the path ascends steep pastures to the left, and at the top of the hill traverses a beautiful wood of stone-pines. It then descends, crosses another bridge, and ascends to (1 hr.) *Cresta* (6397'; modest entertainment at the schoolmaster's; guides, Simon Heinz, Peter Stoffel), the principal village in the Averser-Thal, which expands here and is carpeted with beautiful pastures. This is one of the highest inhabited valleys among the Alps, and lies in a **sunny** situation. To the N. rises the *Weissberg* (9990').

The path then ascends slightly, passing the **handsome** house of the *Podestat*, or chief magistrate, and the mouth of the *Val Bregalga*, which is enclosed by fine glaciers, to (1½ hr.) *Juf* (6685'); **then** to the left **across** pastures and through a desolate rock-strewn valley to the (1½ hr.) pass of the **Stallerberg** (8480'; beautiful view of the Julier Mts., etc.). The path, quite distinct, now descends, keeping to the left, to (2 hrs.) *Stalla* (p. 392). — From Juf through the *Val Faller* to *Molins*, see p. 392.

A path leads from Juf to the S.E. over the Forcellina (8770') to the (3½ hrs.) *Septimer* (p. 392), and thence to (2 hrs.) *Casaccia* in the *Val Bregaglia*, or over the *Lunghino Pass* to the (5 hrs.) *Maloja* (see **pp. 392, 395**). — From the Forcellina Pass we may ascend a peak known **in the** Averser-Thal as the *Forcellina* (9918'; admirable view) in 1½ hr., **and descend to** the S.E. into the *Val Turba*. We then reach the Septimer route 20 min. below the pass, by the second bridge over the Septimerbach (p. 392). — From Cresta through the Val Bregalga and across the **Passo della Duana** (9187'; with guide) to *Soglio* in the Val Bregaglia (p. 424), 7-8 hrs., interesting. The pass, between *Pizzo Marcio* (9534') and *Pizzo della Duana* (p. 423), and also the descent afford a fine view of the Bregaglia Mts., especially of the Val Bondasca with the finely-shaped Piz Badile.

FROM CANICÜL TO PIANAZZO on the Splügen route (4½ hrs.; with guide). The path ascends steeply on the right side of the *Val d'Emet*, through wood, to the (1¼ hr.) *Alp Emet* (6194'), whence the cairn on the pass is visible; then over the soft and uneven soil of the Alp to the top in 1 hr. more. Retrospective view of the Piz Beverin; afterwards the Calanda comes in sight. From the Passo di **Madesimo** (7480'; frontier of Switzerland and Italy) the *Piz Tambo* (see below) is seen to the W., and the *Cima di Lago* (9892') and *Piz Gallegione* (10,285') to the S.E. We descend past the N. side of the pretty *Lago d'Emet*, on the left bank of the *Madesimo*, then across meadows, to the huts of *Al Tecchio* and (1½ hr.) *Madesimo* (p. 388). New road thence to (1½ M.) *Pianazzo* (p. 383).

The Splügen road winds upwards, passes the scanty ruins of the *Bärenburg*, and enters the wooded *Rofna Ravine, in which the Rhine forms a series of waterfalls. Near the entrance (2¼ M. from Andeer) the road crosses the *Averser-Rhein* (*Melchior's Inn), which here issues from the *Val Ferrera* (p. 381) and forms a fine waterfall a little way up the valley. Walkers may make a short-cut.

Towards the end of the gorge (2½ M.), we pass an old bridge over the Rhine on the right. The valley expands. The road crosses (¾ M.) a torrent which drains the *Suretta-Thal* on the left. We next (½ M.) pass through a rocky gateway (*Sassa Plana*; 4390'). At (1 M.) the prettily situated **Hôt.-Pens. Hinterrhein* a bridge leads to the village of *Sufers* (4673') on the left bank of the Rhine. Farther on we enter a wooded ravine and cross (1¼ M.) the wild stream in its profound gorge by a bold bridge (4727'). After a short ascent we obtain a survey of the broad *Val Rhein (Rheinwald-Thal)*; on the right the barren *Kalkberg* (9763'); opposite, the *Einshorn* (9650'); in the background, the *Rheinwaldhorn* (11,150'); to the left, adjoining the *Guggernüll* (9472'), is the *Piz Tambo* (10,748'); farther back the *Piz Curvèr* (p. 380). — Then (1¼ M.) —

32½ M. **Splügen** (4757'; pop. 424; **Hôt. Bodenhaus*, R., L., & A. 3½, D. 3, pens. 7-8 fr.; **Hôt. Splügen*, R., L., & A. 2-2½, B. 1, D. 2½, pens. 5-6 fr.), the capital of the Rheinwald-Thal, enlivened by the traffic on the Splügen and Bernardino routes. The ruined castle on the old road commands a pretty view.

EXCURSIONS. (Guides, *Peter Schwarz* and *Joh. Sprecher*.) Pleasant walks to the *Fluhgründ* (1 hr.) and *Donatzhöhe* (1½ hr.). By carriage to the *Bernardino Pass* (p. 386); the Alp beyond the inn commands a splendid survey of the Rheinwald Glacier. — The **Guggernüll** (9472'; 4½ hrs.; guide 6 fr.), by the *Tambo Alp*, and the Einshorn (9650'; from Nufenen, 4-5 hrs.; 8 fr.) are two fine points, and not difficult. — The **Pizzo Tambo** (*Tambohorn* or *Schneehorn*, 10,748'; 14 fr.), ascended from the Splügen Pass in 3½ hrs., is fatiguing, but for experts free from danger. Most extensive view, N. to Swabia, and S. to Milan, whence the peak is visible.

Excursion to the *Source of the Hinter-Rhein*, p. 385. — Over the *Löchliberg* to the *Safier-Thal*, see p. 367.

Molins
Sur
Marmorera
Cresta
Silvaplana
Sils
Casaccia
Vicosoprano
Soglio
Stampa
Bondo
Castasegna

96. From Splügen to the Lake of Como.

41 M. DILIGENCE twice daily to Chiavenna (10 fr., coupé 12 fr.) in 5 hrs.; from Chiavenna to Colico, 17 M., RAILWAY in 1 hr. (3 fr. 10, 2 fr. 15, 1 fr. 40 c.) corresponding with the steamboats to Como.

The road divides at the village of *Splügen* (p. 382). The Bernardino route leads straight on (p. 385), while the SPLÜGEN ROAD, which was constructed by the Austrian government in 1819-21, crosses the Rhine to the left by an iron bridge, ascends in windings (avoided by short-cuts), and farther up passes through a tunnel 93 yds. long, beyond which we see the head of the pass. The road crosses the *Häusernbach* twice in a bleak valley, at the end of which the old bridle-path ascends direct to the pass. The road ascends on the W. slope in numberless zigzags, past the lonely *Berghaus* (6677'), and through a long gallery of masonry, to the (2½ hrs.; 5 M.) **Splügen Pass** (*Colmo dell' Orso*, 6945'), between the *Piz Tambo* (10,748'; see p. 382) on the right, and the *Surettahorn* (8925') on the left, the boundary between Switzerland and Italy.

Beyond the pass and the first *Cantoniera*, we reach (1½ M.) the **Dogana** (6247'), or Italian custom-house, a group of houses at the head of a bleak valley (*Monte Splnga Inn, plain; Post). In winter the snow here sometimes reaches to the windows of the upper story. During snow-storms, bells are rung in the four highest houses of refuge as a guide to travellers.

The old bridle-path turned to the right by the second **wooden bridge**, and led through the *Cardinell* gorge direct to Isola, a **route** much exposed to avalanches. In traversing this ravine in Dec., 1800, the French under Gen. Macdonald sustained severe losses, **whole columns** being precipitated into the abyss. The new road descends the E. slope in numerous zigzags, being protected at places against avalanches by long galleries of solid masonry (first 249 yds. long, second 228, third 550 yds.), with sloping roofs to enable the snow to slide off, and openings at the sides for light.

On quitting the third gallery, we obtain a fine view of the old road, which was destroyed by an inundation in 1834, and the village of *Isola*. At the end of the gallery is a copious spring. The new road avoids the dangerous *Liro Gorge* between Isola and Campo Dolcino. Near *Pianazzo* (Inn, dear) a road descends to the right to Isola. Just beyond Pianazzo, near the entrance to a short tunnel, the *Madesimo* descends into the valley, forming a *Fall 650' high (best viewed from a small platform by the road, where the conductor stops the diligence).

From Pianazzo a road (two-horse carr. from Splügen and back 40 fr., fee 4 fr.) ascends to the hamlet of (1½ M.) **Madesimo** (4920'), with a chalybeate spring and a hydropathic *Curhaus (pens. 7½ fr.), recommended as a health-resort. — To *Cavrell* over the *Passo di Madesimo*, see p. 381.

The part of the road which we now enter upon is the boldest in point of construction, with numerous tunnels, and terraces rising perpendicularly one above the other.

15½ M. **Campodolcino** (3457'; *Croce d'Oro; Posta* or *Corona*, mediocre) consists of four groups of houses; the second contains the church and the 'campo santo' or burial-ground. A Latin inscription on the rock, beyond one of the galleries, is in honour of the Emp. Francis, who made this road from '*Clavenna ad Rhenum*'. The *Liro Valley*, or *Valle S. Giacomo*, is strewn with fragments of rock, chiefly of brittle white gneiss, which reddens on exposure to the air. The wildness of the scene is somewhat softened by the rich foliage of the chestnuts visible lower down, from among which rises the slender white campanile of the church of *Gallivaggio*. Near *S. Giacomo* are whole forests of chestnuts, which extend far up the steep slopes. We soon reach the vineyards of Chiavenna, where the luxuriance of Italian vegetation is fully displayed.

25 M. **Chiavenna.** — Hotels. *HÔTEL CONRADI, 5 min. from the station, with railway-ticket and luggage office, R., L., & A. 3½-5, B. 1¼, D. 3-4½, S. with wine 3, pens. 6½-8, omn. ½-¾ fr.; *ALBERGO SPECOLA, at the station, R., L., & A. 2½, B. 1 fr.; *CHIAVE D'ORO, on the Promenade, in the Italian style.

The Station (*Café-Restaurant, lunch 2½ fr., beer) lies to the E. of the town. Through-tickets are here issued to the steamboat-stations on the Lago di Como, with coupon for the omnibus between the railway-station and the quay at Colico.

Chiavenna (1090'), the *Clavenna* of the Romans, an ancient town with 4086 inhab., is charmingly situated on the *Mera*, at the mouth of the Val Bregaglia (p 416). Opposite the Hôtel Conradi are the ruins of an unfinished château of De Salis, the last governor appointed by the Grisons. Picturesque view from the '*Paradiso*' or garden of the ruin (adm. 50 c.). *S. Lorenzo*, the principal church, has an elegant detached *campanile* or clock-tower, rising from the former burial-ground. In the octagonal baptistery (closed, fee 15-20 c.) is an ancient font of 1206 with reliefs. The neighbouring hills of *Val Capiola* contain numerous giant cauldrons ('Marmitte dei Giganti') of varying size (guides at the hotels).

The RAILWAY TO COLICO (fares, see p. 375) traverses three tunnels soon after starting, beyond which we enjoy a fine retrospect of Chiavenna. The line runs through a rich vine-bearing country, the lower parts of which, however, are exposed to the inundations of the Liro and Mera. The valley *(Piano di Chiavenna)* is enclosed on both sides by lofty mountains. On the right bank of the Mera lies *Gordona*, at the mouth of the *Val della* **Forcola** (p. 387), beyond which the *Boggia* forms a pretty waterfall in its precipitous descent from the narrow *Val Bodengo*. — 6 M. *Samolaco* is the station for the large village of that name on the opposite (right) bank of the Mera, at the mouth of the *Val Mengasia*. Before (8½ M.) *Novate*, the railway reaches the *Lago di Mezzola*. This lake was originally the N. bay of the Lake of Como, from which it has been almost separated by the deposits of the *Adda;* but the shallow channel which connects the lakes has again been rendered navigable. To the S. appears the pyramidal **Mte. Legnone (p. 458)**. The railway crosses the diluvial land formed by the mountain-stream issuing from the *Val Codera* on the left, and, supported by masonry and traversing tunnels, skirts the E. bank of the lake viâ *Campo* and *Verzeia*. It crosses the Adda beyond (12½ M.) *Dubino*. The Valtellina railway (p. 422) joins ours from the left; on a hill to the right the ruined castle of *Fuentes*, once the key of the Valtellina, erected by the Spaniards in 1603, and destroyed by the French in 1796.

17 M. Colico (722'), see p. 458. The station is nearly ½ M. from the quay. The omnibus-coupons are collected at the exit from the station. There is abundant time to permit of passengers walking to the quay.

97. From Splügen to Bellinzona. S. Bernardino.
Comp Map, p. 382.

45½ M. DILIGENCE daily (between S. Bernardino and Bellinzona twice daily) in 8¼ ., returning in 11 hrs. (15 fr. 25, coupé 18 fr. 95 c.). EXTRA-POST with two horses from Coire to Bellinzona 171 fr. 20 c., with three horses 240 fr. 50 c.; from Splügen to Bellinzona with two horses 95 fr. 80 c. CARRIAGE AND PAIR from Coire to Bellinzona 180 fr., from Splügen to Bellinzona (in 3 days) 115 fr.; fee 10 per cent of the fare.

Splügen (4757'), see p. 382. We traverse the upper *Val Rhein*, passing below (1 M.) *Medels* (5030'). On the left bank, ¾ M. farther on, lies the pasture of *Ebi*, now partly covered with débris, where the 'Landsgemeinde' used to assemble biennially on the first Sunday in May. Then (2 M.) *Nufenen* (5145'), at the mouth of the *Areue-Thal*, at the head of which appears the *Curciusa Glacier*. On the left are the huge rocky *Guggernüll* (p. 382), concealing the *Piz Tambo* (p. 382), and the *Einshorn* (9650') Near (2¼ M.) —

6 M. **Hinterrhein** (5300'; *Post, plain), the highest village in the valley, the Rheinwald Mts., the Marscholhorn, Rheinquellhorn, Rheinwaldhorn, Hochberghorn, and Kirchalphorn come in sight.

Source of the Hinter-Rhein. From Hinterrhein to the Zapport Chalet 2¼ hrs., thence to the club-hut ¾ hr., rough, and hardly repaying (guide advisable, 6 fr.; *G. Trepp, Joh. Lorez*). The path, damaged annually by inundations and landslips, diverges to the right from the Bernardino road, beyond the Rhine bridge (see below), and at first traverses the level floor of the valley. After ½ hr. the valley narrows. The path loses itself in a stony chaos on the right slope of the valley, while the steep N. side is partly covered with poor pastures. The wild infant Rhine is in many places covered with avalanche-snow which lies here the whole year. By one of these snow-bridges we cross to the left bank, where a narrow path, kept in order by the shepherds in summer, leads to the (1¼ hr.) *Zapport Chalet* (6420'), occupied in July and August by the Bergamasque shepherds, who pasture their flocks on the sunny *Zapport-Alp*. The route to the club-hut (¾ hr.) next passes the *Hölle*, a wild cliff on the right bank, at the foot of which the Rhine forms a small fall; and on the same bank higher up is a poor rock-strewn Alpine pasture, called by way of antithesis the *Paradies*. The Zapport *Club-Hut* (7613'), with room for 10-12 persons, is also occupied in summer by the shepherds. The narrow valley is terminated by the Rheinwald Glacier, the lower part of which is called the *Paradies Glacier*. The Hinter-Rhein issues from an aperture in the glacier (7270'), in shape resembling a cow's mouth, immediately below the chalet. This chief source of the river (*Sprung* or *Ursprung*) is soon augmented by numerous small tributaries from crevasses of the glacier. From the club-hut we may ascend the Rheinwald Glacier in order to survey the vast *Adula* or *Rheinwald Mts.*: the *Zapporthorn* (10,330'), *Rheinquellhorn* (10,500'), *Vogelberg* (10,565'), *Rheinwaldhorn*, *Güferhorn* (11,130'), etc. — The Rheinwaldhorn (*Piz Valrhein* or *Adulahorn*, 11,150') may be ascended from the club-hut in 4 hrs. by the *Lentalücke* (9692') and the N.E. arête (not difficult for experts). The *Güferhorn* (3½-4 hrs. from the club-hut, by the Lentalücke and the S.W. arête); the *Vogelberg* and *Rheinquellhorn* (each 3½-4 hrs. from the club-hut, over the *Rheinwald Glacier*); and the *Zapporthorn* (3½-4 hrs. from the Bernardino Pass, over the *Muccia Glacier*) are all tolerably easy ascents.

From Hinterrhein over the *Valser Berg* to the Lugnetz Valley and Ilanz, see p. 370; over the *Zapportgrat* or the *Lentalücke* to Zervreila, p. 370. Trying passes (*Vogeljoch*, 9640'; *Passo del Cadabbi*, 9680'; *Zapport Pass*, 10,140') lead to the S. from the Rheinwald and Zapport glaciers to *Malvaglia* (p. 377).

The BERNARDINO ROAD crosses the Rhine by a bridge (5300') of three arches, ½ M. beyond Hinterrhein, and ascends the steep bush-clad slope in windings. (A good short-cut diverges to the right from the second winding.) Looking back, we have a fine view of the Rhine Valley and the Kirchalphorn, Lorenzhorn, Schwarzhorn, and Hochbergborn, which bound it on the north. On the left, before (2½ M.) we cross the *Masek-Bach* (5680'), is the solitary *Dürrenbühl Chalet*. Traversing a bleak valley, and passing the *Thäli-Alp* on the left, we reach the (3 M.) S. Bernardino Pass (6770'; Inn, poor), at the N. end of the little *Lago Moësola*, from which three rocks project. This pass was known to the Romans, and down to the 15th cent. it was called the *Vogelberg*. When St. Bernardin of Siena preached the gospel here at that period, a chapel was erected on the S. slope of the mountain, and the pass has since been named after him. On the left rise the *Pizzo Uccello* (8910') and *Mittaghorn* (8560'); on the right the *Marscholhorn* (or *Piz Moësola*; 9520'). Magnificent view from a large white boulder, ¾ hr. above the hotel to the N.W. (guide unnecessary).

We descend in numerous windings on the left bank of the *Moësa*, which issues from the lake, and pass a Cantoniera. On the W. rises the *Zapporthorn* (10,330') with the *Stabbio-Grat* (8995'), from which the *Muccia Glacier* descends. To the E. are the *Piz Lumbreda* (9770'), *Piz Mutun* (9360'), and *Piz Curciusa* (9423'). Lower down, we cross the Moësa by a handsome bridge, and descend in a wide bend to (5 M.) —

17 M. S. **Bernardino** (5335'; *Hôt. Victoria*, R., L., & A. 2-5, D. 4, pens. 10½ fr.; *Hôt. Brocco*, *Hôt. Ravizza*, board 7½-9½ fr.; *Albergo Menghetti*), the highest village in the *Val Mesocco* or *Mesolcina*, with a mineral spring which attracts many invalids in summer. The valley, especially the lower part, contrasts strongly with the Val Rhein in language, culture, and climate. Everything here is Italian, and the inhabitants are Roman Catholics, Cardinal Borromeo (p. 451) having successfully crushed the germs of the Reformation. — Over the *Passetti Pass* **to the** *Val Calanca*, see p. 387.

To the N., above the Bernardino Pass, towers the sharp tooth of the *Piz Uccello* (see above). The road ascends a little, and then descends in numerous zigzags (which footpaths cut off). A fine fall of the Moësa, in the gorge to the right, is well seen only if we follow the path leading from S. Bernardino to S. Giacomo, first on **the left**, and then on the right bank of the stream. At (4½ M.) *S. Giacomo* (3760'; **Alb.** Toscano) the road crosses the Moësa (pleasing view), and then descends rapidly to (**4 M.**) —

26 M. **Mesocco** or *Cremeo* (2560'; *Posta*, well spoken of; *Hôt. Toscani*, dirty), where walnut-trees, chestnuts, **vines, and fields of** maize proclaim the Italian climate. On a rocky height to the left of the road, ½ M. below **the village, rises the grand ruined castle of** *Mesocco* (or *Misox*), with its four **towers, which was** destroyed by natives of the Grisons in 1526. From **the slopes descend numerous** brooks, and between Mesocco **and Lostallo there are eight water**falls, some of them considerable.

Beyond (1½ M.) **Soazza** (2067') we reach the bottom of the valley. Near the second bridge **below Soazza the** *Buffalora* forms a fine cascade. Then (2½ M.) *Cabbiolo* (1475'), (1 M.) *Lostallo*, with extensive vineyards **and the first fig-trees,** and (4½ M.) —

35½ M. **Cama** (1260'), with a Capuchin monastery.

FROM CAMA TO CHIAVENNA a fatiguing, but interesting route (14-15 hrs., guide necessary to the summit of the pass only, 5 fr.) ascends the steep *Val Cama*, containing the little lake of that name (4058'), crosses the (5½ hrs.) Bocchetta di Val Cama (6780'), and descends through the *Val Bodengo* to (3½ hrs.) *Bodengo* (rustic inn) and by a steep path, with steps, through the gorge of the *Boggia* to *Gordona* and (5 hrs.) *Chiavenna*. — A somewhat easier, but less interesting path from Soazza (see above) crosses the **Passo della Forcola** (7270') and leads through the valley **of the same name** to Chiavenna (12-13 hrs.; with guide).

Then (¾ M.) *Leggia* (1125') and (1¼ M.) **Grono** (1000'; *Hôtel Calancasca*), a thriving village at the mouth **of the Val Calanca**, with the *Florentina* tower, and near it a chapel with old frescoes.

The picturesque **Val Calanca** is traversed by **a road,** first **on the left,** then on the right bank of the *Calancasca*, leading by *Molina*, *Arvigo*, *S. Domenica*, and *Augio* to (10 M.) *Rossa* (3570'; Inn), the chief village in the valley. (Toilsome route hence to the W., over the *Giumella Pass*, 6955', to *Malvaglia* in the *Val Blenio*, p. 377.) Bridle-path hence to (1 hr.) *Valbella* (4383'), the highest hamlet in the valley, from which an easy route to the E. crosses the *Passo di Trescolmine* (7064') to (5 hrs.) *Mesocco*; then (1 hr.) *Alp Alogna* (4695'), whence we may cross the *Passo di Passetti* (6808') to the E. to *S. Bernardino* (p. 386) in 4-5 hrs. (guide). At the head of the Val Calanca, but difficult of access thence, lies the grand mountain-basin of the *Stabbio Alps* (6590'), which may be reached in 4-5 hrs. from S. Bernardino by crossing the *Passo Tre Uomini* (8704').

39 M. **Roveredo** (975'; pop. 1065; *Angelo; Croce*), the capital of the lower Val Mesocco, with the ruined castle **of the once power**ful Trivulzio family.

S. Vittore (880') is the last village **of the Grisons,** *Lumino* the first in Canton Ticino. The Bernardino route passes *Castione*, on the right, a station on the St. Gotthard Railway (p. 117), joins **the St.** Gotthard road, and crosses the Moësa. Below the confluence of the Moësa and the *Ticino* lies *Arbedo* (813'), a village of sad memory in Swiss history. On 30th July, 1422, a battle took place here between

3000 Swiss and 24,000 Milanese, in which 2000 of the former fell. They were interred beneath several mounds of earth near the church of St. Paul, which is called *Chiesa Rossa* from its red colour.

45½ M. **Bellinzona**, see p. 433.

98. From Coire to the Engadine over the Albula Pass.

Comp. Map, p. 360.

DILIGENCE twice daily in summer: viâ Churwalden and Lenz to Samaden, 45½ M., in 12½ hrs. (18 fr. 25 c., coupé 21 fr. 90 c.; to Bergün, where passengers dine, in 7 hrs.; from Bergün to Ponte 4 hrs.); from Samaden to St. Moritz, 5 M., in 1 hr. 10 min. (in immediate correspondence with the preceding); from Samaden to Pontresina, 3½ M., in 55 minutes. — EXTRAPOST and pair from Coire to Samaden 105 fr. 80 c., or by the Schyn and Albula passes 121 fr. 20 c.; to St. Moritz or Pontresina 117 fr. 20 or 132 fr. 80 c. — CARRIAGE and pair from Coire to Bergün 70, over the Albula Pass to Samaden 120, Pontresina or St. Moritz 110, Tarasp 170 fr (viâ Schyn and Albula 80, 110, 120, or 180 fr.) and driver's fee of 10 per cent of the fare (to Samaden 1½-2 days). — A most interesting route; fine mountain-scenery.

From Coire either viâ *Churwalden* to **Lenz** in 3¼ hrs., or viâ *Thusis* and *Schyn* to *Tiefenkasten* in 5¼ hrs., see RR. 99, 95. The Albula road diverges at Lenz (or Tiefenkasten) to the left from the Julier road, passes (16½ M.) *Brienz* (p. 364; a direct path to *Surava* and Bad Alvaneu diverges to the right at the last house of Brienz), then turns twice to the left at intervals of 5 min.) and below the ruined château of *Belfort*, and winds down the *Crapanaira Ravine* to —

20 M. **Bad Alvaneu** (3115') in the *Albula-Thal*, where the roads from Lenz and Tiefenkasten unite. The sulphur-springs are of repute for rheumatism, etc. (*Hotel*, R., L., & A. 3-5, D. 4, pension 6½-11 fr.; *Pens. Schuler*, unpretending). On the opposite bank is a picturesque waterfall.

The **Piz Michèl** (10,375'; 6-7 hrs.; with guide) may be ascended by experts without much difficulty from Bad Alvaneu through the *Schaftobel*. View of striking grandeur. — In the *Val Spadlatscha*, 4 hrs. above Bad Alvaneu or Filisur, and 3 hrs. from Bergün (see below), is the *Aela Club-Hut* (7020'), from which the **Tinzenhorn** (10,430') may be ascended in 4 hrs., and the Piz d'Aela (10,960') in 4½-5 hrs. (the latter difficult, and both requiring experience). Difficult descent from the Tinzenhorn on the steep W. side to the *Tinzenthor Pass* (p. 392) and by the *Tigiel Alp* to *Tinzen* (p. 392).

Above **Alvaneu** (1 M.) the road crosses the *Landwasser*, which falls into the Albula here, and ascends to the right to (1 M.) **Filisur** (3410'; *Hôt. Schönthal*, *Weisses Kreuz*, both plain), a pleasant village, commanded by the scanty ruins of *Greifenstein* (3985'). We then **descend** to the *Albula* and gradually ascend the thickly wooded valley on the **right bank**. Walkers should prefer the old road on the left bank of **the Albula**, which rejoins the road on the right bank above (2¼ M.) *Ballalūna* (3615'), a saw-mill (Inn, with a few beds). We then cross the *Stulser Bach*, ascend in a curve through wood, and enter the (1¼ M.) ***Bergüner Stein** (*Il Crap*, 4280'), a profound gorge with perpendicular sides. For 800 paces the road, constructed in 1696 and originally 4-6' wide, is hewn through the solid rock, being protected at places by a wall. The brawling stream at the bottom of the gorge is visible at one point only. At the end of the gorge, on the right, tower the *Tinzenhorn* (10,430') and the *Piz d'Aela* (10,960'), and we enter the green basin, enclosed by wooded hills, of (1½ M.) —

27 M. **Bergün**, Roman. *Bravuogn* (4475'; pop. 435; *Hôt. Piz Aela* or *Post*, R., L., & A. 2½-5, D. 3, pens. from 7 fr.; *Weisses Kreuz*, R. 2½, B. 1¼, D. 2½, pens. 6½-7½ fr.; *Edelweiss*; *Sonne*), a village with a mineral spring (small bath-house), an old Romanesque church, and a handsome prison-tower.

EXCURSIONS (guides, *P. Mettier* and *Albert Rauch*). Above Bergün, to the N.E., is the village of **Latsch** (5215'), on the slope of the *Latscher Kulm* (or *Cuolm da Latsch*, 7515'; ascent repaying, 2 hrs.). — Over the *Sertig Pass* to *Davos*, see p. 362. — Over the **Fuorcla Pischa** (9193') to *Madulein*, fatiguing, 9-10 hrs., with guide, through the *Val Tuors* and the *Val Plazbi*. From the pass, between Piz Kesch and Piz Blaisun, adepts may ascend the **Piz Kesch** (11,230') in 2 hrs. (but better from the *Kesch-Hütte*, over the *Porchabella Glacier*, in 2½-3 hrs.; comp. p. 362). — *Piz d'Aela* and *Tinzenhorn*, see p. 388. (The *Aela Club-Hut* may be reached from Bergün by the *Alp Uglix* in 3 hrs.) — Over the **Aela Pass** (9588'), between Piz d'Aela and Piz Val-Lung, to the *Val d'Err* and *Tinzen* (p. 392), viâ *Naz* (see below), 5 hrs. (with guide), interesting and not difficult.

We now ascend the beautifully wooded valley, passing the *Val Tisch* on the left. The Albula forms several small waterfalls and one of some size above the (3½ M.) Alpine hamlet of *Naz* (5725'). On the bold pinnacles to the right *(Piz d'Aela, Piz Val-Lung, Piz Salteras)* are seen patches of snow at places. The road ascends in long windings, past the chalets of *Preda* and *Palpuogna*, and on the right, below the road, the pale-green *Lake of Palpuogna*, to the (2¾ M.) **Inn** (D. 3 fr.) on the **Weissenstein**, Roman *Crap Alv* (6660'). It next describes a wide curve (footpath to the left much shorter) at the base of the two rocky horns of the *Giumels* (9137'), avoiding a marshy basin in which the Albula rises, and ascends the rock-strewn *Teufels-Thal* to the (2¼ M.) **Albula Pass** (7595'; *Hospice*, plain), lying between the summits of the Albulastock, the *Crasta Mora* (9635') on the right, consisting of granite, and the *Piz Uertsch* or *Albulahorn* (10,738'), on the left, being limestone.

The road now proceeds straight on through a dreary valley. Before us rises the Piz Mezaun, a fine pyramid; adjoining it on the right, at the head of the Val Chamuera, are the Piz Lavirum and Piz Cotschen; farther to the right are the Piz Muraigl and Piz Languard. We then begin to descend past several chalets and finally by seven long bends commanding fine views of the Piz Quatervals and Piz del Diavel, and afterwards of Ponte and Camogasc, with Madulein and Guardaval on the hill to the left. [The former bridle-path, first on the right, then on the left bank, is much shorter.] Traversing a larch-wood, we reach (2½ hrs., or 1½ hr. by the bridle-path) —

42 M. **Ponte** (5548'). Thence to *Samaden*, see p. 412; to *Schuls* and *Nauders*, see R. **102**.

99. From Coire to the Engadine over the Julier.

Comp. Maps, pp. 360, 382, 402.

DILIGENCE to Samaden in summer daily by Churwalden in 13 hrs. (20 fr. 85, coupé 25 fr. 5 c.), by the Schyn in 14¼ hrs. (22 fr. 70, coupé 27 fr. 45 c.). — EXTRA-POST and pair from Coire to St. Moritz 120 fr. 10 c., to Samaden 126 fr. 10 c. (or by the Schyn and Julier, 133 fr. 30 c. and 139 fr. 30 c.). — CARRIAGE and pair from Coire to St. Moritz over the Julier 100, to Pontresina or Samaden 110 fr. (by the Schyn and Julier 110 or 120 fr.); driver's fee 10 per cent of the fare.

Coire (1935′), see p. 354. By the Steinbock Hotel the road crosses the *Plessur* and ascends in windings (several short-cuts), with views of the town, the Rhine Valley, and the Calanda. To the E. opens the *Schanfigg-Thal* (p. 364), watered by the Plessur in its deep channel. A finger-post 1¼ M. from Coire indicates the route to the left to *Bad Passugg* (p. 356), and another, ¾ M. farther on, the way to the *Künzeli* (p. 356). We ascend the valley of the *Rabiusa*, which falls into the Plessur far below, and pass *Malix* (3800′; with a mineral spring) and the ruin of *Strassberg*.

6 M. **Churwalden** (4120′; **Krone*, R., L., & A. 3, D. 4, pens. 7-11 fr.; **Hôt. Gengel*, R. & A. 2½ fr.; **Hôt.-Pens. Mettier & Schweizerhaus*; *Pens. Hemmi*; **Hôt.-Pens. Rothhorn*, R. 1-2, L. ½, B. 1, D. 2½, board 4 fr.; **Weisses Kreuz*, R., L., & A. 1½-2½, B. 1, D. 2½, pens. 4-6 fr.), a health and whey-cure resort, with an old church and the former monastery of *Aschera*, lies picturesquely in a narrow valley.

The road ascends more rapidly; a pleasant path through wood runs parallel with it, on the left bank of the stream, which it crosses immediately before —

8½ M. **Parpān** (4957′; **Curhaus & Post*, R., L., & A. from 2½, pens. 7-9 fr.; **Hôt. Stätzerhorn*, pens. 6-8 fr.), a pleasant Alpine village in an open situation. The ancestral mansion of the Buol family, built at the end of the 16th cent., contains rooms in the mediæval style and various relics.

Pleasant walk to the (2 hrs.) Churer Joch (6656′), at the foot of the *Gürgaletsch*: view of Coire, the Rhine Valley as far as the Sentis, etc. The *Stätzer Horn* (*Piz Raschil*, 8160′; 3 hrs., without guide), a favourite point of view, the highest peak of the range between the valley of Churwalden and the Domleschg (see p. 378), is ascended from Parpan by the S. A. C.'s new bridle-path. Beyond the hamlet of *Sartans* straight on, avoiding the path to the right. Inn closed and falling to decay. Grand panorama of the valleys of Schanfigg, Churwalden, Oberhalbstein, Schams, Domleschg, and the Vorder-Rhein as far as Ilanz; of the entire Rhætikon Chain, Calanda, Tödi, St. Gotthard, Piz Beverin, Rheinwald Glacier, Piz Tambo, Bernina, Albula, etc. (Panorama by *A. Heim*.) Beautiful pastures and rare plants on the slopes. The descent on the Domleschg side is longer, and the last part is fatiguing, but cannot be mistaken; this route leads by the alps of *Raschil* and *Schall* to the chalets of *Almens*, and then to the left to *Scharans* and Thusis in the Rhine Valley (4 hrs. in all). Mountaineers may also descend by *Obervatz* to the *Solis Bridge* (p. 379).

From Parpan to *Arosa*, 4½ hrs., see p. 366.

On the top of the pass (5090′) we obtain a fine view of the Oberhalbstein Mts. to the S., the pyramid of the *Lenzer Horn*

(9548') and *Piz Michēl* (10,375') on the left, and the *Calanda* (p. 366) to the N. We descend to *Valbella* and *Canols*, pass several tarns and the *Heidsee* (4880'), surrounded by forest (*Chalet-Restaurant on an island, pens. 4-5 fr.), cross the wooded *Lenzer Heide*, Rom. *Planeira*, a region justly dreaded during snow-storms, to *Lai* (Post, pens. 4-5 fr.) and the (2³/₄ M.) *Curhaus Lenzer Heide* (4775'; pens. 5-6 fr.).

The *Aroser Rothhorn (9790'), a splendid point of view, is ascended hence by a new club-path in 3½ hrs. (guide 10 fr.; comp. p. 366). — The *Lenzerhorn (9550'; 3½ hrs.; guide) is also easy and attractive.

Travellers bound for the *Schyn Road* take the road diverging to the right at *Lai* (1/2 M. to the N. of the Curhaus), which leads over the *Heidbach* to (50 min.) *Obervatz* (4015'). We keep to the left before reaching Obervatz, so as to avoid the roads leading to *Lain*, which lies higher. Beyond Obervatz we descend abruptly viâ *Zorten* and *Nivaigl* to (40 min.) the *Solis Bridge* (p. 379).

14½ M. **Lenz**, Rom. *Lansch* (4285'; *Krone* or *Post*), an important military point before the construction of the Splügen route. The Duc de Rohan in 1635, and Lecourbe in 1799 took up a position here against the Austrians. *Albula Road to Bad Alvaneu* and *Bergün*, see p. 388.

Our road descends in numerous windings (avoided by short-cuts) to the Albula, overlooking the picturesque Oberhalbstein; in the foreground is the village of Alvaschein (p. 379); beyond the Schyn Pass lies Stürvis (p. 379); and far below is Tiefenkasten. Near the farm of *Vazerols*, to the right, below the road, is a small monument marking the spot where the Three Leagues took the oath of eternal union in 1471 (comp. p. 355).

17½ M. **Tiefenkasten**, more correctly *Tiefenkastell*, Roman. *Casti* (2790'; *Hôt. Julier*, R., L., & A. 1½-4, B. 1¼, D. 3, pens. 8-10 fr.; *Hôt. Albula*, R., L., & A. 3, B. 1¼, D. 3 fr.; *Rhätia*, plain), almost entirely rebuilt after the fire of 1890, lies picturesquely in the deep valley, with its church on a hill (2917') above the confluence of the *Julia* and the Albula. (To *Surava* and *Bad Alvaneu*, see p. 388; *Schyn Road to Thusis*, see p. 379.)

The road again ascends rapidly, and skirts the *Stein*, a bold limestone cliff (rock-gallery and tunnel). Far below flows the *Julia* or *Oberhalbstein Rhine*. (The Romanic word *Rhein* means 'flowing water'.) We next enter (4½ M.) the broad and populous part of the valley called the *Oberhalbstein (Sur Seissa)*, 5 M. in length, and pass the villages of *Burvein*, (1¼ M.) *Conters* (Post), and (³/₄ M.) Savognin or *Schweiningen* (4060'; *Hôt.-Pens. Pianta*, post and telegraph office, pens. 6 fr.; *Hôt. Piz Michel*, pens. 6 fr.; *Rhätia*). On the W. slope lie *Salūx*, *Präsāns*, *Reāms* (with a handsome castle, now a prison), and other villages.

EXCURSIONS. Piz Curver (9760'; 5 hrs.; guide), from Savognin by *Ziteil*, not difficult, a very fine point (see p. 380; descent to Zillis or Andeer). — FROM SAVOGNIN TO AUSSER-FERRERA OVER THE FIANELL PASS, 5½ hrs., easy and pleasant. A narrow road leads through the smiling *Val Nandrò* to the (2 hrs.) *Alp Curtins* (6400'); here we ascend to the right to the (1 hr.) *Alp Schmorras* (7500') and the (1 hr.) **Fianell** or **Schmorras Pass** (8850'), opposite the *Piz Grisch* (*Piz Fianell*, 10,000'); then descend by the *Alp Moos* and *Sutt Foina* to (1½ hr.) *Ausser-Ferrera* (p. 381). We next reach (1¼ M.) **Tinzen**, Rom. *Tinizung* (4070'; Hôt. *Tinzenhorn*), prettily situated at the mouth of the *Val d'Err*. In the background rise *Piz Val-Lung* and *Piz d'Aela* (p. 388).

From **Tinzen** to *Bergün* over the *Aela Pass*, 4 hrs., see p. 389. To the N. a trying route (5 hrs.; with guide) crosses the **Tinzenthor Pass (8465')**, between the Piz Michel and the Tinzenhorn, to *Bad Alvaneu* (p. 388). — Piz Michel (10,375'; 6 hrs.; with guide), more difficult from here than from Alvaneu (p. 388). — To Samaden over the Errjoch (10,270'), 9 hrs., with guide, laborious, but repaying. Ascent through the picturesque *Val d'Err* and over the *Err Glacier* to **the pass**, lying to the N.E. of the *Piz d'Err* (see below); descent through the *Val Bever* (p. 401).

Above Tinzen the Julia forms several fine waterfalls. The road leads alternately through curious rounded basins, probably formed by erosion, and picturesque rocky ravines. We next reach (1½ M.) *Roffna* (4760'; Löwe, plain), and (2¾ M.) —

29½ M. **Molins**, Ger. *Mühlen* (4793'; *Löwe*, R. 2½, D., incl. wine, 3½-4 fr.), beautifully situated, where the diligence halts for dinner.

From the **Val da Faller**, which debouches here and divides into the *Val Gronda* and the *Val Bercla* ¾ hr. farther up, routes little used (guide) cross the *Val Gronda Joch* (9193'), on the E. of the *Weissberg*, to (6 hrs.) *Cresta* (p. 381), and the *Fallerjoch* (about 9090'), past the *Flüh Lakes*, to (5½ hrs.) *Juf* in the Averser-Thal (p. 384). — The **Piz Platta** (11,110'), ascended through the *Val Faller* and *Val Bercla* in 5½ hrs. (guide), commands a splendid view. — *Piz d'Err* (11,138'), *Piz d'Arblatsch* (10,512'), and *Piz Forbisch* (10,690'), for experts (guides at the 'Löwe').

The route from this point to Stalla, skirting the rapid Julia, presents a succession of grand rocky landscapes. One of the finest points is near the bridge before (¾ M.) *Sur* is reached. On a beautiful wooded hill, in the middle of the valley, stands the square watch-tower of *Splüdatsch* (5260'; path to it beyond Sur; fine view). On the right, ¾ M. farther on, appears the ruined castle of *Marmorera*, partly built in a rocky cavity halfway up the hill. The next villages are (1¼ M.) *Marmorera* (*Marmels*, 5360'), **at the mouth of the** *Val Natons; Stalvedro* (5613'); and (3 M.) —

34½ M. **Stalla** (5827'; *Post*), or *Bivio*, the Roman *Bivium*, where the Julier and Septimer routes separate.

The SEPTIMER ROUTE, a bridle-path (to **Casaccia** 4 hrs.; guide unnecessary in fine weather), one of the oldest **Alpine** routes, anciently traversed by Roman and German emperors **with their** armies, diverges to the right from the road above Stalla, and ascends the *Val Cavreccia*. At the chalets of (1 hr.) *Cadnal* it crosses the brook, enters a defile, and ascends the somewhat marshy meadows of *Pian Canfer*, to the (1 hr.) Septimer Pass (*Passo di Sett*; 7582'), with a dilapidated hospice. (Over the *Forcellina* to *Juf*, and by *Lunghino* to the *Maloja*, see p. 395.) A height to the left of the pass, indicated by two stones, commands a magnificent view of the mountains of the Maloja, Piz della Margna (10,355'), Monte dell' Oro (10,544'), etc. Descent by a rough paved path, crossing the *Septimer Bach* (*Acqua di Settimo*) three times, to the valley of the *Mera*, and on its left bank, the latter part very steep and stony, to (2 hrs.) *Casaccia* (p. 422).

From **Stalla** to *Andeer*, over the *Stallerberg* and through the *Averser Thal* and *Val Ferrera*, see p. 381. — To SILS over the Fuorcla di Gravasalvas (3806', with guide), 5½ hrs., interesting. Below the Julier Pass we ascend to the right, past the small *Gravasalvas Lake*, to the pass, on the W. side of the *Piz Lagrev*, with a fine view of the Bernina, etc.; then a steep descent to the *Lake of Sils* (p. 396).

The road, completed in 1827, ascends the stony slopes of the Julier *(Giulio)* in numerous windings (carriages ascend in 2 and descend in 1 hr.). Walkers reach the pass in 1¾ hr. From November to the middle of May the mountain is usually crossed by sledges, though the Julier is clear of snow before any other pass of equal height, and the least exposed to avalanches. A little on this side of the summit are a few houses (7360') including a rustic inn. On the (38½ M.) pass (7500') are two round milestones of mica-slate, 5' in height, without inscription, erected in the time of Augustus, who constructed a military road from Clavenna (p. 384) to the Curia Rætorum (Coire) over the Maloja and the Julier. Roman coins have also been found here. Near the milestones, to the right, is a small clear lake, which contains trout notwithstanding its great height.

On the E. slope of the Julier, 1 M. from the top, lies the small *Julier Alp*, with two chalets. On the left rise *Piz Julier* and *Piz d'Albana*, and on the right *Piz Pulaschin* (p. 398). In descending we soon obtain a superb view of the snow and ice mountains of the Bernina (p. 403). In the foreground rise Piz Surlej and Mt. Arlas, above which tower Piz Tschierva, Piz Morteratsch, Piz Bernina, and Piz Corvatsch and Piz della Margna on the right. The Upper Engadine, with its green lakes, comes gradually into full view.

44½ M. *Silvaplana* (5958'), and thence to —
52 M. *Samaden* (5670'), see pp. 397-401.

ENGADINE.

The *Engadine (Rom. *Engiadina*), a valley 60 M. long, and seldom more than 1 M. broad, descending from S.W. to N.E., and watered by the *Inn*, is bounded by lofty mountains, partly covered with glaciers and snow. The *Upper Engadine*, between the Maloja and Samaden, with its numerous lakes and the valley of Pontresina, is the most attractive part of the valley, while the *Lower Engadine* (R. 102), below Samaden, is less picturesque. The strong and bracing air of the Upper Engadine makes that region one of the most famous health-resorts in the world.

The temperature rises in summer to 66-76° Fahr. in the shade; in winter the thermometer frequently falls to 30-40° below zero. 'Nine months winter and three months cold', is the laconic, but rather exaggerated account the natives give of their climate. Very abrupt changes in the temperature, and even white frosts and snow are by no means uncommon in August, so that winter-wraps should not be forgotten by those who purpose to spend even a few weeks here. As the Upper Engadine is crowded in summer, rooms had better be ordered beforehand. — **Heavy luggage** may be forwarded through a goods-agent, e.g. *Messrs. Bavier, Kient, & Co.*, of Coire and Silvaplana.

At first sight the bottom of the Upper Engadine resembles a vast and almost treeless meadow. The lower slopes of the mountains are chiefly clothed with the larch and the *pinus cembra*, or Swiss stone-pine (Ger. *Arve*), a stately tree, sometimes called the 'cedar of the Alps', but commoner in the Pyrenees, the Carpathians, and the south of Siberia than in Switzerland. Its light, close-grained wood, which is white in colour and has a pleasant fragrance, is extremely durable, and is much esteemed for cabinet-work. The kernels (30 to 40) of the cones, enclosed in a very hard triangular shell, have a pleasant flavour, not unlike that of the pine-apple.

The Engadiners, a sober, industrious, and frugal race, are with few exceptions, Protestants. The Romanic mother-tongue renders all the Romance languages comparatively easy to them, while they are taught German in the schools from the age of ten. They frequently emigrate in early life to different parts of Europe, where they earn their living as confectioners, coffee-house keepers, makers of liqueurs and chocolate, etc.; and when they have amassed a competency they usually return to their native valleys to spend the evening of a busy and active life. To persons of this class belong many of the comfortably furnished and neat white houses in the Engadine. The windows are made small to exclude the cold. The pasturage is excellent, but is seldom in the hands of the inhabitants, being let by them to Bergamasque shepherds, who spend the summer here with their flocks (paying 1 fr. for each sheep) and sell in autumn the long wool to the manufacturers of Bergamo. The hay in the meadows is also collected by Italian reapers.

100. The Upper Engadine, from the Maloja to Samaden.

Comp. Map, p. 402.

15 M. DILIGENCE twice daily in 3 hrs., comp. p. 422. OMNIBUS from Maloja to Sils in 1 hr., on Mon., Wed., and Frid. at 6 p.m.; to St. Moritz daily in 1½ hr., at 6.30 p.m. and 5.30 a.m. (3 fr.; there and back 5 fr.).

The *Engadine* (comp. p. 393) begins at the summit of the **Maloja**, or *Maloggia* (5960′), the lowest pass between Switzerland and Italy, which ascends gently from the Engadine, and descends suddenly on the W. side to the *Val Bregaglia* (p. 422). A little before the summit, on the S. side, is the *Hôtel Maloja-Kulm* (R. 2 fr.), opposite a projecting rock commanding a beautiful view of the Val Bregaglia, and beyond it is the (1/4 M.) *Hôt. Osteria Vecchia*, in the Swiss style (R. from 2¹/₂, pens. 7-9 fr.). To the left, higher up, is the unfinished *Château of Count Renesse* (6128′), with extensive grounds commanding splendid views of the Val Bregaglia; the finest walk is the 'Chemin des Artistes' (from the Cursaal and back in 1½ hr.). Farther on are some private villas in the Swiss style, and the *Hôtel Longhin* (pens. 6-6¹/₂ fr.). To the right of the road, at the upper end of the Lake of Sils, is the large *Hôtel Cursaal-Maloja* (R., L., & A. 6-9, déj. 4, D. 5, board 8, music ½ fr.; open from June 1st to Sept. 30th), a first-class establishment owned by a Belgian company. English Church Service. The view comprises to the E. the Lake of Sils, the Piz Mortel, and Piz Lagrev; to the N. the Piz Gravasalvas, N.W. Piz Lunghino, W. the mountains of the Septimer, S.W. Piz Grande, Piz Cacciabella, and Mte. Zocca; to the S., between Pizzo Salecina and Piz della Margna, in the Val Cavloccio, the Monte del Forno and behind it the beautiful white Cima di Rosso.

EXCURSIONS. Below the pass, a little to the W., a footpath, and ½ M. farther on a cart-road diverge to the left from the Maloja road, cross the *Orlegna* (waterfall, see below) near the lowest houses of the hamlet of *Ordeno*, and ascend on the left bank through meadows and wood to the

Engadine. MALOJA. *VI. Route 100.* 395

(50 min.) sequestered, dark-blue *'Cavloccio Lake* (6243'), surrounded by lofty mountains; to the S. the finely shaped *Monte del Forno* (10,545'); to the left of it the snowy Muretto Pass (see below). The large *Cavloccio Alp*, at the S. end of the lake, is occupied in spring and autumn only; in the height of summer the cattle are pastured on the higher alps. From this point to the Forno Glacier and back, 2 hrs. (see below). — On the other side of the Orlegna bridge (see above) a path (finger-post) leads to the left to the (40 min.) little *Lago di Bitabergo* (6110'), and thence to the (3/4 hr.) **Motta Salecina** (7055'), at the foot of the *Pizzo Salecina* (8500'), with a fine view of the Bregaglia and the Upper Engadine.

The **Orlegna Fall** is reached by descending the windings of the Maloja road to a (1 M.) finger-post, and diverging by a path to the left, which leads to a (2 min.) rocky plateau above the chief fall.

A pretty walk leads to the E. on the S. bank of the lake, diverging to the right (finger-post, 'Pian Cunchetta') from the **path to Isola**, to (3/4 hr.) *Aira della Palsa* (6645') and thence to (20 min.) *L'Ala* (7000'), with fine view.

To the **Forno Glacier** (guide advisable, to the Forno Hut 10, to the glacier circus 15 fr.; Jac. Uffer, Agost. Clalüna), also interesting. We follow the Muretto route (p. 388) to the (1½ hr.) *Alp Piancanino* (6520'); then ascend to the right (before the bridge) for 3/4 hr. over turf and moraine to the *Forno Glacier*, which we cross to (1¼ hr.) the *Forno Hut* (about 8200'), on a projecting rock on the W. side of the glacier, opposite the Mte. del Forno. Imposing amphitheatre of glaciers, commanded by the Piz Bacone, Cima di Cantone, Cima di Castello, Pizzo Torrone, Mte. Sissone, Cima di Rosso, and Monte del Forno. — The *Piz Bacone* (10,637'; 2½-3 hrs.; 35 fr.), *Cima di Castello* (11,158'; 3½ hrs.; 35-40 fr.), *Pizzo Torrone* (10,825'; 3-3½ hrs.; very difficult; 70 fr.), *Monte Sissone* (11,030'; 3-3½ hrs.; 30 fr.), and *Cima di Rosso* (11,045'; 3 hrs.; 30 fr.) may be ascended hence. *Monte della Disgrazia* (12,050'; 75-80 fr.), see p. 421. — Over the **Forno Pass** (about 10,500'), between the *Pizzo Torrone Orientale* (10,825') and the *Monte Sissone*, to the *Val di Mello* and the *Bagni del Masino*, 11 hrs. from Maloja, for experts only, with good guides (50 fr.), see p. 422.

Piz Lunghino (9120'; 3 hrs., guide 10 fr.), remunerative. From the Hôtel Longhin a bridle-path ascends to the left over pastures to the (2 hrs.) blue *Lunghino* or *Longhin Lake* (8136'), from which the Inn emerges; footpath thence over rocks and stones (guide necessary for novices) to the top. Splendid view.

From the **Maloja to Cresta in the Averser-Thal**, 7½ hrs., an attractive route (guide unnecessary for adepts provided with Siegfried's map). From the (2 hrs.) Lunghino Lake (see above) a footpath leads to the W. to the (½ hr.) *Longhin Pass* or *Forcletta di Lunghino* (8645'), whence we descend (no path), leaving the *Motta da Sett* or *Septimerberg* (8645') to the left, to the (1 hr.) *Septimer Pass* (p. 392), where we cross the old Septimer route. A footpath ascends hence, at first keeping somewhat to the left and then following the right side of the ravine (cairns), to the (1½ hr.) Forcellina **Pass** (8790'), where we obtain a view of the Averser-Thal. We descend to the right to the floor of the valley, at first gradually, then in rapid zigzags; 1¼ hr. *Juf*; 1 hr. *Cresta* (p. 384).

From the **Maloja over the Muretto Pass to Chiesa** in the *Val Malenco*, 7 hrs. (guide 20 fr.), a toilsome but repaying route. To the (1 hr.) *Cavloccio Alp*, see above. Hence a new path ascends to the (25 min.) *Piancanino Alp* (6520'), situated at the confluence of the Forno and the Muretto (foot-bridge over the former). A steep and fatiguing ascent over rocky debris, along the small *Muretto Glacier*, and over snow leads hence to the (1½ hr.) Muretto Pass (8390'), between the *Mte. del Forno* (10,545') and the *Mte. Muretto* (10,197'), where we get a fine survey of the grand *Mte. della Disgrazia* (p. 421). Descent over snow, then by a rough path over stony and grassy slopes on the left bank of the wild *Matero*, with admirable views of the Mte. della Disgrazia, the Mte. Sissone, Cima di Rosso, etc., to the *Chiareggio Alp* (5478'; quarters), and by a road passing numerous slate-quarries to (4 hrs.) *Chiesa* (3287') in the *Val Malenco* (p. 421).

FROM THE MALOJA TO PROMONTOGNO OVER THE CASNILE AND CACCIABELLA PASSES (14 hrs., guide 35 fr.), most interesting, traversing the grand Bregaglia Mts. (fatiguing, but for experts not difficult). To the (8½ hrs.) *Forno Hut* (about 8200'), where the night may be spent, see p. 395; thence to the right to the (1½ hr.) **Passo di Casnile** (9744'; superb view). Descent across snow, through a couloir, and over rock, to the foot of the *Cantone Glacier*, and then across two moraines to the (1½ hr.) *Albigna Glacier* (through the Val Albigna to Vicosoprano, see p. 423). We next ascend the steep and stony slope of *Cacciabella* ('fine hunting'; a resort of chamois) to the (2 hrs.) **Passo di Cacciabella** (9444'), another fine point of view, and descend to the (2 hrs.) *Alp di Sciora* (6785') and through the wild *Val Bondasca* (p. 423) to the (2½-3 hrs.) *Hôtel Bregaglia* (p. 423). — In coming from Promontogno (14-15 hrs. to the Maloja) it is advisable to spend the night at the (4 hrs.) *Alp Sciora* or, if that be empty, at the *Alp Naravedro*, 3 hrs. from Promontogno; comp. p. 423.

At the Cursaal we cross the infant *Inn*, here called *Ova d'Oen*, which descends in cascades from the *Piz Lunghino* (p. 395) to the W., and at the chalets of *Capolago* reach the pale-green **Lake of Sils**, Rom. *Lej da Segl* (5890'), 4½ M. long and 240' deep, the N.W. bank of which we follow. Walkers may take the path (fingerpost 'Pian Curtinatsch') leading along the S.E. bank, passing the hamlet of *Isola*, which lies on a green plateau at the mouth of the *Fedoz* (from the Cursaal ¾ hr.; to Sils-Maria 1½ hr.). In the gorge near Isola the Fedoz descends in a beautiful fall. Above Isola appears the beautiful Piz Corvatsch (p. 408), beyond the *Crap da Chüern*, a rocky promontory which divides the lake into two basins. As we approach the peninsula of Chastè (see below), the rifted Fedoz Glacier, at the head of the Val Fedoz, appears to the S., above Isola. At the E. end of the lake lies (1½ hr.) —

4½ M. **Sils** (5895'), Rom. *Segl*, embracing the hamlets of *Sils-Baseglia* (with the diligence-office), immediately to the right of the road, at the foot of the precipitous *Piz Lagrev* (10,400'), and *Sils-Maria*, ½ M. to the S., pleasantly situated among low larch-covered hills, through which the Fex flows. The wooded peninsula of *Chastè* (castle), which stretches into the lake between the hamlets, bears traces of an ancient castle. Sils-Maria (*Alpenrose, R., L., & A. 3½-6, D. 4, S. 3, board 7 fr.; **Hôt. Edelweiss*, R., L., & A. 2½-5½, D. 4, pens. 8½-12 fr.) is well adapted, on acount of the numerous shady walks in the vicinity, for a residence of some time.

OMNIBUS from Sils-Maria to St. Moritz daily at 7 a.m., returning at 10.30 a.m. (on Tues., Thurs., Sat., and Sun. also at 2 p.m., returning at 5.30 a.m.), in 1 hr.; to the Maloja Hotel on Mon., Wed., and Frid. at 2 p.m., returning at 6 p.m., also in 1 hr. Fare for each route 1½, there and back 2½ fr. — CARRIAGE with one horse from Sils to St. Moritz 10, to Pontresina 15 fr.

WALKS. Immediately to the E. of the Hôtel Alpenrose is the *Muot Maria*, a small hill with view. The three chief points of 'View among the low larch-covered hills (behind and to the W. of the hotel), over which passes the narrow road to the valley of Fex, beginning at the bridge over that stream, are the *Laret-Höhe* (¼ hr., in the direction of Silvaplana), the *Bellavista* (20 min., in the direction of the Maloja), and a bench on the Fex road (20 min.; view of a fall on the Fex and over the wooded hills in the foreground to the snow and ice-covered mountains of the Fex valley beyond). — The ascent of the *Muot Marmoré* (about 7220'), a rounded

Engadine. SILVAPLANA. *VI. Route 100.* 397

spur of the rugged *Furtschellas* (9620'), forms an attractive and easy excursion (1¼ hr. from the Hôtel Edelweiss). The Piz Corvatsch adjoins the Furtschellas on the E. — Pleasant walks lead eastward from the Hôtel Edelweiss along the wooded slopes to a saw-mill, and thence to (1¼ hr.) Surlej. — Another fine view may be enjoyed from the *Plaz* (6240'), a projection on the **slope** of the Piz Lagrev, to which a path, nearly opposite the bridge **over the** Inn at Sils-Baseglia, ascends in 20 min. The view towards the Maloja is best in the morning, towards the Fex Valley and the Piz Corvatsch in the evening.

The "Fex Valley *(Val Fex* or *Schafthal)* may be visited from Sils-Maria in 4-5 hrs. (there and back). The narrow carriage-road ascends the left bank of the Fex, while a shorter footpath follows the right bank. Beyond the bench mentioned above the road descends to the farm of *Vaüglia*, but re-ascends, leaving the houses of *Platta* on the left, to the little church of (50 min. from Sils) *Crasta*, shortly before which it is joined on the left by the above-mentioned footpath (recommended as a return-route). A *View-Bench, about 3 min. beyond the church, affords on fine evenings perhaps the most satisfactory view of the mountain-amphitheatre forming the background of the valley. Those who are pressed for time may turn here. The road crosses the stream, and reaches (¼ hr.) the *Restaurant zur Edelweisshalde*, and ½ M. farther on, beyond the hamlet of *Curtins* (6480'), the *Restaurant Philipp*. Beyond (10 min.) a ruined house, we recross the Fex, and in 20 min. (ground marshy at places) reach the top of the *Muot Selvas*, an old moraine-hill, projecting obliquely into the valley, and affording an excellent survey of the beautiful *Fex Glacier*, surrounded by the Chapütschin, Piz Tremoggia, Chapütsch, Piz Fora, Piz Güz, and Piz Led. Below us the Fex emerges from its broad stony bed. In the opposite direction is the green Fex Valley, with the indented chain of Piz Lagrev and Piz Pulaschin in the background.

A path (guide unnecessary) ascends to the right from the church of Crasta to an alp, then leads to the left through larch-wood to the **(1½ hr.)** *Muot Ota* (8065'), which commands a view of the Fex and Fedoz Glaciers. The view is still better higher up, on the way to the *Plaun Grand* (8200'). — The path to the *Fedoz Valley* diverges to the S.E. from the carriage-road to the Fex Valley, at a point about 100 paces to the S. of Vaüglia; to the Fedoz Chalet, ¾ hr.

MOUNTAIN ASCENTS (guides. *Chr. Klucker, J.* and *A. Eggenberger*). The *Piz Led* (10,135'; 4 hrs.; guide 10 fr.), *Piz della Margna* (10,855'; 4½-5 hrs.; 16 fr.), *Piz Chapütschin* (11,180'; 4½-5 hrs.; 15, returning by Pontresina 25 fr.), and *Piz Tremoggia* (11,322'; 5-6 hrs.; 13 fr.) may be ascended from Sils by adepts without difficulty. More toilsome ascents are those of the *Piz Glüschaint* (11,800'; 5½-6 hrs.; 30 fr.), *Piz Fora* (11,053'; 6-7 hrs.; 20 fr.), and *Piz Corvatsch* (5 hrs.; 14, returning by Pontresina 16 fr.; more trying from Sils than from Pontresina. see p. 408).

FROM SILS TO PONTRESINA over the *Fuorcla Fex-Roseg* (18 fr.), the *Fuorcla Chapütschin* (30 fr.), or the *Fuorcla Glüschaint* (35 fr.), see p. 409. — To MALENCO over the *Fex Glacier* and the **Tremoggia Pass** (9910'; 25 fr.), between the Chapütsch and Piz Tremoggia, or over the **Fuorcla** Fex-Scerscen (10,236'; 40 fr.), between Piz Tremoggia and Piz Glüschaint, both suited only for mountaineers (9-10 hrs.); descent over the *Scerscen Glacier*; then steeply, to the W. of Mte. Nero, to the *Val Entova* and *Chiesa* (p. 421).

Beyond Sils-Baseglia the road (in shade in the afternoon), skirting the foot of the *Piz Pulaschin* (9900'), follows the left bank of the artificial channel of the Inn and that of the *Lake of Silvaplana* (5885'), 2 M. long, to (2¾ M.) Silvaplana. Walkers may leave Sils-Maria to the N., and follow the path over the meadows, then skirt the larch-clad hill, crossing several brooks, and finally passing a pretty *Waterfall* of the Surlej brook, to (1¼ hr.) Surlej; thence they may proceed to St. Moritz, viâ Crestalta.

7¼ M. **Silvaplana** (5958'; **Wilder Mann & Post*, R., L., & A.

2½-4, D. 8½, pens. 7-10 fr.; *Hôt. Corvatsch, to the W. of the village, pens. from 7 fr.; *Sonne, plain), where we reach the Julier road (R. 99), lies pleasantly on a green pasture, on the alluvial deposits of the brook descending from the Julier, which separate the lakes of Silvaplana and Campfèr. Opposite, on the E. side of the valley, is the village of *Surlej* ('above the lake'), destroyed by a torrent in 1834. It possesses a chalybeate spring.

To PONTRESINA OVER THE FUORCLA SURLEJ, 7-8 hrs. (guide, not required by adepts, 10, horse 20 fr.), a bridle-path, very attractive. Beyond the church of Surlej (see above), we do not turn to the left (route to Crestalta) but keep straight on, soon cross the brook to the right, and ascend into the wood; 1 hr. *Alp Surlej* (6976'); then to the S. over a pasture, towards the Piz Corvatsch. Above a second chalet the path turns to the left, and, near the *Corvatsch Glacier*, reaches the (2½ hrs.) *Fuorcla Surlej (9010'; splendid view), between *Piz Corvatsch* (p. 408; ascended from the pass in 2½ hrs.) and *Mt. Arlas*. Descent by a good path to the (½-¾ hr.) *Alp Surovel* (7424'; milk) and the (½ hr.) *Restaurant du Glacier*, in the Roseg-Thal, 1¾ hr. from Pontresina (p. 405). — From the Baths of St. Moritz a good bridle-path leads to (3½ hrs.) the Fuorcla Surlej, either viâ the Quellenhügel or the Johannisberg and the (1¼ hr.) *Hahnensee* (Restaurant, high charges). Comp. p. 399.

Piz Julier (11,105'), 5 hrs. from Silvaplana (guide 20 fr.), trying. An interesting descent (for adepts only) may be made to the S.E. viâ the *Julier-Scharte* (between the Piz Julier and Piz d'Albana) to the *Val Suvretta* (p. 401; to St. Moritz 4 hrs.). — Easier, but less interesting, is *Piz Pulaschin* (9900'; 3½ hrs., with guide).

The Silvalana Lake is connected by a hannel with the small *Lake of Campfèr*, which is bisected by a promontory. The road skirts the W. bank of the latter Opposite rises the wooded cheight of *Crestalta* (6250'; Restaurant, mediocre), 1 M. from Silvaplana, which affords an admirable view of the lakes and mountains. (Footpath to St. Moritz, ¾ hr.) Below the Campfèr Lake the Inn takes the name of *Sela* until it enters the Lake of St. Moritz.

8½ M. Campfèr, Rom. *Chamfèr* (6000'; *Hôt. Julierhof; *Hôt. d'Angleterre; Pens. Cazin; Engl. Ch. Serv.). The road divides here. The S. branch, on which the diligence runs in summer, crosses the Inn and leads by *Bad St. Moritz* (station) to (3 M.) *Dorf St. Moritz*, while the N. road, ½ M. shorter, runs high above the Inn, on its left bank, and below the Lower Alpina (p. 399), to the village.

10 M. Baths of St. Moritz. — °CURHAUS (*Grand Hôtel des Bains*), with upwards of 250 beds; board 8 fr., R. for 1-2 pers. usually 10 fr. per day; °NEUES STAHLBAD (*Gr. Hôtel des Nouveaux Bains*), with 250 rooms, handsomely fitted up, with covered promenade, etc.; from both of these visitors can go to the baths and the spring under cover in bad weather. °HÔTEL VICTORIA, opposite, with *Villa Beausite*, R., L., & A. 7-8 fr. and upwards. A few paces farther on, on the left bank of the Inn, °HÔTEL DU LAC, R. 5-12, L. ½, A. 1, B. 1¼, D. 6, pens. from 15 fr.; these four first-class. — HOF ST. MORITZ; °ENGADINER HOF, Nearer the village: *HÔTEL & GRAND CAFÉ CENTRAL (Munich beer), pens. 8-12 fr.; *HÔTEL BELLEVUE AU LAC, with *Villa Monplaisir*, R., L., & A. 10-11 fr. — PENSIONS. Near the Curhaus, *Villa Pidermann-Brugger; near the Hôtel Central, *Edelweiss*, *Flütsch*. — Band several times daily; soirées dansantes for the guests of the first four hotels twice weekly.

BATHS. In the *Curhaus* (in the long wing) 7-10 a.m. 2 fr., 10 a.m. to noon 2½ fr., 12-6 p.m. 1½ fr.; in the new filled baths, 7 to 9.30 a.m. 3, 9.30 to 1 p.m. 4 fr.; Turkish bath 2 fr.; tickets at the post-office in the Cur-

Engadine.   ST. MORITZ.   *VI. Route 100.* **399**

haus. In the *Neues Stahlbad*: 7-9 a.m. 2½ fr.; 9-10 a.m. 3 fr., 10-1 p.m. ¾ fr.; subscription for mineral water 15 fr. — PHYSICIANS: *Dr. Holland* (p. 400), *Dr. Barnard* (English), *Drs. Berry*, *Nolda*, *Hössli*, *Zangger*, and *Veraguth*.

CARRIAGES. To the *Meierei*, with one horse for 1-2 pers. 5, 3 pers. 7 fr.; to the *Village of St. Moritz* or *Campfèr* with one horse 2-3, with two horses for 4 pers. 4, 5 pers. 5, 6 pers. 6 fr.; to *Pontresina* one-horse 9-11, two-horse 20, 24, 28 fr.; to the *Morteratsch Glacier* one-horse 12-15, two-horse 25, 29, 38 fr.; to the *Roseg Glacier* one-horse 18-22 fr.; *Silvaplana*, in the forenoon, one-horse 5-6, two-horse 16, 20, 24 fr.; afternoon, one-horse 7-9, two-horse 20, 24, 28 fr.; *Sils* one-horse 9-11, two-horse 20, 24, 28 fr.; *Fex Valley* one-horse 16-19 fr.; the *Maloja* one-horse 12-15, two-horse 24, 29, 38 fr. Fee 10 per cent of the tariff.

OMNIBUS to *Sils-Maria*, see p. 396; to the *Maloja*, see p. 394; to *Samaden* at 11 a.m., in 1 hr.; to *Pontresina* at 2 p.m. in 1 hr., 2 fr., there and back 3 fr.; to the *Morteratsch Glacier* in 1½ hr., 2½ fr., there and back 4 fr.

ENGLISH CHURCH (see below).

The *Baths of St. Moritz* (5805') owe their importance to the mineral springs rising at the foot of Piz Rosatsch, strongly impregnated with carbonic acid and alkaline salts, pronounced the best of its kind in Europe by Paracelsus as early as 1539, and annually resorted to by numerous patients of all nations. The water is used for drinking as well as bathing. The season is from the middle of June to the middle of September. Patients will find warm clothing necessary; comp. p. 393.

The scanty grounds in front of the Curhaus are adjoined by a broad street, with several fine shops, which leads past the Hôtel Victoria and the *Post Office* to the lake and the village. To the right, at the foot of Piz Rosatsch, is the *Neues Stahlbad*. On the lake lies the *Casino St. Moritz*, with concert, reading, and conversation rooms, café-restaurant, etc. To the right, beyond the Inn, rises a new *Roman Catholic Church*.

Behind the E. wing of the Curhaus, promenades, passing the *French Protestant Church*, ascend the (20 min.) pine-clad *Quellenhügel*, and lead thence to the (¾ hr.) *Johannisberg*, commanding a pretty view of St. Moritz (the bridle-path proceeds to the Fuorcla Surlej, p. 398). — Another walk leads on the S. bank of the *Lake of St. Moritz*, or over the hill at the foot of the Rosatsch, to the (½ hr.) *Meierei* (dairy) or *Acla Silva* (Restaurant, an afternoon resort), on the way to Pontresina. — To the (35 min.) *Lower Alpina* (Restaurant, dear) a path ascends to the right just beyond the upper Inn bridge, ¼ M. to the S.W. of the Curhaus. Higher up is the (20 min.) *Upper Alpina* (Restaurant). — A guide-post above the Curhaus, to the N., indicates the way to the 'Wald Promenade', which follows the slope above the road (see p. 393) between the Alpina and the village. — To the (¾ hr.) *Crestalta* (p. 398) a pleasant wood-walk ascends to the S.W. from the Curhaus on the right bank of the Inn.

On the road from the Baths to the Village of St. Moritz is the little *English Church*, in the round-arch style.

11¼ M. **Village of St. Moritz.** — HÔT.-PENS. ENGADINER KULM, an extensive pile of buildings at the upper end of the village, with a fine view and every convenience for both summer and winter, patronized by the English and Americans; high charges, board from 10½ fr. R. in summer

8-10, in winter 1-7 fr. — At the end next the Baths, *Hôt. Bavier zum Belvedere. R. L., & A. 3½, lunch 3, D. 4½, pens. from 10½ fr. In the village: *Hôt.-Pens. Caspar Badrutt, R. from 3, L. & A. 1, lunch 3, D. 5, pens. from 12 fr.; Steffani, R. 1½-3½, L. & A. 1, lunch 2½, D. 3, pens. 8-10 fr.; Hôtel-Pens. Suisse; Hôt.-Pens. Veraguth, pens. 7 fr.; Hôt.-Pens. National; Hôt.-Pens. Helvetia, with restaurant and confectioner's; Hôt.-Pens. Wettstein; Hôt.-Pens. Rosatsch; Hôt. Petersburg, a little below the Kulm Hotel, with good view, R. 3-4, L. & A. 1, lunch 3, D. 5, pens. from 12 fr.; Hôt.-Pens. Beaurivage, in an open situation, overlooking the lake, R. 3-4, L. ½, A. 1, D. 5, pens. from 12 fr.; adjacent, Sonnenegg, an unpretending inn. — Pensions, beginning from the lower end: *Rhaetia*, *Villa Berry*, *Joos. Flugi, Schmidt, Gartmann, Villa Grünberg, Pidermann; Villa Languard*, next the Kulm Hotel; *Tagnoni-Badrutt* (private hotel), finely situated above the lake. Outside the village, on the Samaden road, *Zum Bären & Pens. Stecher*. — English Physician: *Dr. Holland* (in summer and winter). — *English Church*, see p. 399.

Carriages. With one horse to the Curhaus for 1-2 pers. 2, 3-4 pers. fr.; with two horses for 3-4 pers. 4. for 5 pers. 5 fr.; to *Campfèr* 5-6 or 10-12 fr.; to the *Alpina* viâ Campfèr 6-7 or 13-14 fr.; to *Samaden* 6-8 or 11-15 fr.; to *Pontresina* 8-10 or 15-18 fr.; to the *Rosegg Glacier*, one-horse 16-22 fr.; to the *Morteratsch Glacier* 12-14 or 22-25 fr.; to the *Bernina Houses* 14-16 or 25-28 fr.; to the *Bernina Hospice* 20-24 or 34-36 fr.; to *Poschiavo* 40 or 70-80 fr.; to the *Maloja* 12-16 or 21-27 fr.; to *Chiavenna* 45 or 70-90 fr.; to *Coire* 70 or 110-120 fr.; fee 10 per cent of the tariff. — Omnibus to the Maloja daily at 10 a.m. in 1¾ hr. (8 fr., there and back 5 fr.). Omnibus for patients in the forenoon between the village and the baths every hour.

Guides' Tariff given in the different excursions. *Wieland Wieland, Dan. Schlegel, Alex., Abr., and Stephan Wieland, Flor. Grass, Joh. Luzi, Barth. Schocher*, etc., may be recommended as **guides**. — Trespassing on the meadows before hay-harvest is punishable by a fine.

St. Moritz, Rom. *San Murezzan* (6090'; **pop. 822**), the highest village in the Engadine, 130' higher than the Maloja, lies on a slope to the N. of the *Lake of St. Moritz*, which abounds in trout, and commands a fine view of the mountains, from the Piz Languard westwards to the Piz Julier, particularly of the Piz Surlej, with its glacier, the Piz Corvatsch and (farther distant) the Piz della Margna. The majority of visitors are English or American; Italians are also numerous. Several hundred patients usually spend the winter here, which they enliven with skating and tobogganing.

A guide-post at the W. end of the village indicates the way to the 'Wald-Promenade', which leads in 25 min. to the *Alpina* (p. 399). — From the centre of the village a road descends to the S.E. past the Hotel Beaurivage and (8 min.) crosses the Inn, which forms a fine waterfall 100 yds. below the bridge. On the right bank is the *Restaurant & Pens. Waldhaus*, with a view-terrace. Hence we may either proceed on the hillside along the forest, or take the footpath which begins at the bridge and skirts the lake to the (20 min.) *Meierei* (p. 399). — From the Inn bridge (finger-post) a very attractive path leads on the right bank through the gorge of *Charnadüra* to (½ hr.) **Celerina** and **Pontresina**. — From the E. extremity of the village opposite the Hotel Kulm a good path ascends to the N.E. to the (¾ hr.) *Alp Laret* (6893'); another past the new town hall to the N.W. to the (1 hr.) *Alp Glop* (7100').

From the Alp Laret we may proceed to the (¾ hr.) *Sass da Muottas* (7766'), with fine view of the Bernina chain and Inn valley; descent through

Engadine. SAMADEN. *VI. Route 100.* 401

the *Val Saluver* to (³/₄ hr.) Celerina. — From the Alp Glop a path leads to the (2¹/₂-3 hrs.) top of the *Piz Nair* (10,040'; guide advisable, 7 fr., with descent to the Val Suvretta 10 fr.); superb view.

To SAMADEN THROUGH THE VAL SUVRETTA, AND THE VAL BEVER, 7 hrs., interesting, especially for botanists (guide unnecessary). The route from the baths leads by the *Lower Alpina*, and that from the village by the *Alp Glop*. We then ascend past the *Alp Suvretta* to the small *Suvretta Lake* (8563') and the (3 hrs.) pass (8590') which separates the S. *Val Suvretta da St. Moritz* from the N. *Val Suvretta da Samaden*. We descend the latter, to the (³/₄ hr.) *Alp Suvretta - Samaden* (7024'), where the Val Suvretta opens into the *Val Bever*, and reach the (1¹/₄ hr.) *Alp Prasüratsch*, where a road begins. Thence back to St. Moritz by carriage previously ordered (16-20 fr.), viâ *Bevers* and **Samaden** in 2 hrs.

The *Piz Rosatsch* (9825'; guide 8 fr.) and the *Piz Surlej* (10,455'; guide 10 fr.) may be ascended from the Acla Silva (p. 399) viâ the *Stats Alp*; both rather fatiguing (comp. p. 408).

An *Excursion on the Bernina Road as far as the *Hospice* (p. 411), including a visit to the *Morteratsch Glacier* (p. 404) or the *Alp Grüm* (p. 411), takes 10 hrs. by carriage (p. 400).

The FOOTPATH TO PONTRESINA, ³/₄ hr., is shorter than the carriageroad viâ Celerina. From the *Dairy* (*Acla Silva*, p. 399) it passes the N. end of the *Statzer See* (where the road to Celerina leads straight on), turns to the right, and then to the left after a few paces, and traverses a wood, rounding the base of the **Rosatsch**. Below (¹/₂ hr.) Pontresina we either cross the Berninabach to the (¹/₄ hr.) Hôtel Roseg; or we may cross the Roseg, to the right, and the Punt Ota to the Hôtel Saratz.

The Samaden road ascends for a short distance, and then descends in a long bend through larch-wood (short-cut for walkers by the old road). On quitting the wood we enjoy an admirable survey of the Inn Valley, extending nearly in a straight line to the *Munt Baseglia* near Zernetz (p. 413) which apparently closes the valley. Passing Cresta, Rom. *Crasta* (5690'; *Pens. Misani*, with restaurant), we cross the *Schlatteinbach*, descending from the Val Saluver (see p. 402), to —

13 M. **Celerina**, Rom. *Schlarigna* (*Hôt.-Pens. Murail*, pens. from 8 fr.). The road divides here. The branch to the right, to (1 hr.) **Pontresina** (p. 402), crosses the Inn and passes the dilapidated chapel of *St. Gian*, crosses the *Berninabach*, and joins the Samaden road (see below). The left branch leads to Samaden.

Footpath through the *Charnadüra* to the *Acla*, see p. 400. It diverges to the right before the Inn bridge, leads through a meadow on the bank of the Inn, crosses to the right bank, and ascends gradually through wood.

Near Samaden the *Flatzbach* or *Berninabach*, descending from the Bernina, falls into the Inn.

15 M. **Samäden**. — *HÔTEL BERNINA, R., L., & A. from 5¹/₂, B. 1¹/₂, lunch 3¹/₂, D. 5 fr., at the lower end of the village; HÔT.-PENS. DES ALPES, moderate; *HÔTEL BELLEVUE (*J. Lis*), near the Inn bridge on the Pontresina road, pens. from 7 fr.; KRONE, unpretending, well spoken of for single gentlemen, R. 2¹/₂, B. 1 fr. — CARRIAGE with one horse to the *Village of St. Moritz* for 2 pers. 5, 3 pers. 7, with two horses for 4, 5, or 6 pers. 10, 12, or 15 fr.; to the *Baths of St. Moritz* one-horse 7 or 9, two-horse 14, 16, or 18 fr., there and back, or with luggage 8 or 11, or 16, 18, 20 fr.; to *Pontresina* one-horse 4 or 5¹/₂, two-horse 8, 10, or 13 fr., there and back, or with luggage one-horse 5 or 7, two-horse 10, 12, or 15 fr.; *Morteratsch Glacier* one-horse 8 or 10, two-horse 16, 20, 25 fr.; *Roseg Glacier* one-horse 15 or 20 fr.; *Bernina Houses* one-horse 12 or 14 fr., two-horse 22, 27, 35 fr.; *Bernina Hospice* one-horse 15 or 20 fr., two-horse 30, 35, 45 fr. —

BAEDEKER, Switzerland. 16th Edition. 26

Omnibus daily 7 a.m. from the Hôtel Bernina to the Baths of St. Moritz (in 1¹/₄ hr.; returning at 11 a.m.) and to Pontresina and the Morteratsch Glacier.

Samaden, Rom. *Samēdan* (5670'; pop. 842), the chief village of the Upper Engadine, with handsome houses and a new *English Church*, is another summer-resort, beautifully situated on the W. side of the Inn Valley. The principal old house is that of the *Planta* family, a name intimately connected with the history of the country for nearly 1000 years. Splendid view to the S.W. (finest from the Bevers road below the village) of the imposing Bernina Chain, culminating in the beautiful white Piz Palü, the lofty Piz Bernina, Piz Tschierva, and farther to the right Piz Roseg; in the foreground the Piz Rosatsch and to the extreme right, above the hills of St. Moritz, the Piz della Margna.

WALKS. To the N., past the English church, to the (¹/₂ hr.) *Munterätsch*, a larch-clad hill, with a fine view of the Bernina group. Thence to the right, by a pleasant wood-walk, to the (¹/₂ hr.) saw-mill of *Resgia* in the Val Bever. — To the W. in 20 min. to the hill of *Salvasplanas*, above the church of *St. Peter* (5895'), with tombstones of the Planta, Salis, Juvalta, and other families, and the (1 hr.) *Alpetta*. — To the S. to the (¹/₂ hr.) wooded hill of *Christolais*, between Samaden and Celerina.

The *Muottas Muraigl (p. 406; 2¹/₂ hrs.) is a very fine point. The new path (steep and somewhat sunny; horse or mule 10 fr.) diverging to the left from the Pontresina road at the bridge over the Inn, descends along the right bank and after 25 min. turns to the right into the *Val Champagna*, through which it rapidly ascends to the (2 hrs.) summit. From the Muottas Muraigl to *Pontresina* (1¹/₂ hr.), see p. 406; to the top of the *Schafberg* (1¹/₂ hr.), see p. 406.

To the W. above **Samaden rises Piz Padella** (9460'), a grotesquely cleft limestone rock, which may be ascended by a good path in 3 hrs., diverging from the Piz Ot route at the point where a small valley begins at the back of the Padella. *View of the Inn Valley, from Silvaplana to Zernetz. Rich flora. A rocky ridge with three peaks (*Trais Fluors*, 'three flowers'; 9700'; an interesting climb, for experts only) connects the Piz Padella with the massive *Piz Ot (10,660'; 'lofty peak'; guide 10 fr.). This granite peak, rising abruptly in a pyramidal form, and formerly accessible to experts only, is now ascended without danger in 4-4¹/₂ hrs. from Samaden. Bridle-path to the (2¹/₂ hr.) *Fontauna Fraida* ('cold spring'; 8840'), where it is joined by the direct path from St. Moritz and Celerina through the *Val Saluver* and the *Fuorcla da Trais Fluors*. Finally, the path ascends for ³/₄ hr. in zigzags, iron rods being attached to the rock at awkward places. Imposing view, little inferior to that from the Piz Languard (p. 407).

FROM SAMADEN TO PONTRESINA (3¹/₄ M). The road (Bernina Road, R. 103) soon crosses the *Inn*, traverses the bottom of the valley, and at the point where it reaches the *Flatzbach* is joined by the road from Celerina (p. 401). It then crosses the *Muraigl* (p. 406) Near Pontresina, to the right, appears the beautiful *Roseg Glacier* (p. 405); in the background rise *Piz Morteratsch*, *Piz Tschierva*, *La Sella*, and *Piz Glüschaint*.

101. Pontresina and Environs.
Comp. also Map, p. 394.

Hotels (frequently so full from the middle of July to the middle of August as to render engagement of rooms in advance prudent); At *Lower Pontresina*: *HÔTEL ROSEG, at the N. end of the village (largely patronized by English and Americans), with a large 'dépendance' and full view of

PONTRESINA. VI. Route 101. 403

the Roseg Valley, R., L., & A. from 5, lunch 3½, D. 5, pens. from 12 fr.; *Hôt. Enderlin, E., L., & A. from 5, lunch 3½, D. 5, pens. from 11 fr.; °Weisses Kreuz (*Enderlin Sen.*), R., L., & A. from 3¼, B. 1½, lunch 3, D. 3½, pens. from 9 fr.; °Kronenhof & Bellavista, with fine view (patronized by English travellers), R., L., & A. 3-7, lunch 3, D. 4½, pens. 10-14 fr.; *Hôt. Languard, R. 2-6, L. & A. 1, lunch 3, D. 4½, pens. 10-14 fr.; °Hôt. Saratz, R. from 3, L. & A. 1, lunch 3½, D. 5, pens. 12 fr.; *Hôt.-Pens. Pontresina, R., L., & A. from 4, B. 1½, lunch 3½, D. 5, pens. 12 fr.; *Hôt. Müller, R., L., & A. from 2½, B. 1¼, lunch 2½, D. 3½, pens. from 8 fr.; Hôt. Bernina, R., L., & A. 4-5, D. 3, pens. 7-10 fr. — At *Upper Pontresina*: *Steinbock, R., L., & A. from 2½, D. 3½, S. 3, pens. from 9 fr. — Private Apartments at *Villa Jenny*, *Villa Ludwig*, *Villa Carduff*, Mme. *Gross* (R., L., & A. 2½ fr.), *M. Lina*, etc. — Beer at the *Hôt. Enderlin*, the *Kronenhof*, and the *Hôt. Pontresina*. — *Café Casino*; *Café Ma Campagne*, above the Hôtel Pontresina (rooms to let); *Chalet Sanssouci* (p. 404).
Guides. *Martin Schocher*, *Hans Grass the Younger*, *Andr. Rauch*, *Benedict Cadonau*, *L. Caflisch*, *Hermann Freimann*, *Paul Müller*, *Chr. Schnitzler*, *Peter Beeli*, etc. On all excursions for which the tariff is 50 fr. or over, two guides, or one guide and a porter, are prescribed. The charges for the excursions are given in each case. Smaller excursions, not fixed in the tariff, 10 fr. daily; if more than three persons, each 2 fr. extra. The guide carries luggage not exceeding 15 lbs.
Photographs, etc., at *Fluri's*, near the Hôtel Pontresina. — Physicians: Dr. *Barnard* (p. 399); Dr. *Stuart Tidey*; Dr. P. *Gredig*; Dr. *Bernhardt*. — Alpine plants at *Caviezel's*.
Post & Telegraph Office, below the Hôtel Pontresina.
Carriages. The fares here given are the return-fares, and in each case include waiting for 1 hr., each additional hr. 1 fr. for one-horse, 2 fr. for two-horse carriages. To *Morteratsch*, with one horse, 1-2 pers. 5, 3 pers. 6 fr., with two horses, 4 pers., 10 fr.; *Roseg*, with one horse 9 or 11½ fr. (there or back only, 8 or 10 fr.); *Bernina Houses* one-horse 6 or 7½, two-horse 12 fr.; *Val del Fain* one-horse 10 or 12 fr.; *Bernina Hospice* one-horse 13 or 16, two-horse 25 fr.; *Village of St. Moritz* one-horse 7 or 9, two-horse 14 fr.; *Baths of St. Moritz* one-horse 8 or 10, two-horse 15 fr.; *Samaden* 5 and 6, or 10 fr.; *Maloja* 17 and 21, or 32 fr. Fee for driver of one horse, half-day 50 c., whole day 1 fr.; for longer excursions 10 per cent of the fare. For each day of rest, 10 fr. per horse. Detailed tariff for longer journeys at the hotels. — Omnibus from Samaden by Pontresina to the Morteratsch Glacier daily (comp. p. 402).
English Church (*Holy Trinity*); service during the season (lending library under the care of the chaplain).

Pontresina (5915'; pop. 500), a considerable village, extending along the right bank of the *Berninabach* or *Flatzbach* on both sides of the Bernina road for more than ½ M., consists of *Lower Pontresina* (Rom. *Laret*), with the large church, and *Upper Pontresina* (Rom. *Spiert*), about ¼ M. apart, between which lies a group of houses called *Bellavita*, including the *English Church*. Above Spiert are the houses of *Giarsun* and *Carlihof*, with the loftily situated little church of *S. Maria* (adjoined by the small churchyard) and the ruined tower of *La Spaniola*. Pontresina owes its importance as a mountaineering station to the proximity of the Bernina Chain, which separates the Upper Engadine and the Val Bregaglia from the Valtellina, and vies in the grandeur of its snowclad peaks and glaciers (Rom. *Vadret*, Ital. *Vedretta*) with the Mte. Rosa group. The highest summit, the Piz Bernina (p. 408), is not visible from Pontresina. A splendid view, however, opens from Lower Pontresina between the Piz Rosatsch and Piz Chalchagn of the Roseg Valley, with the Roseg Glacier and the Piz Tschierva, Sella, Glüschaint, and Chapütschin in the background. From Upper Pontresina, the top of Piz Palü only is visible above the Morteratsch Valley, to the right of Munt Pers. To the N.W. rises the rocky pyramid of Piz Ot. — The majority of visitors formerly used to be English; now about one half of them are German.

Pretty *WOOD PROMENADES skirt the slope of Piz Chalchagn on the left bank of the Berninabach and extend up the Roseg Valley. They are usually entered by descending past the Hotel Saratz and crossing the bridge Punt Ota, beyond which the Schlucht-Promenade leads to the left through wood along the narrow gorge of the Berninabach, into which we may descend at two points (the second the easier), to the (¼ hr.) prettily situated Chalet Sanssouci (café). A few paces before we reach it, paths diverge to the left and right : to the left, we may descend to the bridge over the Berninabach, and re-ascend to Upper Pontresina; to the right is the Tais Promenade, which after ¼ hr. joins the Rusellas Promenade in the Roseg Valley (see below). Straight on, the Schlucht-Promenade continues on the left bank of the Berninabach to Morteratsch (see below); after ¼ hr. we may diverge to the left, cross the bridge opposite the Languard Fall, and return by the road. — Ascending straight on from the Punt Ota (to the right, the carriage-road to the Roseg Valley, p. 405) and then following the shady walk on nearly the same level, passing the finger-post to the 'Muottas' on the left, we reach the Rusellas Promenade in the Roseg Valley, where (¼ hr.) a bench commands a fine view of the Roseg Glacier. About 20 min farther on, we join the carriage-road (p. 405).

A path leading straight on from the Punt Ota (finger-post, see above) ascends in 1¼ hr. to the 'Signal' on the Muottas da Pontresina (7690'; fine view).

About ¼ M. from the Punt Ota on the Roseg road, beyond the bridge (see below), a finger-post shows the path to the (1½ hr.) Muottas da Celerina, a N.E. spur of Piz Rosatsch (fine view; best from the second signal). The same finger-post also shows the path to ST. MORITZ, which is (10 min.) joined by the path crossing the Berninabach below the Hôtel Roseg; it then ascends through wood to the (½ hr.) Lake of Statz and the (¼ hr.) Acla Silva on the Lake of St. Moritz (p. 399).

The *Morteratsch Glacier (Vadret da Morteratsch; guide unnecessary; carriage in ½ hr., see p. 403) is 3 M. to the S. of Upper Pontresina. Pedestrians follow the Schlucht-Promenade, or ⅔ M. from Upper Pontresina opposite the Languard Fall diverge to the right from the Bernina road, cross the Berninabach, and then proceed by a shady path on the left bank of the brook to the restaurant. The Road diverges to the right from the Bernina road about 1½ M. farther on (see p. 410), and crosses (1 M.) the Berninabach, which forms pretty falls above and below the bridge. Then across the Morteratsch Brook to the (¼ M.) Restaurant-Pension Morteratsch (6260'; R. 2-3, pens. 9-10 fr.), situated 10 min. from the foot of the glacier (view of the Piz Palü, Bellavista, Crast'agüzza, and Piz Bernina).

In the glacier is an artificial grotto (½ fr.), the way to which is shown by a flag; thence to the top of the glacier 10 min. (guide indispensable, 5 fr.). — To the right of the flag a path ascends, at first through wood, past a chalet, and about 10 min. beyond it to the right, to the (25 min.) *Chünetta*, a point of view affording a complete survey of the glacier and its grand environment (from the Munt Pers towards the right: Piz Palü, Bellavista, Zupò, Crast'- agüzza, Bernina, part of the Roseg, Morteratsch, Boval, Tschierva).

A closer survey is obtained from the *Boval Hut* (8070'), 1½ hr. higher up (2 hrs. from the inn), on the W. side of the glacier. The path (guide convenient) ascends the slope of the valley from a point 5 min. below the Chünetta, finally through a chimney, to the hut, maintained by the S. A. C., the starting-point for the Bernina, Morteratsch, Palü, etc. (p. 403). Less ambitious travellers should at least walk hence across the glacier to the fall of the *Pers Glacier* (there and back 3½ hrs.; with guide only, 12 fr.; comp. p. 408).

*Roseg Glacier (road to the inn 6 M., carriage in ¾ hr., see p. 403; thence to the glacier ½ hr.). From the Punt Ota (p. 404), we keep to the right, cross the *Roseg Brook*, and ascend its left bank between the wooded *Piz Chalchagn* on the left and the *Piz Rosatsch* on the right. After 2½ M., by the *Alp Prüma*, we cross the stream; beyond the bridge, on the right bank, the path coming from the Rusellas Promenade (p. 404) emerges from the wood on the left. A little farther on, there is a good spring on the right. After 1½ M. more, beyond the wooded *Muot da Cresta*, we again cross the brook, and soon reach the (¼ M.) small *Restaurant du Glacier* (6560'; rooms), ¾ hr. from the *Roseg Glacier*, which has receded greatly of late, but is well surveyed from the inn, with the peaks surrounding it (Piz Tschierva, Bernina, Roseg, La Sella, Piz Glüschaint, La Monschia, and Piz Chapütschin). The glacier consists of two large ice-cataracts (E. the *Vadret da Roseg*, and W. the *Vadret da Tschierva*), which unite below. Between them rises the isolated green height of *Aguagliouls*, where sheep graze in summer. Through the telescope at the inn, grazing chamois may generally be discovered in the afternoon high up on the slopes of Piz Misaun. — A more extensive view of the imposing amphitheatre is obtained from the *Alp Ota (7385'); the path leads from the inn for 20 min. at the same level, and ascends past a projecting rock on the right to the (½ hr.) two chalets of the Alp. Passing to the right of the chalets, we reach the best point (on the Mortel path) in 20 min. more, where, in addition to the above-named peaks to the left and right of Piz Bernina, the Piz Morteratsch and Monte Scerscen are visible; between the Morteratsch and Bernina the Fuorcla Prievlusa, between the Scerscen and Roseg the Porta Roseg, and between the Roseg and Sella the Sella Pass.

For the glacier itself a guide is necessary (7 fr.; to be had at the Restaurant du Glacier); a path on the right side of the Roseg brook leads by the *Alp Misaun* to the (1 hr.) *Margum Misaum* (7398') and thence across the glacier to the rocky hill of Aguagliouls (nearest point, 8780'), 1½ hr.; view grander and more complete than from the Alp Ota.

An admirable survey is also obtained from the Alp Surovèl (7425'; milk), ³/₄ hr. from the Roseg Restaurant, on the way to the *Fuorcla Surlej* (p. 898). — A path, commanding splendid views, leads from the Alp Ota along the slope to the (1 hr.) **Mortèl Club-Hut** (7840'), grandly situated, the starting-point for Piz Roseg, the Sella Pass, etc. From the hut across the Roseg Glacier to the rock of *Aguagliouls* 1¹/₄ hr.; thence back to the Roseg Restaurant 2 hrs.; a very fine round, with guide (15 fr.).

A most interesting excursion is the ascent of the *Schafberg (*Munt della Bes-cha*, 8965'; bridle-path in 2¹/₂ hrs.; guide unnecessary). Good paths lead from the Hôtel Roseg, passing the picturesque chalet of Herr Nitzschner and to the left of the large church of Lower Pontresina, to the (20 min.) hill *Crast' Ota* (fine views), where they unite. We then ascend through wood to a (50 min.) *Chalet Restaurant* (7320'), an admirable point of view. At our feet lie Pontresina and the snow-girt Roseg valley, bounded by the Piz Rosatsch on the right and the Piz Chalchagn on the left, with the glistening peaks of the Sella, Piz Glüschaint, the Monschia, and the Chapütschin in the background; adjoining the Piz Chalchagn on the right is the Piz Morteratsch, on the left the Bellavista, Piz Palü, Piz Cambrena, Munt Pers, and Sassal Masone; then the Languard valley with the Paradies and the Piz Albris; to the right, below us, at the foot of the Rosatsch, are the sombre little Lake of Statz and the blue Lake of St. Moritz; above these rise the mountains on the N. side of the Inn, Piz Lunghino, Lagrev, Albana, Julier, Nair, Ot, and the serrated Crasta Mora near the Albula Pass. — From this point a bridle-path ascends in 1¹/₄ hr. to the top of the saddle between *Las Sruors* (see below) and the summit of the *Schafberg* (8965'), to the left, reached in 10 min. more. On the summit is an unpretending *Restaurant*. The *View embraces the whole Bernina group (beside the peaks already mentioned we see, beginning at the Bellavista, the Piz Zupò, Argient, Crast'agüzza, Piz Bernina, Piz Bianco, Mte. di Scerscen, Piz Morteratsch, Piz Roseg; on the other side of the Roseg valley, Piz Corvatsch and Piz Surlej), Piz Uertsch, Piz Kesch, to the right of the Albula, the valley of the Inn as far as the Maloja (with the lakes of Campfèr and Sils).

The W. peak (9783') of **Las Sruors** ('the sisters') is easily ascended from the Schafberg in ³/₄ hr. and commands a grand view of the Bernina group and the Ortler. The two other peaks are difficult and for experts only (guide 20 fr.).

A path descends the N. side of the Schafberg in zigzags into the bleak *Muraigl Valley*, affording a view of the *Piz Vadret* (10,400'), to the right. In ¹/₂ hr. we reach the bridge over the Muraigl, the right bank of which we follow, passing the chalets of Muraigl, to a second bridge, by which we regain the left bank. We skirt the N. slope of the Schafberg, through fine wood, and reach the Hôtel Roseg in Pontresina in ¹/₂ hr. more.

Another very fine view is afforded by the *Muottas Muraigl (8270'; guide unnecessary; horse 10 fr.), easily ascended from Pontresina in 2 hrs. We follow the path above described, diverge to the left by a guide-post near the chalet above the Hôtel

Roseg, skirt the W. and N. slope of the Schafberg by a shady path, and after 1¼ hr. cross the bridge to the *Lower Muraigl Alp* (7216'), where the path divides. The shorter but worse branch ascends very steeply (½ hr.); the right branch goes straight on for some distance, then turns to the left beyond a ruined hut, and reaches the (¾ hr.) *Upper Alp* (2100') and the *Inn of J. Lis* (4 beds). The best point of view is beside a stone man, a few minutes farther on, where the path from Samaden ends (p. 402). From this point we survey the glaciers of the Bernina (the Roseg Valley, with the Piz Morteratsch, Piz Bernina, etc., being particularly striking), the green Upper Engadine with its lakes, from Ponte to the Maloja, and the mountains on the N. side of the Inn Valley from Piz Lunghino to Piz Kesch. — Descent to Samaden, see p. 402; over the Schafberg to Pontresina, 3 hrs., see above.

*Piz Languard (10,715'; 4 hrs., way not to be mistaken; guide, advisable after fresh snow, 10 fr. for 3 pers., each additional pers. 2 fr.; horse to the foot of the peak 10 fr.), fatiguing, but in fine weather deservedly a favourite point of view. We start early, in order to avoid the mists which often rise about 8 a.m.; and in this case the path is in shade as far as the foot of the peak. From Lower Pontresina the route is indicated by a guide-post near the Hôtel Languard; from Upper Pontresina we follow the path to the left near the Hôtel Steinbock, passing above the small burial-chapel, and ascend the stony slope in zigzags, to the (1 hr.) *Alp Languard* (7872'; rfmts., moderate). Beyond the Alp we ascend the bleak Languard Valley to the (1¼ hr.) base (9090') of the Languard peak, where the bridle-path ends. A steep zigzag path leads hence to the (1½ hr.) summit, on which is a trigonometrical signal (*Inn, moderate). The *VIEW (comp. Panorama) extends to the S.W. as far as Mte. Rosa, to the S.E. to the Adamello, to the N.W. to the Tödi, and to the N.E. to the Zugspitze.

Mountaineers may descend across the *Albris Glacier* and past the little *Pischa Lake* (9121'), which is sometimes frozen over until late in summer, to the *Val del Fain* (p. 410) and the (2½ hrs.) Bernina houses (guide 12 fr.). In descending it is advisable to keep several hundred paces to the right of the waterfall which issues from the lake, as all the other descents are very steep and difficult. — From the Languard Alp we may ascend the Paun da Zücher (*pain de sucre;* 9495'; 2½ hrs., guide 15 fr.), and Piz Albris (10,387'; 3 hrs., guide 20 fr.); both fatiguing.

The *Diavolezza Tour (9-10 hrs., which may be distributed over two days since the erection of the inn on the pass; guide 15, including night's stay 20 fr.) is one of the finest and least fatiguing of glacier-excursions, and is often made. The path diverges to the right from the Bernina road at the *Bernina Houses* (6720'; 5 M. from Pontresina; carriages, see p. 403) and ascends grassy and stony slopes to the (1½ hr.) picturesque little *Diavolezza Lake* (8463'; bridle-path thus far, horse from Pontresina 10 fr. and fee); then over loose stones and snow, to the S.E. of *Munt Pers* ('lost mountain'; 10,533'), to the (1½ hr.) *Diavolezza Pass* (9767'; Inn, plain

408 *VI. Route 101.*   PONTRESINA.   *Diavolezza.*

but not cheap), commanding a grand *View (of overwhelming beauty in the early morning) of the neighbouring Bernina group: from left to right, Piz Cambrena, Palü, Bellavista, Crast'agüzza, Bernina, Morteratsch, and Tschierva; below us lie the Pers and Morteratsch glaciers. Steep descent (rope desirable for novices) over débris to the moraine of the *Pers Glacier*; then across the glacier to the (1 hr.) rocky *Isla Persa*; lastly over the middle moraine to the *Morteratsch Glacier*, and down the latter to the (2½ hrs.) *Restaurant Morteratsch* (p. 404).

Those on their way from the Bernina Hospice to the Diavolezza need not descend to the Bernina Houses, but diverge to the left from the road below *Lej Pitschen* (p. 410) and soon strike a narrow path, which ascends through the *Val d'Arlas* and joins the ordinary route near the Diavolezza Lake.

Piz Rosatsch (9825'; 4-5 hrs.; guide 12 fr.) and Piz Chalchagn (10,350'; 5-6 hrs.; guide 15 fr.), without special interest. — *Piz Surlej (10,455'), in 5-6 hrs. (guide 14 fr.), an admirable point, is best ascended from the *Acla Silva* on the Lake of St. Moritz, over the *Statz Alp*, or from *Silvaplana* (4-4½ hrs.). — *Piz Corvatsch (11,345'; 6 hrs.; guide 16, back by Silvaplana 18, by Sils 20 fr.), somewhat laborious. From the (4½ M.) *Roseg Inn* (p. 405) we ascend to the (¾ hr.) *Alp Surovel* and follow the Surlej route to the (1½ hr.) highest chalet (*Margum Sura*, 8000'); then turn to the left towards a snow-peak visible to the S.W., and ascend grassy and stony slopes to the (1 hr.) *Corvatsch* or *Alp Ota Glacier*. Lastly up the glacier, the crevasses of which require caution, to the (2 hrs.) summit, covered with rocks, and generally free from snow. The guides usually halt on the *Piz Mortèl* (11,293'), but it is preferable to go on to the (¼ hr.) highest peak, where the view to the S.W. is far more picturesque. The great attraction of the view consists in the double survey, to the E. and S.E., of the imposing Bernina amphitheatre, and, to the W., of the green Engadine with its villages and lakes immediately below us. Distant view very extensive, like that from Piz Languard; on the S.W. it extends to the Monte Viso. Descent by the *Fuorcla Surlej* to *Silvaplana* (comp. p. 398). The descent on the W. side by *Marmoré* to (3 hrs.) *Sils*, for experts only, is steep and trying.

The *Piz Morteratsch (12,315'; 5-6 hrs. from the Roseg Restaurant, p. 405), though requiring a steady head, is the easiest of the higher peaks, but difficult when there is little snow. Descent to the Boval Hut fatiguing (guide 30, including the transit from the Roseg-Thal to the Morteratsch-Thal 35 fr.) — Chapütschin (11,133'), 8-9 hrs., or from the Mortèl Hut 4 hrs.; guide 25, with descent to Fex 30 fr. — Piz Tschierva (11,713'; 5-6 hrs. from the Roseg Restaurant; 25 fr.), fatiguing, but repaying. — La Sella (11,770'; 8-9 hrs.; from the Mortèl Hut 4 hrs.; guide 30 fr.) and Piz Glüschaint (11,805'; 8-9 hrs.; guide 35 fr.) are not difficult, but require experience. — *Piz Palü (12,835'), conspicuous for the beauty of its form and the purity of its snow, from the Diavolezza Inn 4½-5½ hrs., from Boval 7, or from the Capanna Marinelli (p. 409) 5 hrs., trying, but with good guides (50 fr., for all three peaks 60 fr.) free from danger. From the first (E.) peak (12,755) a narrow arête, descending perpendicularly on the S. side (steady head necessary), leads to the double-peaked second (12,835') and the third peak (12,545'). The descent may be made by the *Bellavista Saddle* and the *Festung* to the *Morteratsch Glacier* (to the Hôtel Morteratsch 5-6 hrs., guide 60 fr.). — *Piz Zupò ('*Verborgne Horn*', 13,120'), from the Boval Hut by the *Fortezza* (see below), or under favourable conditions of the snow, direct by the Morteratsch Glacier in 6-8, or from the Capanna Marinelli by the *Crast'agüzza Saddle* in 4-5 hrs., toilsome (guide 50 fr.); panorama of surpassing grandeur. — Crast'agüzza (12,705'), a ridge between Piz Bernina and Piz Zupò, rising almost perpendicularly from the glacier, 14 hrs. from Boval, very difficult, but most interesting (guide 80 fr.).

The *Piz Bernina (13,295'; 8-10 hrs. from the Boval Hut; guide 70 fr.), the highest peak of the group, first ascended in 1850, is highly interesting, but should be attempted by none but thorough experts. The route ascends, according to the state of the snow, either direct through the central icefall of the Morteratsch Glacier (the '*Labyrinth*'), and over rock and glacier to the right; or by the so-called *Festung* or *Fortezza* to a basin of snow between Piz Bernina and Crast'agüzza, and thence by the arête from the S. E. side to the top. The ascent is shorter from the S. side (from the *Capanna Marinelli*, see below, over the *Crast'agüzza Saddle*, 6-7 hrs.). A much more difficult route ascends from the *Tschierva Glacier* and up the W. slope, and then on the N. side by the *Fuorcla Prievlusa* (11,325'), the *Pizzo Bianco* (13,117'), and the *Bernina Scharte* (accomplished for the first time in 1878 by Dr. Güssfeldt; 9-10 hrs. from the Roseg Restaurant). — More difficult are Piz Roseg (12,935', 9-10 hrs. from the Mortèl Hut; guide 80 fr.), first ascended in 1865, and Monte di Scerscen (13,015'; guide 150 fr.), ascended for the first time in 1877 by Dr. Güssfeldt. In 1894 a safe, though not easy route up and down the Scerscen, by one of the rocky ribs on the E. face, was discovered and taken thrice in one week by the guide Roman Imboden of St. Nikolaus. Between Monte di Scerscen and Piz Roseg lies the difficult Porta Roseg (*Fuorcla Tschierva-Scerscen* or *Güssfeldt-Sattel*; 11,578'), first crossed by Dr. Güssfeldt in 1872 (ascent in 9-12 hrs. from the Roseg Restaurant).

Passes. From Pontresina to Sils, several routes. The easiest (but rather trying; 9 hrs., guide 20 fr.) crosses the Fuorcla da Fex-Roseg (10,110'). From the Roseg Restaurant we ascend, viâ the *Ota Alp*, along the slope of the Mortel-Thal, finally over rubble and past the little *Chapütschin Glacier*, to the (3½ hrs.) pass (splendid view); then a steep and toilsome descent to the *Lej Sgrischus*, well stocked with trout, and either into the *Fex Valley* and to *Curtins* (p. 397), or to the right by *Marmoré* to (2½ hrs.) *Sils-Maria* (p. 396). — From Pontresina to Sils over the Fuorcla Chapütschin (10,590'), between the Chapütschin and Monschia, or over the Fuorcla Glüschaint (about 11,000'), between the Monschia and Piz Glüschaint, both for experts only (guide 35 fr.).

Over the Sella Pass to the Val Malenco, grand and interesting, but trying (from the Mortel Hut to Fellaria 8-9, to Chiesa 12-13 hrs.; guide to Poschiavo or Chiesa, or back to Pontresina by the Cambrena or Bellavista Pass, 65 fr.). From the Mortèl Hut we ascend behind the Aguagliouls rock and over the *Roseg Glacier* and the crevassed *Sella Glacier* to the (3-3½ hrs.) Sella Pass (*Fuorcla Sella*, 10,843'), lying to the S.W of the huge rock and ice precipices of *Piz Roseg* (12,935'). Descent over the *Scerscen Glacier*, with splendid views of the S. side of the Bernina group (Mte. di Scerscen, Piz Bernina, Crast'agüzza, Zupò, and Mte. Nero and Disgrazia to the right), and across a snow-saddle running out from Piz Zupò (to the left, higher up, the *Capanna Marinelli*, see below) to the névé of the *Fellaria Glacier*; then down the right side of the glacier, over rock and débris to the (4-5 hrs. from the pass) *Fellaria Chalets* in the *Val Campo Moro* (7335'; poor, occupied in the height of summer only). Thence down the *Val Lanterna* to *Lansada* and (4 hrs.) *Chiesa* in the *Val Malenco* (p. 421). — Instead of going to Chiesa, the traveller may prefer to complete the Circuit of Piz Bernina and return to Pontresina. In this case we do not descend to the Fellaria Chalets. On the upper part of the Scerscen Glacier we keep to the left, again ascend, and reach (1½-2 hrs. from the Sella Pass) the *Capanna Marinelli*, a club-hut of the C. A. I., situated on the rocks running out from the Piz Zupò (about 9840'), between the Scerscen and Fellaria glaciers, 3 hrs. above the Fellaria Chalets. This is the starting-point for the Piz Bernina, Palü, etc. (see p. 408). The direct route hence Back to Pontresina, over the Fuorcla Bellavista (12,080'), between the Bellavista and Piz Palü, and down by the *Fortezza* (see above) and the *Morteratsch Glacier*, is laborious (9-10 hrs.; guide 50 fr.). — To the Bernina Hospice over the Cambrena Pass, 8-9 hrs., fatiguing, but repaying (guide 50 fr.). From the Fellaria Glacier we cross a saddle of névé on the S. side of Piz Palü, to the *Palü Glacier*, skirt the slopes of Piz Palü and *Piz Cambrena* (11,885'), and reach the Cambrena Pass (11,250'), between Piz Cambrena and *Piz Caralo*. Descent

410 *VI. Route 101.*   PONTRESINA.   *Val del Fain.*

over the *Cambrena Glacier* to the Lago Nero (see below) and the Bernina Hospice. During the ascent a view extending from Mte. della Disgrazia to the snow-mountains of the Oetzthal is gradually revealed. This route is easier in the reverse direction, a night being spent at the Bernina houses or the hospice. In this case, too, the place exposed to falls of ice is passed early in the morning. — Experts may, without difficulty, descend from the snow-saddle on the side of Piz Palü (p. 408) direct to the *Palü Glacier*, avoiding the crevasses by keeping to the left, and then over turf and rock, past the *Sassal Masone*, to the *Bernina Hospice* (7-8 hrs. from the Capanna Marinelli, 12-13 hrs. from the Mortèl Hut; guide 50 fr.).

To POSCHIAVO a route leads from Fellaria to the E. over the Passo Rovano or Confinale (8590'), and through the *Val Orse*, in 3¹/₂ hrs.; another crosses the Canciano Pass (8360'; comp. p. 421), lying farther S. (also 3¹/₂ hrs.). To reach the latter pass from the Fellaria Chalets we descend a little over old moraines of the Fellaria Glacier, and then ascend to the left through the *Val Poschiavina* to the (1¹/₂ hr.) pass, where we have a fine survey of the Fellaria and Verona Glaciers, of the Piz Zupò and Piz Roseg, and of the Canciano Glacier to the S. Descent by the *Alp d'Ur* (6350') and through the *Val di Gola* to (2 hrs.) *Poschiavo* (p. 420).

FROM PONTRESINA TO MALENCO OVER THE CHAPÜTSCHIN PASS AND THE FUORCLA FEX-SCERSCEN, 12-13 hrs. from the Mortèl Hut (guide 65 fr.), a toilsome route, for experts only. Over the *Fuorcla Chapütschin* or the *Fuorcla Glüschaint* to the *Fex Glacier* (difficult descent), see p. 409. Instead of descending to the right to the Fex Valley, we turn to the left to the snowy saddle of the *Fuorcla Fex-Scerscen* and then descend the *Scerscen Glacier* to the *Val Malenco* (p. 421).

FROM PONTRESINA TO THE BERNINA HOSPICE, 9¹/₂ M., a beautiful day's excursion (carriages, see p. 402), including a visit to the Sassal Masone or the Alp Grüm. — From Pontresina to the point where the road to the Morteratsch Glacier diverges, see p. 404. The Bernina road begins to ascend. To the right a splendid *View of the Morteratsch Glacier, **with its huge** medial moraine, overshadowed by the dazzling Piz Palü, **Bellavista**, Zupò, Argient, Crast'agüzza, the Piz Bernina, Morteratsch, and Tschierva. (From one of the windings of the road, by a horse-trough, a path diverges to the Bernina Falls and the Morteratsch Glacier.) About 5 M. from Pontresina are the solitary *Bernina Houses* (6723'; Inn), near the entrance to the *Val del Fain*. — *Diavolezza Route*, see p. 407.

The **Val del Fain**, or *Heuthal*, 5 M. long, is interesting to botanists; edelweiss grows on the slopes at the head of the valley. A bridle-path (practicable for light vehicles for 2¹/₂ M.; carriages, see p. 402) ascends the valley and crosses the *Alp La Stretta* to the **Passo Fieno** (8145'), between the *Piz Stretta* (10,195') and the *Piz dels Lejs* (10,015'), whence a steep and stony footpath descends into the *Spöl Valley* to (6 hrs.) *Livigno* (p. 414). — Ascent of *Piz Languard* by *La Pischa*, see p. 407.

Beyond the Bernina Houses (³/₄ M.) the old bridle-path diverges to the right, and leads on the left side of the brook over the *Alp Bregaglia* to the pass. The road crosses the brook and ascends gradually on the E. side of the valley, passing the mouth of the *Val Minor*. (To the left rise *Piz Alv* and *Piz Lagalb*, to the right the stony slopes of the *Diavolezza*, p. 407.) The zone of trees is now quitted. The road passes the small *Lago Minore* (Rom. *Lej Pitschen*) and *Lago Nero* (Rom. *Lej Nair*), leads to the left above the light green *Lago Bianco* (Rom. *Lej Alv*; 7316'), describes a sharp bend, and crosses a brook descending to the left from Piz Lagalb. The narrow barrier between the Lago Nero and the Lago Bianco forms the watershed between the Black Sea and the Adriatic, the waters of the former descending to the Inn, and those of the latter to the Adda. To the right lies the *Cambrena Glacier*, commanded by *Piz Cambrena* (11,835') and *Piz Carale* (11,247'); to the left *Sassal Masone* (9970'). Before us rises *Piz Campascio* (see below); to the left of it is the conical *Pizzo di Teo*, to the right the *Pizzo di Sena*. Pedestrians ascend from the Bernina Houses in 1½ hr. to the —

12½ M. **Bernina Hospice** (7575'; *Hotel*, R. 2-2½, déj. at 11 a.m. 2½, D. at 1 p.m. 4-4½ fr.), finely situated above the Lago Bianco and opposite the Cambrena Glacier. To the E., at the back of the hospice, is the little *Lago della Crocetta*.

EXCURSIONS. (Guides and horses at the hospice.) Piz Campascio (8585'; guide 4 fr.), to the S. of the hospice, rising perpendicularly on the E. side, ascended by a good path in 1½ hr., commands a very striking view. — Piz Lagalb (9718'), to the N. (see above), also affords a fine view (2 hrs.; 4 fr.).

From the hospice to the SASSAL MASONE or the ALP GRÜM (1¼-1½, there and back 3-4 hrs.; guide 4 fr., unnecessary; donkey or mule 7 fr.; chaise-à-porteurs, with 2 porters, 25 fr.!), very interesting. A few paces to the S. of the hospice the bridle-path diverges from the road to the right and skirts the E. bank of the Lago Bianco. It crosses (¼ hr.) the brook issuing from the S. end of the lake, and follows the right slope of the valley, skirting the little *Lago della Scala*. A finger-post (¼ hr.) indicates the path to the right to the Sassal Masone (½ hr.); the path straight on leads to the Alp Grüm. The *Sassal Masone Alp (7800'; rfmts.), with its two round huts, lies at the foot of the *Sassal Masone* (9970'), and commands a grand view of the Palü Glacier, Pizzo di Verona, Piz Palü, the Poschiavo Valley, and the Val Viola Mts. — The view of the glacier is still more imposing from the Alp Grüm. Where the path to the Sassal Mason diverges (see above), we go straight on; then, where the path divides and Piz Palü appears to the right, to the left at the same level; and (½ hr.) reach the *Alp Grüm* (7182'; *Restaurant*), where the superb Palü Glacier, separated from us by a narrow valley only, and the Poschiavo Valley far below, with its lake and the villages of Le Prese, Prada, and St. Antonio, are suddenly revealed. To the S.E., in the distance, rise the Adamello and Presanella.

FROM THE ALP GRÜM TO POSCHIAVO (2¾ hrs.). The path descends steeply to the right, and afterwards widens into a stony cart-track; ½ hr., *Alp La Dotta*; ¼ hr., hamlet of *Cavaglia* (5680'), in a wider part of the valley. We cross (¼ hr.) the *Cavagliasco*, descending from the Palü Glacier through a wild rocky gorge; then skirt the slope to the right by a very rough and stony path (often the bed of a torrent), and descend rapidly to (1¾ hr.) *Poschiavo* (p. 420). Fine view of the valley and the opposite heights, on which runs the Bernina road. Travellers intending to visit the Alp Grüm from Poschiavo (advisable only in dry weather) should have the beginning of the route pointed out (boy from the hotel for a small fee).

Over the *Cambrena Pass* to *Fellaria*, grand but toilsome, see p. 410.

A few paces to the E. of the hospice is the top of the **Bernina Pass** (7658'). Beyond it the road passes through two galleries and descends rapidly in windings (avoidable by short-cuts), past *La Motta* (6510'), to (3¾ M.) —

16½ M. **La Rösa** (6162'; poor Inn).

To the N. of La Motta opens the *Val Lagone*, containing strata of gypsum and alabaster, through which a narrow road leads over the Forcola di Livigno (7638') to (6 hrs.) *Livigno* (p. 414).

THROUGH THE VAL VIOLA TO BORMIO (p. 425) 10 hrs., interesting; guide unnecessary in fine weather (from Pontresina to Bormio 45 fr.). Provisions should be taken. The bridle-path diverges from the Bernina road to the left at *Sfazzu* (p. 419) and ascends the *Val di Campo*, past the chalets of *Saiba*, *La Tonta*, and *Plan Sena* (6500′), to (2 hrs.) *Longacqua*, the highest chalet or 'malga'. To the N. lies the *Val Mera*, with the beautiful *Corno di Campo* (10,395′), whence a fatiguing route crosses the *Colle di Campo* (8776′) to Livigno. From this point through the *Val Viola Poschiavina* to the (1½ hr.) **Val Viola Pass** (8070′) the path is ill-defined at places (guide desirable for the inexperienced; keep to the left before the summit), leading at first through woods of stone-pines, in which several pretty little blue lakes lie to the right. Fine retrospective view of the Bernina Mts.; to the S. the precipices of the *Cima Saoseo*. Beyond the pass the path, again distinct, gradually descends to (¾ hr.) the first chalet in the *Val Viola Bormina*, on the little *Val Viola Lake* (7480′). It then leads high along the N. slope of the valley, affording beautiful views of the *Val di Dosdè* to the right, with the *Pizzo di Dosdà* (10,760′) and the *Cima Lago Spalmo* (10,820′), and then descends rapidly through wood to the (1½ hr.) *Ponte Minestra* (6490′; below which is a waterfall) and the (¾ hr.) hamlet of *Campo*. Then across pastures and through wood at places, past several houses and barns, to (1½ hr.) *S. Carlo* (5185′), a village with a church. On the right rise the *Cima di Piazzi* (11,280′), with the *Piazzi Glacier*, and the *Corno di S. Colombano* (9915′). Descent to the *Val di Deutro* and *Semogo* (route to Livigno by Foscagno, see p. 425), and by *Isolaccia* (Osteria by the bridge) and *Pedenosso* to (2 hrs.) *Premadio*. We now cross the Adda, and reach *Bormio* in ½ hr. by the road to the right, or the *New Baths* (p. 426) in ¼ hr. by that to the left.

From Bormio we may ascend through the *Val di Dosdè* (see above), finally crossing a small glacier, to (7-8 hrs.) the **Capanna di Dosdè**, built by the Italian Alpine Club on the *Passo di Dosdè* (9850′). This is the starting-point for the ascents of the *Cima Saoseo* (10,715′; 3 hrs., with descent to Poschiavo 8 hrs.), the *Cima Viola* (11,100′; 3 hrs.), and the *Corno di Dosdè* (10,608′; 4 hrs.). Beyond the hut we descend through the *Val Vermolera* to the *Val Grosina* and (6 hrs.) *Grosio* (p. 426).

An easy and attractive route leads through the *Val Verva*, which diverges to the S. from the Val Viola, and over the **Passo di Verva** (7590′), between the *Cima di Piazzi* and the *Pizzo di Dosdà*. It then descends to *Eita* (accommodation **in the** 'Casa d'Eita' of the Italian Alpine Club) and through the pretty **Val Grosina** to (10 hrs.) *Grosio* (p. 426).

102. From Samaden to Nauders. Lower Engadine.

50 M. DILIGENCE from Samaden to Schuls thrice daily in 5½ hrs. (13 fr. 60, coupé 16 fr. 35 c.); from Tarasp to Nauders twice daily in 4 hrs. (7 fr. 20, coupé 8 fr. 65 c.). (Diligence in connection to Landeck, p. 420.) The scenery is pretty at places, but may be sufficiently surveyed from an open carriage. The road is very dusty in dry weather. — CARRIAGE with one horse from Tarasp to Samaden 30, to Pontresina 40 fr.; EXTRA-POST and pair from Samaden to Landeck in two days 150 fr. and 15 fr. fee, to Meran viâ Martinsbruck in three days 200 fr. and 20 fr. fee.

Below *Samaden*, we enjoy a grand panorama of the Bernina range (comp. p. 402). 1½ M. **Bevers** (5610′; *Schmid's Inn*), a thriving village, lies at the foot of the indented *Crasta Mora* (p. 389). **Hr. Krättli**, a botanist, sells dried plants here. (Through the *Val Bever* and *Val Suvretta* to *St. Moritz*, see p. 401.) The road passes the (¾ M.) *Agnas Inn*, and leads along the canalized *Inn* to (1¾ M.) —

4 M. **Ponte** (5548′; **Hôtel Albula*, R., L., & A. 1½-2, B. 1 fr.; **Krone*, beyond the bridge, plain), at the beginning of the *Albula Route* (R. 98). On the **opposite** bank lies *Campovasto* or *Camogasc*, at the entrance to the narrow *Val Chamuera* (p. 413).

°**Munt Müsella** (8632′), on the right bank of the Inn, to the S.E. of Ponte, is easily ascended in 2½ hrs. (guide desirable); beautiful view. —

Piz Uertsch (*Albulahorn*, 10,738') is ascended from the Albula Pass in 3 hrs.; a fatiguing climb, requiring a steady head; splendid view (guide 35 fr.).

From Ponte to Livigno (6 hrs.) a bridle-path; guide desirable. We ascend the *Val Chamuera* to the (1½ hr.) chalets of *Serlas* (6634'), where the *Val Lavirum* diverges; then rapidly through the latter to the (2½ hrs.) **Fuorcla Lavirum** (*Passo dell' Everone*; 9250'), between (r.) *Piz Lavirum* (*Pizzo dell' Everone*; 10,030'; ¾ hr. from the pass; splendid view of the Ortler) and (l.) *Piz Casanella* (9616'). Then a steep descent into the *Val Federia*. After 1 hr. the path descending from the Casana Pass (see below) on the left joins our route; 1 hr., *Livigno* (p. 414).

The road follows the left bank of the Inn to (¾ M.) **Madulein** (5515'; *Restaurant Guardaval*), with the ruin of *Guardaval* on a steep rock to the left (5873'; ascent ¼ hr.), erected in 1251 by Bishop Volkard to 'guard the valley'. Then (1½ M.) —

6 M. **Zuoz**, or *Zutz* (5548'; pop. 429; *Hôt. Concordia & Post*, with hydropathic, R. 2½–3½, D. 4, pens. from 7 fr.; *Schweizerbund*; *Pens. Poult*, 5½ fr.), a prosperous village, in a sheltered situation about 300' above the bottom of the valley, visited as a summer-resort. Pretty walks, affording fine views, lead hence to the hill of *Crasta* (¼ hr.) and up the valley of the Inn, through meadows and wood, to the (1 hr.) ruin of *Guardaval* (see above); to the *Schivera Gorge* (½ hr.); to the *Arpiglia Gorge* (½ hr.); *Acla Perini* (1 hr.), etc.

*Piz Griatschouls (9755'; 4 hrs.), not difficult; extensive view. Descent by the Val Sulsanna to Capella (see below). — Piz Mezaun or Mezzem (9727', 5 hrs.; guide), easy; very fine view. — *Piz Kesch (11,230'), not difficult for experts, 5½–6 hrs., with guide (Flury Claradetscher, Jacob Gyr, Christ. Jud; 30 fr.). Cart-road to the Alp Eschia, near which a club-hut is to be erected. Superb view from the top (comp. p. 362).

Near (1 M.) **Scanfs** (5413'; *Scaletta, Post*, both plain) the Inn is crossed by a handsome bridge, but the road follows the left bank.

To the right opens the *Val Casana*, whence a bridle-path crosses the *Casana Pass* (8832'; splendid view) to (7 hrs.) *Livigno* (p. 414). The pass lies between *Punta Casana* (9870') and *Punta Casanella* (9616'), both easily ascended, the former better from the *Val Trupchum*, on the N. side.

On the right rises *Piz d'Esen* (10,270'). Below (1½ M.) *Capella* the road crosses the *Sulsanna*. (Through the *Val Sulsanna* and over the *Scaletta* and *Sertig* passes to Davos, see p. 362.) We next skirt a pine-clad gorge of the Inn. Below *Cinuskel* (5300'; Post), near *Brail* (Kreuz), the *Punt Ota*, a bridge over a brook emerging from the *Val Puntota*, separates the Upper from the Lower Engadine. At the end of the gorge we have a fine view of the river and the covered wooden bridge (4890') which carries the road to the right bank. Through the opening of the valley we see the *Munt Baselgia* and the *Piz Nüna* (see below). Near (8½ M.) *Zernetz* the valley expands into a wide and partially cultivated basin. To the N. appears the snow-furrowed head of *Piz Linard* (p. 415).

17 M. **Zernetz** (4910'; pop. 570; *Bär*, R., L., & A. 4 fr.), at the influx of the dark *Spöl* into the Inn, with a handsome church of 1623, has been almost entirely rebuilt since a fire in 1872.

*Munt Baselgia (9780'; 4 hrs.; guide 5 fr.), Piz d'Arpiglia (9945'; 5 hrs.; 6 fr.), and Piz Nüna (10,260'; 6 hrs.; 8 fr.) are ascended from Zernetz (all

414   *VI. Route 102.*   ZERNETZ.   *From Samaden*

rather trying). — Piz Sursura (10,420'; 6-7 hrs.; 12 fr.), through the *Val Sursura* and over the glacier of that name, fatiguing.

FROM ZERNETZ TO MÜNSTER (24½ M.; diligence daily in 6 hrs., 9 fr. 80, coupé 11 fr. 80 c.). The road, attractive even for walkers, gradually ascends on the right bank of the *Spöl* through the wild and wooded defile of *La Serra*, crossing several ravines (*Val da Barcli, Val Laschadura*) and the wooded plateau of *Champ Sech* to the (5½ M.) bridge over the *Ova d'Spin* (5997'). Beyond the bridge the shorter old bridle-path ascends in a straight direction over the hill of *Champ Lông* and through the *Val Flur* to the Ofen Inn, while the new road makes a long circuit to the right, skirting the wooded hill of *Crastatscha*. We cross the (2¾ M.) *Ova del Fuorn* (5610'), in its wild ravine (bridle-path to the right to Livigno, see below). The road skirts the left bank of the Ova del Fuorn, crosses it, and reaches (1½ M.) the *Inn* on the Ofenberg (*Il Fuorn*, 5920'). It next passes the mouth of the *Val del Botsch*, the *Val da Stavelchod*, and *Val Nüglia*, and ascends the marshy *Alp Buffalora* to the (5 M.) Ofen Pass (*Sü Som*, 7070'), with fine view of the Ortler. (Thence across the *Buffalora Pass* to the *Fraele Valley* and *Bormio*, p. 425.) We descend through stone-pines to (8 M.) *Cierfs* (5460'; *Alpenrose; Weisses Kreuz*), in the Münster-Thal, or *Val Mustair*, watered by the *Rambach*. Then (1½ M.) *Fuldera* (to the left above which lies *Lü*, p. 418), (2 M.) *Valcava* (4632'; *Post), and (1½ M.) *St. Maria* (p. 428). Thence to (2 M.) *Münster* and (9½ M.) *Mals*, see p. 428. Over the *Wormser Joch* to *Bormio*, see p. 426; through the *Val da Scarla* to *Schuls*, see p. 418.

FROM ZERNETZ TO LIVIGNO, 8 hrs. Road to the (9 M.) bridge over the Ova del Fuorn (1½ M. before the Ofenberg Inn, see above); then a bridle-path (4½-5 hrs.; finger-post on the left bank by the bridge), crossing the hill and ascending the *Spöl Valley* alternately on the right and on the left bank of the torrent. At the bridge over the *Acqua del Gallo* is the Italian frontier (the boundary district is exempt from custom duties); comp. the Map, p. 412. — *Livigno* (5940') is a scattered village in the wide open valley of the Spöl, with several churches; near the church of *S. Antonio* is the plain, but well managed *Pension Alpina* (R. 2 fr.). To the S., the valley is closed by the *Vedretta del Vogo*. — From Livigno to the Bernina road by the Forcola (5 hrs.; narrow road, mountain vehicle with one horse 15-20 fr.) or the Passo Fieno, see pp. 411, 410; to Ponte by the Lavirum Pass, see p. 413; to Scanfs by the Casana Pass, see p. 413; to Zernetz by the *Passo del Diavel* (9285'), to the W. of the *Piz dell' Acqua* (10,260'), a fatiguing glacier-pass, little frequented (9-10 hrs.; guide 20 fr.). — From Livigno to *Bormio* (7 hrs.), see p. 425; the path begins at the church of S. Antonio.

FROM ZERNETZ TO BORMIO viâ *Buffalora*, see p. 425. A shorter route (9½-10 hrs. to Bormio, guide necessary) diverges to the right from the Münster-Thal road beyond the bridge over the *Fuorn*, and leads by the *La Schera Alp* and *S. Giacomo di Fraele* to the *Scale di Fraele* and *Bormio*.

Below Zernetz the road recrosses the Inn (behind rises the *Piz Quatervals*, 10,855'), and enters a narrow, pine-clad gorge, extending as far as (3½ M.) —

20½ M. **Süs**, Rom. *Susch* (4689'; *Schweizerhof; *Rhätia & Post*, R., L., & A. 2, B. 1, D. 3, pens. 5-7 fr.; *Hôt. Flüela*, plain; brewery by the bridge), surmounted by the ruins of a castle (*Fortezza*), perhaps of Roman origin. To the E. rise *Piz Mezdi* and *Piz d'Arpiglia*. (*Flüela Road to Davos*, see p. 360.) Then over the *Sagliains* brook, through the valley of which runs the route over the Vereina Pass to Klosters (p. 359), to (2 M.) —

22½ M. Lavin (4690'; *Piz Linard*, R., L., & A. 2½ fr.; *Steinbock*), at the mouth of the *Val Lavinuoz*. To the S.W. is the large *Sursura Glacier* (p. 406).

EXCURSIONS. (Guides, the schoolmaster *Clagüna, Joh. Paravicini, J. S. Bonifazi*, and others.) *Sass Aula* (2 hrs.) and *Murtèra* (3 hrs.), both easy

to Nauders. ARDETZ. *VI. Route 102.* 415

and interesting. — Through the *Val Lavinuoz* to the *Tiatscha Glacier*, 3 hrs., also attractive. — Piz **Mezdi** (9593'; guide 10 fr.) is ascended through the *Val Zernina* in 5 hrs., the last part rather steep. Splendid view of the Engadine, the Silvretta, etc. The Val Zeznina ends, 4 hrs. from Lavin, in the mountain-basin of *Macun* (8645'), with its small glaciers and six little lakes, environed by Piz d'Arpiglia, Munt della Baseglia, and Piz Macun. — Piz **Linard** (11,207'; 6-8 hrs.; guide 20 fr.), the highest peak of the *Silvretta* group, affording a most superb panorama, is trying and fit for experts only. Bridle-path to the (3 hrs.) *Alp Glims*, with a poor refuge-hut; thence to the top 3-4 hrs. (the last 1½ hr. steep and toilsome). — From Lavin to Klosters over the *Vernela Pass* or the *Verstanklathor*, see p. 359.

The right bank of the Inn, generally steep, affords few sites for villages, while on the left bank, on broad, sunny heights, lie *Lavin*, *Guarda*, and *Ardetz*, said to be of Etruscan origin, picturesquely commanded by towers and ruined castles. The Inn flows through a deep gorge, swelled by many brooks descending from lateral valleys.

Beyond **Lavin** the road leads through a rocky gateway, and near (2 M.) *Giarsun* crosses the mouth of the *Val Tuoi* (p. 431).

A road to the left ascends to (1¼ M.) **Guarda** (5413'; *Hôt.-Pens. Meisser*, with 'dépendance' *Zur Sonne*, R. 1½, B. 1, pens. 5 fr.; *Osteria Silvretta*), prettily situated, which is reached (1 hr.) more pleasantly by the old road gradually ascending from Lavin. The ascents of *Piz Cotschen* (9838'; 4 hrs.; guide 10 fr.) and of *Piz Buin* (10,870'; 6 hrs.; 25 fr.), a magnificent point, are recommended (guide, B. Padrun). — To *Klosters* over the *Silvretta Pass*, see p. 359; to the *Montafon* over the *Vermunt Pass*, see p. 431. — From Guarda the old road descends to *Boschia* and *Ardetz* in 1 hr. Walkers bound for Schuls will find it better to follow the old road by Fetan (keep up to the left, at a point ¼ hr. beyond Boschia), which bends into the Val Tasna at the ruined houses of *Canova*, and shortly afterwards joins the new road from Ardetz. From Guarda to Fetan 2½ hrs.

The road skirts a stony slope high above the Inn, enters a pleasant larch-wood, and then traverses meadows and fields to (3 M.) —

27½ M. **Ardetz**, Ger. *Steinsberg* (4826'; pop. 628; *Post*; *Pens. Alpina*), picturesquely situated, and commanded by the ruin of *Steinsberg*, with its well-preserved tower.

A road (diligence every afternoon in 1 hr.) commanding fine views leads from Ardetz, across the *Val Tasna* and up the sunny pastures on the N. side of the valley, to (4½ M.) Fetan (5405'; *Victoria*, pens. from 7 fr.; *Restaurant zur Alten Post*), largely rebuilt since a fire in 1885, and commanding a charming view of the mountains on the S. side of the valley (finest from the *Paradies* pavilion, near a grove ¼ hr. to the W. of the village). — *Muotta Naluns* and *Piz Clüna*, see p. 417. — From Fetan to Schuls, 3 M., carriage-road (omnibus twice a day from the Hôtel Victoria to the springs at Tarasp). A direct footpath to Tarasp diverges to the right from the road after the last wide curve, beyond the stream.

The wild **Val Tasna**, with its woods and pastures, ascends between (l.) *Piz Cotschen* (9940') and (r.) *Piz Minschun* (10,080') for 3 hrs., and then divides into (l.) the *Val d'Urezzas* and (r.) the *Val Urschai*. From the latter a difficult path crosses the ice-clad *Futschöl Pass* (9060'), with fine views of the huge *Fluchthorn* (11,140'), to the Tyrolese *Jamthal* and (8-9 hrs.) *Galtür* in the *Patznaun*; see *Baedeker's Eastern Alps*.

Beyond Ardetz the road traverses stony slopes, and is hewn in the rock at places. From a bend we obtain a most picturesque view of Schloss Tarasp; to the right, on the S. bank of the Inn, rise Piz Plavna, Piz Pisóc, Lischanna, and Ayutz. The road then describes a wide curve, enters the deep *Val Tasna* (see above), and crosses it by a stone bridge. The road leads high above the deep wooded gorge of the Inn. To the right a fine view of the sombre, pine-clad *Val Plavna*, with the **Piz Plavna** Dadaint (p. 417) in the background. In the foreground, on the right bank of the Inn, is Schloss Tarasp. The road then descends to the Inn, passes at the back of the Curhaus, and reaches the post-station of —

31½ M. **Bad Tarasp** (3946'; *Curhaus*, R., L., & A. from 5 fr., B. 1½, D. 5, pens. from 13, visitors' tax 17, baths 1½-2½ fr.; Eng. Church, see below), situated in a small expansion of the deep valley of the Inn, with celebrated mineral springs resembling those of Carlsbad. The *Lucius* and *Emerita* springs, both containing salt and carbonate of soda, are those chiefly used for drinking. The baths are supplied with chalybeate water from the Carolaquelle. Physician, Dr. **Leva**. A covered wooden bridge leads from the Curhaus to the springs on the right bank, with the *Trinkhalle* (concerts in the morning at the Trinkhalle, afternoon and evening in the garden of the Curhaus). A good road ascends thence in zigzags to the (½ M.) health-resort of **Vulpera** (4160'), prettily situated on a sunny plateau near the wood, and also frequented by patients (*Waldhaus*, farthest to the E., with dépendances, R., L., & A. from 3, B. 1½, D. 3½, pens. 9 fr.; *Bellevue*, R. from 2½, A. ½, pens. from 9 fr.; *Tell & Alpenrose*, pens. from 8½ fr.; all of these belong to the same hotel company; *Conradin, 7½ fr.).

Beyond the **Curhaus** the road re-ascends, past the **English Church** (on the left), to —

34 M. **Schuls**. — *Hôt. Belvedere*, with view-terrace (pens. from 9 fr.), with the dépendance *Hôt. du Parc & Alt-Belvedere* in Unter-Schuls (pens. from 7 fr.). — *Post*, R., L., & A. 4, D. 4, S. 2½ fr.; *Quellenhof*, R. 2, B. 1, S. 2½ fr.; *Hôt. König 'Zum Piz Chiampatsch'*, R. 2, D. 2 fr. 80, S. 2 fr. 20 c., B. 1, board 5 fr.; *Krone*, plain; all these at Upper Schuls. At Lower Schuls, *Helvetia*, moderate (R. and B. only); *Hôtel Central*, plain.

Omnibuses of the innkeepers from Schuls to Tarasp between 6 and 8 a.m. every 10 min., between 8 and 12 every hour; fare there and back 30 c. — Extra-Post to *Davos*, with 2, 3, or 4 horses, 73 fr. 70, 101 fr. 75, 129 fr. 80 c.; to *Nauders*, 37 fr. 20, 51 fr. 50, 65 fr. 80 c. — Carriage from Schuls to Tarasp Curhaus and back, with stay of ½ hr., 3 fr., with two horses 5 fr.; to Vulpera and back (½ day) 7 or 12, to Sent 7 or 12, to Fetan 9 or 15, Süs 15 or 25, St. Moritz or Pontresina 38 or 70, Landeck 50 or 85, **Meran** 90 or 170 fr.

Physicians. *Dr. Dorta; Dr. Vogelsang*. — Weekly subscription to the 'Verschönerungsverein' of Schuls, Tarasp, and Vulpera 75 c.

Schuls (3980'; pop. 940), Rom. *Scuol*, the capital of the Lower Engadine, picturesquely situated opposite a noble range of mountains extending from Piz Lat to Piz Plavna, consists of *Upper* and *Lower Schuls*, between which the high-road runs. On this road is the *Badehalle Schuls*, with chalybeate and ordinary baths (1½-2½ fr.). In the vicinity are several chalybeate springs. The most important are the *Vihquelle*, with an interesting hill of iron-ore, ½ M. to the N., and the carbonic *Sotsass-Quelle*, a little to the E., on the road to Sent (p. 418). Many visitors of the Baths of Tarasp live at Schuls. — The direct path from Schuls to (½ hr.) Vulpera and Tarasp diverges to

to Nauders. SCHULS. *VI. Route 102.* 417

the left from the road at the W. end of Schuls, crosses the Inn above the junction of the *Clemgia* and then divides, the right branch skirting the Inn by the Cur-Promenade to (1/2 hr.) Tarasp, the left branch ascending through wood to (1/2 hr.) Vulpera.

ENVIRONS. The handsome **Castle of Tarasp** (4935'), 1 hr. from Vulpera, now dilapidated, was the residence of the Austrian governors down to 1803. A good road leads round its N. base to the hamlets of *Florins* (Restaurant) and (1 hr.) *Fontana* (4690'; rfmts. at the former Hôtel Tarasp, now a nunnery), at the S.W. base, with a Capuchin monastery and a small lake. Pleasant walk thence to the (1 1/2 hr.) *Alp Laisch* (5995'; milk), at the entrance to the picturesque Val Plavna. — Beautiful view from the *Kreuzberg* (4860'), especially by evening light (from Fontana past the castle of Tarasp and via *Sparsels*, 1/4 hr.; from Vulpera direct, 1 hr.). — Pretty walk from Vulpera to the farm of (35 min.) *Avrona* (4790'), situated above the deep *Clemgia Gorge* and at the base of the Piz Pisoc, and to the small dark-green *Schwarze See* (5050'), 20 min. higher, where we obtain a fine view of Piz Linard.

Road from Schuls to (4 1/2 M.) *Fetan*, see p. 415; a footpath leads past the Vihquelle, and along the edge of the wood in 1 1/4 hr. — Road to (2 3/4 M.) *Sent*, see p. 418 (diligence in summer twice daily). — Beautiful walk from Sent (2 1/2 hrs. there and back) to the wild *Val Sinestra* (p. 410), which may be ascended to a point opposite *Manas;* splendid forest; far below, between limestone rocks, the brawling torrent.

To the Val d'Uina: a picturesque footpath follows the right bank of the Inn, passing *Pradella* to (4 1/2 M.) *Sur En* (3650'; Bär, with sign painted by Paul Meyerheim), situated at the mouth of the valley, opposite Crusch (p. 418). Driving is also practicable to this point, via Crusch. A tolerable path ascends hence through the richly-wooded valley, passing several waterfalls and through a romantic rocky gorge, to the chalets of (1 1/2 hr.) *Ausser-Uina* (4980') and (1 hr.) *Inner-Uina*. An attractive pass (guide unnecessary for the expert) leads hence over (1 1/2 hr.) *Sursass* (7735') and through the pleasant *Val Schlinga* to (3 hrs.) *Mals*, see p. 428.

ASCENTS (guides, *Joh. Rauch, Jak. Bischoff, Jak.* and *Ed. Truog, Jak. Widal*, and *Brunett*). To the N. of Schuls rises the grassy **Muotta Naluns** (7015'; guide, not indispensable, 6-8 fr.), ascended in 2 1/2 hrs. (or from Fetan in 1 1/2 hr.). View limited; better from the *Piz Glüna* (9175'; from the Muotta Naluns in 2 hrs.; from Fetan by the *Alp Laret* in 3 1/2-4 hrs.; guide 10 fr.). — More extensive panorama from **Piz Champatsch** (9596'; 5 hrs. from Schuls; guide 12 fr.), by the *Alp Champatsch*, and thence round the summit, ascending finally on the E. side. The direct ascent from the S. is steep, stony, and tiring.

°**Piz Lischanna** (10,200'; 6 hrs.; guide 15 fr.) is perhaps the finest point of view near Schuls. From the Scarl road (see p. 418), at the second bend, we diverge to the left by a steep forest-path to *St. Jon* (4820'), with the ruins of a house. Here we turn to the left and skirt the base of the *Piz St. Jon*, then ascend through pastures and wood in the *Val Lischanna*, to (3 hrs.) the *Schafalp* (6760'; no accommodation). The path then ascends a stony slope in long zigzags, passing the *Lischanna Glacier* on the right, above us, and skirting steep rocks at places, to the (3 hrs.) iron vane on the top. The view is superb: immediately in the foreground rise the bare and riven peaks of the Piz St. Jon, Ayutz, and Pisoc; far below lies the green Engadine from Lavin to Martinsbruck; to the S. are the Ortler, the Valtellina Alps, and the Bernina; in the distance, to the W., the Bernese Alps, the Tödi, and nearer us Piz Linard and Piz Buin; to the N. the Augstenberg, Fluchthorn, and the distant Zugspitze; to the E. the Oetzthal Mts. with the Wildspitze and Weisskugel, and farther distant the fantastic Dolomites. — Adepts (with guide, 25-30 fr.) descend the *Lischanna Glacier* to the *Val Seesvenna* and *Scarl* (13 hrs. from Schuls; see p. 418).

Piz Pisoc (10,427'; 7 hrs.; guide 25 fr.), Piz Plavna Dadaint (10,413'; 8 hrs.; 30 fr.), and Piz Seesvenna (10,565'; 8 hrs.; 25 fr.; night spent at Scarl), all difficult, for experts only. Piz St. Jon (9980'; 8 hrs.; 15 fr.), Piz Cot-

BAEDEKER, Switzerland. 16th Edition. 27

schen (p. 415), *Piz Minschun* (10,080'; from Fetan 5 hrs.; 10 fr.), and *Piz Foraz* (10,150'; 7 hrs.; 15 fr.) are less difficult.

FROM SCHULS TO ST. MARIA IN THE MÜNSTER-THAL, through the Scarlthal (*Val da Scarla*), 8 hrs., interesting (guide 25 fr., unnecessary). We ascend the road to the S. from the Inn bridge, soon enter a larch-wood, and reach the plateau on which St. Jon (see above) lies farther to the left. Opposite, high up on the left side of the deep gorge of the *Clemgia*, lies the farm of *Avrona* (p. 417). The road, bad at places, gradually descends through wood into the valley, enclosed by the huge furrowed slopes of *Piz Pisoc* on the right and *Piz St. Jon* and *Piz Madlain* on the left, and frequently crosses the Clemgia, the inundations of which are often very destructive. After 2 hrs. the sequestered *Val Minger* diverges to the right, with *Piz Foraz* (see above) in the background. To the left is the *Val del Poch*. Passing a deserted foundry, we next reach (1 hr.) Scarl (5948'; *Adler*, *Edelweiss*, *Pens. Feuerstein*, all plain but not cheap), a hamlet at the mouth of the *Val Seesvenna*, whence *Piz Cornet* (9951'), *Piz Cristannes* (10,237'), and *Piz Seesvenna* (see above) may be ascended. To the left, ¹/₂ hr. above Scarl, a bridle-path leads over the *Cruschetta Pass* (*Scarljöchl*, 7600'), and through the pretty *Val Avigna*, in 3 hrs. to *Taufers* (p. 428). The road ends here. The bridle-path crosses the valley, which expands here (beautiful stone-pines); it passes the chalets of *Astras Dadora* (i.e. outer), and *Dadaint* (i.e. inner), and, bearing to the left, leads between (r.) *Piz d'Astras* (9836') and (l.) *Piz Murtera* (9836') to the Costainas Pass (7885'), 2 hrs. from Scarl. It then descends to the extensive dairy of *Champatsch* (7034'), in the parish of Valcava, rounds the rock of *La Durezza*, and leads through wood (avoid steep path to Cierfs, descending to the right) to *Lü* (6293'), a sunny and sheltered hamlet; then by a narrow road to *Lüssai*, and across the *Rambach* to *Furom*, a solitary house on the Ofenberg road halfway between *Fuldera* and *Valcava*. Thence to *St. Maria* (2 hrs. from the Cotainas Pass), see p. 414.

The OFENBERG INN (p. 414) may be reached from Schuls by the *Val da Scarl*, the *Costainas Pass*, and *Cierfs* (thence by road) in about 10 hrs. (see above). A shorter route ascends the wild *Val Plavna* from *Fontana* (p. 417) and crosses the *Fuorcletta* (8785') to the *Val del Botsch*, which opens about 1 M. before the *Ofenberg Inn* (*Osteria del Fuorn*), on the road described at p. 406 (about 8¹/₂ hrs.; guide desirable). From the Ofenberg to Livigno (5-5¹/₂ hrs.; guide unnecessary), see p. 414.

The road to Nauders leads along the slope above the river. About ¹/₄ M. from Schuls a road ascends to the left, passing *Sotsass*, with its mineral spring (carbonic acid gas), to (2¹/₂ M.) Sent (4724'; Rhätia), a handsome village (1000 inhab.), with the ruins of the Romanesque church of *St. Peter* on a rocky point. Farther on we pass *Pradella*, on the right bank of the Inn. At (4¹/₂ M.) Crusch (4075'; Kreuz) the Sent road (see above) rejoins ours. About ¹/₂ M. farther on a road descends to the right to (1 M.) *Sur En*, at the mouth of the *Val d'Uina* (p. 417). Above us on the left, beyond the deep ravine of the *Val Sinestra*, lies (2¹/₄ M.) *Remüs*, Rom. *Ramosch* (4022'), with the ruined castle of *Tschanuff*

Piz Arina (9452), from Remüs 4 hrs., with guide, somewhat laborious; the view of the Oetzthal and Arlberg Alps is scarcely inferior to that from Piz Lischanna. — An easy and attractive route leads through the *Val Sinestra*, with chalybeate springs containing arsenic, and over the Fimber Pass (8694') to *Ischgl* in the Patznaun (8¹/₂ hrs.; guide 20 fr.). The bridle-path ascends on the left bank of the *Sinestra* by *Manas*, past the mouth of the *Val Laver* on the left and the farm of *Suort*, to the (2 hrs.) chalets of *Griosch* (5943'), at the foot of the huge *Stammerspitze* (10,683'; highest peak first ascended in 1884 by Prof. Schulz of Leipzig). On the right opens the *Val Tiatscha*, with the *Muttler* (10,827') in the background. Then through the *Val Chöglias* to the alp of that name, and to the left to the (2¹/₂ hrs.) pass, where we have a striking survey of the Fluchthorn.

Descent through the *Fimber-Thal* to (4 hrs.) *Ischgl* in the *Patznaun*; see Baedeker's *Eastern Alps*.

The valley contracts; to the right are the ruin of *Serviezel* and a bridge over the Inn. In the narrow *Val d'Assa* on the right (fine waterfall at the entrance) is the (2 hrs.) intermittent *Fontana Chistaina*, which flows once in 3 hrs. only. Near it is an interesting stalactite cavern. A fine view of the loftily situated *Schleins* is soon revealed; above it to the left rise the *Muttler* and the indented *Stammerspitze* (p. 418); to the right *Piz Lat* (9190').

The next village (4 1/4 M.) is *Strada*. Near (1 1/4 M.) **Martinsbruck** (3343'; *Hôt. Denoth zur Post*) the scenery becomes grander. The Inn Bridge is the boundary between Switzerland and Tyrol (Austrian custom-house). On the left are the ruins of another castle named *Serviezel*. (Path on the left bank of the Inn viâ the *Novellerhof* in 1 1/2 hr. to *Old Finstermünz*, see p. 429, and on to *Pfunds*; guide advisable for novices.) The new road to Nauders, on the Tyrolese side, winds up the wooded hill which separates the Inn Valley from that of the *Stille Bach*. (The old road, preferable and shorter, ascends to the right by the custom-house, past the small houses.) At the top of the hill we enjoy an admirable retrospective view of the Lower Engadine; to the N. rises *Piz Mondin* (10,375'). Lastly a slight descent to (5 1/2 M.) —

50 M. *Nauders* (4468'), see p. 428.

103. From Samaden-Pontresina over the Bernina to Tirano and through the Valtellina to Colico.

Comp. Maps, pp. 394, 412.

76 M. DILIGENCE in summer from Samaden to (24 1/2 M.) Poschiavo twice daily in 5 1/2 hrs. (9 fr. 80, coupé 11 fr. 80 c.); thence to (11 M.) Tirano in 1 3/4 hr. (4 fr. 30, coupé 5 fr. 15 c.); from Tirano to (16 M.) Sondrio in 2 3/4 hrs. RAILWAY from Sondrio to (25 1/2 M.) Colico in 1 hr. 35 min. (4 fr. 65, 3 fr. 25, 2 fr. 10 c.). — EXTRA-POST and pair from Samaden to Poschiavo 60 fr.; CARRIAGE with one horse from Pontresina to Poschiavo 35, with two horses 70, to Tirano 50 and 90 fr.; one-horse carriage from Poschiavo to Tirano 12, with two horses 22 frs., to Sondrio 30 and 45, to Bormio 40 and 65, to Pontresina 30 and 50, to St. Moritz 40 and 60 fr. This route will even repay walkers, as far as Tirano, but should not be preferred to the Val Bregaglia (p. 422).

The **Bernina Pass**, the only carriage-road over the Bernina chain (p. 403), is the chief route between the Engadine and the Valtellina, and is frequented even in winter. — The journey through the VALTELLINA has been much facilitated by the new railway from Sondrio to Colico, which also affords an excellent route to the North Italian lakes from the Engadine.

From *Samaden* to (3 1/4 M.) *Pontresina*, see p. 402; from Pontresina over the *Bernina Pass* to (16 1/2 M.) *La Rösa*, see p. 411. — The road soon passes to the E. slope, where we obtain a passing view of the upper part of the narrow *Poschiavino Valley*, down to Poschiavo. Below (1/2 hr.) *Sfazzu* (where a direct, but bad and stony footpath from La Rösa debouches; bridle-path to the *Val Viola Pass* see p. 412) we cross the brook descending from the *Val di Campo*, pass *Pisciadella* (4910') on the right, and descend in a wide

27*

curve on the E. side of the valley. The road reaches the bottom of the valley at (4½ M.) *S. Carlo* (3590'), where it passes through a gateway. On the hill to the right appears a glacier descending from the *Pizzo di Verona* (11,360').

24 M. **Poschiavo**, Ger. *Puschlav* (3315'; pop. **2953**; *Croce Bianca; Hôt. Albricci*, in the principal piazza, R., L., & A. 3½ fr.), a busy little town, with several factories and handsome houses. The language is Italian, and one-third of the inhabitants are Protestant. The *Roman Catholic Church* dates from 1494, but the tower is much older; good wood-carving in the interior. The charnel-house behind the church contains numerous skulls and bones. The town-hall bears the arms of the town. The *Protestant Church* is modern.

Sassalbo (9375'; 6 hrs.; with guide), tiring, but very attractive. From Poschiavo we ascend to the E. to the (3 hrs.) *Alp Sassiglione* (6310'; spend night), and mount by the *Forcola di Sassiglione* (8330') on the S. side to the (3 hrs.) summit. Grand panorama: W. the Bernina, E. the Ortler, S.E. the Adamello. — To the *Val Malenco* over the *Canciano* or the *Confinale Pass*, see p. 410.

Omnibus from Poschiavo to (3 M.) Le Prese 4 times **daily (6 and** 10 a. m., 2 and 6.30 p. m.) in 1½ hr. (1 fr.; carr. with one **horse 4,** two horses 7 fr.). The road crosses the Poschiavino, traverses a **pleasant level** valley, and passes *S. Antonio*.

27 M. **Le Prese** (3155'; *Curhaus*, R., L., & A. from 5½, D, 4½, pens. **8-12** fr., Engl. **Ch. Serv.** in summer; *Inn*, fair), a wateringplace at the N.W. end of the *Lago di Poschiavo*, well stocked with trout, is suitable for some stay. The alkaline and sulphureous spring (46° Fahr.; baths heated by steam, 2 fr.) rises 100 paces from the bath-house.

The road skirts the W. bank of the lake, passing old fortifications, destroyed in 1814. At the S. end is the (2 M.) village of *Meschino*, with a beautiful view of the lake with the snow-mountains in the background. We now descend a narrow, rocky gorge, accompanied by a series of waterfalls all the way, and reach (30½ M.) **Brusio,** Ger. *Brüs* (2477'; *Post*, poor), the last large Swiss village (pop. **1160;** ⅓ Prot.), with a Roman Catholic and a Protestant church, the latter built at the beginning of the 17th century.

The road descends through **walnut and** chestnut-plantations (pretty fall of the *Sajento* on the right) to *Campascio* and —

31½ M. **Campo Cologno** (1835'; *Albergo Rezia*, near the postoffice, R., L., & A. 2, B. 1, D. 3, pens. 5 fr.), where vineyards begin. The Italian custom-house is near the old fort *Piatta mala*.

34 M. **Madonna di Tirano** (1500'; *Albergo S. Michele*, R. 3, B. 1 fr.) is a small village built **around** an imposing pilgrimage-church of the 16th century. We **here reach the** *Valtellina*, Ger. **Veltlin, the** broad valley of the **Adda**, which belonged to the Grisons down to 1797. The floor of the valley is frequently devastated by inundations. The fertile slopes yield excellent red wine (p. 354). The road unites here with the Stelvio route (p. 425), on which lies —

35 M. **Tirano** (1475'; pop. 6000; *Italia*, dear; *Hôt. Stelvio*, by the lower bridge, well spoken of, *Posta*, well spoken of), a small town with old mansions of the Visconti, Pallavicini, and Salis families. In the background, to the E., rises *Monte Mortirolo*.

The road to Colico leads back to Madonna di Tirano, and crosses the *Poschiavino*. At (42½ M.) *Tresenda* (1235') a bridge crosses the Adda to the road which leads by the *Passo d'Aprica* (4040') to *Edòlo* and *Brescia* (see *Baedeker's Northern Italy*; a footpath to the Passo d'Aprica, ½ hr. shorter, leads to the left from Madonna, viâ the hamlet of *Staziona*). The old watch-tower of *Teglio* on the hill to the right gives its name to the valley *(Val Teglino)*.

51½ M. **Sondrio** (1140'; pop. 6900; *Posta*, R., L., & A. 4½, D. 4 fr.; *Maddalena*; *Restaurant Briolini Marino*, in the Piazza Vittorio Emanuele, with beds, well spoken of), the capital of the Valtellina, grows excellent wine (Sassella, Grumello, Inferno, Montagna). The wild *Malero*, descending from the Val Malenco (see below), which has frequently endangered the town, now flows through a broad artificial channel. The old castle is used as a barrack.

The *Corno Stella* (8665'; very attractive and not difficult) may be ascended in 7-8 hrs. from Sondrio viâ the *Val del Livrio*.

In the *Val Malenco* a good road on the right bank of the Malero (diligence in 2¾ hrs., down in 1½ hr.) ascends by *Torre* to (10 M.) *Chiesa* (3297'; *Hôt. Olivo*), the principal village in the valley, finely situated. (Guides, Mich. and Silvio Schenatti, G. Olivo.) Interesting asbestos-mines in the neighbourhood. Pleasant walks from Chiesa: to the *Palù Lake* (6920'), beautifully situated; by *Lanzada* to the waterfall at the head of the *Val Lanterna*; to the *Pirlo Lakes* (6890'), etc. — From Chiesa over the *Muretto Pass* (8890') to the *Maloja* (8 hrs.), see p. 395; over the *Tremoggia* or the *Scerscen Pass* to *Sils* (9-10 hrs.), see p. 397; over the *Sella Pass*, the *Bellavista Saddle*, or the *Cambrena Pass* to *Pontresina* (16-17 hrs.), see p. 409; over the *Canciano* or *Confinale Pass* to *Poschiavo* (8-9 hrs.), see p. 410. The *Fellaria Chalets* (p. 409) may be reached from Chiesa in 4½ hrs., through the *Val Lanterna* (guide advisable, as there is no path; from Fellaria to the *Capanna Marinelli* 3 hrs.). — Monte della Disgrazia (12,050'), 11 hrs. from Chiesa, not difficult for adepts. We spend the night in the (7-8 hrs. from Chiesa) *Capanna della Disgrazia* of the I. A. C. on the *Cornarossa Pass* (9180'), between the Val Malenco and the Val di Sasso Bissolo; and thence ascend to the (4 hrs.) summit (small hut), which commands a splendid view. A shorter ascent leads from the Val Masino (see below). From *Cataeggio* (1½ hr. from the Bagni del Masino) we proceed through the *Val di Sasso Bissolo* viâ the *Preda Rossa Alp* to the (4½ hrs.) *Capanna Cecilia* of the I. A. C. (8280'), 5 hrs. below the summit (guide 30 fr.). The descent through the *Val di Mello* to the Bagni (about 7 hrs.) is not difficult and highly picturesque.

The RAILWAY STATION lies about ½ M. to the S. of the town (omnibus 50 c.). As the train leaves it we have a glimpse into the Val Malenco and cross the *Malero*. To the right, on a rocky height and supported by galleries, rises the church of *Sassella*. 3½ M. *Castione*; the village lies on the slope to the right. 7 M. *S. Pietro-Berbenno*. — 11 M. *Ardenno-Masino*, at the mouth of the *Val Masino*.

Val Masino. The road (carr. from the Curhaus at the station, 7 fr. each pers., return 5 fr.) leads viâ *Masino*, *Pioda*, and *Cataeggio*, at the mouth of the *Val di Sasso Bissolo* (see above), to (7 M.) *S. Martino* (3724'), where the valley divides: to the right the *Valle di Mello* (route over the

Passo di Zocca or the *Forno Pass* to the Val Bregaglia, see pp. 423, 395), so the left the *Valle dei Bagni*. In the latter lie the (1¹/₄ M.) **Bagni del Matino**, with a good *Curhaus* (4330'). This valley, called *Val Porcellizza* above this point, turns to the N.; at its head towers the fine *Badile* group. The E. peak (*Piz Cengalo*, 11,060') is fatiguing, but without danger for mountaineers with good guides; the night is spent in the (4 hrs.) *Capanna Badile*, whence the top is reached in 3 hrs. The central *Piz Badile* (10,850') is very difficult (guide 25 fr.). Easier and at the same time attractive ascents are those of the *Piz Porcellizzo* (10,000'; 5¹/₂ hrs. viâ the Baita di Porcellizzo), *Monte Spluga* (9335'; 7 hrs., viâ the *Alp* and the *Bocchetta di Merdarola*), *Cavalcorto* (9070'; 4 hrs., viâ *Alp Scione*), etc. — *Monte della Disgrazia*, see p. 421. — Over the *Bondo Pass* (10,200') to the *Val Bondasca* (trying, for experts only), see p. 424. — Guides, **Ant. Baroni**, **Giul.** and **Giov. Fiorelli**, and **Fed. Cotta**, of S. Martino.

The train crosses the Adda, the right bank of which is here precipitous; the road runs high above. To the right, in the Val Masino, appears the *Mte. della Disgrazia* (p. 421). 14 M. *Talamona*. 16 M. **Morbegno** (853'; *Ancora*), noted for its silk-culture, is situated at the mouth of the *Val del Bitto*, through which a bridle-path leads over the *Passo di S. Marco* (5996') to *Piazza S. Martino* in the *Val Brembana* and to *Bergamo* (see Baedeker's *Northern Italy*). 18 M. *Cosio-Traona*. Beyond (21 M.) *Delebio*, on the *Lesina* (p. 458), which descends from Mte. Legnone, the railway unites with the line from Chiavenna to Colico (p. 385). On a crag to the right is the ruin of *Fuentes*. — 25¹/₂ M. *Colico*, see p. 458.

104. From the Maloja to Chiavenna. Val Bregaglia.
Comp. Map, pp. 382, 394.

19¹/₂ M. DILIGENCE from Samaden to (34 M.) Chiavenna over the Maloja, twice daily in 7 hrs. (from St. Moritz 5¹/₂, Silvaplana 5, Maloja Caveaal 4 hrs.); fares 13 fr. 65 c., coupé or banquette 16 fr. 40 c. — CARRIAGE with one horse from St. Moritz 45, with two horses 75-90 fr. — EXTRA-POST with two horses from Samaden 69 fr. 20 c. — Railway from Chiavenna to Colico, see p. 384. — This is the finest route from the Engadine to the Italian lakes. The change in the vegetation is very striking.

Maloja (5960'), see p. 394. The road, which was constructed in 1835-39, descends the precipitous slope of the Maloja (about 820' in height) in 12 curves, which may be avoided by walkers (from the sixth bend a footpath leads to the left to the Orlegna Waterfall, see p. 395). The pines and other coniferous trees immediately below the summit of the pass are very luxuriant. We then pass, on the right bank of the *Orlegna*, the ruins of the church of *S Gaudenzio* (on the right), where we have our last retrospect of the château of Count Renesse on the Maloja.

3 M. **Casaccia**, Rom. *Casätsch* (4790'; *Hôt.-Pens. *Stampa*), the highest village in the Val Bregaglia, is commanded by the ruin of *Turratsch*. Bridle-path to Stalla over the *Septimer*, p. 392.

The ***Val Bregaglia** (perhaps 'Prægallia', 'in front of' Cisalpine Gaul), Ger. *Bergeller-Thal*, is watered by the *Mera* or *Maira* and for the first two-thirds of its extent belongs to Switzerland. The inhabitants speak Italian, though some of the communities are exclusively Protestant. Nowhere else is the transition from the scanty vegetation of the higher Alps to the luxuriant flora of Italy so abrupt as in this valley. — The road intersects the open valley to the S. and ¾ M. below Casaccia crosses the Oriegna before its junction with the Maira. Beyond the hamlet of (½ M.) *Lobbia* (4720') we see to the left the *Cascata dell' Albigna* (see below) and other fine waterfalls descending from the mountains. The road now descends in windings (cut off by the old road, being the paved Roman road, following the telegraph-line) to *Asarina* (4435') Then past the mouth of the Val Albigna and the *Grotta di Albigna* (a beer-cellar) we reach —

7½ M. Vicosoprano, Rom. *Vespran* (3565'; pop. 339; **Couronne et Poste*, kept by *Maurizio*), the capital of the Val Bregaglia, with a handsome church, at the influx of the *Albigna* into the Mera. Curious rock-formations in the vicinity.

The Val Albigna deserves a visit. About ¾ M. above Vicosoprano we diverge to the right from the road, and ascend through wood to the (3 hrs.) *Cascata dell' Albigna*, a fine fall of the Albigna in a wild ravine, near the foot of the *Albigna Glacier*. The adjoining chalet (6773') is not always occupied. From this point over the *Cacciabella Pass* to *Bondo*, and over the *Casnile Pass* to the *Maloja*, see p. 396. — A trying route, to the S., crosses the *Albigna Glacier* and the *Forcella di S. Martino* (*Passo di Zocca*, 9000'), between the *Cima di Castello* (11,160'; ascended from the pass without difficulty in 2 hrs.) and the *Mte. di Zocca* (10,890'), to the *Val di Mello* and *S. Martino* (p. 421).

The Pizzo della Duana (10,280'; 6-7 hrs.; guide, the forester *Giov. Stampa* at Stampa, etc.), a magnificent point of view, is not difficult for adepts. The route leads from Vicosoprano to the N., by the *Alp Zocchetta* and *Pianlò*, to the small *Lago di Val Campo*, and ascends the arête from the E. side to the top. We may descend by the *Alp Pianaccio* to *Soglio*.

The next villages are *Borgonuovo*, Rom. *Bornöv* (3470') and Stampa (*Albergo Piz Duan, moderate). Picturesquely situated on the hill to the right lies *Coltura*, with the modern red château of Baron Castelmur and the white church of *S. Pietro*. The tower of Castelmur and the church of Promontogno are visible in front of us. Walnut trees and chestnuts now begin to appear in considerable numbers; but we do not find ourselves amidst the full luxuriance of the S. Alpine Italian flora until after we have passed the rocky gate of *La Porta*, at —

11 M. Promontogno (2685'), commanded by the handsome church of Our Lady and the ruined castle of *Castelmur*, from which old walls stretch down into the valley. — In an open situation below the village, to the left, is the **Hôt.-Pens. Bregaglia* (R., L., & A. 4-5, D. 4½, luncheon 3½, pens. 9 fr.). Behind the latter, at the entrance of the *Val Bondasca*, of which we obtain an attractive glimpse, lies the large village of Bondo, with a château of the Salis family. For three months in the year this village never sees the sun. Chestnuts and rhododendrons flourish here side by side.

Pleasant excursion (guide desirable; *Andrea Picenoni* of Bondo) to the Val Bondasca, and over the *Lombardoi*, *Laretto*, and *Naravedro Alps* to the (4 hrs.) highest *Alp di Sciora* (6785'), grandly situated. To the E. rise the Piz Cacciabella (9745') and the Pizzi di Sciora; to the S. are the Bondasca

Glacier and the bold Badile group (p. 422). — Over the crevassed *Bondasca Glacier* and the *Forcella di Bondo* (10,200') a hazardous route leads to the *Val Porcellizza* and *Bagni del Masino* (p. 422; 10 hrs. from Bondo). — Over the *Cacciabella Pass* to the *Albigna Glacier*, or over the *Casnile Pass* to the *Maloja* (15 hrs. from Promontogno), see p. 396.

The road now crosses the Mera, here swollen by the wild *Bondasca*, and passes the houses of *Spino* (2630'). A carriage-road diverges to the right to Soglio (see below). Mulberries, figs, and vines flourish here in luxuriant abundance.

13½ M. Castasegna (2235'; *Restaurant Schumacher*; *Alb. Svizzero*), a closely-built but pleasant village, is the last Swiss place.

Pleasant walk through a beautiful chestnut-wood, past the waterfall of the *Acqua di Stoll*, to (1 hr.) Soglio, Ger. *Sils* (3570', *Hôt.-Pens. Giovanoli*, in an old mansion of the Salis family). In the garden of the hotel the stone-pine or Alpine cedar is seen in curious juxtaposition with the chestnut. Fine view of the Bondasca Glacier. Descent by a new road to Spino (see above; carr. to Vicosoprano 10 fr.). — Over the *Duana Pass* to the Averser-Thal, see p. 382. — The Piz Gallegione (10,285'), 5 hrs. from Soglio, is not difficult (guide necessary). From Soglio in 3½ hrs. to the saddle (*Forcella*, 8924'), between the Gallegione and the *Cima di Cavlo*; then to the left in 1½ hr. to the top (splendid view).

Immediately below Castasegna, on the other side of the *Lovere*, which descends from the right, is the Italian Dogana or custom-house.

15 M. Villa, called *Villa di Chiavenna* to distinguish it from other places of the same name, is a large and picturesquely-situated village, with a conspicuous pilgrimage-church. 1¼ M. farther down we pass the village of *S. Croce*.

Near S. Croce (to the left), but on the opposite bank of the *Mera*, formerly stood the prosperous little town of *Plurs*, with 2430 inhab., which was entirely destroyed by a landslip from Mte. Conto in 1618. The mass of earth and rock which buried the town is 60' thick, and is now richly clothed with chestnuts. In 1861 one of the town-bells was found. — Near *Curtinaccio*, ¾ M. from the road and 4½ M. from Chiavenna, is the old baronial *Villa Roncalia*, with a fine panelled hall.

A little to the right of *S. Abbondio* is the fine double waterfall of the *Acqua Fraggia*. The road now leads through *Campedello* and a suburb of Chiavenna, the name of which, *Borgo Nuovo Piuro*, recalls the buried town of Plurs, to —

19½ M. *Chiavenna*, see p. 385. The railway-station is on the opposite side of the town.

105. From Tirano to Nauders over the Stelvio.
Comp. Maps, pp. 394, 412.

79 M. Messagerie from Tirano to Bormio daily in 6 hrs. (9 fr. 20 c.). Diligence daily (from middle of June to end of Sept.) from the Baths of Bormio over the Stelvio to Eyrs in 10½ hrs. (coupé 7 fl. 35 kr.; also open carriages), leaving the Baths at 6.30 a.m., arriving at S. Maria at 10.30, Franzenshöhe at 1, Trafoi at 3, Prad at 4.30, and Eyrs at 5.20 p.m.; from Eyrs to Nauders daily in 5, to Landeck (p. 429) in 10½ hrs. (7 fl. 14 kr.). — Extra-Post with two horses from Tirano to the Baths of Bormio 50 fr. — Return-carriages to Tirano and Bormio are frequently met with at Poschiavo (p. 420). One-horse carr. from Pontresina to Bormio 80, two-horse 120 fr., a night being spent at Le Prese (to Le Prese 5¾ hrs.; thence to Bormio 8 hrs.). Carriage and pair from Samaden to Meran over the Stelvio in 3½ days, 250 fr. and 25 fr. fee. Extra-post and pair from the Baths of Bormio to Trafoi in 6½ hrs., 60 fr.

WALKING. The scenery will reward walkers. From the Baths of Bormio to S. Maria 4½–5, to the Stelvio Pass 1, Franzenshöhe 1½, Trafoi 1½, Prad 2 hrs.; so also from S. Maria over the Wormser Joch to S. Maria in the Münster-Thal in 3, Münster ¾, Taufers ⅝, and Mals in 1½ hr.

For fuller details as to RR. 105, 106, see *Baedeker's Eastern Alps*.

Tirano (1475'), see p. 421. The road ascends along vine-clad hills to the *Sernio* (2080') region of the valley. To the N. rises the precipitous *Mte. Masuccio* (9240'), a landslip from which in 1807 blocked the bed of the *Adda*, and converted the valley as far as *Tovo* into a lake. At (5 M.) *Mazzo* the road crosses the Adda, and at the large village of *Grosotto* (Albergo Pini) the *Roasco*, which descends from the *Val Grosina* (p. 412). On the left are the conspicuous ruins of the fortress of *Venosta*. We cross the Adda again beyond *Grosio*.

12 M. **Bolladore** (2820'; *Posta or Angelo; Hôt. des Alpes*). On the hillside to the N. stands the pretty church of *Sondalo*. The valley contracts; the vegetation becomes poorer; below us dashes the grey glacier-water of the Adda. At *Le Prese*, about 1½ M. beyond (½ M.) *Mondadizza* we again cross the Adda. The defile of *La Serra di Morignone*, 3¾ M. long, separates the Valtellina from the region of Bormio; at the entrance to it, on the right, are remains of old fortifications. The *Ponte del Diavolo* was the scene of a sharp skirmish between the Austrians and Garibaldians on 26th June, 1859. At the end of the defile is the hamlet of *Morignone* in a green dale *(Valle di Sotto)*; the church stands on the hill far above it. The next group of houses is *S. Antonio*, with brick-works.

Beyond (3¾ M.) **Ceppina** opens the broad green valley *(Piano)* of Bormio, enclosed by lofty mountains, which are partly covered with snow. At *S. Lucia* the road crosses the *Frodolfo*, which unites with the Adda below the bridge, and turns to the N.E. to (3¼ M.) —

25½ M. **Bormio**, Ger. *Worms* (4020'; *Posta or Leone d'Oro; *Torre or Cola, in the Piazza Cavour), at the entrance to the *Val Furva*, an old-fashioned little Italian town, with dilapidated towers.

FROM BORMIO TO LIVIGNO a bridle-path (7 hrs.; without guide; better in the reverse direction; narrow road under construction). At *Premadio* it crosses the Adda and ascends the Val di Dentro to (1½ hr.) *Isolaccia* (p. 412). On the slope to the right lies the hamlet of *Pedenosso*, above which, on the saddle of the *Monte delle Scale*, rise two towers which once defended that pass *(Scale di Fraele;* 6370'). [Over the Scale di Fraele to *S. Giacomo di Fraele* (6390') and over the *Val Mora Pass* and the *Giufplan* (7723') to the *Buffalora Alp*, near the *Ofen Pass* (p. 414), and *Zernetz*, 12 hrs.; guide desirable, 20 fr.] Beyond Isolaccia the path ascends on the left bank of the brook; ½ hr., *Semogo* (4673'; Martinelli); above us, opposite, at the mouth of the *Val Viola*, is the church of *S. Carlo*. (Val Viola Pass to the Bernina, see p. 412.) From the (2½ hrs.) **Foscagno Pass** (7556'), with two little lakes, we have a retrospect of the Val Viola and the S. Ortler Mts. Descent to (1 hr.) *Trepalle* (6850'); then to the W., over the hill, to (1½ hr.) *Livigno* (p. 414).

At Bormio the windings of the Stelvio road begin. (The diligence starts from the New Baths, 2 M. from Bormio; carriage thither from Bormio in the morning, if ordered previously.)

27½ M. **Baths of Bormio.** The *New Baths (Bagni Nuovi, 4370'; with post-office; R., L., & A. 3½-4, B. 1½, D. 4, S. 3 fr.), on a terrace, with a fine view of the valley of Bormio and the surrounding mountains, are much frequented in July and August, and remain open till the middle of October. The water (92-100°) is conveyed by pipes from the springs, ³/₄ M. higher, **at the Old Baths** (see below), **which may be reached by a footpath** as well as by the road.

The Stelvio road, constructed in 1820-25, ascends in a long curve, with beautiful **retrospects of the valley** from Bormio to Ceppina; to the S.W. the *Corno di S. Colombano* (9915'), *Cima di Piazzi* (11,280'), and *Cima Redasco* (10,300'), to the S.E. the *Mte. Valaccetta* (10,425') and **the** icy pyramid of *Piz Tresero* (11,820'), at the upper end of the *Val Furva;* to the W. the *Val Viola* (p. 412). We cross an iron bridge, and pass through a short tunnel *(Galleria dei Bagni)*, beyond which the *Old Baths (Bagni Vecchi;* 4757') lie below the road **on the left**. Beyond the deep gorge of the Adda rises the precipitous *Mte. delle Scale* (p. 425).

To the left, farther on, the *Adda* emerges from the wild *Val Fraele*. A copious brook, which flows from the cliffs below the mouth of the Val Fraele, is sometimes erroneously described as the source of the Adda. A succession of galleries, partly of wood, and partly hewn in the rocks, for protection against avalanches, carry the road through a defile *(Il Diroccamento)* to the *Iᵃ Cantoniera di Piatta Martina* (5585'), a hospice for travellers, and the *IIᵃ Cantoniera al piede di Spondalonga* (6495'), which was destroyed by Garibaldians in 1859. On the W. side of the valley rises the abrupt *Mte. Braulio* (9777'). The road crosses the brook issuing from the *Val Vitelli* by the *Ponte Alto*, and ascends in windings (short-cuts). In a gorge to the left are the *Falls of the Braulio*. We next pass the *Casino dei Rotteri di Spondalonga* (7510') and the *IIIᵃ Cantoniera al Piano del Braulio* (7875'; Inn, tolerable), with a chapel.

34 M. **S. Maria** (8150'; *Gobbi's Inn*), the *IVᵃ Cantoniera* and the Italian custom-house.

A bridle-path, formerly the only route between the Vintschgau and Valtellina, diverges to the left from the Cantoniera S. Maria to the **Wormser Joch** or *Giogo di S. Maria* (8240'), and descends through the *Muranza Valley* to (2½-3 hrs.) the Swiss village of *S. Maria* in the Münster-Thal (p. 428); thence by *Taufers* in 3½ hrs. to *Mals* (p. 428) in the Adige valley.

*Piz Umbrail (9950'), the E. and highest peak of the bold serrated mountains which bound the Val Braulio on the N., is a remarkably fine point (1¾ hr.; guide, for novices, 5-6 fr.). Turning to the left by the dogana, we ascend a grassy slope and then a stony zigzag path to the summit. Superb view (see Panorama by Faller). Travellers from Bormio may ascend this peak by diverging to the left from the road ¼ hr. beyond the Third Cantoniera (see above) and ascending to (1 hr.) a small lake, whence they mount over rocks to the (1 hr.) top. Descent to the Fourth Cantoniera.

Beyond S. Maria the road affords glimpses of the Münster-Thal to the left. On the right rises the huge **Eben Glacier**. The **pass is never free from snow except in** warm summers; in June heaps of snow, 6-8' deep, are often seen on the roadside. On the (1½ M.)

Stelvio Pass (*Stilfser Joch* or *Ferdinandshöhe*, 9055') stands a road-menders' house (rfmts. and beds at the *Dreisprachen-Hütte*). A column to the right marks the boundary between Italy and Tyrol. About 1/2 M. to the N. is the frontier of Switzerland (Grisons).

A path ascends to the left in 10 min. to the *Dreisprachenspitze* (9325'), a rocky height affording an admirable view, particularly of the Ortler, whose snowy dome rises immediately opposite. The bare, reddish *Monte Pressura* (*Röthelspitze*; 9940'), towards the N.W., intercepts the view of the Münster-Thal.

The road now descends the **talc-slate** slopes in long windings (to Trafoi 33 in all). To the right, high above the snowy slopes, rise the *Geisterspitze* (11,355') and *Tuckettspitze* (11,400'). As the road affords the finest views, the short-cuts should be avoided.

40 1/2 M. **Franzenshöhe** (7180'; *Inn*, R. 70 kr.), formerly a post-station. To the S. the huge *Madatsch Glacier* extends far into the valley. About 2 M. farther on, just beyond the 18th kilomètre-stone, is the spot where Madeleine de Tourville, an English lady, was murdered by her husband, a Walloon, in 1876. The *Weisse Knott*, a small platform a little farther on, is an excellent point of view: before us rises the sombre Madatschspitz; to the right the Madatsch Glacier, with its splendid ice-fall; to the left the Trafoier Ferner, and above it the Pleisshorn and Ortler; in the valley far below, amidst pines, is the chapel of the Three Holy Springs (see below). In the back-ground to the N. rises the snowy pyramid of the Weisskugel.

45 M. **Trafoi** (5080'; *Post*; *Zur Schönen Aussicht*), a small hamlet, finely situated at the foot of the *Ortler* (12,800'), the highest of the Eastern Alps, which may be ascended hence in 8-9 hrs., or from Sulden in the Sulden Valley in 7-8 hrs. (spending the night in the Payer Hut; comp. *Baedeker's Eastern Alps*). One-horse carr. to Prad 3 1/2 fl. Austrian custom-house.

Pleasant walk (3/4 hr.) to the *Three Holy Springs* (5263'), which rise in the valley below, at the foot of the Ortler. The path (guide unnecessary) diverges from the road to the left, 3 min. above the Post, and leads at the same level through meadows and wood, and over moraine. At the end of the valley are figures of Christ, Mary, and St. John, under a roof, from whose breasts flows the very cold 'holy water'. Adjacent are a chapel and an auberge for pilgrims. Opposite rises the huge and abrupt Madatsch, over the dark limestone rocks of which two brooks are precipitated. Above, to the left, are the ice-masses of the Trafoier and Lower Ortler-Ferner, overtopped by the Trafoier Eiswand. The scene is interesting and impressive.

We now follow the impetuous *Trafoi-Bach*, the inundations of which are sometimes very destructive, and pass (3 1/2 M.) *Gomagoi*, Ger. *Beidewasser* (4265'; *Reinstadler*), with a small fort erected in 1860. To the right opens the picturesque *Sulden Valley*, a great resort of mountaineers, with the (2 1/4 hrs.) village of *St. Gertrud* or *Sulden* (6050'; *Sulden*, first-class; *Eller*; comp. *Baedeker's Eastern Alps*).

The narrow valley barely affords room for the road and river. The latter forms several falls. On the hill to the left lies the village of *Stilfs*, Ital *Stelvio*, which gives its name to this route.

53 M. Prad (2940'; *Neue Post; *Alte Post*), or *Brad*, lies at the foot of the Stelvio route. The road intersects the broad valley of the *Etsch* or *Adige*, crosses a marsh and the river by a long viaduct, and reaches (2 M.) *Neu-Spondinig* (2855'; *Hirsch*), on the highroad from Botzen and Meran to Landeck, 1½ M. to the W. of *Eyrs*.

WALKERS may avoid the glaring and fatiguing road from Prad by Spondinig to Mals by diverging to the left at Prad, on the right bank of the Adige, and following the foot of the mountains, to *Aguins*, *Lichtenberg* ("Iun), charmingly situated amidst fruit-trees, with a ruined castle (see below), *Glurns* (3260'; Krone), a small fortified town with an old church, and (2½ hrs.) *Mals* (see below).

To THE MÜNSTER-THAL a narrow road leads from Glurns to the W., on the right bank of the *Rambach*, which here falls into the Adige. After 2½ M. it crosses the brook. (Route on the right bank by *Riffair* not recommended.) The (5 M.) loftily situated village of *Taufers* (4042'), with its three churches, is commanded by three ruined castles. (Over *La Cruschetta* to *Scarl*, see p. 418.) A broad road leads hence to the (1/2 M.) Swiss frontier and (1/2 M.) **Münster**, Rom. *Mustair* (3765'; *Hôt.-Pens. Münsterhof; Piz Ciavalatsch; Hirsch*), the first village in the Grisons, with a large Benedictine church. The road descends, crosses the Rambach (passing the *Ava da Pisch*, a fine waterfall in a wooded gorge to the left), and ascends gradually by *Sielva* to (2 M.) **S. Maria** (4553'; *Piz Umbrail; *Weisses Kreuz*), a large village at the mouth of the *Val Muranza*. Over the Wormser Joch to Bormio, see p. 426; over the Ofen Pass to Zernetz, see p. 414; through the Val da Scarla to Schuls, see p. 418.

The road to Nauders skirts the hillside at some distance from the Adige. The valley is called the *Upper Vintschgau*, after the *Venosti*, its ancient inhabitants. To the left, on the opposite bank, rises the half-ruined castle of *Lichtenberg*. On the right is the *Churburg*, a château of Count Trapp. To the left lies *Glurns* (see above), near which the *Rambach* (see above) flows into the Adige. On the road lies *Tartsch*. Near Mals is the ancient tower of the *Frölichsburg*.

61½ M. **Mals** (3445'; *Post* or *Adler; Bär; Hirsch*) is a village of Roman origin (viâ *Sursass* to the *Val d' Uina*, see p. 417). On the opposite bank of the Adige rises the large Benedictine Abbey of *Marienberg*. To the left, farther on, are the village of *Burgeis* and the castle of *Fürstenburg*. This monotonous part of the valley is called the *Malser Heide*. The road ascends and soon reaches the E. bank of the *Heider-See*, and beyond it —

69 M. **St. Valentin** *auf der Heide* (4695'; *Post*). Magnificent retrospective *VIEW (most striking when approached from Nauders) of the ice-clad Ortler range, which forms the entire background. Skirting the E. bank of the *Mitter-See*, the road leads to (3½ M.) *Graun*, at the entrance of the *Langtauferer Thal*. To the left is the green *Reschen-See*, the source of the Adige. Beyond (2 M.) *Reschen* (4888'; *Villa Fischersheim; Stern*), at the N end of the lake, we reach the **Reschen-Scheideck** (4898'), the watershed between the Black Sea and the Adriatic, and then descend by the *Stille Bach* to (4½ M.) —

79 M. **Nauders** (4468'; *Post; Löwe; Mondschein*). The old castle of *Naudersberg* contains the district courts of justice.

From Nauders to the **Lower Engadine** (diligence to *Schuls* daily), see p. 419.

106. From Nauders to Bregenz over the Arlberg.

Comp. Maps, pp. 356, 54.

103 M. DILIGENCE from Nauders to Landeck (27 M.) daily in 5¼ hrs. (also an omnibus). RAILWAY from Landeck to Bregenz, 76½ M., in 4¼-6 hrs.; fares 3 fl. 83, 2 fl. 55, 1 fl. 28 kr. (express 5 fl. 75, 3 fl. 83, 1 fl. 92 kr.).

The road through the *Finstermünz Pass* runs high above the river, being hewn at places in the perpendicular slate-rock (three tunnels, two avalanche-galleries). At the entrance to the pass is a small fort, and beyond it a pretty waterfall. The finest point on the route is **Hoch-Finstermünz** (3630'), a group of houses with a *Hotel*. Far below, on the Inn, is *Alt-Finstermünz* (3203'), with its old tower. The ravine of the Inn, with the Engadine Mts. in the background, is very picturesque.

The road descends gradually, passing through three short tunnels, and crosses the Inn near —

8 M. **Pfunds** (3185'), consisting of two villages, separated by the Inn: on the right bank *Pfunds* (*Post), on the left bank *Stuben* (*Traube), through which the road runs. To the S.W. towers *Piz Mondin* (10,375'), a peak of the N. Engadine chain; to the S.E. the *Glockthurm* (11,010') and other peaks of the Oetzthaler Ferner. The road again crosses the Inn near (4 M.) *Tösens*.

17½ M. **Ried** (2875'; *Post; Maass*), a thriving village, with the castle of *Siegmundsried*. The road crosses to the left bank at (2 M.) *Prutz* (Rose), at the mouth of the *Kaunser-Thal*, in which, farther on, the Grieskogel is visible. On a precipice to the left is the ruined castle of *Laudegg*; near it lies the village of *Ladis* (3900'), with sulphur-baths (moderate), 1 hr. from Prutz; ½ hr. higher is *Obladis* (4545'), a bath-house with mineral springs, well fitted up and finely situated, but not accessible by carriage.

The road recrosses the Inn by the (2½ M.) *Pontlatz Bridge* (2820'), 6 M. from Landeck, where the Bavarian invaders of Tyrol were signally defeated by the Tyrolese 'Landsturm' in 1703 and 1809. To the right *Flies*, with *Schloss Piedenegg*. To the left, on the opposite bank, a fall of the *Urgbach*, high above which is the village of *Hochgallmig*. The Inn dashes through a narrow gorge and forms several series of cataracts.

27 M. **Landeck** (2670'; **Post; Schwarzer Adler; Goldner Adler*), a large village on both banks of the Inn, is commanded by an ancient castle, now tenanted by poor families. The *Railway Station* (*Restaurant, R. 1 fl. 20 kr.) lies 1¼ M. to the E.

The *ARLBERG RAILWAY crosses the Inn. Looking back, we get a glimpse of the picturesque Landeck to the left, and of the huge *Parseierspitze* (9955') to the right. The train ascends on the right bank of the deep *Sanna-Thal* to (30½ M. from Nauders) Stat. *Pians* (2990'), opposite the village of that name (Alte and Neue Post), above which lies *Grins*. After crossing several viaducts we reach (32 M.) **Wiesberg**, with an old château (recently restored), beyond which we cross the *Trisanna*, which emerges from the *Patznaun-Thal* and unites with the *Rosanna* to form the Sanna, by means of a bold bridge, 286 yds. long and 282' high. Then a tunnel, 221 yds. long.

34 M. *Strengen* (3355') lies at the N. base of the *Petziner Spitze* (8353'). To the W. rises the *Riffler* (9880'), with its glacier. We follow the right bank of the Rosanna to (36 M.) **Flirsch** (3795'; *Post*), at the foot of the *Eisenspitze* (9400'), prettily situated. The valley expands. The train crosses the Rosanna three times. 40 M. *Pettneu.* Crossing the stream twice more, we next reach (44 M.) **St. Anton** (4270'; *Post; Adler*), the highest village in the Rosanna Valley or *Stanzer-Thal*, at the E. base of the *Arlberg*.

Immediately beyond St. Anton the train enters the great *Arlberg Tunnel, nearly 6⅓ M. long (St. Gotthard Tunnel 9¼ M.), ascends slightly to the middle of it (4300' above the sea-level; 1600' below the Arlberg Pass), and then descends rapidly to the *Kloster-Thal*, watered by the *Alfenz*. 50½ M. *Langen* (3990'; Buffet), on the right bank of the stream. At first running high up on the N. side of the valley, the train descends to (54 M.) *Danöfen* and —

58 M. **Dalaas** (3055'); the village (2750'; *Post*) lies far below in the valley. Several more viaducts and tunnels. 59½ M. *Hintergasse* (2700'). At (62 M.) *Bratz* (2315'; *Löwe*) the train reaches the bottom of the valley. To the right a considerable fall of the *Fallbach*. The train then traverses the broad valley of the *Ill* to —

66½ M. **Bludenz** (1905'; *Bludenzer Hof, *Scesaplana*, **Hôt. Arlberg*, at the station; in the town, *Post, *Kreuz, *Krone*), prettily situated. To the S. is the picturesque ravine of the *Brandner-Thal*, with the ice-clad Scesaplana in the background.

EXCURSION TO THE LÜNERSEE AND ASCENT OF THE SCESAPLANA, very interesting. (To the lake 6-6½ hrs., an easy route.) We descend and cross the Ill to *Bürs*, and ascend the charming *Brandner-Thal* to (3 hrs.) *Brand* (3375; °Beck; °Kegele). The path mounts on the right bank of the *Alvierbach* to the *Alp Lagant*, and ascends thence in zigzags over grass, débris, and rock. To the right rises the *Seekopf*, with its huge stony slopes; over the rocks to the left falls a fine cascade of the brook issuing from the Lünersee. We next reach the depression on the N.W. side of the beautiful, blue °Lünersee (6310'), the largest lake among the Rhætian Alps, 4 M. in circumference. On the W. bank is the (3-3½ hrs.) *Douglas Hut* (Inn).

The ascent of the °Scesaplana (9740'), the highest peak of the Rhætikon chain, is toilsome, but safe (4 hrs.; guide from Bludenz 9½, from Brand 7 fl.). Leaving the Douglas Hut, we skirt the lake for a little way and then ascend to the right, at first on turf, and then over loose stones and the dreary, rock-strewn *Todte Alp*. Lastly we pass through a steep 'cheminée' to the arête and to the top without difficulty. The imposing prospect embraces the whole of Swabia towards the N., as far as Ulm; the Vorarlberg and Algäu Alps to the N.E.; the Oetzthal, Stubai, and Zillerthal Alps to the E.; the Swiss Alps from the Silvretta and Bernina to the St. Gotthard and the Bernese Mts., and the Rhine Valley, Canton of Appenzell, and Lake of Constance to the S. and W. — Descent to the *Schamella Club-Hut* and by the *Alp Palus* to (4 hrs.) *Seewis* in the Prätigau, see p. 357. From the Douglas Hut to (7 hrs.) Schruns (p. 431) an attractive route leads past the grand °*Schweizer-Thor* (7055'; pass to the Prätigau, p. 357) to the *Oefen Pass*, and descends to the finely situated *Sporer*

Alp and through the *Gauer-Thal* (see p. 431; adepts may dispense with a guide).

The **Montafon** (see Map, p. 356, and comp. *Baedeker's Eastern Alps*), to the S.E. of Bludenz, is a beautiful and populous valley, watered by the Ill, and separated from the Prätigau on the S. by the *Rhaetikon Mts.* (for a fuller description, see *Baedeker's Eastern Alps*). The road (omnibus to Schruns several times daily, 80kr.), leads by *St. Peter* to (4 M.) *St. Anton*, a hamlet on a mound of débris at the base of the *Schwartzhorn*; then, following the right bank of the Ill, to (3½ M.) Schruns (2280'; pop. 1710; *Löwe; *Taube), the chief place in the valley, charmingly situated at the base of the *Bartholomäberg* (4880'; a fine point of view). On the opposite bank of the Ill lies *Tschaguns*, at the entrance to the *Gauer-Thal*, from which a path crosses the *Drusenthor* (7220'), between the *Drusenfluh* (9298') and the *Sulzfluh* (9265'), to (8 hrs.) *Schiers* (p. 357) in the Prätigau. (To the Lünersee, see p. 430.) Over the *Partnun* or *Gruben Pass*, or over the *Plasseggen Pass*, to (7-8 hrs.) *Küblis* see p. 358. — The *Sulzfluh* (9265'; 7 hrs.; guide 9 fl.) is a splendid point, hardly inferior to the Scesaplana, and not difficult; to the *Tilisuna Hut* (Inn) 5 hrs., thence to the top 2 hrs. more.

Above Schruns the valley contracts. At (2 hrs.) *Gallenkirch* (2730'; Adler, Rössle) the *Gargellen-Thal* opens to the S., through which tolerable routes cross the *St. Antönien-Joch* (7665') to the W. to (6 hrs.) *Küblis* (p. 351), and the *Schlappina Joch* (7100') to the E. to (6 hrs.) *Klosters* in the Prätigau (p. 358). Passing *Gortepohl*, we next reach (2 hrs.) *Gaschurn* (3120'; *Rössl; *Krone), prettily situated at the mouth of the *Gannera-Thal*, and (1 hr.) *Patenen* (3435'; Sonne), the last village in the Montafon. (Passes into the *Patznaun*, see *Baedeker's Eastern Alps*.)

FROM PATENEN OVER THE VERMUNT PASS TO GUARDA in the Lower Engadine (10 hrs.; with guide), trying, but attractive. We ascend the *Gross-Vermunt-Thal* to the right, passing the grand *Stüber Fall* or *Höllen Fall* to the (3½ hrs.) *Madlener-Haus* (6495'; Inn), on the *Gross-Vermunt-Alp*, on the W. side of the *Bieler Höhe*. We next ascend to the S. to the source of the Ill (7140') at the foot of the great *Vermunt Glacier*, and toil up the moraine and the glacier to the *Vermunt Pass* (9180'), between the *Dreiländerspitze* (10,350') on the E. and *Piz Buin* (10,900'), the highest of the Vorarlberg Mts., on the W. (ascended by adepts without difficulty from the Madlener-Haus in 6 hrs.). Steep descent to the *Val Tuoi* and *Guarda* (p. 415). — Over the *Kloster Pass* to *Klosters*, see p. 359.

The line crosses the Ill beyond (70 M.) *Strassenhaus*, and the *Mänkbach*, descending from the *Gamperton-Thal* on the left, near (73 M.) *Nenzing*. 77½ M. *Frastanz*, at the mouth of the *Samina-Thal*. The Illthal, below Bludenz called the *Wallgau*, contracts. At Feldkirch the river forces a deep passage (*Obere* and *Untere Ill-klamm*) through the limestone rocks before emptying itself into the broad Rhine Valley. The train crosses the Ill, enters the Upper Klamm, and passes through a short tunnel.

80 M. **Feldkirch** (1510'; pop. 3800; *Vorarlberger Hof*, at the rail. station, R. from 80 kr., pens. from 2½ fl.; *Englischer Hof*, R. from 80 kr., D. 1½ fl.; *Bär; *Löwe; Schäfle, well spoken of; beer at the *Rössl; Rail. Restaurant*), a natural fortress, hemmed in by mountains, and once the key to Tyrol, is a pleasant little town, above which rises the ruined *Schattenburg*. A large Jesuit school here is called the *Stella Matutina*. The *Parish Church*, erected in 1487, contains a 'Descent from the Cross' attributed to Holbein; and the *Capuchin Church* has another good painting of the same subject. By the Gymnasium is a small botanical garden.

Fine view of the Rhine Valley, from the Falknis to Lake Constance, and of the gorge of the Ill, from the *Margarethenkapf (1880'), a hill 20 min. to the W., on the left bank of the Ill, with the villa and pleasant park of Hr. v. Tschavoll (tickets at the hotels; visiting-card also sufficient).

FROM FELDKIRCH TO BUCHS (1¹/₂ M.) railway in ³/₄ hr. (fares 60, 40, 20 kr.). It sweeps round the *Ardetzenberg*, crosses the Ill at *Nofels*, and intersects the broad Rhine Valley. Stations *Nendeln* and *Schaan*. (*Vaduz*, 2 M. to the S., p. 61.) Near *Buchs* (p. 61) it crosses the Rhine.

The train now skirts the E. side of the wooded and vine-clad *Ardetzenberg*. 83 M. *Rankweil*, at the mouth of the *Laternser-Thal*, with a picturesquely situated church. Above the alluvial plain of the Rhine rise several wooded knolls, the chief of which is the *Kummenberg* (2186'), to the left. Near (88 M.) *Götzis*, with its modern Romanesque church, are two ruined castles of the Montforts.

91 M. Hohenems (1420'; *Post; Krone*) lies at the foot of bold rocks, crowned with the castles of *Neu-Hohenems* and *Alt-Hohenems*. In the village is a château of Count Waldburg-Zeil. Crossing the *Dornbirner Ach*, we next reach —

95 M. Dornbirn (1435'; pop. 10,700; *Hôt. Weiss*, at the station; *Dornbirner Hof; *Hirsch; Mohr*), a thriving little town, upwards of 2 M. in length. The S.W background is formed by the Appenzell Mts., the Kamor and Hohekasten, the snow-clad Sentis, and the serrated Curfirsten.

98 M. *Schwarzach*; 100¹/₂ M. *Lautrach*. (Junction-line to the left to St. Margrethen, p. 60.) The train then crosses the *Bregenzer Ach* to —

103 M. Bregenz. — OESTERREICHISCHER HOF, on the quay; HÔT. EUROPA, *MONTFORT, HABSBURGER HOF, at the station; *WEISSER KREUZ, SCHWEIZERHOF, Römer-Str.; KRONE; ADLER. — Wine at *F. King's*, on the road to the Gebhardsberg. Beer at *Forster's Brewery* and at the *Hirsch*, on the same road.

Bregenz (1260'; pop. 7000), the capital of the *Vorarlberg*, the *Brigantium* of the Romans, is beautifully situated at the E. end of the *Lake of Constance*. The *Old*, or *Upper Town*, on a height, occupies the site of the *Roman Camp*, and formerly had two gates, the southern of which has been removed. Fine survey from the Pier.

The *Gebhardsberg (1963'; ascent ¹/₂ hr., the last half through wood), with the ruined castle of *Hohen-Bregenz*, an auberge, and a pilgrimage-church, commands the Lake of Constance, the valley of the Bregenzer Ach and the Rhine, and the Alps of Appenzell and Glarus. Picturesque foreground, formed by precipitous pine-clad hills.

The *Pfänder (3465'), to the E. of Bregenz, commands a far more extensive prospect. The path (1¹/₂ br.) ascends to the right by the old barracks at the N. end of Bregenz, traverses wood, passes the (50 min.) 'Halbstation-Pfänder' auberge, and follows the telegraph-wires to the large *Hotel* (pens. 3¹/₂ fl.), 5 min. from the top. The longer carriage-road (2-2¹/₂ hrs.) leads through the upper part of the town to the 'Berg-Isel' (rifle-range), then chiefly through wood to the hamlet of *Fluh* (Krone) and thence to the hotel.

Railway to *Lindau* (6 M.; p. 59) by *Lochau* in 22 min. (60, 42, 30 kr.). Steamboats on the *Lake of Constance*, see p. 28.

VII. THE ITALIAN LAKES.

107. From Bellinzona to Lugano and Como (Milan) . . . 433
  Excursions from Lugano: Monte S. Salvatore; Monte Brè; Monte Caprino; S. Bernardo; Bigorio; Monte Boglia; Monte Camoghè; Monte Tamaro, 436-438. — Monte Generoso, 438.
108. From Bellinzona to Locarno. Val Maggia 440
  Val Verzasca, 440. — From Locarno to Domodossola through the Centovalli and Vigezzo valleys. Val Onsernone, 441. — Val Rovana. Val di Campo. Val di Bosco, 442. Excursions from Bignasco. Through the Val Bavona to the Tosa Falls or to Airolo; Basodino, 442, 443. — Val Prato; Campo Tencia. From Fusio to Airolo or to Fiesso, 443.
109. Lago Maggiore 443
  Railway from Bellinzona to Luino and Novara, 443. — Val Cannobbina; viâ Finero to S. Maria Maggiore, 445. — Sasso del Ferro; Monte Nudo; S. Caterina del Sasso. From Laveno to Como and to Milan, 446. — From Intra to Bee and to Premeno, 447. — Borromean Islands, 448, 449. — Monte Mottarone, 450. — From Arona to Milan, 451.
110. From Domodossola to Novara. Lake of Orta . . . 451
  From Gravellona to Pallanza or to Baveno-Stresa, 452. — Excursions from Orta, 452. — From Varallo to Ponte Grande and to Alagna; Val Sermenza, 454.
111. From Luino on Lago Maggiore to Menaggio on the Lake of Como. Lake of Lugano 455
  From Ponte Tresa to Lugano by land, 455. — Grottoes of Osteno and Rescia; Lanzo d'Intelvi, 456.
112. Lake of Como 457
  Monte Legnone, 458. — Monte Grigna, 459. — Monte S. Primo; Monte Crocione; Monte Galbiga, 461. — Lake of Lecco, 463. — From Lecco to Milan and to Bergamo, 463. — From Como to Erba and Bellagio by land, 464.
113. From Como to Milan 465

107. From Bellinzona to Lugano and Como *(Milan)*.
Comp. *Maps, pp. 434, 456.*

RAILWAY (comp. p. 108) from Bellinzona to *Lugano*, 18½ M., in 56-68 min. (4 fr. 70, 3 fr. 30, 2 fr. 35 c.); from Lugano to *Como*, 19½ M., in 2 hrs. (3 fr. 20, 2 fr. 25, 1 fr. 50 c.); from Lugano to *Milan*, 48½ M., in 3-3¾ hrs. (8 fr. 55, 6 fr. 5, 4 fr. 30 c.).

Bellinzona, Ger. *Bellenz* (760'; pop. 3360; *Rail. Restaurant; Hôt. Suisse & Poste*, R., L., & A. 3, D. 4 fr.; *Cervo; Alb. Ristor. Ferrari*), a town of quite Italian character, **with a handsome abbeychurch, is the capital of Canton Ticino. It is commanded on the W. by the** *Castello Grande*, **on an isolated hill; on the E. by the** *Castello di Mezzo*, or *di Svitto*, and the *Castello Corbario* or *Corbè*, **the highest of the three (1500'). In the middle ages Bellinzona**

was strongly fortified and was regarded as the key to the route from Lombardy to Germany. In the 16-18th cent., the three castles were the residences of the bailiffs of Uri, Schwyz, and Unterwalden (comp. p. 116). The Castello Grande is now used as a prison and arsenal (visitors admitted; fee); the other two are in ruins. — Bellinzona is the junction of the St. Gotthard line (to Lugano-Como, see below) and the lines to Locarno (p. 440) and Luino (p. 443).

Beautiful walk (1½ hr. in all) from the station to the S. through the town (10 min.), then ascend to the **left** to the highest castle by a stony path in numerous zigzags, affording constantly improving views. After about 40 min., the path to the castle (12 min.) diverges to the left; the main path leads straight on to *Daro* and the conspicuous chapel of *S. Maria della Salute*, commanding a picturesque view of the two lower castles. We may descend to the left of the chapel by a path enclosed with vineyard walls and regain the railway-station in 20-25 minutes.

Ascent of *Monte Camoghè* (7308'; from Bellinzona 7-8 hrs., with guide), see p. 438; over the *Passo di S. Jorio* to the **Lake** *of Como*, see p. 458.

A tunnel of 313 yds. carries the train under the *Castello di Mezzo* (p. 433). At (2½ M.) **Giubiasco** (765') the lines to the Lago **Maggiore** (p. 443) diverge to the **right**.

Trending to the left, the Lugano line approaches the foot of the mountains near **Camorino**, and begins to ascend the *Monte Cenere*, through walnut and chestnut-trees. *S. Antonio* lies below on the right; then *Cadenazzo* (p. 440). Two tunnels (the *Precassino*, 435 yds.; and the *Meggiagra*, 111 yds.). View of the Ticino Valley, the influx of the Ticino into the Lago Maggiore, Locarno, and the Val Maggia Mts., improving as we ascend. We pass under **Monte Cenere**, the top of which is 370' above, by means of a tunnel, 1840 yds. long (1437' above the sea-level; shut the windows), to —

9 M. *Rivera-Bironico* (1560'), in the bleak valley of the *Leguana*, which soon joins the *Vedeggio*, descending from the *Mte. Camoghè* (p. 438), to form the *Agno*. Beyond the short *Molincero Tunnel* is (15 M.) Taverne (1105'; *Inn at Taverne Inferiori*). At **Lamone** (1033') the train leaves the Agno and ascends past *Cadempino* and **Vezia** to the *Massagno Tunnel* (1135'; 1016 yds. long).

18½ M. **Lugano**. — The Railway Station (1110'; Pl. C, 2; *Restaurant*) lies on the hill above the town, of which, as well as of the lake, it commands a fine view. Besides the road there are a shorter footpath and a CABLE TRAMWAY (*Funicolare*; comp. Pl. C, 2, 3) to the town (fares up 40 or 20 c., down 20 or 10 c.). — The STEAMBOAT (p. 456) has three piers *Lugano-Città*, at the Piazza Giardino; *Lugano-Parco*, near the Hôtel du Parc; and *Lugano-Paradiso*, for Paradiso and the Mte. Salvatore.

Hotels (the chief of which send omnibuses to meet the trains and steamers). *On the lake*: *HÔTEL DU PARC (Pl. a; B, C, 4), in an old monastery at the S. end of the town, with garden (band thrice a day) and the dépendances of *Belvedere*, **Villa Ceresio**, and *Villa Beau-Séjour* (Pl. b, B 4; the last, with fine garden, alone open in winter), R., L., & A. 4-6, B. 1½, lunch 3, D. 5, omnibus 1½, pens. 9-11 fr.; *GRAND HÔTEL SPLENDIDE (Pl. c; B, 5), ¼ M. farther to the W., on the Paradiso road (p. 436), patronized by English and Americans, R., L., & A. from 5, D. 5, pens. from 12 fr. — HÔT.-PENS. LUGANO (Pl. e; C, 3), with a small garden, well spoken of; ALBERGO-TRATTORIA AMERICANA (Pl. f; D. 3), Piazza Giardino, pens. 6 fr., with a good restaurant. — *In the town*: HÔT. SUISSE (Pl. g; D, 3), near the Piazza Giardino, R. & A. 2½-4, B. 1¼, lunch 2½, D. 3½ fr.; PENSION ZWEIFEL,

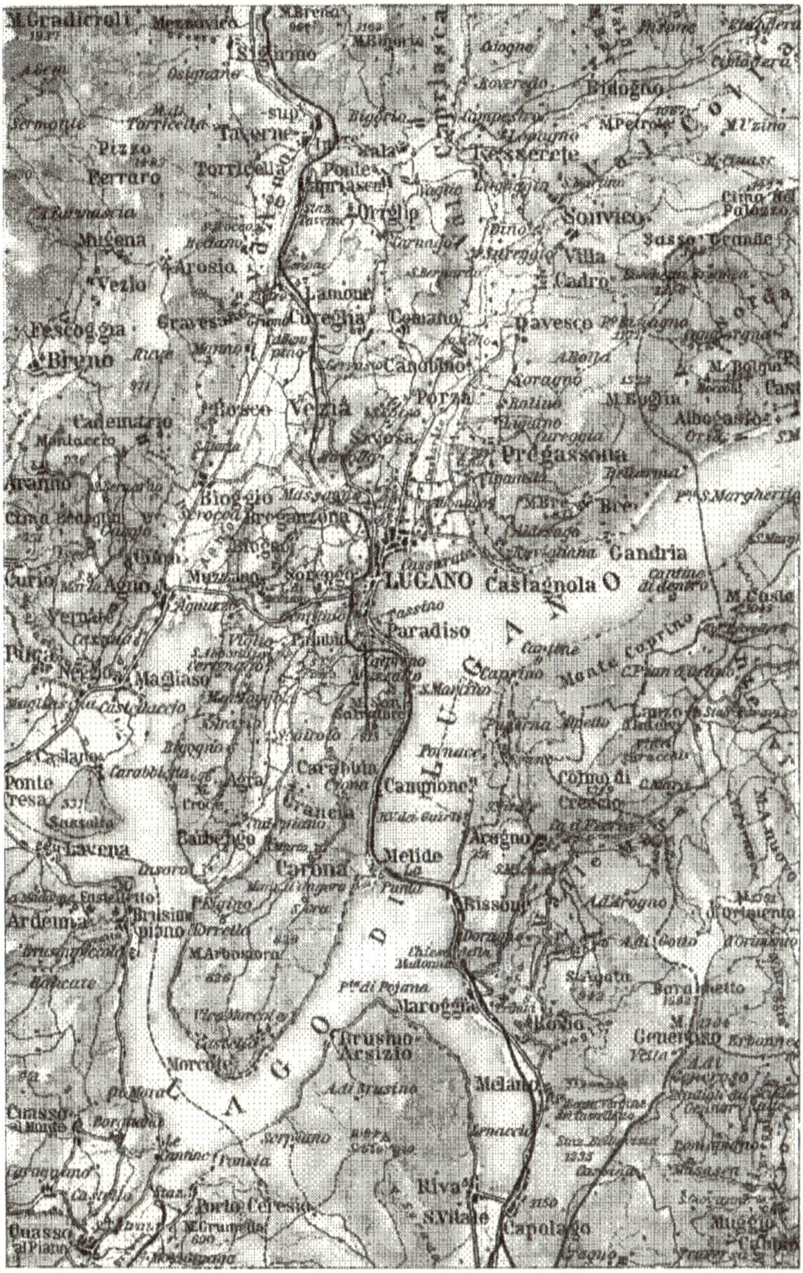

Tesserete
Sonvico
Villa
Cadro
Davesco
Lamone
Canobbio
Comano
Porza
Vezia
Pregassona
Gandria
Bioggio
Agno
Magliaso
LUGANO
Castagnola
Paradiso
Monte Caprino
Campione
Carona
Melide
Bissone
Maroggia
Rovio
Brusino
Arsizio
Morcote
Melano
Riva
S.Vitale
Capolago
Porto Ceresio
Ponte
Tresa
Lavena
Muggio

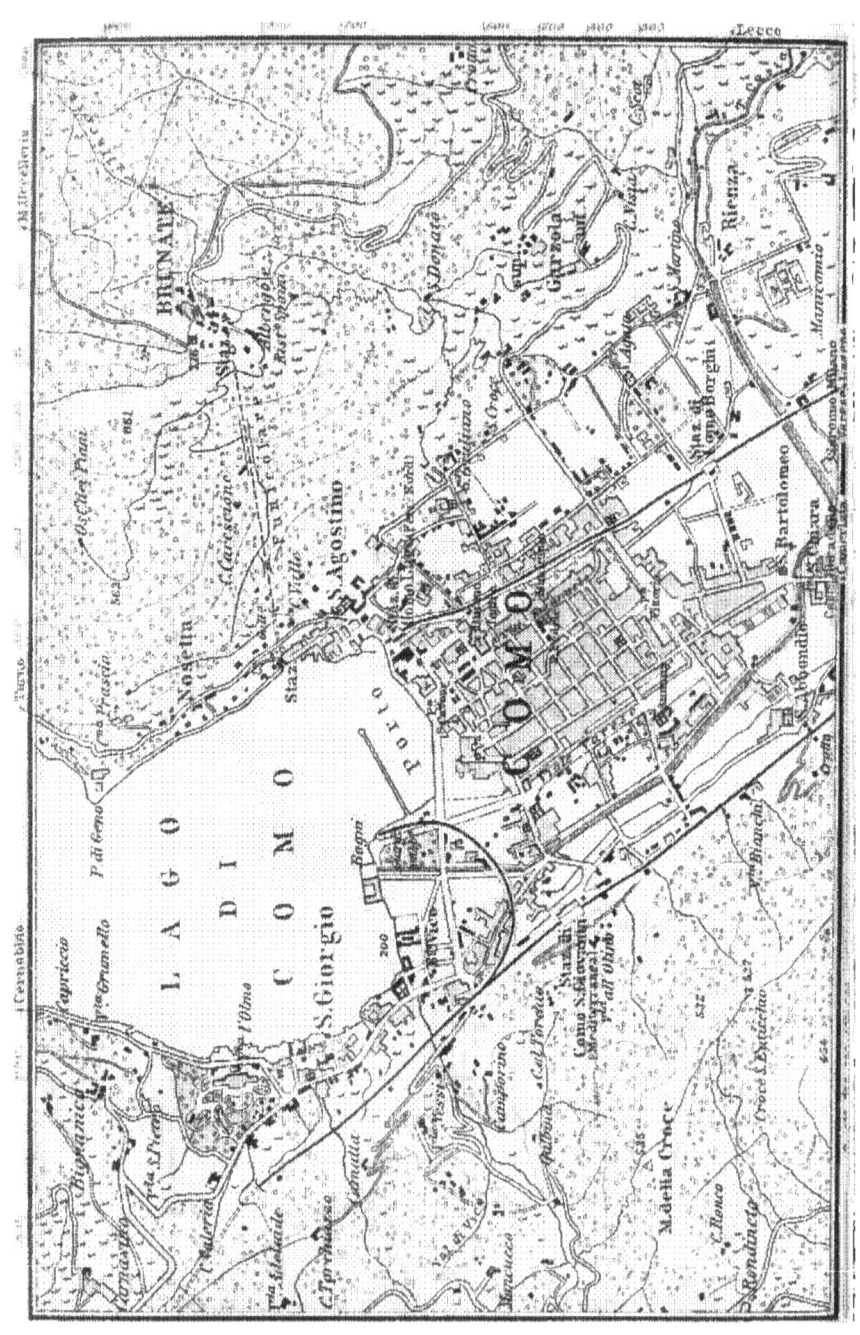

4-5 fr. — *Near the station:* to the S., *HÔT.-PENS. BEAU-REGARD (Pl. 1; B, 3), R., L., & A. 2¹/₂-4, B. 1¹/₄, lunch 2¹/₂, D. 3¹/₂, pens. 7¹/₂-10 fr.; *HÔT. ST. GOTTHARD (Pl. k; C, 3); still farther on, PENS. PASKAY, well spoken of; to the N., *HÔT. WASHINGTON (Pl. d; C, 1), in an elevated and open situation, R., L., & A. 3-3¹/₂, pens. 6-8 fr. Below the station: *HÔT.-PENS. ERICA, with dépendance *Villa Clarita* (Pl. 1; C, 2), R., L., & A. 3, D. 3¹/₂ fr.; *HÔT. DE LA VILLE & PENS. BON-AIR (Pl. o; C, 2), pens. 5-7 fr.; PENS. INDUNI, moderate. — At *Paradiso* (p. 436): *HÔT.-PENS. REICHMANN (Pl. n; B, 6), on the lake, R., L., & A. 2¹-4, D. 3¹/₂, pens. 7-9 fr.; HÔT. DU LAC, also on the lake; PENS. RUFIBACH-STALDER, unpretending; *HÔT.-PENS. SALVADOR (Pl. m; A, B, 6), from 6 fr.; *BELLEVUE (Pl. A, 6), pens. 6-8 fr. — At *Cassarate*, 1 M. to the E., in a sheltered position, with a S. aspect: *PENS. VILLA CASTAGNOLA (Pl. G, 3), with pretty garden, R., L., & A. 2¹/₂-3, pens. 6-8 fr.; PENS. VILLA ETOILE, 5-6 fr.; PENS. VILLA DU MIDI (Pl. G, 6), ¹/₄ M. farther on, 4¹/₂-5 fr.; *PENS. VILLA MORITZ, higher up on the hillside, pens. 5-6 fr.

Restaurants. At the Hotels; *Trattoria Biaggi (also rooms and pens.), to the W. of the Piazza della Riforma, on the way to the cable-tramway, thoroughly Italian; *American Bar*, Piazza Giardino. — Beer at the *Deutsches Brauhaus*, at the N.E. corner of the Piazza Giardino; *Walter* (rooms), *Straub*, both on the quay, near the Hôtel Lugano. — *Café Centrale*, Piazza Giardino. — Confectioners. *Meister*, a little to the S.W. of the Palazzo Civico; *Forster*, Via Canova, by the post-office.*

Lake Baths on the Paradiso road (30 c., towels 20 c., cabine 60 c.). WARM BATHS at *Anastasi's*, near the Hôtel du Parc.

Post & Telegraph Office (Pl. D, 3), Via Canova, near the Hôt. Suisse. — Physicians, *Dr. Cornils, Dr. Zbinden, Dr. Reali*, etc. — Bookseller, *Schmid, Francke, & Co. (Libreria Dalp)*, Piazza Giardino.

Carriage to or from the St. Gotthard Railway Station and the town, with one horse, 1 pers. 1, 2 pers. 1¹/₂, 3 pers. 2 fr., two horses, 1-2 pers. 2, 3-5 pers. 3 fr.; same fares from the station to Paradiso, and from the town to Cassarate. From the town to *Castagnola*, or from the St. Gotthard or the Salvatore station to *Cassarate*, 1¹/₂, 2, 2¹/₂, 3, or 4 fr.; from these stations to *Castagnola* 2, 2¹/₂, 3, 4, 5, and 6 fr. — *Circuit of Mts. S. Salvatore* (p. 436; viâ Pambio, Figino, Morcote, Melide, in 2¹/₂ hrs. repaying) with one horse 7, with two horses 12 fr.; to *Capolago* 7 and 12, *Luino* 12 and 20, *Varese* 16 and 30 fr. Fee 10 per cent of the fare.

Boats, with 1 rower 1³/₄, 2 rowers 3 fr. for the first hour; each additional ¹/₂ hour 1 rower ¹/₂, two rowers 1 fr., and fee. Sailing-boats 3¹/₂ fr. for the first hour, each additional ¹/₂ hr. 1¹/₂ fr.

English Church Service in a chapel beside the Hôtel du Parc.

Lugano (932'; pop. 7000), the largest town in Canton Ticino, charmingly situated on the lake of the same name, with quite an Italian climate, is a pleasant place for a prolonged stay. The scenery is Italian in character; numerous villages and country-seats are scattered along the banks of the lake, and the lower hills are covered with vineyards and gardens, contrasting beautifully with the dark foliage of the chestnuts and walnuts. Immediately to the S. rises the dolomitic *Monte S. Salvatore*, wooded to its summit; to the E., beyond the lake, is the *Monte Caprino*, with the *Monte Generoso* to its right. To the left the *Monte Brè* and the beautiful *Monte Boylia*. On the N. opens the broad valley of the *Cassarate*, backed by an amphitheatre of mountains among which the double peak of *Monte Camoghè* is conspicuous.

A broad *Quay*, planted with trees, and frequented as an evening promenade, extends along the lake. Opposite the steamboat-pier is the handsome *Palazzo Civico* (Pl. C, 3), with a fine colonnaded

28*

court. To the E. is the spacious *Piazza Giardino*, at the harbour, with gardens and a column with barometer, thermometer, etc. The *Piazza della Riforma* lies farther back. At the S. end of the quay rises a Fountain Statue of Tell, by Vela. — The church of *S. Maria degli Angioli* (adjoining the Hôtel du Parc) contains a fresco on the rood-loft by *Luini*, the *Passion, one of his finest works, with numerous figures. On the wall to the left is the Last Supper, in three sections, and in the 1st Chapel on the right a fine Madonna, both on panel by Luini.

The interior of the town, with its arcades, workshops in the open air, and granite-paved streets, is also quite Italian in its character. *S. Lorenzo* (Pl C, 2), the principal church, on a height below the station, probably erected by Tommaso **Rodari at the** close of the 15th cent., has a tasteful marble façade. The terrace in front of the railway-station commands an extensive *Prospect over land and lake.

Pleasant WALK to the S., on the high-road past the Hôtel du Parc and Hôtel Splendide, through the suburb of *Paradiso* (Pl. A, B, 6) and along the foot of Mte. S. Salvatore to the (1¼ M.) headland of *S. Martino*, a charming point of view. To Melide, see p. 438. At Paradiso a path diverges to the right to the (5 min.) *Belvedere*, with view of the lake and town. — To the W. by the winding road to **Ponte** Tresa (Pl. A, B, 4, 5), which diverges to the S. at the Villa Beauséjour (short-cuts for walkers), to the (1¾ M.) hill on which lies the frequented *Restaurant du Jardin*. The village of *Sorengo* is situated on a hill to the right (fine view from the church); to the W. is the Lake of Muzzano. A carriage-road leads from the Restaurant du Jardin, to the left, viâ *Gentilino*, to (1½ M.) the conspicuous church of *S. Abbondio* (1345), in the graveyard of which are several monuments by Vela. A very attractive walk (3 M. longer) leads from Gentilino to the right through fine chestnut-woods to *Montagnola*, and then back to S. Abbondio. — To the E., from the Piazza **Castello** [where to the right (No. 227) is the entrance to the shady park of the *Villa Ciani*, now *Gabrini* (Pl. D, E, 3), with a marble figure of a mourning woman ('la Desolazione') by Vinc. Vela; ½-1 fr. to the gardener], we may follow the Via Carlo Cattaneo, which crosses the (¼ M.) *Cassarate*, to (¾ M.) *Cassarate* (Pl. G, 3), and thence proceed by the sunny road skirting the foot of the Mte. Brè to (1 M.) *Castagnola*, where we obtain a fine view of the Mte. S. Salvatore. Thence we may skirt the lake up and down hill to (1-1¼ hr.) *Gandria* (p. 456).

The most interesting excursion is the ascent of the *Mte. S. Salvatore (2980'), by CABLE RAILWAY (1 M. long) from Paradiso in 25 min. (fare 3, down 2, up and down 4 fr.; half-fare on Sun.). The lower station (1245'; Restaurant; Pl. A, 6) lies ¼ M. from the steamboat-pier *Lugano-Paradiso* (steamboat from Lugano-Città in 10 min.). The line, with an initial gradient of 17:100, crosses the St. Gotthard Railway, traverses a viaduct (110 yds. long; 38:100) supported by iron pillars, and reaches the halfway station of *Pazzallo* (1007), where carriages are changed. Here are the machine-house for the electric motor and the steam-engine. The line now ascends over granite rock, at an increasing gradient (finally 60 : 100), to the terminus (2900'; Restaurant). Thence we ascend on foot to the (7 min.) summit (*Vetta*), on which there is a pilgrimage-chapel. The *View embraces all the arms of the Lake of Lugano, the mountains and their wooded slopes, and the beautiful villas and gardens above Lugano. To the E. above Porlezza is Monte Legnone (p. 458); to the N., above Lugano, rises the double peak of Monte Camoghè; to the left of this are the distant Rheinwald Mts.; towards the W. is the Monte Rosa chain, with the Matterhorn and other Valaisian Alps to the right. (Morning light most favourable.) — Those who prefer to make the whole ascent on foot follow the road passing the Hôtel Bellevue (comp. Pl. A, 6) and under the St. Gotthard Railway At (1½ M.) Pazzalo they follow the lane named 'Al **Monte**' towards the E. and cross (12 min.) the cable-railway. The top is reached from Lugano in 2 hrs. — The beautiful and fragrant Daphne Cneorum and the Helleborus niger, or 'Christmas Rose', both adapted for transplantation to gardens, are found on this mountain.

The *Monte Brè (3050'; ascent 2½-3, descent 1¾ hr. from Lugano; guide needless; mule 10 fr.) affords another beautiful walk. From the Piazza Castello to the iron bridge over the *Cassarate*, see above. Beyond the bridge we turn to the left, then after about 130 paces to the right, and ascend the winding road between low walls to the large mill, *Molinazzo* (Pl. G, 2), where mules may be hired. Thence by the same road to (1 M.) *Viganello*; below the hill crowned by the church of *Pazzolino* we turn to the right to (1½ M.) *Albonago*. Thence the road again ascends, partly between walls, and among chestnuts, figs, and vines, to (¾ hr.) *Aldesago*, on the mountain-slope, the highest village visible from Lugano. Aldesago may also be reached in ¾-1 hr. from *Castagnola* (see above), via *Ruvigliano*. Above Aldesago the path divides: both branches lead round to the (½-¾ hr.) village of *Brè* (2630'; 2 hrs. from Lugano; Restaurant & Pension Forni), at the back of the hill. From the church of Brè a narrow path ascends to the W. through brushwood to the (½ hr.) top of the hill. This path also divides, both branches being attractive: that to the right ascends at once; that to the left first leads to a spur in the direction of Lugano, and then ascends at the back of the hill. Beautiful view of the different bays of the Lake of Lugano, especially towards Porlezza, and of the surrounding mountains. Lugano is visible from the above-mentioned spur, but not from the top.

Opposite Lugano, on the E. bank of the lake, rises **Monte Caprino**, the 'Cantine' in the cool grottoes of which are much frequented on Sun. and holidays (in winter on Mon. & Frid.; closed in the evening). Good 'Asti' and other wines of icy coolness are sold here. Another favourite resort is the *Cavallino Garden Restaurant*, to the S. of the Cantine, near which is a fine waterfall. Rowing boat there and back in 2½ hrs. incl. stay; also steamboat on Sun. and holidays. — From the cellars a path ascends the Mte. Caprino and follows the ridge to the S.W. to the (3 hrs.) *Colmo di Creccio* (4298'), with a picturesque view.

To S. Bernardo and Bigorio (to the station of Taverne, 3½-4 hrs.). A cart-track on the fertile slopes to the N. of Lugano leads by *Massagno*, *Savosa*, *Porza*, and *Comano* to the (1½ hr.) church of S. Bernardo (2310'), on a rocky plateau, with a picturesque view. (At the S.E. base of the hill are the village of *Canobbio* and the château of *Trevano*.) Thence (at first following the top of the hill to the N.; no path) to *Sala* and the (1¼ hr.) monastery of Bigorio (2360'; rfmts.), charmingly situated on the wooded hill of that name. (The church contains a Madonna attributed to Guercino or Perino del Vaga.) Back by (1 M.) *Ponte Capriasca* (1425'; with a church containing a good old copy of Leonardo da Vinci's Last Supper; best light 11-1) to the (1½ M.) railway-station of *Taverne* (p. 434).

*Monte Boglia (4960', 4-4½ hrs.), a hill visible from Lugano to the left of Mte. Brè (guide desirable). Ascent by *Sorayno* and the *Alp Bolla* or from *Brè* (see above) in 1¾ hr. View little inferior to that from Mte. Generoso. Descent on the E. side through the grassy *Val Solda* to *Castello* and *S. Mamette* (a steamboat-station; p. 456) or *Oria* (p. 456).

Monte Camoghè (7303'; 7-8 hrs. from Lugano; guide from Colla), famous point of view, is fatiguing. Road viâ *Canobbio* and *Tesserete* (*Ser. Antonini*); then to the right into the *Val Colla*, to (12 M.; carr. in 2½ hrs.) *Scareglia* or *Lower Colla* (3205'; *Osteria Garzirola*). We then (with guide) ascend on foot by Colla and the *Alp Pietrarossa*, leaving the *Mte. Garzirola* (see below) to the right, to the (3 hrs.) *Alp Sertena* (5922') and the (1½ hr.) top. — The descent may be made to the N., by the *Rivolte* and *Levene* Alps, to the *Val Morobbia*, *Giubiasco*, and (5 hrs.) *Bellinzona* (p. 426; ascent of the Camoghè from Bellinzona, 7-8 hrs.). — *Monte Garzirola* (6912'), 3 hrs. from Colla, also repaying. — From the Val Colla an interesting walk over the pass of *S. Lucio* (5960') to *Porlezza*, or over the *Cima dell' Aratione* (5928'; fine view) to the *Val Solda* (p. 456), or to the Val Solda by a path passing the curious dolomite pinnacles of the *Denti di Vecchia*.

Monte Tamaro (6433'; 4 hrs.; guide) from *Taverne* (p. 434) or *Bironico* (p. 434), not difficult. Splendid view of Lago Maggiore, etc.

A pleasant excursion may be made in a light mountain-carriage (16-17 fr.) viâ *Boggio* (1053') to (2 hrs.) *Cademario* (2407), whence the carriage is sent to Agno. From Cademario we ascend on foot to (20 min.) San Bernardo (2955'; view of Lago Maggiore, etc.). We next proceed to the Aronno-Iseo road and follow it to the left to *Iseo* (1254'), *Cimo*, *Vernate*, and (2 hrs.) *Agno* (p. 455). The chapel of *S. Maria* (2566') lies near the road, between Iseo and Cimo.

To the *Grotto of Osteno*, see p. 456.

FROM LUGANO TO COMO (Milan). The train crosses the *Tassino Valley*, by means of a viaduct, 120' high (charming view to the left), and passes through the *Paradiso Tunnel* (833 yds.) under the N.E. spur of *Monte S. Salvatore* (p. 436). It then skirts the lake, with views (to the left) of the wooded slopes of the E. bank and the villages upon it. 23 M. *Melide* (905'), 1½ M. from the promontory of S. Martino (p. **436**), with the Grotto Demicheli Restaurant and the Grotto Civelli (cold meats). The train and the road cross the lake to *Bissone* by a stone viaduct ½ M. long, which sadly mars the scenery. At each end there is an arch for the passage of boats. Fine views in both directions. Two tunnels. Then (26 M.) *Maroggia* (Elvezia), at the W. base of the *Mte. Generoso*; continuous view of the lake on the right.

27½ M. Capolago (**Hôt.-Pens. du Lac*, with garden; *Rail. Restaurant*), at the head of the S.E. arm of the lake, the station for the *Generoso Railway* (steamboat from Lugano twice daily in ¾ hr.).

The *Monte Generoso, owing to its isolated situation, opposite the principal chain of the Swiss Alps, and to its elevation above the Italian lakes and the plains of Lombardy, commands perhaps the most magnificent view to the S. of the Alps, and may justly be compared with the Rigi. A RACK-AND-PINION RAILWAY, now, like the hotels, the property of the Pasta family, leads from Capolago to within 200' of the summit. The line, on Abt's system, is 5½ M. long and has a maximum gradient of 22 : 100 (Rigi Railway 25 : 100). Four trains ascend daily to the summit (Vetta) in 1¼ hrs., to Bellavista (Hôtel Generoso) in 56 min.; fare to Bellavista 5 fr. 85, to Vetta 7 fr. 50 c., from Vetta to Capolago 5 fr.; return-fare to the top 10 fr. (half-fares on Sun.); tickets including the railway journey, and R., D., and B. at the Hôtel Kulm 18 fr — The trains start from the steamboat-pier at *Capolago* (p. 438), where the toothed rail begins, and halt at (2 min.) the St. Gotthard Railway station (p. 438). The train crosses the road and the St. Gotthard Railway and ascends the slope of the Generoso (gradient 20:100, afterwards 22:100), with a continuous view, on the right, of the fertile Val di Laveggio, girt with wooded hills, the little town of Mendrisio, and behind, of the Lake of Lugano with S. Vitale on the W. bank, and N. to the Mte. Salvatore. Then we skirt abrupt cliffs and pass through a curved tunnel (150 yds. long), immediately before which the summit of Monte Rosa is visible, to the station of (1³/₄ M.) **S. Nicolao** (2820′), in the wooded *Val di Solarino*. The line next describes a wide curve, threads a tunnel 50 yds. long, and proceeds high up on the mountain-slope, with views of the plain as far as Milan and Varese, and of the wooded valleys of the Generoso (to the right appears Monte Bisbino, with its pilgrimage-church). — 3¹/₂ M. **Bellavista** (4010′; *Restaurant*). A promenade leads from the station along the mountain-slope (benches) to the (10 min.) *Perron*, a mountain-spur (railings) immediately above Capolago, with a beautiful view (best in the morning) of the Lake of Lugano and the surrounding heights, backed by the line of snow-peaks stretching from the Gran Paradiso to the St. Gotthard. About ¹/₂ M. to the E. of the station (hotel-porter meets the trains) is the *Hôtel Monte Generoso (3960′; R., L., & A. 4-5, B. 1¹/₂, déj. 3, D. 5, pens. 9-12 fr.; Engl. Church Service in summer)*, situated on a mountain-terrace commanding a view over the plain of Lombardy as far as Mte. Viso. A bridle-path leads hence to the summit in 1¹/₄ hr. — Beyond Bellavista the railway ascends through another tunnel (90 yds. long) and closely skirts the barren ridge, affording occasional views to the left of the lake and town of Lugano, and to the right, below, of the villages of Muggio and Cabbio. Beyond two short tunnels we reach the station of (5¹/₂ M.) **Vetta** (5355′; *Hôtel Kulm*, R. 5, B. 1¹/₂, déj. 3¹/₂, D. 5 fr., connected by view-terraces with the *Restaurant Vetta*). A new path provided with railings leads hence in 10 min. to the summit of *Monte Generoso (5560′). The *View, no less striking than picturesque (comp. the panorama), embraces the lakes of Lugano, Como, Varese, and Maggiore, the entire Alpine chain from the Monte Viso to the Pizzo dei Tre Signori, and to the S. the plain of Lombardy, watered by the Po and backed by the Apennines, with the towns of Milan, Lodi, Crema, and Cremona. — From the station of Vetta the descent on foot to the Hôtel du Generoso or to Bellavista station may be made in ³/₄ hr.

Monte Generoso may also be ascended from *Maroggia* (p. 438) by *Rovio* (Hôt.-Pens. Mte. Generoso, open in winter also; pens. 5-6 fr.), or from *Balerna* (p. 440) by *Muggio* in 4-4¹/₂ hrs. (roads to Rovio and Muggio, beyond which the ascent is fatiguing). — From *Lanzo d'Intelvi* (bridle-path, 5¹/₂ hrs.), see p. 456 (better for returning; to Osteno 6 hrs.).

30 M. **Mendrisio** (1180'; *Alb. dell' Angelo*, Italian, R. & A. 2½ fr.), with 2872 inhab., ½ M. from the station, **lies at the beginning of the bridle-path to the Hôt. du Generoso** (see above; 3 hrs., mule 6 fr.). The short *Coldrerio Tunnel* carries us through the watershed between the **Laveggio** and the *Breggia*. 33 M. *Balerna*.

35 M. **Chiasso** (764'; *Rail. Restaurant;* *Alb. S. Michele*, by the station), the last Swiss village (custom-house; usually a long halt). The line pierces the *Monte Olimpino* by means of a tunnel 3190 yds. long (view of the Lago di Como to the left), and passes *Borgo Vico*, a suburb of Como, on the left.

38 M. **Como** (p. 463); thence to (30 M.) *Milan*, see R. 113.

108. From Bellinzona to Locarno. Val Maggia.

RAILWAY to Locarno, 14 M., in ¾ hr. (2 fr. 30, 1 fr. 60, 1 fr. 15 c.). — DILIGENCE from Locarno to *Bignasco* twice daily in 3½ hrs., coupé 4 fr. 60 c.; diligence from Bignasco to Fusio in summer daily in 3 hrs. Carriage with one horse from Locarno to Bignasco 19, with two horses 30 fr., back 18 and 25 fr.; from Bignasco to Fusio and back 18 or 35 fr.

To (5½ M.) *Cadenazzo*, see p. 434. The Locarno line (change of carriages) diverges to the right and below (r.) *Cugnasco* crosses the *Ticino*. — 10 M. *Gordola*, with productive vineyards, at the mouth of the **Val Verzasca**.

Val Verzasca. A road (diligence from Locarno to Sonogno daily in 4½ hrs.) ascends the deep and picturesque valley, watered by the beautiful *Verzasca* with its countless falls. This stream and its tributaries abound in fish and are often of an exquisite transparent green. The lover of nature should descend into the ravine and explore some of the delicious rocky pools. The road leads by (r.) *Vogorno* and (l.) *Corippo* to (8 M.) *Lavertezzo* (Osteria della Posta) and (4 M.) *Brione* (2497'; *Inn*), the chief village in the valley, at the mouth of the *Val d'Osola*, through which a path (with guide) leads to the *Forcarella Cocco* (7010'), the *Val Cocco*, and (8 hrs.) *Bignasco* (p. 442). Ascending to the N. through the main valley, we next come to *Gerra, Frasco*, and (4½ M.) Sonogno (2982'; *Inn*), the last village, where the valley again divides. Thence to the W. over the *Passo di Redorta* (7140'), between the Corona di Redorta and Mte. Zucchero, to the *Val Pertusio* and (8 hrs.; guide) *Prato* (p. 443), interesting. Another attractive route leads to the N. by *Cabione* and the *Alp Bedeglia* to the *Passo di Laghetto* (6920'), to the W. of the Cima Bianca; it then descends to the *Alp del Lago* (6046'), with its little lake ('laghetto') and through the *Val Chironico* to (8 hrs.) *Giornico* (p. 117).

The train crosses the brawling *Verzasca* and runs on the bank of the *Lago Maggiore* to —

14 M. **Locarno**. — °GRAND HÔTEL LOCARNO, with garden, view of the lake, and English Chapel, R., L., & A. 5-6, B. 1½, lunch 3, D. 5 fr.; pens. 8-12½ fr.; °HÔT.-PENS. DU PARC, with garden and view, R. 2-5, B. 1½ déj. 3, D. 4, pens. 6-10 fr.; °HÔT.-PENS. BELVEDERE; PENS. BEAU-RIVAGE; °HÔT.-PENS. REBER, with garden on the lake, moderate, pens. 6-7 fr.; °CORONA, Italian, R., L., & A. 2-3, B. 1¼ fr.; HÔT. SUISSE, in the chief piazza, moderate; HÔT. DU LAC; PENS. VILLA RIGHETTI, on the way to the Madonna; °PENS. VILLA MURALTO, pens. 5 fr.; ALBERGO S. GOTTARDO, near the station, R. from 1½, B. 1, D. with wine 3, pens. 4 fr.; furnished rooms at *Giul. Borghetti's*. — *Rail. Restaurant*.

STEAMBOATS on *Lago Maggiore*, see p. 441; departure according to Roman time, which is 20 min. in advance of Swiss time.

Locarno (680'; pop. 3353, Rom. Cath.), a busy little town of thoroughly Italian character, is beautifully situated on the Lago Maggiore at the mouth of the *Maggia*. Since 1513 it has belonged to Switzerland. In the 15th cent. the town is said to have contained 5000 inhab.; but by an intolerant decree in 1553 several of the most industrious Protestant families were banished for refusing to conform to the Roman Catholic ritual. A number of these (the Orelli, Muralto, and others) repaired to Zürich, where they founded the silk-manufactories which still flourish. The market-place, with the old *Government Buildings* and the *Post Office*, lies to the W. of the harbour. The houses have arcades on the groundfloor. In front of the church of *S. Antonio* is a memorial fountain to the *Marchese Marcacci* (d. 1854). Another monument commemorates the deputy *Mordasini* (d. 1888). At the market held at Locarno on alternate Thursdays the picturesque costumes of the neighbouring peasantry are seen to advantage. The greatest gala-day is 8th Sept., the Nativity of the Virgin.

Fine view from the *Madonna del Sasso* (1168'), a pilgrimage-church on a wooded rock above the town (½ hr.; steep paved path, with 14 'stations'). The church contains an Entombment by Ciseri and a Flight into Egypt by Bramantino. Ascending to the left at the back of the church, we reach (10 min.) a *Chapel* containing a painted terracotta group of the Resurrection by Rossi (1887), and affording a most picturesque retrospect of the Madonna del Sasso. The chapel of the *Trinità del Monte*, still farther up, commands a view of the upper part of Lago Maggiore. The whole walk may be accomplished in 1½ hr. (evening light favourable).

WALKS. Pleasant walks may be taken to the W. viâ *Solduno* to the (3 M.) *Ponte Brolla* (p. 442); to the S.W. across the bridge over the Maggia to (2 M.) *Losone*, with cool wine-cellars (wine good and cheap), or to (2½ M.) *Ascona* (p. 444), and thence along the bank of the Lago Maggiore to (6 M.) *Brissago* (p. 444); preferable is the walk 'over the hill' to (2 hrs.) *Ronco* and thence down to (1 hr.) Brissago; — to the E. to (1½ M.) *Minusio* and into the (¾ M.) romantic *Navegna Gorge*, with a chalybeate spring; to the N to the mountain-villages of *Orselina* and *Brione* (each 3 M.), with charming views; or to (6 M.) *Mergoscia* in the *Val Verzasca* (p. 440), etc. — About 2 hrs. above Locarno is the *Pens. Alpenheim* (L. Borghetti), visited as a health and whey-cure resort.

FROM LOCARNO TO DOMODOSSOLA, 12 hrs., a beautiful route, through the Val Centovalli and the Val di Vigezzo. Road to *Losone* and (6 M.) *Intragna* (1300'; Inn), picturesquely situated at the confluence of the *Meleza* and the *Onsernone*. Then a new road leads along the left bank of the Meleza, passing below *Borgnone* (r.) and (6½ M.) an *Osteria* (l.; fair), to (1 M.) *Camedo*, the last Swiss village. Thence a fatiguing and hilly path, crossing the Italian frontier and passing the villages of *Oglio* and *Dissimo*, leads to (2½ hrs.) *Rè* (several inns), a resort of pilgrims, with a large new hospital. Road thence by (3 M.) *Malesco* (Leon d'Oro), where the new road from the Val Cannobbina joins ours on the left (p. 445), to (1½ M.) S. Maria Maggiore (2713'; *Croce di Malta*), the capital of the populous *Val Vigezzo*, and on viâ *Druogno* and *Riva* and down through a pretty valley to (9½ M.) *Domodossola* (p. 314). [The road to *Crevola* (p. 313) diverges to the right about 2 M. before we reach Domodossola.]

Val Onsernone. Road (diligences from Locarno to Comologno and Vergeletto daily in 3½ hrs.) across the *Ponte Brolla* (p. 442) to (1½ M.)

Cavigliano, where a road to *Intragna* (p. 441) diverges to the left. We then ascend to the N.W., through the picturesque *Val Onsernone*, in numerous windings, to *Loco* (Inn) and (6½ M.) *Russo* (2638'), where the valley divides. The road bends into the W. branch of the valley and at the picturesque *Ponte Oscuro*, where the road to Vergeletto diverges to the right, it turns and ascends the S. branch past *Crana* to (1½ hr.) *Comologno* (3503'; no tolerable inn). From (1¼ hr.) *Spruga*, where the road stops, a bridle-path crosses the Italian frontier to the (1¾ M.) rustic *Bagni di Craveggia*, with a sulphur-spring, whence an easy route crosses the *Bocchetta di S. Antonio* to (5 hrs.) *S. Maria Maggiore* (p. 441). — In the N. branch of the valley, 3 M. from Russo, lies *Vergeletto* (2990'; °Osteria Domenigone). Thence to Cimalmotto (see below) over the *Passo di Porcareccio*, or to Cevio by the *Lago di Alzasca*, interesting (with guide).

The *Val Maggia, 25 M. long, with its bold rock scenery, its rich vegetation, and its pretty villages **and grand** waterfalls, deserves a visit, particularly in spring **or** autumn. **The road** (diligence and carriages, see p. 440) leads on the left bank of the *Maggia*, with its numerous falls, past the picturesque (3 M.) *Ponte Brolla* (820'; route to the Val Onsernone, see above), to *Avegno*, where the snow-covered **summit of** the Basodino is visible for a short time, and to (4½ M.) **Maggia** (1138'), a considerable village. To the right is the fine *Cascata della Pozzaccia*. Then by *Coglio, Giumaglio, Someo* (Osteria al Ponte; Ristor. del Soladino), with its handsome houses, and *Riveo* (passing the beautiful *Soladino Fall, 330' high, on the left) to *Visletto*, at the foot of massive cliffs, and over the Maggia **to** (7½ M.) **Cevio** (1380'; *Ristor. del Basodino*, with a few rooms; *Ristor. della Posta*), the capital of the valley (514 inhab.), with fine groups of trees and an old church, at the mouth of the *Val Rovana*.

The steep Val Rovana divides at (3½ M.) *Collognasca* (2640) into (l.) the *Val di Campo* and (r.) the *Val di Bosco*. In the former lie (3½ M.) *Campo* (4430'; Inn) and (1½ M.) *Cimalmotto* (Inn), the church of which has a porch with interesting frescoes. Thence over the *Porcareccio Pass* to *Vergeletto*, see above; over the *Passo di Bosa* (7405') and through the *Val Isorno* to (6 hrs.) *Crevola*, easy; over the *Passo di Craverola* (*Scatta del Forno*, 8290') to *Premia*, or over the *Passo della Scatta* (8420') and the *Passo di Comella* to *Crodo* in the *Val Antigorio* (p. 321), both easy (guide). — In the **Val di Bosco**, 5 M. from Collognasca, lies *Bosco* (4030'; Inn), called also **Crin** or **Gurin**, the only German village in Canton Ticino. Thence over the *Criner Furka* to the **Val Formazza**, see p. 320.

1¼ M. (18 M. from Locarno) *Bignasco* (1424'; *Hôt. du Glacier*, R., L., & A. 3-3½, D. 4, pens. 7-9 fr.; pop. 202), is charmingly situated at the mouth of the *Val Bavona*, and is well adapted for a stay (English Church Service in summer). About ½ M. to the S.E. is the pretty *Waterfall of Bignasco*.

Pleasant walk to the (¾ hr.) *Madonna dei Monti* (2360'), a fine point of view (ascend to the left after crossing the Maggia below the hotel). Beyond the chapel we proceed still **farther** into the valley, passing several chalets and ascending on the other **side of** the brook to (20 min.) the *Incino Alp*, whence we descend again past two fine waterfalls (*Bagni di Nerone* and *Piccolo Niagara*) to (40 min.) Bignasco. — We may also follow the Fusio road (guide-post) to the (¾ M.) *Pontelatto*, cross the Maggia, and return on the left bank. — Other walks to *Brontallo* **and** (3 M.) *Menzonio* (2380'; fine view); to (3 hrs.) *S. Carlo*, (3½ hrs.) *Fusio*, etc. (see p. 443).

FROM BIGNASCO TO THE (10 hrs.) TOSA FALLS, or TO (11 hrs.) AIROLO. Through the picturesque *Val Bavona, which opens to the N.W. of Bignasco, a road shaded by walnut and chestnut trees, leads viâ *Cavergno*,

Fontana, Foroglio (with a fine waterfall), *Fontanella*, and *Sonlerto* to (3 hrs.) *S. Carlo* (3150'; *Albergo Basodino, unpretending), whence the *Basodino* (10,750') may be ascended with guide (G. Padovani) in 5-6 hrs. (laborious; descent to Auf der Frut, 3¼ hrs., see p. 320). From S. Carlo we ascend rapidly viâ *Campo*, with guide, to the (2½ hrs.) *Alp Robiei* (6566'; accommodation) and to the W. through the *Val Fiorina* to the (3-3½ hrs.) Bocchetta di Val Maggia (8608') and (2½ hrs.) *Auf der Frut* (p. 319). — Travellers bound for Airolo, instead of crossing the bridge leading to the Alp Robiei, follow the left bank of the stream (with guide) and ascend by the *Alp Lielpe* and *Pioda*, past the little *Lago Sciundrau* (7720'), to the (5 hrs.) Forcla di Cristallina (8474'), to the W. of the *Cristallina* (9547'); then descend over snow into the *Val Torta* and through the *Val Cristallina* to *Ossasco* (p. 319) and (3 hrs.) *Airolo* (p. 114).

The road in the Val Maggia, called *Val Broglio* above this point, next leads to *Broglio* and (4½ M.) Prato (2460'; *Inn*, rustic), at the mouth of the *Val Prato*, which ascends to the E. to the *Campo Tencia*. (Over the *Redorta Pass* to the *Val Verzasca*, see p. 410.) The Campo Tencia (N. summit, 10,038'; 8-9 hrs. from Prato, with guide), a magnificent point of view, is trying. Through the Val Prato to the highest chalets of the *Corte di Campo Tencia* (7250') 5 hrs.; then, on the E. side, up the crest of the *Crozlina Glacier* to the (3-4 hrs.) summit. Experts may descend across the glacier to the E. to the *Alp Crozlina* and by *Dalpe* to *Faido* (p. 116).

At (1¼ M.) Peccia (2785'; Inn, rustic) the *Val Peccia* opens to the left, with the pyramidal *Poncione di Braga* (9405') in the background. The highest portion of the Val Maggia is named the *Val Lavizzara* from the 'lavezzo' stone found there. The road ascends in windings (short-cuts for walkers) to the flat upper part of the valley, crosses (4 M.) the wild gorge of the Maggia, and leads past (right) *Mogno*, still ascending in windings (shorter footpath to the right), to (2 M.) Fusio (4202'; *Hôt. Dazio*), the last village in the Val Maggia, most picturesquely situated.

A picturesque walk may be taken, through wood, from Fusio to (1½ M.) the pretty hamlet of *Sambucco* (4485'), with a fine waterfall. — Passes from Fusio (with guide): to the N. by *Corte*, and either the Sassello Pass (7697') or the Passo dei Sassi (8200'; for experts only), to (5½-6 hrs.) *Airolo*; to the W. over the Passo di Naret (8015') and past the small lake of the same name, to (7 hrs.) *Ossasco* in the *Val Bedretto*; to the N.E. by *Colla* and the *Alp Pianascio* to the (2½-3 hrs.) Campolungo Pass (7595'; fine view); descent either to the right by the *Alp Cadonighino* and *Dalpe* (see above) to (3 hrs.) *Faido* (p. 116), or (very steep) to the left past the little *Lago Tremorgio* (5997') to (2 hrs.) *Rodi-Fiesso* (p. 116). — The *Poncione Tremorgio* (8786'), a splendid point of view, may be scaled from the Campolungo Pass in 1½ hr.

109. Lago Maggiore.

Railway FROM BELLINZONA TO NOVARA VIÂ LUINO, 67 M., in 4-5 hrs. (fares 12 fr., 8 fr. 45 c., 6 fr.); TO LUINO in 1¼-1½ hr. (fares 4 fr. 50, 3 fr. 20, 2 fr. 10 c.). — Intermediate stations: 2½ M. *Giubiasco*; 5½ M. *Cadenazzo*; 10½ M. *Magadino*; 12½ M. *S. Nazzaro*; 14½ M. *Ranzo-Gero*; 17 M. *Pino*, the first Italian station; 21 M. *Maccagno*; 25 M. Luino, with both the Italian and the Swiss custom-houses; 29 M. *Porto Valtravaglia*; 34 M. *Laveno*; 36½ M. *Leggiuno-Monvalle*; 40½ M. *Ispra*; 43½ M. *Taino-Angera*; 47 M. *Sesto-Calende* (see *Baedeker's Northern Italy*). — FROM BELLINZONA TO LOCARNO, see p. 440.

Steamboat twice daily in summer from Locarno to Laveno, and seven or eight times daily from Laveno to Intra, Pallanza, the Borromean Islands, Stresa, and Arona. From Locarno to Arona 5½ hrs., from Luino to Isola Bella 2¾ (from Laveno 1¼) hrs.; from Isola Bella to Arona 1¼ hr. (fare from Locarno to Arona 5 fr. 85 or 3 fr. 20 c., from Luino to Isola Bella 2 fr. 15 or 1 fr. 80 c., from Isola Bella to Arona 1 fr. 70 c. or 1 fr., *landing and embarking included*). Strict punctuality (Roman time, p. 432) is not always observed. Some of the boats are saloon-steamers, with restaurants on board (lunch 3, D. 4½ fr.). — The names of the STATIONS are printed below in bolder type; those always touched at are *Locarno, Brissago, Cannobbio, Luino, Laveno, Intra, Pallanza, Baveno, Isola Bella, Stresa, Belgirate, Lesa, Meina, Arona*.

The *Lago Maggiore (646', greatest depth 2800'), the *Lacus Verbanus* of the Romans, is about 37 M. long, and averages 1½-3 M. in width. The N. end for a distance of 9 M., sometimes called the *Lake of Locarno*, belongs to Canton Ticino. The W. bank beyond the brook *Valmara*, and the E. bank from the *Dirinella* belong to Italy. The chief tributaries of the lake are on the N. the *Ticino* and the *Maggia*, and on the W. the *Tosa*. The river emerging from the S. end retains the name of *Ticino*. At the N. end the lake is enclosed by lofty mountains, for the most part wooded, while the E. bank towards the lower end slopes gradually down to the plains of Lombardy. The water is green in its N. arm, and deep blue at the S. end.

Locarno, see p. 440. Opposite, in the N.E. corner of the lake, at the mouth of the *Ticino*, lies **Magadino** (railway-station; *Hôt. Bellevue, Pens. Viviani*, 5 fr., both on the lake), consisting of two villages, *Magadino Inferiore* and *Superiore*, at the foot of *Monte Tamaro* (6433').

To the S. of Locarno, where the deposits of the *Maggia* have formed a large delta, the Val Maggia (p. 442) opens, with its numerous villages. Farther on the W. bank is covered with villages, country-houses, and campanili. In an angle lies **Ascona** (small-boat stat.), with a ruined castle and several villas; then *Ronco*, higher up the hillside. Passing the two small *Isole de Brissago* the steamer reaches **Gera** (railway-station) on the E. bank; and then, on the W. bank, **Brissago** *(Hôt.-Pens. Beau-Séjour; Hôt. Suisse; Pens. Köhler)*, the last Swiss village, a delightful spot, with a fine group of cypress-trees near the church. The slopes above are covered with fig-trees, olives, and pomegranates; even the myrtle flourishes in the open air. — To the S. of Brissago is a large 'international tobacco manufactury'. Italian custom-house examination on board the steamer.

Opposite Brissago, on the E. bank, lies the Italian village of *Pino* (railway-station).

The next Italian villages are *S. Agăta* and **Cannobbio** (*Hôt. Cannobbio*, on the lake, R. 2½-3, pens. 6 fr.; *Albergo delle Alpi*, moderate; *Pens. Villa Badia*, 1¼ M. to the S., 260' above the lake, quiet, pens. 6-7 fr.). Cannobbio (pop. 2600), one of the oldest and most important places on the lake, lies at the entrance of the *Val Cannobbina*, and is overshadowed by wooded mountains. The church *Della Pietà*, the dome of which is attributed to Bramante, contains a Bearing of the Cross by Gaud. Ferrari.

Pleasant walk (also omnibus) up the picturesque **Val Cannobbina** to (1½ M.) *La Salute* (Hydropathic), and to the (20 min.) *Orrido*, a wild rocky gorge, where there is a waterfall in spring (best viewed from a boat, ½-1 fr.). — A new road ascends the beautiful valley, frequently crossing the river, and passing the villages of *Spoccia* (Osteria Americana, on the roadside), *Orasso*, *Cursolo*, and *Gurro*, on the heights on each side. It then crosses a low pass to *Finero* (Inn) and *Malesco*, in the *Val Vigezzo*, and descends to (19 M.) *S. Maria Maggiore* (p. 411). One-horse carriage from Cannobbio to S. Maria in 5 hrs., 15 fr., two-horse carr. 30 fr.

The steamer now steers to the E. bank, and stops at **Maccagno** (railway-station; *Alb. della Torre*), with a picturesque church and an ancient tower. The viaducts and tunnels of the railway from Bellinzona to Genoa are seen skirting the lake. Passing *Casneda* in a wooded ravine, we next reach —

Luino (railway-station). — The STEAMBOAT PIER adjoins the waiting room (déj. incl. wine 2½, D. 4½ fr.) of the *Steam Tramway to Ponte Tresa* (Lugano; see p. 155). By passing to the left of this station and the statue of Garibaldi and following the broad 'Via Principe di Napoli' we reach the (10 min.) STAZIONE INTERNAZIONALE, the station of the Bellinzona and Genoa line, where the Italian and Swiss custom-house examinations take place (*Restaurant*, déj. 3 fr.). Omnibus from the steamboat-pier 40 c., small trunk 25, large 50 c.

Hotels. *GRAND HÔTEL SIMPLON ET TERMINUS, on the lake, to the S. of the town, with a garden; HÔTEL POSTE & SUISSE, R., L., & A. 2½-3½ fr.; VITTORIA, well spoken of, these two near the pier. — Near the Staz. Internazionale: MILANO, déj. 2, D. 3 fr., incl. wine; ANCORA. — *Café Clerici*.

Luino or *Luvino*, a busy little town with 1800 inhab., is situated at the base and on the slopes of the hills, a little to the N. of the mouth of the *Tresa*. It affords good headquarters for a stay on account of its ample railway and steamer facilities. The *Statue of Garibaldi*, near the pier, commemorates his brave but futile attempt to continue the contest here with his devoted guerilla band after the conclusion of the armistice between Piedmont and Austria on Aug. 15th, 1848. The church of *S. Pietro* is adorned with frescoes by Bernardino Luini, a native of the place (ca. 1470-1530). At the mouth of the *Margorabbia*, ½ M. to the S., lies Germignaga, with the large silk-spinning *(filanda)* and winding *(filatoja)* factories of E. Stehli-Hirt of Zürich. (Admission by application to the manager.)

Near the W. bank, on rocks rising from the lake, are the two grotesque-looking *Castelli di Cannero*, half in ruins, the property of Count Borromeo. In the 15th cent. they harboured the five brothers Mazzarda, notorious brigands, the terror of the district. Cannero *(Albergo Nizza; Alb. Cannero)* is beautifully situated amidst vineyards and orchards. We next pass *Barbé*, with its graceful spire, **Oggebbio** and **Ghiffa** (small-boat station; *Hôt. Ghiffa*) on the W. bank, and **Porto-Valtravaglia** (railway-station; *Osteria Antica*) on the E. bank, villages at which the steamers do not always stop. In a wooded bay lies *Caldè*, with the old tower of *Castello di Caldè* on a hill. To the S. rises the green *Sasso del Ferro*, the most beautiful mountain on the lake (see p. 446). To the W., Monte Rosa and the Simplon group are visible. To the E. lies —

Laveno (railway-station; *Posta, Italian), a large village, beautifully situated in a bay at the mouth of the *Boesio*, once an Austrian war-harbour. The steamboat-pier adjoins the station of the *Varese and Milan* line; the station of the *St. Gotthard Railway* (p. 443) is about 1/2 M. farther on (omnibus). Near the quay is a monument for Garibaldians killed in 1859. Large pottery *(Società Ceramica Italiana)* on the site of the former Fort St. Michele (to the left); above it is the *Villa Pullè*, with a view-tower containing several relics of 1859.

The **Sasso del Ferro** (3485'), easily ascended from Laveno in 2 1/2 hrs., affords a magnificent view of the lake, the plain as far as Milan, and the huge snow-peaks of the Monte Rosa chain. About 6 M. to the N.E. of Laveno, at the back of the Sasso del Ferro, is the hamlet of *Vararo* (about 2600'), whence the *Monte Nudo (4050') may be ascended in 1 1/2 hr. Splendid view (surpassing that of Mte. Mottarone) of the Lago Maggiore, the lakes of Lugano and Varese, and the chain of the High Alps. — Pleasant excursion from Laveno via *Cerro* (which is reached by a road diverging to the right beyond the bridge over the Boesio, a few min. before the St. Gotthard station), to the (1 1/4 hr.) monastery of **S. Caterina del Sasso**, situated on the hillside high above the lake. Imbedded in the roof of the church is a mass of rock which fell upon it in the 17th century and has remained there ever since.

From Laveno to the *Borromean Islands* and *Pallanza* (p. 447), boat, with 3 rowers, 10-12 fr. (to Isola Bella 1 1/2 hr., thence to Isola Madre 20 min., Pallanza 20 min.).

FROM LAVENO VIA VARESE TO COMO (32 M.; railway in 2 1/4 hrs.) or MILAN (45 1/2 M.; railway in 2 1/4-3 hrs.). The line leads along the S. base of the Sasso del Ferro through the *Val Cuvio*, watered by the Boesio, via *Cittiglio, Gemonio, Cocquio, Gavirate, Barasso,* and *Casbeno*, to (14 M.) **Varese** (1250'; *Gr.-Hôtel Varese*, 1 M. to the W., near Casbeno, R., L., & A. 5 1/2, D. 5 fr.; *Italia, Europa, Angelo,* etc.), a town of 5800 inhab., charmingly situated near the lake of that name, with numerous villas. Splendid view from the pilgrimage-church of the *Madonna del Monte*, 2 1/2 hrs. to the N.W. A branch-line runs hence to *Induno* and (6 1/2 M.; 1/2 hr.) *Porto Ceresio*, on the Lake of Lugano (p. 450). — 16 1/2 M. *Malnate*, the junction of the lines to Milan via *Saronno*, and to Como via *Solbiate, Olgiate* (the highest point on the line, with numerous villas), *Lurate-Caccivio, Civello, Grandate,* and *Camerlata* (p. 465).

FROM LAVENO TO MILAN VIA GALLARATE, 45 M., railway in 2 1/2 hrs. — 2 1/2 M. *Sangiano*. The line diverges to the left from the line to Sesto (to the right is Monvalle, p. 443), and passes through a tunnel, to (5 M.) **Besozzo**. 10 M. *Ternate-Varano*, on the lovely Lago di Comabbio. Beyond a long tunnel are (13 1/2 M.) *Cragnola-Cimbro* and (16 1/2 M.) *Besnate*. — 20 M. *Gallarate*. Thence to (45 M.) Milan, see p. 451.

As we approach Intra, a valley opening **to the W. discloses a** passing **survey** of the N. neighbours of Monte Rosa: **first the Strahlhorn**, then the Mischabelhörner and the Simplon.

Intra (*Hôtels Vitello d'Oro, Leon d'Oro*, and *de la Ville*, all three united, R. & A. 2 1/2-3 1/2, B. 1 1/4 fr.; *Hôt. Intra; Agnello*), a thriving town of 5700 inhab., **lies** on alluvial soil between the mouths of two torrents, the *S. Giovanni* and *S. Bernardino*, which supply the numerous factories of the town with water-power. On the quay is a marble statue of *Garibaldi*; in the Piazza del Teatro a bronze statue of *Victor Emmanuel II.*, **by** Barsaglia. — On the lake, 1/2 M. to the N., is the *Villa Barbò*, with beautiful gardens, and 3/4 M. beyond it is the *Villa Ada* of Sign. Ceriani, also noteworthy for its luxuriant vegetation (palms, huge eucalypti, etc.); to the S., the

Villa S. Remigio; farther on, on the promontory of Castagnola, the small and ancient church and the *Villa Ashburner* (a red building in the Gothic style).

Pleasant walk from Intra to the N. by the new road (shaded short-cuts for walkers; carr. with two or three horses 25 fr.), viâ *Arizzano* to (3³/₄ M.) Bee (1935'; **Alb. Bee*), with a fine view of Lago Maggiore, and to (3 M.) Premeno (2600'; **Hôt.-Pens. Premeno*, finely situated). Above it (10 min.) is the *Tornico*, a platform laid out in honour of Garibaldi, with a good spring and a beautiful view of the Alps. About ¹/₄ hr. higher is the *Bellavista*, commanding the lake to the E. and the beautiful and fertile Val Intragna to the W., with its numerous villages.

To the S. of Intra the *Punta della Castagnola*, with the Grand Hôtel Eden (see below), stretches far into the lake. When the steamer has rounded the promontory and enters the wide W. bay of Lago Maggiore, the *Borromean Islands are disclosed to view: near the S. bank the Isola Bella, W. of it the Isola dei Pescatori, in the foreground Isola Madre. The little *Isola S. Giovanni*, near Pallanza, with its chapel, house, and gardens, is also one of the Borromean Islands. Beyond the Isola dei Pescatori rises the blunt pyramid of Mte. Mottarone, with the hotel near the top; farther to the W. the white quarries near Baveno are visible, while the picture is closed by the snow-covered mountains between the Simplon and Monte Rosa.

Pallanza. — Hotels (omn. at the pier, 1 fr.). *GRAND HÔTEL PALLANZA, a large house, finely situated about ¹/₂ M. to the E., with several dépendances and the *Villa Montebello*, R., L., & A. 3¹/₂-12, B. 1¹/₂, déj. 3, D. 5, warm bath 2¹/₂, lake-bath 1¹/₂, pens. in summer 7¹/₂-12¹/₂, in winter 7-10¹/₂ fr. *GRAND HÔTEL EDEN, 3 min. farther on, splendidly situated on the promontory of Castagnola, R., L., & A. 3¹/₂-7, B. 1¹/₂, déj. 3, D. 5, pens. 7-12 fr. — POSTA, R., L., & A. 3, B. 1¹/₄, D. 4, pens. 5-8 fr.; S. GOTTARDO, R., L., & A. from 2, D. 3¹/₂, pens. from 5 fr.; well spoken of; HÔT. MILANO, R. 2, D. incl. wine 3¹/₂ fr., these three near the pier; *PENS. VILLA MAGGIORE, R. 2, D. 3, pens. 5-6 fr. — *Café Bolongaro*, near the pier, Munich beer.

Diligence (office opposite the Alb. S. Gottardo) to *Gravellona* (p. 452; 6 M.), 4 times daily, in 1 hr. (1 fr. 65 c., banquette 2¹/₂ fr., incl. 33 lbs. of baggage); connecting thrice with an omnibus to Intra (p. 446; 25 min.; 50 c.). The Hôtel Pallanza also sends a private omnibus to Gravellona (1¹/₂ fr.).

Boats. With one rower to Isola Madre and back 2¹/₂, with two 4¹/₂ fr.; to Isola Bella and back 3¹/₂ or 6 fr.; to both islands and back 4 or 7 fr.; to Stresa and back 3¹/₂ or 6 fr.; to Laveno and back 3¹/₂ or 7 fr., etc.; boat without rower usually 1 fr. per hour. The hirer should ascertain the charge before embarking. The hotels have boats of their own at similar charges. Small gratuity usual.

ENGLISH CHURCH SERVICE in the Grand Hôt. Pallanza (April-Oct.).

Pallanza, a busy little town with 3200 inhab., is beautifully situated opposite the Borromean Islands and commands a fine view. Opposite the steamboat-pier is the market-place, with the town hall (*Municipio*) and the church of *S. Leonardo* (lower portion of campanile of Roman origin). The street to the right leads past the villas *Branca, Bozzotti* (right), *Montebello* (left), the nursery-garden of *Rovelli* (worth visiting; left) and the large hotels mentioned above, then round the promontory of Castagnola to Intra. — In the street leading to the N. from the market-place is the *Post Office* (right) and at its end (left) the church of *Santo Stefano* (to the left of the gate a Roman inscription). Following straight on the broad 'Viale Principe Umberto', past the bathing establishment of *Caprera* (alkaline spring), we reach (¹/₄ hr.) the domed church of the *Madonna della Campagna*, at the base of *Monte Rosso* (2273').

WALK ROUND MONTE ROSSO (3¹/₂-4 hrs., fatiguing; no rfmts. except bread and wine). From the Madonna della Campagna we follow the road straight on and (¹/₄ hr.) cross the *S. Bernardino* (p. 446; footpath up the left bank); 20 min., road from Intra, where we keep to the left. In (6 min.) the village of *Trobaso* we turn to the left; 12 min., cross-way; to the right to Unchio (see below); to the left by a handsome bridge to the right bank of the S. Bernardino and (¹/₄ hr.) *Santino*. Thence by a rough and stony path viâ (¹/₂ hr.) *Brieno* to (¹/₂ hr.) *Cavendone*, passing the pilgrimage-church below the village; shortly afterwards opens the view of the lake; 1¹/₂ hr. *Suna* (see below). — Following beyond Trobaso the road to the right, to (¹/₄ hr.) *Unchio* and (40 min.) *Cossogno* (here to the left by the 'Via Solferino'), we reach by stony paths the (¹/₄ hr.) *'Roman Bridge'* across the picturesque gorge of the S. Bernardino. We may now ascend by flights of steps to the (¹/₄ hr.) church of *Rovegro*, where we turn to the right to the village and ascend to the left (boy to show the way desirable) by **stony** paths over the hill **to** (1 hr.) *Santino* (see above).

To the W. of Pallanza the road skirts the **lake to** (1 M.) **Suna** (small-boat station; *Pens. Camenisch; Alb. Pesce*) and (3 M.) *Fondo Toce*, situated at the mouth of the impetuous *Tosa (Toce)*, where a road to the little *Lago di Mergozzo* diverges to the right; thence past the granite quarries of *Mont' Orfano* and by a five-arched bridge over the Tosa **to the** railway-station of *Gravellona* (p. 452; 6 M. from Pallanza; omnibus, see p. 447).

The next steamboat station (seldom visited; small-boat stat.) is Feriolo, 2¹/₂ M. from Gravellona (p. 452; omnibus **from** Stresa, see p. 449). The large granite quarries which border **the** hillside from Feriolo to Baveno furnished the material for the **columns** in Milan cathedral, the church of San Paolo fuori le Mura at Rome, and other buildings. The *Stabilimento Nic. Della Casa*, about ³/₄ M. to the W. of Baveno, where the **granite is** worked, deserves a visit. — Then —

Baveno. — GRAND HÔTEL BELLEVUE, R., L., & A. 5-7, D. 5 fr., with a beautiful garden; *BEAURIVAGE, also with garden; *HÔT.-PENS. SUISSE, R. from 1¹/₂, B. 1, lunch 2, D. 3, pens. from 5 fr. — DILIGENCE to *Gravellona* (5 M.) thrice daily in 40 min., 1 fr. 15, coupé or banquette 1 fr. 75 c.

ROWING BOATS to the Borromean Islands, same charges as from Stresa (p. 449). Halfway between Baveno and Stresa is **a** ferry, where the charge for the short crossing (10 min.) is 1-2 fr.

ENGLISH CHURCH in the garden of the **Villa Clara.**

Baveno, with 700 inhab. and a picturesque view of the lake, is well adapted for a stay **of some** time. The handsome *Villa Clara*, on the E. side of the village, formerly the property of Mr. Henfrey, was occupied by Queen Victoria for three weeks in April, 1879, and for a month by the invalid Crown Prince of Germany in Oct.-Nov., 1887 (no admission).

The steamer now approaches the —

*****Borromean Islands**, and touches (on some trips only) at the westernmost, the *Isola Superiore* or *dei Pescatori*, and then (always)

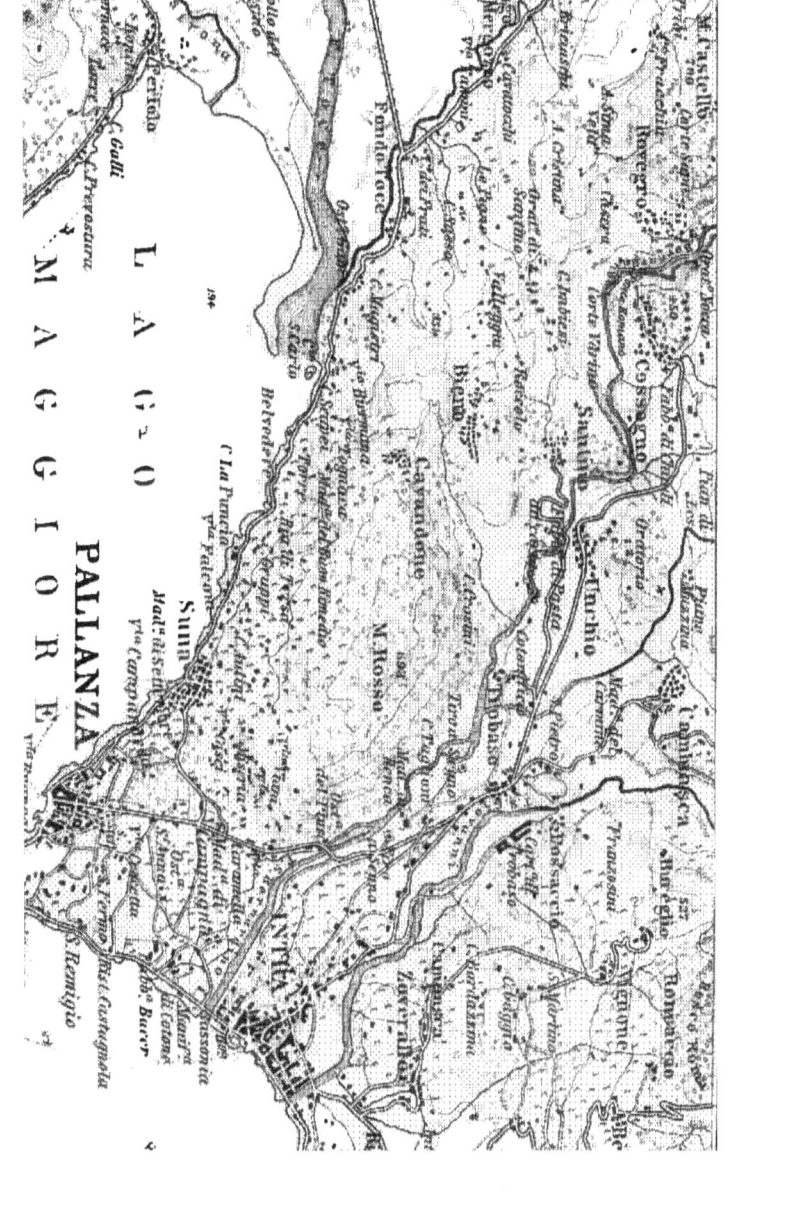

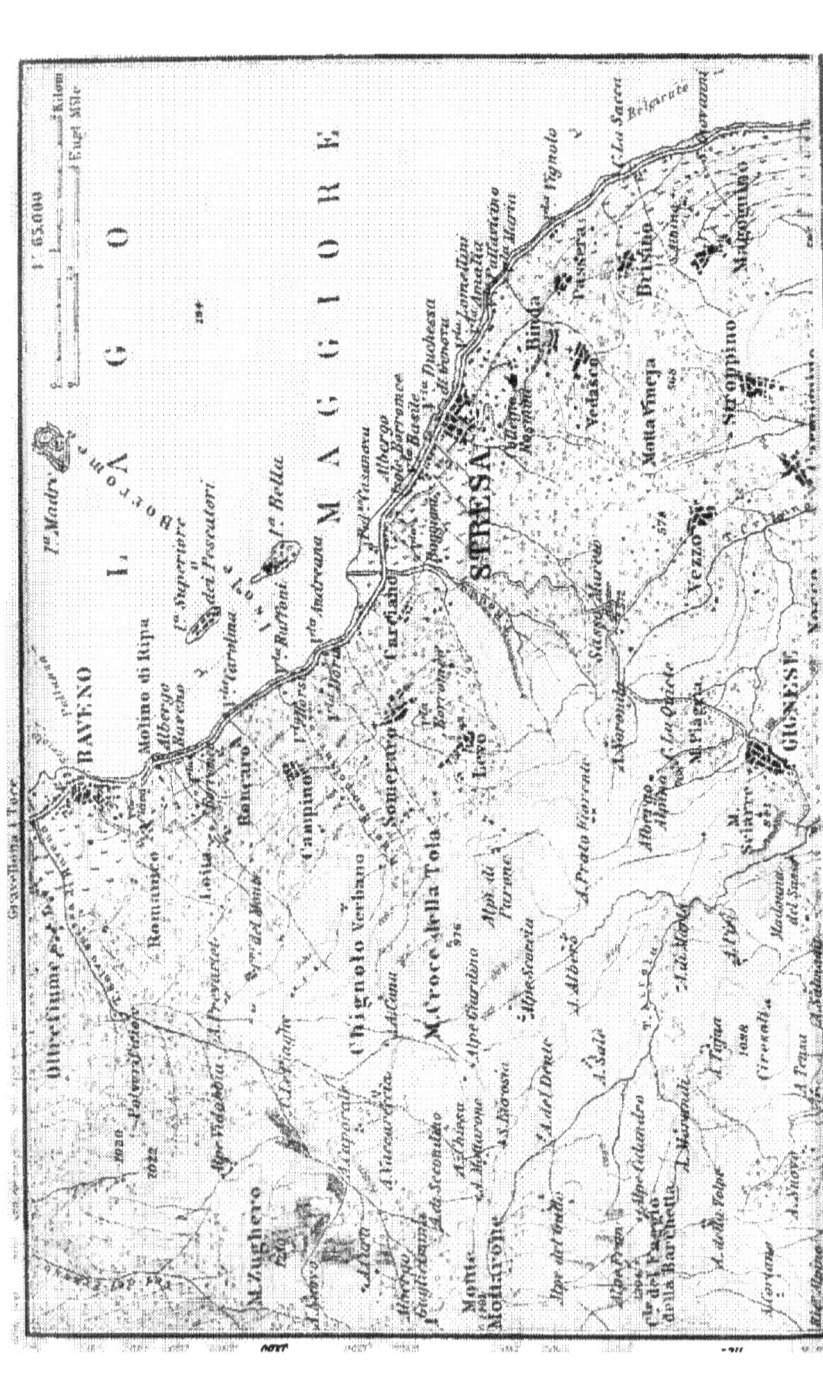

at the *Isola Bella*, the southernmost, which, with the *Isola Madre*, belongs to the Borromeo family. The scenery around the Borromean Islands rivals that of the Lake of Como in grandeur, and perhaps surpasses it in richness. Visitors are admitted to Isola Bella and Isola Madre from 15th March to 15th Nov. daily, except Mon., from 9 a.m. to 3, 4, or 5 p.m., according to the season.

In the splendour-loving, but tasteless 17th cent. *Count Vitaliano Borromeo* (d. 1690) erected a château on *Isola Bella, and converted the barren mica-slate rock into beautiful gardens, rising on ten terraces 100' above the lake, and displaying all the wealth of Italian vegetation: lemon-trees, cedars, magnolias, cypresses, orange-trees, laurels, magnificent camellias and oleanders, etc. (evening light best for the beautiful view). The grounds are disfigured with shell-grottoes, fountains (dry), mosaics, and statues in the style of the period. Travellers from the north cannot fail to be struck with the loveliness of the banks, studded with innumerable dwellings, and clothed with luxuriant vegetation (chestnuts, mulberries, vines, figs, olives), and of the deep-blue lake, enhanced by the snow-mountains in the background. The large *Château*, the N. wing of which is unfinished, contains handsome saloons, a collection of pictures, a chapel with tombs of the Borromeo family of 1485 and 1575, transferred hither from Milan, a gallery with valuable tapestry, etc. The view through the arches of the long galleries under the château is curious. A servant attends visitors in the château (fee 50 c., 1 fr. for a party), and the well-informed gardener shows the grounds for a similar fee. Adjoining the château are the *Hôt. du Dauphin* or *Delfino* (R., L., & A. 3, B. 1¼, D. 4, pens. 7 fr.), and the *Ristor. del Vapore*. Boat to Isola Madre and back with two rowers 3 fr.

The *Isola Madre (not a steamboat-station) is on its S. side laid out in seven terraces, with lemon and orange-trellises. On the highest terrace is an uninhabited Palazzo, with a beautiful view. On the N. side are delightful grounds, with luxuriant vegetation (gardener 1 fr.). — The **Isola dei Pescatori** or **Superiore** (*Hot.-Ristor. d'Italia*, pens. 5-6 fr., well spoken of) is also worth visiting for the sake of the picturesque views it commands. The island is almost entirely occupied by a fishing-village (300 inhab.).

Opposite Isola Bella, on the W. bank lies —

Stresa. — *Hôtel des Iles Borromées (Omarini's)*, ½ M. from the landing-place, comfortable, with a fine garden, R., L., & A. from 4, B. 1½, déj. 3, D. 5, pens. 9-12; omnibus 1 fr.; *Hôt.-Pension Beau-Séjour* (same proprietors), with a large garden; Hôtel Milan, with a small garden on the lake, near the pier, R. from 2, B. 1½, déj. 3, D. 4, pens. from 6 fr.; Albergo Reale Bolongaro, on the lake, Italian, well spoken of, R., L., & A. 2½, B. 1, déj 3, D. 4, pens. 7 fr.; Italia & Pension Suisse, R., L., & A. from 2, B. 1¼, déj. 2½, D. 3½, board 5 fr.; Albergo S. Gottardo, R. from 1½, pens. 5-6 fr.; these three second-class, but very fair.

Boat (*barca*) with one rower 2 fr. for the first hour, and 50 c. for each additional half-hour (comp. p. 448). — Diligence to *Gravellona* (7½ M.; p. 452) thrice daily in 1¼ hr., 1 fr. 80, coupé or banquette 2 fr. 70 c.

English Church Service at the Hôtel des Iles Borromées (April-Sept.).

Stresa, with 1300 inhab. and a picturesque view of the Borromean Islands, lies in a cooler and airier situation than the other places on Lago Maggiore and is therefore preferable for a stay during the hot season. In the vicinity are numerous villas of the Italian **aristocracy**. The *Villa Ducale,* adjoining the Alb. Milano to the W., **is the property** of the Duchess of Genoa; a new mansion in the park **is the** summer residence of her son, the Duke of Genoa. — The **handsome** *Rosminian Monastery* (875'), on the hillside 1/2 M. to the S. of the town, is now a school. The church contains the monument of Ant. Rosmini (d. 1855), **with an** admirable statue by Vela. On the lake, 1/2 M. to the S., are **the** beautifully situated *Villa Pallavicino* and (1/4 M. farther on) **the** *Villa Vignolo,* with fine gardens (visitors admitted).

From Baveno or Stresa to the Monte Mottarone, 3 1/2-4 hrs.; **guide** (convenient for the last third of the ascent) 5 fr.; **mule** 5 fr., with **guide** 3 fr.; mountain-vehicle with one horse from Stresa to **the** Albergo Alpino 10 fr. — The route from Baveno ascends mostly through wood by *Romanico, Campino,* and *Someraro* (1500'), where it is joined by a route diverging from the Baveno and Stresa road opposite the Isola Bella, to the (1 3/4-2 hrs.) hamlet of *Levo* (1915'; *Hôt. Levo, pens. 6-7 fr.). A path leads hence to the left to the Albergo Alpino (25 min.; see below). The path to the Mottarone steadily ascends over pastures, past the *Alpe Giardino* (3055'), to the (1 hr.) chapel of *S. Eurosia* (3685'), without a tower, where we turn to the right; 20 min. *Alpe del Mottarone,* surrounded by fine beeches and elms; 1/2 hr. Albergo Mottarone (see below). — From Stresa **we** follow a carriage-road diverging from the high-road to the E of the Hôtel des Iles Borromées; 1 hr. *Ristorante Zanini,* a chalet on an open meadow, beyond which a finger-post directs us to the right to Levo (see above). The carriage-road proceeds straight on to *Gignese,* but before reaching this village (25 min. from the Ristor. Zanini) a narrow road diverges **to the** right to the (3/4 hr.) *Albergo Alpino* (2525'; pens. 7 1/2-8 fr.), with **view of** Pallanza, Intra, and Baveno. Hence we ascend over pastures, **past** the *Alpe del Mottarone* (see above), to the (1 3/4 hr.) °*Albergo Mottarone,* **kept by** the brothers *Guiglielmina* (4675'; R., L., & A. 3, D. 1 1/2, lunch 3 1/2, pens. incl. wine 9 fr.), 10 min. below the grassy summit of the *Monte Mottarone (4892'), the highest elevation of the *Margozzolo* group of hills. The **view** from the top (panorama by Bossoli, in the hotel) embraces the Alps, **from the** Col di Tenda and Monte Viso on the W. to the Ortler and Adamello on the E. The most conspicuous feature is the Mte. Rosa group; to the right **of** it appear the Cima di Jazzi, Strahlhorn, Rimpfischhorn, Allalinhorn, Alphubel, Mischabel (Täschhorn, Dom, Nadelhorn), Pizzo Bottarello, Portjengrat, Bletschhorn, Mte. Leone, Jungfrau, Helsenhorn, Flescherhörner; **then** **more** distant, to the E. of the peak of Mte. Zeda, the mountains of the Rheinwald group, Bernina, Disgrazia, Mte. Legnone, Mte. Generoso, Mte. Grigna. At our feet lie seven lakes, those of Orta, Mergozzo, Maggiore, Biandrone, Varese, Monate, and Comabbio; farther to the right stretch the great **plains** of Lombardy and Piedmont, with Milan and its lofty cathedral in the centre. The silvery Ticino and Sesia meander through the plains, **and** by a singular optical delusion seem to traverse a lofty table-land. **The** Mottarone consists of a number of bare peaks, studded with a few chalets among tall trees; its base is encircled with chestnut-trees, and the surrounding plain is also well wooded.

On the W. side a path, rather steep at places (guide advisable), descends direct to (2 hrs.) *Omegna* (rail. stat., see p. 452). — Travellers bound for Orta (4 1/4 hrs.) soon reach a broad bridle-path on the S. side of the hill (guide unnecessary), descending by *Cheggino* (2100') to (2 1/2 hrs.) *Armeno* (1720'; Alb. al Mottarone), on the high-road, which they follow to the S. From (12 min.) the point where the road forks, the left branch leads to

Miasino (p. 452), the right by *Carcegna*, crossing the railway to Gravellona (the station of Orta-Miasino lying to the left), to (4 M.) *Orta* (p. 452).

The banks now become flatter; to the W appears Monte Rosa. On the W. bank is **Belgirate** (*Gr. Hôt. Belgirate*; 700 inhab.), with the villas *Fontana*, *Principessa Matilda*, etc. Then **Lesa** and **Meina** *(Alb. Zanetta)* on the W., and **Angera** (railway-station) on the E. bank, with a handsome château of Count Borromeo.

Arona (738'; pop. 3300; *Albergo Reale d'Italia & Posta*; *Alb. S. Gottardo*, both on the quay; *Ancora*, behind the S. Gottardo; *Caffè della Stazione*; *Café*, next the Alb. Reale; *Café du Lac*, at the harbour), an old town on the W. bank of the lake, about 3 M. from its S. end, extends up the slope of the hill. In the principal church, *S. Maria*, is the chapel of the Borromeo family, to the right of the high-altar, containing an *Altar-piece, the Holy Family, by *Gaudenzio Ferrari*, of 1511.

On a commanding height $1/2$ hr. to the N. is a colossal *Statue of S. Carlo*, 70' high, resting on a pedestal 43' high, erected in 1697 in honour of the famous cardinal, Count Carlo Borromeo, Archbishop of Milan, who was born here in 1538 (d. 1584, canonised 1610). The head, hands, and feet of the statue are of bronze, the robe of wrought copper. Ascent in the interior disagreeable (50 c.). Relics of S. Carlo are preserved in the neighbouring *Church*, near which is a large *Seminary for Priests*.

FROM ARONA TO MILAN, 42 M., railway in $2^1/_4$ - $2^1/_2$ hrs. (6 fr. 80, 4 fr. 55, 2 fr. 65 c.). The most important stations are *Sesto-Calende* and (17 M. *Gallarate*, the junction of the lines from Varese and Laveno (p. 446).

110. From Domodossola to Novara. Lake of Orta.

Comp. Map, p. 448.

56 M. RAILWAY in $3^1/_2$ hrs. (fares 10 fr. 30, 7 fr. 15, 4 fr. 60 c.); to *Gravellona*, the station for the Lago Maggiore (omnibus to Pallanza and Stresa, see pp. 447, 449), 20 M., in $1^1/_4$ hr. (fares 3 fr. 70, 2 fr. 55, 1 fr. 65 c.).

Domodossola, see p. 314. — The line runs straight along the foot of the cliffs bounding the *Val d'Ossola* to the W., on the right bank of the *Tosa* or *Toce*, which divides into numerous rivulets and fills with its broad gravelly bed the whole bottom of the valley. — At ($3^3/_4$ M.) *Villa*, or *Villadossola*, the *Val Antrona* (p. 314) opens on the right. — Near (5 M.) *Pallanzeno* (750') the railway skirts the Tosa for a short distance, and then traverses a broad grassy expanse. At ($6^3/_4$ M.) *Piedimulera* (800'; *Corona*; *Alb. Piedimulera) the *Val Anzasca* opens on the right (to *Macugnaga*, see p. 343). The railway crosses the *Anza* to (8 M.) *Rumianca* and the Tosa to ($8^1/_2$ M.) **Vogogna** (740'; *Corona*), a small town at the foot of steep rocks, with a ruined castle. — $10^1/_2$ M. *Premosello*. Beyond (13 M.) *Cuzzago* we cross the Tosa by a bridge, 515' yds. in length, and reach ($15^1/_2$ M.) **Ornavasso** *(Italia; Croce Bianca)*. The marble quarries on the hill to the left belong to the cathedral-chapter of Milan. —

452 *VII. Route 110.*   ORTA.   *From Domodossola*

At (18½ M.) **Gravellona**, or *Gravellona-Toce* (*Railway Restaurant*; Inns indifferent), a place with large cotton-mills, the *Strona* falls into the Tosa. Travellers bound for the Lago Maggiore descend here (road to *Pallanza*, 6¼ M., viâ Fondo Toce and Suna, see p. 448; omnibus, see p. 447; road to *Baveno*, 5 M., viâ Feriolo, and to *Stresa*, see p. 448; omnibus, see p. 449).

The railway to Novara runs to the S. through the fertile valley of the Strona. Beyond (21 M.) *Crusinallo* it crosses the river and immediately afterwards the *Nigulia Canal*, which drains the Lake of Orta.

23 M. **Omegna** *(Alb. Manin; Croce Bianca)*, a thriving little town with a large paper-mill, at the N. extremity of the **Lago d'Orta** (950′), a charming lake 7½ M. in length, now called *Lago Cusio* after its dubious ancient name. — The train skirts the lake, commanding beautiful views. Beyond (26½ M.) *Pettenasco* it crosses the *Pescone* and then the imposing **Sassina Viaduct**.

28½ M. **Orta**, also the station for *Miasino*. — The STATION is about 1 M. above Orta; at the exit, we keep to the left, pass below the railway, and walk straight on, past the (½ M.) *Villa Crespi*, in the Moorish style, beyond which, to the right, a finger-post shows the way to the Monte d'Orta and the (¼ hr.) Alb. Belvedere.

Hotels. *ALBERGO BELVEDERE, in a commanding position on the W. brow of the Monte d'Orta, R. & A. 3, D. 4 fr. — ALB. S. GIULIO, ALB. ORTA, both 1¼ M. from the railway-station, in the market-place on the lake.

The little town of *Orta*, essentially consisting of a small Piazza opening towards the lake and a long and narrow street, with a number of pretty villas extending towards the station, lies opposite the little *Isola S. Giulio*, at the foot of the finely wooded **Monte d'Orta** (1315′) which protrudes far into the lake. Ascent halfway between the town and the station (see above), or from the Piazza through the garden of the *Villa Natta* (50 c.). In the 16th cent. 20 chapels were erected on this hill in honour of St. Francis of Assisi, each containing a scene from his history in painted life-size figures of terracotta (the best in the 13th, 16th, and 20th chapels; in the last is represented the canonisation of the saint). The hill, which is also called the *Sacro Monte* or *Santuario*, is laid out as a park. Various points command charming views of the lake, and from the Campanile on the highest point we enjoy a panorama, which is dominated to the W. by the snowy head of Monte Rosa (50 c.).

Boat to the **Isola S. Giulio**, there and back 1 fr. 50 c. The church, founded by St. Julius, who came from Greece in 379 to convert the natives, and frequently restored, contains several good reliefs, old frescoes, and a fine Romanesque pulpit; in the sacristy is a Madonna by Gaudenzio Ferrari, and in the crypt below the high-altar a shrine of crystal and silver containing the body of St. Julius. On the hill is a seminary for priests, with a garden affording charming views of the lake.

Picturesque excursions may be made from Orta to the (1 hr.) *Madonna della Bocciola* (1565′), situated on the hill above the station; to the S. to the (1¼ hr.) *Torre di Buccione* (1500′; an ancient watch-tower dating from Emp. Frederick Barbarossa) at the S. end of the lake (1¼ hr.; boat to Buccione 1½ fr.), both points commanding good views. By *Pella* (see below) to (½ hr.) *Alzo*, with extensive granite-quarries (branch-railway from Gozzano, see p. 453) and to (1 hr.) the *Madonna del Sasso* (2090′), the pretty church of the hamlet of *Boletto*, on a lofty cliff, commanding a fine survey of the entire lake.

From Orta to the *Mottarone* (4-5 hrs.), viâ *Carcegno*, *Armeno*, and *Cheggino*, see p. 450 (on the houses, arrows pointing 'al Mottarone' or 'al Mergozzolo'); guide 6 fr., mule 10 fr.

Beautiful view of the island of *S. Giulio* (p. 452) as we proceed; on the steep cliffs of the W. bank is the church of *Madonna del Sasso* (see above). Beyond (30 M.) *Corconio* the train traverses a cutting on the W. side of the *Castello di Buccione* (see above) and quits the Lake of Orta. 32 M. *Bolzano*. 33½ M. *Gozzano*, a place of considerable size, is the junction for *Alzo* (p. 452). We now traverse the fertile *Val d'Agogna*. 37 M. *Borgomanero* (Alb. del Ramo Secco), 7½ M. to the S.W. of Arona (p. 451). — 41 M. *Cressa-Fontaneto*; 43 M. *Suno*; 46½ M. *Momo*; 51 M. *Callignaga*; 54 M. *Vignale*; 56 M. *Novara*. From Novara to *Milan* (p. 465), railway in 1¼ hr.; to *Laveno* (p. 446) in 1½ hr. Comp. *Baedeker's Northern Italy*.

FROM ORTA OVER THE COLMA TO VARALLO, 4½ hrs., a beautiful walk (donkey 6, to the Colma 3 fr.; guide, 5 fr., unnecessary). On the W. bank of the lake, opposite Orta, peep the white houses of **Pella** (*Pesce d'Oro*, unpretending) from amidst vines, chestnuts, and walnuts. (Boat from Orta in 20 min., fare 1½ fr.) At Pella we strike the new road leading from Alzo (see above) along the slope to (3 M.) *Arola* (2020'), which commands a lovely view in the direction of the Lake of Orta. We turn to the left 5 min. beyond the village, descend a little, and then keep on for ½ hr. on the same level, skirting the gorge of the Pellino, which here forms a pretty waterfall. We next ascend through wood, over crumbling blocks of granite, to the (¾ hr.) wooded Col di Colma (3090'). The eminence to the left commands a splendid view, embracing Monte Rosa, the lakes of Orta and Varese, and the plain of Lombardy. In descending (to the right), we overlook the fertile *Val Sesia*, with its numerous villages. The path leads through groves of chestnuts and walnuts to (¾ hr.) *Civiasco* (2350'; several taverns), whence a new road (short-cut by the old path descending to the left), the first part of which affords a view of Mte. Rosa, leads to (¾ hr.) —

Varallo (1480'; pop. 2300; *Italia*, R. & A. 3½, D. 4 fr.; *Croce Bianca*, good cuisine; *Posta*), the capital of the *Val Sesia* and terminus of the line Novara-Borgosesia-Varallo, at the mouth of the *Mastallone*. The Piazza Vitt. Emanuele, at the entrance to the town from the station, is embellished by a monument to Victor Emmanuel. Over the high-altar of the collegiate church of *S. Gaudenzio* is a Marriage of St. Catharine by *Gaudenzio Ferrari* (1484-1549), a native of the neighbouring Val Duggia. The church of *S. Maria delle Grazie* contains frescoes by the same master, in the choir. His statue in marble, by Della Vedova, stands near the church. Beyond the bridge over the Mastallone is a large new *Stabilimento Idroterapico*, open from June to the end of Sept. (pens.

9-11 fr.). — A paved road, shaded by fine chestnut-trees, ascends from S. Maria delle Grazie past the church of *S. Maria di Loreto* (above the portal, a fresco by Gaud. Ferrari, the Adoration of the Child) in 20 min. to the **Sacro Monte** (*Santuario di Varallo*; 1995'), a great resort of pilgrims and sight-seers. On the top of the hill and on its slopes are a church and 46 chapels, or oratories, containing scenes from the life of the Saviour in painted lifesize figures of terracotta, beginning with the Fall in the 1st chapel, and ending with the Entombment of the Virgin in the 46th, dating mostly from the end of the 16th century. The hill now belongs to the town (Inn and Café at the top).

Varallo is a capital starting-point for excursions into the very attractive and easily accessible valleys in the vicinity.

FROM VARALLO VIÂ FOBELLO TO PONTEGRANDE (and Macugnaga), 9 hrs., guide hardly necessary. A road ascends the pretty *Val Mastallone*, passing the (3 M.) picturesque *Ponte della Gula*, *Cravagliana*, and *Ferrera*, and crossing the *Landwasser* (see below) by the (8¹/₄ M.) *Ponte delle Due Acque* to (10¹/₂ M.) Fobello (2887'; *Posta*; *Italia*). Thence a bridle-path leads viâ *Boco*, *Piana*, *S. Maria*, and *Giavino* to the (3 hrs.) Col di Baranca (5970'), with a chapel and an *Inn. Steep descent, with pretty views of the Val Anzasca, through the *Vall' Ollocchia* to *Bannio* and (3 hrs.) *Pontegrande* (p. 344). — From the Ponte delle Due Acque (see above) a road ascends along the Landwasser to (6 M.) Rimella (8874'; *Alb. Fontana*), a German community of about a dozen hamlets, with 1100 inhabitants. The situation is magnificent. A fine but toilsome route leads hence over the *Colle Drochetta* to *Bannio* and *Pontegrande* (p. 344) in 5 hrs. (guide).

FROM VARALLO THROUGH THE VAL SESIA TO ALAGNA (23 M.). Omnibus daily at 1 p.m. in 5 hrs. The road ascends the fertile valley, on the left bank of the Sesia, by *Valmaggia* and *Vocca*, to (5 M.) Balmuccia (1900'), at the influx of the *Sermenza*.

[A road ascends the picturesque **Val Sermenza** (*Valle Piccola*) by (1¹/₂ M.) *Boccioleto* (2188'; *Pens.-Restaurant della Fenice*) and *Ferrera*, to (¹/₂ hr.) *Fervento* (Restaurant Valle Sermenza), and a bridle-path leads thence to (1 hr.) *Rimasco* (2970'; two inns, the upper the better), where the valley divides. In the branch on the right (E.; *Val d'Egua*) lies (2 hrs.) *Carcoforo* (4280'; Monte Moro, plain), and in the *Val Piccola* to the left (W.) are *Rima S. Giuseppe* and (2 hrs.) Rima (4650; *Albergo Tagliaferro*), the last belonging to the German communities at the S. base of Mte. Rosa (comp. pp. 346, 347). FROM CARCOFORO TO PONTEGRANDE over the *Col d'Egua* (7836') and *Col di Baranca* (see above), 6-7 hrs., with guide, interesting; to PESTARENA over the *Passo della Mariana* (about 8180'), 6 hrs., with guide, fatiguing; to MACUGNAGA over the *Col della Bottiglia* (8765'), 7 hrs., with guide, also fatiguing (descent through the *Val Quarazza*, p. 346); to RIMA over the *Bocchetta del Temu* (7700'), 4¹/₂ hrs., with guide, easy. — FROM RIMA (see above) TO MACUGNAGA over the **Passo di Rima** (*Col del Piccolo Altare*; 8630'), 5 hrs., bridle-path; to ALAGNA over the *Colle Mond* or the *Bocchetta Moanda*, see p. 347.]

The road, following the left bank of the Sesia, next leads by *Scopa* (Albergo Topini), *Scopello* (Alb. Deblasi; Valsesia), *Pila*, *Piode*, and *Campertogno* to (10 M.) Mollia (2887'; *Alb. Valsesiano*). Thence through the narrowing valley to (5¹/₂ M.) *Riva Valdobbia* (3628'; *Hôt. delle Alpi*), beautifully situated, where several peaks of Monte Rosa become visible to the N., and (2¹/₂ M.) *Alagna* (p. 346). The façade of the church of Riva Valdobbia is adorned with a large fresco of the Last Judgment, of the school of Gaud. Ferrari.

111. From Luino on Lago Maggiore to Menaggio on the Lake of Como. Lake of Lugano.

Comp. Maps, pp. 448, 456.

RAILWAY (STEAM TRAMWAY) from Luino to (8 M.) *Ponte Tresa* in 1 hr. (2 fr. 65, 1 fr. 45 c.). STEAMER from Ponte Tresa to (15 M.) *Lugano* in 1³/₄ to (26 M.) *Porlezza* in 2³/₄ hrs. (4 fr. 50, 2 fr. 70 c.). STEAM TRAMWAY from Porlezza to (8 M.) *Menaggio* in 1 hr. (2 fr. 65, 1 fr. 45 c.). Through-tickets 9 fr. 80, 5 fr. 60 c.; return, Sunday, and circular tickets at reduced rates (to be had on board any of the steamers). — Swiss custom-house examination on board the steamers on the Lake of Lugano; Italian custom-house at Porlezza or Ponte Tresa.

Luino, see p. 445. The station of the steam-tramway adjoins the landing-place. The train crosses the St. Gotthard railway near the Luino station, and at (2 M.) *Creva*, a manufacturing place, reaches the *Tresa*, the river descending from the Lake of Lugano, which falls into the Lago Maggiore at *Germignaga* (p. 445). After winding up the abrupt right bank of the Tresa, the train crosses the river, which here forms the boundary between Switzerland and Italy, passes through two tunnels, and stops at (4¹/₂ M.) *Cremenaga* (833'). It then follows the left bank, affording pretty views of picturesque villages and churches, to (8 M.) **Ponte Tresa**, on the Italian side of the river. The village of that name, on the Swiss side, lies on a bay of the **Lake of Lugano** which is so enclosed by mountains that it looks like a complete little lake in itself.

The LUGANO ROAD (6 M.), which may also be recommended to pedestrians, crosses the *Vallestna* to (1¹/₂ M.) *Magliaso*, and leads, with the Monte S. Salvatore on the right, to (1¹/₂ M.) Agno (968'). Crossing the stream of that name, and passing the little *Lake of Muzzano*, we gradually ascend to the Restaurant du Jardin in Sorengo (p. 436), and descend to (2¹/₂ M.) *Lugano* (p. 434).

STEAMBOAT JOURNEY. The vessel steers through the *Stretto* or strait of *Lavena* (with the abrupt *Mte. Caslano*, 1710', on the left), and enters the W. arm of the **Lake of Lugano** (890'; Ital. *Lago Ceresio*), where the wooded banks are somewhat monotonous. To the N. we soon obtain a fine view of the bay of *Agno* (see above), with high mountains behind it (Mte. Tamaro, Mte. Bigorio, etc.). The steamer turns to the S., passing *Figino* on the left (with Mte. S. Salvatore and its chapel in the distance, p. 436), and touches at *Brusimpiano* on the right. Farther on we skirt the wooded slopes of the *Mte. Arbostora* (2750') on the left, at the foot of which runs a road to Lugano (p. 435). In a bay of the S. bank lies **Porto** or *Porto Ceresio*. (Railway to *Varese* in ¹/₂ hr., see p. 446.) On a distant hill is seen the *Madonna del Monte* (p. 446).

The steamer turns to the N., to **Morcōte** (*Hôtel-Pension Raygi-Kauffmann*, pens. 5 fr.), with a picturesque church, prettily situated on the S. angle of Mte. Arbostora. We follow the W. bank; *Brusin-Arsizio* lies on the right, and the long indented crest of *Mte. Generoso* soon appears (p. 439). The vessel touches at *Melide* on the W. and at *Bissone* on the E. bank, and passes through the railway viaduct (picturesque view through the arch). Then *Campione*, on the

456 *VII. Route 111.* OSTENO.

E. bank (interesting old frescoes in the church of the Madonna dell'Annunziata). To the left rises *Mte. S. Salvatore* (p. 436), to the right *Mte. Caprino* (p. 437).

Lugano, see p. 434 (the station of the St. Gotthard Railway lies high above the town, 1 M. from the steamboat-pier).

Between Lugano and S. Mamette is the finest part of the lake. On the N. bank is *Castagnola*, most picturesquely situated at the foot of *Mte. Brè* (p. 437); then **Gandria** *(Pens.)*, with its gardens borne by lofty arcades and its vine-terraces. Beyond this point the lake assumes a wilder and lonelier character. The next villages are (1.) *Bellarma* (frontier); **Oria** with the *Villa Bianci; Albogasio;* and **S. Mamette** *(Stella d'Italia)*, beautifully situated at the mouth of the picturesque *Val Solda*, with *Castello* high above it (p. 437). The S. bank is wooded and abrupt. To the left *Loggio, Cresogno,* and *Cima*, opposite which (S.) lies **Osteno** *(Hôt. du Bateau; Restaurant della Grotta)*, much frequented on account of its gorge (return-ticket 2 fr. 35 c ; tickets for the grotto sold on board the steamer, 75 c.).

The *Grotto of Osteno (*Orrido* or *Pescara*, 'fisherman's gorge') is 7 min. from the landing-place. We pass through the village; outside the gate we descend to the right before the stone bridge, and cross the brook. The mouth of the gorge, in which there is a small waterfall, is near a projecting rock (Restaurant). Visitors embark in a small boat and enter the grotto, the bottom of which is occupied by the brook. The narrow ravine through which we thread our way is curiously hollowed out by the water. Far above, the roof is formed by overhanging bushes, between which glimpses of blue sky are obtained. The gorge is terminated by a waterfall. — The Tufa Grottoes of Rescia, though much less interesting, may also be visited if time allows (1 hr. there and back). Boat (with two rowers, there and back 2 fr. each) round the promontory to the E. of Osteno in 1/4 hr. to the hamlet of *Rescia;* thence by a narrow path to the grottoes in 5 min. (adm. and torches 1/2 fr.). The dome-shaped grottoes, encrusted with calcareous sinter and stalactites, are connected by a low passage (caution necessary). From the second is seen a small waterfall in a gorge. In the vicinity are tufa-quarries, containing interesting fossils.

A new road leads from Osteno to the S.W. to (6 M.) Lanzo d'Intelvi (3417; *Caffè Centrale*, moderate, déj. 2 fr.); 11/4 M. above is situated the '*Hôt. Belvedere* (pens. 8-10 fr.), a pleasant spot for some stay, with a fine view of the Lake of Lugano and the Alps with Mte. Rosa (English Church Service in summer; English physician). [Those whose destination is the Hôtel Belvedere take the footpath to the right, about 3/4 M. before reaching Lanzo, which soon joins the road ascending to the hotel.] A road also leads to Lanzo from *Maroggia* (9 M.; see p. 438), and another from *Argegno* on the Lake of Como (121/2 M.; see p. 461). Near Lanzo (20 min.) are the baths of *Paraviso*. Bridle-path to *Mte. Generoso* (p. 439), 51/2 hrs.

The N. bank of the lake now becomes rocky and precipitous. At the N. end of this bay lies **Porlezza** (*Alb. del Lago,* indifferent), with the Italian custom-house. Boat to Lugano 10-12 fr.

From Porlezza to Menaggio. The station of the tramway (comp. p. 455) is close to the landing-place. The train runs through the broad valley of the *Cuccione*, by *Tavordo, S. Pietro* (last view of the Lake of Lugano), and (21/2 M.) *Piano*, on the little *Lago del Piano*. It then ascends more rapidly (4 : 100) viâ *Bene-Grona* to (5 M.) *Grandola* (1260'), the highest point on the line, 610' above the Lake of Como. It next descends on the lofty right bank of the

Val Sanagra in numerous curves, the line being hewn in the rock in many places and supported by buttresses of masonry. Beyond a tunnel 110 yds. long the line takes a long bend towards the S., affording a delightful *View of the Lake of Como, with its luxuriantly fertile banks, sprinkled with towns, villages, and villas, and enclosed by high mountains. To the right are the beautiful peninsula of Bellagio and the bay of Lecco. After running towards the S for about 1/2 M., the train turns back at a sharp angle and descends rapidly (5:100) to (8 M.) *Menaggio* (p. 459), where the terminus is close to the steamboat-pier and the Hôtel Menaggio. (The village of Menaggio has a pier of its own, see below.)

112. The Lake of Como.

Steamboat thrice daily from Colico to Como in $3^1/_2$-5 hrs. (5 times from Bellagio to Como, and 8 times from Torriggia to Como); thrice between Colico and Lecco ($3^3/_4$-$4^1/_2$ hrs.), and thrice between Como and Lecco ($3^1/_2$ hrs.). Stations: *Colico* (pier), *Domaso*, *Gravedona* (pier), *Dongo* (pier), *Musso*, *Cremia*, *Dervio*, *Rezzonico*, *Acquaseria*, *Bellano* (pier), *Varenna* (pier), *Menaggio-Bridge* (pier), *Menaggio-Station* (pier), *Bellagio* (pier), *Cadenabbia* (pier), *S. Giovanni & Tremezzo* (pier), *Azzano*, *Lenno*, *Lezzeno & Campo*, *Sala*, *Argegno* (pier), *Nesso*, *Torriggia*, *Pognana*, *Palanzo*, *Carate* (pier), *Urio*, *Torno*, *Moltrasio*, *Blevio*, *Cernobbio* (pier), *Como* (pier). Embarkation and landing free (the tickets have a coupon which is given to the boatman). Those who embark at intermediate stations must procure a ticket at the pier; otherwise they are liable to be charged for the whole distance from Como or Colico. Some of these stations are often passed without stopping, and the advertised hours are not rigidly adhered to. Some of the boats are handsome saloon steamers, with good restaurants. In the following description the stations with piers are marked P, those where the landing is effected by boats with B.

Railway on the E. bank from *Bellano* (p. 458) to *Lecco* (15 M.), of no particular interest for tourists. Numerous tunnels and viaducts.

Boats *(barche)*. First hour $1^1/_2$ fr., for each additional hour 1 fr. per rower. From Bellagio to Cadenabbia and back, or vice versâ, each rower $2^1/_2$ fr.; Bellagio to Tremezzo, Bellagio to Menaggio, and Bellagio to Varenna also $2^1/_2$ fr. each rower; Bellaggio to Villa Melzi, Villa Carlotta, and back, each rower 3 fr. — One rower generally suffices, unless time is limited. If a second proffers his services, he may be dismissed with: *'basta uno'* (one is enough). The boatmen reduce their fares when customers are not numerous. The following phrases may be useful: *'Quanto volete per una corsa d'un ora (di due ore)? Siamo due (tre, quattro) persone. È troppo, vi darò un franco (due franchi)'*, etc. — The boatmen generally expect a fee *(buonamano)* of $^1/_2$-1 fr. in addition to the fare.

The **Lake of Como (700')**, Ital. *Lago di Como*, or *Il Lario*, the *Lacus Larius* of the Romans, extolled by Virgil (Georg. ii. 159), is by many considered the most beautiful lake in N. Italy. From the N. end to Como it is 30 M. in length; between Menaggio and Varenna, its broadest part, it is nearly $2^1/_2$ M. in breadth; and its greatest depth is 1930'. At Bellagio (p. 460) the lake divides into two arms, the *Lake of Como* (W.) and that of *Lecco* (E.). The *Adda* falls into the lake at the N. end and emerges from it again at Lecco.

Numerous gay villas of the Milanese aristocracy, with luxuriant gardens and vineyards, are scattered along the banks of the lake, and above these extend groves of chestnuts and walnuts of brilliant green, contrasting strongly with the dull-gray tint of the olive, which to the unaccustomed eye resembles the willow. The gay and fragrant oleanders add a great charm to summer. Among smaller botanical specimens are numerous varieties of saxifrage, rare orchids, the very uncommon Ceterach Maranthæ, the maiden hair (Adianthum Capillus), and other ferns. The mountains rise to a height of 7000'. The scenery of the lake, viewed from the steamboat, somewhat resembles that of a vast river, the banks on both sides being distinguishable. The lake is well stocked with fish, such as the palatable little '*Agoni*', and trout ('*Trote*') of 20 lbs. weight are occasionally captured.

EASTERN BANK.

Colico (P; *Isola Bella*; *Hôt. Risi*), at the N. extremity of the lake, see p. 385.

Ogliasca, *Dorio*, and **Corenno**, with a ruined castle.

Dervio (B) lies at the mouth of the *Varrone*, and at the foot of *Monte Legnone* and its spur, the *Monte Legnoncino* (5680').

Monte Legnone (8565'), the highest mountain of Lombardy, may be ascended hence in 7 hrs., with guide (not difficult for adepts, and very attractive). A bridle-path leads to (2 hrs.) *Suaglio* (2580'; *Osteria Pinzetta*, plain), and via *Introzzo* and *Valle Lavade* to the (2 hrs.) *Rifugio* of the Italian Alpine Club near the *Roccoli Lorla* (4460'; good accomodation and beds), on the saddle between Legnone and Legnoncino. Thence in 2¼ hrs. to the *Capanna Alpina* (7010'; no night-quarters) and in 1 hr. more to the summit, with magnificent view. The ascent on the N. side, from *Delebio* (p. 123), is easier: bridle-path through the *Val della Lesina* to the (4 hrs.) *Alp Cappello*, and over the *Bocchetta di Legnone* in 3 hrs. to the summit.

Bellano (P; *Alb. *Bellano*) with 1400 inhab. and important manufactories and iron-works, the temporary terminus of the Lecco and Colico line (p. 457), lies at the mouth of the industrious *Val Sassina*, through which a bridle-path leads to *Taceno* (road thence via *Introbbio* to Lecco). A little above its influx into the lake the *Pioverna* forms a waterfall 200' high (*Orrido di Bellano*, ½ fr.). By the pier is a monument to the poet *Tom. Grossi* (d. 1853).

WESTERN BANK.

Gera (B). — *Domaso* (P), charmingly situated, with the *Villas Venini*, *Miani*, and others.

Gravedōna (P; *Alb. Gravedona; Alb. del Lauro*), with 1600 inhab., is picturesquely situated at the entrance of a ravine. At the upper end of the village rises the *Palazzo del Pero* with its four towers, erected at the end of the 16th century. Adjoining the old church of *S. Vincenzo* is the *Baptisterium S. Maria del Tiglio* of the 12th century, containing two Christian inscriptions of the 5th century.

To the W. opens the *Val di Gravedona*, through which a bridle-path leads to (9 hrs.) Bellinzona (p. 433), crossing the *Passo di S. Jorio* (6415').

Dongo (P; *Alb. Dongo*), a large village in a sheltered situation, at the mouth of the valley of the same name.

On a precipitous rock above *Musso* (B) are the three ruined castles of *Rocca di Musso*, where the condottiere Giov. Giac. de' Medici resided in 1525-31 and held sway over the whole lake.

Pianello and *Cremia* (B), with the handsome church of S. Michele (altar-piece, *St. Michael, by Paolo Veronese).

MENAGGIO.

EASTERN BANK.

Through the *Val Sassina*, which opens at Bellano, a narrow road leads viâ *Taceno* to (6 M.) *Cortenova*, and thence viâ *Introbbio* to Lecco.

Gittana is the landing-place for the hydropathic establishment of *Regoledo*, beautifully situated 500′ above the lake (cable-railway from the pier to the hotel).

Varenna (P; *Hôtel Royal Marcioni*), with beautiful gardens, is charmingly situated on a promontory at the mouth of the *Val Esino*. Admirable view from the ruin of *Torre di Vezio*, near the hamlet of *Vezio*, high above the town (1/2 hr.). Near Varenna several tunnels are hewn in the rock for the passage of the road and the railway (p. 457). The marble from the neighbouring quarries is cut and polished in the town.

About 3/4 M. to the S. of the town the *Fiume Latte* ('milk-stream') falls in several leaps from a height of 1000′ (copious between March and May and in autumn; dry in summer).

The *Monte Grigna* (7907′; 8 hrs.) is a very fine point. From Varenna a bridle-path leads on the right bank of the *Esino* by *Perledo* to (21/2 hrs.) *Esino* (*Alb. Monte Codeno, moderate), prettily situated. Thence (guide desirable; to the club-hut 4, to the summit 7 fr.) to the *Alp Cainallo* 11/2 hr., *Alp Prada* 11/2 hr., *Club Hut* of the C.A.I. (5933′) 1/2 hr., and to the top (*Grigna di Moncodine*) in 2 hrs. more (the last part rather trying). Superb view of the whole Alpine chain from the Mte. Viso to the Ortler (the Mte. Rosa group particularly fine), and of the plains of Lombardy to the distant Apennines. We may descend to the W. (steep) to the club-hut *Capanna di Releggio* (5840′) in the *Val Neria* and to *Mandello*, or to the E. to *Pasturo* in the *Val Sassina* (p. 458).

WESTERN BANK.

Rezzonico (B), with the *Villa Litta*; on the castle-hill a restored fortress of the 13th century. Then *S. Abbondio*. A dangerous footpath crosses the wild precipice of *Il Sasso Rancio* ('the orange rock'), traversed by the Russians under Bellegarde in 1799, when many lives were lost.

Acquaseria (P), the chief place in the parish of *S. Abbondio*.

Menaggio (P). — Two *Steamboat Piers*: one, to the N., near the Hôtel Victoria and the Corona; the other, near the Hôtel Menaggio, for the *Steam Tramway* to Porlezza (Lugano, p. 457). Omnibuses of the hotels at both piers. — *Hotels:* *GRAND HÔTEL VICTORIA, R., L., & A. from 41/2, B. 11/2, déj. 31/2, D. 5, pens. 8-11 fr., Engl. Ch. Service; *HÔT. MENAGGIO, R., L., & A. 31/2-51/2, B. 11/2, déj. 3, D. 5, pens. 8-11 fr., both with gardens on the lake; CORONA, Italian, second-class.

Menaggio (1000 inhab.), with a large silk factory, offers a fine view of Bellagio. A little to the S., on the lake, is the palatial *Villa Mylius*. — To the N. of Menaggio, near the church of *Loveno* (*Inn), is the (11/4 M.) **Villa Vigoni** (fee of 1 fr. to the gardener), with a superb view of Bellagio, Menaggio, and the three arms of the lake (finest from the Chalet Suisse 1/4 hr. farther up). A summer-house contains two admirable reliefs by *Thorwaldsen* (Nemesis) and a marble group by *Argenti*. — Adjacent are the *Villa Massimo d'Azeglio*, containing paintings by the well-known author (d. 1866), and the *Villa Garoviglio*.

The finest point of view is the church of *Madonna della Breglia*, reached by a climb of 11/2 hr. from the Villa Vigoni.

The lake is divided here by the *Punta di Bellagio* into two arms, the BAY OF COMO to the S.W., and the BAY OF LECCO to the S.E.

460 VII. Route 112.   BELLAGIO.   Lake of

Bay of Como.

EASTERN BANK.

Bellagio (P). — Hotels (good, when not overcrowded; omn. at the pier). °GRANDE BRETAGNE, °GRAND HÔTEL BELLAGIO, two large first-class hotels, beautifully situated on the lake, R., L., & A. from 5½, B. 1½, déj. 3½, D. 5, pens. 10-16 fr.; HÔT.-PENS. VILLA SERBELLONI, a dépendance of the Gr. Hôt. Bellagio (pens. 12-14 fr.), with less comfortable rooms, in the fine park mentioned below. — °GENAZZINI, also well situated on the lake, R., L., & A. 3½-5, B. 1½, déj. 2½, D. 4½, pens. 7-10 fr. — Plainer: *HÔT.-PENS. FLORENCE, R., L., & A. 2½-4, B. 1½, déj. 2½, D. 4. pens. 7-9 fr.; PENS. SUISSE, 6-7 fr.; HÔT.-PENS. DES ETRANGERS, 7-9 fr.; ALB. DEL VAPORE, all on the lake. — Beer at the *Café des Etrangers*, on the quay, and in the *Restaurant* of the Hôtel de Florence. — *Boats*; see p. 457. — Objects in olive-wood, silk goods, lace, and antiquities in numerous shops. — Chemist, *Laviazari*. — *English Church* (services April-Oct.).

Bellagio (708'; pop. 800), situated at the W. base of the promontory separating the two arms of the lake, is perhaps the most delightful spot in the lake-district of N. Italy. The church of *S. Giovanni* contains an altar-piece by Gaud. Ferrari.

On the height above Bellagio (25 min. from Genazzini's Hotel to the highest point) stands the *Villa Serbelloni (adm. 1 fr., free for guests of the Grand Hôt. Bellagio). The park extends to the end of the wooded promontory, and affords charming views of Varenna, the Villa Arconati, the Villa Carlotta, etc.

The *Villa Belmonte* (adm. 50 c.), belonging to an Englishman, also commands a charming view.

About 1 M. to the S. of the lower entrance to the Villa Serbelloni, beyond the cemetery,

WESTERN BANK.

Cadenabbia (P). — °BELLEVUE, next the Villa Carlotta, with shady grounds on the lake (closed in Dec., Jan., & Feb.); °BELLE-ILE, R., L., & A. 2-4, B. 1½, déj. 2½, D. 4, pens. 7-10 fr.; *BRITANNIA, R. 2-4, L. ³⁄₄, A. ½, B. 1½, déj. 3, D. 4½, pens. 7-12 fr.; HÔT.-PENS. CADENABBIA, 7-8 fr. — *Café Lavecari*. — *English Church* (services April-Nov.).

Cadenabbia lies about halfway between Como and Colico, in a very sheltered situation. — A little to the S.W., in a garden sloping down to the lake, stands the famous *Villa Carlotta* (formerly *Sommariva*, after the count to whom it once belonged). It was purchased by the Princess Albert of Prussia in 1843, and named **after her** daughter Charlotte (d. 1855), and now belongs to the **Duke of** Saxe-Meiningen. Visitors ring at the entrance to **the garden and ascend the steps (adm. 8-5; 1 fr. and fee).**

The MARBLE SALOON has a frieze adorned with celebrated °Reliefs by *Thorwaldsen*, representing the Triumph of Alexander (for which Count Sommariva paid a sum equal to 14,285*l.* sterling); it also contains several sculptures: °Cupid and Psyche, Magdalene, Palamedes, and Venus, all by *Canova*; Mars and Venus, by *Acquisti*; Cupid offering water to doves, by *Bienaimé*, etc. The BILLIARD ROOM contains casts, and a chimney-piece with sculptured frieze, representing a Bacchanalian procession, said to be one of *Thorwaldsen's* early works. In the GARDEN SALOON are several modern pictures: Romeo and Juliet by *Hayez*; Atala by *Lordon*; also a marble relief of Napoleon, as consul, by *Lavarini*.

The *GARDEN, extending on the S. to Tremezzo, and on the N. to the Hôtel Bellevue, displays the most luxuriant vegetation. Near the S. side of the villa is a magnolia 1½ ft. in diameter. Striking view of Bellagio from under the trees on the S. side of

EASTERN BANK.

we reach a small blue gate on the left, leading to the *Villa Giulia, the property of Count Blome of Vienna, with beautiful *Gardens (open on Sun. and feast days; fee 1/2-1 fr.).

To *Civenna (p. 464) a delightful excursion (carr. with one horse 8 fr.; there and back in 3 hrs.), with which a visit to the Villa Giulia is easily combined.

*Monte S. Primo (5555'), from Bellagio 4½ hrs. with guide (10 fr.), interesting. We ascend past the Villa Giulia and Casate to a (2 hrs.) chapel, where the road forks; thence by the cart-road to the right to the chalets of Villa and Borzo and by a good path to the (2½ hrs.) top, which commands a splendid view of the Lake of Como, the Brianza, and the range of the High Alps.

To the S. of Bellagio is the (½ M.) *Villa Melzi, belonging to the Duchess of Melzi, with numerous art-treasures and a fine garden (adm. on Thurs. and Sat.; 1 fr.).

The Villa Trivulzio, formerly Poldi (the family name of the Gonzagas) contains the mausoleum of the last Gonzaga, a round Romanesque tower (charming view).

S. Giovanni (B) and Villa Trotti. Villa Besana.

Near Lezzeno (B) is one of the deepest parts of the lake.

WESTERN BANK.

the garden. The mortuary chapel of the Sommariva family, at the end of the garden-wall, contains numerous works in marble.

Halfway up the Sasso S. Martino, a rock behind Cadenabbia, is the little church of Madonna di S. Martino, with a beautiful view (1½ hr.; rough path).

The Monte Crocione (5365'), a higher hill to the W. (3½-4 hrs., with guide, 5 fr.; fatiguing; to avoid heat, start at 2 a.m.), commands a striking view of the Lake of Como and Bellagio. A more extensive view of the Valaisian Alps, etc., is enjoyed from the *Monte Galbiga (5600'), adjoining the Crocione to the W. and reached from it by the arête in 50 min. Descent by the Alp Penna to (3 hrs.) Osteno (p. 456).

Tremezzo (P; *Alb. Bazzoni) is almost a continuation of Cadenabbia. Between them is the Villa Carlotta. This district, the Tremezzina, is justly called the garden of Lombardy.

Interesting excursion (3-4 hrs. there and back) by Lenno (Ristor. Brentani) to *S. Maria del Soccorso (1374'), a 'Mt. Calvary' with a superb view (rfmts. at the sacristan's); back by Mezzegra.

At the end of the long peninsula of Lavedo is the handsome Villa Arcomati, formerly Balbianello, with its colonnade (splendid view). In the bay lie Azzano (B) and Lenno (B). To the S. of the promontory is Campo (B), charmingly situated, and beyond it Sala (B); between these lies the islet of Comacina, with the little church of S. Giovanni. Then Colonno (B).

Argegno (P; Alb. & Ristor. Telo; Alb. Barchetta, at the mouth of the fertile Val Intelvi.

A road leads hence viâ Castiglione and S. Fedele d'Intelvi (2522'; Alb. S. Rocco) to (12½ M.) Lanzo (see p. 456).

VII. *Route 112.* CERNOBBIO. *Lake of Como.*

EASTERN BANK.

Nesso (B), at the mouth of the *Val di Nesso*, which ascends to the *Piano del Tivano* (3800'), with a high waterfall in a narrow gorge, often dry in summer.

Careno and *Quarsano*; then *Pognana* (B) and *Riva di Palanzo* (P).

The *Villa Pliniana*, in the bay of *Molina*, at the entrance of a narrow gorge, erected in 1570 by Count Anguissola, is now the property of Marchesa Trotti. It derives its name from a spring near it which daily changes its level, a peculiarity observed by both the Plinys. The quotations are inscribed on the walls of the court.

To the S. of *Torno* (P; *Bella Venezia*) both banks are dotted with villas.

Villa Taverna, formerly *Tanzi*, with beautiful gardens; *Villa Ferranti* or *Pasta*, formerly the property of the celebrated singer (d. 1865); *Villa Taglioni*, once the property of the famous danseuse.

Blevio, with numerous villas (*Mylius, Ricordi*); then, beyond the promontory (with the *Villa Cornaggia*), *Borgo S. Agostino*, the N.E. suburb of Como.

A new road (carr. with one horse 8 fr., with two horses 15 fr.) and a cable-railway (*funicolare*) lead hence to (4½ M.) Brunate (2405'; *Albergo-Ristorante Spaini*, with electric light; *Bellavista*, R. 1½, B. 1¼, déj. 2½, D. 4 fr.), enjoying a beautiful view towards the W., as far as Monte Rosa.

WESTERN BANK.

Brienno (B), embosomed in laurels.

Torrigia (P; *Ristorante Casarico*); on the promontory the *Villa Elisa*. To the S., on the lake, rises a lofty *Pyramid*, erected by Joseph Frank, a professor at Pavia (d. 1851).

Germanello, Laglio, Carate (P; Alb. Lario), *Urio* (B), all with fine villas.

Moltrasio (P; Alb. Caramazza), in a beautiful situation, with the large *Palazzo Passalacqua*, rising above terraced gardens.

Villa Volpi, formerly *Pizzo*, on a promontory extending far into the lake. High above it is the church of *Rovenna*.

Cernobbio (P; **Grand Hôtel Villa d'Este & Reine d'Angleterre*, with fine park, R., L., & A. 3, pens. 10-13 fr.; *Hôtel Reine Olga & Cernobbio*, R., L., & A. 3-5, B. 1½, déj. 3, D. 4, pens. 8-12 fr.; **Alb. Milano*, Italian), with the villas *Belinzaghi, Baroggi*, etc. Steam-tramway to Como.

The Monte Bisbino (4385'), with a pilgrimage-church (fine view), may easily be ascended from Cernobbio or Brienno (see above) in 3 hrs.

Farther on are the *Villa Cima*, with fine grounds, the *Villa Gonzalez*, and the *Villa Tavernola*, below the mouth of the *Breggia*.

*Villa dell' Olmo (shown to visitors), formerly *Raimondi*, the largest on the lake, now the property of Duke Visconti-Modrone, with gorgeous rooms and fine park, is at *Borgo Vico*, the N.W. suburb of Como.

Como, see p. 463. Omnibus to the station 30 c.

COMO. *VII. Route 112.* 463

Lake of Lecco.

The S.E. arm of the Lake of Como, 12 M. long, is less charming than the S.W. arm, but offers grander mountain-scenery. The E. bank is skirted by the railway mentioned at p. 457. Steamers thrice daily from Bellagio (Como) to Lecco, and thrice daily from Colico to Lecco and vice versâ (comp. p. 457).

The steamer rounds the *Punta di Bellagio* (p. 459); on the height above are the grounds of the Villa Serbelloni, and farther on the Villa Giulia. Then (l.) *Lierna* (B), at the foot of the steep *Cima Pelaggia*, with a fine retrospect to the N.; (r.) *Limonta* (B), **Vassena** (B), *Onno* (B), opposite the rocky *Mte. Grigna* (p. 459). Farther on (l.) *Tonzanico* and *Mandello* (P; Corona), at the foot of the *Mte. Campione*; then *Abbadia* (B), on a promontory stretching far into the lake, at the mouth of the *Val Gerona*. On the W. bank we see a row of cement-kilns, at the foot of the *Corni di Canzo* (4510'). Opposite Lecco lies (r.) *Parè*, separated from *Malgrate* by the promontory of *S. Dionigio*. Malgrate lies at the entrance of the *Val Madrera*, through which a road leads to Como viâ Erba (p. 454). The lake contracts to the river *Adda*, flowing out of it, which is crossed by the *Ponte Grande*, a stone bridge of ten arches, built in 1335, and by the new *Railway Bridge* of the Lecco and Como line (p. 457).

Lecco (P; *Alb. Mazzoleni*, at the steamboat-pier; *Croce di Malta; Corona*, all in the Italian style), a busy town with silk and cotton factories and iron-works (6100 inhab.) at the foot of *Monte Resegone* (6160'), is admirably described in Manzoni's 'I Promessi Sposi'. In the Piazza are statues of Manzoni (1785-1873) and Garibaldi, by Confalonieri. Pleasant walks to the hill of *S. Gerolamo* and the pilgrimage-church on *Mte. Baro* (3150'; *View).

From Lecco *to Milan* (31½ M.), railway viâ *Monza* in 2-2½ hrs., and to *Bergamo* (20½ M.) in 1¼ hr., see *Baedeker's Northern Italy*.

Como. — *Hôtel Volta*, with café-restaurant, R., L., & A. 4-6, B. 1½, D. 5 fr.; *Hôt. Cavour*, with lift and electric light; Italia; Hôt.-Pens. Suisse, with café-restaurant; Hôt.-Pens. Bellevue, with the *Café-Restaurant Marinoni*; all in the Piazza Cavour, near the quay. — Restaurant: *Trattoria Frasconi*, in a recess of the Piazza Cavour, at the end of a street at right angles to the harbour. — *Caffè Plinio*, next door to the Hôt. Volta. — *Baths* in the lake, near the *Giardino Pubblico* (to the left, beyond the pier). — *Station* of the St. Gotthard Railway *(Stazione Como S. Giovanni* or *Mediterranea*, for Milan and Lugano), to the right from the steamboat-pier and across the piazza past the hotels (½ M.; omnibus 30 c.; gratuitous for holders of through-tickets). A second station *(Stazione Como Lago* or *Ferrovia Nord)*, for the lines to Milan viâ Saronno and for Varese and Laveno (p. 446), is 4 min. to the left from the pier. — Books, photographs, etc., at the *Libreria Dalp* (Schmid, Francke, & Co.), in the Hôtel Volta.

Como (705'; pop. 11,000), the birthplace of the younger Pliny and of Volta, the physicist (whose *Statue* by P. Marchesi is on the W. side of the town near the harbour), lies at the S. end of the S.W. arm of the Lake of Como, amidst an amphitheatre of mountains. The *Cathedral*, begun in the Lombard Gothic style in 1396, and altered in the Renaissance style by Tommaso and Jacopo Rodari after 1486, is built entirely of marble and is one of the finest in N. Italy.

464 *VII. Route 112.*   CANZO.

Adjacent is the **Town Hall (Municipio)**, completed in 1215, the walls of which are curiously built of stones of different colours. The *Porta del Torre*, a massive five-storied structure, is also noteworthy. In the Piazza Vittoria a bronze *Statue *of* **Garibaldi**, by Vela, was erected in 1889. Large silk-manufactories. — Outside the town, on the promenade, is the *Chiesa del Crocefisso*, of the 17th cent., and ½ M. beyond it, on the slope to the left, is the handsome *Basilica S. Abbondio*, of the 8th and 11th centuries.

EXCURSIONS. ON THE E. BANK a beautiful new road leads along the hillside, high above the lake, affording a variety of charming views, to (5 M.) *Torno* (p. 462). — *Brunate*, see p. 462; the station of the cable-tramway (*Funicolare*; opened in 1894) is near the 'Stazione Ferrovia Nord', on the lake (fare 2, down 1½, there and back **3 fr.**). The line is about ⅔ M. long, and the steepest gradient is 54:100.

FROM COMO TO ERBA AND BELLAGIO (about 28 M.), a very pleasant drive or walk (one-horse carr. in 5-6 hrs.; fare 25 fr.; and gratuity of 3 fr.). We follow the Lecco road from the Porta Milanese, and gradually ascend the hills to the E. The lake is hidden by the wooded *Monte S. Maurizio*; to the S. we overlook the country in the direction of Milan; on the S. E. lies the Brianza, an undulating and very fertile tract, 12 M. long, 6 M. wide, lying between the Lambro and the Adda, and stretching N.E. to Lecco. This is the 'Garden of Lombardy', and a favourite resort of the Milanese, who have numerous villas here. The church of the village of *Camnago*, to the N. of the road, contains Volta's tomb (p. 453). Farther on, to the S. of the road, near a small lake, is the sharp ridge of *Montorfano*. Near *Cassano* we observe a curious leaning tower. Beyond *Albesio* the view embraces the valley of Erba (*Pian d'Erba*) and the lakes of *Alserio*, *Pusiano*, and *Annone*, above which on the E. rise the *Corni di Canzo* (4540) and the serrated *Resegone di Lecco* (6160').

Near (9 M.) **Erba** (1017; *Inn*), a small town in a very fertile district, are several villas, the finest being *Villa Amalia*, on the N.W. side, with a splendid view of the Brianza. Near *Incino*, with its lofty Lombard tower, ¼ hr. to the S.E. of Erba, stood the *Licinaforum* of the Romans, mentioned by Pliny. — Railway from Erba to *Milan*, by *S. Pietro*, in 1½ hr.

Beyond Erba we cross the *Lambro*, which is here artificially conducted into the *Lago di Pusiano*, a little to the S.E. Just beyond it the Bellagio road diverges to the left from the Lecco road, and passes through *Longone*, on the W. bank of the narrow *Lago del Segrino*. The next place is (6 M.) **Canzo** (*Croce di Malta*), which extends almost to (1 M.) *Asso* (joint population 3200). At the entrance to Asso is a large silk-factory (*Casa Versa*).

The road now gradually ascends the pretty VALL' ASSINA, the valley of the *Lambro*, with wooded slopes. It passes through several villages, (2 M.) *Lasnigo*, (2 M.) *Barni*, and *Magréglio*, where it becomes steeper. First view of both arms of the Lake of Como from the top of the hill near the (1 M.) *Chapel*. A charming *Survey of the whole W. arm to Lecco, and beyond it, is obtained beyond the first church of (1 M.) **Civenna** (*Angelo*, unpretending, R. 1 fr.), with its graceful tower. The road now runs for 2 M. on the top of the wooded hill which extends into the lake as far as Bellagio. Beyond the chapel we obtain striking views of the Bay of Como, the Tremezzina with the Villa Carlotta and Cadenabbia (p. 460), the E. arm (Lake of Lecco), a large portion of the road on the E. shore resting on masonry and embankments, the entire lake from the promontory of Bellagio to Domaso (p. 454), the promontory itself, and far below us the hill with the Serbelloni park (p. 460).

The road winds downwards for nearly 8 M., passing the *Villa Giulia* (p. 461) and the cemetery of Bellagio. From Civenna to the hotels at *Bellagio* on the lake (p. 460) about 6 M.

An interesting but rather fatiguing circuit (path very stony at places) may be made by ascending the **Monte S. Primo** (p. 461; 5055'; 4-5 hrs., with guide) from Canzo, and descending to (2½ hrs.) Nesso or (3 hrs.) Bellagio.

113. From Como to Milan.

30 M. RAILWAY viâ *Monza* (comp. p. 463) in 1¼-1¾ hr.; 5 fr. 50, 3 fr. 85, 2 fr. 75 c. (another line by *Camnago* and *S. Pietro*, 27 M., in 1½ hr.). TRAMWAY from Como to Milan (station at Como near the steamboat-pier) by *Lomazzo*, *Saronno*, *Bollate*, *Novate*, and *Borisa* (29 M., in 2-2½ hrs.).

Como, see p. 463. — On a hill near (3 M.) *Albate-Camerlata* rises the lofty old **tower** of the *Castello Baradello*, where Frederick Barbarossa occasionally resided. 5½ M. *Cucciago*; 8½ M. *Cantù-Asnago*; 9½ M. *Carimate*; 12 M. *Camnago*. The hilly district to the right and left is the fertile *Brianza*, with its numerous villas (p. 454); in the background the long, indented *Mte. Resegone* (p. 463). 15½ M. *Seregno*; 17½ M. *Desio*. Several tunnels.

21 M. **Monza** (pop. 18,500; *Albergo del Castello* at the station; *Falcone*), an old town. The *Cathedral*, founded in 595 by Queen Theodolinde, and rebuilt in the 14th cent., contains the '*Iron Crown*' of the Lombard kings and a rich treasury (5 fr.). The royal *Summer Palace* near Monza has a fine large park. — 25 M. *Sesto S. Giovanni*.

30 M. **Milan**. — The STATION (Pl. F, G, 1; *Restaurant*) is a handsome building adorned with frescoes and sculptures. Cab into the town (by day or night) 1¼ fr.; each trunk 25 c.; hotel-omnibus 1-1½ fr. — Tramway from the station to the town 10 c. — Porter for luggage under 100 lbs. 50 c.

Hotels. *HÔTEL DE LA VILLE (Pl. a; F, 5), Corso Vittorio Emanuele; *HÔTEL CAVOUR (Pl. b; F, 3), Piazza Cavour; *GRAND HÔTEL MILAN (Pl. c; F, 3, 4), Via Al. Manzoni 29, R., L., & A. from 4½ fr.; *HÔT. CONTINENTAL (Pl. e; E, 4), Via Al. Manzoni; all these of the first class: R., L., & A. from 5, D. 5, B. 1½, omnibus 1½ fr. — Less expensive: *GRANDE BRETAGNE & REICHMANN (Pl. d; D, E, 6), Via Torino 45; *METROPOLE, in the Piazza del Duomo; *REBECCHINO (Pl. p; E, 6), Via S. Margherita; *EUROPA (Pl. f; F, 5), Corso Vitt. Emanuele 9; *MANIN (Pl. k; E, 2), Via Manin, near the Giardini Pubblici; *ROMA (Pl. g; F, 5), Corso Vitt. Emanuele 7; *POZZO (Pl. l; F, 6), Via Torino, D. 4½ fr.; *FRANCIA (Pl. m; F, 5), Corso Vitt. Eman. 19. — BISCIONE & BELLEVUE, in the Piazza Fontana, next the Piazza del Duomo, R., L., & A. 3, D. incl. wine 4 fr.; *CENTRAL ST. MARC (Pl. h; E, 6), Via del Pesce; *BELLA VENEZIA (Pl. i; E, F, 5), Piazza S. Fedele; ANCORA (Pl. n; F, 5), Via Agnello; *LION & TROIS SUISSES (Pl. o; G, 4, 5), Corso Vitt. Emanuele and Via Durini; HÔTEL-PENSION SUISSE, Via Visconti, commercial; *TERMINUS HOTEL, unpretending, conveniently situated near the central railway-station.

Restaurants (*Trattorie*). *BIFFI, *Gambrinus*, see below; *COVA, with garden, Via S. Giuseppe; *Rebecchino*, see above; *Guffanti*, Via S. Giuseppe. — **Cafés**. *BIFFI, Gambrinus, in the Galleria Vitt. Emanuele; *Cova*, see above; *Accademia*, Piazza della Scala; *Delle Colonne*, Corso Venezia 1; several in the Giardini Pubblici (p. 468). *Caffè latte*, coffee with milk; *caffè nero*, black coffee. — Beer ('*birra*') at the cafés (30 c. per 'tazza'). Also at the *Birreria Nazionale*, opposite the cathedral; *Birreria Svizzera*, next door to the Hôtel Métropole.

Baths. *Bagno di Diana* (Pl. H, 2), outside the Porta Venezia; *Bagno Nazionale* (Pl. D, 8), outside the Porta Ticinese; *Bagno dell' Annunziata*, Via Annunziata 11, etc.

Cabs ('*Broughams*') 1 fr. per drive, at night 1¼ fr.; per ½ hr. 1, per hr. 1½ fr.; from the station to the town, 1¼ fr. (comp. above).

Tramway every 5 min. (10 c.) from the Piazza del Duomo to most of the city gates and to the Cimitero. — Steam Tramway to *Monza* (see above) in 1 hr. (60-80 c.); also to *Saronno-Como* (p. 463), *Gussano*, *Vaprio*, etc.

BAEDEKER, Switzerland. 16th Edition. 30

Post Office (Pl. E, 6), near the cathedral, Via Rastrelli 20, behind the Palazzo Reale, open from 8 a.m. to 9 p.m. — **Telegraph Office** (Pl. E, 5), Piazza dei Mercanti 19, on the N.W. side of the Piazza del Duomo.

Theatres. *Teatro della Scala* (Pl. E, 4), the largest in Italy next to S. Carlo at Naples, open during the Carnival only. *Teatro Manzoni* (Pl. E, 5), Piazza della Scala, comedies. *Teatro dal Verme* (Pl. D, 4), operas and ballet. *Teatro Filodrammatico* (Pl. E, 4), operas.

English Church Service, Via Andegari 8.

Milan (more fully described in *Baedeker's N. Italy*), which was rebuilt after its total destruction in 1162 by the Emp. Frederick Barbarossa, is the capital of Lombardy, and one of the wealthiest manufacturing cities in Italy, silk being the staple commodity. The city is upwards of 7 M. in circumference, and has a population of 315,000, exclusive of the garrison, or 418,000 including the suburbs.

The business-centre and also the most attractive part of Milan is the *PIAZZA DEL DUOMO (Pl. E, F, 5), recently much extended, and flanked with palatial edifices, designed by *Gius. Mengoni*, which, with the majestic 'Duomo', present a striking appearance. This is also the focus of the tramway and omnibus system.

The **Cathedral (Pl. E, F, 5), in the Gothic style, one of the largest churches in Europe, built entirely of white marble, and decorated with 98 turrets and 2000 marble statues, was begun in 1386 by the splendour-loving Giangaleazzo Visconti, and completed by Napoleon I.

The **INTERIOR, with its double aisles, borne by 52 pillars, and its beautiful stained windows, is very impressive. In the S. transept a °*Monument to Giacomo and Gabriele de' Medici*, by Leoni, erected in 1564 by Pope Pius IV. to the memory of his brothers. °*Stained-Glass Windows* in the choir. An ancient sarcophagus of St. Dionysius, in porphyry, now serves as a *Font*. The subterranean *Cappella S. Carlo* contains the tomb of S. Carlo Borromeo (in summer 5-10, in winter 7-10 a.m.; at other times, fee of 1 fr.).

The ascent (in the corner of the right transept) of the °ROOF and TOWER (354'; ticket 25 c.; 157 steps to the roof) is recommended, as the visitor is thus enabled to inspect the architecture of the exterior more closely, and obtains a noble prospect of the Alps and Apennines (Panorama by Bossoli at Pirola's, Piazza della Scala 6, 1 fr.).

The *Galleria Vittorio Emanuele (Pl. E, 5), a fine arcade with tempting shops, built by *Gius. Mengoni* in 1865-72, and adorned with statues of 24 celebrated Italians, connects the Piazza del Duomo with the Scala. — In the *Piazza della Scala* (Pl. E, 4) rises a marble statue of *Leonardo da Vinci* (d. 1519) by Magni. The great master is surrounded by his pupils Cesare da Sesto, Marco da Oggiono, Salaino, and Boltraffio.

Of the other eighty churches of Milan, the following are noteworthy. — *S. Ambrogio (Pl. C, 6), founded by St. Ambrose in the 4th cent., and re-erected in the 12th cent., contains an 'Ecce Homo' by *Luini*, and several ancient monuments. — *S. Maria delle Grazie (Pl. B, 5), of the 15th cent., attributed to *Bramante*, contains pictures by *Ferrari*, *Caravaggio*, and *Luini*. In the refectory of the monastery is *Leonardo da Vinci's* far-famed **LAST SUPPER, painted on the wall in oils, and now almost obliterated (daily 9-4, Sun. 12-3:

1 fr., on Sun. gratis). — **S. Maria presso S. Celso** (Pl. E, 8), by *Bramante*, also contains good pictures. — **S. Maurizio** (Pl. C, 5) has fine frescoes by *Luini*. — *****S. Lorenzo** (Pl. D, 7) once formed part of a Roman palace; the isolated *Colonnade* is borne by sixteen Corinthian columns. — **S. Carlo Borromeo** (Pl. F, 4), completed in 1847, contains two groups in marble by *Marchesi*.

The *****Brera** (Pl. E, 3), or *Palazzo di Scienze, Lettere ed Arti*, formerly the Jesuits' College, contains the *Public Library* (300,000 vols., 1000 MSS.), a *Collection of Coins* (50,000), the *Observatory*, *Casts* from the antique, an *Archaeological Museum* and a most interesting *Picture Gallery* (*Pinacoteca*; open daily 9-4, Sun. and holidays 12-4; adm. 1 fr., on Sun. and holidays gratis). In the court are marble statues.

<small>PICTURE GALLERY. Antechambers I. and II.: Frescoes by *Luini*, *Ferrari*, *Bramantino*, and *Marco da Oggionno*, the finest being *Luini's* Angels (Nos. 14, 26, 45, 49, 54, 68), works of a 'genre' character (2, 11, 13), and scenes from the life of Mary (5, 19, 42, 43, 51, 63, 69, 73); *47. Madonna with SS. Anthony and Barbara; 25. *Gaud. Ferrari*, Adoration of the Magi. — Oil-paintings. 1st Room: 87. *Bernardino de' Conti*, Madonna; 88. *Solaino*, Madonna. — 2nd R.: 159. *Gentile da Fabriano*, Mary in glory; 167. *Bartol. Montagna*, Madonna; 168. *Gent. Bellini*, Preaching of St. Mark; 179. *Ercole di Roberti*, Madonna; 191. *Cima da Conegliano*, Saints; 193. *Crivelli*, Madonna. — 3rd R.: 206. *Moretto*, Madonna with SS. Jerome, Anthony, and Francis; 209. *Bonifacio I.*, Finding of Moses; *P. Veroness*, 219. SS. Gregory and Jerome, 220. Adoration of the Magi, 221. SS. Ambrose and Augustine, 227. SS. Antonius Abbas, Cornelius, and Cyprian. — 4th R.: 248. *Titian*, St. Jerome. — 5th R.: 288 bis. *Titian*, Portrait; 261. *Giov. Bellini*, Madonna; 264. *And. Mantegna*, Large altar-piece; 265. *Bern. Luini*, Madonna; *267. *Leonardo da Vinci* (?), Head of Christ; **270. *Raphael's* famous 'Sposalizio', or Marriage of the Virgin, painted in 1504; *Mantegna*, 273. Pietà, *282. Madonna; 106. *A. Solario*, Madonna. — 6th R.: 283. *Crivelli*, Madonna and Saints; *Giov. Bellini*, *284. Pietà, *297. Madonna; *300. *Cima da Conegliano*, SS. Peter, Paul, and John the Baptist. — 7th R.: 253, 254, 255. *Lorenzo Lotto*, Portraits. — 8th R.: 331. *Guercino*, Expulsion of Hagar; 333. *Dossi*, St. Sebastian; 334. *Fr. Francia*, Annunciation. — 9th R.: Dutch and Flemish works; 449. *Rembrandt*, The artist's sister. — 10th R.: 390. *Velazquez* (?), Dead monk; 442. *Van Dyck*, Madonna and Child with St. Anthony of Padua. — 11th R.: 456. *Domenichino*, Madonna and saints. — To the left is a suite of rooms with modern pictures, sketches by academicians, casts, etc. — On the groundfloor is the Museo Archeologico (daily 12-3, adm. 1/2 fr.; Sun. 2-4, gratis), a collection of antique, mediæval, and Renaissance sculptures and old frescoes, most of them found at Milan. Among the best sculptures are those by *Agostino Busti*, surnamed *Il Bambaja*.</small>

The famous *****Bibliotheca Ambrosiana** (Pl. D, E, 5), open daily, except Wed. & Sun., 10-3 (fee 1/2-1 fr.; pictures on Wed. 10-12 1/2, gratis), founded in 1609 by Card. Fed. Borromeo, contains 160,000 vols. and 8000 MSS. Among the pictures is *Raphael's* cartoon for his School of Athens.

The *****Museo Poldi-Pezzòli** (Pl. F, 4), Via Moroni 10, contains an admirable collection of weapons, pictures, sculptures in marble, bronze, and terracotta, furniture, tapestry, trinkets, etc., exhibited in the house of the founder Cavaliere Poldi-Pezzoli (d. 1879). Adm. daily 9-4, Sun. and holidays 11-3; 1 fr.; catalogue 1 fr.

The *****Ospedale Maggiore** (Pl. F, 6), a remarkably fine brick

468 VII. Route 113.   MILAN.   Castello.

edifice, begun by *Ant. Filarete* of Florence in 1457, contains nine
different courts. The external terracotta incrustation is observed
on other Milanese buildings, but the façade of the Ospedale with its
rich and beautiful windows is probably unsurpassed.

The **Castello** (Pl. D, 3, 4), adjoining the *Piazza d'Armi*, once
the seat of the Visconti and Sforza, is now a barrack. Behind it
lies the *Arena*, a kind of circus for 30,000 pers., founded by
Napoleon I. (fee ½ fr.). On the N.W. side of the Piazza d'Armi
rises the *Arco del Sempione (Pl. B, 2), a triumphal arch of marble,
founded by Napoleon in 1804 by way of termination to the Simplon
road, and completed in 1838 (107 steps to the top).

The **Giardini Pubblici** (Pl. F, G, 2, 3), between the Porta Venezia and Porta Nuova, are the chief promenade of the Milanese. In
the older part is the so-called *Salone*, containing the *Museo Artistico*
(daily 11-4; 50 c.; Sun. and Thurs. gratis). Adjoining the W.
side is the *Museo Civico*, containing natural history collections (same
hours of admission). At the W. entrance to the new Giardino Pubblico is the Piazza Cavour, embellished with a *Statue of Cavour* in
bronze, by Tabacchi, erected in 1865.

The new *Cemetery (*Cimitero*; 50 acres in area), outside the
Porta Tenaglia (Pl. C, D, 1), contains many fine monuments (several
with marble statues of the surviving mourners) and a 'Tempio di
Cremazione' for burning the dead. *View of the Alps.

INDEX.

Aaberli Alp 43.
Aadorf 49.
Aarau 22.
Aarberg 220.
Aarburg 17.
Aare, the 12. 14. 20. 146.
151 etc.
Aare Glaciers, the 189.
—, Gorge of the 182.
Aathal 44.
Abbadia 463.
Abbaye, L' 221.
S. Abbondio on the Lake of Como 459.
— near Lugano 436.
Abendberg, the 162.
Abgschütz 132.
Abläntschen 203.
Abondance, Vallée d' 259.
Abschwung, Im 189.
Achtelsassgrat, the 136.
Acla Silva 399. 400.
Acletta 124. 373.
—, Piz d' 124.
Acqua, Hospice all' 315.
—, Piz dell' 414.
Acquarossa 377.
Acquaseria 459.
Adda, the 384. 420. 425. 457. 463. etc.
Adelboden 196.
Adige, the 428.
Adler Glacier 338. 343.
— Pass 343.
St. Adrian 104.
Adula Mts. 386.
Adulahorn 386.
Aela Hut 388. 389.
— Pass 382.
—, Piz d' 388.
Ælpeltispitz 358.
Ælplisee 365.
Ærnen 317. 318.
Æsch 10. 141.
—, Im 72.
Æschach 53.
Æschi 156.
Affoltern 21. 79.
Agassizhorn 190.
Agassiz-Joch 190.
St. Agata 444.
Ägeri 80. 107.
Ägerisee 80.
Agittes, Aux 247.

Agno 438. 445.
Agogna, Val d' 458.
Aguagliouls 405.
Agums 428.
Ai, Tour d' 247.
—, Lac d' 252.
Aigle 246.
Aiguille Grise 281.
— Verte 277. 283.
—, Plan de l' 281.
Aiguilles Dorées 300.
— Marbrées 282.
— Rouges (Chamonix) 275.
— — (Argentière) 288.
— — (Val d'Hérens) 323.
— —, Glacier des 323.
Airolo 114.
Aix-les-Bains 265.
Alagna 346.
Albana, Piz 393.
Albate 465.
Albbruck 24.
Albert-Hauenstein 28.
Albertville 268.
Albesio 464.
Albeuve 253.
Albigna, Cascata dell' 423.
— Glacier 396. 423.
—, Val 423.
Albinen 195.
Albis, the 40.
— Hochwacht 40.
Albisbrunn 79.
Albogasio 456.
Albris, Piz 407.
Albrist 197.
Albrunhorn 318.
Albrun Pass 318.
Albula, the 363. 378. 379. 388.
Albula Pass 389.
Albulahorn 389. 413.
Aletschbord 317.
Aletsch Glacier, the Great 175. 178. 190. 316. 317.
—, the Upper 309. 310.
Aletschhorn 310. 316.
Algaby 312.
Allalin Glacier 342. 343.
Allalinhorn 341.
Allalin Pass 343.
Allaman 236. 245.
Allamans, Les 260.

Allée, Alp de l' 328.
—, Col de l' 325. 329.
—, Pigne de l' 325. 328.
Allée Blanche 291.
— —, Glacier de l' 289. 291.
Allenbach-Thal 196.
Allèves 300.
Alliaz, Bains de l' 211.
Allières 253.
Allinges, Les 257.
Allmannshöhe 31.
Allmendhubel 168.
Allweg 101.
Almagell 314. 342.
Alphubel 341.
Alphubel-Joch 342.
Alpien 312.
Alpienbach, the 313.
Alpligen 165.
— Glacier 118.
— Lücke 118.
Alpnach 131.
—, Lake of 101. 131.
Alpnach-Stad 102. 131.
Alpschelenbubel 198.
Alpthal 106. 107.
Alserio, Lago d' 464.
Altdorf 110.
Alteingrat 362.
Altels 193.
Alten-Alp 57. 58.
Altendorf 43.
Altenoren Alp 124.
Altmann 59.
Altmatt 107.
Altorf 110.
Alt-St. Johann 63.
Altstad, islet 87.
Altstaffel 315. 319.
Altstätten (Rhine Valley) 61.
Altstetten (near Zürich) 21. 78.
Alv, Piz 410.
Alvaneu 364.
Alvaneu, Bad 388.
Alvaschein 379.
Alvascheiner Alp 362.
Alvier 47. 61.
Alzasca, Lago d' 442.
Alzo 452. 453.
Ambri 116.
Amden 45.
Amdener Berg 45.

INDEX.

Amianthe 305.
Amisbühel 158.
Ammerten Glacier 201.
Ammertengrat 201.
Ammertenhorn 200.
Ammerten Pass 201.
Ammon 45.
Amphion, Bath 257.
Amselflub 366.
Amsigen-Alp 102.
Amsoldingen 153. 202.
Amsteg 112.
Andeer 381.
Andelfingen 33.
Andermatt 119.
Andermatten 320.
Andey, Pointe d' 272.
Andolla, Pizzo d' 313.
Anengrat 317.
Angera 451.
Anières 256.
St. Anna, Castle of 52.
—, Glacier of 120.
Annecy 269.
—, Lac d' 269.
Annemasse 271.
Annes, Col des 270.
Anniviers, Val d' 327
Annone, Lago d' 464.
Anterne, Col d' 275.
Antey-St-André 349.
St. Anthony, Chapel 55.
Anthy 257.
Antigine, Passo d' 314.
—, Pizzo d' 314.
Antigorio, Val 313. 320. 442.
St. Anton, on the Arlberg 430.
St. Antönien 358.
St. Antönier-Joch 358. 481.
S. Antonio, near Poschiavo 420.
—, Bocchetta di 442.
Antrona Piana 314.
— Pass 314.
Anzasca, Val d' 344.
Anzeindaz 255.
Aosta 294.
Appenzell 56.
—, Canton 53.
Aprica, Passo d' 421.
Aquila 377.
Arabione, Cima dell' 438.
Aravis, Col des 270.
Arbedo 387. 117.
Arbenhorn 329.
Arblatsch, Piz 392.
Arbola, Bocchetta d' 318.
—, Punta d' 317.
Arbole, Chalets d' 256.

Arbon 81.
Arbostora, Mte. 455.
Ardenno 421.
Ardetz 415.
Ardon 256. 307.
Arenaberg 26. 32.
Areue Valley 385.
Areuse, Gorges de l' 213.
Argegno 456. 461.
Argentière 283.
—, Aiguille d' 283.
—, Col d' 283.
—, Glacier d' 283.
Argentine 248. 255.
Arina, Piz 418.
Arlas, Mt. 398.
Arlberg 430.
Arlen Glacier 188.
Arlesheim 10.
Arly, the 263. 269.
Armeno 450.
Armillon 201.
Arnaz 297.
Arnen Lake 251.
Arnex 220.
Arni Alp 129.
Arola 453.
Arolla 323.
— Glacier 324. 323.
—, Pigne d' 305. 306. 322. 323.
Arona 451.
Arosa 365.
Arpette, Vallée & Fenêtre d' 286. 296.
Arpiglia, Piz d' 413. 414.
Arpille 250.
Arpitetta, Alp 328.
—, Pointe d' 328.
Arren, Pointe d' 273.
Arth 104. 96.
Arth-Goldau 96. 109.
Arve, the 224. 270. 271. 277. 282 etc.
Arveye 247. 251.
Arveyron, Source of the 278.
Arvier 294.
Arvigo 387.
Arvigrat 182.
Arzinol, Pic d' 323.
Ascona 441. 444.
Asnago 465.
Assa, Val d' 419.
Assina, Valle 464.
Astras dadaint 418.
— dadora 418.
—, Piz 418.
Attinghausen 111.
Au, Convent near Einsiedeln 107.
—, in the Rhine Valley 60.

Au, on the Lake of Zürich 42.
Auberig, the Grosse 43.
St. Aubin 214.
Aubonne 236. 245.
Au Devant 254.
Audon, Becca d' 251.
Audoz, Crête d' 233.
Auengüter 70.
Augst 19.
Augstbord Pass 331.
Augstholz 141.
Augstmatthorn 163.
Aul, Piz 370.
Ault, Piz 372.
Aurona, Punta d' 311.
— Glacier 311.
Ausserbinn 317.
Ausser-Ferrera 381.
Auvernier 211.
Avançon, the 247. 255.
Avants, Les 244. 254.
Aven 256.
Avenches 219.
Aventina Glacier 348.
Averser-Thal 381.
Avigna, Val 418.
Avril, Mont 305. 306.
Avrona 417. 418.
Axalp 184.
Axalphorn 174.
Axenberg or Axenfluh 92. 110.
Axenfels 91.
Axenstein 91.
Axenstrasse 92.
Ayas, Val d' 348.
Ayent 201.
Ayer 328. 329.
Ayerne, Roc d' 259.
Aymaville 294.
Ayutz, Piz 415.
Azzano 461.

Baar 81.
Baceno 321.
Bachalp 179. 180.
Bächistock 74.
Bachtel 44.
Bächtelen 151.
Bacone, Piz 395.
Baden 20.
Badile, Piz 422.
Badus 120. 374.
Bageschwand-Höhe 140.
Bagnes, Val de 301.
Baldegg 141.
Baldegger See 141.
Bâle 2.
Balen 340.
Balenfirn 341.
Balerna 439. 440.
Balfrinhorn 341. 342.

INDEX. 471

Ballaigues 221.
Ballalüna 388.
Ballenbühl 140.
Ballswyl 215.
Ballwyl 141.
Balme 272. 306.
—, Aig. de 287.
—, Col de 287.
—, Chalet à la 290.
—, La 287. 293.
Balmenhorn 333.
Balmeregghorn 132.
Balmhorn 193.
Balmuccia 454.
Balmwand 72.
Balsthal 13.
Baltschieder-Joch 198.
Balzers 61.
Bannio 344. 454.
Baradello, Castello 465.
Baranca, Col di 344. 454.
Barberine, the 275. 260. 284.
—, Col de 260. 286.
Bard 297.
Bardonnèche 268.
Bäregg 177.
Bärengrube 130.
Bärenhorn 370.
Baretta Balma 359.
Bargis 368.
Barma, La 304. 322.
Barnaz, La 349.
Barni 464.
Baro, Monte 463.
Barrhorn 380.
Barrjoch 331.
Bärschis 47.
Bärschwyl 10.
Barthélemy, Val St. 296.
Baseglia, Mt. 401. 413.
Basel, see Bâle.
Basel-Augst 19. 3.
Basodino, the 320. 443.
Basset, Le 241.
Batiaz, La, Castle 249.
Bâtie, Bois de la 231.
Bätten-Alp 184.
Batzenheid 62.
Bäuchlen 139.
Bauen 92. 90.
Baugy 241.
Bauma 49.
Baumgarten-Alp 70.
Baveno 448.
Bavona, Val 320. 412.
Bäzberg 119.
St. Beatenberg 158.
Beatenbucht 157.
St. Beatusbad 157.
Beatushöhle 157.
Beaufort 268.
Beaulmes, Aig. de 215.

Bechburg 13.
Beckenried 88.
Bedretto 315.
—, Val 315.
Bee 447.
Beichbirn 309. 310.
Beichgrat 309.
Beinwyl 141.
Belachat, Plan 279.
Belalp 309.
Belalphorn 309.
Belfort, ruin 364. 388.
Belfoux 217.
Belgirate 451.
Bellagio 460.
—, Punta di 459. 463.
Bellano 458.
Bellarma 456.
Bella Tola 330. 328.
Bellavista 439. 447.
—, Fuorela 409.
Bellegarde on the Rhone 264.
— in the Jaunthal 208.
Bellenhöchst 162.
Bellerive 10.
Belleville 268.
Bellevue 234. 245.
—, Pavillon de 288.
Bellinzona 433. 117.
Belmeten 111.
Belmistock 111.
Bel Oiseau 285.
Belotte 256. 233.
Belp 151. 153.
Belpberg 151.
Belvedere (Macugnaga) 345.
— (Little St. Bernard) 293.
Bendlikon 42.
Bene-Grona 456.
Béranger, Col de 289.
—, Pierre à 278.
Bérard, Vallée de 275. 284.
—, Cascade à 284.
—, Pierre à 275.
Bercher 289.
Bercla, Val 392.
Bergell, Valley of 422.
Bergli (Engelberg) 129.
— (Sigriswyl) 156.
Bergli Hut 178. 173.
Bergli-Joch 187.
Berglistock (Grindelwald) 177. 187.
— (Oberalp) 374.
Bergue, La 274.
Bergün 389.
Berguner Stein 388.
Beringen 24.
Berisal 311.

Berlingen 26. 32.
Bern 144.
St. Bernard, Great 301.
—, Little 293. 290.
S. Bernardino 386.
— Pass 386.
S. Bernardo 437. 438.
Bernegg 51. 60.
Bernese Oberland 143.
Bernetsmatt 123.
Bernex 231.
Bernhalden, Alp 63.
Bernina 401. 408.
— Hospice 411.
— Houses 407.
— Pass 411.
—, Piz 408.
Beroldingen 90.
Berra 217.
Bertol, Col de 324. 340.
—, Dents de 324.
—, Glacier de 323. 324.
Besso 328.
Bétemps Hut 338.
Betlis 45.
Betschwanden 69.
Bettaforca 348.
Bettelmatt 319.
Bettelried 199.
Bettlihorn 311. 317.
Bettliner Pass 348.
Bettmer-Alp 317.
Bettmerhorn 317.
Bettmersee 317.
Bettolina 348.
Beuggen 28.
Bevaix 214.
Bever, Val 401. 412.
Beverin, Piz 380.
Bevers 412.
Bevieux 247.
Bévilard 11.
Bex 247.
Bianco, Corno 347.
—, Pizzo (Macugnaga) 345.
— (Bernina) 409.
Biasca 117.
Biaschina Ravine 116.
Biberbrücke 105.
Biberlikopf 45.
Biberstein, Castle 22.
Bider Glacier 341.
Bief d'Etoz 210.
Biel in the Canton of Bern 12.
— in the Valais 316.
Bielenstock 125.
Bieler Höhe 431.
Bienenberg 13.
Bienne 12.
—, Lake of 12. 206.
Bieno 448.

Bies Glacier 333.
— Joch 331.
Bietschhorn 198.
Bietschjoch 198.
Bietschthal 308.
Biferten Glacier 69.
— Stock 71. 371.
Biglenalp 172.
Bignasco 442.
Bigorio 437.
Bilten 43.
Binn 317.
Binnenthal 317.
Binningen 9.
Bionaz 321.
Bionnassay 288.
—, Aig. de 282. 289.
Birmensdorf 78.
Bironico 434.
Birrenberg 217.
Birrwyl 142.
Birs, the 4. 10. 13.
Birseck, Château 10.
Birsig-Thal 9.
Bisbino, Mte. 462.
Bischofzell 49.
Bise, Cornettes de 258.
Bisi-Thal 73.
Bissone 438. 455.
Bistenen Pass 332.
Bististaffel 332.
Bitto, Val del 422.
Bitzistock 136.
Bivio 392.
Blackenstock 130.
St. Blaise 206.
Blaitière, Aig. de 278. 280.
—, Cascade de 277.
Blanchard 258.
Blankenburg 199. 204.
Blas, Piz 115. 378.
Blauberg 126.
Blaue Gletscher 181.
— Schnee 58.
— See 192.
Blanen 9.
Blenio, Val 377.
Blevio 462.
Blindenhorn 316.
Blinden-Thal 316.
Blitzingen 316.
Blonay, near Vevey 241.
—, near Evian 258.
Blonnière, La 270.
Bludenz 430.
Blume 156.
Blumenstein, Baths 153.
Blümlisalp 192.
Blümlisalpfirn (Urirothstock) 130. 93.
Blümlisalp Glacier (Kienthal) 193. 191.

Blümlisalphorn 192.
Blümlisalp-Rothhorn 192.
Blümlisalpstock 192.
Boccareccio, Passo del 318.
Boccioleto 454.
Bochard, Aig. du 277. 278.
Bocken 61.
Bockli 111.
Bocktschingel 124.
Boden Glacier 335.
Bodengo 387.
—, Val 384. 387.
Bodensee 28.
Bodio 117.
Bodmen 312. 363.
Bodmer, Castle 356.
Bodmer Glacier 313.
Bödmer Alp 72.
Bœuf, Pas du 330.
Boganggen, Alp 170.
Boglia, Mte. 457.
Bognanco, Val 314.
Bois, Les 278. 283.
—, Glacier des 279. 283.
—, Tête de 300.
Bolladore 425.
Bollingen 44.
Boltigen 203.
Bolzano 458.
Bommen-Alp 57.
Bonaduz 367. 378.
Bonaveau, Chalets of 259.
Bondasca Glacier 424.
—, Val 423.
Bonder-Krinden 197.
Bonderlen-Thal 197.
Bonderspitz 197.
Bondo 423.
—, Forcella di 424.
Bonhomme, Col du 290. 268.
—, Croix du 290.
Bönigen 184. 163.
Boniswyl 142.
Bon-Nant, the 273. 289.
Bonne 274.
Bonneville 271.
Bonport 242.
Bons-St-Didier 233. 265.
Bonstetten 78.
Borca 344.
Bordon, Garde de 328.
Borgofranco 298.
Borgomanero 453.
Borgonuovo 423.
Bormio 425.
—, Baths of 426.
Bornand, Petit and Grand 270.
Bornengo, Passo Pian 374.

Borromean Islands 448.
Bors, Alp 347.
—, Glacier 347.
Bortelhorn 311.
Borterthal 330.
Bosa, Passo di 442.
Bösalgäu, Alp 163.
Bosco 320. 442.
—, Val di 442.
Bosses, Val des 303.
— du Dromadaire 278. 281.
Bossey 264. 233.
Bosson, Becs de 323. 326.
Bossons, Glacier des 280.
—, Les 280.
Boswyl 22.
Bottarello, Pizzo 341.
Bottiglia, Col della 454.
Bottmingen 9.
Bötzberg, the 20.
Bötzenegg 20.
Boudry 213.
Bougy 236.
—, Signal de 236.
Boujean 12.
Bouquetin 328. 325.
Bonquetins, Col des 325. 326.
—, Dent des 324. 326.
Bourdeau 266.
Bourg-St-Maurice 294.
— St-Pierre 300.
Bourget, Le 266.
—, Lac du 265. 266.
Boussine 305.
—, Tour de 305.
Bouveret 258.
Boval Hut 405.
Boveresse 212.
Bovernier 299.
Boveyre, Glacier de 300.
Bözingen 12.
Brail 413.
Bramegg 138.
Bramois 321.
Brand 100. 430.
Brändlisberg 153.
Brandner-Thal 430.
Branson 250.
Brasses, Pointe des 274.
Brassus, Le 221. 235.
Bratz 430.
Braunwaldberg 69.
Brè 437.
—, Monte 437.
Bregaglia, Val 422.
Bregalga, Val 381.
Bregenz 432.
Breil 350.
Breithorn, near Zermatt 337.
—, the Lauterbrunnen 198.

Breithorn, the Lötschen-thaler 310.
Breithorn Pass 311.
Breitlauenen 164.
Brembana, Val 422.
Bremgarten 22.
Brenet, Lac 221.
Brenets, Les 211.
—, Lac des 211.
Breney, Col de 306.
—, Glac. de 305.
Brennet 23.
Brenno, the 117. 376. 377.
Brenva, Glac. de 291.
Bréonna, Col de 325. 329.
—, Couronne de 323. 329.
Brestenberg 142.
Bretayes 247. 252.
Breuil 350.
Brévent 279.
—, Col du 275.
Breya, la 299.
—, Col de la 299.
Brianza, the 464. 465.
Bricolla, Alp 325.
—, Glacier de 325.
—, Pointe de 325.
— —, Col de la 325.
Brides-les-Bains 268.
Brieg 309.
Brienno 462.
Brienz in the Canton of Bern 182.
— (Grisons) 364.
—, Lake of 183.
Brienzer Rothhorn 182. 133.
Brienzwyler 182.
Brig 309.
Brigels 375. 71.
Brigelser Horn 371.
Brione 440. 441.
Brisi 45.
Brissago 441. 444.
Bristen 122.
Bristenstock 112.
Britterhöhe 46.
Brizon 272.
Broc 203.
Brodbüsi 202.
Broglio 443.
Brolla, Ponte 441. 442.
Brot, Saut de 212.
Brouillard, Glacier du 282.
Brozet, Col du 250.
—, Glacier du 200.
Brugg 20. 23
Brügg 12.
Bruggen 50. 51.
Brûlé, Mont 300. 304. 324.
Brülisau 53. 59.
Brunate 462.

Bründlen-Alp 86. 103.
Brünig 134.
— Pass 133.
Brunneggborn 330.
Brunnegg-Joch 331.
Brunnen 90. 110.
Brunnenstock 137.
Brunni 103.
Brunni Glacier 124.
— Pass 124.
Brunnistock 93.
Brunni-Thal 72. 123. 124.
Brusimpiano 455.
Brusin-Arsizio 455.
Brusio 420.
Brusson 348.
Bubendorf, Bad 13.
Bubikon 44.
Buccione, Torre di 452.
Buchberg in the Rhine Valley 60.
— on the Lint Canal 43. 44.
Buchenthal 50.
Buchs 61. 432.
Buchs-Dællikon 21.
Budden, Capanna 296.
Budri, Roc de 331.
Buet 275. 284.
Buffalora Pass 414.
Bühlalp 128. 93.
Bühlbad 192.
Bühler 56. 59.
Buin, Piz 359. 415. 431.
Bülach 48. 21.
Bulle 252.
Büls, Alp 47.
Bundalp 170.
Bunderbach 192.
Bündnerbergfirn 368.
Bünzen 22.
Buochs 88.
Buochser Horn 89. 128.
Buonas 104.
Büren 17. 89. 128.
Burg (near Bâle) 9.
— (on the Rhine) 26.
— (Grindelwald) 180.
Burgdorf 17.
Bürgeln 133.
Bürgenstock 101.
Burgfeldstand 158.
Burgfluh (Lenk) 200.
— (Wimmis) 154. 202.
Burghalden 105.
Burgistein 153.
Burglauenen 171.
Bürglen (Thurgau) 49.
— (Uri) 111.
Bürglen-Sattel 203.
Burier 246.
Bursinel 245.
Burvein 391.

Buscagna Pass 311. 318.
Büsingen 25.
Bussalp 179.
Busseralies, Casc. de 350.
Busswyl 12.
Buthier, the 295. 296. 303.
Bütscheggen 196.
Bütschelegg 151.
Bütlassen 170. 191.

Cabbiolo 387.
Cacciabella Pass 396. 424.
—, Piz di 424.
Cadabbi, Passo del 336.
Cademario 438.
Cadenabbia 460.
Cadenazzo 434. 440.
Cadlimo, Val 115. 373. 374. 376.
—, Bocca di 115.
Cadval 392.
Caille, La 270.
Cairasca, Val 313. 318.
Calanca, Val 387.
Calanda 62. 356.
Caldè 445.
Calfreisen 364.
Calmot 374.
Cama 387.
—, Bocchetta di Val 387.
Camadra, Cima 376.
—, Val 370.
Camana Alp 367.
Cambrena, Piz 409. 411.
— Pass 409.
Cambriales, Piz 123.
Camerlata 446. 465.
Camnago 464. 465.
Camoghè, Cima di (Val Piora) 115.
—, Mte. (near Lugano) 438. 434.
Camona Alp 370.
Camoscio, Corno del 347.
Campascio, Piz 411.
Camperio, Hospice 376.
Campfèr 398.
—, Lake of 398.
Campi, Ruin 879.
Campione 455.
—, Mte. 463.
Campo (Val di Campo) 376. 383. 442.
— (Lake of Como) 461.
— (Val Viola) 412.
—, Colle di 412.
—, Corno di 412.
—, Lago di Val 423.
—, Val di 376. 412. 442.
Campo Cologno 420.
— Dolcino 384.
— Moro, Val 409.
— Tencia 443.

Campolungo Pass 443.
Campovasto 412.
Campsut 381.
Canardhorn 358.
Canaria Valley 115. 374.
Canciano Glacier 140.
— Pass 410.
Canfèr, Pian 392.
Canicül 381.
Cannero 445.
Cannobbino, Val 445.
Cannobbio 441.
Cantone Glacier 396.
Cantù 465.
Canzo 461.
—, Corni di 463. 464.
Capella 362. 413.
Capolago (Lake of Sils) 396.
— (Lake of Lugano) 438. 439.
Caprino, Monte 437.
Carale, Piz 409. 411.
Carate 462.
Carcoforo 454.
Careno 462.
Carimate 465.
S. Carlo (Val Bavona) 442. 443.
— (Val Poschiavina) 420.
— (Val Viola) 412. 425.
Carmenna Pass 306.
Carouge 232.
Carrel, Capanna 350.
Casaccia (Val Bregaglia) 422.
—, Hospice on the Lukmanier 376.
Casana, Val 413.
—, Pass and Piz 413.
Casanella, Piz 413.
Casanna 358.
Casneda 445.
Casnile Pass 396, 423.
Cassano 464.
Cassarate 435. 436. 437.
Castagnola 436. 437. 456.
Castasegna 424.
Castel, Château 26.
Castelfranco Glacier 310. 315.
Castello 437. 456.
—, Cima di 395. 423.
Castelmur, Ruin 423.
Castiel 364.
Castione 117. 387.
Castor 348.
S. Caterina del Sasso 446.
St. Catharinenthal 25.
Catogne, Mont 299.
Catscharauls 124. 372.
Cauma Lake 368.

Caux, Mont 243.
Cavaglia 411.
Cavalcorto 422.
Cavandone 448.
Cavanna Pass 125.
Cavardiras, Piz 124.
Cavel, Piz 370. 371.
Cavel-Joch (Somvixer Thal) 370. 371.
Cavell-Joch (Prätigau) 357.
Cavigliano 442.
Cavloccio Lake 395.
Cavorgia 373.
Cavreccia, Val 392.
Celerina 401.
—, Muottas da 404.
Céligny 235. 245.
Cenere, Monte 434.
Cengalo, Piz 422.
Cenis, Mont 268.
Centovalli, Valley 441.
Centrale, Pizzo 121.
Ceppomorelli 344.
St. Cergue 285.
Cerlier 206.
Cerniat 203.
Cernobbio 462.
Cervin, Mont 337. 338.
Cery 239.
Cevio 412.
Chablais, the 257.
Chable 301.
Chablettes, Les 280.
Chailly 241.
Chalame, Val 297.
Chalchagn, Piz 405. 408.
Challant, Val 297. 348.
Challaz, La 258.
Challes 267.
Cham 80. 104.
Chambave 296.
Chambéry 267.
Chambésy 245.
Chambotte, La 266.
Chambrelien 209.
Champfèr 398.
Chamois, Col des 300.
Chamonix 276.
Chamossaire 246. 252.
Champagna, Val 402.
Champ du Moulin 212, 214.
Champatsch, Piz 417. 418.
Champéry 259.
Champex 286. 299.
—, Lac de 299.
Champlan 202. 348.
Champorcher, Val 297.
Champsec 304.
Chamuera, Val 412. 413.
Chancy 264.
Chandolin (Val d'Anniviers) 327.

Chandolin (near Sion) 251.
Chanélaz 213.
Chanrion 305.
Chapeau 278.
Chapieux, Les, or Chapiu 290.
Chapütschin, Piz 397. 408.
—, Fuorcla 397. 409.
Chardonnet, Aig. du 283.
—, Col du 283.
—, Pav. du 283.
Charlanoz, Chal. de 280.
Charmettes, Les 267.
Charmey (Galmis) in the Jaunthal 203.
— (Galmitz) near Aarberg 220.
Charmilles, Les 231.
Charmontel 219.
Charmoz, Aiguille de 278.
Charnadüra 400.
Charpex 242. 244.
Charvensod 295. 296.
Chasseral, the 12. 206.
Chasseron, the 214. 212.
Chat, Col du 266.
—, Dent du 266.
—, Mont du 266.
Château des Dames 350.
Château d'Oex 254.
Châtel (Jaunthal) 203.
— (Drance Valley) 259.
Châtel St. Denis 258.
Châtelaine 231.
Châtelard, Le (Arve Valley) 273.
— (Eau Noire) 285.
—, Château 241. 242.
Châtelet 250.
Châtillens 219.
Châtillon (Aosta Valley) 297.
— (Arve Valley) 272.
— (Lac du Bourget) 265.
Chauderon, Gorge du 244.
Chaulin 244.
Chaumont 209.
Chaussy, Pic de 252. 255.
Chaux-de-Fonds, La 210.
Chavans, Les 286.
Chavonnes, Lac des 247.
Chavornay 215.
Chécouri, Col de 292.
Cheggino 450.
Chemin, Mont 299.
Chenalette 302.
Chêne 271. 255.
Chermignon 195.
Chermontane, Grande 305.
—, Col de 324.
Chésalette 203.
Chesières 246.

Chessel 258.
Chétif, Mont 292.
Cheville, Pas de 256.
Chèvres, Pas de 324.
Chexbres 218.
—, Signal de 218.
Chiaclavuot 362.
Chiamut 374.
Chiareggio 396.
Chiasso 440.
Chiavenna 384.
Chiesa 395.
Chiésaz, La 241.
Chignin-les-Marches 267. 268.
Chillon, Castle 244.
Chindrieux 265.
Chippis 327.
Chöglias, Val 418.
Chosalets, Les 283.
Chougny 238.
Chünetta 405.
Chur 354.
Churer Joch 390.
Churwalden 390.
Cierfs 414. 418.
Cima 456.
Cimalmotto 442.
Cimes Blanches, Col des 349.
Cingino, Pizzo del 314.
Cinuskel 413.
Ciprianspitz 357.
Civenna 461. 464.
Civiasco 453.
Clarens 241.
Clariden 71.
Claridenfirn 124.
Clariden Pass 124. 71.
Claridenstock 123. 124.
Claro 117.
—, Pizzo di 117.
Clavadel 362. 363.
Clavalité, Val 296.
Cleuson, Col de 305.
Clusanfe, Col de 259. 285.
—, Alp 260.
Cluse, La, near Pontarlier 213.
—, Montagne de la 206.
Cluses 272.
Coblenz 23. 48.
Cocco, Forcarella 440.
Codelago, Lake of 318.
Coire 354.
Colico 385. 458.
Colla 438. 443.
Collon, Col de 324.
—, Mont 323. 324.
—, Petit Mt. 324.
Collonge 256.
Collonges 264.
Colma, Col di 453.

Cologny 233. 256.
S. Colombano, Corno di 412. 420.
Colombey 259.
Colombier 218.
—, the 265.
Colonno 461.
Coltura 423.
Columbe, Passo 376. 115.
Commabbio, Lago di 446.
Comacina, Isola 461.
Gombal Lake 289. 291.
Comballaz 255.
Combin, Grand 300. 304. 305.
— de Corbassière 30
Combloux 269.
Comboé 296.
Comella, Passo di 442.
Commugny 234.
Como 463.
—, Lake of 457.
Comologno 442.
Concise 244.
Concordia Hut 171. 173. 178.
Confin, Fuorcla del 359.
Confinale, Passo 410.
Constance 29.
—, Lake of 28. 432.
Constantia Hut 328.
Contamines, Les 289.
Conters 358. 365. 391.
Conthey 256. 307.
Convers, Les 210.
Conversion, La 218.
Coppet 234. 245.
Corandoni 115.
Corbassière, Glac. de 304.
—, Combin de 304.
Corbeyrier 247.
Corcelles, near Grandson 214.
—, near Neuchâtel 209.
—, near Payerne 217. 219.
Corconio 453.
Corenno 458.
Corgémont 11. 211.
Corjeon, Dent de 253.
Corna Rossa Pass 425.
Cornaux 206.
Cornera, Passo 318.
—, Val 373. 374.
Cornier, Grand 325. 329.
—, Col du Grand 325. 329.
Corno, Val 319.
Coroi, Piz 370.
Corsier 256.
Cortaillod 214.
Cortenova 459.
Curtlys 347. 348.
Corvatsch, Piz 408. 397. 398.

Corvatsch Glacier 395. 408.
Corvo, Piz 376.
Cosio-Traona 422.
Cossogno 418.
Cossonay 215.
Costainas Pass 418.
Côte, La 235. 245.
Cotschen, Piz 415.
Courmayeur 292.
Cournère, Col du Val 350.
Couronne, Col de 325. 329.
Courrendlin 10.
Court 11.
Couvet 212.
Coux, Col de 260. 257.
Crammont, the 293.
Crampiolo 318.
Crans, Château de 235.
— sur Sierre 308.
Crap Alv 372.
Crasta (near Celerina) 401.
— (Fex Valley) 397.
Crastagüzza, the 408.
— Saddle 408. 409.
Crasta Mora 389. 412.
Cravate, the 339.
Craveggia 442.
Craverola, Passo di 442.
Cray, Mont 254
Creccio, Colmo di 437.
Cremenaga 455.
Cremia 458.
Crémine 11.
Cresogno 456.
Cressier 206.
Cresta in the Averser-Thal 381.
— in the Engadine 401.
—, Lake 368.
Crestalla 398. 399.
Crésus 203.
Crêt, Col du 304. 322.
Crête Sèche, Col de 306.
— —, Glacier de 306.
Crêtes, Chât. des 241.
Creux de Champ 251.
— du Van 212.
Creva 455.
Crevola 313. 321. 442.
Crin 442.
Criner Furca 320. 442.
Crispalt 373.
Crispausa 373. 374.
Cristallina (Grisons) 375.
—, Piz (Tessin) 443.
—, Forcellina 376.
—, Forcla di 443.
—, Passo 375.
—, Val (near Airolo) 443.
— — (Grisons) 375.
S. Croce 424.
Crocione, Monte 461.
Crodo 321. 442.

Croix, La 251. 286.
—, Col de la 247. 251.
Ste. Croix 214.
Crot 334.
Croy 220.
Crozlina, Alp and Glacier 443.
Crugnolo 446.
Crusch 418.
Cruschetta Pass 418. 428.
Crusinallo 452.
Cubli, Mont 244.
Cucciago 465.
Cudrefin 220.
Culet 259.
Cully 218. 239. 246.
Culoz 265.
Cumbels 370.
Curaglia 375.
Curciusa Glacier 385.
—, Piz 386.
Curfirsten 45.
Curtinaccio 421.
Curtins 397.
—, Alp 391.
Curtnätscherhof 367.
Curvèr, Piz 380. 391.
Cusio, Lago 452.
Cuvio, Val 446.
Cuzzago 451.

Dächli, Unteres and Oberes 97.
Dachsen 32. 27.
Daenikon 21. 22.
Dala, the 195. 303.
— Glacier 198.
Dalaas 430.
Dallenwyl 128.
Dalley, Casc. du 285.
Dalpe 443.
Dalvazza 358.
Dammafirn 118. 114.
Damma Pass 118. 137.
Dammastock 137. 118.
Dangio 377.
Danöfen 430.
Dard, Casc. du 251. 280.
Därligen 155.
Darreï, Le 299.
Dartgas, Piz 70.
Daube, the (Gemmi) 194.
— (Schynige Platte) 164
Daubenhorn 194.
Daubensee 194.
Davos Platz 361.
— Dorf 361. 360.
Davoser See 360.
Day, Le 221.
Dazio Grande 116.
Delebio 422. 458.
Delémont 10.
Délices, Les 231.

Delle 10.
Delsberg 10.
Dent Blanche 325. 339.
— —, Col de la 325.
— —, Glacier de la 325.
Dent, Grande & Petite 324.
Dentro, Val di 412.
Dents Blanches 260.
Derborence, Lac de 256.
Dervio 458.
Désert, Grand 305.
Desio 465.
Dévens 247.
Devero-Alp 318.
Diableret, the 251. 255.
Diablons 328. 330.
—, Col des 330.
Diavel, Passo del 414.
Diavolezza, La 407.
Diechterhorn 137.
Dielsdorf 48.
Diemoz 296.
Diemtig-Thal 202.
Diesbach 69. 151.
Diesrut, Pass 370.
Diessenhofen 25. 32.
Dietikon 21.
Digg 368.
Dintikon 20. 22.
Diosaz, Gorges de la 273.
Dischma-Thal 360. 362.
Disentis 372.
Disgrazia, Monte della 421. 395.
Distel Glacier 310.
Divonne 234. 235.
Dix, Val des 304. 322.
Dixenze, the 322.
Doire, the 291. 293.
Doldenhorn 193.
Dôle 235.
Dolent, Col du Mont 283. 292.
—, Mont 293. 283.
Dolf, Piz 368.
Dollfus, Pavillon 189. 178.
Dollone 292.
Dom 338. 341.
Domaso 458.
Dôme, Cabane du (Mont Blanc) 281.
—, Glacier du 281.
Dom-Joch 342.
Domleschg, the 378.
Domodossola 314.
Dongio 377.
Dongo 458.
Donnas 298.
Dora Baltea 291. 295.
Dorio 458.
Dornach 10.
Dornbirn 432.
Doron, the 268. 269.

Dosdè, Capanna di 412.
—, Pizzo di 412.
Dossen 95. 88. 100.
Dossenhorn 185. 186.
Dossenhütte 177. 186.
Dottikon 20. 22.
Döttingen 23.
Doubs, Côtes du 210.
—, Saut du 211.
Douglas Hut 430.
Doussard 269.
Donvaine 233.
Drance, the, in the Chablais 257.
—, in the Valais 249. 293. 298.
Dreckloch-Alp 69. 74.
Dreibündenstein 356.
Dreiländerspitze 431.
Dreiländerstein 105.
Drei Schwestern 61.
Dreisprachenspitze 427.
Drochetta, Colle 454.
Dronaz, Pic de 302.
Dru, Aiguille du 277. 279.
Drusenfluh 431.
Drusenthor 357. 431.
Duana, Passo della 382. 424.
—, Pizzo della 382. 423.
Dubino 385.
Ducan, Hoch- 362.
— Pass 362.
Düdingen 215.
Dufour-Spitze 338. 345. 348.
Duin, Tour de 247.
Duingt, Château 269.
Dündenhorn 193.
Dünden Pass 170.
Dungel Glacier 200. 250.
Durand, Glacier de (Val des Dix) 306. 324.
— — (Val de Zinal) 325. 328.
—, Col 329. 340.
—, Mont 329.
Duranna Pass 365.
Durgin, Piz 71.
Durnant, Gorges du 299.
Dürrboden 360. 362.
Dürrenäsch 142.
Dürrenberg 170. 168.
Düssistock 123.
Dza, Col de 350.

Eau-Morte 289.
— Noire, the 260. 284.
Ebenalp 57.
Ebihorn 329.
Ebikon 81.
Ebnat 63.
Ebnefluh 171.

Ebneflub-Joch 171.
Ecandies, Col des 299.
—, Pointe des 299.
Echallens 239.
Echelle, Pas de l' 232.
Echevenoz, Les 303.
Eclépens 215.
Ecluse, Fort de l' 264.
Ecoulaies, Glacier des 304.
Ecublens 219.
Effingen 20.
Effretikon 44. 48.
Egerkingen 14.
Eggfluh 341.
Egginerhorn 341.
Egginer Pass 342.
Eggishorn 316.
Eginen Valley 315. 319.
Eglisau 48.
Egua, Col d' 454.
Ehrenfels 378. 379.
Ehrlose 141.
Eierhals 80.
Eigenthal 86. 138.
Eiger 178.
Eiger Glacier 174.
Eigerjoch 178.
Einfisch-Thal 327.
Einshorn 382.
Einsiedeln 106.
Eisboden 176.
Eismeer 177.
Eita 412.
Elgg 49.
Elm 75.
Elsighorn 197.
Elslücke 317.
Emaney, Col d' 260. 285.
Emd 382.
Emdthal 154.
Emet, Alp 362.
—, Lago di 382.
Emilius, Mt. 296.
Emmen 140.
Emmenbrücke 19.
Emmenmatt 140.
Emmen-Thal 140.
Emmetten 88.
Emosson 275. 285.
Ems 367.
Encel, Pas d' 259. 260.
Enclaves, Col d' 291.
Engadine, the 393.
Enge, near Bern 150.
—, near the Giessbach 184.
Engelberg 128.
Engelberger Rothstock 129. 93.
Engelhorn 185.
Enggistein 140.
Engi 75. 47.
Engstlen-Alp 135.

Engstlen-See 136.
Engstlig Alp and Falls (Adelboden) 197. 201.
Engstligengrat 197. 195.
Ennenda 63.
Ennethühl 63.
Ennetlint 68. 69.
Enney 253.
Entfelden 18.
Entlebuch 138. 139.
Entlen-Thal 139.
Entova, Val 397.
Entremont Valley 299.
Entreroches, Canal d' 215.
Enziswiler 53.
Epagny 253.
Ependes 215.
Eptingen 14.
Erba 464.
Erde 256.
Eringer-Thal 322.
Erlach 206.
Erlen 49.
Erlenbach in the Simmen-Thal 202.
—, Lake of Zürich 41.
Erlimoos 14.
Erlisbach 13. 22.
Ermatingen 26. 31.
Ermensee 141.
Erschenz 26. 32.
Escher Canal 46. 67.
Eschia, Val d' 413.
—, Alp 413.
—, Fuorcla d' 362.
Eschlikon 49.
Escholzmatt 139.
Esel 102. 103.
Esen, Piz d' 413.
Esino 459.
Essets, Col des 243.
Estavayer 217.
Etablons, Col des 307.
Etivaz 293.
Etoile, Mont de l' 323.
Etremblères 271. 282.
Etroubles 303.
Etsch, see Adige.
Ettingen 9.
Etzel, the 103. 106.
Etzli-Alp 373.
Etzli-Thal 112. 122.
Etzweilen 32. 25.
Eugensberg 26. 32.

Euseigne 322.
Euthal 107.
Evançon 297. 348.
Evêque 324.
—, Col de l' 306. 324. 340.
Evian-les-Bains 257.
Evilard 12.
Evionnaz 249.
Evires 270.
Evolena 322.
Ewig - Schneehorn 189. 186. 187.
Excenevrex 257.
Fadera 357.
Faderhorn 344.
Fafler-Alp 171. 310. 317.
Fahlensee 58. 61.
Fahrwangen 141.
Faido 116.
Fain, Val del 407. 410.
Faldum Pass 198.
Faldum-Rothhorn 198.
Falkenburg 54.
Falkenfluh 151. 163.
Falkenstein 13.
Falknis 62.
Fallbodenhubel 174.
Faller, Val 392.
Fallère, Mt. 296.
Fallerhorn 346.
Fallerjoch 392.
Fang 327.
Faong 219.
Fardün, Ruin 380.
Farnbühl, Baths 138.
Färnigen 138.
Fätschbach, the 69. 71.
Faucille, Col de la 285.
Faulen, the 69. 73. 74.
—, the Böse 69.
—, the Hohe 111. 112.
Faulensee (Lake of Thun) 157.
— (Erstfeld Valley) 112.
—, Bad 157.
Faulhorn 179.
Faverges 269.
Fayet, Le 269. 273.
Fedoz, Vadret da 396.
—, Val 397.
Fee 311.
— Glacier 341. 342.
— Pass 342.
Feldbach 26. 32. 41. 317.
Feldkirch 431.
Feldmeilen 41.
Felik Glacier 348.
Felikjoch 339.
Fellaria Glacier 409.
— Chalets 409. 421.
Fellers 369.
Fellücke 412.

Felli-Thal 112.
Felsberg 367.
Felsenbach 357.
Felsenegg 79.
Felsenhorn 197.
Fenêtre, Col de (Gr. St. Bernard) 303.
— (Val de Bagnes) 306.
—, Glacier de 306.
Fer-à-Cheval, Vallée du 275.
Ferden 198.
— Pass 198. 195.
— Rothhorn 198.
Feriolo 448.
Ferney 231.
Ferpècle 325.
—, Glacier de 325.
Ferrera Valley 381. 382.
Ferret 293. 303.
—, Col 293. 303.
—, Valley of 291. 292. 299.
Ferro, Sasso del 446.
Fervento 454.
Festi Glacier 332. 338.
Fetan 415.
Feuerstein 133. 139.
Feuerthalen 24. 32.
Feuillerette-Alp 195.
Feusisberg 105.
Fex Glacier 397.
—, **Valley** of 397. 409.
— Roseg, Fuorcla 409. 397.
— Scerscen, Fuorcla 397. 410.
Fianell Pass 392.
—, Piz 392.
Fibbia 121.
Fidaz 368.
St. Fiden 51.
Fideris 357.
—, Baths of 357.
Fideriser Alp 365.
Fieno, Passo 410.
Fier, Défilé du 267.
—, Gorges du 267.
Fiery 339. 348.
Fiesch 316.
Fiescher-Alp 316.
Fiescher Glacier (Grindelwald) 177. 178.
— — (Valais) 190. 316.
Fiescherhorn, the Kleine 178.
Fiescher-Joch 178.
Fiesso 116.
Figino 455.
Filisur 388.
Fillar-Alp 345.
Fillarkuppe 339. 344.
Fillar Pass 339.

Fillinges, Pont de 274.
Filzbach 46.
Fimber Pass 418.
Findelen 335. 337.
— Glacier 335. 337. 343.
Finero 445.
Finge 308.
Finhaut, or
Fins-Hauts 285.
Finsteraar Glacier 178. 189. Finsteraarhorn 189. 177.
Finsteraar-Joch 178. 190.
Finstermünz Pass 429.
Fionney 304.
Fiorina, Val 320. 443.
Fisistöcke 199.
Fiume Latte 459.
Flamatt 215.
Fläscherberg 61. 62.
Flawyl 50.
Fleckistock 118.
Flégère 279.
Flendruz 204.
Fless Pass 359.
—, Val 359. 360.
Fletschhorn 312. 341.
Fleurier 212.
Flies 429.
Fliesbordkamm 58.
Flims 368. 75.
Flimser Alpen 75.
— See 368.
— Stein 363. 75.
Flirsch 430.
Flis Alp 58.
Floria, Aig. de la 279.
Florissant 232.
Fluchthorn 415.
Flüela Pass 360.
— Valley 360.
Flüelen 92. 110.
Fluh-Alp (Loëche) 195. 198.
— (Zermatt) 337. 343.
Fluhberg 43.
Flühen 9. 118.
Flühli (Entlebuch) 139.
— (Melchthal) 135.
Flühmatt 129.
Fluhseeli 200.
Flühseen 392.
Flumet 268. 270.
Flums 47.
Flurins 417.
Fobello 454.
Follaterres, Les 250.
Folly, La 293. 283.
Fond de la Combe 275.
Fondei 365.
Fondo Toce 448.
Fonds, Vallée des 275.
—, Col des 275.
Fongio 115.

Fontana (Bedretto) 122. 315.
— (Tarasp) 417.
Fontauna, Alp 362.
Foo Pass 76.
Foppa Pass 66.
Foppiano 320.
Fora, Piz 397.
Foraz, Piz 418.
Forbisch, Piz 392.
Forcellina, the 381. 395.
Forchetta, Passo di 311.
Forclaz, La 247.
—, Col de la, near St. Gervais 273.
— —, near Martigny 284. 286.
Forcletta, Pas de la 323. 331.
Forcola, Passo della 387.
Formazza Valley 318. 320.
Forno Glacier 395.
— Hut 395. 396.
—, Mte. 395.
— Pass 395. 422.
—, Scatta del 442.
Foron, the 271.
Fort, Mont 305.
Foscagno Pass 425.
Fouilly, Le 273.
Fours, Col des 290.
—, Pointe des 290.
Ste. Foy 294.
Fracle, Val 414. 425. 426.
Frakmünd, Alp 103.
Franzenshöhe 427.
Frastanz 431.
Frau, the 192.
—, the Weisse 192.
—, the Wilde 191. 170.
Frauenbalmhütte 191. 193.
Frauenfeld 49.
Frauenkirch 363.
Frauenthor 369.
Freibergen (Rigi) 95.
Freiburg 215.
Fremd-Vereina 359.
Frenières 247.
Frenkendorf 13.
Fresnay, Glac. du 282.
Fréty, Mont 292.
Freudenberg 51. 61.
—, Ruin 64.
Fribourg 216.
Frick 20.
Fridolin Hut 70.
Friedau 14.
Friedliswart 12.
Friedrichshafen 28.
Frieswylhubel 151.
Frinvillier 12.
Frisal, Piz 371.

INDEX. 479

Frisal, Val 371.
Frohburg 14.
Fröllchsegg 59. 51.
Fronalp 68.
Fronalpstock (near Glarus) 47. 68.
— (near Brunnen) 91. 89.
Fronscha, La 371.
Frontenex 268.
Frudiera, Becca di 348.
Frümsel 45.
Fründenhorn 193. 192.
Fründenjoch 193.
Frut, Auf der 318. 319. 443.
Frutberg 75.
Frutigen 191.
Frutt (Melchsee) 132.
Frutwald 320.
Fuentes, Ruin 385. 422.
Fuldera 414. 418.
Fuorcla Prievlusa 409.
Fuorcletta 418.
Fuorn 414.
Furcla, Piz 373.
Furcletta 366.
Furgg Alp 341.
— Glacier 314. 336. 339.
— Joch 339.
— Valley 314.
Furggenbaumhorn 311.
Furggenbaum Pass 311.
Furka, the 125. 315.
— (Criner) 320. 442.
— (Maienfelder) 336.
— (Rieder) 317.
Furkahorn (Furka) 126.
— (Arosa) 366.
Furna 357.
Furom 418.
Fürren Alp 129.
Furth 370.
Furtwang-Sattel 137.
Furva, Val 425. 426.
Fusio 442. 443.
Fusshörner 309. 310.
Futschöl Pass 415.

Gabelhorn, Ober- 329. 335. 339.
—, Unter- 337.
Gäbris 55. 56.
Gadmen 137. 397.
Gadmen-Thal 137.
Gadmer Flühe 135.
Gagenhaupt 335.
Gaglianera, Piz 370. 371. 372.
Gais 56.
Galbiga, Monte 461.
Galenstock 126.
St. Gall, Hospice 376.
Gallarate 440. 451.

Galleglone, Piz 424.
St. Gallen 50.
Gallenkirch 431.
Gallina, Piz 315.
Gallivaggio 384.
Galmhorn 195.
Galmis 203.
Galtür 415.
Gamchi Glacier 191.
Gamchilücke 170. 191.
Gämmerrhein, the 374.
Gampel 198. 306.
Gams 61. 63.
Gamser Glacier 312.
— Joch 312.
Gamslücke 190.
Gamstock 120.
Gandegg Hut 336. 387.
Gandria 436. 456.
Ganter Bridge 511.
Gantrist Pass 203. 153.
Gargellen 431.
Garina, Cima 376.
Garstelet Glacier 339. 348.
Garzirola, Mte. 438.
Gaschurn 431.
Gasterndorf 199.
Gasternholz 199.
Gastern-Thal 170. 193. 198.
Gastlose 203.
Gatschiefer 358.
Gätterli 100.
Gauli Glacier 186. 187.
— Pass 186. 187.
Géant, Col du 282. 292.
—, Dent du 292. 282.
—, Glacier du 277. 282.
—, Grotte du 350.
Gebhardsberg 432.
Geissholz 186. 187.
Geisspfad Pass 318.
Geisterspitze 427.
Geló, Mont 805. 806.
Gelfingen 141.
Gellihorn 173.
Gelmerhorn 188. 314.
Gelmersee 188.
Geltenbach 199.
Gelten Glacier 250.
— Pass 250.
Gelterkinden 13.
Gemeinen Wesen, Alp 58. 63.
Gemmenalphorn 158. 163.
Gemmi 194.
Gemsfayrenstock 89. 124.
Gemshörn 342.
Gemsmättli 102.
Gemsstein 317.
Generoso, Monte 438.
Geneva 221.
—, Lake of 233.
Geneveys, Les Hauts 210.

Geneveys, sur-Coffrane 210.
Genf, see Geneva.
Genthal-Alp 135.
Genthod 234. 245.
Gentilino 436.
St. Georges 236.
Gera 444. 458.
Geren Pass 315.
Geren-Thal 315.
Gerihorn 191.
St. Germain 294. 297.
Germanello 462.
Germignaga 445. 455.
Gerona, Val 463.
Gerra 440.
Gers, Lac de 275.
Gersau 89. 87.
Gerschni-Alp 136.
Gerstenhorn 314.
St. Gervais 278.
—, Baths of 278.
Geschenen (Reuss Valley), see Göschenen.
— (Rhone Valley) 315.
Gessenay 204.
Geta, Les 257. 274.
Gex 235. 231.
—, Pays de 234.
Ghiffa 445.
Ghirone 370. 376.
S. Giacomo on the Bernardino 387.
— in the Liro Valley 384.
— near Sedrun 378.
— d'Ayas 348.
— di Fraele 414. 425.
— Pass 329.
Giarsun 415.
Gibloux 217.
Gibswyl 49.
Giebel 72. 134. 151.
Gierm, Val 378.
Giessbach, the 183.
Giessen 317.
Giétroz, Glacier de 304. 305. 306. 324.
—, Casc. du 305.
Giettaz, La 270.
Giglistock 137.
Gignod 303.
Gigot, Mont 266.
Gilly 236. 245.
Gimel 236.
Gimmelwald 169.
Gingins 235.
St. Gingolph 258. 265.
St. Gion, Hospice 376.
Giop, Alp 400. 401.
Giornico 117. 440.
S. Giovanni, Island, in the Lake of Como 461.

S. Giovanni, Island in
 the Lago Maggiore 446.
 447.
Girespitz 58.
Giselafluh 22.
Gisikon 80.
Giswil 133.
Giswiler Stock 133.
Glt, Piz (Maderaner-
 Thal) 123.
— (Val Nalps) 373.
Gitschen 111.
Gittana 459.
Gittaz, La 290.
Gitte, the 288. 290.
Gitzi-Furgge 195. 198.
Giubiasco 434. 433.
Giuf, Val 373.
S. Giulio, Island 452.
Giumella Pass 387.
Giumels 389.
Glacier, Aig. du 290. 291.
Glaciers, Glacier des 289.
 290.
—, Val des 290.
Gland 245.
Glaris 363.
Glärnisch 74.
—, Vorder 68. 74.
Glarus 67.
Glas Pass 367.
Glat thugg 48.
Glattenfirn 112. 130.
Glattensee 73.
Gleckstein Hut 177. 178.
 186.
Gléresse 206.
Gletsch 315.
Gletscher-Alp 341.
Gletscherhorn (Lauter-
 brunnen) 171.
— (Susten) 137.
— (Wildstrubel) 200.
Gletscherstaffel 310.
Gletschhorn 125.
Gliems Glacier 71.
Gliemspforte 71.
Glims, Alp 415.
Glion 242. 243.
Glis 309. 310.
Glishorn 309. 310.
Glockhaus 132.
Glockthurm 429.
Gloggeren 57. 58.
Glovelier 10.
Glüna, Piz 417.
Gluringen 316.
Glurns 428.
Glüschaint, Piz 397. 408.
—, Fuorcla 397. 408.
Gnepfstein 102.
Gnifetti, Cap. 339. 347.
—, Punta 348. 333. 344.

Guippen 109.
Gnof, Alp 123.
Goldau 109.
Goldenberg 33.
Golderen 184. 182.
Goldiwyl 163.
Goldswyl 163. 184.
Golèse, Col de la 257. 260.
 274.
Golzern Alp 123.
Gomagoi 427.
Gondo 313.
—, Ravine of 313.
Gonten 50.
Gontenbad 50.
Gonzen 48. 61.
Gordola 440.
Gordona 384.
Gorgier 214.
Gorner Glacier 333. 335.
 336.
— Gorge 334.
— Grat 334.
Göschenen 114. 118.
— Thal 114. 118.
Göschener-Alp 118.
Gossau 50. 49.
Gotschna 358.
S. Gottardo, Sasso di 120.
Gotteron, Pont de 216.
St. Gotthard, the 120. 121.
— Hospice 121.
— Railway 108.
— Road 112. 118.
— Tunnel 114.
Gottlieben 26. 31.
Gottschalkenberg 106. 80.
Götzis 432.
Gouille, Mt. de la 301.
Goumois 210.
Goûter, Aiguille du 278.
 281.
—, Dôme du 282.
Gozzano 453.
Grabs 63.
Grafenort 128.
Graggi-Hütte 137.
Grammont, the(Chablais)
 253.
Grand Bornand 270.
— Combin 300. 304. 305.
— Cornier 325. 329.
—, Col du 325. 329.
— Désert 305.
— Mœveran 248. 255.
— Plan 299.
— Plateau 281.
— Saconnex 231.
— Salève 232.
— Tournalin 350.
— Villard 253.
Grande Dent 324.
— Fourche 299.

Grande Gorge 233.
Grandes Roches 239.
Grandola 456.
Grands Montets, Col des
 283.
Grands Mulets 281.
Grandson 214.
Grandval 11.
Grandvaux 216.
Granfelden 11.
Granges 219. 258. 307.
— Neuves 304.
Granier, Mont 267.
Gran-Semetta 349.
Grapillon, Pas du 293.
Gräplang, Ruins 47.
Grasonet 283.
Grassen Pass 130.
Gratschlucht Glacier 126.
Grau-Haupt 348.
Grauhörner 68.
Graun 428.
Graustock 135.
Gravasalvas, Fuorcla di
 392.
Gravedona 458.
Gravellona 448. 452.
Greifensee, the 42.
Greina Pass 370.
Grellingen 10.
Grenchen 17.
Grengiols, Bridge of 318.
Grenzach 23.
Grenz Glacier 338. 339.
Greppen 104. 87.
Gressoney-St-Jean 347.
— la-Trinité 339. 347.
Grésy, Casc. de 266.
— sur-Aix 266.
— sur-Isère 266.
Grialetsch Glacier 360.
Griatschouls, Piz 413.
Griaz, La 288.
—, Glacier de 273.
Gries Glacier (Valais)
 319.
— — (Clariden) 72. 124.
— — (Tödi) 71.
— — Pass 316.
Gries Pass 319.
Grieset 69. 74.
Griessen Glacier 129.
Grigna, Monte 459. 463.
Grimbach-Thal 202.
Grimentz 327.
Grimisuat 202.
Grimmenstein, Ruins 60.
Grimmi 202.
Grimsel, Hospice 188.
— Pass 190.
Grindel-Alp 180.
Grindelwald 175.
— Glaciers 176. 186.

Grins 429.
Grisch, Piz 381. 392.
Grisons, Canton 353.
Grivola 294.
Groisy-le-Plot 270.
Grond, Piz 369.
Gronda, Val 392.
Grono 387.
Grosina, Val 412. 425.
Grosio 412. 425.
Gros-Jean, Tête du 255.
Grosotto 425.
Grosshorn 198.
Gross-Lauteraarhorn 178.
— Litzner 359.
— Lohner 197.
— Schreckhorn 177.
— **Seehorn** 359.
— Spannort 128.
— Strubel 197. 200.
Grossthal (Uri) 93.
— (Glarus) 68.
Grotte aux Fées 248.
— du Géant 350.
Grub 55.
Gruben 330.
Grubenberg 203.
Gruben Pass 358.
Grüm, Alp 411.
Grünenberg 163.
Grünhorn Hut 70.
Grünhornlücke 190.
Grünsee 335.
Grüsch 357.
Grüsisberg 153.
Grütsch-Alp 165. 167.
Gruyères 253.
Grynau, Castle **44**.
Gryon 255.
Gschwandenmad - Alp 185.
Gspaltenhorn 191.
Gstad 250.
Gsteig, near Interlaken 163. 165.
—, Sarine Valley 250.
Gsteigthal 250.
Gsteigwyler 163. **164**.
Gstein 312.
Gsür 197.
Guarda 415.
Guardavall, Ruins **413**.
Güda Glacier 370.
Gueula, Col de la 285. 275.
Gueuroz 249. 286.
Güferhorn 386.
Gugel, the 334.
Guggerloch 56.
Guggernüll 382.
Guggi Glacier 178.
— Hut 174. 178.
Güggisgrat 158.
Guiu 215.

Gula, Ponte della 454.
Gumfluh 204. 254.
Gümlingen 151. 140.
Gummegg 140.
Gummihorn 164.
Gündlischwand 164.
Gunten 156.
Gunz 46. 318.
Guppen-Alp 68.
Gürbenbach, the 162.
Gurf 320.
Gürgaletsch 380.
Gurnigel, Upper 153. 203.
Gurnigelbad 153.
Gurschenstock 120.
Gurten 151.
Gurtnellen 112. 113.
Guschenkopf 66.
Güschihorn 318.
Guspis Valley 120.
Güssfeldt-Sattel 409.
Gütsch (Lucerne) 85. 81.
— (Brunnen) 19. 90.
Guttannen 188.
Güttingen 31.
Gwächtenhorn 138.
Gwärtler 135.
Gwatt 155. 199.
Gydisdorf 176.
Gyrenbad 49.

Haag 61. 63.
Habkern 163.
Habsburg, Ruins 23.
—, Neu-, Château 101.
Hacken 107.
Hägendorf 14.
Hagleren 139.
Hagnau 29.
Hahnenmoos 197.
Hahnenschritthorn 250.
Hahnensee 393.
Haibützli 76.
Haldenstein 62. **356**.
Hallau 24.
Hallwyl 142.
—, Lake of 141.
Haltenegg 153.
Hammetschwand 101.
Handegg Falls 188.
Hangbaum-Alp 93.
Hangendgletscherhorn 186.
Hanghorn **129**.
Hannig Alp **341. 342**.
Hard, Schloss **26. 32**.
Harder 162.
Haselmatt 80.
Hasenmatt 16.
Hasle 18. 139.
Haseleberg 182. 181.
Haslen 68.
Haslen-See 67.

Hasli, Valley of 181. 187.
Hasli-Grund 187.
Hasli-Jungfrau 177.
Hasli-Scheidegg 186.
Hätzingen 68. 69.
Haudères 323.
Hauenstein, the Obere 13. 14.
— Tunnel 14.
Hausen 79. 134.
Hausstock 70. 75.
Haute-Combe 266.
Haute-Luce 268.
Hauteville 267.
—, Château 241.
Hauts-Geneveys, **Les** 210.
Hedingen 79.
Heidegg 141.
Heidel Pass **47**.
Heiden 54.
Heider See 391. 428.
Heiligenschwendi 153.
Heiligkreuz 96. 139. 318.
Heimwehfluh 162.
Heinrichsbad 50.
Heinzenberg 367. 378.
Heldsberg 60.
Helsenhorn 318.
Hemishofen 25. **26**.
Hendschikon 20. 22.
Henggart 33.
Henniez 219.
Herbagères, Mont des 287.
Herbrigen 332.
Hérémence 322.
—, Val d' 322.
Hérens, Col d' 324. **325.** 340.
—, Dent d' 339.
—, Val d' 307. 322.
Hergiswyl 101. 103. 131.
Herisau 50.
Hermance 256.
Hermättje 336.
Herrenrüti 129.
Herrgottswald 86.
Herrliberg 41.
Hertenstein 87.
Herthen 23.
Herzogenbuchsee 17.
Hettlingen 33.
Heuboden Alp 47. 68.
Heustric bad 154.
Heuthal 410.
Hildisrieden 141.
Hilfikon 141.
Hilterfingen 156.
Hindelbank 18.
Hinterburg-See 184.
Hinter-Clärnisch 74.
Hinter-Meggen 104.
Hinterrhein, Village 358.
Hinter-Rhein, the 367. 370. 385.

Hinterrück 45.
Hinweil 44.
Hirli 380.
Hirondelles, Col des 282.
Hirzel 81.
Hirzelhöhe 81.
Hitzkirch 141.
Hochbalen Glacier 342.
Hochdorf 141.
Hoch-Ducan 362.
Hoch-Finstermünz 429.
Hochfluh 25. 318.
Hoch-Ryalt, Ruin 378.
Höchst 197.
Hochstuckli 107.
Hochwacht (Albis) 40.
—, near Regensberg 48.
—, on the Zugerberg 79.
Hoch-Wülflingen 48.
Hockenhorn 199.
Hofers-Alpe 312.
Hoffnungsau 363.
Hohberg Glacier 333.
Hohe Brisen 127. 93.
Hohe Faulen 112.
Hohe Kasten 59.
Hohe Licht 347.
Hohenems 432.
Hohenklingen 26. 32.
Hohenrain 141.
Hohen-Rhätien 378.
Hohenstollen 132. 182.
Höhentwiel 25.
Hohe Rhonen 105.
Hohe Thurm 69.
Hohfluh 182. 184.
Hohgant 163.
Hohgleifen 198.
Hohle Gasse 105.
Hohlicht 129. 339.
— Glacier 338.
Hohsaas Hnt 312.
Hohsand, Alp 319.
— Glacier & Pass 317. 318.
Hohthäligrat 335.
Hohtbürli Pass 170.
Hohwäng Glacier 329.
Hoierberg 58.
Holderbank 13.
Hölle, the 110. 365.
Höllenschlund 375.
Hölstein 13.
Homberg 141. 142.
Hombrechtikon 41.
Hône-Bard 297.
Honegg 101.
Hôpitaux, Les 221.
Horbachgütsch 79.
Horben, Schloss 22. 141.
Hörbisthal 129.
Horgen 41. 42. 81.
Horger Egg 81.

Horn 81. 62.
Hörnli (Adelboden) 196.
— (Arosa) 366.
— (Zermatt) 386.
Hornussen 20.
Horw 131.
Hospenthal 120.
Hospitalet, L' 301.
Houches, Les 288. 273.
Hüfi Alp 124.
— Glacier 123. 124.
— Pass 124.
Hugisattel 190.
Hühnerstock 121. 125.
Hüllehorn 318.
Hundschüpfi 135.
Hundsfluh 170.
Hünegg, Château 156.
Hüngigütsch 79.
Hunzenschwyl 22.
Hurden 42.
Hürnberg 140.
Hutegg 340.
Hutstock 129. 132.
Hütten 105.
Huttwil 17.

Ibach 110.
Iberg 107.
Iberger Egg 107.
Iffigen-Alp 200. 201.
Iffigen Fall 201.
Ignes, Casc. des 323.
Ilanz 369.
Iles, Les 283.
Ilfingen 11.
Illgraben 308.
Illhorn 327.
Illiez, Val d' 259.
Illklamm 431.
Im Feld 317.
Im Hof 185. 187.
St. Imier 310.
Immensee 104.
Immenstaad 29.
Incino 461.
Inden 195.
Indren Glacier 339.
Ingenbohl 110.
Inn, the 396. 402. 412.
Inner-Ferrera 381.
— Rothhorn 341.
Innerthal 43.
Innertkirchen **187.**
Inschi 112.
Intelvi Valley 461.
Interlaken 158.
Intra 446.
Intragna 441. 442.
Introbbio 453. 459.
Introd, Château 294.
Introzzo 453.
Ischa, Alp 362.

Iselle 313.
Iselten-Alp 164. 180.
Iseltwald 184.
Isenaux 251.
Isenfluh 165.
Isenthal 92. 93.
Isère, the 268.
Isleten 92.
Isola in the Engadine 396.
— on the Splügen 383.
Isola Bella 449.
— Madre 449.
— dei Pescatori, or Superiore 448. 449.
Isolaccia 412. 425.
Isorno, Val 442.
Ispra 443.
Issime 343.
Ivrea 298.

Jacobsbad 50.
Jägerhorn 340. **344.**
Jäger-Joch 340.
Jägernstöcke 71.
Jägerrücken 345.
Jagihorn 341.
Jaillet, Col 270.
St. Jakob in Unterwalden 101.
— in Uri 93.
Jakobshübeli 152.
Jaman 243.
—, Dent de 248. 254.
—, Col de la Dent de 254.
—, Plan de 254.
Jamthal, the 415.
Janzo, Casa di 347.
Jardin (Chamonix) 278.
— (Argentière) 283.
Jätzalp 75.
Jaun 203.
Jaun, Valley of 203.
Javernaz, Croix de 248.
Jazzi-Alp 345.
—, Cima di 337. 344.
— Glacier, the 340.
Jäzzihorn 344.
Jazzi Pass 340.
St. Jean-d'Aulph 257. 274.
— de-Sixt 270.
— in the **Val d'Anniviers** 327.
Jenatz 357.
Jenins 62.
Jenisberg 363.
St. Jeoire 274.
Jochli 93. 128.
Joch Pass 136.
Joderhorn 345.
St. Johann am Platz 361.
—, Alt 63.
—, **Neu** 63.
Johannisburg 43.

Joli, Col 268.
—, Mont 289.
Jolimont, the 206.
St. Jon 417.
—, Piz 417. 418.
Jona 44.
Jonswyl 62.
Jorasses, Grandes 292.
Jorat, Mont 287.
Jordils 286.
Jörifless Pass 359.
Jöri Glacier 359.
— Lakes 359.
S. Jorio, Passo di 458.
St. Joseph am Gänsbrunnen 11.
Jougne 221.
Jouplane, Col de 257. 274.
Joux, La 283.
—, Col de 348.
—, Fort de 213. 221.
—, Lac de 221.
Jouxtens 239.
Jovet, Plan 290.
Juchli 129. 132.
Juchlistock 132.
Juf 381. 392.
St. Julien 264. 232.
Julier, the 398.
—, Piz 398.
Jumeaux, Les, see Zwillinge.
— du Vallon 350.
Jungen 331.
Jungfrau 173. 316.
—, Hasli- 177.
Jungfrau-Joch 178. 316.
Jung Pass 331.
Jupiter, Plan de 303.
Jura, the 10. 16. 209. 235. etc.
Justis-Thal 157.
Juvalta 378.

Käferberg 48.
Kägiswyl 132.
Kaien 55.
Kaiser-Augst 19.
Kaisereggschloss 217.
Kaiserstock 92.
Kaiserstuhl 93. 133.
— near Lungern 133.
— on the Rhine 43.
Kalchthal 133.
Kalfeisen-Thal 47. 76.
Kalkberg 382.
Kalkstock 111. 123.
Kalli, the 177. 178.
Kalpetran 382.
Kaltbad (Rigi) 95.
Kaltwasser Glacier 311.
— Pass 311.
Kammerstock 69.

Kammli-Alp 124.
Kammlilücke 124.
Kammlistock 123.
Kamor 59.
Kanal Glacier 370.
Kanalthal, the 370.
Kander, the 191. 202.
Kanderfirn 170.
Kandergrund 192.
Kandersteg 192.
Kander-Thal 199.
Kändle, the 202.
Kappel on the Albis 79.
— in the Toggenburg 63.
Kärpf or Kärpfstock 69. 75.
Karrenalp 73.
Karrholen 163.
Kärstelenbach, the 112. 122.
Käserruck 47.
Kastanienbaum 101.
Kastelen-Alp 86.
Kastelhorn 320.
Kästris 367. 369.
Katzensee 21.
Katzenzagel 73. 92.
Katzis 378.
Kaunserthal 429.
Kehle Glacier 118. 137.
Kehlen-Alp 137.
Kehrbüchi 319.
Kehrsatz 153.
Kehrsiten 101.
Kemptthal 45.
Kerenzen-Berg 46. 47.
Kerns 132. 101.
Kernwald 101.
Kesch, Piz 362. 389. 413.
— Hut 362. 389.
Kessiloch 139.
Kesswyl 81.
Kienthal 191. 170.
Kiesen 151.
Kilchberg 42.
Kinzig Kulm 73.
Kippel 198.
Kirchberg 29.
—, the 69. 73.
Kirchenthurnen 153.
Kirchet 185. 187.
Kirchspalt Glacier 169.
Kisten Pass 71.
Kistenstein 365.
Kistenstöckli 71.
Klausen Pass 72. 111.
Klein-Basel 4.
— Buin 359.
— Fiescherhorn 178.
— Schreckhorn 177.
— Spannort 130.
Klein-Thal, in the Canton of Glarus 68. 75.

Klein-Thal, in the Canton of Uri 93.
Kleinthalfirn 130.
Klenenhorn 310.
Klimsenhorn 102. 103.
Klönthal 74. 68.
Klönthaler See 74. 68.
Klösterli (Rigi) 96.
Kloster Pass 359.
Klosters 358.
Klus, Oensinger 13.
— in the Kanderthal 199.
— in the Prätigau 356.
— in the Simmenthal 203.
Knonau 79.
Knörihubel 140.
Kohleren Ravine (Thun) 152.
— — (Adelboden) 196.
Kollbrunn 49.
Kölliken 18.
Königsfelden 20.
Konolfingen 140.
Krabelwand 96.
Kralalp 58.
Kranzbergfirn 171.
Krattigen 155. 156.
Krauchthal 75.
Kräzern Pass 68. 50.
Kreuz, the 357.
Kreuzberg, the 417.
Kreuzlingen 81.
Kreuzli Pass 112. 373.
Kriegalp Pass 318.
Kriens 86.
Kriesiloch 103.
Krinnen 250. 201.
Krinnenfirn 177.
Kronbühl 51.
Krönlet, or Krönte 112.
Krönte-Hütte 112.
Küblibad, the 157.
Küblis 358.
Kühalphorn 362.
Kühalp Thal 362.
Kühboden Glacier 315.
Kühbodenhorn 121. 315.
Kühlauenen Glacier 172.
Kunisbergli 197.
Kunkels Pass 66.
Kurfirsten, the 45.
Kurzegg 51. 55.
Kurzenburg 51.
Küsnacht on the Lake of Zürich 41.
Küssnacht on the Lake of Lucerne 104.
Kyburg 49.

Laax 369.
Lac Noir 217. 247.
Lacerandes, Pointe des 302.

31*

Lachat, Mont 288.
Lachaud, Mont 308.
Lachen 48.
Ladis, Baths 429.
Ladt, Im 319.
Lagalb, Piz 410. 411.
Lägerngebirge 21. 48.
Laghetto, Passo di 440.
Laglio 462.
Lago, Cima di 381. 382.
Lago Bianco 410.
— Maggiore 440. 443.
— Minore 410.
— Nero 410.
— Spalmo, Cima 412.
Lagone, Val 411.
Lagrev, Piz 396.
Laiblau, Piz 373.
Laisch, Alp 417.
Lämmern Glacier 194. 197. 201.
Lampertsch-Alp 370.
Lancebranlette 294.
Lancettes, Glacier des 289.
Lancey 305.
Landeck 429.
Landenberg 132.
Landeron 206.
Landmark 61. 55.
Landquart 62. 356.
Landskron 9.
Landwasser Road 361. 362.
Lange Fluh 341. 342.
Langen 430.
Langenberg, the 40.
Langenbruck 13.
Langenegg 55.
Langenfluh Glacier 335.
Langenthal 17.
Langnau 130. 13. 40.
Langtauferer-Thal 428.
Läng-Thal 318.
Languard, Piz 407.
— Alp 407.
Langwies 365.
Lanterna, Val 409. 421.
Lanzada 409. 421.
Lanzo d'Intelvi 439. 456.
Laquinhorn 341.
Laquin-Joch 312. 341.
Laret, Ober and Unter- 358. 359.
Laret Alp (Fettan) 417.
— — (St. Moritz) 400.
Lasa Alp 66.
Lasnigo 464.
Latelhorn 314. 341.
Latsch 389.
Latscher Kulm 389.
Lattenfirn 71.
Latterbach 202.

Lauberhorn 174.
Laucherhorn 130.
Laucherspitze 198.
Lauenen (Genthal) 135.
— (Lauenenthal) 250.
— See 250.
— Thal 250.
Laufbodenhorn 200.
Läufelfingen 14.
Laufen, Schloss 27.
— on the Birs 10.
Laufenburg 23. 20.
Lauinenthor 171.
Laupen 215.
St. Laurent 270. 300.
Laurenzenbad 22.
Lausanne 236. 245.
Lauteraar Glacier 189.
— Horn 178.
— Sattel 178.
Lauterbrunnen 165.
— Scheidegg 174.
Lautersee 375.
Lantrach 432.
Lavancher 278. 283.
Lavaz, Val 371. 372.
Lavaz-Joch 372.
Laveigrat 197.
Laveno 446. 453.
Lavertezzo 440.
Lavey, Baths of 248.
Lavin 414. 415.
Laviner Joch 359.
Lavirum, Fuorcla, Piz, and Val 318.
Lavizzara, Val 443.
Lavorgo 116.
Lax 318.
Layaz, La 251.
Lebendun Lake 318.
Lecco 463.
—, Lago di 463.
Lechaud, Col 275.
Lécherette, La 255.
Leckihorn 120. 121.
Lecki Pass 121.
Led, Piz 397.
Leggiuno 443.
St. Légier 241.
Legnoncino, Monte 458.
Legnone, Monte 458.
Lei, Valle di 384.
Leidensee Pass 112.
Leissigen 155. 156.
Leistkamm 45.
Lejs, Piz dels 410.
Lemene 267.
Lenk 200.
Lenno 461.
Lens, Pas du 299.
Lenta-Lücke 370. 386.
— Valley 370.
Lenz 364. 391.

Lenzburg 22. 142.
Lenzer Heide 391.
— Horn 390. 391.
Lenzjoch 341. 342.
St. Léonard 307.
Leone, Monte 311.
Lerow 157
Lesa 451.
Leschaux, Col de 270.
—, Glacier de 277.
Leubringen 12.
Leuerfall, the 57.
Leuggelbach 68.
Leuk 196. 308.
—, Baths of 194.
— Susten 303.
Leutschach-Thal 112.
Leventina 116. 122.
Levo 450.
Leysin 252.
Lezzeno 461.
Liappey 322. 304.
Lichtenberg, Castle 428.
Lichtensteig 63.
Liddes 300.
Liechtenstein 61. 62.
Liedernen 92.
Liestal 13.
Ligerz 206.
Lignerolles 220.
Lignières 206.
Lillianes 348.
Limmat, the 20. 35. 48 etc.
Limmern Glacier 71.
Limmern-Thal 71.
Limonta 463.
Linard, Piz 415.
Lindau 53.
Lindenberg 141.
Lindenhof 53.
Lint, the 40. 44. etc.
— Canal 44. 49.
—, Colony of the 45.
Linthal 69.
Linty, Capanna 339. 347.
Lion, Col du 338.
Liro Valley 384.
Lischanna Glacier 417.
—, Piz 417.
Littau 188.
Litzner 359.
Liverogne 294.
Livigno 414.
—, Forcola di 411.
Livournea, Col de 296.
Livrio, Val 421.
Lizerne, the 256.
Lobhörner 172.
Locarno 440. 444.
Loccie, Col delle 346.
—, Punta delle 346.
Lochau 432.
Lochberg 118.

Löchliberg 367.
Löchli Pass 367.
Locle, Le 211.
Loco 442.
Luderio 377.
Loëche 308.
— les-Bains 194.
— Ville 196.
Löffelhorn 315.
Loges, Col des 210.
Loggio 456.
Lognan, Pav. de 283.
Lohlen Pass 374.
Lohner 197.
Lombard 299.
Lona, Pas de 326.
Longeborgne 307.
Longhin Pass 395.
—, Piz 395.
—, Lake 395.
Longone 464.
Lorze, the 79. 80. 81.
Losone 441.
Lostallo 387.
Lostorf 14. 22.
Lothenbach 104.
Lötschen Glacier 199. 317.
— Lücke 198. 317.
— **Pass** 199.
— **Thal** 198. 308.
Lottigna 377.
Lourtier 304.
Louvie, Col de 305.
Lovagny 267. 270.
Lovenex, Lake of 258.
Loveno 453.
Lowerz 109.
Lowerzer See 109.
Lü 414. 418.
St. Luc 329.
St. Luce 336.
Lucel 323.
Lucendro Pass 120. 121.
—, Piz 120. 121.
—, Lake of 120.
Lucerne 81.
—, Lake of 83. 86.
Luchsingen 69.
S. Lucio, Pass of 438.
Lucomagno, Pizzo 115.
Lüderen-Gässli 140.
Ludiano 377.
Ludwigshöhe 338. 339.
Luette 322.
—, La 305.
Lugano 434.
—, Lake of 455.
Lugeten 43. 106.
Lugnetz Valley 369.
Lugrin 258. 265.
Luino 445.
Luisettes 300.
Luisin 286.

Lukmanier 376.
Lumbreda, Piz 386.
Lumbrein 370.
Lampegnia, Alp 372.
Lüner See 430.
Lungern 133.
—, Lake of 133.
Lunghino, Piz 395. 396.
—, Fuorcla di 395.
Lürlibad 355.
Lüsch, Lake of 367.
Lüscherz 206.
Luseney, Mont 296. 297.
Lüsgen-Alp 309.
Lüssai 418.
Lütisburg 62.
Lutry 239. 218. 245.
Lütschen-Thal 171.
Lütschine, the 160. 165.
184, etc.
—, the Black 165. 171.
175. 176.
—, the White 165.
Lützelau 87.
Lützelflüh 18.
Luvino 445.
Luvis 369.
St. Luziensteig 62.
Luzzone, Val 370.
Lyrerose, Glacier de 306.
324.
Lys Glacier 339.
Lys Joch 339.
Lyskamm 338. 348.
Lys Valley 298. 347. 348.
Lyss 12. 220.
Lyssach 18.

Maasplank-Joch 118. 137.
Maccagno 445.
Macolin 12.
Macugnaga 344.
— Glacier 345.
Macun 415.
Madatsch 427.
— Glacier 427.
Maderaner-Thal 122.
Madesimo 383.
—, Passo di 382.
Madlain, Piz 418.
Madlenerhaus 359. 431.
Madonna del Monte 446.
455.
— della Bocciola 452.
— della Campagna 448.
— di S. Martino 461.
— del Sasso, on the Lago
Maggiore 441.
— —, near the Lake of
Orta 452.
Madrera, Val 463.
Madriser Thal 381.

Madulein 413.
Magadino 444.
Magenhorn 312.
Mägenwyl 22.
Maggia 442.
—, the 441. 442. 444.
—, Val 320. 442.
Maggiore, Lago 443.
Magglingen 12.
Magland 272.
Magreglio 464.
Maienfeld 61.
Maienwang 190.
Maierhof 369.
Maigels Pass 347.
—, Val 374.
Mainau, Island of 31.
Maira, the 422.
Maisons Blanches 300.
— —, Col des 300. 305.
Majing Glacier 195.
Majinghorn 198.
Maladers 364.
Malans 61. 62. 356.
Malenco, Val 395. 409. 421.
Malero, the 395. 421.
Malesco 441. 445.
Mallet, Mont 278.
Malnate 446.
Maloggia, or
Maloja 394. 421.
Mals 428. 426.
Malser Heide **428.**
Malters 138.
Malvaglia 377. 387.
S. Mamette 437. 456.
Mammern 26. 32.
Mandello 459. 463.
Männedorf 41.
Mannenbach 26. **32.**
Männlichen 174.
Männliflüh 197.
Maran 365.
Marbach 26. 32. 61.
St. Marcel 296.
Marcellaz 267.
Marcelly, Pointe de **274.**
Marchairuz, Col de 221.
286.
Marchhorn **320.**
Marcio, Pizzo **382.**
S. **Marco** 313.
—, Passo di 422.
Märenberge 71.
Margna, Piz 397.
Margozzolo, Mte. 450.
St. Margretenberg 66.
St. Margrethen 60. 432.
S. Maria del Soccorso
461.
— Maggiore 314. 441. 442.
St. **Maria** der Engeln,
Monastery 63.

S. Maria, Hospice on the Lukmanier 373. 376.
— in the Münsterthal 418. 426. 428.
— on the Stelvio 426.
—, Muot 396.
—, Sils- 396.
Marinstein 9.
Marienberg 52.
Marienhorn 348.
Marienthal 133. 139.
Marignier 272. 274.
Marin 206.
Marinelli, Capanna (Mte. Rosa) 315.
— (Bernina) 400. 421.
Märjelen Alp 316.
— See 316.
Markelfingen 25.
Marlioz 266.
Marly 217.
Marmels 392.
Marmorè, Muot 396.
Marmorera 392.
Marnein 357.
Maroggia 433.
Marscholhorn 386.
Martigny 249.
— Bourg 250.
St. Martin in the Kalfeisen-Thal 47. 76.
— in the Lugnetz Valley 369. 370.
— Charvonnex 270.
Martinet, Glacier de 248.
S. Martino (Val Masino) 421.
— (near Lugano) 436.
—, Forcella di 423.
—, Madonna di 461.
—, Piazza 423.
—, Sasso 461.
Martinsbruck 419.
Martinsloch 75. 368.
Martinstobel 52.
Mary, Mt. 304.
Masein 378. 379.
Masino 421.
—, Bagni del 395. 422. 424.
—, Val 421.
Massa, the 318.
Massongex 260.
Mastallone, Val 453. 454.
Masuccio, Piz 425.
Matan 357.
Matmoire, the 297. 349.
Matt 75. 76.
Matten near Interlaken 160.
— in the Upper Simmenthal 199.
Matterhorn 338. 350.

Matterhorn, the Little 353. 357.
Matterjoch 336.
Mattervisp, the 332. 335. 340.
Matthorn 102. 103.
Mattmark Alp 314. 343.
Mattwald Glacier 312.
Mattwaldhorn 312.
Maudit, Mont 291.
Mauensee 18.
St. Maurice 248.
Maurienne Valley 268.
Mauvais Pas 278.
Mauvoisin 305.
Mayen, Tour de 247.
Medel, Piz 376.
Medels 385.
—, Mompè 372.
Medelser Glacier 372. 375. 376.
— Thal, the 375.
Meeralp 75.
Meerenalp 46.
Meersburg 29.
Mégève 283.
Meggen 104.
Meggenhorn 87. 104.
Megglisalp 57. 58.
Melden 330.
Meidenhorn 330.
Meiden Pass 328. 330.
Meien 138.
Meienreuss, the 112. 113.
Meienschanz 138.
Melcrhof 361.
Meilen 41.
Meilleret, Pointe de 251.
Meillerie 258. 265.
Meina 451.
—, Col de la 322. 323.
Meiringen 181.
Meisterschwanden 141. 142.
Meitschlingen 112.
Melchsee 132. 135.
Melchthal 132.
—, the Kleine 133.
Meldegg 60.
Melide 438. 455.
Mellen, Piz 71.
Mellichen Valley 342. 343.
Mellig 341. 342.
Mello, Val di 395. 421. 423.
Mels 47. 76.
Menaggio 459.
Mendrisio 440.
Menthon, Château 269.
Menzberg 133.
Menzikon 141.
Menzingen 80.
Mer de Glace, near Chamonix 278.

Mer de Glace, near Argentière 283.
Mera, the 384. 392. 422.
Merdarola, Bocchetta di 422.
Mergoscia 441.
Mergozzo, Lago di 448.
Méribé 322.
Merignier 300.
Merlen-Alp 46. 68.
Merligen 157.
Mesocco 387.
Mesolcina, Val 386.
Messernalp 318.
Mettelhorn 337.
Mettenberg 176. 177.
Mettlen-Alp 172.
Mettmenstetten 79.
Meyrin 264.
Mezaun, Piz 413.
Mezdi, Piz 415.
Mezzaselva 353.
Mezzem, Piz 413.
Mezzola, Lago di 384.
Miage, Col de 282.
—, Glacier de 281. 289. 291.
St. Michaelskreuz 105.
Michel, Piz 388. 392.
St. Michel 268.
Midi, Aiguille du 280. 281.
—, Col du 280.
—, Dent du 259. 285.
Mieussy 274.
Miax 253.
Milan 465.
Milar, Val 373.
Milchbach 176.
Mind, Mont 325. 326.
— —, Glacier du 324.
Minger, Val 418.
Minor, Val 410.
Minschun, Piz 415. 418.
Miolans, Castle 268.
Misaum, Alp 405.
Mischabelhörner 635.
Mischabel-Joch 340. 341. 342.
Misox, Ruins 387.
Misoxer Thal 387.
Mission 328.
Mitlödi 68.
Mittagfluh 203.
Mittaggüpfi 102.
Mittaghorn (Lauterbrunnen Valley) 171.
— (Bernardino) 386.
— (Binnenthal) 316.
— (Rawyl) 201.
— (Saas Valley) 341.
Mittagjoch 171.
Mittelhorn 177.
Mittelzell 25.
Mittenberg 355. 364.

Mittler-Glärnisch 74.
Moanda, Bocchetta 347. 454.
Modane 268.
Moësa, the 117. 386. 387.
Moësola, Lago 386.
—, Piz 386.
Mœveran, Grand 248. 255.
Mogno 443.
Möhlin 19.
Moine, Aiguille du 277.
Moiry, Glac. de 325. 326. 328. 329.
—, Val 326. 328.
Môle 272. 274.
Moléson 253.
Mollia 454.
Mollis 67.
Mols 47.
Moltrasio 462.
Moming 329. 339.
— Pass 329. 340.
Mompé-Medel 372.
— Tavetsch 373.
Monbaron, Colma di 298.
Mönch, the 175. 178.
Mönchalp-Thal 359.
Mönchenstein 10.
Mönchjoch 178. 179. 317.
Moncodine 459.
Mondelli Pass 344. 346.
Mondin, Piz 419. 429.
Monnetier 232. 271.
Monstein 60. 361.
Montafon 431.
Montagna s. Heinzenberg.
Montagnaia, Colle 296.
Montalto Dora 298.
Montana 308.
Montanvert 277.
Montbarry 252.
Mont Blanc 281.
— — de Seilon 305.
— — du Tacul 280. 282.
— —, Rocher du 281. 291.
Montbovon 253.
Mont Brûlé, Col du 325. 340.
Montbrun 251.
Mont Durand, Glacier du 305.
Montées, Les 273.
Monteluna 66.
Montenvers 277.
Monterascio Pass 370.
Montets, Col des 283.
Monthey 259.
Montjoie Valley 273. 289.
Montjovet 297.
Montmélian 268.
Montorfano 464.
Montoz 11.
Montreux 242. 246.

Montriond 257.
Mont Rouge, Col du 306. 324.
Mont Ruan, Glacier du 260.
Montsalvens, Ruins 203.
Monvalie 443.
Monza 463. 465.
Morast 319.
Morat 219.
—, Lake of 219.
Morbegno 422.
Morcles 248.
—, Dent de 248.
Morcote 455.
Mörel 317. 318.
Morez 235.
Morgarten 80. 107.
Morge, the, in Savoy 258.
—, the, in the Cant. of Valais 251. 256. 307.
Morgenberghorn 157. 162.
Morgenhorn 192.
Morges 236. 245.
Morgetenbachfall 208.
Morgex 294.
Morgin 259.
Moriana, Passo della 454.
Morignone 425.
Möringen 206.
Morissen 369.
St. Moritz (Engad.) 398.
—, Baths of 398.
—, Lake of 399. 400.
Mornex 232. 271.
Moro, Monte 345.
— Pass 345. 348.
Morobbia, Val 433.
Morschach 91.
Mörschwil 52.
Mort, Mont 302.
Morteau 211.
Mortèl, Piz 408.
—, Club Hut 406.
Morteratsch, Piz 402. 408.
— Glacier 404. 408. 409.
Mortirolo, Monte 421.
Morzine 257.
Mosses, Les 255.
Môtiers 212.
—, Grotte de 212.
Motta, La 411.
Mottarone, Monte 450.
Möttelischloss 52.
Mottelon 256.
Mottets 268. 290.
Motto 377.
Moud, Colle 347. 454.
Moudon 219.
Moulins, Les 254.
Mountet Hut 328. 325.
Moutier in the Jura 11.

Moutier, Val 10.
Moûtiers en Tarentaise 268.
Mouxy 266.
Muccia Glacier 386.
Muggio 439.
Mühlebach-Thal 47. 75.
Mühlehorn 46.
Mühlen 392.
Mühlestalden 136.
Mühlethal 134. 136.
Mülenen 191.
Mulets, the Grands 281.
Mulins 368.
Mülkerblatt 200.
Mumpf 19.
München-Buchsee 12.
Mundaun, Piz 369.
Münsingen 151.
Munster on the Birs 11.
—, Grisons 414. 423.
—, Valais 315.
Münsterlingen 31.
Münster-Thal (Grisons) 414.
— in the Jura 10.
Muota, the 72. 110. etc.
Muotathal 78.
Muottas, Sass da 400.
Muraigl, Alp 407.
—, Muottas 402. 406.
Muranza, Val 426. 428.
Muraun, Piz 372. 375.
Muraz 256. 259.
Muretto, Monte 395.
— Pass 395. 421.
Murg on the Rhine 23.
— on the Walensee 46.
Murgenthal 17.
Murgsee-Furkel 47. 68.
Murgseen 46. 48.
Murgthal 46.
Muri 22.
Mürren 168.
Murten 219.
Murtèra 414.
—, Piz 418.
Mürtschen-Alp 46. 47. 68.
Mürtschenstock 46.
Müsella, Munt 412.
Musso 458.
Mustail 380.
Mustair, Val 414.
Mutten 379.
— Glacier 121.
Muttenhorn 121. 126.
Muttensee 70.
Mutenstock 70.
Muttenthaler Grat 76.
Muttenz 13.
Mutthorn 170.
Muttler 418. 419.
Muttnerhorn 379.

488 INDEX.

Mutan, Piz 386.
Muzzano, Lake of 455.
Myten 110.
Mytenstein 91.

Nadelhorn 341.
Nadeljoch 342.
Näfels 67.
Nägelis-Grätli **126. 190.**
Nair, Piz 401.
Naira, Piz 366.
Nalps, Pass 373.
—, Val 373.
Naluns, Muotta 417.
Nandro, Val 391.
Nant 243.
—, Pont de 248.
Nant-Borrant 289.
Nant-Bride 260.
Nanzer-Thal 309. 332.
Napf, the 139.
Napoléon, Pont 310.
Naret, Passo di 443.
Naters 309. 318.
Natons, Val 392.
Nauders 428.
Nava, Pointe de 328.
Navegna, the 441.
Navigenze, the 326. 327.
Naye, Rochers de 243. 254.
Naz 389.
S. Nazzaro 443.
Nebikon 18.
Neftenbach 33. 48.
Nendaz 305.
—, Val de 305.
Nendeln 432.
Neria, Val 459.
Nernier 257.
Nessel 309.
Nessen-Thal 136.
Nesslau 63.
Nesslern, Alp 162.
Nesso 462.
Nesthorn 310.
Netstall 67.
Neuchâtel 207.
—, Lake of 207.
Neuenburg 207.
Neuenkamm 46.
Neuenstadt 206.
Neu-Habsburg 104.
Neuhaus 157.
Neuhausen 24. 27.
Neumünster 41.
Neu St. Johann 63.
Neuschels 203.
Neuva, Glacier de la 283. 293.
Neuveville 206.
Neu-Wartburg, Castle 14.
S. Nicolao 439.

St. Nicolas de Veroce 289.
Nidelbad 42.
Nidfurn 68.
Nieder-Bauen 88.
Niederbipp 15.
Niedergestelen 308.
Niederglatt 21. 48.
Niederhallwyl 142.
Niederhorn 158.
Nieder-Rawyl 201.
Niederrickenbach 128.
Niederried 184.
Nieder-Schönthal 13.
Niederschwörstadt 23.
Nieder-Surenen 129. 130.
Nieder-Tagstein 378.
Nieder-Urnen 67.
Niederwald 316.
Niesen 154.
Niesenhorn 200.
St. Niklaus (Göschenen-Thal) 118.
— (Melchthal) 132.
— (Visp Valley) 332.
Niouc 327.
Niva, Alp 323.
Niven 198.
— Pass 198.
Nivolet, Dent du 267.
Nofels 432.
Noiraigue 212.
Noir-Mont 235.
Nolla, the 378.
Nona, Becca di 295.
Nordend (Monte Rosa) 338. 340. 344.
Notkersegg 51. 55.
Notre-Dame de la Gorge 289.
— de Briançon 268.
— de Guérison 291.
— du Sex 243.
Nottwyl 19.
Novai 359.
Novara 453.
Novate 384.
Novel 211. 258.
Novena, Passo di 315.
Nudo, Monte 446.
Nufenen 385.
— Pass 315. 115.
Nufenenstock 315. 319.
Nüna, Piz 413.
Nünalphorn 132.
Nuolen, Baths of 43.
Nurschallas, Piz 374.
Nus 296.
Nüschen-Alp 70.
Nyon 235. 245.

Oberaar Glacier 189. 190.
Oberaarhorn 190.

Oberaar-Joch 190.
— Hut 190. 189.
Oberaar-Rothjoch 190.
Ober-Aegeri 80.
Ober-Aletsch Glacier 309.
— Hut 309.
Oberalp 375.
— Pass 374.
Oberalpstock **123. 373.**
Ober-Arth 96.
Ober-Bauen 95.
Oberbipp 15.
Oberblegisee 69.
Oberbuchen 141.
Ober-Gabelhorn 339.
Obergestelen 315.
Oberglatt 48.
Oberhalbstein-Thal 398.
Oberhaupt, the 102. 103.
Oberhofen 156.
Oberhornsee **167.**
Oberkäsern, **Alp 45.**
Oberkastels **370.**
Oberland, the **Bernese** 143.
—, the Bündner 371.
Oberlaubhorn, the 200.
Oberlauchringen 24.
Obermeilen 41.
Ober-Mutten 379.
Oberreinach 141.
Ober-Rickenbach 93. 128.
Oberried on the Lake of Brienz 184.
— in the Simmenthal 200.
Oberrieden 42.
Oberriet 361.
Ober-Rothhorn 338.
Obersaxen 369.
Ober-Schönenbuch 72. 91.
Obersee (Wiggis) 67.
— (Erstfeld Valley) 112.
Oberstaad 26. 32.
Ober-Stocken 202.
Oberurmi 88. 89.
Ober-Urnen 67.
Obervatz 379. 390. 391.
Oberwald 315.
Ober-Winterthur **32.**
Oberwyl 9. 104.
Obladis **429.**
Obort 70.
Obrist, Alp 380.
Obstalden 46.
Obwalden 132.
Oche, Dent d' 258.
Ochsen 203.
Ochsenhorn 332.
Ochsenjoch 178.
Ochsenkopf 73.
Oehningen 26. 82.
Oensingen 15.

Oerlikon 48. 21.
Oeschinen, Alp 170. 192.
—, Lake of 192.
Oeschinengrat 170.
Oeschinenhorn 192.
Oeschinen-Joch 193.
Ofenberg 414. 413.
Ofenhorn 317.
Ofen Pass 414.
Ofenthal 314. 343.
Ofenthal Pass 314.
Oggebbio 445.
Oira 321.
Oldenhorn 250. 251. 322.
Olen, Col d' 347.
Olgiasca 458.
Olgiate 446.
Olivone 377.
Ollocchia, Val 454.
Ollomont 306.
—, Val 306.
Ollon 246. 247.
Olten 14.
Oltingen 13.
Oltschibach 182.
Oltschikopf 184.
Omegna 450. 452.
Onno 463.
Onsernone, Val 441. 442.
Orbe 220. 215.
Oren, Col d' 306.
—, Combe d' 321.
Oriano, Monte 448.
Oria 437. 456.
Orlegna-Fall 394. 395.
Ormelune 294.
Ormont-dessous 252.
— -dessus 251.
Ornavasso 451.
Orny, Cabane d' 299. 287.
—, Glacier d' 287.
—, Pointe d' 299.
Oron 218.
Orsia 347. 348.
Orsières 299.
Orsino, Piz 120.
Orsino Pass 121.
Orsirora Lake 121.
Orso, Colmo dell' 383.
Orta 452.
—, Lago d' 452.
— Miasino 452.
Ortler 427.
Ortstock 69. 71. 73.
Orvin 11.
Osogna 117.
Osola, Val d' 440.
Ossasco 315. 443.
Ossola, Val d' 313. 451.
Osteno 456.
Ostermundingen 151.
Ot, Piz 402.
Ota. Alp 405. 409.

Otanes, Col des 304.
Otelfingen 21.
Otemma, Col d' 306.
—, Glacier d' 306. 324.
—, Pointe d' 305.
Othmarsingen 20. 22.
Ouches, Les 273. 288.
Ouchy 236.
Oyace 324.
St. Oyen 303.

Padella, Piz 402.
Pain de Sucre 303.
Painsec 327.
Palesieux 293.
Palette, the 251.
Palézieux 218. 253.
Palfries, Alp 47.
Pallanza 447.
Pallanzeno 451.
Palpuogna 389.
Palü, Piz 408.
— Glacier 409. 410.
— Lake 421.
Paneyrossaz, Glac. de 255.
Panix 75.
— Pass 75.
Panossière, Cabane de 300. 304.
Pantenbrücke 70.
Paradies 385.
— Glacier 385.
Paradis, Piz 373.
—, Fuorcla da 373.
Paradiso 436.
Paraviso 456.
Pardisla 357.
Parè 463.
Parmelan 270.
Parpan 390.
Parrain, Mt. 304.
Parrot-Spitze 338. 348.
Parseier-Spitze 429.
Part-Dieu, Convent 252.
Partnun Pass 358.
Partnuner Staffel 358.
Pass Mal 379.
Passetti Pass 387.
Passugg 356.
Pasturo 459.
Patenen 359. 431.
Patnaul, Fuorcla da 370.
Patznaun Valley 415.
Paudèze, the 218. 239. 245.
Paun da Zücher 407.
Payerne 219.
Pazzola, Piz 373.
Pecetto 344. 345.
Peccia (Val Maggia) 443.
— (Val Vogna) 347.
Pedenosso 412. 425.
Peiden, Baths 370.
Peiler Thal 370.

Peilz, Ile de 245.
—, Tour de 240. 246.
Peist 364.
Pelaggia, Cima 463.
Pèlerins, Nant des 280.
—, Glacier des 281.
Fella 452. 453.
Pellina, Val 296. 303. 306.
Pelouse, Pointe 275.
Pendant 283.
Pennine Alps 303.
Penthalaz 215.
Percée, Pointe 272. 273.
Perdatsch 375.
Perralotaz, Pont 273. 280.
Perroc, Dent 324.
Perron, the 285.
Pers, Munt 407
— Glacier 405. 408.
Persa, Isla 408
Pesciora, Piz 121.
Pestarena 344.
St. Peter (Lugnetz) 370.
— (Schanfigg-Thal) 364.
—, Isle of 206.
Peter and Paul 51.
Petersgrat 171. 191. 193. 198.
Petriolo Alp 345.
— Glacier 345.
Pettenasco 452.
Pettnen 430.
Peulaz, Col de la 293.
—, Chalets de la 293.
Peuteret, Aiguille de 291.
Pfäfers, Bad 64.
—, Village 65.
Pfaffen Glacier 136.
Pfaffensprung 112. 113.
Pfaffenwand 136.
Pfäffikon (Lake of Zürich 42. 43.
— (near Wetzikon) 44.
Pfänder 432.
Pfannenstiel 41.
Pfannenstock 69.
Pfin 308.
Pfunds 419. 429.
Pian Canfèr 392.
Pianazzo 383.
Piancanino 395.
Pianello 458.
Piano 456.
—, Lago del 456.
Pians 429
Piazzi, Cima di 412. 426.
Piccola, Valle 347. 454.
Piccolo Altare, Col 454.
Pièce, Glacier de 323. 324.
Piedimulera 451. 344.
Pierre a dzo, La 259.
— à Béranger 278.

INDEX.

Pierre à Bérard 275.
— à l'Echelle 280.
— Grept, Tête à 248. 265.
— Joseph, Col de 252.
— Pertuis 11.
— Pointue, Pav. de la 280.
— à Vire 805.
— à Voir 250. 299. 304.
307.
St. Pierre d'Albigny 263.
— Mont-Joux 300.
Pieterlen 17.
Piglimohorn 346.
Pignieu, Baths of 381.
Pilatus, the 102.
— Lake 86.
Pilo Alp 347.
Piller-Gletscher 359.
Pillon, Col de 251.
Pino 444.
Pinterjoch 348.
Piode Joch 339.
Piora, Hôtel and Val 115.
Piotta 116.
Piottino, Monte 116.
Pirlo Lakes 421.
Pischa, Lake 407.
—, Fuorcla 389.
Pischahorn 359. 361.
Pisciadella 420.
Pisoc, Piz 417. 418.
Pisse, Col delle 339. 347.
Pissevache, the 249.
Pitons, Les 232.
Pitschen, Lej 408. 410.
Piuro 424.
Pizalun 66.
Pizol 66.
Plaine Morte, Glacier de la 200. 197.
Plan (Ormont) 251.
— de l'Aiguille 281.
— Bel Achat 279.
— des Dames 290.
— Lachat 279.
— Névé, Glacier de 248.
—, Aig. du 281.
Planches, Les 242.
Planchettes, Les 210.
Plangolin, Col de 304.
Planken-Alp 93. 129.
Plaupraz 275. 280.
Plan **Rai**, Glacier 359.
Plans, Les (Bex) 247.
—, Vallée des 248.
Plantour, the 246.
Planura Pass 124.
Plasseggen Pass 358.
Platé, Désert de 275.
Platifer 116.
Platta (Fexthal) 397.
— (Val Medel) 375.
—, Piz 392.

Plattas, Val 372. 375.
Platten 309.
Plattenhörner 359.
Plattenschlucht 370.
Plattje, Unteres & Oberes (Monte Rosa) 338.
— (near Saas) 341.
Plana, Piz 417.
—, Val 416. 418.
Pleiades, Les 241.
Plessur, the 354. 364. 390.
Pletschen 165. 330.
Pleureur, Mont 304. 305.
Plines, Col des 299.
Plurs 424.
Poch, Val del 418.
Pochtenbach Fall 191.
Pochtenkessel 196.
Pognana 462.
St. Point, Lake of 213.
Polleggio 117.
Pollux 335.
Pommat, the 320.
Pont, Le 221.
— de Nant 248.
— de Risse 274.
— Pélissier 273.
— St. Martin 298. 348.
— Ste. Marie 273.
— Serrand 293.
Pontarlier 213.
Ponte (Engadine) 412.
— Capriasca 457.
— del Diavolo 425.
— Grande 344. 454. 463.
— Tresa 455.
Ponti, Ai 311. 318.
Pontlatz Bridge 429.
Pontresina 402.
—, Muottas da 404.
Porcareccio, Passo di 442.
Porcellizza, Val 422. 424.
Porchabella Glacier 362. 389.
Porlezza 456.
Porrentruy 10.
Port Valais 258.
Porta, La 423.
Portalet 299.
Porte du Sex, La 258.
Portjengrat 313.
Porto Ceresio 446. 455.
— Valtravaglia 445.
Poschiavo 420.
—, Lago di 420.
Pougny 264.
Pourri, Mont 290.
Pozzolo, Pizzo 344.
Prad 428.
Prada, Alp 459.
Pradella 417. 418.
Pragel 73.
Pralaire 233. 274.

Pralong 270.
Prangins 235. 245.
Pra-Rayé 296. 324. 350.
Prarion 273.
Prasüratsch 401.
Prätigau 356.
Prato 440. 443.
—, Val 443.
Prätsch 365.
Pratteln 13.
Praz, Les 278. 279. 283.
— de Fort 283. 293.
— Sec 292.
Präzerhöhe 379.
Prazfleuri 325.
—, Col de 305.
Prazlong 322.
Pré, Le 254.
— de Bar 292.
Preda 359.
— Rossa, Alp **421.**
Predelp Pass 116. 376.
Pregny 231.
Premadjo 412. 426.
Premeno 447.
Premia 321. 442.
Premosello 451.
Prequartero 344.
Pré-St-Didier 293.
Prese, Le 420. 425.
Pressura, Monte 427.
St. Prex 236. 245.
S. Primo, Monte 461. 464.
Priausch 46.
Pringy-la-Caille 270.
Promenthoux 235.
Promontogno 423.
Prosa, Mte. 120. 121.
Proz, Cantine de 301.
—, Glacier de 301.
Pruntrut 10.
Prutz 429.
Pulaschin, Piz 398.
Pully 239.
Puntaiglas, Val **371.**
Puntota, Bridge near Pontresina 404. 405.
— —, near Zernetz 413.
Puschlav 420.
Pusiano, Lago di 464.
Pyrimont 265.

Quarazza, Val di 344. 346. 451.
Quarsano 462.
Quart, Château 296. 300.
Quart-Villefranche 296.
Quarten 46. 47.
Quatervals, Piz 414.
Quincinetto 298.
Quinten 45. 46.
Quintino Sella Hut (Mont Blanc) 281.

INDEX. 491

Quintino (Lysjoch) 348. 339.
Quinto 116.

Rabenfluh 153.
Rabiosa, the (near Coire) 356. 360.
Rabius 371.
Rabiusa, the (Safier-Thal) 367.
Rachisberg 18.
Radolfzell 25.
Rafrüti 140.
Ragatz 64. 61.
Ragol 66.
Rain 141.
Raisse, Ravine of the 212.
Ralligstöcke 157.
Ramin Pass 76.
Ramisfluh 132.
Ramosa, Fuorcla da 370.
Ramsey 18.
Ranasca Alp 75.
Rancio, Sasso 459.
Randa 333.
Randen, Beringer 25.
—, Hohe 25.
Rang, Tête de 210.
Rankweil 432.
Ranzo-Gera 443.
Ranzola, Colle 348.
Rappenhorn 317.
Rapperswil 42. 41.
Raron 308.
Raschil, Alp 390.
—, Piz 390.
Räterichsboden 188.
Rathhausen 140.
Rauthorn 312.
Rautifelder 67.
Rautispitz 67.
Raveisch Lakes 362.
Ravins, Les 201.
Rawyl 201.
Rawylhorn 201.
Räzlberg 200.
Räzli Glacier 201.
Rò 441.
Realp 125.
Realta, Ruin 378.
Rebarmaz, La 285.
Rebbio, Punta del 311.
—, Forca del 311.
Reckingen 48. 316.
Réclère 10.
Redasco, Bima 426.
Redorta, Corona di 440.
—, Passo di 440. 443.
Regenbolshorn 197.
Regensberg 48.
Regensdorf 21.
Regina, Piz 370.

Regina Margherita, Capanna 296. 348.
Regoledo 459.
Rehetobel 55.
Reichenau 367. 377.
—, Island of 25.
Reichenbach 191.
—, Falls of the 185.
Reichenburg 43.
Reichenstein 204.
Reiden 18.
Reidenbach 208.
Reignier 271.
Reinach 141.
Reisen 14.
Releccio, Capanna di 459.
Remüs 418.
St. Rémy 303.
Renens 215. 246.
Renfenhorn 186.
Rentiert, Alp 372.
Reposoir 270.
—, Rocher du 293.
Reschen 428.
— Scheideck 428.
Rescia 456.
Resegone di Lecco 463. 464. 465.
Resti Pass 198.
— Rothhorn 193.
— Tschingel Glacier 123.
Résy 348.
Retico, Lago 375.
Rettan Lake 251.
Reuchenette 11.
Reulissenberg 201.
Reuse, the 209. 212.
Reuse d'Arolla, Col de la 306.
Reuss, the 19. 20. 80. 83. etc.
Reuti 182.
Reutigen 202.
Revard 266.
Rezzonico 459.
Rhäticon 431.
Rhäztins 378.
Rhein, Averser 381. 382.
—, Hinter 367. 370.
—, — (Source of the) 385.
—, Medelser, or
—, Mittel 372. 375.
—, Oberhalbstein 391.
—, Val 382. 385.
—, Valser 370.
—, Vorder 367. 372. 374. 375.
—, Vriner 370.
Rheinegg 60.
Rheinfelden 19. 23.
Rheingau, Upper 60.
Rheinklingen 25.
Rheinquellhorn 386.

Rheinthal, Vorder- 368. 377.
Rheinwald Glacier 385.
Rheinwaldhorn 386.
Rheinwald-Thal 382. 385.
Rhine, the 3. 23. 29.
—, Falls of the 27. 24.
Rhodan, the 315.
Rhone, the 126. 224. 246. 253. 306. 314 etc.
— Glacier 126. 314.
—, Perte du 261.
Rhonestock 118. 315.
Richensee 141.
Richetli Pass 75.
Richisau 73.
Richtersweil 43.
Rickenbach 107. 110.
—, Ober- 98.
—, Nieder- 128.
Riddes 307.
Ried on the Inn 429.
— (Lötschenthal) 198.
— (Muotathal) 72.
— (Valais) 317.
Rieden 44.
Rieder Alp 317.
— Furka 317.
Riederhorn 317.
Riedern 74.
Riedmatten, Col de 306. 322. 324.
Ried Pass 341. 342.
Riedwyl 17.
Riein 369.
—, Piz 368.
Riemenstalden-Thal 73. 92. 110. 111.
Rienzer Stock 113.
Rieseten Pass 75.
Riffelalp 334.
Riffelberg 334.
Riffelbord 335.
Riffelhaus 334.
Riffelhorn 335. 337.
Riffler 430.
Riggisberg 153.
Rigi, the 94.
Rigidalstock, the 129.
Rigi-Felsenthor 95. 96.
— First 94. 99.
— Hochfluh 89. 100.
— Kaltbad 95.
— Klösterli 96.
— Kulm 96. 97.
— Railways 94.
— Rothstock 95.
— Scheidegg 94. 100.
— Staffel 95. 96.
Rikon 40.
Rima 454. 347.
—, Passo di 45
Rimasco 454.

Rimella 454.
Rimpfischhorn 338. 343.
Rimpfischwänge 343.
Rinderhörner 193.
Ringelspitz 378.
Ringgenberg 168. 184.
Rinkenberg 371.
Rinkenkopf 75.
Rionda, La 285.
Ripaille, Castle 257.
Riseten Fall 88.
Ritom, Lake 115.
Ritter Pass 318.
Ritzengrätli 179. 180.
Ritzingen 316.
Riva Valdobbia 454.
— di Palanzo 462.
Rivasco 320.
Riva z St. Saphorin 218. 239. 246.
Rivera-Bironico 431.
Riviera 117.
Robiei, Alp 320. 443.
Roc Noir 328. 329.
S. Rocco 321.
Roche in the Jura 11.
— on the Rhone 246.
— Percée, La 284.
— sur Foron 270. 271.
Roches, Col des 211.
Rodi-Fiesso 116.
Rodont-Bridge 120.
Rofelhörner 344.
Rofelstaffel 345.
Roffna 392.
Rofna-Ravine 382.
Roggenhorn 358.
Rohrbachstein 200. 201.
Rohren 101.
Roisetta 350.
Rolle 236. 245.
Romainmotier 220.
Romanshorn 49. 31.
Römerswyl 141.
Romiti 95. 97.
Romont 218.
Romoos 131.
Ronco 441. 444.
Rondadura Pass 373.
—, Piz 373. 376.
Rongellen 380.
Rophaien 92.
Rorschach 52.
Rorschacher Berg 52.
Rösa, La 411.
Rosa Blanche, Pte. de 304. 305.
Rosa, Monte 338. 345. 348.
Rosatsch, Piz 401. 403.
Roseg, Piz 409.
— Glacier 405. **409**.
—, Porta 409.
—, Vadret da 405.

Roselette, **Mt.** 289.
Rosenberg 51.
Rosenhorn 187.
Rosenlaui, Baths of 185.
— Glacier 185. 186.
Rossa 387.
—, Bocca 318.
Rossberg 108. 79. 80. 109.
Rossboden Glacier 312.
Rosebodenhorn 312.
Rossboden Pass 312.
Rossbühel 52. 55.
Rossinière 254.
Rosso, Mte. 448.
—, Cima di 395.
Rossstock 92. 111.
Rothe Boden 335. 337.
— Kumme 197.
— Kummen 335.
Röthelspitze 427.
Rothenbrunnen 378.
Rothendossen 102.
Rothenegg 162.
Rothenthurm 107.
Rothfluh 88.
Rothgrätli 93. 129.
Rothhorn, Brienzer 182. 133.
—, Aroser 366. 361.
—, Blümlisalp 192.
—, Faldum 198.
—, Ferden 193.
—, Resti 198.
— (Macugnaga) 344.
— (Saas) 312. 338.
— (Sigriswyl) 156.
— (Zermatt) 333. 335.
— (Zinal) 329. 339.
Röthi 16.
Röthihorn 179. 180.
Rothkreuz 22. 80. 109.
Roth-See 81.
Rothstock, Rigi- 95.
—, Uri- 129. 93.
—, Engelberger 129. 93.
— Lücke 93.
Rothtbal Hut 171. **173**.
— Sattel 171. 173.
Rothtbor, the 46.
Rotondo, Pizzo 114. 121.
—, Passo 114. 121.
Rotten, the 315.
Rotzberg 101.
Rotzloch 102.
Rougemont 204.
Rouges, Aiguilles (Chamonix) 275. 283.
—, — (Evolena) 323.
Rousses, Les 235.
Rovana, Val 442.
Rovano, Passo 410.
Rovenna 462.
Roveredo 387.

Rovio 439.
Ruan, Mont 260.
Rubigen 151.
Rühlihorn 204. 251.
Ruch-Eptingen 14.
Ruchen, Grosse 123. 72.
Ruchenglärnisch 74.
Ruchkehlen Pass 72. 124.
Ruchi 67. 70.
Rüchi 70.
Rudenz, Château 93. 133.
Rue 218. 219.
Rüegsau 18.
Rueras 373.
Rugen, the **Kleine** 161.
Ruinaz 294.
Ruinette 305.
Ruis 75. 371.
Rumianca 451.
Rumilly 266.
Ruosalp 73.
Ruosalper Kulm 73.
Ruppersweil 22.
Ruschlikon 42.
Rusein, Piz 70.
—, Val 71. 372.
Russo 442.
Ruth, Dent de 208.
Ruti in the Rhine Valley 61.
— **near** Arosa 365.
— near Rapperswil 41.
— near Stachelberg 69.
Rüth 91.
Rutor 293. 294. 304.
— Falls 293.
Rüttiftrn 138.
Rüttihubelbad 140.
Ruz, Val de 210.
Ryalt, Hoch-, Ruin 378.

Saane, see **Sarine.**
Saanen 204.
— Möser 204.
Saas im Grund 340.
— -Fee 341.
— in the Prätigau 358.
Saasberg 69. 74.
Saasgrat 309. 332.
Saas Pass 311.
Sacconnex 231.
Sachseln 133.
Säckingen 23. 20.
Safenwyl 18.
Safien-Platz 367.
Safierberg 367.
Safier-Thal 367.
Saflisch Pass 311.
Sage, La 323. 325. 326.
Sagerou, Col de 260.
Sägisthal 180.
Sagne, Mont 210.
Sala 437. 461.

Saland 49.
Salanfe, the 249. 260. 285.
— Alp 260. 285.
—, Col de 260.
Salay 325.
Salecina, Motta 895.
Saleinaz, Fenêtre de 299.
—, Cabane & Glacier de 289. 300.
Salenstein 26. 32.
Sâles 243.
Saletz 61. 59.
Salève, Mont 232.
Salgesch 196. 303.
Salins 268.
Sallanches 272.
Sallières, Tour 260. 286.
Salquenen 308.
Salteras, Piz 389.
Saltine, the 309.
Saluver, Val 401. 402.
Salux 391.
Salvagny 275.
Salvan 285.
S. Salvatore, Monte 436.
Samaden 401.
Sämbtis-See 58. 61.
Sambucco 443.
Samoëns 274.
Samolaco 384.
Samstagern 105. 43.
Sand-Alp 70. 372.
Sandalp Pass 71. 372.
Sandfirn 71. 124. 372.
Sandgrat 124. 372.
Sandhubel 363. 366.
Sanetschhorn 250.
Sanetsch Pass 250.
Santino 443.
Saoseo, Cima 412.
Sapün 365.
Sardasca Alp 359.
Sardona Alp 76.
— Glacier 76.
— Pass 76.
Sargans 47. 61.
Sarina Alp 62.
Sarine, the 203. 204. 215. 216. 253 etc.
Sarnen 132.
—, Lake of 133.
Saronno 446.
Sarraz, La 220. 215.
Sartuns 390.
Sassalbo 420.
Sassal Masone 410. 411.
— —, Alp 411.
Sass Auta 414.
Sassella 421.
Sassello Pass 443. 115.
Sasseneire 323. 326.
Sassi, Passo dei 115. 443.

Sassiglione, Forcola di 420.
Sassina, Val 458. 459.
Sasso Bissolo, Val di 425.
Satarma 323.
Satigny 264.
Sattel 108.
Sattelhorn 310. 317.
Sätteli 136.
Satteltelücke 370.
Saugern 10.
Sauren Glacier 76.
— Pass 76.
Saurenstock 75. 76.
Saussure, Pavillon 293.
—, Aig. de 282.
Sauterot 322.
Savaranche, Val 294.
Savognin 380. 391.
Sax 61.
Saxe, La, Baths 291.
—, Mont de 292.
Saxer Lucke 61.
Saxeten 162.
Saxeten-Thal 162.
Saxon, Baths of 306.
Scai, Piz 115. 376.
Scala, Lago della 411.
Scale, Mte. delle 425. 426.
Scaletta Pass 362.
— Glacier 362.
Scalettahorn 362.
Scanfs 413.
Scara Orell 121.
Scaradra Pass 370.
Scareglia 438.
Scarl 417. 418.
Scarljöchl 418.
Scarlthal 418.
Scatta, Passo della 442.
— Minojo 318.
Scerscen Glacier 397. 409.
—, Monte di 409.
— Pass 421.
Scesaplana 357. 430.
Schaan 432.
Schachen (near Lindau) 53.
— (in the Entlebuch) 138.
Schachenbad 53.
Schächen-Thal 72. 111.
Schadau, Château 153. 155.
Schadburg 163.
Schafberg (near Wildhaus) 63.
— (Lötschenthal) 198.
— (Melchthal) 135.
— (Oeschinenthal) 170.
— (Pontresina) 406.
Schafboden 58.
Schaffhausen 24.
Schäfle's Egg 59.

Schafjoch 156.
Schafmatt 13. 139.
Schaftobel 388.
Schalliberg Alp 339.
Schallihorn 329.
Schalli-Joch 329. 340.
Schamella Hut 357. 430.
Schams, Valley of 380.
Schanfiggthal 364.
Schangnau 163.
Schänis 44.
Schäniser Berg 45.
Scharans 378. 390.
Schättorf 111.
Schatzalp 366.
Schaubhorn 188.
Schauenberg 49.
Schauenburg, Bad 13.
Schauensee 86. 103.
Scheerhorn 72. 123.
Scheerhorn Griggeli Pass 124.
Scheerjoch 124.
Scheibe, Grosse 76.
Scheibe Pass 76.
Scheibenstoll 45.
Scheidegg, Great 186.
—, Hasli 186.
—, Lauterbrunnen 174.
—, Little 171.
—, Reschen 428.
—, Rigi 100.
Scheidstock 70.
Schera Alp 414.
Scherzligen 151. 155.
Scheye 67.
Schiahorn 361. 362. 366.
Schienhorn 310.
Schiers 357.
Schiesshorn 366.
Schiffli 162.
Schild 67 68.
Schilt 99.
Schiltwald 168.
Schilthorn (Lötschen Pass) 199.
— (near Mürren) 168.
Schimberg 139.
—, Bad 139.
Schindellegi 105. 43.
Schinznach 23. 20.
Schlagstrasse 108.
Schlans 371.
Schlapina Joch 359. 431.
Schleins 419.
Schleuis 369.
Schlieren 21.
Schlieren-Thal 133.
Schlinga, Val 417.
Schlossberg 130.
— Glacier 111.
Schlossberglücke 112. 130.
Schlösslikopf 66.

Schlossstock 130.
Schlossstock-Lücke 93. 130.
Schlündi 204.
Schmadribach Fall 167.
Schmadri-Joch 171.
Schmerikon 44.
Schmidhäuser 317.
Schmitten 215. 363.
Schmorras Pass 392.
Schnaus 371.
Schneehorn 173. 382.
Schneestock 137.
Schneidehorn 201.
Schnittweyer Bad 153.
Schnurtobel 95.
Schöllenen 118.
Schönboden 106.
Schönbrunn 80.
Schöneck 88.
Schönegg 163. 164. 356.
Schönegg Pass 93.
Schönenwerth 22.
Schönfels 79.
Schönhorn 311.
Schrattenflühe 139.
Schrättern, Alp 187.
Schreckhorn 177.
Schreienbach, the 70.
Schrinen, Alp 47.
Schruns 431.
Schuls 416.
Schüpfheim 139.
Schüss, the 11. 210.
Schwabhorn 180.
Schwaldis, Alp 47.
Schwalmern 157. 162.
Schwalmis 128.
Schwanau, Island of 109.
Schwand 129.
Schwandegg 80.
Schwanden 88.
Schwandfeldspitze 197.
Schwändi 68. 103.
Schwarenbach 193.
Schwarzach 432.
Schwarzberg Glacier 343.
— -Weissthor 343.
Schwärze Glacier (Furka) 121.
— (Monte Rosa) 339.
Schwarzegg Hut 177. 178. 190.
Schwarzenbach 50. 73.
Schwarzenberg 138.
Schwarze See, near Schuls 417.
— near Freiburg 217.
Schwarz-Gletscher 193.
Schwarzgrat 111.
Schwarzgrätli 197.
Schwarzhorn (Augstbord Pass) 331.

Schwarzhorn (near Grindelwald) 181. 184.
— (Flüela Pass) 360. 362.
— (Kienthal) 170.
— (Monte Rosa) 338.
— (Parpan) 366.
Schwarz-See (near Zermatt) 336.
— (near Klosters) 359.
— (Arosa) 365.
Schwarzsee-Bad 217.
Schwarzthor 339.
Schwarzwald Glacier 186.
— Huts 186.
Schwefelberg 153. 203.
Schwein-Alp 43. 74.
Schweiningen 391.
Schweizerhalle 13.
Schweizer-Thor 357. 430.
Schwellisee 365.
Schwendi 54. 57. 58. 133.
Schwendifluh 90.
Schwendi-Kaltbad 133.
Schwendienbad 140.
Schwyz 109.
Schyn Road, the 379.
Schyngrat 93.
Schynige Platte 164.
Sciez 257.
Sciora, Alp 396. 423.
Sciundrau, Lago 443.
Scopa 454.
Scopello 454.
Scopi 376.
Sedrun 373.
Séchex 257.
Seealpsee, the 57.
Seeboden-Alp 97.
Seedorf 93.
Seehörner (Silvretta) 359
Seelibühl 153.
Seelisberg 89.
Seelisberger Kulm 88.
— See 88. 90.
Seengen 142.
Seerüti 74.
Seesvenna, Val 417. 418.
—, Piz 417. 418.
Seethal 141.
Seewen 109.
Seewenalp 133. 139.
Seewenegg 133. 139.
Seewinen Glacier 346.
Seewis 357.
Séez 294.
Seezberg 2.
Seezthal 47. 76.
Sefinen-Alp 168.
Sefinen-Furgge 170.
Sefinen-Thal 169. 168.
Segl, Lej da 396.
Segnas 373.
Segnes Glacier 75. 76. 368.

Segnes Pass 75. 368.
— Piz 75. 76.
Segrino, Lago del 464.
Seigne, Col de la 291.
Seignelégier 210.
Seilon, Alp 306. 322.
—, Col de 306. 324.
—, Glacier de 306. 324.
—, Mont Blanc de 305.
Seiloz, La 293.
Selbsanft 69. 70.
Selkingen 316.
Sella, La 402. 403.
— Pass 409.
— Glacier 409.
— Lake 121. 122.
—, Rifugio (Mont Blanc) 281.
—, — (Lyskamm) 338.
—, — (Weissthor) 339. 345.
Selun 45.
Selva 374.
Selzach 17.
Sembrancher 299.
Semione 377.
Semnoz 269.
Semogo 412. 425.
Sempach 19.
Sena, Pizzo di 411.
Sengla, la 324.
Sennhof 48.
Sennthum 330.
Sennwald 59. 61.
Sent 418.
Sentier, Le 221.
Sentis 57.
Seon 142.
Sepey, Le (Ormont) 252.
— (Val d'Hérens) 325.
Septimer 392. 395.
Seregno 465.
Serena, Col de la 303.
Serenbach, the 46.
Serengia, Piz 378.
Sergnement 265.
Sermenza, Val 454.
Serneus 358.
Sernf-Thal 68. 75.
Serpentine 305.
Serra Neire 329.
Serrières 211.
Sertig-Dörfli 362.
— Pass 362.
Sertig-Thal 362. 363.
Servaplana 256.
Serviezel, Ruin 419.
Servoz 273.
Sesia Joch 339.
— Valley 346. 453.
Sesto-Calende 451.
— S. Giovanni 465.
Sether Furka 75.
Sett, Passo di 392.

Settimo Vittone 296.
Sevelen 61.
Sevenen, Alp 312.
St. Séverin 256.
Sevreu, Alp 304.
—, Col de 304.
Sévrier 269. 270.
Sexblanc, Col de 304. 300.
Sex Rouge 251.
Seyon, the 207. 209.
Seyssel 265.
Sfazzu 419.
Sgrischus, Lej 409.
Sichellauenen 166.
Siders 307.
Sieben Brunnen, the 200.
Siebnen 43.
Siedelhorn, the Little 189.
Siedeln Glacier 125.
Sierre 307.
Sierroz, Gorges du 266.
Signalhorn 369.
Signalkuppe 348.
Signau 140.
Signayes 304.
Sigriswyl 156.
— Grat 156. 157.
— Rothhorn 156. 157.
Sihl, the 21. 35. 42. 48. 78. 81 etc.
Sihl-Brücke 81.
Sihlseeli 74.
Sihlthal 40. 107.
Sihlwald 40.
Silberhorn 173.
Silbern 73. 74.
Silberstock 69. 73.
Silenen 112.
Sillerngrat 197.
Sils (Engadine) 396.
— in the Rheinthal 379.
—, Lake of 396.
Silvaplana 397.
Silvretta-Hut 358. 359.
— Glacier 358. 359.
Silvrettahorn 359.
Silvretta Pass 359.
Simano 377.
Simelihorn 179. 180.
Simeli Pass 312.
Simme, the 199. 200. 202 etc.
Simmenegg 203.
Simmenfluh 202.
Simmenthal 199. 202.
St. Simon 266.
Simplon 311. 312.
—, Pass 311.
Sinestra, Val 417. 418.
Singen 25. 32.
Sion 307.
—, Mayens de 324.
—, Monastery of 44.

Sirnach 49.
Sirvoltenhorn 312.
Sirvolten Pass 312.
Sislkon 92. 110.
Sismonda, Signal 296.
Sissach 13.
Sissacher Fluh 13.
Sisseln 20.
Sissone, Monte 395.
Sitter, the 49. 50. 56.
Six-Madun 120. 374.
Sixt 275.
Soazza 387.
Soglio 382. 423. 424.
Soja, Val 377.
Soladino Fall 442.
Solalex 255.
Solda, Val 437. 438. 456.
Solis Bridge 379.
Soleure or Solothurn 15.
Som la Proz 298. 299.
Someo 442.
Someraro 450.
Sommerau 14.
Sommerikopf 63.
Somvix 371.
—, Val 370. 371.
Sonadon, Col du 300. 305.
—, Glacier du 300. 305.
Sonceboz 11. 211.
Sondrio 421.
Sonnenberg, near Lucerne 86. 138.
— near Zürich 34.
— near Seelisberg 89.
Sonnighorn 341.
Sonogno 440.
Sonvillier 210.
Sonzier 248. 254.
Sorebois, Col de 326.
—, Corne de 326. 328.
Sörenberg 139.
Sorengo 436.
Sorescia 121.
Sorreda Pass 370.
Sosto, Mt. 376.
Souste, La 308.
Soyhières 10.
Spadlatscha, Val 388.
Spannegg 68.
Spannort, the Great and Little 130. 112.
Spannort-Hütte 130.
Spannort-Joch 112. 130.
Sparrhorn 309.
Speer 45. 63.
Speicher 55.
Spescha, Porta da 71. 372.
Spicherfluh 132. 135.
Spiellaui-See 373.
Spiez 155. 156.
Spiezwyler 154.

Spinabad 363.
Spino 424.
Spiringen 72. 111.
Spitalmatte 193.
Spitelrüti 71.
Spitzberg 125.
Splüdatsch, Castle 392.
Splüga, Monte 422.
Splügen 382.
— Pass 383.
Spoccia 445.
Spöl, the 413. 414.
Spondinig 428.
Spontisköpfe 356.
Spruga 442.
Sruors, Las 406.
Staad 60. 52.
Stabbio Alps 387.
Stabbio-Grat 386.
Stachelberg, Baths of 69.
Stäfa 41.
Stäfel-Alp 123.
Stafeln, the 123.
Staffel-Alp 320. 337.
Staffelwald 320.
Stalden in the Visp Valley 332.
—, on the Pragel 73.
—, in Unterwalden 138.
Staldenried 332. 340.
Stalla 392.
Stallerberg 381.
Stalvedro 392.
—, Stretto di 116.
Stammerspitz 418. 419.
Stammheim 32.
Stampa 423.
Stans 127.
Stanser Horn 127.
Stansstad 101.
Stanz 127.
Starkenbach 83. 45.
Starlera, Val 381.
Statz Alp 401. 408.
—, Lake of 401. 404.
Stätzer Horn 390.
Staubbach, the 166.
Stäuberfall 72.
Staufberg 22. 142.
Stavelatsch, Fuorcla 372.
Stechelberg 166.
Steckborn 32. 26.
Steffisburg 153.
Steigli-Egg 102.
Stein, Zum 137.
— zu Baden 21.
— on the Rhine 20. 26. 32.
— (Toggenburg) 63. 45.
Steinach, Castle 52.
Steinalp-Brisen 128.
Steinberg, the 137.
—, the Lower 167.
—, the Upper 167.

Steinen 109.
Steinenberg Alp 170.
Steinerberg 108.
Steinerne Tisch 52. 60.
Stein Glacier 137.
Stein-Limmi 137.
Steinsberg 415.
Steinthalhorn 381.
Steje, Becco delle 298.
Stella, Corno 421.
Stellihorn 343.
Stelvio Pass 427.
St. Stephan 199.
Stilfs 427.
Stilfser Joch 427.
Stock 120. 296. 375.
Stockalp 190. 316.
Stock Glacier 325. 326.
Stockgron 71.
Stockhorn (Simmenthal) 202.
— (Zermatt) 337.
Stockje 324. 326.
Stockknubel 337.
Stöckle 120. 375.
Stoos, near Brunnen 91.
Storegg 132.
Stoss, near Gais 56.
Strahlegg 178. 357.
Strahlhorn 338. 343.
Strassberg, Ruin 365. 390.
Strättligen 153. 155.
Strela Pass 366.
Strengen 430.
Stresa 449.
Stretta, La 410.
—, Piz della 410.
Strich, Zum 344.
Strim Glacier 123.
Strimthal 123. 373.
Strubelegg 197.
Stuben 429.
Stücklistock 138.
Studerhorn 190.
Studerjoch 190.
Stufenstein-Alp 171.
Stulsergrat 363. 364.
Sturnaboden 357.
Stürvis 379.
Stutz 101. 357.
Sublage 251.
Suchet, Mont 215. 220.
Süd-Lenzspitze 341.
Sueglio 458.
Sufers 382.
Suggithurm 163.
Suhr 18. 22.
Suldalp, Untere 157.
Sulden-Thal 427.
Suldthal 157. 162.
Sulegg 162. 165.
Sulgen 49.
St. Sulpice 212. 236.

Suls, Alp 162.
Sulsanna, Val 362. 413.
Sulzfluh 358. 431.
Sumiswald 18.
Suna 448.
Sundgraben 157. 158.
Sundlauenen 157.
Sur 392.
Sura, Alp 368. 372.
Surava 384. 388.
Sur En 417. 418.
Surenen Pass 130.
Suretta, Val 382.
Surettahorn 381. 383.
Surlej 398.
—, Fuorcla da 398. 406.
—, Piz 401. 403.
Surovel, Alp 398. 406. 408.
Surpalix, Val 374.
Surrhein 371. 373.
Sur Sass 417.
Sursee 13.
Sursura, Piz 414.
— Glacier 414.
Sü Som 414.
Süs 414.
Susasca Valley 359.
Süser Thal 359.
Sussillon 327.
Susten 308.
Susten Alp 129.
— Hörner 137.
— Joch 118.
— Limmi 137. 118.
Suvoroff Bridge 72.
Suvretta, Alp 401.
—, Val 308. 401. 412.
Suzanfe, see Clusanfe.

Tabor, Mont 339.
Taceno 458. 459.
Taconay, Glacier de 273.
Tacul, Glacier du 277. 282.
—, Montblanc du 280.
Tägertschi 140.
Tagliaferro 347.
Taille, Mont 348.
Taillères, Lac des 213.
Taino 443.
Talamona 422.
Talèfre, Aig. de 282.
—, Col de 282.
—, Glacier de 277. 278. 282.
Talloires 269.
Tamaro, Monte 438. 444.
Tambohorn 382.
Tamić, Col de 268.
Tamina, the 61. 64. 65.
Tamins 368.
Taneda 115.
Taney, Lake of 258.
Taninges 274. 272.

Tannenalp 132. 135.
Tannenberg 68.
Tanneverge, Pic and Col de 275. 286.
Tanzbödeli 157. 162.
Tapiaz, La 281.
Tarasp, Castle 417.
—, Baths of 416.
Tarentaise 263. 294.
Tartsch 428.
Täsch 333. 342.
Täsch-Alp 342.
Täschhorn 341. 335.
Täsch Pass 343.
Tasna, Val 415.
Tatlishorn 199.
Tätschbach Fall 129.
Taubenloch, Gorge 12.
Taufers 426. 428.
Tavagnasco 293.
Tavanasa 371.
Tavannes 11.
Tavel 241.
Taverne 434. 437.
Tavetsch 373.
—, Mompè 373.
Tavordo 456.
Tecknau 13.
Teglio 421.
Tell's Chapel (near Küssnacht) 105.
— (near Bürglen) 111.
— (Lake of Uri) 92.
Tell's Platte 92.
Telli, the 168.
Telli-Thal 171.
Temu, Bocchetta del 454.
Tencia, Campo 443.
Tendre, Mont 221.
Tène, La 206.
Teniger Bad 371.
Teo, Pizzo di 411.
Termine, Val 115.
Terrarossa, Punta di 311.
Terri, Piz 370.
Territet 244. 246.
Tesserete 438.
Tessin, see Ticino.
Tête Blanche 324. 325.
— de Bois 300.
— Noire (near the Col de la Forclaz) 284.
— (near St. Gervais)273.
— Rouge, Glacier de 281.
Teufelsbrücke, in the Reussthal 119.
— near Mürren 168.
— in the Sihlthal 106.
Teufelsmünster 90.
Teufen 59.
Tgietschen, Piz (Oberalpstock) 123. 373.
— (Pass Diesrut) 370.

Thal 60.
Thalacker 80.
Thalalpsee 46. 68.
Thaleggli 137.
Thälliboden 345.
Thältistock 137.
Thalweil 42.
Théodule Glacier 336. 337.
— Pass 336. 339. 349. 350.
Theodulhorn 336.
Therwil 9.
Thièle or Toile, the 214.
Thièle or Zihl, the 207.
Thiengen 24.
Thierachern 153.
Thieralplistock 137.
Thierberg 121. 137. 200.
— Gletscher 200.
— Limmi 118. 137.
Thierbergli 137.
Thierfehd 70.
Thierwies 58.
Thônes 270.
Thonon 257.
Thuile, La 293.
Thun 151.
—, Lake of 155.
Thur, the 33. 49. 62. etc.
Thurgau, Canton 49.
Thurm, the Aeussere and Innere 343.
Thusis 378.
Tiarms, Pass da 374.
—, Piz, Val 374.
Tiatscha Pass 359.
—, Val 418.
— Glacier 415.
Ticino, the 115. 116. 122. 315. 387. 440. 444.
—, the Canton of 433.
Tiefenbach 125.
Tiefengletscher 125.
Tiefenkasten 380. 391.
Tiefenmatten Glacier 325.
— Joch 339.
Tiefensattel 125. 137.
Tiejerfluh 366.
Tignes 294.
Tine, La 254.
Tines, Les 278. 283.
Tinière, Col de la 245.
Tinzen 392.
Tinzenhorn 363. 388. 389.
Tinzenthor-Pass 392.
Tirano 421.
—, Madonna di 420.
Tisch, Val 389.
Titlis 130. 135.
Tivano, Piano del 462.
Toce, see Tosa.
Tödi 70. 372.
—, Bündner 371.
—, Lesser 372.

Toggenburg 62.
Toggia, Valle 320.
Toma Lake 120. 374.
Tomlishorn 103.
Tondu, Col du Mt. 289.
Torgnon Glacier 324.
Torno 462. 464.
Torre 377. 421.
Torrent, Col de 326.
Torrentalp (Leuk) 195.
— (Einfisch-Thal) 326.
Torrenthorn 195.
Torrigia 462.
Torrone, Pizzo 395.
Torta, Val (Ticino) 443.
— (near Klosters) 359.
Tosa, the 313. 319. 444. 448. 451. etc.
—, Falls of the 319.
Tösens 429.
Töss 48.
—, the 33. 48.
Tongues 257
Tounot 328.
—, Alp 330.
Tour, Le 288.
Tour, Aig. de la 280.
—, Aig. du 284. 287. 299.
—, Cabane de la 338.
—, Col du 287.
— Glacier du 287. 288.
— Noire 283.
— de Peilz, La 240. 246.
— de Trême, La 253.
Tournalin, Grand 350.
Tournanche, Col de 339.
Tournelon Blanc 304. 305.
Tournette, Mont (Isère Valley) 268.
— (near Annecy) 269. 270.
Tour-Ronde 258.
Tourtemagne 308.
Trachsellauenen 166.
Tracht 182.
Tracuit, Alp 328.
—, Col de 330.
Trafoi 427
Trais Fluors 402.
Trasquera 311. 313. 318.
Travers 212.
—, Val de 212.
Treib 89.
Trelatête, Aig. de 291.
—, Col de 289.
—, Glacier de 289.
—, Pavillon de 289.
Trélechamp 283.
Trélex 235.
Tremezzina, the 461.
Tremezzo 461.
Tremoggia, Piz 397.
— Pass 397.
Tremola, Val 122.

Tremorgio, Poncione 443.
Trepalle 425.
Tresa, Ponte 455.
Trescuhnine Pass 387.
Tresenda 421.
Tresero, Piz 426.
Trévelin 236.
Tre Uomini, Passo 387.
Tricot, Aig. du 289.
Triège, Falls of the 285.
Trient 284.
—, the 284.
—, Col de 284.
—, Glacier de 284. 286. 299.
—, Gorges du 249.
Trift Alp (Saas) 341.
— Glacier (Triftthal) 136.
— (near Saas) 341.
— (near Zermatt) 329.
Triftgrätli 341.
Trifthorn 329. 341.
Trifthütte 137.
Triftjoch 329. 340.
Triftlimmi 137
Trift Pass 341.
Trift Valley 136. 137. 337.
Trins 368.
Trinserhorn 378.
Triolet, Aig. de 292. 282. 283.
—, Cabane de 283. 292.
— Col de 282.
—, Glacier de 292.
St. Triphon 247.
Triquent 285.
Trobaso 448.
Trogen 55.
Troistorrents 259.
Trub 140.
Trübbach 61.
Trubschachen 139.
Trübsee 136.
— -Alp 136.
Trudelingen 72.
Trümleten-Thal 166. 172. 173.
Trümmelbach Fall 166.
Truns 371.
Trupchum, Val 413.
Trüttlisberg 250. 201.
Tschamut 374.
Tschanuff, Ruin 418.
Tschappina 367. 379.
Tschera, Piz la 381.
Tschiertschen 366.
Tschierva, Piz 402. 408.
—, Vadret da 405.
Tschingelalp 191.
Tschingel Glacier 167. 170.
Tschingelhorn (Lauterbrunnen) 171. 198.

BAEDEKER, Switzerland. 16th Edition.

Tschingelhörner (Sernfthal) 75.
Tschingellochtighorn 197.
Tschingeln-Alp (near Elm) 75.
— (near Walenstadt) 47.
Tschingel Pass 170. 191. 193.
Tschingeltritt 170.
Tschuggen (Grindelwald) 174.
— (Arosa) 365.
— (Fluela Pass) 360.
Tübach 52.
Tubang, Mont 380.
Tuckettspitze 427.
Tummenen 831.
Tuoi, Val 359. 415.
Tuors, Val 362. 389.
Turbach Valley 201.
Turbenthal 49. 382.
Turgi 20.
Turlo Pass 346.
Turtig 308.
Turtmann 308. 30.
— Glacier 330.
— Valley 308. 330.
Twann 206.
Twannberg 206.
Tyndall, Col and Pic 339.
Tzeudet, Glacier 300.

Uccello, Piz 386.
Ueberlinger See 29.
Ueblenberg 191.
Ueli Alp 70.
Uerikon 41.
Uertsch, Piz 364. 389. 413.
Ueschinen-Thal 193.
Ueschinen-Thäli 197.
Uetikon 41.
Uetliberg 39.
Uffern, Val 375.
— Pass 375. 376.
—, Piz dell' 373. 376.
— Hut 373.
Ufnau, Island of 41.
Ugines 269.
Ugines, Fontaines d' 268. 269.
Uina, Val 417. 418.
Ulrichen 315.
Ulrichshorn 341. 342.
Umbrail, Piz 426.
Unchio 443.
Ungeheuerhorn 359.
Unspunnen, Ruins 162.
Unter-Aar Glacier 178. 189.
Unter-Aegeri 80.
Unteralp Pass 115.
Unter-Gabelhorn 337.
Unterhorn 368.

Unter-Laret 359.
Unter-Mustail 380.
Unter-Mutten 379.
Unterschächen 72.
Untersee 25. 26. 31.
Unterseen 160. 161.
Unter-Sihlwald 40.
Unter-Solis 379.
Unterstetten 100.
Unterterzen 46. 47.
Unterwald 320.
Unterwalden, Canton 127.
Unterwasser 58. 63.
Uomo, Piz dell' (Lukmanier) 376.
— — (St. Gotthard) 115. 120.
— Pass 115.
Uratstöcke 137.
Urbach-Thal 186. 187.
Urden-Fürkli 366.
Urdorf 78.
Urezas, Val 415.
Uri, Canton 91. 98.
—, Lake of 91. 110.
—, Rothstock 93. 129.
Urio 462.
Urlaun, Piz 71. 371.
Urlichen 315.
Urmein 367.
Urnäsch 50.
Urnenalp 187.
Urner Boden 71.
— Loch 119.
— See 91.
Ste. Ursanne 10.
Urschai, Val 415.
Urseren Valley 119.
Urweid, Innere 187.
Ussin 350.
Uster 44.
Uttigen 151.
Uttwyl 31.
Utzensdorf 17.
Utznach 44.
Utzwyl 50.
Vache, Roc de la 328.
Vadalles, Les 253.
Vadret, Piz (Dischma Valley) 360. 362.
— (near Pontresina) 403. 406.
Vadura 66.
Vaduz 61. 432.
Vaira, Val 313.
Valais, Upper 315.
Valaisan, Mt. 293.
Valbella 387. 391.
Valcava 414. 418.
Valcournère, Col de 850.
Valdobbia, Col di 347.
Valendas 367.

Valens 66.
St. Valentin 369. 428.
Valettes, Les 299.
Valgronda-Joch (Val Faller) 392.
— (Val Somvix) 371.
Vallatsch 370.
Valletta, Pizzo la 120.
Vallorbes 221.
Vallung, Piz 389. 392.
Valmaggia 454.
—, Bocchetta di 443.
Valorcine 284.
Valpelline 306.
—, Col de 296. 325. 340.
Valrhein, Piz 386.
Vals am Platz 370.
Valsainte 208.
Valser Berg 370. 386.
Valsorey, the 300.
—, Col du 300.
—, Glacier du 300.
Valtellina, the 420.
Valtendra, Passo di 311. 318.
Valtenigia, Alp 371.
Valtournanche 350.
—, Glacier 349. 350.
Valtravaglia, Porto 445.
Valurgut 65.
Valzeina 357.
Valzeinerspitz 357.
Van d'en haut 260. 285
Vandœuvres 233.
Vanescha Pass 370.
Vanzone 344.
Varallo 453.
Varenna 459.
Varens, Aig. de 272. 273.
Varese 446.
Varzo 313.
Vasanenkopf 66.
Vasevey, Col de 306.
Vasön 66.
Vassena 463.
Vättis 66.
Vaud, Canton de 237.
Vauderens 218.
Vaulion 221.
—, Dent de 221.
Vaulruz 253. 218.
Vaumarcus, Castle 214.
Vaux, La 239. 245.
Vazerols 391.
Vecchio, Passo 374.
Vedro, Val di 313.
Veglia, Alp 311. 313. 318.
Veislvi, Dents de 324.
Velan, Mont 301.
Veltlin, see Valtellina.
Venay 270.
Veni, Val 291.
Vercorins 327.

Vereina Pass 359.
St. Verenathal 16.
Vergeletto 442.
Vermolera, Val 412.
Vermunt Pass 431.
— Glacier 431.
Vernayaz 249. 260. 286.
Vernaz, Col de 258.
Vernela Pass 358.
Vernex 241. 242. 246.
Vernier 231.
Vernok Pass 370.
Verona, Pizzo di 420.
Verra Glacier 339.
— Pass 339.
Verrès 297. 348.
Verrières Suisse 213.
Versam 367.
Versegère 304.
Vers l'Église 251.
Versoix 234. 245.
Verstanklahorn 359.
Verstanklathor 359.
Verva, Passo di 412.
Verzasca, Val 440.
Vésenaz 233.
Vespran 423.
Vessona, Col de 296.
Vétroz 307.
Vevey 239. 246.
Vex 321.
Veyrier 232. 264. 270.
Veytaux 242. 246.
Vezio, Torre di 459.
Via Mala 380.
Vial, Piz 370. 371. 372.
Vicosoprano 423.
Vierwaldstätter See 86.
Viesch, see Fiesch.
Viganello 437.
Vigens 370.
Vigezzo Valley 441. 445.
Vigne Glacier 339. 346.
Vilan 357.
Villa near Airolo 315.
— (Val Bregaglia) 424.
— (Val d'Hérens) 322. 326.
— (Vrinthal) 370.
Villadossola 314. 451.
Villard-sous-Mont 258.
Villars (near Aigle) 246. 251.
— (near Lausanne) 220.
Ville d' Issert 293.
Villeneuve in the Aosta Valley 294.
—, Lake of Geneva 245. 246.
Villers-le-Lac 211.
Villette, La 289. 232.
Villmergen 20. 22. 141.
Vilters 61.

St. Vincent 297. 348.
Vincenthütte 347.
Vincent Pyramid 348.
Vindels, Alp 66.
Vindonissa 20.
Vintschgau, the 428.
Viola, Cima, Val and Pass 412. 425. 426.
Vionnaz 259.
Viou, Becca di 297.
Visaille, Cant. de la 291.
Visp 309.
—, the 308. 331.
—, the Matter 332. 335. 340.
—, the Saaser 332. 340.
Vispach 309.
Visperterbinen 332.
Vissoye 327.
S. Vittore 387.
Vitznau 87.
Vitznauer Stock 88. 89.
Viviers 267.
—, Grotto 258.
Vizan, Piz 381.
Vocca 454.
Vogelberg 386.
Vögelisegg 50. 51.
Vogeljoch 386.
Vogna, Val 347.
Vogogna 451.
Voirons 233. 274.
Vorab 75. 368.
Voralpthal 118.
— Hut 118.
Voranen 74.
Vorder-Glärnisch 68. 74.
— Meggen 104.
— Thierberg 137.
Vonasson, Pointe de 323.
Vouvry 258.
Voza, Col de 288.
Vreneligärtli 74.
Vrin 370.
V'iache, Mont 264.
Vufflens, Castle 236. 245.
Vuibez, Glacier de 323. 324.
—, Serra de 324.
Vully, Mont 219. 220.
Vulpera 416.

Wabern 151. 153.
Wädenswell 42.
Wagenlucke 58.
Wäggithal 43.
Wahlalp 202.
Waid, near Zürich 34.
—, near St. Gallen 51.
Walchwyl 104. 109.
Wald near Rüti 49.
— near Trogen 55.
Waldenburg 13.

Waldhäuser 368.
Waldibruck 141.
Waldisbalm, Grotto 88.
Waldshut 24. 20.
Waldspitz 179.
Waldstatt 50.
Walen-See 45.
Walenstadt 46. 47.
—, Lake of 45.
Walkringen 140.
Wallisellen 48.
Waltensburg 371.
Waltersfirren Alp 123.
Walzenhausen 60.
Wandfluh 308. 325.
Wand Glacier 342. 343.
Wangen (Aare) 14. 15.
— (Untersee) 26. 32.
— (Lake of Zürich) 43.
Wannenstock 73.
Wartburg, Neu-, Ruins 14.
Wartegg 52. 54. 60.
Wartensee 54. 19.
Wartenstein, Pens. 65.
Wasen 112. 113.
Wasenhorn 311.
Wasserauen 57. 58.
Wasserfluh 22.
Wasserwendi 134. 182.
Watt 21.
Wattenwyl 153.
Wattingen 113.
Wattwyl 63. 44.
Wauwyl 18.
Weesen 45.
Weggis 87.
Weiach 48.
Weinburg, Castle 52. 60.
Weinfelden 49.
Weissbad 56.
Weissberg 391. 392.
Weisse Frau 192.
Weissenau, Ruins 155.
Weissenburg 202.
—, Baths of 202.
Weissenfluh 88.
Weissenstein, in the Grisons 389.
— near Soleure 16.
Weissensteinhorn 370.
Weissfluh 358. 361. 365.
Weisshorn (Rawyl) 201.
— (near Zermatt) 339.
— (Fluela Pass) 360.
— (Arosa) 366.
— (Parpan) 366.
— Hotel 328.
— Hut 339.
Weismies 341. 313.
Weisstannen 47. 75. 76.
Weissthor, Old 339. 345.
—, New 339. 345.
—, Schwarzberg 340. 343.

32*

Weiss-Wasserstelz 48.
Weitenalpstock 123. 373.
Weiterschwanden 72.111.
Weit-Riss 132.
Wellenkuppe 338.
Wellhorn 185. 186.
Welschtobel 365. 366.
Wenden Glacier 137.
Wenden-Joch 130.
Wendenstöcke 135. 136.
Wengen 172.
Wengern-Alp 173.
Wengi, Baths of 79. 191.
Wengistein 16.
Wenslingen 13.
Werdenberg 63.
—, Castle 61.
Werthenstein 138.
West-Lenzspitze 341.
Wetterhorn 177. 186.
Wetterlimmi 186.
Wetterlücke 171.
Wettingen 21.
Wettschwyl 78.
Wetzikon 44.
Wetzsteinhorn 201.
Wichlen Alp 75.
Wichtrach 151
Widderegg 123.
Widderfeld 86. 102. 129.
Widerstein-Furkel 47.
Wienachten 54.
Wiesberg 429.
Wiesen 363.
Wiggen 139.
Wiggerthal 18.
Wiggis 67 74.
Wild-Andrist 169.
Wilde Frau 191. 192. 170.
Wildegg 22.
Wilderswyl 163. 165.
Wildgeissberg 132. 135.
Wildgerst 181.
Wildhaus 63.
Wildhorn 200. 201. 251.
— Hut 200. 250.
Wildkirchli 57.
Wildspitz 108. 79. 80.
Wildstrubel 194. 197. 200.
Willigenbrücke 182. 185. 187.
Willisau 18.
Wimmis 154. 202.
Windegg Hut 137.
Windgällen (Maderaner-Thal) 123. 124.
Windgälle, the Schächenthaler 72.
Windjoch 342.

Winkel 101.
Winkelmatten 334. 335.
Winkeln 50.
Winter Glacier 125.
Winterberg 118. 137.
Winterjoch 118. 137.
Winterlücke 125.
Winterthur 48.
—, Ober 32.
Wittwe 193.
Wohlen 20. 22.
Wohlhausen 138.
Wolfenschiessen 128.
Wolfgang 360.
Wolfhalden 55.
Wolfsberg 26. 32.
Wollerau 43.
Wollishofen 42.
Worb 140.
Wormser Joch 426.
Wörth, Chât. 28. 27.
Wülflingen 48.
Wülpelsberg 22. 23.
Wurmspach, Convent 44.
Wyhlen 23.
Wyl 49.
Wyla 49.
Wylen 52. 50.
Wyler 112. 135. 136.
Wyler-Alp 134.
Wylerhorn 134.
Wynigen 17.
Wyttenwasser Glacier 121.

Yverdon 214.
Yvoire 257.
Yvonand 217.
Yvorne 246.
Ywerberhörner 120.

Za, Aiguille de la 323.
Za de l'Ano 329.
Za-de-Zan, Col de 324.
—, Glacier de 296. 323. 324.
Zadrell, Fuorcla 359.
Zanfleuron 200. 250.
— Glacier 250. 256. 322.
Zapportgrat 370. 386.
Zapporthorn 386.
Zapport-Hut 385.
— Pass 386.
Zarmine, Col de 324.
Zäsenberg 177.
Zäsenberghorn 177.
Zaté, Col du 325. 329.
—, Pointe de 329.
Zatelet Praz 326.
Zavragia Ravine 371.

Zäziwyl 140.
Zell 49.
Zerbion, Mt. 297.
Zermatt 333.
Zermeigeren 342.
Zermettje 333.
Zernetz 413.
Zertannen 344. 345.
Zervreila 370.
Ziegelbrücke 45. 43.
Zigiorenove, Glacier de 323. 324.
Zihl, the 12. 207.
Zihlistock-Alp 89. 100.
Zillis 380.
Zimmerberg 81.
Zimmerwald 151.
Zinal 328.
—, Glacier de 325. 328.
—, Pointe de 329.
— Rothhorn 329. 339.
Zinkenstöcke 189.
Ziteil 380. 391.
Zizers 62.
Zmutt 337.
— Glacier 296. 326. 329.
— Valley 336. 337.
Zocca, Passo di 422. 423.
Zofingen 18.
Zollikofen 12. 18.
Zollikon 41.
Zozanne, Lac 326.
Zug 75.
—, Lake of 103.
Züge, the 363.
Zuger Berg 79.
Zum See 334. 336. 337.
— Steg 318. 320.
— Strich 344.
Zumsteinspitze 348.
Zuoz 413.
Zupò, Piz 408.
Zürchersmühle 50.
Zürich 33.
—, Lake of 40.
Zürichberg 41.
Zürich-Letten 41.
Zustoll, the 45.
Zweilütschinen 165. 164.
Zweisimmen 203.
Zwillinge 335.
Zwillings Pass 339. 348.
Zwingen, Schloss 10.
Zwingli Pass 58.
Zwing-Uri 112.
Zwischbergen Pass 313.
Zwischen-Thierbergen 137.
Zwitzer Egg 201.

Leipsic: Printed by Breitkopf & Härtel.

www.ingramcontent.com/pod-product-compliance
Lightning Source LLC
Chambersburg PA
CBHW030259010526
44108CB00038B/624